Review question ~~X~~ (2-3) ~~X~~ page 11-16
do the work sheet

4.1 - 4.2 11 - 16 RQ 2-3 ✓

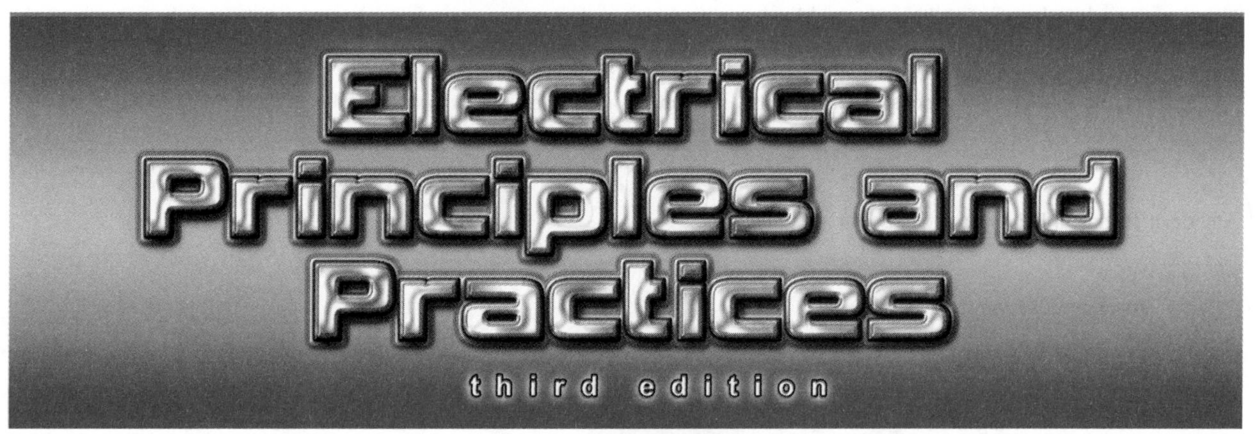

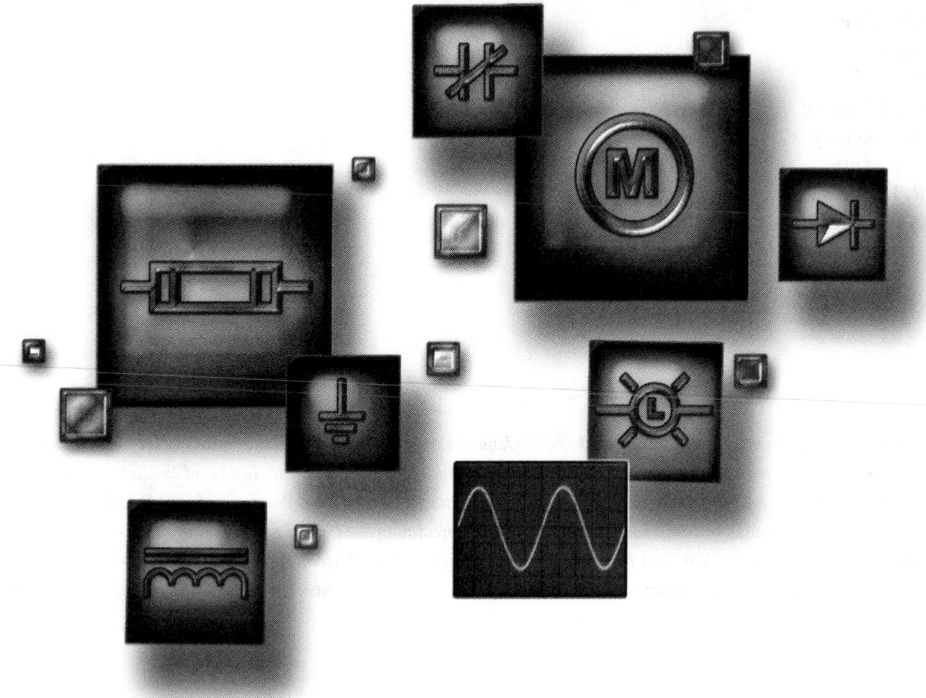

AMERICAN TECHNICAL PUBLISHERS, INC.
HOMEWOOD, ILLINOIS 60430-4600

Glen A. Mazur
Peter A. Zurlis

American Technical Publishers, Inc., Editorial Staff

Editor in Chief:
Jonathan F. Gosse
Art Manager:
James M. Clarke
Copy Editor:
Valerie A. Deisinger
Cover Design:
James M. Clarke
Illustration/Layout:
Jennifer M. Hines
William J. Sinclair
Lauren A. Gately
CD-ROM Development:
Carl R. Hansen

3 4 5 6 7 8 9 – 07 – 9 8 7 6 5 4 3

Printed in the United States of America

ISBN 978-0-8269-1803-1

This book is printed on 30% recycled paper.

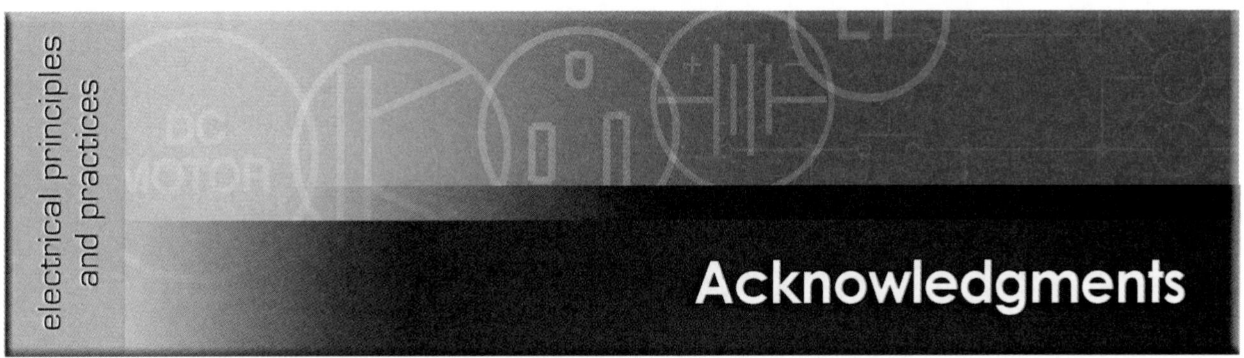

Acknowledgments

The authors and publisher are grateful to the following companies, organizations, and individuals for providing photographs, information, and technical assistance:

Baldor Electric Co.
B & K Precision
Bunn-O-Matic Corporation
Calculated Industries, Inc.
Cooper Bussman, Inc.
Fluke Corporation
GE Lighting
Gould Fiber Optics
Lab Safety Supply, Inc.
Leviton Manufacturing Co., Inc.
Motorola, Inc.
Philips Lighting
Rockwell Automation, Inc.
Ruud Lighting, Inc.
Saftronics, Inc.
Salisbury
SPG Solar
Square D Company
The Stanley Works

Contents

CD-ROM Contents

Introduction

Electrical Principles and Practices, 3rd Edition, is an introduction to electrical and electronic principles and practices and their uses in residential, commercial, and industrial applications. This edition has been reorganized to provide coverage of key electrical concepts at the beginning of the text. Other chapters have been expanded to include greater coverage on personal protective equipment, NFPA 70E, basic first aid, test instruments, printreading, and programmable logic relays.

The *Electrical Principles and Practices* CD-ROM in the back of the book is designed as a self-study aid to enhance text content. The CD-ROM includes Quick Quizzes®, an Illustrated Glossary, Math Formulas, and Reference Material. The Quick Quizzes® provide an interactive review of the topics covered in the chapters. The Illustrated Glossary provides a helpful reference to key terms included in the text. Selected terms are linked to illustrations, video clips, and animated graphics that augment the definition provided. The Math Formulas provide formulas for interactive calculation and application of Ohm's Law, the Power Formula, series circuits, parallel circuits, electric motor operation, and RCL circuits. The Reference Material button accesses links to Internet sites that contain related electrical-product manufacturer and reference information. Instructions for using the CD-ROM can be found on the last page of this book. A video icon appears on selected question screens and provides access to a video clip or animated graphic related to the question topic.

Electrical Principles and Practices, 3rd Edition, is one of several products published by American Technical Publishers, Inc. To obtain information on related educational products, please visit the American Tech web site at www.go2atp.com.

The Publisher

electrical principles and practices

Features

Detailed drawings illustrate common electrical principles.

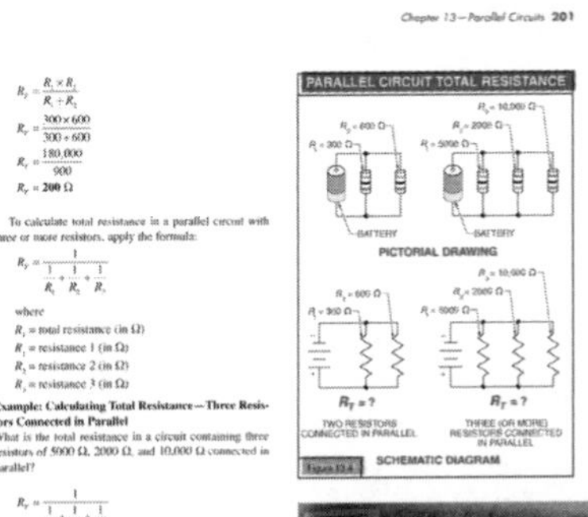

Industrial application photos supplement text and illustrations.

Tech Facts provide supplemental facts related to the topic discussed.

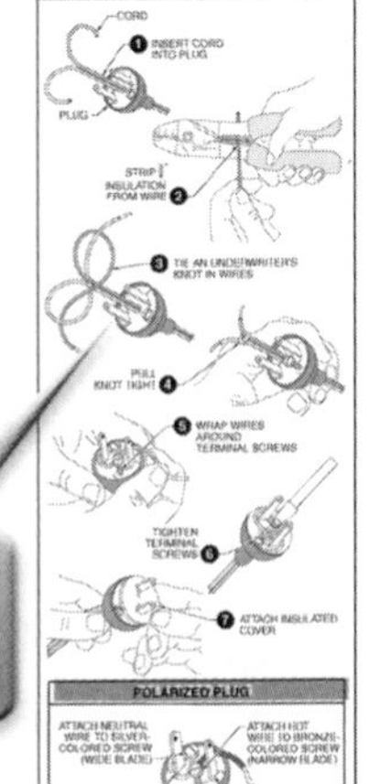

Drawings illustrate proper wiring practices to ensure safety.

Electricity

In 600 BC, the Greeks discovered a substance, amber (fossilized sap), that when rubbed against wool, caused other substances to be attracted to it. The word electricity is derived from the Greek word for amber. Today, this attraction is referred to as static electricity. One of the two basic forms of electricity, *static electricity* is an electrical charge at rest. The electrical shock a human encounters by touching an object is an example of the transfer of a static charge by contact. Lightning is an example of the transfer of a static charge by a spark. Static electricity has some limited practical uses, such as in electrostatic air filters and electrostatic spray-painting. **See Figure 1-1.**

The other form of electricity is electricity that is generated. *Produced electricity* is the electricity produced either by magnetism (generator), chemicals (battery), light (photocell), heat (thermocouple), or pressure (crystals). Produced electricity is used in circuits that are specifically designed to carry electrical charges through a controlled path to operate specific loads.

Of the basic forms of energy (electrical, chemical, nuclear, light, heat, and mechanical), electricity is the most widely used form of energy. Electricity is used to provide energy for traditional electrical applications such as lighting, heating, cooling, cooking, communication, and transportation. In addition to improvements and increases in traditional applications, new uses of electricity are increasing on a daily basis in the medical, aviation, aerospace, military, computing, networking, business, marketing, and automotive industries.

Electricity is the movement of electrons from atom to atom. An understanding of the electrical properties of matter is required because the movement of electrons from atom to atom produces electrical energy. This energy is used to produce work.

Matter

Matter is anything that has mass and occupies space. All objects consist of matter. Matter can exist in the state of a solid, liquid, or gas. **See Figure 1-2.** A *solid* is a state of matter that has a definite volume and shape. A *liquid* is a state of matter that has a definite volume but not a definite shape. A *gas* is a state of matter that is fluid, has a relatively low density, and is highly compressible.

Metal, wood, and plastic are examples of matter in a solid state. Water and gasoline are examples of matter in a liquid state. Hydrogen and carbon dioxide are examples of matter in a gas state. Some matter, such as water, can exist in a solid, liquid, or gas state. The state that water exists in is changed by the addition or removal of heat.

All matter has electrical properties. The electrical behavior of matter varies according to the physical makeup of the matter. Some types of matter, such as copper, allow electricity to easily move through them and can act as conductors. A *conductor* is material that has very little resistance and permits electrons to move through it easily.

Other types of matter, such as rubber, do not allow electricity to easily move through them and can act as insulators. An *insulator* is a material that has a very high resistance and resists the flow of electrons. The properties of different matter must be understood when designing electrical components and circuits, working around electrical equipment, and troubleshooting electrical circuits.

STATIC ELECTRICITY

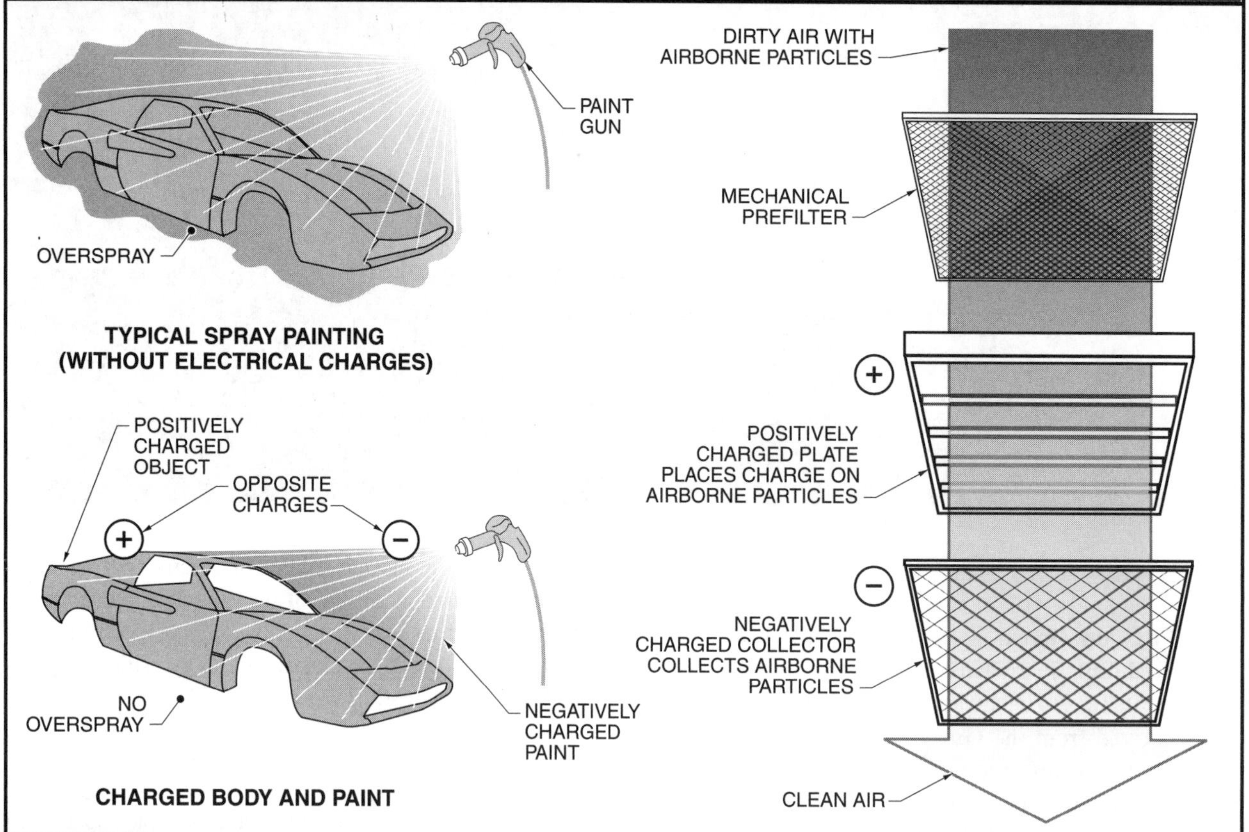

PAINT GUN
OVERSPRAY
TYPICAL SPRAY PAINTING (WITHOUT ELECTRICAL CHARGES)
POSITIVELY CHARGED OBJECT
OPPOSITE CHARGES
NO OVERSPRAY
NEGATIVELY CHARGED PAINT
CHARGED BODY AND PAINT
DIRTY AIR WITH AIRBORNE PARTICLES
MECHANICAL PREFILTER
POSITIVELY CHARGED PLATE PLACES CHARGE ON AIRBORNE PARTICLES
NEGATIVELY CHARGED COLLECTOR COLLECTS AIRBORNE PARTICLES
CLEAN AIR

Figure 1-1 ELECTROSTATIC SPRAY-PAINTING ELECTROSTATIC AIR FILTER

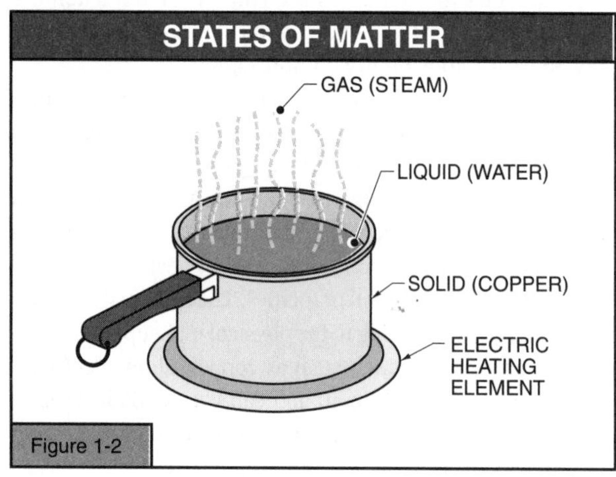

Figure 1-2

Atoms

An *atom* is the smallest particle that an element can be reduced to and still maintain the properties of that element. The three principle parts of an atom are the electron, neutron, and proton.

An *electron* is a negatively charged particle in an atom. A *neutron* is a neutral particle, with a mass approximately the same as a proton, that exists in the nucleus of an atom. A *proton* is a particle with a positive electrical charge of 1 unit that exists in the nucleus of an atom. Every atom has a definite number of protons, neutrons, and electrons. **See Figure 1-3.** The number of protons in an atom determines the atom's weight (mass) and its atomic number. For example, a hydrogen (H) atom has the fewest protons of all atoms, has the least amount of mass, and is assigned the atomic number of one.

In the 18th century, Benjamin Franklin used the terms positive (+) and negative (–) to describe an electrical charge. A *positive charge* is an electrical charge produced when there are fewer electrons than normal. A *negative charge* is an electrical charge produced when there are more electrons than normal. A proton of an atom has a positive (+) charge, the electron has a negative (–) charge, and the neutron has no charge. The neutron and proton combine to form the nucleus of an atom. Since the neutron has no charge, the nucleus has a positive (+) charge.

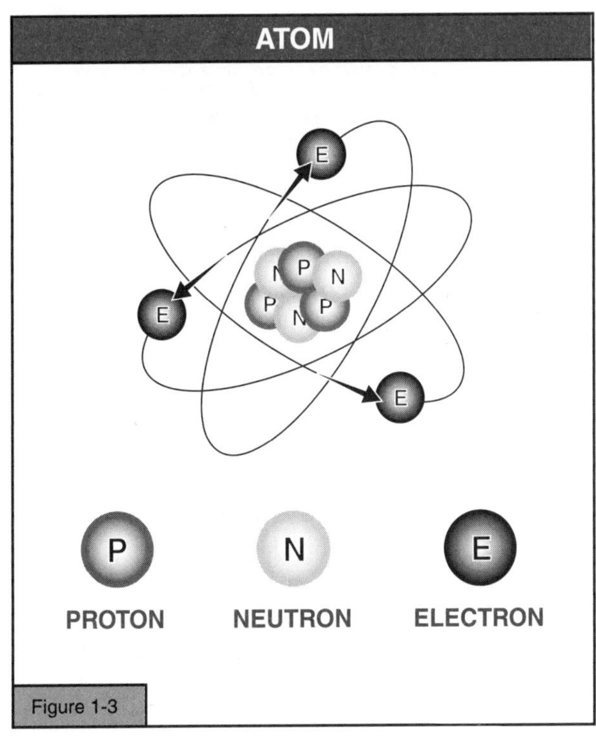

Figure 1-3

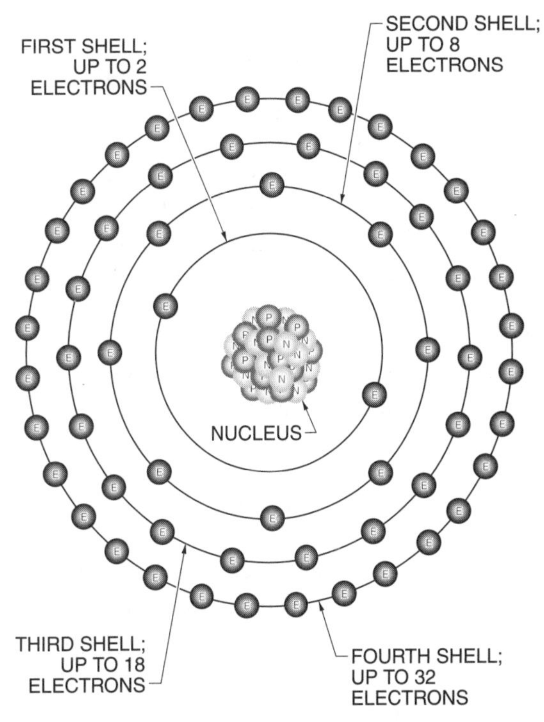

ELECTRON CONFIGURATION		
Shell Number	**Shell Letter**	**Maximum Number of Electrons**
1	K	2
2	L	8
3	M	18
4	N	32
5	O	50
6	P	72
7	Q	98

Figure 1-4

The electrons orbit the nucleus of an atom, completing billions of trips around the nucleus each millionth of a second. In all atoms the electrons are arranged in shells at various distances from the nucleus according to the amount of energy they have. The total number of shells an atom has varies from one to seven shells. The shells are numbered innermost to outermost 1, 2, 3, 4, 5, 6, and 7 or are lettered K, L, M, N, O, P, and Q. **See Figure 1-4.**

Each shell can hold only a specific number of electrons. The innermost shell can hold two electrons, the second shell can hold eight electrons, the third shell can hold 18 electrons, the fourth shell can hold 32 electrons, the fifth shell can hold 50 electrons, the sixth shell can hold 72 electrons, and the seventh shell can hold 98 electrons. The shells are filled starting with the inner shells and working outward, so that when the inner shell is filled with as many electrons as it can hold, the next shell is started. However, the outer shells are never completely filled. Most materials used in the electrical/electronic field contain four shells or less.

TECH FACT *An atomic symbol of an element is derived from the first one or two letters of its English name, though in some cases it is based on a Latin name.*

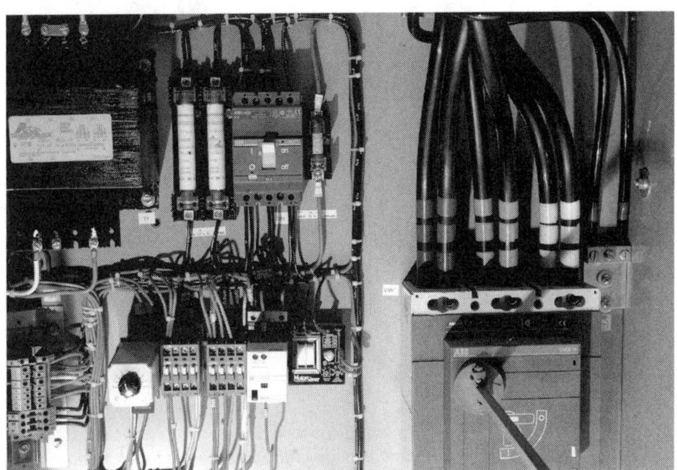

Insulation used on electrical conductors is an example of a material that does not allow valence electrons to move from atom to atom.

Valence Electrons

The outer shell of an atom includes electrons and is called the valence shell. A *valence shell* is the outermost shell of an atom and contains the electrons that form new compounds. For example, two hydrogen atoms combine with the electrons in the outer shell of an oxygen atom to form water. The electrons in the valence shell are important because they can be used to produce an electric current flow.

Most elements do not have a completed valence shell containing the maximum allowable number of electrons. The number of electrons in the valence shell determines whether an element allows electrons to easily move from atom to atom or resists the flow of electrons. A material that allows electrons to easily move from atom to atom can be used as a conductor. For example, copper is a good conductor because copper atoms allow their valence electrons to move from atom to atom. **See Figure 1-5.**

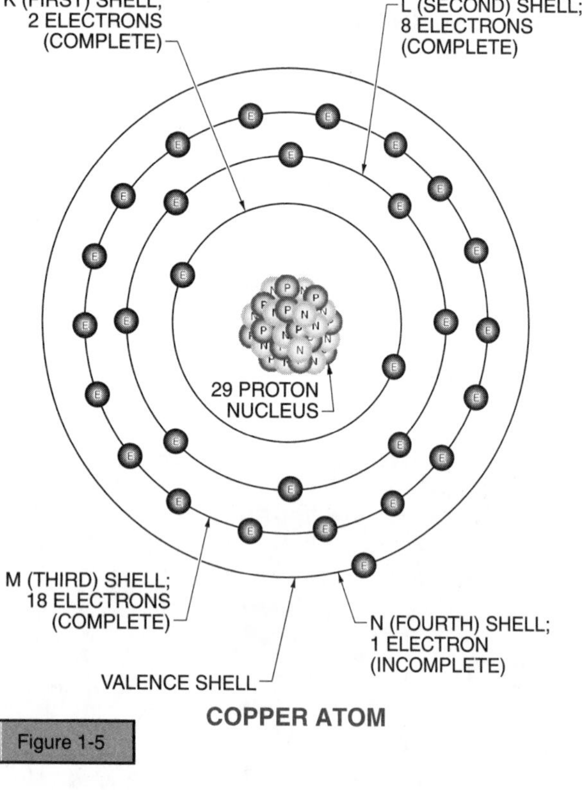

Figure 1-5

TECH TIP *Individual conductors must be insulated and protected per NEC® requirements for maximum safety.*

TECH FACT *In 1869, Dmitri Mendeleev developed the periodic table by arranging known elements in a particular way, with elements in the same column displaying similar properties.*

A material that does not allow electrons to easily move from atom to atom can be used as an insulator. For example, chlorine is a good insulator because chlorine atoms do not allow their valence electrons to easily move from atom to atom. However, most insulators used in the electrical field are made from compounds such as glass, rubber, plastic, or paper. Insulators have five or more valence electrons.

Conductors have three or less valence electrons. Most conductors have only one or two valence electrons. Most metals are good conductors. Silver is the best conductor, followed by copper, gold, aluminum, and iron.

A *semiconductor* is an electronic device that has electrical conductivity between that of a conductor (high conductivity) and that of an insulator (low conductivity). Semiconductors (carbon, germanium, and silicon) are made from materials that have exactly four valence electrons. Semiconductor materials are not conductors and not insulators. Semiconductor materials do not conduct electricity easily and are not good insulators.

In a conductor atom, an outside force can be applied to force the atom to lose or gain valence electrons. Electrons can be forced to move by chemical reaction, friction, pressure, heat, light, or a magnetic field. These forces can occur naturally, as with lightning and static electricity, or they can be produced, as in a battery or generator. **See Figure 1-6.**

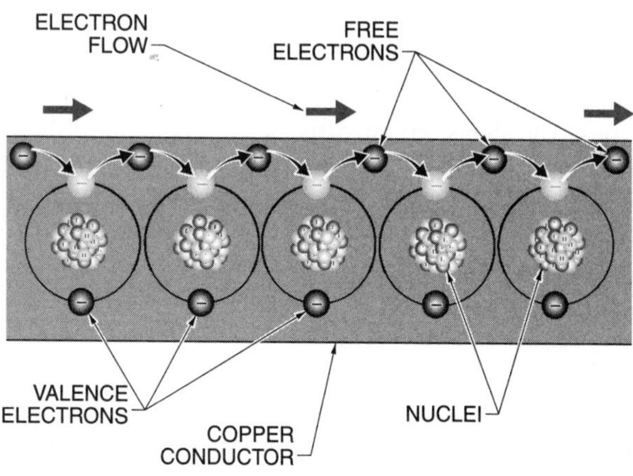

Figure 1-6

Elements

Elements are the basic materials that make up all matter. An *element* is a substance that cannot be chemically broken down and contains atoms of only one variety. All solids, liquids, and gases are made up of elements. Some matter may be made up of only one element, but most matter is made up of more then one element. For example, a pure copper kettle is made up of the copper (Cu) element and the water inside the kettle is made up of hydrogen (H) and oxygen (O) elements. Oxygen is the most plentiful element on Earth. **See Figure 1-7.**

There are 109 elements, 92 of which are natural. The others are artificial (manufactured). The number of elements increases as more elements are discovered and developed. Each element has a symbol that is an abbreviation for the element. The abbreviations for elements are internationally recognized and commonly used. The symbols of elements are used to identify material composition and to write formulas. **See Figure 1-8. See Appendix.**

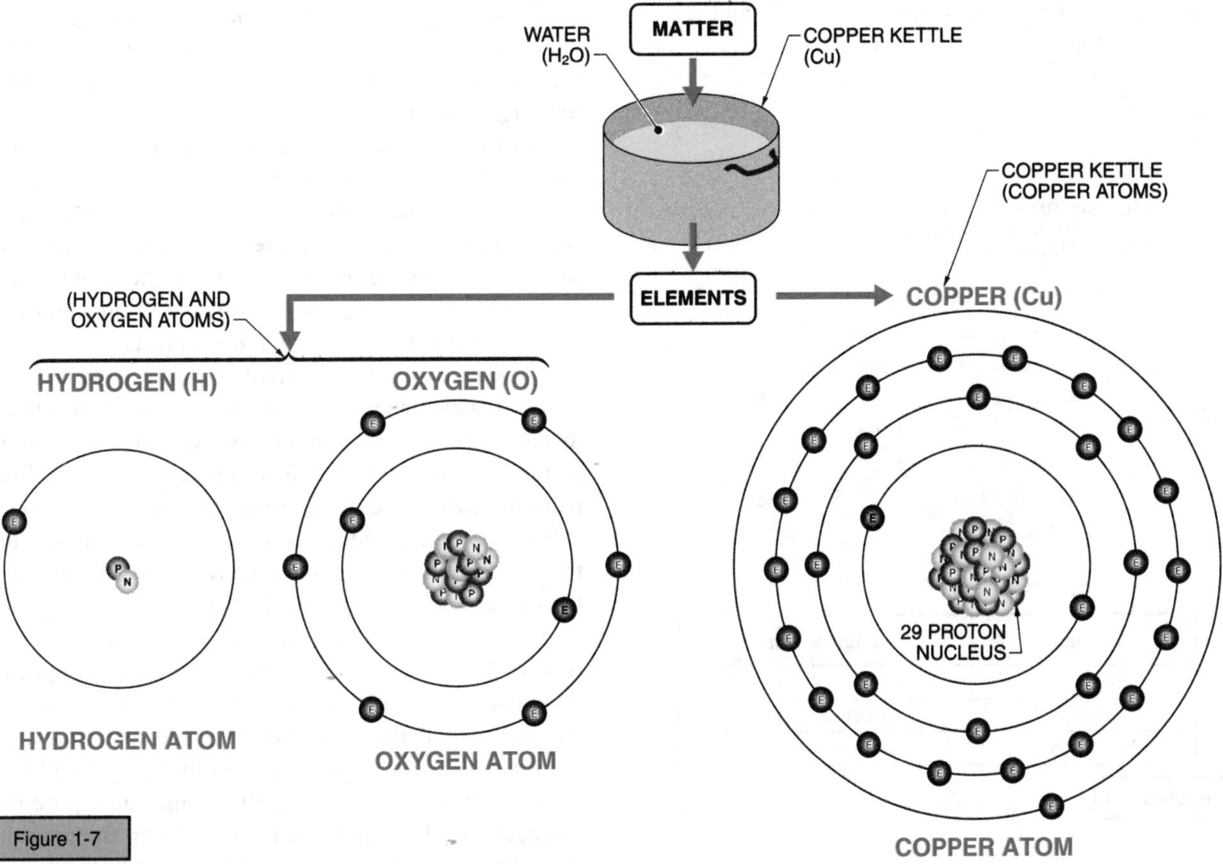

Figure 1-7

CHEMICAL ELEMENTS							
Name	Symbol	Atomic Weight*	Atomic Number	Name	Symbol	Atomic Weight*	Atomic Number
Actinium	Ac	[227]	89	Neon	Ne	20.183	10
Aluminum	Al	26.9815	13	Neptunium	Np	[237]	93
Americium	Am	[243]	95	Nickel	Ni	58.71	28
Antimony	Sb	121.75	51	Niobium	Nb	92.906	41
Argon	Ar	39.948	18	Nitrogen	N	14.0067	7
Arsenic	As	74.9216	33	Nobelium	No	[255]	102
Astatine	At	[210]	85	Osmium	Os	190.2	76
[illegible]	[illegible]	137.34	56	[illegible]	[illegible]	[illegible]	8

Figure 1-8

Solid elements, such as iron (Fe), copper (Cu), aluminum (Al), silver (Ag), and gold (Au), are of special interest in the electrical field because they are conductors. Gaseous elements, such as oxygen (O), helium (He), mercury-vapor (Hg), scandium (Sc), argon (Ar), and neon (Ne), are also of special interest because they are used in specific applications. For example, neon lights are lamps that contain neon, helium, or argon gas under low pressure. Depending upon which gas is used and the color of the gas tube, many different-colored light outputs are possible. Neon is used to produce red and orange light. Argon is used to produce blue and green light. Helium is used to produce gold and white light. **See Figure 1-9.**

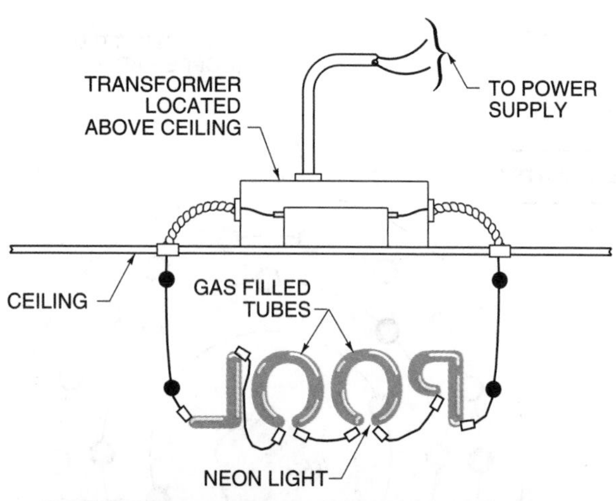

Element	Color Output
Neon (Ne)	Reds and Oranges
Argon (Ar)	Blues and Greens
Helium (He)	Golds and Whites

Figure 1-9

The ability to distinguish symbols in the electrical field is important because symbols are used on equipment and for ordering components. For example, conductors are commonly made of copper, aluminum, and copper-clad aluminum. Copper and aluminum are the most commonly used materials. Copper-clad aluminum is a conductor that has copper bonded to an aluminum core. Some electrical equipment is not designed to be used with an aluminum conductor. This equipment is marked "CU" and "CU-CLAD." Verifying that a conductor type matches equipment type assures safe and reliable electrical installation. **See Figure 1-10.**

In addition to using symbols when installing equipment, symbols are also used when ordering equipment, such as switching equipment. Electrical contacts are used to switch (start and stop) the flow of electricity in devices such as switches and relays. The material the contacts are made of determines how effectively the contacts operate.

Relay contacts are available in several different materials. Material such as silver, silver-cadmium, gold-plated silver, and tungsten are typical elements used to make electrical contacts. The material used is determined by the application. If the wrong material is used, the equipment wears prematurely, causing equipment malfunction. The appropriate material should be specified when ordering a contact for a given application. **See Figure 1-11.**

Most contacts include silver, because silver has the highest electrical conductive property of all materials. However, pure silver sticks, welds (from arcing), and is subject to sulfidation when used for many applications. *Sulfidation* is the formation of film on the contact surface. Sulfidation increases the resistance (opposition) to the flow of electricity through the contacts.

To reduce sulfidation, silver is alloyed with other metals. For example, silver is alloyed with cadmium (Cd) to produce a silver-cadmium contact. Silver-cadmium contacts have good electrical characteristics and low resistance, which helps the contact resist arcing but not sulfidation. Silver-cadmium contacts are used in circuits that switch higher currents at more than 12 V, because the higher current burns off the sulfidation.

Some electrical contacts are used in applications that require frequent switching; others are used in applications that require very little switching. When contacts are seldom switched, sulfidation can build up and damage the silver contacts. To prevent this problem, gold-plated silver contacts are used in applications that require infrequent switching to minimize sulfidation and provide a good electrical connection. However, gold-plated contacts are not used in high-current applications because the gold burns off quickly at high currents.

Tungsten contacts are used in high-voltage applications because tungsten has a high melting temperature and is less affected by arcing. Tungsten contacts are also used when high repetitive switching is required.

TECH TIP

Gold-plated silver contacts should be used in applications that require intermittent switching to provide electrical conduction and reduce sulfidation.

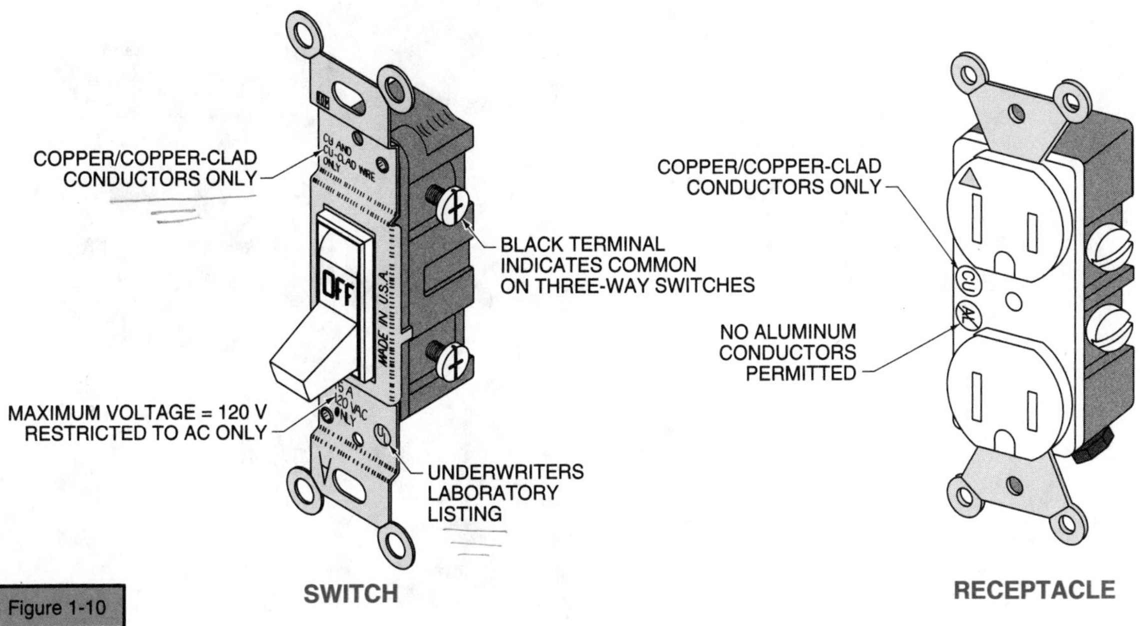

Figure 1-10

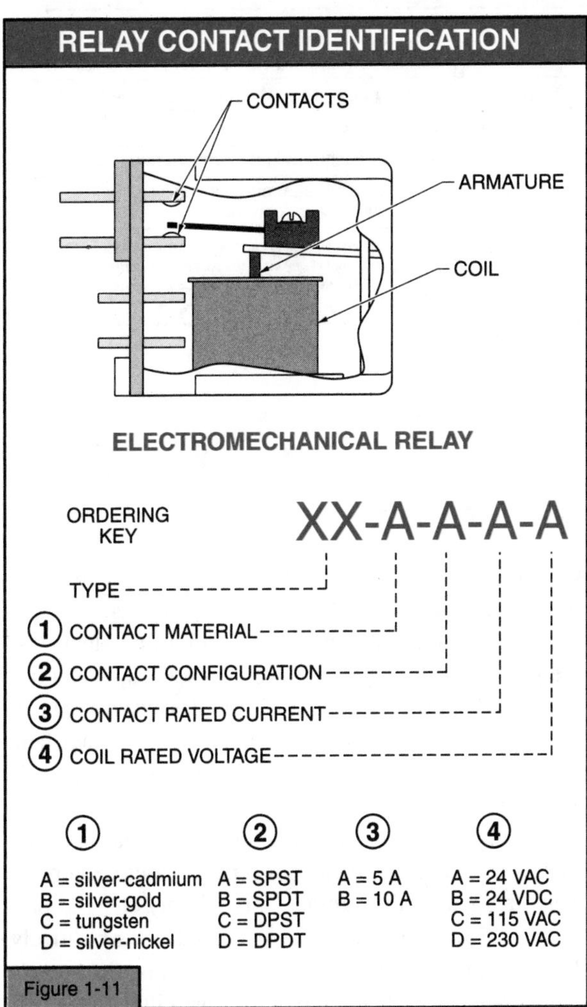

Figure 1-11

Compounds

A *molecule* is the smallest particle that a compound can be reduced to and still possess the chemical properties of the original compound. If the compound were divided further, only atoms of elements would remain. For example, a drop of water contains billions of water molecules. If the drop of water were divided until only a single water molecule remained, that final molecule would still possess all the chemical properties of water. However, if the one molecule of water were further divided, only two hydrogen atoms and one oxygen atom would remain.

A *compound* is a combination of the atoms of two or more elements. Every compound has a definite composition described by a chemical formula. There are many more compounds than there are elements. This is because elements can be combined to produce materials having characteristics that are completely different from the original elements. Water (H_2O) is an example of a simple compound composed of hydrogen and oxygen elements. Water consists of two atoms of hydrogen linked to one atom of oxygen. The chemical formula for water is written H_2O because there are exactly twice as many hydrogen atoms as oxygen atoms in any sample of water. **See Figure 1-12.**

There are only 109 elements, but these elements can be combined in many different ways to form millions of compounds. Elements can even be combined in different proportions to produce different compounds. For example, the elements of carbon and hydrogen can combine to form methane (CH_4) or propane (C_3H_8).

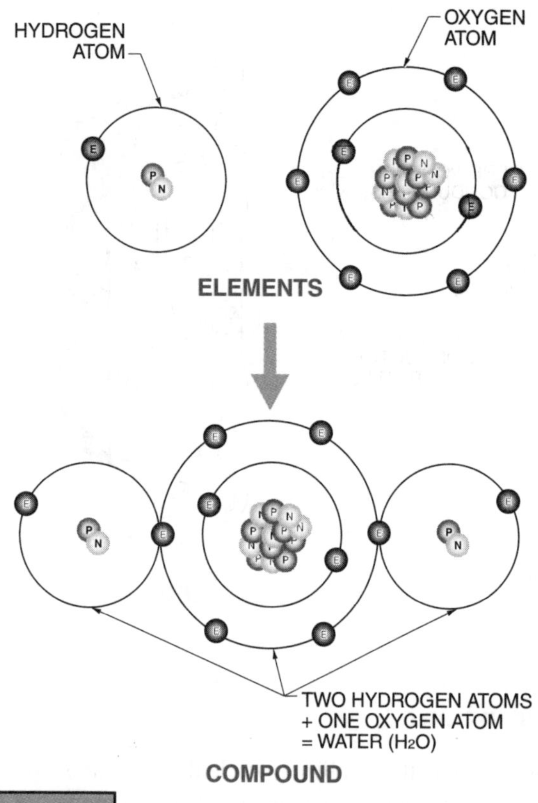

Figure 1-12

When some elements are combined, the properties of the compound may differ considerably from that of the individual elements. For example, the element sodium (Na) is a soft metal that reacts violently with water, and the element chlorine (Cl) is a poisonous gas. Common table salt (NaCl) is formed when the elements sodium and chlorine are combined.

Elements such as carbon (C), silicon (Si), and germanium (Ge) are combined with other elements to produce specific compounds that have unique electrical characteristics. For example, carbon is used to make electrical resistors. Germanium and silicon are used to make electronic components such as diodes and transistors.

The electrical properties of compounds are important in water treatment, gas and oil refining, food processing, chemical and pharmaceutical manufacturing, medical applications, and research. In such areas, low-voltage electricity is passed through the compounds and used to take measurements that provide information about the product being tested. For example, a carbon monoxide sensor can be used to troubleshoot a malfunctioning furnace. The carbon monoxide sensor displays the amount of carbon monoxide (CO) in the air in parts per million. **See Figure 1-13.**

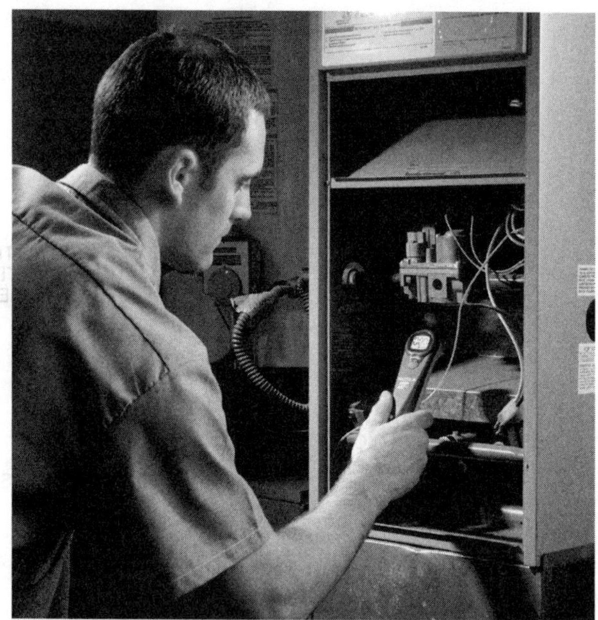

Fluke Corporation

CARBON MONOXIDE (CO) LEVEL STANDARDS*	
0–1 PPM	Normal background levels
9 PPM	ASHRAE Standard 62-1989 for living areas
50 PPM	OSHA enclosed space 8-hour average level†
100 PPM	OSHA exposure limit†
200 PPM	Mild headache, fatigue, nausea, and dizziness
800 PPM	Dizziness, nausea, and convulsions. Death within 2 to 3 hours

† U.S. Department of Labor, Occupational Safety and Health Administration (OSHA) Regulation 1917.24: The CO content in any enclosed space shall be maintained at not more than 50 PPM (0.005%). Remove employees from enclosed space if the CO concentration exceeds 100 PPM (0.01%).

* in parts per million (ppm)

Figure 1-13

Law of Electric Charges

Although an atom is normally electrically neutral because there are as many electrons (–) in it as there are protons (+), an atom can lose or gain a few electrons. The gaining or losing of electrons produces an electric charge in the atom. All charged particles exert forces on one another, even if they are not in physical contact. The exerted force is due to the electric field that surrounds all charged particles.

The *law of charges* is a law that states that opposite charges attract and like charges repel. Therefore, a positively charged particle attracts a negatively charged particle, a positively charged particle repels another positively charged particle, and a negatively charged particle repels another negatively charged particle. **See Figure 1-14.**

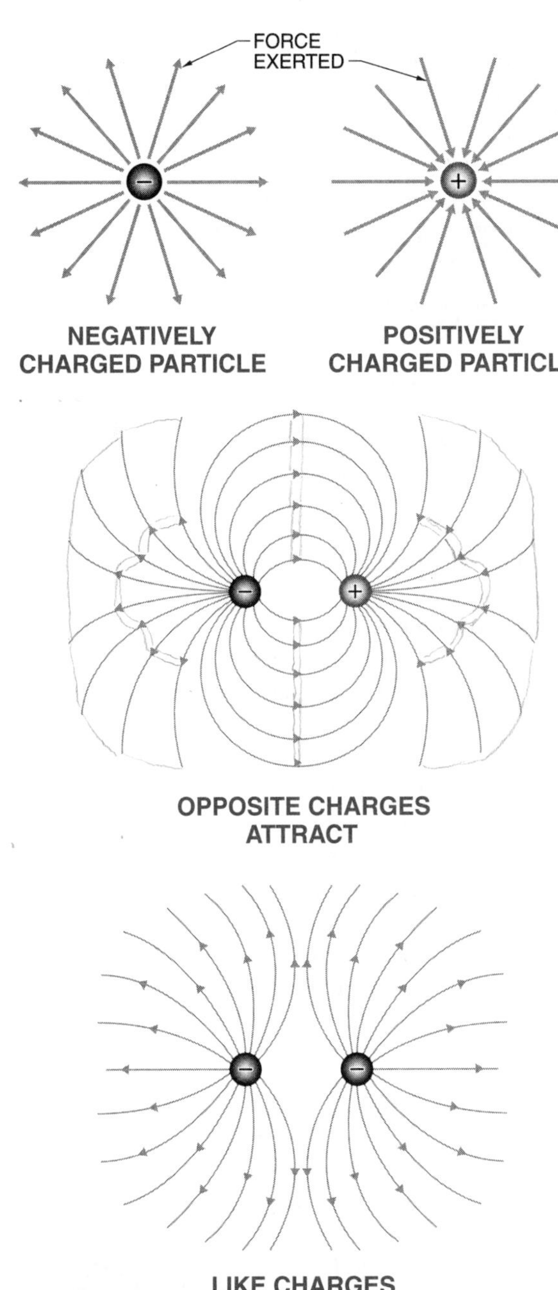

Figure 1-14

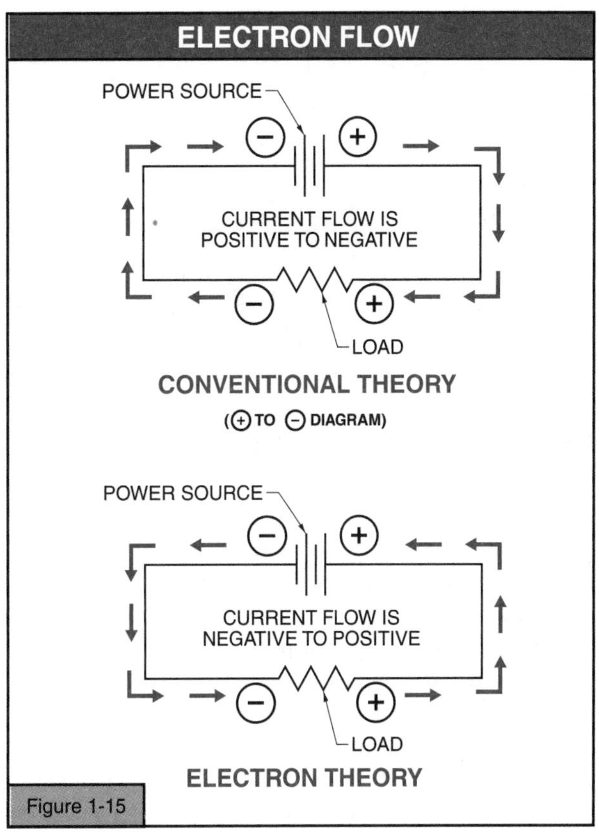

Figure 1-15

Electron Flow

Early scientists believed that electrons flowed from positive to negative. Later, when atomic structure was studied, electron flow from negative to positive was introduced. The two different theories are called conventional current flow and electron current flow. *Conventional current flow* is current flow from positive to negative. *Electron current flow* is current flow from negative to positive. **See Figure 1-15.**

Both conventional current flow and electron current flow are still used. Conventional current flow is used more in the electrical field and by electrical engineers to aid in explaining electrical circuit properties. Conventional current flow is also widely used in the small engine, automotive, marina, and aviation fields. Electron current flow is used more in the electronic semiconductor field to assist in explaining the operation of solid-state electronic components.

Electricity Production

Potential energy is stored energy a body has due to its position, chemical state, or condition. Solar cells can convert solar power directly into electricity. However, converting other potential energy into a force that turns a generator produces the vast majority of electricity. The major forms of energy used to produce electricity through a generator include fossil fuels (coal, natural gas, and oil), nuclear power, and hydroelectric (water) power. **See Figure 1-16.**

A *fossil fuel* is a fuel formed by plant and animal remains taken from the ground. Available quantities of fossil fuel are limited. These natural resources cannot be formed as quickly as they are being depleted.

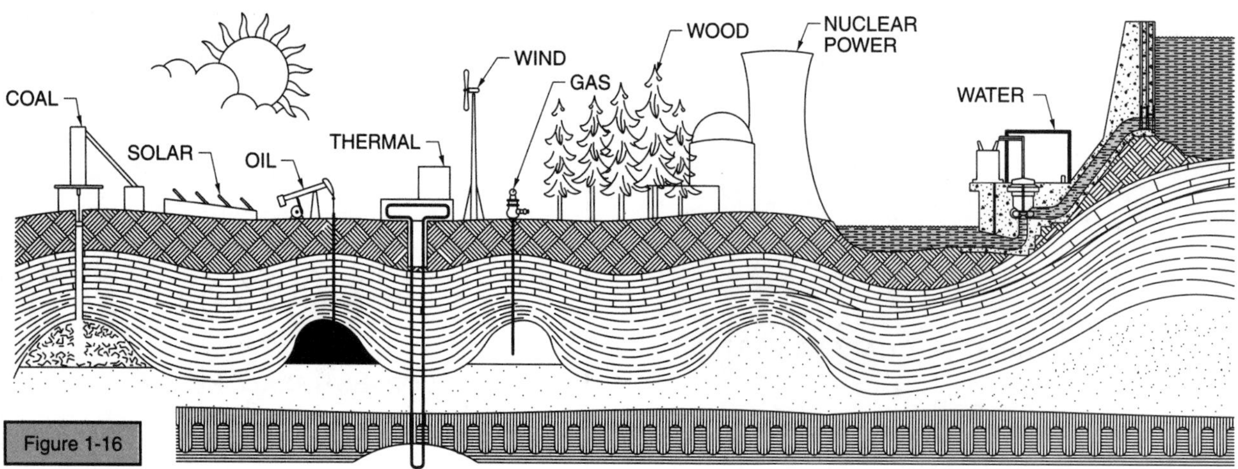

Figure 1-16

Coal is the most widely used fossil fuel. It is burned to produce the steam necessary to drive the generators that produce electricity. The United States produces about 600,000,000 t (tons) of coal per year. The production of electricity requires approximately 350,000,000 t per year. The balance of the coal is used to produce heat.

The cost of coal is relatively lower than oil, based on Btu per unit and the fact that the coal consumed in the United States is mined in the United States. A *British thermal unit (Btu)* is the amount of heat required to raise the temperature of 1 lb of water 1°F. Most of the oil used in the United States is transported from other countries, which increases its price.

Natural gas is found in the earth along with oil. Natural gas is an efficient form of energy used to produce heat. Natural gas and oil are transported overland by pipeline. Oil is transported overseas in huge tankers. Since natural gas and oil are not as abundant or as inexpensive as coal, the generators that use them to produce electricity are used only when the demand for electricity is great. Electricity is in greatest demand between 4:00 PM and 6:00 PM.

Nuclear power is created through a carefully controlled nuclear fission process. *Nuclear fission* is the splitting of atomic nuclei, resulting in the release of large amounts of energy. Nuclear fission creates heat that is used to transform water into steam. The steam is used to drive a generator and produce electricity.

The primary use of nuclear power is for the production of electricity. Uranium 235 (^{235}U) is the major element used for nuclear fission. The extraction of uranium is expensive because of the precautions taken to protect the miners from radiation. The United States has a large reserve of uranium that is estimated to be approximately one-fourth of the world's supply.

Approximately 40% of the electricity generated in the United States is used in industry, 34% in residences, and 26% in commercial applications. The increased demand for electricity has resulted in a greater reliance on nuclear power for generating electricity.

Hydroelectric power uses the force of water created by a dam in the production of electricity. The force behind the moving water drives generators that produce electricity. The large amounts of water in reservoirs behind dams put some restrictions on locations for power stations. The use of hydroelectric power has increased because little waste is created and the process is fairly clean. **See Figure 1-17.**

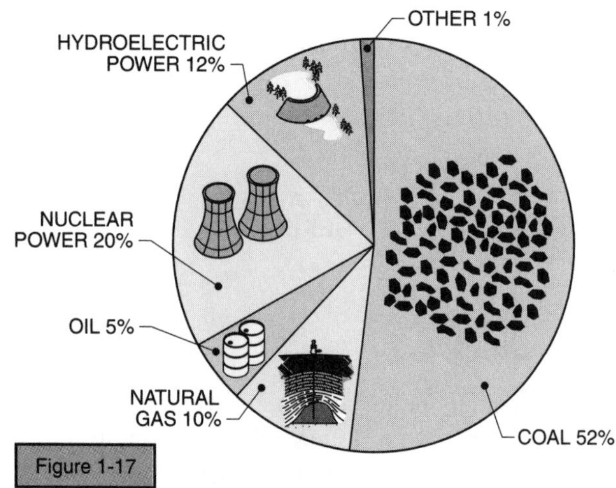

Figure 1-17

Refer to the CD-ROM in the back of the book for Chapter 1 Quick Quiz® and related information.

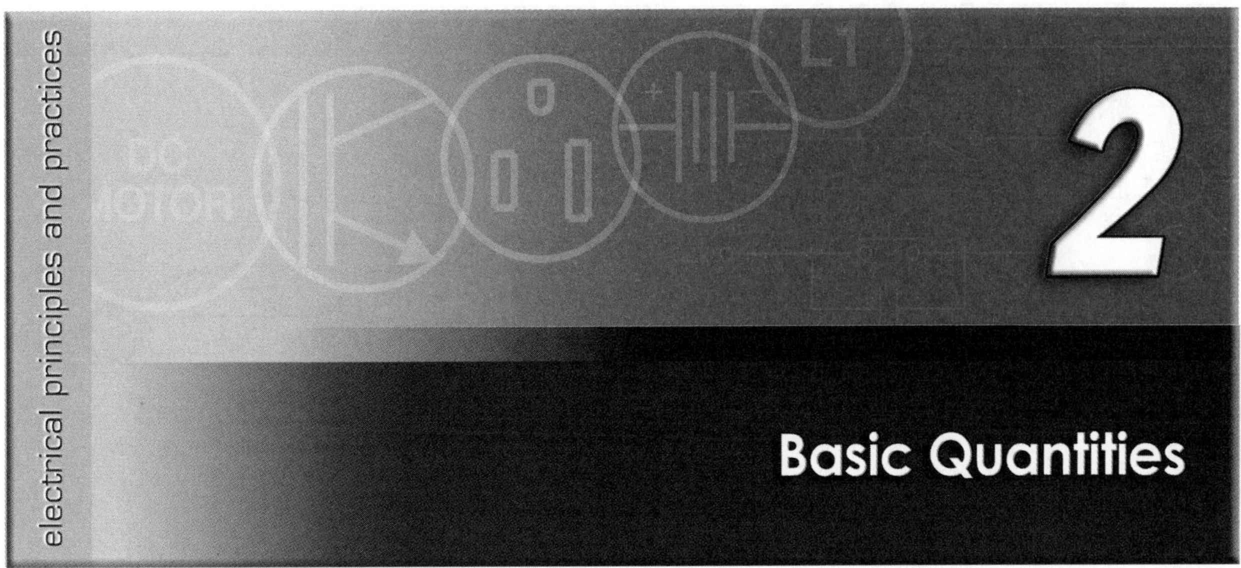

2 Basic Quantities

Energy

Energy is the capacity to do work. The two forms of energy are potential energy and kinetic energy. *Potential energy* is stored energy a body has due to its position, chemical state, or condition. For example, water behind a dam has potential energy because of the position of the water. A compressed spring has potential energy because of its condition. Gasoline has potential energy based on its chemical state. **See Figure 2-1.**

Kinetic energy is the energy of motion. Examples of kinetic energy include falling water, a released spring, a speeding automobile, etc. Kinetic energy is released potential energy. Energy released when water falls through a dam is used to generate electricity for residential, commercial, and industrial use. Energy released by a compressed spring is used to apply a braking force on a motor shaft. Energy released by the burning of gasoline is used to propel a vehicle forward and drive the alternator that produces the electrical power.

The major sources of energy are coal, oil, gas, wood, and nuclear power. Solar, wind, water, and thermal sources also provide energy. These energy forms are used to produce work when converted to electricity, steam, heat, and mechanical force. Some energy sources, such as coal, oil, and gas, cannot be replaced. They are consumed when used. Other energy sources, such as solar power, wind, and water, are not consumed when used.

Electricity is produced by converting potential energy directly or indirectly into electricity. Solar cells convert solar power directly into electricity. Most electricity is produced by converting potential energy into a force that turns a generator. Energy used to produce electricity through a generator include fossil fuels (coal, natural gas, and oil), nuclear power, and hydroelectric power.

Voltage

All electrical circuits must have a source of power to produce work. The source of power used depends on the application and the amount of power required. All sources of power produce a set voltage level or voltage range.

Voltage (E) is the amount of electrical pressure in a circuit. Voltage is measured in volts (V). Voltage is also known as electromotive force (EMF) or potential difference. Voltage may be produced by electromagnetism (generators), chemicals (batteries), light (photocells), heat (thermocouples), pressure (piezoelectricity), or friction (static electricity). **See Figure 2-2.**

Voltage is produced any time there is an excess of electrons at one terminal of a voltage source and a deficiency of electrons at the other terminal. The greater the difference in electrons between the terminals, the higher the voltage.

Electromagnets use a combination of voltage and magnetism to move large amounts of ferrous materials.

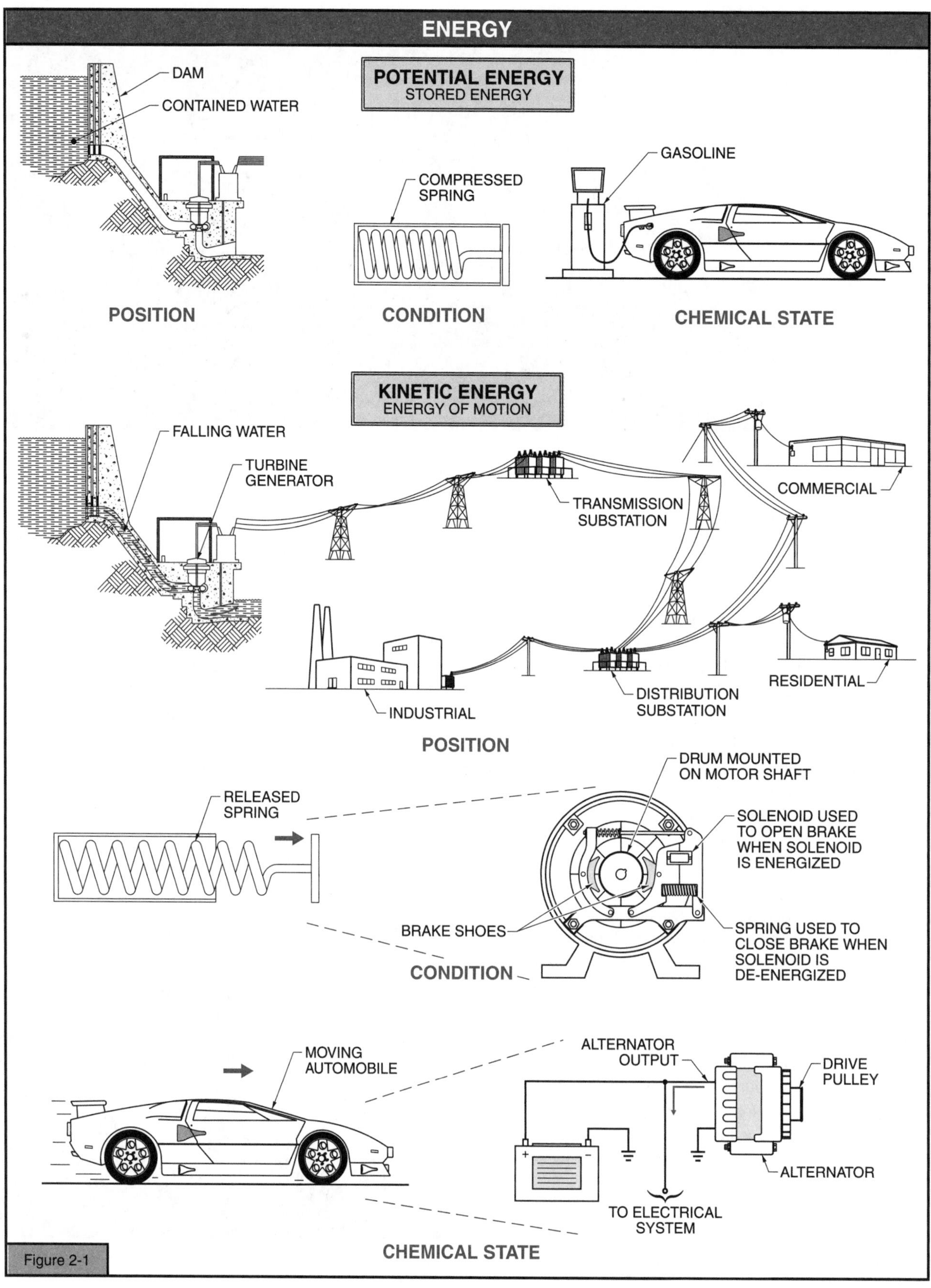
ENERGY
POTENTIAL ENERGY
STORED ENERGY
DAM
CONTAINED WATER
POSITION
COMPRESSED SPRING
CONDITION
GASOLINE
CHEMICAL STATE
KINETIC ENERGY
ENERGY OF MOTION
FALLING WATER
TURBINE GENERATOR
TRANSMISSION SUBSTATION
COMMERCIAL
INDUSTRIAL
DISTRIBUTION SUBSTATION
RESIDENTIAL
POSITION
RELEASED SPRING
DRUM MOUNTED ON MOTOR SHAFT
SOLENOID USED TO OPEN BRAKE WHEN SOLENOID IS ENERGIZED
BRAKE SHOES
SPRING USED TO CLOSE BRAKE WHEN SOLENOID IS DE-ENERGIZED
CONDITION
MOVING AUTOMOBILE
ALTERNATOR OUTPUT
DRIVE PULLEY
ALTERNATOR
TO ELECTRICAL SYSTEM
CHEMICAL STATE
Figure 2-1

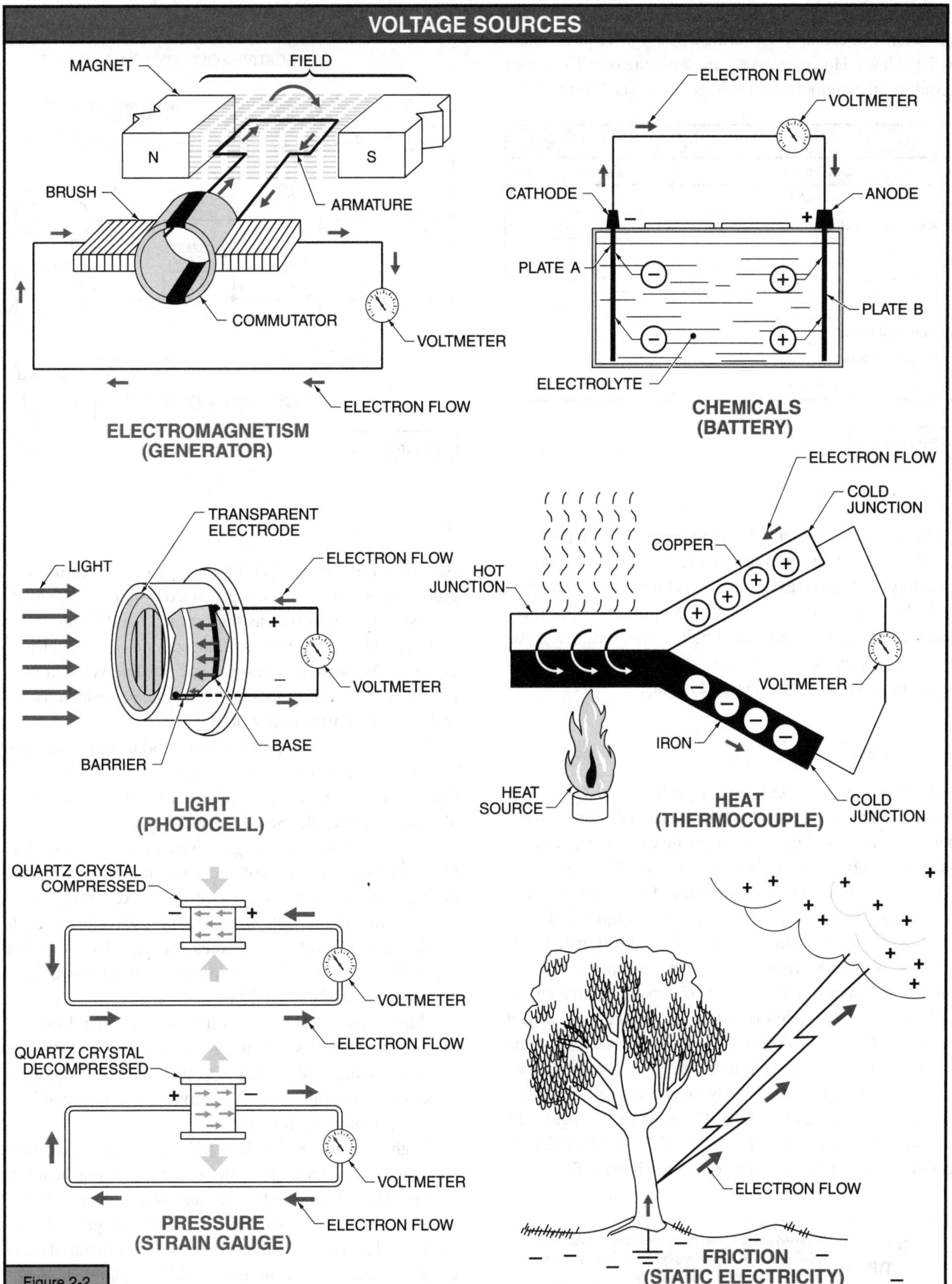

VOLTAGE SOURCES

MAGNET
FIELD
N
S
BRUSH
ARMATURE
COMMUTATOR
VOLTMETER
ELECTRON FLOW

ELECTROMAGNETISM (GENERATOR)

ELECTRON FLOW
VOLTMETER
CATHODE
ANODE
PLATE A
PLATE B
ELECTROLYTE

CHEMICALS (BATTERY)

TRANSPARENT ELECTRODE
LIGHT
ELECTRON FLOW
VOLTMETER
BASE
BARRIER

LIGHT (PHOTOCELL)

ELECTRON FLOW
COLD JUNCTION
COPPER
HOT JUNCTION
VOLTMETER
IRON
HEAT SOURCE
COLD JUNCTION

HEAT (THERMOCOUPLE)

QUARTZ CRYSTAL COMPRESSED
VOLTMETER
ELECTRON FLOW
QUARTZ CRYSTAL DECOMPRESSED
VOLTMETER
ELECTRON FLOW

PRESSURE (STRAIN GAUGE)

ELECTRON FLOW

FRICTION (STATIC ELECTRICITY)

Figure 2-2

The amount of voltage in a circuit depends on the application. For example, powering a flashlight requires a low voltage level. However, producing a picture on a TV screen requires an extremely high voltage level. **See Figure 2-3.**

COMMON VOLTAGE LEVELS	
Device	**Level***
Flashlight battery (AAA, AA, C, D)	1.5
Automobile battery	12
Golf cart	36
Refrigerator, TV, VCR	115
Central air conditioner	230
Industrial motor	460
TV picture tube	25,000
High-tension power line	up to 500,000

* in V

Figure 2-3

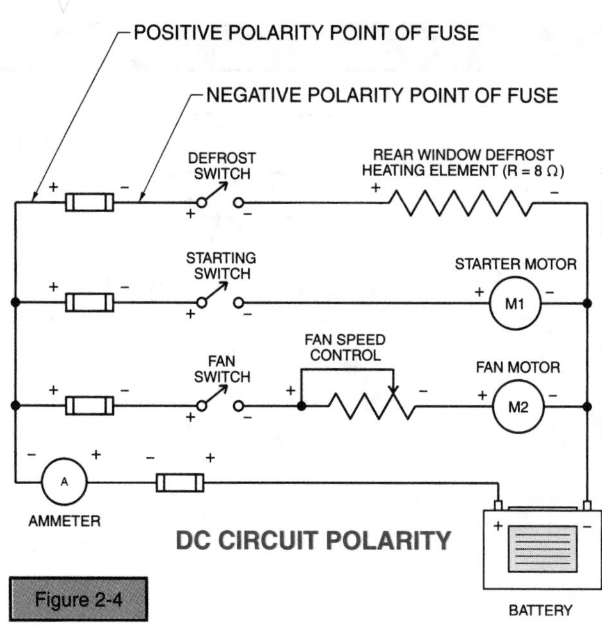

Figure 2-4

Voltage Types

Voltage is either direct (DC) or alternating (AC). *DC voltage* is voltage that flows in one direction only. *AC voltage* is voltage that reverses its direction of flow at regular intervals. DC voltage is used in almost all portable equipment (automobiles, golf carts, flashlights, cameras, etc.). AC voltage is used in residential, commercial, and industrial lighting and power distribution systems.

DC Voltage

All DC voltage sources have a positive and a negative terminal. The positive and negative terminals establish polarity in a circuit. *Polarity* is the positive (+) or negative (–) state of an object. All points in a DC circuit have polarity. The most common power sources that directly produce DC voltage are batteries and photocells. **See Figure 2-4.**

In addition to obtaining DC voltage directly from batteries and photocells, DC voltage is also obtained from a rectified AC voltage supply. DC voltage is obtained any time an AC voltage is passed through a rectifier. A *rectifier* is a device that converts AC voltage to DC voltage by allowing the voltage and current to move in only one direction. DC voltage obtained from a rectified AC voltage supply varies from almost pure DC voltage to half-wave DC voltage. Common DC voltage levels include 1.5 V, 3 V, 6 V, 9 V, 12 V, 24 V, 36 V, 125 V, 250 V, 600 V, 1200 V, 1500 V, and 3000 V. **See Figure 2-5.**

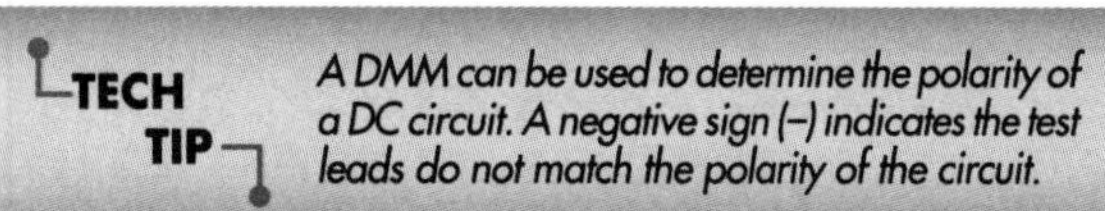

A DMM can be used to determine the polarity of a DC circuit. A negative sign (–) indicates the test leads do not match the polarity of the circuit.

AC Voltage

AC voltage is the most common type of voltage used to produce work. AC voltage is produced by generators which produce AC sine waves as they rotate. An *AC sine wave* is a symmetrical waveform that contains 360 electrical degrees. The wave reaches its peak positive value at 90°, returns to 0 V at 180°, increases to its peak negative value at 270°, and returns to 0 V at 360°.

A *cycle* is one complete positive and negative alternation of a wave form. An *alternation* is half of a cycle. A sine wave has one positive alternation and one negative alternation per cycle. **See Figure 2-6.**

AC voltage is either single-phase (1ϕ) or three-phase (3ϕ). Single-phase AC voltage contains only one alternating voltage waveform. Three-phase AC voltage is a combination of three alternating voltage waveforms, each displaced 120 electrical degrees (one-third of a cycle) apart. Three-phase voltage is produced when three coils are simultaneously rotated in a generator.

Almost any level of AC voltage is available. Low AC voltages (6 V to 24 V) are used for doorbells and security systems. Medium AC voltages (110 V to 120 V) are used in residential applications for lighting, heating, cooling, cooking, running motors, etc.

High AC voltages (208 V to 480 V) are used in commercial applications for large-scale cooking, lighting, heating, cooling, etc., as required by restaurants, office buildings, shopping malls, schools, etc. High AC voltages are also used in industrial applications to convert raw materials into usable products, in addition to providing lighting, heating, and cooling for plant personnel.

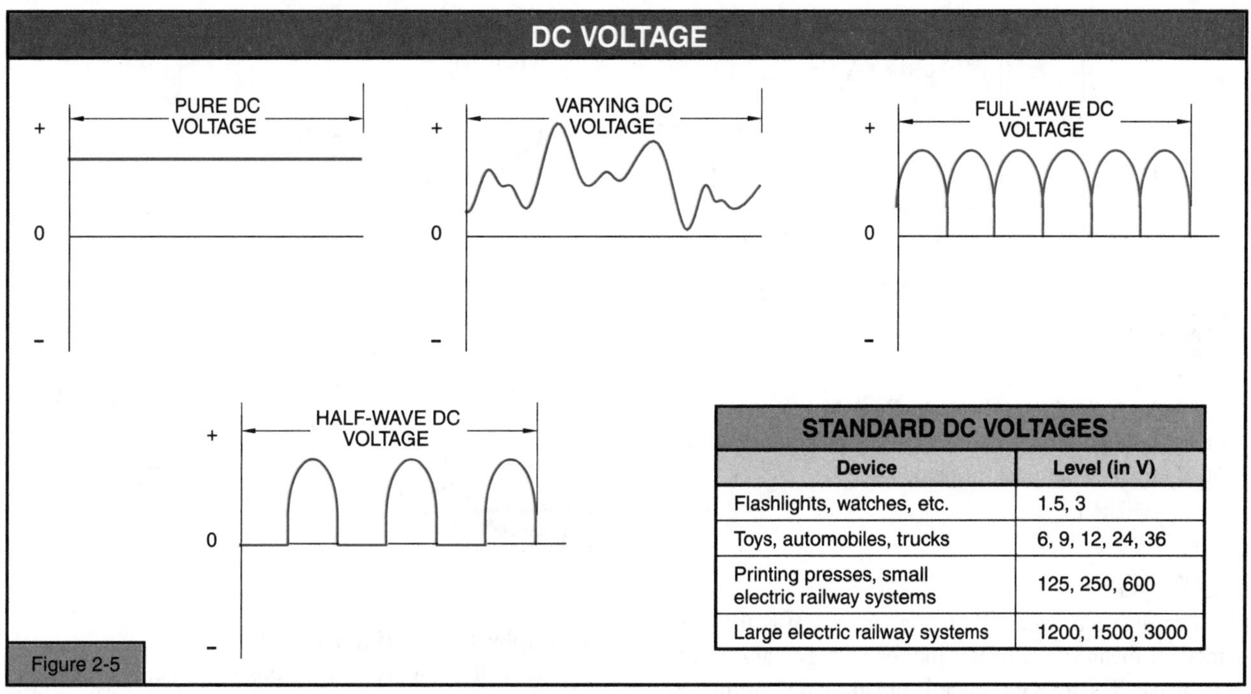

STANDARD DC VOLTAGES	
Device	**Level (in V)**
Flashlights, watches, etc.	1.5, 3
Toys, automobiles, trucks	6, 9, 12, 24, 36
Printing presses, small electric railway systems	125, 250, 600
Large electric railway systems	1200, 1500, 3000

Figure 2-5

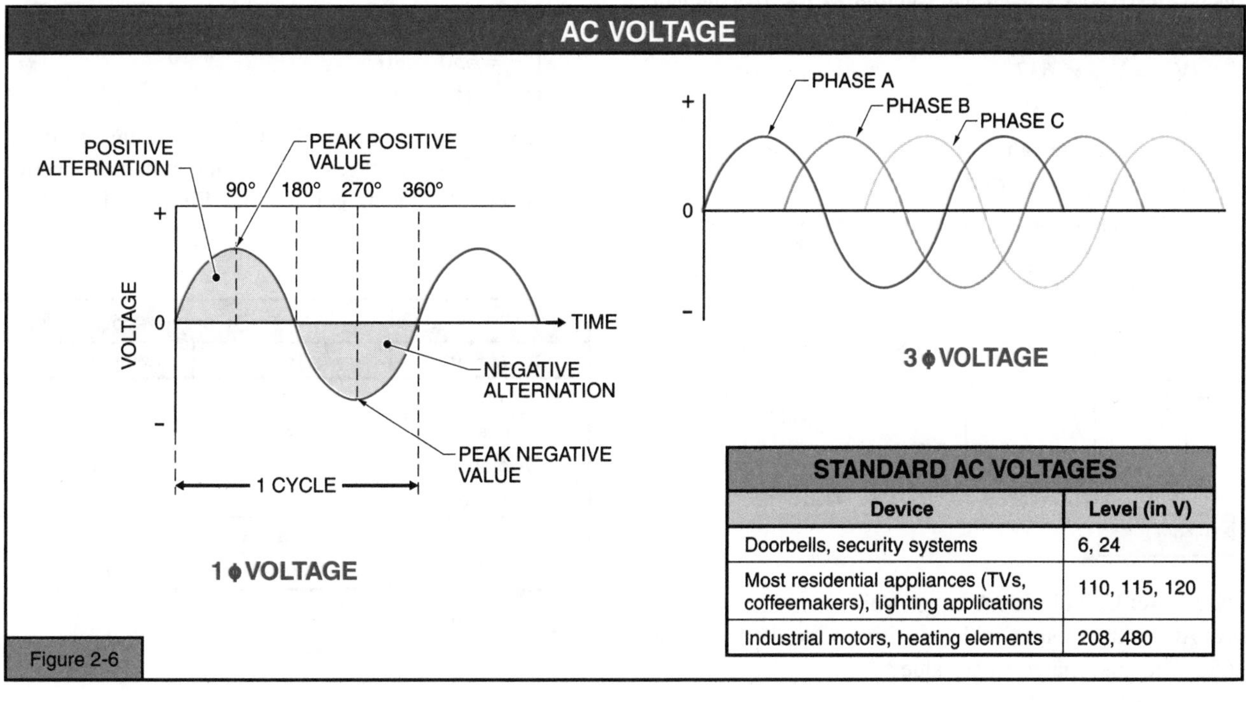

STANDARD AC VOLTAGES	
Device	**Level (in V)**
Doorbells, security systems	6, 24
Most residential appliances (TVs, coffeemakers), lighting applications	110, 115, 120
Industrial motors, heating elements	208, 480

Figure 2-6

AC Voltage Values

AC voltage is stated and measured as peak, average, or rms values. The peak value (V_{max}) of a sine wave is the maximum value of either the positive or negative alternation. The positive and negative alternation are equal in a sine wave. The peak-to-peak value (V_{p-p}) is the value measured from the maximum positive alternation to the maximum negative alternation. **See Figure 2-7.** To calculate peak-to-peak value, apply the formula:

$$V_{p\text{-}p} = 2 \times V_{max}$$

where

$V_{p\text{-}p}$ = peak-to-peak value (in V)

2 = constant (to double peak value)

V_{max} = peak value (in V)

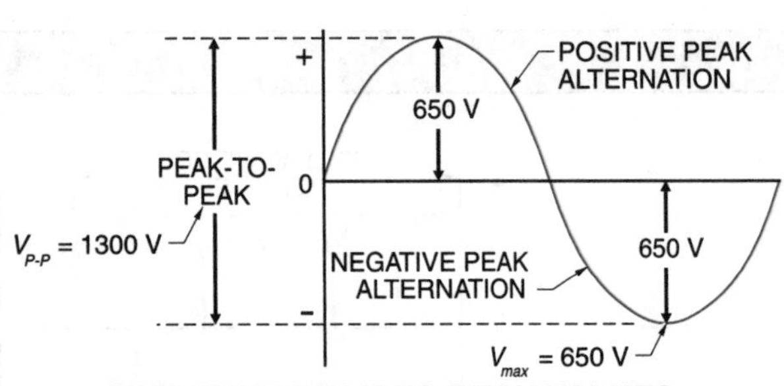

Figure 2-7

Example: Calculating Peak-to-Peak Value

An AC voltage source with a positive peak alternation of 650 V equals what peak-to-peak value?

$V_{p\text{-}p} = 2 \times V_{max}$

$V_{p\text{-}p} = 2 \times 650$

$V_{p\text{-}p} = 1300$ V

The *average value* (V_{avg}) of a sine wave is the mathematical mean of all instantaneous voltage values in the sine wave. The average value is equal to .637 of the peak value of a standard sine wave. **See Figure 2-8.** To calculate average value, apply the formula:

$V_{avg} = V_{max} \times .637$

where

V_{avg} = average value (in V)

V_{max} = peak value (in V)

.637 = constant (mean of instantaneous values)

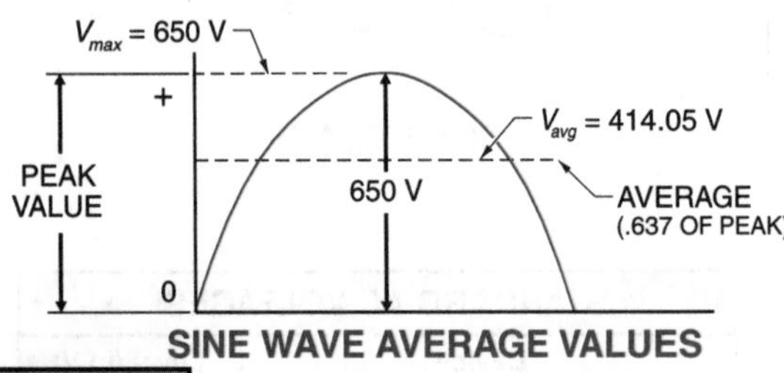

Figure 2-8

Example: Calculating Average Value

An AC voltage source with a positive peak alternation of 650 V equals what average value?

$V_{avg} = V_{max} \times .637$

$V_{avg} = 650 \times .637$

$V_{avg} =$ **414.05 V**

The *root-mean-square (effective) value* (V_{rms}) of a sine wave is the value that produces the same amount of heat in a pure resistive circuit as a DC current of the same value. The rms value is equal to .707 of the peak value in a sine wave. **See Figure 2-9.** To calculate rms value, apply the formula:

$V_{rms} = V_{max} \times .707$

where

V_{rms} = rms value (in V)

V_{max} = peak value (in V)

.707 = constant

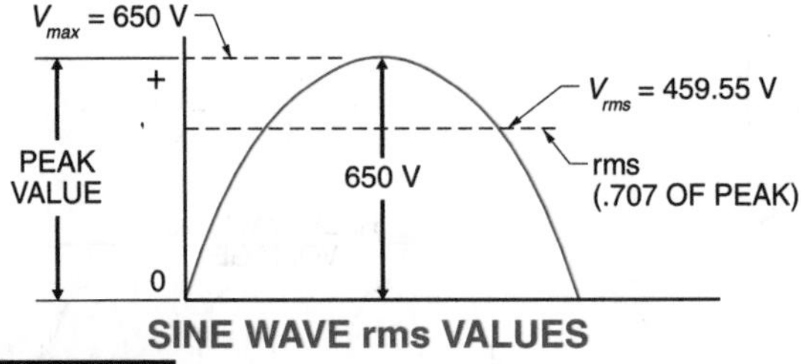

Figure 2-9

Example: Calculating rms Value

An AC voltage source with a positive peak alternation of 650 V equals what rms value?

$V_{rms} = V_{max} \times .707$

$V_{rms} = 650 \times .707$

$V_{rms} =$ **459.55 V**

AC voltages may be listed or measured in any one of the basic values. AC voltages of one value may be converted to another value when required. **See Figure 2-10.**

VOLTAGE CONVERSIONS		
To Convert	**To**	**Multiply By**
rms	Average	.9
rms	Peak	1.414
Average	rms	1.111
Average	Peak	1.567
Peak	rms	.707
Peak	Average	.637

Figure 2-10

Current

Current flows through a circuit when a source of power is connected to a device that uses electricity. *Current (I)* is the amount of electrons flowing through an electrical circuit. Current is measured in amperes (A). An *ampere* is the number of electrons passing a given point in one second. The more power a load requires, the larger the amount of current flow. **See Figure 2-11.**

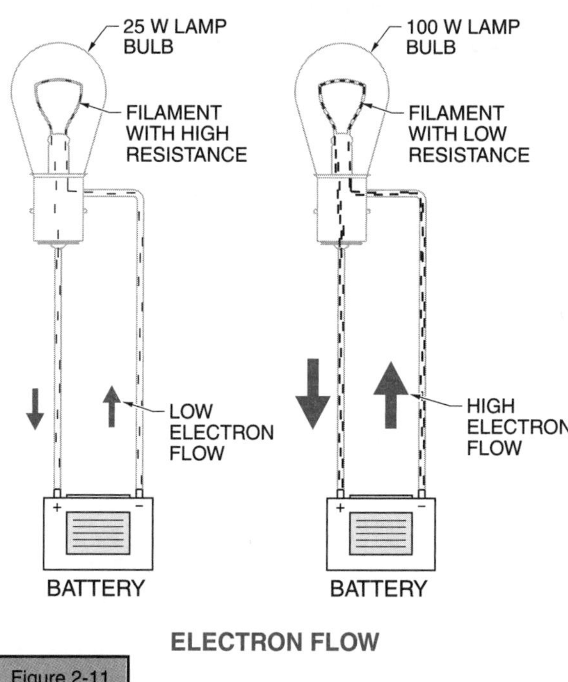

Figure 2-11

Current Levels

Different voltage sources produce different amounts of current. For example, standard AA, A, C, and D size batteries all produce 1.5 V, but each size is capable of delivering different amounts of current. Size AA batteries are capable of delivering the least amount of current, and size D batteries are capable of delivering the most amount of current. For this reason, a load connected to a size D battery lasts longer than a load connected to a size A battery. **See Figure 2-12.**

COMMON LOAD CURRENT LEVELS

Load	Level (in A)
25 W light bulb	.22
100 W light bulb	.88
VCR	.35
25″ color TV	2
½ HP motor	9.8
2 HP motor	24

* when connected to a 115 VAC supply

Figure 2-12

Current may be direct or alternating. Direct current (DC) is current that flows in only one direction. Direct current flows in any circuit connected to a power supply producing a DC voltage. Alternating current (AC) is current that reverses its direction of flow at regular intervals. Alternating current flows in any circuit connected to a power supply producing an AC voltage.

Current Flow

Early scientists believed that electrons flowed from positive (+) to negative (–). Later, when atomic structure was studied, electron flow from negative to positive was introduced. *Conventional current flow* is current flow from positive to negative. *Electron current flow* is current flow from negative to positive. **See Figure 2-13.**

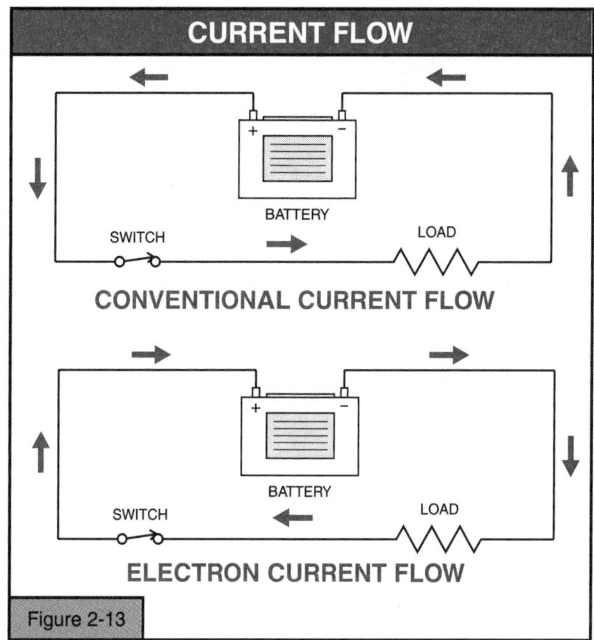

Figure 2-13

Power

Electrical energy is converted into another form of energy any time current flows in a circuit. Electrical energy is converted into sound (speakers), rotary motion (motors), light (lamps), linear motion (solenoids), and heat (heating elements). **See Figure 2-14.**

Power (P) is the rate of doing work or using energy. Power may be expressed as either true power or apparent power. *True power* (P_T) is the actual power used in an electrical circuit. True power is expressed in watts (W). *Apparent power* (P_A) is the product of the voltage and current in a circuit calculated without considering the phase shift that may be present between the voltage and current in the circuit. Apparent power is expressed in volt amps (VA). True power is always less than apparent power in any circuit in which there is a phase shift between voltage and current. **See Figure 2-15.**

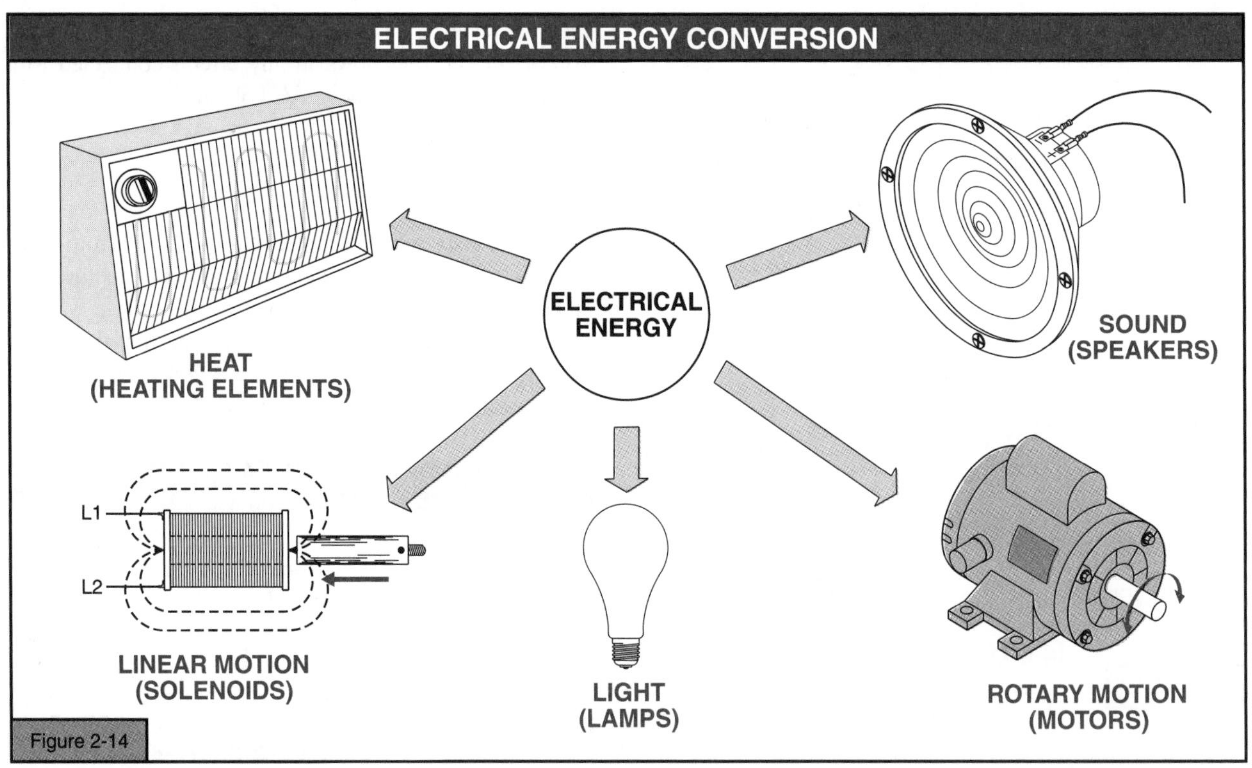

Figure 2-14

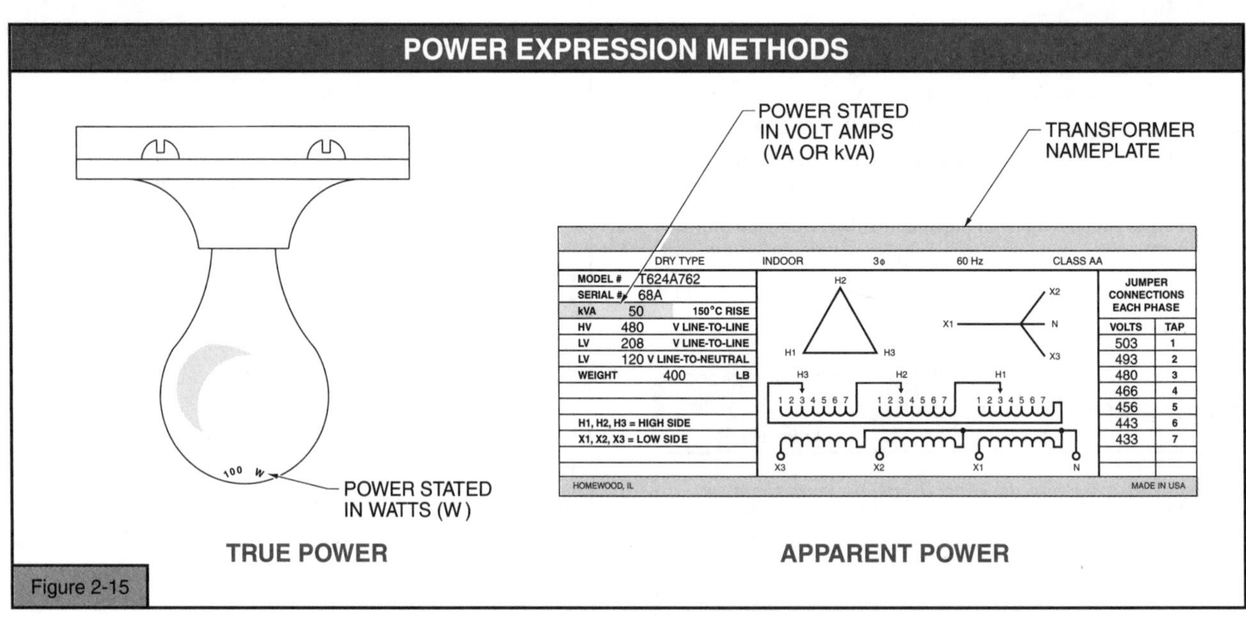

Figure 2-15

Phase Shift

Phase shift is the state when voltage and current in a circuit do not reach their maximum amplitude and zero level simultaneously. There is little to no phase shift in AC circuits that contain resistive devices. There is a phase shift in AC circuits that contain devices causing inductance or capacitance.

Resistive Circuits

A *resistive circuit* is a circuit that contains only resistance. Resistance may be produced by heating elements and incandescent lamps. Alternating voltage and current are in-phase in resistive circuits. *In-phase* is the state when voltage and current reach their maximum amplitude and zero level simultaneously. **See Figure 2-16.**

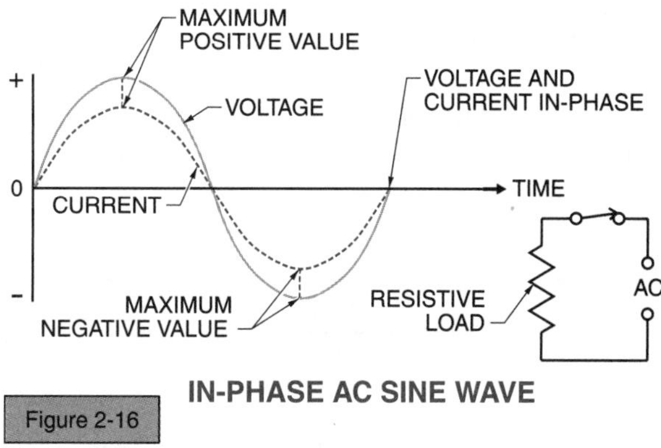

IN-PHASE AC SINE WAVE

Figure 2-16

Inductive Circuits

Inductance is the property of an electric device that opposes a change in current due to its ability to store energy in a magnetic field. All coils (motor windings, transformers, solenoids, etc.) create inductance in an electrical circuit. A phase shift exists between alternating voltage and current in an inductive circuit. An *inductive circuit* is a circuit in which current lags voltage. The greater the inductance (compared to resistance) in a circuit, the closer the phase shift to 90°. **See Figure 2-17.**

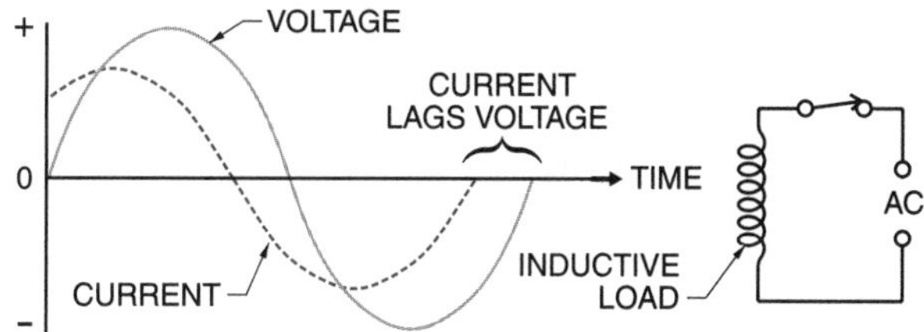

INDUCTIVE CIRCUIT AC SINE WAVE

Figure 2-17

TECH FACT *Capacitance is added to a power circuit to increase true power when a circuit includes resistive and inductive loads.*

Capacitive Circuits

Capacitance is the ability to store energy in the form of an electrical charge. Capacitors create capacitance in an electrical circuit. A phase shift exists between voltage and current in a capacitive circuit. A *capacitive circuit* is a circuit in which current leads voltage. The greater the capacitance (compared to resistance) in a circuit, the closer the phase shift to –90°. **See Figure 2-18.**

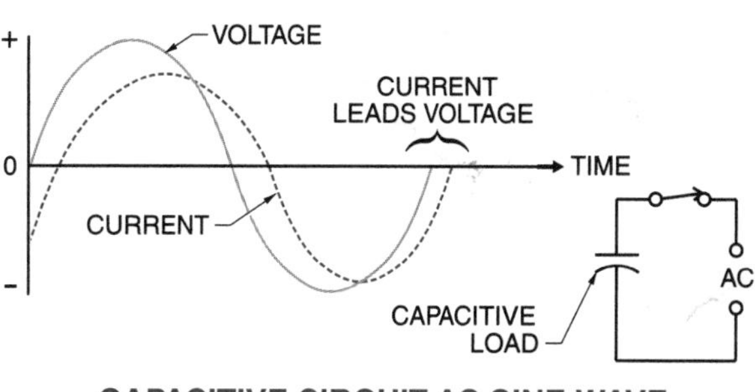

CAPACITIVE CIRCUIT AC SINE WAVE

Figure 2-18

Power Factor

Power factor (PF) is the ratio of true power used in an AC circuit to apparent power delivered to the circuit. Power factor is expressed as a percentage. True power equals apparent power when the power factor is 100%. When the power factor is less than 100%, the circuit is less efficient and has higher operating cost. To calculate power factor, apply the formula:

$$PF = \frac{P_T}{P_A} \times 100$$

where

PF = power factor (percentage)

P_T = true power (in W)

P_A = apparent power (in VA)

100 = constant (to convert decimal to percent)

Example: Calculating Power Factor

What is the power factor when the true power of an electrical circuit is 1500 W and the apparent power is 1700 VA?

$$PF = \frac{P_T}{P_A} \times 100$$

$$PF = \frac{1500}{1700} \times 100$$

$$PF = .882 \times 100$$

$$PF = \mathbf{88.2\%}$$

Resistance

Resistance (R) is the opposition to current flow. Resistance is measured in ohms. The Greek symbol Ω (omega) is used to represent ohms. Resistance limits the flow of current in an electrical circuit. The higher the resistance, the lower the current flow. Likewise, the lower the resistance, the higher the current flow. **See Figure 2-19.**

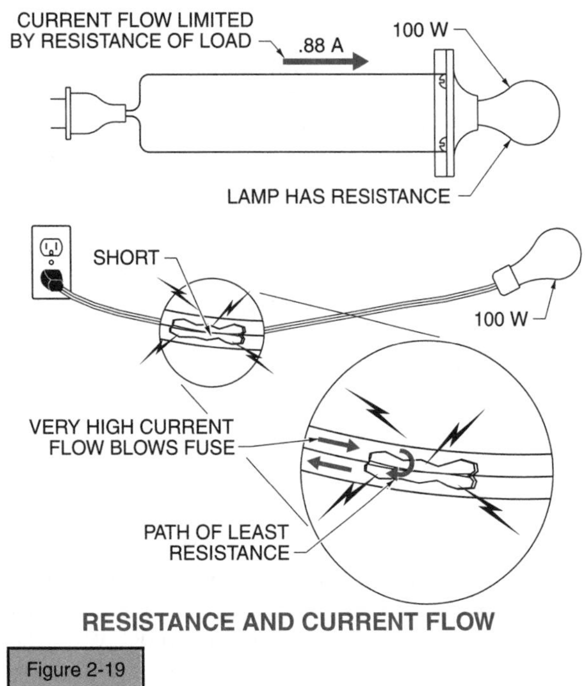

Figure 2-19

Conductors and Insulators

All materials have some amount of resistance to current flow. A conductor is a material that has very little resistance and permits electrons to move through it easily. Most metals are good conductors. Copper is the most commonly used conductor. Silver is a better conductor than copper, but is too expensive for most applications. Aluminum is not as good of a conductor as copper, but costs less and is lighter. For this reason, aluminum is also commonly used as a conductor. Conductors are available as wire, cord, or cable. Conductors may be solid or stranded.

Copper conductors are often used to connect ground circuits to metal such as piping to safely conduct any current flow through the path to ground.

An *insulator* is a material that has a very high resistance and resists the flow of electrons. Common insulators include rubber, plastic, air, glass, and paper. Electrical insulations are classified by their temperature ratings as 60°C (140°F), 75°C (167°F), and 90°C (194°F). **See Figure 2-20.**

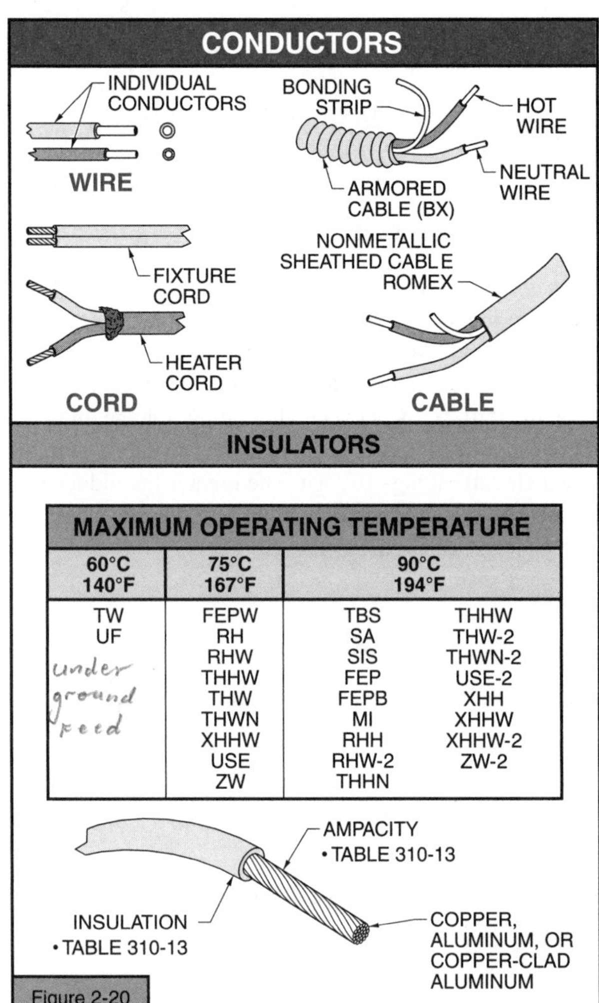

MAXIMUM OPERATING TEMPERATURE			
60°C 140°F	75°C 167°F	90°C 194°F	
TW	FEPW	TBS	THHW
UF	RH	SA	THW-2
	RHW	SIS	THWN-2
	THHW	FEP	USE-2
	THW	FEPB	XHH
	THWN	MI	XHHW
	XHHW	RHH	XHHW-2
	USE	RHW-2	ZW-2
	ZW	THHN	

Figure 2-20

Heat

Heat is thermal energy. Anything that gives off heat is a heat source. Electrical energy may be converted into heat. Electricity may be used to heat almost any gas, liquid, or solid. Heat produced from electricity is used in residential and commercial applications to heat rooms and water, cook food, dry clothes, etc. Most industrial processes require the heating of a variety of liquids, solids, and gases. Heat is used to separate metals from their ores, refine crude oil, process foods, remove moisture, shape metals and plastics, join metals and plastics, harden metals, and to cure finishes such as paint, enamels, and varnishes.

Heat is measured in British thermal units (Btu) or calories (cal). British thermal units are used to measure the amount of heat in the U.S. system of measurements. A *British thermal unit (Btu)* is the amount of heat required to raise the temperature of 1 lb of water 1°F. Calories are used to measure the amount of heat in the Metric system of measurements. A *calorie* is the amount of heat required to raise 1 g of water 1°C. One Btu is equivalent to 252 calories or .252 kilocalories. **See Figure 2-21.**

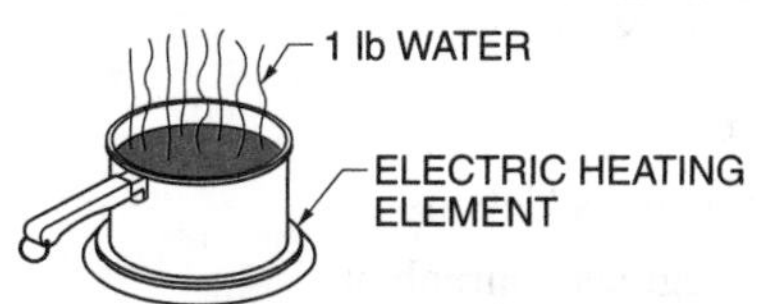

1 Btu = 1°F TEMPERATURE CHANGE
1 Btu = 252 CALORIES

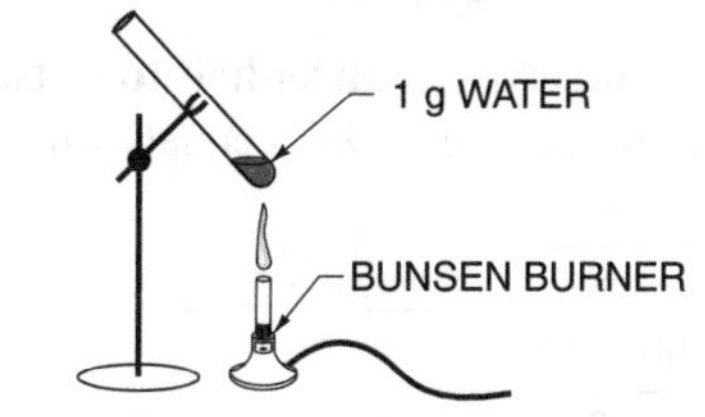

1 CALORIE = 1°C TEMPERATURE CHANGE
252 CALORIES = 1 Btu

Figure 2-21 UNITS OF HEAT

Most electric heating elements are rated in watts and/or Btu. The amount of electrical energy used by a heating element depends on the power rating of the heating element (in watts) and the amount of time (in hours) the heating element is ON. To calculate the amount of electrical energy used by a heating element, apply the formula:

$E = P \times T$

where

E = energy (in Wh)

P = power (in W)

T = time (in hr)

Example: Calculating Electrical Energy

A heater rated at 3000 W uses how much energy in 1 hr, and how much energy in 10 hr?

One Hour

$E = P \times T$

$E = 3000 \times 1$

E = **3000 Wh (3 kWh)**

Ten Hours

$E = P \times T$

$E = 3000 \times 10$

E = **30,000 Wh (30 kWh)**

Converting watts to Btu is required because heating elements are normally rated in watts or Btu. One Wh is equal to 3.412 Btu. Likewise, 1 Btu is equal to .293 Wh. **See Figure 2-22.**

HEAT CONVERSIONS		
To Obtain	**Multiply**	**By**
Btu	kcal	3.968
Btu	Wh	3.412
Btu	kWh	3412
Cal	Btu	252
Kcal	Btu	.252
Wh	kWh	1000
Wh	Btu	2.93
kWh	Btu	.0002929

Figure 2-22

Using formulas or conversion factors, a power rating in watts may be converted to Btu, or a power rating in Btu may be converted to watts.

Example: Converting Wh to Btu

A heater rated at 3000 Wh (3 kWh) is equal to how many Btu?

$Btu = kWh \times 3412$

$Btu = 3 \times 3412$

Btu = **10,236 Btu**

Example: Converting Btu to kWh

A heater rated at 30,000 Btu is equal to how many kWh?

$kWh = Btu \times .0002929$

$kWh = 30{,}000 \times .0002929$

kWh = **8.787 kWh**

Temperature

All heating elements produce heat. Temperature increases when heat is produced. *Temperature* is the measurement of the intensity of heat. The amount of increase in temperature depends on the amount of heat produced, the mass of the body being heated, and the material of which the heated body is made. **See Figure 2-23.**

The higher the amount of heat produced, the faster the temperature increases in the body being heated. The larger the body to be heated, the slower the temperature rise of the body. The better conductor of heat the material being heated is, the faster the heat transfer. The rate a material conducts heat depends on the thermal conductivity rating of the material.

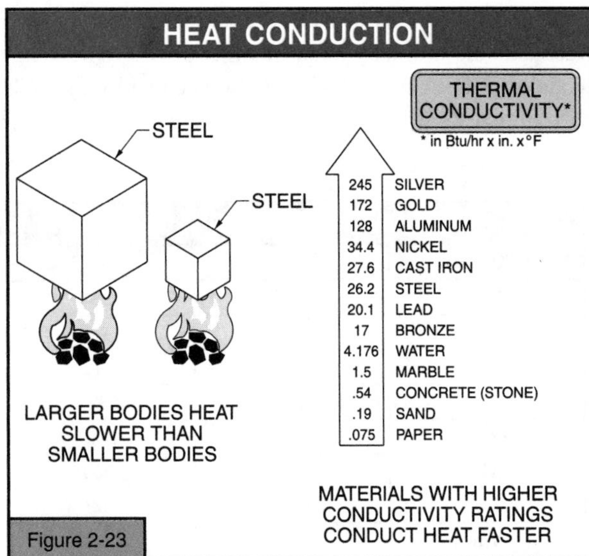

Figure 2-23

Thermal conductivity is the property of a material to conduct heat in the form of thermal energy. The higher the thermal conductivity rating, the faster the material conducts (transfers) heat. The lower the thermal conductivity rating, the slower the material conducts (transfers) heat. The thermal conductivity rating (number) assigned to a given material is based on that material's ability to transfer heat through 1 sq ft of the surface area for a given thickness in Btu/hr, per 1°F temperature difference.

Gases and liquids such as air and water have poor thermal conductivity. Solids such as aluminum and copper have good thermal conductivity. The thermal conductivity number assigned to a given material is based on the ability of the material to transfer heat. The number is based on the amount of heat transferred through 1 sq ft of the surface area for a given thickness in Btu per hour, per 1°F difference through the material. Aluminum (thermal conductivity = 128) is a much better thermal conductor than steel (thermal conductivity = 26.2). Since aluminum is a much better conductor of heat, it is used in most heat sink applications. A *heat sink* is a device that conducts and dissipates heat away from a component. **See Figure 2-24.**

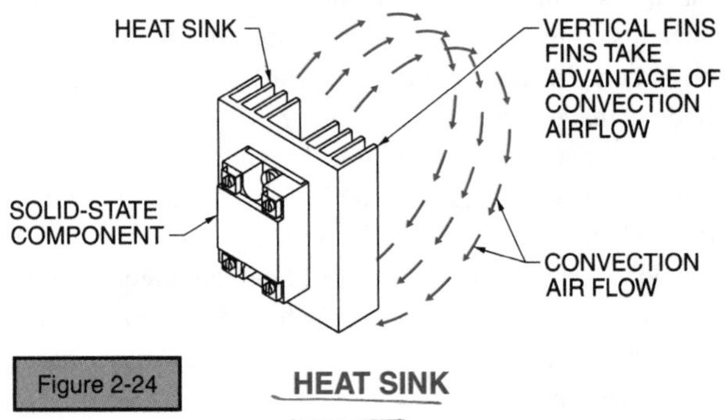

Figure 2-24

Temperature Conversion

Temperature is normally measured in degrees Fahrenheit (°F) or degrees Celsius (°C). Converting one unit to the other is required because both Fahrenheit and Celsius are commonly used in the electrical field. **See Figure 2-25.**

To convert a Fahrenheit temperature reading to Celsius, subtract 32 from the Fahrenheit reading and divide by 1.8. To convert Fahrenheit to Celsius, apply the formula:

$$°C = \frac{°F - 32}{1.8}$$

where

$°C$ = degrees Celsius

$°F$ = degrees Fahrenheit

32 = difference between bases

1.8 = ratio between bases

Example: Converting Fahrenheit to Celsius

A temperature reading of 90°F equals what °C?

$$°C = \frac{°F - 32}{1.8}$$

$$°C = \frac{90 - 32}{1.8}$$

$$°C = \frac{58}{1.8}$$

$$°C = \mathbf{32.2\bar{2}°C}$$

To convert a Celsius temperature reading to Fahrenheit, multiply 1.8 by the Celsius reading and add 32. To convert Celsius to Fahrenheit, apply the formula:

$$°F = (1.8 \times °C) + 32$$

where

$°F$ = degrees Fahrenheit

1.8 = ratio between bases

$°C$ = degrees Celsius

32 = difference between bases

Example: Converting Celsius to Fahrenheit

A temperature reading of 150°C equals what °F?

$$°F = (1.8 \times °C) + 32$$

$$°F = (1.8 \times 150) + 32$$

$$°F = 270 + 32$$

$$°F = \mathbf{302°F}$$

Light

Light is that portion of the electromagnetic spectrum which produces radiant energy. The electromagnetic spectrum ranges from cosmic rays with extremely short wavelengths to electric power frequencies with extremely long wavelengths. **See Figure 2-26.**

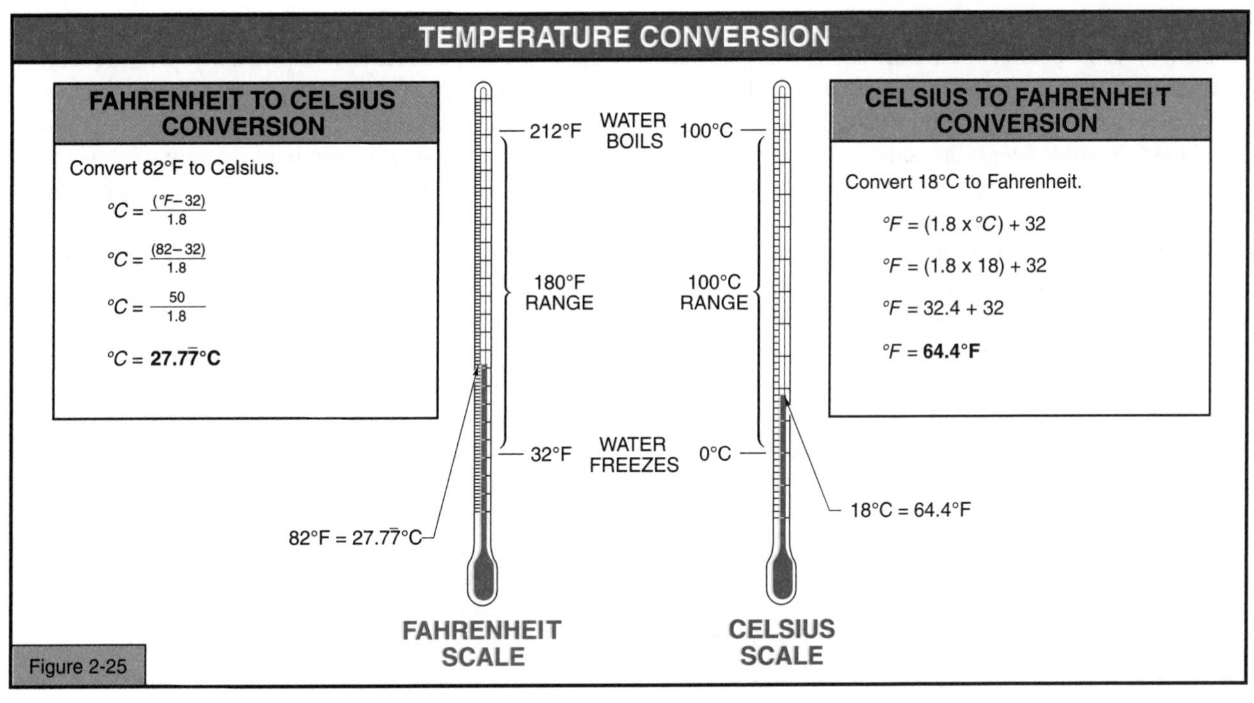

Figure 2-25

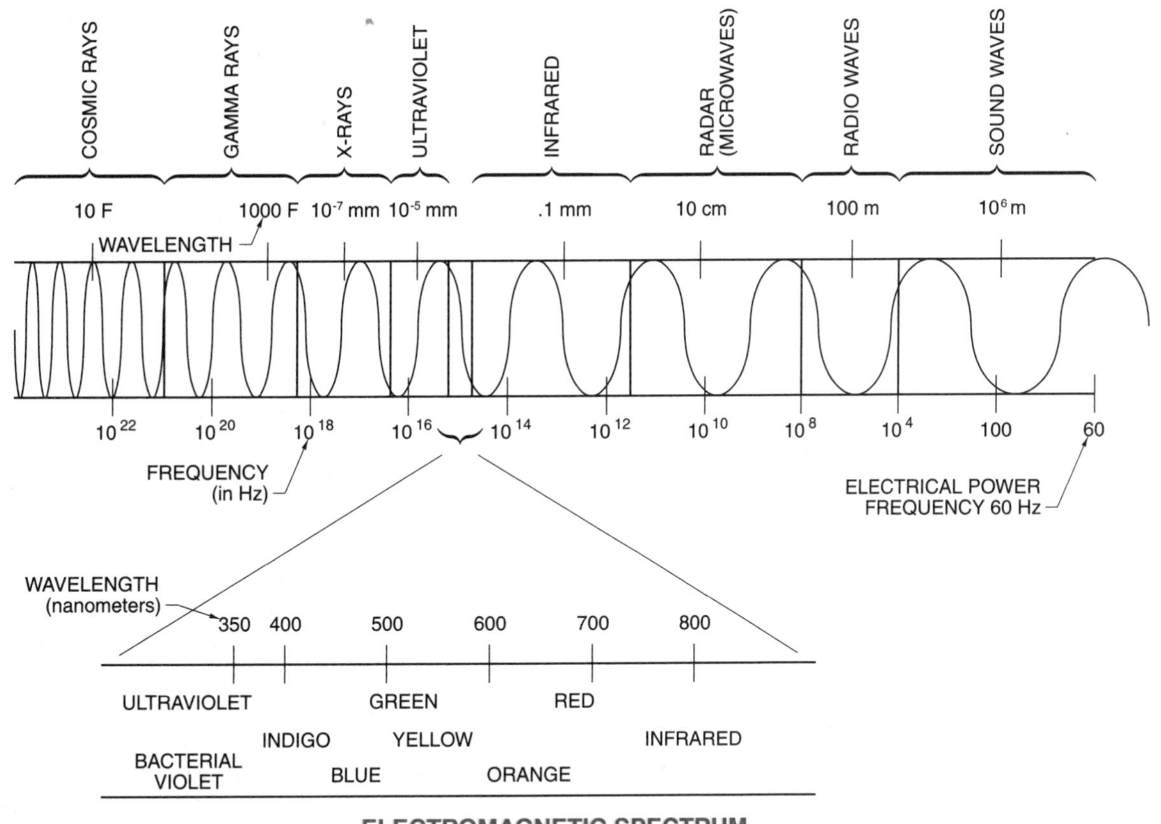

Figure 2-26

Light can be in the form of visible light or invisible light. *Visible light* is the portion of the electromagnetic spectrum to which the human eye responds. Visible light includes the part of the electromagnetic spectrum that ranges from violet to red light. *Invisible light* is the portion of the electromagnetic spectrum on either side of the visible light spectrum. Invisible light includes ultraviolet and infrared light.

The color of light is determined by its wavelength. Visible light with the shortest wavelengths produce the color violet. Visible light with the longest wavelengths produce red. Wavelengths between violet and red produce blue, green, yellow, and orange. The combination of colored light produces white light when a light source, such as the sun, produces energy over the entire visible spectrum in approximately equal quantities. A non-white light is produced when a light source, such as a low-pressure sodium lamp, produces energy that mostly lies in a narrow band of the spectrum (yellow-orange). **See Figure 2-27.**

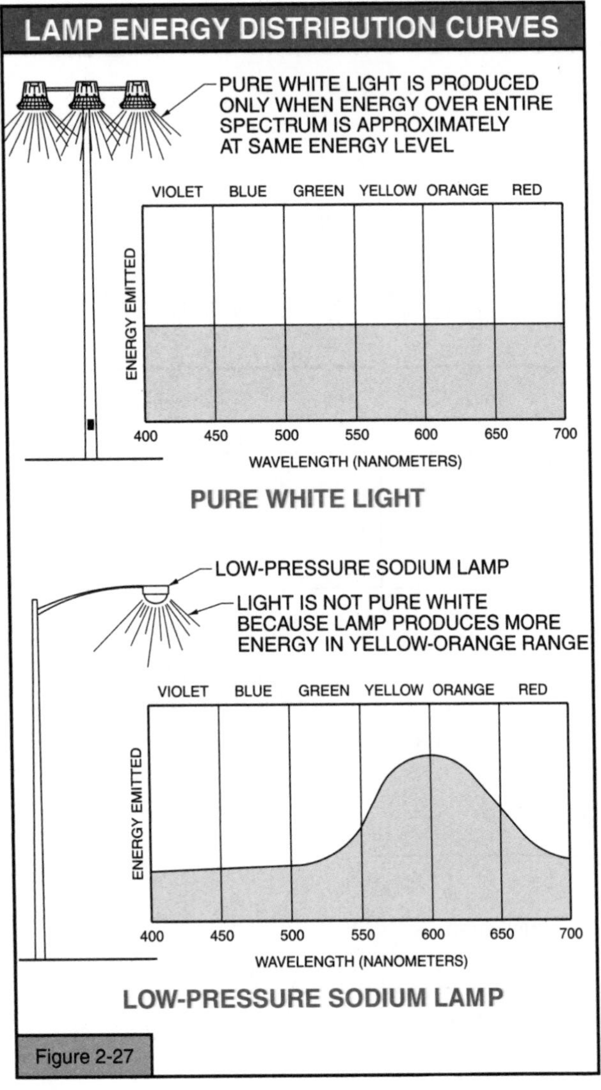

Figure 2-27

Color distortion of light-colored objects viewed under the lamp may be extreme because the low-pressure sodium lamp produces most of its energy in the yellow-orange area of the light spectrum. However, since the low-pressure sodium lamp produces more light per watt of power than any other lamp type, it is used in applications such as street lighting in which color distortion may be tolerated.

The *ultraviolet region* is the region of the spectrum with wavelengths just short of the color violet. The *infrared region* is the region of the spectrum with wavelengths slightly longer than the color red. Both ultraviolet and infrared light are invisible to the human eye. However, ultraviolet and infrared light is used in special applications. Depending on the wavelength, ultraviolet light is used in germicidal (bacteria killing) lamps, photochemical lamps, black lights, and sun lamps. Infrared light is used in heat lamps.

A *lamp* is an output device that converts electrical energy into light. The amount of light a lamp produces is expressed in lumens. A *lumen (lm)* is the unit used to measure the total amount of light produced by a light source. For example, a standard 40 W incandescent lamp produces about 480 lm, and a standard 40 W fluorescent lamp produces about 3100 lm. Manufacturers rate lamps (light bulbs) in the total amount of light (lumen) produced by the lamps. Since the lumen is the total amount of light, it is comparable to the amount of current (amperes) in an electrical circuit or the amount of flow (gpm) in a hydraulic system.

The light produced by a light source causes illumination. *Illumination* is the effect that occurs when light falls on a surface. The unit of measure of illumination is the footcandle. A *footcandle (fc)* is the amount of light produced by a lamp (lumens) divided by the area that is illuminated.

Light spreads as it travels farther from the light-producing source. The relationship between the amount of light produced at the source and the amount of illumination at different distances from the light source is expressed by the inverse square law. The *inverse square law* states that the amount of illumination on a surface varies inversely with the square of the distance from the light source. **See Figure 2-28.**

One lm produces an illumination of 1 fc on a 1 sq ft area. Thus, if the amount of light falling on a surface 1′ away from the light source produces 1 fc, then the light source produces ¼ fc at a 2′ distance, and ⅑ fc at a 3′ distance. This is because the light produced at the source covers four times the area at a 2′ distance and nine times the area at a 3′ distance, than it does at a 1′distance. **See Figure 2-29.**

The amount of light produced on a surface is dependent upon the amount of lumens produced by the light source and the distance the surface is from the light source. The light level required on a surface varies widely. For example, an operating table in a hospital requires much more light than a teller's station in a bank.

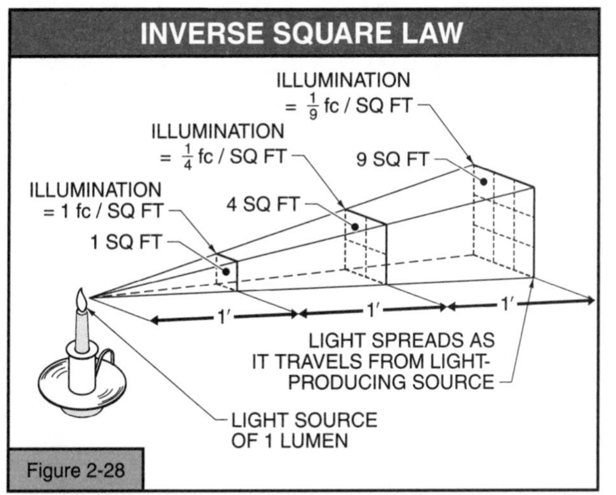

Figure 2-28

G E Lighting

The appropriate amount of light (in footcandles) must be considered when lighting areas such as outdoor sports stadiums.

RECOMMENDED LIGHT LEVELS

Interior Lighting		Exterior Lighting		Sports Lighting	
Area	**fc**	**Area**	**fc**	**Area**	**fc**
Assembly Rough, easy seeing Medium Fine	 30 100 500	**Airports** Terminal Apron – Loading	 2	**Baseball** Outfield Infield	 100 150
Auditorium Exhibitions	30	**Building Construction** General Excavation	 10 2	**Basketball** College and Professional Recreational	 50 10
Banks Lobby, general Waiting areas Teller station	 50 70 150	**Buildings** Light surface Dark surface	 15 50	**Billiards** Recreational Tournament	 30 50
Clothing Manufacturing Pattern making Shops	 50 100	**Loading areas**	20	Boxing Professional Championship	 200 500
Hospital/Medical Lobby Dental chair Operating table	 30 1000 2500	**Parking areas** Industrial Shopping	 2 5	**Golf** Tee Fairway Green Miniature	 5 2 5 10
Machine Shop Rough bench Medium bench	 50 100	**Piers** Freight/Passenger	 20	**Racing** Auto, Horse Dog	 20 30
Offices Regular office work Accounting Detailed work	 100 150 200	**Railroad Yards** Switch points	 2	Ski Slope	1
Printing Proofreading Color inspecting	 150 200	**Service Station** Pump island area Service areas	 25 5	**Volleyball** Recreational Tournament	 10 20
Warehouses Inactive Active	 5 30	**Streets** Local Expressway	 .9 1.4	**Tennis Courts** Recreational Tournament	 15 30

Figure 2-29

Refer to the CD-ROM in the back of the book for Chapter 2 Quick Quiz® and related information.

Ohm's Law

Experiments performed by the German scientist George Simon Ohm in 1827 proved there was a direct relationship between voltage (E), current (I), and resistance (R) in an electrical circuit. This relationship is known as Ohm's law.

Ohm's law is the relationship between voltage, current, and resistance in a circuit. Ohm's law states that current in a circuit is proportional to the voltage and inversely proportional to the resistance. Any value in this relationship can be found when the other two are known. The relationship between voltage, current, and resistance may be visualized by presenting Ohm's law in pie chart form. **See Figure 3-1.**

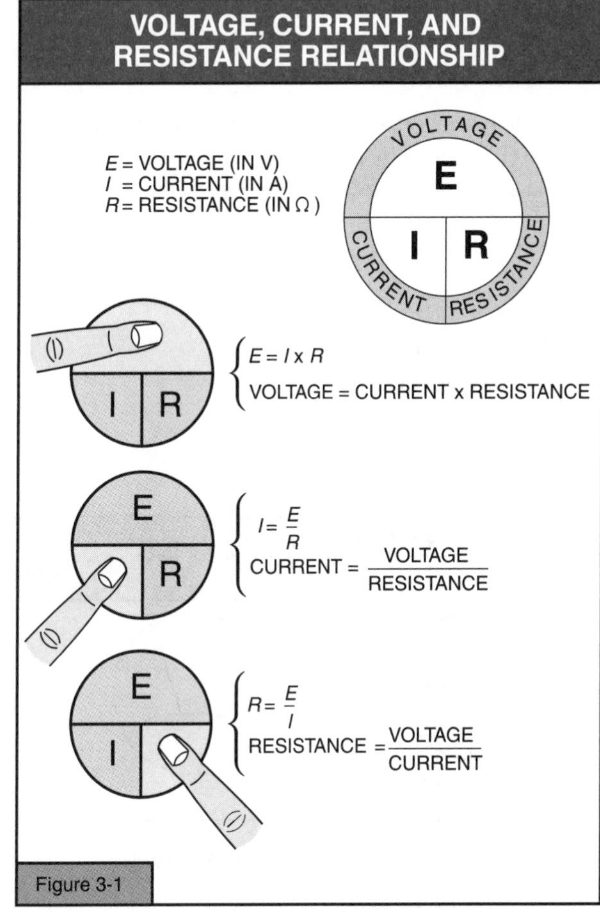

Figure 3-1

Calculating Voltage Using Ohm's Law

Ohm's law states that voltage (E) in a circuit is equal to resistance (R) times current (I). **See Figure 3-2.** To calculate voltage using Ohm's law, apply the formula:

$E = I \times R$

where

E = voltage (in V)

I = current (in A)

R = resistance (in Ω)

Example: Calculating Voltage Using Ohm's Law

What is the voltage in a circuit that includes a 3 Ω heating element that draws 4 A?

$E = I \times R$

$E = 4 \times 3$

$E = \mathbf{12\ V}$

TECH TIP

It is better to oversize a wire than undersize a wire. A larger wire (smaller AWG number) has less resistance than a smaller wire (larger AWG number).

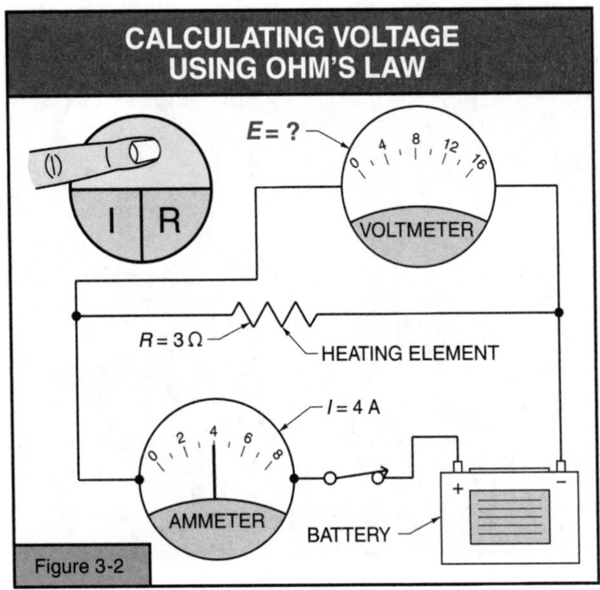

Figure 3-2

Calculating Current Using Ohm's Law

Ohm's law states that current (I) in a circuit is equal to voltage (E) divided by resistance (R). **See Figure 3-3.** To calculate current using Ohm's law, apply the formula:

$$I = \frac{E}{R}$$

where

I = current (in A)

E = voltage (in V)

R = resistance (in Ω)

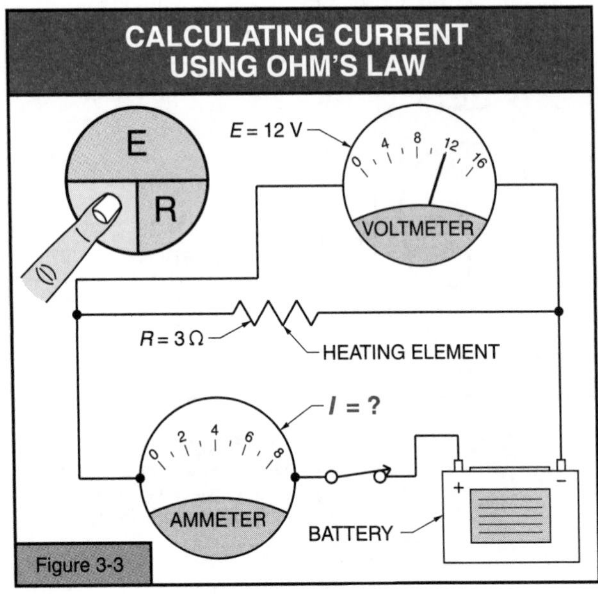

Figure 3-3

Example: Calculating Current Using Ohm's Law
What is the current in a circuit with a 3 Ω heating element connected to a 12 V supply?

$$I = \frac{E}{R}$$

$$I = \frac{12}{3}$$

$$I = \mathbf{4\ A}$$

Calculating Resistance Using Ohm's Law

Ohm's law states that the resistance (R) in a circuit is equal to voltage (E) divided by current (I). **See Figure 3-4.** To calculate resistance using Ohm's law, apply the formula:

$$R = \frac{E}{I}$$

where

R = resistance (in Ω)

E = voltage (in V)

I = current (in A)

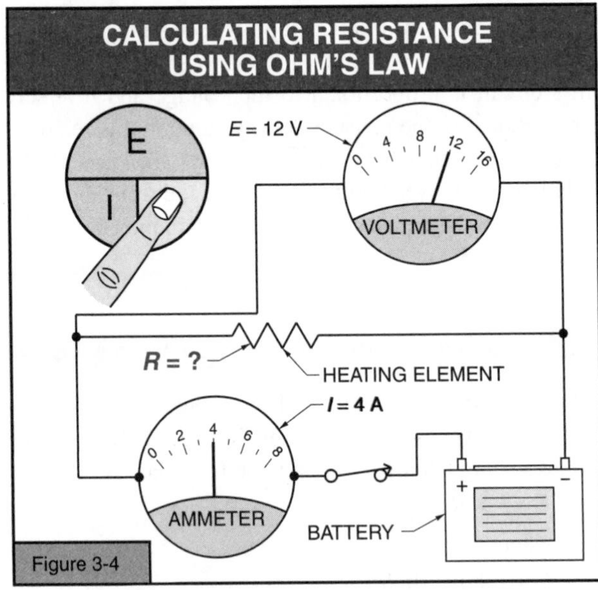

Figure 3-4

Example: Calculating Resistance Using Ohm's Law
What is the resistance of a circuit in which a load that draws 4 A is connected to a 12 V supply?

$$R = \frac{E}{I}$$

$$R = \frac{12}{4}$$

$$R = \mathbf{3\ \Omega}$$

Voltage/Current Relationship

Ohm's law states that if the resistance in a circuit remains constant, a change in current is directly proportional to a change in voltage. This is shown in an application that uses a variable power supply and has a fixed resistance load. An example is a heat shrink sealing gun connected to a variable power supply. **See Figure 3-5.**

In this circuit, a fixed load resistance of 4 Ω is connected to a variable power supply which may be varied from 0 V to 24 V. The current in the circuit may be found for any given voltage by applying Ohm's law. For example, if the voltage in the circuit is set at 8 V, the current equals 2 A ($8/4 = 2$ A).

The graph shows the direct proportional relationship between voltage and current in the circuit. The voltage values are marked on the horizontal axis and the current values are marked on the vertical axis. A linear resistance line is developed when the given values for voltage and current are plotted on the graph. The plotted resistance values show how any change in voltage results in a proportional change in current.

The direct proportional relationship may also be shown in a circuit that uses a defrost heating element as the load. **See Figure 3-6.** In this circuit, the supply voltage is 12 V, and the resistance of the circuit load is 12 Ω.

Using Ohm's law, the current is equal to 1 A ($12/12 = 1$ A). If the voltage in the circuit increases from 12 V to 24 V, the current in the circuit increases to 2 A ($24/12 = 2$ A). If the voltage in the circuit decreases from 12 V to 6 V, the current in the circuit decreases to .5 A ($6/12 = .5$ A).

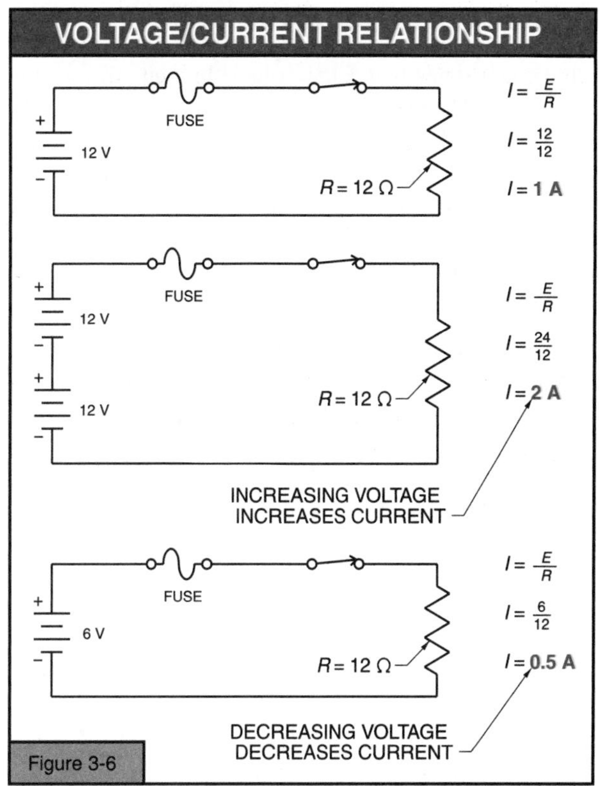

Figure 3-6

TECH FACT *Connecting batteries in series increases total voltage. Connecting batteries in parallel increases total current.*

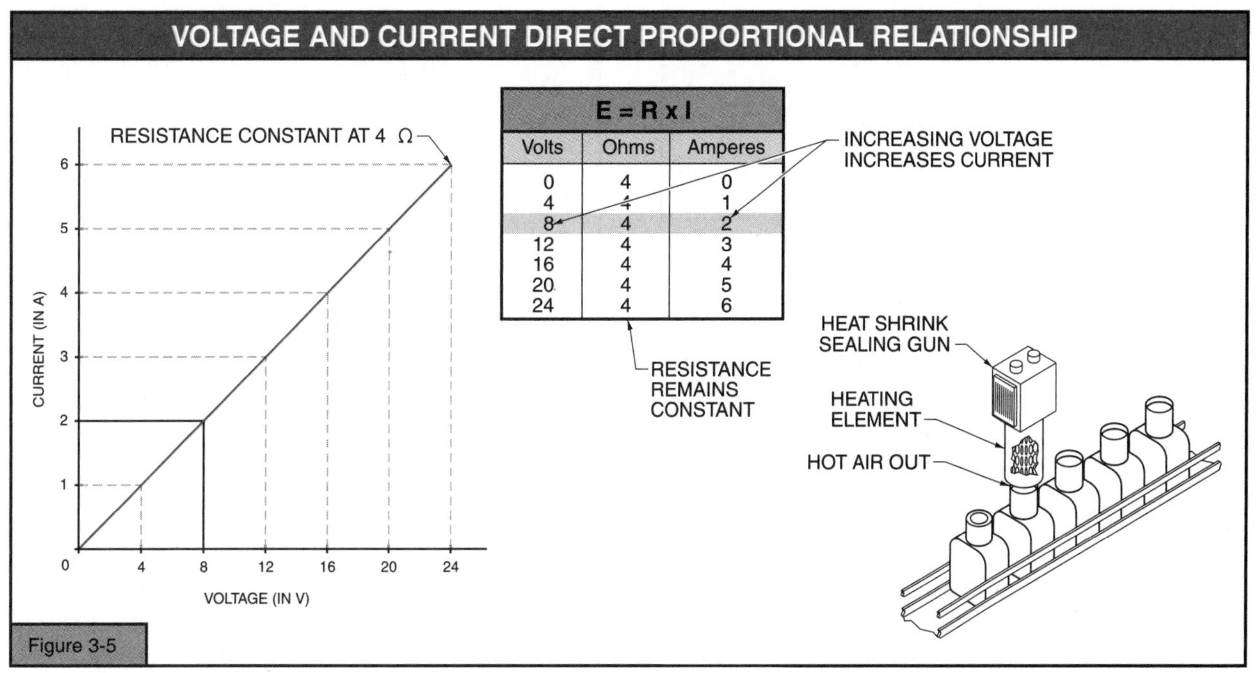

E = R x I		
Volts	Ohms	Amperes
0	4	0
4	4	1
8	4	2
12	4	3
16	4	4
20	4	5
24	4	6

Figure 3-5

Current/Resistance Relationship

Ohm's law states that if the voltage in a circuit remains constant, a change in resistance produces an inversely proportional change in current. The current in a circuit decreases with an increase in resistance, and the current in the circuit increases with a decrease in resistance. This is shown in any application that uses a variable resistance load. **See Figure 3-7.**

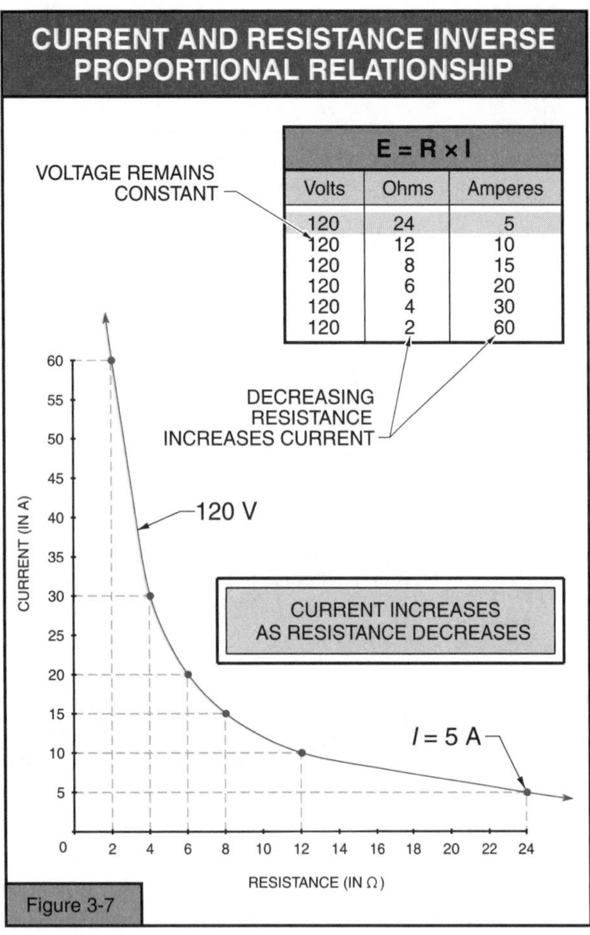

Figure 3-7

In this circuit, a variable resistance of between 2 Ω and 24 Ω is connected to a fixed power supply. The current in the circuit may be found for any given resistance by applying Ohm's law. For example, if the resistance in the circuit is set at 24 Ω, the current equals 5 A ($^{120}/_{24} = 5$ A).

The graph shows the inverse proportional relationship between current and resistance in the circuit. The changing resistance values are marked on the horizontal axis. The changing current values are marked on the vertical axis. The voltage is plotted on the graph. The graph shows that current in a circuit is inversely proportional to the resistance.

The inversely proportional relationship may also be shown in a circuit that uses a three element heater. Each heating element is connected by a four-position selector switch. **See Figure 3-8.**

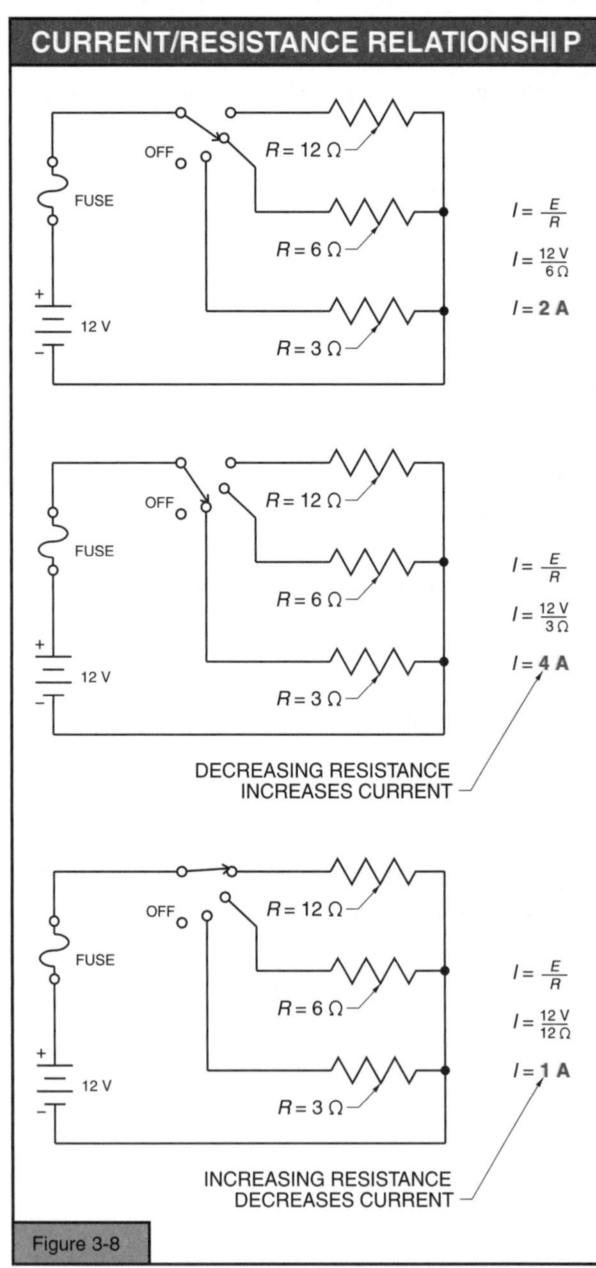

Figure 3-8

In this circuit, the supply voltage is 12 V, and the load has a resistance of 6 Ω. Using Ohm's law, the current is equal to 2 A ($^{12}/_{6} = 2$ A). If the resistance decreases from 6 Ω to 3 Ω, the current in the circuit increases to 4 A ($^{12}/_{3} = 4$ A). If the resistance increases from 6 Ω to 12 Ω, the current in the circuit decreases to 1 A ($^{12}/_{12} = 1$ A).

Applications Using Ohm's Law

Ohm's law is used in engineering applications and troubleshooting applications. In engineering applications, Ohm's law is used to solve for the proper values of voltage, current, or resistance during circuit design and to predict circuit characteristics before power is applied to a circuit when two of the three electrical values are known. Calculating these values helps determine the correct size of conductors and components and the voltage level that must be used for proper circuit operation.

In troubleshooting applications, Ohm's law is used to determine how a circuit should operate and how it is operating under power. When troubleshooting a circuit, the circuit is already designed. To determine the problem, voltage and current measurements are taken. These measured values are used to calculate the circuit's resistance and help locate the problem.

For example, when the insulation of equipment or conductors breaks down, resistance in the circuit decreases and current increases. By measuring a current level higher than normal, the troubleshooter knows that the circuit resistance has decreased or the circuit voltage has increased. This information may be used to determine potential problems such as insulation breakdown or a high-voltage condition. Likewise, if in the same circuit the measured current level is lower than normal, the troubleshooter knows that the circuit resistance has increased or the circuit voltage has decreased. Circuit resistance typically increases due to poor connections, loose connections, corrosion, or damaged components.

Using Ohm's Law When Designing Circuits

Changing a circuit's resistance changes the amount of current flowing though the circuit. This principle is applied in any circuit that uses variable resistor (potentiometer or rheostat) to control a load. For example, variable resistors are used to control the sound output of speakers (volume controls), to control the brightness level of lamps (dimmer switches), and to control the amount of heat output of heating elements (temperature controls). **See Figure 3-9.**

In this circuit, the dimmer control that adjusts the brightness of the panel lamps uses a rheostat as a dimmer control to adjust the brightness of the lamps. The dimmer control increases or decreases the resistance of the instrument panel lighting circuit. Adjusting the dimmer control changes the amount of current through the circuit and the voltage applied to the panel lamps.

When designing a circuit like the automobile lighting circuit, the design engineer uses Ohm's law to determine the resistance range required to control the lamps through their full range of brightness. The design engineer determines the required resistance range based on the known electrical values. In this circuit, the known electrical values are the battery voltage and the current draw of the lamps selected.

For example, if the lamps are rated at .25 A per bulb and four bulbs are used, the total lamp current is 1 A (.25 × 4 = 1 A). The resistance of the circuit at full lamp brightness equals 12 Ω (12 ÷ 1 = 12 Ω).

Knowing that the total resistance is 12 Ω and the total current is 1 A, the design engineer can determine the required variable resistor range to control the lamps. For example, if the midrange of the dimmer switch is to produce a 50% lamp brightness, the current has to be reduced by 50%. A 50% current value of .5 A (1 × .5 = .5 A) is used to find the total circuit resistance required. The total required circuit resistance equals 24 Ω (12 ÷ .5 = 24 Ω). Fifty percent lamp brightness resistance equals 24 Ω.

Thus, 24 Ω total resistance is required to reduce the brightness of the lamps by 50%. Since the four lamps produce 12 Ω of resistance, an additional 12 Ω of resistance must be produced by the dimmer control. In this application, a variable resistor that produced 12 Ω of resistance at midrange is required. By using Ohm's law, the design engineer has determined that a variable resistor that has a full range of 0 Ω to 24 Ω (using 12 Ω as the midrange) could be used as the panel lamp dimmer control.

TECH TIP *Ensure that electrical connections are tight. Loose connections increase circuit resistance.*

Calculated Industries, Inc.

Electrician's calculators are used to calculate electrical parameters such as voltage, resistance, current, power, and power factor.

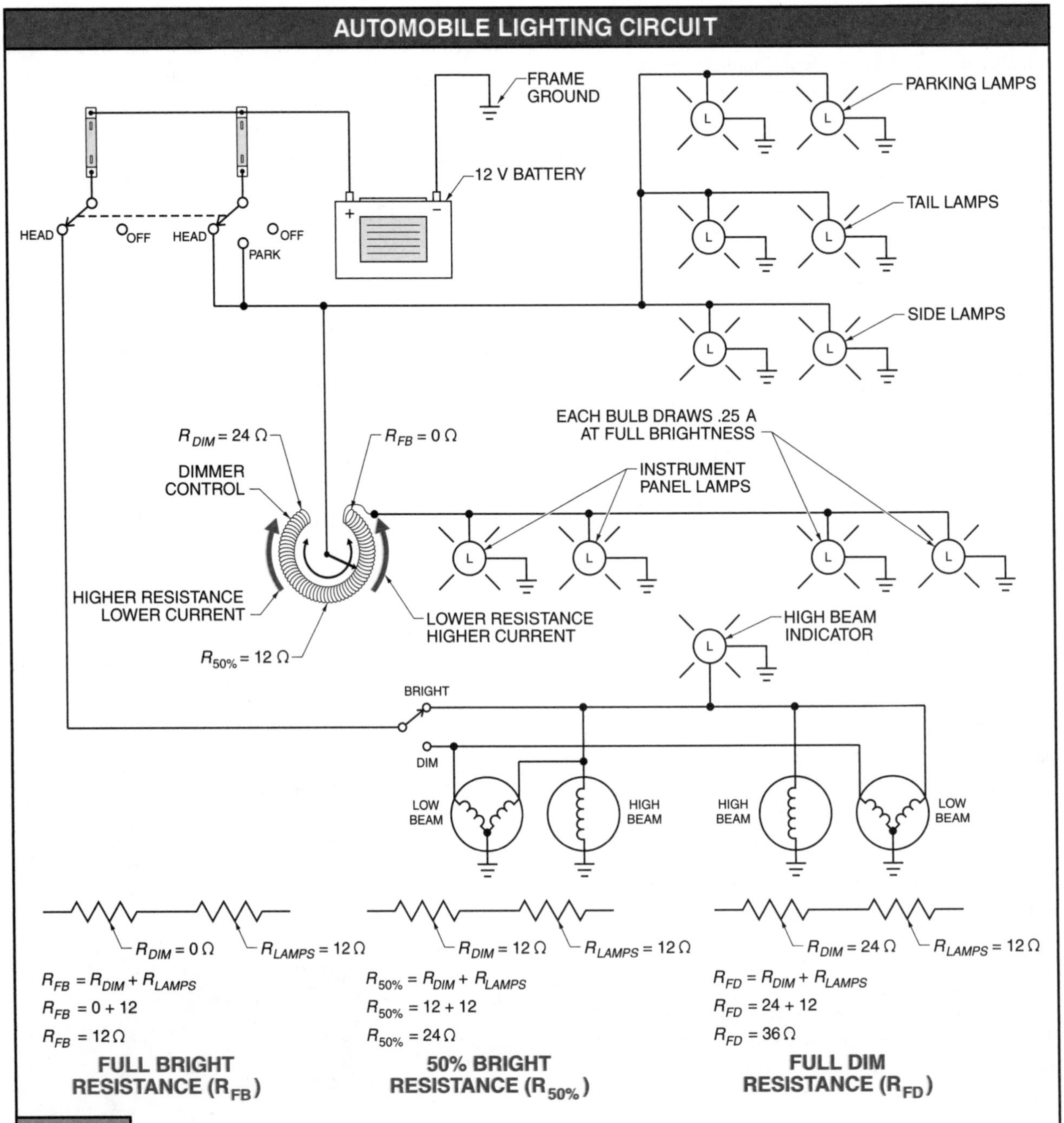

AUTOMOBILE LIGHTING CIRCUIT
FRAME GROUND
12 V BATTERY
PARKING LAMPS
TAIL LAMPS
SIDE LAMPS
HEAD
OFF
HEAD
OFF
PARK
EACH BULB DRAWS .25 A AT FULL BRIGHTNESS
INSTRUMENT PANEL LAMPS
R_DIM = 24 Ω
R_FB = 0 Ω
DIMMER CONTROL
HIGHER RESISTANCE LOWER CURRENT
LOWER RESISTANCE HIGHER CURRENT
R_50% = 12 Ω
HIGH BEAM INDICATOR
BRIGHT
DIM
LOW BEAM
HIGH BEAM
HIGH BEAM
LOW BEAM
R_DIM = 0 Ω
R_LAMPS = 12 Ω
R_FB = R_DIM + R_LAMPS
R_FB = 0 + 12
R_FB = 12 Ω
FULL BRIGHT RESISTANCE (R_FB)
R_DIM = 12 Ω
R_LAMPS = 12 Ω
R_50% = R_DIM + R_LAMPS
R_50% = 12 + 12
R_50% = 24 Ω
50% BRIGHT RESISTANCE (R_50%)
R_DIM = 24 Ω
R_LAMPS = 12 Ω
R_FD = R_DIM + R_LAMPS
R_FD = 24 + 12
R_FD = 36 Ω
FULL DIM RESISTANCE (R_FD)

Figure 3-9

Using Ohm's Law When Troubleshooting

Ohm's law is used during troubleshooting to determine circuit conditions. Voltage and current measurements are taken because resistance cannot be measured on a circuit that is powered. After voltage and current measurements are taken, Ohm's law is applied to determine the resistance of the circuit. **See Figure 3-10.**

In this circuit, the resistance of an electric heater may be determined by measuring the circuit's voltage and current. Using Ohm's law, the normal circuit's resistance is equal to 60 Ω ($^{140}/_{4}$ = 60 Ω).

TECH TIP

When selecting or sizing a switch, potentiometer, or rheostat, make sure the voltage, current, and power ratings are higher than the current in the circuit.

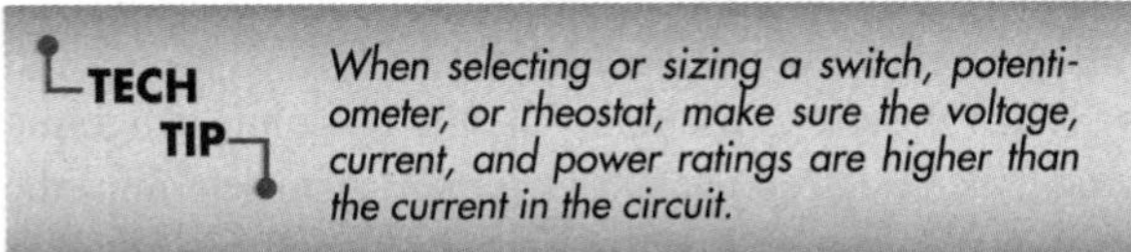

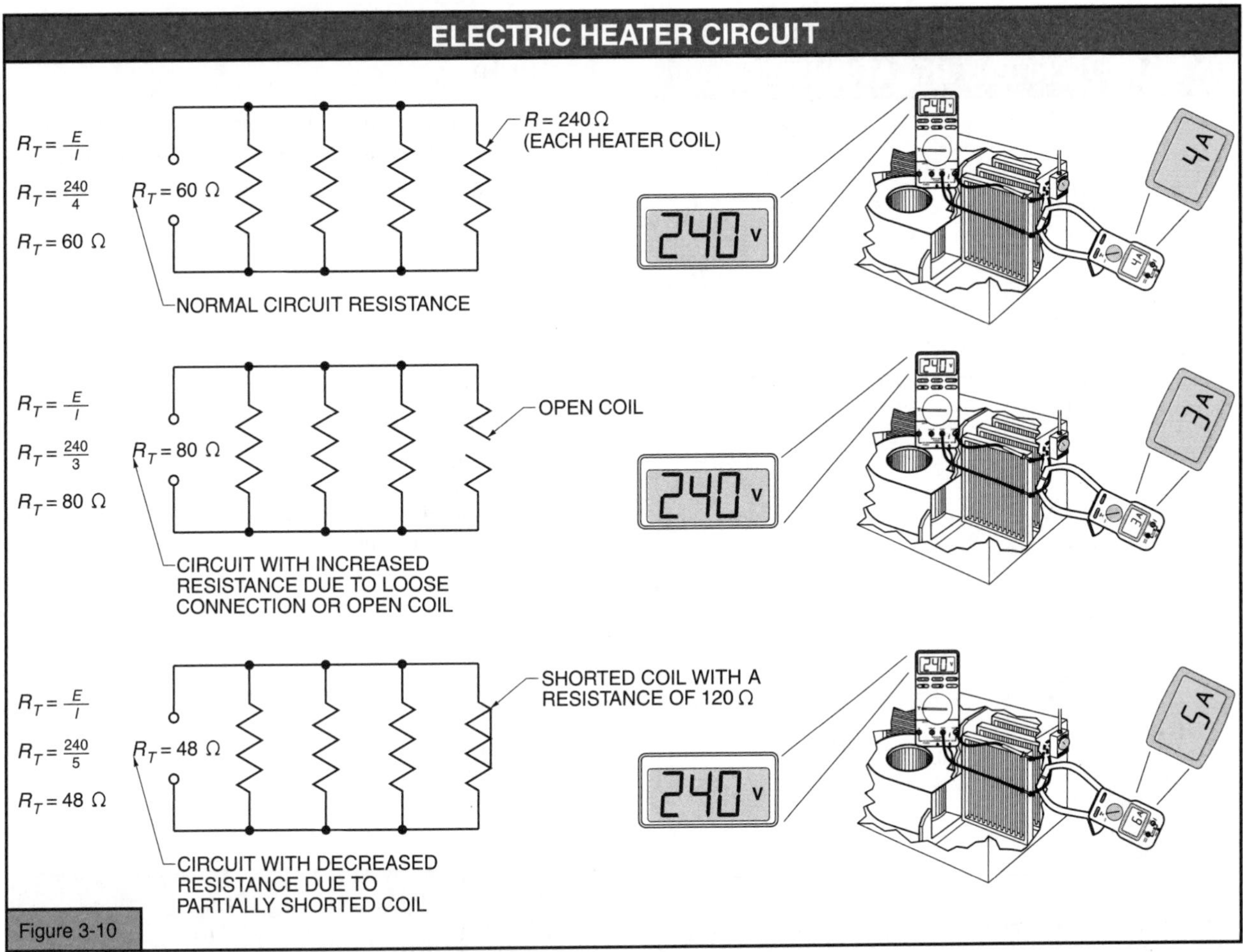

Figure 3-10

A troubleshooter can use the 60 Ω resistance value to help determine the condition of the circuit. For example, if the ammeter measured 3 A instead of 4 A, the troubleshooter knows that the circuit's resistance has increased ($^{140}/_{12}$ = 80 Ω). An additional 20 Ω of circuit resistance is present (80 Ω – 60 Ω = 20 Ω). The troubleshooter knows that the additional 20 Ω could come from a loose (or dirty) connection or open coil sections. Open coil sections increase the total circuit resistance.

Likewise, if the ammeter measured 5 A instead of 4 A, the troubleshooter knows that the circuit's resistance has decreased ($^{240}/_{5}$ = 48 Ω). Because the circuit's resistance has decreased by 12 Ω, the troubleshooter knows that the decreased resistance could be caused by a partially shorted coil or insulation breakdown.

Power Formula

The *power formula* is the relationship between power (P), voltage (E), and current (I) in an electrical circuit. Any value in this relationship may be found using the power formula when the other two are known. Using the formula listed in a chart form, the unknown value is converted. The relationship between power, current, and voltage may be visualized by presenting the power formula in pie chart form. **See Figure 3-11.**

Calculating Power Using Power Formula

The power formula states that power (P) in a circuit is equal to voltage (E) times current (I). **See Figure 3-12.** To calculate power using the power formula, apply the formula:

$P = E \times I$

where

P = power (in W)

E = voltage (in V)

I = current (in A)

Example: Calculating Power Using Power Formula

What is the power of a load that draws .5 A when connected to a 120 V supply?

$P = E \times I$

$P = 120 \times .5$

$P = \mathbf{60}$ W

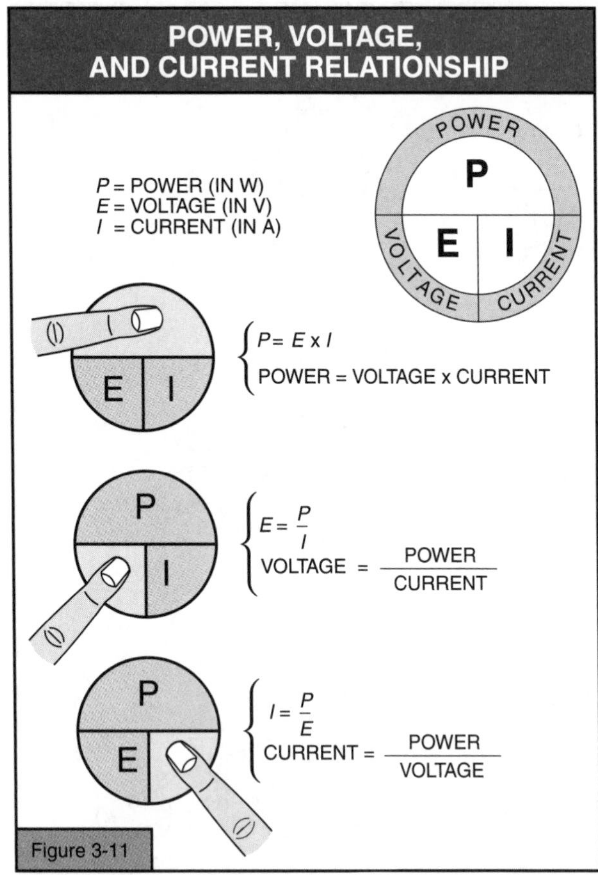

Figure 3-11

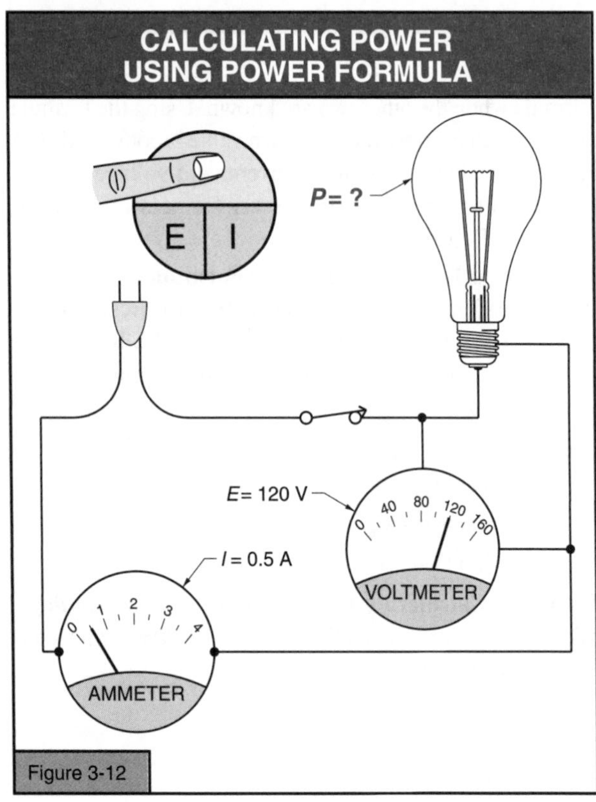

Figure 3-12

Calculating Voltage Using Power Formula

The power formula states that voltage (E) in a circuit is equal to power (P) divided by current (I). **See Figure 3-13.** To calculate voltage using the power formula, apply the formula:

$$E = \frac{P}{I}$$

where

E = voltage (in V)

P = power (in W)

I = current (in A)

Example: Calculating Voltage Using Power Formula
What is the voltage in a circuit in which a 60 W load draws .5 A?

$$E = \frac{P}{I}$$

$$E = \frac{60}{.5}$$

$$E = \mathbf{120\ V}$$

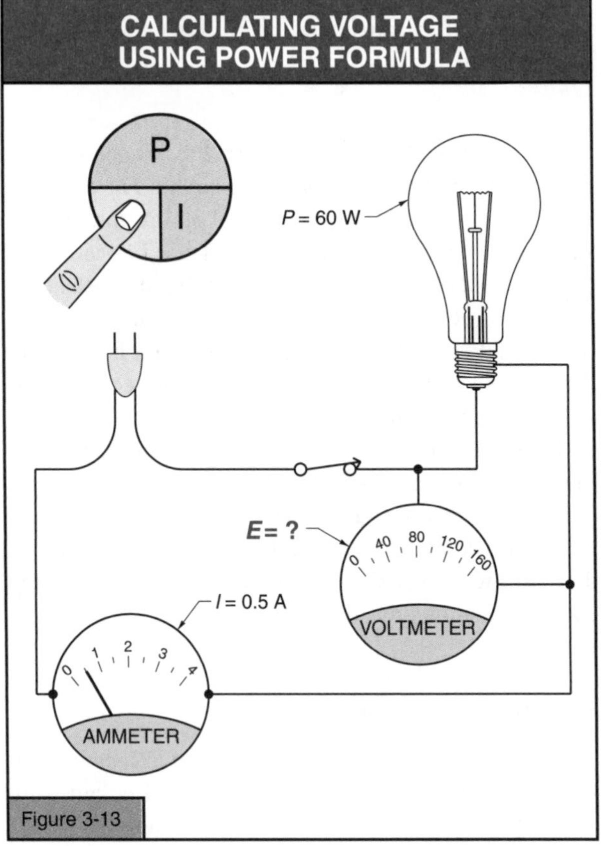

Figure 3-13

Calculating Current Using Power Formula

The power formula states that current (I) in a circuit is equal to power (P) divided by voltage (E). **See Figure 3-14.** To calculate current using the power formula, apply the formula:

$$I = \frac{P}{E}$$

where

I = current (in A)

P = power (in W)

E = voltage (in V)

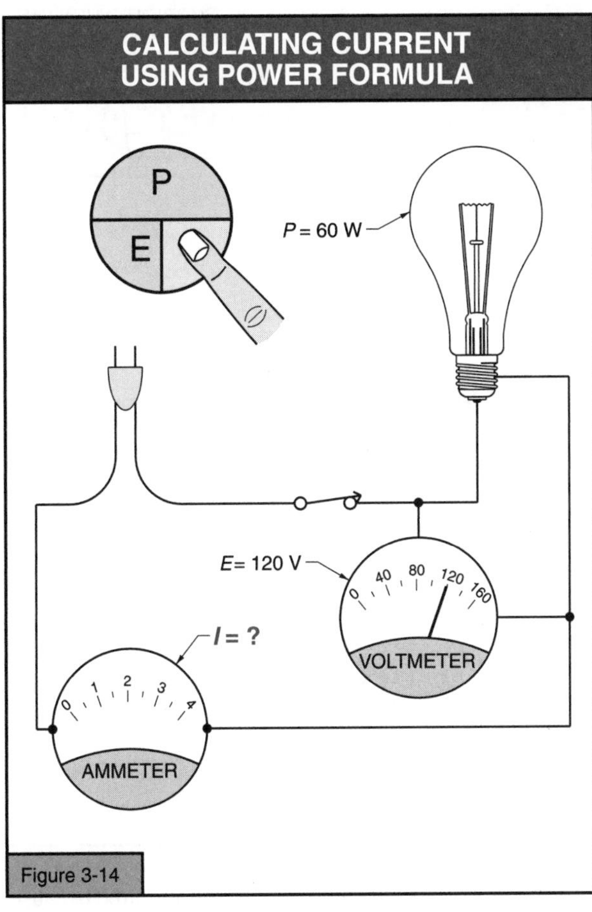

CALCULATING CURRENT USING POWER FORMULA

Figure 3-14

Example: Calculating Current Using Power Formula
What is the current in a circuit in which a 60 W load is connected to a 120 V supply?

$$I = \frac{P}{E}$$

$$I = \frac{60}{120}$$

$$I = .5 \text{ A}$$

Power/Current Relationship

The power formula states that if the voltage in a circuit remains constant and the power required from the circuit changes, the current in the circuit also changes. The power required from a circuit changes any time loads are added (power increase) or removed (power decrease). This principle is shown in any circuit when loads are added. **See Figure 3-15.**

In this circuit, a 240 W fan, a 120 W sign, and a 1200 W motor are connected to a 120 V supply. The total circuit current may be found using the power formula. For example, when the 240 W fan is turned ON, the total circuit current equals 2 A ($^{240}/_{120}$ = 2 A). The current of the fan equals 2 A.

If, in addition to the fan being ON, the sign is also turned ON, the circuit's current increases by 1 A ($^{120}/_{120}$ = 1 A). The current of the sign equals 1 A. The total current with both the fan and the sign ON equals 3 A (2 + 1 = 3 A).

If, in addition to the fan and sign being ON, the motor is also turned ON, the circuit's current increases by 10 A ($^{1200}/_{120}$ = 10 A). The current of the motor equals 10 A. The total circuit current with all three loads ON equals 13 A (2 + 1 + 10 = 13 A).

Applications Using Power Formula

Like Ohm's law, the power formula is used when troubleshooting and to predict circuit characteristics before power is applied. Any electrical value (P, E, or I) can be calculated when the other two values are known or measured. The power formula is useful when determining expected current values because most electrical equipment lists a voltage and power rating. The listed power rating is given in watts for most appliances and heating elements or in horsepower for motors.

Saftronics Inc.

Power and current calculations are required when designing equipment such as overhead bridge cranes.

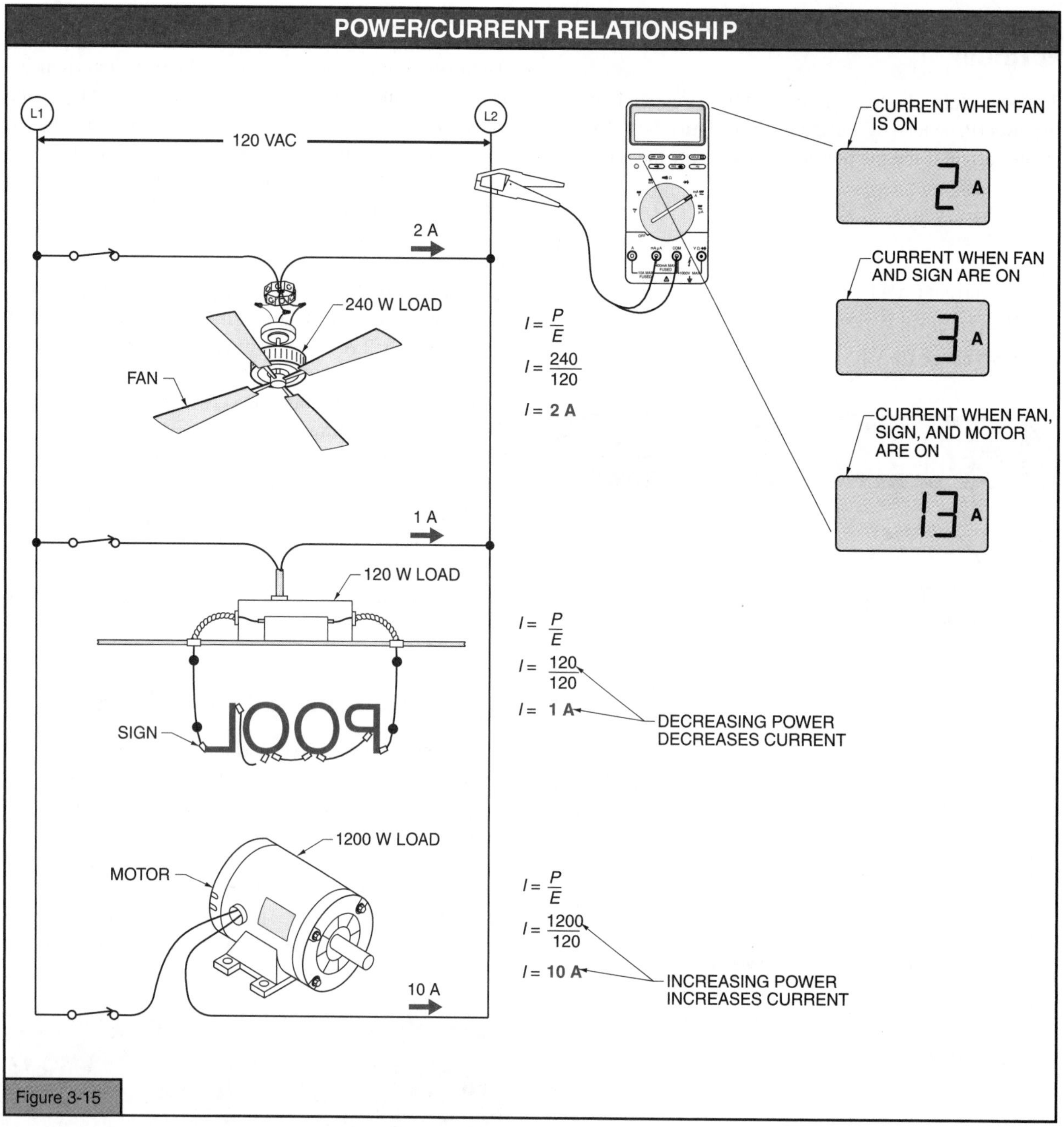

POWER/CURRENT RELATIONSHIP
L1
L2
120 VAC
2 A
240 W LOAD
FAN
I = P/E
I = 240/120
I = 2 A
CURRENT WHEN FAN IS ON
2 A
CURRENT WHEN FAN AND SIGN ARE ON
3 A
CURRENT WHEN FAN, SIGN, AND MOTOR ARE ON
13 A
1 A
120 W LOAD
SIGN
POOL
I = P/E
I = 120/120
I = 1 A
DECREASING POWER DECREASES CURRENT
1200 W LOAD
MOTOR
10 A
I = P/E
I = 1200/120
I = 10 A
INCREASING POWER INCREASES CURRENT

Figure 3-15

Heat is produced any time electricity passes through a wire that has resistance. This method is used to produce heat in such devices as toasters, portable space heaters, hair dryers, and electric water heaters. When selecting a heating element, the size of the heating element is usually stated in watts. The size selected is based on the application. The greater the required heat output, the higher the required wattage. Once a wattage size is selected, the current draw is determined by using the power formula. For example, if a 2000 W heating element is connected to a 115 V supply, the circuit's current equals 17.39 A ($^{2000}/_{115}$ = 17.39 A). **See Figure 3-16.**

Finding the current value (17.39 A) is required because the wire size, switch rating, and fuse (or circuit breaker) rating are all based on the amount of current. Likewise, the conduit size is based on the wire size. Thus, knowing the amount of current a load draws is essential to putting a system together.

TECH TIP

One watt of power equals the work done in 1 sec by 1 V of potential difference in moving 1 C of change. One coulomb per second equals 1 A.

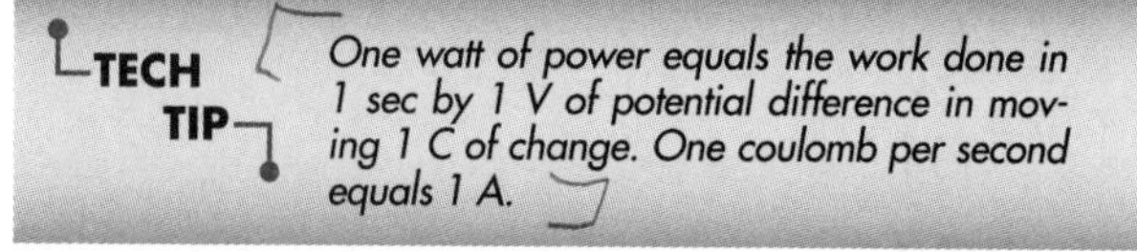

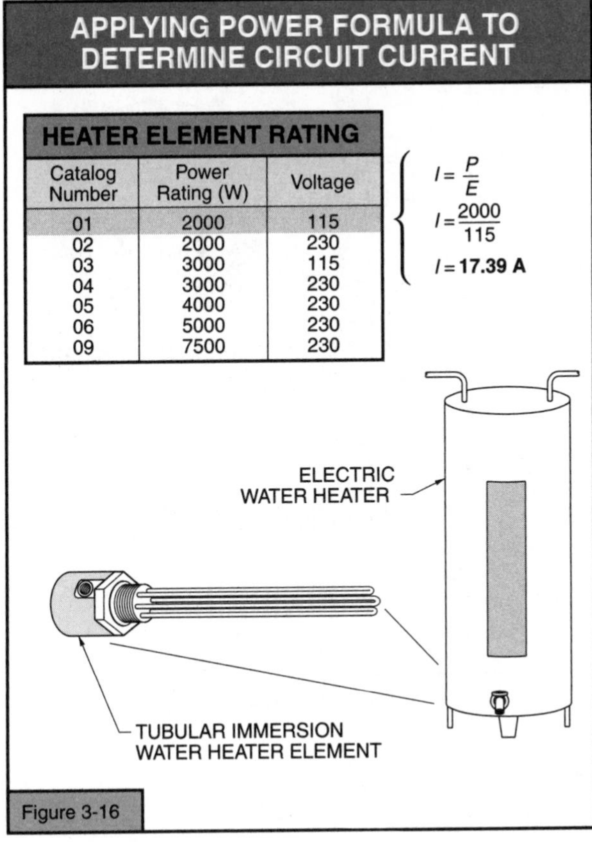

Catalog Number	Power Rating (W)	Voltage
01	2000	115
02	2000	230
03	3000	115
04	3000	230
05	4000	230
06	5000	230
09	7500	230

Figure 3-16

Combining Ohm's Law and Power Formula

Ohm's law and the power formula may be combined mathematically and written as any combination of voltage (E), current (I), resistance (R), or power (P). This combination lists six basic formulas and six rearranged formulas. **See Figure 3-17.**

Ohm's Law and Impedance

Ohm's law and the power formula are limited to circuits in which electrical resistance is the only significant opposition to the flow of current. This limitation includes all direct current (DC) circuits and any alternating current (AC) circuits that do not contain a significant amount of inductance and/or capacitance. AC circuits that do not include inductance and/or capacitance include such devices as heating elements and incandescent lamps. AC circuits that include inductance are any circuits that include a coil as the load. Motors, transformers, and solenoids all include coils. AC circuits that include capacitance are any circuits that include a capacitor(s).

In DC circuits and AC circuits that do not contain a significant amount of inductance and/or capacitance, the opposition to the flow of current is resistance (R). In circuits that contain inductance (X_L) or capacitance (X_C), the opposition to the flow of current is reactance (X). In circuits that contain resistance (R) and reactance (X), the combined opposition to the flow of current is impedance (Z). Impedance is stated in ohms.

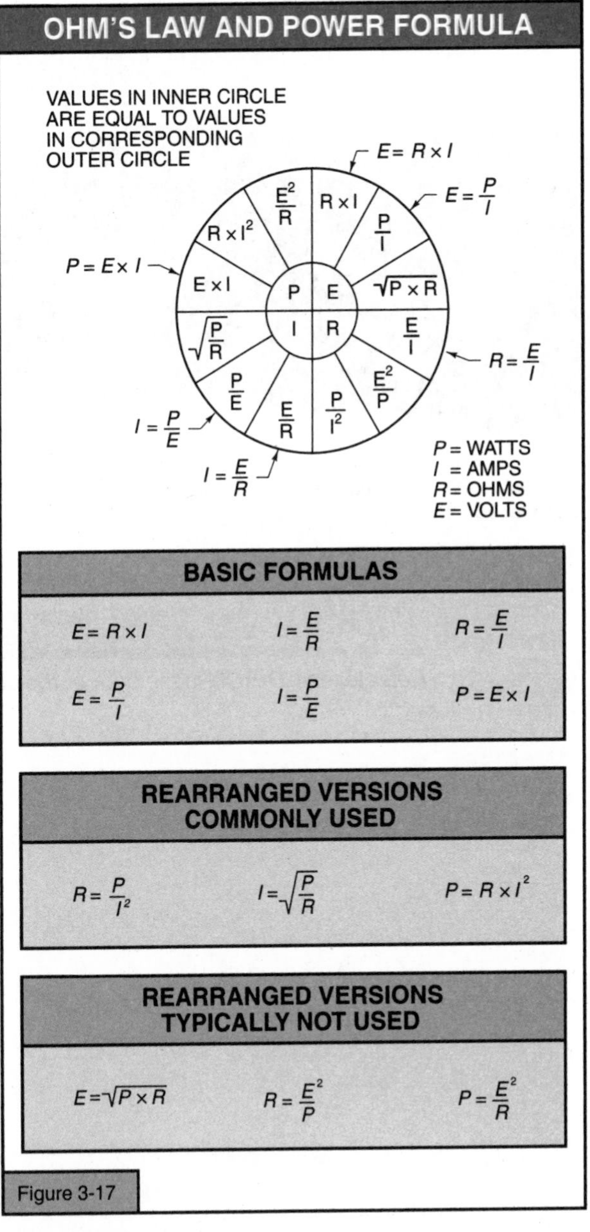

Figure 3-17

TECH FACT

Georg S. Ohm discovered that all electrical quantities are proportional to each other and can therefore be expressed as mathematical formulas.

Ohm's law is used in circuits that contain impedance, however, Z is substituted for R in the formula. Z represents the total resistive force (resistance and reactance) opposing current flow. The relationship between voltage (E), current (I), and impedance (Z) may be visualized by presenting the relationship in pie chart form. **See Figure 3-18.**

Motorola Inc.

Calculations of impedance are required when designing equipment such as digital cable receivers.

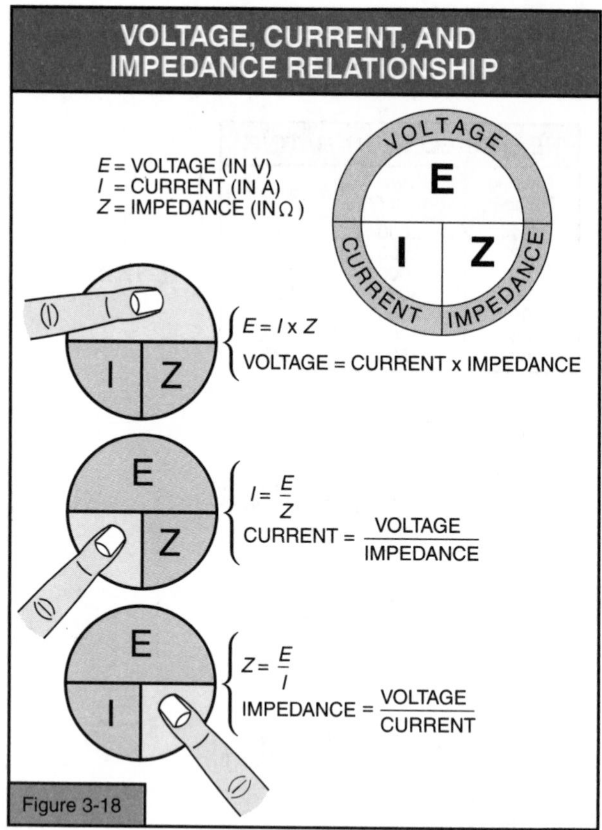

Figure 3-18

Refer to the CD-ROM in the back of the book for Chapter 3 Quick Quiz® and related information.

Codes and Standards

National, state, and local codes and standards are used to protect people and property from electrical dangers. A *code* is a regulation or minimum requirement. A *standard* is an accepted reference or practice. Codes and standards ensure electrical equipment is built and installed safely and every effort is made to protect people from electrical shock. **See Figure 4-1.**

National Fire Protection Association (NFPA)

The *National Fire Protection Association (NFPA)* is a national organization that provides guidance in assessing the hazards of the products of combustion. The NFPA publishes the National Electrical Code (NEC®). The purpose of the NEC® is the practical safeguarding of persons and property from the hazards arising from the use of electricity. The NEC® is updated every three years. Many city, county, state, and federal agencies use the NEC® to set requirements for electrical installations.

American National Standards Institute (ANSI)

The *American National Standards Institute (ANSI)* is a national organization that helps identify industrial and public needs for standards. ANSI coordinates and encourages activities in national standards development.

Occupational Safety and Health Administration (OSHA)

The *Occupational Safety and Health Administration (OSHA)* is a federal agency that requires all employers to provide a safe environment for their employees. All work areas must be free from hazards likely to cause serious harm. The provisions of this act are enforced by federal inspection. With few exceptions, OSHA uses the NEC® to help ensure a safe electrical environment. OSHA has developed color codes to help ensure a safe environment. **See Figure 4-2.** The color codes help quickly identify fire protection equipment, physical hazards, dangerous parts of machines, radiation hazards, and locations of first aid equipment.

National Electrical Manufacturers Association (NEMA)

The *National Electrical Manufacturers Association (NEMA)* is a national organization that assists with information and standards concerning proper selection, ratings, construction, testing, and performance of electrical equipment. NEMA standards are used as guidelines for the manufacture and use of electrical equipment.

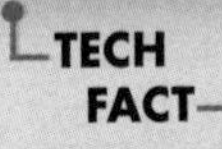

Per OSHA regulations, it is an employer's duty to furnish each employee a place of employment that is free from recognized hazards that cause or are likely to cause death or serious physical harm.

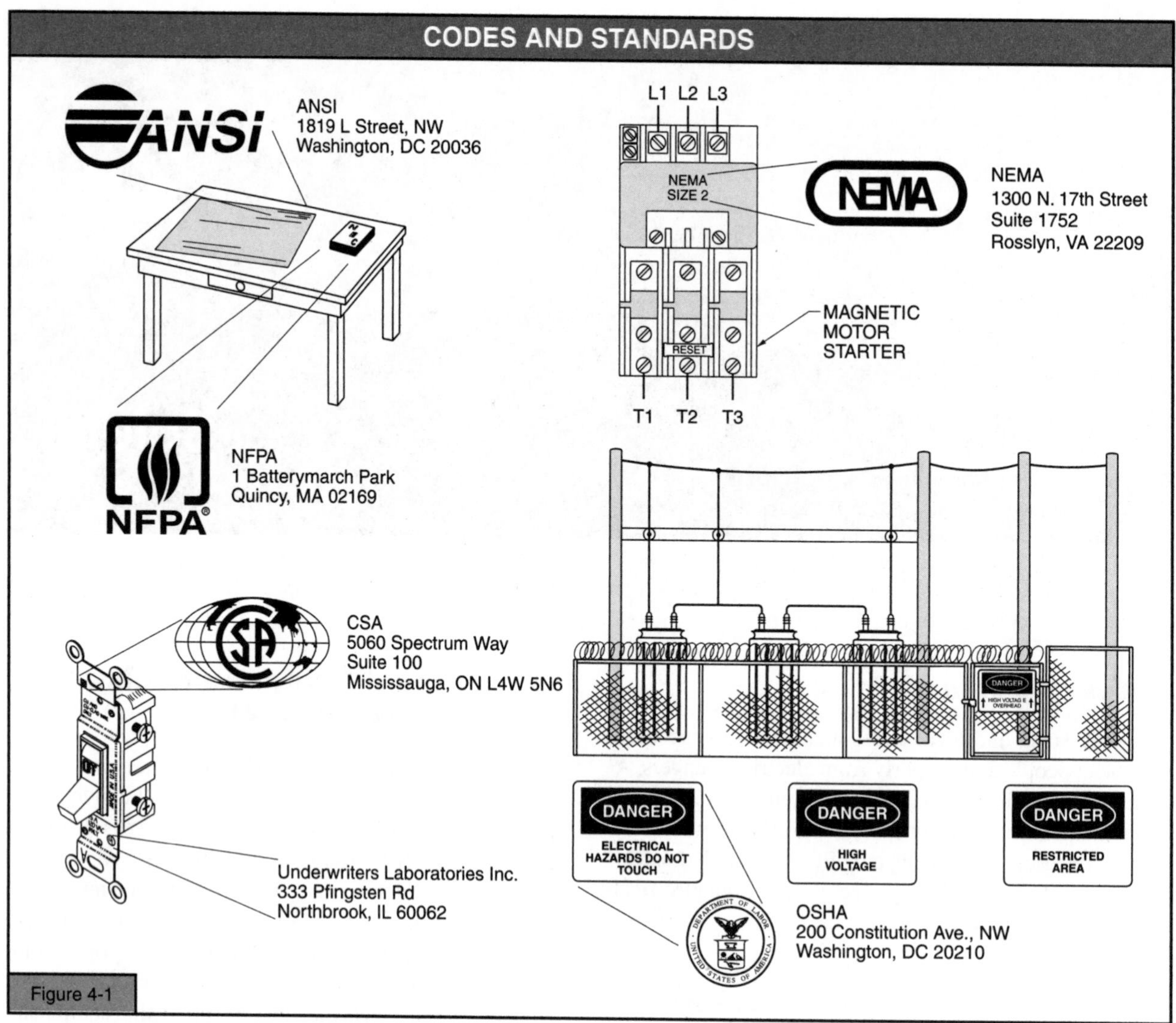

Figure 4-1

OSHA SAFETY COLOR CODES

Color	Examples
Red	Fire protection equipment and apparatus, portable containers of flammable liquids, emergency stop pushbuttons and switches
Yellow	Caution and for marking physical hazards, waste containers for explosive or combustible materials, caution against starting, using, or moving equipment under repair, identification of the starting point or power source of machinery
Orange	Dangerous parts of machines, safety starter buttons, the exposed parts (edges) of pulleys, gears, rollers, cutting devices, power jaws
Purple	Radiation hazards
Green	Safety, location of first aid equipment (other than fire fighting equipment)

Figure 4-2

Underwriters Laboratories Inc. (UL®)

Underwriters Laboratories Inc. (UL®) is an independent organization that tests equipment and products to see if they conform to national codes and standards. Equipment tested and approved by UL® carries the UL® label. UL®-approved equipment and products are listed in its annual publication.

Canadian Standards Association (CSA)

The *Canadian Standards Association (CSA)* is a Canadian organization similar to UL® that tests equipment and products to ensure they meet national standards.

Fire Extinguisher Classes

The five classes of fires are Class A, Class B, Class C, Class D, and Class K. Class A fires include burning wood, paper, textiles, and other ordinary combustible materials containing carbon. Class B fires include burning oil, gas, grease, paint, and other liquids that convert to a gas when heated. Class C fires include burning electrical devices, motors, and transformers. Class D is a specialized class of fires including burning metals such as zirconium, titanium, magnesium, sodium, and potassium. Class K fires include grease fires in commercial cooking equipment. **See Figure 4-3.**

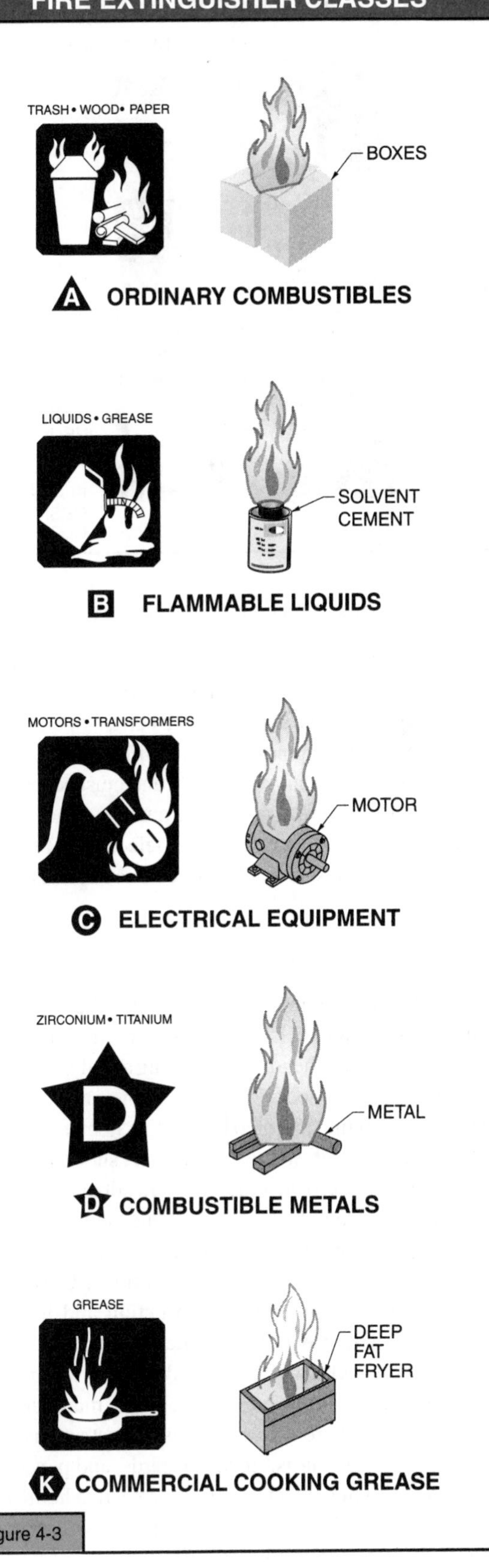

Figure 4-3

Personal Protective Equipment

Personal protective equipment (PPE) is clothing and/or equipment used by a technician to reduce the possibility of injury in the work area. The use of personal protective equipment is required whenever work may occur on or near energized exposed electrical circuits. The National Fire Protection Association standard *NFPA 70E, Standard for Electrical Safety in the Workplace,* addresses "electrical safety requirements for employee workplaces that are necessary for the safeguarding of employees in pursuit of gainful employment."

For maximum safety, personal protective equipment and safety requirements must be followed as specified in NFPA 70E, OSHA Standard Part 1910 *Subpart 1—Personal Protective Equipment* (1910.132 through 1910.138), and other applicable safety mandates. Personal protective equipment includes flame-resistant clothing, head protection, eye protection, ear protection, hand protection, foot protection, back protection, knee protection, rubber insulating matting, and arc blast protection. **See Figure 4-4.**

Per NFPA 70E, "Only qualified persons shall perform testing on or near live parts operating at 50 V or more." All personal protective equipment and tools are selected to be appropriate to the operating voltage (or higher) of the equipment or circuits being worked on or near. Equipment, devices, tools, and test instruments must be suited for the work to be performed. In most cases, voltage-rated gloves and tools are required. Voltage-rated gloves and tools are rated and tested for the maximum line-to-line voltage upon which work will be performed. Protective gloves must be inspected or tested as required for maximum safety before each task.

A facility fire protection program prevents personal injury and loss of life, protects property, prevents the inception of accidental fires, and gives early warning of their occurrence.

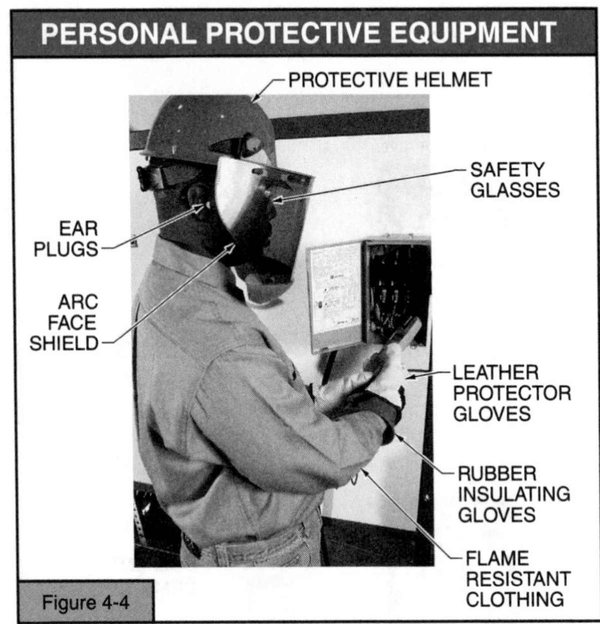

Figure 4-4

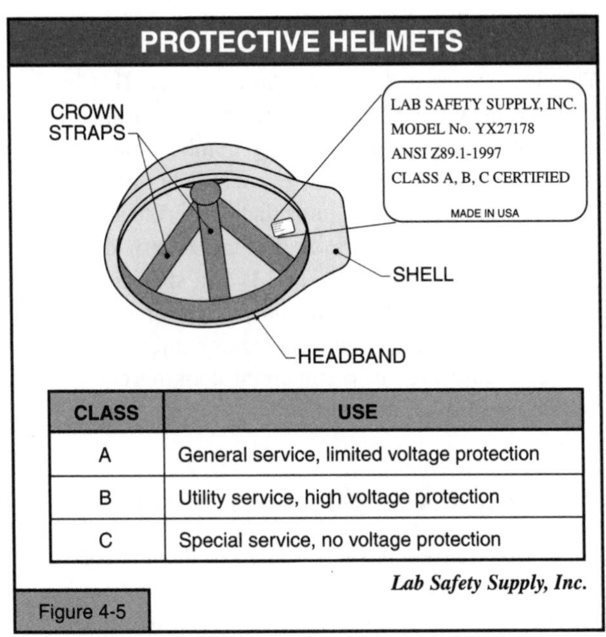

CLASS	USE
A	General service, limited voltage protection
B	Utility service, high voltage protection
C	Special service, no voltage protection

Lab Safety Supply, Inc.

Figure 4-5

Flame-Resistant (FR) Clothing

Sparks from an electrical circuit can cause a fire. Approved flame-resistant (FR) clothing must be worn for protection from electrical arcs when performing certain operations on or near energized equipment or circuits. FR clothing must be kept as clean and sanitary as practical and must be inspected prior to each use. Defective clothing must be removed from service immediately and replaced. Defective FR clothing must be tagged "unsafe" and returned to a supervisor.

Head Protection

Head protection requires using a protective helmet. A *protective helmet* is a hard hat that is used in the workplace to prevent injury from the impact of falling and flying objects, and from electrical shock. Protective helmets resist penetration and absorb impact force. Protective helmet shells are made of durable, lightweight materials. A shock-absorbing lining consists of crown straps and a headband that keeps the shell away from the head to provide ventilation.

Protective helmets are identified by class of protection against specific hazardous conditions. Class A, B, and C helmets are used for construction and industrial applications. Class A protective helmets protect against low-voltage shocks and burns and impact hazards and are commonly used in construction and manufacturing facilities. Class B protective helmets protect against high-voltage shock and burns, impact hazards, and penetration by falling or flying objects. Class C protective helmets are manufactured with lighter materials, yet provide adequate impact protection. **See Figure 4-5.**

Eye Protection

Eye protection must be worn to prevent eye or face injuries caused by flying particles, contact arcing, and radiant energy. Eye protection must comply with OSHA 29 CFR 1910.133, *Eye and Face Protection*. Eye protection standards are specified in ANSI Z87.1, *Occupational and Educational Eye and Face Protection*. Eye protection includes safety glasses, face shields, and arc blast hoods. **See Figure 4-6.**

Safety glasses are an eye protection device with special impact-resistant glass or plastic lenses, reinforced frames, and side shields. Plastic frames are designed to keep the lenses secured in the frame if an impact occurs and to minimize the shock hazard when working with electrical equipment. Side shields provide additional protection from flying objects. Tinted-lens safety glasses protect against low-voltage arc hazards.

A *face shield* is an eye and face protection device that covers the entire face with a plastic shield and is used for protection from flying objects. Tinted face shields protect against low-voltage arc hazards. An *arc blast hood* is an eye and face protection device that consists of a flame-resistant hood and face shield.

Safety glasses, face shields, and arc blast hoods must be properly maintained to provide protection and clear visibility. Lens cleaners are available that clean without risk of lens damage. Pitted, scratched, and crazed lenses (crazing is a defect caused by exposure to aggressive solvents, chemicals, or heat that leaves microscopic cracks within the lenses) reduce vision and may cause lenses to fail on impact.

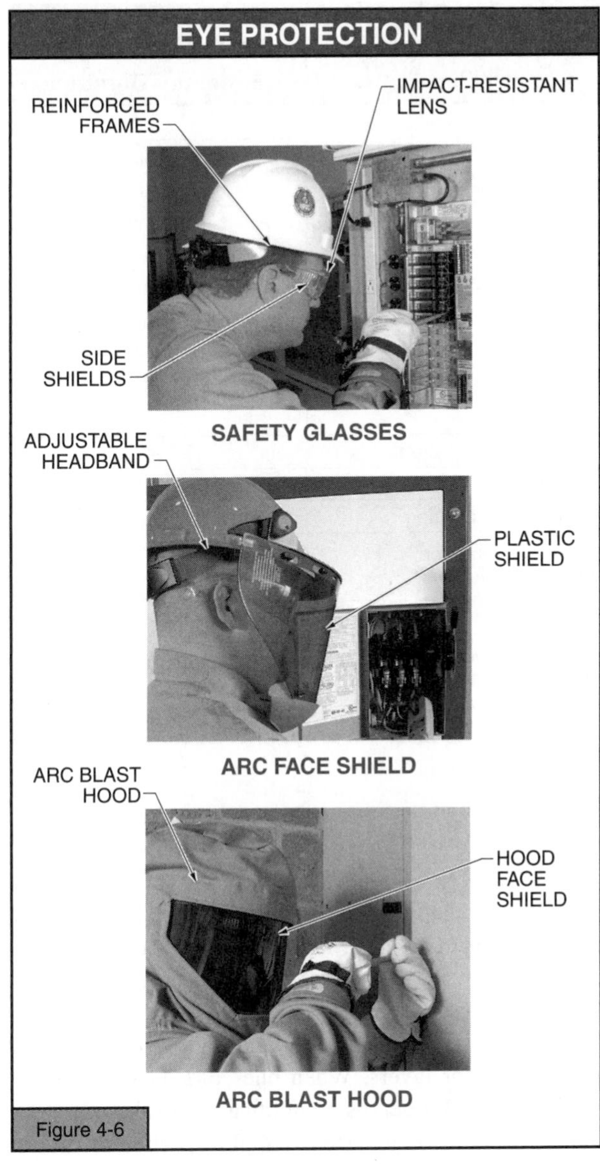

Figure 4-6

Ear Protection

Ear protection devices are worn to limit the noise entering the ear and include earplugs and earmuffs. An *earplug* is an ear protection device made of moldable rubber, foam, or plastic and inserted into the ear canal. An *earmuff* is an ear protection device worn over the ears. A tight seal around an earmuff is required for proper protection.

Power tools and equipment can produce excessive noise levels. Technicians subjected to excessive noise levels may develop hearing loss over time. The severity of hearing loss depends on the intensity and duration of exposure. Noise intensity is expressed in decibels. A *decibel (dB)* is a unit of measure used to express the relative intensity of sound. **See Figure 4-7.** Ear protection is worn to prevent hearing loss.

Ear protection devices are assigned a noise reduction rating (NRR) number based on the noise level reduced. For example, an NRR of 27 means that the noise level is reduced by 27 dB when the device is tested at the factory. To determine approximate noise reduction in the field, 7 dB is subtracted from the NRR. For example, an NRR of 27 provides a noise reduction of approximately 20 dB in the field.

Hand Protection

Hand protection consists of rubber insulating gloves and leather protectors worn to prevent injuries to hands caused by cuts or electrical shock. The primary purpose of rubber insulating gloves and leather protectors is to insulate hands and lower arms from possible contact with live conductors. Rubber insulating gloves offer a high resistance to current flow to help prevent an electrical shock and the leather protectors protect the rubber glove and add additional insulation. Rubber insulating gloves are rated, labeled, and color coded as to the maximum voltage allowed. **See Figure 4-8.**

Warning: Rubber insulating gloves are designed for specific applications. Leather protector gloves are required for protecting rubber insulating gloves. Rubber insulating gloves must not be used alone. Rubber insulating gloves offer the highest resistance and greatest insulation. Serious injury or death can result from improper use of rubber insulating gloves, or from using outdated and/or the wrong type of rubber insulating gloves for an application.

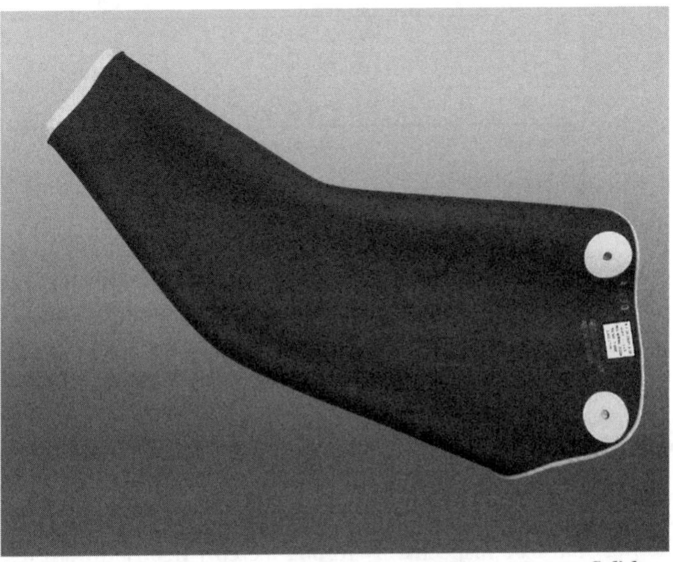
Salisbury

Insulating sleeves can be used with rubber insulating gloves for additional protection from contact with electrical lines and equipment.

SOUND LEVELS			
Average Decibel (dB)	**Loudness**	**Examples**	**Exposure Duration**
140	Deafening	Jet airplane taking off, air raid siren, locomotive horn	
130	Pain threshold		2 min
120	Feeling threshold		7 min
110	Uncomfortable		30 min
100	Very loud	Chain saw	2 hr
90	Noisy	Shouting, auto horn	4 hr
85			8 hr
80	Moderately loud	Vacuum cleaner	25.5 hr
70	Loud	Telephone ringing, loud talking	—
60	Moderate	Normal conversation	—
50	Quiet	Hair dryer	—
40	Moderately quiet	Refrigerator running	—
30	Very quiet	Quiet conversation, broadcast studio	—
20	Faint	Whispering	—
10	Barely audible	Rustling leaves, soundproof room, human breathing	—
0	Hearing threshold	Intolerably quiet	—

Figure 4-7

RUBBER INSULATING GLOVE RATING*		
Class	**Maximum Use**	**Label Color**
00	500 V	Beige
0	1 kV (1000 V)	Red
1	7.5 kV (7500 V)	White
2	17 kV (17,000 V)	Yellow
3	26.5 kV (26,500 V)	Green
4	36 kV (36,000 V)	Orange

* Refer to ASTM D 120-02 Standard Specification for Rubber Insulating Gloves.

Figure 4-8

Inspection of Rubber Insulating Gloves

Rubber insulating gloves must be field tested (by visual inspection and air test) prior to each use. Rubber insulating gloves must also be laboratory tested by an approved laboratory every six months.

Rubber insulating gloves must also be air-tested when there is cause to suspect any damage. The entire surface must be inspected by rolling the cuff tightly toward the palm in such a manner that air is trapped inside the glove, or by using a mechanical inflation device. When using a mechanical inflation device, care must be taken to avoid over-inflation. The glove is examined for punctures and other defects. Puncture detection may be enhanced by listening for escaping air.

Some brands of rubber insulating gloves are available with two color layers. When one color layer becomes visible, the color serves as notification to the user that the gloves must be replaced. Visual inspection of rubber insulating gloves is performed by stretching a small area (particularly fingertips) and checking for defects such as punctures or pin holes, embedded or foreign material, deep scratches or cracks, cuts or snags, and deterioration caused by oil, heat, grease, insulating compounds, or any other substances that may harm rubber.

When visual inspection is completed, an air test is performed using the following procedure:

1. Grasp the gauntlet of the glove between thumb and forefinger.
2. Spin the glove around rapidly to fill with air.
3. Trap the air by squeezing the gauntlet with one hand, while using the other hand to squeeze the palm, fingers, and thumb while looking for defects.
4. Hold the glove to the face or ear to detect escaping air.

Proper care of leather protector gloves is essential for user safety. Leather protector gloves are checked for cuts, tears, holes, abrasions, defective or worn stitching, oil contamination, and any other condition that might prevent the glove from adequately protecting rubber insulating gloves. Rubber insulating gloves or leather protector gloves found to be defective shall not be discarded or destroyed in the field, but shall be tagged "unsafe" and returned to a supervisor.

Foot Protection

Foot protection consists of shoes worn to prevent foot injuries that are typically caused by objects falling less than 4′ and having an average weight of less than 65 lb. Safety shoes with reinforced steel toes protect against injuries caused by compression and impact. Insulated rubber-soled shoes are commonly worn during electrical work to prevent electrical shock. Protective footwear must comply with ANSI Z41, *Personal Protection—Protective Footwear.*

Back Protection

A back injury is one of the most common injuries resulting in lost time in the workplace. Back injuries are the result of improper lifting procedures. Back injuries can be avoided through good planning and by following work procedures. Assistance should be sought when moving heavy objects. When lifting objects from the ground, ensure the path is clear of obstacles and free of hazards. Bend the knees and grasp the object firmly. Next, lift the object, straightening the legs and keeping the back as straight as possible. Finally, move forward after the whole body is in the vertical position. Keep the load close to the body and keep the load steady. **See Figure 4-9.**

Knee Protection

A *knee pad* is a rubber, leather, or plastic pad strapped onto the knees for protection. A knee pad is worn by technicians who spend considerable time working on their knees or who work in close areas and must kneel for proper access to electrical equipment. Buckle straps or Velcro™ closures secure knee pads in position. **See Figure 4-10.**

TECH FACT *While back braces and lifting belts provide extra support for back muscles when performing lifting tasks, they do not prevent injury. Injury is prevented only by using proper lifting techniques.*

Figure 4-9

KNEE PROTECTION

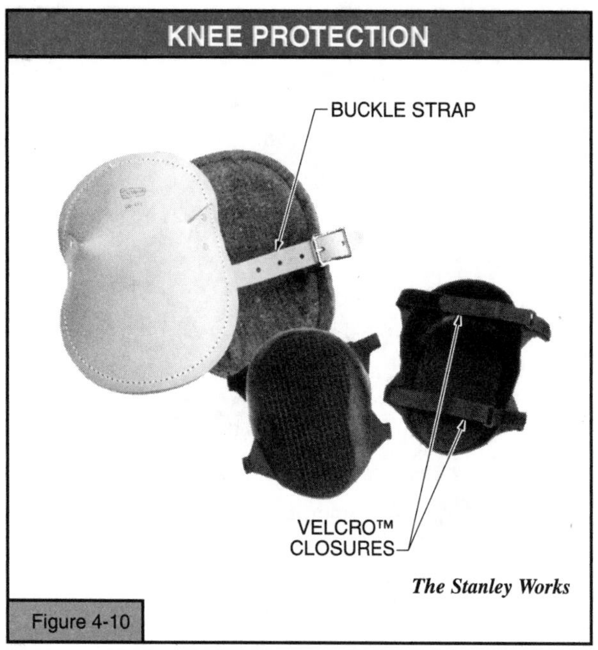

The Stanley Works

Figure 4-10

Rubber Insulating Matting

Rubber insulating matting is a floor covering that provides technicians protection from electrical shock when working on live electrical circuits. Dielectric black fluted rubber matting is specifically designed for use in front of open cabinets or high-voltage equipment. Matting is used to protect technicians when voltages are over 50 V. Two types of matting that differ in chemical and physical characteristics are designated as Type I natural rubber and Type II elastomeric compound matting. **See Figure 4-11.**

Arc Blast

An *electric arc* is a discharge of electric current across an air gap. Arcs are caused by excessive voltage ionizing an air gap between two conductors, or by accidental contact between two conductors and followed by reseparation. When an electric arc occurs, there is the possibility of an arc flash or an arc blast. An *arc flash* is an extremely high-temperature discharge produced by an electrical fault in the air. Arc flash temperatures reach 35,000°F. An *arc blast* is an explosion that occurs when the air surrounding electrical equipment becomes ionized and conductive. The threat of arc blast is greatest from electrical systems of 480 V and higher. Arc blasts are possible in systems of lesser voltage, but are not likely to be as destructive as in a high-voltage system.

Arc flash and arc blast are always a possibility when working with electrical equipment. Arc flash can occur when using a voltmeter or DMM to measure voltage in a 480 V or higher electrical system when a power line transient occurs, such as a lightning strike or power surge. A potential cause for arc flash and arc blast is improper test instrument and meter use. For example, an arc blast can occur by connecting an ammeter across two points of a circuit that is energized with a voltage higher than the rating of the meter. To avoid causing arc blast or arc flash, an electrical system needs to be de-energized, locked out, and tagged out prior to performing work. Only qualified electricians are allowed to work on energized circuits of 50 V or higher.

TECH FACT *Matting used for prevention of electrical shock is available in vinyl composition for better oil, chemical, ozone, and flame resistance.*

RUBBER INSULATING MATTING RATINGS

Safety Standard	Material Thickness		Material Width (in in.)	Test Voltage	Maximum Working Voltage
	Inches	Millimeters			
BS921*	.236	6	36	11,000	450
BS921*	.236	6	48	11,000	450
BS921*	.354	9	36	15,000	650
BS921*	.354	9	48	15,000	650
VDE0680†	.118	3	39	10,000	1000
ASTM D178‡	.236	6	24	25,000	17,000
ASTM D178‡	.236	6	30	25,000	17,000
ASTM D178‡	.236	6	36	25,000	17,000
ASTM D178‡	.236	6	48	25,000	17,000

* BSI–British Standards Institiute
† VDE–Verband Deutscher Elektrotechniker Testing and Certification Institute
‡ ASTM International–American Society for Testing and Materials

Figure 4-11

Flash Protection Boundary

A *flash protection boundary* is the distance at which PPE is required to prevent burns when an arc occurs. **See Figure 4-12.** Per NFPA 70E, systems of 600 V and less require a flash protection boundary of 4′, based on the clearing time of the circuit breaker to act. While a circuit that is being worked on should always be de-energized, the possibility exists that nearby circuits are still energized and within the flash protection boundary. Barriers such as insulation blankets, along with the proper PPE, must be used to protect against an arc flash from nearby energized circuits.

Personal Protective Equipment for Arc Blast Protection

The proper personal protective equipment (PPE) must always be worn when working with energized electrical circuits for protection against an arc blast or arc flash. Clothing made of synthetic materials such as nylon, polyester, or rayon, alone or combined with cotton, must never be worn, as synthetic materials burn and melt to the skin. Per NFPA 70E, the type of PPE worn depends on the type of work being performed. The minimum PPE requirement for electrical work is an untreated natural material long-sleeve shirt, long pants, safety glasses with side shields, and rubber insulating shoes or boots. Additional PPE includes flame-resistant (FR) coveralls, FR long-sleeve shirts and pants, a hard hat with an FR liner, hearing protection, and double-layer flash suit jacket and pants. Flash suits are similar to firefighter turnout gear and must be worn when working near a Category 4 hazard/risk area. **See Figure 4-13.**

Electrical Shock

According to the National Safety Council, over 1000 people are killed by electrical shock in the United States each year. Electricity is the number one cause of fires. More than 100,000 people are killed in electrical fires each year. Improper electrical wiring or misuse of electricity causes destruction of equipment and fire damage to property.

Safe working habits are required when troubleshooting an electrical circuit or component because the electric parts that are normally enclosed are exposed during troubleshooting. An electrical shock results any time a body becomes part of an electrical circuit. Electrical shock varies from a mild shock to fatal current. The severity of an electrical shock depends on the amount of electric current (in mA) that flows through the body, the length of time the body is exposed to the current flow, the path the current takes through the body, the physical size and condition of the body through which the current passes, and the amount of body area exposed to the electric contact.

The amount of current that passes through a circuit depends on the voltage and resistance of the circuit. During an electrical shock, a person's body becomes part of the electrical circuit. The resistance a person's body offers to the flow of current varies. Sweaty hands have less resistance than dry hands. A wet floor has less resistance than a dry floor. The lower the resistance, the greater the current flow. The greater the current flow, the greater the severity of shock. **See Figure 4-14.**

TECH FACT *Arc-blast-resistant clothing is used to prevent sustained burns to an individual. Most arc-blast burn injuries are a result of the continued burning of non-arc-resistant or non-flame-resistant clothing.*

APPROACH BOUNDARIES TO ENERGIZED PARTS FOR SHOCK PREVENTION

Nominal System (Voltage, Range, Phase to Phase*)	Limited Approach Boundary: Exposed Movable Conductor	Limited Approach Boundary: Exposed Fixed-Circuit Part	Restricted Approach Boundary (Allowing for Accidental Movement)	Prohibited Approach Boundary
0 to 50	N/A	N/A	N/A	N/A
51 to 300	10′-0″	4′-0″	Avoid contact	Avoid contact
301 to 750	10′-0″	4′-0″	1′-0″	0′-1″
751 to 15,000	10′-0″	4′-0″	2′-2″	0′-7″

* in V

Figure 4-12

FLAME-RESISTANT PROTECTIVE EQUIPMENT REQUIREMENTS				
Flame-Retardant Clothing Type	Category Number (1 = Least Hazardous)			
	1	2	3	4
Flash suit jacket				X
Flash suit pants				X
Head protection				
Hard hat	X	X	X	X
Flame-retardant hard hat liner			X	X
Safety glasses w/side shields or goggles	X	X	X	X
Face protection (2-layer hood)			X	X
Hearing protection (ear canal inserts)			X	X
Rubber gloves w/leather protectors	X	X	X	X
Leather shoes w/rubber soles	X	X	X	X

* in V

Figure 4-13

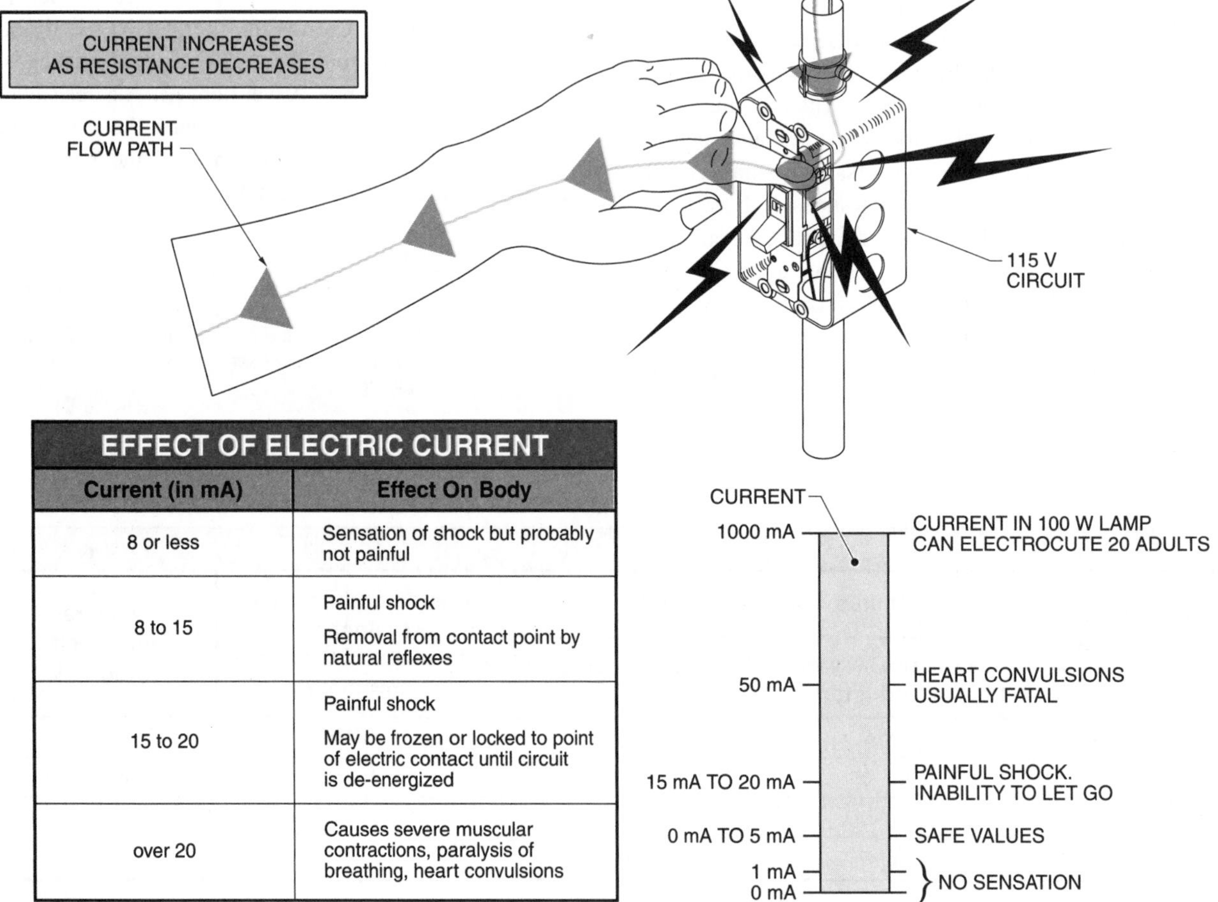

EFFECT OF ELECTRIC CURRENT	
Current (in mA)	Effect On Body
8 or less	Sensation of shock but probably not painful
8 to 15	Painful shock Removal from contact point by natural reflexes
15 to 20	Painful shock May be frozen or locked to point of electric contact until circuit is de-energized
over 20	Causes severe muscular contractions, paralysis of breathing, heart convulsions

Figure 4-14

When handling a victim of an electrical shock accident, apply the following procedures:

1. Break the circuit to free the person immediately and safely. Never touch any part of a victim's body when in contact with the circuit. When the circuit cannot be turned OFF, use any nonconducting device to free the person. Resist the temptation to touch the person when power is not turned OFF.
2. After the person is free from the circuit, send for help and determine if the person is breathing. When there is no breathing or pulse, start CPR if trained to do so. Always get medical attention for a victim of electrical shock.
3. When the person is breathing and has a pulse, check for burns and cuts. Burns are caused by contact with the live circuit, and are found at the points that the electricity entered and exited the body. Treat the entrance and exit burns as thermal burns and get medical help immediately.

Basic First Aid

Accidents that cause injury to individuals can happen at any time and at any place. Immediate medical treatment is required for the victim, regardless of the extent of the injury. Often, first aid given immediately at the scene of an accident can improve the victim's chances of survival and recovery. *First aid* is help for a victim immediately after an injury and before professional medical help arrives. First aid includes procedures performed to help save a person's life before professional help arrives. If someone is injured, steps taken to keep that person as safe as possible until professional help arrives include the following:

- Remain calm.
- Call 911 or the workplace emergency number immediately if an individual is seriously injured.
- Never move an injured person unless a fire or explosives are involved. Moving an injured person may make the injury worse.
- Assess the injured person carefully and perform basic first aid procedures.
- Maintain first aid procedures until professional medical help arrives.
- Report all injuries to the supervisor.

A conscious accident victim must provide consent before care can be administered. To obtain consent, the victim must be asked if help can be provided. Once the victim gives consent, provide the appropriate care. Do not give care if the victim does not give consent. In such a case, call 911 or the workplace emergency number. Consent is implied in cases where an accident victim is unconscious, confused, or seriously injured. Implied consent allows an individual to provide care to an accident victim because the victim would agree to the care if he or she could.

Various states have enacted Good Samaritan laws to give legal protection to individuals who provide emergency care to accident victims. These laws are designed to encourage individuals to help others in emergencies. Good Samaritan laws vary from state to state. A legal professional or the local library should be consulted for information regarding the Good Samaritan laws in a particular state.

Do not attempt to provide care for which you have not been trained. The American Red Cross may be consulted for training programs covering standard first aid and CPR. Basic first aid procedures can be used to help treat shock, bleeding, burns and scalds, choking, heat exhaustion and heat stroke, and poisoning.

Shock

Shock usually accompanies severe injury. Shock can threaten the life of an injured person if not treated quickly. Shock occurs when the body's vital functions are threatened by lack of blood or when the major organs and tissues do not receive enough oxygen. Symptoms of shock include cold, clammy, or pale skin, chills, confusion, nausea or vomiting, shallow breathing, and unusual thirst. To treat shock, the victim should lie down with his or her legs elevated if there is no sign of broken bones or spinal injury. The victim should also be covered to prevent chills or loss of body heat. Any obvious signs of bleeding should be controlled. A victim who is unconscious or bleeding from the mouth should lie on one side so breathing is easier. Get professional medical help as soon a possible.

Bleeding

Bleeding is the most visible result of an injury. Most people can lose a small amount of blood without major problems. However, if a quart or more of blood is lost quickly, shock and/or death is possible. Bleeding is controlled by placing a clean, dry cloth on the wound and applying pressure with a finger or palm of the hand, depending on the severity of the wound. The pressure should be kept on the wound until the bleeding stops. If there is no sign of broken bones, the wound should be elevated above the victim's heart to slow the bleeding at the wound. Once the bleeding stops, the cloth against the wound should not be removed, as it could disturb the blood clotting and restart the bleeding.

If possible, rubber or latex gloves should be worn before touching any blood, because touching blood involves health risks. If rubber or latex gloves are not available, a clean plastic bag may be used to cover the hands. If the injury is extensive, the victim may go into shock and should be treated for it. Get professional medical help as soon as possible.

Burns and Scalds

Burns can be caused by heat, chemicals, or electricity. For burns caused by heat or chemicals, the burn should be immediately flushed with cool water for a minimum of 30 min. Do not apply ice, because it may cause further damage to the burned area. Maintain this treatment until the pain or burning stops. After flushing the burn with cool water, cover the burn with a clean cotton cloth. If a clean cotton cloth is not available, do not cover the burn.

Do not try to remove clothing that is stuck to a burn. Do not scrub the burn or apply any soap, ointments, greases, or powders. Avoid breaking any blisters that appear. Also, do not offer the burn victim anything to drink or eat. Keep the victim covered with a blanket to maintain a normal body temperature. Treat all burn victims for shock. Get professional medical help as soon as possible.

A burn caused by electricity requires first ensuring that the victim is removed from the power source. Once the victim is clear of the power source, check the victim for any airway obstruction, and check his or her breathing and circulation. Administer CPR if necessary. Once the victim is stable, flush the burn with cool water for a minimum of 30 min. Do not move the victim, and do not scrub the burn or apply any soap, ointments, greases, or powders. After flushing the burn with cool water for 30 min, cover the burn with a clean cotton cloth. If a clean cotton cloth is not available, do not cover the burn. Treat the victim for shock and try to maintain a normal body temperature. Get professional medical help as soon as possible.

Choking

Choking occurs when food or a foreign object obstructs the throat and interferes with normal breathing. Permission must be obtained from a conscious victim before providing care. The following steps are advised if a choking victim is unable to speak or cough:

1. Ask the victim if they are choking.
2. Shout for help if the victim cannot cough, speak, or breath.
3. Call 911 or the workplace emergency number.
4. Perform abdominal thrusts. Abdominal thrusts are performed by wrapping the arms around the victim's waist, making a fist, and placing the thumb of the fist on the middle of the victim's abdomen just above the navel. The fist is grasped with the other hand and both fists are pressed into the abdomen with a quick upward thrust.

Heat Exhaustion and Heat Stroke

Heat exhaustion can occur wherever there is poor air circulation, such as around machinery or an industrial furnace. Heat exhaustion can also occur if a person has not adjusted to very warm temperatures and may be due to loss of body fluids and salts. In a person suffering from heat exhaustion, the body reacts by increasing the heart rate and strengthening blood circulation. The symptoms of heat exhaustion include excessive fatigue, dizziness, disorientation, and skin with a normal temperature but a damp and clammy feeling. A person with heat exhaustion is treated by moving the victim to a cool location. The victim should be encouraged to drink water and rest.

Heat stroke is much more serious than heat exhaustion. Heat stroke occurs when the body's sweat glands shut down. Symptoms of heat stroke include confusion, collapse, unconsciousness, and fever with dry, mottled skin. A heat stroke victim can die quickly. A person suffering from heat stroke should be moved immediately to a cool area and cool water should be poured over the victim. The victim should be fanned to provide good air circulation. Get professional medical help as soon as possible.

Poisoning

Poisons may be in solid, liquid, or gas form. If the poison has been ingested in a solid form, such as pills, remove any poison that is in the victim's mouth using a clean cloth wrapped around a finger. If the poison is a corrosive liquid on the skin, remove any clothing from the affected area and flush the skin with water for 30 min. If the poison is in contact with the eyes, flush the victim's eyes for a minimum of 15 min with clean water. When calling for medical help, the poisonous product's container or label should be available. This will enable the caller to answer questions about the poison.

When the poison is in the environment, the victim must be removed immediately from the poison source. Once the victim is removed from the poison source, the appropriate treatment should be administered, based on the form of the poisoning. If the poison is a gas, a respirator may be required for protection when entering the area. Move the victim to fresh air and get professional medical help as quickly as possible.

First Aid Kits

To administer effective first aid, adequate supplies must be maintained in a first aid kit. A first aid kit can be purchased stocked with the necessary supplies, or one can be made by assembling a kit containing adhesive bandages, butterfly closures, rolled gauze, nonstick sterile pads, and various first aid tapes. Additional items can be included as required, such as tweezers, aspirin, an additional analgesic, first aid cream, a thermometer, and an ice pack.

Grounding

Electrical circuits are grounded to safeguard equipment and personnel against the hazards of electrical shock. Proper grounding of electrical tools, machines, equipment, and delivery systems is one of the most important factors in preventing hazardous conditions.

Grounding is the connection of all exposed non-current-carrying metal parts to the earth. Grounding provides a direct path for unwanted (fault) current to the earth without causing harm to persons or equipment. Grounding is accomplished by connecting the circuit to a metal underground pipe, a metal frame of a building, a concrete-encased electrode, or a ground ring. **See Figure 4-15.**

Non-current-carrying metal parts that are connected to ground include all metal boxes, raceways, enclosures, and equipment. Unwanted current exists because of insulation failure or because a current-carrying conductor makes contact with a non-current-carrying part of the system. In a properly grounded system, the unwanted current flow blows fuses or trips circuit breakers. Once the fuse is blown or circuit breaker is tripped, the circuit is open and no additional current flows.

Lockout/Tagout

Electrical power must be removed when electrical equipment is inspected, serviced, repaired, or replaced. Power is removed and the equipment must be locked out and tagged out to ensure the safety of personnel working with the equipment.

Per OSHA standards, equipment is locked out and tagged out before any preventive maintenance or servicing is performed. *Lockout* is the process of removing the source of electrical power and installing a lock that prevents the power from being turned ON. *Tagout* is the process of placing a danger tag on the source of electrical power that indicates that the equipment may not be operated until the danger tag is removed.

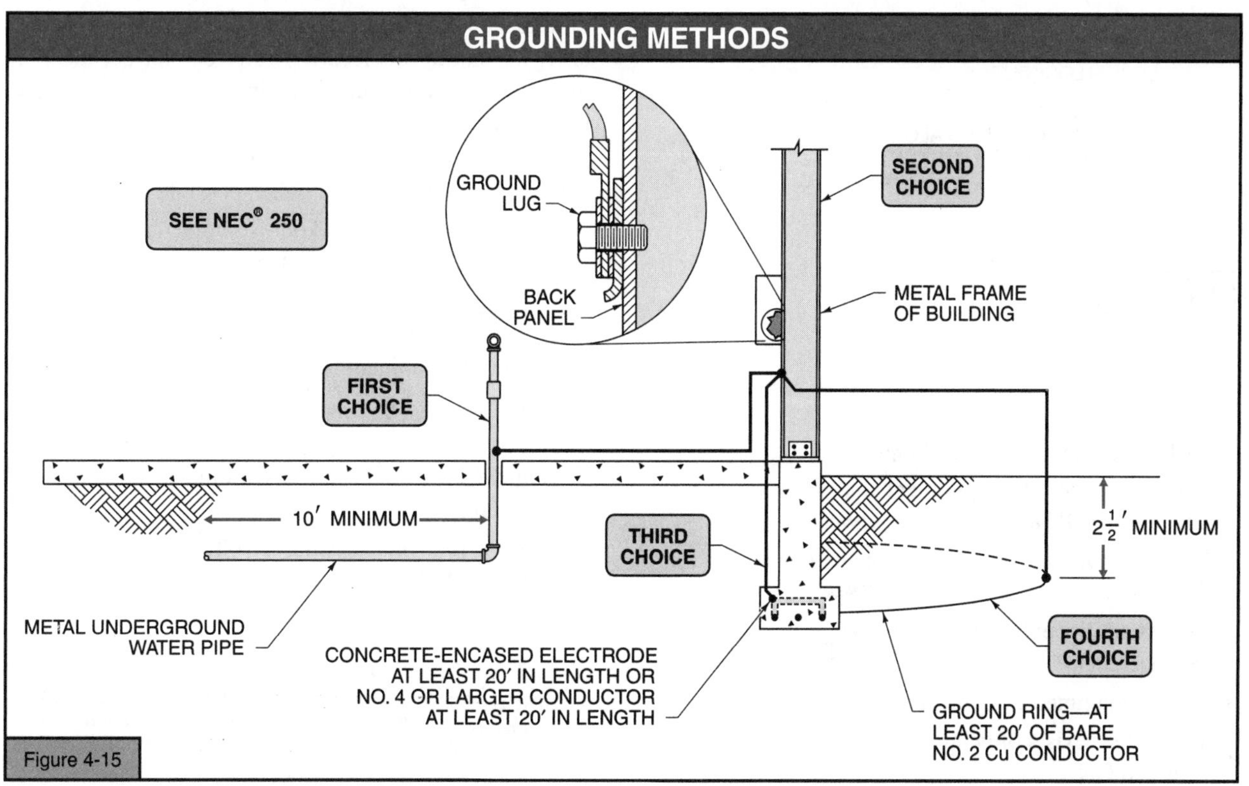

Figure 4-15

A danger tag has the same importance and purpose as a lock and is used alone only when a lock does not fit the disconnect device. The danger tag shall be attached at the disconnect device with a tag tie or equivalent and shall have space for the worker's name, craft, and other required information. A danger tag must withstand the elements and expected atmosphere for as long as the tag remains in place. **See Figure 4-16.**

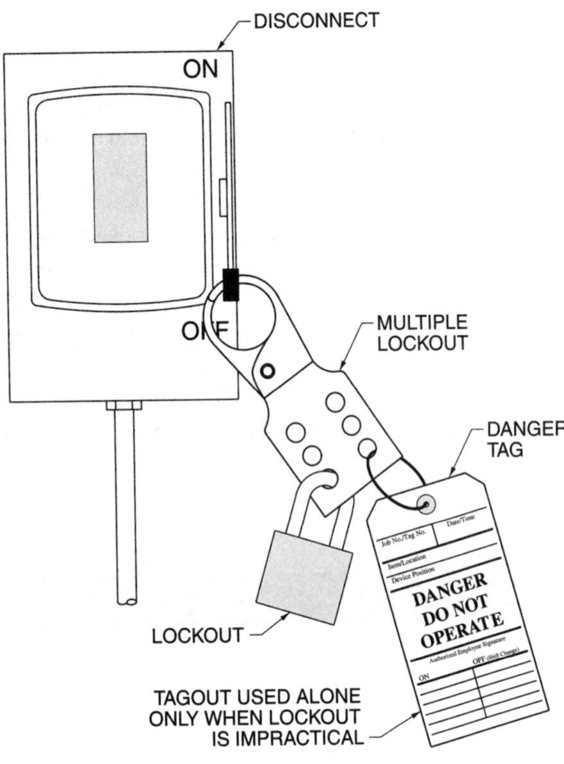

Figure 4-16 **LOCKOUT/TAGOUT**

A lockout/tagout is used when:

- Servicing electrical equipment that does not require power to be ON to perform the service
- Removing or bypassing a machine guard or other safety device
- The possibility exists of being injured or caught in moving machinery
- Clearing jammed equipment
- The danger exists of being injured if equipment power is turned ON

Lockouts and tagouts do not by themselves remove power from a circuit. An approved procedure is followed when applying a lockout/tagout. Lockouts and tagouts are attached only after the equipment is turned OFF and tested to ensure that power is OFF. The lockout/tagout procedure is required for the safety of workers due to modern equipment hazards. OSHA provides a standard procedure for equipment lockout/tagout, as follows:

1. Prepare for machinery shutdown
2. Shut down machinery or equipment
3. Isolate machinery or equipment
4. Apply lockout and/or tagout
5. Release stored energy
6. Verify isolation of machinery or equipment

Warning: Personnel should consult OSHA Standard 1910.147—*The Control of Hazardous Energy (Lockout/Tagout)* for industry standards on lockout/tagout.

A lockout/tagout must not be removed by any person other than the person who installed it, except in an emergency. In an emergency, the lockout/tagout may be removed only by authorized personnel. The authorized personnel must follow approved procedures. A list of company rules and procedures are given to any person who may use a lockout/tagout.

When more than one electrician is required to perform a task on a piece of equipment, each electrician shall place a lockout /tagout on the energy-isolating device(s). A multiple-lockout/tagout device (hasp) must be used because energy-isolating devices typically cannot accept more than one lockout/tagout at one time. A *hasp* is a multiple-lockout/tagout device.

Always remember the following:

- Use a lockout and tagout when possible.
- Use a tagout when a lockout is impractical. A tagout is used alone only when a lock does not fit the disconnect device.
- Use a multiple-lockout device when individual employee lockout of equipment is impractical.
- Notify all employees affected before using a lockout/tagout.
- Remove all power sources including primary and secondary.
- Measure for voltage using a voltmeter to ensure that power is OFF.

TECH FACT *Failure to properly control hazardous energy sources during maintenance operations accounts for over 65,000 injuries per year.*

Lockout Devices

A *lockout device* is a lightweight enclosure that allows the lockout of standard control devices. Lockout devices are available in various shapes and sizes that allow for the lockout of ball valves, gate valves, and electrical plugs.

Lockout devices resist chemicals, cracking, abrasion, temperature changes, and are available in colors to match ANSI pipe colors. Lockout devices are sized to fit standard industry control-device sizes. **See Figure 4-17.**

Locks used to lock out a device may be color-coded and individually keyed. The locks are rust resistant and are available with various-size shackles.

Danger tags provide additional lockout and warning information. Various danger tags are available. Danger tags may include warnings such as "Do Not Start" or "Do Not Operate," or may provide space to enter worker, date, and lockout reason information. Tag ties must be strong enough to prevent accidental removal and must be self-locking and nonreusable.

Lockout/tagout kits are also available. A lockout/tagout kit contains items required to comply with the OSHA lockout/tagout standards. Lockout/tagout kits contain reusable danger tags, tag ties, multiple-lockout devices, locks, magnetic signs, and information on lockout/tagout procedures. **See Figure 4-18.**

Electrical Enclosures

An *electrical enclosure* is a housing that protects wires and equipment and prevents personnel injury from accidental contact with a live circuit. Enclosures are used to protect electrical equipment such as programmable controllers, motor starters, fuses/breakers, and control circuits. An enclosure provides the main protection from atmospheric conditions. Using the proper enclosure helps prevent problems caused by contamination, moisture, and physical damage. Enclosures are categorized by the protection they provide. An enclosure is selected based on the location of the equipment and NEC® requirements. **See Figure 4-19.**

LOCKOUT DEVICES

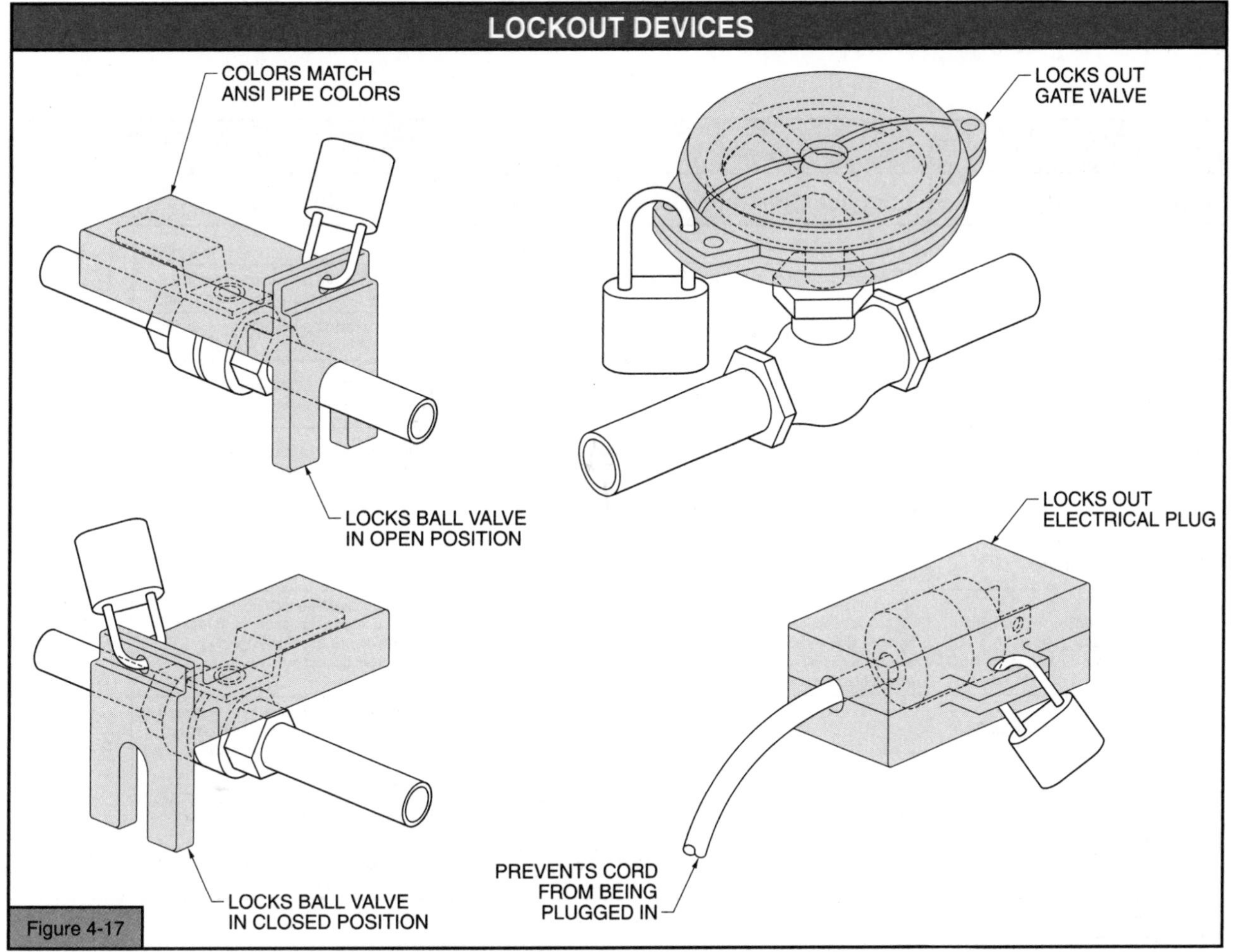

Figure 4-17

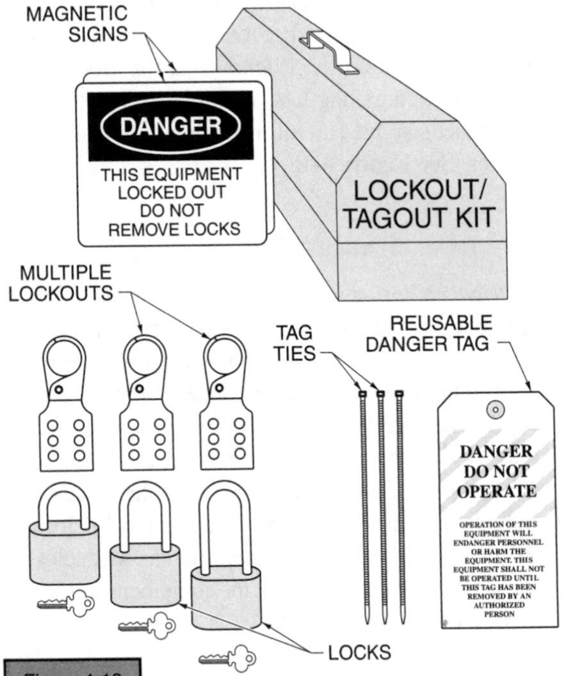

Figure 4-18 LOCKOUT/TAGOUT KIT

The NEC® classifies hazardous locations according to the properties and quantities of the hazardous material that may be present. Hazardous locations are divided into three classes, two divisions, and seven groups.

Class is the hazardous location category that refers to the generic hazardous material present. Class I applies to locations where flammable gases or vapors may be present in the air in quantities sufficient to produce an explosive or ignitable mixture. Class II applies to locations where combustible dusts may be present in sufficient quantities to cause an explosion. Class III applies to locations where the hazardous material consists of easily ignitable fibers or airborne materials that are not normally in suspension in the air in quantities large enough to produce an ignitable mixture.

Division is the hazardous location category that refers to the probability that a hazardous material is present. Division 1 applies to locations where ignitable mixtures exist under normal operating conditions found in the process or operation or during periodic maintenance. Division 2 applies to locations where ignitable mixtures exist only in abnormal situations. Abnormal situations occur as a result of accidents or when equipment fails.

ENCLOSURES

Type	Use	Service Conditions	Tests	Comments	Type
1	Indoor	No unusual	Rod entry, rust resistance		1
3	Outdoor	Windblown dust, rain, sleet, and ice on enclosure	Rain, external icing, dust, and rust resistance	Do not provide protection against internal condensation or internal icing	
3R	Outdoor	Falling rain and ice on enclosure	Rod entry, rain, external icing, and rust resistance	Do not provide protection against dust, internal condensation, or internal icing	
4	Indoor/outdoor	Windblown dust and rain, splashing water, hose-directed water, and ice on enclosure	Hosedown, external icing, and rust resistance	Do not provide protection against internal condensation or internal icing	4
4X	Indoor/outdoor	Corrosion, windblown dust and rain, splashing water, hose-directed water, and ice on enclosure	Hosedown, external icing, and corrosion resistance	Do not provide protection against internal condensation or internal icing	4X
6	Indoor/outdoor	Occasional temporary submersion at a limited depth			
6P	Indoor/outdoor	Prolonged submersion at a limited depth			
7	Indoor locations classified as Class I, Groups A, B, C, or D, as defined in the NEC®	Withstand and contain an internal explosion of specified gases, contain an explosion sufficiently so an explosive gas-air mixture in the atmosphere is not ignited	Explosion, hydrostatic, and temperature	Enclosed heat-generating devices shall not cause external surfaces to reach temperatures capable of igniting explosive gas-air mixtures in the atmosphere	7
9	Indoor locations classified as Class II, Groups E or G, as defined in the NEC®	Dust	Dust penetration, temperature, and gasket aging	Enclosed heat-generating devices shall not cause external surfaces to reach temperatures capable of igniting explosive gas-air mixtures in the atmosphere	9
12	Indoor	Dust, falling dirt, and dripping noncorrosive liquids	Drip, dust, and rust resistance	Do not provide protection against internal condensation	12
13	Indoor	Dust, spraying water, oil, and noncorrosive coolant	Oil explosion and rust resistance	Do not provide protection against internal condensation	

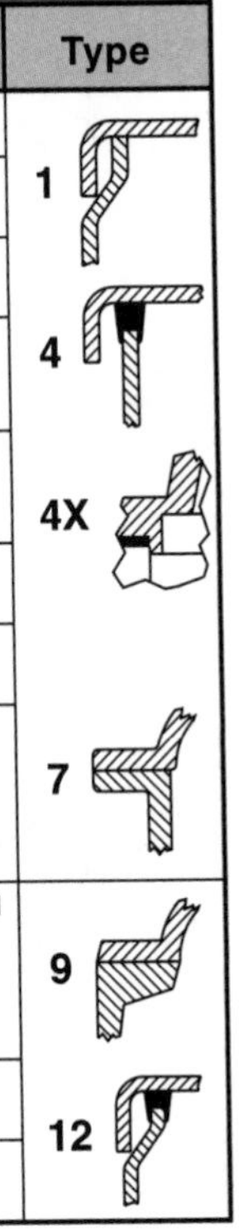

Figure 4-19

Air mixtures of gases, vapors, and dusts are grouped according to their similar characteristics. The NEC® classifies gases and vapors in Groups A, B, C, and D for Class I locations, and combustible dusts in Groups E, F, and G for Class II locations. **See Figure 4-20.** For example, a Type 7 enclosure is required for an indoor application where gasoline is stored (Class I, Group D).

Electrical Safety

Electrical workers must exercise caution and work safely at all times. Basic safety rules must be followed when working on or around electrical equipment. Electrical safety rules help prevent injuries from electrical energy sources. **See Figure 4-21.**

HAZARDOUS LOCATIONS

Hazardous Location – A location where there is an increased risk of fire or explosion due to the presence of flammable gases, vapors, liquids, combustible dusts, or easily-ignitable fibers or flyings.

Location – A position or site.

Flammable – Capable of being easily ignited and of burning quickly.

Gas – A fluid (such as air) that has no independent shape or volume but tends to expand indefinitely.

Vapor – A substance in the gaseous state as distinguished from the solid or liquid state.

Liquid – A fluid (such as water) that has no independent shape but has a definite volume. A liquid does not expand indefinitely and is only slightly compressible.

Combustible – Capable of burning.

Ignitable – Capable of being set on fire.

Fiber – A thread or piece of material.

Flyings – Small particles of material.

Dust – Fine particles of matter.

Classes	Likelihood that a flammable or combustible concentration is present
I	Sufficient quantities of flammable gases and vapors present in air to cause an explosion or ignite hazardous materials
II	Sufficient quantities of combustible dust are present in air to cause an explosion or ignite hazardous materials
III	Easily ignitable fibers or flyings are present in air, but not in a sufficient quantity to cause an explosion of ignite hazardous materials

Divisions	Location containing hazardous substances
1	Hazardous location in which hazardous substance is normally present in air in sufficient quantities to cause an explosion or ignite hazardous materials
2	Hazardous location in which hazardous substance is not normally present in air in sufficient quantities to cause an explosion or ignite hazardous materials

Groups: Atmosphere containing flammable gases or vapors or combustible dust		
Class I	Class II	Class III
A B C D	E F G	none

DIVISION I EXAMPLES

Class I:

- Spray booth interiors
- Areas adjacent to spraying or painting operations using volatile flammable solvents
- Open tanks or vats of volatile flammable liquids
- Drying or evaporation rooms for flammable vents
- Areas where fats and oil extraction equipment using flammable solvents are operated
- Cleaning and dyeing plant rooms that use flammable liquids that do not contain adequate ventilation
- Refrigeration or freezer interiors that store flammable materials
- All other locations where sufficient ignitable quantities of flammable gases or vapors are likely to occur during routine operations

Class II:

- Grain and grain products
- Pulverized sugar and cocoa
- Dried egg and milk powders
- Pulverized spices
- Starch and pastes
- Potato and wood flour
- Oil meal from beans and seeds
- Dried hay
- Any other organic material that may produce combustible dusts during their use or handling

Class III:

- Portions of rayon, cotton, or other textile mills
- Manufacturing and processing plants for combustible fibers, cotton gins, and cotton seed mills
- Flax processing plants
- Clothing manufacturing plants
- Woodworking plants
- Other establishments involving similar hazardous processes or conditions

Figure 4-20

ELECTRICAL SAFETY

Always comply with the NEC®.

Use UL® approved appliances, components, and equipment.

Keep electrical grounding circuits in good condition. Ground any conductive component or element that does not have to be energized. The grounding connection must be a low-resistance conductor heavy enough to carry the largest fault current that may occur.

Turn OFF, lock out, and tag disconnect switches when working on any electrical circuit or equipment. Test all circuits after they are turned OFF. Insulators may not insulate, grounding circuits may not ground, and switches may not open the circuit. ①

Use double-insulated power tools or power tools that include a third conductor grounding terminal which provides a path for fault current. Never use a power tool that has the third conductor grounding terminal removed.

Always use protective and safety equipment. ②

Know what to do in an emergency.

Check conductors, cords, components, and equipment for signs of wear or damage. Replace any equipment that is not safe. ③

Never throw water on an electrical fire. Turn OFF the power and use a Class C rated fire extinguisher. ④

Work with another individual when working in a dangerous area or with dangerous equipment.

Learn CPR and first aid. ⑤

Do not work when tired or taking medication that causes drowsiness.

Do not work in poorly lighted areas.

Always use nonconductive ladders. Never use a metal ladder when working around electrical equipment.

Ensure there are no atmospheric hazards such as flammable dust or vapor in the area. A live electrical circuit may emit a spark at any time.

Use one hand when working on a live circuit to reduce the chance of an electrical shock passing through the heart and lungs.

Never bypass or disable fuses or circuit breakers.

Extra care must be taken in an electrical fire because burning insulation produces toxic fumes.

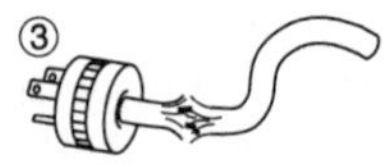

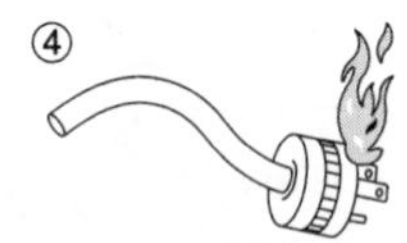

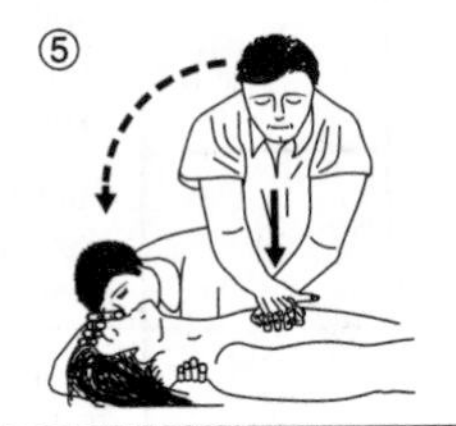

Figure 4-21

Electric Motor Safety

Two areas requiring attention when working with electric motors are the electrical circuit and rotating shaft. Basic electric motor safety rules include:

- Connecting a motor to the correct grounding system.
- Ensuring that guards or housings are connected to the rotating parts of a motor or anything connected to the motor. **See Figure 4-22.**
- Using the correct motor type for the location. For example, a DC or universal motor must never be used in a hazardous location that contains flammable materials because the sparking at the brushes can ignite the material.
- Connecting a motor to the correct voltage and power source.
- Providing the motor with the correct overload and overcurrent protection to protect the motor when starting or shorted (overcurrent protection) and when running (overload protection).

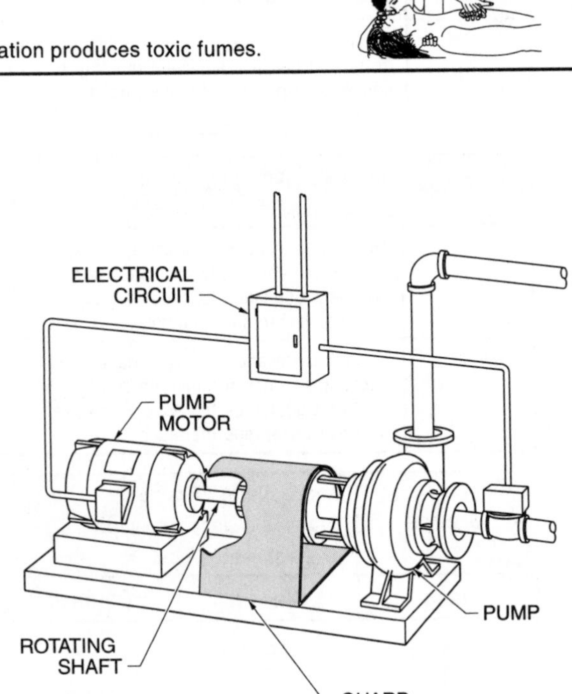

Figure 4-22

Pressure Relief Valves

A *pressure relief valve* is a valve that limits the amount of pressure in a fluid power system. A pressure relief valve is used as a safety valve to prevent damage that may be caused by exceedingly high pressure. Pressure relief valves are two-way, normally closed valves that may be set to open at a given pressure. The valve is closed and no fluid is allowed through the valve when the pressure in the system is below the setting of the relief valve. The valve opens and relieves the pressure in the system when the pressure in the system increases to the setting of the valve (or higher). When the pressure in the system drops below the control knob setting, the relief valve closes. **See Figure 4-23.**

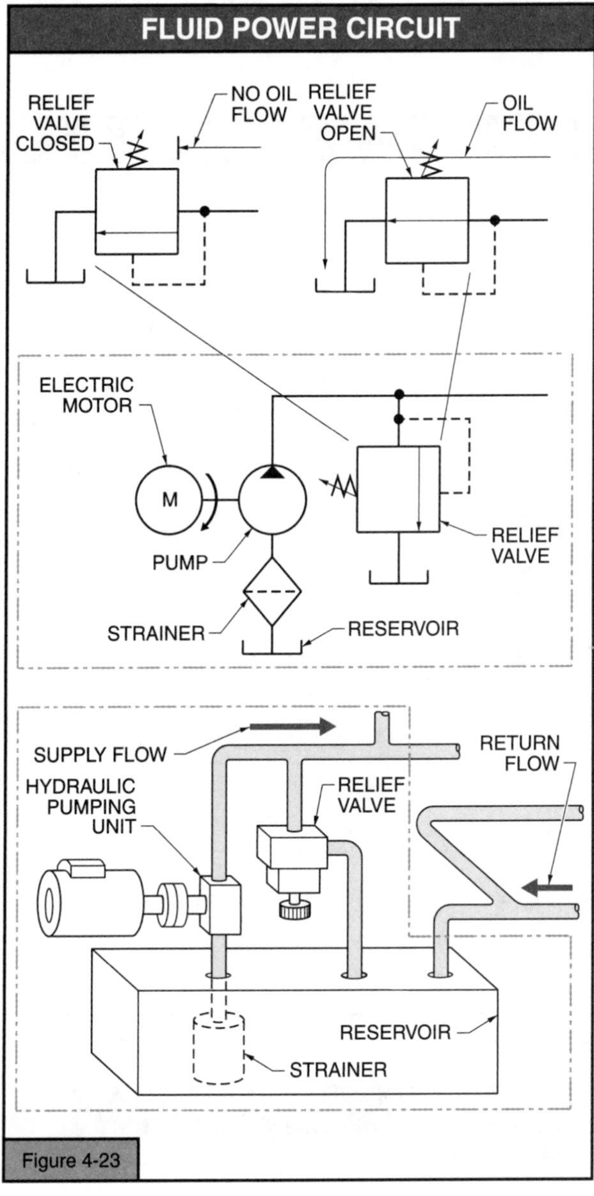

Figure 4-23

Pressure relief valves are also used to relieve excessive pressure in residential, commercial, and industrial heating systems. Relieving the pressure is required when water is heated because water is not compressible. Water expands approximately 4% when heated from 60°F to 212°F. A relief valve is installed to relieve the pressure caused by this expansion in case there is no expansion room and/or the temperature control switch fails to turn OFF the heater. **See Figure 4-24.**

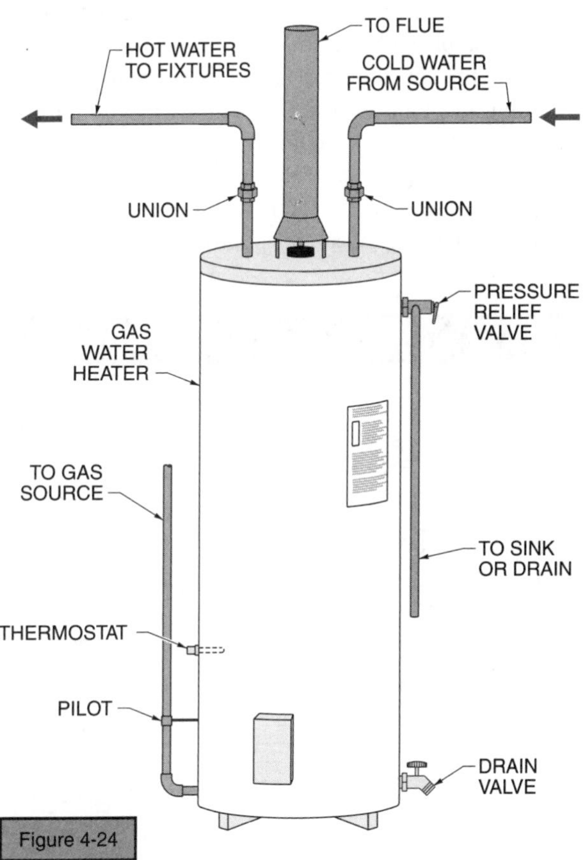

Figure 4-24

Pressure relief valves are installed on the cold water inlet side of the tank. The relief valve's outlet should be piped to the nearest drain. Local codes normally specify which type (diaphragm, etc.) of relief valve must be used for the given application.

Care must be taken to ensure that unauthorized persons do not adjust the setting of the relief valve higher than the system design pressure. Leaks may occur, hoses may break, and components may fail if the setting is set too high.

Pressure relief valves may be used anywhere in a system where it is necessary to prevent pressure from exceeding a maximum safe level. The most common use of a pressure relief valve is to prevent a fluid power pump from developing excessively high pressure. This condition occurs whenever the system becomes overloaded.

An overloaded system occurs whenever the fluid in the system has no place to go and the pump is forcing additional fluid into the system. The pump or other components may be damaged if this condition is allowed to continue. By relieving pressure at the pump, the relief protects the pump motor from overloading.

Documentation Forms

Documentation forms provide a written record of the steps taken to comply with industry safety standards as well as a method of tracking progress. Setting safety goals and rules, providing safe tools, working conditions, and training are all important factors in reducing accidents in the workplace. Safety steps must be documented to ensure proper compliance with industry safety requirements and protect employees and employers. **See Figure 4-25.**

Different documentation forms may be used depending on the product (or process) being manufactured, company policy, and applicable codes and standards. Documentation generally requires several different steps and forms for record keeping. Documentation forms include: Safety Program Goals, Safety Program Personnel, Danger/Equipment Identification, Protection Device Location Identification, Protection Device Log, and Safety Program Training Log.

TECH FACT

OSHA requires that personnel be trained in-house in areas such as lockout/tagout, hazardous material handling, hazardous waste disposal, bonding and grounding techniques, and understanding material safety data sheets (MSDS). All in-house training programs must be documented by the company and signed by each participant.

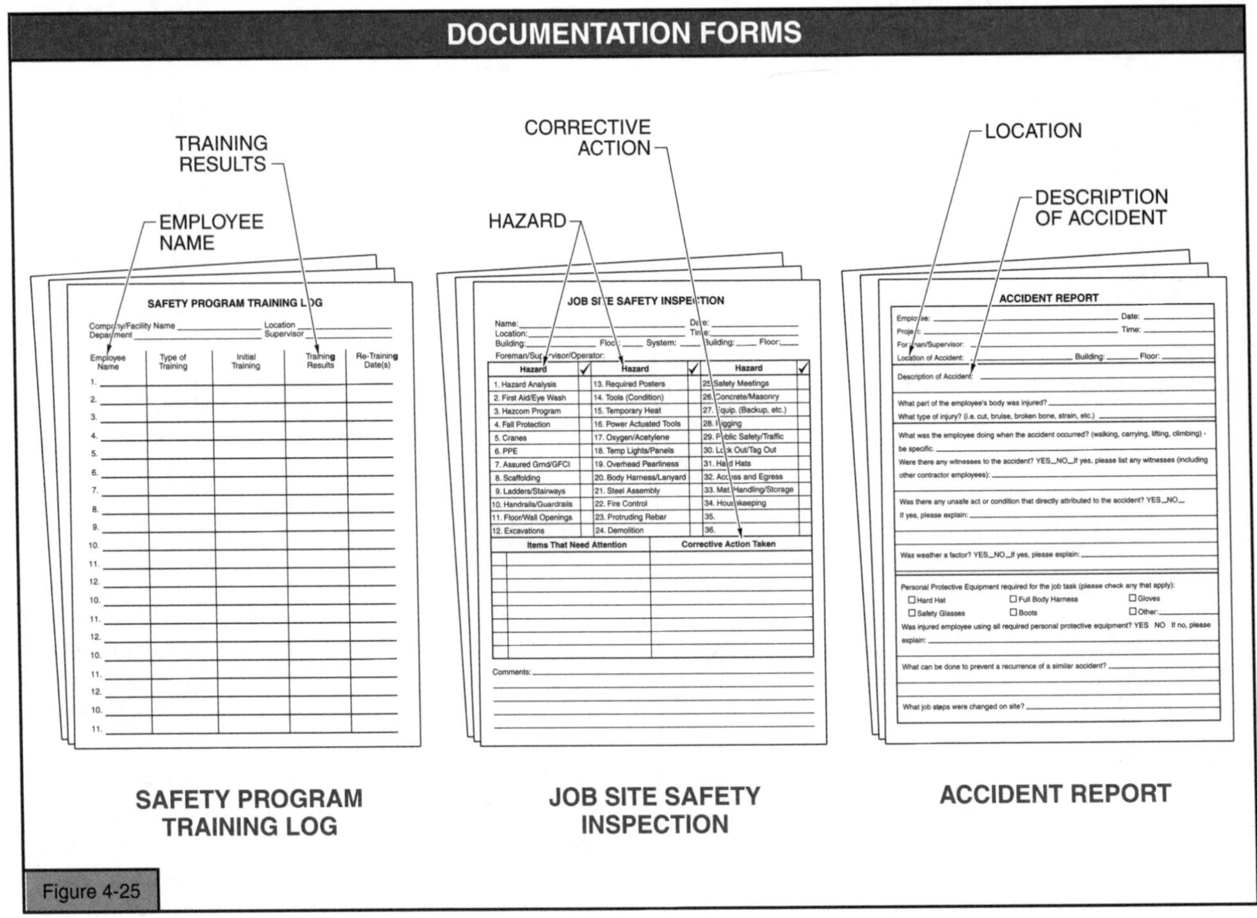

Figure 4-25

Refer to the CD-ROM in the back of the book for Chapter 4 Quick Quiz® and related information.

Mathematics

Mathematics is the science of numbers and their operations, interrelations, and combinations. *Arithmetic* is the branch of mathematics that involves the computation (addition, subtraction, multiplication, and division) of positive real numbers. A *real number* is any number associated with a point on a line.

Electricians must be competent in addition, subtraction, multiplication, and division, as well as in converting fractions to decimals, calculating percentages, converting decimals to fractions, metric conversions, etc. These operations are used when calculating service entrance capacity, conduit fill, conduit bends, wire ampacity, etc.

Whole Numbers

Whole numbers (integers) are numbers that have no fractional or decimal parts. For example, 1, 2, 8, 34, 1845, etc. are whole numbers. Whole numbers are used for counting all things that can be counted as separate objects. Whole numbers are odd or even. **See Figure 5-1.** An *odd number* is any number that cannot be divided by 2 an exact number of times. Odd numbers always end in 1, 3, 5, 7, or 9. For example, numbers such as 3, 35, 125, 239, etc. are odd numbers. An *even number* is any number that can be divided by 2 an exact number of times. An even number is any number ending in 2, 4, 6, 8, or 0. For example, the numbers 6, 22, 180, 1904, etc. are even numbers.

A *prime number* is a number that can be divided an exact number of times, only by itself and the number 1. For example, the numbers 1, 2, 3, 5, 7, 11, 13, 19, 23, etc. are prime numbers.

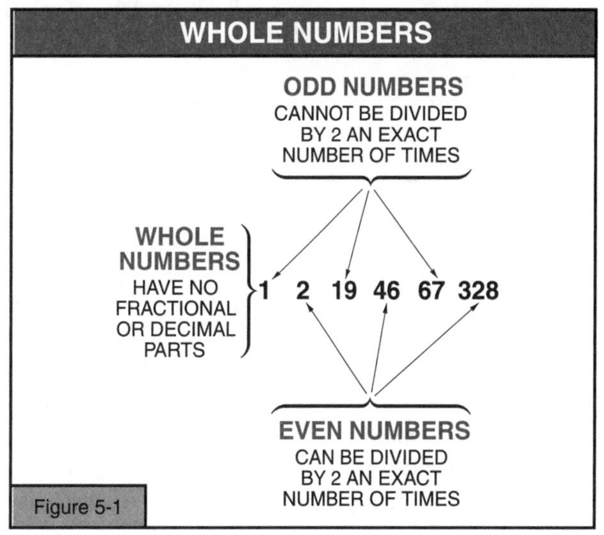

Figure 5-1

Many more numbers exist that are partially fractional or decimal than there are integers because most quantities cannot be directly counted, but must be measured in some way. *Continuous quantities* are the numbers resulting from measurements that cannot be directly counted.

Fractions

A *fraction* is a part of a whole unit or number. Fractions are shorthand methods of describing a part of a total amount. Any number or item smaller than one is a fraction and can be divided into a number of fractional parts. **See Figure 5-2.**

The top number of a fraction is the numerator, and the bottom number is the denominator. A *denominator* is a number in a fraction that indicates the number of equal parts into which a unit is divided. A *numerator* is a number in a fraction that indicates how many of these parts are used. For example, if a circle is divided into four equal parts, each part is 1 part out of the total of four parts. Each part may be written as the fraction ¼. The circle may be divided into any number of equal parts with each part being 1 part out of the total number of parts.

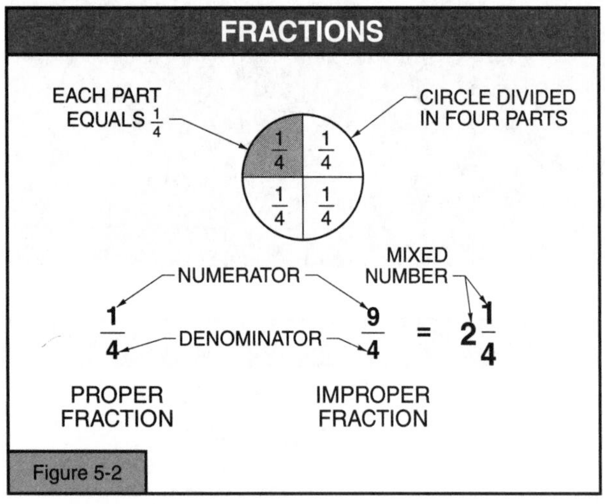

Figure 5-2

A *proper fraction* is a fraction that has a numerator smaller than its denominator. The value of a proper fraction is less than one unit. An *improper fraction* is a fraction that has a numerator larger than its denominator. An improper fraction contains one or more than one whole unit. An improper fraction is changed into a mixed number (a whole number and a proper fraction) by dividing the numerator by the denominator. For example, when dividing the numerator 9 of the improper fraction 9⁄4 by its denominator 4, the result is 2¼. The improper fraction 9⁄4 contains two whole units and one fourth of a unit.

Reducing Fractions

Reducing fractions is the process of changing a fraction from one fractional form to another. Fractions are reduced when they are changed to lower terms. Fractions are reduced because it is easier to use and perform calculations with smaller fractions. For example, the fractions 72⁄144, 36⁄72, 18⁄36, and ½ are all equivalent fractions. In calculations, it is easier to work with ½ which is 72⁄144, 36⁄72, and 18⁄36 reduced to their lowest terms.

TECH FACT *A fraction is reduced to lowest terms when the numerator and denominator have no common divisor except 1.*

To reduce a fraction to lower terms, divide the numerator and denominator by the same terms. The value of a fraction is not changed when the numerator and denominator are multiplied or divided by the same number. For example, the fraction 3⁄9 can be reduced to lower terms by dividing 3 (numerator) and 9 (denominator) by the same number (3). **See Figure 5-3.** Reducing 3⁄9 to lower terms does not change its value.

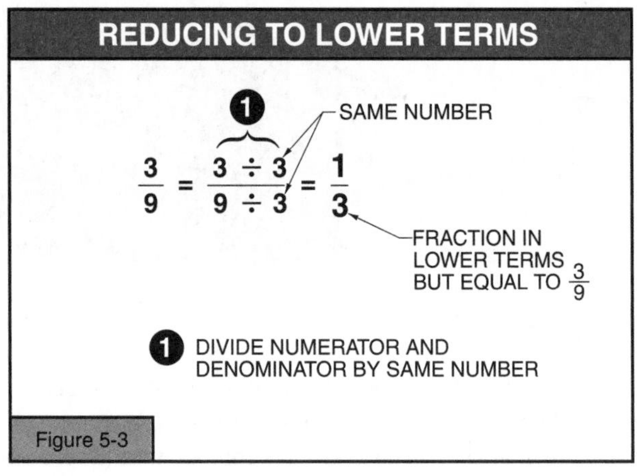

Figure 5-3

Adding Fractions

Fractions that have the same denominator are added or subtracted by adding or subtracting the numerators and putting the result over the common denominator. **See Figure 5-4.** For example, to add ⅛ and ⅛, add the numerators (1 + 1 = 2) and place the sum over the denominator (2⁄8). The fraction 2⁄8 can be reduced to ¼.

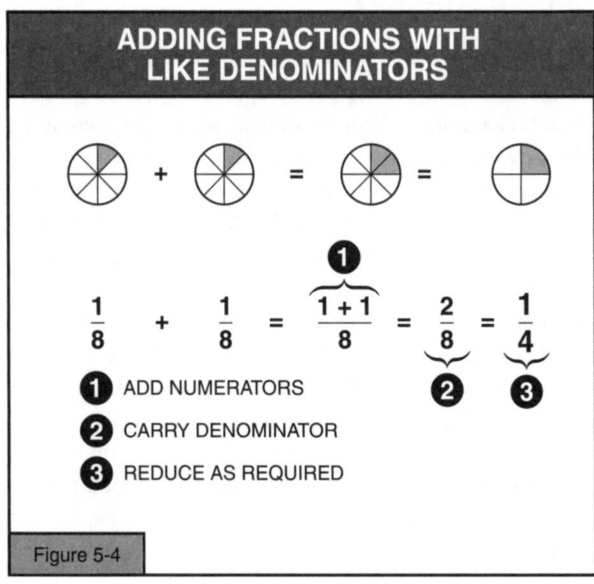

Figure 5-4

Fractions with unlike denominators must be changed to have a common denominator (lowest common denominator) before they can be added. The numerators are added or subtracted after the fractions are changed to equivalent fractions with the same denominator.

The *lowest common denominator (LCD)* is the smallest number into which the denominators of a group of two or more fractions divides an exact number of times. To find the LCD for a group of fractions, arrange the denominators in a horizontal row. A division sign is drawn under the row. The lowest possible number that divides an exact number of times into two or more of the denominators is found. **See Figure 5-5.**

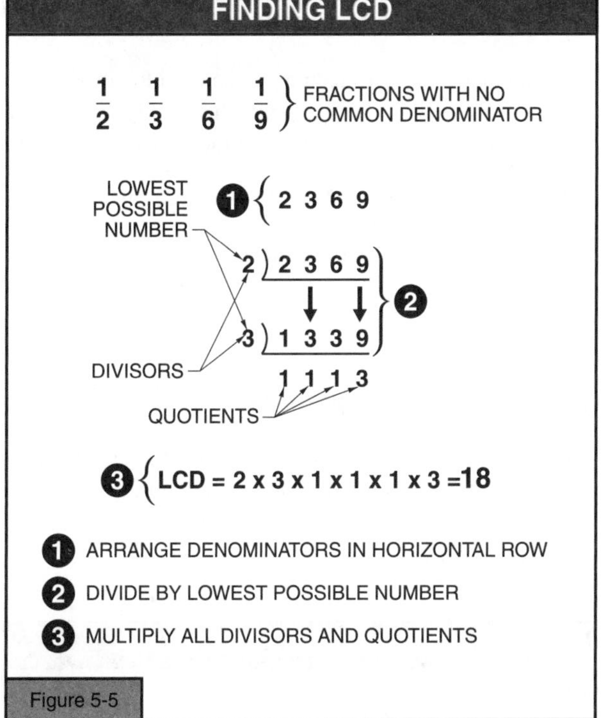

Figure 5-5

The denominators are divided by this number and the quotients are placed under the denominators that are divisible an exact number of times. The denominators that are not divisible by this number are brought down. The steps are repeated until division is no longer possible. Multiply all the divisors and quotients (numbers left in the last line). The product is the LCD.

For example, to add $\frac{1}{2}$ and $\frac{3}{4}$, the fractions are reduced to common terms (LCD) or changed to have common denominators. **See Figure 5-6.** The fractions have an LCD of 4. The equivalent fractions are $\frac{1}{2} = \frac{2}{4}$ and $\frac{3}{4} = \frac{3}{4}$. Add the numerators (2 + 3 = 5). Place the 5 over 4 ($\frac{5}{4}$). The fraction $\frac{5}{4}$ is an improper fraction which reduces to the mixed number $1\frac{1}{4}$.

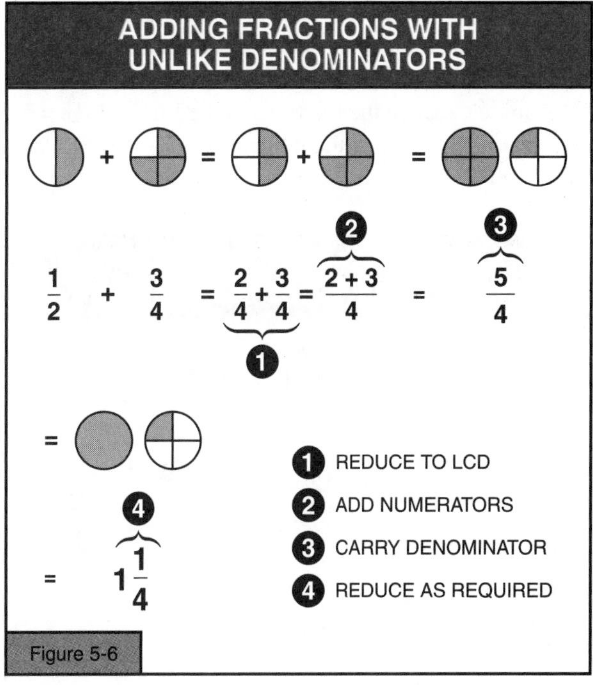

Figure 5-6

Subtracting Fractions

Fractions must have common denominators before they can be subtracted. All fractions to be subtracted must be reduced to their LCD. To subtract mixed numbers, whole numbers and fractions are subtracted separately and the two differences are added.

To subtract fractions that have common denominators, subtract the numerators and place the difference over the denominator. **See Figure 5-7.** For example, to subtract $\frac{5}{16}$ from $\frac{9}{16}$, 5 is subtracted from 9 (9 – 5 = 4). The 4 is placed over 16 ($\frac{4}{16}$). Reduce $\frac{4}{16}$ by dividing the numerator and denominator by 4 ($\frac{4 \div 4}{16 \div 4} = \frac{1}{4}$).

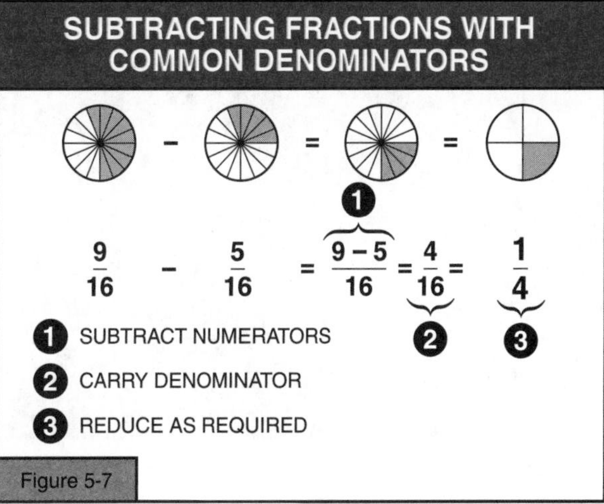

Figure 5-7

To subtract fractions having unlike denominators, reduce the fractions to the LCD. Subtract one numerator from the other. **See Figure 5-8.** For example, to subtract $\frac{1}{6}$ from $\frac{7}{12}$, reduce the fractions to the LCD 12 ($\frac{1}{6} = \frac{2}{12}$ and $\frac{7}{12} = \frac{7}{12}$). Subtract the numerators ($7 - 2 = 5$). Place the 5 over 12 ($\frac{7}{12} - \frac{2}{12} = \frac{5}{12}$).

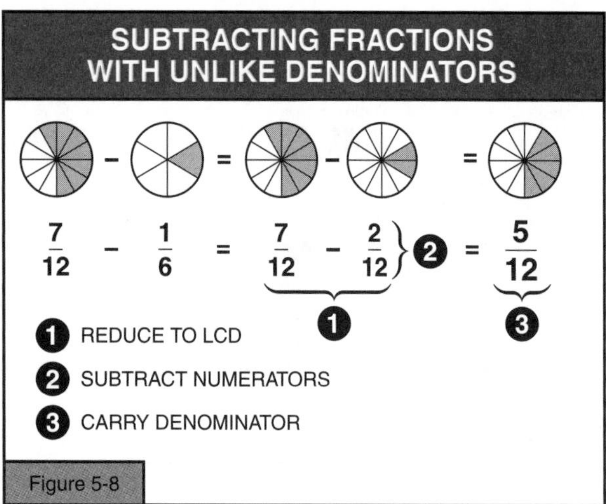

Figure 5-8

Multiplying Two Fractions

To multiply two fractions, multiply the numerator of one fraction by the numerator of the other fraction, and multiply the denominator of one fraction by the denominator of the other fraction. Reduce the answer as required. **See Figure 5-9.** For example, to multiply $\frac{1}{4}$ by $\frac{1}{2}$ ($\frac{1}{4} \times \frac{1}{2}$), multiply the numerators and multiply the denominators ($\frac{1 \times 1}{4 \times 2} = \frac{1}{8}$).

Fractional measurements are used when sizing electrical metal tubing (EMT) for electrical installations.

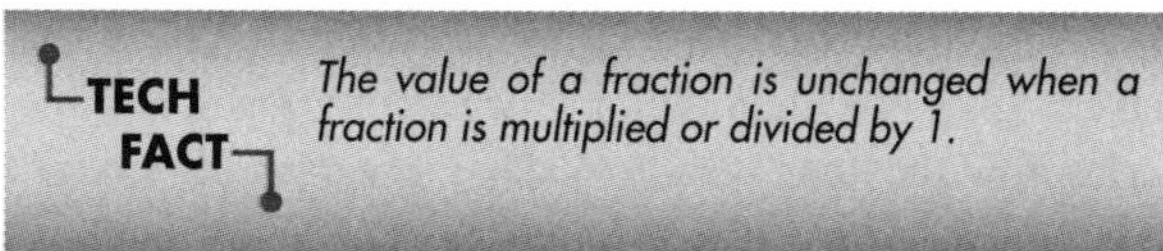

The value of a fraction is unchanged when a fraction is multiplied or divided by 1.

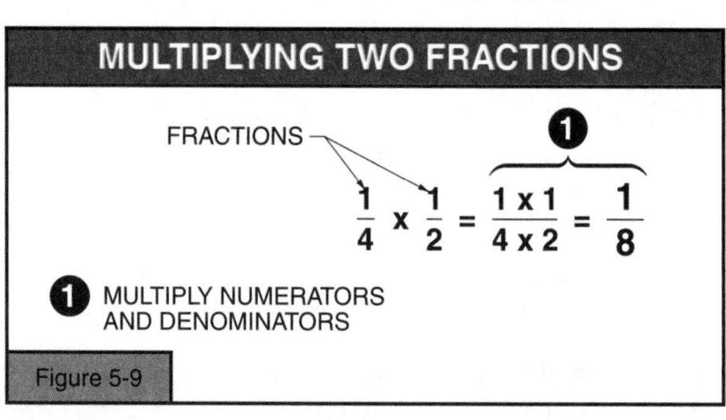

Figure 5-9

In many cases, cancellation can be used to simplify the multiplication process. For example, when multiplying $\frac{2}{5} \times \frac{1}{8}$, the 2 divides into 2 of the $\frac{2}{5}$ and 8 of the $\frac{1}{8}$ an exact number of times ($\frac{2 \div 2}{5} \times \frac{1}{8 \div 2} = \frac{1}{5} \times \frac{1}{4}$). No more cancellation is possible so the numerators and denominators are multiplied ($\frac{1}{5} \times \frac{1}{4} = \frac{1 \times 1}{5 \times 4} = \frac{1}{20}$).

Dividing Two Fractions

To divide one fraction by another, invert the divisor fraction. Use cancellation if possible. Then multiply the numerators and denominators. Reduce the answer as required. **See Figure 5-10.**

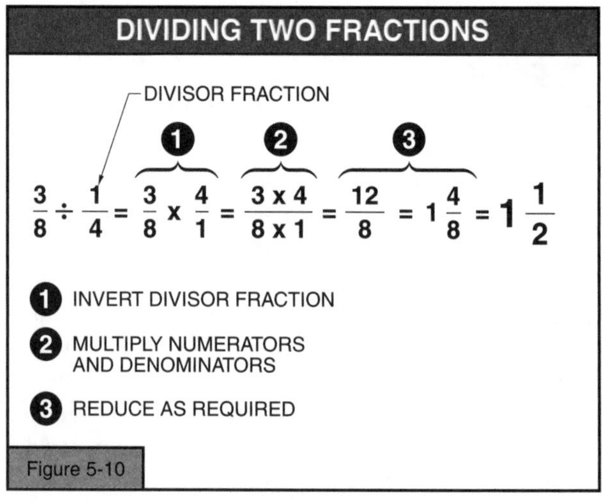

Figure 5-10

For example, to divide $\frac{3}{8}$ by $\frac{1}{4}$, invert $\frac{1}{4}$ (divisor fraction) and multiply by $\frac{3}{8}$ ($\frac{3}{8} \times \frac{4}{1} = \frac{12}{8}$). Reduce as required ($\frac{12}{8} = 12 \div 8 = 1\frac{4}{8} = 1\frac{1}{2}$).

Decimals

A *decimal fraction* is a fraction with a denominator of 10, 100, 1000, etc. The number 1 is the smallest whole number. Anything smaller than 1 is a decimal and can be divided into any number of parts. For example, .85 indicates that the whole number 1 is divided into 100 parts, and 85 parts are present.

Any fraction with 10, 100, 1000, or another multiple of 10 for its denominator may be written as a decimal. For example, $\frac{7}{10}$ is .7, $\frac{75}{100}$ is .75, and $\frac{725}{1000}$ is .725 in the decimal system.

A *decimal* is a number expressed in base 10. The two types of decimal numbers are proper and mixed decimal numbers. A *proper decimal number* is a decimal number that has no whole numbers. For example, .482 is a proper decimal number. A *mixed decimal number* is a decimal number that has a whole number and a decimal number separated by a decimal point. For example, 6.75 is a mixed decimal number.

A *decimal point* is the period at the left of a proper decimal number or the period that separates the parts of a mixed decimal number. All numbers to the left of the decimal point are whole numbers. All numbers to the right of the decimal point are less than whole numbers. **See Figure 5-11.**

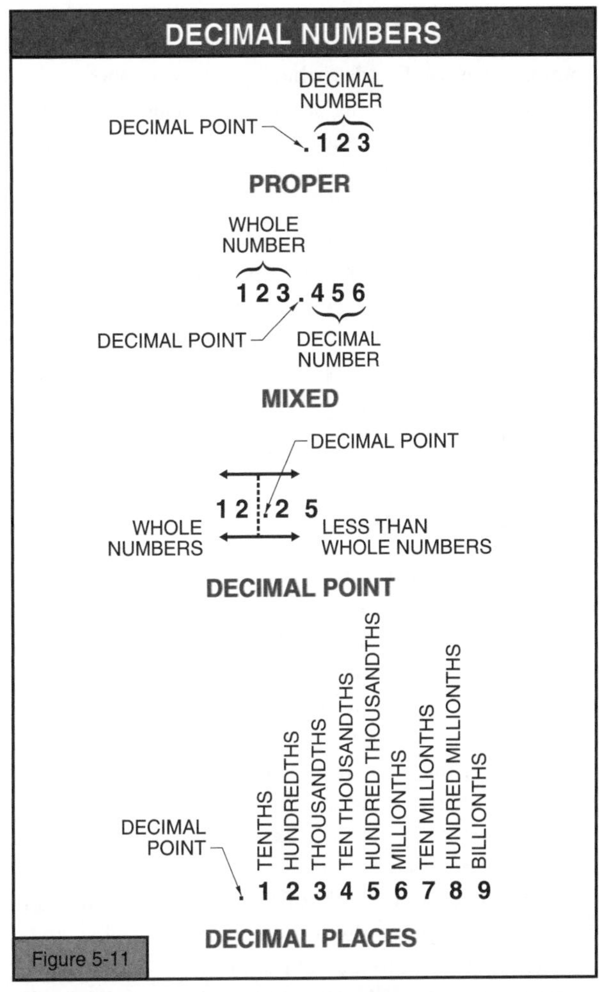

Figure 5-11

Converting Fractions to Decimals

Electricians are constantly taking measurements and reading specifications. Many measurements and specifications are in fractions or decimals. To change a fraction to a decimal, divide the numerator by the denominator. **See Figure 5-12.**

For example, to convert the fraction $\frac{7}{16}$ to a decimal, divide 7 (numerator) by 16 (denominator). Seven divided by 16 equals .4375. The fraction $\frac{7}{16}$ has a decimal equivalent of .4375.

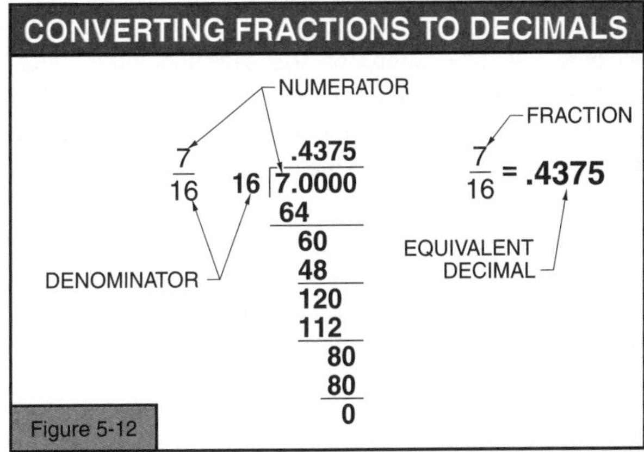

Figure 5-12

Converting Decimals to Fractions

To convert a decimal to a fraction, place the decimal number over 10, 100, 1000, or 10,000 depending on the number of digits (decimal places) in the decimal. Reduce this number to the smallest fraction. **See Figure 5-13.**

For example, to convert .75 to a fraction, place the 75 over 100. The fraction $\frac{75}{100}$ is equivalent to the decimal number .75. The fraction $\frac{75}{100}$ can be reduced to $\frac{3}{4}$.

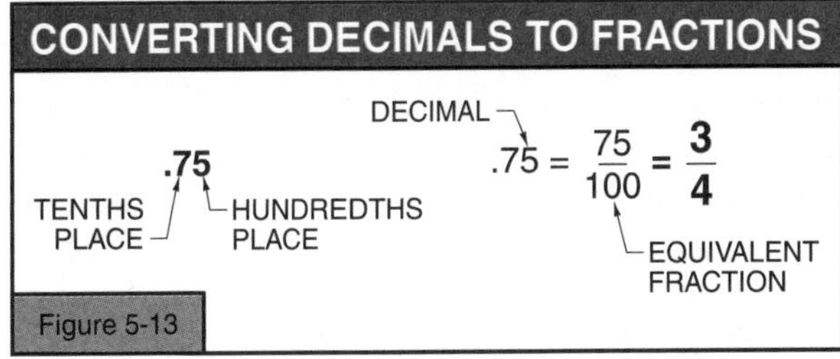

Figure 5-13

TECH FACT — *The number of decimal places in a product is equal to the sum of the number of decimal places in the numbers being multiplied.*

Significant Digits

Numbers are used to describe quantities by showing an exact value as in counting or an approximate value which is derived from measuring. An exact number includes a motor that costs $357.89. An approximate number includes a circuit that averages 120 V.

When a voltmeter measures a value of 120 V, the reading has three significant digits and is precise to the nearest 1 V (smallest quantity). However, if the meter shows 120.0 V, the measurement is accurate to four significant digits and the degree of accuracy is a tenth of a volt. The accuracy of a number is often specified by the number of significant digits (smallest quantity).

Rounding Numbers

Rounding is the process of obtaining a number to a required degree of accuracy. In measurements, more places in a decimal number indicate a higher degree of accuracy. For example, when measuring tolerances on a machined part, the tolerance may be measured in hundredths or thousandths (1.25″ + .01″ or 1.250″ + .005″). The overall size of the parts is the same but the degree of accuracy is more critical for the part measured in thousandths.

To round a number, locate the number of places to which the number is to be rounded. Add 1 to the last retained digit if the digit to the right of the place to be rounded is 5 or more. Discard all digits to the right of the place to be rounded if the digit to the right of the place to be rounded is less than 5. The NEC® rounds down below 5; leaves the 5; and rounds up above 5. **See Figure 5-14.** For example, when rounding the number 65.263 to the hundredths place, the three to the right of the place being rounded is less than 5, so the numbers to the right of the 6 are dropped. The number 65.263 rounded to the hundredths place is 65.26. When rounding the number 35.7168 to the thousandths place, add 1 to the 6 (thousandths) because the 8 occupying the ten thousandths place is 5 or greater. Drop the rest of the decimal. The number 35.7168 rounded to the thousandths place is 35.717.

Percentages

A *percentage* is a fraction that has a denominator of 100. The symbol for percent is %. Percentages are used throughout the electrical and electronic fields to describe and compare electrical effects and indicate the efficiency of electric machines. A number itself is not a fraction, but becomes a fraction when the percent symbol is attached to it. For example, the number 44 is not a fraction. However, when the percent sign is added, it becomes 44⁄100 or 44%.

Percentages may be converted to decimal numbers. To convert a percentage to a decimal number, move the decimal point two places to the left and drop the percent sign. **See Figure 5-15.** For example, to convert 32% to a decimal number, move the decimal point two places to the left and drop the percent sign (.32). To change a decimal number to a percentage, move the decimal point two places to the right and add the percent sign. For example, to convert .66 to a percentage, move the decimal point two places to the right and add a percent sign (66%). To change a common fraction to a percentage, express the fraction as a decimal number then move the decimal point two places to the right and add the percent sign.

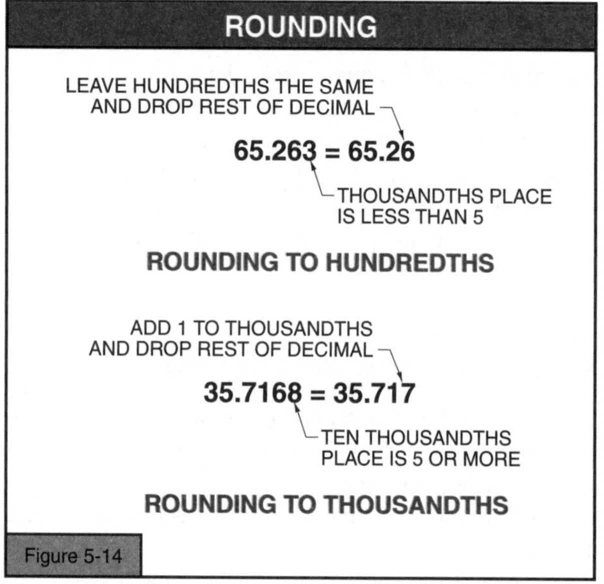

Figure 5-14

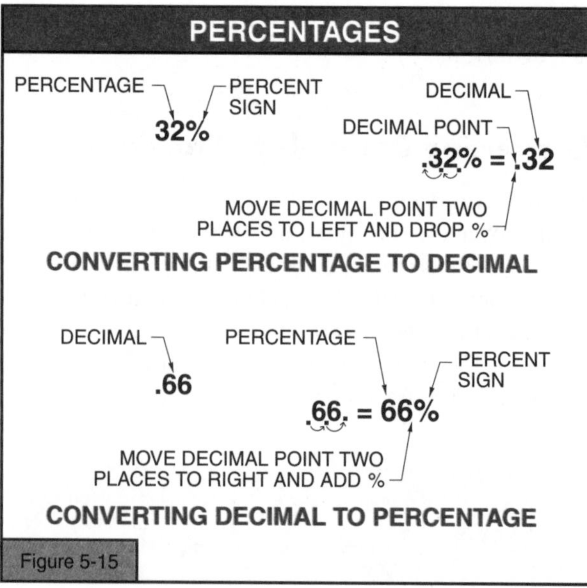

Figure 5-15

To calculate the percentage one number is of another, divide the larger number into the smaller number and multiply the result by 100%. **See Figure 5-16.** For example, to calculate what percentage 3 is of 12, divide 3 by 12 ($3 \div 12 = .25$). Multiply $.25 \times 100$ ($.25 \times 100 = 25\%$). The number 3 is 25% of the number 12.

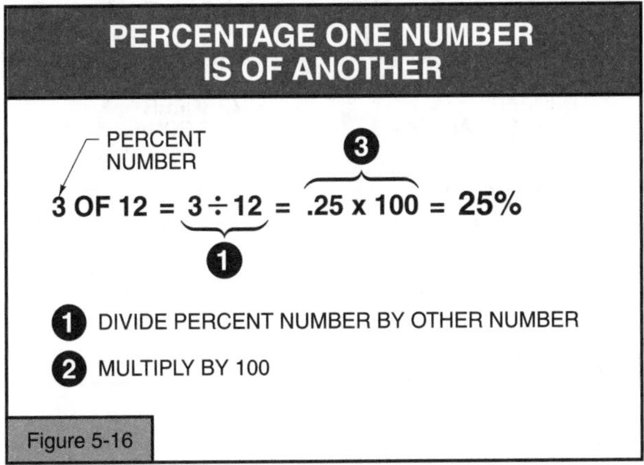

Figure 5-16

Powers

Factors are two or more numbers multiplied together to give a product. For example, in the problem $3 \times 4 = 12$, the 3 and 4 are factors of 12. A *power* is the product of a repeated factor. A *base* is the repeated factor that is multiplied to obtain a power.

For example, in the problem $3 \times 3 \times 3 \times 3 = 81$, the product (81) is the power and the factor (3) is the base. The number 3 is the base of the power 81, and 81 is the power of the base 3.

The product is to the second power or square when the base is used twice as a factor. For example, the square of 4 is 4×4 or 16. The product is to the third power or cube when the base is used three times as a factor. For example, the cube of 5 is $5 \times 5 \times 5$ or 125. The product is to the fourth power, fifth power, sixth power, etc. and is indicated by an exponent when the base is used as a factor more than three times. **See Figure 5-17.**

POWERS

BASE — EXPONENT (SUPERSCRIPT)

$10^3 = 10 \times 10 \times 10 = 1000$

BASE — FACTORS — POWER (PRODUCT)

$10^6 = 10 \times 10 \times 10 \times 10 \times 10 \times 10 = 1{,}000{,}000$

EXPONENT

Figure 5-17

An *exponent* is a number that indicates the number of times a base is raised to a power (number of times used as a factor). Exponents are indicated with a superscript. A *superscript* is a small number above and to the right of the base. For example, 2^4 (two raised to the fourth power) indicates that the base (2) is multiplied four times as a factor ($2 \times 2 \times 2 \times 2 = 16$), and 16 is the fourth power of 2.

Exponents may be positive, negative, or zero. A base number raised to a positive power equals the base number multiplied by itself as many times as indicated by the exponent. For example, 3^3 equals $3 \times 3 \times 3$ or 27. Any base raised to the power of zero is equal to 1. For example, $1^0 = 1$, $14^0 = 1$, and $829^0 = 1$. The first power of any number is equal to the number itself. Thus, $2^1 = 2$ and $10^1 = 10$. A base raised to a negative power is the reciprocal of the base with a positive exponent. To find the value of a base raised to a negative power, invert the base to find the reciprocal and solve the denominator. **See Figure 5-18.** For example, to find the value of 10^{-4}, the base number (10) is reciprocated ($\frac{1}{10}$) and the exponent is made positive ($\frac{1}{10^4}$). The denominator is then solved by multiplying by the number of times indicated by the exponent ($\frac{1}{10} \times \frac{1}{10} \times \frac{1}{10} \times \frac{1}{10} = \frac{1}{10{,}000}$).

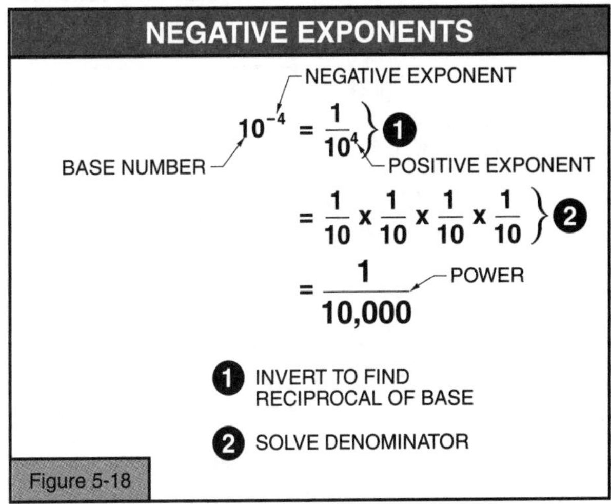

Figure 5-18

Positive Base Numbers

To raise a positive base number to a power, multiply the positive base number by itself as many times as indicated by the exponent. **See Figure 5-19.** For example, to solve 4^3, multiply 4 by 4 by 4 ($4 \times 4 \times 4 = 64$). The third power of 4, or 4^3, = 64.

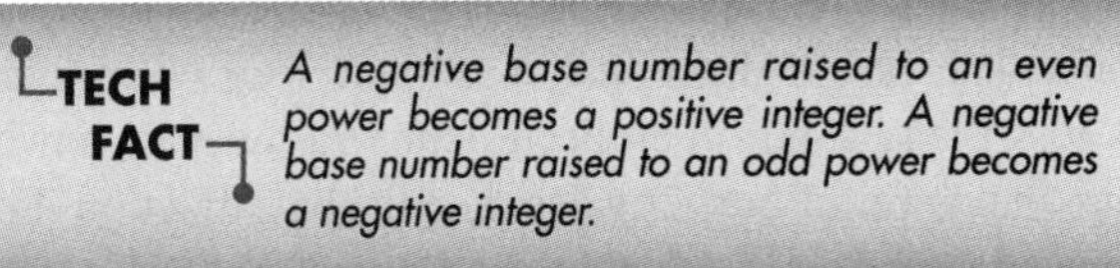

A negative base number raised to an even power becomes a positive integer. A negative base number raised to an odd power becomes a negative integer.

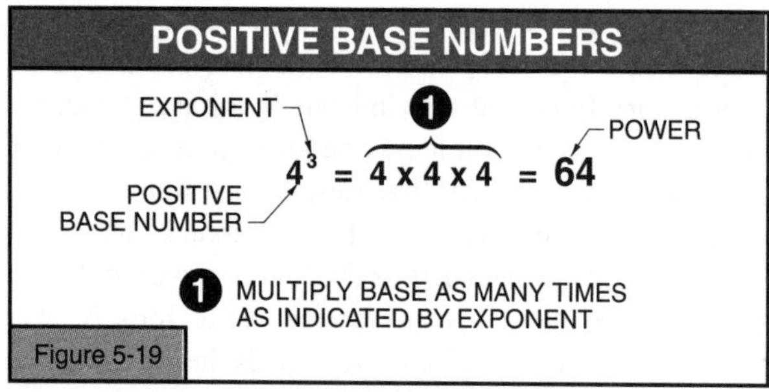

Figure 5-19

Negative Base Numbers

To raise a negative base number to a power, multiply the negative base number by itself as many times as indicated by the exponent. **See Figure 5-20.** The answer is a positive number if the negative base number is raised to an even power. The answer is a negative number if the negative base number is raised to an odd power. Parentheses determine to what base the exponent applies. For example, the difference between $(-4)^2$ and $-(4)^2$ is that $(-4)^2$ is $-4 \times -4 = 16$, and $-(4)^2$ is $-(4 \times 4) = -16$.

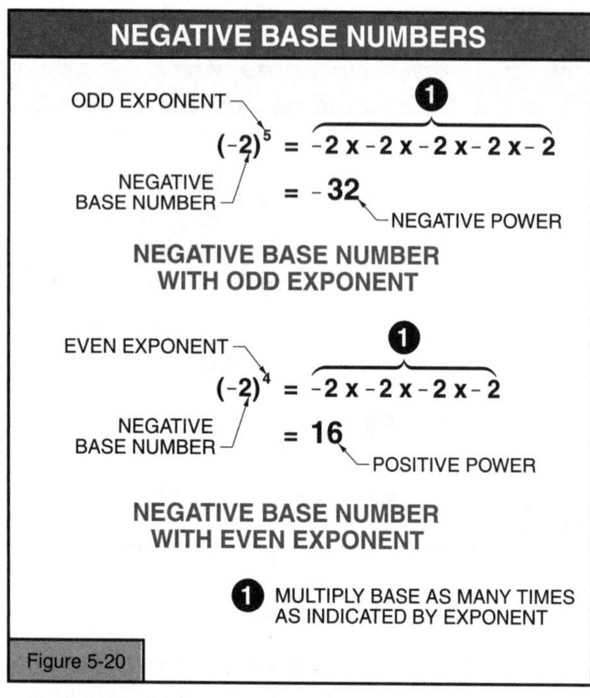

Figure 5-20

Fractional Base Numbers

To raise a fraction to a power, multiply the numerator and denominator by the power separately. **See Figure 5-21.** A proper fraction raised to a power has an answer of lesser value than the base fraction. For example, $(3/5)^2 = 9/25$ which is less than $3/5$. An improper fraction raised to a power has an answer of greater value than the base fraction. For example, $(4/3)^2 = 16/9$, which is greater than $4/3$.

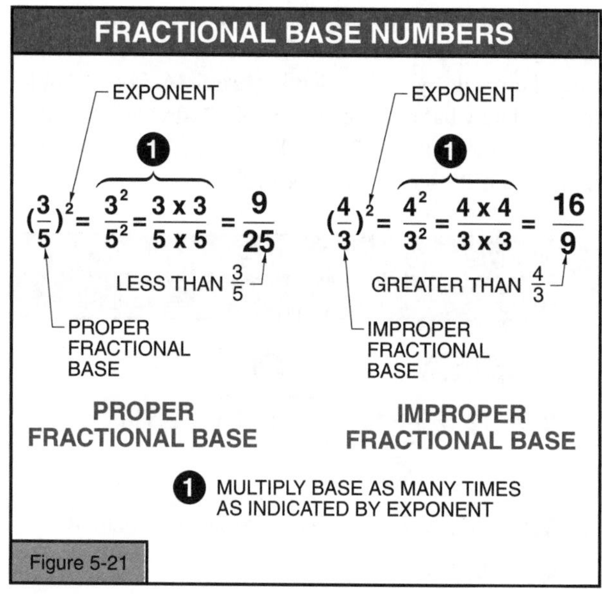

Figure 5-21

Decimal Base Numbers

To raise decimals or mixed decimals to a power, multiply the decimal by itself as many times as indicated by the exponent. **See Figure 5-22.** For example, to solve $(.25)^3$, multiply .25 by .25 by .25 ($.25 \times .25 \times .25 = .015625$).

Decimals are similar to proper fractions in that decimals raised to a higher power have an answer of lesser value. For example, $.2^2 = .2 \times .2 = .04$, which is less than .2, and $.2^3 = .2 \times .2 \times .2 = .008$ which is less than .2 and $.2^2$.

To find the number of decimal places in the answer, multiply the number of decimal places in the base by the exponent. For example, $.3^6 = .3 \times .3 \times .3 \times .3 \times .3 \times .3 = .000729$ which has six decimal places because one (decimal place) multiplied by 6 (exponent) equals 6.

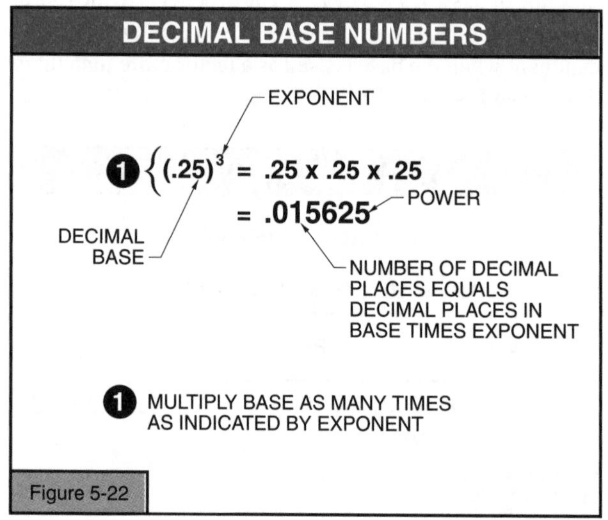

Figure 5-22

Power of a Power

The power of a power is found by multiplying the two exponents to find the new exponent which indicates the number of times the base is used as a factor. **See Figure 5-23.** For example, to raise 2^2 to the third power or $(2^2)^3$, multiply 2 (base exponent) by 3 (second exponent) to get 6 (new exponent). The new problem is 2 to the sixth power (2^6). To solve 2^6, multiply 2 by 2 by 2 by 2 by 2 by 2 ($2 \times 2 \times 2 \times 2 \times 2 \times 2 = 64$).

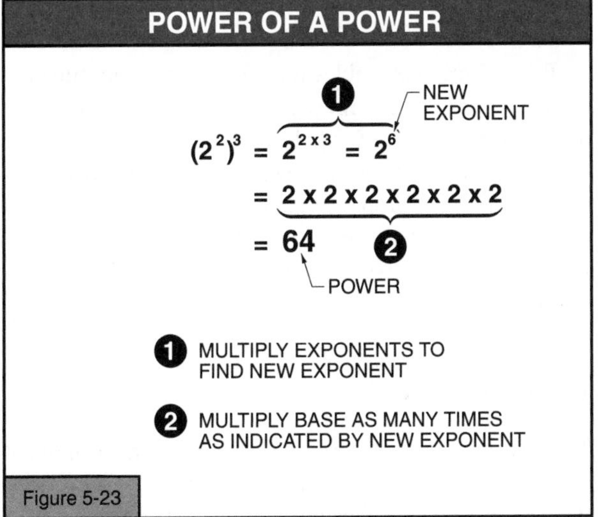

Figure 5-23

Multiplication of Powers

To multiply two or more like bases with exponents, add the exponents to find the new exponent for the same base. **See Figure 5-24.** For example, to multiply 5^2 and 5^3, add the 2 and 3 for the new exponent 5 ($2 + 3 = 5$). Place the 5 as the exponent next to the base and solve ($5^2 \times 5^3 = 5^{2+3} = 5^5 = 3125$).

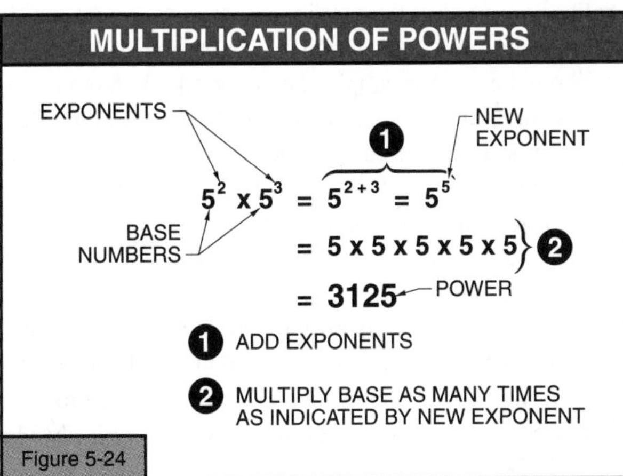

Figure 5-24

Division of Powers

To divide two like bases with exponents, subtract the exponent of the divisor from the exponent of the dividend to find the new exponent for the same base. **See Figure 5-25.** For example, to divide 5^6 by 5^4, subtract 4 from 6 to find the new exponent 2 ($6 - 4 = 2$). Place the 2 as the exponent next to the base and solve ($5^6 \div 5^4 = 5^{6-4} = 5^2 = 5 \times 5 = 25$).

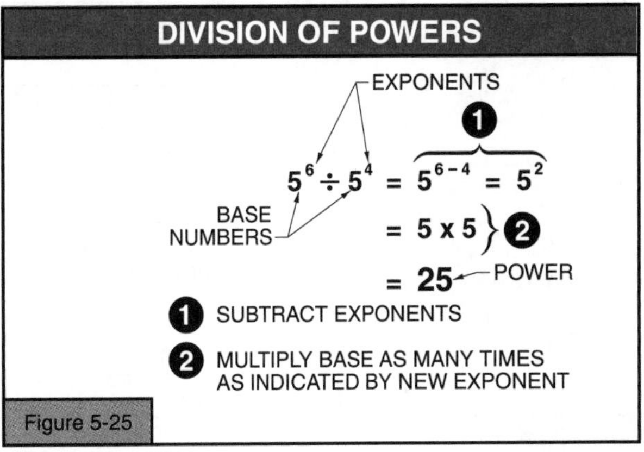

Figure 5-25

Scientific Notation

In the electrical and electronic fields, calculations may involve very large or very small numbers. For example, a capacitor may have a value of .000002 F (farads) or an electrical power transmission line may have a voltage of 345,000 V. Electrical problems may be solved with formulas which use exponents.

A *coefficient* is a factor that is multiplied by the base number. For example, in the number 2×10^3, the number 2 is the coefficient. To express a large number as a small number times a power of ten, move the decimal point to the left and count the number of places to the original decimal point. The number of places counted gives the positive power of ten. For example, $152 = 15.2 \times 10^1$, or 1.52×10^2, or $.152 \times 10^3$.

To express a decimal fraction as a whole number times a power of 10, move the decimal point to the right and count the number of places to the original point. The number of places counted is the proper negative power of 10. For example, $.005 = 5 \times 10^{-3}$, or $.5 \times 10^{-2}$, or $.05 \times 10^{-1}$. **See Figure 5-26.**

Roots

A *root* of a number is the number which must be multiplied by itself a given number of times to obtain the original number. The original number results when the roots of the number are multiplied together. For example, the two equal factors of 25 are 5 and 5 because $25 = 5 \times 5$. Therefore, 5 is a root of 25.

POWERS OF TEN

	Scientific Notation	Decimal Notation
Whole Numbers	10^6	1,000,000 – mega (M)
	10^5	100,000
	10^4	10,000
	10^3	1000 – kilo (K)
	10^2	100
	10^1	10
	10^0	1 – base
Fractions	10^{-1}	0.1
	10^{-2}	0.01
	10^{-3}	0.001 – milli (m)
	10^{-4}	0.0001
	10^{-5}	0.00001
	10^{-6}	0.000001 – micro (μ)

Figure 5-26

A *square root* of a number is a number which, when multiplied by itself, gives the original number. For example, the square root of 16 is 4 (4 × 4 = 16). The mathematical symbol for square root is the radical sign ($\sqrt{\ }$). The *index* is the small figure to the left of the radical sign ($\sqrt[x]{\ }$). The *radicand* is the number that appears under the radical sign. A *cube root* is a number which has three equal roots. For example, the cube root of 27 is 3 (3 × 3 × 3 = 27). When the radical sign is used without an index it is understood to mean square root. **See Figure 5-27.**

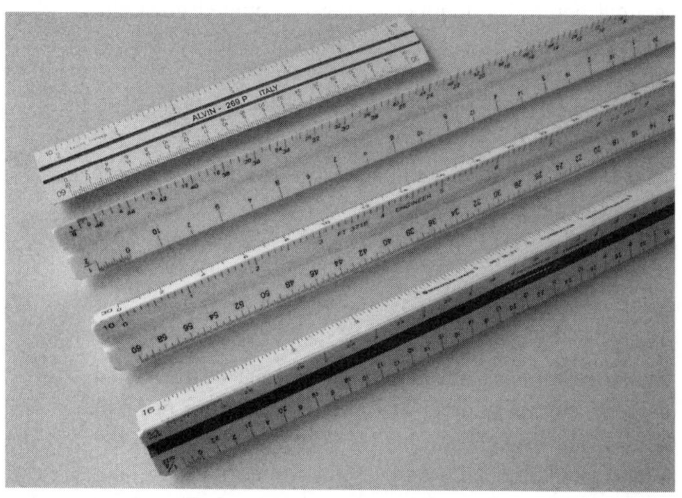

Architect's and engineer's scales are used to draw objects to a specific size and to reduce or enlarge them proportionately.

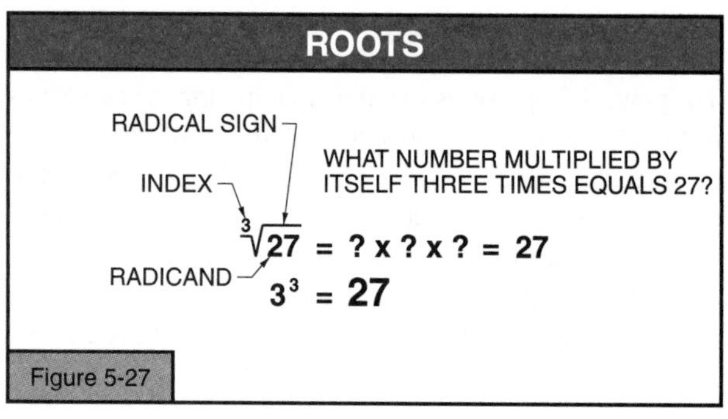

Figure 5-27

Powers and roots tables may be used to determine the square, cubes, square root, cube root, and reciprocal of numbers. **See Figure 5-28.** To find the square, cube, etc., find the number in the Number column and move horizontally to the appropriate column. For example, to find the square root of 11, find 11 in the Number column and move horizontally to the Square Root column. The square root of 11 is 3.31662.

Measurement

The two most commonly used systems of measurement are the English system and the metric system. Conversion factors are used to change measurements from one system to the other. The English system is primarily used in the United States. This system uses the inch (in. or ″), foot (ft or ′), pint (pt), quart (qt), gallon (gal.), ounce (oz), pound (lb), etc. as basic units of measure. **See Figure 5-29.**

The metric system is the most common measurement system used in the world. The metric system is based on the meter (m), liter (l), and gram (g). **See Figure 5-30.**

Metric Prefixes

Prefixes are used in the metric system to represent multipliers. Metric prefixes simplify the notation of large and small numbers. For example, the prefix kilo (k) has a prefix equivalent of 1000, so 1 kilometer (1 km) = 1000 m. Metric prefixes may be used with electrical quantities such as 1,000,000 Ω (1 MΩ), 1000 V (1 kV), .001 W (1 mW), and .000001 A (1 μA). The most commonly used prefixes in the electrical/electronic field are mega (M), kilo (k), milli (m), and micro (μ).

To change from a quantity to another prefix, multiply the quantity by the number of units that equals one of the original metric units (conversion factor). For example, to change 544 m to kilometers (1 m = $\frac{1}{1000}$ km), multiply 544 by .001 (544 × .001 = .544 km).

POWERS, ROOTS, AND RECIPROCALS						
Number	**Square**	**Cube**	**Square Root**	**Cube Root**	**Reciprocal**	**Number**
1	1	1	1.00000	1.00000	1.0000000	**1**
2	4	8	1.41421	1.25990	.5000000	**2**
3	9	27	1.73205	1.44225	.3333333	**3**
4	16	64	2.00000	1.58740	.2500000	**4**
5	25	125	2.23607	1.70998	.2000000	**5**
6	36	216	2.44949	1.81712	.1666667	**6**
7	49	343	2.64575	1.91293	.1428571	**7**
8	64	512	2.82843	2.00000	.1250000	**8**
9	81	729	3.00000	2.08008	.1111111	**9**
10	100	1000	3.16228	2.15443	.1000000	**10**
11	121	1331	3.31662	2.22398	.0909091	**11**
12	144	1728	3.46410	2.28943	.0833333	**12**
13	169	2197	3.60555	2.35133	.0769231	**13**
14	196	2744	3.74166	2.41014	0.714286	**14**

Figure 5-28

The decimal point in the base is moved to the left or right and a prefix is added to convert a unit without a prefix (base) to a unit with a prefix. The decimal point is moved to the left and a prefix is added to convert a large value to a simpler term. The decimal point is moved to the right and a prefix is added to convert a small value to a simpler term. **See Figure 5-31.**

A conversion table may be used to change a metric quantity to another prefix. For example, to change 150,000 V to kV, move the decimal point three places to the left and add the prefix k (150,000 V = 150 kV).

Conversions

Conversions can be performed in all units of measure. English and metric measurements are converted from one system of measurement to the other by multiplying by the number of units of one system that equals one unit of measurement of the other system. The number of units is the conversion factor. Conversions are performed by applying the appropriate conversion factor. Equivalent tables are used to convert between measurement systems. **See Appendix.**

To convert a metric measurement to an English measurement, multiply the measurement by the number of English units that equal one of the metric units (conversion factor). For example, to convert 50 mm to inches, multiply 50 by .039 (50 × .039 = 1.95″).

To convert an English measurement to a metric measurement, multiply the measurement by the number of metric units that equals one of the English units (conversion factor). For example, to convert 3 mi to kilometers, multiply 3 by 1.609 (3 × 1.609 = 4.827 km).

Calculators are often used on a job site to perform mathematical functions.

ENGLISH SYSTEM				
LENGTH		**Unit**	**Abbr**	**Equivalents**
		mile	mi	5280′, 320 rd, 1760 yd
		rod	rd	5.50 yd, 16.5′
		yard	yd	3′, 36″
		foot	ft *or* ′	12″, .333 yd
		inch	in. *or* ″	.083′, .028 yd
AREA A = l x w		square mile	sq mi *or* mi^2	640 A, 102,400 sq rd
		acre	A	4840 sq yd, 43,560 sq ft
		square rod	sq rd *or* rd^2	30.25 sq yd, .00625 A
		square yard	sq yd *or* yd^2	1296 sq in., 9 sq ft
		square foot	sq ft *or* ft^2	144 sq in., .111 sq yd
		square inch	sq in. *or* in^2	.0069 sq ft, .00077 sq yd
VOLUME V = l x w x t		cubic yard	cu yd *or* yd^3	27 cu ft, 46,656 cu in.
		cubic foot	cu ft *or* ft^3	1728 cu in., .0370 cu yd
		cubic inch	cu in. *or* in^3	.00058 cu ft, .000021 cu yd
CAPACITY	*U.S. liquid measure*	gallon	gal.	4 qt (231 cu in.)
WATER, FUEL, ETC.		quart	qt	2 pt (57.75 cu in.)
		pint	pt	4 gi (28.875 cu in.)
		gill	gi	4 fl oz (7.219 cu in.)
		fluidounce	fl oz	8 fl dr (1.805 cu in.)
		fluidram	fl dr	60 min (.226 cu in.)
		minim	min	⅙ fl dr (.003760 cu in.)
VEGETABLES, GRAIN, ETC.	*U.S. dry measure*	bushel	bu	4 pk (2150.42 cu in.)
		peck	pk	8 qt (537.605 cu in.)
		quart	qt	2 pt (67.201 cu in.)
		pint	pt	½ qt (33.600 cu in.)
DRUGS	*British imperial liquid and dry measure*	bushel	bu	4 pk (2219.36 cu in.)
		peck	pk	2 gal. (554.84 cu in.)
		gallon	gal.	4 qt (277.420 cu in.)
		quart	qt	2 pt (69.355 cu in.)
		pint	pt	4 gi (34.678 cu in.)
		gill	gi	5 fl oz (8.669 cu in.)
		fluidounce	fl oz	8 fl dr (1.7339 cu in.)
		fluidram	fl dr	60 min (.216734 cu in.)
		minim	min	1/60 fl dr (.003612 cu in.)
MASS AND WEIGHT	*avoirdupois*	ton	t	2000 lb
COAL, GRAIN, ETC.		short ton		2000 lb
		long ton		2240 lb
		pound	lb *or* #	16 oz, 7000 gr
		ounce	oz	16 dr, 437.5 gr
		dram	dr	27.344 gr, .0625 oz
		grain	gr	.037 dr, .002286 oz
GOLD, SILVER, ETC.	*troy*	pound	lb	12 oz, 240 dwt, 5760 gr
		ounce	oz	20 dwt, 480 gr
		pennyweight	dwt *or* pwt	24 gr, .05 oz
		grain	gr	.042 dwt, .002083 oz
DRUGS	*apothecaries'*	pound	lb ap	12 oz, 5760 gr
		ounce	oz ap	8 dr ap, 480 gr
		dram	dr ap	3 s ap, 60 gr
		scruple	s ap	20 gr, .333 dr ap
		grain	gr	.05 s, .002083 oz, .0166 dr ap

Figure 5-29

METRIC SYSTEM			
LENGTH	**Unit**	**Abbreviation**	**Number of Base Units**
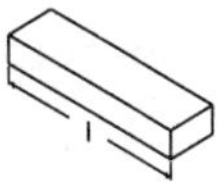	kilometer	km	1000
	hectometer	hm	100
	dekameter	dam	10
	meter*	m	1
	decimeter	dm	.1
	centimeter	cm	.01
	millimeter	mm	.001
AREA A = l x w 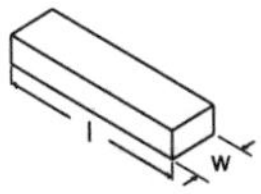	square kilometer	sq km *or* km^2	1,000,000
	hectare	ha	10,000
	are	a	100
	square centimeter	sq cm *or* cm^2	.0001
VOLUME V = l x w x t 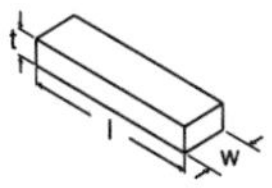	cubic centimeter	cu cm, cm^3, *or* cc	.000001
	cubic decimeter	dm^3	.001
	cubic meter*	m^3	1
CAPACITY WATER, FUEL, ETC. 	kiloliter	kl	1000
	hectoliter	hl	100
	dekaliter	dal	10
VEGETABLES, GRAIN, ETC. 	liter*	l	1
	cubic decimeter	dm^3	1
	deciliter	dl	.10
DRUGS 	centiliter	cl	.01
	milliliter	ml	.001
MASS AND WEIGHT	metric ton	t	1,000,000
COAL, GRAIN, ETC. 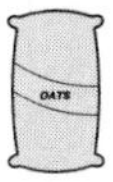	kilogram	kg	1000
	hectogram	hg	100
	dekagram	dag	10
GOLD, SILVER, ETC. 	gram*	g	1
	decigram	dg	.10
DRUGS 	centigram	cg	.01
	milligram	mg	.001

* base units

Figure 5-30

Graphs

A *graph* is a diagram that shows the continuous relationship between two or more variables. Graphs represent information in a simple form and are commonly used by component manufacturers to illustrate data and specifications. On a graph, one known variable is plotted horizontally and another is plotted vertically. The relationship between the two variables is represented by a straight or curved line. The point at which either variable line intersects the straight or curved line represents the value of the unknown variable.

544 m = ? km

544 x .001 = .544 km

CONVERSION FACTOR

OR

MOVE DECIMAL POINT THREE PLACES TO LEFT

.544 m = ? km

544 m = .544 km

PREFIXES			
Multiples and Submultiples	**Prefixes**	**Symbols**	**Meaning**
1,000,000,000,000 = 10^{12}	tera	T	trillion
1,000,000,000 = 10^{9}	giga	G	billion
1,000,000 = 10^{6}	mega	M	million
1000 = 10^{3}	kilo	k	thousand
100 = 10^{2}	hecto	h	hundred
10 = 10^{1}	deka	d	ten
Unit 1 = 10^{0}			
.1 = 10^{-1}	deci	d	tenth
.01 = 10^{-2}	centi	c	hundredth
.001 = 10^{-3}	milli	m	thousandth
.000001 = 10^{-6}	micro	μ	millionth
.000000001 = 10^{-9}	nano	n	billionth

CONVERSION TABLE												
Initial Units	**Final Units**											
	giga	**mega**	**kilo**	**hecto**	**deka**	**base**	**deci**	**centi**	**milli**	**micro**	**nano**	**pico**
giga		3R	6R	7R	8R	9R	10R	11R	12R	15R	18R	21R
mega	3L		3R	4R	5R	6R	7R	8R	9R	12R	15R	18R
kilo	6L	3L		1R	2R	3R	4R	5R	6R	9R	12R	15R
hecto	7L	4L	1L		1R	2R	3R	4R	5R	8R	11R	14R
deka	8L	5L	2L	1L		1R	2R	3R	4R	7R	10R	13R
base	9L	6L	3L	2L	1L		1R	2R	3R	6R	9R	12R
deci	10L	7L	4L	3L	2L	1L		1R	2R	5R	8R	11R
centi	11L	8L	5L	4L	3L	2L	1L		1R	4R	7R	10R
milli	12L	9L	6L	5L	4L	3L	2L	1L		3R	6R	9R
micro	15L	12L	9L	8L	7L	6L	5L	4L	3L		3R	6R
nano	18L	15L	12L	11L	10L	9L	8L	7L	6L	3L		3R
pico	21L	18L	15L	14L	13L	12L	11L	10L	6L	6L	3L	

R = move the decimal point to the right

L = move the decimal point to the left

Figure 5-31

For example, a graph may be used to illustrate the effect of the ambient temperature on the operating characteristics of a fuse. *Ambient temperature* is the temperature of the air surrounding a piece of equipment. As the ambient temperature increases, the opening time and capacity rating of the fuse decreases. As the ambient temperature decreases, the opening time and capacity rating of the fuse increases. At 140°F, the opening time of a fuse is 70% of the standard rated opening time and the capacity is 90% of the standard rated carrying capacity. Likewise, at –40°F, the opening time of the fuse is approximately 133% of the standard rated opening time and the capacity is approximately 118% of the standard rated carrying capacity. **See Figure 5-32.**

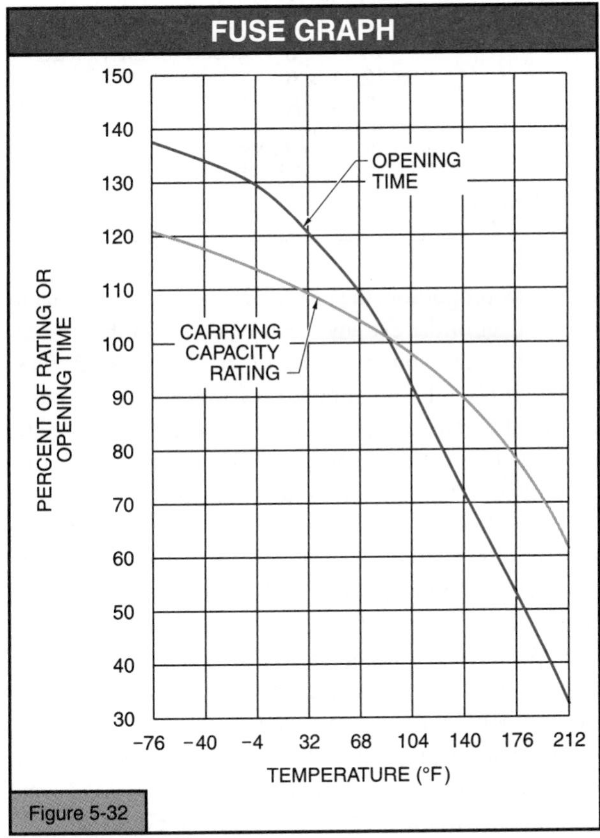

Figure 5-32

Relationships between variables are often given in tables. Tables present the relationship between several variables. **See Figure 5-33.** A table may be converted to a graph to show the continuous relationship between the variables. To convert a table to a graph, apply the procedure:

1. Draw two axes at right angles.
2. Label the two axes.
3. Select scales based on the values given.
4. Plot the given points from the table.
5. Draw a line or arc to connect the points.

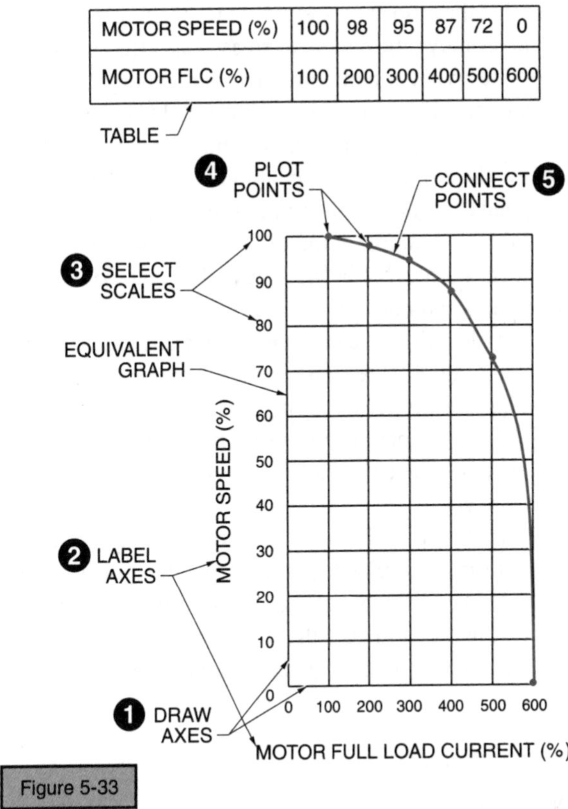

Figure 5-33

Ratios

A *ratio* is the relationship between two quantities or terms. Ratios are used in the electrical and electronic fields when determining the ratio of the primary winding to the secondary winding in a transformer. A ratio is the mathematical method of making a comparison. *Relation* is how much larger or smaller one term is in comparison with another term. Terms must be changed into common units of measure before a comparison can be made. No comparison is possible if they cannot be reduced to a common unit.

Two terms are known in all ratio calculations. A colon (:) is the symbol used to indicate a relation between terms. **See Figure 5-34.** For example, the ratio of 8 to 2 is written as the ratio expression 8:2.

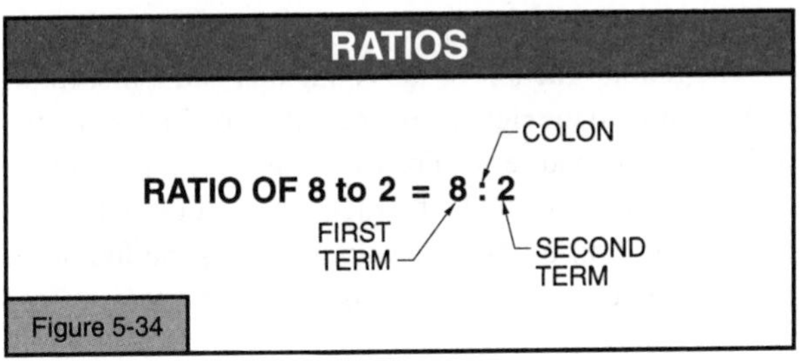

Figure 5-34

To find the ratio of two terms, divide the first term by the second term. A ratio can be expressed as a fraction where the first term is the numerator and the second term is the denominator. For example, the ratio of 15 to 5 = 15:5 = 15 ÷ 5 = $^{15}/_{5}$ = 3. Because a ratio may be expressed as a fraction, the numerator and denominator of a ratio may be multiplied or divided by the same number without changing its value. For example, a ratio of 20:4 can be reduce to an equivalent ratio of 5:1 because 20 and 4 can be divided by 4 five times and one time respectively.

A simple ratio is the ratio between two terms. A simple ratio where the first term is smaller than the second term is similar to a proper fraction with the answer being less than 1. For example, in the ratio 5:10, the first term (5) is smaller than the second term (10) and equals 5 divided by 10, or ½.

A simple ratio where the first term is larger than the second term is similar to an improper fraction with the answer being either a whole number or mixed number. For example, in the ratio 10:5 the first term (10) is larger than the second term (5) and equals 10 divided by 5, or 2.

Proportions

A *proportion* is an expression indicating that two ratios are equal. For example, the ratio expressions 8:4 and 12:6 both have a ratio of 2. All proportions are composed of two equal ratio expressions. **See Figure 5-35.** An equal sign is used between the two ratio expressions to indicate that one expression is equal to the other. The equal proportion is written as 8:4 = 12:6.

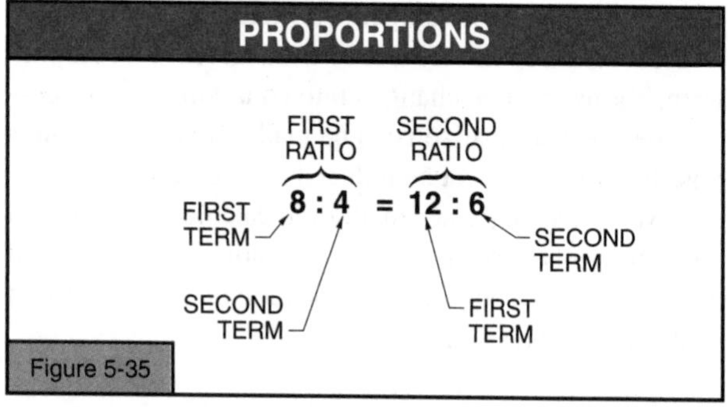

Figure 5-35

The ratio to the left of the equal sign is the first ratio, and the ratio to the right is the second ratio. Each ratio has a first and second term. For example, in the proportion 8:4 = 12:6, the 8 is the first term of the first ratio, the 4 is the second term of the first ratio, the 12 is the first term of the second ratio, and the 6 is the second term of the second ratio.

Means and Extremes

Means are the two inner numbers of a proportion. *Extremes* are the two outer numbers of a proportion. **See Figure 5-36.** For example, in the proportion 8:4 = 12:6, the 4 and 12 (two inner numbers) are the means. The 8 and 6 (two outer numbers) are the extremes.

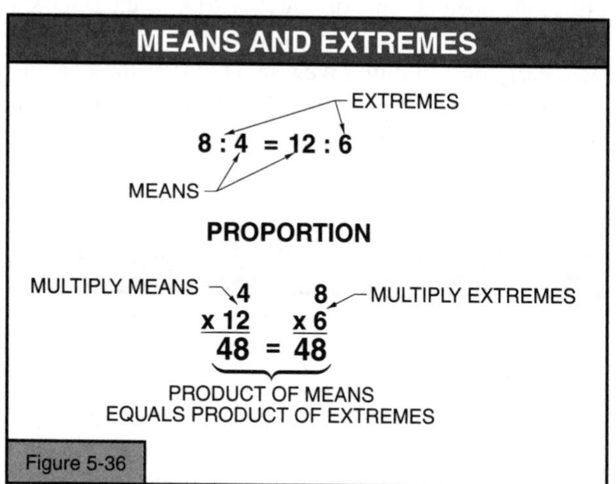

Figure 5-36

In any proportion, the product of the means is equal to the product of the extremes. For example, in the proportion 8:4 = 12:6, multiply 4 by 12 (means) (4 × 12 = 48) and multiply 8 by 6 (extremes) (8 × 6 = 48). The ratios are equal because the product of the means is equal to the product of the extremes. No proportion expression is a true proportion unless the two ratios are equal.

Proportions are either direct proportions or inverse proportions. A *direct proportion* is a statement of equality between two ratios in which the first of four terms divided by the second equals the third divided by the fourth. For example, in the proportion 8:2 = 12:3, both ratios equal 4 (8 ÷ 2 = 4 and 12 ÷ 3 = 4). An increase in one term results in a proportional increase in the other related term. For example, in the proportion 8:2 = 12:3, if the 8 is increased, then the 12 must be proportionally increased by the same amount to keep the proportion equal. **See Figure 5-37.**

An *inverse proportion* is a statement of equality between two ratios in which an increase in one quantity results in a proportional decrease in the other related quantity. For example, in a two-pulley system, if one pulley is 10″ in diameter and the other is 2″ in diameter, the smaller pulley rotates five times when the larger pulley rotates one time (10 ÷ 2 = 5). **See Figure 5-38.** The number of rotations of a pulley is inversely proportional to the diameter of the pulley. If the diameter of the small pulley is increased to 5″, then it rotates twice as the large pulley rotates once (10:5 = 2:1). The diameter of the small pulley is increased and the number of rotations decreased.

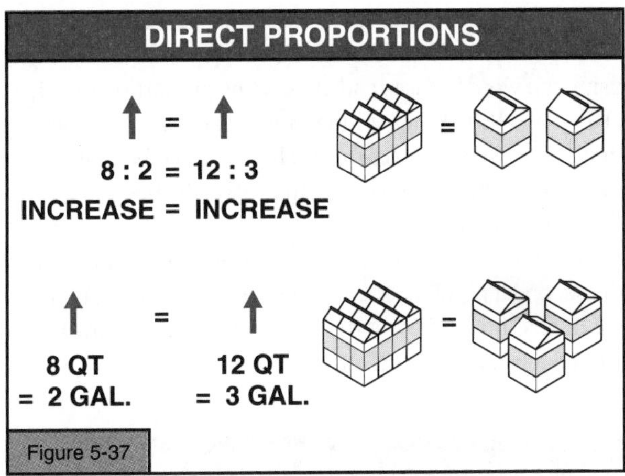

Figure 5-37

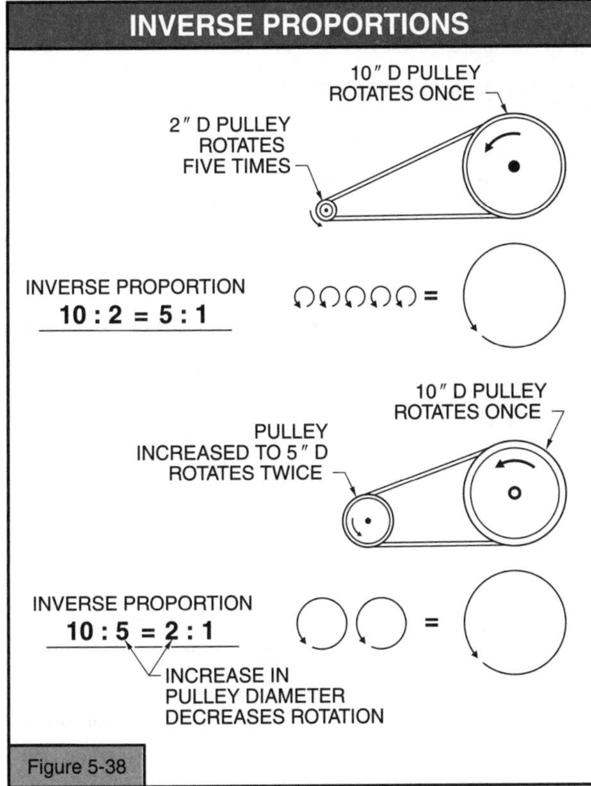

Figure 5-38

Applications of proportions are presented as problems that give a statement of certain conditions and an indication of the required answer. In all proportion problems, three of the four numbers, which become terms of the two ratios, are given in the statement of the problem.

To solve a proportion problem it must be determined if the problem is a direct or inverse proportion, then the unknown number is found. For example, if a metal joint 16′ long requires 80 rivets, how many rivets are required for a joint 11′ long?

To solve the problem, determine which of the quantities or numbers is the first term in the first ratio, the second term in the first ratio, etc. The first two terms in a proportion form a ratio, the last two terms form a ratio, and a ratio can only be made of numbers with common measures.

The first ratio in the proportion consists of 16 and 11 which are in feet. The second ratio must also have numbers representing common units of measure. The number 80 represents rivets and the unknown number also represents rivets. Therefore, the 80 and the unknown number form the second ratio. **See Figure 5-39.**

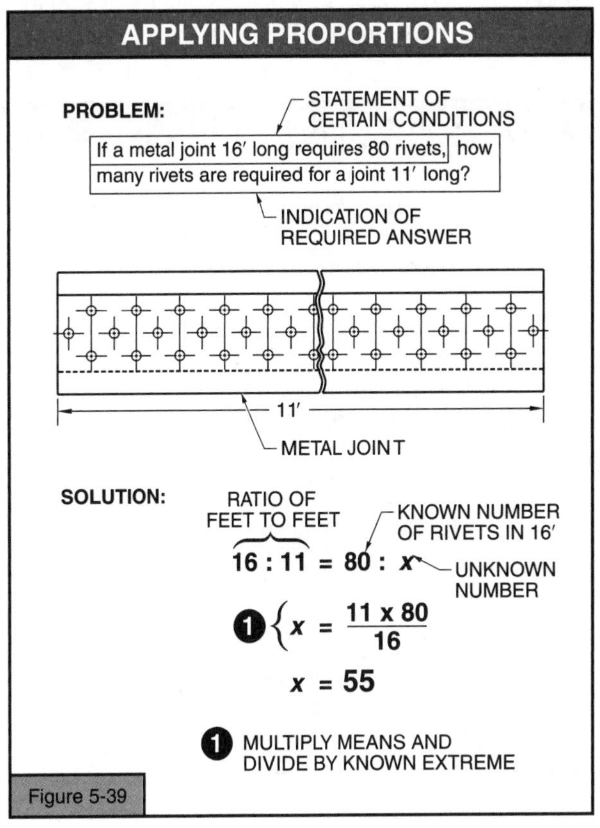

Figure 5-39

To decide where to place the three numbers given in any problem, always make the unknown number the second term in the second ratio. The unknown number is indicated by the lowercase letter x. The first terms of both ratios are the terms that complete a statement. For example, “a metal joint 16′ long requires 80 rivets” is a complete statement, and 16 and 80 are the first terms of the ratios. The second term of the first ratio is 11′. Thus, the ratio is $16:11 = 80:x$ because terms with common units of measure are compared in ratios. If this were an inverse proportion, the second ratio would be inverted to get $16:11 = x:80$.

In a direct proportion, the second term of the second ratio is an extreme of a proportion. To solve for the unknown number, find the missing extreme by multiplying 11 and 80 (11 × 80 = 880) and divide 880 by 16 (880 ÷ 16 = 55). The number of rivets in the 11′ long joint is 55.

Compound Proportions

A *compound proportion* is a proportion in which some terms are products of two variables. A *variable* is a quantity that may change. **See Figure 5-40.** For example, "If five welders fabricate 130 joint assemblies in 12 hours, then 10 welders can fabricate 195 joint assemblies in 9 hours" is a statement of a compound proportion.

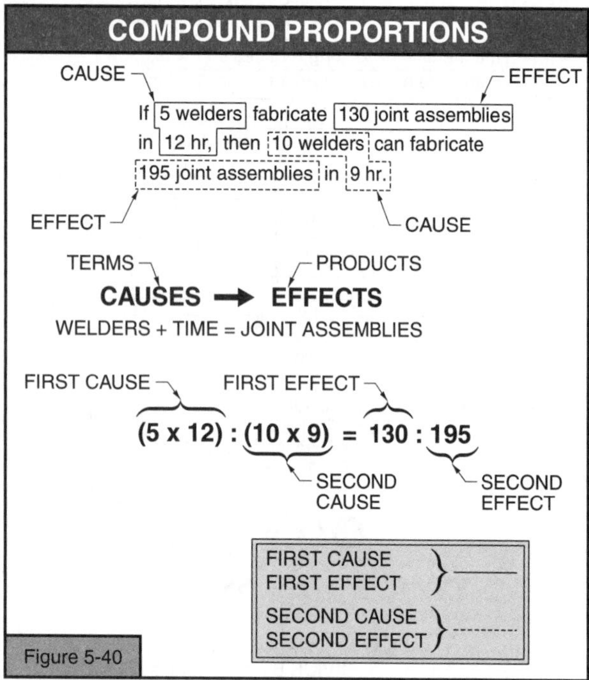

Figure 5-40

To simplify the procedure of solving compound proportions, each term of the ratios is renamed as either a cause or an effect. Every question in any proportion, especially in a compound proportion, may be considered as a comparison between two causes and two effects.

Causes are actions that produce effects. *Effects* are the results of actions or causes. For example, if a carpenter builds a deck, the action of building is the cause and the deck is the effect.

In the statement, "If five welders fabricate 130 joint assemblies in 12 hr, then 10 welders can fabricate 195 joint assemblies in 9 hours," the welders (5 and 10) and the time (12 hr and 9 hr) are the causes, and the number of joint assemblies (130 and 195) are the effects.

To simplify the causes, multiply the related terms. The proportion is written (5 × 12):(10 × 9) = 130:195. When using the words causes and effects, the proportion (5 × 12):(10 × 9) = 130:195 is written in first cause:second cause = first effect:second effect form. The 5 × 12 is the first cause, the 10 × 9 is the second cause, the 130 is the first effect, and the 195 is the second effect.

The unknown term (answer required) in a proportion is the second term of the second ratio. **See Figure 5-41.** When solving compound proportion problems, the unknown term can be a cause or an effect.

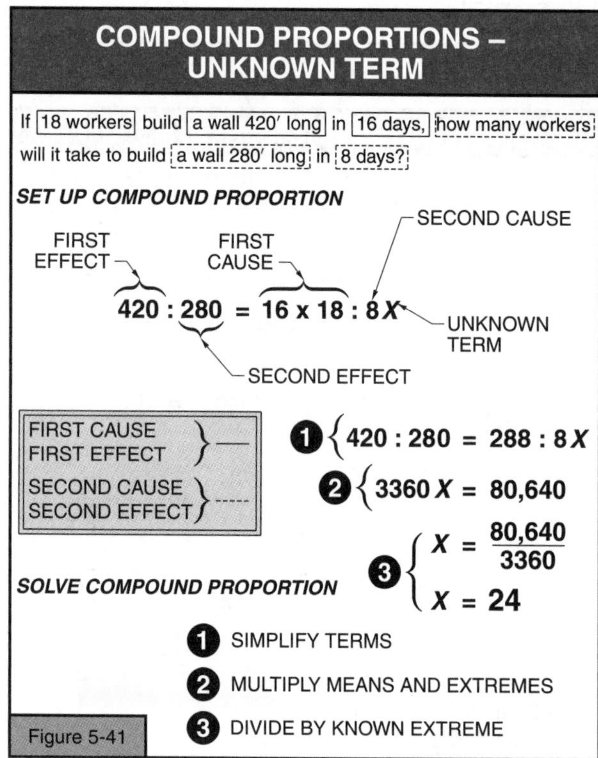

Figure 5-41

If the unknown term is an effect, write the compound proportion in first cause:second cause = first effect:second effect form. If the unknown term is a cause, write the proportion in first effect:second effect = first cause:second cause form. In both cases, the unknown term is the second term of the second ratio.

For example, in the problem "If 18 workers build a wall 420′ long in 16 days, how many workers can build a wall 280′ long in eight days?" The "workers" and "days" are the causes, and "feet" are the effects. The workers build the wall over a number of days, and the number of feet the wall is built is an effect, or the result of action. The unknown term is the number of workers. Effects (feet) form the first ratio, and causes (workers and days) form the second ratio.

To determine the number of workers, the compound proportion is set up by placing an x as the second term of the second ratio. The first term of the second ratio must be workers because ratios only include common measures.

The terms of the second ratio must be a product of days and workers (16 × 18 and 8 × x) because causes are workers and days. Place 420 (feet related to the first cause) as the first effect, and 280 (feet related to the second cause) as the second effect.

Calculate the first cause by multiplying 16 (days) by 18 (workers) to get 288 (16 × 18 = 288). Calculate the second cause by multiplying 8 (days) by x (workers) to get $8x$ ($8 \times x = 8x$). Multiply 280 by 288 (means) to get 80,640. Divide 80,640 by 3360 (known extreme) to get 24. Thus, 24 workers are needed to build a wall 280′ long in eight days.

Area

Area is the number of unit squares equal to the surface of an object. Area is used in the electrical and electronic fields to determine conduit fill and material use. For example, a standard size piece of drywall contains 32 sq ft (4 x 8 = 32 sq ft). Area is expressed in square inches, square feet, and other units of measure. A *square inch* is a measurement of 1″ × 1″. A *square foot* is a measurement that contains 144 sq in. (12″ × 12″ = 144 sq in.). The area of any plane figure can be determined by applying the proper formula. **See Figure 5-42.**

Circumference is the boundary of a circle. A *circle* is a plane figure generated about a centerpoint. *Diameter* is the distance from circumference to circumference through the centerpoint of a circle. To calculate the circumference of a circle when diameter is known, apply the formula:

$C = \pi D$

where

C = circumference

π = 3.1416

D = diameter

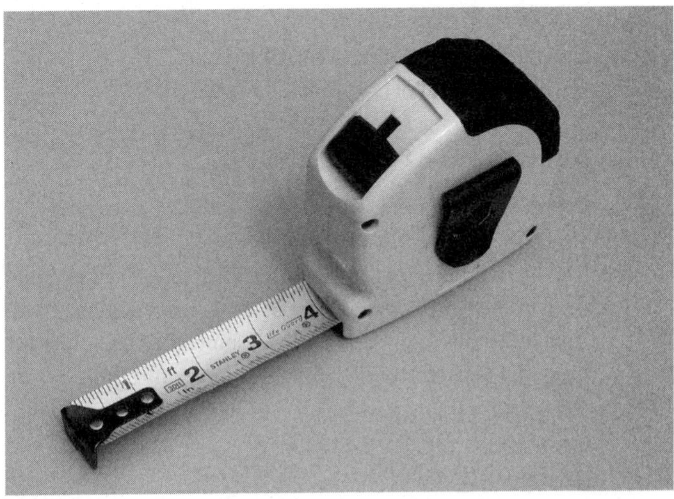

A tape measure can be used to take measurements such as length, width, and height.

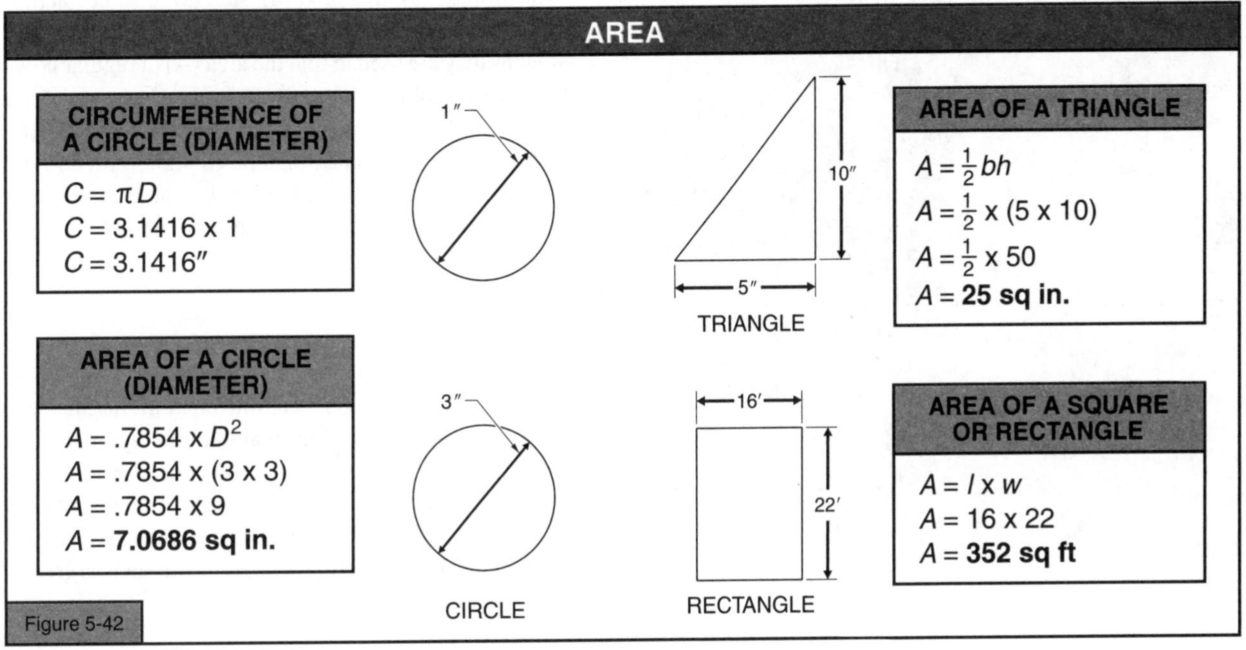

Figure 5-42

Example: Calculating Circumference

What is the circumference of a 1″ OD pipe?

$C = \pi D$

$C = 3.1416 \times 1$

$C =$ **3.1416″**

To calculate the area of a circle when the diameter is known, apply the formula:

$A = .7854 \times D^2$

where

A = area

.7854 = constant ($\pi \div 4$)

D^2 = diameter squared

Example: Calculating Area of Circle

What is the area of a 3″ diameter pipe?

$A = .7854 \times D^2$

$A = .7854 \times (3 \times 3)$

$A = .7854 \times 9$

$A =$ **7.0686 sq in.**

A *square* is a four-sided figure with all sides equal and four 90° angles. A *rectangle* is a four-sided figure with opposite sides equal and four 90° angles. To calculate the area of a square or a rectangle, apply the formula:

$A = l \times w$

where

A = area

l = length

w = width

Fluke Corporation

Power quality meters use mathematical values to calculate parameters such as power and power factor.

TECH FACT

A vector is a quantity involving direction and magnitude, represented by a straight line with an arrowhead on one end. A vector is used to represent the changing value of a sine wave. The length of the line represents the magnitude, or quantity, of the vector. Vectors are used to represent factors such as force, direction, and velocity.

Example: Calculating Area of Rectangle

What is the area of a 16′ × 22′ storage room?

$A = l \times w$

$A = 16 \times 22$

$A =$ **352 sq ft**

A *triangle* is a three-sided figure with three interior angles. To calculate the area of a triangle, apply the formula:

$A = \frac{1}{2}bh$

where

A = area

½ = constant (180° ÷ 360° = ½)

b = base

h = height

Example: Calculating Area of Triangle

What is the area of a triangle with a 5″ base and a 10″ height?

$A = \frac{1}{2}bh$

$A = \frac{1}{2} \times (5 \times 10)$

$A = \frac{1}{2} \times 50$

$A =$ **25 sq in.**

The altitude (a), diagonal (d), side (s), distance across corners (c), and distance across flats (f) are related to one another as they are used to find the area (A) of regular polygons. For example, the altitude and side of an equilateral triangle are related in that the altitude is 0.866s and the side is 1.155a. The area of an equilateral triangle is then found by applying the constant for the known value. A *constant* is a number with a fixed value. **See Figure 5-43.**

Math Symbols

A *symbol* is a graphic element that represents an operation, a quantity, or unit. Symbols are universal in that they are not language dependent. For example, the symbol for a triangle is shown the same whether it is used in the USA, France, China, etc. The symbol is a graphic representation of a triangle. Math symbols are used to show an operation such as adding or subtracting; a relationship such as greater than or equals; or a figure such as a square or rectangle. **See Figure 5-44.**

CONSTANTS*				
a = altitude	c = distance across corners	d = diagonal	f = distance across flats	s = side

EQUILATERAL TRIANGLE $A = 0.866s$ $s = 1.155a$ $A = 0.433s^2$ $A = 0.577a^2$	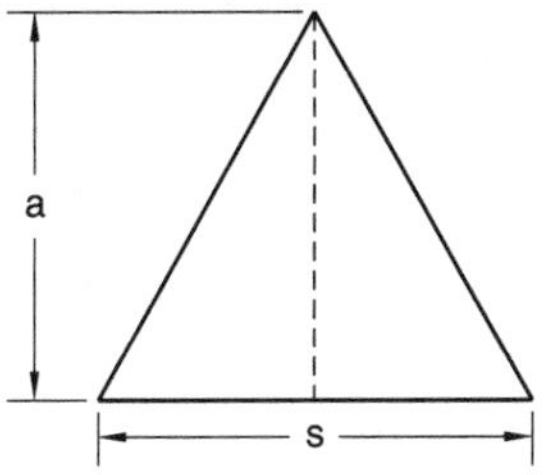 $A = 0.433s^2$ $A = 0.433\ (3^2)$ $A = 0.433\ (9)$ $A =$ **3.897 sq in.** (side 3″) 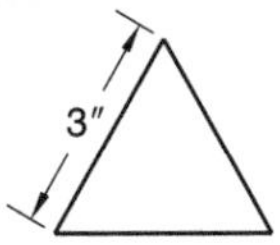
	$A = 0.577a^2$ $A = 0.577\ (3^2)$ $A = 0.577\ (9)$ $A =$ **5.193 sq in.** (altitude 3″) 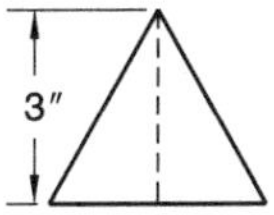
SQUARE $d = 1.414s$ $a = 0.707d$ $A = s^2$ $A = 0.5d^2$	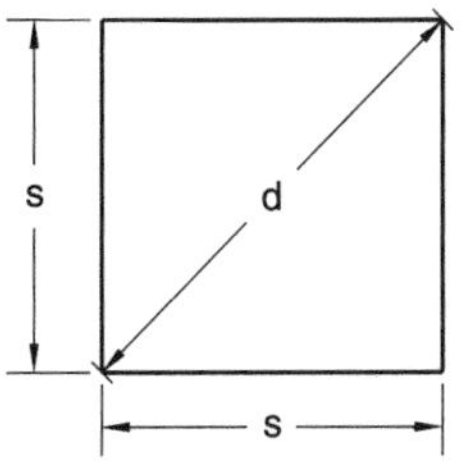 $A = s^2$ $A = 3^2$ $A = 3 \times 3$ $A =$ **9 sq in.** (3″ × 3″) 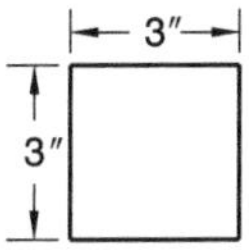
	$A = 0.5d^2$ $A = 0.5\ (3^2)$ $A = 0.5\ (9)$ $A =$ **4.5 sq in.** (diagonal 3″) 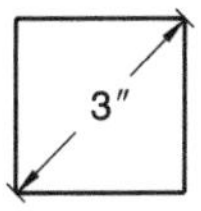
REGULAR HEXAGON $c = 2s$ or $c = 1.155f$ $f = 1.732s$ or $f = 0.866c$ $s = 0.5c$ or $s = 0.577f$ $A = 0.650c^2$ $A = 0.866f^2$ $A = 2.598s^2$	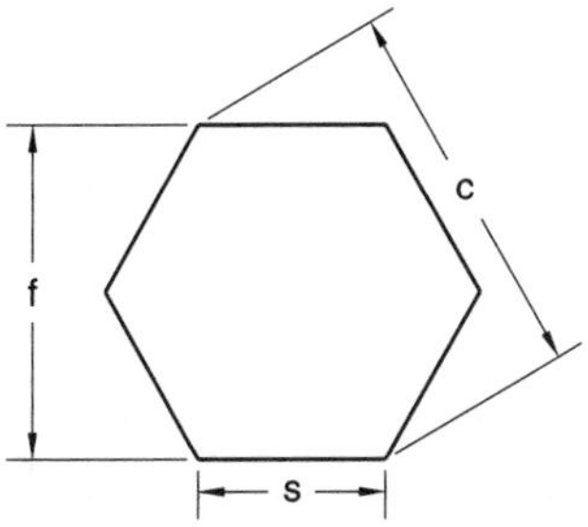 $A = 0.650c^2$ $A = 0.650\ (3^2)$ $A = 0.650\ (9)$ $A =$ **5.85 sq in.** (3″ across corners) 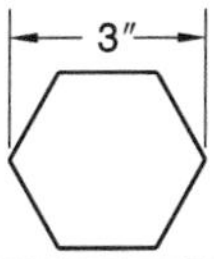
	$A = 0.866f^2$ $A = 0.866\ (3^2)$ $A = 0.866\ (9)$ $A =$ **7.794 sq in.** (3″ across flats) 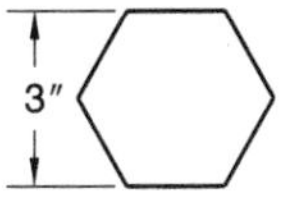
	$A = 2.598s^2$ $A = 2.598\ (3^2)$ $A = 2.598\ (9)$ $A =$ **23.382 sq in.** (side 3″) 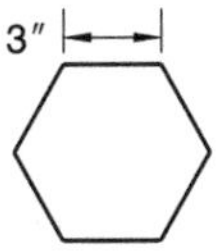
REGULAR OCTAGON $c = 2.613s$ or $c = 1.083f$ $f = 2.414s$ or $f = 0.924c$ $s = 0.383c$ or $s = 0.414f$ $A = 0.707c^2$ $A = 0.828f^2$ $A = 4.828s^2$	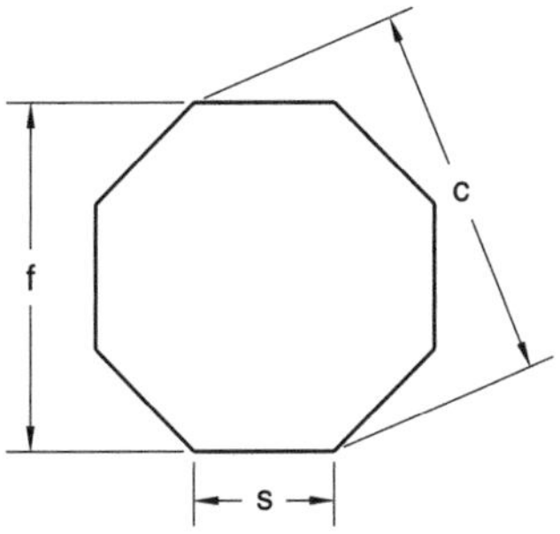 $A = 0.707c^2$ $A = 0.707\ (3^2)$ $A = 0.707\ (9)$ $A =$ **6.363 sq in.** (3″ across corners) 
	$A = 0.828f^2$ $A = 0.828\ (3^2)$ $A = 0.828\ (9)$ $A =$ **7.452 sq in.** (3″ across flats) 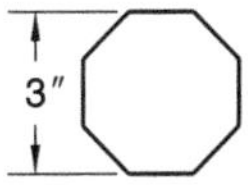
	$A = 4.828s^2$ $A = 4.828\ (3^2)$ $A = 4.828\ (9)$ $A =$ **43.452 sq in.** (side 3″) 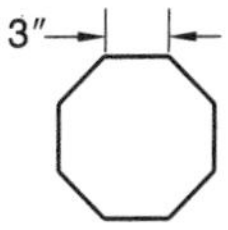

*A = area

Figure 5-43

MATH SYMBOLS					
Symbol	**Read**	**Symbol**	**Read**	**Symbol**	**Read**
+	Plus	≈	Is approximately equal to	∥	Parallel
-	Minus	≡	Is identical with	°	Degree(s)
±	Plus or minus	≅	Is congruent to	′	Minute(s)
×	Multiplied by; times	:	Is to; ratio of	″	Second(s)
÷	Divided by	∴	Therefore	′	Foot (feet)
=	Equals	∞	Infinity	″	Inch (inches)
≠ or	Not equal to	⌒	Arc	√	Square root
>	Is greater than	○	Circle	α	Alpha
<	Is less than	∠	Angle	β	Beta
≥	Is greater than or equal to	∟	Right angle	γ	Gamma
≤	Is less than or equal to	△	Triangle	θ	Theta
≯	Is not greater than	□	Square	ϕ	Phi
≮	Is not less than	▭	Rectangle	π	Pi
~	Is similar	⊥	Perpendicular	Σ	Sigma

Figure 5-44

Refer to the CD-ROM in the back of the book for Chapter 5 Quick Quiz® and related information.

Problem Solving

An electrical, electronic, fluid power, mechanical, or combination system must be correctly designed and maintained to operate properly. Basic rules, laws, and principles must be applied and followed when designing a system. These same rules, laws, and principles are also applied to keep the system operating. Understanding how to solve a problem requires an understanding of the system, circuits, components, and formulas that apply. Formulas must be applied and used to solve practical application problems.

Heating Air Required Power

Air is heated for general room temperature control (heating), oven control (baking and cooking food), and material processing (heat shrinking, drying, baking a finish, etc.). Stainless steel sheath resistor heating elements are commonly used for heating air. The electric resistance heating elements produce heat and a circulating fan is generally used to move air over the heating elements. **See Figure 6-1.**

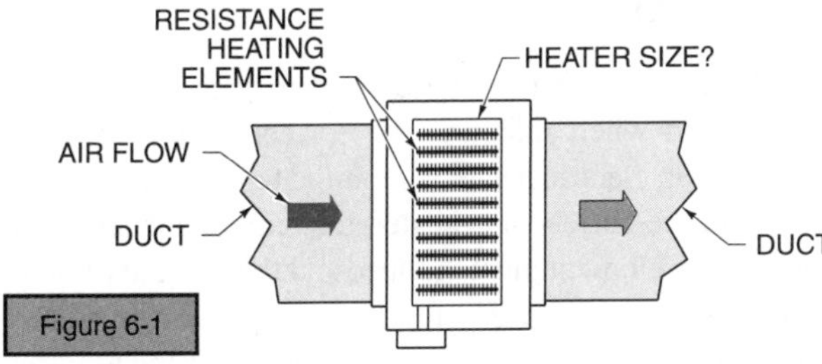

Figure 6-1

The amount of power required to heat air must be determined when selecting the correct heater size. For example, an electric heater is required for an application. To select the correct size heater, the temperature rise that the heating element must produce and the rate of air flow over the heating elements must be known. The higher the required temperature rise and/or the faster the air flow, the larger the required heater size. To calculate the required heater size for normal air heating applications, apply the formula:

$$P_a = \frac{Q \times TR}{3000}$$

where

P_a = required power to heat air (in kW)

Q = air flow rate (in cfm)

TR = temperature rise (in °F)

3000 = constant

Example: Calculating Heating Air Required Power

How many kilowatts of electrical power is required by an electric heater to raise the temperature of air by 100° with an air flow rate of 300 cfm?

$$P_a = \frac{Q \times TR}{3000}$$

$$P_a = \frac{300 \times 100}{3000}$$

$$P_a = \frac{30{,}000}{3000}$$

$$P_a = \mathbf{10\ kW}$$

Heating Water Required Power

Water is heated for cooking, cleaning, processing (material separation, etc.), and conversion to steam (generating electricity, etc.). Copper sheath resistor heating elements are commonly used for heating water. The electric resistance heating elements are heated and natural convection occurs (heated water rises), or circulating pumps are used to move the water over the heating elements. **See Figure 6-2.**

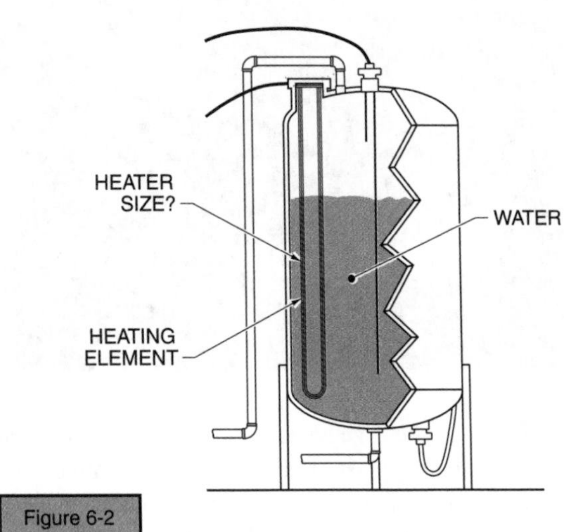

Figure 6-2

The amount of power required to heat water must be determined when selecting the correct heater size. For example, an electric heater is to be used to heat water in a tank. To select the correct size heater, the temperature rise that the heating element must produce and the amount of water to be heated must be determined. The higher the required temperature rise and/or the more gallons of water, the larger the required heater size. Additional power is required if there is a high heat loss. To calculate the required heater size for normal water heating applications, apply the formula:

$$P_w = \frac{G \times TR}{325 \times HT}$$

where

P_w = required power to heat water (in kW)

G = amount of water (in gal.)

TR = temperature rise (in °F)

325 = constant

HT = heat-up time (in hr)

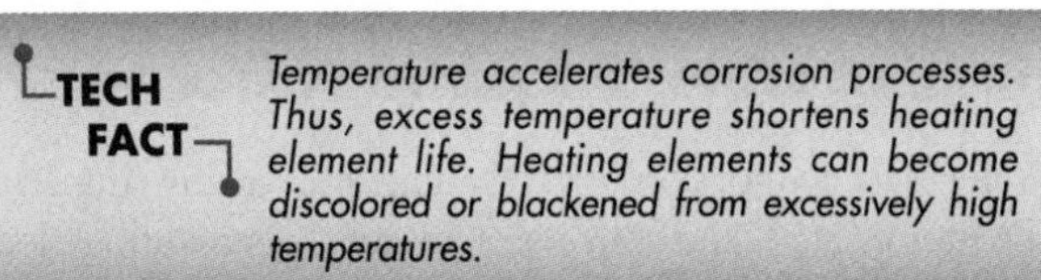
TECH FACT — *Temperature accelerates corrosion processes. Thus, excess temperature shortens heating element life. Heating elements can become discolored or blackened from excessively high temperatures.*

Example: Calculating Heating Water Required Power

How many kilowatts of electrical power is required by an electric heater to raise the temperature of 100 gal. of water in a tank by 100°F in 1 hr?

$$P_w = \frac{G \times TR}{325 \times HT}$$

$$P_w = \frac{100 \times 100}{325 \times 1}$$

$$P_w = \frac{10,000}{325}$$

$$P_w = \mathbf{30.77\ kW}$$

Heating Oil Required Power

Oil is heated for cooking (olive oil, soybean oil, corn oil, cotton seed oil, etc.), processing fuel oils (motor oil, kerosene, etc.), and refining other oils (linseed oil, etc.). Steel sheath resistor heating elements are commonly used for heating oil. The electric resistance heating elements are heated and natural convection occurs (heated oil rises), or circulating pumps are used to move the oil over the heating elements. **See Figure 6-3.**

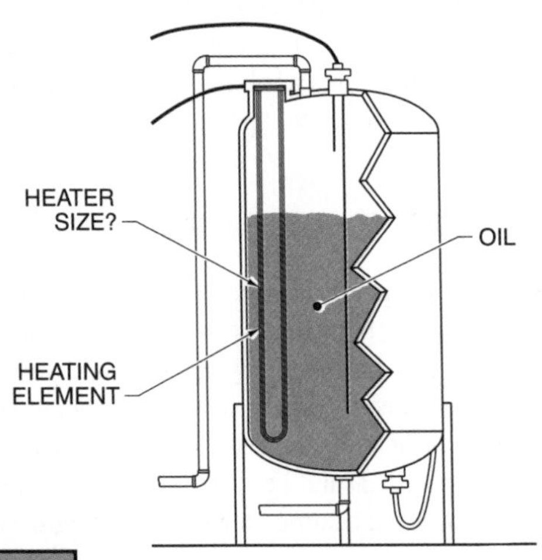

Figure 6-3

The amount of power required to heat oil must be determined when selecting the correct heater size. For example, an electric heater is required to heat oil in a tank. To select the correct size heater, the temperature rise that the heating element must produce and the amount of oil to be heated are required. The higher the required temperature rise and/or the more gallons of oil, the larger the required heater size. Additional power is required if there is a high heat loss. To calculate the required heater size for normal oil heating applications, apply the formula:

$$P_o = \frac{G \times TR}{800 \times HT}$$

where

P_o = required power to heat oil (in kW)

G = amount of oil (in gal.)

TR = temperature rise (in °F)

800 = constant

HT = heat-up time (in hr)

Example: Calculating Heating Oil Required Power
How many kilowatts of electrical power is required by an electric heater to raise the temperature of 200 gal. of oil in a tank by 50° in 2 hr?

$$P_o = \frac{G \times TR}{800 \times HT}$$

$$P_o = \frac{200 \times 50}{800 \times 2}$$

$$P_o = \frac{10,000}{1650}$$

$$P_o = \mathbf{6.25\ kW}$$

Electric Heater Output

Electric heating elements are rated in wattage for a given applied voltage. However, the applied voltage is not always the same as the actual voltage. The heating element's output is different for any applied voltage other than the rated voltage. For example, the heating elements in an electric forced-air heating unit produce a different heat output if connected to a voltage supply greater or less than their rated voltage. **See Figure 6-4.**

The voltage supply to the forced-air heating unit may vary because of the difference between the heating element's rated voltage and the supply voltage. For example, a 220 V rated heating element may be connected to a 208 V power supply. In such a case, the 5.45% voltage reduction would result in an 11% heating power reduction. The voltage may also be intentionally varied through a control circuit to control the heat output. To calculate the amount of actual power (in watts) delivered by an electric heating element (heater), apply the formula:

$$P_A = P_R \times \frac{V_A^2}{V_R^2}$$

where

P_A = actual heater wattage delivered (in W)

P_R = rated heater wattage (in W)

V_A = voltage applied to heating element (in V)

V_R = rated voltage of heating element (in V)

Fluke Corporation
Infrared thermometers are used to spot-check temperature readings and verify that heating and cooling systems are operating properly.

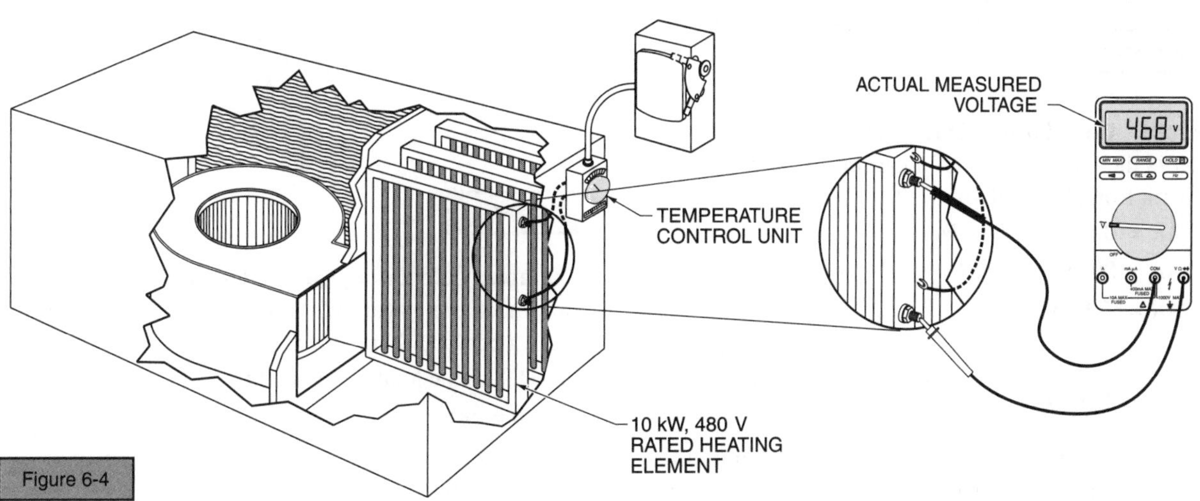

Figure 6-4

Example: Calculating Electric Heater Output
What is the actual heater power output when a 10,000 W, 480 V heating element is connected to a 468 V supply?

$$P_A = P_R \times \frac{V_A^2}{V_R^2}$$

$$P_A = 10,000 \times \frac{468^2}{480^2}$$

$$P_A = 10,000 \times \frac{219,024}{230,400}$$

$$P_A = \mathbf{9500\ W\ (9.5\ kW)}$$

Air Resistance (Aerodynamics)

Air produces resistance against all moving objects. For example, as a car travels, the resistance of the air slows the movement of the car. Air resistance must be kept to a minimum in some applications. Air resistance is reduced by reducing the frontal area of the moving object and/or reducing the speed of the moving object. **See Figure 6-5.** To determine air resistance, apply the formula:

$AR = .0025 \times mph^2 \times FA$

where

AR = resistance of air (in lb)

.0025 = constant

mph = speed of object (in mph)

FA = frontal area of moving object (in sq ft)

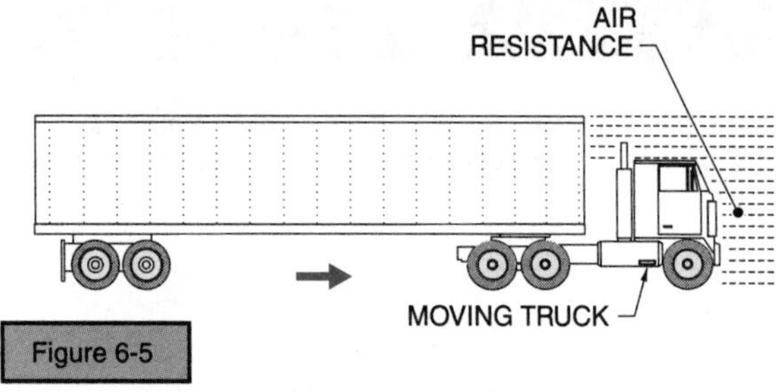

Figure 6-5

Example: Calculating Air Resistance
What is the air resistance of a car traveling 55 mph and having a frontal area of 76 sq ft?

$AR = .0025 \times mph^2 \times FA$

$AR = .0025 \times 55^2 \times 76$

$AR = \mathbf{574.75\ lb}$

Energy Usage

The unit of electric power is the watt. The amount of energy used by an electric device and the amount charged for that use by a power company is expressed in watt-hours. Watt-hours is the amount of power being used over time. One watt-hour equals one watt used in 1 hr. In most power measurements, the kilowatt-hour (kWh) is used instead of the watt-hour. One kilowatt-hour equals 1000 Wh. **See Figure 6-6.** To find the total amount of energy (in kilowatt-hours) used, apply the procedure:

1. Determine total operating time.

   $op = hr \times d$

   where

   op = operating time (in hr)

   hr = hours

   d = days

2. Determine total energy used.

   $$W_T = \frac{P \times op}{1000}$$

   where

   W_T = total energy used (in kWh)

   P = amount of power (in W)

   op = operating time (in hr)

   1000 = constant (to convert W to kW)

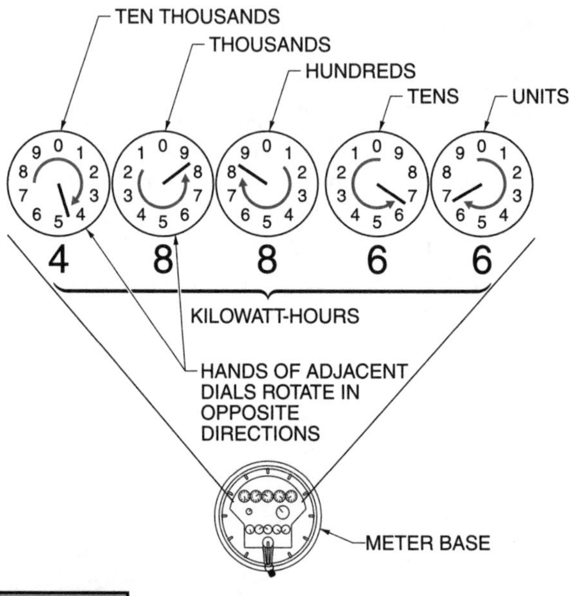

Figure 6-6

Example: Calculating Energy Usage
How much energy (in kWh) is used by a 10,000 W electric heater operating 8 hr a day for 30 days?

1. Determine total operating time.

   $op = hr \times d$

   $op = 8 \times 30$

   $op = 240$ hr

2. Determine total energy used.

$$W_T = \frac{P \times op}{1000}$$

$$W_T = \frac{10{,}000 \times 240}{1000}$$

$$W_T = \frac{2{,}400{,}000}{1000}$$

$$W_T = \mathbf{2400\ kWh}$$

Branch Circuit Voltage Drop

All conductors (wires) have resistance. The longer a conductor run, the higher the total resistance. The higher the resistance, the greater the voltage drop. Branch circuit conductors must be sized properly for safe equipment operation. Conductors are sized large enough to prevent no more than a 3% voltage drop per NEC® 210.19(A) FPN No. 4 at the farthest point. **See Figure 6-7.** To determine the approximate allowable length a conductor can be run, apply the procedure:

1. Determine voltage drop.

   $VD = VC \times 3\%$

   where

   VD = voltage drop (in V)

   VC = circuit voltage (in V)

   3% = constant (210.19 (A) FPN No. 4)

2. Determine maximum length.

   $$L = \frac{VD \times CM}{2 \times K \times I}$$

   where

   L = maximum length (in ft)

   VD = voltage drop (in V)

   CM = wire circular mils (from Copper Conductor Ratings chart)

   2 = constant

   K = conductor resistivity constant (13 for copper, 19 for aluminum)

   I = current of load (in A)

Example: Calculating Conductor Maximum Length

How far can a 120 V, 14 A load be placed from the power panel using an AWG #12 Cu conductor?

1. Determine voltage drop.

   $VD = VC \times 3\%$

   $VD = 120 \times .03$

   $VD = 3.6\text{ V}$

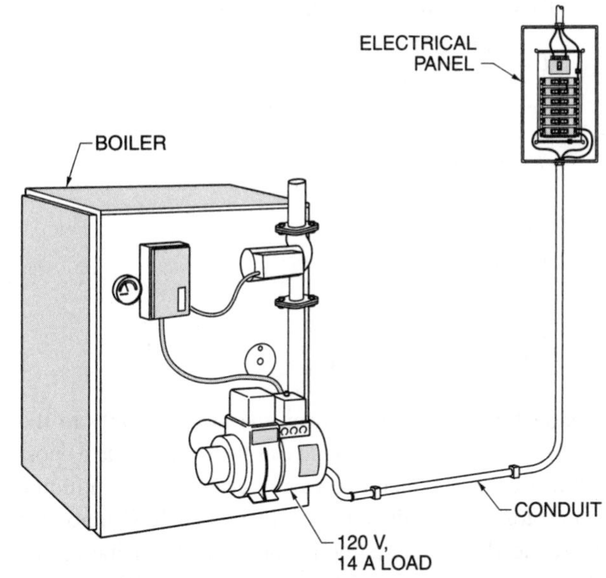

COPPER CONDUCTOR RATINGS*

SIZES INCREASE / NUMBERS DECREASE

AWG	Ampacity	DIA Mils	CIR Mils
18	----	40.0	1624
17	----	45.0	2048
16	----	51.0	2583
15	----	57.0	3257
14	20	64.0	4107
12	25	81.0	6530
10	30	102.0	10,380
8	40	128.0	16,510
6	55	162.0	26,250
4	70	204.0	41,740
3	85	229.0	52,630
2	95	258.0	66,370
1	110	289.0	83,690
0	125	325.0	105,500
00	145	365.0	133,100
000	165	410.0	167,800

* @ 60°C

Figure 6-7

2. Determine maximum length.

$$L = \frac{VD \times CM}{2 \times K \times I}$$

$$L = \frac{3.6 \times 6530}{2 \times 13 \times 14}$$

$$L = \frac{23,508}{364}$$

$$L = \mathbf{64.6'}$$

Conductor Resistance

Conductors carry current from the power supply to the loads. A conductor should have as little resistance as possible for the conductor to carry current with a minimal power loss. All conductors have resistance. Resistance can be reduced by using a good conductor and/or increasing the size (cross-sectional area) of the conductor. For any given conductor, the larger the wire's diameter, the less the resistance. The cross-sectional area of round conductors is measured in circular mils (cmil). **See Figure 6-8.** To find the resistance of a conductor, apply the formula:

$$R_T = L \times \frac{R_C}{1000}$$

where

R_T = total resistance of conductor (in Ω)

L = length of conductor (in ft)

R_C = resistance of 1000′ of conductor (from Copper Conductor Ratings chart)

1000 = constant (to determine resistance per foot of conductor)

COPPER CONDUCTOR RATINGS*

AWG	Ampacity	DIA Mils	CIR Mils	OHMS (Ω)**
18	----	40.0	1624	6.510
17	----	45.0	2048	5.163
16	----	51.0	2583	4.094
15	----	57.0	3257	3.247
14	20	64.0	4107	2.575
12	25	81.0	6530	1.619
10	30	102.0	10,380	1.018
8	40	128.0	16,510	.6405
6	55	162.0	26,250	.4028
4	70	204.0	41,740	.2533

* 60°C temperature rating

** per 1000′ of copper wire at room temperature (72°F)

Figure 6-8

Example: Calculating Conductor Resistance

What is the total resistance of 250′ of AWG #14 conductor?

$$R_T = L \times \frac{R_C}{1000}$$

$$R_T = 250 \times \frac{2.575}{1000}$$

$$R_T = 250 \times .002575$$

$$R_T = \mathbf{.6437\ \Omega}$$

TECH TIP *As a general rule, a motor develops a 3 lb-ft of torque per HP at 1800 rpm and 4.5 lb-ft of torque per HP at 1200 rpm.*

Volts Per Hertz (V/Hz) Ratio

One of the primary advantages of a variable frequency drive (VFD) is it can be applied to standard 3ϕ induction motors. Three-phase induction motors are the standard motor used in most industrial applications. A VFD can control the speed of a standard induction motor. However, the voltage applied to the stator must be boosted when a motor is operating below approximately 15 Hz. The extra voltage is needed to make up for the large power losses all AC motors have at low speed. The VFD must be programmed to apply a boost voltage at low motor speeds to compensate for power losses at low speeds. The additional boost gives the motor more rotor torque at very low speeds. The applied motor's V/Hz ratio must be calculated to program the boost. The V/Hz ratio is the ratio between the voltage applied to the stator and the frequency of the voltage applied to the stator of the motor. This calculated value is used to program the VFD. **See Figure 6-9.** To calculate the V/Hz ratio for an induction motor, apply the formula:

$$V/Hz = \frac{V}{Hz}$$

where

V/Hz = volts per Hertz ratio

V = rated motor nameplate voltage (in V)

Hz = rated motor nameplate frequency (in Hz)

Example: Calculating V/Hz Ratio

What is the V/Hz ratio for a motor with a nameplate rating of 230 V and 60 Hz?

$$V/Hz = \frac{V}{Hz}$$

$$V/Hz = \frac{230}{60}$$

$$V/Hz = \mathbf{3.83}$$

Note: 3.83 is the value used to program the VFD boost parameter.

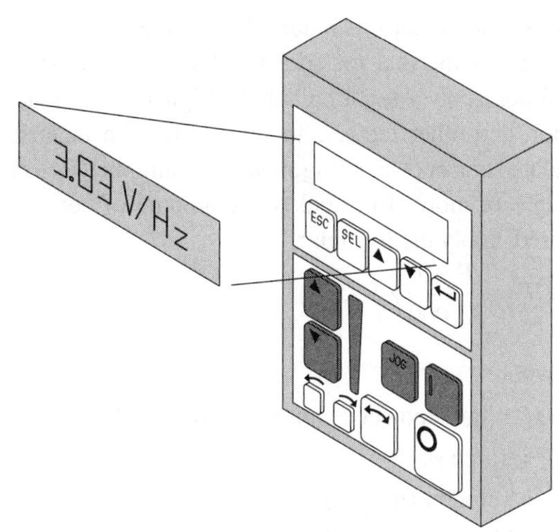

Figure 6-9

Moving Object Activating Speed

Proximity and photoelectric switches are used to detect moving objects without touching them. Proximity and photoelectric switches are used in applications in which the object to be detected is excessively light, heavy, hot, or untouchable. Proximity and photoelectric switches have a rated activating frequency which is often stated in operations per second (ops). For example, a switch with an ops of 1000 means that the switch can detect 1000 objects per second passing by it. **See Figure 6-10.**

Strobe tachometers are often used to measure the speed of moving objects. Strobe tachometers normally display the speed of the moving objects in feet per minute (fpm). A strobe tachometer's reading (in fpm) must be converted to the proximity or photoelectric switch's specified activating frequency (in ops) when using a strobe tachometer to determine a proximity or photoelectric switch's compatibility to an application. To convert fpm to ops, apply the formula:

$$ops = \frac{npf \times v}{60}$$

where

ops = operations per second

npf = objects per foot

v = object speed (in fpm)

60 = constant (sec per min)

Example: Calculating Operations Per Second

What is the number of operations per second when a strobe tachometer measures moving cans at a rate of 120 fpm when the cans are spaced 6 per foot?

$$ops = \frac{npf \times v}{60}$$

$$ops = \frac{6 \times 120}{60}$$

$$ops = \frac{720}{60}$$

$$ops = \mathbf{12}$$

A proximity or photoelectric switch with a specified operating frequency of at least 12 ops (or more) may be used. For proper operation and future speed changes, a switch with an operating frequency of at least 20% greater should be used.

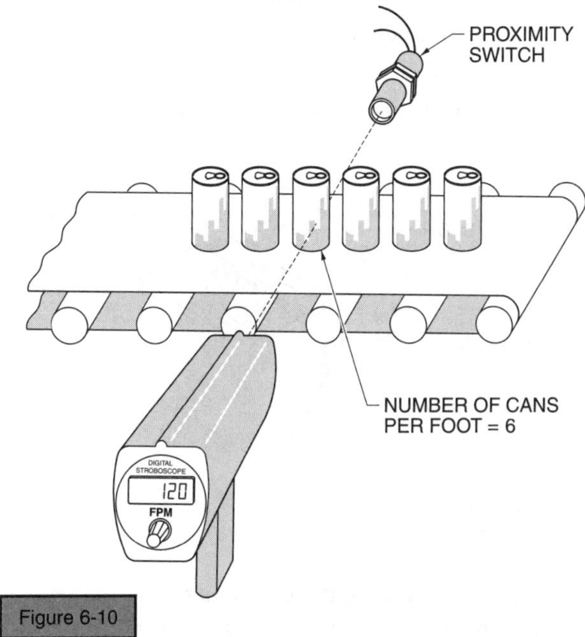

Figure 6-10

Baldor Electric Co.

Three-phase induction motors are used in applications such as outdoor industrial conveyor systems.

Motor Pulley Size

AC induction motors operate at a constant speed for a given frequency and number of poles. However, the load's motor drive must often operate at different speeds than the motor. For example, a fan motor may operate at 1725 rpm, but the blower unit the motor is driving may have to operate at 345 rpm. A pulley can be used to change the output speed of a motor. **See Figure 6-11.** To calculate the driven machine pulley diameter, apply the formula:

$$PD_m = \frac{PD_d \times N_d}{N_m}$$

where

PD_m = driven machine pulley diameter (in in.)

PD_d = drive pulley diameter (in in.)

N_d = motor drive speed (in rpm)

N_m = driven machine speed (in rpm)

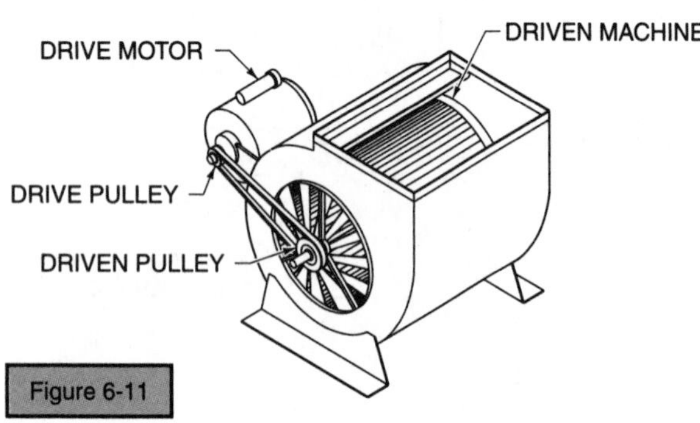

Figure 6-11

Example: Calculating Pulley Diameter

What is the required driven machine pulley diameter if a motor running at 1725 rpm has a 1″ pulley and the driven machine is to be operated at 345 rpm?

$$PD_m = \frac{PD_d \times N_d}{N_m}$$

$$PD_m = \frac{1 \times 1725}{345}$$

$$PD_m = \frac{1725}{345}$$

$$PD_m = \mathbf{5''}$$

Vertical Conveyor Required Horsepower

An application's horsepower requirements must be determined when selecting a motor or motor drive (VFD) because motors and drives are horsepower rated. **See Figure 6-12.** For example, a motor is required to lift a load vertically in elevator, mining and material handling applications. The actual horsepower required may be higher due to machine friction and other factors. To overcome such factors, the calculated horsepower rating should be multiplied by a factor of 125% or more. To calculate the amount of horsepower required, apply the formula:

$$HP = \frac{F \times v}{33{,}000} \times 1.25$$

where

HP = horsepower required for vertical lift

F = force (in lb)

v = velocity (in fpm)

33,000 = constant (lb-ft per minute)

1.25 = constant (for increased HP)

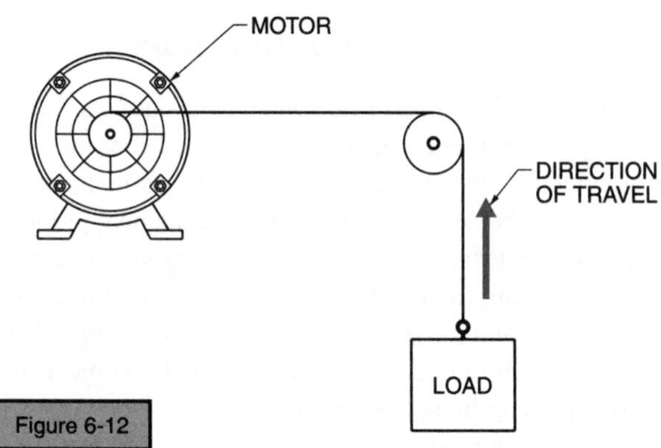

Figure 6-12

Example: Calculating Vertical Conveyor Required Horsepower

What is the motor horsepower required to lift 2200 lb vertically at a rate of 75 fpm?

$$HP = \frac{F \times v}{33{,}000} \times 1.25$$

$$HP = \frac{2200 \times 75}{33{,}000} \times 1.25$$

$$HP = \frac{165{,}000}{33{,}000} \times 1.25$$

$$HP = 5 \times 1.25$$

$$HP = \mathbf{6.25\ HP}$$

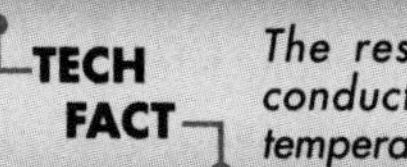

TECH FACT — *The resistance of copper and aluminum conductors increases with an increase in temperature.*

Energy-Efficient Motor Savings

Electric motors are used to produce work. To produce work, motors convert electrical energy (power) into rotating mechanical energy. Motors use more electrical power compared to all other loads (lamps, solenoids, etc.). In most residential, commercial, and industrial locations, motors account for more than 50% of all power used. To reduce energy consumption, motor manufacturers produce energy-efficient motors. Energy-efficient motors cost more, but save energy. The more the motor is used, the greater the energy saving and the shorter the payback period. **See Figure 6-13.** To determine the payback period of selecting an energy-efficient motor over a standard motor, apply the procedure:

1. Calculate power saved.

$$P_S = .746 \times HP \times \left(\frac{1}{Eff^1} - \frac{1}{Eff^2} \right)$$

where

P_S = power saved (in kW)

.746 = constant

HP = motor horsepower rating (in HP)

Eff_1 = efficiency rating of standard motor (in %)

Eff_2 = efficiency rating of energy-efficient motor (in %)

2. Determine the amount of money saved per year.

$$D_{SY} = P_S \times D_{kWh} \times H_y$$

where

D_{SY} = amount saved per year (in $)

P_S = power saved (in kW)

D_{kWh} = cost of power per kWh (in $)

H_y = hours motor operates per year

3. Determine the payback period.

$$P_P = \frac{P_D}{D_{SY}}$$

where

P_P = payback period (in yr)

P_D = price difference between standard motor and energy-efficient motor (in $)

D_{SY} = amount saved per year (in $)

Example: Calculating Motor Payback Period

What is the payback period when selecting a 25 HP energy-efficient motor for an application as compared to a standard motor? *Note:* The motor is operated 40 hr per week (2080 hr/yr), the cost of power is $0.085 per kWh, and the energy-efficient motor costs $350.00 more than a standard motor.

1. Calculate power saved.

$$P_S = .746 \times HP \times \left(\frac{1}{Eff^1} - \frac{1}{Eff^2} \right)$$

$$P_S = .746 \times 25 \times \left(\frac{1}{.88} - \frac{1}{.93} \right)$$

$$P_S = 18.65 \times (1.136 - 1.075)$$

$$P_S = 18.65 \times .61$$

$$P_S = 1.1376$$

2. Determine the amount of money saved per year.

$$D_{SY} = P_S \times D_{kWh} \times H_y$$

$$D_{SY} = 1.1376 \times .085 \times 2080$$

D_{SY} = $201.13 (amount saved by using an energy-efficient motor)

TYPICAL MOTOR EFFICIENCIES					
HP	Standard Motor (%)	Energy-Efficient Motor (%)	HP	Standard Motor (%)	Energy-Efficient Motor (%)
1	76.5	84.0	30	88.1	93.1
1.5	78.5	85.5	40	89.3	93.6
2	79.9	86.5	50	90.4	93.7
3	80.8	88.5	75	90.8	95.0
5	83.1	88.6	100	91.6	95.4
7.5	83.8	90.2	125	91.8	95.8
10	85.0	90.3	150	92.3	96.0
15	86.5	91.7	200	93.3	96.1
20	87.5	92.4	250	93.6	96.2
25	88.0	93.0	300	93.8	96.5

Figure 6-13

3. Determine the payback period.

$$P_P = \frac{P_D}{D_{SY}}$$

$$P_P = \frac{350}{201.13}$$

$$P_P = \mathbf{1.74\ yr}$$

Sizing Fans and Blowers

Fans and blowers are used in numerous applications, such as forced air furnaces, air conditioners, air purifiers, food warmers, ventilation systems, medical equipment, air curtains, vending machines, dryers, etc. Fans and blowers are driven by electric motors. In any application, the fan or blower must move a required amount of air if the system is to operate properly.

The required air flow rate (Q) is normally given in cubic feet per minute (cfm). The major factors that affect the air flow rate include the fan's operating speed (in rpm), the fan's impeller diameter (in in.), and the fan's impeller width (in in.). All of these factors must be considered because they are all interrelated and affect the unit's physical size, operating sound level, and electrical requirements.

An increase (or decrease) in air flow rate may be achieved without a blower or impeller size change by changing the blower operating speed (N). Air flow rate varies directly as the speed. **See Figure 6-14.** To calculate air flow rate by increasing the blower speed, apply the formula:

$$Q_2 = Q_1 \times \frac{N_2}{N_1}$$

where

Q_2 = desired air flow (in cfm)

Q_1 = known air flow (in cfm)

N_2 = new fan speed (in rpm)

N_1 = present fan speed (in rpm)

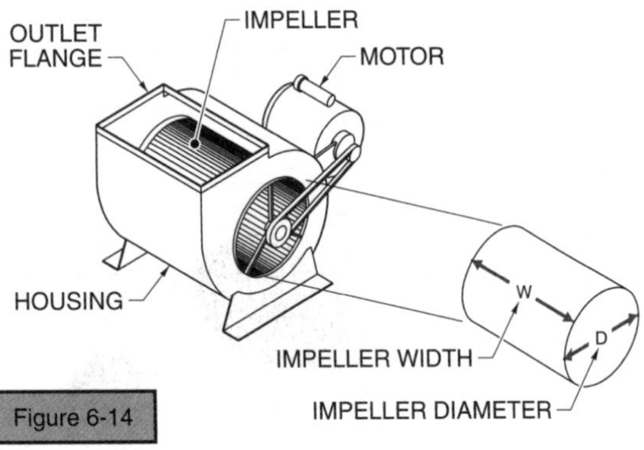

Figure 6-14

Example: Calculating Air Flow Rate—Increasing Speed

How much air does a blower deliver if the operating speed is increased to 1800 rpm if the blower is rated to deliver 250 cfm when operating at 1200 rpm?

$$Q_2 = Q_1 \times \frac{N_2}{N_1}$$

$$Q_2 = 250 \times \frac{1800}{1200}$$

$$Q_2 = 250 \times 1.5$$

$$Q_2 = \mathbf{375\ cfm}$$

Although increasing speed increases air flow, normally an increase in speed also increases the operating sound level. To reduce the sound level, the speed can be reduced and the impeller size increased. However, the slower the speed and larger the impeller size, the larger the total housing space required. The impeller diameter may be increased if additional air flow is required. Air flow rate varies directly with the cube of the blower impeller diameter. To calculate air flow rate by increasing the blower diameter, apply the formula:

$$Q_2 = Q_1 \times \left(\frac{D_2}{D_1}\right)^3$$

where

Q_2 = desired air flow (in cfm)

Q_1 = known air flow (in cfm)

D_2 = new fan diameter (in in.)

D_1 = present fan diameter (in in.)

Example: Calculating Air Flow Rate—Increasing Diameter

How much air flow is produced by a blower with a 6″ diameter impeller if a blower with a 3″ diameter impeller produces 250 cfm?

$$Q_2 = Q_1 \times \left(\frac{D_2}{D_1}\right)^3$$

$$Q_2 = 250 \times \left(\frac{6}{3}\right)^3$$

$$Q_2 = 250 \times 2^3$$

$$Q_2 = 250 \times 8$$

$$Q_2 = \mathbf{2000\ cfm}$$

Although increasing the diameter of the blower increases air flow, increasing the diameter also requires more power to drive the larger diameter impeller. Air flow may also be increased by increasing the width of the impeller. Air flow rate varies directly with the width of the impeller.

To calculate air flow rate by increasing the blower impeller width, apply the formula:

$$Q_2 = Q_1 \times \frac{W_2}{W_1}$$

where

Q_2 = desired air flow (in cfm)

Q_1 = known air flow (in cfm)

W_2 = new impeller width (in in.)

W_1 = present impeller width (in in.)

Example: Calculating Air Flow Rate—Increasing Width

How much air flow is delivered by a 12″ wide blower when a 6″ wide blower delivers 250 cfm?

$$Q_2 = Q_1 \times \frac{W_2}{W_1}$$

$$Q_2 = 250 \times \frac{12}{6}$$

$$Q_2 = 250 \times 2$$

$$Q_2 = \mathbf{500\ cfm}$$

Motor Operation at 50 Hz

A motor may be operated at different frequencies as long as the motor can produce enough torque to drive the load and there is enough air flow over the motor to prevent overheating the motor windings. Operating a motor at different frequencies is normally the result of using a VFD or operating a 60 Hz rated motor on a 50 Hz power supply or a 50 Hz rated motor on a 60 Hz power supply.

Follow the VFD manufacturer's recommendations for programming the VFD to match the motor specifications. Operating a motor at 50 Hz reduces the speed of the motor by 20%. **See Figure 6-15.** Reducing the speed of the motor may reduce the air flow output if the motor is connected to a fan or blower. To calculate the operating speed of a 60 Hz rated motor when powered by a 50 Hz power supply, apply the formula:

$$S_{50} = S_{60} \times .8$$

where

S_{50} = motor speed when 60 Hz rated motor is operated on 50 Hz power (in rpm)

S_{60} = motor speed when 60 Hz rated motor is operated on 60 Hz (in rpm)

.8 = constant (for 20% reduction in speed)

TECH FACT — *To prevent electrical shock from metal-sheathed heaters, ground the metal sheath and use a ground-fault circuit interrupter (GFCI).*

Example: Calculating Motor Speed—60 Hz Motor Operated on 50 Hz

What is the operating speed of a 1740 rpm, 60 Hz rated motor powered by a 50 Hz power supply?

$$S_{50} = S_{60} \times .8$$

$$S_{50} = 1740 \times .8$$

$$S_{50} = \mathbf{1392\ rpm}$$

A 50 Hz rated motor operates at a 20% increase in speed when operated on 60 Hz power. To calculate the operating speed of a 50 Hz rated motor when operated by a 60 Hz power supply, apply the formula:

$$S_{60} = S_{50} \times 1.2$$

where

S_{60} = motor speed when 50 Hz rated motor is operated on 60 Hz power (in rpm)

S_{50} = motor speed when 50 Hz rated motor is operated on 50 Hz power (in rpm)

1.2 = constant (for 20% increase in speed)

Example: Calculating Motor Speed–50 Hz Motor Operated on 60 Hz

What is the operating speed of a 1740 rpm, 50 Hz rated motor powered by a 60 Hz power supply?

$$S_{60} = S_{50} \times 1.2$$

$$S_{60} = 1740 \times 1.2$$

$$S_{60} = \mathbf{2088\ rpm}$$

Horsepower values of diesel freight locomotives range between 3000 HP and 6000 HP.

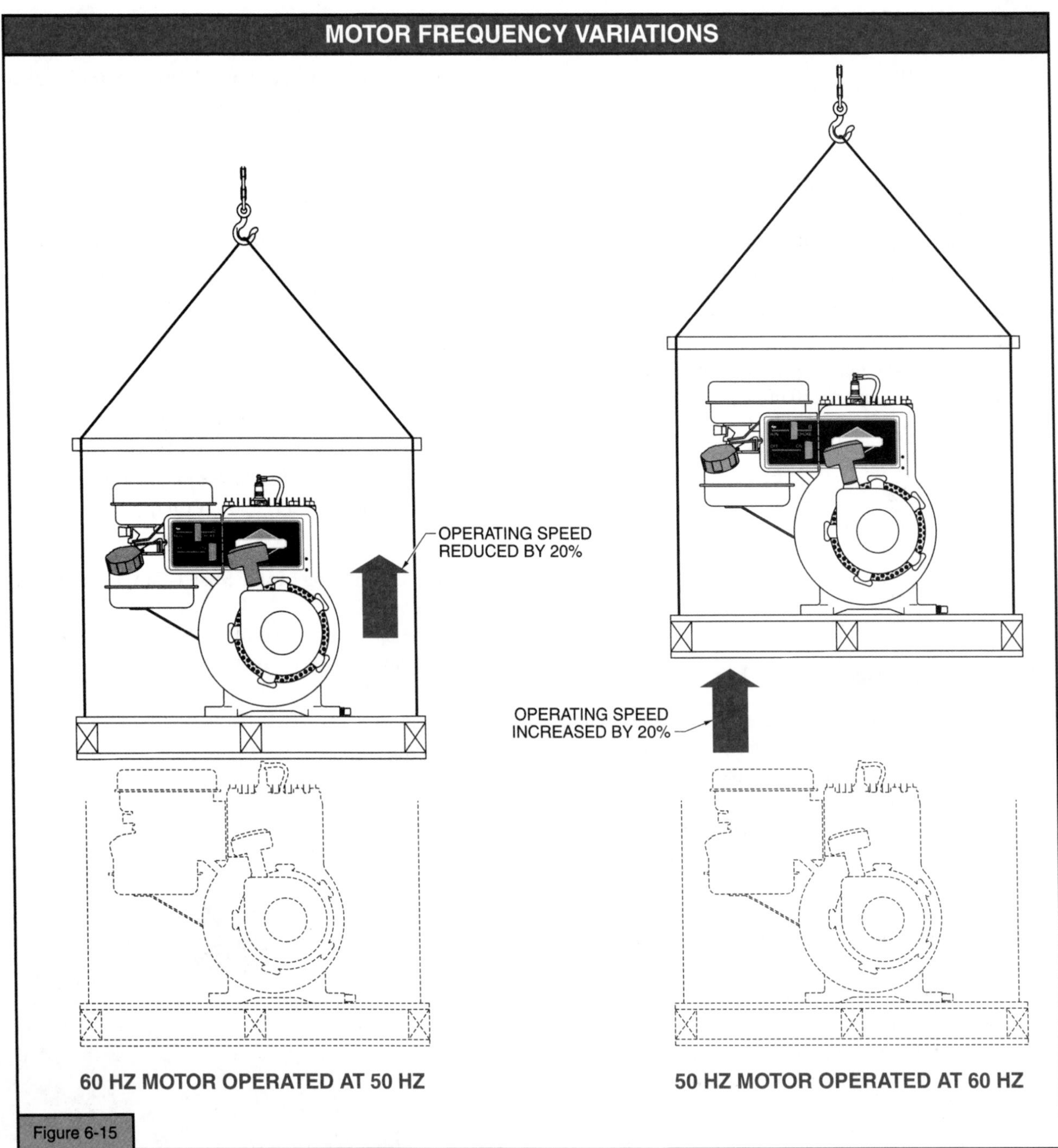

Figure 6-15

Amplifier and Speaker Resistance

Amplifiers and speakers have impedance. *Impedance (Z)* is the total opposition of any combination of resistance, inductive reactance, or capacitive reactance offered to the flow of alternating current. Impedance is measured in ohms (Ω). Most speakers are 8 Ω but smaller speakers are usually 4 Ω, and some larger speakers are 16 Ω. Most amplifiers are designed for connection to an 8 Ω speaker or speakers connected so their total resistance is 8 Ω. In addition to an 8 Ω speaker connection, some amplifiers may include a 4 Ω or 16 Ω speaker connection. When connecting speakers to amplifiers, the speakers may be connected in parallel or series. A *parallel connection* is a connection that has two or more components connected so that there is more than one path for current flow. A *series connection* is a connection that has two or more components connected so there is only one path for current flow. **See Figure 6-16.**

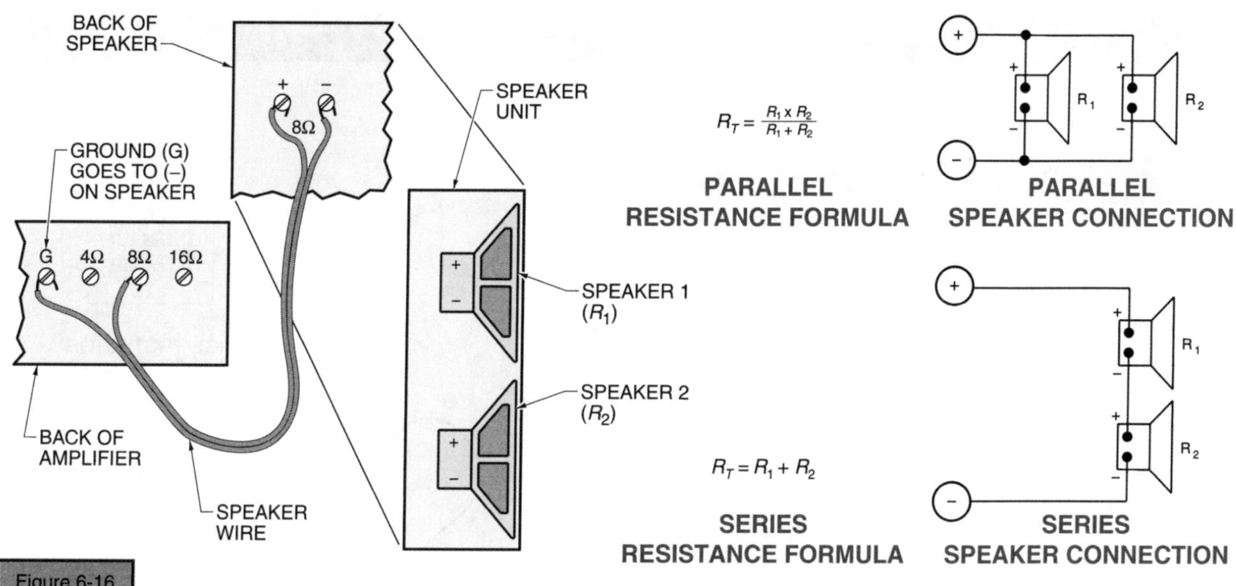

Figure 6-16

When two speakers are connected in parallel, their total resistance decreases. To find the total resistance of two speakers connected in parallel, apply the formula:

$$R_T = \frac{R_1 \times R_2}{R_1 + R_2}$$

where

R_T = total speaker resistance (in Ω)

R_1 = resistance of speaker 1 (in Ω)

R_2 = resistance of speaker 2 (in Ω)

Example: Calculating Total Resistance of Speakers Connected in Parallel

What is the total resistance of two 8 Ω speakers connected in parallel?

$$R_T = \frac{R_1 \times R_2}{R_1 + R_2}$$

$$R_T = \frac{8 \times 8}{8 + 8}$$

$$R_T = \frac{64}{16}$$

$$R_T = \mathbf{4\ \Omega}$$

When two speakers are connected in series, their total resistance increases. To find the total resistance of two speakers connected in series, apply the formula:

$$R_T = R_1 + R_2$$

where:

R_T = total speaker resistance (in Ω)

R_1 = resistance of speaker 1 (in Ω)

R_2 = resistance of speaker 2 (in Ω)

Example: Calculating Total Resistance of Speakers Connected in Series

What is the total resistance of two 8 Ω speakers connected in series?

$$R_T = R_1 + R_2$$

$$R_T = 8 + 8$$

$$R_T = \mathbf{16\ \Omega}$$

Surge Protectors

Computers, video and audio equipment, most electronic equipment, and small appliances are extremely sensitive to damaging voltage surges and spikes. *Surge protectors* are electrical devices that continuously monitor incoming power and automatically clamp any excess voltage that could cause damage. Surge protectors are rated for the maximum voltage for which they are designed. An understanding of peak and rms voltage is necessary because surge protector voltage ratings are listed as peak values, not as rms values. **See Figure 6-17.**

To convert an rms voltage (V_{RMS}) to peak voltage (V_P), multiply the rms voltage by 1.414. To convert a peak voltage to an rms voltage, multiply the peak voltage by .707. For example, what is the normal peak voltage of a standard 120 VAC receptacle outlet?

$$V_{RMS} = V_P \times 1.414$$

$$V_{RMS} = 120 \times 1.414$$

$$V_{RMS} = \mathbf{169.68\ (170\ VAC)}$$

The normal peak voltage of a standard 120 VAC receptacle is 170 VAC. When selecting a surge protector, look for the lowest limiting peak voltage rating.

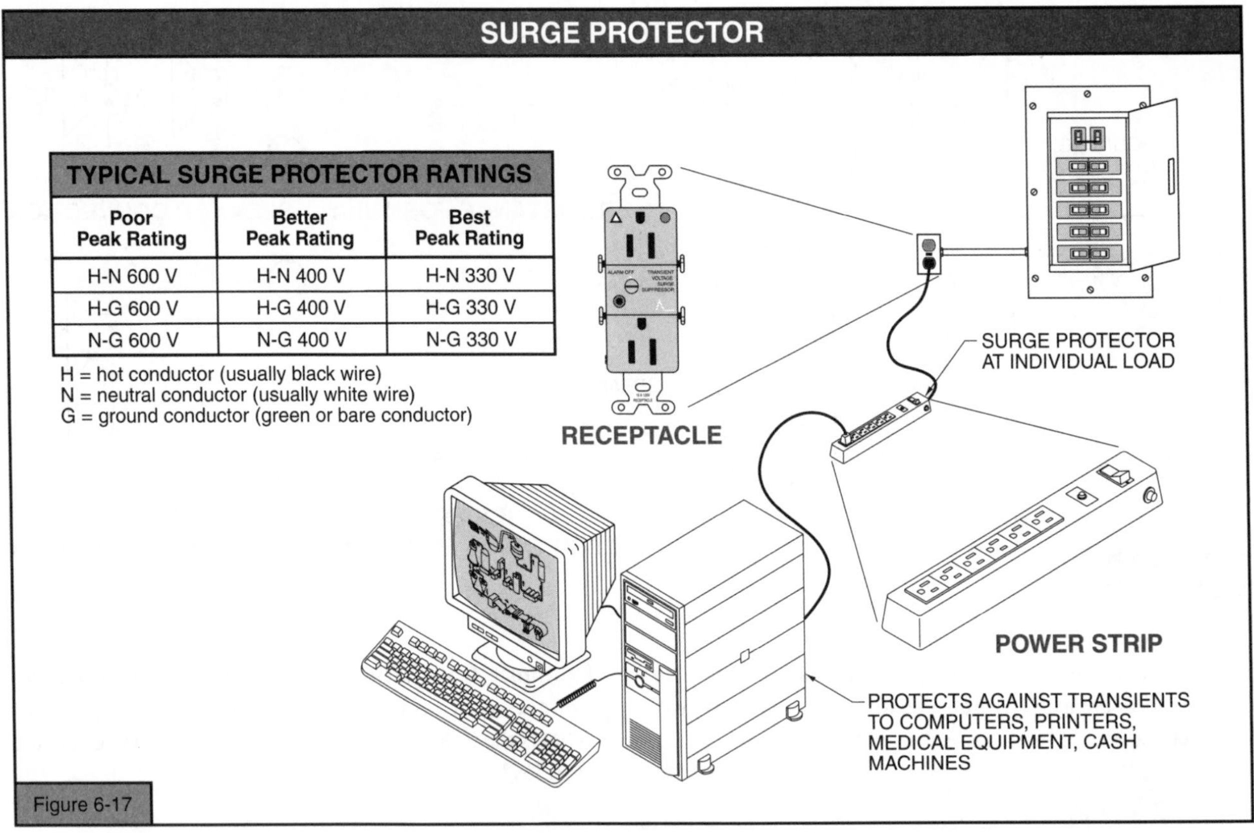

TYPICAL SURGE PROTECTOR RATINGS		
Poor Peak Rating	**Better Peak Rating**	**Best Peak Rating**
H-N 600 V	H-N 400 V	H-N 330 V
H-G 600 V	H-G 400 V	H-G 330 V
N-G 600 V	N-G 400 V	N-G 330 V

H = hot conductor (usually black wire)
N = neutral conductor (usually white wire)
G = ground conductor (green or bare conductor)

Figure 6-17

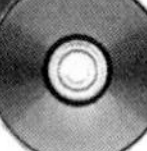

Refer to the CD-ROM in the back of the book for Chapter 6 Quick Quiz® and related information.

7 Numbering Systems and Codes

Decimal Numbering System

A *decimal* is a number expressed in base 10. The decimal numbering system is the most commonly used numbering system. The 10 digits used in the decimal numbering system are 0, 1, 2, 3, 4, 5, 6, 7, 8, and 9.

Each digit in a decimal number occupies a position that has a certain weight. The position farthest to the right has the least weight and the position farthest to the left has the most weight. The position weights are 1, 10, 100, 1000, etc.

Prefixes and exponents may be used to simplify the expression of large decimal numbers. For example, 1000 may be expressed as 1 k or 10^3 and 1,000,000 may be expressed as 1 M or 10^6, **See Figure 7-1.**

To evaluate the decimal number 17,684, the digit in each position is multiplied by the appropriate weighted factor. Thus, the number 17,684 represents 1 ten-thousand, 7 thousands, 6 hundreds, 8 tens, and 4 ones. To evaluate the decimal number 527.231, the digit in each position is multiplied by the appropriate weighted factor. Thus, the number 527.231 represents 5 hundreds, 2 tens, 7 ones, 2 tenths, 3 hundredths, and 1 thousandths. **See Figure 7-2.**

> **TECH FACT** — *The United States monetary system is based on the decimal numbering system.*

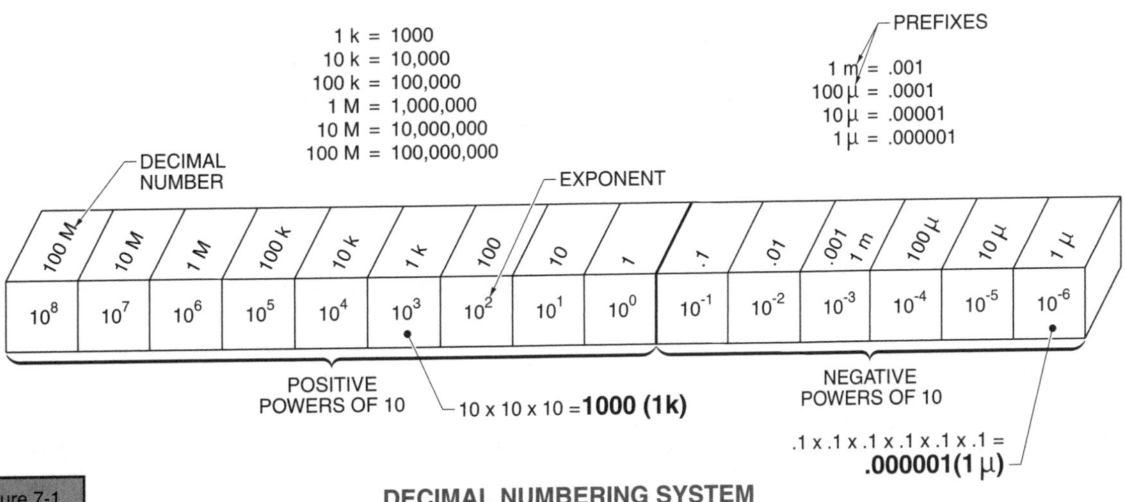

Figure 7-1 DECIMAL NUMBERING SYSTEM

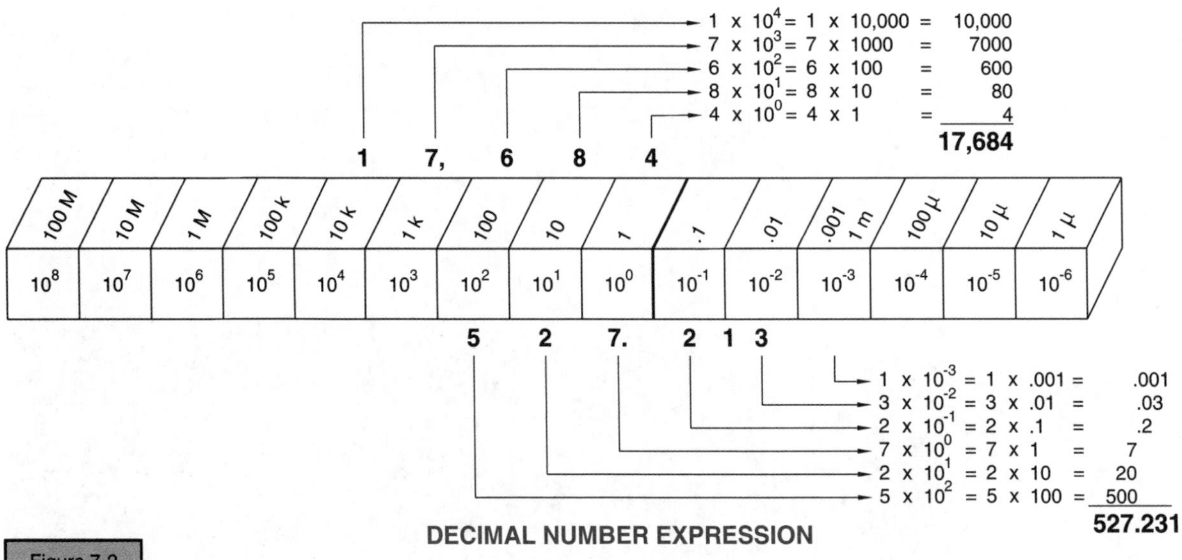

Figure 7-2

Binary Numbering System

A *binary numbering system* is a base two numbering system that uses only two digits, 0 and 1. There are no 2s, 3s, 4s, etc. Digital electronic systems use the binary numbering system because it uses the digits 0 and 1, which are used to represent the two physical states that most electrical elements have (ON/OFF, open/closed, etc.). Electronic systems process information quickly because they can manipulate millions of 0s and 1s per second. **See Figure 7-3.**

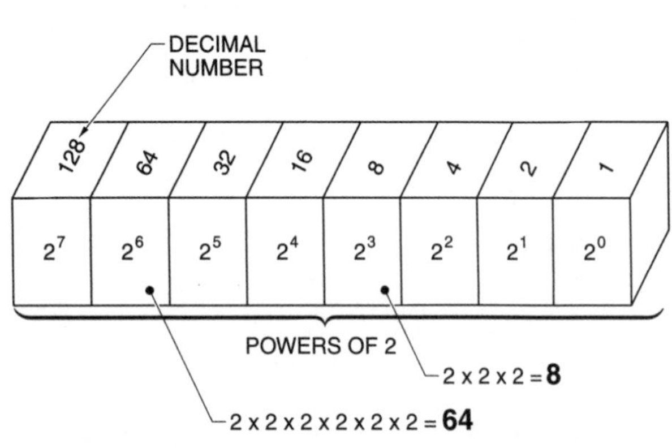

Figure 7-3

TECH FACT *An algebraic expression containing one term is monomial, two terms is binomial, and three terms is trinomial. An algebraic expression containing two or more terms is also multinomial.*

A *bit* is each digit (0 or 1) of a binary number. The binary number 11 (decimal number 3) has 2 bits, binary number 101 (decimal number 5) has 3 bits, and 10000001 (decimal number 129) has 8 bits. A *nibble* is a group of 4 bits (1001, 0011, etc.). A *byte* is a group of 8 bits (10001101, etc.). A *word* is a group of one or more bytes (1111000011110000, etc.) that is transmitted or stored as a unit. **See Figure 7-4.**

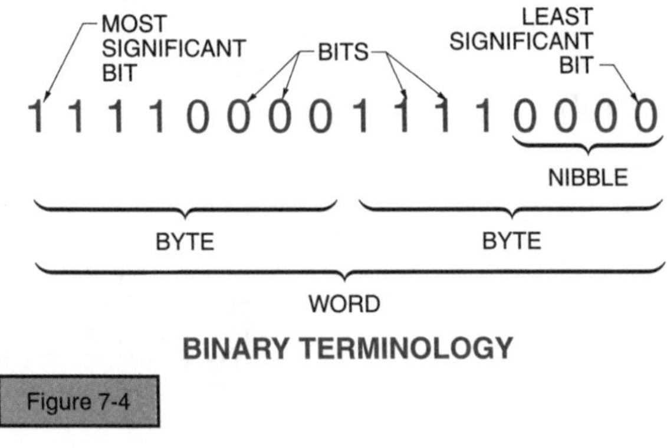

Figure 7-4

Binary-to-Decimal Conversion

To convert a binary number to a decimal number, write the decimal value of each position where a 1 occurs in the binary number. Leave a blank space for each 0 in the binary number. Multiply each 1 by the weight for the position where it occurs. Add all the numbers to determine the decimal equivalent. **See Figure 7-5.**

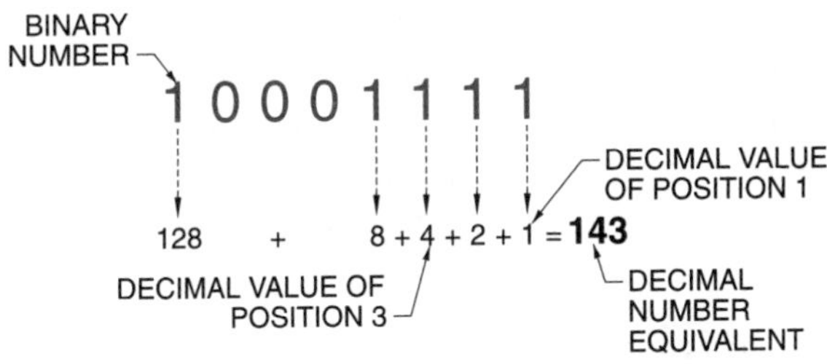

Figure 7-5

Example: Binary-to-Decimal Conversion
What is the decimal number equivalent to the binary number 10001111?

A 1 occurs in positions weighted 1, 2, 4, 8, and 128 in the binary number 10001111. Adding these values equals 143. The decimal number equivalent to the binary number 10001111 is 143.

Decimal-to-Binary Conversion

To convert a decimal number to a binary number, the decimal number is divided by 2 in repetitive steps. A 1 is recorded if there is a remainder after each step. A 0 is recorded if there is no remainder after each step. The division process is repeated until the quotient is 0. **See Figure 7-6.**

Example: Decimal-to-Binary Conversion
What is the binary number equivalent to the decimal number 50?

The decimal number 50 is divided by 2 a total of six times. There is a remainder of 1 after dividing 50 by 2 the second, fifth, and sixth times. These values represent the weighted positions 2, 16, and 32. Each of these positions receive a binary number 1.

There is a remainder of 0 after dividing 50 by 2 the first, third, and fourth times. These values represent the weighted positions 1, 4, and 8. Each of these positions receive a binary number 0. These values are combined starting at the most significant bit (32) through the least significant bit (1). The binary number equivalent to the decimal number 50 is 110010.

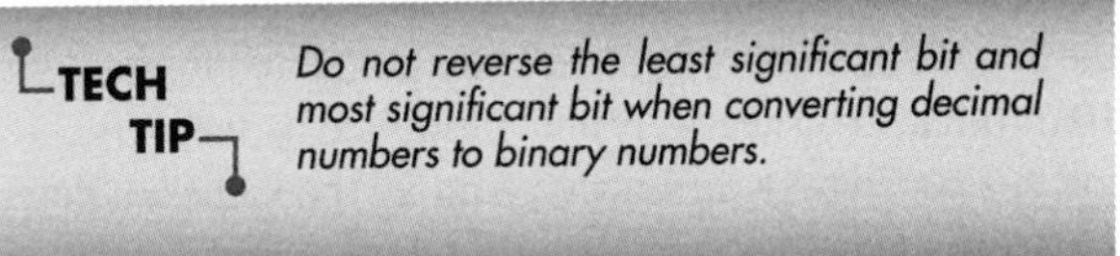

Do not reverse the least significant bit and most significant bit when converting decimal numbers to binary numbers.

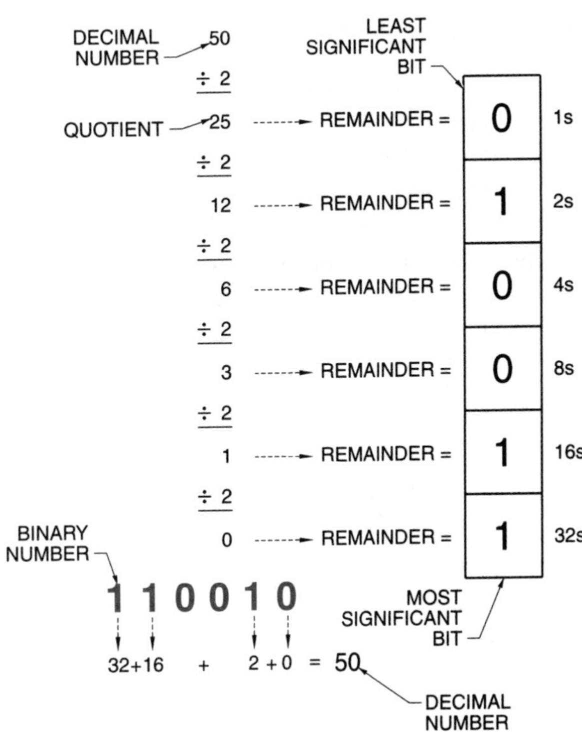

Figure 7-6

Octal Numbering System

The *octal numbering system* is a base eight numbering system that uses only digits 0, 1, 2, 3, 4, 5, 6, and 7. The octal number system is used to process large numbers by digital circuits and to simplify the expression of binary numbers. The octal numbering system uses the same basic principle as the decimal and binary numbering systems. Like the decimal and binary numbering systems, each number occupies a position that has a certain weight. The position weights are 1, 8, 64, 512, 4096, 32,768, 262,144, etc. **See Figure 7-7.**

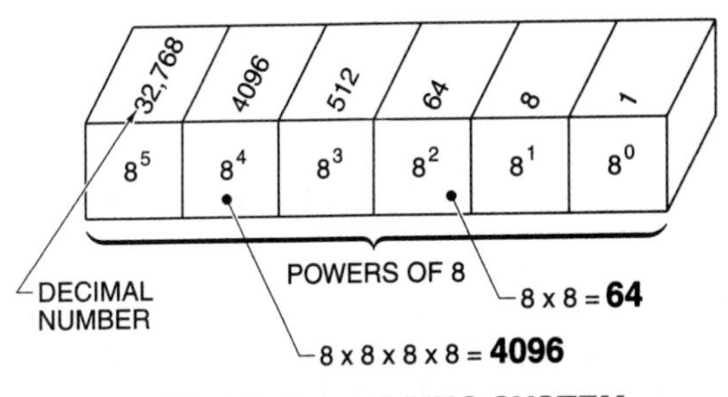

Figure 7-7

Octal-to-Decimal Conversion

To convert an octal number to a decimal number, each position where a number occurs in the octal number is multiplied by the weight for the position. Leave a blank space for each 0 in the octal number. Add all the values to determine the decimal equivalent number. **See Figure 7-8.**

Example: Octal-to-Decimal Conversion
What is the decimal number equivalent of the octal number 114?

A 1 occurs in positions weighted 64 and 8 in the octal number 114. A 4 occurs in the position weighted 1 in the octal number 114. The two number 1s are multiplied by 64 and 8. The number 4 is multiplied by 1. Adding the products equals 76. The decimal number equivalent to the octal number 114 is 76.

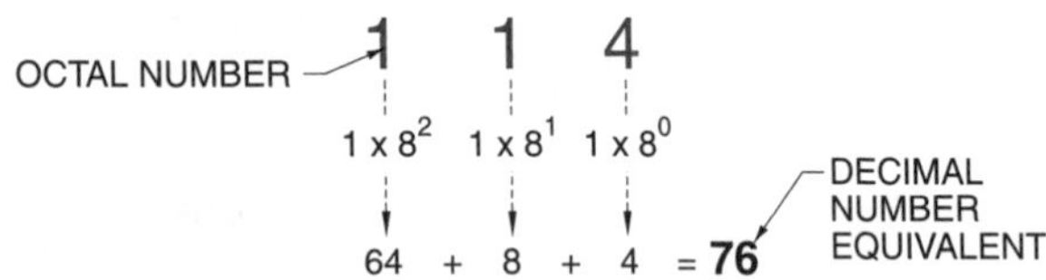

Figure 7-8

Decimal-to-Octal Conversion

To convert a decimal number to an octal number, the decimal number is divided by 8 in repetitive steps. Any remainder is recorded. This process is repeated until the quotient is 0. **See Figure 7-9.**

Example: Decimal-to-Octal Conversion
What is the octal number equivalent of the decimal number 180?

The decimal number 180 is divided by 8 a total of three times. There are remainders of 4, 6, and 2. These values represent the weighted positions 1, 8, and 16. These values are combined starting at the most significant bit (16) through the least significant bit (1). The octal number equivalent to the decimal number 180 is 264.

Hexadecimal Numbering System

The *hexadecimal numbering system* is a base 16 numbering system that uses digits 0, 1, 2, 3, 4, 5, 6, 7, 8, 9, and the first six letters of the alphabet, A, B, C, D, E, and F. The letters A through F are used to denote the digits 10 through 15 respectively. Like the octal numbering system, the hexadecimal numbering system is used by digital circuits to process large numbers and simplify the expression of binary numbers. Just as in the decimal, binary, and octal numbering systems, each number occupies a position that has a certain weight. The position weights are 1, 16, 256, 4096, 65,536, etc. **See Figure 7-10.**

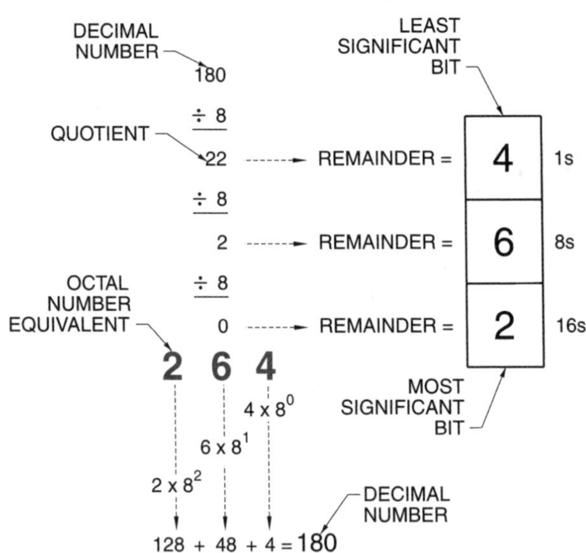

Figure 7-9

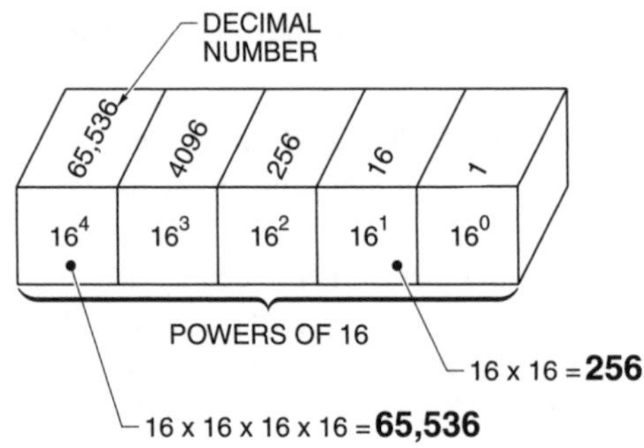

Figure 7-10

Hexadecimal-to-Decimal Conversion

To convert a hexadecimal number to a decimal number, each position where a number occurs in the hexadecimal number is multiplied by the weight for the position where they occur. Letters A through F are converted to the digits 10 through 15 respectively and multiplied by the weight for the position where they occur. Add all the values to determine the decimal equivalent number. **See Figure 7-11.**

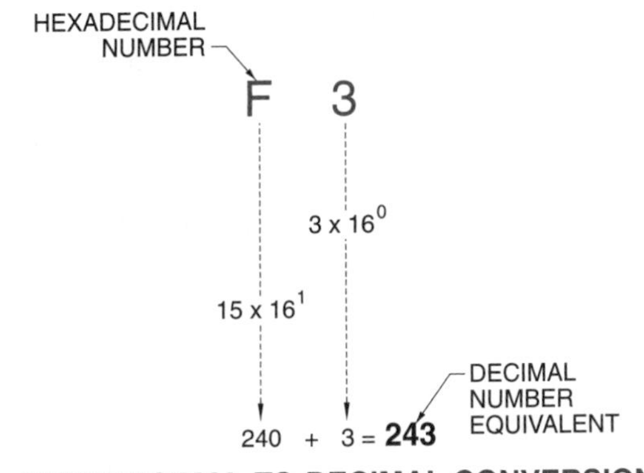

Figure 7-11

Example: Hexadecimal-to-Decimal Conversion
What is the decimal number equivalent of the hexadecimal number F3?

An F occurs in the position weighted 16, and a 3 occurs in the position weighted 1 in the hexadecimal number F3. The letter F represents 15 in the hexadecimal numbering system. The number 15 is multiplied by 16 and the number 3 is multiplied by 1. Adding the products equals 243. The decimal number equivalent to the hexadecimal number F3 is 243.

Decimal-to-Hexadecimal Conversion

To convert a decimal number to a hexadecimal number, the decimal number is divided by 16 in repetitive steps. Any remainder is recorded. This process is repeated until the quotient is 0. In the hexadecimal numbering system, A = 10, B = 11, etc. **See Figure 7-12.**

Example: Decimal-to-Hexadecimal Conversion
What is the hexadecimal number equivalent of the decimal number 213?

The decimal number 213 is divided by 16 twice. There are remainders of 5 and 13. The number 13 is equivalent to the hexadecimal letter D. These values represent the weighted positions 1 and 16. These values are combined starting at the most significant bit (16) through the least significant bit (1). The hexadecimal number equivalent to the decimal number 213 is D5.

Binary Coded Decimal (BCD) System

A *binary coded decimal (BCD) system* is a coding system that represents each digit from 0 through 9 as a four-bit binary number. For example, the decimal number 103 has the digits 1, 0, and 3. Converting each digit to binary gives the BCD number 0001-0000-0011. The space between each group of digits is important when displaying BCD numbers. BCD is used to provide a means of converting a numbering system easily handled by people (decimal) to a code easily handled by machines (binary). A BCD number is easier to read than the identical binary number because of the spacing between each four-bit group. **See Appendix.**

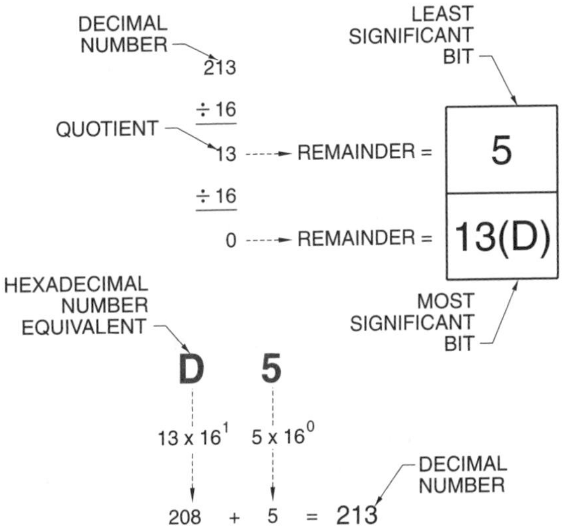

Figure 7-12

Binary numbering systems are used when programming devices such as programmable logic controllers (PLCs).

BCD-to-Decimal Conversion

To convert a BCD number to a decimal number, write the decimal value of each eight-bit group. Continue writing the decimal equivalent of each group of eight bits until there are no more numbers. Combine these values to obtain the decimal number equivalent. **See Figure 7-13.**

Figure 7-13

Example: BCD-to-Decimal Conversion
What is the decimal number equivalent of the BCD number 0000 0100 0000 0101 0000 0110?

The decimal number equivalent of the eight-bit set 0000 0100 equals 4. The decimal number equivalent of the eight-bit set 0000 0101 equals 5. The decimal number equivalent of the eight-bit set 0000 0110 equals 6. Combining these values equals 456. The decimal number equivalent of the BCD number 0000 0100 0000 0101 0000 0110 is 456.

Decimal-to-BCD Conversion

To convert a decimal number to a BCD number, write the BCD value of each decimal number. Continue writing the BCD equivalent of each decimal number until there are no more numbers. Combine these values to obtain the BCD number equivalent. **See Figure 7-14.**

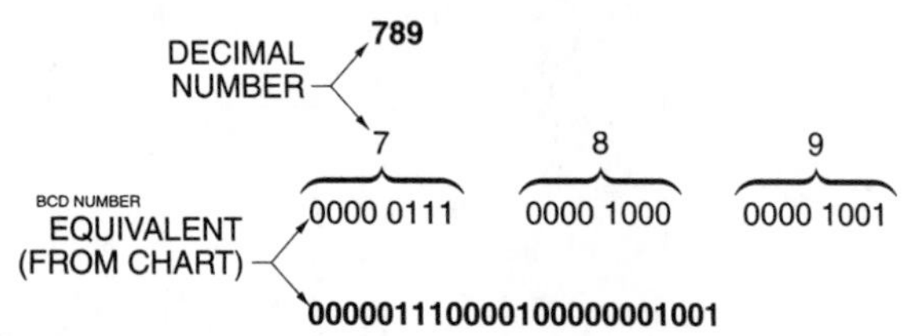

Figure 7-14

TECH FACT — *The analog signal of music is converted to a digital format and stored on a CD. A CD player converts the digital signal back to an analog signal and sends it to the speakers.*

Example: Decimal-to-BCD Conversion
What is the BCD number equivalent of the decimal number 789?

The BCD number equivalent of 7 equals 0000 0111. The BCD number equivalent of 8 equals 0000 1000. The BCD number equivalent of 9 equals 0000 1001. The BCD number equivalent of 789 equals 0000 0111 0000 1000 0000 1001. **See Appendix.**

Gray Code

As many as four digits could change for a single count when using the binary numbering system. For example, 0111 (decimal number 7) changes all four bits at the next larger number 1000 (decimal number 8). This may cause problems when using the binary numbering system to input data on fast moving machines. The Gray code is used to solve this problem. In the Gray code, only one digit changes for each count from one number to the next. **See Figure 7-15.**

FOUR-BIT GRAY CODE		
Decimal	**Binary**	**Gray**
0	0000	0000
1	0001	0001
2	0010	0011
3	0011	0010
4	0100	0110
5	0101	0111
6	0110	0101
7	0111	0100
8	1000	1100
9	1001	1101
10	1010	1111
11	1011	1110
12	1100	1010
13	1101	1011
14	1110	1001
15	1111	1000

Figure 7-15

The Gray code is used in such applications as indicating the angular position of a shaft on an automated machine. In the Gray code shaft application, only one digit changes for each ⅛ increment of shaft. A shaft divided into 16 increments could be used if higher accuracy is required. Gray code numbers from 0 through 15 are used on a shaft divided into 16 increments. **See Figure 7-16.**

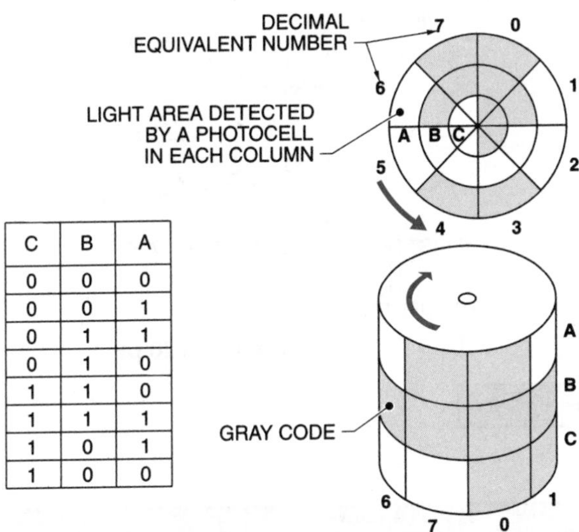

C	B	A
0	0	0
0	0	1
0	1	1
0	1	0
1	1	0
1	1	1
1	0	1
1	0	0

Figure 7-16

Resistor Color Codes

Resistors are devices that limit the current flowing in an electronic circuit. Small resistors use color bands to represent their resistance value. The first two color bands represent the first two digits in the value of the resistor. The third color band (multiplier) indicates the number of zeros that must be added to the first two digits. The fourth band (tolerance) indicates how far the actual measured value can be from the coded value. **See Figure 7-17.**

RESISTOR COLOR CODES				
Color	Digit		Multiplier	Toler-ance*
	1st	2nd		
Black (BK)	0	0	1	0
Brown (BR)	1	1	10	—
Red (R)	2	2	100	—
Orange (O)	3	3	1000	—
Yellow (Y)	4	4	10,000	—
Green (G)	5	5	100,000	—
Blue (BL)	6	6	1,000,000	—
Violet (V)	7	7	10,000,000	—
Gray (GY)	8	8	100,000,000	—
White (W)	9	9	1,000,000,000	—
Gold (Au)	—	—	0.1	5
Silver (Ag)	—	—	0.01	10
None	—	—	0	20
	BAND 1	BAND 2	BAND 3	BAND 4

* in %

Figure 7-17

Example: Determining Resistance
What is the resistance of a resistor that has red, black, orange, and silver colored bands?

A resistor with a red, black, orange, and silver colored band has a resistance value of 20,000 Ω (20 kΩ) ±10%. **See Figure 7-18.**

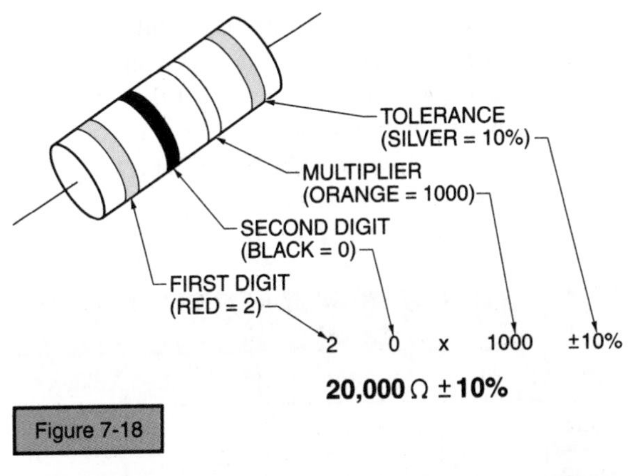

Figure 7-18

TECH TIP
Measure the value of old or dark resistors. The value of an old resistor may exceed its rated tolerance because normal resistor heating may change the value of the resistor.

Capacitor Color Codes

Capacitors are color-coded to indicate their capacitance value. Capacitor color codes are standardized by the Electronic Industries Association (EIA). The color-coded capacitance value is always in picofarad (pF) units. The colors used are the same as those used for resistors.

Mica capacitors normally use a six-dot system. **See Figure 7-19.** The six-dot system is read starting from the top row from left to right and then the bottom row from right to left. A white or silver first dot indicates the new EIA coding. The first dot may also be black for a joint Army-Navy (JAN) code. In both cases, the capacitance value is read from the second, third, and fourth dots. Dot five specifies tolerance and dot six gives the EIA classification. Seven classes (from A to G) specify temperature coefficient, leakage resistance, and additional variable factors. **See Figure 7-20.**

Example: Determining Capacitance—Mica Capacitors
What is the capacitance if the colors on a mica capacitor are violet, red, and brown for dots two, three, and four?

The capacitance of a capacitor with a color code of violet, red, and brown for dots two, three, and four is 720 pF. Ceramic capacitors have stripes or dots with three or five colors. They may be tubular with axial leads or disc-shaped with radial leads. **See Figure 7-21.**

For tubular ceramic capacitors, the wide color band specifying temperature coefficient indicates the left end. The last color indicates the tolerance. The middle three colors give the picofarad capacitance value. For example, a ceramic capacitor with second, third, and fourth bands of brown, black, and brown indicates a 100 pF capacitor.

Gray and white are used as decimal multipliers for very small value capacitors. Gray indicates a multiplier of 0.01 and white indicates a multiplier of 0.1. For example, a ceramic capacitor with second, third, and fourth bands of green, black, and white indicates a 5 pF capacitor (50 × 0.1 = 5 pF).

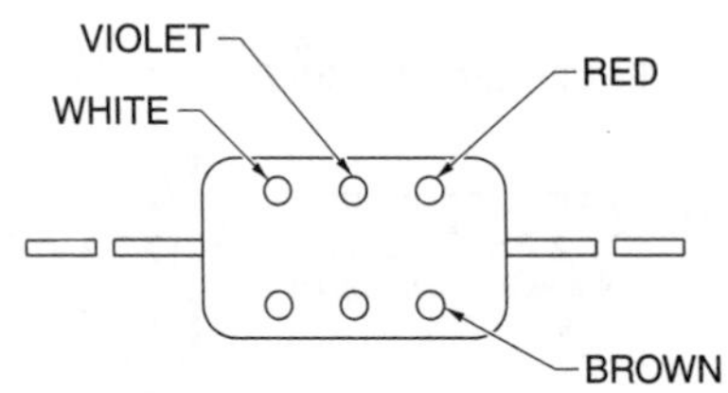

(FIRST DOT) WHITE = EIA STANDARD
(SECOND DOT) VIOLET = 7
(THIRD DOT) RED = 2
(MULTIPLIER) BROWN = 10

72 x 10 = 720 pF

Figure 7-20

MICA CAPACITOR COLOR CODES

Color	Number	Multiplier	Tolerance*	Voltage**
None			20	500
Black	0	1		
Brown	1	10	1	100
Red	2	100	2	200
Orange	3	1000	3	300
Yellow	4	10,000	4	400
Green	5	100,000	5 (EIA)	500
Blue	6	1,000,000	6	600
Violet	7	10,000,000	7	700
Gray	8	100,000,000	8	800
White	9	1,000,000,000	9	900
Gold		0.1	5 (JAN)	1000
Silver		0.01	10	2000

* in %
** in V

WHITE (EIA)
BLACK (JAN)
FIRST DIGIT
SECOND DIGIT
MULTIPLIER
CLASSIFICATION
TOLERANCE

MICA

Figure 7-19

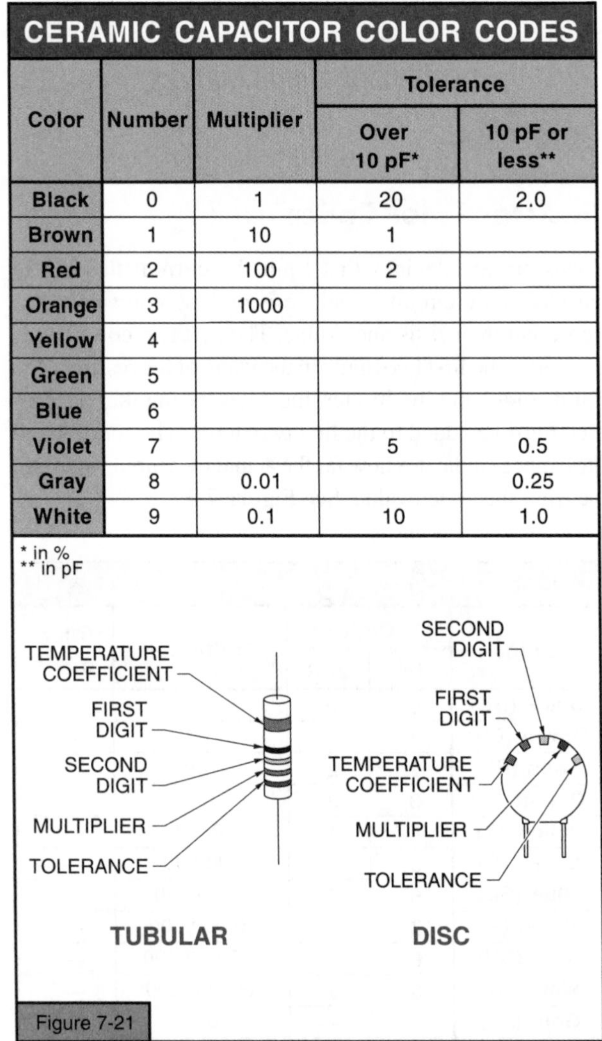

CERAMIC CAPACITOR COLOR CODES

Color	Number	Multiplier	Tolerance	
			Over 10 pF*	10 pF or less**
Black	0	1	20	2.0
Brown	1	10	1	
Red	2	100	2	
Orange	3	1000		
Yellow	4			
Green	5			
Blue	6			
Violet	7		5	0.5
Gray	8	0.01		0.25
White	9	0.1	10	1.0

* in %
** in pF

Figure 7-21

Refer to the CD-ROM in the back of the book for Chapter 7 Quick Quiz® and related information.

8 Meter Abbreviations and Displays

Meter Types

Meters measure quantities such as voltage, current, speed, pressure, counts, etc. Meters may be permanent or portable. Permanent meters are installed to constantly measure and display quantities. Portable meters are used to take momentary measurements. Portable meters are normally used when setting up or troubleshooting a circuit.

Meters are either single-function meters or multimeters. A *single-function meter* is a meter capable of measuring and displaying only one quantity. A *multimeter* is a meter that is capable of measuring two or more quantities. Quantities measured by meters are either electrical quantities or nonelectrical quantities. The meter used depends on the quantity measured and the application. **See Figure 8-1.**

MEASURABLE QUANTITIES . . .

ELECTRICAL

Device		Measures	Unit of Measure	Typical Uses
VOLTMETER	14.7 V	Amount of electrical pressure in a circuit	Volts (V)	Gives a voltage reading in applications such as battery chargers, power supplies, and power distribution systems
AMMETER	1.510 A	Amount of electron flow in a circuit	Amperes (A)	Indicates amount of current a load, circuit, or process is using
WATTMETER	16.8 W	Amount of electrical power in a circuit	Watts (W)	Gives a power reading in such applications as amplifiers, heating elements, and power distribution systems
FREQUENCY METER	57.1 Hz	Number of electrical cycles per second	Hertz (Hz)	Indicates AC power line frequency in such applications as variable-speed motor drives
OHMMETER	10.08 kΩ	Resistance to flow of electricity	Ohms (Ω)	Indicates a load, circuit, or component resistance before power is applied to a circuit
CONDUCTIVITY METER		Ability to conduct electricity	Mhos (usually Mhos/cm) or microsiemens (usually μS/cm)	Measures the ability of a solution (usually water) to conduct electricity

. . . MEASURABLE QUANTITIES

NONELECTRICAL				
Device		**Measures**	**Unit of Measure**	**Typical Uses**
PRESSURE GAUGE	1000 500 1500 0 2000 PSI	Amount of fluid (air or liquid) pressure in a system	Pounds per square inch (psi), kilograms per centimeter (kg/cm), or bars (bar)	Monitors HVAC systems, pollution control systems, fluid systems, and machine conditions
TACHOMETER	1000 RPM	Speed of a moving object	Revolutions per minute (rpm)	Monitors speed of moving objects, motors, gears, engines, and machine parts
TEMPERATURE METER	478 °C	Intensity of heat in an object or area	Degrees Fahrenheit (°F) or degrees Celsius (°C)	Monitors temperature of products, machines, fluids, processes, and areas
ANEMOMETER		Air velocity (distance traveled per unit of time)	Feet per minute (fpm) or meters per second (mps)	Monitors flow of air in HVAC systems
MANOMETER		Pressure differential between two points in a system	Inches water column (in in. wc) pounds per square inch (psi) or centimeters (cm)	Monitors pressure drop across air filters, dampers, and refrigeration coils
HYGROMETER	% RELATIVE HUMIDITY	Relative humidity	Percent (0% to 100%) relative humidity (% RH)	Monitors freezers, HVAC systems, storage bins, computer rooms, libraries, and warehouses
pH METER	7.08 pH	Acidity or alkalinity of a solution	0 pH to 14 pH	Monitors cooling towers, process steam, feedwater, pulp and paper operations, and wastewater treatment
VIBRATION METER	2.19 IN/SEC	Amount of imbalance in a machine or system	Velocity (in/s or cm/s) or displacement (in. or mm)	Monitors machines and motors for excess vibration. Indicates when a machine is not properly loaded or when alignment is required
FLOWMETER	36.7	Amount of fluid moving in a system	Gallons per minute (gpm) or standard cubic feet per minute (scfm)	Indicates that gas or liquid is moving and monitors rate of movement
COUNTER	654321	Number of devices moving past a given location	Numerical	Maintains production values, parts used, and inventory

Figure 8-1

Meter Abbreviations

Meter switches, scales, displays, and manuals use abbreviations and symbols to convey information, such as settings and units of measure. An *abbreviation* is a letter or combination of letters that represents a word. Abbreviations depend on a particular language. A *symbol* is a graphic element that represents an operation, a quantity, or unit. Symbols are independent of language because a symbol can be recognized regardless of the language a person speaks. All meters use standard abbreviations to represent a quantity or term. Abbreviations can be used individually (100 V) or in combinations (100 kV). **See Figure 8-2.**

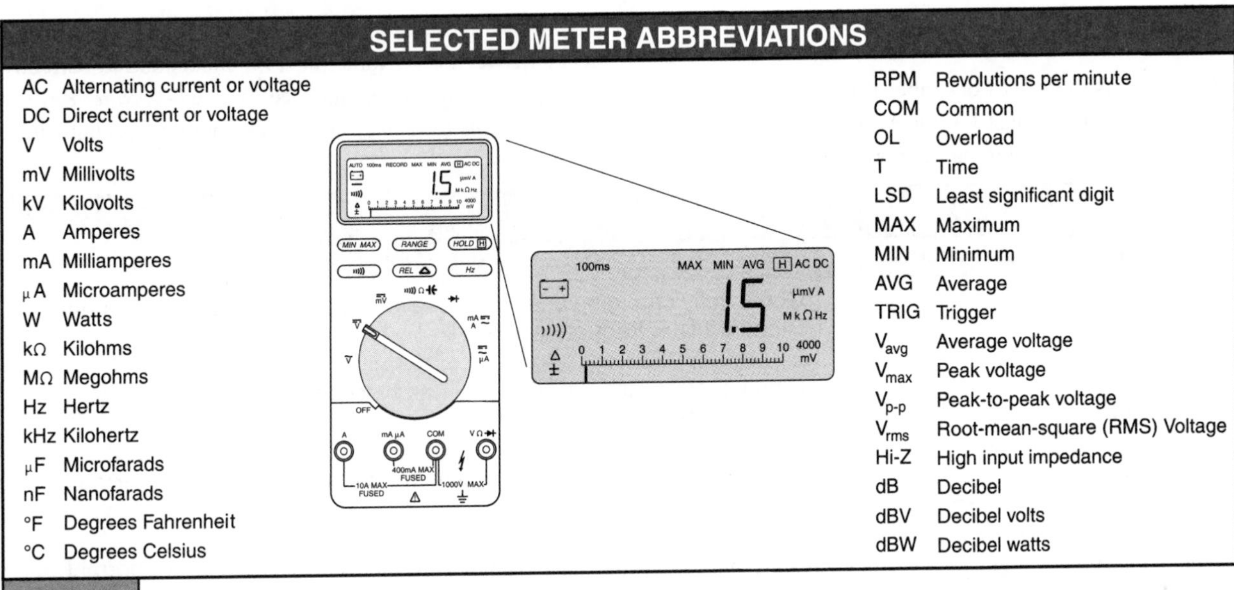

Figure 8-2

Meter Symbols

Most meters use standard symbols to represent an electrical component (battery, etc.), term (ground, etc.), or message to the user (warning, etc.). **See Figure 8-3.**

TECH FACT *The higher the current in a circuit, the higher the heat produced and more likely that damage may occur.*

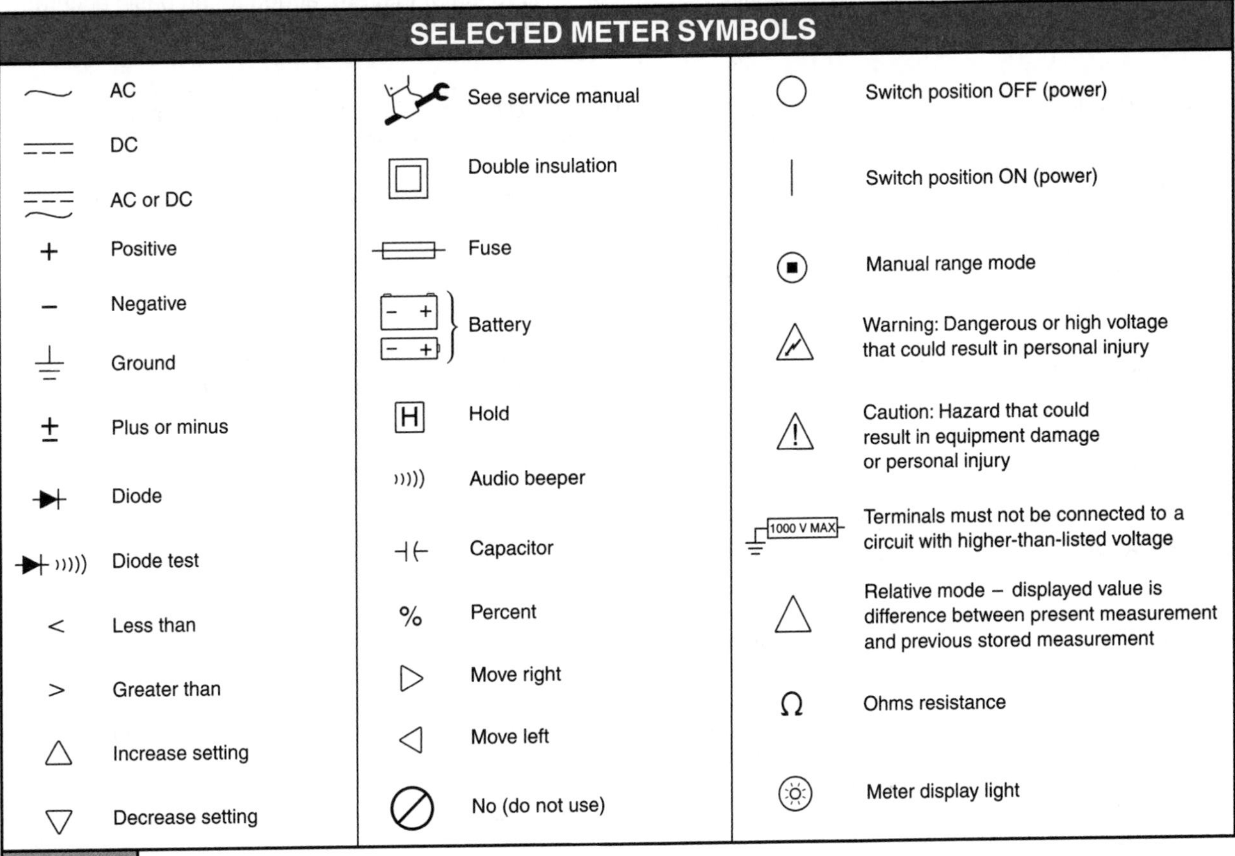

Figure 8-3

Meter Terminology

All meters use terms to describe displayed information. Meter terminology must be understood for proper meter use. For example, a meter may display an electrical quantity as an average, effective, peak, or rms value. The difference between each value must be understood to correctly interpret the measurements. **See Figure 8-4.** *Note:* Review all operating and safety precautions before working with a new meter.

METER TERMINOLOGY . . .		
TERM	**SYMBOL**	**DEFINITION**
AC		Continually changing current that reverses direction at regular intervals. Standard U.S. frequency is 60 Hz
AC COUPLING		Signal that passes an AC signal and blocks a DC signal. Used to measure AC signals that are riding on a DC signal
AC/DC		Indicates ability to read or operate on alternating and direct current
ACCURACY ANALOG METER		Largest allowable error (in percent of full scale) made under normal operating conditions. The reading of a meter set on the 250 V range with an accuracy rating of ± 2% could vary ± 5 V. Analog meters have greater accuracy when readings are taken on the upper half of the scale
ACCURACY DIGITAL METER	100.0 AC V	Largest allowable error (in percent of reading) made under normal operating conditions. A reading of 100.0 V on a meter with an accuracy of ±2% is between 98.0 V and 102.0 V. Accuracy may also include a specified amount of digits (counts) that are added to the basic accuracy rating. For example, an accuracy of ± 2% (± 2 digits) means that a display reading of 100.0 V on the meter is between 97.8 V and 102.2 V
ALLIGATOR CLIP		Long-jawed, spring-loaded clamp connected to the end of a test lead. Used to make temporary electrical connections
AMBIENT TEMPERATURE		Temperature of air surrounding a meter or equipment to which the meter is connected
AMMETER		Meter that measures electric current
AMMETER SHUNT		Low-resistance conductor that is connected in parallel with the terminals of an ammeter to extend the range of current values measured by the ammeter
AMPLITUDE		Highest value reached by a quantity under test
ATTENUATION		Decrease in amplitude of a signal
AUDIBLE		Sound that can be heard
AUTORANGING		Function that automatically selects a meter's range based on signals received
AVERAGE VALUE		Value equal to .637 times the amplitude of a measured value

. . . METER TERMINOLOGY . . .		
TERM	**SYMBOL**	**DEFINITION**
BACKLIGHT		Light that brightens the meter display
BANANA JACK		Meter jack that accepts a banana plug
BANANA PLUG		Long, thick terminal connection on one end of a test lead used to make a connection to a meter
BATTERY SAVE		Feature that enables a meter to shut down when battery level is too low or no key is pressed within a set time
BNC		Coaxial-type input connector used on some meters
CAPTURE		Function that records and displays measured values
CELSIUS	°C	Temperature measured on a scale for which the freezing point of water is 0° and the boiling point is 100°
CLOSED CIRCUIT		Circuit in which two or more points allow a predesigned current to flow
COUNTS		Unit of measure of meter resolution. A 1999 count meter cannot display a measurement of 1⁄10 of a volt when measuring 200 V or more. A 3200 count meter can display a measurement of 1⁄10 of a volt up to 320 V
DC		Current that constantly flows in one direction
DECIBEL (dB)		Measurement that indicates voltage or power comparison in a logarithmic scale
DIGITS		Indication of the resolution of a meter. A 3½ digit meter can display three full digits and one half digit. The three full digits display a number from 0 to 9. The half digit displays a 1 or is left blank. A 3½ digit meter displays readings up to 1999 counts of resolution. A 4½ digit meter displays readings up to 19,999 counts of resolution
DIODE		Semiconductor that allows current to flow in only one direction
DISCHARGE		Removal of an electric charge
DUAL TRACE		Feature that allows two separate waveforms to be displayed simultaneously
EARTH GROUND		Reference point that is directly connected to ground

... METER TERMINOLOGY ...		
TERM	**SYMBOL**	**DEFINITION**
EFFECTIVE VALUE		Value equal to .707 of the amplitude of a measured quantity
FAHRENHEIT	°F	Temperature measured on a scale for which the freezing point of water is 32° and the boiling point is 212°
FREEZE		Function that holds a waveform (or measurement) for closer examination
FREQUENCY		Number of complete cycles occurring per unit of time
FUNCTION SWITCH		Switch that selects the function (AC voltage, DC voltage, etc.) that a meter is to measure
GLITCH		Momentary spike in a waveform
GLITCH DETECT		Function that increases the meter sampling rate to maximize the detection of the glitch(es)
GROUND		Common connection to a point in a circuit whose potential is taken as zero
HARD COPY		Function that allows a printed copy of the displayed measurement
HOLD BUTTON	HOLD H	Button that allows a meter to capture and hold a stable measurement
LIQUID CRYSTAL DISPLAY (LCD)		Display that uses liquid crystals to display waveforms, measurements, and text on its screen
MEASURING RANGE		Minimum and maximum quantity that a meter can safely and accurately measure
NOISE		Unwanted extraneous electrical signals
OPEN CIRCUIT		Circuit in which two (or more) points do not provide a path for current flow
OVERFLOW		Condition of a meter that occurs when a quantity to be measured is greater than the quantity the meter can display
OVERLOAD	OL	Condition of a meter that occurs when a quantity to be measured is greater than the quantity the meter can safely handle for the meter range setting
PEAK		Highest value reached when measuring
PEAK-TO-PEAK		Highest and lowest voltage value of a waveform

. . . METER TERMINOLOGY		
TERM	**SYMBOL**	**DEFINITION**
POLARITY		Orientation of the positive (+) and negative (−) side of direct current or voltage
PROBE		Pointed metal tip of a test lead used to make contact with the circuit under test
PULSE		Waveform that increases from a constant value, then decreases to its original value
PULSE TRAIN		Repetitive series of pulses
RANGE		Quantities between two points or levels
RECALL		Function that allows stored information (or measurements) to be displayed
RESOLUTION		Sensitivity of a meter. A meter may have a resolution of 1 V or 1 mV
RISING SLOPE		Part of a waveform displaying a rise in voltage
ROOT-MEAN-SQUARE	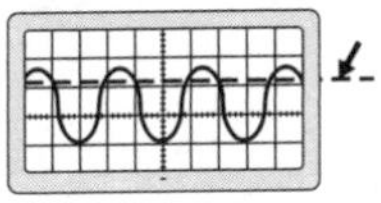	Value equal to .707 of the amplitude of a measured value
SAMPLE		Momentary reading taken from an input signal
SAMPLING RATE		Number of readings taken from a signal every second
SHORT CIRCUIT		Two or more points in a circuit that allow an unplanned current flow
TERMINAL	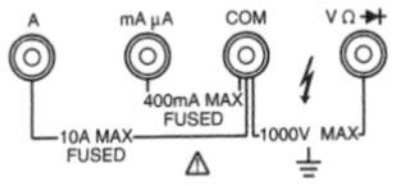 A mA µA COM VΩ 400mA MAX FUSED 10A MAX FUSED 1000V MAX	Point to which meter test leads are connected
TERMINAL VOLTAGE		Voltage level that meter terminals can safely handle
TRACE		Displayed waveform that shows the voltage variations of the input signal as a function of time
TRIGGER		Device which determines the beginning point of a wavelength
WAVEFORM		Pattern defined by an electrical signal
ZOOM	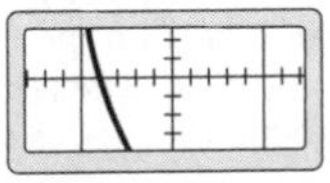	Function that allows a waveform (or part of waveform) to be magnified

Figure 8-4

Reading Analog Displays

An *analog display* is an electromechanical device that indicates readings by the mechanical motion of a pointer. Analog displays use scales to display measured values. Analog scales may be linear or nonlinear. A *linear scale* is a scale that is divided into equally spaced segments. A *nonlinear scale* is a scale that is divided into unequally spaced segments. **See Figure 8-5.**

Analog scales are divided using primary divisions, secondary divisions, and subdivisions. A *primary division* is a division with a listed value. A *secondary division* is a division that divides primary divisions in halves, thirds, fourths, fifths, etc. A *subdivision* is a division that divides secondary divisions in halves, thirds, fourths, fifths, etc. Secondary divisions and subdivisions do not have listed numerical values. When reading an analog scale, add the primary, secondary, and subdivision readings. **See Figure 8-6.**

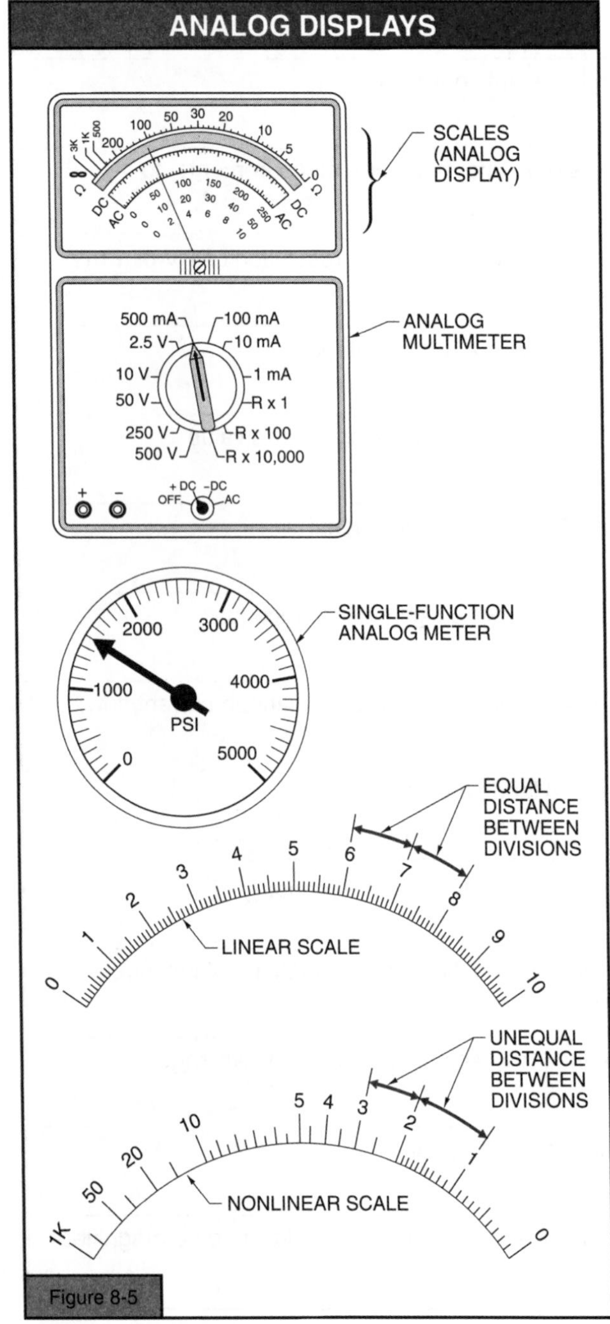

Figure 8-5

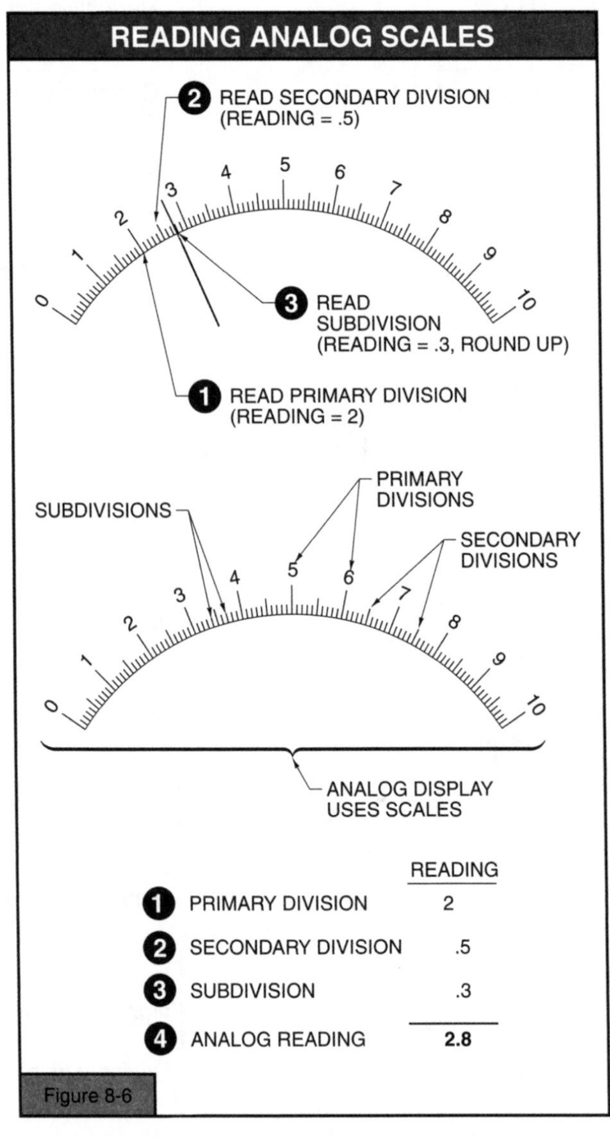

Figure 8-6

TECH TIP

Always start on the highest possible setting and use a meter that has a fused current setting when measuring current in a circuit.

To read an analog scale, apply the procedure:

1. Read the primary division.
2. Read the secondary division if the pointer moves past a secondary division. *Note:* This may not occur with very low readings.
3. Read the subdivision if the pointer is not directly on a primary or secondary division. Round the reading to the nearest subdivision if the pointer is not directly on a subdivision. Round the reading to the next highest subdivision if rounding to the nearest subdivision is unclear.
4. Add the primary division, secondary division, and subdivision readings to obtain the analog reading.

Digital Displays

A *digital display* is an electronic device that displays readings as numerical values. Digital displays help eliminate human error when taking readings by displaying exact values measured. Errors occur when reading a digital display if the displayed prefixes, symbols, and decimal points are not properly applied.

Digital displays display values using either a light-emitting diode (LED) or a liquid crystal display (LCD). LED displays are easier to read and use more power than LCD displays. Most portable digital meters use an LCD.

The exact value on a digital display is determined from the numbers displayed and the position of the decimal point. A range switch determines the placement of the decimal point.

Typical voltage ranges on a digital display are 3 V, 30 V, and 300 V. The highest possible reading with the range switch on 3 V is 2.999 V. The highest possible reading with the range switch on 30 V is 29.99 V. The highest possible reading with the range switch on 300 V is 299.9 V. Accurate readings are obtained by using the range that gives the best resolution without overloading. **See Figure 8-7.**

Bar Graphs

Most digital displays include a bar graph to show changes and trends in a circuit. A *bar graph* is a graph composed of segments that function as an analog pointer. The displayed bar graph segments increase as the measured value increases and decrease as the measured value decreases. Reverse the polarity of test leads if a negative sign is displayed at the beginning of the bar graph. **See Figure 8-8.**

Wrap-Around Bar Graphs

A *wrap-around bar graph* is a bar graph that displays a fraction of the full range on the graph. The pointer wraps around and starts over when the limit of the bar graph is reached. **See Figure 8-9.**

Fluke Corporation

Individual segments of a wraparound bar graph provide a graphic representation of measurement changes.

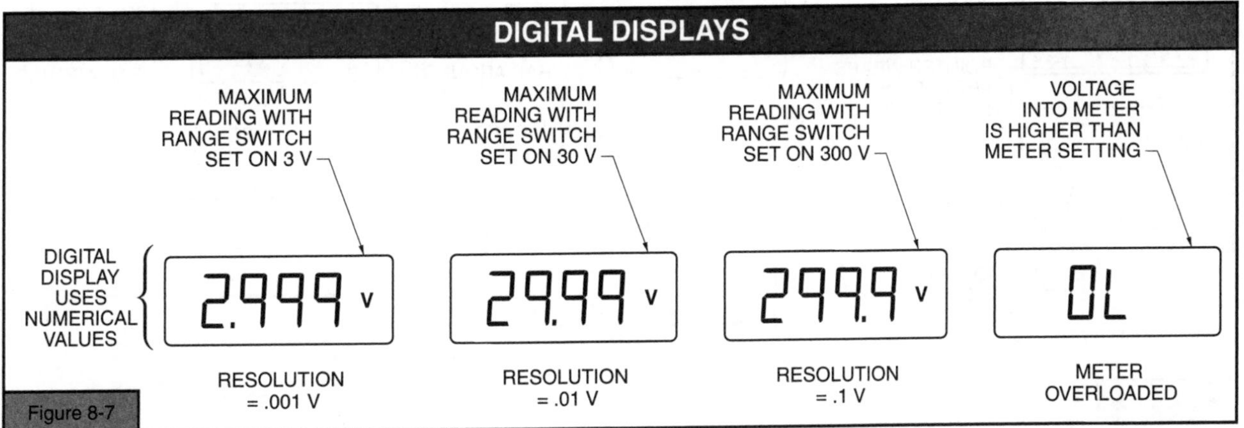

Figure 8-7

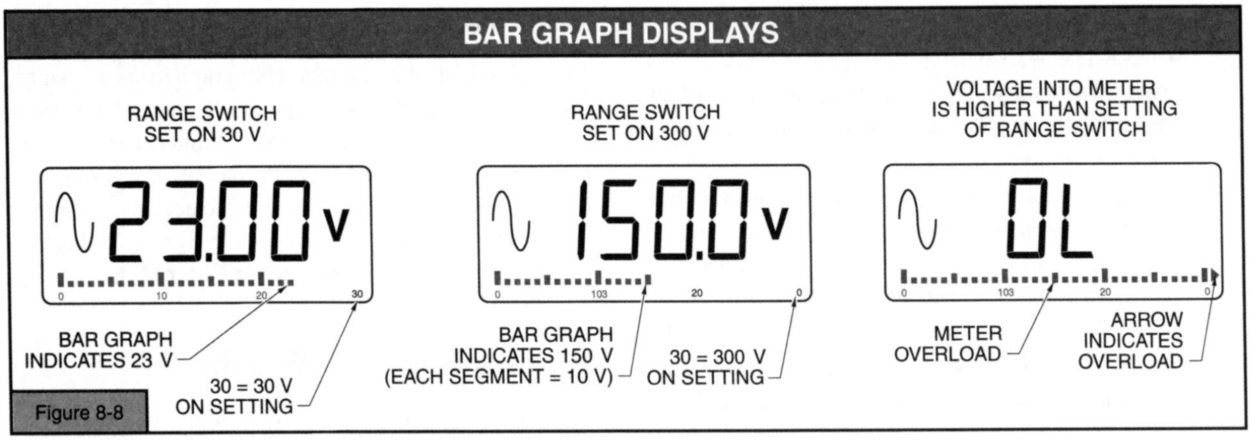

BAR GRAPH DISPLAYS

RANGE SWITCH SET ON 30 V
23.00 V
BAR GRAPH INDICATES 23 V
30 = 30 V ON SETTING

RANGE SWITCH SET ON 300 V
150.0 V
BAR GRAPH INDICATES 150 V (EACH SEGMENT = 10 V)
30 = 300 V ON SETTING

VOLTAGE INTO METER IS HIGHER THAN SETTING OF RANGE SWITCH
OL
METER OVERLOAD
ARROW INDICATES OVERLOAD

Figure 8-8

Bar Graph Use

A bar graph reading is updated 30 times per second. A numerical reading is updated 4 times per second. The bar graph is used when quickly changing signals cause the numerical display to flash or when there is a change in the circuit that is too rapid for the numerical display to detect.

For example, mechanical relay contacts may bounce open when exposed to vibration. Contact bounce causes intermittent problems in electrical equipment. Frequency and severity of contact bounce increases as a relay ages.

A contact's resistance changes momentarily from zero to infinity and back when a contact bounces open. A numerical digital display cannot indicate contact bounce because most numerical digital displays require more than 250 ms to update their displays. The quick response of bar graphs enables detection of most contact bounce problems. The contact bounce is displayed by the movement of one or more segments the moment the contact opens.

Ghost Voltages

A meter set to measure voltage may display a reading before the meter is connected to a powered circuit. The displayed voltage is a ghost voltage that appears as changing numbers on a digital display or as a vibrating analog display. A *ghost voltage* is a voltage that appears on a meter that is not connected to a circuit.

Ghost voltages are produced by the magnetic fields generated by current-carrying conductors, fluorescent lighting, and operating electrical equipment. Ghost voltages enter a meter through the test leads because test leads not connected to a circuit act as antennae for stray voltages. **See Figure 8-10.**

Ghost voltages do not damage a meter. Ghost voltages may be misread as circuit voltages when a meter is connected to a circuit that is believed to be powered. A circuit that is not powered can also act as an antenna for stray voltages. To ensure true circuit voltage readings, connect a meter to a circuit long enough so that the meter displays a constant reading.

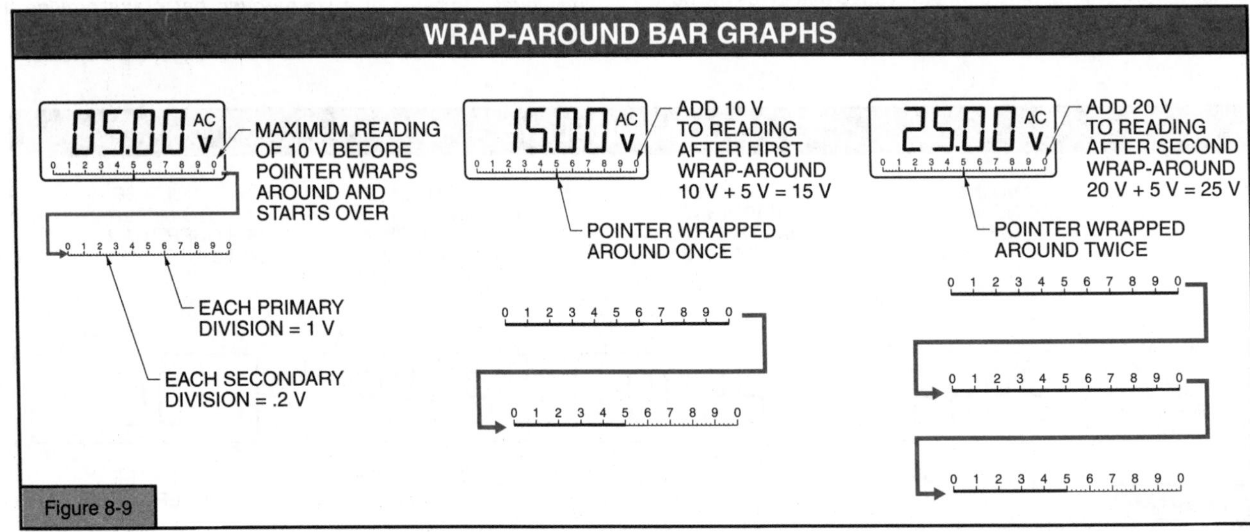

Figure 8-9

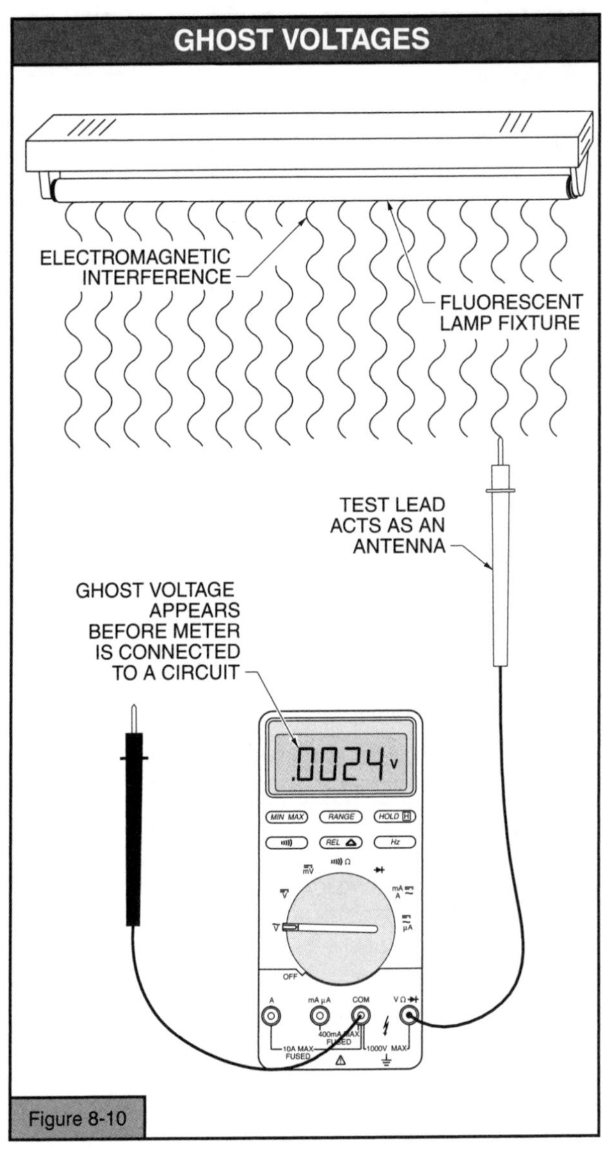

Figure 8-10

Meter Precautions

A variety of analog and digital meters measure electrical quantities such as voltage, current, resistance, power, speed, temperature, etc. Each meter includes an instruction manual. The instruction manual shows and explains individual meter specifications and features, proper operating procedure and safety, and specific warnings and applications.

TECH TIP

In areas of high electromagnetic interference, always use a meter that has a shielded case to reduce the effects of electromagnetic interference.

Read the manufacturer's instruction manual before using any test equipment. Always refer to the instruction manual for information concerning specific use, safety, and meter capacity. Safety precautions are required when using some meters because of the high voltage and currents in some circuits. Safety precautions for using meters include:

- Ensuring that the meter leads are connected to the correct meter jacks.
- Ensuring that the selector switch is set to the proper range and function before applying test leads to a circuit.
- Knowing meter accuracy limits. Any reading taken varies from the true reading by the accuracy limits of the meter.
- Averaging several readings when accuracy is critical. This reduces the effect of stray alternating fields on some meters.
- Starting with the highest range when measuring unknown values.
- Checking meter leads for frayed or broken insulation.
- Avoiding taking measurements under humid or damp conditions.
- Ensuring that hands, shoes, and work areas are dry.

Meter Characteristics

Meters are available in a wide range of prices. Meter cost is increased based on added features, safety, warranties, and improved specifications. Any hand-held digital multimeter (DMM) should be able to take basic voltage (AC and DC), DC current (less than 10 A), and resistance measurements. Additional features may be added to improve safety, usage, and accuracy. **See Figure 8-11.**

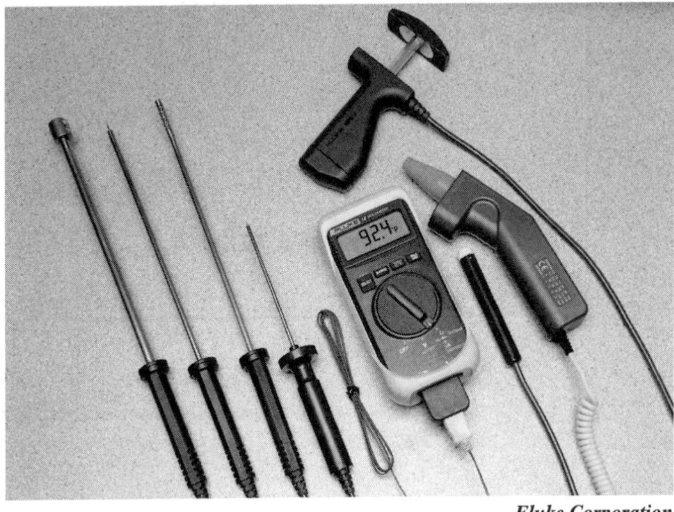

Fluke Corporation

Most digital test instruments can be used with attachments to perform additional tests.

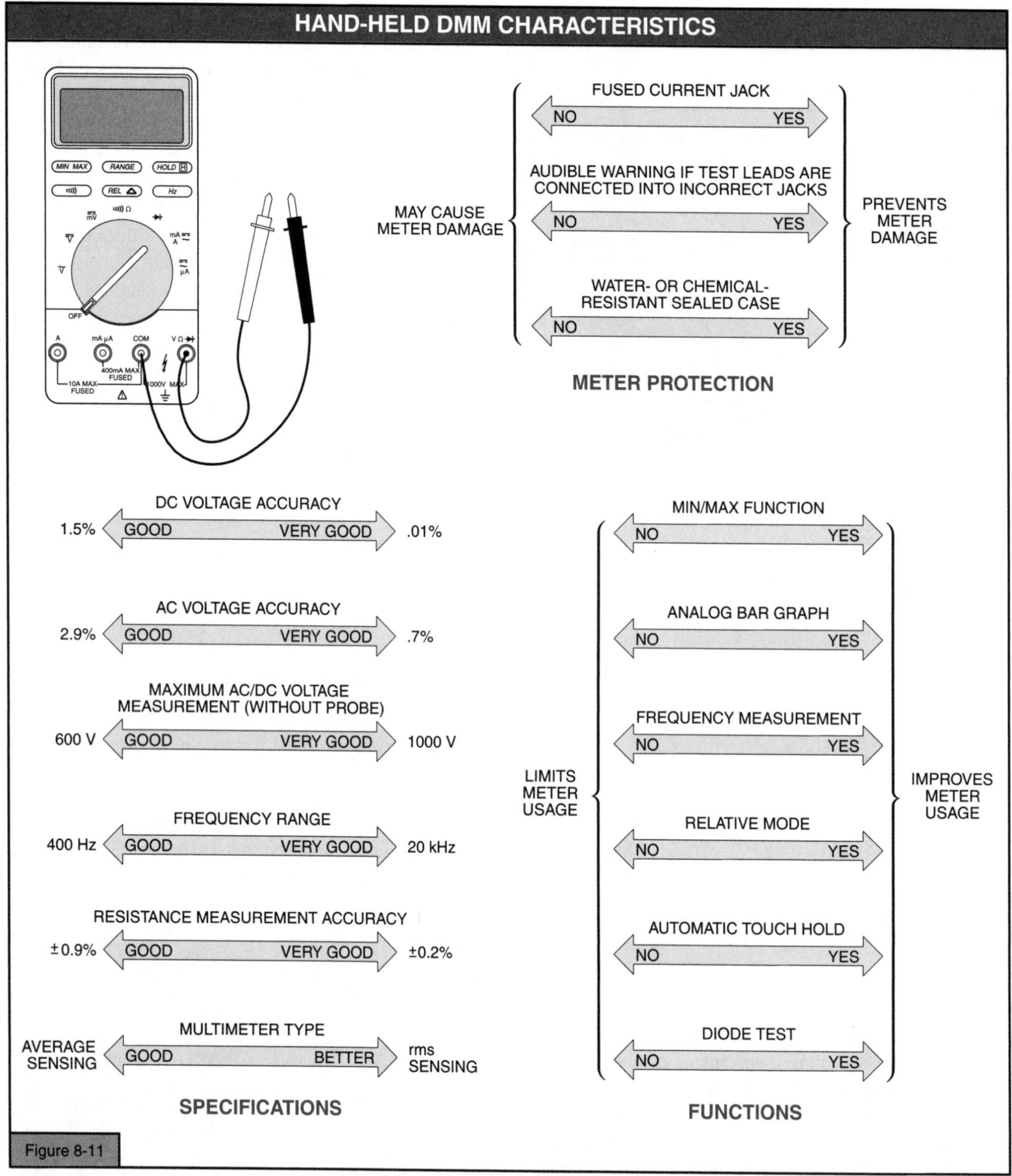

Figure 8-11

Refer to the CD-ROM in the back of the book for Chapter 8 Quick Quiz® and related information.

Taking Measurements

In most electrical, electronic, and fluid power circuits, measurements are taken when there is a problem (troubleshooting), or to display the presence and level of a measurable quantity, such as voltage, pressure, speed, or temperature during normal operation. Measurements are taken using portable equipment such as a voltmeter when troubleshooting. Built-in panel meters are used when taking measurements for normal display indication. For example, some automobile dashboards include a voltmeter to show the amount of system voltage during normal operation. A hand-held voltmeter is used to troubleshoot an electrical problem in the system.

Taking and displaying measurements during normal operation gives a visual indication of circuit performance. Seeing the level of pressure, temperature, voltage, current, speed, etc. at any time, allows for a safe and productive system.

Taking measurements during troubleshooting gives a clear picture of any problems in a circuit, circuit operation, and possible future problems. Always take measurements when troubleshooting, even if the problem appears obvious. For example, the obvious problem may be that a motor is burned out. Replacing the motor may get the system running, but the reason why the motor burned out is not determined. Taking current measurements indicates whether the motor is overloaded and the fuses are properly sized. Taking voltage measurements indicates whether there is a poor electrical connection upstream (low voltage) from the motor. Taking temperature measurements may indicate improper ventilation or bad bearings.

Voltage

Voltage (E) is the amount of electrical pressure in a circuit. Voltage is measured in volts (V). The amount of voltage in a circuit can be measured using test lights, voltage testers, voltage indicators, and voltmeters. Safety precautions must be taken to ensure that no part of the body comes in contact with a live part of the circuit. Safety precautions must also be taken to ensure that voltage test instruments are properly used because voltage measurements are taken when a circuit is powered.

Voltage measurements are normally taken to establish that there is voltage at a given point in a circuit and to determine if the voltage is at the proper level. For example, if there is voltage present on the input side of a fuse but not on the output side of the fuse, the fuse is bad (open). Establishing that there is voltage at a given point in a circuit enables the troubleshooter to know which direction (upstream or downstream from the test point) to troubleshoot.

The second reason voltage measurements are taken is to determine if the voltage is at the proper level. Most electrical devices, circuits, and power sources operate at a rated voltage level (1.5 V, 9 V, 24 V, 115 V, 230 V, etc.). The exact voltage level in a circuit may vary. For most electrical circuits, the voltage level may vary ±10% from the rated voltage. Lower-than-normal voltage is a more common problem than higher-than-normal voltage. Low voltage occurs due to loose connections, corrosion at connections, undersized feeder conductors, damaged conductors, or when additional loads are added to a circuit.

AC Voltage Measurement — Test Lights

A *test light* is a test instrument with a bulb that is connected to two test leads to give a visual indication when voltage is present in a circuit. The most common test light is a neon test light. **See Figure 9-1.** A *neon test light* has a bulb that is filled with neon gas and uses two electrodes to ionize the gas (excite the atoms). Neon test lights are preferred because neon bulbs have extended lifetimes compared to other bulbs. The long life of neon bulbs is attributed to the bulbs having a very high resistance so neon bulbs draw very little current when taking a measurement. The bulb of a test light illuminates when voltage is present in the circuit being tested.

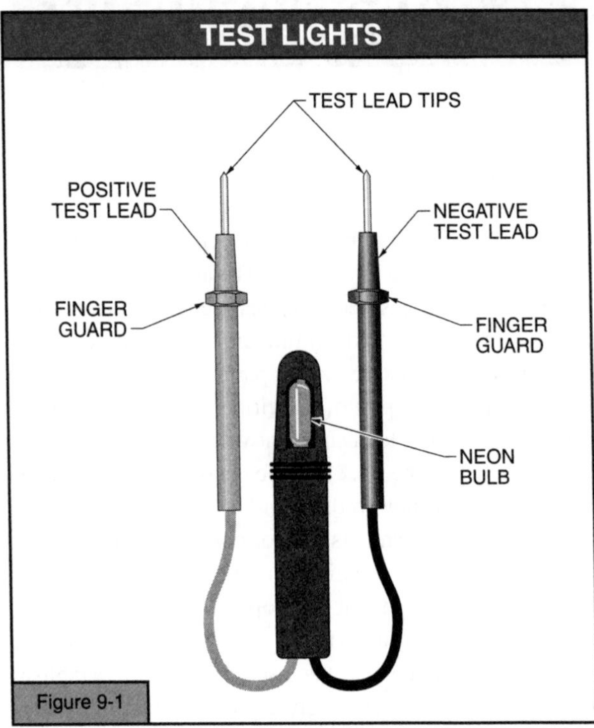

Figure 9-1

Test lights are inexpensive, small enough to carry in a pocket, and easy to use. Test lights however, have a limited voltage-indicating range and cannot determine the actual voltage of a circuit, only that voltage is present in a circuit. Test lights that have a wider voltage range are better than test lights that have only one voltage rating. For example, a neon test light rated for 90 VAC to 600 VAC is better than a test light rated for only 120 VAC. Neon test lights must not be used to test GFCIs or GFIs (ground fault circuit interrupters or ground fault interrupters), because the neon bulb does not draw enough current to trip the GFCI when connected between the hot side of the receptacle and ground.

Test lights are primarily used to determine when voltage is present in a circuit (the circuit is energized), such as when testing receptacles. When testing a receptacle, the test light bulb illuminates when the receptacle is properly wired and energized. **See Figure 9-2.**

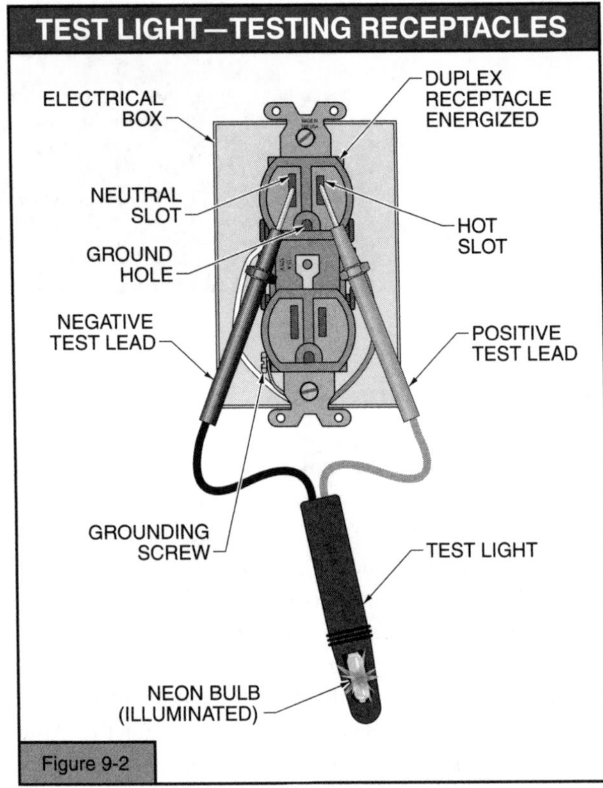

Figure 9-2

When a receptacle is properly wired, a test light bulb illuminates when the test light leads are connected from the neutral slot to the hot slot. A test light bulb also illuminates when the leads are connected from the ground slot to the hot slot. If the test light illuminates when the leads are connected from the neutral slot to the ground slot, the hot (black) and neutral (white) wires are reverse wired. The situation of having the hot and neutral wires reversed is a safety hazard and must be corrected.

If a test light illuminates when the test leads are connected to the neutral slot and hot slot but does not light when connected to the ground slot and the hot slot, the receptacle is not grounded. When a test light illuminates, but is dimmer than when connected between the neutral slot and hot slot, the receptacle has an improper ground (having higher resistance). Improper grounds are also a safety hazard and must be corrected.

Testing receptacles is also possible with a receptacle tester. A *receptacle tester* is a device that is plugged into a standard receptacle to determine if the receptacle is properly wired and energized. **See Figure 9-3.** Some receptacle tester models include a ground fault circuit interrupter (GFCI) or ground fault interrupter (GFI) test button that allows the receptacle tester to be used on GFCI or GFI receptacles.

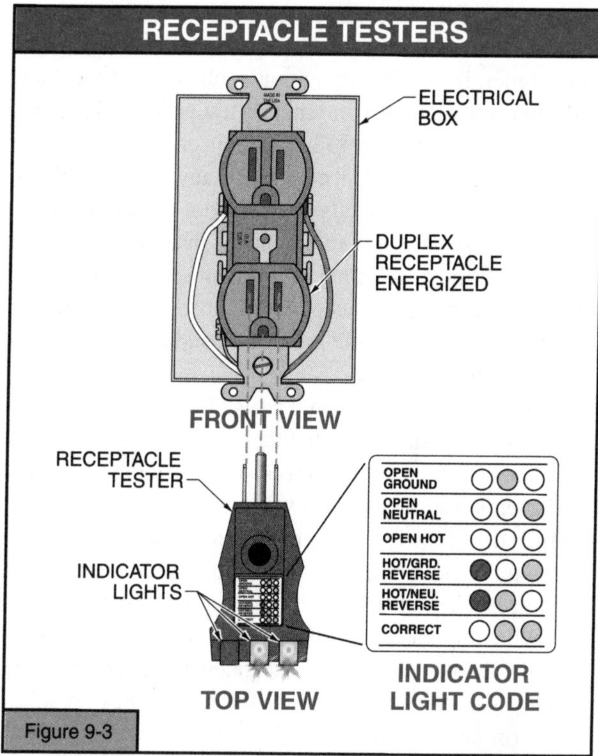

Figure 9-3

Before using a test light or any voltage measuring instrument, always check the test light on a known energized circuit that is within the test light's rating to ensure that the test light is operating correctly. **See Figure 9-4.**

Warning: Always wear proper protective equipment when working around energized circuits. Exercise caution when testing voltages over 24 V.

Before taking any measurements using a test light, ensure the test light is designed to take measurements on the circuit being tested. Refer to the operating manual of the test instrument for all measuring precautions, limitations, and procedures. To test for voltage using a test light, apply the procedure:

1. Verify that the test light has a voltage rating higher than the highest potential voltage in the circuit. Care must be taken to guarantee that the exposed metal tips of the test light leads do not touch fingers or any metal parts not being tested.
2. Connect one test lead of the test light to one side of the circuit or ground. When testing a circuit that has a neutral or ground, connect to the neutral or ground side of the circuit first.
3. Connect the other test lead of the test light to the other side (hot side) of the circuit. Voltage is present when the test light bulb illuminates. Voltage is less than the rating of the test light when the test light is dimly lit and is higher than the rating of the test light when the test light glows brighter than normal. Voltage is not present or present at a very low level when a test light does not illuminate.
4. Remove the test light from the circuit.

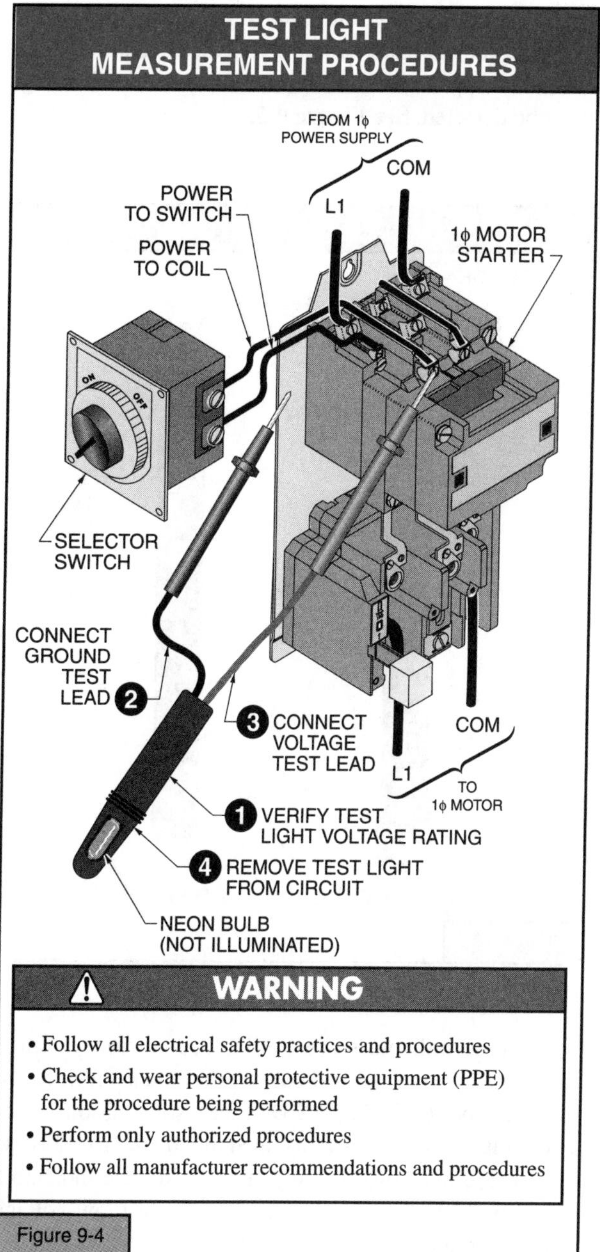

WARNING

- Follow all electrical safety practices and procedures
- Check and wear personal protective equipment (PPE) for the procedure being performed
- Perform only authorized procedures
- Follow all manufacturer recommendations and procedures

Figure 9-4

Warning: When a test light does not illuminate, a voltage can still be present that could cause an electrical shock. A test light can be damaged during testing by too high a voltage, so always retest a test light on a known energized circuit to verify that the bulb of a test light that indicated no voltage is still operating correctly.

Voltage Indicators

A *voltage indicator* is a test instrument that indicates the presence of voltage when the test tip touches, or is near, an energized hot conductor or energized metal part. The tip glows and/or the device creates a sound when voltage is present at the test point. Voltage indicators are used to test receptacles, fuses/circuit breakers, breaks in cables, and other applications in which the presence of voltage must be detected. **See Figure 9-5.**

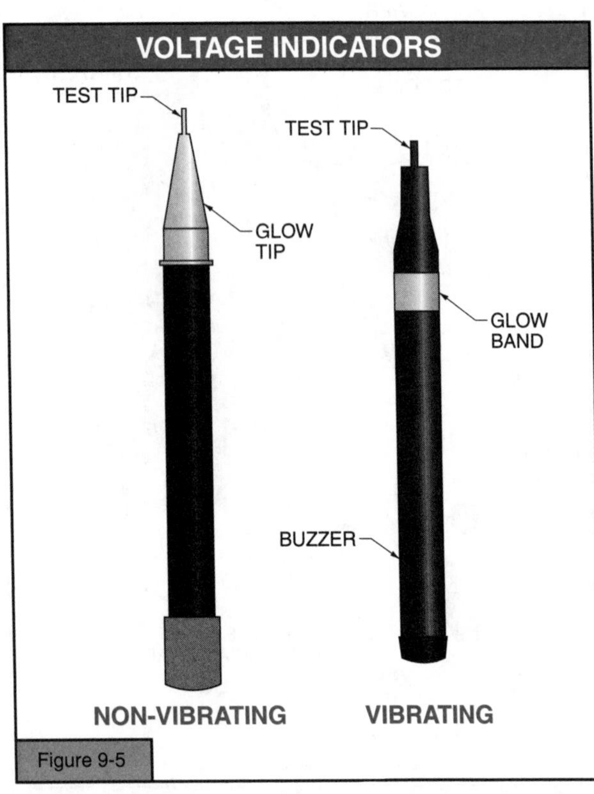

Figure 9-5

Voltage indicators are available in various voltage ranges (a few volts to hundreds of volts) and in the different voltage types (AC, DC, AC/DC) for testing various types of circuits. Voltage indicators rated for 90 VAC to 600 VAC are used to test power-supply circuits. Voltage indicators rated for 24 VAC to 90 VAC are used to test low-voltage control circuits.

Voltage indicators are inexpensive, are small enough to carry in a pocket, are easy to use, are nonconductive, and indicate a voltage without touching any live parts of the circuit, even through conductor insulation. Voltage indicators, however, only indicate that voltage is present but do not indicate the actual voltage amount. In addition, voltage indicators may not provide an indication that voltage is present, even when voltage is present, as when the wire being tested is shielded.

Voltage indicators are used for finding a break in an extension cord or wire, or determining when a receptacle is hot (energized). One important application of a voltage indicator is in making a preliminary test to determine if any metal parts are hot before beginning service on a circuit or component. A fault occurs when any hot (energized) conductor touches a metal part that is not grounded. Faults occur because of a nick in conductor insulation, or when an exposed metal terminal screw contacts another metal part when a switch or receptacle is loose. An ungrounded situation occurs any time a ground conductor is loose or not connected properly.

Before touching any metal parts of a circuit, use a voltage indicator to indicate whether any conductors or metal components are hot (energized). Voltage indicators glow and/or emit sounds when a wire or metal component is energized. The wire or component must be tested with a voltmeter to determine the actual voltage amount. When a voltage indicator does not glow and/or sound, the wire or metal component is probably not hot (energized), but it may be energized when shielded conductors are being used. Voltage indicators must only be used as a quick precheck to determine if a wire or metal component is energized.

Warning: Always follow a voltage indicator test with a test using a voltmeter to verify the voltage indicator measurement any time the voltage indicator indicates no voltage.

Before using a voltage indicator, always check the voltage indicator on a known energized circuit that is within the voltage rating of the voltage indicator to verify proper operation. A voltage indicator is operating properly if there is a steady glow when the tip is placed next to a hot (energized) conductor. The hot side (short slot or black wire) of a standard receptacle is typically used to test voltage indicators. **See Figure 9-6.**

Warning: Always wear proper protective equipment when working around energized circuits. Exercise caution when testing voltages over 24 V.

Before taking any measurements using a voltage indicator, ensure the indicator is designed to take measurements on the circuit being tested. Refer to the operating manual of the test instrument for all measuring precautions, limitations, and procedures. Always wear required personal protective equipment and follow all safety rules when taking the measurement. To test for voltage using a voltage indicator, apply the procedure:

1. Verify that the voltage indicator has a voltage rating higher than the highest potential voltage in the circuit being tested. When circuit voltage is unknown, slowly bring the voltage indicator near the conductor being tested. The voltage indicator glows and/or sounds when voltage is present. The brighter a voltage indicator glows, the higher the voltage or the closer the voltage indicator is to the voltage source.
2. Place the tip of the voltage indicator on or near the wire or device being tested. When testing an extension cord for a break, test several points along the wire. Expect the voltage tester to turn ON and OFF when moved along a cord that has twisted wire conductors, because the hot wire changes position along the cord.
3. When the voltage indicator does not indicate the presence of voltage by glowing or making a sound, do not assume that there is no voltage in a circuit. Always take a second test instrument (voltmeter or multimeter) and measure for the presence of voltage before working around or on exposed wires and electrical components.
4. Remove voltage indicator from circuit.

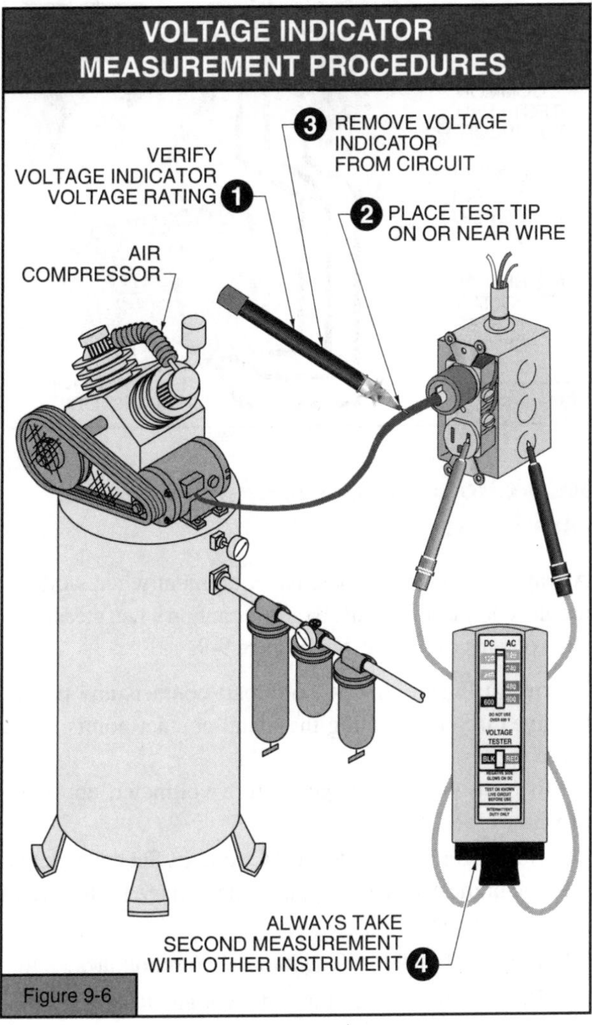

Figure 9-6

Warning: When a voltage indicator does not glow or sound, a voltage that can cause an electrical shock may still be present. Always retest a voltage indicator on a known energized circuit after use, to verify that the voltage indicator is operating properly.

Voltage Measurement — Voltage Testers

A *voltage tester* is an electrical test instrument that indicates the approximate voltage amount and type of voltage (AC or DC) in a circuit by the movement of a pointer (and vibration on some models). When a voltage tester includes a solenoid, the solenoid vibrates when the tester is connected to AC voltage. Some voltage testers include a colored plunger or other indicator such as a light that indicates the polarity of the test leads as positive or negative when measuring a DC circuit. **See Figure 9-7.**

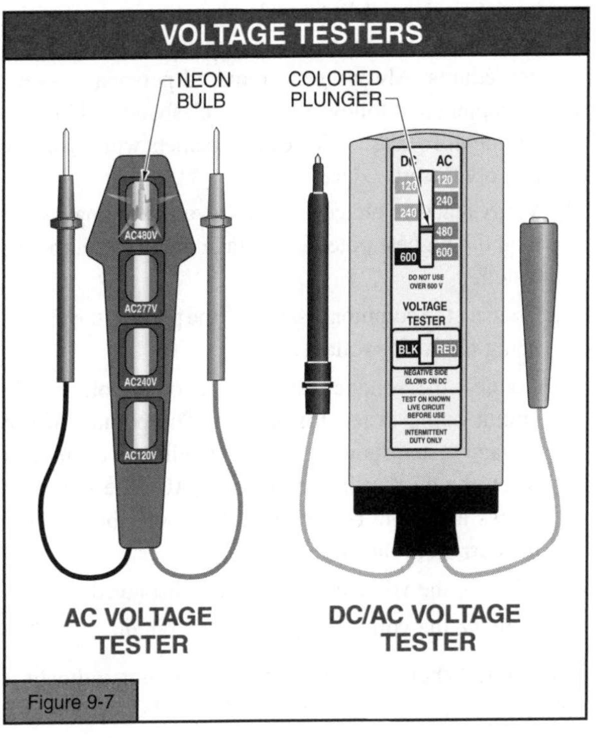

Figure 9-7

Voltage testers have a lower impedance (resistance) than voltage indicators or digital multimeters. GFCIs are designed to trip at approximately 6 mA (.006 A). Because voltage indicators and multimeters draw less than 6 mA, voltage indicators and multimeters cannot be used to trip a GFCI during a test (when connected from the hot receptacle slot to ground). The lower impedance of voltage testers allows voltage testers to be used for testing GFCIs.

Voltage testers are used to take voltage measurements anytime the voltage of a circuit being tested is within the rating of the tester and an exact voltage measurement is not required. Exact voltage measurements are not required to determine when a receptacle is energized (hot), a system is grounded, fuses or circuit breakers are good or bad, or when a system is a 115 VAC, 230 VAC, or 460 VAC circuit. Because test lights, voltage indicators, and voltmeters do not draw enough current to trip a GFCI receptacle, voltage testers are considered the best test instrument for testing GFCI receptacles. Properly wired GFCI receptacles trip when the test button on the receptacle is pressed.

Before using a voltage tester or any voltage measuring instrument, always check the voltage tester on a known energized circuit that is within the voltage rating of the voltage tester to verify proper operation. **See Figure 9-8.**

Warning: Always wear proper protective equipment when working around energized circuits. Exercise caution when testing voltages over 24 V.

Before taking any voltage measurements using a voltage tester, ensure the tester is designed to take measurements on the circuit being tested. Refer to the operating manual of the test instrument for all measuring precautions, limitations, and procedures. Always wear required personal protective equipment and follow all safety rules when taking the measurement. To take a voltage measurement with a voltage tester, apply the procedure:

1. Verify that the voltage tester has a voltage rating higher than the highest potential voltage in the circuit being tested.
2. Connect the common test lead to the point of the circuit being tested for voltage.
3. Connect the voltage test lead to another point in the circuit being tested for voltage. The pointer of the voltage tester shows a voltage reading and vibrates when the current in the circuit is AC. The indicator shows a voltage reading and does not vibrate when the current in the circuit is DC.
4. Observe the voltage measurement displayed.
5. Remove the voltage tester from the circuit.

Warning: When a voltage tester does not indicate a voltage, a voltage that can cause an electrical shock may still be present. A voltage tester can be damaged during testing by excessive high voltages, so always retest a voltage tester on a known energized circuit to verify that the tester that did not indicate a voltage is still operating correctly.

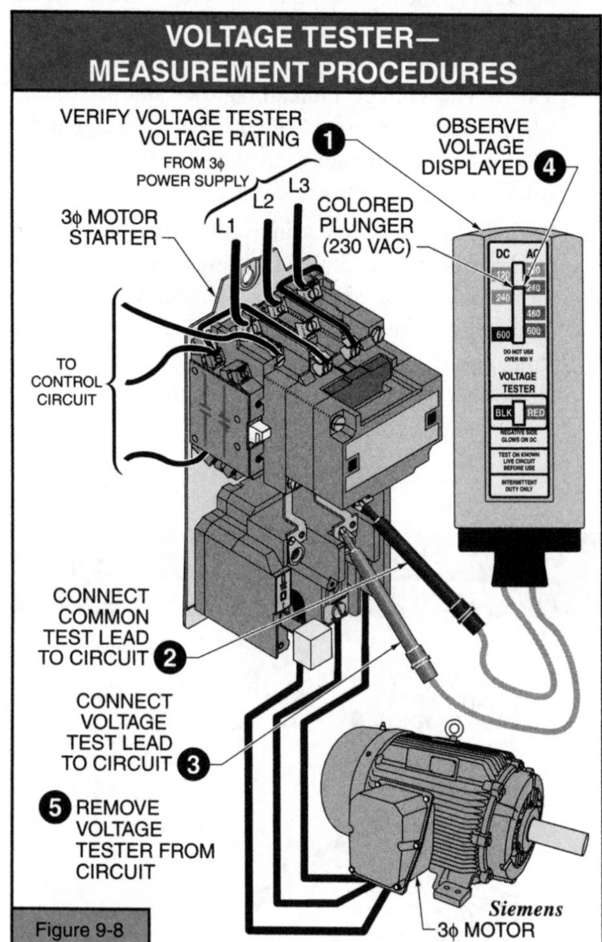

Figure 9-8

AC Voltage Measurement — Digital Multimeters

Always wear proper protective equipment when working around energized circuits. Exercise caution when measuring AC voltages over 24 V. **See Figure 9-9.**

Warning: Ensure that no body part contacts any part of the live circuit, including the metal contact points at the tip of the test leads.

To measure AC voltages with a voltmeter, apply the procedure:

1. Set the function switch to AC voltage. Set the meter on the highest voltage setting if the voltage in the circuit is unknown.
2. Plug the black test lead into the common jack.
3. Plug the red test lead into the voltage jack.

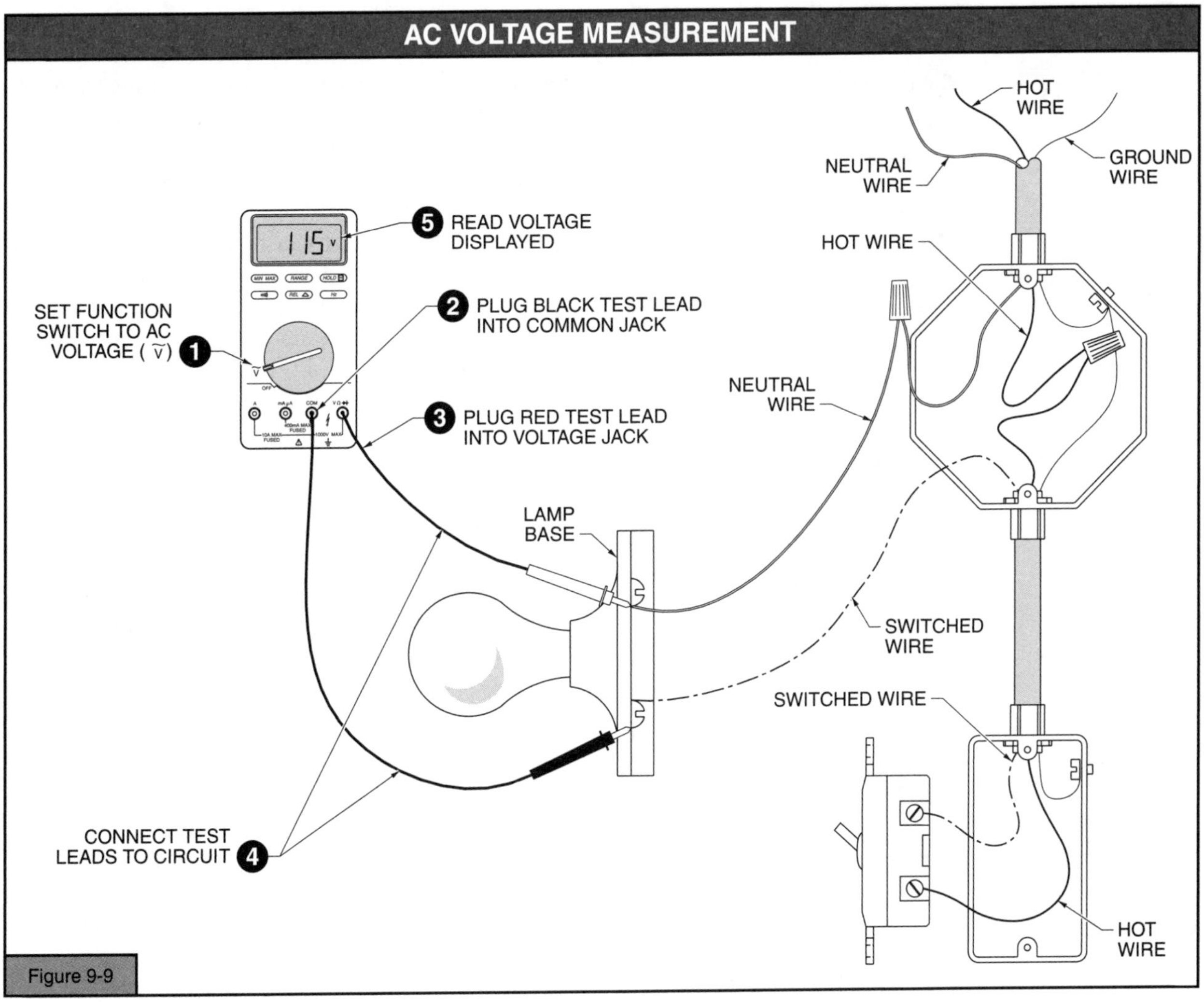

AC VOLTAGE MEASUREMENT

Figure 9-9

4. Connect the test leads to the circuit. The position of the test leads is arbitrary. Common industrial practice is to connect the black test lead to the grounded (neutral) side of the AC voltage.
5. Read the voltage displayed on the meter.

DC Voltage Measurement – Digital Multimeters

Exercise caution when measuring DC voltages over 60 V. **See Figure 9-10.**

Warning: Ensure that no body part contacts any part of a live circuit, including the metal contact points at the tip of the test leads.

To measure DC voltages with a voltmeter, apply the procedure:

1. Set the function switch to DC voltage. Select a setting high enough to measure the highest possible circuit voltage if the meter has more than one voltage position or if the circuit voltage is unknown.
2. Plug the black test lead into the common jack.
3. Plug the red test lead into the voltage jack.
4. Connect the meter test leads in the circuit. Connect the black test lead to circuit ground and the red test lead to the point at which the voltage is under test. Reverse the black and red test leads if a negative sign appears in front of the reading on a digital meter.
5. Read the voltage displayed on the meter.

Exceeding the limits of a meter can damage the meter and may also cause personal injury. A meter that is connected to higher-than-rated voltages and/or misused may explode.

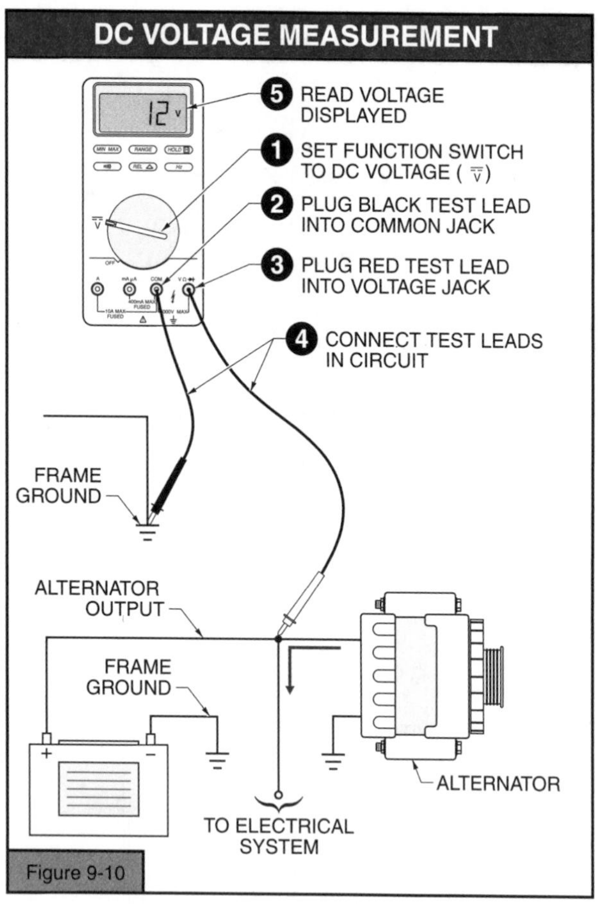

Figure 9-10

Current

Current (I) is the amount of electrons flowing through an electrical circuit. Current is measured in amperes (A). The amount of current in a circuit can be measured using an ammeter. An ammeter is used to measure the amount of current flowing in a circuit that is powered. Safety precautions must be taken to ensure that no part of the body comes in contact with a live part of the circuit. Safety precautions must also be taken to ensure that the ammeter is properly used because current measurements are taken when a circuit is powered.

Amperage measurements are normally taken to indicate the amount of circuit loading or the condition of a load. Every component (lamp, motor, heating element, speaker, etc.) that converts electrical energy into some other form of energy (light, rotating motion, heat, sound, etc.) uses current. The more energy required, the higher the current usage. Thus, every time a load is added to a circuit (or switched ON), the circuit must deliver more current. Current limits must be set and monitored because all electrical components (conductors, switches, motors, etc.) have a rated current limit. Current limits are normally set by the proper sizing of fuses, circuit breakers, overloads, and other types of current monitors.

Measuring the amount of current in a circuit is the best indication of the amount of loading of a circuit or component. Circuit current is measured using an in-line ammeter or a clamp-on ammeter. An *in-line ammeter* is a meter that measures current in a circuit by inserting the meter in series with the component(s) under test. In-line ammeter readings require the circuit to be opened so the ammeter can be inserted. A *clamp-on ammeter* is a meter that measures current in a circuit by measuring the strength of the magnetic field around a conductor. Clamp-on ammeters take current readings without opening the circuit. In-line ammeters are used to measure very small amounts of current (milli and microamperes) and are normally limited to measuring currents of less than 10 A. Clamp-on ammeters are used to measure currents from .01 A or less to 1000 A or more. In-line ammeters are normally used to measure DC currents, but some models allow for DC or AC current measurements. Clamp-on ammeters are normally used to measure AC currents, but some models allow for AC or DC current measurements. Always check the type (AC and/or DC) and limits (maximum measurable amount) of the meter being used.

In-Line Ammeter — DC Measurement

Care must be taken to protect the meter, circuit, and person using the meter when measuring DC with an in-line ammeter. Always apply the following rules when using an in-line ammeter:

- Always wear proper protective equipment when working around energized circuits.
- Check that the power to the test circuit is OFF before connecting and disconnecting test leads.
- Do not change the position of any switches or settings on the meter while the circuit under test is energized.
- Turn the power to the meter and circuit OFF before any meter settings are changed.
- Connect the ammeter in series with the component(s) to be tested.
- Do not take current readings from any circuit in which the current may exceed the limit of the meter.

Many meters include a fuse in the low-ampere range to prevent meter damage caused by excessive current. Before using a meter, check to see if the meter is fused on the current range being used. The meter is marked as fused or not fused at the test lead terminals. An external fuse may be connected in series with the meter test leads if the meter is not fused. To protect the meter, the fuse rating should not exceed the current range of the meter. **See Figure 9-11.**

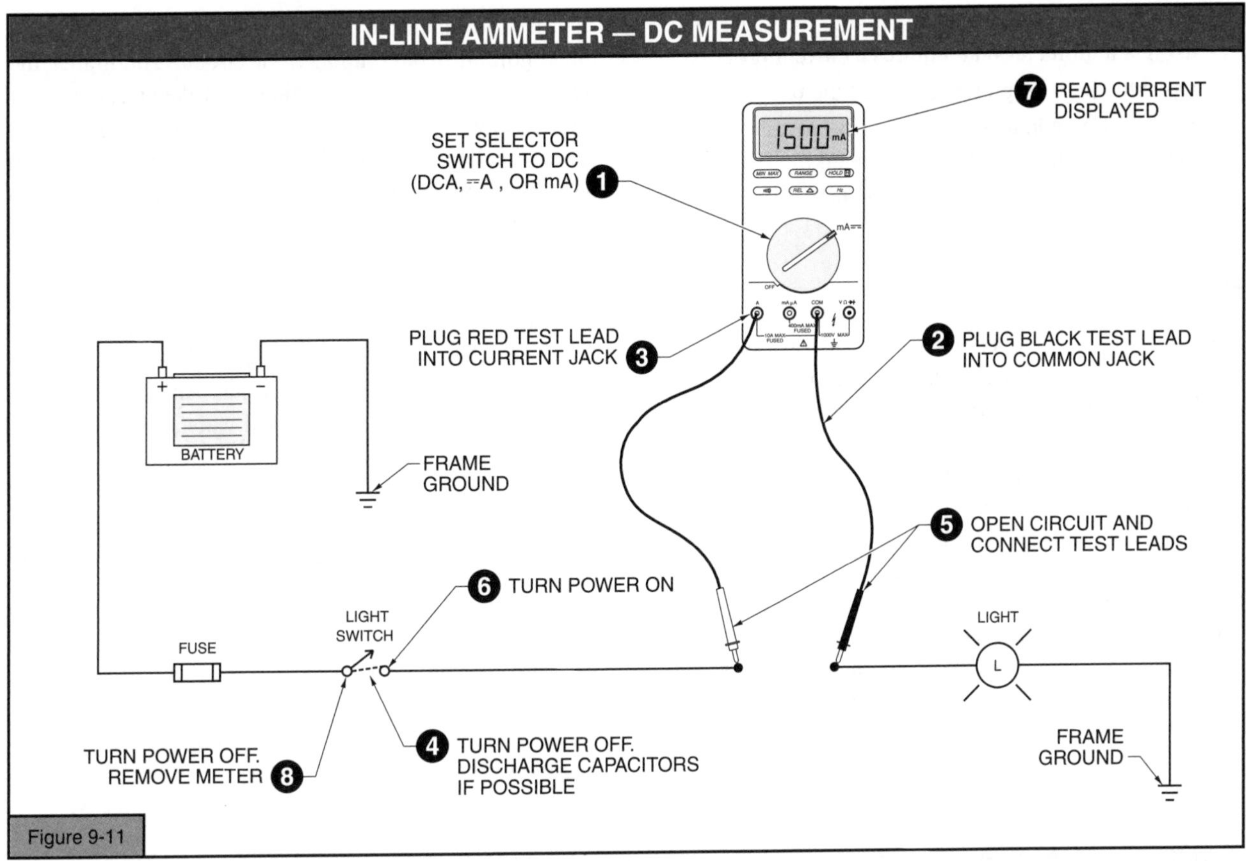

Figure 9-11

Warning: Ensure that no body parts contact any part of the live circuit, including the metal contact points at the tip of the test leads.

To measure DC using an in-line ammeter, apply the procedure:

1. Set the selector switch to DC. Select a setting high enough to measure the highest possible circuit current if the meter has more than one DC position.
2. Plug the black test lead into the common jack. The common jack may be marked com (common), – (negative), or lo (low).
3. Plug the red test lead into the current jack. The current jack may be marked + (positive), mA (milliamps), or hi (high).
4. Turn the power to the circuit or device under test OFF and discharge all capacitors if possible.
5. Open the circuit and connect the test leads to each side of the opening. The black (negative) test lead is connected to the negative side of the opening and the red (positive) test lead is connected to the positive side of the opening. Reverse the black and red test leads if a negative sign appears in front of the displayed reading.
6. Turn the power to the circuit under test ON.
7. Read the current displayed on the meter.
8. Turn the power OFF and remove the meter from the circuit.

The same procedure is used to measure AC with an in-line ammeter, except that the selector switch is set on AC current.

Clamp-On Ammeter — AC Measurement

Clamp-on ammeters measure the current in a circuit by measuring the strength of the magnetic field around a conductor. **See Figure 9-12.**

Warning: Care must be taken to ensure that the meter does not pick up stray magnetic fields. Whenever possible, separate the conductors under test from other surrounding conductors by a few inches.

To measure AC using a clamp-on ammeter, apply the procedure:

1. Set the function switch to AC current. Select the proper setting to measure the highest possible circuit current if the meter has more than one current position or if the circuit current is unknown.

2. Plug the current probe accessory into the meter when using a multimeter that requires a current probe.
3. Open the jaws by pressing against the trigger.
4. Enclose one conductor in the jaws. Ensure that the jaws are completely closed before taking readings.
5. Read the current displayed on the meter.

Resistance

Resistance (R) is the opposition to current flow. Resistance is measured in ohms (Ω). Resistance can be measured using an ohmmeter or megohmmeter. An *ohmmeter* is a device that is used to measure the amount of resistance in a component (or circuit) that is not powered. A *megohmmeter* is a device that detects insulation deterioration by measuring high resistance values under high test voltage conditions. Megohmmeter test voltages range from 50 V to 5000 V. Precautions must be taken to ensure that all circuit power is turned OFF before any measurements are taken because resistance readings are taken on circuits that are not powered. This can be done by removing the component to be tested from the circuit or using a voltmeter to ensure the power is OFF.

Resistance measurements are normally taken to indicate the condition of a component or circuit. The higher the resistance, the lower the current flow. Likewise, the lower the resistance, the higher the current flow. Components that are designed to insulate (rubber, paper, plastic, etc.) should have a very high resistance. Components that are designed to conduct (switch contacts, conductors, etc.) should have a very low resistance. As insulators are damaged by moisture, overheating, etc., their resistance decreases. As conductors are damaged by burning, corrosion, etc., their resistance increases. Other components, such as heating elements, should have a fixed resistance value. Any major change in the resistance value of a fixed resistance component normally indicates a problem.

Ohmmeter — Resistance Measurement

An ohmmeter measures resistance with all power to the circuit OFF. Low voltage applied to a meter set to measure resistance causes inaccurate readings. High voltage applied to a meter set to measure resistance causes meter damage. Check for voltage using a voltmeter. **See Figure 9-13.**

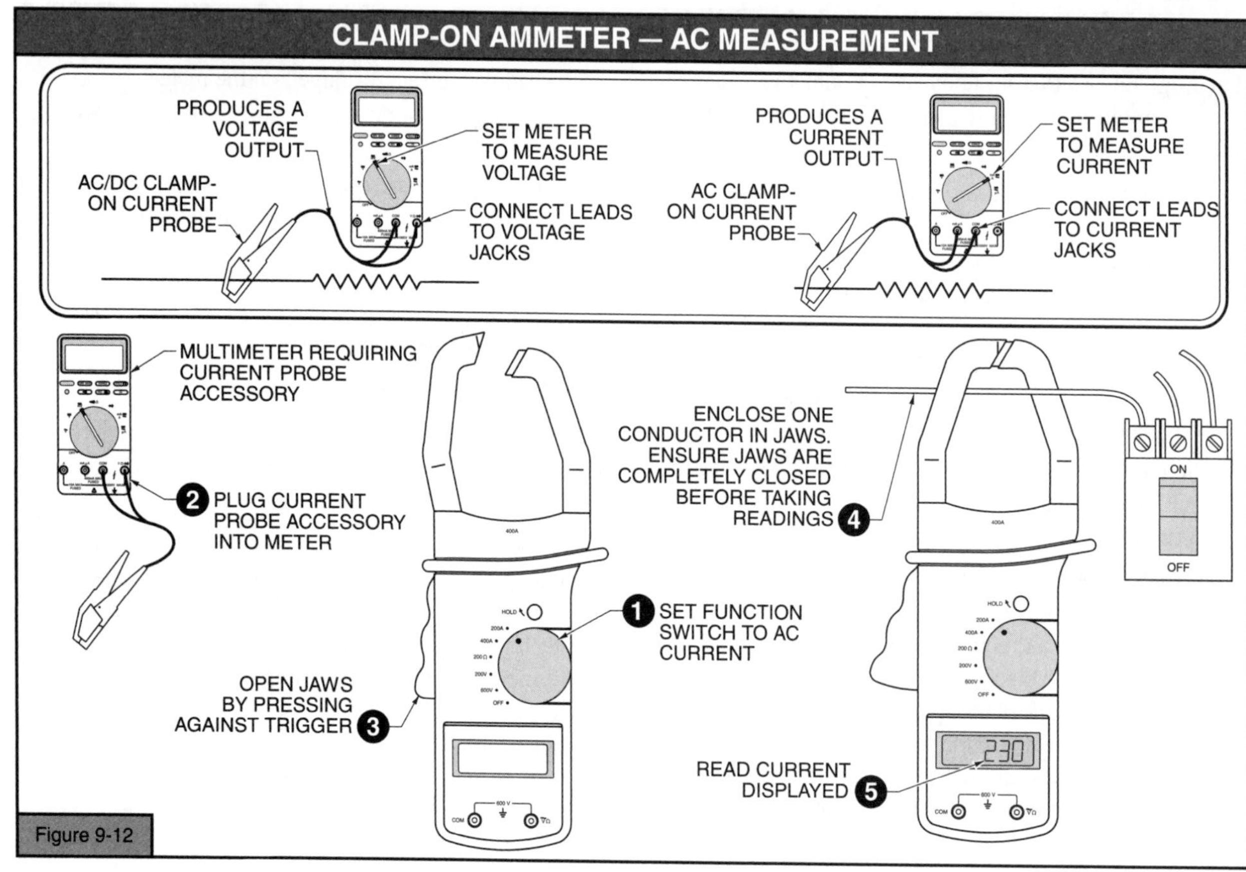

Figure 9-12

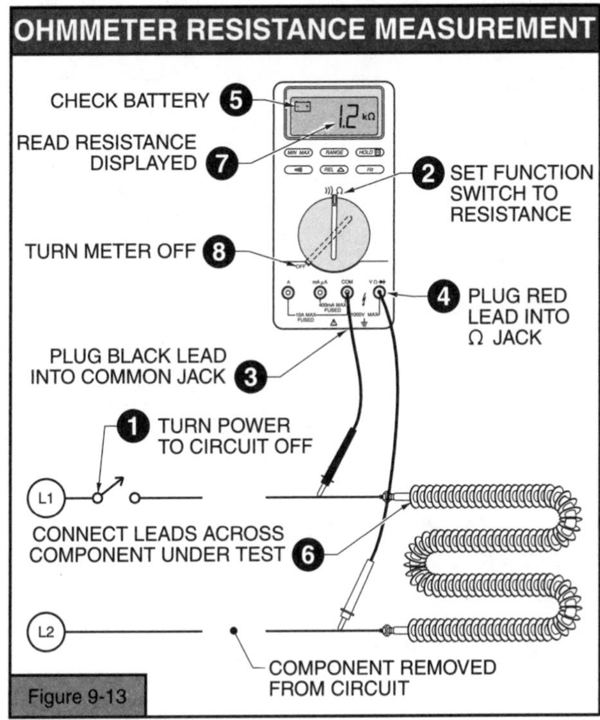

Figure 9-13

Warning: Ensure that no voltage is present in the circuit or component under test before taking resistance measurements.

To measure resistance using an ohmmeter, apply the procedure:

1. Check to ensure that all power is OFF in the circuit or component under test and disconnect the component from the circuit.
2. Set the function switch to the resistance position, which is marked Ω on digital meters.
3. Plug the black test lead into the common jack.
4. Plug the red test lead into the resistance jack.
5. Ensure that the meter batteries are in good condition. The battery symbol is displayed when the batteries are low. Digital meters are zeroed by an internal circuit.
6. Connect the meter test leads across the component under test. Ensure that contact between the test leads and the circuit is good. Dirt, solder flux, oil, and other foreign substances greatly affect resistance readings.
7. Read the resistance displayed on the meter. Check the circuit schematic for parallel paths. Parallel paths with the resistance under test cause reading errors. Do not touch exposed metal parts of the test leads during the test. The resistance of a person's body can cause reading errors.
8. Turn the meter OFF after measurements are taken to save battery life.

Megohmmeter — Resistance Measurement

Megohmmeters deliver a high voltage to the circuit under test. A megohmmeter detects insulation failure or potential failure of insulation caused by excessive moisture, dirt, heat, cold, corrosive substances, vibration, and aging. **See Figure 9-14.**

Warning: Ensure that no voltage is present in a circuit or component under test before taking any resistance measurements. Ensure that no body part contacts the high voltage.

To use a megohmmeter, apply the procedure:

1. Ensure that all power is OFF in the circuit or component under test. Test for voltage using a voltmeter.
2. Set the selector switch to the voltage at which the circuit is to be tested. The test voltage should be as high or higher than the highest voltage to which the circuit under test is exposed.
3. Plug the black test lead into the negative (earth) jack.
4. Plug the red test lead into the positive (line) jack.
5. Ensure that the batteries are in good condition. The meter contains no batteries if the meter includes a crank. The meter contains no batteries or crank if the meter plugs into a standard outlet.
6. Connect the line test lead to the conductor under test.
7. Connect the earth test lead to a second conductor in the circuit or earth ground.
8. Press the test button or turn the crank and read the resistance displayed on the meter. Change the meter resistance or voltage range if required.
9. Consult the equipment manufacturer or meter manufacturer for the minimum recommended resistance values. The insulation is good if the meter reading is equal to or higher than the minimum value.

Temperature

Temperature is the measurement of the intensity of heat. Temperature is normally measured in degrees Fahrenheit (°F) or degrees Celsius (°C). The amount of temperature at a location can be measured using a contact thermometer or infrared meter. A *contact thermometer* is an instrument that measures temperature at a single point. An *infrared meter* is a meter that measures heat energy by measuring the infrared energy that a material emits. All materials emit infrared energy in proportion to their temperature. **See Figure 9-15.**

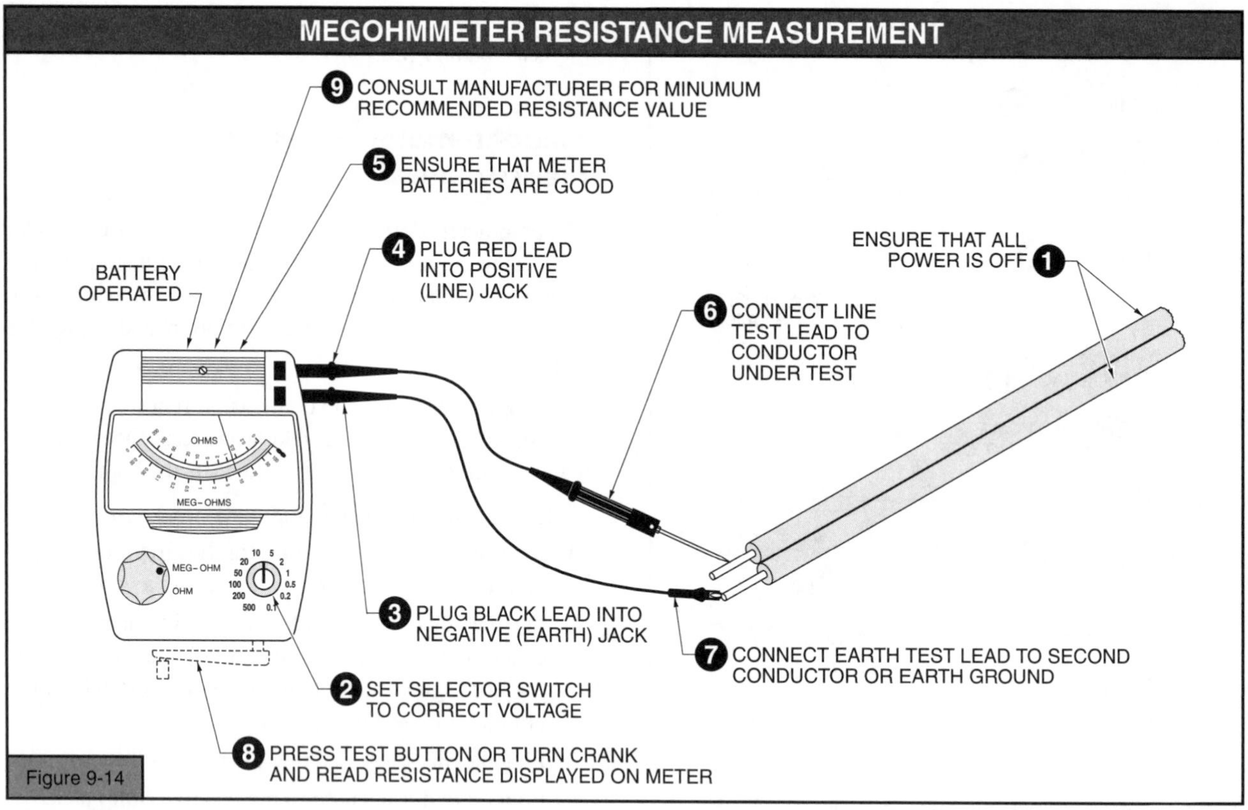

Figure 9-14

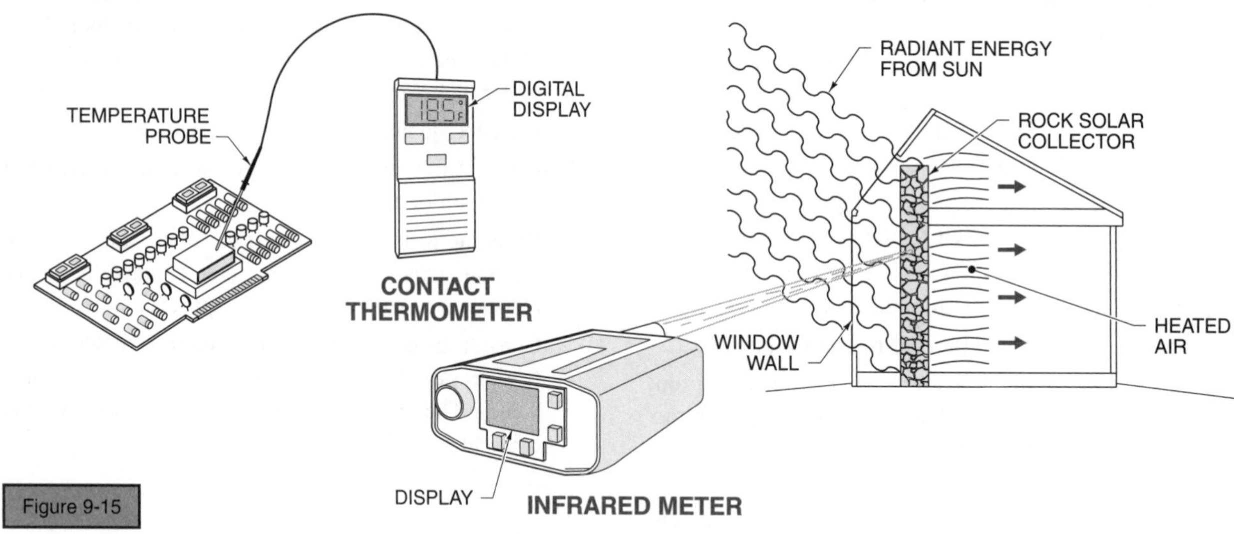

Figure 9-15

Contact Thermometer Temperature Measurement

Measuring hot devices presents a safety hazard. Ensure that proper equipment is used and that safety precautions are taken. Use a temperature probe that is rated higher than the highest possible temperature and always wear safety glasses. **See Figure 9-16.**

Warning: Avoid contact with any material that can cause burns.

To measure temperature using a contact thermometer, apply the procedure:

1. Select a temperature probe. A probe must have a higher temperature rating than the highest temperature it may contact and should have a shape that allows good contact with the device under test.

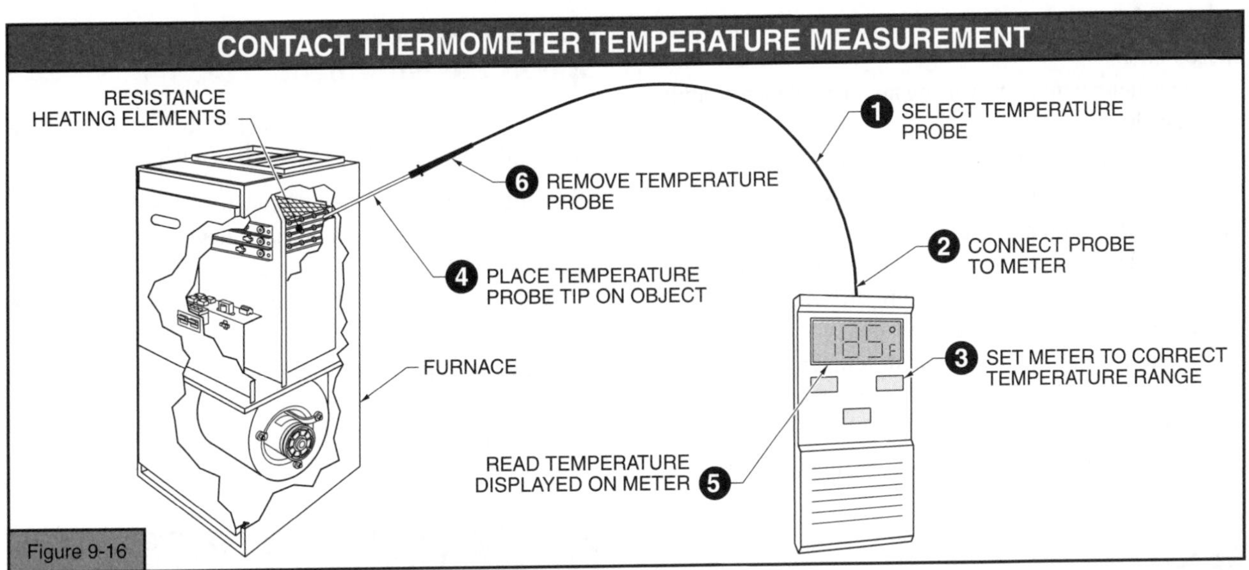

Figure 9-16

2. Connect the temperature probe to the meter.
3. Set the meter to the correct temperature range. Select the highest range if the temperature is unknown.
4. Place the temperature probe tip on the object or in the area to be measured.
5. Read the temperature displayed on the meter.
6. Remove the temperature probe from the object or area under test.

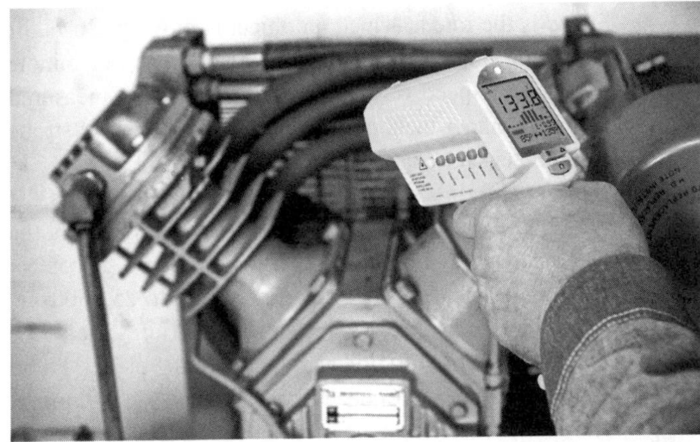
Fluke Corporation

Noncontact thermometers are used by maintenance technicians to locate problems such as misaligned or overheated equipment, lubrication problems, and overloads.

Infrared Meter Temperature Measurement

An infrared meter displays an image on a screen that shows different temperatures indicated by different colors or gives a direct digital temperature readout. **See Figure 9-17.**

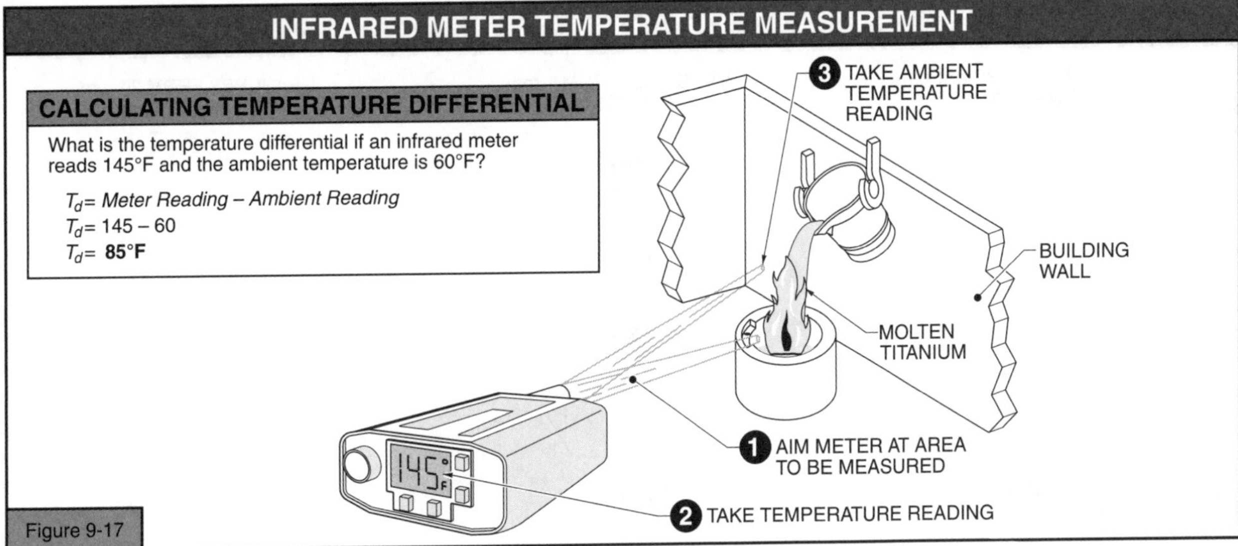

Figure 9-17

Warning: Avoid contact with any material that can cause burns.

To measure temperature using an infrared meter, apply the procedure:

1. Aim meter at area to be measured. Focus meter based on the distance the object is from the meter.
2. Take the temperature reading of any areas suspected to have temperatures above ambient temperature. To determine temperature differential between ambient temperature and the area of increased temperature, subtract the ambient temperature reading from the reading obtained for the area of increased temperature.
3. Take the ambient temperature reading for reference.

Speed

Speed is the rate at which an object is moving. Speed of a rotating object is measured in revolutions per minute (rpm). The speed of a rotating object can be measured using a contact tachometer, photo tachometer, or strobe tachometer. **See Figure 9-18.**

A *contact tachometer* is a device that measures the rotational speed of an object through direct contact of the tachometer tip with the object to be measured. A contact tachometer measures speeds from .1 rpm to 25,000 rpm. A *photo tachometer* is a device that measures the speed of an object without direct contact with the object. Photo tachometers measure speeds from 1 rpm to 99,999 rpm. A *strobe tachometer* is a device that uses a flashing light to measure the speed of a moving object. Strobe tachometers measure speeds from 20 rpm to 1,000,000 rpm.

Contact Tachometer Speed Measurement

A contact tachometer must be placed in contact with the shaft of the rotating object. **See Figure 9-19.**

Warning: Exercise caution when working around moving objects. Use a photo or strobe tachometer if there is danger of contact with a moving object.

To measure speed with a contact tachometer, apply the procedure:

1. Place the tip of the tachometer in direct contact with the moving object.
2. Read the speed displayed on the meter.

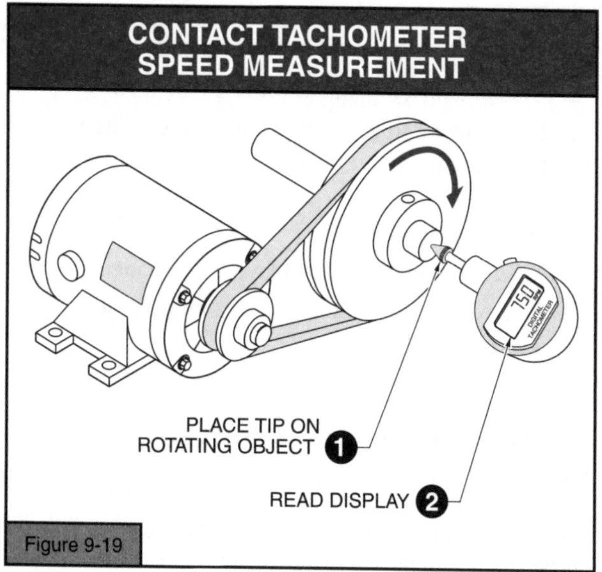

Figure 9-19

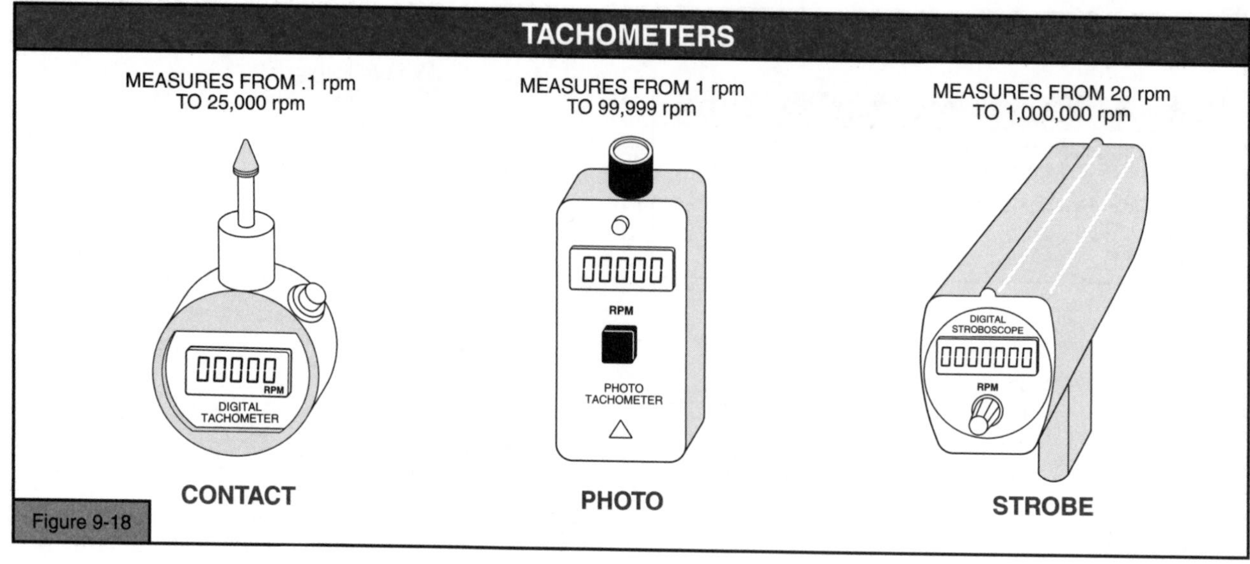

Figure 9-18

Photo Tachometer Speed Measurement

A photo tachometer measures speed by focusing a light beam on a reflective area or a section of reflective tape. **See Figure 9-20.**

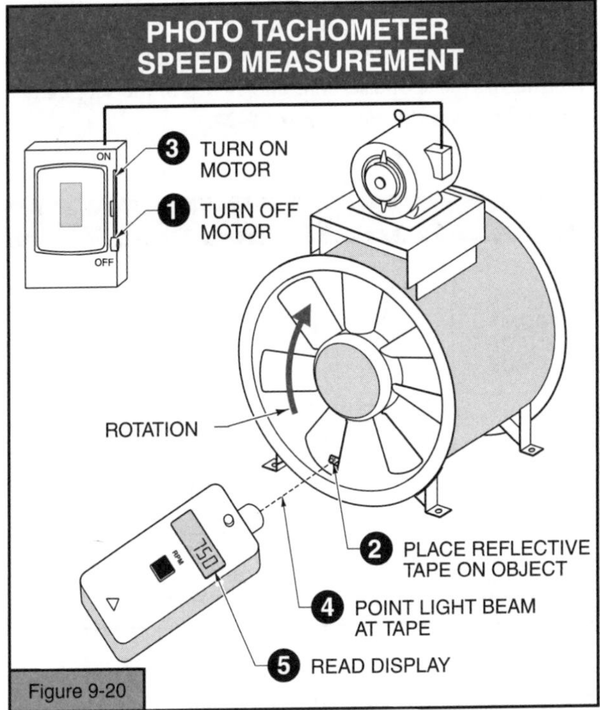

Figure 9-20

Warning: Exercise caution when working around moving objects. Turn OFF, lock out, and tag disconnect switches of motors and rotating machinery as required. Use a photo or strobe tachometer if there is danger of contact with a moving object.

To measure speed with a photo tachometer, apply the procedure:

1. Turn OFF, lock out, and tag disconnect switches of motors or other rotating machinery.
2. Place reflective tape on the object to be measured.
3. Turn ON the motor of machinery.
4. Point the light beam of the photo tachometer at the reflective tape.
5. Read the speed displayed on the meter.

Strobe Tachometer Speed Measurement

A strobe tachometer measures speed by synchronizing its light's flash rate with the speed of the moving object. **See Figure 9-21.**

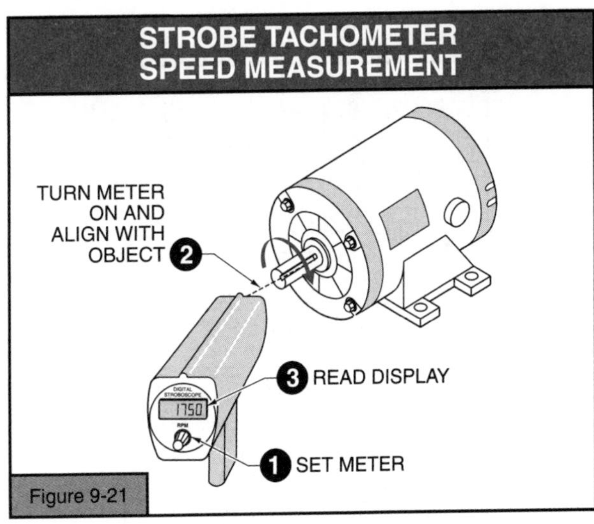

Figure 9-21

Warning: Exercise caution when working around moving objects. Use a photo or strobe tachometer if there is danger of contact with a moving object.

To measure speed with a strobe tachometer, apply the procedure:

1. Set the meter for the best speed range for the application.
2. Turn the tachometer ON and align the visible light beam with the object to be measured.
3. Read the speed displayed on the meter.

Scopes

A graphical display of circuit voltage has more meaning than a numerical value in some troubleshooting situations. Circuits that include rapidly fluctuating signals, stray signals, and phase shifts are easily detected on a graphical display.

A *scope* is a device that gives a visual display of voltages. A scope shows the shape of a circuit's voltage and allows the voltage level, frequency, and phase to be measured. The two basic types of scopes used in troubleshooting are oscilloscopes and scopemeters. **See Figure 9-22.**

Oscilloscopes

An *oscilloscope* is an instrument that displays an instantaneous voltage. An oscilloscope is used to display the shape of a voltage waveform when bench testing electronic circuits. *Bench testing* is testing performed when equipment under test is brought to a designated service area. Oscilloscopes are used to troubleshoot digital circuits, communication circuits, TVs, VCRs, and computers. Oscilloscopes are available in basic and specialized types that can display different waveforms simultaneously.

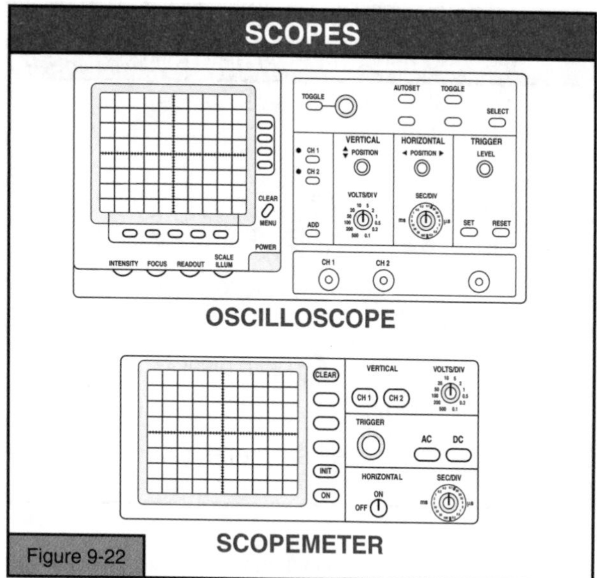

Figure 9-22

Scopemeters

A *scopemeter* is a combination oscilloscope and digital multimeter. A scopemeter is used to display the shape of a voltage waveform when troubleshooting circuits in the field. A scopemeter is portable and can be used as a multimeter and a scope. A scopemeter does not have all the features of a specialized oscilloscope.

Scope Displays

A scope displays the voltage under test on the scope screen. The scope screen contains horizontal and vertical axes. The horizontal (x) axis represents time. The vertical (y) axis represents the amplitude of the voltage waveform.

The horizontal and vertical lines divide the screen into equal divisions. The divisions help to measure the voltage level and frequency of the displayed waveforms. **See Figure 9-23.**

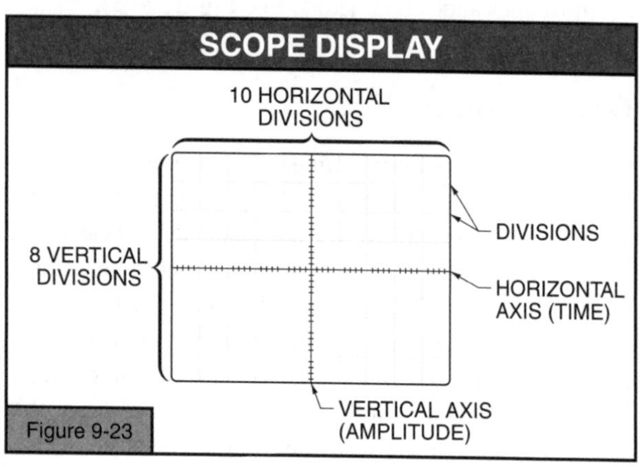

Figure 9-23

Scope Trace

A trace is established on the screen before a circuit under test is connected. A *trace* is a reference point/line that is visually displayed on the face of the scope screen. The trace is normally positioned over the horizontal center line on the screen.

The starting point of the trace is located near the left side of the screen. *Sweep* is the movement of the displayed trace across the scope screen. The sweep of the scope trace is from left to right. **See Figure 9-24.**

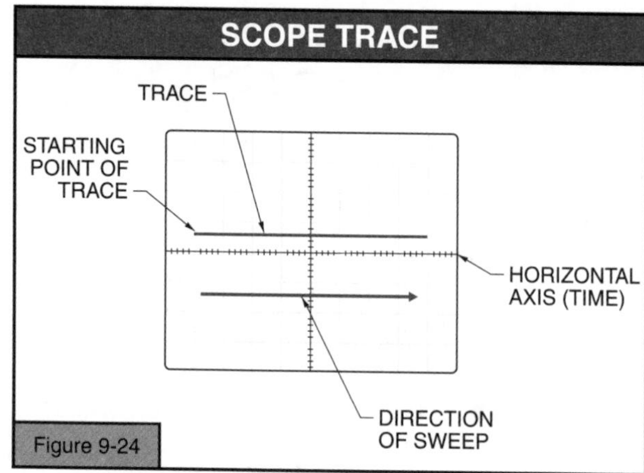

Figure 9-24

Manually-Operated Controls

Manually-operated controls are adjusted to view a waveform. Typical manually-operated scope adjustment controls include intensity, focus, horizontal positioning, vertical positioning, volts/division, and time/division.

Intensity is the level of brightness. The intensity control sets the level of brightness on the displayed voltage trace. The intensity level is kept as low as possible to keep the trace in focus. The focus control adjusts the sharpness of the displayed voltage trace.

The horizontal control adjusts the left and right positions of the displayed voltage trace. The horizontal control sets the starting point of the trace. The vertical control adjusts the up and down positions of the displayed voltage trace. **See Figure 9-25.**

The volts/division (volts per division) control selects the height of the displayed waveform. The setting determines the number of volts each horizontal screen division represents. For example, if a waveform occupies 4 divisions and the volts/division control is set at 20, the peak-to-peak voltage (V_{p-p}) equals 80 V (4 × 20 = 80 V). Eighty volts peak-to-peak equals 40 V peak (V_{max}) (80 ÷ 2 = 40 V). Forty volts peak equals 28.28 V_{rms} (40 ×.707 = 28.28 V). **See Figure 9-26.**

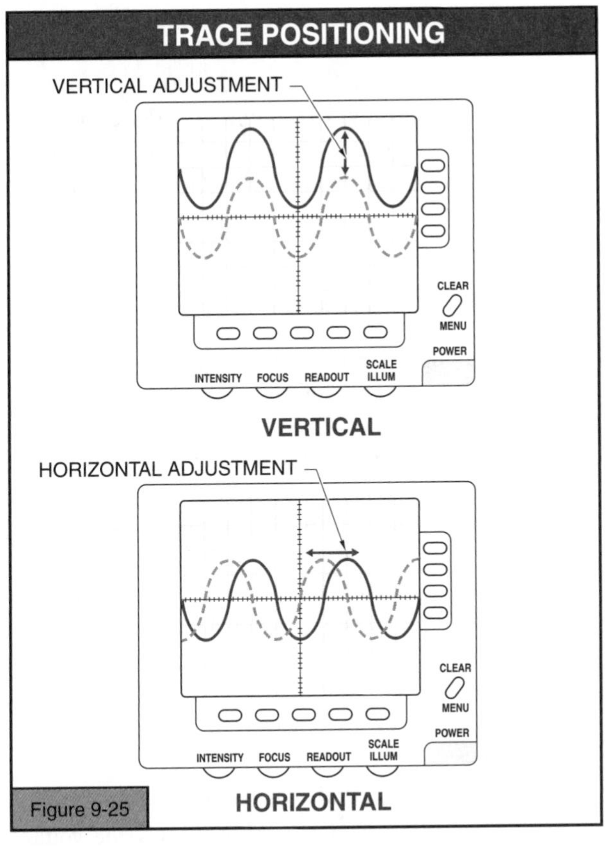

Figure 9-25

The time/division (time per division) control selects the width of the displayed waveform. The setting determines the length of time each cycle takes to move across the screen. For example, if the time/division control is set at 10, each vertical screen division equals 10 ms (milliseconds). If one cycle of a waveform equals 4 divisions, the displayed time equals 40 ms (4 × 10 = 40 ms). **See Figure 9-27.** To determine the frequency of the displayed waveform, first calculate the period of the waveform and then calculate the frequency. To calculate the frequency of a waveform, apply the procedure:

1. Calculate time period.

   To calculate the time period, multiply the number of divisions by the time/division setting.

2. Calculate frequency. To calculate frequency, apply the formula:

   $$f = \frac{1}{T}$$

   where

   f = frequency (in Hz)

   1 = constant (for reciprocal relation between f and T)

   T = time period (in sec)

VOLTS PER DIVISION CONTROL

CALCULATING PEAK-TO-PEAK VOLTAGE

What is the peak-to-peak voltage when a waveform occupies 4 divisions and the volts/division control is set on 20?

$V_{p\text{-}p}$ = *Divisions* × *Volts/Division*

$V_{p\text{-}p}$ = 4 × 20

$V_{p\text{-}p}$ = **80 V**

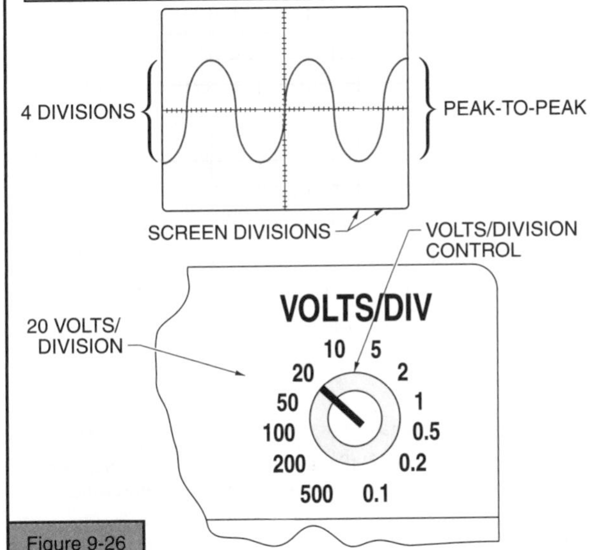

4 DIVISIONS

PEAK-TO-PEAK

SCREEN DIVISIONS

VOLTS/DIVISION CONTROL

20 VOLTS/ DIVISION

VOLTS/DIV

10 5 2 1 0.5 0.2 0.1 500 200 100 50 20

Figure 9-26

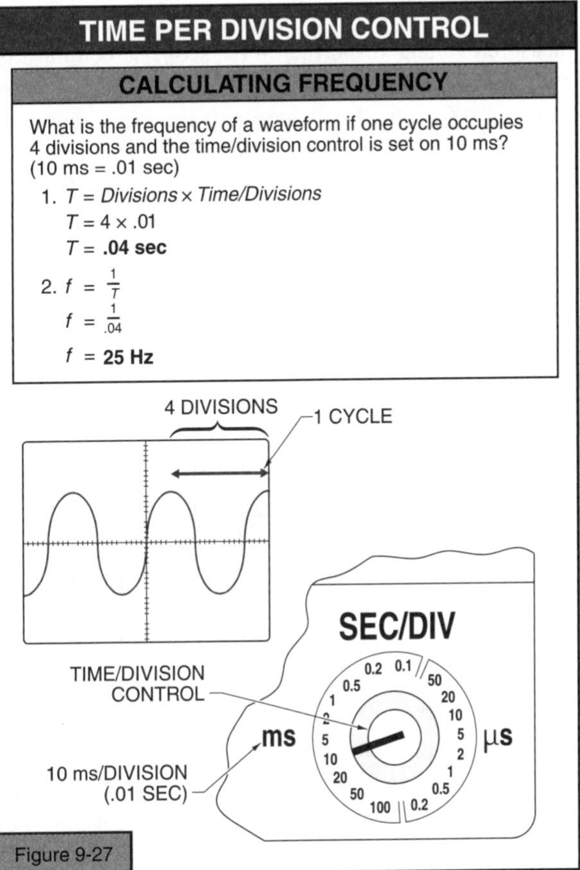

Figure 9-27

Example: Calculating Frequency

What is the frequency of a waveform if one cycle of the waveform occupies 4 divisions and the time/division control is set on 10 ms (10 ms = .01 sec).

1. Calculate time period.

   $T = 4 \times .01$

   $T = .04$ sec

2. Calculate frequency.

   $f = \frac{1}{T}$

   $f = \frac{1}{.04}$

   $f = \mathbf{25\ Hz}$

Scope Measurement — AC Voltage

A scope is connected in parallel with a circuit or component under test. The scope is connected by a probe on the end of a test lead. A 1X probe (1 to 1) is used to connect the input of the scope to the circuit under test when the test voltage is lower than the voltage limit of the scope.

A 10X probe (10 to 1) is used to divide the input voltage by 10. The scope voltage limit equals 10 times the normal rated voltage when a 10X probe is used. The amount of measured voltage displayed on the scope screen must be multiplied by 10 to obtain the actual circuit voltage when using a 10X probe. For example, if the measured scope voltage is 25 V when using a 10X probe, the actual circuit voltage is 250 V (25 V × 10 = 250 V). **See Figure 9-28.** To use a scope to measure AC voltage, apply the procedure:

1. Turn the power switch ON and adjust the trace brightness on the screen.
2. Set the AC/DC control switch to AC.
3. Set the volts/division control to display the voltage level under test. Set the control to the highest value if the voltage level is unknown.
4. Connect the scope probe to the AC voltage under test.
5. Adjust the volts/division control to display the full waveform of the voltage under test.
6. Set the time/division control to display several cycles of the voltage under test.
7. Adjust the waveform by using the vertical control to set the lower edge of the waveform on one of the lower lines.
8. Measure the vertical amplitude of the waveform by counting the number of divisions displayed (Vp-p).

To calculate V_{rms}, first calculate V_{p-p}. V_{p-p} is calculated by multiplying the number of divisions by the volts/division setting. For example, if a waveform occupies 4 divisions and the volts/division setting is 10, V_{p-p} equals 40 V (4 × 10 = 40 V). V_{max} equals 20 V (40 ÷ 2 = 20 V).

V_{max} is multiplied by .707 to find V_{rms}. For example, if V_{max} is 20 V, V_{rms} equals 14.14 V (20 V × .707 = 14.14 V). V_{rms} is the value of the voltage under test as measured on a voltmeter.

Scope Measurement — Frequency

In AC applications such as variable frequency motor drives, it may be necessary to measure the frequency in the circuit. A frequency meter should be used in applications where high accuracy is required. A scopemeter gives an adequate reading in most frequency measurement applications. A scopemeter also shows any distortion present in the circuit under test.

To measure frequency, the scope probes are connected in parallel with the circuit or component under test. **See Figure 9-29.** To use a scope to measure frequency, apply the procedure:

1. Turn the power switch ON and adjust the trace brightness.
2. Set the AC/DC control switch to AC.
3. Set the volts/division control to display the voltage level under test. Set the control to the highest value if the voltage level is unknown.
4. Connect the scope probe to the AC voltage under test.
5. Adjust the volts/division control to display the vertical amplitude of the waveform under test.
6. Set the time/division control to display approximately two cycles of the waveform under test.
7. Set the vertical control so that the center of the waveform is on the centerline of the scope screen.
8. Set the horizontal control so that the start of one cycle of the waveform begins at the vertical centerline on the scope screen.
9. Measure the number of divisions between the start point and end point of one cycle.

To determine frequency, multiply the number of divisions by the time/division setting. This value is the time period for one cycle. To determine the frequency of the waveform, divide the time period by 1.

TECH FACT — *A true RMS meter is the only type of digital meter that can provide an accurate measurement of nonsinusoidal waveforms.*

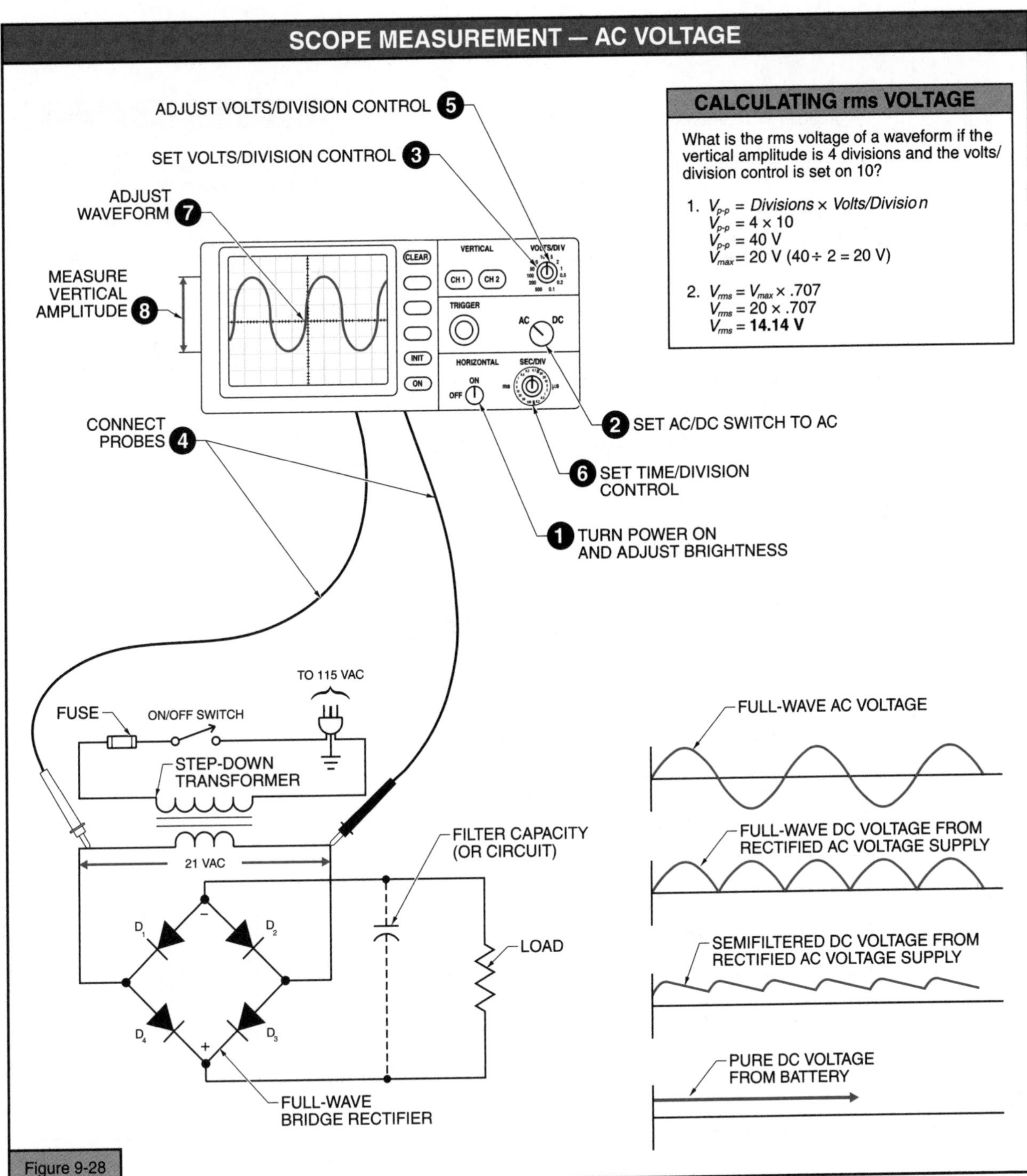

Figure 9-28

Example: Calculating Frequency
What is the frequency if one cycle of a waveform occupies 5 divisions and the time/division setting is 20 ms (microseconds).

1. Calculate time period.

$T = 5 \times .00002$

$T = .0001$ sec

2. Calculate frequency.

$$f = \frac{1}{T}$$

$$f = \frac{1}{.0001}$$

$f =$ **10,000 Hz**

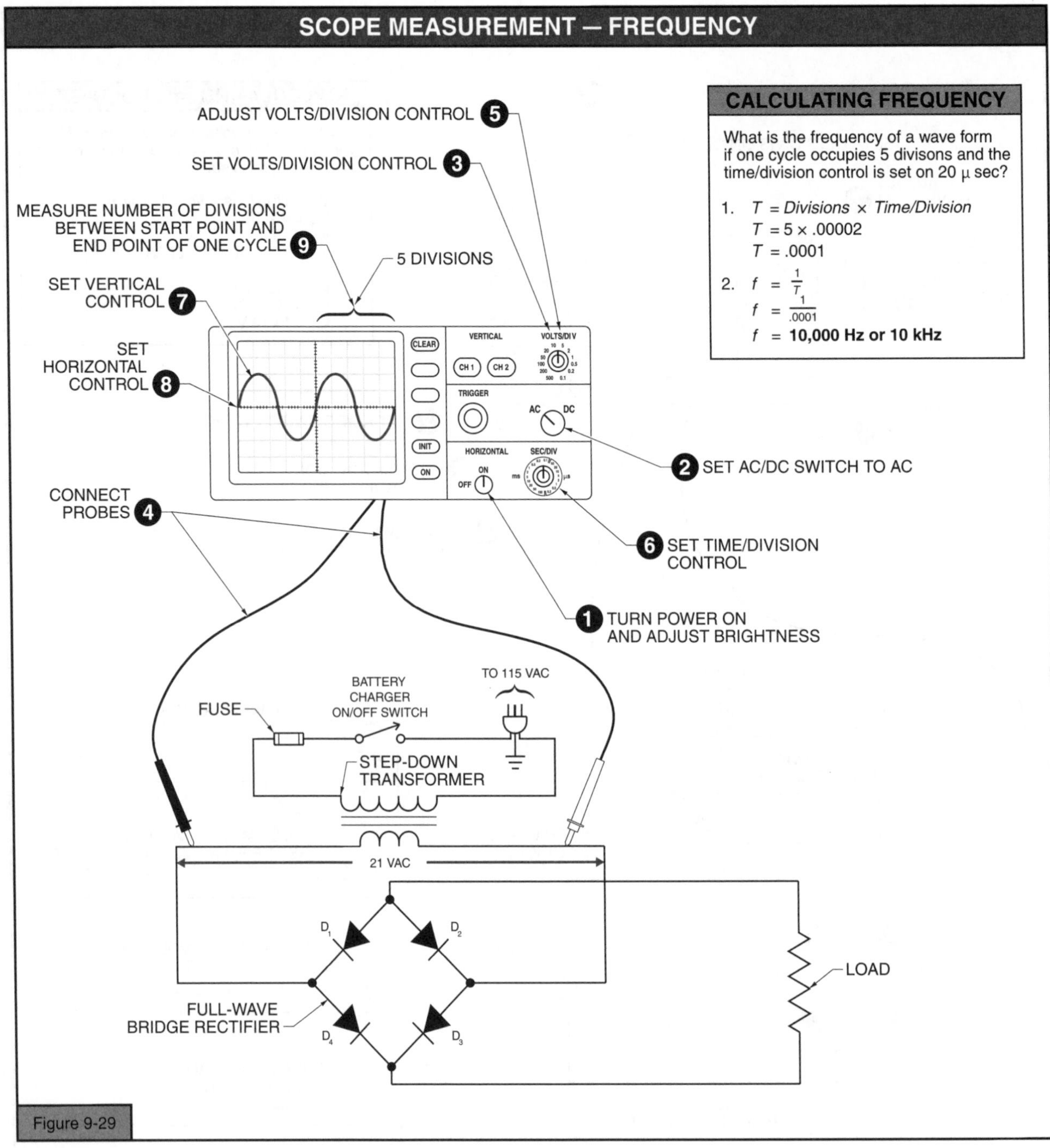

Figure 9-29

Scope Measurement — DC Voltage

To measure AC voltage using a scope, a test probe is connected to the point in the circuit where the DC voltage is to be measured. The ground lead of the scope is connected to the ground of the circuit. The voltage is positive if the trace moves above the center line. The voltage is negative if the trace moves below the center line. **See Figure 9-30.** To use a scope to measure DC voltage, apply the procedure:

1. Turn the power switch ON and adjust the trace brightness.
2. Set the AC/DC control switch to DC.
3. Set the volts/division control to display the voltage level under test. Set the control to the highest value possible if the voltage level under test is unknown.
4. Connect the scope probe to the ground point of the circuit under test.

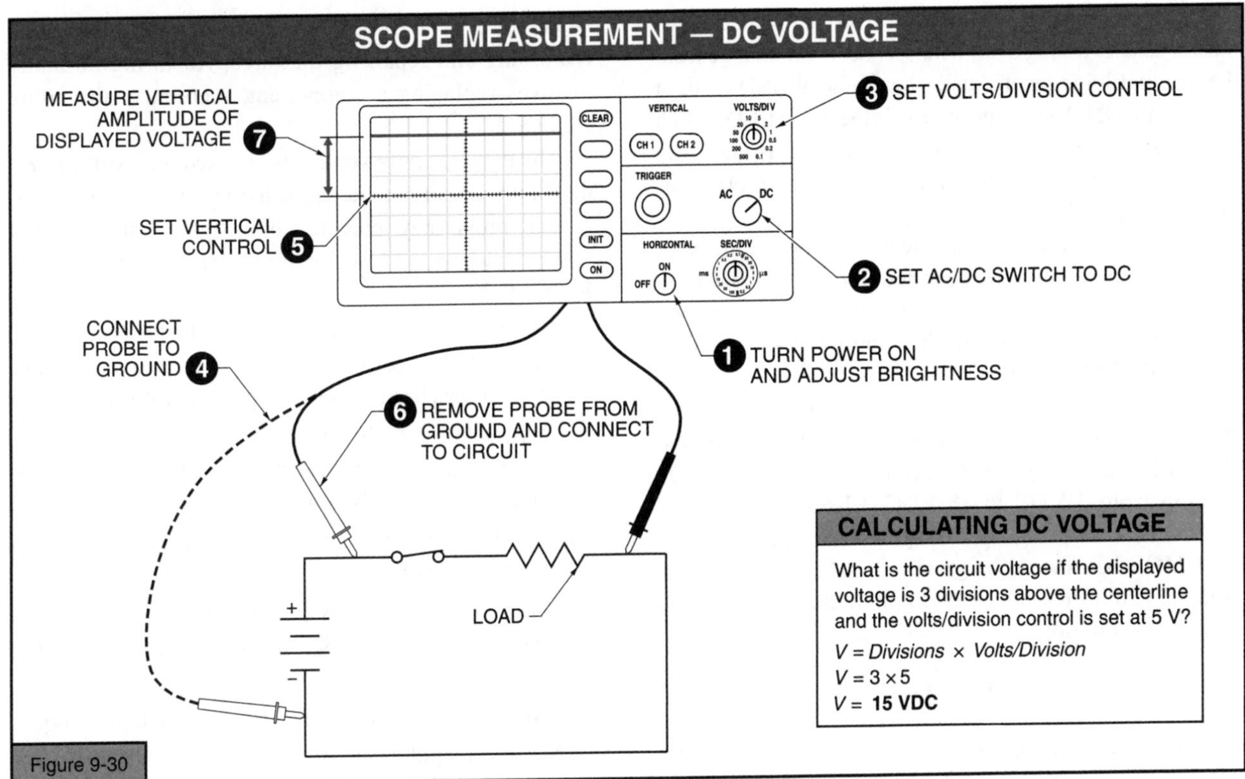

Figure 9-30

5. Set the vertical control so that the displayed line is in the center of the screen. The displayed line represents 0 VDC.
6. Remove the scope probe from ground point and connect it to the DC voltage under test. The displayed voltage moves above or below the scope centerline depending on the polarity of the DC voltage under test.
7. Measure the vertical amplitude of the voltage from the centerline by counting the number of divisions from the centerline.

The number of displayed divisions is multiplied by the volts/division setting to determine the DC voltage under test. For example, if a waveform is located at 3 divisions above the centerline and the volts/division control is set at 5 V, the voltage equals 15 VDC (3 × 5 = 15 V).

Digital Logic Probes

A *digital logic probe* is a special DC voltmeter that detects the presence or absence of a signal. Displays on a digital logic probe include logic high, logic low, pulse light, memory, and TTL/CMOS. **See Figure 9-31.**

The high light-emitting diode (LED) lights when the logic probe detects a high logic level (1). The low LED lights when the logic probe detects a low logic level (0). The pulse LED flashes relatively slowly when the probe detects logic activity in a circuit. Logic activity indicates that the circuit is changing between logic levels. The pulse light displays the changes between logic levels because the changes are usually too fast for the high and low LEDs to display.

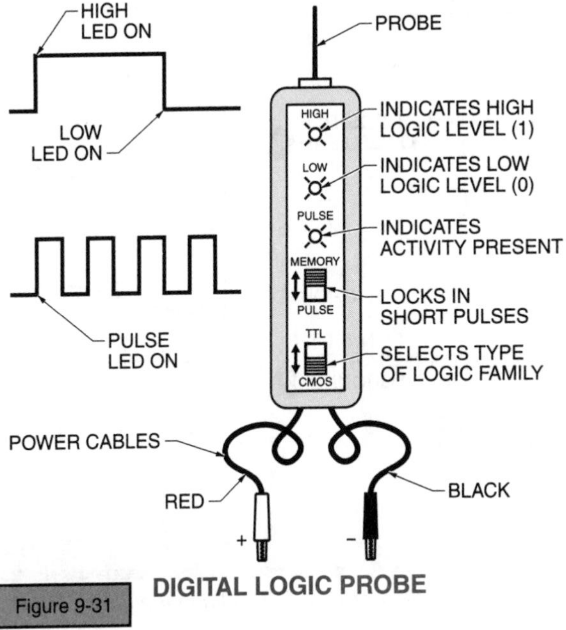

Figure 9-31

The memory switch sets the logic probe to detect short pulses, usually lasting a few nanoseconds. Any change from the original logic level causes the memory LED to light and remain ON. The memory LED is the pulse LED switch in the memory position. The memory switch is manually moved to the pulse position and back to the memory position to reset the logic probe.

The TTL/CMOS switch selects the logic family of integrated circuits (ICs) to be tested. *Transistor-transistor logic (TTL) ICs* are a broad family of ICs that employ a two-transistor arrangement. The supply voltage for TTL ICs is 5.0 VDC ± .25 V.

Complementary metal-oxide semiconductor (CMOS) ICs are a group of ICs that employ MOS transistors. CMOS ICs are designed to operate on a supply voltage ranging from 3 VDC to 18 VDC. Check circuit schematics for CMOS circuit voltages. The supply voltage for CMOS ICs should be greater than –5% of the rated voltage. CMOS ICs are noted for their exceptionally low power consumption.

Digital Logic Probe Use

Digital circuits fail because the signal is lost somewhere between the circuit input and output stages. The problem is usually solved by finding the point where the signal is missing and repairing that area. Repairing normally involves replacing a component, section, or the entire PC board.

The power supply voltage is checked with voltmeter if the digital circuit or logic probe has intermittent problems. A logic probe may indicate a high signal, but the supply voltage may be too low for proper circuit operation. **See Figure 9-32.**

To use a digital logic probe, apply the procedure:

1. Connect the positive (red) power lead to the positive side of the circuit's power supply. The positive power supply is + 5 VDC for TTL circuits.
2. Connect the negative (black) power lead to the ground side of the circuit's power supply.
3. Set the selector switch to the logic family (TTL or CMOS) under test.
4. Touch the probe tip to the point being tested. Start at the input side of the circuit and move to the output of the circuit.
5. Note the condition of the LEDs on the logic probe.

Single-shot pulses are stored indefinitely by placing the switch in the memory position.

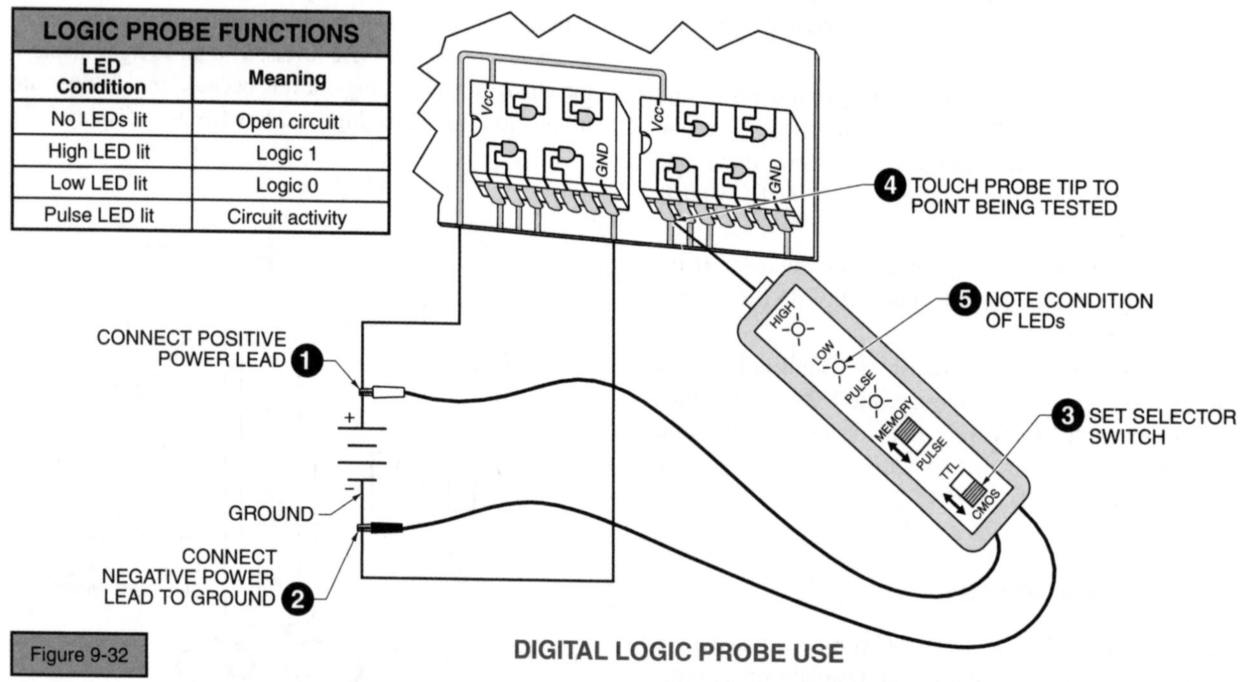

LOGIC PROBE FUNCTIONS	
LED Condition	Meaning
No LEDs lit	Open circuit
High LED lit	Logic 1
Low LED lit	Logic 0
Pulse LED lit	Circuit activity

Figure 9-32

Refer to the CD-ROM in the back of the book for Chapter 9 Quick Quiz® and related information.

Symbols and Printreading

Using Abbreviations and Prints

Electricity and fluid power are used to produce work. To produce work, individual components must be interconnected to control and convert electricity and fluid power into heat, light, sound, rotary force, and linear force. The interconnected components form circuits. Interconnected circuits form systems. Interconnected systems form networks. Circuits, systems, and networks must be properly designed, installed, and maintained to perform safely.

Engineers use standard abbreviations, acronyms, and symbols to develop prints when designing circuits and systems. Prints are used to build and install the individual components after a circuit or system is designed. Prints are also used when troubleshooting and making circuit changes after the circuit or system is placed in operation. Electrical, electronic, and fluid power components are interconnected to perform their required operation(s). For this reason, the meaning of all abbreviations, acronyms, and symbols must be understood when working with electrical, electronic, and fluid power circuits. **See Figure 10-1.**

Abbreviations and Acronyms

An *abbreviation* is a letter or combination of letters that represents a word. Abbreviations are used on prints, in service manuals, on components, and on test equipment. Abbreviations are used to shorten a word or phrase. For example, the letter V is used on prints to represent volts.

An *acronym* is a word formed from the first letter(s) of a compound term. For example, the letters SCR are used to represent a silicon controlled rectifier. Abbreviations and acronyms may have the same meaning in most (or all) technical fields or may have different meanings in different technical fields. For example, the acronym AC represents alternating current in the electrical, electronic, fluid power, and other technical fields. However, the abbreviation T represents transformer in the electrical/electronic field and torque, thrust, or tank in the fluid power field.

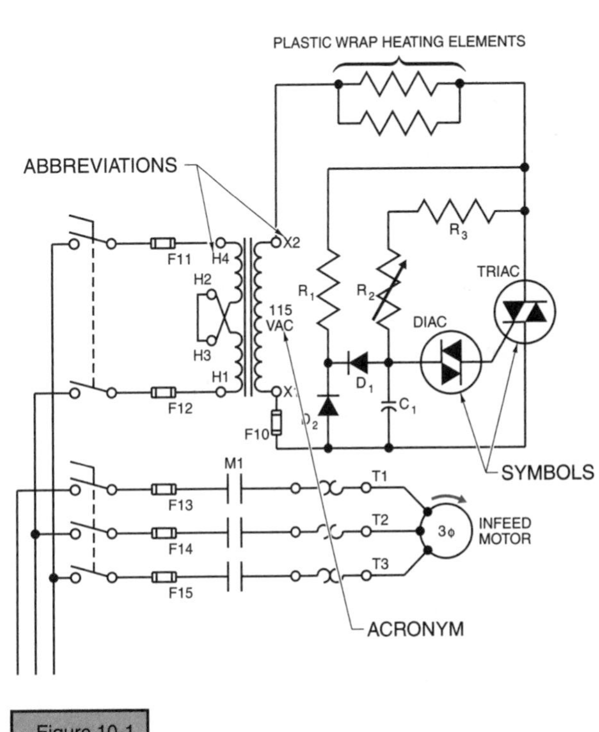

Figure 10-1

The actual meaning may not be the same even when the meaning of an abbreviation or acronym appears to be the same in more than one technical field. For example, the acronym NO represents normally open and the acronym NC represents normally closed in the electrical and fluid power fields. However, the term NO in the electrical field represents a circuit that does not allow power to flow in the normal position. In the fluid power field, NO represents a circuit that allows power to flow in the normal position. For this reason, the meaning of an abbreviation or acronym must be understood as it is used in each field. **See Figure 10-2. See Appendix.**

Symbols

A *symbol* is a graphic element that represents an operation, a quantity, or unit. Symbols, like abbreviations and acronyms, are used on prints, in service manuals, on components, and on test equipment. Symbols used in electrical, electronic, and fluid power diagrams show a simplified picture of the various components in the circuit. Symbol use enables easier understanding of circuit operation and logic. To troubleshoot, repair, or replace components, the actual component that the symbol represents must be known. For example, on a print, the symbol for a battery is a series of long and short lines. This is much different than the actual appearance of a battery that is installed in a circuit. **See Figure 10-3.**

Power Sources

A *power source* is a device that converts various forms of energy into electricity. All electrical circuits must have a source of power. A power source may be chemical (battery), magnetic (generator), heat (thermocouple), or solar (photovoltaic cell). The power source used depends on the application and the amount of power required. The power source in a circuit is normally the point at which to start when reading or troubleshooting a print. **See Figure 10-4.**

Chemical power sources such as batteries convert chemical energy into electrical energy. On the symbol, the long lines represent positive terminals. Plus and minus signs are optional. Cells are connected in series to increase voltage or are connected in parallel to increase current.

Electrical power such as alternating current (AC) is supplied by a generator. Direct current (DC) is rectified from AC. AC supply voltage reverses its direction of flow twice per cycle and does not have a definite positive and negative side. DC supply voltage flows in one direction only and has a definite positive and negative side.

SELECTED ELECTRICAL/ELECTRONIC ABBREVIATIONS/ACRONYMS

Abbreviation/Acronym	Meaning
AC	Alternating current
ALM	Alarm
AM	Ammeter
ARM	Armature
AU	Automatic
BAT	Battery (electric)
BR	Brake relay
CAP	Capacitor
CB	Circuit breaker
CEMF	Counter electromotive force
CKT	Circuit
CONT	Control
CR	Control relay
CRM	Control relay master
CT	Current transformer
D	Down (or diode)

SELECTED FLUID POWER ABBREVIATIONS/ACRONYMS

Abbreviation/Acronym	Meaning
Btu	British thermal unit
C	Degrees Celsius
CC	Closed center (valves)
CCW	Counterclockwise
CFM	Cubic feet per minute
CFS	Cubic feet per second
CIM	Cubic inches per minute
COM	Common
CPM	Cycles per minute
CPS	Cycles per second
CW	Clockwise
CYL	Cylinder
D	Drain

Figure 10-2

SYMBOLS

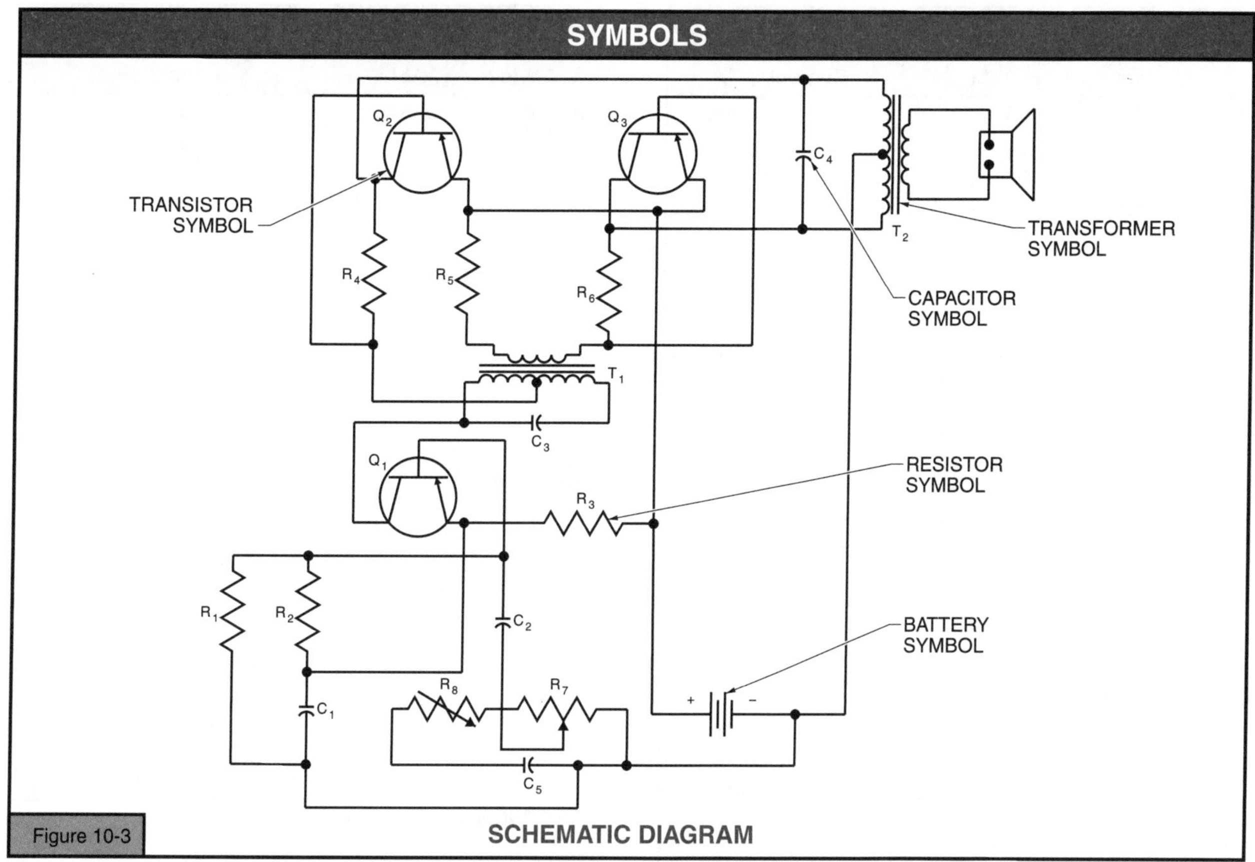

Figure 10-3

Heat power sources such as thermocouples convert heat energy into electrical energy. They are composed of two metals joined at the junction. Electrons flow from one metal to the other as heat is applied.

Solar power sources such as photovoltaic cells convert light energy into electricity. They produce DC directly proportional to the amount of light energy striking the cell surface.

Disconnects and OCPDs

A *disconnect switch (disconnect)* is a switch that disconnects electrical circuits from motors and machines. The NEC® requires a means of disconnecting a motor and controller from the circuit. An *overcurrent protection device (OCPD)* is a disconnect switch with circuit breakers (CBs) or fuses added to provide overcurrent protection of the switched circuit. Disconnects and OCPDs apply and remove circuit power and provide overcurrent protection. **See Figure 10-5.**

Disconnects are used to manually remove or apply power to a circuit. With few exceptions, they are required in circuits that control large loads (over 10 A). Overcurrent protection is added to protect the circuit conductors, controls, and loads from short circuits, grounds, and excessive current levels. Disconnects are used in most motor, high-intensity discharge (HID) lighting, and heating applications.

CBs interrupt (disconnect) circuit power when a predetermined value of current (in amperes) has been exceeded. A CB is reset after the fault is cleared. CBs provide short-circuit and overcurrent protection. A single-pole CB is used for DC and low-voltage, 1ϕ AC circuit protection. A two-pole CB is used for high-voltage, 1ϕ AC circuit protection. A three-pole CB is used for 3ϕ AC circuits. A disconnect with breakers interrupts circuit power when a predetermined value of current (in amperes) has been exceeded. The disconnect switch is placed before the overcurrent protection device.

Fuses and disconnects with fuses interrupt circuit power when a predetermined value of current (in amperes) has been exceeded. The disconnect with fuses is replaced after it has blown and the fault is cleared. Disconnects with fuses provide overcurrent protection and disconnect the circuit.

TECH FACT

An arrow placed across a symbol indicates that the component is variable (variable resistor, variable capacitor, variable flow control valve, etc.).

POWER SOURCES		
Device	Abbr	Symbol
BATTERIES (CHEMICAL)	BAT Pos = (+) Neg = (−)	OPTIONAL; POSITIVE TERMINAL; NEGATIVE TERMINAL SINGLE CELL
	BAT	MULTIPLE CELL
DIRECT CURRENT (MAGNETIC)	DC	(+) (−) STANDARD
	DC	(+) (−) ON METERS
ALTERNATING CURRENT (MAGNETIC)	AC VAC	WAVEFORM REPRESENTS AC
THERMOCOUPLES (HEAT)	TC	OPEN REPRESENTS TERMINAL CONNECTIONS; SOLID REPRESENTS THERMOCOUPLE ELEMENT
PHOTOVOLTAIC CELLS (LIGHT)	PV	LIGHT ENERGY; POSITIVE TERMINAL; NEGATIVE TERMINAL

Figure 10-4

Thermal overloads detect the amount of current flowing in a circuit by sensing heat. The more the current flow, the greater the heat. The heat operates contacts at a preset temperature. Thermal overloads are also known as heaters.

Contacts

Contacts are the conducting part of a switch that operates with another conducting part to make or break a circuit. All switches including relays, contactors, and starters use contacts to start or stop the flow of current in a circuit. The position of the contacts, number of poles, number of throws, and type of break are used to describe contacts. **See Figure 10-6.**

Normally open contacts make or break a circuit. Contacts are designed for low-power switching such as relays, or for high-power switching such as contactors or motor starters. Normally closed contacts switch low currents. Control contacts are used in devices such as pushbuttons, limit switches, relays, etc.

DISCONNECTS AND OCPDs		
Device	Abbr	Symbol
DISCONNECT SWITCHES (ON / OFF)	SPST	TERMINALS FOR CONNECTING WIRE; KNIFE SWITCH SINGLE-POLE, SINGLE-THROW
	DPST	MECHANICALLY TIED TOGETHER, BUT NOT ELECTRICALLY DOUBLE-POLE, SINGLE-THROW
	3PST	THREE-POLE, SINGLE-THROW
CIRCUIT BREAKERS	SPCB	CIRCUIT BREAKER ELEMENT SINGLE-POLE, CIRCUIT BREAKER
	DPCB	TWO-POLE, CIRCUIT BREAKER
	3PCB	THREE-POLE, CIRCUIT BREAKER
DISCONNECT WITH BREAKERS	DISC SW/CB	POWER LINE SIDE; LOAD SIDE THREE-PHASE
FUSES	FU	OR; FUSE ELEMENT SINGLE
DISCONNECT WITH FUSES	DISC SW/FU	POWER LINE SIDE; LOAD SIDE THREE-PHASE
THERMAL OVERLOADS	OL	THERMAL ELEMENT (HEATER) THREE-PHASE

Figure 10-5

CONTACTS		
Device	Abbr	Symbol
CONTACTS	NO	MECHANICAL CONTACTS; SOLID-STATE CONTACTS — NORMALLY OPEN
SMALL	NC	CONTACT CLOSED — NORMALLY CLOSED
LARGE	SPST or S	POLE; TERMINALS — SINGLE-POLE, SINGLE-THROW, SINGLE-BREAK
	SPST	BREAK BOTH SIDES — SINGLE-POLE, SINGLE-THROW, DOUBLE-BREAK
	SPDT or S_3	SINGLE-POLE, DOUBLE-THROW, SINGLE-BREAK
SWITCH	SPDT	CLOSED CONTACT — SINGLE-POLE, DOUBLE-THROW, DOUBLE-BREAK
	DPST	MECHANICALLY CONNECTED — DOUBLE-POLE, SINGLE-THROW, SINGLE-BREAK
	DPDT	COMMON TERMINALS — DOUBLE-POLE, DOUBLE-THROW, SINGLE-BREAK
	DPDT	DOUBLE-POLE, DOUBLE-THROW, DOUBLE-BREAK
		COMMON TERMINAL — MULTIPLE CONTACT RELAY

Figure 10-6

Single-pole, single-throw, single-break contacts make or break one circuit. Single-pole, double-throw, single-break contacts divert power to one of two circuit paths. Single-pole, double-throw, double-break contacts make or break two circuits. Double-pole, single-throw, single-break contacts make or break two circuits. Double-pole, double-throw, single-break contacts divert power from two different circuits. Double-pole, double-throw, double-break contacts make or break four circuits. Multiple contact rotary contacts divert power from one circuit to multiple circuits.

Control Switches

A *control switch* is a switch that controls the flow of current in a circuit. Switches can be activated manually, mechanically, or automatically. Once activated, the switch operates contacts. The contacts are used to start or stop the flow of current. Control switches include selector, limit, proximity, temperature, pressure, flow, level, pushbuttons, and foot switches. **See Figure 10-7.**

A *pushbutton* is a switch that makes (NO) or breaks (NC) a circuit when manually pressed. Single-circuit normally open pushbuttons are one of the simplest and most common forms of control. A pushbutton normally consists of one or more contact blocks, an operator, and a legend plate.

Single-circuit, normally closed pushbuttons break a circuit. Double-circuit, normally open and closed pushbuttons break and/or make a circuit. Mushroom head pushbuttons are used as emergency stop or start pushbuttons. Maintained pushbuttons operate a second (interlocking) circuit when one pushbutton is pressed. Illuminated pushbuttons give a visual indication of circuit condition.

A *selector switch* is a switch with an operator that is rotated (instead of pushed) to activate the electrical contacts. Two-position selector switches make (NO) or break (NC) a circuit when manually rotated. The basic difference between a pushbutton and a selector switch is the method of operation. The operator is rotated to open and close the contacts with a selector switch. Selector switches are used to select one of several different circuit possibilities. Typical machine operation includes manual or automatic, low or high, up or down, and stop or run. A selector switch can be spring-loaded or have maintained positions.

A *limit switch* is a switch that converts mechanical motion into an electrical signal. Limit switches accomplish this conversion by using a lever to open or close contacts within the limit switch enclosure. An NC limit switch breaks a circuit. An NO, held closed limit switch breaks a circuit when an object holding the switch closed is removed.

CONTROL SWITCHES

Device	Abbr	Symbol
PUSHBUTTONS 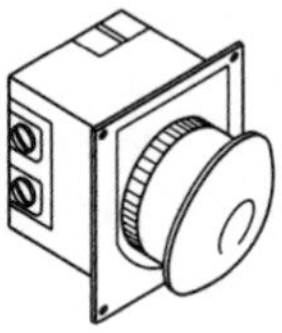	PB or PB-NO	MANUAL OPERATION TERMINALS **SINGLE CIRCUIT, NORMALLY OPEN**
	PB or PB-NC	**SINGLE CIRCUIT, NORMALLY CLOSED**
	PB or PB-NO/NC	**DOUBLE-CIRCUIT, NORMALLY OPEN AND CLOSED**
	PB or PB-NO/NC	MUSHROOM OPERATOR **MUSHROOM HEAD**
	PB	MECHANICAL LINK **MAINTAINED**
	PB/LT	R R = red G = green A = amber B = blue LAMP INSIDE OPERATOR **ILLUMINATED**
SELECTOR SWITCHES 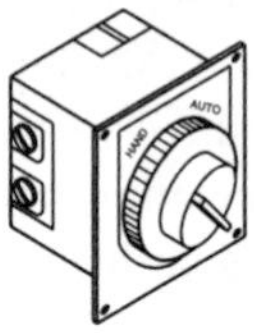	SELT SW or SS	RUN POSITION JOG POSITION JOG RUN A1 A2 Truth table: columns J, R; A1: X under J; A2: X under R CONTACTS OPEN "X" INDICATES CONTACTS CLOSED **TWO-POSITION**
	SELT SW or SS	J = jog R = run or reverse S = stop U = up D = down F = forward A = automatic M = manual STOP POSITION JOG STOP RUN A1 A2 Truth table: columns J, S, R; A1: X under J; A2: X under R TRUTH TABLE FOR SWITCH **THREE-POSITION**
FOOT SWITCHES 	FTS	FOOT OPERATOR **NORMALLY OPEN**
	FTS	**NORMALLY CLOSED**
LIMIT SWITCHES (MECHANICAL) 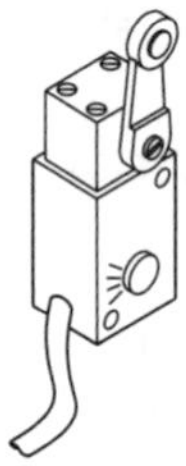	LS	MECHANICAL OPERATOR **NORMALLY OPEN**
	LS	OPERATOR IN CLOSED POSITION **NORMALLY OPEN, HELD CLOSED**
	LS	**NORMALLY CLOSED**
	LS	OPERATOR IN OPEN POSITION **NORMALLY CLOSED, HELD OPEN**
PROXIMITY SWITCHES 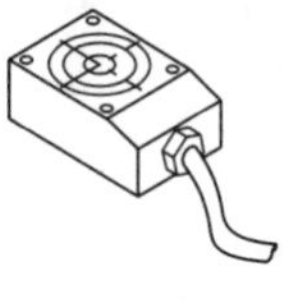	PROX	SOLID-STATE OR **NORMALLY OPEN**
	PROX	OR **NORMALLY CLOSED**
TEMPERATURE SWITCHES 	TEMP SW	TEMPERATURE OPERATOR **NORMALLY OPEN**
	TEMP SW	**NORMALLY CLOSED**
PRESSURE SWITCHES 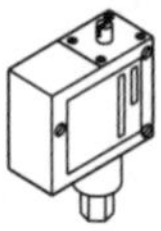	PS	PRESSURE OPERATOR **NORMALLY OPEN**
	PS	**NORMALLY CLOSED**
FLOW SWITCHES 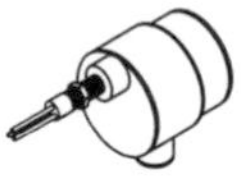	FLS	FLOW OPERATOR **NORMALLY OPEN**
	FLS	**NORMALLY CLOSED**
LEVEL SWITCHES 	LS	LEVEL OPERATOR **NORMALLY OPEN**
	LS	**NORMALLY CLOSED**

Figure 10-7

A *proximity switch* is a switch that detects the presence or absence of an object without touching the object. Proximity switches use solid-state components to start or stop the flow of current. The three basic types of proximity switches are inductive, capacitive, and magnetic (Hall effect).

A *temperature switch* is a switch that responds to temperature changes. A temperature switch may be used to maintain a specified temperature within a process, or to protect against overtemperature conditions. Temperature switches are available to monitor the temperature of liquids, air, or surfaces.

A *pressure switch* is a switch that opens or closes contacts in response to pressure changes in media such as air, water, or oil. The electrical contacts may be used to stop motors or fans, open or close dampers or louvers, or signal a warning light or alarm.

A *flow switch* is a switch that senses the movement of a fluid. The fluid may be air, water, oil, or other gases or liquids. A flow switch is normally inserted into a pipe or duct. The sensing element is normally a valve or paddle that extends into the pipe or duct. This element moves and activates contacts whenever the fluid flow is sufficient to overcome a spring tension.

A *level switch* is a switch that measures and responds to the level of a material. The material may be water, oil, paint, granules, or other solids. Common detection methods include a ball float switch, electrodes, photoelectric, and other types of transmitter and receiver devices.

A *foot switch* is a control switch that is operated by a person's foot. A foot switch allows free use of hands while providing for manual control of a machine. Many foot switches include a guard to prevent accidental operation.

Relays and Timers

A *relay* is a device that controls one electrical circuit by opening and closing contacts in another circuit. Relays multiply the number of available output contacts or permit a small current to switch a high current. The two basic types of relays are electromechanical and solid-state relays. An *electromechanical relay* is a switching device that has sets of contacts which are closed by a magnetic force. A *solid-state relay* is a relay that uses electronic switching devices in place of mechanical contacts. **See Figure 10-8.**

Electromechanical relays switch small currents in a control circuit. The control relay consists of an electromagnet that moves an armature which operates one or more sets of contacts. Electromechanical relays may include all NO, all NC, or any combination of contacts. Industrial relays allow contacts to be changed to either NO or NC.

RELAYS AND TIMERS

Device	Abbr	Symbol
CONTROL RELAYS (ELECTRO-MECHANICAL)	CR	CR — COIL
	NO	NORMALLY OPEN CONTACT
	NC	NORMALLY CLOSED CONTACT
CONTROL RELAYS (SOLID-STATE)	SSR	SSR — INPUT (COIL)
	NO	SSR — NORMALLY OPEN OUTPUT
	NC	SSR — NORMALLY CLOSED OUTPUT
TIMERS, ON-DELAY	TR	TR — COIL
	NOTC	TC OR — TIMED CLOSED — NORMALLY OPEN, TIMED CLOSED
	NCTO	TO OR — POINTS IN THE DIRECTION OF TIME DELAY — TIMED OPEN — NORMALLY CLOSED, TIMED OPEN
TIMERS, OFF-DELAY	TR	TR — COIL
	NOTO	TO OR — NORMALLY OPEN, TIMED OPEN
	NCTC	TC OR — POINTS IN THE DIRECTION OF TIME DELAY — NORMALLY CLOSED, TIMED CLOSED

Figure 10-8

Solid-state relays interface electronic circuits to high power electrical circuits. The control relay depends on electronic devices such as SCRs and triacs to switch the current flow. An NO solid-state relay makes a circuit. An NC solid-state relay breaks a circuit.

A *timer* is a control device that uses a preset time period as part of the control function. Timers have mechanical contacts or solid-state switching devices. The two most common timers are ON-delay and OFF-delay timers.

An *ON-delay timer* is a timer that delays for a predetermined time after receiving a signal to activate or turn ON before the timer's contacts are changed from their normal condition. NO contacts close and NC contacts open after the predetermined time period. NO timed closed ON-delay timers make a circuit after time delay. NC timed open ON-delay timers break a circuit after time delay.

An *OFF-delay timer* is a timer that delays for a predetermined time period when the timer no longer receives a signal to stay activated or is turned OFF before the timer contacts return to their normal condition. NO contacts open and NC contacts close after the predetermined time period. NO timed open OFF-delay timers make a circuit and delay the breaking of a circuit. NC timed closed OFF-delay timers break a circuit and delay the making of a circuit.

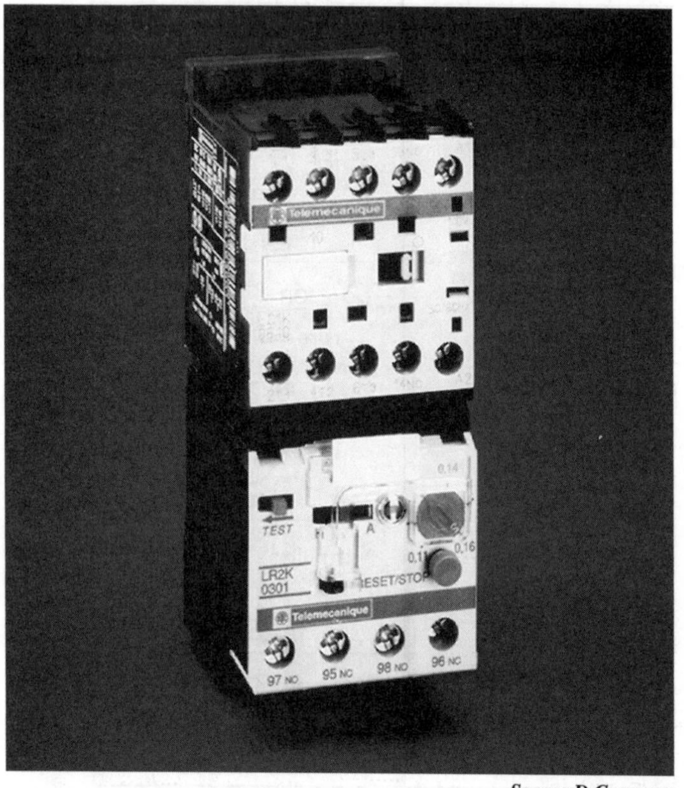

Square D Company

Small contactors are used in applications that have size restrictions.

Contactors and Motor Starters

A *contactor* is a control device that uses a small control current to energize or de-energize the load connected to it. Contactors turn ON and OFF high current loads that do not require overload protection. They are used to switch lights, heating elements, and solenoids. These loads require overcurrent protection, but not overload protection. **See Figure 10-9.**

A contactor includes a coil, NO contacts, and NC contacts. Contactors switch loads ON or OFF when a control switch energizes the coil. The coil operates the high-current contacts. NO contacts make a circuit. NC contacts break a circuit.

A motor starter is an electrically-operated switch (contactor) that includes motor overload protection. Motor starters may be mechanical or solid-state. Motor starters include a coil, which may be single-voltage or dual-voltage, overload protection, and contacts.

The coil switches loads ON or OFF when energized by a control switch. The coil operates high-current contacts. Overload protection, in addition to overcurrent protection, is required with motor loads. Overload protection is provided by thermal or magnetic overload elements that operate overload contacts. The overload contacts are used in the control circuit to remove the motor from power during an overload. Overload protection protects the motor when running. Overcurrent protection is provided by fuses and circuit breakers. Overcurrent protection protects the motor when starting.

Motor starters include high-power contacts for switching the motor and normally include additional low-power relay contacts that can be used in the control circuit. The high-power NC contacts in a motor starter turn a motor OFF when opened. The high-power NO contacts in a motor starter turn the motor ON when closed. The low-power NO contacts may be used to sound an alarm. The low-power NC contacts may be used to run additional loads ON any time the motor starter is OFF.

Solenoids

A *solenoid* is an electric output device that converts electrical energy into a linear mechanical force. A solenoid has an electromagnet that moves an iron plunger when powered. The plunger transmits the force created by the solenoid into useful work.

Solenoids produce linear motion whenever power is applied (or removed if spring-loaded) from the coil. The moving plunger is used to control electrical contacts, as in a contactor, or can be used directly to move nonelectrical devices, such as the spool of a valve. **See Figure 10-10.**

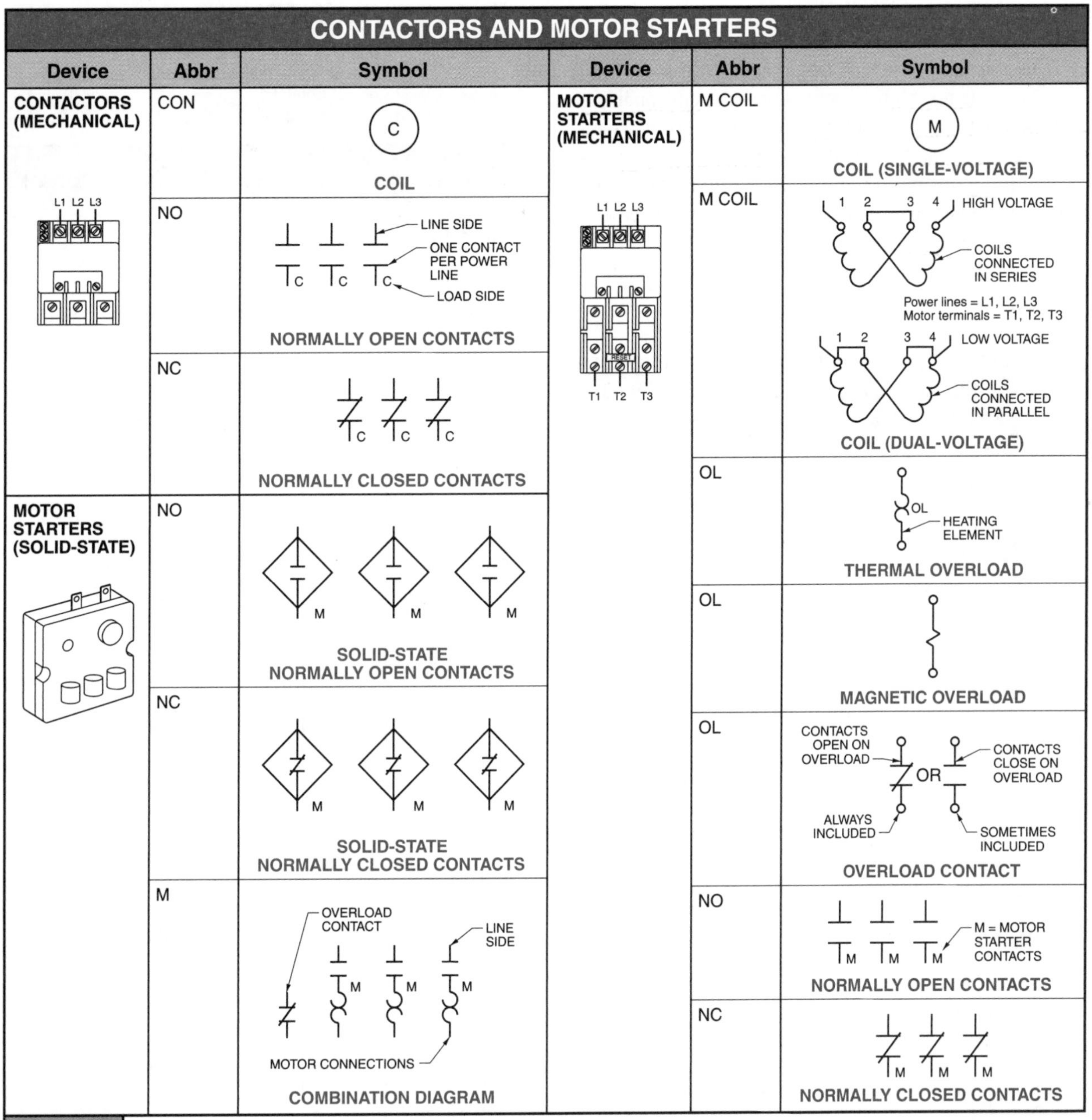

CONTACTORS AND MOTOR STARTERS

Device	Abbr	Symbol
CONTACTORS (MECHANICAL) L1 L2 L3	CON	C COIL
	NO	LINE SIDE ONE CONTACT PER POWER LINE C C C LOAD SIDE NORMALLY OPEN CONTACTS
	NC	C C C NORMALLY CLOSED CONTACTS
MOTOR STARTERS (SOLID-STATE)	NO	M M M SOLID-STATE NORMALLY OPEN CONTACTS
	NC	M M M SOLID-STATE NORMALLY CLOSED CONTACTS
	M	OVERLOAD CONTACT LINE SIDE M M M MOTOR CONNECTIONS COMBINATION DIAGRAM

Device	Abbr	Symbol
MOTOR STARTERS (MECHANICAL) L1 L2 L3 RESET T1 T2 T3	M COIL	M COIL (SINGLE-VOLTAGE)
	M COIL	1 2 3 4 HIGH VOLTAGE COILS CONNECTED IN SERIES Power lines = L1, L2, L3 Motor terminals = T1, T2, T3 1 2 3 4 LOW VOLTAGE COILS CONNECTED IN PARALLEL COIL (DUAL-VOLTAGE)
	OL	OL HEATING ELEMENT THERMAL OVERLOAD
	OL	MAGNETIC OVERLOAD
	OL	CONTACTS OPEN ON OVERLOAD OR CONTACTS CLOSE ON OVERLOAD ALWAYS INCLUDED SOMETIMES INCLUDED OVERLOAD CONTACT
	NO	M M M M = MOTOR STARTER CONTACTS NORMALLY OPEN CONTACTS
	NC	M M M NORMALLY CLOSED CONTACTS

Figure 10-9

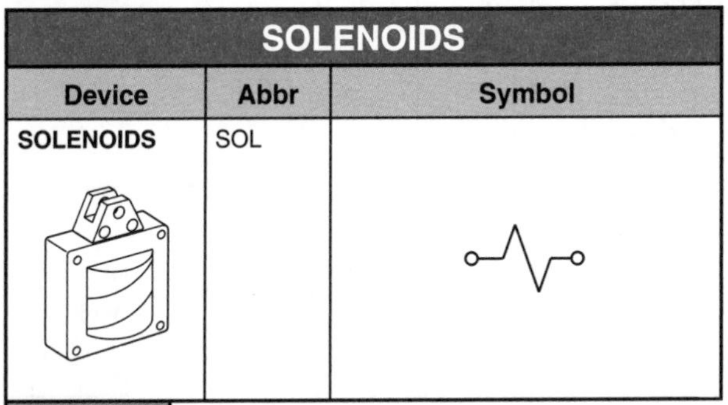

SOLENOIDS

Device	Abbr	Symbol
SOLENOIDS	SOL	

Figure 10-10

Resistors

Electrical resistance is opposition to the flow of electrons through any material. A *resistor* is a device that limits the current flowing in an electronic circuit. Resistors are classified by their resistance value (in Ω) and their power dissipation (in W). **See Figure 10-11.**

Resistors are used for dividing voltage, reducing voltage, developing heat, and limiting current. Resistors may be fixed, variable, or tapped. A *fixed resistor* is a resistor with a set value, such as 100 Ω. A *variable (adjustable) resistor* is a resistor with a set range of values, such as 0 Ω to 1000 Ω.

A *tapped resistor* is a resistor that contains fixed tap points of different resistances.

A *thermistor* is a device that changes resistance with a change in temperature. Thermistors are semiconductor devices. The two types of thermistors are the positive temperature coefficient (PTC) and the negative temperature coefficient (NTC) thermistors. A *PTC thermistor* is a thermistor that has an increasing resistance value with an increase in temperature. An *NTC thermistor* is a thermistor that has a decreasing resistance value with an increase in temperature. The NTC is the most common type of thermistor.

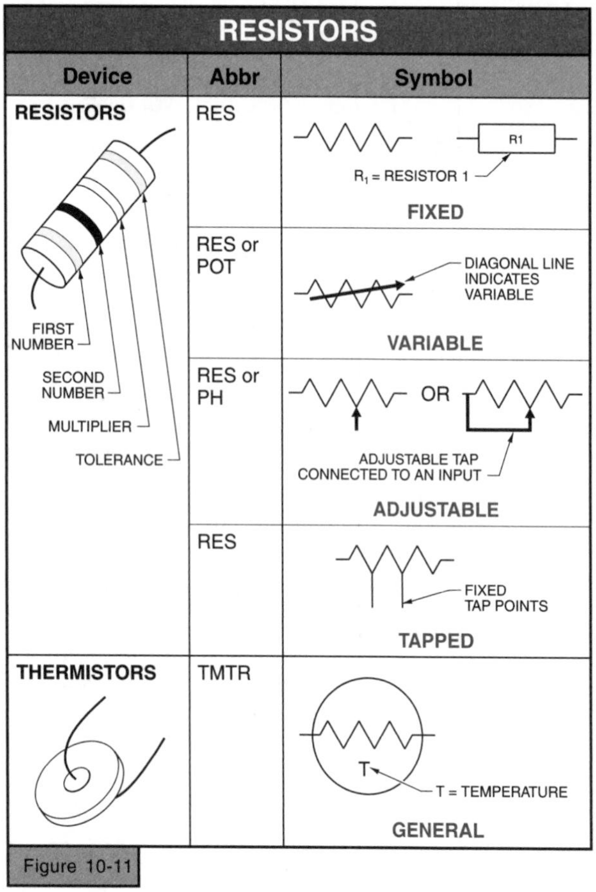

RESISTORS

Device	Abbr	Symbol
RESISTORS (FIRST NUMBER, SECOND NUMBER, MULTIPLIER, TOLERANCE)	RES	R1; R_1 = RESISTOR 1; FIXED
	RES or POT	DIAGONAL LINE INDICATES VARIABLE; VARIABLE
	RES or PH	OR; ADJUSTABLE TAP CONNECTED TO AN INPUT; ADJUSTABLE
	RES	FIXED TAP POINTS; TAPPED
THERMISTORS	TMTR	T; T = TEMPERATURE; GENERAL

Figure 10-11

Capacitors and Diodes

Capacitance is the ability to store energy in the form of an electrical charge. Capacitance is the property of an electric device that permits the storage of electrically separated charges when potential differences exist between the conductors. A *capacitor* is an electric device designed to store electrical energy by means of an electrostatic field. Capacitors are used as filters in AC circuits, to block DC voltages in electronic circuits, and to improve torque in motors. **See Figure 10-12.**

TECH TIP *Wiring diagrams are used during the design and assembly of electrical devices and circuits to show the location and relationships of components.*

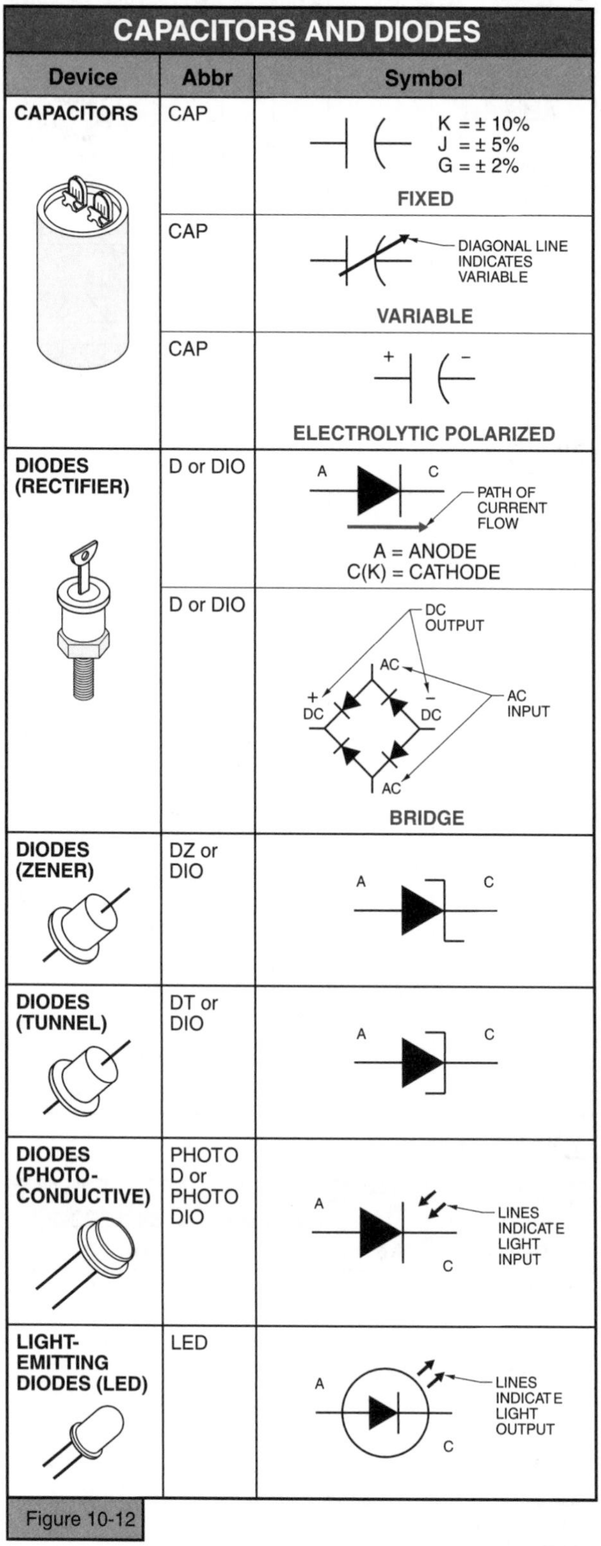

CAPACITORS AND DIODES

Device	Abbr	Symbol
CAPACITORS	CAP	K = ± 10%; J = ± 5%; G = ± 2%; FIXED
	CAP	DIAGONAL LINE INDICATES VARIABLE; VARIABLE
	CAP	+ −; ELECTROLYTIC POLARIZED
DIODES (RECTIFIER)	D or DIO	A C; PATH OF CURRENT FLOW; A = ANODE; C(K) = CATHODE
	D or DIO	DC OUTPUT; AC; + DC; − DC; AC INPUT; AC; BRIDGE
DIODES (ZENER)	DZ or DIO	A C
DIODES (TUNNEL)	DT or DIO	A C
DIODES (PHOTO-CONDUCTIVE)	PHOTO D or PHOTO DIO	A C; LINES INDICATE LIGHT INPUT
LIGHT-EMITTING DIODES (LED)	LED	A C; LINES INDICATE LIGHT OUTPUT

Figure 10-12

Capacitors include fixed, variable, and electrolytic polarized. Once the capacitors are discharged, they must be recharged. Tolerance of some capacitors is indicated by a letter code following the number.

A *diode* is a semiconductor device that offers very high opposition to current flow in one direction and very low opposition to current flow in the opposite direction. Diodes are also known as rectifiers because they change AC into pulsating DC. Diodes are rated according to their type, voltage, and current capacity. The most common types of diodes include zener, tunnel, photoconductive, and light-emitting diodes (LEDs). Diodes are available in current ranges of a few milliamps to over 1000 A. A diode bridge rectifier provides a full-wave DC output.

A *zener diode* is a silicon PN junction that differs from a standard diode in that it operates in the reverse breakdown region. Zener diodes operate as a voltage regulator in an electronic circuit. Zener diodes allow varying amounts of reverse current to flow through them and continue to maintain a relatively constant voltage drop when reverse biased.

A *tunnel diode* is a diode designed so that the current flowing through the diode decreases with an increase in applied voltage for a specified range of forward voltage. Tunnel diodes operate as an amplifier or oscillator in an electronic circuit. Tunnel diodes exhibit negative resistance when operated within a specific range of forward voltage. Tunnel diodes are used in logic circuits and for level sensing.

A *photoconductive diode* is a diode that conducts current when energized by light. Photoconductive diodes decrease resistance as light increases. The diode is made of a photosensitive material that decreases in resistance with an increase in light.

A *light emitting diode (LED)* is a diode that emits light when forward current is applied. In an LED, light is produced when current is passed through the diode. LEDs are commonly used as visual indicators because they have a very long life.

Thyristors

A *thyristor* is a solid-state switching device that switches current ON by a quick pulse of control current. Thyristors are used as solid-state switches. They are designed to be either ON or OFF. In the ON position, they allow current to flow in the circuit. In the OFF position, they block the flow of current in the circuit. **See Figure 10-13.**

Once thyristors are switched ON, no further control current is required to keep them ON. A thyristor is switched OFF only when the current flowing through the thyristor reaches zero. Thyristors can switch currents from a few milliamperes to over 1000 A. The three basic types of thyristors are silicon controlled rectifiers (SCRs), diacs, and triacs.

THYRISTORS

Device	Abbr	Symbol
SILICON CONTROLLED RECTIFIERS	SCR	A, G, K(C); DIRECTION OF CURRENT FLOW; A = Anode, G = Gate, K(C) = Cathode
	SCR	USED IN AC CIRCUITS; ANTI-PARALLEL SCR
DIACS		T, T; T = Terminal
TRIACS		MT2, G, MT1; MT1 = Main Terminal 1, MT2 = Main Terminal 2, G = Gate

Figure 10-13

A *silicon controlled rectifier (SCR)* is a three-terminal semiconductor thyristor that is normally an open circuit until a signal applied to the gate switches it to the conducting state in one direction. SCRs switch direct current. If two SCRs are placed in an anti-parallel configuration, they can be used to switch alternating current. Anti-parallel SCRs are often used when switching high currents.

A *diac* is a thyristor that triggers in either direction when its breakover voltage is exceeded. Diacs allow current to flow in either direction once it reaches a rated breakover voltage. The diac functions like an open switch (in both directions) until a set value of voltage is reached. At this voltage level, the diac conducts current. Once the diac starts conducting current, it continues until the current is reduced to a minimum level.

A *triac* is a three-terminal semiconductor thyristor that is triggered into conduction in either direction by a small current to its gate. Triacs switch alternating current.

TECH FACT *Standard practices for drawing electrical and electronic diagrams are detailed in Electrical and Electronic Diagrams, ANSI Y14.15.*

Transistors

A *transistor* is a three-terminal device that controls current through the device depending on the amount of voltage applied to the base. Transistors are made of semiconductor material. They are part of almost all electronic circuits. Transistors are used to start or stop (switch) current flow or to increase (amplify) current flow in DC circuits. **See Figure 10-14.**

TRANSISTORS		
Device	**Abbr**	**Symbol**
TRANSISTORS	Q or NPN	E, B, C; E = EMITTER, C = COLLECTOR, B = BASE; NPN
	Q or PNP	E, B, C; PNP
	UJT	E, B_2, B_1; E = EMITTER, B_1 = BASE 1, B_2 = BASE 2; P-TYPE; E, B_2, B_1; N-TYPE; UNIJUNCTION TRANSISTOR
	JFET	G, D, S; G = GATE, D = DRAIN, S = SOURCE; P-CHANNEL; G, D, S; N-CHANNEL; JUNCTION FIELD-EFFECT TRANSISTOR
PHOTO-TRANSISTORS		E, C, B; THREE LEAD; E, C; TWO LEAD; NPN
		E, C, B; PNP

Figure 10-14

Transistors may be NPN, PNP, unijunction, and junction field-effect transistors and phototransistors. An *NPN transistor* is a transistor that is formed by sandwiching a thin layer of P-type material between two layers of N-type material. A *PNP transistor* is a transistor formed by sandwiching a thin layer of N-type material between two layers of P-type material. NPN and PNP transistors function as a switch. They start, stop, or increase the flow of current in a DC circuit. They may also function as amplifiers in allowing a very small current to control a much larger current. A *unijunction transistor (UJT)* is a transistor consisting of N-type material with a region of P-type material doped within the N-type material. A UJT is used in timer circuits and gate control for SCRs. A *junction field-effect transistor (JFET)* is a device in which output current is controlled by the voltage on the input.

A *phototransistor* is a transistor that controls the amount of current flowing through the emitter/base junction based on the amount of light. Phototransistors include NPN and PNP transistors. There is an increase of current flow with an increase in light. A phototransistor is a combination of a photodiode and transistor.

Digital Logic Gates

A *digital logic gate* is a circuit that performs a special logic operation such as AND, OR, NOT, NOR, NAND, and exclusive OR. Digital logic gates are used in most electronic circuits. **See Figure 10-15.**

Digital circuits operate on binary signals. Binary signals have two states. The signal is either high (1) or low (0). A high signal is normally 5 V, but can range from 2.4 V to 5 V. A low signal is normally 0 V, but can range from 0 V to .8 V.

An *AND gate* is a logic gate that provides a logic level 1 only if all inputs are at logic level 1. An *OR gate* is a logic gate that provides a logic level 1 if one or more inputs are at logic level 1. A *NOT gate* is a logic gate that provides an output that is the opposite of the input. A *NOR gate* is a logic gate that provides a logic level 0 if one or more inputs are at logic level 1. A *NAND gate* is a logic gate that provides a logic level 0 only if all inputs are at logic level 1. An *exclusive OR gate* is a logic gate that provides a logic level 1 only if one, but not both of the inputs are at logic level 1.

TECH FACT *Color has meaning on prints. For example, black represents negative (ground) in electronic prints and meters, and hot (ungrounded) in residential and industrial wiring.*

DIGITAL LOGIC GATES		
Device	**Boolean Expression**	**Symbol**
AND GATES	Y = A • B (• = AND)	INPUT A, B; Y OUTPUT
OR GATES	Y = A + B (+ = OR)	INPUT A, B; Y OUTPUT
NOT (INVERTER) GATES	$Y = \overline{A}$ (– = NOT)	INPUT A; Y OUTPUT; BUBBLE MEANS INVERTED
NOR GATES	$Y = \overline{A+B}$	INPUT A, B; Y OUTPUT
NAND GATES	$Y = \overline{A \bullet B}$	INPUT A, B; Y OUTPUT
EXCLUSIVE OR GATES	Y = A⊕B (⊕ = EXCLUSIVE)	INPUT A, B; Y OUTPUT; MEANS EXCLUSIVE

Figure 10-15

COILS		
Device	**Abbr**	**Symbol**
COILS WOUND COIL AIR CORE	L or INDT	OR **AIR CORE**
	L or INDT	OR **IRON CORE**
WOUND COIL IRON CORE	L or INDT	TAPS CAN BE AT ANY POINT **TAPPED CORE**
	L or INDT	DIAGONAL LINE MEANS ADJUSTABLE **ADJUSTABLE CORE**

Figure 10-16

Coils

A *coil* is a winding consisting of insulated conductors arranged to produce magnetic flux. Coils are also known as inductors or chokes. Coils are made by winding a length of conductor around a core. The conductor is normally copper wire coated with a thin layer of enamel insulation. When current flows through the coil, the magnetic field developed by the coil opposes a change of current in the circuit in which the coil is used. **See Figure 10-16.**

Coils may be air core, iron core, tapped core, and adjustable core coils. An *air core coil* is a coil with an air core that develops a magnetic field when current passes through the coil. An *iron core coil* is a coil with an iron core added to increase the strength of the magnetic field. A *tapped core coil* is a coil with taps for selecting different magnetic field strengths. An *adjustable core coil* is a coil with a means for varying the magnetic field. Adjustable core coils are used in tuning circuits.

Transformers

A *transformer* is an electric device that uses electromagnetism to change voltage from one level to another or to isolate one voltage from another. Transformers have no moving parts. Voltage is changed through the interaction of the coils' magnetic fields. The voltage is applied to the primary coil and the loads are connected to the secondary coil. **See Figure 10-17.**

Transformers may have an air core or iron core. An iron core in a transformer increases the strength of the electromagnetic induction. A *control circuit transformer* is a transformer used to provide power for a machine or system control circuit. Control circuit transformers have a dual-voltage input, normally 230/460 VAC. Output is normally 115 VAC.

An *autotransformer* is a transformer that changes voltage level using the same common coil for both the primary and secondary sides. Taps allow for voltage adjustments.

A *current transformer* is a transformer that measures the amount of current flowing through a wire by measuring the strength of the magnetic field around the wire. Current transformers are used in solid-state overloads and instruments.

Transformers have the ability to prevent short circuits between the primary and secondary windings. They are designed to withstand voltage higher than the nameplate rating, allowing them to survive faults such as short circuits or lightning strikes. Insulation deterioration causes premature failure from exposure to these faults.

TRANSFORMERS		
Device	**Abbr**	**Symbol**
TRANSFORMERS	T or XFMR	PRIMARY SIDE SECONDARY SIDE **AIR CORE**
	T or XFMR	PRIMARY SIDE SECONDARY SIDE **IRON CORE**
	CONT or XFMR	H1 H3 H2 H4 TWO COILS USED ON PRIMARY SIDE H = HIGH SIDE X = LOW SIDE X1 X2 SECONDARY SIDE **CONTROL CIRCUIT**
AUTO-TRANSFORMERS TAPS	AT or AUTO XFMR	TRANSFORMER TAPS
CURRENT TRANSFORMERS	CT	ONE COIL PER LINE SINGLE-PHASE THREE-PHASE

Figure 10-17

Motors

A *motor* is a machine that develops torque (rotating mechanical force) on a shaft which is used to produce work. Electric motors convert electrical energy into mechanical energy. Motors accomplish this by producing rotary motion. The rotary motion is developed by the opposition of magnetic fields within the motor. **See Figure 10-18.**

The two types of magnetic fields in a motor that produce rotation are the stationary field and the rotating field. The stationary field is known as the stator in AC motors and the field in DC motors. The rotating field is known as the rotor in AC motors and the armature in DC motors. The shaft of the motor is connected to the rotating magnetic field.

Direct current motors include an armature, series field, shunt field, and interpoles. The armature is the rotating field. The series field is the stationary field. Interpoles are used to reduce sparking at the brushes of large DC motors.

A *DC series motor* is a DC motor that has the series field coils connected in series with the armature. It is used as a traction motor because it produces the highest torque of all DC motors. A *DC shunt motor* is a DC motor that has the field connected in parallel with the armature. It is used where constant or adjustable speed is required and starting conditions are moderate. A *DC compound motor* is a DC motor that has the series field connected in series with the armature and the shunt field connected in parallel with the series field and armature. It is used when a high starting torque and constant speed are required. A *DC permanent-magnet motor* is a DC motor that has only armature connections and no field connections. It is used in applications in which the motor is only required to run for a short time, such as in automobile power windows and power seats.

An *alternating current motor* is a motor that operates on AC. Alternating current motors include 1ϕ single- or dual-voltage, 1ϕ two-speed single-voltage, and 3ϕ single- or dual-voltage motors. Single-phase two-speed single-voltage motors are used when more than one fixed speed is required. Three-phase single- or dual-voltage motors are the standard industrial motors used in sizes from fractional HP to over 500 HP.

Lights, Alarms, and Meters

In some electrical circuits, a visual or audible indication of the circuit condition is required. A light can be used to indicate the presence or absence of voltage in part of a circuit, but not the amount. Lights produce visual signals. **See Figure 10-19.**

Lights include pilot, indicator, neon AC, and neon DC lights. A *pilot light* is a light that provides illumination or gives a visual indication of the presence or absence of power in part of a circuit. When used in control circuits, they are known as indicating or pilot lights. Pilot lights generally include a nameplate that tells the meaning of the light such as "motor ON," "gate open," "tank filled," etc.

Alarms are used if the application calls for an audible indication of circuit condition. Alarms include bells, horns, and buzzers. Alarms are generally used to signal a problem. They can be connected to recycle timers to give different signals, such as long or short blasts from the same alarm.

A meter is used if a visual indication of the amount of an electrical property is required. Meters include voltmeters, ammeters, wattmeters, and frequency meters. Meters measure or monitor the condition and properties of an electrical circuit. In some applications, meters are connected as a permanent part of the circuit. In other applications, they are connected temporarily for troubleshooting or measuring purposes.

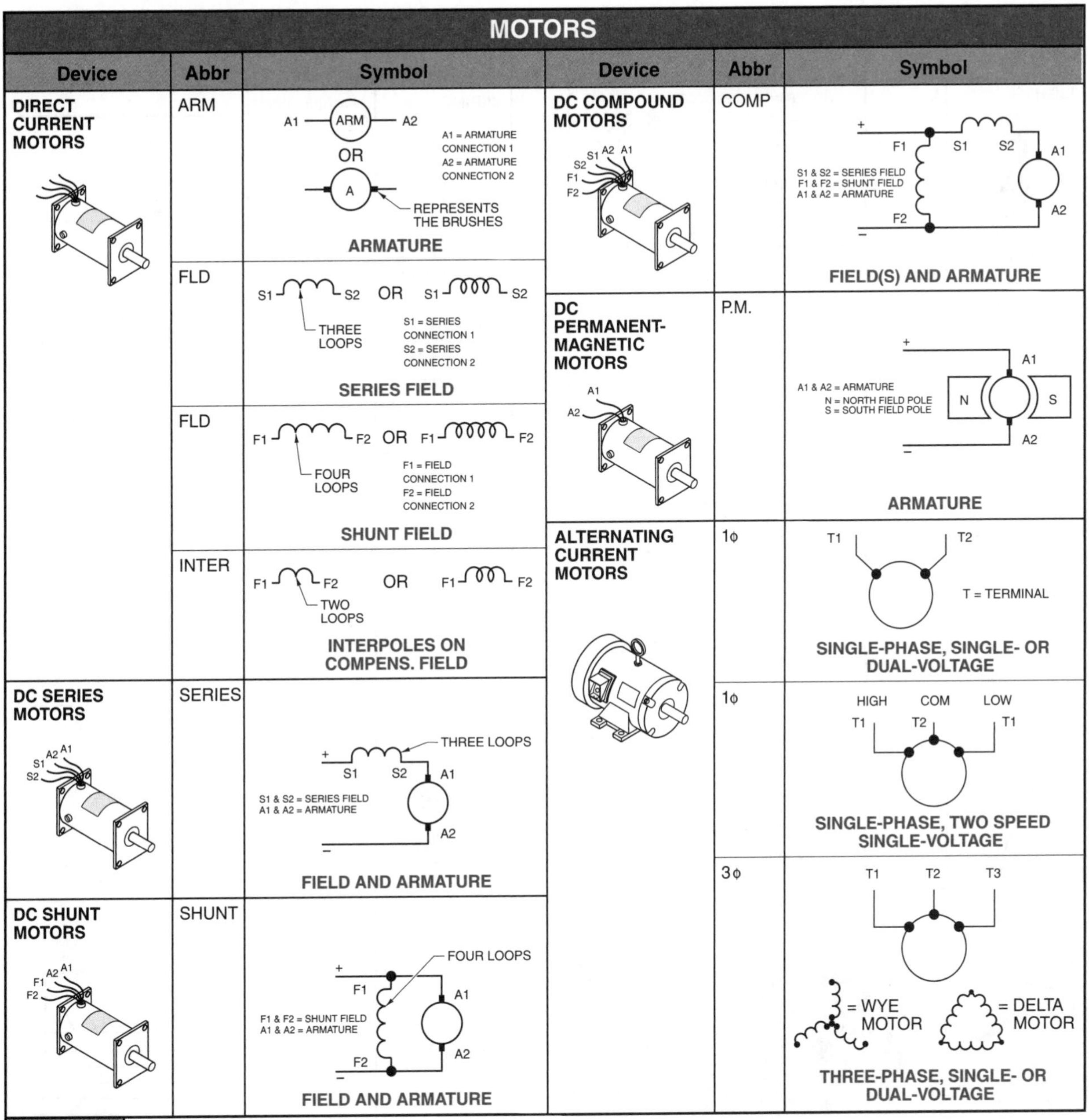

MOTORS					
Device	**Abbr**	**Symbol**	**Device**	**Abbr**	**Symbol**
DIRECT CURRENT MOTORS	ARM	A1 ARM A2 A1 = ARMATURE CONNECTION 1 A2 = ARMATURE CONNECTION 2 OR A REPRESENTS THE BRUSHES **ARMATURE**	**DC COMPOUND MOTORS** S1 A2 A1 S2 F1 F2	COMP	+ F1 S1 S2 A1 S1 & S2 = SERIES FIELD F1 & F2 = SHUNT FIELD A1 & A2 = ARMATURE A2 F2 − **FIELD(S) AND ARMATURE**
	FLD	S1 S2 OR S1 S2 THREE LOOPS S1 = SERIES CONNECTION 1 S2 = SERIES CONNECTION 2 **SERIES FIELD**	**DC PERMANENT-MAGNETIC MOTORS** A1 A2	P.M.	+ A1 A1 & A2 = ARMATURE N = NORTH FIELD POLE S = SOUTH FIELD POLE N S A2 − **ARMATURE**
	FLD	F1 F2 OR F1 F2 FOUR LOOPS F1 = FIELD CONNECTION 1 F2 = FIELD CONNECTION 2 **SHUNT FIELD**	**ALTERNATING CURRENT MOTORS**	1ϕ	T1 T2 T = TERMINAL **SINGLE-PHASE, SINGLE- OR DUAL-VOLTAGE**
	INTER	F1 F2 OR F1 F2 TWO LOOPS **INTERPOLES ON COMPENS. FIELD**		1ϕ	HIGH COM LOW T1 T2 T1 **SINGLE-PHASE, TWO SPEED SINGLE-VOLTAGE**
DC SERIES MOTORS A2 A1 S1 S2	SERIES	THREE LOOPS + S1 S2 A1 S1 & S2 = SERIES FIELD A1 & A2 = ARMATURE A2 − **FIELD AND ARMATURE**		3ϕ	T1 T2 T3 = WYE MOTOR = DELTA MOTOR **THREE-PHASE, SINGLE- OR DUAL-VOLTAGE**
DC SHUNT MOTORS A2 A1 F1 F2	SHUNT	FOUR LOOPS + F1 A1 F1 & F2 = SHUNT FIELD A1 & A2 = ARMATURE F2 A2 − **FIELD AND ARMATURE**			

Figure 10-18

General Wiring

General wiring is the wiring used to connect electrical components in a circuit. Electrical circuits are made of interconnected components. Wire is generally used to connect the components. Using and tracing the wiring circuit is required when installing new equipment, troubleshooting, and when trying to understand the circuit. **See Figure 10-20.**

The wire size determines the amount of current that the wire may safely carry. The larger the wire, the greater the current-carrying capacity. Copper wire may carry a higher current than aluminum wire for a given wire size. Wire is normally coated with an insulating material. The insulating material comes in many different colors. The different colors help in monitoring the wires' usage. For example, in residential wiring, black is used as a hot wire, white as a neutral wire, red as a switched wire, and green as a ground wire.

LIGHTS, ALARMS, AND METERS		
Device	**Abbr**	**Symbol**
LIGHTS	L	LETTER INDICATES COLOR; L OR R; A = AMBER, G = GREEN, R = RED, B = BLUE; **PILOT LIGHT**
ALARMS	AL or BELL	**BELL**
	AL or HORN	**HORN**
	BUZZ	**BUZZER**
METERS ANALOG	VM	VM OR V; **VOLTMETER**
	AMM	AM OR A; **AMMETER**
DIGITAL	WM	W OR W; **WATTMETER**
	FREQ M	F OR F; **FREQUENCY METER**

Figure 10-19

An electric motor drive can be connected to a 3φ motor to control the flow of water to a process.

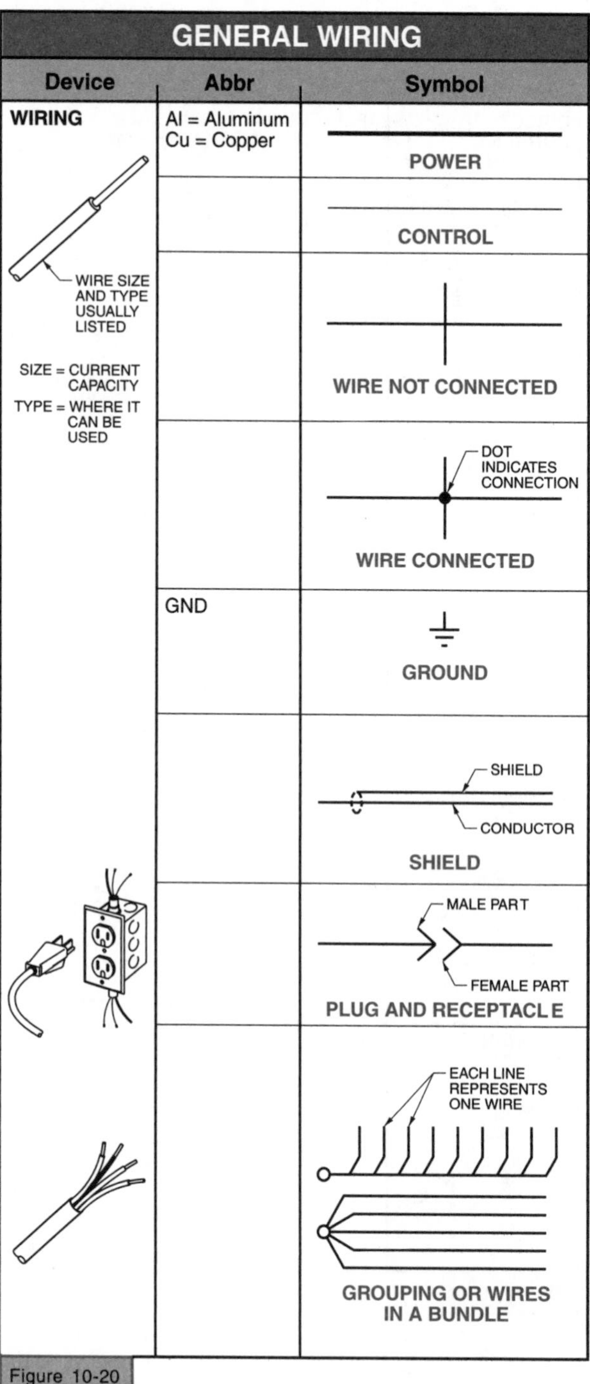

GENERAL WIRING		
Device	**Abbr**	**Symbol**
WIRING	Al = Aluminum Cu = Copper	**POWER**
WIRE SIZE AND TYPE USUALLY LISTED		**CONTROL**
SIZE = CURRENT CAPACITY TYPE = WHERE IT CAN BE USED		**WIRE NOT CONNECTED**
		DOT INDICATES CONNECTION; **WIRE CONNECTED**
	GND	**GROUND**
		SHIELD; CONDUCTOR; **SHIELD**
		MALE PART; FEMALE PART; **PLUG AND RECEPTACLE**
		EACH LINE REPRESENTS ONE WIRE; **GROUPING OR WIRES IN A BUNDLE**

Figure 10-20

Fluid Pumps and Conditioners

A *fluid pump* is a pump that converts mechanical power to hydraulic or pneumatic power by pushing the fluid into the system. Pumps are classified as positive-displacement or non-positive displacement. A *positive displacement pump* is a pump that delivers a definite quantity of fluid for every stroke, revolution, or cycle. A *non-positive displacement pump* is a pump that circulates or transfers fluid. **See Figure 10-21.**

Fluid pumps may be fixed displacement pumps or variable displacement pumps. A *fixed displacement pump* is a pump that develops a fixed amount of flow for a given pump speed. Displacement is only changed by changing the drive speed of the pump. A *variable displacement pump* is a pump in which the amount of flow is manually adjusted or automatically adjusted without changing the pump speed.

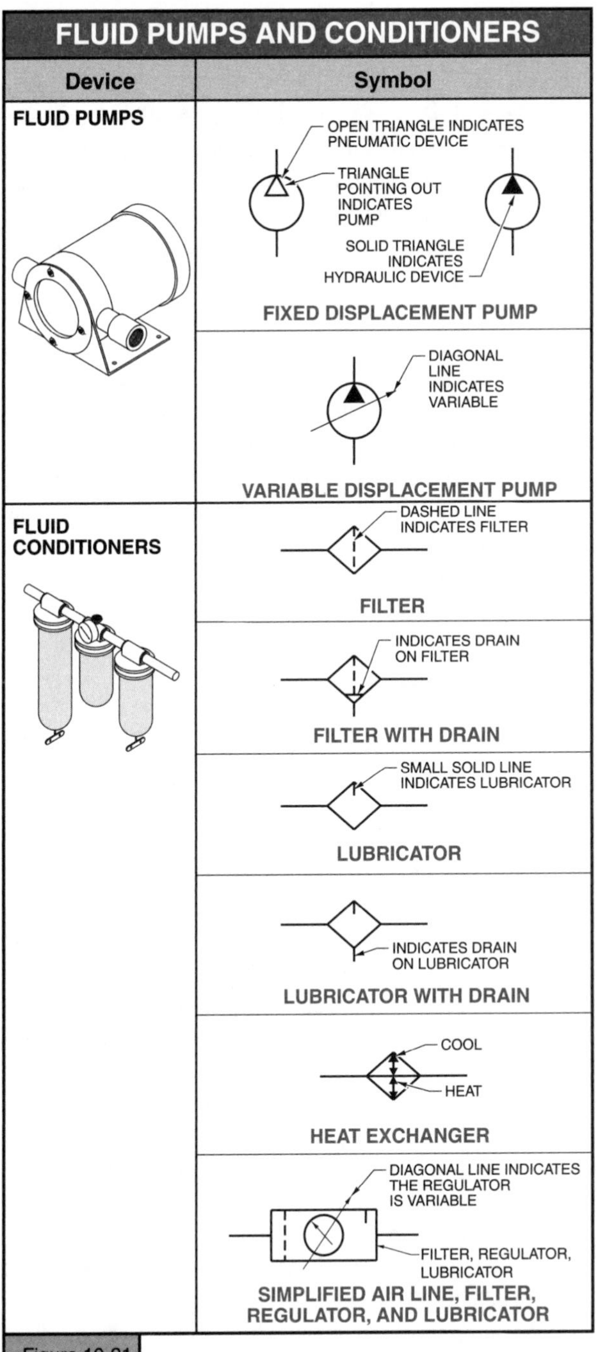

Figure 10-21

A *fluid conditioner* is a device that maintains clean fluid in the system. Fluid systems and components require clean fluid (air or oil) to operate properly. Fluid conditioners include filters, filters with drain, lubricators, lubricators with drain, and heat exchangers. A *filter* is a device that removes solid contaminants from a fluid power system. A *filter with drain* is a device that removes contaminants and moisture from a fluid power system. A *lubricator* is a device that adds lubrication to pneumatic systems. A *lubricator with drain* is a device that adds lubrication and removes moisture from a pneumatic system. A *heat exchanger* is a device that removes heat from the fluid in a hydraulic system.

Fluid Actuators

The pressurized fluid in a hydraulic or pneumatic system must be converted back to mechanical power after it is transmitted to the point where it is to be used. A *fluid actuator* is a device that converts fluid power to mechanical power. Fluid actuators include cylinders, fluid power motors, and oscillators. **See Figure 10-22.**

A *cylinder* is a device that converts fluid energy into a linear mechanical force. Cylinders include single-acting, double-acting, and double-end rod cylinders. Single-acting cylinders provide a high force in one direction. Double-acting cylinders provide a high force in two directions. Double-end rod cylinders provide a pushing and pulling force when the cylinder rod is moving in one direction.

Fluid power motors include fixed displacement, variable displacement, and reversible motors. A *fixed displacement fluid power motor* is a fluid power motor that provides rotary motion and drives a load in one direction. A *variable displacement fluid power motor* is a fluid power motor that drives a load in one direction and provides a variable output. A *reversible fluid power motor* is a fluid power motor that can drive a load in two directions.

A *fluid power oscillator* is a device that moves in fixed increments each time fluid pressure is applied. Typical fluid power oscillator movement increments are 15°, 30°, 45°, 90°, 120°, and 180°.

Directional Control Valves

A *directional control valve* is a valve that connects, disconnects, or directs the fluid pressure from one part of a circuit to another. A directional control valve is commonly connected immediately before an actuator to control that actuator's movement. **See Figure 10-23.**

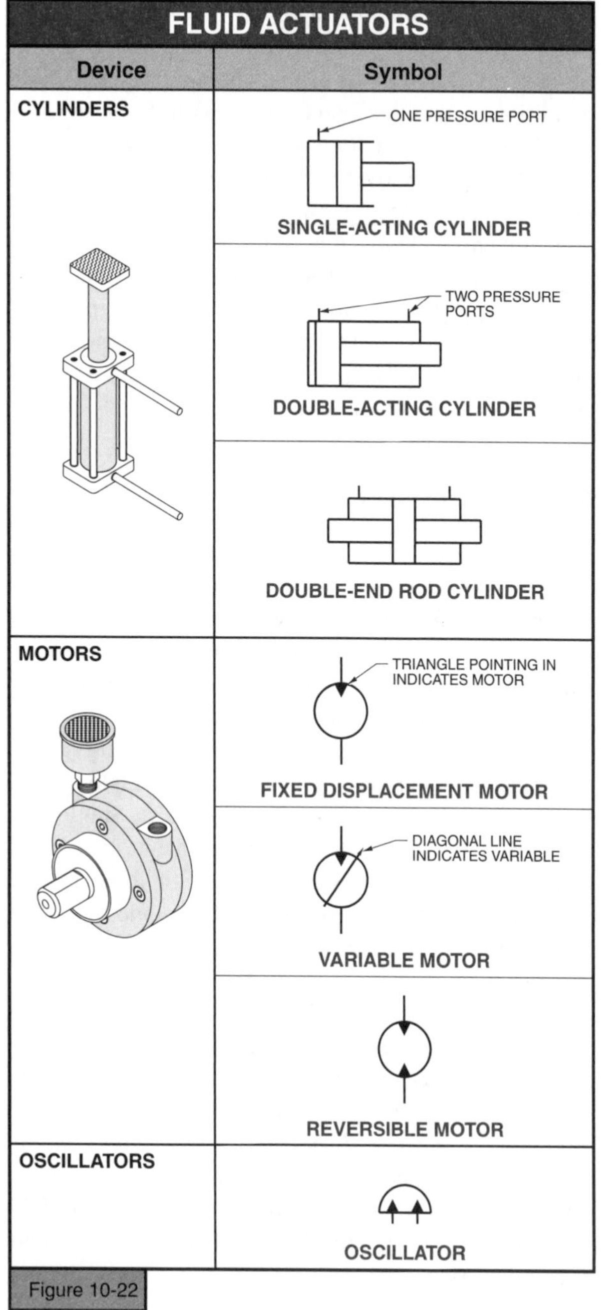

Figure 10-22

Valve symbols are composed of one or more envelopes containing lines that represent flow paths and flow conditions between the ports. An *envelope* is a pictorial representation of what is occurring inside a valve for one given position. A valve has one envelope for each valve position. The number of valve positions is determined by the design of the valve.

Directional control valves include two-way, two-position; three-way, three-position; four-way, two-position; two-way, NO; two-way, NC; three-way, NC; and four-way, three-position valves.

Two-way, two-position valves allow or stop fluid flow. They are used as shutoff, check, or quick exhaust valves. Three-way, two-position valves allow or stop fluid flow and provide an exhaust. They are used to control single-acting cylinders and nonreversible motors. Four-way, two-position valves direct fluid flow to alternating ports. They are used to control double-acting cylinders and reversible motors. Two-way, NO valves allow fluid flow in the spring position. Two-way, NC valves stop fluid flow in the spring position. Three-way, NC valves stop fluid flow in the spring position and provide a path to exhaust. Four-way, three-position valves provide two working positions and one neutral (center) position.

An *actuator* is a device that moves the spool of a valve. Valve actuators may be manual or automatic. Valve actuators include solenoids, palm buttons, levers, foot actuators, springs, detents, and pilots. Fluid power valves can have one or more actuators per valve. Solenoids are used to electrically change the valve position. A *palm button actuator* is a device that is manually operated by a person's hand to change the valve position. A *lever actuator* is an actuator that changes the valve position using a lever. A *foot actuator* is an actuator that changes the valve position manually by using a person's foot. A *spring actuator* is an actuator that automatically returns a valve to a set position. A *detent actuator* is a device that automatically holds a valve in its last position. A *pilot actuator* is an actuator that changes the valve position with air or oil force.

Control Valves

A *control valve* is a valve that sets the flow or pressure of fluid or air in the circuit. A *flow control valve* is a valve used to control the volume of fluid that flows in a part of the circuit. The speed of cylinder and motors is controlled by controlling the rate of flow in the system. Pressure control valves control the pressure in the fluid power system. Their function is to relieve pressure (pressure relief), begin another function (sequence) in the system, or reduce pressure (pressure reducing). **See Figure 10-24.**

Flow control valves include fixed control valves, variable control valves, and flow control with check valve. Fixed valves control the volume of fluid flowing in that part of the circuit. The flow rate of variable control valves can normally be changed from 0% to 100%. The flow rate of a flow control with check valve can be changed but the flow is allowed in only one direction.

A *pressure relief valve* is a valve that limits the amount of pressure in a fluid power system. A pressure relief valve protects the pump and its driver (normally an electric motor) from overloading. The valve opens at a high pressure setting.

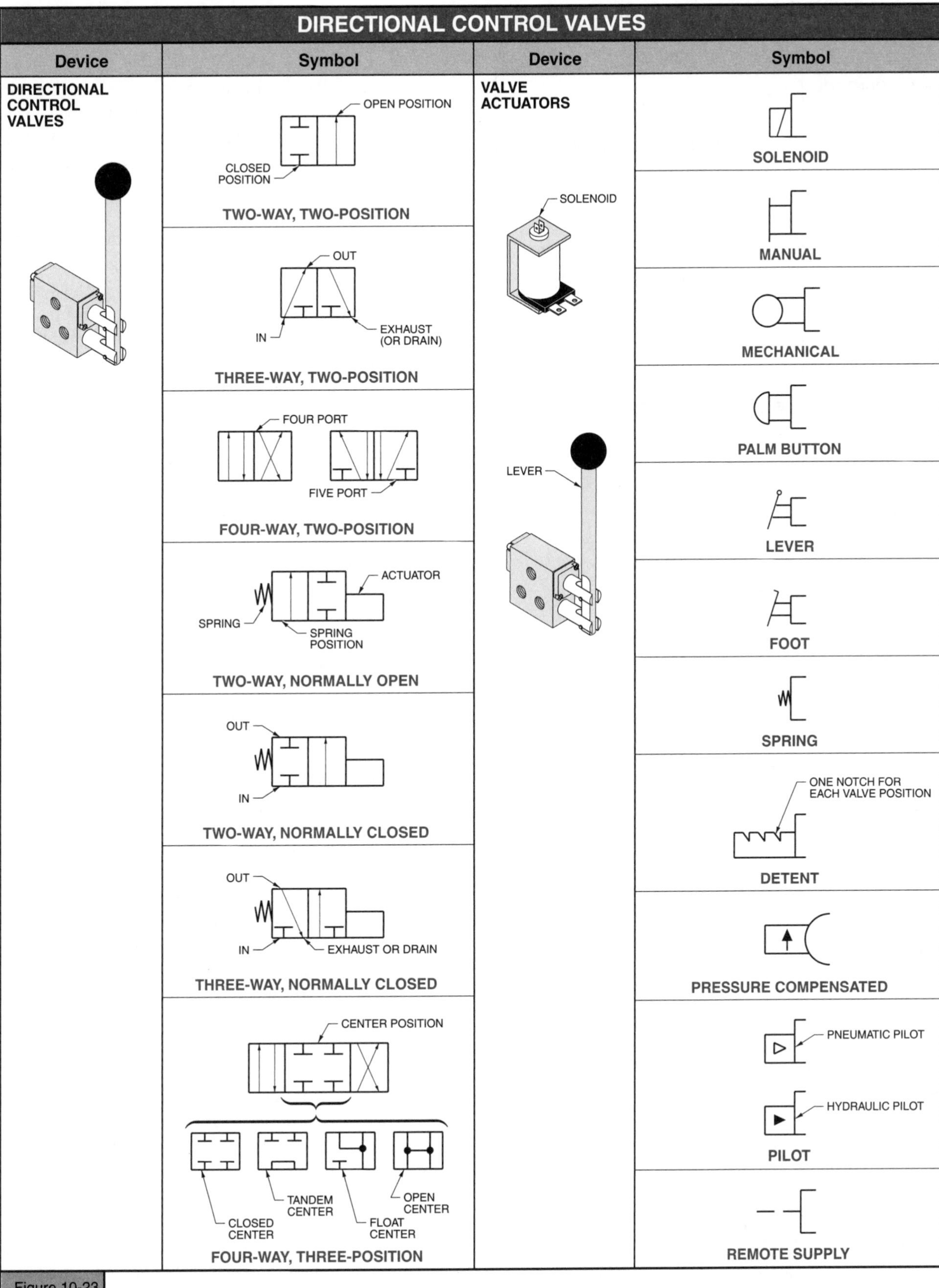
DIRECTIONAL CONTROL VALVES
Device
Symbol
Device
Symbol
DIRECTIONAL CONTROL VALVES
OPEN POSITION
CLOSED POSITION
TWO-WAY, TWO-POSITION
OUT
IN
EXHAUST (OR DRAIN)
THREE-WAY, TWO-POSITION
FOUR PORT
FIVE PORT
FOUR-WAY, TWO-POSITION
ACTUATOR
SPRING
SPRING POSITION
TWO-WAY, NORMALLY OPEN
OUT
IN
TWO-WAY, NORMALLY CLOSED
OUT
IN
EXHAUST OR DRAIN
THREE-WAY, NORMALLY CLOSED
CENTER POSITION
TANDEM CENTER
OPEN CENTER
CLOSED CENTER
FLOAT CENTER
FOUR-WAY, THREE-POSITION
VALVE ACTUATORS
SOLENOID
LEVER
SOLENOID
MANUAL
MECHANICAL
PALM BUTTON
LEVER
FOOT
SPRING
ONE NOTCH FOR EACH VALVE POSITION
DETENT
PRESSURE COMPENSATED
PNEUMATIC PILOT
HYDRAULIC PILOT
PILOT
REMOTE SUPPLY
Figure 10-23

CONTROL VALVES

Device	Symbol
FLOW CONTROL	ORIFICE FIXED FLOW CONTROL
	DIAGONAL LINE INDICATES VARIABLE VARIABLE FLOW CONTROL
	FLOW CONTROL WITH CHECK VALVE
PRESSURE RELIEF	INPUT SIDE SPRING LOADED TO DRAIN
SEQUENCE	INPUT SIDE EXTERNAL DRAIN OUTPUT SIDE
PRESSURE-REDUCING	INPUT SIDE OUTPUT SIDE

Figure 10-24

A *sequence valve* is a valve that directs fluid from one part of a circuit to another part only after the pressure has reached a set value. Sequence valves sequence the movement of different cylinders.

A *pressure-reducing valve* is a valve that reduces the pressure in one part of a circuit. Pressure on the output side can normally be adjusted from 0% to 100% of input pressure.

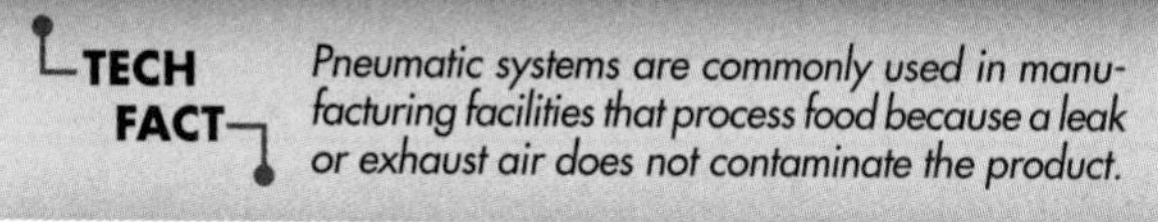

Miscellaneous Fluid Power Components

In addition to the main components in fluid power circuits, additional components are required for specialized functions. Miscellaneous fluid power components include check valves, pressure gauges, air mufflers, manual shutoff valves, pressure accumulators, and connecting lines. **See Figure 10-25.**

A *check valve* is a valve that allows the fluid to flow in one direction only. A *pressure gauge* is a device used to indicate pressure at given points in the system. Pressure gauges monitor and display system pressure. An *air muffler* is a device that reduces the noise that is made when air passes through a restriction. Air mufflers control noise from the air valve exhaust ports. A *manual shutoff valve* is a valve that is manually operated to shut OFF or apply fluid power. An *accumulator* is a device that stores fluid under pressure. Accumulators reduce the required pump size, provide additional volume in part of a circuit when required, and act as a shock absorber. Pressure can be developed from a spring, air charge, or weight.

Connecting lines include main (solid), pilot (dashed), exhaust or drain (dotted), and flexible lines. They interconnect various components in the circuit. Hose (or tubing) used must be rated at a high enough pressure, temperature, and flow rating. Hose must also be rated to work in the environment in which it is used.

Process Control and Instrumentation Standard Graphic Symbols

The Instrument Society of America (ISA) has developed standard symbols and abbreviations for industrial process and instrumentation equipment. In process control and instrumentation drawings, connecting lines with symbols are used to show the kind of signal being transmitted from one part of the system to another. Transmission lines may carry an electrical, pneumatic, hydraulic, mechanical, or other type of signal. The shape of the line and/or the abbreviations used indicate the exact type of signal. **See Figure 10-26.**

The signal lines may be drawn on a diagram entering or leaving the appropriate part of a symbol at any angle. Directional arrowheads shall be added to signal lines when needed to clarify the direction of flow. The electrical, pneumatic, or other power supply to an instrument is not shown unless it is essential to an understanding of the operation of the process or instrument. In process control and instrumentation drawings, abbreviations are used to denote the type of supply.

MISCELLANEOUS FLUID POWER COMPONENTS

Device	Abbr	Symbol
CHECK VALVES 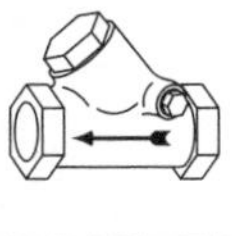		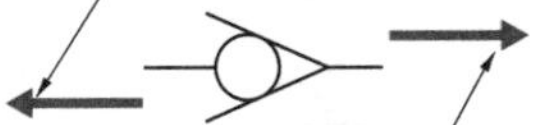 DIRECTION OF FREE FLOW DIRECTION OF NO FLOW
PRESSURE GAUGES 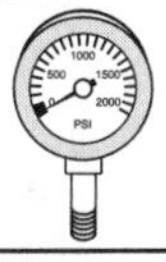		
AIR MUFFLERS 		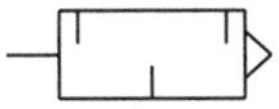
MANUAL SHUT-OFF VALVES 		
ACCUMULATORS 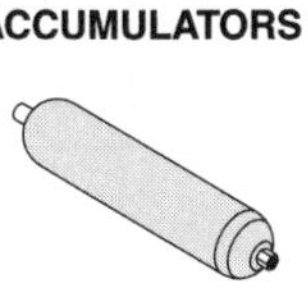		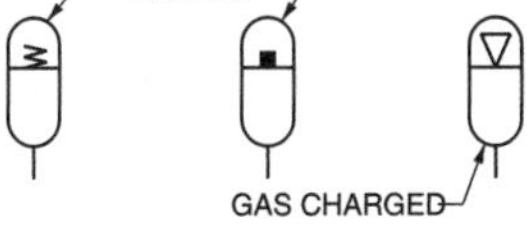 SPRING-LOADED WEIGHTED GAS CHARGED
CONNECTING LINES	R (RED)	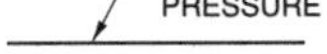 OPERATING PRESSURE MAIN LINE (SOLID)
	O (ORANGE)	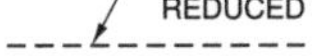 PILOT OR REDUCED PRESSURE PILOT LINE (DASHED)
	B (BLUE) G (GREEN)	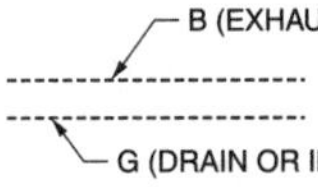 B (EXHAUST) G (DRAIN OR INTAKE) EXHAUST OR DRAIN (DOTTED LINE)
		 FLEXIBLE LINE

Figure 10-25

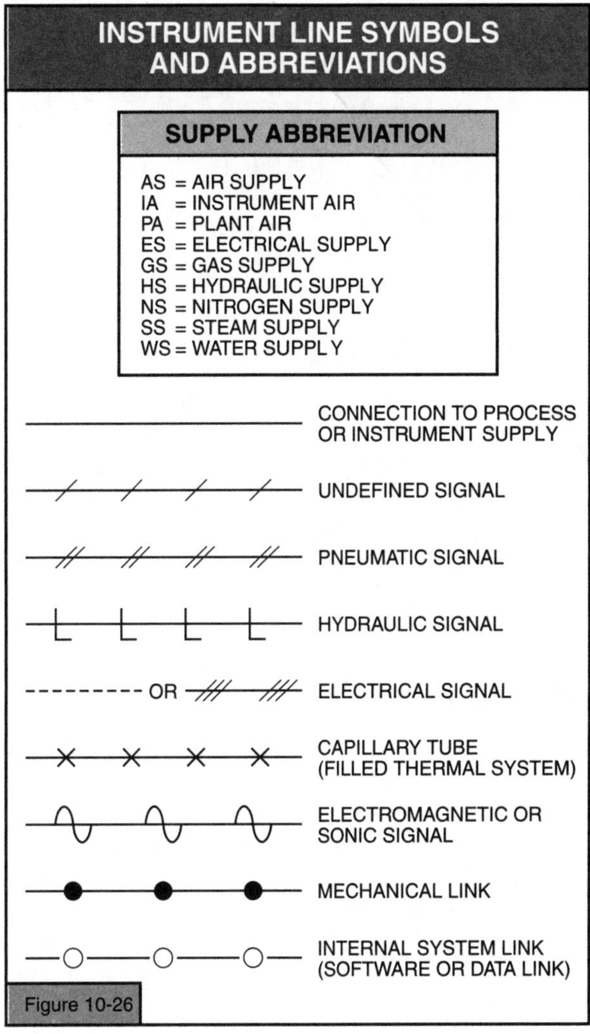

Figure 10-26

Architectural Symbols

An *architectural symbol* is a symbol used on floor plans, site (plot) plans, and detail drawings to show the location of electrical and other devices. Architectural symbols shown on plans and drawings are used for location purposes only and are not intended to show how a device is wired or connected. The actual wiring and connection is determined by the trade (electrical, plumbing, HVAC, etc.) that installs the device in accordance with the National Electrical Code® (NEC®), state codes, local codes, and accepted practices.

Electrical architectural symbols are used to show the location of lights, switches, receptacles, smoke detectors, power panels, and other electrical devices in residential, commercial, and industrial buildings. Electricians use prints (floor plans, etc.) to determine the location, type, and number of required devices to be installed in each area. **See Figure 10-27.**

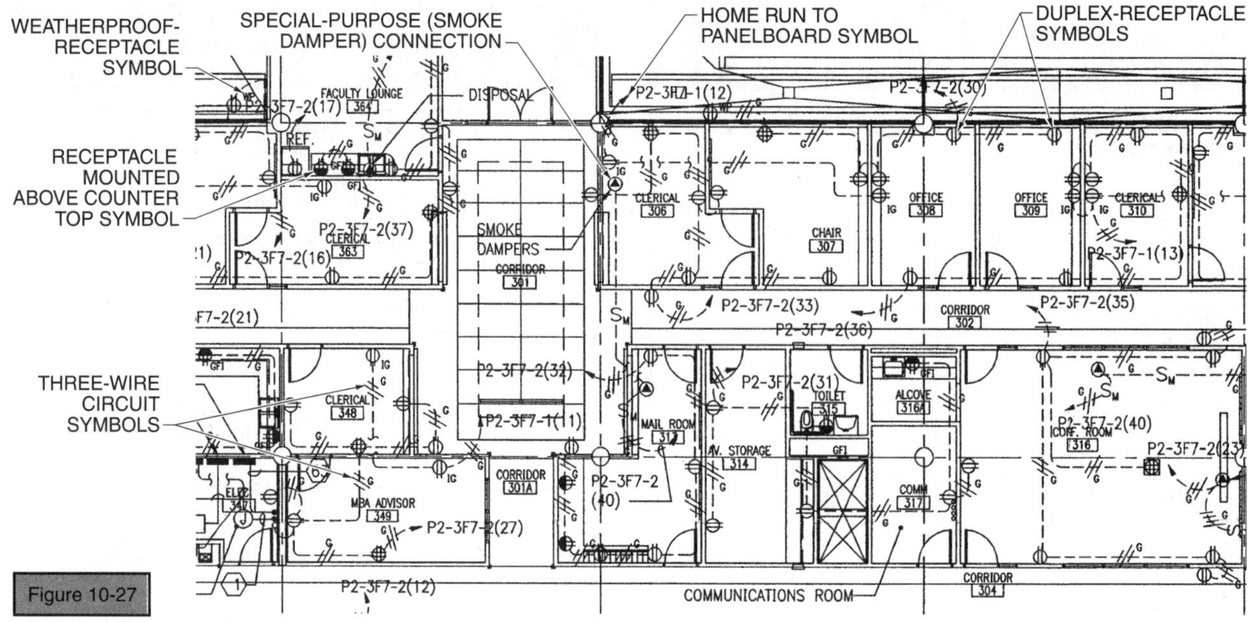

Figure 10-27

Lighting Symbols

Lighting symbols are used on residential and commercial prints to show the location, type of lamp bulb, and type of fixture. In addition to showing the location and type (incandescent, fluorescent, or HID), the symbol can also show the type of required light fixture mounting, such as ceiling or wall. **See Figure 10-28.**

An *incandescent lamp* is an electric lamp that produces light by the flow of current through a tungsten filament inside a gas-filled, sealed glass bulb. Incandescent lamp bulbs are available in a variety of shapes and sizes. The outer glass of the bulb is available in a variety of colors for applications in which a color other than white light is desired. Most commercial and industrial lighting applications use fluorescent and high- intensity discharge (HID) lighting.

Fluorescent lamps are used for large indoor lighting requirements because fluorescent lamps provide more light per watt than incandescent lamps and operate at lower temperatures. A *fluorescent lamp* is a low-pressure discharge lamp in which ionization of mercury-vapor transforms ultraviolet energy generated by the discharge into light. Standard fluorescent lamps vary in diameter from ⅝″ to 2⅛″ and in length from 6″ to 96″. Fluorescent lamp bulbs that produce different shades of color are available.

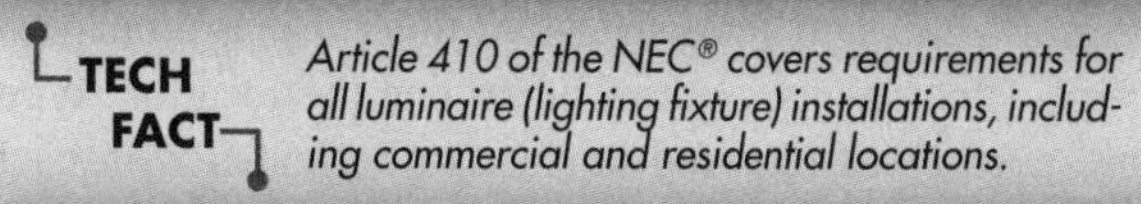
TECH FACT Article 410 of the NEC® covers requirements for all luminaire (lighting fixture) installations, including commercial and residential locations.

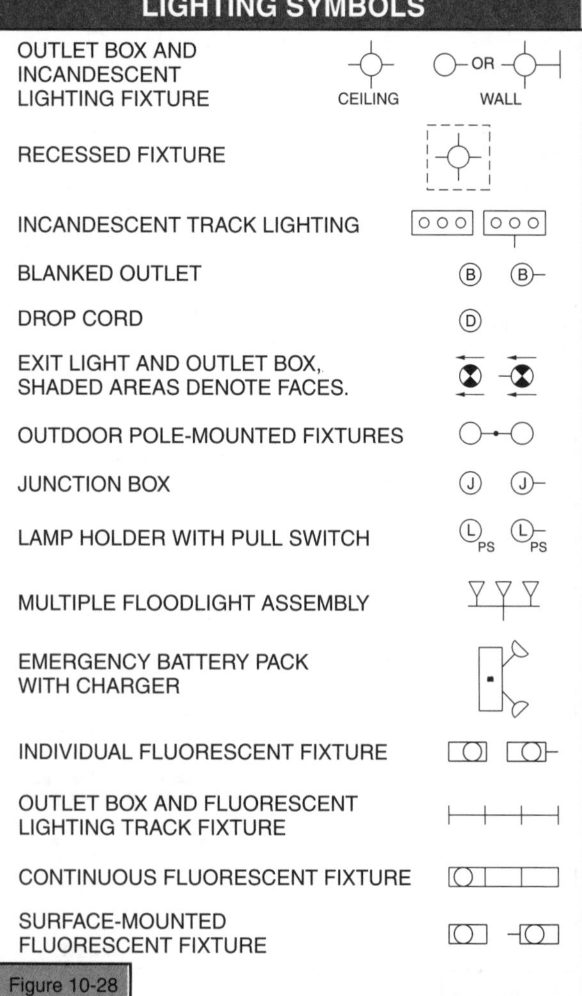

Figure 10-28

A *high-intensity discharge (HID) lamp* is a lamp that produces light from an arc tube. An *arc tube* is the light-producing element of an HID lamp. High-intensity discharge lamps produce more light per watt than any other lamps. The disadvantage of most HID lamps is that they have poor color rendering. *Color rendering* is the appearance of a color when illuminated by a light source. Low-pressure sodium HID lamps produce a yellow to yellow-orange light. Mercury-vapor and metal-halide HID lamps produce a light yellow light. High-pressure sodium HID lamps produce a golden white light. In spite of the poor color rendering of HID lamps, they are still preferred for outdoor lighting and large-area lighting, such as for sports stadiums and warehouses, because they produce far more light per watt than incandescent or fluorescent lamps.

Switch Symbols

All electrical circuits must be controlled. A *switch* is a device that is used to start, stop, or redirect the flow of current in an electrical circuit. The most common types of switches used in lighting circuits are two-way, three-way, and four-way switches. A *two-way switch* is a single-pole, single-throw (SPST) switch that has an ON (closed) position and an OFF (open) position. A *three-way switch* is single-pole, double-throw (SPDT) switch that is used to control lamps from two locations. A *four-way switch* is double-pole, double-throw (DPDT) switch that is used between two three-way switches when controlling a lamp from three or more switch locations. Other electrical switches (dimmers) are used for specialized functions, such as controlling the brightness of a lamp. **See Figure 10-29.**

Receptacle Symbols

Electrical energy can be transformed to other forms of energy, such as heat, light, sound, and mechanical (linear or rotating) energy to produce work. A *receptacle* is a device used to connect equipment with a cord and plug to an electrical system. Receptacle symbols are used on a floor plan to show the location of general-purpose receptacles, dedicated receptacles, duplex receptacles, and special-purpose receptacles. Standard duplex receptacles are the most common type of receptacles used for general power distribution throughout a building. **See Figure 10-30.**

Ruud Lighting, Inc.

High-intensity discharge lighting is commonly used in large industrial buildings.

SWITCH SYMBOLS

SINGLE-POLE SWITCH (TWO-WAY)	S	DIMMER	S_{DM}
DOUBLE-POLE SWITCH	S_2	REMOTE CONTROL SWITCH	S_{RC}
THREE-WAY SWITCH	S_3	WEATHERPROOF SWITCH	S_{WP}
FOUR-WAY SWITCH	S_4	FUSED SWITCH	S_F
AUTOMATIC DOOR SWITCH	S_D	WEATHERPROOF FUSED SWITCH	S_{WP}
KEY-OPERATED SWITCH	S_K	TIME SWITCH	S_T
CIRCUIT BREAKER	S_{CB}	CEILING PULL SWITCH	Ⓢ
WEATHERPROOF CIRCUIT BREAKER	S_{WCB}	SWITCH AND SINGLE RECEPTACLE	$-\ominus_S$
		SWITCH AND DOUBLE RECEPTACLE	$=\ominus_S$

Figure 10-29

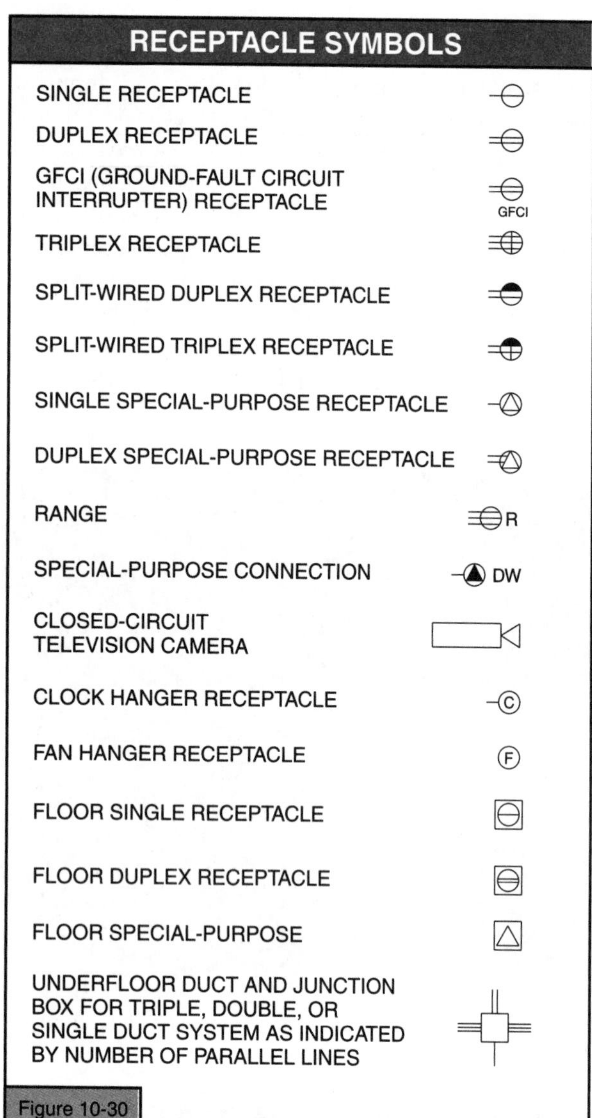

Figure 10-30

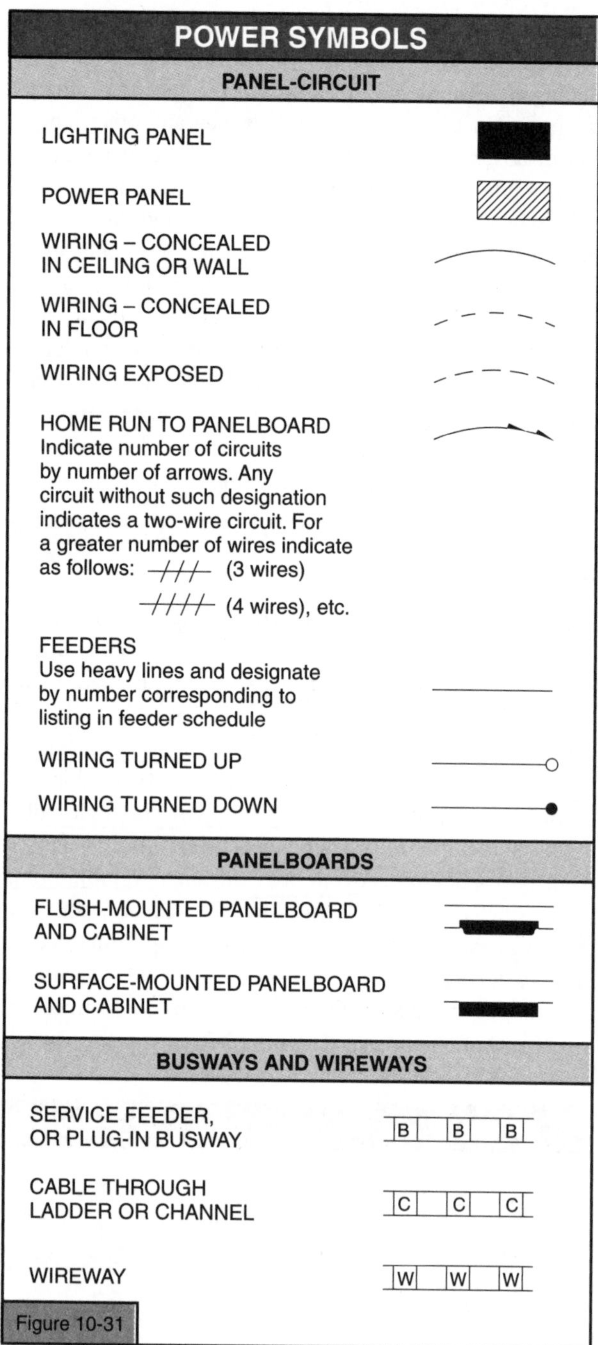

Figure 10-31

Power Symbols

Electrical power must be distributed throughout a building in order to be used by loads such as lamps, computers, TVs, washing machines, and stereos. Power is normally brought into a building from a utility company step-down transformer to a main power and lighting panel. Power and lighting panels are also referred to as panelboards. The main panelboard in a building includes a main multiphase circuit breaker or fuses to protect the entire electrical system. Smaller circuit breakers or fuses protect each individual branch circuit connected to the main panelboard. Various line types (solid or dashed) are used on prints to indicate how the conductors are run to connect the individual receptacles, loads, and switches in each circuit. **See Figure 10-31.**

A power panel is a wall-mounted distribution cabinet in commercial and industrial buildings. A *switchboard* is a freestanding distribution cabinet in commercial and industrial buildings. The individual circuit breakers or fuses in a power panel and lighting panelboard supply power to individual branch circuits. A *branch circuit* is the portion of an electrical circuit between the circuit breaker or fuse and the receptacles, lamps, and loads connected in the circuit.

A *busway* is a metal-enclosed distribution system of busbars available in prefabricated sections. By bolting busway sections together, electrical power is distributed throughout an area and easily available when needed.

Signal Symbols

In addition to lighting and power (receptacle) circuits, electrical circuits can also include specialized circuits for doorbells, sound systems, fire alarms, and other signaling devices. Along with lamps and receptacles, signaling devices are included on electrical prints to show location and usage. **See Figure 10-32.**

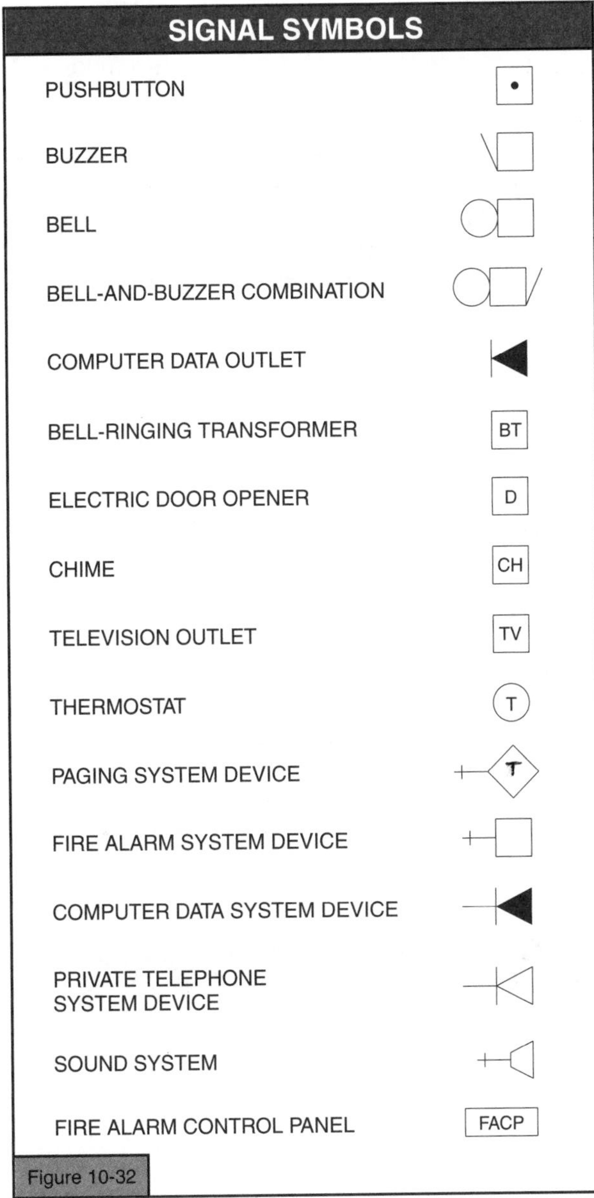

Figure 10-32

The main purpose of signaling devices is to provide a visual and/or audible signal. Alarms, horns, sirens, and bells are typically used to produce a loud signal (warning) that is used to draw attention to an abnormal condition. Audible signals are used in applications that typically require a faster response than a visual signal alone could provide.

Plot-Plan Symbols

When a building is designed, the building floor plan is used to show the location of all electrical devices (lamps, receptacles, power and lighting panels) inside a structure and on its outside walls. Buildings are located at fixed positions on a plot of ground. A *plot plan* is an aerial view of one building lot and provides specific information about the lot.

The building property and adjacent properties include objects such as trees, fences, and other items that must be considered when installing or servicing electrical systems. Symbols are used on plot plans to show the location and type of objects and electrical devices located around a building, such as light standards, fire hydrants, traffic signals, natural gas lines, and manholes. **See Figure 10-33.**

Distribution and Lighting Symbols

Electrical power is distributed to a building using aboveground, underground, or a combination of aboveground and underground devices and equipment. In the past, electrical-distribution and lighting-distribution systems were run aboveground to reduce cost and provide easier access for repairs and expansion. Today, underground electrical-distribution and lighting-distribution systems are preferred to help prevent power outages caused by wind, rain, and ice storms. Aboveground and underground electrical devices and equipment are shown using standard symbols. **See Figure 10-34.**

Electrical prints are used when installing hardwired signaling devices such as smoke alarms.

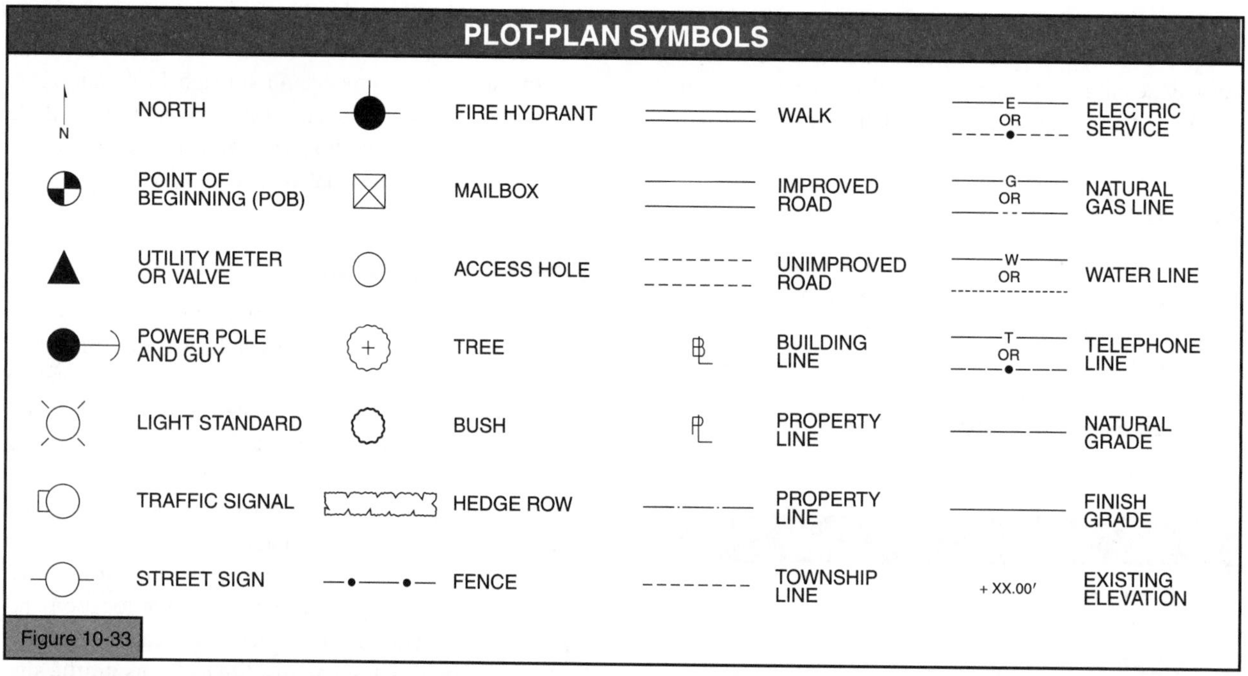

PLOT-PLAN SYMBOLS

Symbol	Description	Symbol	Description	Symbol	Description	Symbol	Description
N	NORTH		FIRE HYDRANT		WALK	E OR	ELECTRIC SERVICE
	POINT OF BEGINNING (POB)		MAILBOX		IMPROVED ROAD	G OR	NATURAL GAS LINE
	UTILITY METER OR VALVE		ACCESS HOLE		UNIMPROVED ROAD	W OR	WATER LINE
	POWER POLE AND GUY		TREE		BUILDING LINE	T OR	TELEPHONE LINE
	LIGHT STANDARD		BUSH		PROPERTY LINE		NATURAL GRADE
	TRAFFIC SIGNAL		HEDGE ROW		PROPERTY LINE		FINISH GRADE
	STREET SIGN		FENCE		TOWNSHIP LINE	+ XX.00′	EXISTING ELEVATION

Figure 10-33

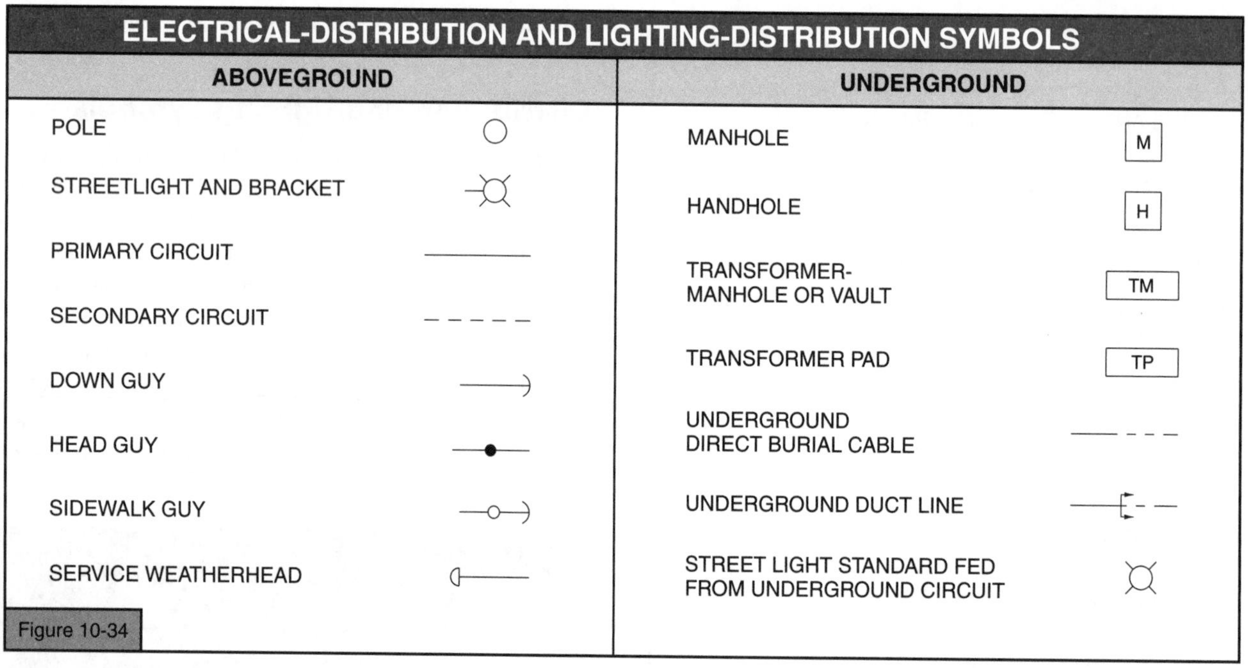

ELECTRICAL-DISTRIBUTION AND LIGHTING-DISTRIBUTION SYMBOLS

ABOVEGROUND		UNDERGROUND	
POLE		MANHOLE	M
STREETLIGHT AND BRACKET		HANDHOLE	H
PRIMARY CIRCUIT		TRANSFORMER-MANHOLE OR VAULT	TM
SECONDARY CIRCUIT		TRANSFORMER PAD	TP
DOWN GUY		UNDERGROUND DIRECT BURIAL CABLE	
HEAD GUY		UNDERGROUND DUCT LINE	
SIDEWALK GUY		STREET LIGHT STANDARD FED FROM UNDERGROUND CIRCUIT	
SERVICE WEATHERHEAD			

Figure 10-34

Refer to the CD-ROM in the back of the book for Chapter 10 Quick Quiz® and related information.

Conductors

Electrical circuits and components are connected using conductors. A *conductor* is a material that has very little resistance and permits electrons to move through it easily. Conductors include wire, cable, and cord. **See Figure 11-1.**

Conductors are available as individual wire or in groups, such as cable and cord. *Wire* is any individual conductor. *Cable* is two or more conductors grouped together within a common protective cover and used to connect individual components. *Cord* is two or more conductors grouped together and used to deliver power to a load by means of a plug.

Most individual conductors are enclosed in an insulated cover to protect the conductor, increase safety, and meet code requirements. Some individual conductors, such as ground wire, may be bare. In coaxial cable, an outer braided conductor is wrapped around an insulated inner conductor. The braided conductor is used to shield the inner conductor from outside electromagnetic interference.

Conductor Materials

Conductor materials include copper, aluminum, copper-clad aluminum, steel, and bronze. Copper (Cu) and aluminum (Al) are the most commonly used materials. Copper is the most common. Copper is preferred because it has a lower resistance than aluminum for any given wire size. For this reason, aluminum conductors must be sized one size larger than copper conductors.

Copper-clad aluminum is a conductor that has copper bonded to an aluminum core. The total amount of copper used is less than 10% of the conductor. Copper is used to counter the disadvantages of aluminum connections. Steel conductors are made from a special high-strength, low-resistance steel. To reduce corrosion, the steel wire is coated with galvanized zinc. Steel wires and cables are used in some aerial and telephone applications. Steel is used because it is more economical in applications that allow its use. Bronze conductors were used on some old trolley systems, but are rarely used today.

Wire Sizes

Wire is sized by using a number, such as No. 12 or No. 14. The wire number is based on the American Wire Gauge (AWG) numbering system. The smaller the AWG number, the larger the conductor and the more current-carrying capacity. For example, a No. 12 copper wire is larger in diameter than a No. 14 copper wire and may carry 5 A more than the No. 14 (No. 12 rated at 20 A; No. 14 rated at 15 A). The wire size used for a circuit depends on the maximum current that the wire must carry and the material of the wire. See NEC® Tables 310.16 through 310.19. **See Figure 11-2.**

Conductors that are smaller than a No. 8 AWG may be either solid or stranded. No. 8 AWG and larger conductors are stranded. Conductors that are No. 18 and No. 16 AWG are normally used for the power cords of small appliances, plug-in lamps, speakers, and bell circuits. No. 14 and No. 12 AWG conductors are used for wiring most lighting circuits and supplying power to standard receptacle outlets. No. 10, No. 8, No. 6, and No. 4 AWG conductors are used for wiring electric appliances, such as ranges, water heaters, furnaces, clothes dryers, and air conditioners. They are also used for supplying power to subpanels and large motors. No. 3, No. 2 AWG, and larger conductors are used for supplying power to main service panels.

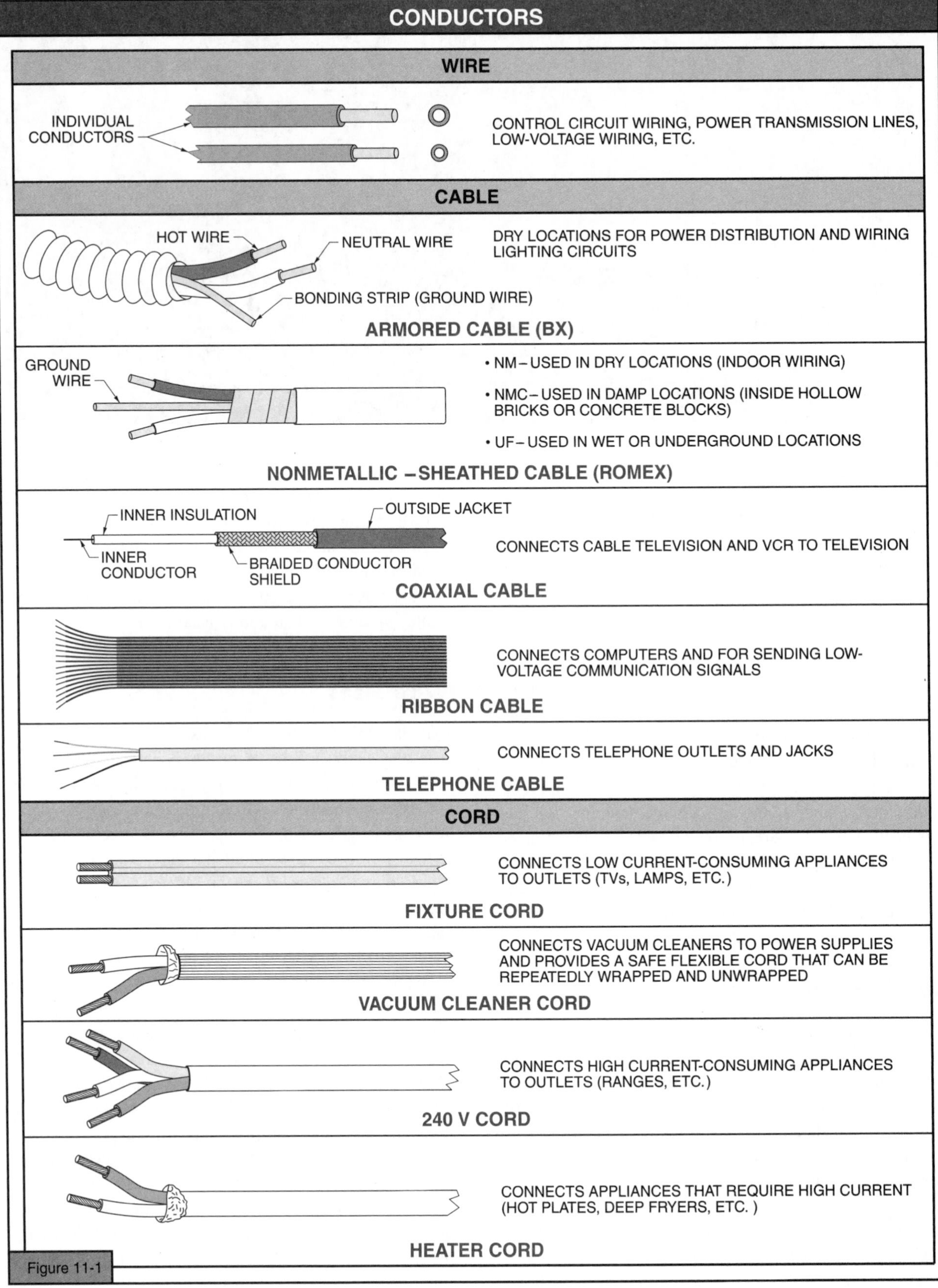
CONDUCTORS
WIRE
INDIVIDUAL CONDUCTORS
CONTROL CIRCUIT WIRING, POWER TRANSMISSION LINES, LOW-VOLTAGE WIRING, ETC.
CABLE
HOT WIRE
NEUTRAL WIRE
BONDING STRIP (GROUND WIRE)
DRY LOCATIONS FOR POWER DISTRIBUTION AND WIRING LIGHTING CIRCUITS
ARMORED CABLE (BX)
GROUND WIRE
• NM – USED IN DRY LOCATIONS (INDOOR WIRING)
• NMC – USED IN DAMP LOCATIONS (INSIDE HOLLOW BRICKS OR CONCRETE BLOCKS)
• UF – USED IN WET OR UNDERGROUND LOCATIONS
NONMETALLIC – SHEATHED CABLE (ROMEX)
INNER INSULATION
OUTSIDE JACKET
INNER CONDUCTOR
BRAIDED CONDUCTOR SHIELD
CONNECTS CABLE TELEVISION AND VCR TO TELEVISION
COAXIAL CABLE
CONNECTS COMPUTERS AND FOR SENDING LOW-VOLTAGE COMMUNICATION SIGNALS
RIBBON CABLE
CONNECTS TELEPHONE OUTLETS AND JACKS
TELEPHONE CABLE
CORD
CONNECTS LOW CURRENT-CONSUMING APPLIANCES TO OUTLETS (TVs, LAMPS, ETC.)
FIXTURE CORD
CONNECTS VACUUM CLEANERS TO POWER SUPPLIES AND PROVIDES A SAFE FLEXIBLE CORD THAT CAN BE REPEATEDLY WRAPPED AND UNWRAPPED
VACUUM CLEANER CORD
CONNECTS HIGH CURRENT-CONSUMING APPLIANCES TO OUTLETS (RANGES, ETC.)
240 V CORD
CONNECTS APPLIANCES THAT REQUIRE HIGH CURRENT (HOT PLATES, DEEP FRYERS, ETC.)
HEATER CORD

Figure 11-1

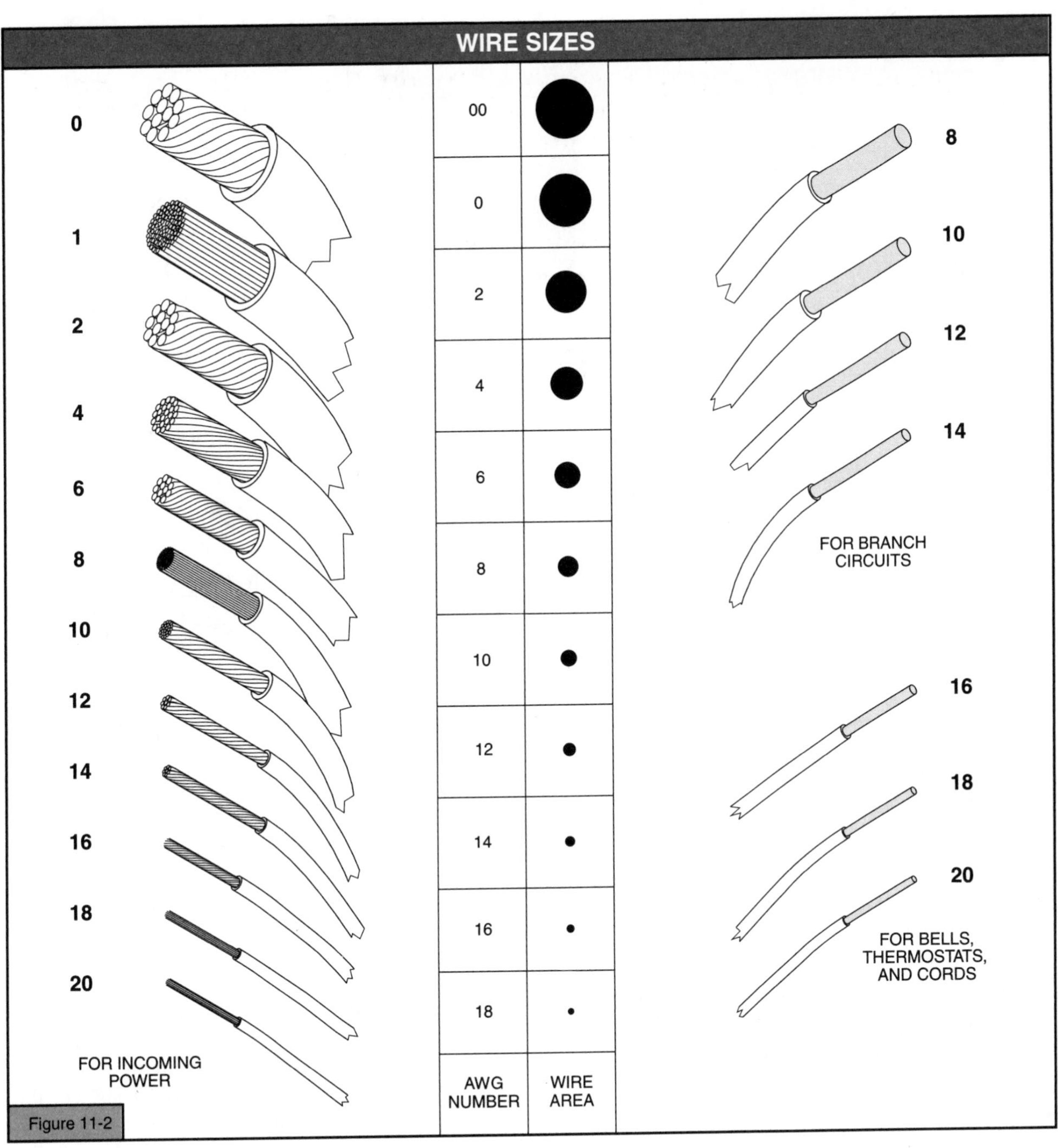

Figure 11-2

Conductor Ratings

Insulated conductors are rated based on the current they may safely carry without damaging the insulation. The current rating depends on the wire size, type of insulation used, conductor's temperature rating, and the type of wire (copper or aluminum). The allowable amount of current a wire may safely carry is listed in NEC® tables. For example, a No. 10 AWG, Cu conductor with Type TW insulation that is designed to operate in an ambient temperature not to exceed 60°C (140°F), is rated at 30 A. See NEC® Table 310.16. **See Figure 11-3.**

The current flowing through a wire must be kept below the wire's rated limit to maintain a safe system. Copper (Cu) is a better conductor (less resistance) than aluminum (Al), and may carry more current for any given AWG size. A large wire is a better conductor (less resistance) than a small wire, and may also carry more current. High temperature-rated insulation may also carry more current because a high temperature is required to break down the insulation material.

COPPER CONDUCTOR RATINGS*				
AWG	AMPACITY	DIA (Mils)	USE	TYPE
18	----	40.0	① Low-voltage signals, alarms, thermostats, bells, buzzers, etc.	SOLID
17	----	45.0		
16	----	51.0		
15	----	57.0		
14	20	64.0	② General lighting and receptacle circuits	SOLID (STRANDED)
12	25	81.0	③ Small appliance circuits	
10	30	102.0	④ Large appliance circuits	
8	40	128.0		STRANDED (SOLID)
6	55	162.0	⑤ Service conductors, very large loads, and power distribution	STRANDED
4	70	204.0		
3	85	229.0		
2	95	258.0		
1	110	289.0		
0	125	325.0		
00	145	365.0		
000	165	410.0		

SIZES INCREASE / NUMBERS DECREASE

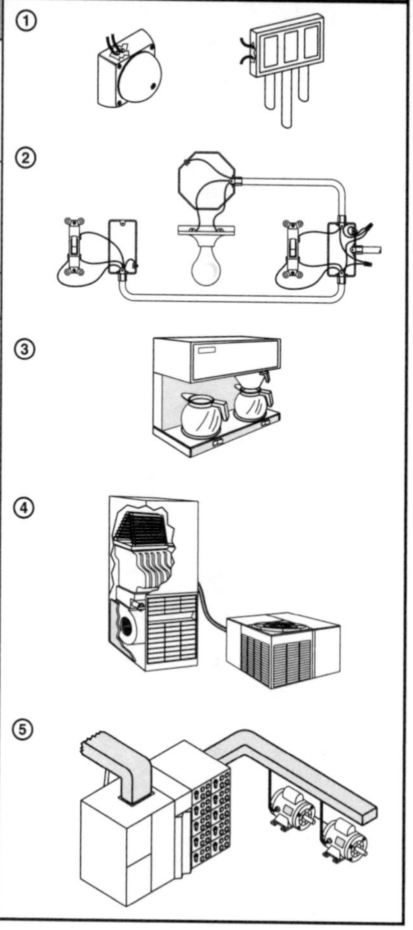

* @ 60° C

Figure 11-3

> **TECH TIP** *Use only cable-pulling lubricant when pulling wire through conduit. Never use oil or grease as a lubricant when pulling wire through conduit, because they can damage the insulation.*

Conduit

Conduit is a hollow pipe used to protect conductors. The conductors are placed inside the conduit. Conduit is either metal or plastic and may be rigid or flexible. Most conduit used in electrical wiring is rigid metal. **See Figure 11-4.**

The two types of metal conduit used in electrical wiring are rigid conduit and electrical metallic tubing (EMT). Rigid conduit uses threaded fittings. EMT uses force-fit fittings. Rigid conduit and EMT conduit may be used indoors or outdoors. Rigid conduit is normally used for underground runs. The exact use of rigid or EMT conduit is determined by local codes and accepted standards.

Plastic conduit is normally referred to as electrical nonmetallic tubing (ENT). ENT is used similarly to metallic conduit, except a plastic bonding cement is used to secure the fittings. ENT is highly resistant to fire, moisture, and many types of chemicals. For this reason, ENT is used in damp and corrosive locations as well as dry locations. ENT may also be used in concrete but is not used in hazardous locations or to support equipment fixtures.

Flexible conduit is a hollow metal or plastic pipe that is easily bent and used to protect conductors. Flexible conduit is used similarly to rigid conduit, in that the conduit is first installed and the wires are added after the conduit is securely in place. Flexible conduit is sometimes used in short runs, such as connecting a central air conditioning unit to the disconnect switch. Flexible conduit is also used in areas that may include vibration, such as connecting a motor to the control box.

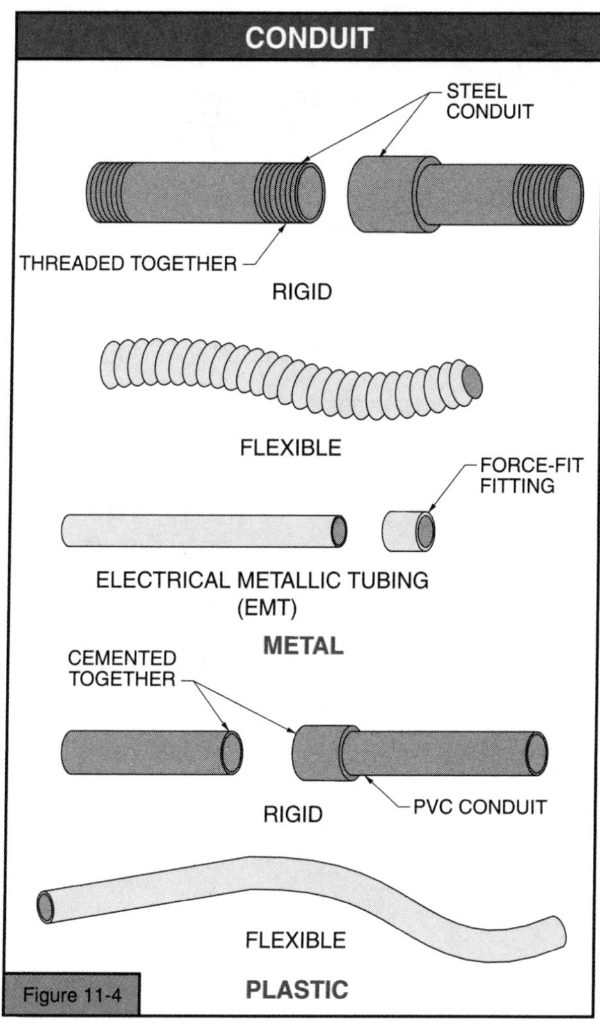

Figure 11-4

Bending Conduit

Conduit is bent to clear obstructions when machinery and equipment are initially installed or relocated. Good techniques and tools are required for making conduit bends. Conduit can be bent properly using hand conduit benders (hickey), mechanical conduit benders, or electric conduit benders. All conduit benders have bending shoes with high supporting sidewalls to prevent flattening or kinking of the conduit and are designed for quick, efficient conduit bending.

A hand conduit bender is used by contractors and maintenance personnel who occasionally need to bend conduit. Hand conduit benders are normally made from heat-treated aluminum or iron. Some hand conduit benders are available with a flared handle that can be used as a conduit straightening tool. A slow, steady hand and foot pressure is used when bending conduit with a hand conduit bender. When bending conduit on the floor, the conduit is secured against the floor by the foot of the bender. When bending conduit in the air, pressure is exerted on the bender as close to the body as possible.

A 30° bend is one of the most commonly used bends because it is the easiest bend to calculate. Conduit is bent using a hand conduit bender by measuring the required length of conduit, checking the required angles, and using the bender sight marks to create accurate bends. In addition to 30° bends, hand conduit benders are also used to make 45° and 90° bends.

Most hand conduit benders are designed so that a 45° angle is achieved by placing the bender on the conduit and raising the handle until it is in the vertical position. **See Figure 11-5.** Hand conduit benders are also used to make 90° bends for corners or formed to a predetermined length for use as stub-ups for floor installation. To make a 90° bend, the required distance is lined up with the edge of the bender. Force is applied to the bender, and the conduit is bent to form a 90° angle. The accuracy of the 90° bend is checked using a level. One leg of the conduit is checked on the horizontal plane, the bent leg is checked on the vertical plane.

Back-to-Back Bends

When running conduit from one outlet box to another box, a back-to-back bend may be required. A *back-to-back bend* is a bend that consists of two 90° bends on one piece of conduit. To make a back-to-back bend, the first bend is made as a regular 90° bend. The distance between boxes is laid out on the tubing. The direction of the bender is reversed, and point *A* of the bender is placed over the correct distance mark on the conduit. The bender handle is raised to form a 90° bend, which finishes the second right angle. Both bends should be parallel with each other.

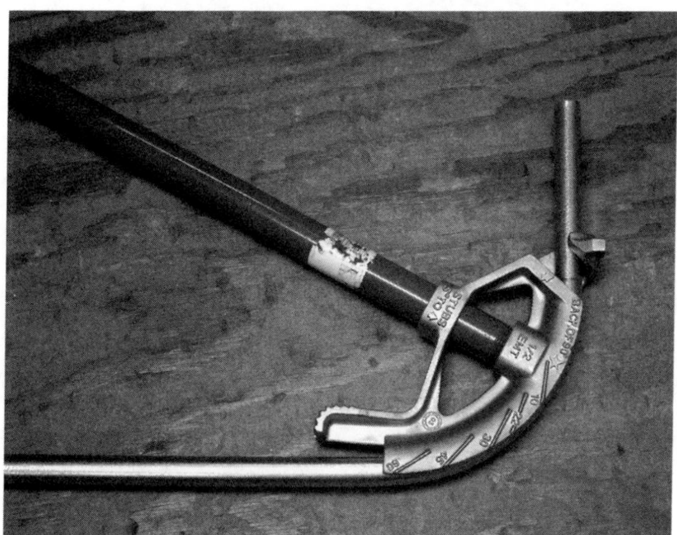
Maintenance personnel and contractors who occasionally need to bend conduit use hand conduit benders.

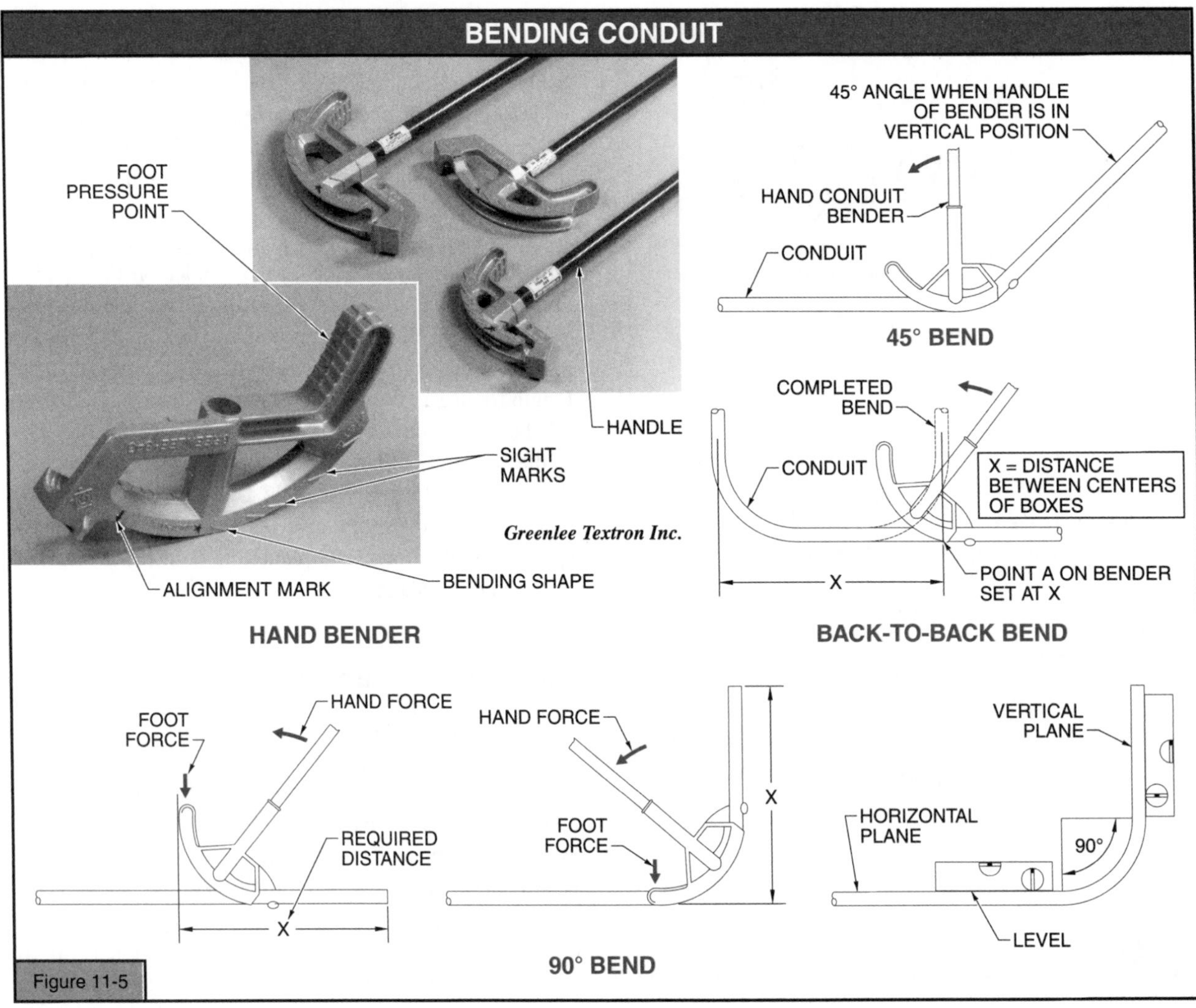

Figure 11-5

Offset Bends

Conduit runs are often interrupted by obstructions and changes in direction that cannot be bypassed with 90° bends. Obstructions are bypassed by using an offset bend. An *offset bend* is a bend that consists of two 45° angles on one piece of conduit. **See Figure 11-6.** To achieve an offset bend, parallel lines are drawn apart on a floor or smooth surface equal to the length (throw) of the offset. The first bend is made at a 45° angle by laying the conduit on the floor and bringing the handle of the bender to the vertical position. The second bend is made by placing the end of the bender at the throw mark and bending until the two legs are parallel. A level is used to check the accuracy of the offset bend.

Double Offset (Saddle) Bends

A double offset (saddle) bend may be necessary to clear obstructions such as beams, piping, or parts of machinery. A *double offset (saddle) bend* is a bend that consists of four 45° offset bends on one piece of conduit. **See Figure 11-7.** To perform a double offset bend, parallel lines are drawn on a floor or smooth surface equal to the saddle distance required for reference. Marking parallel lines on each side of the conduit helps eliminate any deviation from center. To make a double offset bend, a 45° bend is made. The conduit is rotated 180° in the bender and a 45° return bend is made. At the correct distance another 45° bend is made, and finally a second 45° return bend is made.

Mechanical Conduit Benders

A mechanical conduit bender uses a ratchet device to bend conduit. Some mechanical conduit benders are only used on certain types of conduit, while others have changeable bending shoes that allow one bender to be used on all sizes and types of conduit. A mechanical conduit bender is mounted on a frame and may have wheels so it can be moved easily.

As the ratchet is pulled, a series of teeth rotate the bending shoe, bending the conduit. Mechanical conduit benders are generally used by contractors or maintenance personnel who are required to bend conduit frequently but in small volumes. Mechanical conduit benders normally have built-in bending degree indicators and bending charts.

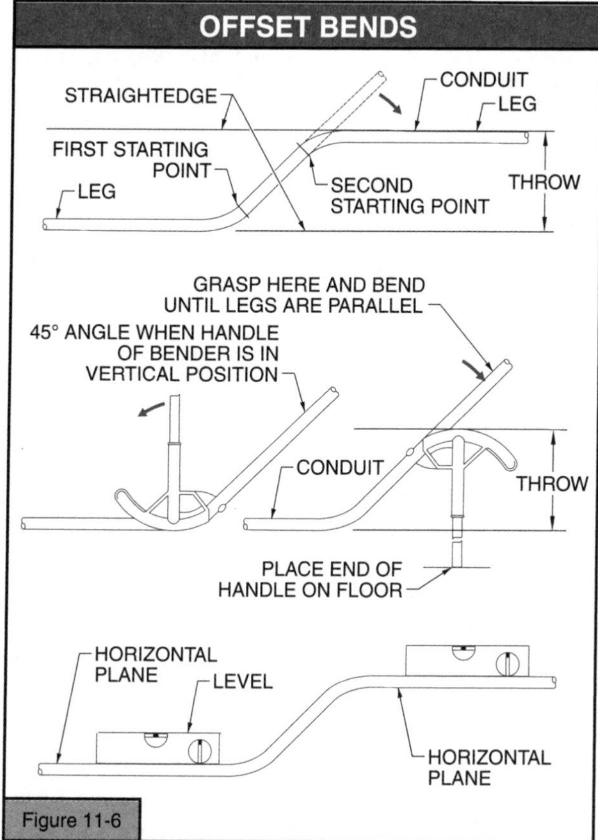

Figure 11-6

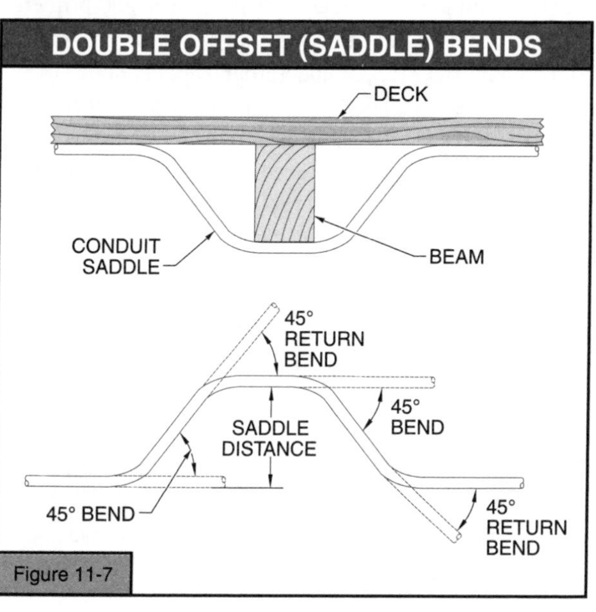

Figure 11-7

Electric Conduit Benders

An electric conduit bender uses an electric motor that powers a chain and sprocket drive. High-volume conduit installers generally use electric conduit benders. Electric conduit benders have interchangeable bending shoes that allow the bender to be used on all types of conduit. Some electric conduit benders are programmable, with digital readouts that help achieve quick, accurate, repetitive bends.

NEC® Conduit Bending Requirements

The NEC® states that bends in conduit shall be made so that the conduit is not damaged and that the internal diameter of the conduit is not effectively reduced. In addition, the NEC® states that a run of electrical metallic tubing from outlet to outlet, from fitting to fitting, or from outlet to fitting shall not contain more than the equivalent of four quarter bends (360° total) including the bends located immediately at the outlet or fitting. NEC® table 344.24 indicates the minimum bend radius acceptable for various sizes of conduit. **See Figure 11-8.**

MINIMUM CONDUIT BENDING RADIUSES

Conduit Size*	One Shot and Full Shoe Benders*	Other Bends*
½	4	4
¾	4½	5
1	5¾	6
1¼	7¼	8
1½	8¼	10
2	9½	12
2½	10½	15
3	13	18
3½	15	21
4	16	24
5	24	30
6	30	36

* in inches

Figure 11-8

Nonmetallic Cable

Nonmetallic cable is a tough, plastic-sheathed cable that is normally made of a moisture-resistant, flame-retardant material. Nonmetallic cable normally includes two or three insulated conductors and a bare ground wire. **See Figure 11-9.** Nonmetallic cable is commonly referred to as Romex (Romex is a trade name). A two-conductor-with-ground nonmetallic cable normally includes one black conductor, one white conductor, and a ground conductor. A three-conductor nonmetallic cable with ground normally includes one black, one white, one red, and a ground conductor. The black conductor is used for the hot line. The white conductor is used for the neutral line. The red conductor is normally used for the switched conductor or as a second hot conductor. The ground conductor (ground wire) is either bare or is insulated with a green covering.

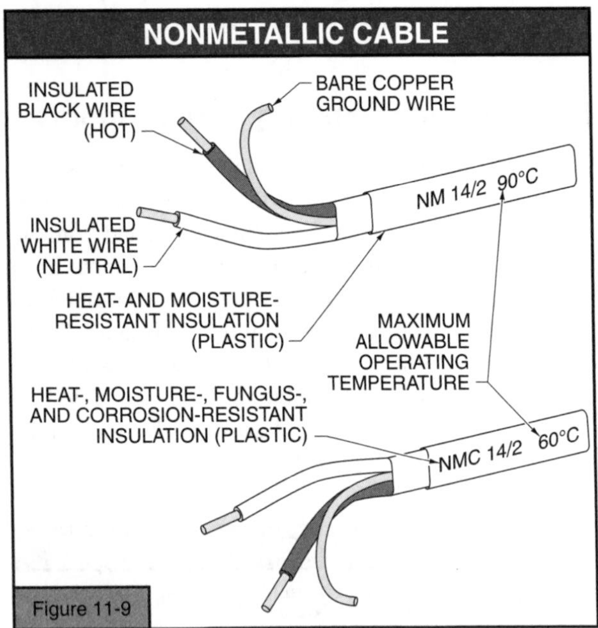

Figure 11-9

NM or NMC nonmetallic cable is normally used for most wiring applications. NM nonmetallic cable has a flame-resistant and moisture-resistant covering. NM nonmetallic cable may be used in dry locations and may be either concealed or exposed. NMC nonmetallic cable may be used in the same locations as NM nonmetallic cable, but is also fungus- and corrosion-resistant. NMC nonmetallic cable is used in damp, moist, or corrosive locations, such as the inside of hollow bricks or concrete blocks. Some local codes allow the use of nonmetallic cable only in certain locations.

Armor-Clad Cable

Armor-clad cable is a cable that has an outer cover made of flexible galvanized steel. **See Figure 11-10.** Armor-clad cable is referred to as sheathed cable or BX. BX was a trademark name, but now it is a generic name for all flexible armored cable. Inside the flexible steel cover are individually wrapped and insulated conductors, as well as a bare conductor which is used for grounding. Armor-clad cable is used in dry locations only. Some local codes prohibit the use of armor-clad cable. For this reason, most armor-clad cable has been replaced by nonmetallic cable.

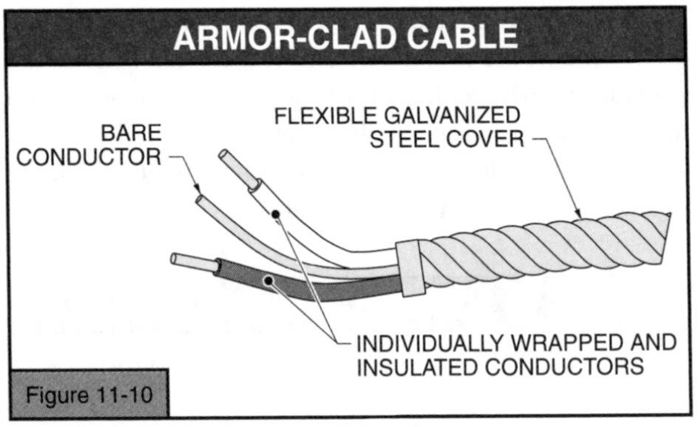

Figure 11-10

Communication Cable

Cable is used to deliver electrical power or signals from one location to another location within an electrical system. Power cable is used to deliver higher powers within the system. Power cables commonly deliver 90 VDC, 115 VAC, 180 VDC, 230 VAC, or 460 VAC to residential, commercial, and industrial loads. However, power cables also deliver higher power to lower voltage loads, such as in automobile, marina, or aviation applications.

Cable is also used to deliver communication signals within an electrical system. *Communication cable* is cable (usually copper or fiber optic) used to transmit data from one location (or device) within a system to another location. Communication cable is also commonly referred to as low voltage cable. Communication cable can be used to interconnect systems that include the following:

- Audio/voice – telephones, speaker systems, intercoms, and digital surround sound
- Video – security systems, television, cable television, digital cable, DVD, projection screens, and multimedia displays
- Data – computer networks, fax machines, and printers

Communication cable can be designed for aerial, protected (behind walls, etc.), or underground applications. *Aerial cable* is cable suspended in the air on poles or other overhead structures. Aerial cable usually includes a steel cable used for strength. Cable designed to be buried underground includes additional outer layers for moisture protection. Underground communication cable is extensively used to interconnect stations, servers, terminals, telecommunication rooms, entrance facilities, and backbone cables. *Backbone cables* are conductors (copper or fiber optic) used between telecommunications closets, or floor distribution terminals and equipment rooms within buildings.

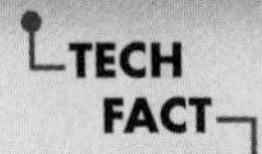

TECH FACT *Aluminum expands and contracts considerably more than does copper. Expansion and contraction loosen a connection that has not been properly tightened.*

Conductor Color Codes and Pin Designations

Communication cable is arranged in twisted pairs of conductors in a common housing. *Twisted conductors* are conductors that are intertwined at a constant rate. The number of pairs usually ranges from four-pair to hundreds of pairs. A standard color code identification system identifies each pair of conductors, regardless of the number of pairs in the cable. **See Figure 11-11.**

COMMUNICATION CABLE COLOR CODE

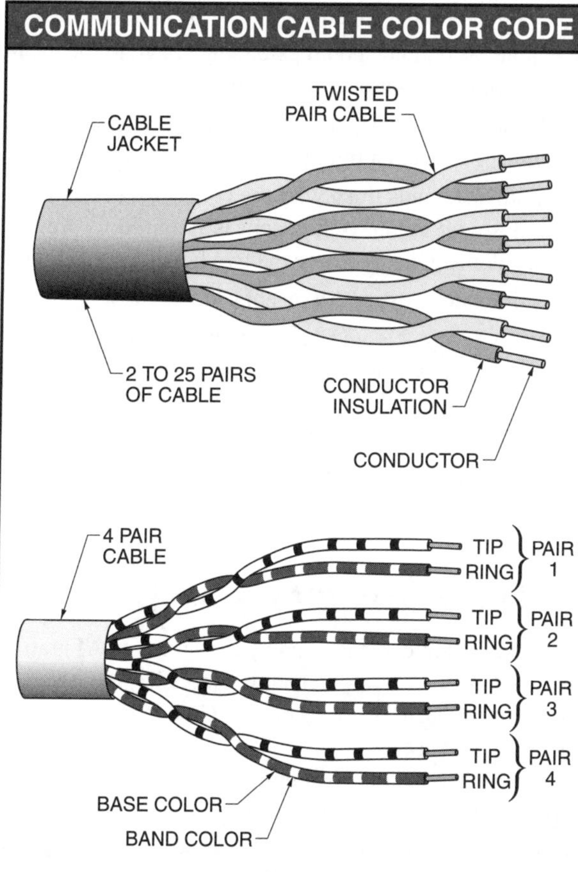

CONDUCTOR COLOR CODING				
Pair No.	Ring Conductor		Tip Conductor	
	Insulation Color	Band Mark	Insulation Color	Band Mark
1	Blue	White	White	Blue
2	Orange	White	White	Orange
3	Green	White	White	Green
4	Brown	White	White	Brown
5	Slate	White	White	Slate
6	Blue	Red	Red	Blue

Figure 11-11

Each conductor pair has a tip conductor and a ring conductor. A *tip conductor* is the first wire in a pair of wires. A *ring conductor* is the second wire in a pair of wires. Tip and ring conductor pair colors complement each other for easy identification. For example, the first pair of conductors is colored white with a blue strip (tip conductor) and blue with a white strip (ring conductor).

Cable Bundle Color Code Designations

Video, voice, and data systems often require communication cable that is bundled into hundreds of individual pairs of conductors. Communication cable, including large amounts of conductor pairs, is arranged so conductor pairs are grouped in sets of 25. Each set of 25 pairs is wrapped with a color-coded string that identifies the grouping. The string color code system follows the same color code used for the 25 pair conductor color code (1 = blue, 2 = orange, 3 = green, etc.). Strings are color-coded using one or two colors. When a one-color string is used, the color of the string follows the color of the tip conductor band color code (blue, orange, green, brown, etc.). When a two-color string is used, the color of the string follows the color of the tip conductor insulation/band color code (white-blue, white-orange, white-green, etc.). **See Figure 11-12.**

In addition to the color-coded strings used to band and identify conductor sets, a cable may also include a ripcord. A *ripcord* is a cord included in a cable that aids in removing the outer jacket.

Mechanical Arrangement Numbering System

Conductor pairs are terminated using connection blocks, cable outlets, plugs, couplers, and other types of quick-connect, quick-disconnect connections. Connectors are designed to provide a good mechanical and electrical connection. The wiring of the connector must also separate and identify the conductors for identification of each pair of conductors.

Conductors are terminated using connectors designed to hold one pair of conductors or multiple pairs of conductors. Standard connectors are designed for holding one, four, and 25 pairs of conductors. Connectors follow a numbering system that allows standardization when wiring conductor pairs. Some recommended communication cable installation tips include the following:

- Use as few connections as possible because excessive connections lower system performance.
- Wire to the highest projected data-rated speed.

COMMUNICATION CABLE STRING COLOR CODE

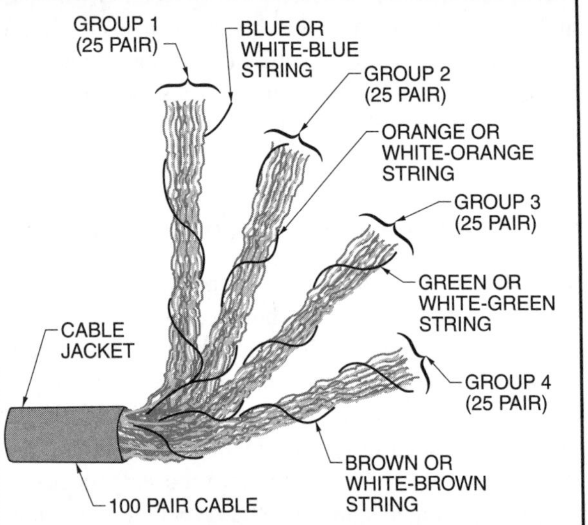

GROUP 1 (25 PAIR)
BLUE OR WHITE-BLUE STRING
GROUP 2 (25 PAIR)
ORANGE OR WHITE-ORANGE STRING
GROUP 3 (25 PAIR)
GREEN OR WHITE-GREEN STRING
CABLE JACKET
GROUP 4 (25 PAIR)
BROWN OR WHITE-BROWN STRING
100 PAIR CABLE

STRING COLOR CODING

Binder Group Color		Pair Count
One Color Coding	Two Color Coding	
Blue	White-Blue	001-025
Orange	White-Orange	026-050
Green	White-Green	051-075
Brown	White-Brown	076-100
Slate	White-Slate	101-125

Figure 11-12

- Allow at least 18″ of spare wire at all connection points.
- Never splice wires on communication cable runs.
- Never run communication cables in parallel with power conductors.
- Keep cable and wires away from heat sources.
- Use nonmetallic staples when securing cable or wiring.
- Place communication outlets at the same height as electrical power outlets.
- Install communication outlets near electrical power outlets, since both powers are usually required.
- Use inner walls for cable runs to avoid problems with insulation.
- Test connections for opens, shorts, and grounds.

Patch Cord Connections

Patch cords are used to join communication links and circuits within an electrical system. Patch cords are terminated with connectors that allow for making easy connections and reconnections. Patch cords are wired using standard configurations that place each conductor pair in the proper order. **See Figure 11-13.** To perform a patch cord connection, apply the following procedure:

1. Cut the cable to length and strip back the cable jacket using a cable stripper or the ripcord included within the cable.
2. Separate conductor pairs in numerical order. Use the cable color code identification system to identify the conductor pairs and each ring and tip conductor within each pair.
3. Align and tie the cable pairs into the connectors. The cable jacket end should extend just beyond the cable tie point. Do not over tighten the cable tie.
4. Ensure that each pair is in the correct channel and order. Maintain a tight twist on each conductor pair.
5. Trim wire ends flush with the connector housing using a wire cutter. A flush cut is required to prevent problems when pressing the two housings together.
6. Align latch cover with receptacle.
7. Press receptacle and latch cover together until connector snaps closed.
8. Inspect the connector and ensure that it is locked in place. If required, additional identification numbers can be added to the connector to identify patch cord.
9. Connect patch cord to appropriate location.

Cable Category (CAT) Ratings

Communication cable is rated by a category (CAT) rating number. CAT ratings are standards adopted by the telecommunication industry. Cable is rated as CAT 1 through CAT 6. **See Figure 11-14.**

A cable's CAT rating determines the conductor bandwidth rating. *Bandwidth* is the range of frequencies that a device can accept within tolerable limits. The higher the bandwidth, the higher the data transfer capacity. Cable CAT ratings include the following:

- CAT 1 – A rating used for older wire and cable used in telephone systems. It is usually limited to voice transmission only.
- CAT 2 – A rating used for inside wire and cable systems for voice applications and data transmission up to 1 MHz.
- CAT 3 – A rating used for inside wire and cable systems for voice applications and data transmission up to 16 MHz. High-speed data/LAN systems use CAT 3. A *local area network (LAN)* is a communication network that spans a relatively small area.

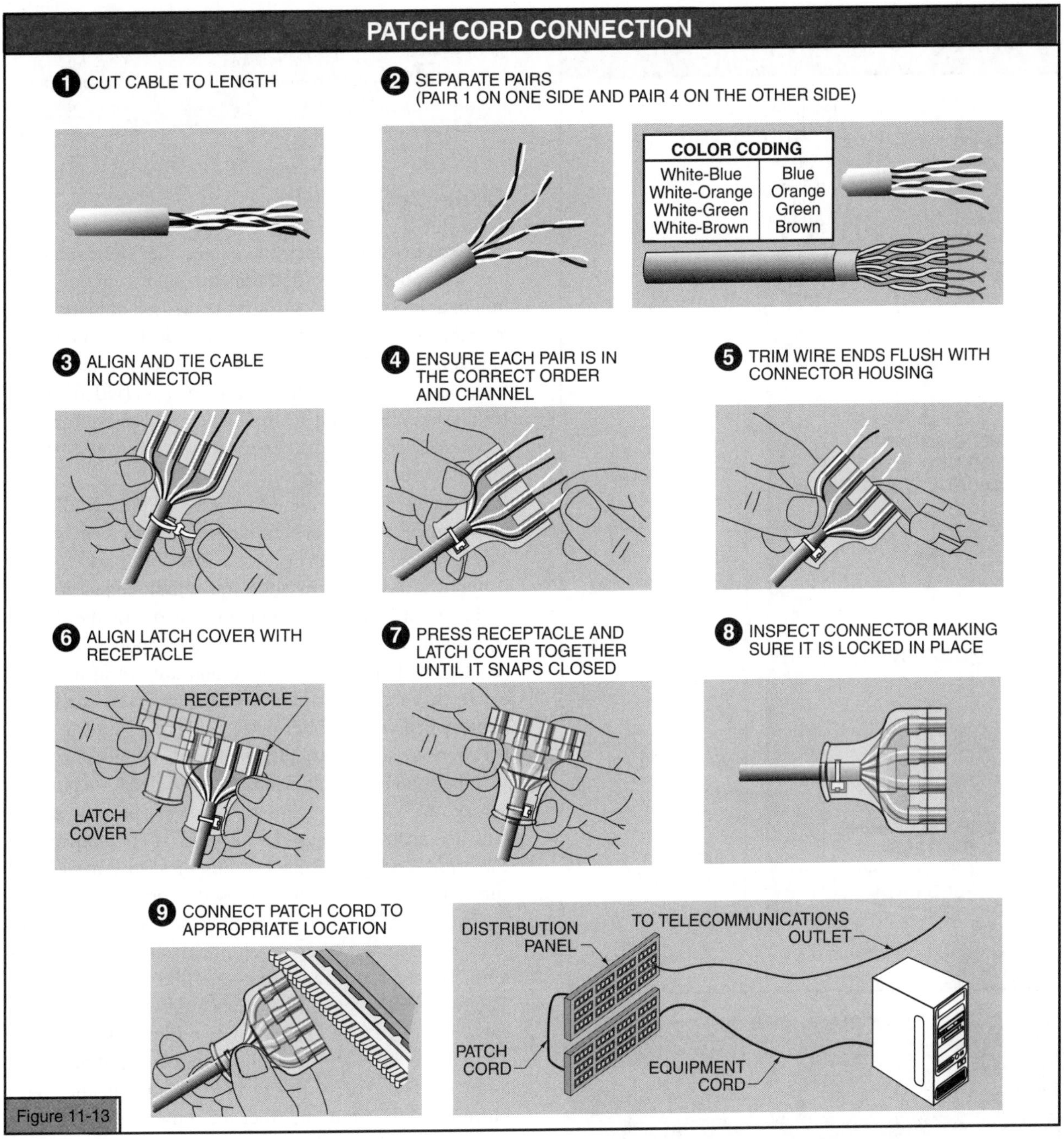

PATCH CORD CONNECTION

1 CUT CABLE TO LENGTH

2 SEPARATE PAIRS (PAIR 1 ON ONE SIDE AND PAIR 4 ON THE OTHER SIDE)

COLOR CODING	
White-Blue	Blue
White-Orange	Orange
White-Green	Green
White-Brown	Brown

3 ALIGN AND TIE CABLE IN CONNECTOR

4 ENSURE EACH PAIR IS IN THE CORRECT ORDER AND CHANNEL

5 TRIM WIRE ENDS FLUSH WITH CONNECTOR HOUSING

6 ALIGN LATCH COVER WITH RECEPTACLE

RECEPTACLE

LATCH COVER

7 PRESS RECEPTACLE AND LATCH COVER TOGETHER UNTIL IT SNAPS CLOSED

8 INSPECT CONNECTOR MAKING SURE IT IS LOCKED IN PLACE

9 CONNECT PATCH CORD TO APPROPRIATE LOCATION

DISTRIBUTION PANEL

TO TELECOMMUNICATIONS OUTLET

PATCH CORD

EQUIPMENT CORD

Figure 11-13

- CAT 4 – A rating used for inside wire and cable systems for voice applications and data transmission up to 20 MHz.
- CAT 5 – A rating used for inside wire and cable systems for voice and data transmission up to 100 MHz. This cable is one of the most commonly used cables in LAN systems because it offers extended frequency transmission (above CAT 3 and CAT 4).
- CAT 5e – A rating used for inside wire and cable systems for voice and data transmission up to 100 MHz. It allows for longer runs because wire is packaged tighter with greater electrical balancing between wire pairs.
- CAT 6 – A rating used for inside wire and cable systems for voice and data transmission up to 250 MHz or more.

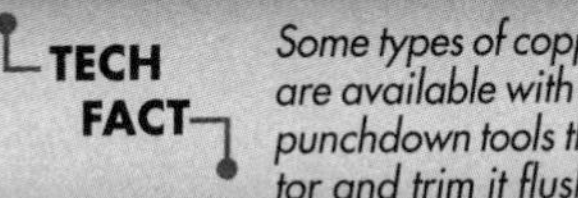

Some types of copper patch-cord termination kits are available with either single-wire or multiwire punchdown tools that seat the wire in the connector and trim it flush.

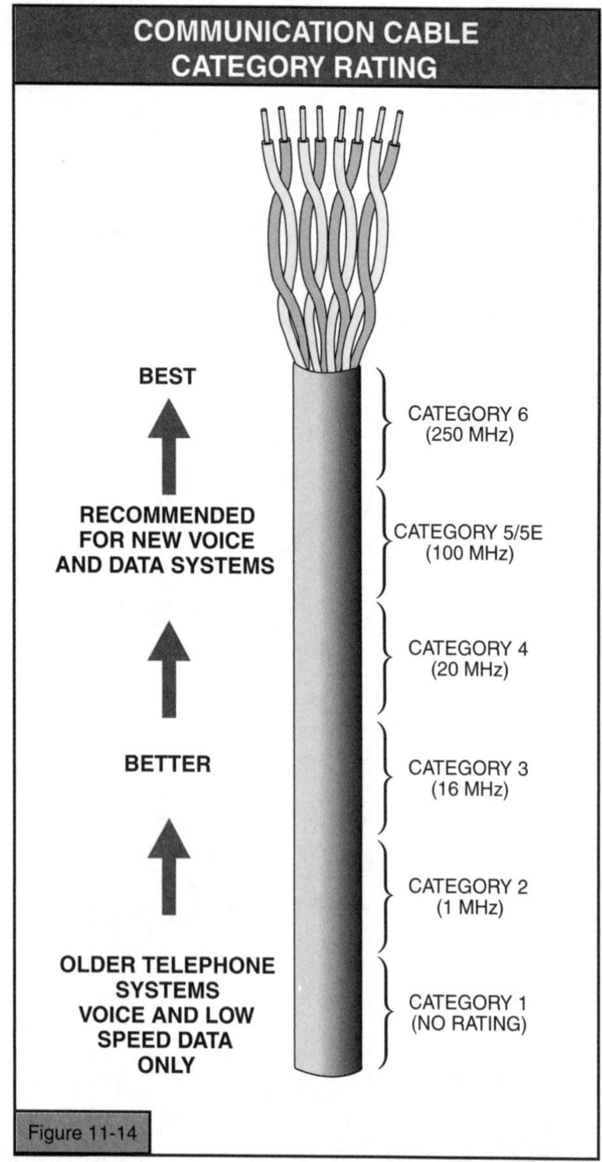

Figure 11-14

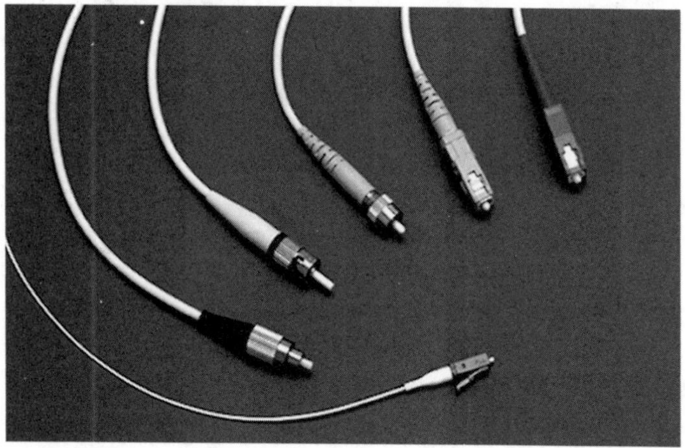
Gould Fiber Optics

Several different types of connectors are used with fiber-optic cable assemblies.

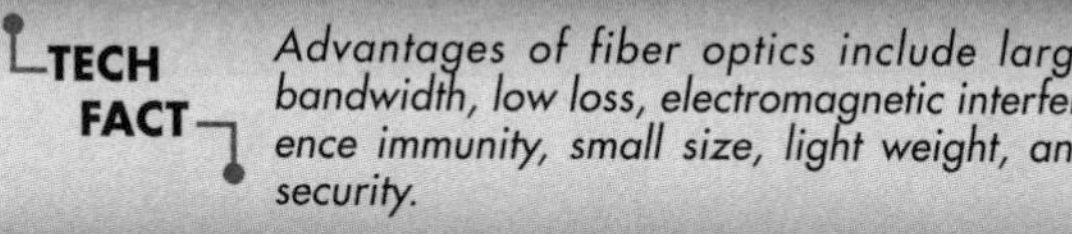

Advantages of fiber optics include large bandwidth, low loss, electromagnetic interference immunity, small size, light weight, and security.

Fiber-Optic Cable

As in copper conductors and cable, fiber-optic conductors and cables are used to transmit data from one location to another. *Fiber optics* is the method of using light to transport information from one location to another location through thin filaments of glass or plastic. In a fiber-optic system, electrical information (video, voice, or data) is converted to light energy, transmitted to a receiver over optical fibers, and then converted back into electrical information.

The optical fibers are either glass or plastic. Glass is more suitable for longer optic data transmission but is more expensive and fragile than plastic. Therefore, plastic is used in applications requiring short transmission lines.

Fiber optic cables are lightweight and offer the highest bandwidth and speeds possible. Fiber optic cables also require smaller diameter cable and are not affected by electromagnetic interference (EMI) or radio frequency interference (RFI). Fiber-optic cables are used in large backbone systems and individual workstations.

Fiber-optic connectors must be protected from physical damage and moisture. To protect the fibers, protective cladding and coatings are used along with strength members. *Cladding* is a layer of glass or other transparent material surrounding the fiber core and used as optical insulation.

There are different types of fiber-optic cables available to meet different cabling requirements. Fiber optic cables normally consist of more than one fiber-carrying strand. The two basic types of fiber cable are tight-buffered fiber cable and loose tube fiber cable. **See Figure 11-15.**

In the loose tube fiber cable, the fiber is not restricted and can move freely inside the housing. In the tight-buffered fiber cable, the fiber movement is much more restricted. Although both types can use air or gel-filled tubes to allow fiber movement, tight-buffered fiber cable usually uses gel-filled tubes. The movement of the fiber cable allows for expansion and contraction with temperature changes. Loose tube fiber cable is stronger and is used in applications requiring long-distance cable runs.

Although signal losses accrue over the length of a fiber cable, losses from splices and connectors often exceed the losses from the actual cable. Common signal losses occur from axial and angular misalignment, excessive separation between cables, and rough ends. **See Figure 11-16.**

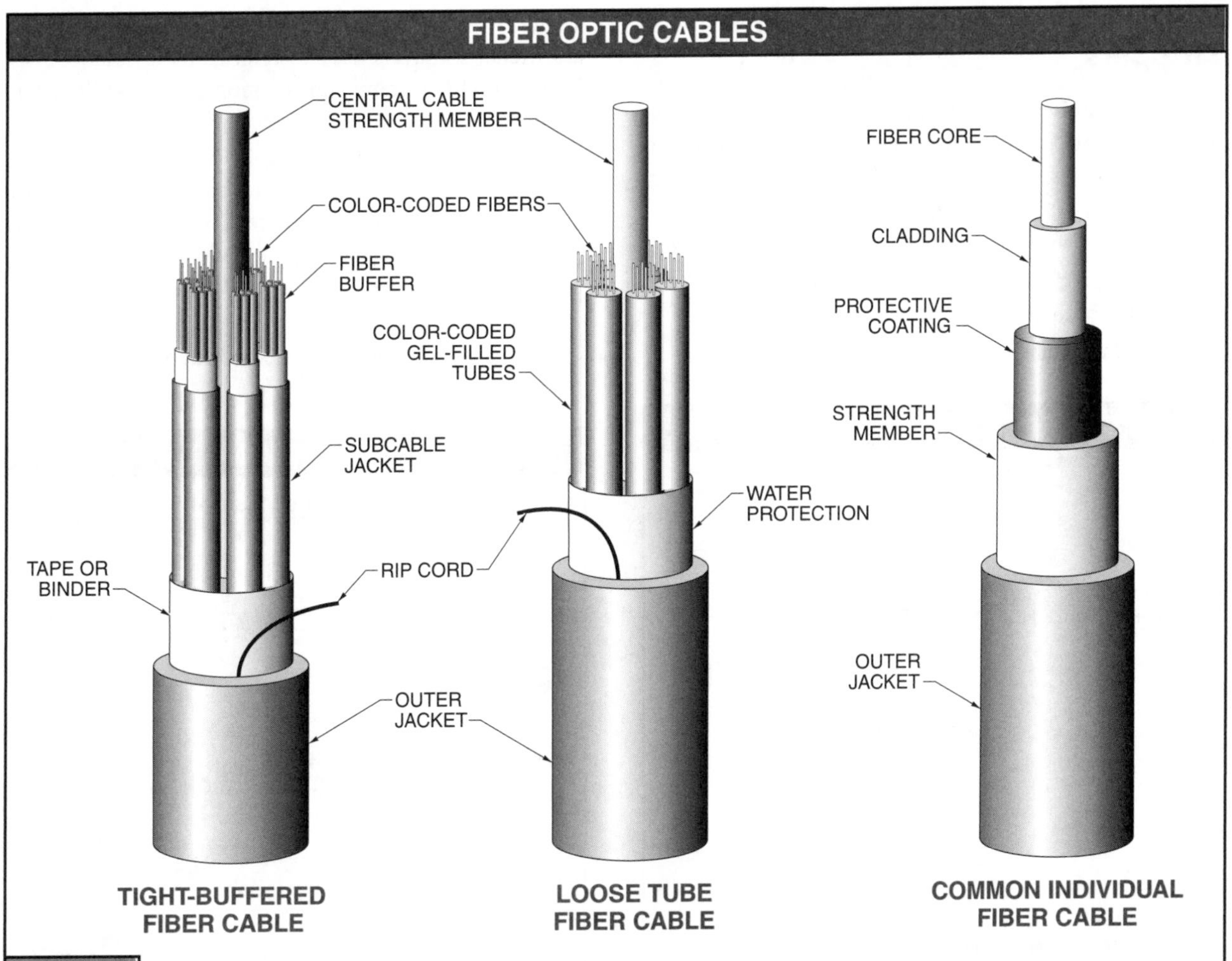

FIBER OPTIC CABLES
CENTRAL CABLE STRENGTH MEMBER
COLOR-CODED FIBERS
FIBER BUFFER
COLOR-CODED GEL-FILLED TUBES
SUBCABLE JACKET
TAPE OR BINDER
RIP CORD
WATER PROTECTION
OUTER JACKET
FIBER CORE
CLADDING
PROTECTIVE COATING
STRENGTH MEMBER
OUTER JACKET
TIGHT-BUFFERED FIBER CABLE
LOOSE TUBE FIBER CABLE
COMMON INDIVIDUAL FIBER CABLE

Figure 11-15

CAUSES OF SIGNAL LOSS
CABLE
FIBER-OPTIC CABLE CONNECTOR
ANGULAR MISALIGNMENT
EXCESSIVE END SEPARATION
AXIAL MISALIGNMENT
ROUGH ENDS

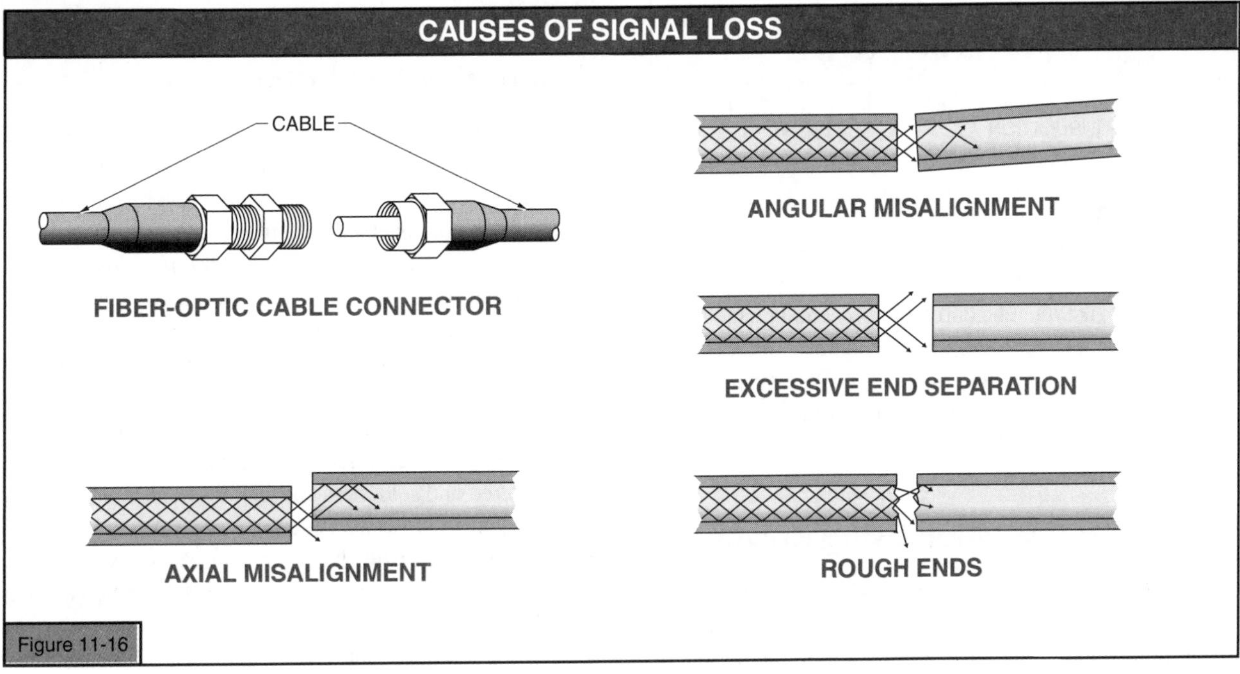

Figure 11-16

Removing Insulation Using Wire Strippers

A *wire stripper* is a tool that is designed to properly remove insulation from small gauge (normally AWG sizes No. 10–No. 22) wires. **See Figure 11-17.** To remove insulation from a wire using a wire stripper, apply the procedure:

1. Insert the wire into the correct wire gauge hole size.
2. Squeeze and rotate the stripper around the wire (or back and forth) to make the cut through the insulation.
3. Strip the insulation off the end of the wire.

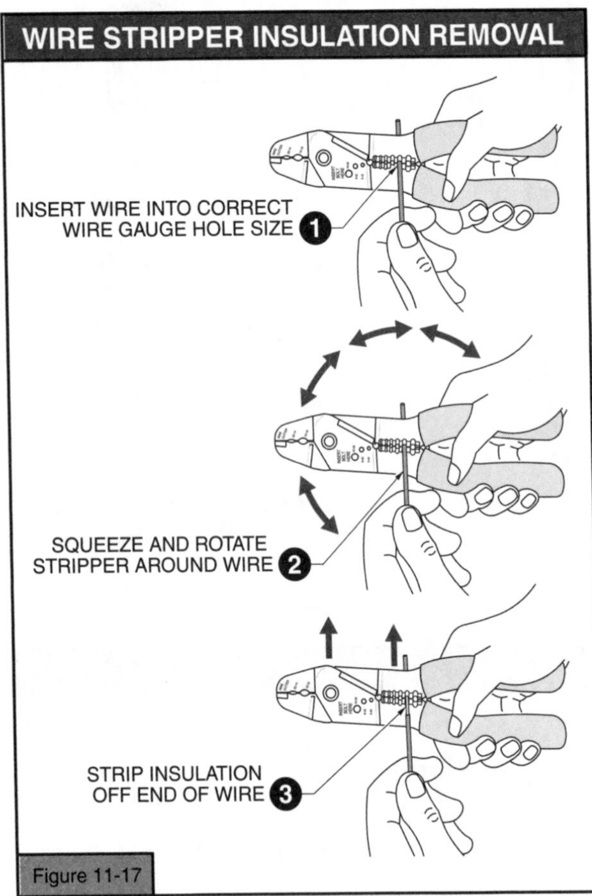

Figure 11-17

The hole size must be correct to prevent damage to the wire. The stripper removes the insulation without cutting into the wire if the correct hole is used. Wire strippers are available that make a square cut and a tapered cut. The tapered cut is preferred.

Removing Insulation Using Knives

A wire stripper is the preferred method of removing insulation from the end of most wire. However, when removing insulation from a large wire or tapping into the middle of a wire, a wire stripper cannot be used. In such cases, an electrician's knife is used to remove the insulation. **See Figure 11-18.** To remove insulation using a knife, apply the procedure:

1. Cut the insulation at an angle to prevent cutting into the wire with the knife. A nick in the wire weakens it and may cause the wire to break when it is bent. Never make a circling cut at right angles to the wire. Keep fingers away from the blade and always cut away from the body.
2. Cut both sides leaving a tapered cut on the insulation.

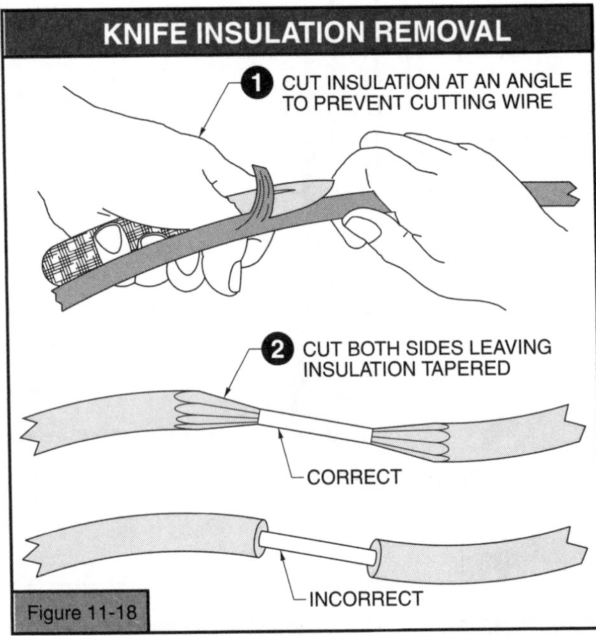

Figure 11-18

Connecting Conductors

Electrical connections are required in almost all electrical and electronic circuits. Electrical connections may be soldered or solderless. A *soldered connection* is a connection that joins conductors by heat to make a strong electrical and mechanical connection. A *solderless connection* is a connection that joins conductors without using heat to make a strong electrical and mechanical connection. Soldered connections are normally used in electronic circuits to connect component leads to printed circuit boards and other connection points. Solderless connections are normally used in electrical circuits to connect conductors.

Soldered and solderless connections must be electrically and mechanically strong. A strong electrical connection ensures that the connection has as low a resistance as possible to the flow of current. A strong mechanical connection ensures that the connection lasts and does not cause a loss of power or possible damage that may be caused if a hot conductor opens and touches some other part of the circuit or a person. Touching some other part of the circuit may cause a spark that could cause a fire or damage additional components.

Soldering

Soldering is the process of joining metals by heat to make a strong electrical and mechanical connection. Solder is used to make the connection. *Solder* is an alloy of tin (Sn) and lead (Pb). The solder that is normally used to make electrical connections is 60/40 solder. This type of solder contains 60% tin and 40% lead. Solder melts at a temperature of about 400°F (204.4°C).

The parts to be soldered must be hot enough to melt the solder for the solder to properly adhere. The parts must be kept clean to allow the solder to flow evenly and adhere. In addition to surface dirt, oil, and corrosion, the metal surfaces must be cleaned of all surface oxides. Oxide is formed on the metal surface by the oxygen in the air. For example, when copper is exposed to air long enough the oxide appears as a green tarnish. Flux is used with solder to remove surface oxide. Flux removes the oxide by making it soluble in the flux and evaporating it as the flux boils off during heating. Rosin flux is contained inside electrical solder. *Rosin-core solder* is electrical solder that contains a rosin core.

A connection must be mechanically strong before it is soldered. Heat is then applied until the materials are hot at which time the solder is applied. The solder must be melted by the heat of the material to be joined, not by the soldering tool.

Only a small amount of solder should be used. The connection should appear smooth and shiny. If it does not, the connection may be a cold solder joint. A *cold solder joint* is a poor electrical and mechanical connection caused by insufficient heat during soldering or connecting parts moving after the solder is applied but before it has cooled.

Care must be taken to not damage surrounding parts that may be damaged by overheating. Semiconductor components, such as transistors and ICs, are very sensitive to heat. A heat sink, such as an alligator clip, is used to help prevent heat damage. The heat sink is placed between the soldered connection point and the component that requires protection. The heat sink absorbs the heat produced during soldering. **See Figure 11-19.** To make a soldered connection, apply the procedure:

1. Select the correct soldering tool and solder. Soldering irons and guns have heating element ratings from about 20 W to 250 W. Most have interchangeable tips that are shaped and designed for most applications. Soldering irons and guns range in temperature from about 500°F to 1000°F.

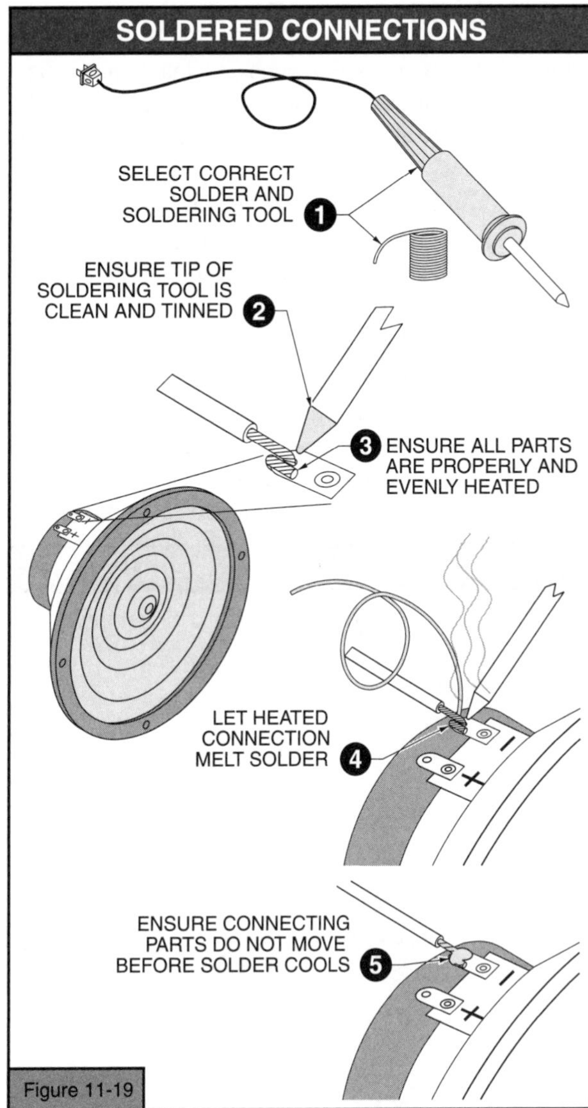

Figure 11-19

2. Ensure that the tip of the soldering tool is clean and tinned. *Tinning* is the process of applying solder to a clean soldering iron tip to prevent corrosion on the tip.
3. Ensure that all the parts to be connected are properly and evenly heated. Do not overheat the connection.
4. Let the heated connection melt the solder and allow it to flow onto the connection. Do not melt the solder on the soldering tip. Use only enough solder to cover the connection.
5. Ensure that the connecting parts do not move before the solder cools.

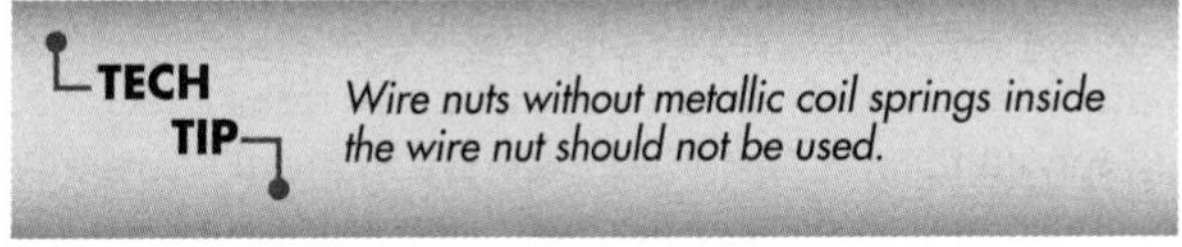

Wire nuts without metallic coil springs inside the wire nut should not be used.

Terminal Screws

Terminal screws are used to connect wires to many electrical components, such as switches and receptacles. **See Figure 11-20.** To make a terminal screw connection, apply the procedure:

1. Strip ¾″ of insulation from the wire ends.
2. Use needle nose pliers to form a loop in the stripped wire. Bend the bare wire around the jaws of the pliers to form a loop.
3. Connect wire to terminal screw by looping the end around the terminal screw in the direction the screw tightens (normally clockwise). As the wire is tightened, the wire is forced under the screw head and clamps. A terminal screw that is not tight, or a wire that is not properly attached is likely to cause problems.

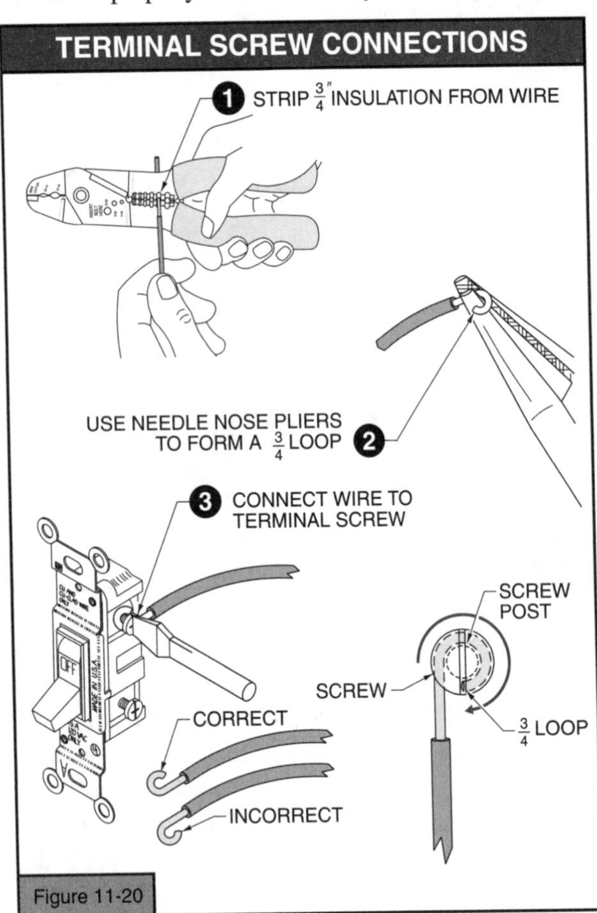

Figure 11-20

Wire Nuts

A *wire nut* is a solderless plastic connector that uses a tapered metal coil spring to twist wires together. Wire nuts are commonly used to connect wire sizes AWG No. 22 through No. 8. **See Figure 11-21.** To make a terminal connection using a wire nut, apply the procedure:

1. Remove enough insulation (normally ½″) from the wire for the connector to cover the bare wire. Although it is normally not necessary to twist the wires together when connecting solid wire, twisting the wires together helps ensure a good connection. Twist the wires together when connecting stranded wires or a stranded to solid wire.
2. Place the conductors inside the wire nut and tighten clockwise. The conductors engage a spiral coil that draws them in and twists them together.
3. Tape the wire nut and wire together with electrical tape.

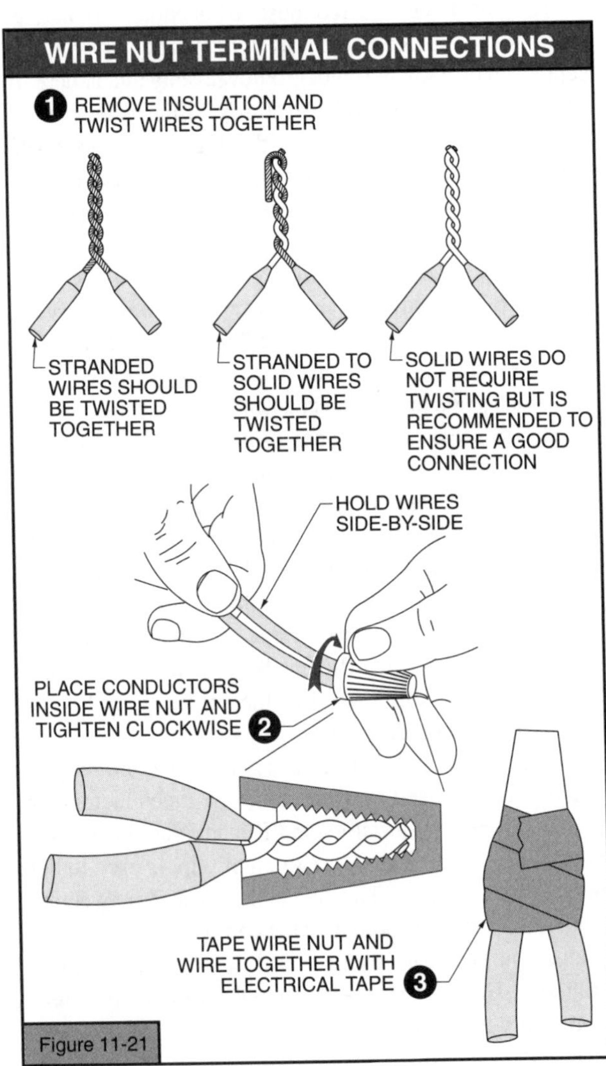

Figure 11-21

All wire nuts have a listing of what size and number of wires that they can safely connect. This listing is normally found on the wire nut box. Manufacturers color code their wire nuts by size, but the standard varies among manufacturers.

Wire Splices

Wire splices are used to connect two wires together to make a good mechanical and electrical connection. The wires are twisted together and then taped, soldered and taped, or connected by a solderless connection. The method used to twist the wires together gives the splice its name, such as pigtail splice, tap splice, knotted tap splice, Western Union splice, or fixture splice. The method used depends on the application. **See Figure 11-22.**

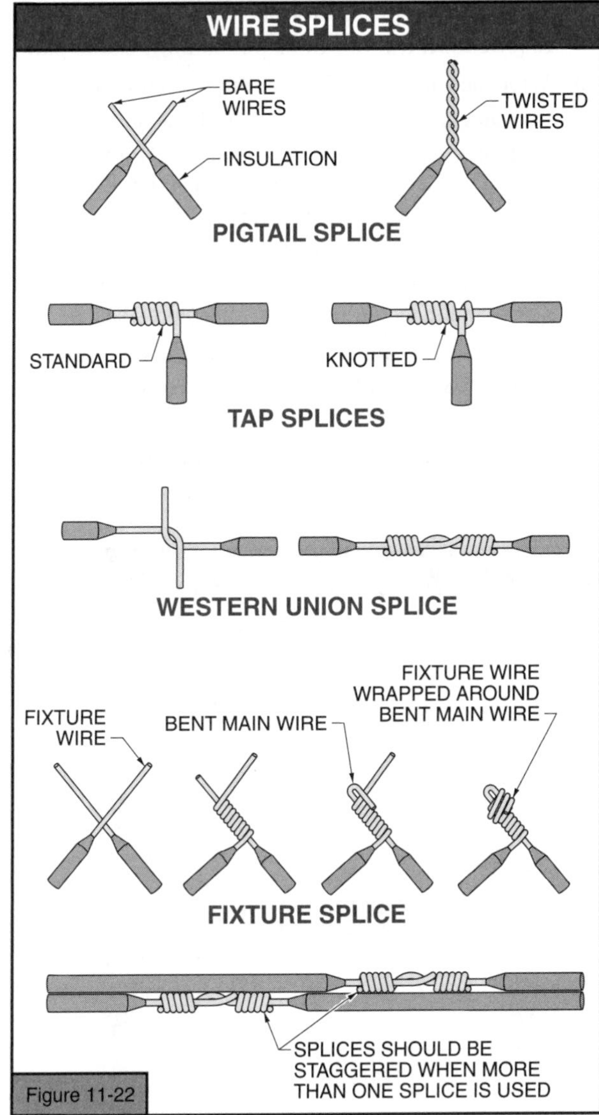

Figure 11-22

Pigtail Splices

A *pigtail splice* is a splice that consists of twisting two wires together. A pigtail splice is used to connect two (or more) wires together by twisting the wires around each other. A pigtail splice is the most common splice used in most residential wiring applications. Ensure that the wires are twisted together and that the splice is not just one wire wrapped around another. A wire nut is normally placed over a pigtail splice.

Tap Splices

A *standard tap splice* is a splice that connects two wires together when one wire is to remain unbroken. A knotted tap splice is a tap splice with a knot made in the splice to increase the mechanical strength of the splice. A wire nut cannot be placed on tap splices, so the splice is taped or soldered and taped.

Western Union Splices

A *Western Union splice* is a splice that connects two wires that may be placed under a lot of mechanical strain. This splice got its name because it was the standard splice used to repair telegraph wire. A wire nut cannot be placed on a Western Union splice, so the splice is taped or soldered and taped.

Fixture Splices

A *fixture splice* is a splice that connects two wires in which one of the wires is smaller than the other, or one wire is stranded and the other is solid. The smaller (or stranded) wire is wrapped around the larger (or solid) wire. The larger (or solid) wire is bent back over the smaller wire. For extra strength, the smaller wire may be wrapped a few extra times after the solid wire is bent over the smaller wire. A wire nut is normally placed over a fixture splice. However, the fixture splice may first have to be soldered if the connection is loose.

Pressure Connections

Pressure connections are used for large wire connections. Pressure connections include split bolt and screw pressure connectors. **See Figure 11-23.** Pressure connections are used for making straight, tap, or terminal connections. The wire ends are stripped, placed in the connector, and the nut (or screw) is tightened. Except when used on grounding wires, the connection is then wrapped with several layers of electrical tape. The tape should be wrapped at least the same thickness as the original insulation.

Crimped Connections

A *crimped connection* is a connection in which two wires are held together by crimping them in a specially designed fitting. Approximately ½″ to ¾″ of insulation is stripped off the wires to be connected. The wires are twisted together if required with pliers. The wire ends are inserted into the connector and the end of the connector is crimped with a crimping tool. A crimped connection may not be approved by some local codes. **See Figure 11-24.**

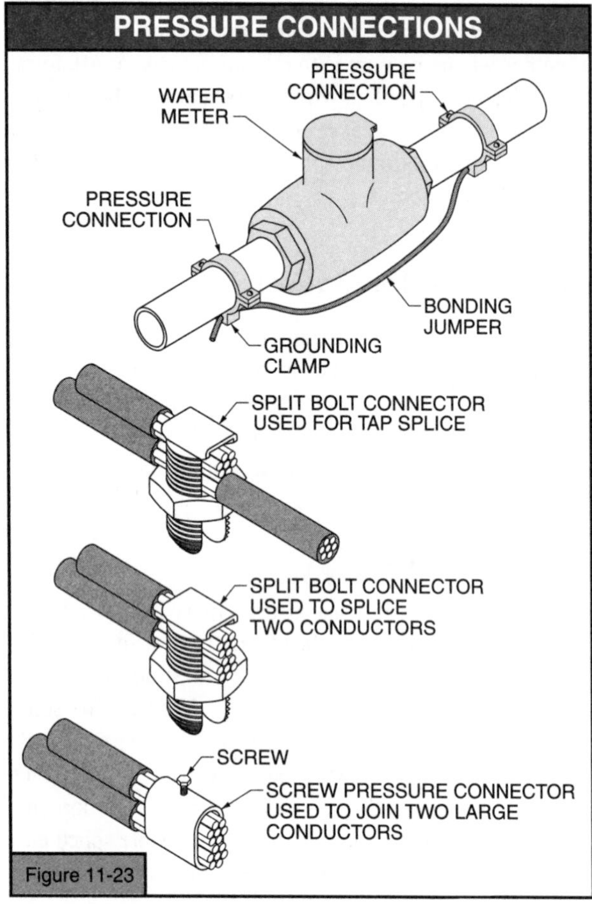

Figure 11-23

Replacing Plugs

A *plug* is the device at the end of a cord that connects the device to the electrical power supply by means of a receptacle. Plugs sometimes crack, break, or lose their electrical connection when the device is disconnected from the receptacle by pulling on the cord instead of the plug. The plug should be replaced when it goes bad. Never try to repair a broken plug. Replace the cord and the plug if the cord is also broken or damaged.

Standard Ungrounded Plugs

Grounded plugs should always be used if the original equipment includes a grounded plug. Always replace a grounded plug with a new grounded plug. **See Figure 11-25.** However, not all electrical devices use a grounded plug. To replace an ungrounded plug, remove the old plug and apply the procedure:

1. Insert the cord into the plug. Pull about 6″ of cord through the plug.
2. Strip about ¾″ insulation from the wire.
3. Tie an underwriter's knot in the two wires. An underwriter's knot is used to relieve the strain on the terminals when the plug is pulled out by pulling on the cord.
4. Pull the underwriter's knot tight. The knot should be large enough so that it cannot be pulled back through the plug. The plug is too large or a double knot is required if the knot can be pulled through the plug.
5. Wrap the wires around the terminal screws. Ensure the hot (black) wire is connected to the smaller blade if the plug is a polarized plug. A *polarized plug* is a plug in which one blade is wider than the other blade. The wider blade is the neutral terminal and is designed to fit only into the neutral side of a polarized receptacle. The hot terminal is normally identified by a bronze-colored screw and the neutral by a silver-colored screw.
6. Tighten the terminal screws. Care must be taken to ensure no strands of wire are exposed on cords that use stranded wire. Solder the stranded ends if required.
7. Attach the insulated cover that comes with the plug. The best plugs include a screw-on cover.

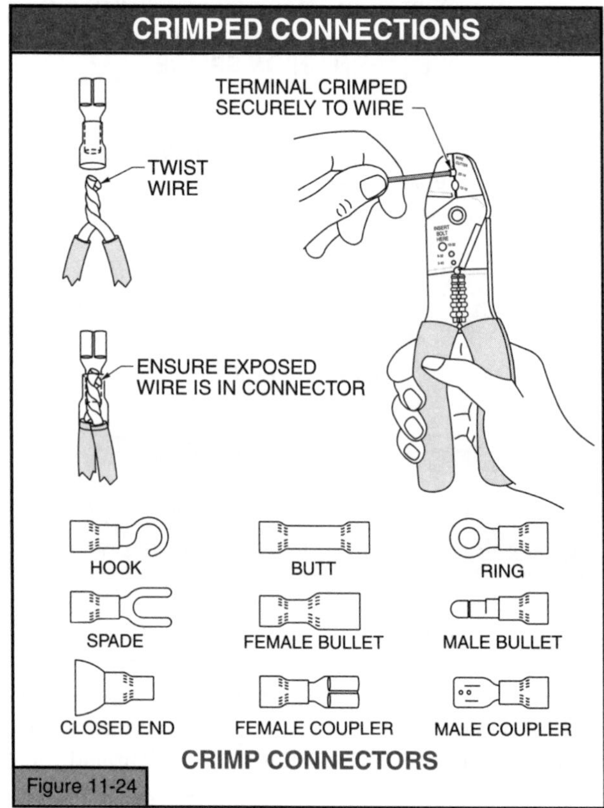

Figure 11-24

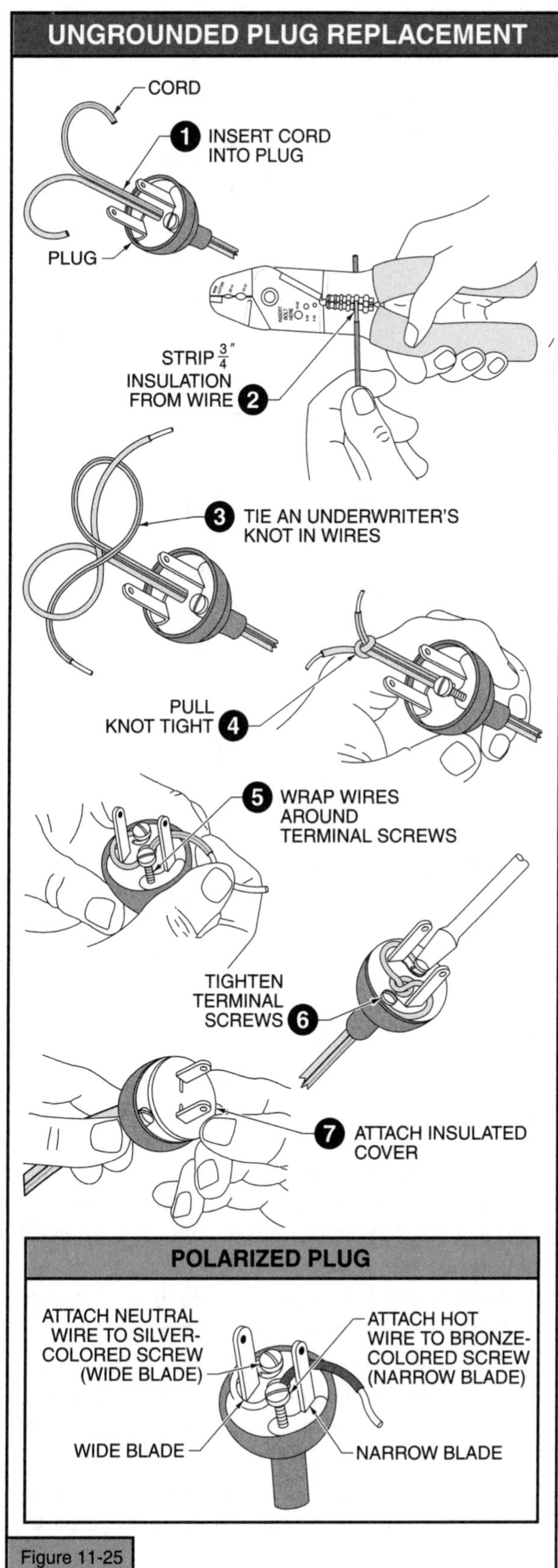

Figure 11-25

Standard Grounded Plugs

Grounded plugs should always be used if the original equipment includes a grounded plug. Always replace a grounded plug with a new grounded plug. The load works but the circuit is not safe if a grounded plug is not used or the ground wire is not connected. Ungrounded equipment can cause a shock or fire. **See Figure 11-26.** To replace a grounded plug, remove the old plug and apply the procedure:

1. Insert the cord into the plug cover and pull about 8″ of cord through the plug.
2. Cut the plastic sheathing away from about 6″ of cord. Peel back and cut off any paper wrapping. Be careful not to cut into the insulation around the individual wires.
3. Strip about ¾″ insulation from the wire.
4. Insert wires into wire pockets. The hot terminal is identified by a bronze colored screw, the neutral by a silver colored screw, and the ground terminal by a green colored screw.
5. Tighten the terminal screws. Care must be taken to ensure no strands are exposed because most cords use stranded wire. Solder the stranded ends if required.
6. Insert plug into cover and tighten assembly screw.

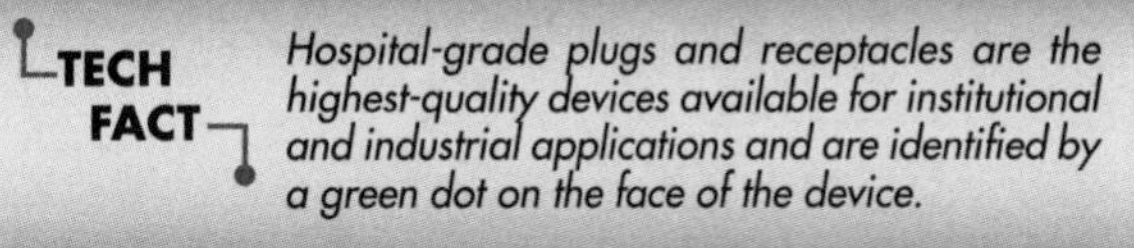

Hospital-grade plugs and receptacles are the highest-quality devices available for institutional and industrial applications and are identified by a green dot on the face of the device.

Appliance (Heater) Plugs

An *appliance (heater) plug* is a plug used to power appliances that produce heat, such as electric grills, roasters, broilers, waffle irons, and large coffeemakers. Appliance plugs fail when the cord is pulled or they are overheated. Replace the cord and plug with a new one if they are bad. **See Figure 11-27.**

To replace a appliance (heater) plug, remove the old plug and apply the procedure:

1. Strip about ¾″ insulation from the wire and twist the stranded ends into a loop. Do not solder because the plug may get hot enough to melt the solder.
2. Wrap the wire around the terminal screws and tighten terminal screws.
3. Screw plug halves together.

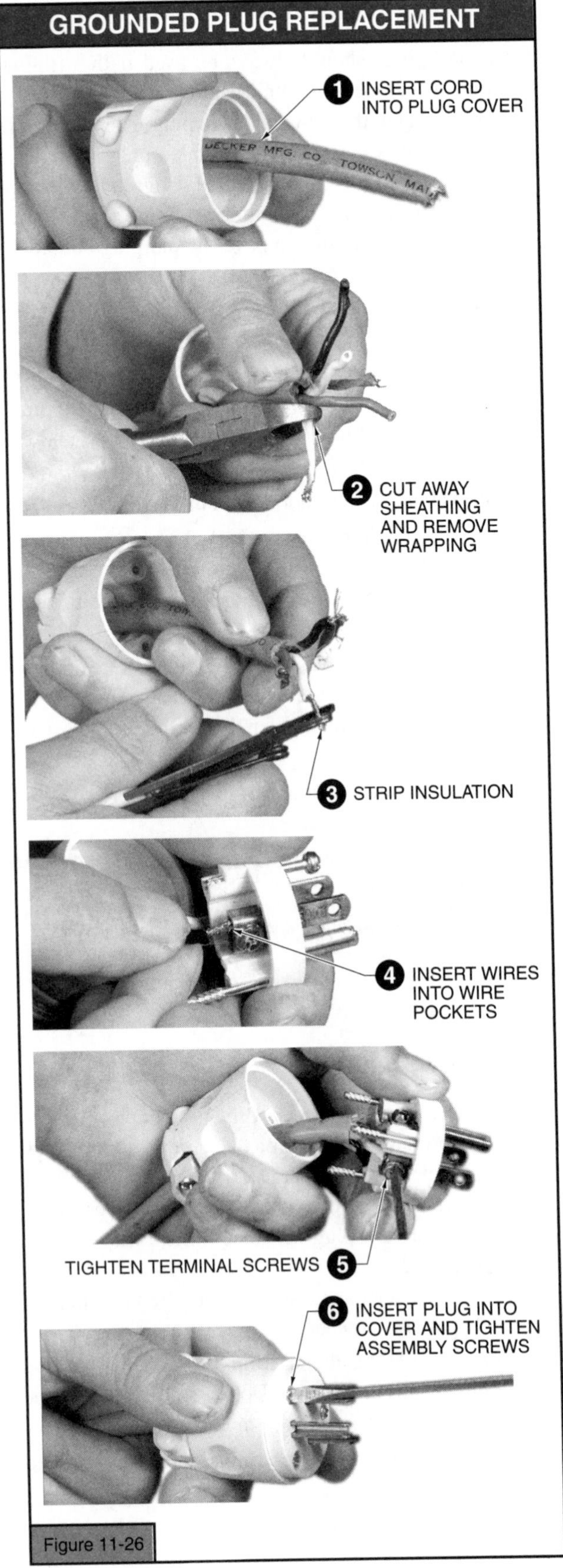

Figure 11-26

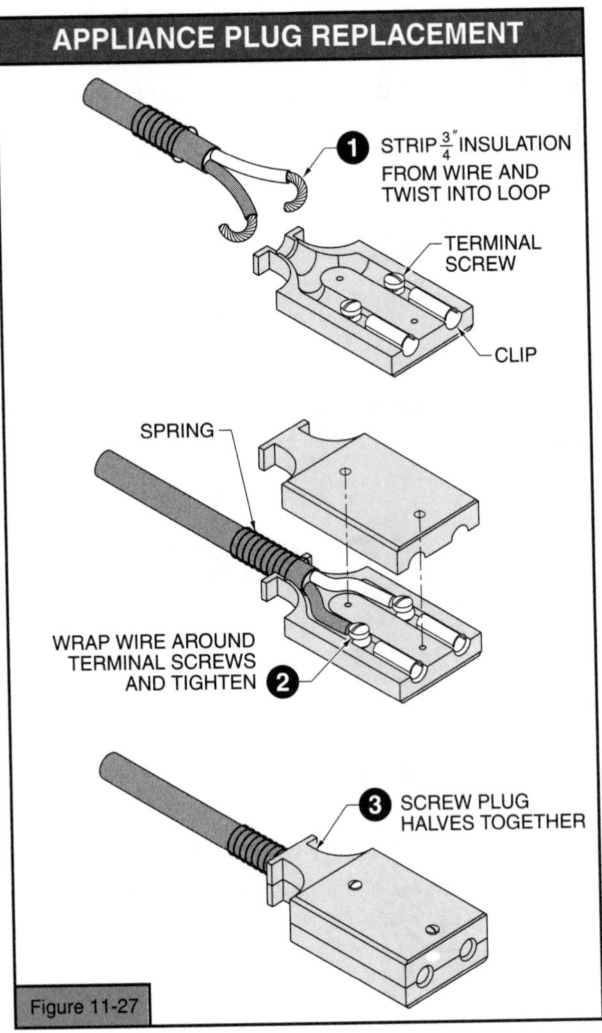

Figure 11-27

Heavy-Duty Plugs

A *heavy-duty plug* is a plug that is used on high-wattage appliances and devices that operate on 230 V or 460 V. These plugs may be used on 1ϕ or 3ϕ powered equipment. The exact size and number of wires required depends on the individual load that the cord is used to power. **See Figure 11-28.** To replace a heavy-duty plug, remove the old plug and apply the procedure:

1. Insert the cord into the plug and strip about ¾" insulation from the wires. Twist the strands and form a loop.
2. Wrap the wires around the correct terminal screws and tighten the terminal screws. The correct terminal screw varies considerably depending on the power supply used.
3. Attach the insulated cover that comes with the plug. The highest quality plugs are manufactured with screw-on covers.
4. Tighten the strain relief screws.

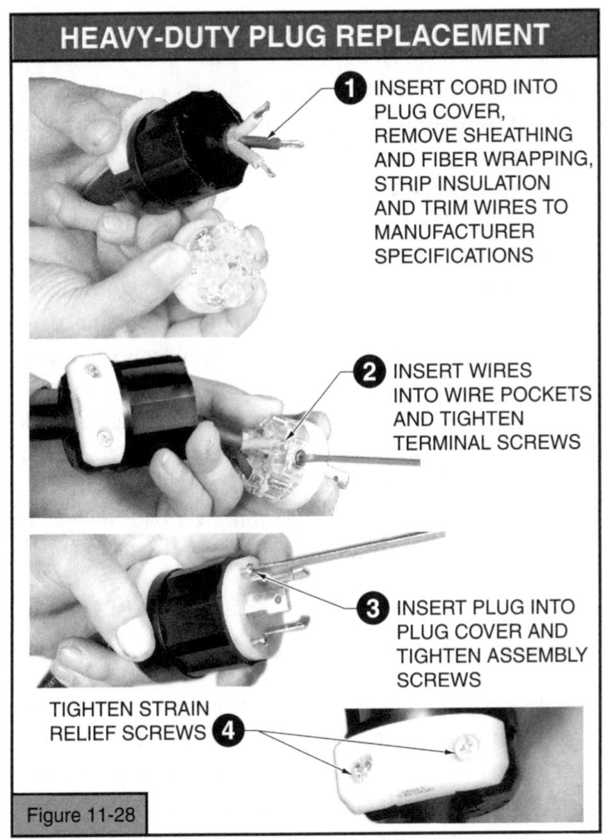

Figure 11-28

Plug and Receptacle Ratings and Configurations

Plugs and receptacles use various configurations to ensure that the current rating of electrical equipment matches the power source to which the equipment is connected. The configuration is designed to prevent a higher current or different voltage plug from being connected to a lesser (or incorrect) voltage receptacle. Plugs and receptacles follow standard configurations. The National Electrical Manufacturers Association (NEMA) has established a set of standard plug and receptacle configurations that identify the type and rating of terminations. **See Appendix.**

Overcurrent

In a properly operating circuit, current flows and is confined to the conductive paths provided by conductors and other components when a load is turned ON. Every load draws a normal amount of current when switched ON. This normal amount of current is the current level for which the load, conductors, switches, and system is designed to safely handle. Under normal operating conditions, the current in the circuit should be equal to or less than the normal current level. However, at times an electrical circuit may have a higher-than-normal current flow (overcurrent).

An *overcurrent* is a condition that exists on an electrical circuit when the normal load current is exceeded. Overcurrents take on two separate characteristics: short circuits and overloads. An overcurrent may be a short circuit current or an overload current.

Short Circuit

A short circuit develops if conductor insulation flakes off enough to expose the bare wire to ground or another wire. A *short circuit* is overcurrent that leaves the normal current-carrying path by going around the load and back to the power source or ground.

A short circuit occurs when the insulation between two wires is broken. The current bypasses the load and returns to the source through the bare wire and not through the load. A short circuit causes the current to rise hundreds of times higher than normal at a very fast rate. Short circuits may cause a fire, shock, explosion, and may damage equipment. All circuits must be protected against a short circuit because short circuits are very dangerous and damaging. **See Figure 11-29.**

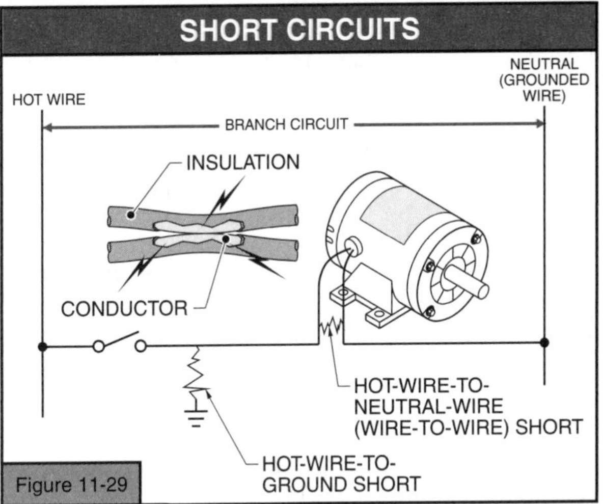

Figure 11-29

Detecting short circuits and removing power from the circuit is not difficult because short circuits draw high current levels. The system's fuses or circuit breakers remove power when a short circuit is detected. However, no matter how quickly power is removed, the danger of the arc that all short circuits produce always exists. When this arc is confined inside conduit, an enclosure, or cable, the chance of fire is greatly reduced. If the arc occurs in the open, a fire may result. Therefore, every effort must be made to minimize the number of places that electrical wire is exposed outside an enclosure. This includes avoiding the use of extension cords and following all local, state, and national codes.

Overloads

An *overload* is the condition that occurs when circuit current rises above the normal current level at which the load and/or circuit is designed to operate. Overloads are caused by defective equipment, overloaded equipment, or too many loads on one circuit. **See Figure 11-30.**

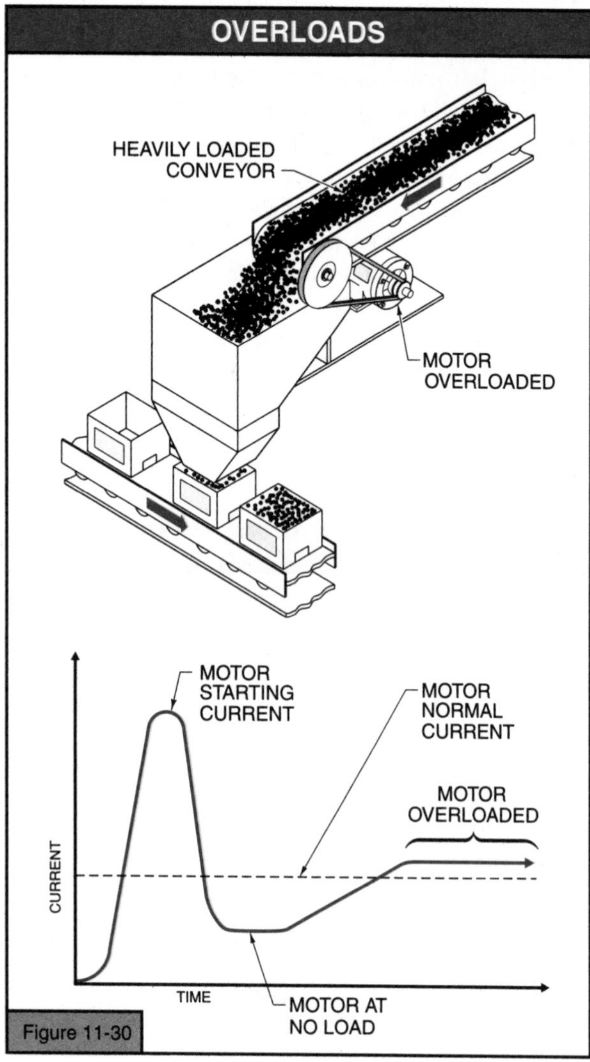

Figure 11-30

Overloads normally range between the normal current level and six times the normal current level, depending on the overload. A normal overload occurs when a motor is started because motors draw more current when starting (up to six times) than when running. An abnormal overload occurs when equipment is overloaded or defective. Overloaded equipment draws a higher-than-normal current based on the degree to which the equipment is overloaded. The more overloaded the equipment, the higher the current draw. When the equipment's load is reduced, the current draw is reduced.

Some motors and equipment are designed to safely handle a small overload for short periods of time. Overloads caused by defective equipment may cause current to increase above the normal current level. Overloads caused by defective equipment do not decrease until the equipment is repaired or replaced.

The temperature rise is minimal and has little or no effect on the equipment or conductors if an overload lasts for a brief time. However, sustained overloads are destructive and must be prevented. A sustained overload results in the overheating of conductors, equipment, and any other components used in the circuit. Like short circuits, sustained overloads must also be removed from the system. Unlike short circuits, overloads do not cause a sudden arc and do not have to be removed immediately. However, if not removed within a short time period, overloads may cause a fire by overheating the equipment and conductors.

Overcurrent Protection

An overcurrent protection device must be used to provide protection from short circuits and overloads. An *overcurrent protection device (OCPD)* is a disconnect switch with circuit breakers (CBs) or fuses added to provide overcurrent protection of the switched circuit. Fuses and circuit breakers are OCPDs designed to automatically stop the flow of current in a circuit that has a short circuit or is overloaded.

A *fuse* is an OCPD with a fusible link that melts and opens the circuit when an overload condition or short circuit occurs. A *circuit breaker* is an OCPD with a mechanical mechanism that may manually or automatically open the circuit when an overload condition or short circuit occurs. **See Figure 11-31.**

Fuses and circuit breakers are widely used in power distribution systems and individual pieces of equipment. Fuses and circuit breakers have their advantages and disadvantages. The OCPD selected is normally determined by the application, economics, and individual preference of the person making the choice.

Fuses and circuit breakers include current and voltage ratings. The current rating is the maximum amount of current the OCPD carries without blowing or tripping. The current rating of the OCPD is determined by the size and type of conductors, control devices used, and loads connected to the circuit.

The voltage rating is the maximum amount of voltage applied to the OCPD. It safely suppresses the internal arc produced when the OCPD stops the current flow. The voltage rating of the OCPD is greater than (or equal to) the voltage in the circuit.

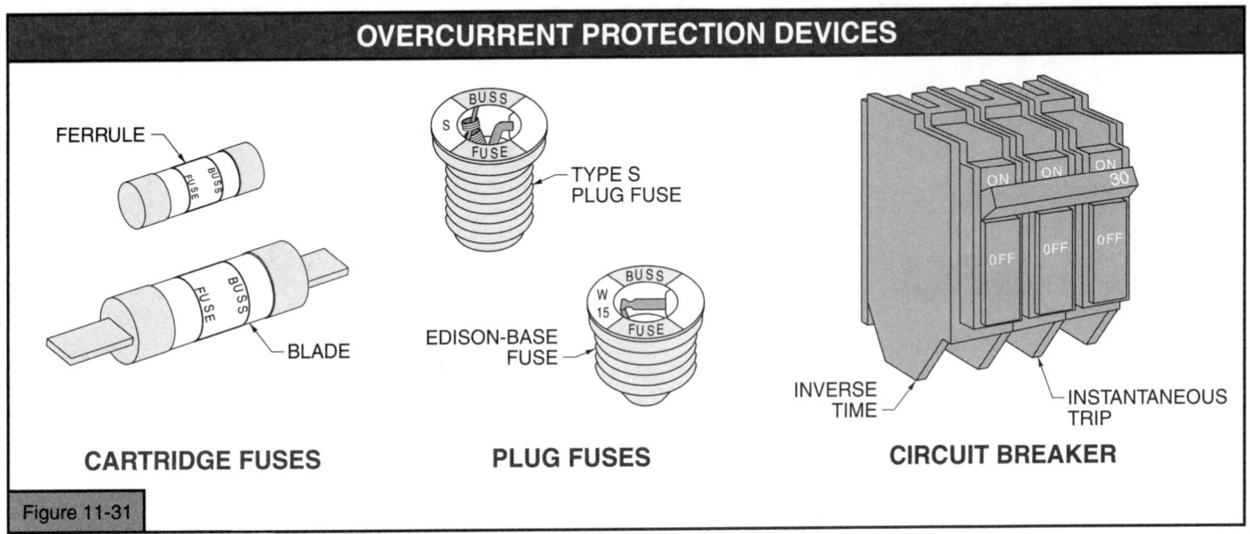

Figure 11-31

Every ungrounded (hot) power line must be protected against short circuits and overloads. A fuse or circuit breaker is installed in every ungrounded power line. One OCPD is required for low voltage, 1ϕ circuits (120 V or less) and all DC circuits. The neutral line (in AC circuits) or the negative line (in DC circuits) does not include an OCPD. Two OCPDs are required for 1ϕ, high-voltage circuits (208 V, 230 V, or 240 V). Both ungrounded power lines include an OCPD. Three OCPDs are required for all 3ϕ circuits (any voltage). All three ungrounded power lines include an OCPD. **See Figure 11-32.**

Fuses

Fuses are very low-resistance devices connected in series with the circuit's conductors. When the circuit current exceeds the rating of the fuse, the fuse opens and prevents current from flowing in that part of the circuit.

The two types of screw-in fuse bases are the Edison base and the Type S base. The Edison base is the oldest type of fuse base. One base size fits several different amperage fuses. The Edison base can no longer be used, except in replacement of existing fuses.

The Type S base is the same as an Edison base, except the Type S base has a fuse and a socket adapter. The socket adapter screws into and becomes part of the socket in the fuse panel. The adapter is constructed so when it is screwed into the socket, it locks in place and cannot easily be removed. The fuse is then screwed into the adapter.

A specific amperage Type S fuse is designed to be screwed into the threads of the same amperage adapter. For example, a 20 A Type S fuse can only be screwed into a 20 A Type S adapter. The screw base (also called a plug fuse) was commonly used to protect branch circuits in many dwellings. Today, circuit breakers are normally used in most new home construction. However, cartridge fuses are still widely used in electrical and electronic circuits.

All fuses contain a short metal strip (fusible link) made of an alloy with a low melting point. When installed in the circuit, the metal strip becomes a conductive link in the circuit.

The fusible link quickly melts and opens the circuit when the current flowing through the strip is greater than the rating of the fuse. The two basic types of fuses are the non-time delay fuse and the time delay fuse.

A *non-time delay fuse (NTDF)* is a fuse that may detect an overcurrent and open the circuit almost instantly. They contain a fusible link that melts and opens the circuit at a set temperature. The fusible link is heated by the current. The higher the current, the higher the temperature. Under normal operation and when the circuit is operating at or below its ampere rating, the fusible link functions as a conductor. However, if an overcurrent occurs in the circuit, the temperature of the fusible link reaches a melting point at the notched section. The notched section melts and burns back, causing an open in the fuse. Only one notched link opens if the overcurrent is caused from an overload. Several notched links open if the overcurrent is caused from a short circuit. NTDFs are also known as single element fuses. **See Figure 11-33.**

Low current rated fuses have only one element, while high current rated fuses have many elements. Very high current rated fuses (several hundred amperes) are developed by using many elements connected in parallel inside the same fuse. The heat and pressure created by the arc(s) are safely contained within the fuse. After removing an overcurrent, the fuse is discarded and replaced. Non-time delay fuses provide a very effective way of removing overloads and short circuits, but may also open on harmless low-level overloads. To prevent this, a time delay may be built into the fuse for use in circuits that have temporary and safe low-level overload currents.

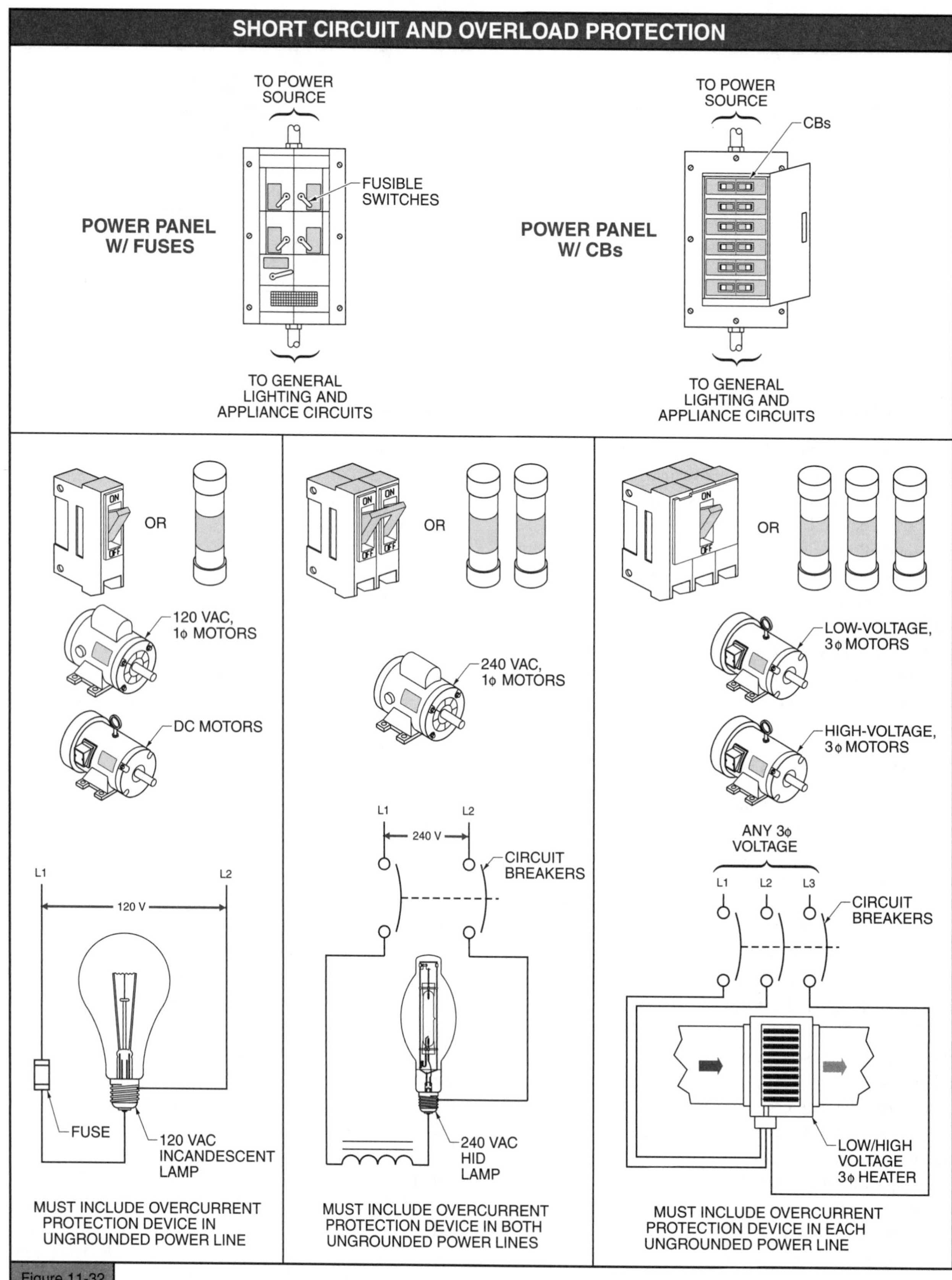
SHORT CIRCUIT AND OVERLOAD PROTECTION
TO POWER SOURCE
FUSIBLE SWITCHES
POWER PANEL W/ FUSES
TO GENERAL LIGHTING AND APPLIANCE CIRCUITS
TO POWER SOURCE
CBs
POWER PANEL W/ CBs
TO GENERAL LIGHTING AND APPLIANCE CIRCUITS
ON
OFF
OR
120 VAC, 1φ MOTORS
DC MOTORS
L1
L2
120 V
FUSE
120 VAC INCANDESCENT LAMP
MUST INCLUDE OVERCURRENT PROTECTION DEVICE IN UNGROUNDED POWER LINE
OR
240 VAC, 1φ MOTORS
L1
L2
240 V
CIRCUIT BREAKERS
240 VAC HID LAMP
MUST INCLUDE OVERCURRENT PROTECTION DEVICE IN BOTH UNGROUNDED POWER LINES
OR
LOW-VOLTAGE, 3φ MOTORS
HIGH-VOLTAGE, 3φ MOTORS
ANY 3φ VOLTAGE
L1
L2
L3
CIRCUIT BREAKERS
LOW/HIGH VOLTAGE 3φ HEATER
MUST INCLUDE OVERCURRENT PROTECTION DEVICE IN EACH UNGROUNDED POWER LINE

Figure 11-32

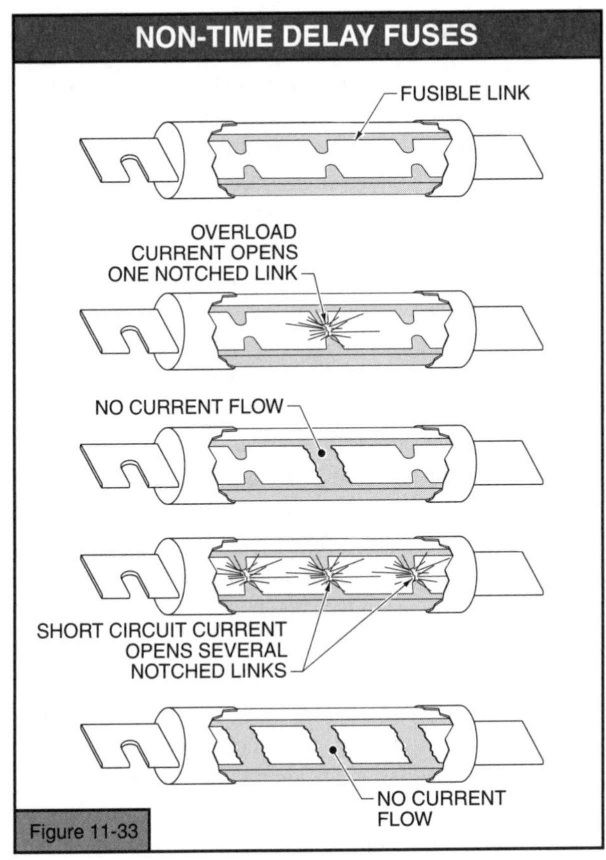

Figure 11-33

A *time delay fuse (TDF)* is a fuse that may detect and remove a short circuit almost instantly, but allow small overloads to exist for a short period of time. Time delay fuses include two elements. One element removes overloads and the other element removes short circuits. The short circuit element is the same type of fuse link element used in the non-time delay fuse. The overload element is a spring-loaded device that opens the circuit when solder holding the spring in position melts. The solder melts after an overload exists for a short period of time (normally several seconds). The overload element protects against temporary overloads up to approximately 800%. TDFs are also known as dual-element fuses. **See Figure 11-34.**

TDFs are used in circuits that have temporary, low-level overloads. Motor circuits are the most common circuit that includes low-level overloads. Motors draw an overload current when starting. As the motor accelerates, the current is reduced to a normal operating level. Therefore, TDFs are always used in motor applications.

Circuit Breakers (CBs)

Circuit breakers are OCPDs which use a mechanical mechanism to protect a circuit from short circuits and overloads. Like fuses, CBs are connected in series with the circuit's conductors. The CB opens and prevents current from flowing in that part of the circuit when the circuit current exceeds the rating of the CB.

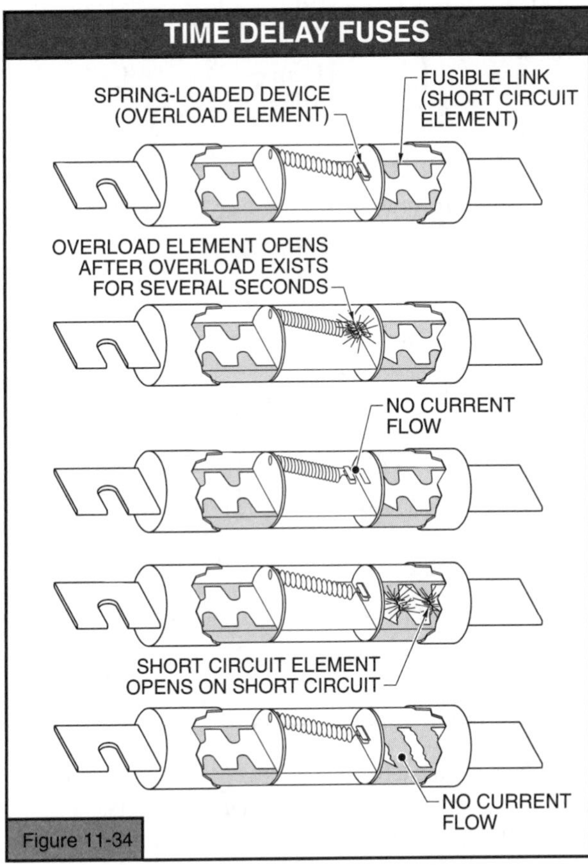

Figure 11-34

CBs contain a spring-loaded electrical contact which opens the circuit. The spring is used to open and close the contacts with a fast snap-action. A handle is added to the contact assembly so the contacts may be manually opened and closed. The contacts are automatically opened on an overcurrent by a bimetal strip and/or an electromagnetic tripping device. The contacts have one stationary contact and one movable contact. The movable contact is attached to a spring that provides a fast snap-action when tripped. **See Figure 11-35.**

The bimetal strip is made of two dissimilar metals that expand at different rates when heated. The strip bends when heated and opens the contacts. The bimetal strip is connected in series with the circuit and is heated by the current flowing through it. The higher the circuit current, the hotter the bimetal strip becomes. Likewise, the higher the current, the shorter the time required to trip the CB.

Like the bimetal strip, the electromagnetic device is connected in series with the circuit. As current passes through the coil, a magnetic field is produced. The higher the circuit current, the stronger the magnetic field. The magnetic field opens the contacts when it becomes strong enough.

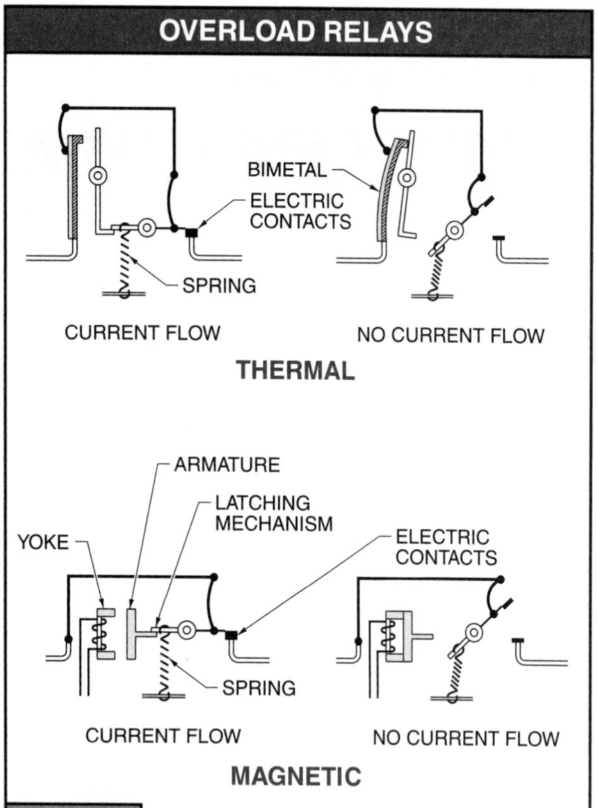

OVERLOAD RELAYS

BIMETAL
ELECTRIC CONTACTS
SPRING
CURRENT FLOW
NO CURRENT FLOW
THERMAL

ARMATURE
LATCHING MECHANISM
YOKE
ELECTRIC CONTACTS
SPRING
CURRENT FLOW
NO CURRENT FLOW
MAGNETIC

Figure 11-35

Whenever the CB contacts are opened, an arc is drawn across the movable and stationary contacts. A short circuit (higher current) produces a much hotter arc than an overload. Therefore, a CB that has tripped numerous times from short circuits may be damaged to the point of requiring replacement.

Data Cable and System Protection

All electrical systems are prone to potentially damaging voltage surges on their power and communication lines. Voltage surges are especially damaging if allowed on communication lines. Voltage surges occur on communication lines by direct or indirect contact with power lines, lightning, or transients. A *transient voltage* is a temporary, unwanted voltage in an electrical circuit. Since these voltage surges can occur anywhere along the electrical system, it is recommended that several levels of protection be provided along the system. **See Figure 11-36.** The levels of protection in an electrical system should include the following:

- Level 1 – The system should be protected from voltage surges entering a building through the use of protection devices. These protection devices should be located at the main power in-feed panel.
- Level 2 – The equipment should be protected from voltage surges introduced between the building entrance and the system equipment. This will protect the system from problems introduced by other electrical loads and faults within the building.
- Level 3 – The equipment should be protected from voltage surges introduced between communication equipment and workstation equipment.

Installing surge suppressors protects equipment and systems. A *surge suppressor* is an electrical device that limits voltage surges that may be present on power lines. Surge suppressors should be installed in the distribution panels, along the building system, and at each workstation of a building. In addition to surge suppressors, filters should also be used. Filters aid in reducing noise introduced on the lines by EMI or RFI.

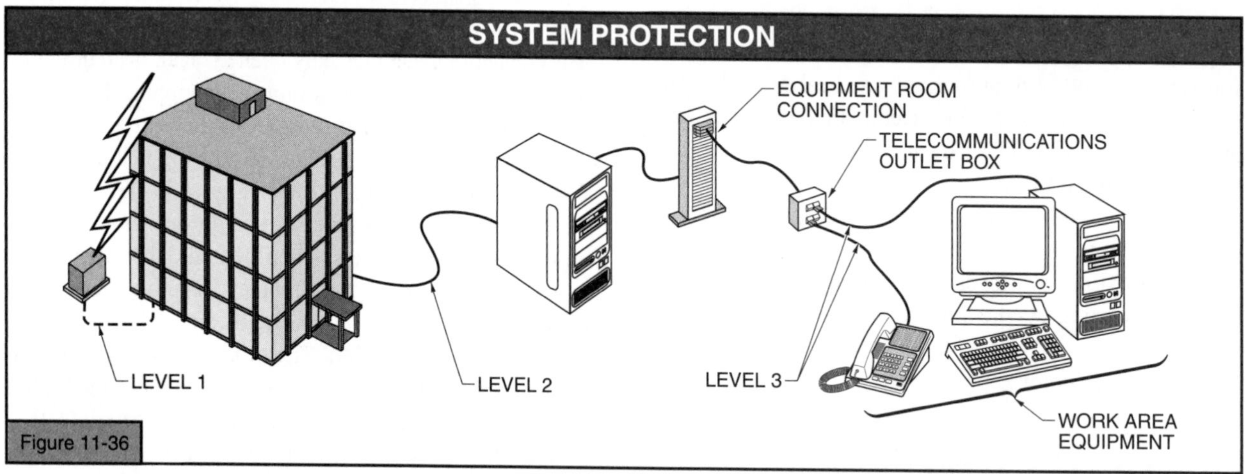

Figure 11-36

Refer to the CD-ROM in the back of the book for Chapter 11 Quick Quiz® and related information.

12 Series Circuits

Series Connections

Switches, loads, meters, fuses, and other electric components can be connected in series. A *series connection* is a connection that has two or more components connected so there is only one path for current flow. Opening the circuit at any point stops the flow of current. Current stops flowing any time a fuse blows, a circuit breaker trips, or a switch or load opens.

A fuse blows or circuit breaker trips when the circuit is overloaded. A switch is opened any time the circuit is turned OFF. The opening of a switch may be manual (toggle switches, etc.), mechanical (limit switches), or automatic (temperature switches). A load opens any time the load's current-consuming component (lamp filament, motor winding, etc.) burns or breaks open. **See Figure 12-1.**

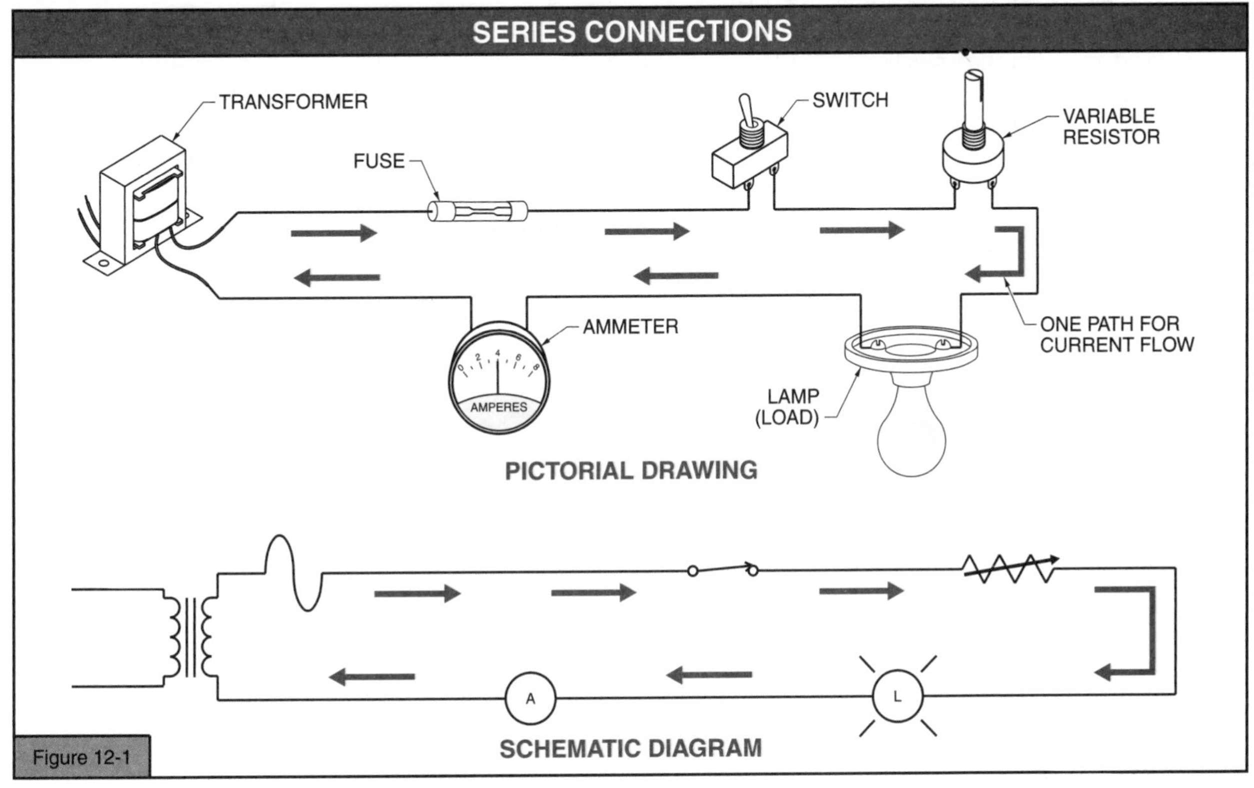

Figure 12-1

Recognizing series-connected components and circuits enables a technician or troubleshooter to take proper measurements, make circuit modifications, and troubleshoot the circuit. A schematic or line diagram is normally used when illustrating a series-connected circuit. A schematic or line diagram shows the circuit in its simplest form.

Series Circuit Polarity

All DC voltage sources have a positive and a negative terminal. The positive and negative terminals establish polarity in a circuit. All points in a DC series circuit have polarity. *Polarity* is the positive (+) or negative (–) state of an object. In a DC series circuit, each component (switch, load, fuse, etc.) has a positive polarity side and a negative polarity side. The side of the component closest to the positive voltage terminal is the positive polarity side. The side of the component closest to the negative voltage terminal is the negative polarity side. **See Figure 12-2.**

Series Switch Connections

All switches connected in series must be closed before current flows in the circuit. Opening any one or more of the switches stops current flow. Series-connected switches include any manual, mechanical, or automatic switch. Switches are usually connected in series to build safety into a circuit. For example, a microwave oven includes a limit switch in the door. The limit switch is connected in series with the oven ON switch. The oven cannot be turned ON unless the limit switch is closed and the oven ON switch is closed.

Manually operated switches are often connected in series with automatic switches. The manual switch provides a point in the circuit for removing power when switched to the OFF position. For example, a manually-operated system ON/OFF switch is included with most thermostats. The manual switch can be used to turn OFF the automatically-operated heating and air conditioning system. **See Figure 12-3.**

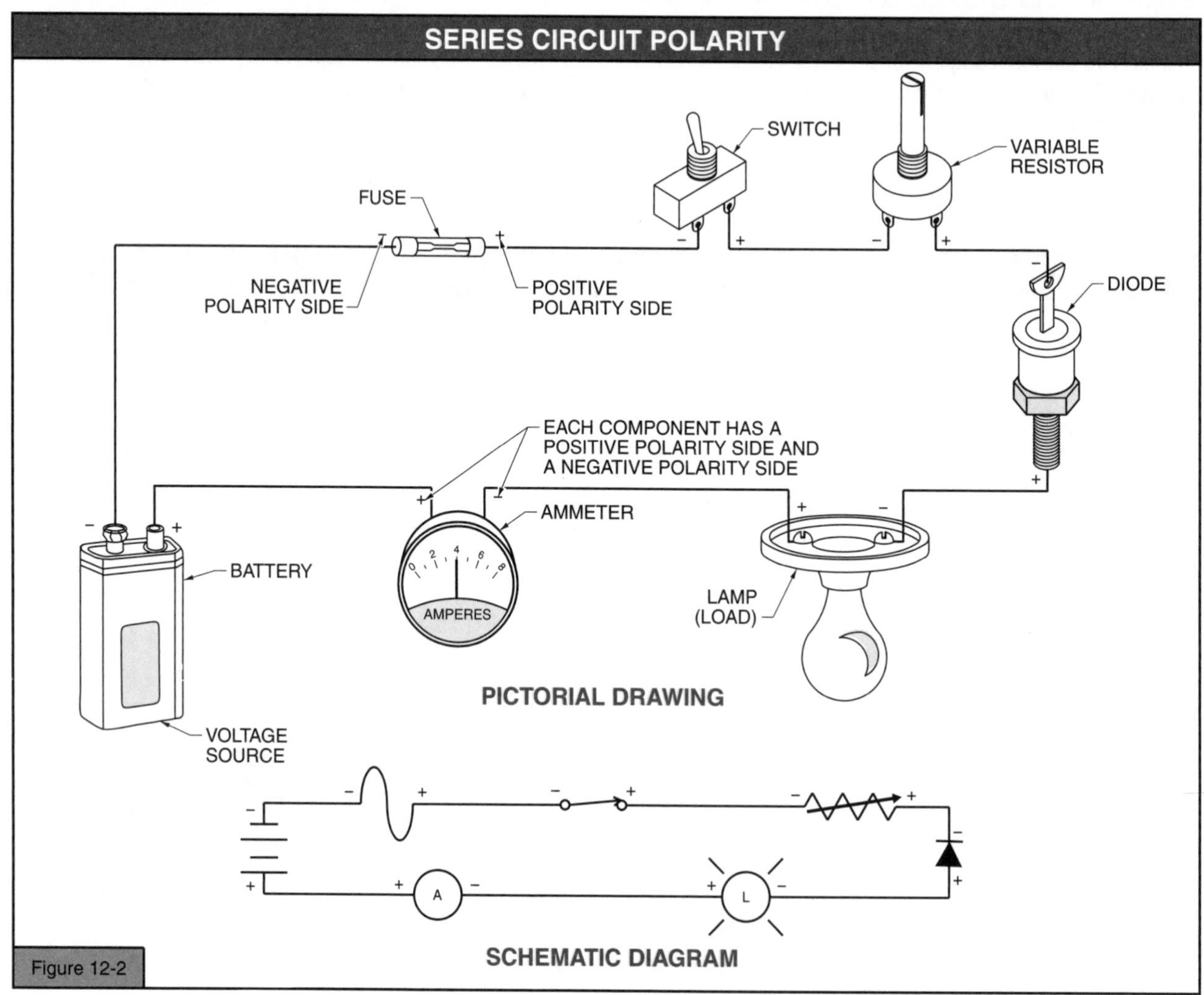

Figure 12-2

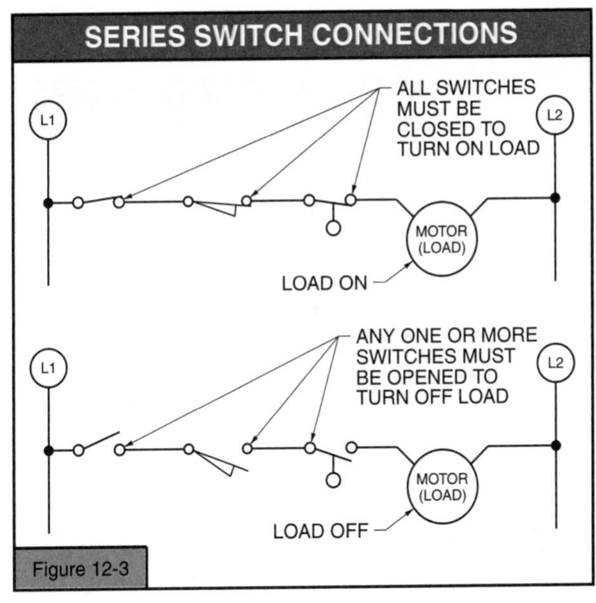

Figure 12-3

Resistance in Series Circuits

Resistance is the opposition to current flow. The total resistance in a circuit containing series-connected loads equals the sum of the resistances of all loads. The resistance in the circuit increases if loads are added in series and decreases if loads are removed. **See Figure 12-4.** To calculate total resistance of a series circuit, apply the formula:

$$R_T = R_1 + R_2 + R_3 + \ldots$$

where

R_T = total resistance (in Ω)

R_1 = resistance 1 (in Ω)

R_2 = resistance 2 (in Ω)

R_3 = resistance 3 (in Ω)

Example: Calculating Total Resistance in Series Circuits

What is the total resistance of a circuit containing resistors of 25 Ω, 50 Ω, and 75 Ω connected in series?

$$R_T = R_1 + R_2 + R_3$$

$$R_T = 25 + 50 + 75$$

$$R_T = \mathbf{150\ \Omega}$$

Voltage in Series Circuits

The voltage applied across loads connected in series is divided across the loads. Each load drops a set percentage of the applied voltage. The exact voltage drop across each load depends on the resistance of each load. The voltage drop across each load is the same if the resistance values are the same. The load with the lowest resistance has the lowest voltage drop and the load with the highest resistance has the highest voltage drop. **See Figure 12-5.**

In an ideal circuit, the load(s) should be the only devices in the circuit that have resistance. For this reason, all of the power supply voltage should be dropped across them. However, any other part of an electrical circuit may also have a voltage drop across it. For example, mechanical switches that are burnt, corroded, or not making good contact have a voltage drop across them. The higher the voltage drop, the more damaged the switch. Likewise, the conductors (wires, etc.) that connect the electrical circuit together also have a voltage drop across them. In theory, the voltage drop across conductors should be zero. In fact, all conductors have some voltage drop across them.

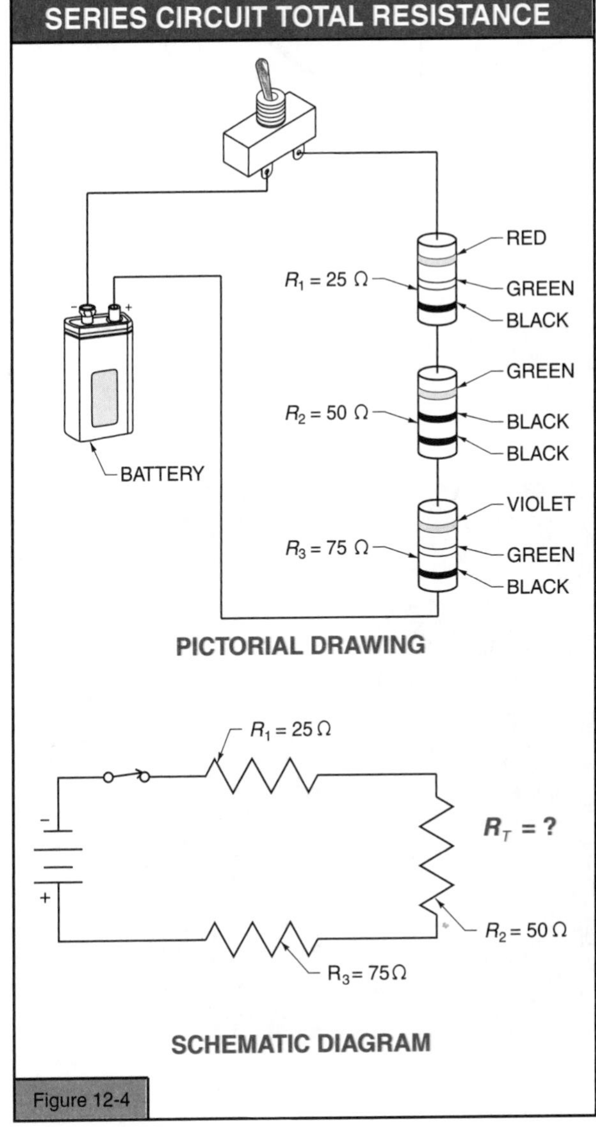

Figure 12-4

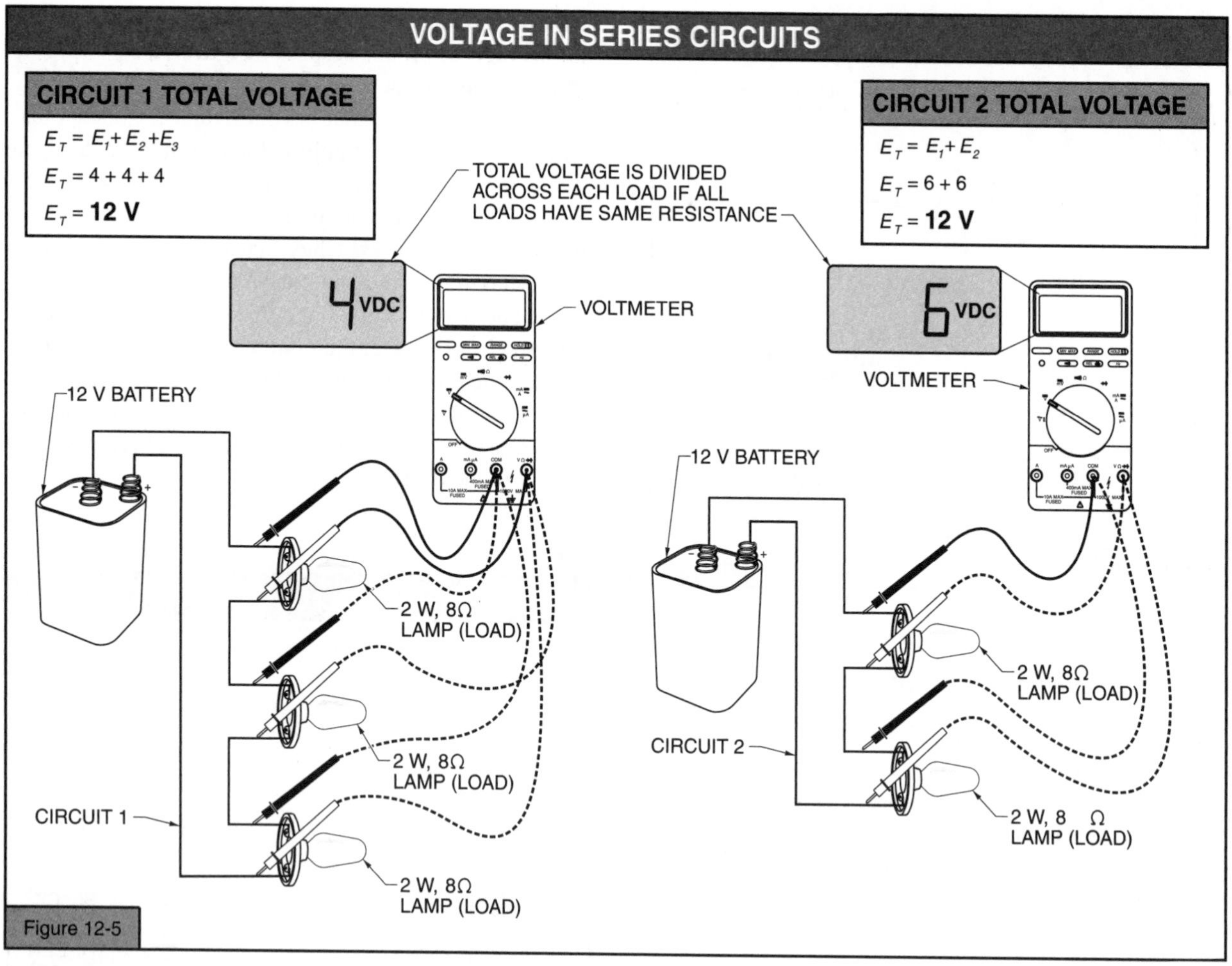

VOLTAGE IN SERIES CIRCUITS

CIRCUIT 1 TOTAL VOLTAGE

$E_T = E_1 + E_2 + E_3$

$E_T = 4 + 4 + 4$

$E_T = \mathbf{12\ V}$

CIRCUIT 2 TOTAL VOLTAGE

$E_T = E_1 + E_2$

$E_T = 6 + 6$

$E_T = \mathbf{12\ V}$

TOTAL VOLTAGE IS DIVIDED ACROSS EACH LOAD IF ALL LOADS HAVE SAME RESISTANCE

4 VDC

VOLTMETER

6 VDC

VOLTMETER

12 V BATTERY

12 V BATTERY

2 W, 8Ω LAMP (LOAD)

2 W, 8Ω LAMP (LOAD)

2 W, 8Ω LAMP (LOAD)

CIRCUIT 1

CIRCUIT 2

2 W, 8Ω LAMP (LOAD)

2 W, 8 Ω LAMP (LOAD)

Figure 12-5

Fluke Corporation

A current clamp attachment is used with a DMM when troubleshooting loads on a power distribution system.

Small conductors have a higher resistance than large conductors made of the same material. Thus, undersized conductors have a higher voltage drop across them. Conductors should be sized large enough to prevent no more than a 3% voltage drop from the power source to the load. The voltage drop across conductors and switches can be measured using a voltmeter.

All loads have some resistance value. The higher the power rating of a load, the lower the resistance value. The lower the power rating of a load, the higher the resistance value. Loads can be lamps, heating elements, solenoids, motors, bells, horns, or any device that uses electrical energy.

Loads are normally represented in a schematic diagram by the resistor symbol. The sum of all the voltage drops in a series circuit is equal to the applied voltage because the voltage in a series circuit is divided across all the loads in the circuit. **See Figure 12-6.** To calculate total voltage in a series circuit when the voltage across each load is known or measured, apply the formula:

$E_T = E_1 + E_2 + E_3 + ...$

where

E_T = total applied voltage (in V)

E_1 = voltage drop across load 1 (in V)

E_2 = voltage drop across load 2 (in V)

E_3 = voltage drop across load 3 (in V)

Example: Calculating Voltage in Series Circuits

What is the total applied voltage of a circuit containing 2 V, 4 V, and 6 V drops across three loads?

$E_T = E_1 + E_2 + E_3 + ...$

$E_T = 2 + 4 + 6$

$E_T = \mathbf{12\ V}$

Current in Series Circuits

The current in a circuit containing series-connected loads is the same throughout the circuit. The current in the circuit decreases if the circuit's resistance increases and the current increases if the circuit's resistance decreases. **See Figure 12-7.** To calculate total current of a series circuit, apply the formula:

$I_T = I_1 = I_2 = I_3$

where

I_T = total circuit current (in A)

I_1 = current through component 1 (in A)

I_2 = current through component 2 (in A)

I_3 = current through component 3 (in A)

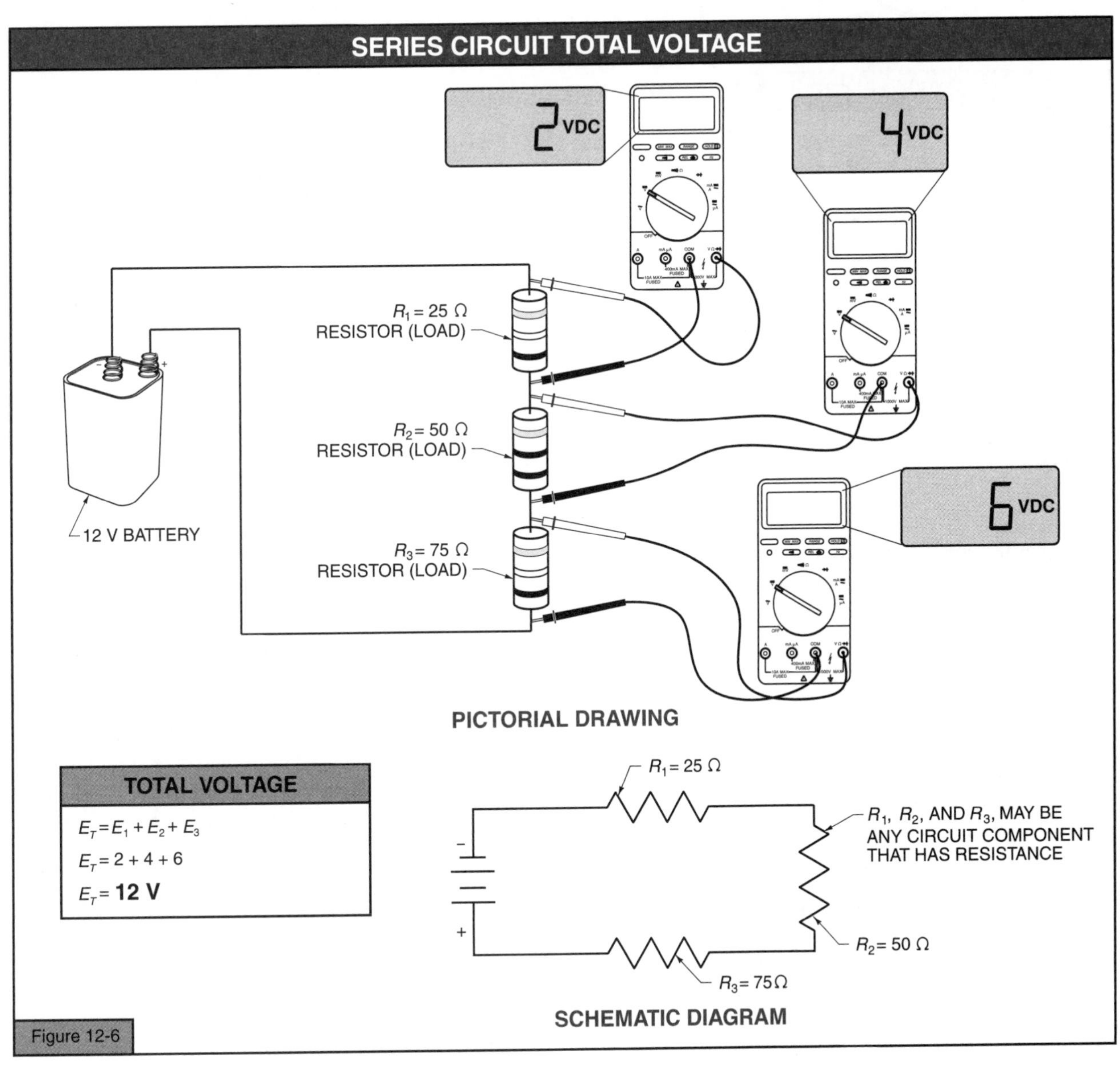

Figure 12-6

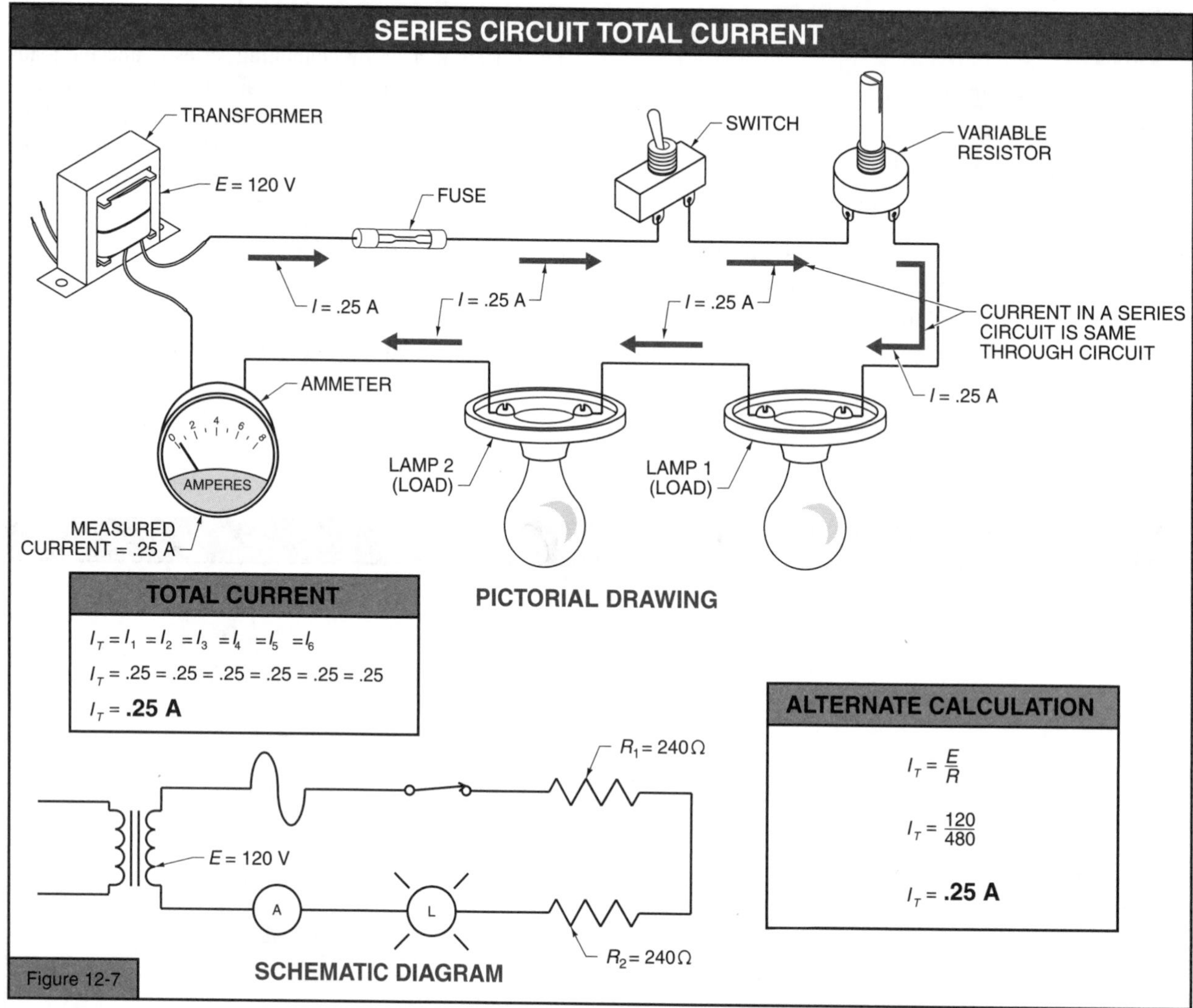

Figure 12-7

Example: Calculating Current in Series Circuits

What is the total current through a series circuit if the current measured at each of six different loads is .25 A (250 mA)?

$I_T = I_1 = I_2 = I_3 = I_4 = I_5 = I_6$

$I_T = .25 = .25 = .25 = .25 = .25 = .25$

$I_T = \mathbf{.25\ A}$

Each component must be sized to safely pass the current of the loads because the current through the switches, fuses, and other components equals the current through each load.

Power in Series Circuits

Power is produced when voltage is applied to a load and current flows through the load. The power produced is used to produce light (lamps), heat (heating elements), rotary motion (motors), or linear motion (solenoids).

The lower the load's resistance or higher the applied voltage, the more power produced. The higher the load's resistance or lower the applied voltage, the less power produced. The amount of power produced is measured in watts (W).

The amount of power produced by a load is equal to the voltage drop across the load times the current through the load. The total power in a circuit is equal to the sum of the individual power produced by each load. **See Figure 12-8.**

To calculate total power in a series circuit when the power across each load is known or measured, apply the formula:

$P_T = P_1 + P_2 + P_3 ...$

where

P_T = total circuit power (in W)

P_1 = power of load 1 (in W)

P_2 = power of load 2 (in W)

P_3 = power of load 3 (in W)

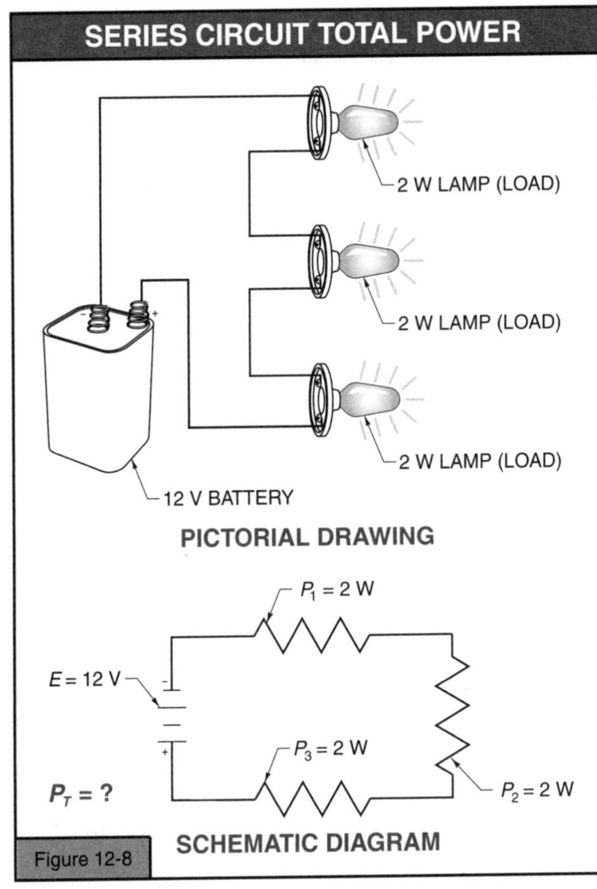

Figure 12-8

Example: Calculating Total Power in Series Circuits

What is the total power in a series circuit if three loads are connected in series and each load produces 2 W?

$P_T = P_1 + P_2 + P_3 + \ldots$

$P_T = 2 + 2 + 2$

$P_T = \mathbf{6\ W}$

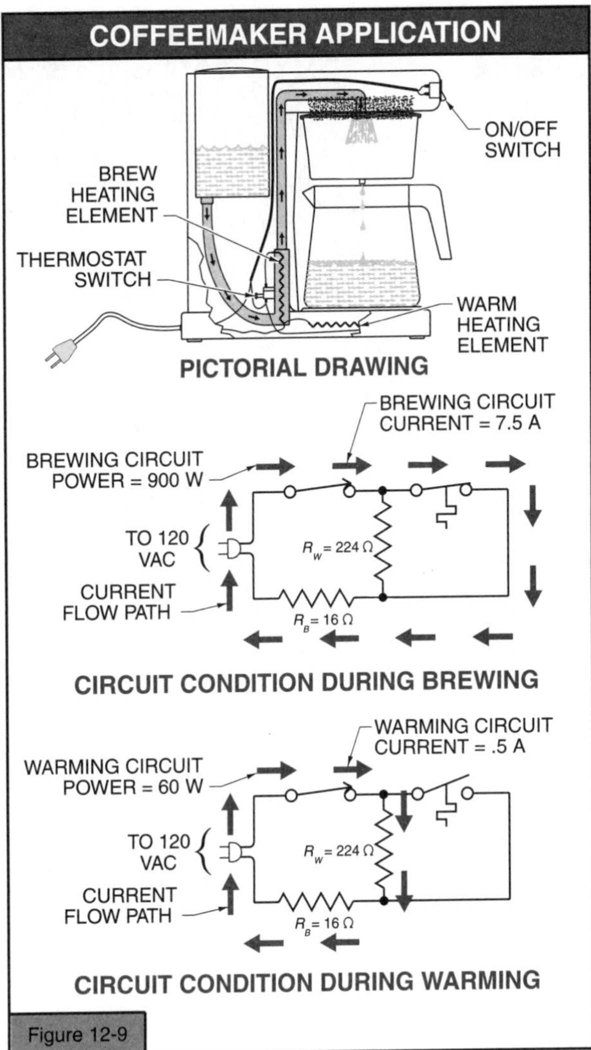

Figure 12-9

Series Circuit Application

The individual concepts and formulas for series circuits can be applied for a total understanding of circuit operation when designing and troubleshooting circuits. For example, a typical coffeemaker applies the concepts of a basic series circuit to brew and keep coffee warm. **See Figure 12-9.**

The coffeemaker includes a brew heating element, a warm heating element, an ON/OFF switch, and a temperature switch. The brew heating element heats the water and forces it over the coffee grounds. The warm heating element keeps the coffee warm after brewing. Since the brew heating element must produce more power than the warm heating element, the brew element has a lower resistance value than the warm heating element. The ON/OFF switch starts and stops the process. The temperature switch changes the circuit's total resistance after the coffee is brewed.

The total resistance of the circuit is equal to the resistance of the brew heating element when the coffee brew cycle is started. The warm heating element is not a part of the electrical circuit at this time because the normally closed temperature switch short circuits the warm heating element during brewing.

The current in the circuit during brewing is found by applying Ohm's law. To calculate circuit current during brewing, apply the formula:

$$I = \frac{E}{R}$$

where

I = current during brewing (in A)

E = applied voltage (in V)

R = brew heating element resistance (in Ω)

Bunn-O-Matic Corporation

The total resistance of the circuit in a coffeemaker is equal to the resistance of the heating element once the brew cycle starts.

Example: Calculating Brewing Current
What is the current in the circuit during brewing when the applied voltage is 120 V and the brew heating element resistance is 16 Ω?

$$I = \frac{E}{R}$$

$$I = \frac{120}{16}$$

$$I = \mathbf{7.5\ A}$$

The power in the circuit during brewing can be found by applying the power formula. To calculate power during brewing, apply the formula:

$$P = E \times I$$

where

P = power during brewing (in W)

E = applied voltage (in V)

I = current during brewing (in A)

Example: Calculating Brewing Power
What is the power in the circuit during brewing when the applied voltage is 120 V and the circuit current is 7.5 A?

$$P = E \times I$$

$$P = 120 \times 7.5$$

$$P = \mathbf{900\ W}$$

The temperature increases after all the water is forced over the brew heating element. The increased temperature opens the temperature switch. The warm heating element and the brew heating element are connected in series when the temperature switch opens. The total resistance of the series circuit is found by adding the resistances of each series-connected component. To calculate the total resistance during warming, apply the formula:

$$R_T = R_1 + R_2$$

where

R_T = total circuit resistance (in Ω)

R_1 = resistance of warm heating element (in Ω)

R_2 = resistance of brew heating element (in Ω)

Example: Calculating Total Resistance During Warming
What is the total resistance in a circuit containing resistances of 224 Ω and 16 Ω?

$$R_T = R_1 + R_2$$

$$R_T = 224 + 16$$

$$R_T = \mathbf{240\ \Omega}$$

The current in the circuit during warming is found by applying Ohm's law.

Example: Calculating Warming Current
What is the current in the circuit during warming when the applied voltage is 120 V and the total circuit resistance is 240 Ω?

$$I = \frac{E}{R}$$

$$I = \frac{120}{240}$$

$$I = \mathbf{.5\ A}$$

The power in the circuit during warming is found by applying the power formula.

Example: Calculating Warming Power
What is the power in the circuit during warming when the applied voltage is 120 V and the circuit current is .5 A?

$$P = E \times I$$

$$P = 120 \times .5$$

$$P = \mathbf{60\ W}$$

Capacitors in Series Circuits

Capacitance (C) is the ability to store energy in the form of an electrical charge. A *capacitor* is an electric device designed to store electrical energy by means of an electrostatic field. **See Figure 12-10.**

A capacitor consists of two metal plates (conductors) separated by a dielectric (insulating material). The dielectric can be air, paper, mica, ceramic, or any other insulator. The capacitor is charged when voltage is applied to its leads and the voltage across the metal plates is equal to the applied voltage. The capacitor holds its charge when the applied voltage is removed. The capacitor discharges whenever there is a path for current flow between the two leads of the capacitor. The stored voltage of a capacitor is used in many electrical and electronic applications.

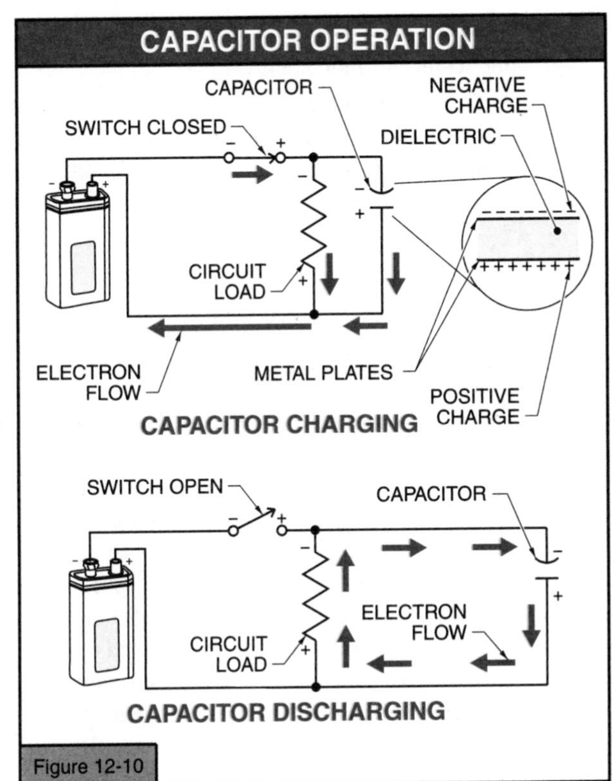

Figure 12-10

The unit used to measure capacitance is the farad (F). The farad is a large unit of capacitance and is not normally used when stating capacitance. The most common unit of capacitance is the microfarad (µF). One µF equals 1/1,000,000 F. **See Figure 12-11.**

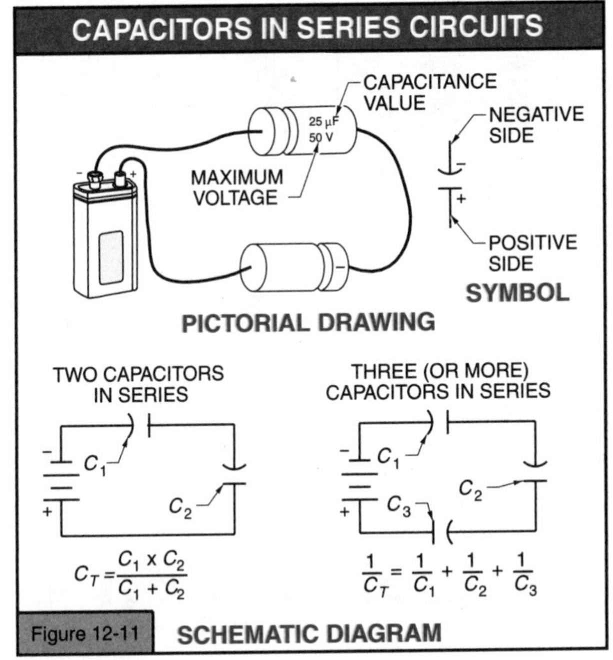

Figure 12-11

To calculate total capacitance of two capacitors connected in series, apply the formula:

$$C_T = \frac{C_1 \times C_2}{C_1 + C_2}$$

where

C_T = total capacitance (in µF)

C_1 = capacitance of capacitor 1 (in µF)

C_2 = capacitance of capacitor 2 (in µF)

Example: Calculating Total Capacitance of Two Capacitors

What is the total capacitance of a 100 mF and 50 mF capacitor connected in series?

$$C_T = \frac{C_1 \times C_2}{C_1 + C_2}$$

$$C_T = \frac{100 \times 50}{100 + 50}$$

$$C_T = \frac{5000}{150}$$

$$C_T = \mathbf{33.33\ \mu F}$$

To calculate total capacitance of three or more capacitors connected in series, apply the formula:

$$\frac{1}{C_T} = \frac{1}{C_1} + \frac{1}{C_2} + \frac{1}{C_3} + \ldots$$

where

C_T = total capacitance (in µF)

C_1 = capacitance of capacitor 1 (in µF)

C_2 = capacitance of capacitor 2 (in µF)

C_3 = capacitance of capacitor 3 (in µF)

Example: Calculating Total Capacitance of Three Capacitors

What is the total capacitance of a 200 µF, 100 µF, and a 300 µF capacitor connected in series?

$$\frac{1}{C_T} = \frac{1}{C_1} + \frac{1}{C_2} + \frac{1}{C_3} + \ldots$$

$$\frac{1}{C_T} = \frac{1}{200} + \frac{1}{100} + \frac{1}{300}$$

$$\frac{1}{C_T} = .005 + .01 + .0033$$

$$\frac{1}{C_T} = .0183 \text{ or } C_T = \mathbf{54.64\ \mu F}$$

TECH FACT

Never touch the terminals of a capacitor, even if the capacitor is not connected to a circuit, because a capacitor can store a charge for a long time.

Inductors in Series Circuits

Inductance (L) is the property of an electric device that opposes a change in current due to its ability to store electrical energy in a magnetic field. An inductor is an electric device designed to store electrical energy by means of a magnetic field.

The three common inductors used in electrical and electronic applications are the air-core, iron-core, and ferrite-core inductors. An *air-core inductor* consists of a coil of wire wrapped around a hollow core. An *iron-core inductor* consists of a coil of wire wrapped around a laminated steel core. A *ferrite-core inductor* consists of a coil of wire wrapped around a ceramic material (ferrite). The ferrite-core inductor consists of a movable core that can be adjusted for different circuit-tuning applications.

A magnetic field builds up around the wire and core when current is applied to an inductor. This magnetic field causes a counter voltage to be induced in the coil. This counter voltage opposes the applied voltage and remains at a constant value when fully energized. This counter voltage opposes any change in current flow. The unit used to measure inductance is the henry (H). **See Figure 12-12.**

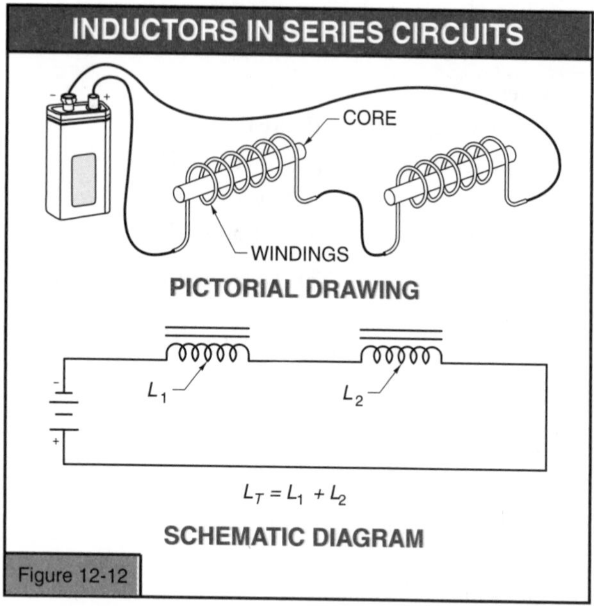

Figure 12-12

To calculate total inductance of inductors connected in series, apply the formula:

$L_T = L_1 + L_2 + ...$

where

L_T = total inductance (in H)

L_1 = inductance of inductor 1 (in H)

L_2 = inductance of inductor 2 (in H)

Example: Calculating Total Inductance

What is the total inductance of a 5 mH and 10 mH inductor connected in series?

$L_T = L_1 + L_2$

$L_T = 5 + 10$

$L_T =$ **15 mH**

Batteries and Solarcells in Series Circuits

A *battery* is a DC voltage source that converts chemical energy to electrical energy. A *cell* is a unit that produces electricity at a fixed voltage and current level. The voltage potential of a battery is increased when cells are connected in series. Cells must have the same voltage rating when connected in series.

A *solarcell* is a voltage source that converts light energy to electrical energy. Solarcells are used in place of batteries in low-current devices or to recharge battery-operated devices. Solarcells are used individually or in combination to increase voltage and current. The voltage potential is increased when solarcells are connected in series. **See Figure 12-13.**

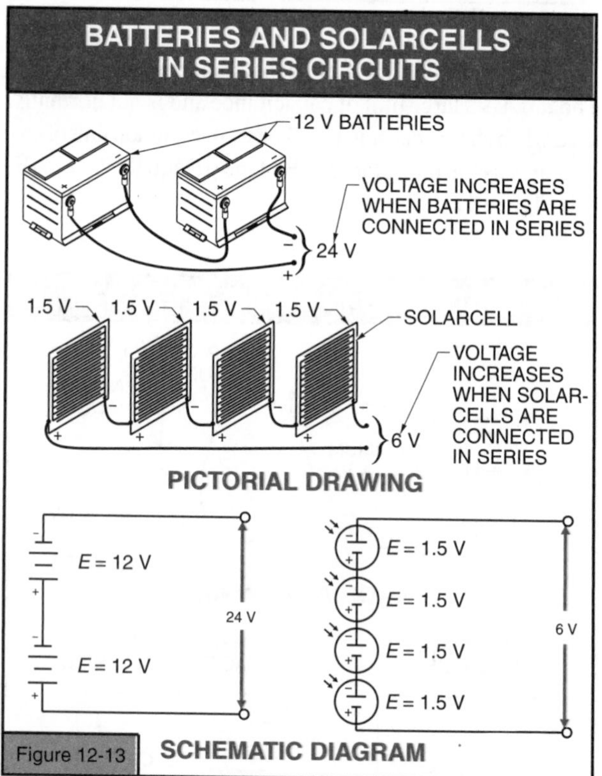

Figure 12-13

Parallel Connections

Switches, loads, meters, and other components can be connected in parallel. A *parallel connection* is a connection that has two or more components connected so that there is more than one path for current flow.

Care must be taken when working with parallel circuits because current can be flowing in one part of the circuit even though another part of the circuit is turned OFF. Understanding and recognizing parallel-connected components and circuits enables a technician or troubleshooter to take proper measurements, make circuit modifications, and troubleshoot the circuit. **See Figure 13-1.**

Parallel Circuit Polarity

All DC voltage sources have a positive and a negative terminal. The positive and negative terminals establish polarity in a circuit. Polarity is the positive (+) or negative (–) state of an object. All points in a DC parallel circuit have polarity. In a parallel DC circuit, each component (meter, load, fuse, etc.) has a positive polarity side and a negative polarity side. The side of the component closest to the positive voltage source terminal is the positive polarity side. The side of the component closest to the voltage source negative terminal is the negative polarity side. **See Figure 13-2.**

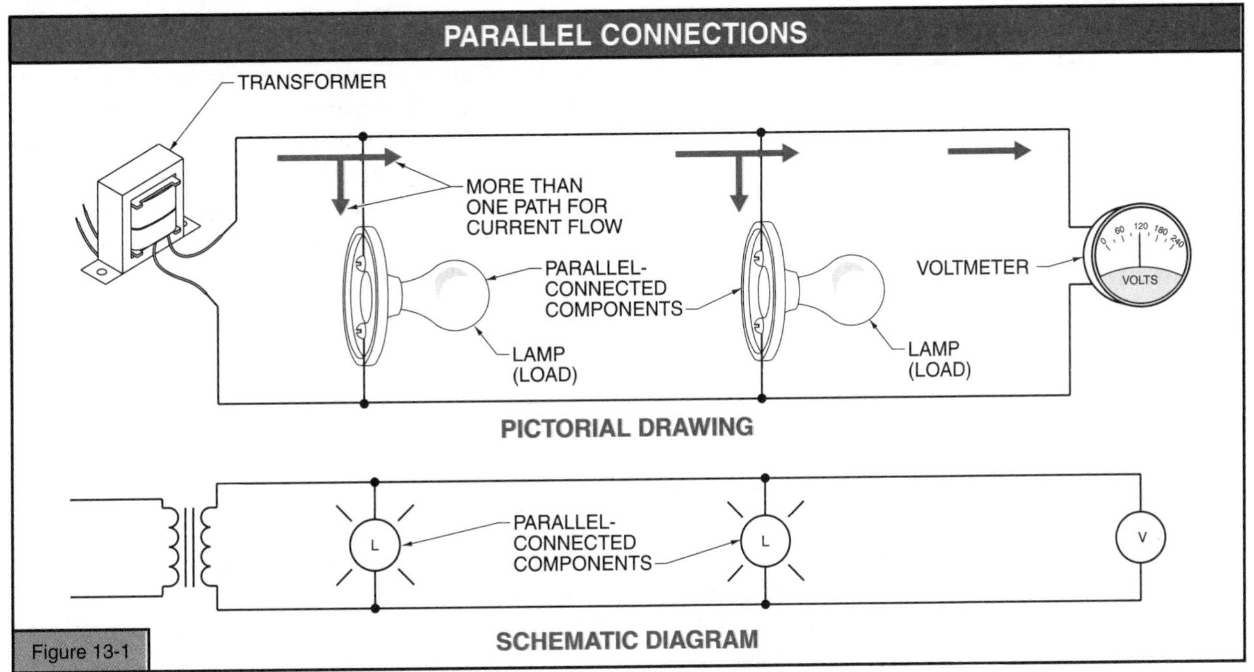

Figure 13-1

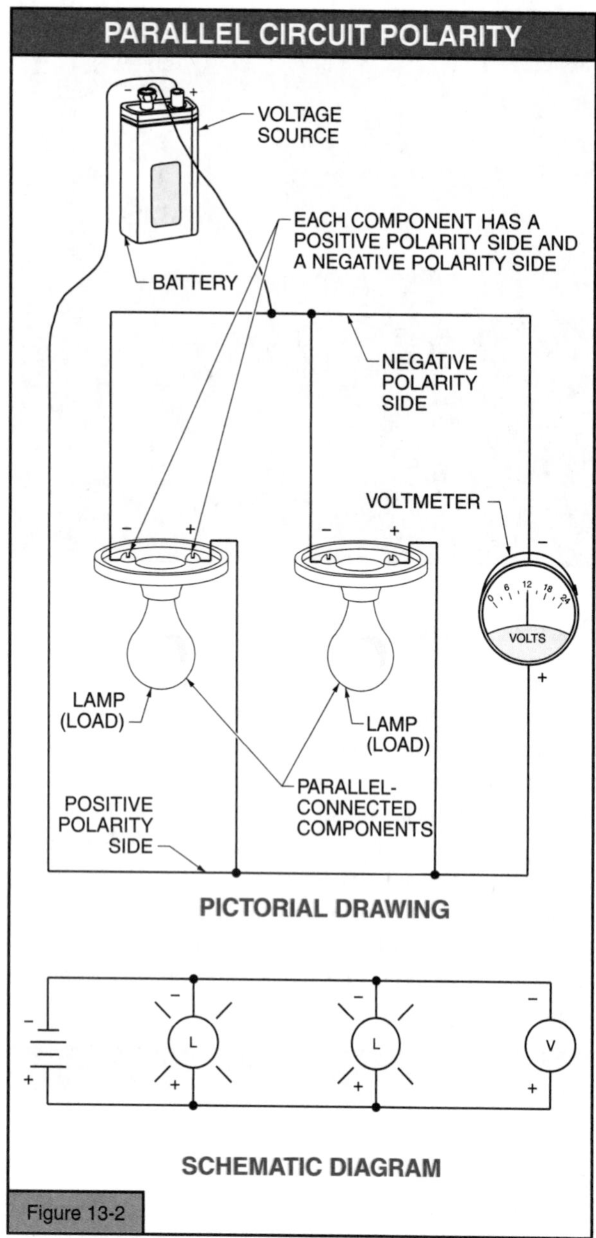

Figure 13-2

Parallel Switch Connections

Parallel-connected switches include any manual, mechanical, or automatic switches. One or more parallel-connected switches must be closed to start current flow. All switches must be opened to stop current flow. Switches are usually connected in parallel to provide additional turn-ON points. For example, several pushbuttons may be connected in parallel to turn ON a doorbell. The switches may be located at the front, back, and side doors. **See Figure 13-3.**

Manually-operated switches are often connected in parallel with mechanical and automatic switches. A manual switch is used to override the mechanical or automatic switch. For example, the dome light in an automobile is automatically turned ON when a door is opened and turned OFF when the door is closed by a limit switch built into the door. A manual switch (usually built into the headlight control switch) is connected in parallel with the door limit switch so that the dome light can be manually turned ON even if all the doors are closed.

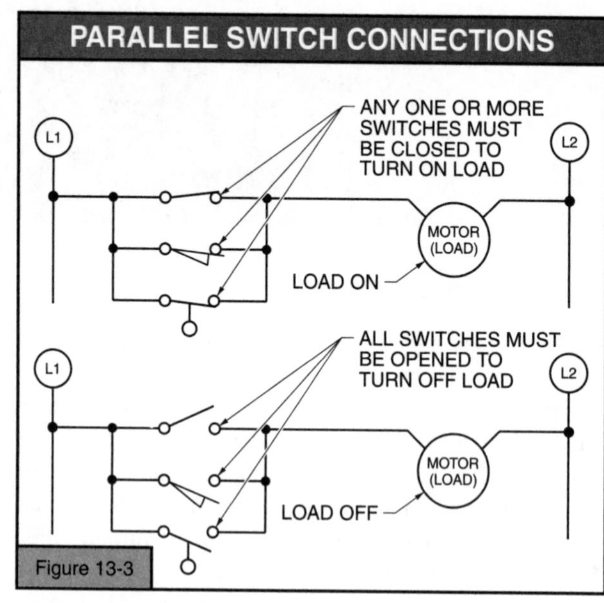

Figure 13-3

Resistance in Parallel Circuits

Resistance is the opposition to current flow. The total resistance in a circuit containing parallel-connected loads is less than the smallest resistance value. The total resistance decreases if loads are added in parallel and increases if loads are removed. **See Figure 13-4.** To calculate total resistance in a parallel circuit containing two resistors, apply the formula:

$$R_T = \frac{R_1 \times R_2}{R_1 + R_2}$$

where

R_T = total resistance (in Ω)

R_1 = resistance 1 (in Ω)

R_2 = resistance 2 (in Ω)

Example: Calculating Total Resistance—Two Resistors Connected in Parallel

What is the total resistance in a circuit containing resistors of 300 Ω and 600 Ω connected in parallel?

$$R_T = \frac{R_1 \times R_2}{R_1 + R_2}$$

$$R_T = \frac{300 \times 600}{300 + 600}$$

$$R_T = \frac{180,000}{900}$$

$$R_T = \mathbf{200\ \Omega}$$

To calculate total resistance in a parallel circuit with three or more resistors, apply the formula:

$$R_T = \frac{1}{\frac{1}{R_1} + \frac{1}{R_2} + \frac{1}{R_3}}$$

where

R_T = total resistance (in Ω)

R_1 = resistance 1 (in Ω)

R_2 = resistance 2 (in Ω)

R_3 = resistance 3 (in Ω)

Example: Calculating Total Resistance—Three Resistors Connected in Parallel

What is the total resistance in a circuit containing three resistors of 5000 Ω, 2000 Ω, and 10,000 Ω connected in parallel?

$$R_T = \frac{1}{\frac{1}{R_1} + \frac{1}{R_2} + \frac{1}{R_3}}$$

$$R_T = \frac{1}{\frac{1}{5000} + \frac{1}{2000} + \frac{1}{10,000}}$$

$$R_T = \frac{1}{.0002 + .0005 + .0001}$$

$$R_T = \frac{1}{.0008}$$

$$R_T = \mathbf{1250\ \Omega}$$

Voltage in Parallel Circuits

The voltage across each load is the same when loads are connected in parallel. The voltage across each load remains the same if parallel loads are added or removed. **See Figure 13-5.** Connecting loads in parallel is the most common method used to connect loads. For example, loads such as lamps, small appliances, fans, TVs, etc., are all connected in parallel when connected to a branch circuit. **See Figure 13-6.**

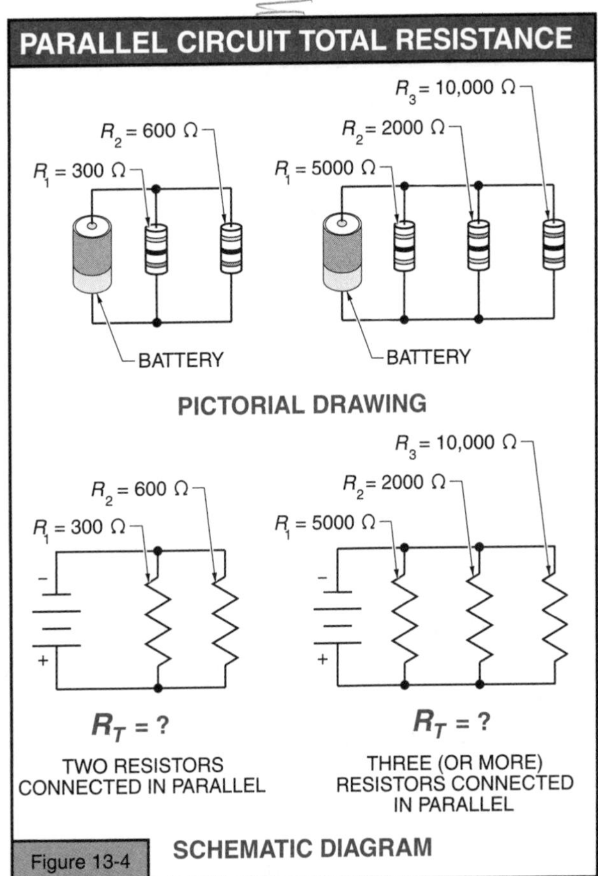

Figure 13-4

Rockwell Automation, Inc.

Programmable logic controllers connected to a communication network are connected in parallel for improved speed of communication.

TECH FACT — *A short circuit has zero resistance. When one branch of a parallel circuit is short circuited, all other paths are also short circuited.*

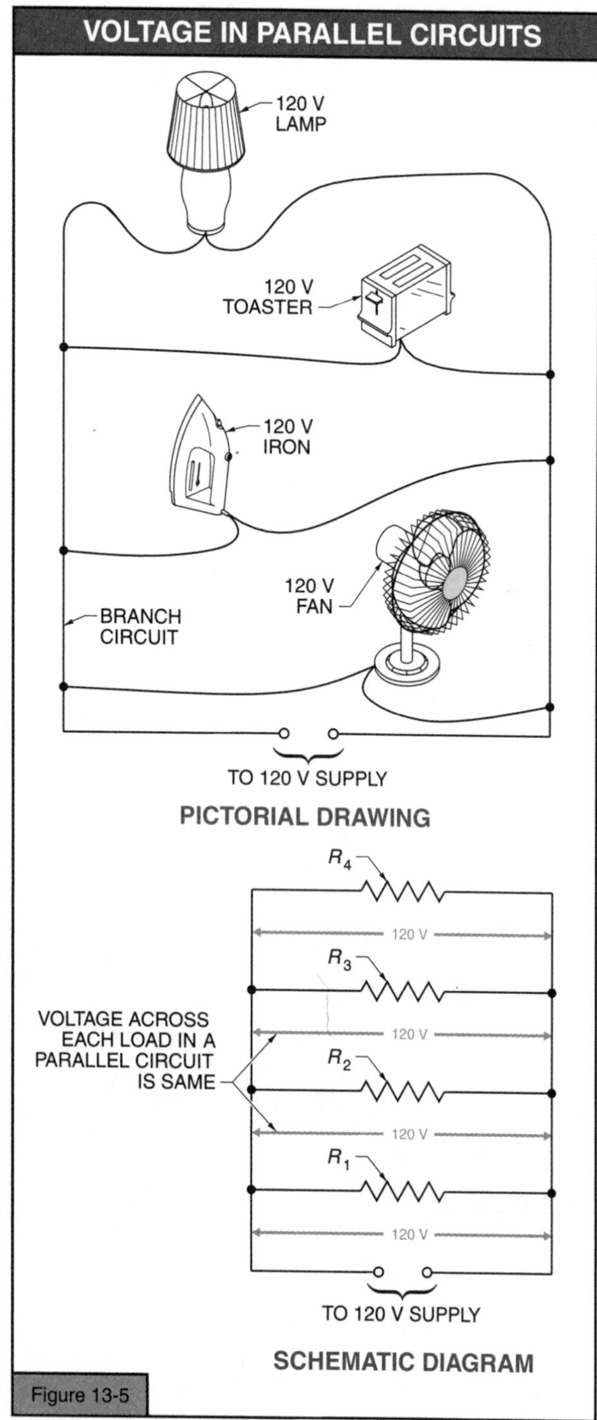

Figure 13-5

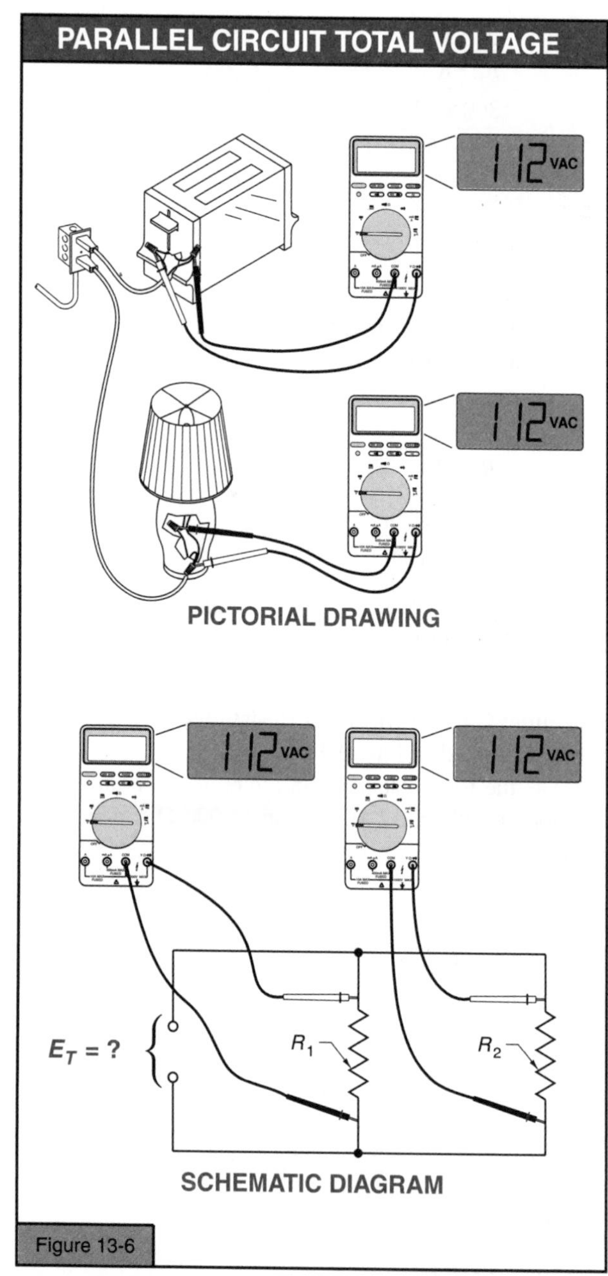

Figure 13-6

TECH TIP *Any parallel-connected switch can energize the circuit and circuit loads. Be careful when working on a live circuit that includes parallel-connected switches.*

To calculate total voltage in a parallel circuit when the voltage across a load is known or measured, apply the formula:

$E_T = E_1 = E_2 = ...$

where

E_T = total applied voltage (in V)

E_1 = voltage across load 1 (in V)

E_2 = voltage across load 2 (in V)

Example: Calculating Voltage in Parallel Circuits

What is the total voltage if the voltage across each load is measured at 112 VAC with a voltmeter?

$E_T = E_1 = E_2$

$E_T = 112 = 112$

E_T = **112 VAC**

Current in Parallel Circuits

Total current in a circuit containing parallel-connected loads equals the sum of the current through all the loads. Total current increases if loads are added in parallel and decreases if loads are removed. **See Figure 13-7.** To calculate total current in a parallel circuit, apply the formula:

$$I_T = I_1 + I_2 + I_3 + \ldots$$

where

I_T = total circuit current (in A)

I_1 = current through load 1 (in A)

I_2 = current through load 2 (in A)

I_3 = current through load 3 (in A)

Example: Calculating Current in Parallel Circuits with Several Loads

What is the total current in a circuit containing four loads connected in parallel if the current through the four loads is .833 A, 6.25 A, 8.33 A, and 1.2 A.

$$I_T = I_1 + I_2 + I_3 + I_4$$

$$I_T = .833 + 6.25 + 8.33 + 1.2$$

$$I_T = \mathbf{16.613\ A}$$

The current through each load and the total circuit current must be known when designing and troubleshooting a circuit. When designing a circuit, the size of wire used is based on the amount of expected current. The greater the current flow, the larger the required wire size. The lesser the current flow, the smaller the required wire size, which reduces circuit cost. When troubleshooting a circuit, an overloaded circuit has a higher-than-normal current reading. A circuit with a lower-than-expected current reading may have a loose connection. A loose connection increases the circuit resistance and decreases the circuit current.

Power in Parallel Circuits

Power is produced when voltage is applied to a load and current flows through the load. The power produced is used to produce light, heat, rotary motion, or linear motion. The lower the load's resistance or higher the applied voltage, the more power produced. The higher the load's resistance or lower the applied voltage, the less power produced. Power produced is measured in watts.

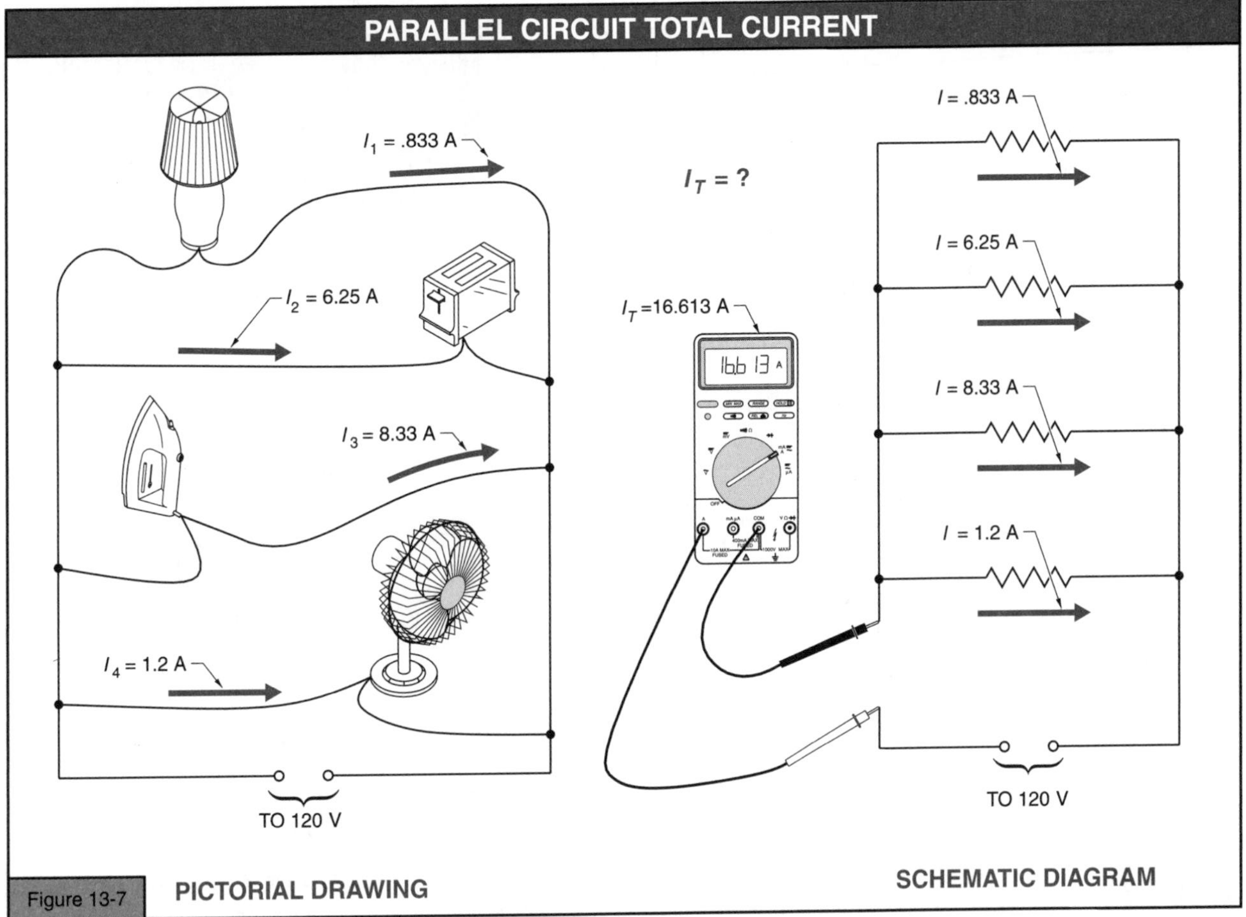

Figure 13-7

Power produced by a load is equal to the voltage drop across the load times the current through the load. Total power is equal to the sum of the power produced by each load. **See Figure 13-8.** To calculate total power in a parallel circuit when the power across each load is known or measured, apply the formula:

$$P_T = P_1 + P_2 + P_3 + \ldots$$

where

P_T = total circuit power (in W)

P_1 = power of load 1 (in W)

P_2 = power of load 2 (in W)

P_3 = power of load 3 (in W)

PARALLEL CIRCUIT TOTAL POWER

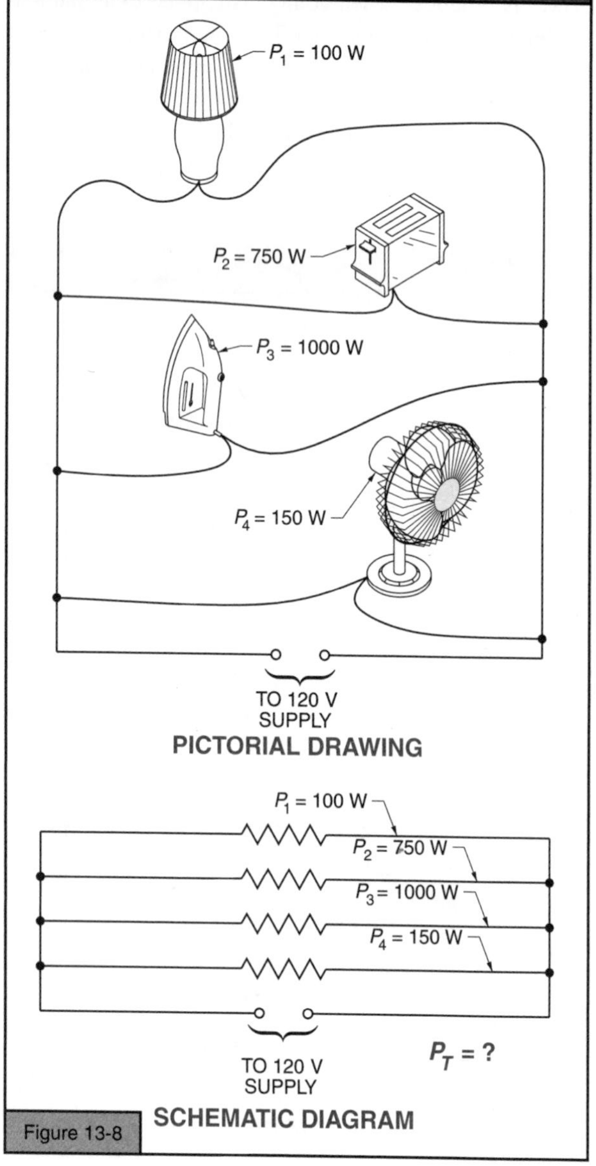

P_1 = 100 W

P_2 = 750 W

P_3 = 1000 W

P_4 = 150 W

TO 120 V SUPPLY

PICTORIAL DRAWING

P_1 = 100 W

P_2 = 750 W

P_3 = 1000 W

P_4 = 150 W

TO 120 V SUPPLY

P_T = ?

SCHEMATIC DIAGRAM

Figure 13-8

Example: Calculating Power in Parallel Circuits with Several Loads

What is the total circuit power if four loads are connected in parallel and the loads produce 100 W, 750 W, 1000 W, and 150 W?

$$P_T = P_1 + P_2 + P_3 + P_4$$

$$P_T = 100 + 750 + 1000 + 150$$

$$P_T = \mathbf{2000\ W}$$

Parallel Circuit Application

The individual concepts and formulas for parallel circuits can be applied for a total understanding of circuit operation when designing and troubleshooting circuits. For example, a waffle iron applies the concepts of a basic parallel circuit to heat the upper and lower heating elements. **See Figure 13-9.**

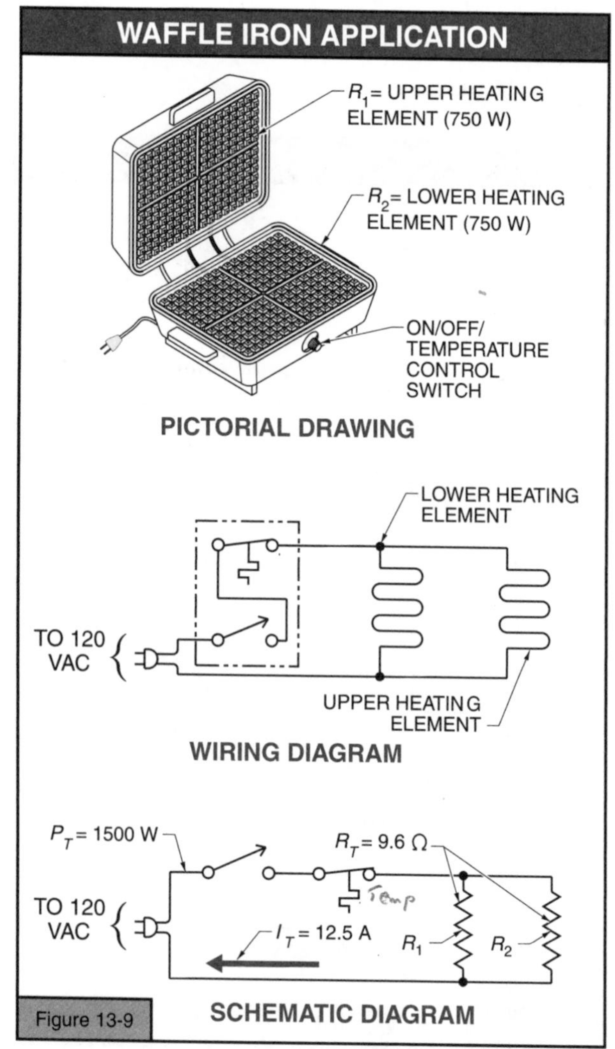

Figure 13-9

The waffle iron includes a heating element that heats the upper section and a heating element that heats the lower section. The two heating elements are connected in parallel. An ON/OFF temperature control switch is used to turn ON the heating elements and control the temperature. A temperature switch is wired in series with the ON/OFF temperature control switch to turn power OFF if overheating occurs in the circuit. To calculate the total power of the waffle iron, apply the formula:

$$P_T = P_1 + P_2$$

where

P_T = waffle iron total power (in W)

P_1 = power of upper heating element (in W)

P_2 = power of lower heating element (in W)

Example: Calculating Waffle Iron Total Power

What is the total power in the waffle iron circuit containing two heating elements with power ratings of 750 W?

$$P_T = P_1 + P_2$$

$$P_T = 750 + 750$$

$$P_T = \mathbf{1500\ W}$$

To calculate the total current used by both waffle iron heating elements, apply the formula:

$$I_T = \frac{P}{E}$$

where

I_T = total current of both heating elements (in A)

P = total power of waffle iron (in W)

E = total voltage (in V)

Example: Calculating Waffle Iron Total Current

What is the total current in the waffle iron if the total power is 1500 W and the total voltage is 120 V?

$$I_T = \frac{P}{E}$$

$$I_T = \frac{1500}{120}$$

$$I_T = \mathbf{12.5\ A}$$

To calculate total resistance of the waffle iron, apply the formula:

$$R_T = \frac{E}{I}$$

where

R_T = waffle iron total resistance (in Ω)

E = waffle iron total voltage (in V)

I = waffle iron total current (in A)

Example: Calculating Waffle Iron Total Resistance

What is the total resistance of the waffle iron if the total voltage is 120 V and the total current is 12.5 A?

$$R_T = \frac{E}{I}$$

$$R_T = \frac{120}{12.5}$$

$$R_T = \mathbf{9.6\ \Omega}$$

Capacitors in Parallel Circuits

Capacitance is the ability to store energy in the form of an electrical charge. A *capacitor* is an electric device designed to store electrical energy by means of an electrostatic field. The equivalent capacitance of capacitors connected in parallel is equal to the sum of the individual capacitors. **See Figure 13-10.** To calculate total capacitance of two capacitors connected in parallel, apply the formula:

$$C_T = C_1 + C_2 + ...$$

where

C_T = total capacitance (in µF)

C_1 = capacitance of capacitor 1 (in µF)

C_2 = capacitance of capacitor 2 (in µF)

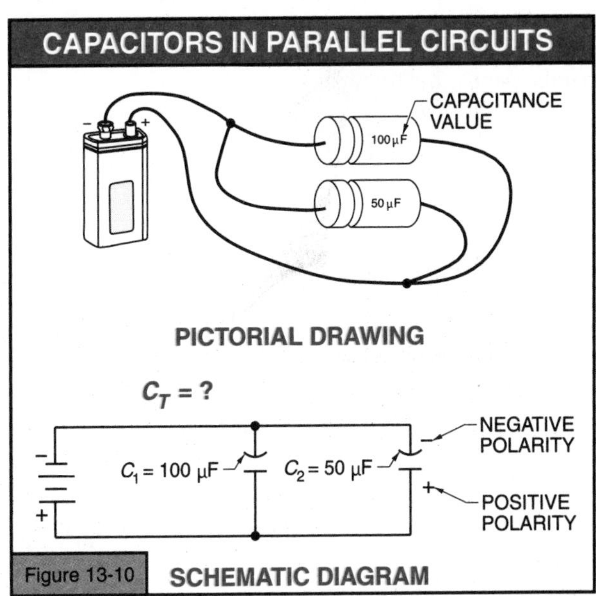

Figure 13-10

Example: Calculating Total Capacitance—Capacitors Connected in Parallel

What is the total capacitance of a 100 µF capacitor connected in parallel with a 50 µF capacitor?

$$C_T = C_1 + C_2$$

$$C_T = 100 + 50$$

$$C_T = \mathbf{150\ \mu F}$$

Inductors in Parallel Circuits

Inductance is the property of an electric device that opposes a change in current due to its ability to store electrical energy in a magnetic field. Inductance is measured in Henrys (H). **See Figure 13-11.**

INDUCTORS IN PARALLEL CIRCUITS

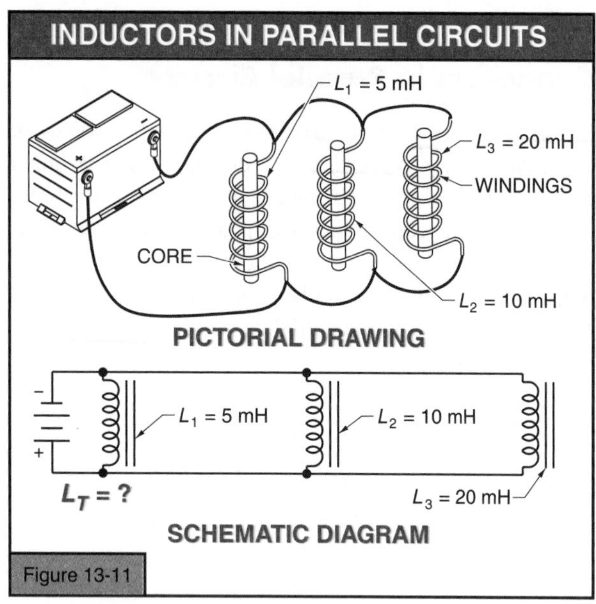

Figure 13-11

To calculate total inductance of inductors connected in parallel, apply the formula:

$$L_T = \frac{1}{\frac{1}{L_1} + \frac{1}{L_2} + \frac{1}{L_3}}$$

where

L_T = total inductance (in H)

L_1 = inductance of inductor 1 (in H)

L_2 = inductance of inductor 2 (in H)

L_3 = inductance of inductor 3 (in H)

Example: Calculating Total Inductance—Inductors Connected in Parallel

What is the total inductance in a circuit containing three inductors of 5 mH, 10 mH, and 20 mH connected in parallel?

$$L_T = \frac{1}{\frac{1}{L_1} + \frac{1}{L_2} + \frac{1}{L_3}}$$

$$L_T = \frac{1}{\frac{1}{5} + \frac{1}{10} + \frac{1}{20}}$$

$$L_T = \frac{1}{.2 + .1 + .05}$$

$$L_T = \frac{1}{.35}$$

$$L_T = \mathbf{2.86\ H}$$

2.86 MH

Batteries and Solarcells in Parallel Circuits

A *battery* is a DC voltage source that converts chemical energy to electrical energy. A *cell* is a unit that produces electricity at a fixed voltage and current level. The current capacity of a battery is increased by connecting cells in parallel.

A *solarcell* is a voltage source that converts light energy to electrical energy. Solarcells are used in place of batteries in low-current devices or are used to recharge battery-operated devices. The current capacity is increased when solarcells are connected in parallel. **See Figure 13-12.**

BATTERIES AND SOLARCELLS IN PARALLEL CIRCUITS

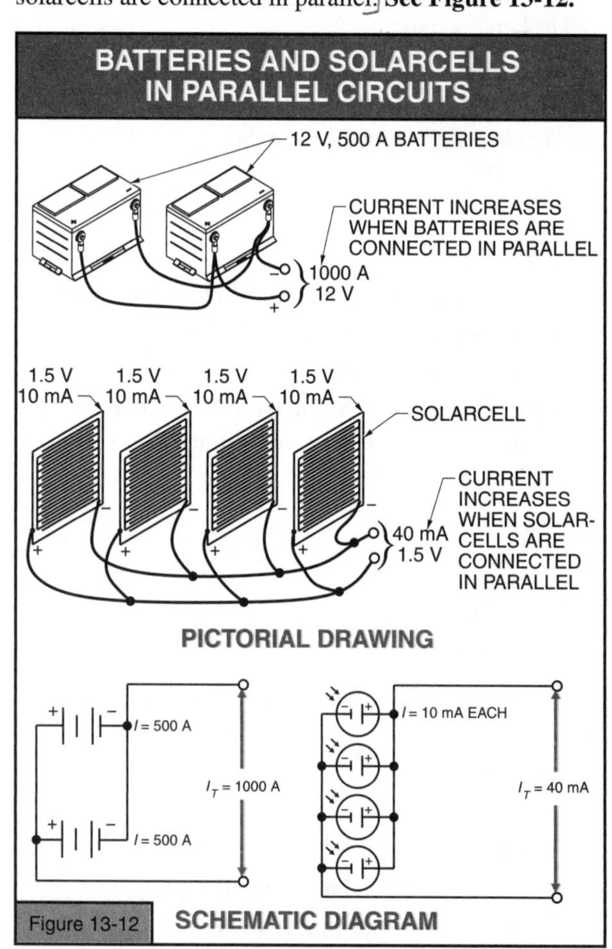

Figure 13-12

Series/Parallel Connections

Switches, loads, meters, fuses, circuit breakers, and other electrical components can be connected in a series/parallel connection. A *series/parallel connection* is a combination of series- and parallel-connected components. The majority of all circuits contain series/parallel-connected components. **See Figure 14-1.**

TECH FACT — *A series/parallel circuit has many practical applications and can be used in a circuit that includes a 12 V lamp to indicate when a 120 V load is ON. In this type of circuit, a resistor used to drop 108 V can be connected in series with the 12 V lamp. The lamp/resistor series combination can then be connected in parallel with the load.*

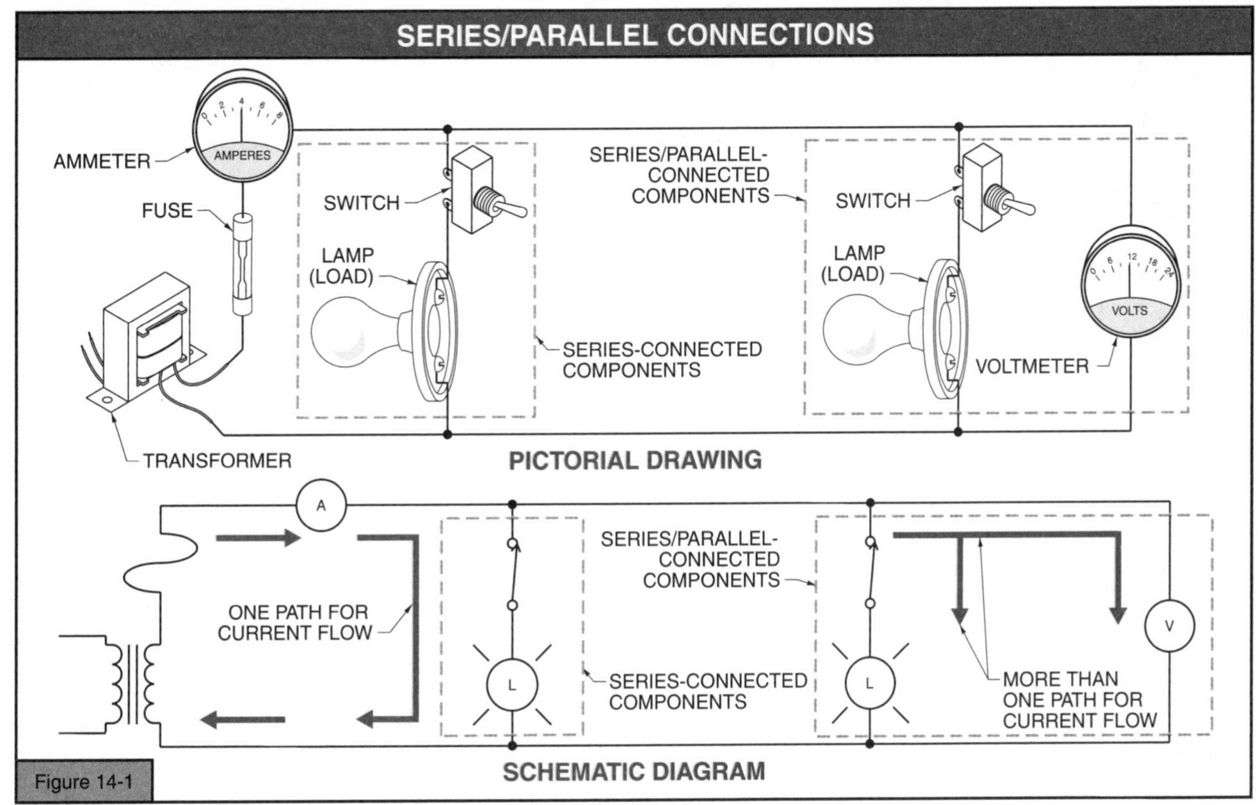

Figure 14-1

Care must be taken when working with series/parallel circuits because current can flow in one part of the parallel circuit even though another part of the circuit is turned OFF. In most circuits, loads such as lamps, motors, and solenoids are normally connected in parallel. Switches, fuses, and circuit breakers that control and monitor the current through the loads are connected in series. Voltmeters are connected in parallel when taking load measurements. Ammeters are connected in series with the load. Understanding and recognizing series- and parallel-connected components and circuits enables a technician or troubleshooter to take proper measurements, make circuit modifications, and troubleshoot the circuit.

TECH TIP *Always install switches that are large enough and have a strong electrical connection (good closing force).*

Series/Parallel Circuit Polarity

All DC voltage sources have a positive and a negative terminal. The positive and negative terminals establish polarity in the circuit. All points in a DC series/parallel circuit have polarity. Each component (switch, load, fuse, etc.) in a DC series/parallel circuit has a positive polarity side and a negative polarity side. The side of the component closest to the positive voltage terminal is the positive polarity side. The side of the component closest to the negative voltage terminal is the negative polarity side. See **Figure 14-2.**

Series/Parallel Switch Connections

Two or more switches must be closed before current flows when the switches are connected in series/parallel. Any one or more series-connected switches or all parallel-connected switches must be opened to stop current flow.

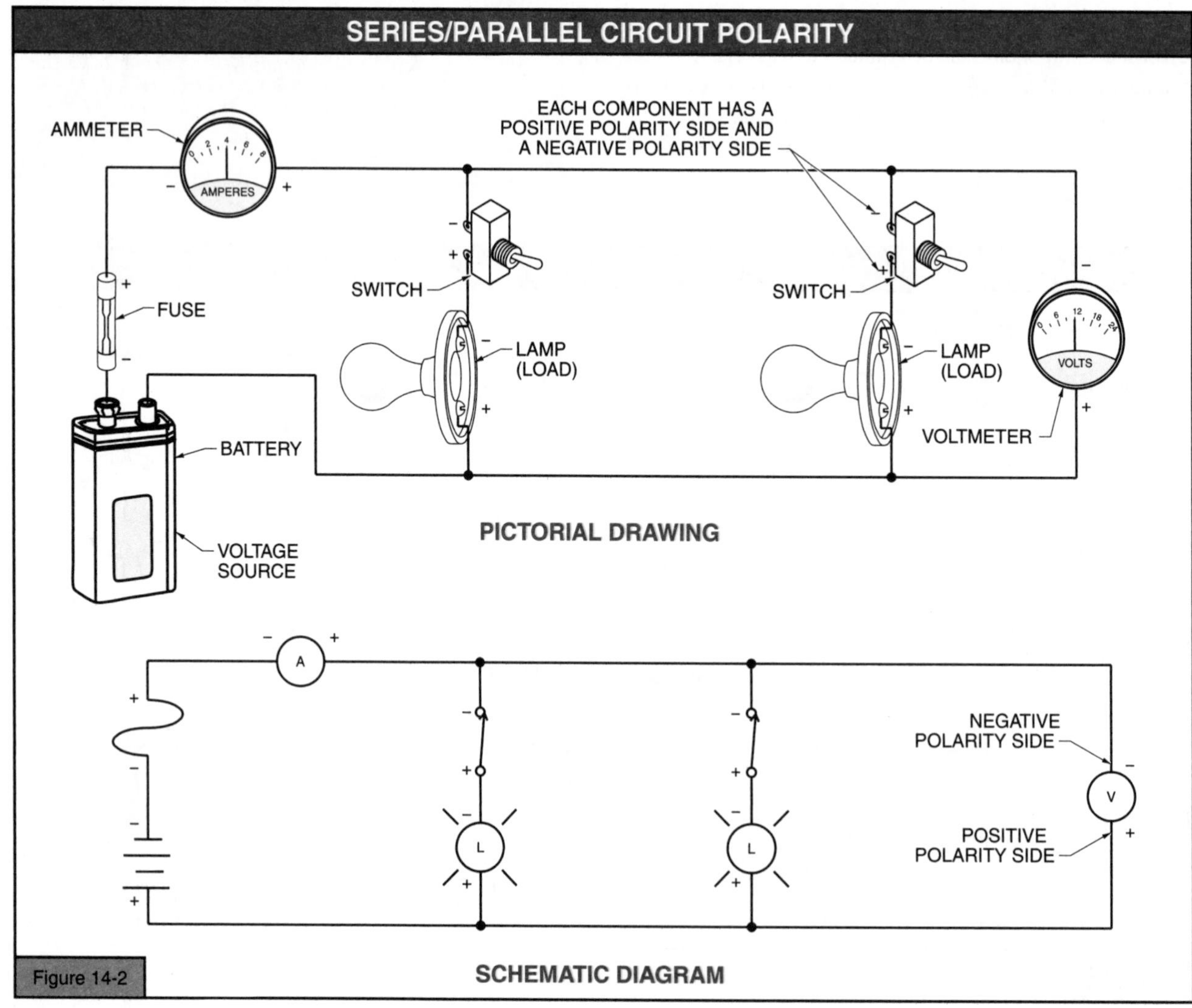

Figure 14-2

Many circuits include switches that are connected in series/parallel combinations. The parallel-connected switches provide different control points and the series-connected switches provide safety. Safety is provided by connecting the series-connected switches to doors and guards. For example, most copiers and printers include built-in limit switches that automatically turn OFF the machine if a machine service door is open. **See Figure 14-3.**

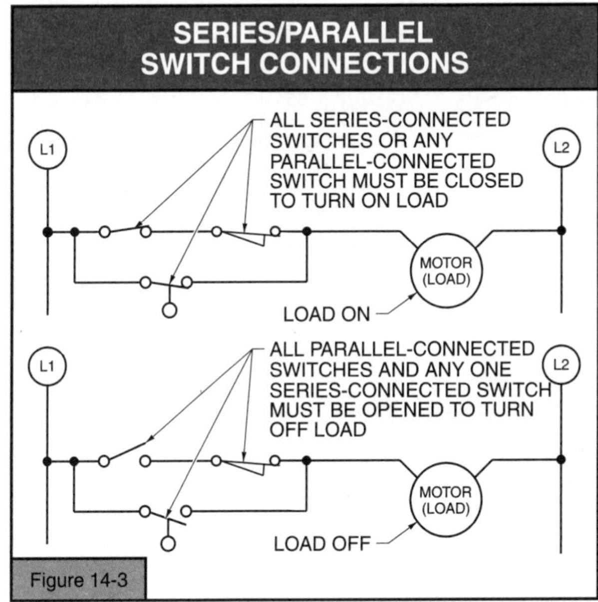

Figure 14-3

Resistance in Series/Parallel Circuits

Resistors and loads, such as heating elements, are often connected in a series/parallel combination. A series/parallel circuit can contain any number of individual resistors (loads) connected in any number of different series/parallel circuit combinations. The series/parallel combination is always equal to one combined total resistance value. The total resistance in a circuit containing series/parallel-connected resistors equals the sum of the series loads and the equivalent resistance of the parallel combinations. **See Figure 14-4.**

To calculate total resistance in a series/parallel circuit that contains two resistors in parallel and series-connected resistors, apply the formula:

$$R_T = \left(\frac{R_{P1} \times R_{P2}}{R_{P1} + R_{P2}}\right) + R_{S1} + R_{S2} + \ldots$$

where

R_T = total resistance (in Ω)

R_{P1} = parallel resistance 1 (in Ω)

R_{P2} = parallel resistance 2 (in Ω)

R_{S1} = series resistance 1 (in Ω)

R_{S2} = series resistance 2 (in Ω)

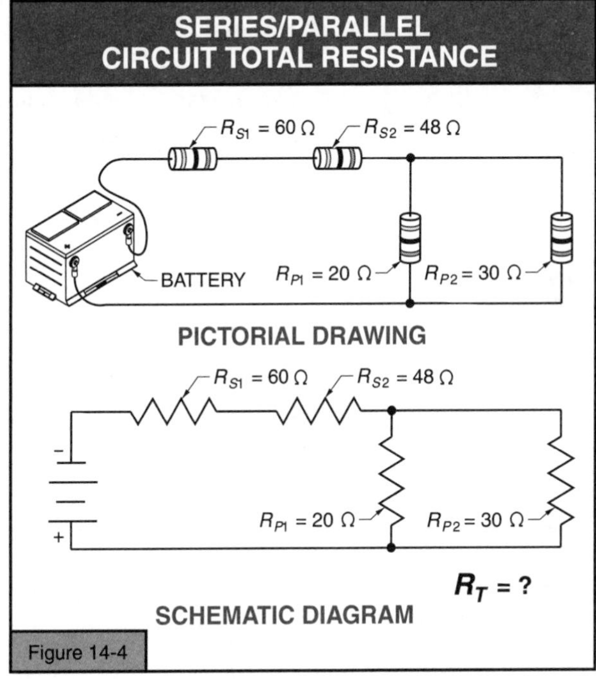

Figure 14-4

Example: Calculating Total Resistance—Series/Parallel Circuit Containing Two Series-Connected and Two Parallel-Connected Resistors

What is the total resistance in a circuit containing a 20 Ω and 30 Ω resistor connected in parallel and a 60 Ω and 48 Ω resistor connected in series?

$$R_T = \left(\frac{R_{P1} \times R_{P2}}{R_{P1} + R_{P2}}\right) + R_{S1} + R_{S2}$$

$$R_T = \left(\frac{20 \times 30}{20 + 30}\right) + 60 + 48$$

$$R_T = \frac{600}{50} + 108$$

$$R_T = 12 + 108$$

$$R_T = \mathbf{120\ \Omega}$$

To find the total resistance in a series/parallel circuit that contains multiple series/parallel combinations, the circuit must be broken down to the basic series and parallel parts. After the circuit is broken down, the rules of series circuits can be applied to the series part(s) and the rules of parallel circuits can be applied to the parallel part(s). **See Figure 14-5.**

To break down a series/parallel circuit into basic series and parallel parts, apply the procedure:

1. Combine the two or more resistance values into a single resistance value if there are two or more resistors (loads) connected in series. This is done by combining the formula for resistors in series.

   $R_T = R_1 + R_2 + R_3 + \ldots$

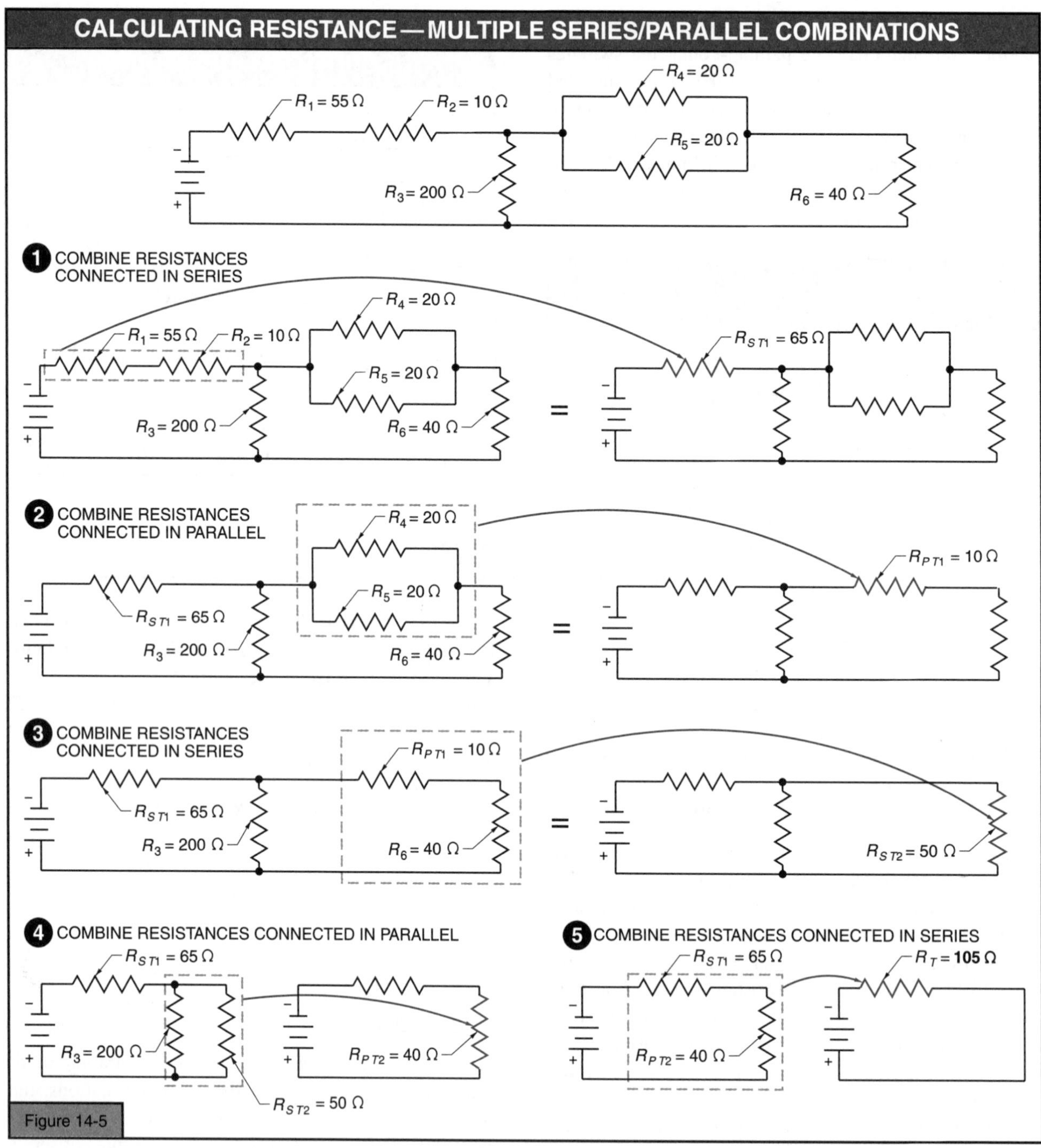

Figure 14-5

2. Combine the two or more resistors (loads) connected in parallel into a single resistance value. This is done by applying the formula for resistors in parallel. Combine each circuit into a single resistance value if there is more than one parallel circuit. For two resistors connected in parallel, apply the formula:

$$R_T = \frac{R_1 \times R_2}{R_1 + R_2}$$

For three or more resistors connected in parallel, apply the formula:

$$R_T = \frac{1}{\frac{1}{R_1} + \frac{1}{R_2} + \frac{1}{R_3} + \ldots}$$

3. Combine the sum of the series resistance values with the sum of each parallel circuit resistance value. In

larger circuits, this combination process may have to be applied several times until all sections of the circuit have been combined.

Example: Calculating Resistance—Series/Parallel Circuit Containing Multiple Series/Parallel Combinations

What is the total resistance of a 55 Ω, 10 Ω, 200 Ω, 20 Ω, 20 Ω, and 40 Ω resistor connected in a series/parallel combination?

1. Combine the two resistors connected in series into a single resistance value.

$R_{ST1} = R_1 + R_2$

$R_{ST1} = 55 + 10$

$R_{ST1} = 65\ \Omega$

2. Combine the two or more resistors connected in parallel into a single resistance value.

$$R_{PT1} = \frac{R_4 \times R_5}{R_4 + R_5}$$

$$R_{PT1} = \frac{20 \times 20}{20 + 20}$$

$$R_{PT1} = \frac{400}{40}$$

$$R_{PT1} = 10\ \Omega$$

3. Combine resistors connected in series into a single resistance.

$R_{ST2} = R_{PT1} + R_6$

$R_{ST2} = 10 + 40$

$R_{ST2} = 50\ \Omega$

4. Combine the two or more resistors connected in parallel into a single resistance value.

$$R_{PT2} = \frac{R_{ST2} \times R_3}{R_{ST2} + R_3}$$

$$R_{PT2} = \frac{50 \times 200}{50 + 200}$$

$$R_{PT2} = \frac{10,000}{250}$$

$$R_{PT2} = 40\ \Omega$$

5. Combine resistors connected in series into a single resistance value.

$R_T = R_{ST1} + R_{PT2}$

$R_T = 65 + 40$

$R_T = \mathbf{105\ \Omega}$

TECH FACT *No material is a perfect conductor or insulator. Resistance is a part of every circuit and must be considered when designing and troubleshooting circuits.*

Current in Series/Parallel Circuits

The total current and current in individual parts of a series/parallel circuit follow the same laws of current in a basic series and a basic parallel circuit. Current is the same in each series part of the series/parallel circuit. Current is equal to the sum of each parallel combination in each parallel part of the series/parallel circuit. **See Figure 14-6.**

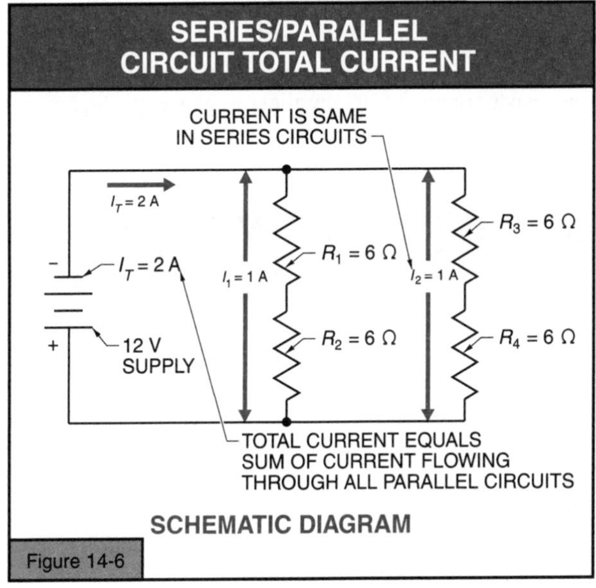

Figure 14-6

Example: Calculating Current in Series/Parallel Circuits

What is the current flowing through two 6 Ω resistors connected across a 12 V power supply?

The current flowing through two 6 Ω resistors connected in series across a 12 V power supply is 1 A at any point in the series circuit. The total current flowing through two parallel-connected branches that each have 1 A flowing through them is 2 A.

To calculate total resistance of two 6 Ω resistors connected in series, apply the formula for resistors in series.

$R_T = R_1 + R_2$

$R_T = 6 + 6$

$R_T = \mathbf{12\ \Omega}$

To calculate current flowing through the 12 Ω series circuit, apply Ohm's law.

$$I = \frac{E}{R}$$

$$I = \frac{12}{12}$$

$$I = \mathbf{1\ A}$$

To calculate total current in the parallel circuit, apply the formula for current in a parallel circuit.

$I_T = I_1 + I_2$

$I_T = 1 + 1$

$I_T = \mathbf{2\ A}$

Voltage in Series/Parallel Circuits

The total voltage applied across resistors (loads) connected in a series/parallel combination is divided across the individual resistors. The higher the resistance of any one resistor or equivalent parallel resistance, the higher the voltage drop. **See Figure 14-7.**

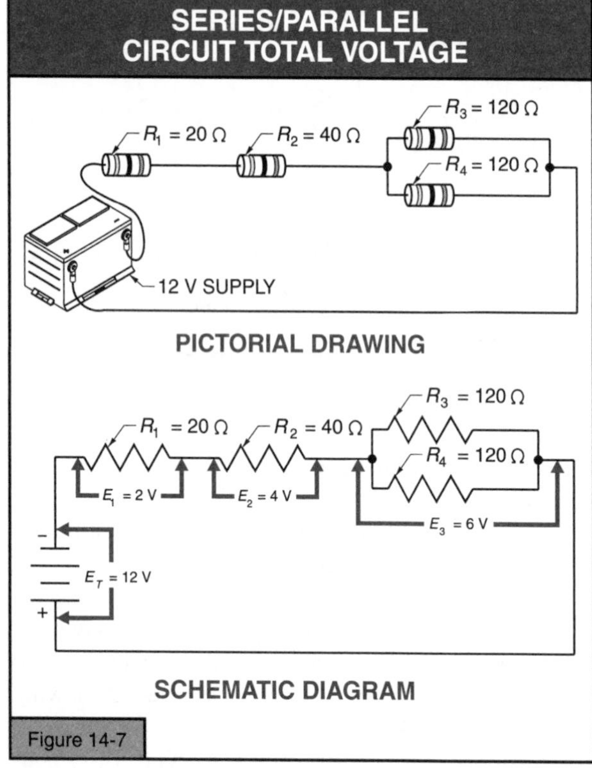

Figure 14-7

Example: Calculating Voltage in Series/Parallel Circuits
What is the voltage drop across each resistor if two 120 Ω resistors are connected in parallel with a 20 Ω, and 40 Ω resistor connected in series that are all connected to a 12 V supply?

To calculate the voltage across each resistor in a series/parallel circuit, the total circuit resistance and total circuit current are calculated first. The total circuit resistance is found by applying the steps used to determine total resistance in a series/parallel circuit.

For example, the total resistance of two 120 Ω resistors connected in parallel with a 20 Ω and a 40 Ω resistor connected in series is 120 Ω. To calculate total resistance, apply the formula for resistance in a series/parallel circuit.

$$R_T = \left(\frac{R_{P1} \times R_{P2}}{R_{P1} + R_{P2}}\right) + R_{S1} + R_{S2}$$

$$R_T = \left(\frac{120 \times 120}{120 + 120}\right) + 20 + 40$$

$$R_T = \frac{14{,}400}{240} + 60$$

$$R_T = 60 + 60$$

$$R_T = \mathbf{120\ \Omega}$$

The total circuit current is found by applying Ohm's law when the total circuit voltage and total circuit resistance is known.

$$I_T = \frac{E_T}{R_T}$$

$$I_T = \frac{12}{120}$$

$$I_T = \mathbf{.1\ A\ (100\ mA)}$$

The voltage drop across each resistor can be calculated once the circuit current is found. To calculate the voltage drop across R_1, apply Ohm's law.

$E_1 = I_1 \times R_1$

$E_1 = .1 \times 20$

$E_1 = \mathbf{2\ V}$

To calculate the voltage drop across R_2, apply Ohm's law.

$E_2 = I_2 \times R_2$

$E_2 = .1 \times 40$

$E_2 = \mathbf{4\ V}$

To calculate the voltage drop across $R_{3\text{-}4}$, apply Ohm's law. The equivalent resistance of two 120 Ω resistors in parallel is 60 Ω.

$E_{3\text{-}4} = I_{3\text{-}4} \times R_{3\text{-}4}$

$E_{3\text{-}4} = .1 \times 60$

$E_{3\text{-}4} = \mathbf{6\ V}$

The voltage across the 20 Ω resistor equals 2 V. The voltage across the 40 Ω resistor equals 4 V. The voltage across the 60 Ω (equivalent resistance of two 120 Ω resistors in parallel) resistor equals 6 V.

TECH FACT

Never place only one conductor (hot wire or neutral wire) in a metal conduit because the magnetic field of the conductor produces heat in the conduit.

Power in Series/Parallel Circuits

Power is produced when current flows through any load or component that has resistance. The lower the resistance or higher the amount of current, the more power produced. The higher the resistance or lower the amount of current, the less power produced. The amount of power produced is measured in watts. See **Figure 14-8.**

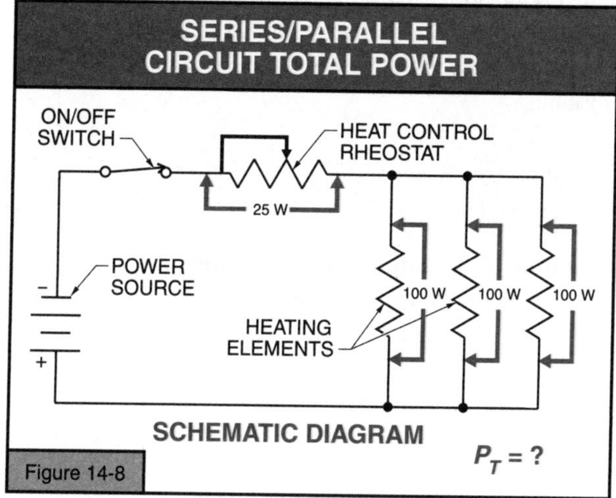

Figure 14-8

The total power in a series/parallel circuit is equal to the sum of the power produced by each load or component. To calculate total power of a series/parallel circuit when the power across each load is known or measured, apply the formula:

$$P_T = P_1 + P_2 + P_3 + \ldots$$

where

P_T = total circuit power (in W)

P_1 = power of load 1 (in W)

P_2 = power of load 2 (in W)

P_3 = power of load 3 (in W)

Example: Calculating Total Power—Series/Parallel Circuits

What is the total power in a circuit containing a load which produces 25 W connected in series and three loads which produce 100 W each connected in parallel?

$$P_T = P_1 + P_2 + P_3 + \ldots$$

$$P_T = 25 + 100 + 100 + 100$$

$$P_T = \mathbf{325\ W}$$

A heating circuit controlled by a rheostat is an example of a series/parallel circuit. The heating elements are connected in parallel and the rheostat is connected in series with the heating elements. The rheostat is used to control the heat output of the circuit. A voltage is dropped across the rheostat because the rheostat has resistance and thus the rheostat dissipates some power.

Series/Parallel Circuit Application

The laws of series/parallel circuits are used in many practical applications. For example, a basic series/parallel circuit is used in any circuit that includes a lamp to indicate when a load is ON. In this circuit, a resistor is connected in series with a lamp, and the load is connected in parallel with the resistor/lamp combination. The load may be any device such as a heating element, motor, or solenoid. See **Figure 14-9.**

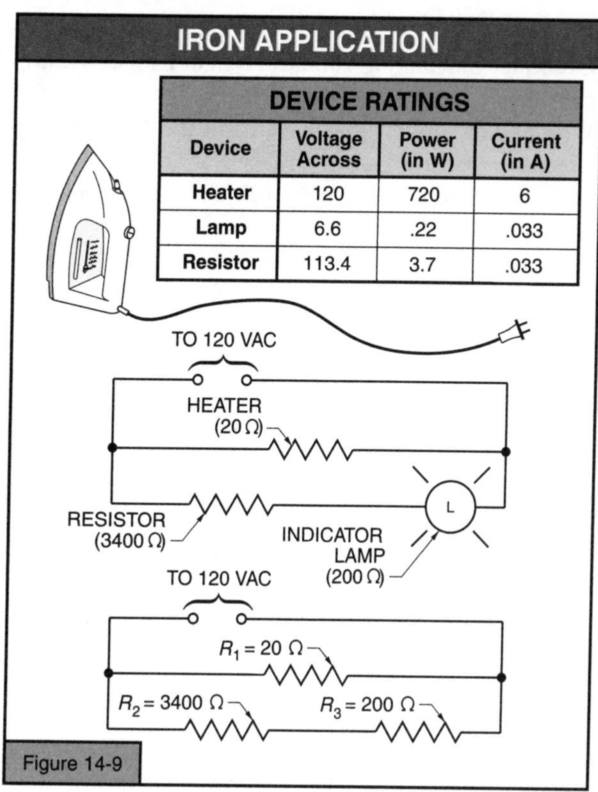

DEVICE RATINGS

Device	Voltage Across	Power (in W)	Current (in A)
Heater	120	720	6
Lamp	6.6	.22	.033
Resistor	113.4	3.7	.033

Figure 14-9

A resistor is connected in series with an indicator lamp so that a lamp that is rated at a low voltage may be used. Using a lamp rated at a low voltage greatly increases the life of the lamp. A resistance value many times the resistance value of the lamp is used so that most of the applied voltage is dropped across the resistor.

Example: Calculating Heater Current

What is the current through a 20 Ω heater connected to a 120 V power supply?

To calculate the current through the heater, apply Ohm's law.

$$I = \frac{E}{R}$$

$$I = \frac{120}{20}$$

$$I = \mathbf{6\ A}$$

Example: Calculating Total Resistance of Resistor/ Lamp Circuit
What is the total resistance of a circuit containing a 200 Ω indicator lamp connected in series with a 3400 Ω resistor?

To calculate the total resistance of the resistor/lamp series circuit, apply the formula for calculating resistances connected in series.

$R_T = R_1 + R_2 + R_3 + ...$

$R_T = 3400 + 200$

$R_T = \mathbf{3600\ \Omega}$

Example: Calculating Total Current of Resistor/Lamp Circuit
What is the current flowing through the resistor/lamp circuit?

To calculate the current through the resistor/lamp circuit, apply Ohm's law.

$$I = \frac{E}{R}$$

$$I = \frac{120}{3600}$$

$I = \mathbf{.033\ A}$

Example: Calculating Voltage Drop Across Lamp
What is the voltage drop across a 200 Ω indicator lamp connected in a circuit that has .033 A of current?

To calculate the voltage drop across the lamp, apply Ohm's law.

$E = I \times R$

$E = .033 \times 200$

$E = \mathbf{6.6\ V}$

Capacitors are connected in series, parallel, and series/parallel combinations in most electronics applications.

TECH FACT *A voltage drop is the amount of reduced voltage that occurs across a component when moving through a circuit from a positive side to the negative side of the power supply.*

Example: Calculating Power of Heater
What is the power in the heater circuit that contains 6 A of current and is connected to a 120 V source?

To calculate the power of the heater, apply the power formula.

$P = E \times I$

$P = 120 \times 6$

$P = \mathbf{720\ W}$

Example: Calculating Power of Lamp
What is the power of the lamp when .033 A of current is flowing in the circuit and it is connected to a 120 V source?

To calculate the power of the lamp, apply the power formula.

$P = E \times I$

$P = 120 \times .033$

$P = \mathbf{.22\ W}$

Capacitors in Series/Parallel Circuits

Capacitors are normally connected in series or in parallel. However, it is possible to connect capacitors in series/parallel combinations. **See Figure 14-10.** To calculate total capacitance of capacitors connected in a series/parallel combination, apply the procedure:

1. Calculate capacitance of parallel capacitors.

   $C_T = C_1 + C_2 + ...$

2. Calculate capacitance of series capacitors.

   $$C_T = \frac{C_1 \times C_2}{C_1 + C_2}$$

Example: Calculating Total Capacitance–Series/Parallel Capacitors
What is the total capacitance of two parallel-connected 60 μF capacitors connected in series with a 30 μF capacitor?

Total capacitance is found by first calculating the parallel branch and then calculating the series combination.

1. Calculate capacitance of parallel capacitors.

   $C_{PT} = C_2 + C_3 + ...$

   $C_{PT} = 60 + 60$

   $C_{PT} = 120\ \mu F$

CAPACITORS IN SERIES/ PARALLEL CIRCUITS

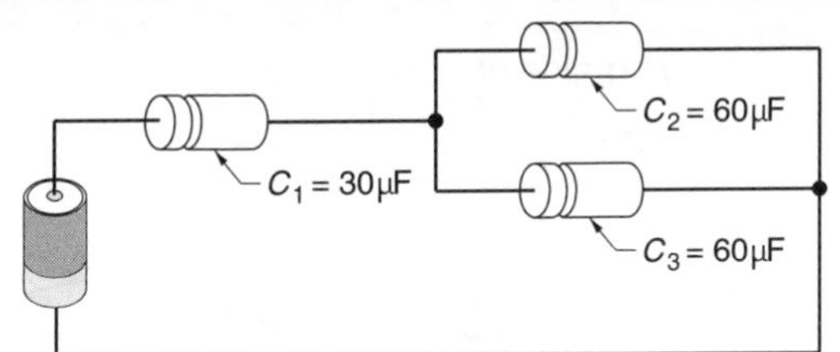

PICTORIAL DRAWING

1 CALCULATE CAPACITANCE OF PARALLEL CAPACITORS

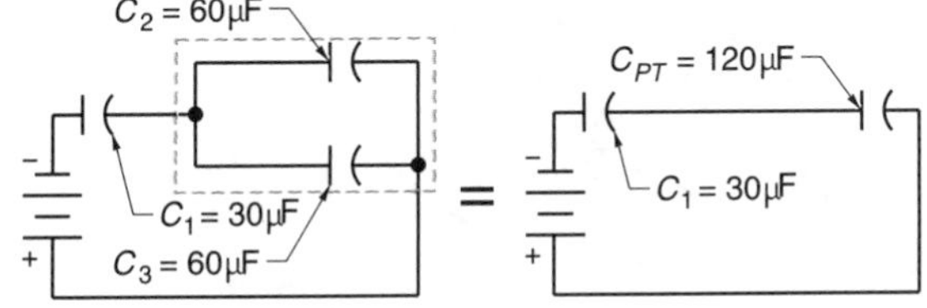

2 CALCULATE CAPACITANCE OF SERIES CAPACITORS

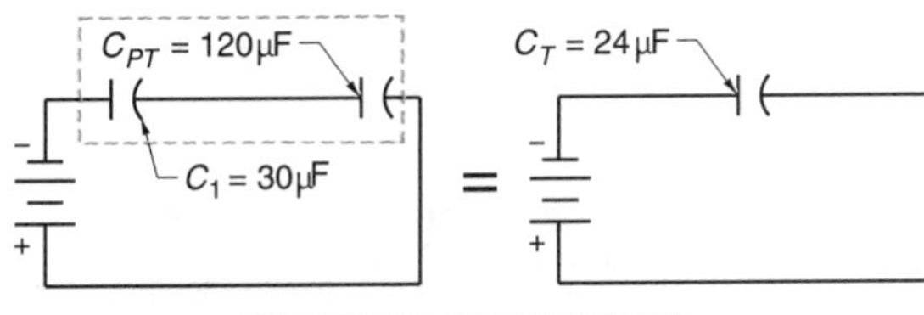

SCHEMATIC DIAGRAM

Figure 14-10

2. Calculate capacitance of series capacitors.

$$C_T = \frac{C_{PT} \times C_1}{C_{PT} + C_1}$$

$$C_T = \frac{120 \times 30}{120 + 30}$$

$$C_T = \frac{3600}{150}$$

$$C_T = \mathbf{24\ \mu F}$$

Inductors in Series/Parallel Circuits

Inductors are normally connected in series or in parallel. However, it is possible to connect inductors in series/parallel combinations. **See Figure 14-11.**

To calculate the total inductance of inductors connected in series/parallel, apply the procedure:

1. Calculate inductance of parallel inductors.

$$L_T = \frac{L_2 \times L_3}{L_2 + L_3}$$

2. Calculate inductance of series inductors.

$$L_T = L_1 + L_2 + \ldots$$

Example: Calculating Total Inductance–Series/Parallel Circuits

What is the total inductance of two parallel-connected 10 H inductors connected in series with a 20 H inductor?

To calculate the total inductance, first calculate the parallel branch and then calculate the series combination.

1. Calculate inductance of parallel inductors.

$$L_{PT} = \frac{L_2 \times L_3}{L_2 + L_3}$$

$$L_{PT} = \frac{10 \times 10}{10 + 10}$$

$$L_{PT} = \frac{100}{20}$$

$$L_{PT} = 5\ \text{H}$$

2. Calculate inductance of series inductors.

$$L_T = L_1 + L_{PT} + \ldots$$

$$L_T = 20 + 5$$

$$L_T = \mathbf{25\ H}$$

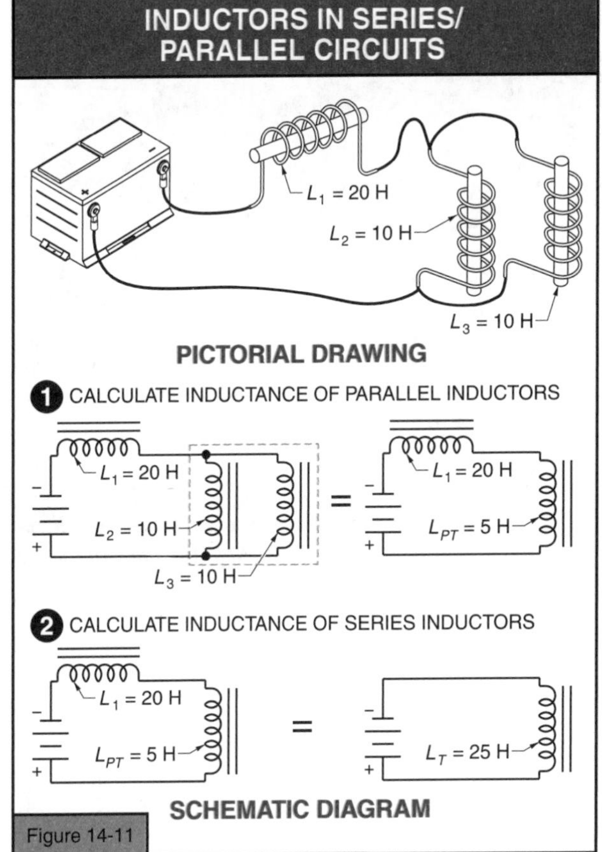

Figure 14-11

Batteries and Solarcells in Series/ Parallel Circuits

Voltage potential is increased when batteries and solarcells are connected in series. Current capacity is increased when batteries and solarcells are connected in parallel. Voltage potential and current capacity are increased when batteries and solarcells are connected in series/parallel combinations. **See Figure 14-12.**

Series/parallel-connected batteries are generally used in large applications such as toys, automobiles, mobile equipment, small trains (used in mines), and small submarines. Batteries are connected in series to increase the voltage to a high level (24 V, 36 V, 90 V, etc.). Batteries are connected in parallel to increase the current.

Series/parallel-connected solarcells are used in small applications (powering calculators) and in large applications (powering satellites). Any number of solarcells can be connected in a series/parallel combination. Many solarcells must be connected in parallel to produce a small amount of current because solarcells produce very little current.

SPG Solar, Inc.

Large solar panels are typically connected in series/parallel combinations.

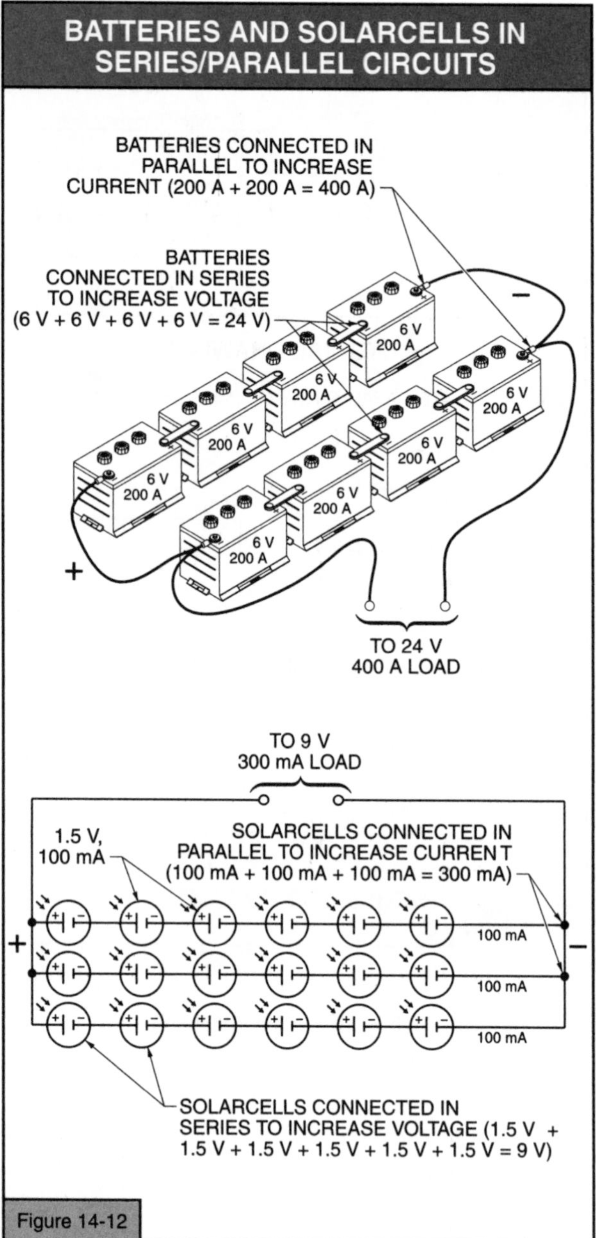

Figure 14-12

Refer to the CD-ROM in the back of the book for Chapter 14 Quick Quiz® and related information.

Magnetism

Magnetism is a force that interacts with other magnets and ferromagnetic materials. *Ferromagnetic materials* are materials, such as soft iron, that are easily magnetized. Magnetism is used to produce most of the electricity consumed, develop rotary motion in motors, and develop linear motion in solenoids.

A *magnet* is a device that attracts iron and steel because of the molecular alignment of its material. All magnets or magnetized material have a north (N) and south (S) pole. The basic law of magnetism states that unlike magnetic poles (N and S) attract each other and like magnetic poles (N and N or S and S) repel each other. The force of attraction between two magnets increases as the distance between the magnets decreases. Likewise, the force of attraction between two magnets decreases as the distance between the magnets increases. **See Figure 15-1.**

Magnetic flux is the invisible lines of force that make up the magnetic field. The more dense the flux, the stronger the magnetic force. Flux is most dense at the ends of a magnet. For this reason, the magnetic force is strongest at the ends of a magnet. The lines of flux leave the north pole and enter the south pole of a magnet or magnetic field.

Magnets may be permanent or temporary. *Permanent magnets* are magnets that hold their magnetism for a long period of time. *Temporary magnets* are magnets that lose their magnetism as soon as the magnetizing force is removed. The most common permanent magnets are bar magnets and horseshoe magnets. Permanent magnets are used in electrical applications, such as in permanent magnet DC motors and reed switches. Temporary magnets are used in most electrical applications, such as motors, transformers, and solenoids.

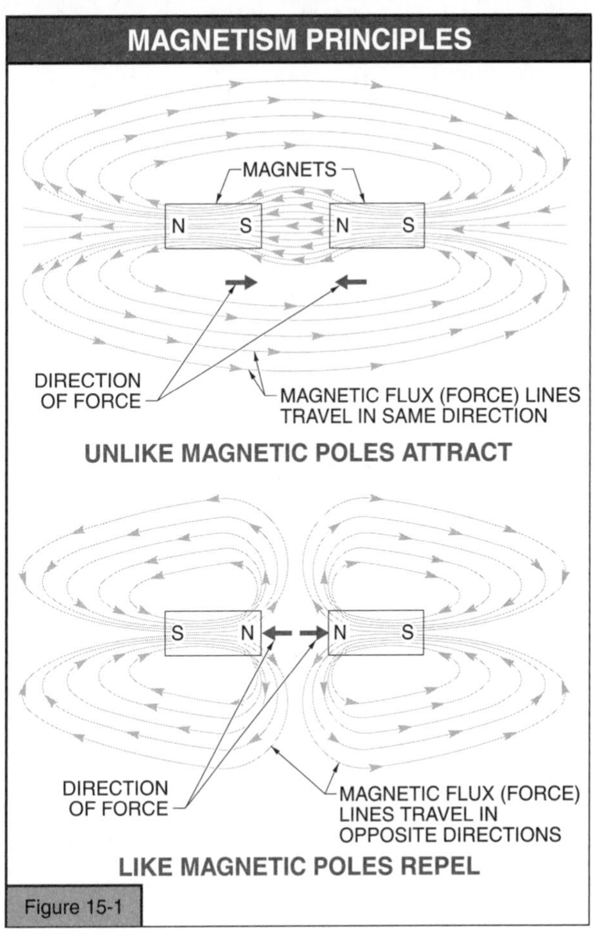

Figure 15-1

Electromagnetism

A magnetic field is produced any time electricity passes through a conductor (wire). *Electromagnetism* is the magnetic field produced when electricity passes through a conductor. Electromagnetism is a temporary magnetic force because the magnetic field is present only as long as current flows. The magnetic field is reduced to zero when the current flow stops.

The magnetic field around a straight conductor is not strong and is of little practical use. The strength of the magnetic field is increased by wrapping the conductor into a coil, increasing the amount of current flowing through the conductor, or wrapping the conductor around an iron core. A strong, concentrated magnetic field is developed when a conductor is wrapped into a coil. The strength of a magnetic field is directly proportional to the number of turns in the coil and the amount of current flowing through the conductor. An iron core increases the strength of the magnetic field by concentrating the field. **See Figure 15-2.**

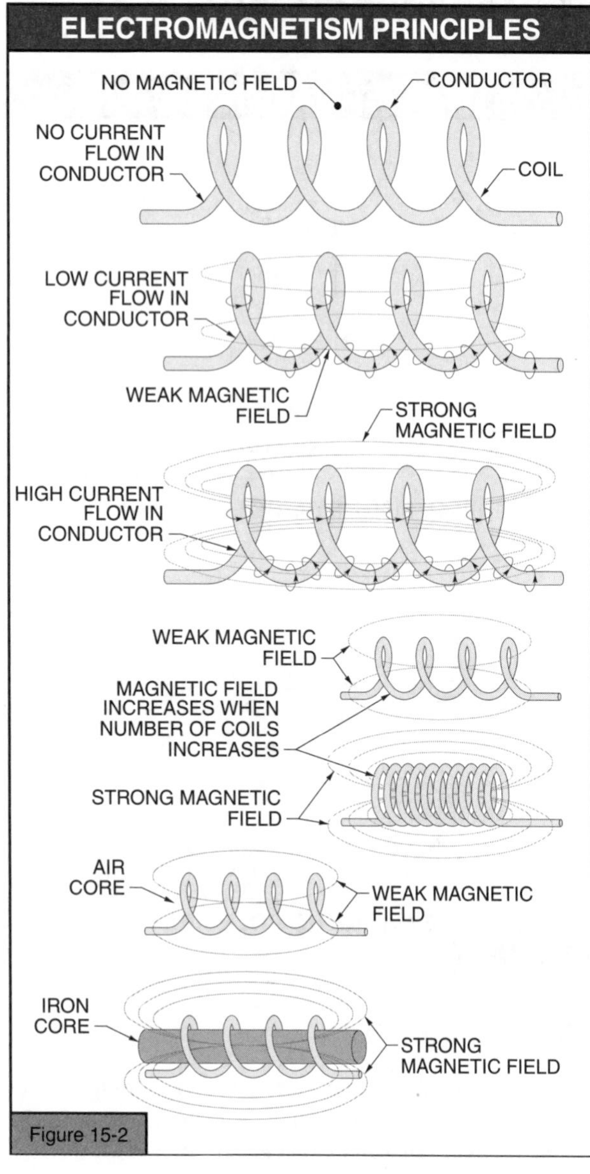

Figure 15-2

Direct current applied to a conductor starts at zero and goes to its maximum value almost instantly. The magnetic field around the conductor also starts at zero and goes to its maximum strength almost instantly. The current and strength of the magnetic field remain at their maximum value as long as the load resistance does not change. The current and the strength of the magnetic field increase if the resistance of the circuit decreases. The current and the magnetic field drop to zero when the direct current is removed.

Alternating current applied to a conductor causes the current to continuously vary in magnitude and the magnetic field to continuously vary in strength. Current flow and magnetic field strength are at their maximum value at the positive and negative peak of the AC sine wave. The current is zero and no magnetic field is produced at the zero points of the AC sine wave. The direction of current flow and polarity of the magnetic field change every time the current passes the zero point of the AC sine wave. On standard 60 Hz (cycle) power frequencies, the current passes the zero point 120 times per second. **See Figure 15-3.**

Solenoids

Electromagnetism is used in solenoids. A *solenoid* is an electric output device that converts electrical energy into a linear, mechanical force. Solenoids produce a linear, mechanical force when electricity is applied to the coil. The coil produces an electromagnetic force when current passes through the coil windings.

The coil windings are wound around an iron frame to increase the magnetic force produced by the coil. An iron core placed near the energized coil causes the magnetic force to draw the iron core into the coil. The magnetic field is not produced when power is removed from the coil. The iron core is removed from the coil by a spring. **See Figure 15-4.**

The amount of linear force a solenoid develops depends on the number of turns of wire in the coil and the amount of applied current. The more coil turns a solenoid has, the greater the linear force. However, as the number of coil turns is increased, the overall size of the solenoid increases. The higher the current flow through the coil, the greater the linear force. The amount of current drawn by the solenoid depends on the applied voltage. For any given size solenoid, the higher the applied voltage, the higher the current. However, if the applied voltage is increased beyond the voltage rating of the solenoid, the high current causes the solenoid to overheat and burn out.

Solenoids develop linear force in applications that use short strokes at low force. Solenoids are used in many electrical applications. Common solenoid applications include:

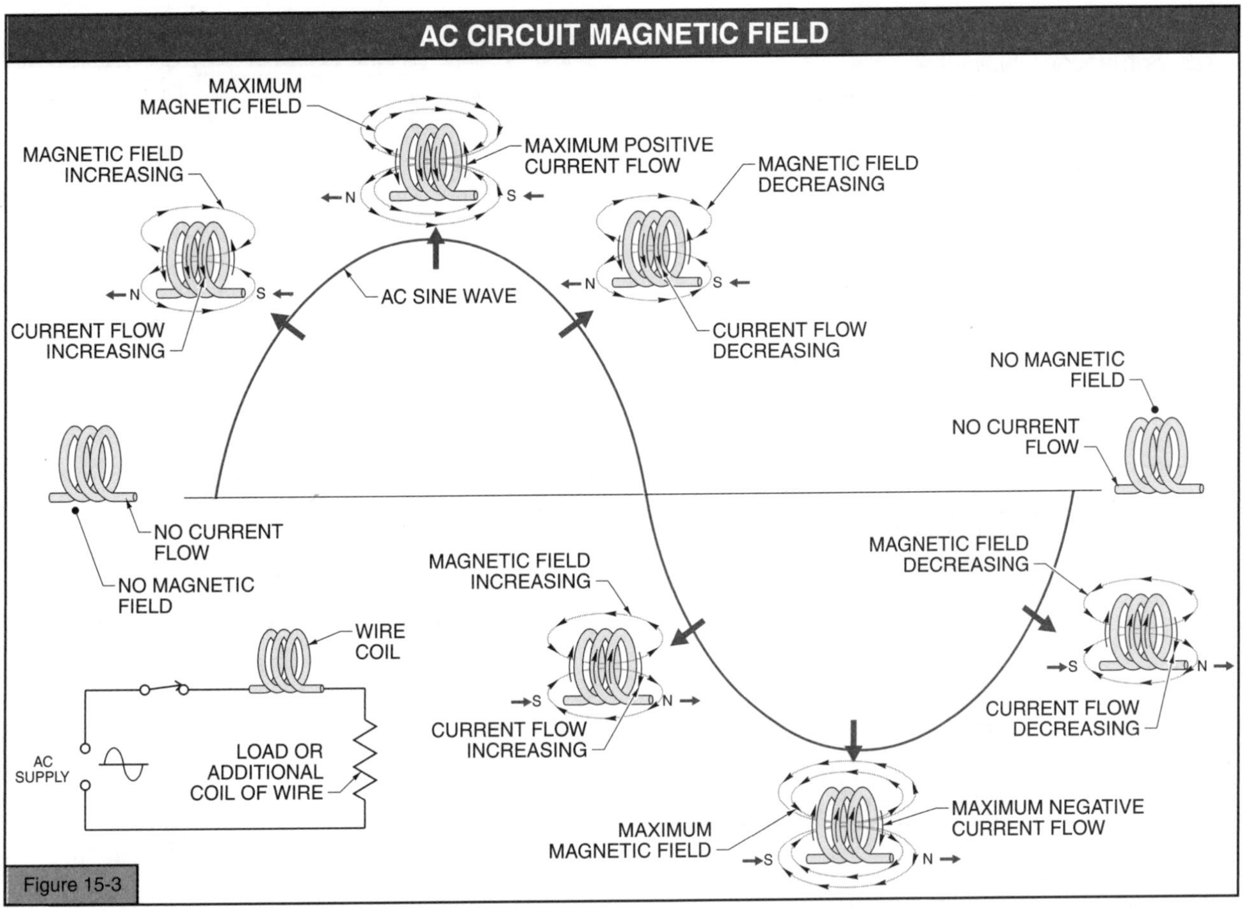

AC CIRCUIT MAGNETIC FIELD

MAXIMUM MAGNETIC FIELD
MAXIMUM POSITIVE CURRENT FLOW
MAGNETIC FIELD INCREASING
MAGNETIC FIELD DECREASING
N
S
AC SINE WAVE
CURRENT FLOW INCREASING
CURRENT FLOW DECREASING
NO MAGNETIC FIELD
NO CURRENT FLOW
NO CURRENT FLOW
NO MAGNETIC FIELD
MAGNETIC FIELD INCREASING
MAGNETIC FIELD DECREASING
WIRE COIL
AC SUPPLY
LOAD OR ADDITIONAL COIL OF WIRE
CURRENT FLOW INCREASING
CURRENT FLOW DECREASING
MAXIMUM MAGNETIC FIELD
MAXIMUM NEGATIVE CURRENT FLOW

Figure 15-3

- Residential—Control the flow of water or gas in washing machines, dishwashers, dryers, furnaces, and automatic sprinkling systems.
- Automobiles—Control locks (doors, trunk, gas tank cover), gas and water flow, antipollution valves, and vacuum valves.
- Boats and Airplanes—Control fuel, coolant, and water flow. Also used for electric control of doors, hatches, missiles, and other devices.
- Commercial and Industrial—Control door locks, open and close valves dispensing product, control movement of parts, stamp information on products, clamp parts, and dispense coins. Also used to operate motor starters, contactors, clutches, brakes, and control fluid power valves.

Transformers

A *transformer* is an electric device that uses electromagnetism to change voltage from one level to another or to isolate one voltage from another. Transformers are primarily used to step up or step down voltage.

Transformers operate on the electromagnetic mutual induction principle. *Mutual inductance* is the effect of one coil inducing a voltage into another coil. **See Figure 15-5.**

AC/DC adapters (wall transformers) are used with small electronic components, accept AC input voltage directly from a wall outlet, and output DC voltage.

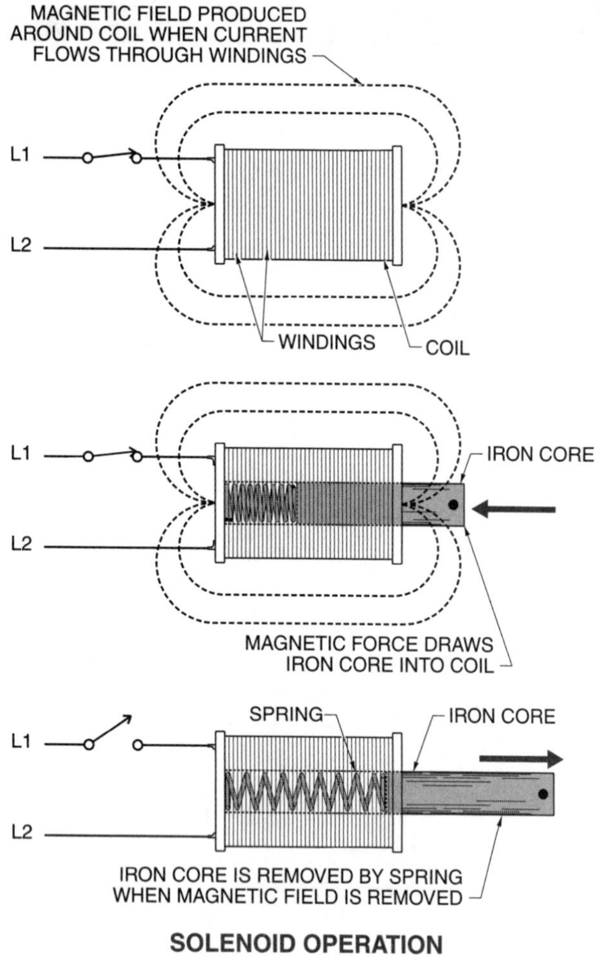

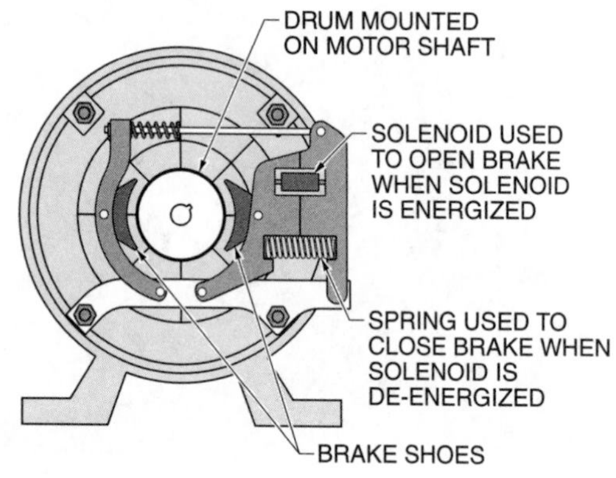

Figure 15-4

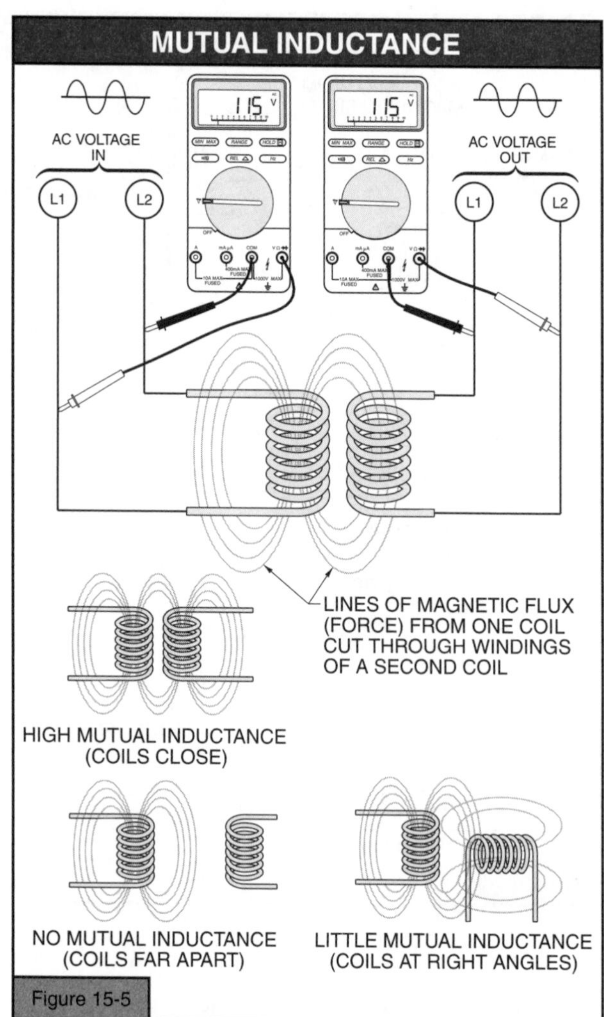

Figure 15-5

The principle of electromagnetic mutual induction states that when the magnetic flux lines from one expanding and contracting magnetic field cut the windings of a second coil, a voltage is induced in the second coil. The amount of voltage induced in the second coil depends on the relative position of the two coils and the number of turns in each coil. The induced voltage in the second coil depends on the relative position of the two coils if the number of turns in each coil are equal. The highest mutual inductance occurs when all the magnetic flux lines from each coil cut through all the turns of wire in the opposite coil. Likewise, no mutual inductance occurs when no magnetic flux lines from one coil cut through any of the turns of wire in the other coil.

Transformers consist of two or more coils of insulated wire wound on a laminated steel core. The steel core is magnetized when a voltage is applied to one coil. The magnetized steel core induces a voltage into the second-ary coil. The *primary coil (input side)* of a transformer is the coil to which the voltage is connected. The *secondary coil (output side)* of a transformer is the coil in which the voltage is induced. The laminated steel core is used so that the magnetic induction between the two coils is as high as possible. **See Figure 15-6.**

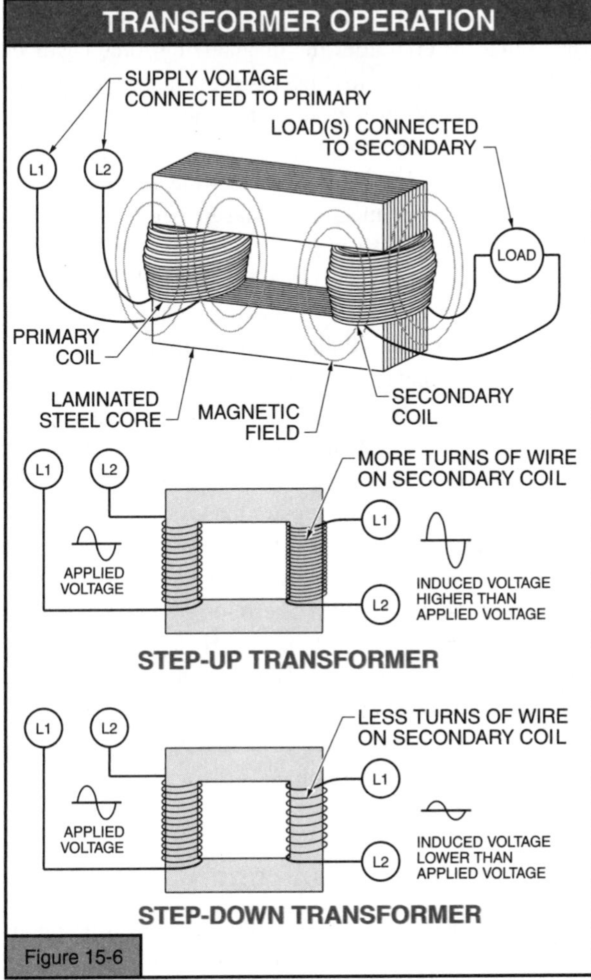

Figure 15-6

The amount of change in voltage between the primary coil and secondary coil depends on the turns ratio (voltage ratio) of the two coils. A *step-up transformer* is a transformer in which the secondary coil has more turns of wire than the primary coil. A step-up transformer produces a higher voltage on the secondary coil than the voltage applied to the primary. A *step-down transformer* is a transformer in which the secondary coil has fewer turns of wire than the primary coil. A step-down transformer produces a lower voltage on the secondary coil than the voltage applied to the primary. To calculate the relationship between the number of turns and the voltage, apply the formula:

$$\frac{N_P}{N_S} = \frac{E_P}{E_S}$$

where

N_P = number of turns in primary coil

N_S = number of turns in secondary coil

E_P = voltage applied to primary coil (in V)

E_S = voltage induced in secondary coil (in V)

This formula may be rearranged to calculate any one value when the other three are known.

Example: Calculating Secondary Induced Voltage

What is the secondary induced voltage of a transformer that has a secondary coil containing 100 turns, a primary coil containing 500 turns, and 25 V applied to the primary coil?

$$E_S = E_P \times \frac{N_S}{N_P}$$

$$E_S = 25 \times \frac{100}{500}$$

$$E_S = \mathbf{5\ V}$$

This transformer is a step-down transformer because the voltage is reduced from 25 V to 5 V. The same formula applies to a step-up transformer. For example, 25 V applied to a transformer primary coil containing 500 turns induces 100 V in the transformer secondary coil containing 2000 turns (25 × (2000 ÷ 500) = 100).

In the process of changing voltage from one level to another, transformers also change the current to a lower or higher level. The ratio of primary current to secondary current is inversely proportional to the voltage ratio. For example, if the voltage ratio is 10:1, the current ratio is 1:10. To calculate the current and voltage relationship, apply the formula:

$$\frac{I_P}{I_S} = \frac{E_S}{E_P}$$

where

I_P = current in primary coil (in A)

I_S = current in secondary coil (in A)

E_S = voltage induced in secondary coil (in V)

E_P = voltage applied to primary coil (in V)

This formula may be rearranged to calculate any one value when the other three are known.

Example: Calculating Transformer Primary Current

What is the current in the primary winding of a transformer when the primary voltage is 240 V, the secondary voltage is 120 V, and a 10 A load is connected to the secondary?

$$I_P = I_S \times \frac{E_S}{E_P}$$

$$I_P = 10 \times \frac{120}{240}$$

$$I_P = \mathbf{5\ A}$$

The current potential of a transformer is stepped down any time a transformer steps up the voltage. Likewise, the current potential is stepped up any time a transformer steps down the voltage.

Transformer Ratings

Transformers are designed to transform power at one voltage level to power at another voltage level. In an ideal transformer, there is no loss or gain. Energy is simply transferred from the primary circuit to the secondary circuit. For example, if the secondary of a transformer requires 500 W of power to operate the loads connected to it, the primary must deliver 500 W.

It is standard practice with transformers to use voltage and current ratings and not wattage ratings. It is also standard practice to rate a transformer for its output capabilities because it is the output of the transformer to which the loads are connected. Thus, transformers are rated by their volt-ampere (VA), or kilovolt-ampere (kVA) output. Small transformers are rated in either VA or kVA. Large transformers are rated in kVA. For example, a 50 VA transformer may be rated as a 50 VA or .05 kVA transformer. A 5000 VA transformer is rated as a 5 kVA transformer.

In an ideal transformer, energy is transferred from the primary circuit to the secondary circuit and there is no power loss. However, all transformers have some power loss. Even though most transformers operate with little power loss (normally .5% to 8%), there is always a power loss in a transformer. The less the power loss, the more efficient the transformer. The efficiency of a transformer is expressed as a percentage. To calculate the efficiency of a transformer, apply the formula:

$$Eff = \frac{P_S}{P_P} \times 100$$

where

Eff = efficiency (in %)

P_S = power of secondary circuit (in W)

P_P = power of primary circuit (in W)

Example: Calculating Transformer Efficiency

What is the efficiency of a transformer that uses 1200 W of primary power to deliver 1110 W of secondary power?

$$Eff = \frac{P_S}{P_P} \times 100$$

$$Eff = \frac{1110}{1200} \times 100$$

$$Eff = .925 \times 100$$

$$Eff = \mathbf{92.5\%}$$

Power loss in a transformer is caused by hysteresis loss, eddy-current loss, and copper loss. The total amount of power loss is a combination of the three.

Transformers are used on AC circuits. In all AC circuits, the voltage (and magnetic field) are constantly changing. *Hysteresis loss* is loss caused by magnetism that remains (lags) in a material after the magnetizing force has been removed. Hysteresis loss occurs every half-cycle of AC when the current reverses direction and some magnetism remains in the iron core.

Eddy-current loss is loss caused by the induced currents that are produced in metal parts that are being magnetized. In a transformer, eddy-current loss occurs in the iron core because the iron core is magnetized. The induced currents in the iron core cause heat in the iron core, primary windings, and secondary windings. Eddy currents are reduced by using a laminated steel core.

Copper is a good conductor of electricity. However, copper has some resistance. *Copper loss* is loss caused by the resistance of the copper wire to the flow of current. The higher the resistance, the greater the loss.

Hysteresis loss, eddy-current loss, and copper loss all produce losses in the transformer due to the heat they develop in the metal parts. These losses may be reduced by proper design and installation. Hysteresis loss is reduced by using a silicon-steel core instead of an iron core. Eddy-current loss is reduced by using a laminated core instead of a solid core. Copper loss is reduced by using as large a conductor as possible.

Transformer Classification

All transformers are used to either step up, step down, or isolate voltage. However, when transformers are used for specific applications, they are normally referred to by the application name, such as a distribution transformer, control transformer, etc. The basic types of transformers include appliance/equipment, control, bell/chime, instrument (current), distribution, isolation, neon sign, and power transformers. **See Figure 15-7.**

Appliance/equipment transformers are transformers specifically designed for the appliance or piece of equipment in which they are used. Electrical and electronic devices, such as TVs, garage door openers, stereos, and computers require different voltage levels to operate their different circuits. The transformer normally has many secondary voltage output levels that can be used to operate several different circuits. Appliance/equipment transformers are rated from 2 VA (or less) to several thousand VA and deliver secondary voltages from 5 V (or less) to several thousand volts.

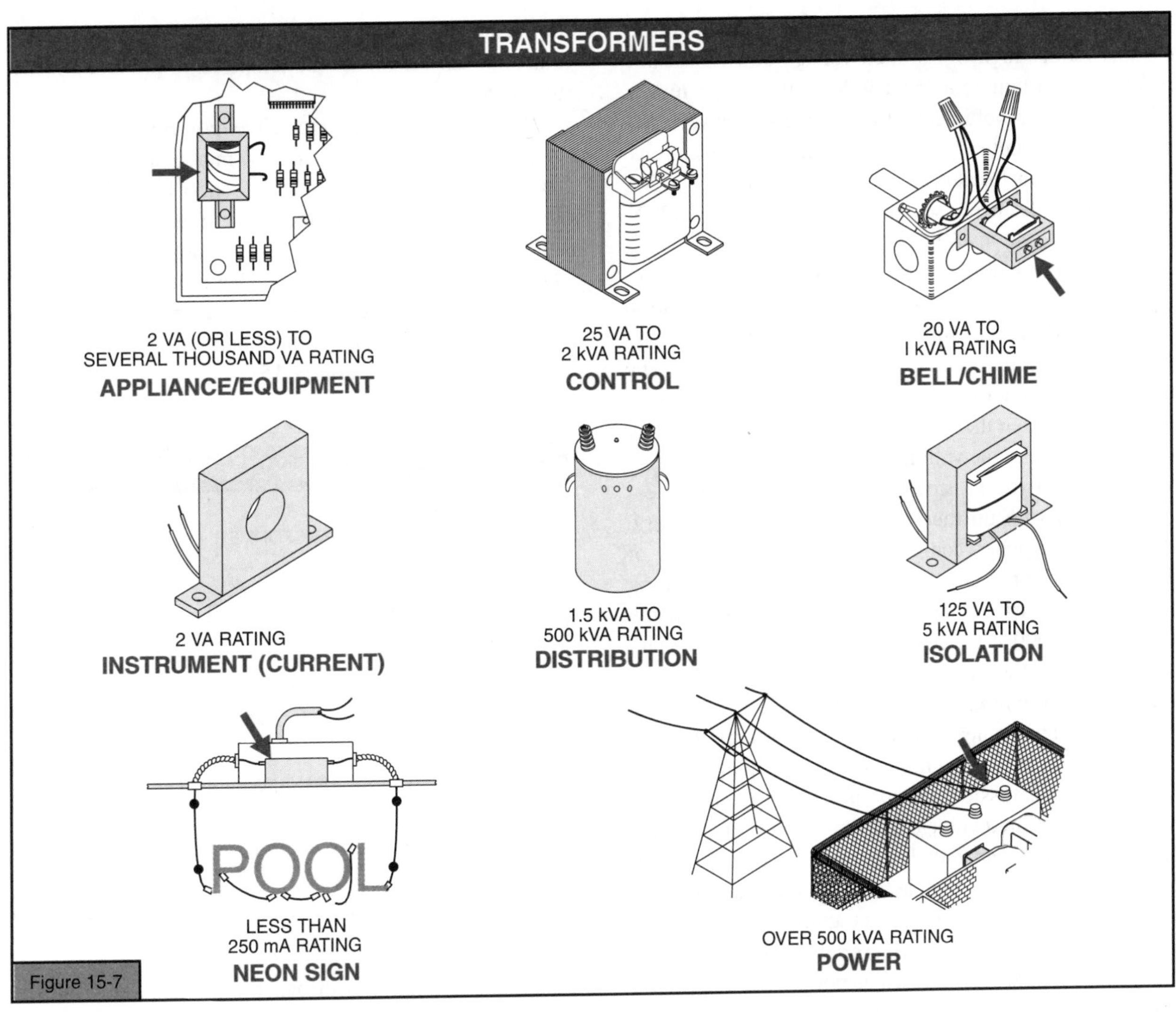

Figure 15-7

Control transformers are step-down transformers that are used to lower the voltage to the control circuit in which the control switches, contactors, and motor starters are connected. Industrial motors and heating elements normally operate on 460/480 V. However, the control circuits used to control the 460/480 V loads normally operate on 115/120 V. Control transformers are normally rated from 25 VA (.025 kVA) to 2000 VA (2 kVA) and deliver a secondary voltage of 115/120 V. Control transformers are mounted inside control cabinets and motor control centers.

Bells, chimes, buzzers, fire systems, and security systems normally operate at voltage levels from 8 V to 24 V. Bell/chime transformers are specifically designed to step down 115/120 V to 8 V, 12 V, 16 V, or 24 V. Bell/chime transformers are normally rated from 20 VA to 1000 VA (1 kVA). The size (VA rating) of the transformer is based on the number and power rating of the loads. The larger the number of loads, the higher the required power rating. Likewise, the higher the power rating of the loads connected to the transformer, the higher the required power rating.

Instrument transformers are transformers specifically designed to step down the voltage or current of a circuit to a lower value that can safely be used to measure the voltage, current, or power in the circuit with a meter (instrument). Instrument transformers are rated according to their primary to secondary ratio, such as 10:1, 50:1, 400:5, etc. For example, a current transformer that has a 10:1 ratio delivers 1 A output on the secondary for every 10 A flowing through the primary. There is 2.5 A flowing through the secondary if there is 25 A flowing through the primary. The ammeter (instrument) that is measuring the circuit current (25 A) is connected to the secondary so that only 2.5 A is flowing into the meter. The meter is calibrated to display 25 A whenever there is a 2.5 A input. Instruments and meters may use an external instrument transformer or include an instrument transformer as an integral part of the meter.

Distribution transformers are step-down transformers that reduce high transmitted voltage down to usable residential, commercial, and industrial voltage levels. Although power is efficiently transmitted at high voltages, high voltage is not safe for use in practical applications. Distribution transformers are normally rated from 1.5 kVA to 500 kVA and deliver a secondary voltage of 115 V, 120 V, 208 V, 230 V, 240 V, 460 V, or 480 V.

Isolation transformers are transformers that are designed to isolate the load from the power source. Isolation transformers have a 1:1 turns (and voltage) ratio. Therefore, the primary voltage is equal to the secondary voltage. The transformer provides isolation between the two sections of the circuit even though the primary and secondary are operating at the same voltage.

Neon sign transformers are step-up transformers designed to step up the secondary voltage to several thousand volts when connected to a primary voltage of 115/230 V. Neon sign transformers are rated according to their secondary voltage. Secondary voltages normally range from 2000 V to 15,000 V. Although the secondary voltage is high, the current rating is relatively low. Current ratings normally range less than 250 mA.

Power transformers are large transformers used by power companies in generating plants to step up voltages for power transmission and in substations to step down voltages. Generated voltage must be stepped up to a high level for efficient transmittal over long distances. The voltage is stepped up to a high level to step the current down to a low level. Low current levels require small wire sizes, reducing cost. Power transformers are normally rated over 500 kVA (500,000 VA) and used on power lines that operate on voltages over 67 kV (67,000 V).

Distribution transformers are the same as power transformers but are smaller and serve a different purpose in the power distribution system. Small distribution transformers are mounted on poles or the ground. Large distribution transformers are mounted on the ground.

Transformer Overloading

All transformers have a power (VA or kVA) rating. The power rating of a transformer indicates the amount of power the transformer can safely deliver. However, like most electrical devices, this rating is not an absolute value. For example, a 100 VA rated transformer is not destroyed if required to deliver 110 VA for a short period of time.

The heat produced by the power destroys the transformer. The heat destroys the transformer by breaking down the insulation, causing short circuits within the windings of the transformer. Thus, temperature is the limiting factor in transformer loading. The more power the transformer must deliver, the higher the temperature produced at the transformer.

Transformers are used to deliver power to a set number of loads. For example, a transformer is used to deliver power to a school. As loads in the school are switched ON and OFF, the power delivered by the transformer changes. At certain times (night), the power output required from the transformer may be low. At other times (during school hours), the power output required from the transformer may be high. *Peak load* is the maximum output required of a transformer. **See Figure 15-8.**

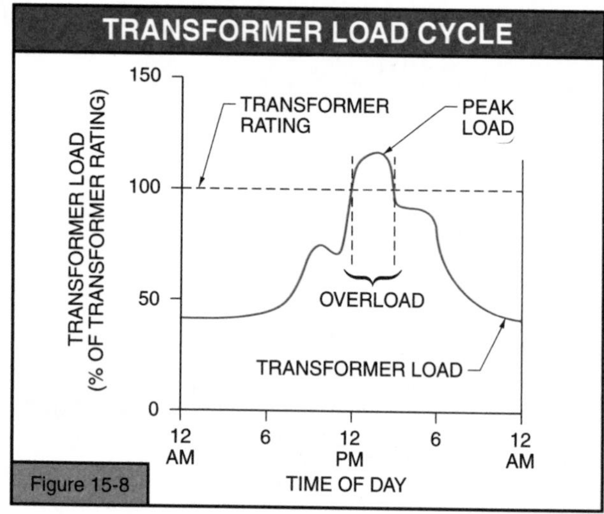

Figure 15-8

A transformer is overloaded when it is required to deliver more power than its rating. A transformer is not damaged when overloaded for a short time period. This is because the heat storage capacity of a transformer ensures a relatively slow increase in internal transformer temperature.

Transformer manufacturers list the length of time a transformer may be safely overloaded at a given peak level. For example, a transformer that is overloaded 3 times its rated current has a permissible overload time of about 6 minutes. **See Figure 15-9.**

Transformer Cooling Methods

The methods used to dissipate heat in a transformer include self-air cooled, forced-air cooled, liquid-immersed/self-air cooled, and liquid-immersed/forced-air cooled. **See Figure 15-10.**

A *self-air cooled transformer* is a transformer that dissipates heat through the air surrounding the transformer.

Heat produced in the windings and core is dissipated into the surrounding air by convection. *Convection* is heat transfer that occurs when currents circulate between warm and cool regions of a fluid (air). Convection heat transfer is increased by adding radiating fins to the transformer.

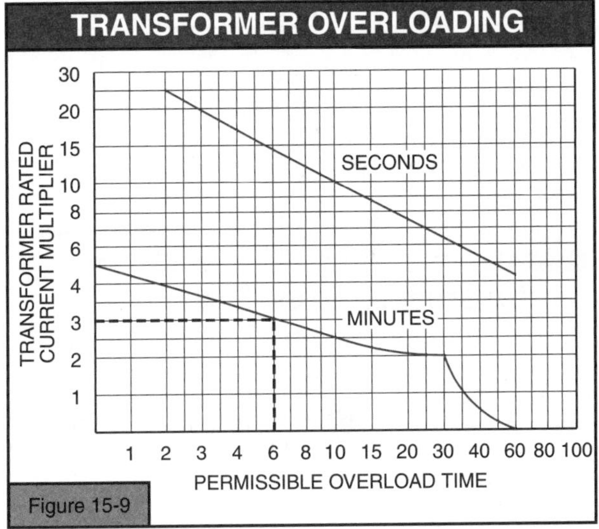

Figure 15-9

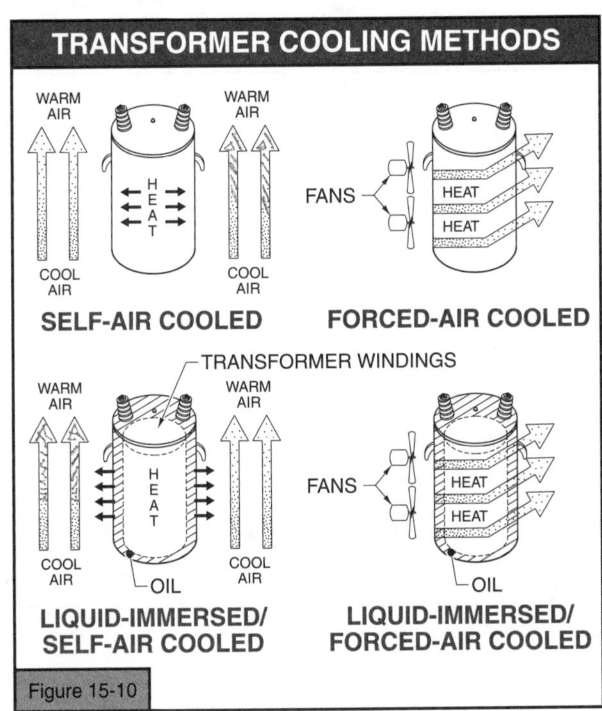

Figure 15-10

Forced-air cooled transformers are transformers that use a fan to move air over the transformer. Using a fan to speed the convection process increases the power that the transformer can deliver by about 30% over the power that can be delivered without a fan. Multiple high-velocity fans are used in some applications to increase the transformer power output by more than 30%. The fans may be designed to remain ON at all times or may be automatically turned ON when the transformer reaches a set temperature.

A *liquid-immersed/self-cooled transformer* is a transformer that uses refined oil or synthetic oil to help cool the transformer windings. The transformer coils and core are enclosed in a metal tank which is immersed in the oil. The oil is used to conduct heat from the windings and core to the outer surface of the transformer. The oil helps slow the heating process by increasing the heat storage capacity of the transformer. This is useful when the transformer is temporarily overloaded during peak usage times.

A *liquid-immersed/forced-air cooled transformer* is a transformer that uses refined oil or synthetic oil and fans to cool the transformer. The oil conducts the heat to the outer surface of the transformer and the fans dissipate the heat to the surrounding air.

Sizing Single-Phase Transformers

A transformer is used any time power is delivered to a residential, commercial, industrial, construction, or other site. The size of the transformer is based on the amount of expected power required. This amount normally takes into consideration present and future needs, peak loading, ambient temperature, load types, and other factors that affect transformer operation and temperature. **See Figure 15-11.** To size a 1ϕ transformer, apply the procedure:

1. Determine the total voltage required by the loads if more than one load is connected. The secondary side of the transformer must have a rating equal to the voltage of the loads.
2. Determine the amperage rating or kVA capacity required by the load(s). Add all loads that are (or may be) ON concurrently.
3. Check load(s) frequency on the nameplate. The frequency of the supply voltage and the electrical load(s) must be the same.
4. Check the supply voltage to the primary side of the transformer. The primary side of a transformer must have a rating equal to the supply voltage. Consider each voltage if there is more than one source voltage available. Use a transformer that has primary taps if there is a variation in the supply voltage. The transformer must have a kVA capacity of at least 10% greater than that required by the loads.

A single-phase full-load currents conversion table may be used to determine proper kVA capacity when the load rating is given in amperes. **See Figure 15-12.**

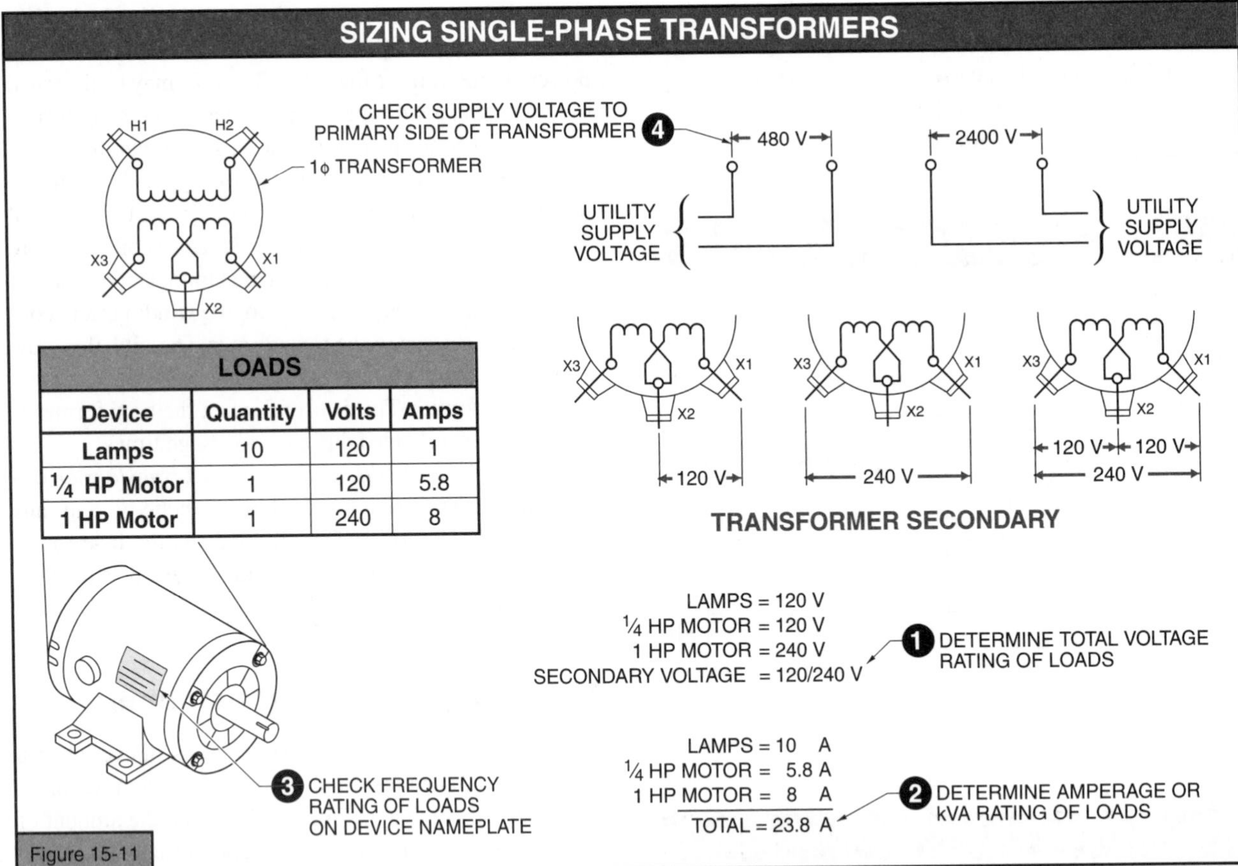

LOADS			
Device	Quantity	Volts	Amps
Lamps	10	120	1
¼ HP Motor	1	120	5.8
1 HP Motor	1	240	8

Figure 15-11

SINGLE-PHASE FULL-LOAD CURRENTS*

kVA	120 V	208 V	240 V	277 V	380V	480 V
.050	.4	.2	.2	.2	.1	.1
.100	.8	.5	.4	.3	.2	.2
.150	1.2	.7	.6	.5	.4	.3
.250	2.0	1.2	1	.9	.6	.5
.500	4.2	2.4	2.1	1.8	1.3	1
.750	6.3	3.6	3.1	2.7	2	1.6
1	8.3	4.8	4.2	3.6	2.6	2.1
1½	12.5	7.2	6.2	5.4	3.9	3.1
2	16.7	9.6	8.3	7.2	5.2	4.2
3	25	14.4	12.5	10.8	7.9	62
5	41	24	20.8	18	13.1	10.4
7½	62	36	31	27	19.7	15.6
10	83	48	41	36	26	20.8
15	125	72	62	54	39	31

* in A

Figure 15-12

TECH TIP

Some transformers are provided with a tap changer to provide a uniform secondary voltage when the primary voltage varies due to line voltage drop.

To calculate kVA capacity of a 1ϕ transformer when voltage and current are known, apply the formula:

$$kVA_{CAP} = E \times \frac{I}{1000}$$

where

kVA_{CAP} = transformer capacity (in kVA)

E = voltage (in V)

I = current (in A)

1000 = constant

Example: Calculating 1ϕ Transformer Capacity

What is the kVA capacity of a 1ϕ transformer used in a 240 V circuit that has loads of 5 A, 3 A, and 8 A connected to the secondary?

$$kVA_{CAP} = E \times \frac{I}{1000}$$

$$kVA_{CAP} = 240 \times \frac{16}{1000}$$

$$kVA_{CAP} = \frac{3840}{1000}$$

$$kVA_{CAP} = \mathbf{3.84\ kVA}$$

A 5 kVA transformer is selected for the application because it is the next size available with a capacity of at least 10% greater than that required by the loads.

Sizing Three-Phase Transformers

The size of a 3ϕ transformer is based on the amount of expected power required in the circuit. **See Figure 15-13.** To size a 3ϕ transformer, apply the procedure:

1. Determine the total voltage required by the loads if more than one load is connected. The secondary side of the transformer must have a rating equal to the voltage of the loads.
2. Determine the amperage rating or kVA capacity required by the load(s). Add all loads that are (or may be) ON concurrently.
3. Check the frequency of the load(s) on the nameplate. The frequency of the supply voltage and the electrical load(s) must be the same.
4. Determine the type of 3ϕ voltage available. This includes three-wire no ground or three-wire with ground (four-wire).
5. Check the supply voltage to the primary side of the transformer.

In emergency situations, most transformers may be kept operational by spraying water on the transformer housing—not exposed electrical parts. This may increase output by up to 60%.

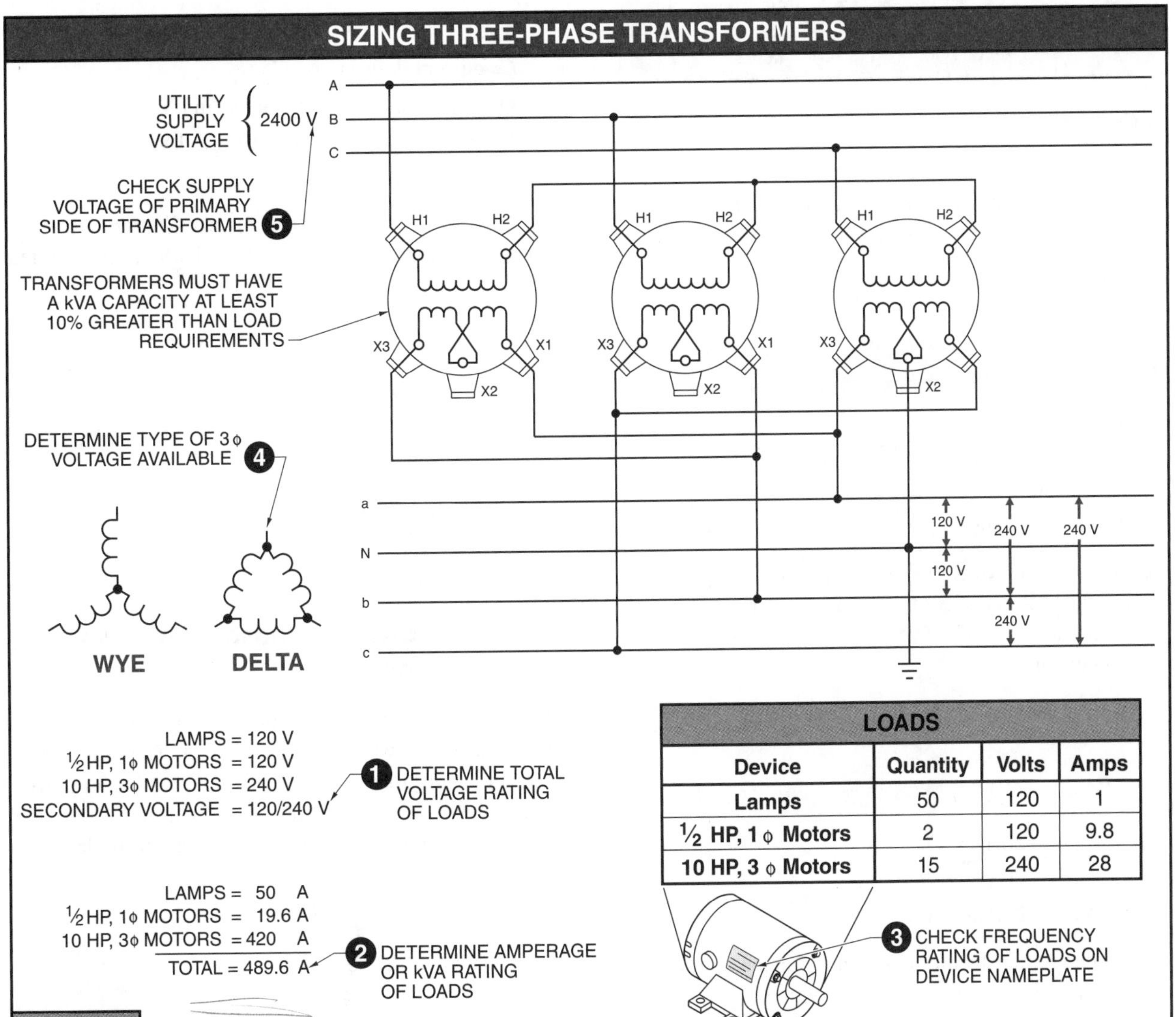

SIZING THREE-PHASE TRANSFORMERS

UTILITY SUPPLY VOLTAGE 2400 V (A, B, C)

5 CHECK SUPPLY VOLTAGE OF PRIMARY SIDE OF TRANSFORMER

TRANSFORMERS MUST HAVE A kVA CAPACITY AT LEAST 10% GREATER THAN LOAD REQUIREMENTS

H1 H2 X3 X1 X2

4 DETERMINE TYPE OF 3ϕ VOLTAGE AVAILABLE

WYE DELTA

a N b c — 120 V, 120 V, 240 V, 240 V, 240 V

LAMPS = 120 V
½HP, 1ϕ MOTORS = 120 V
10 HP, 3ϕ MOTORS = 240 V
SECONDARY VOLTAGE = 120/240 V

1 DETERMINE TOTAL VOLTAGE RATING OF LOADS

LAMPS = 50 A
½HP, 1ϕ MOTORS = 19.6 A
10 HP, 3ϕ MOTORS = 420 A
TOTAL = 489.6 A

2 DETERMINE AMPERAGE OR kVA RATING OF LOADS

LOADS			
Device	**Quantity**	**Volts**	**Amps**
Lamps	50	120	1
½ HP, 1 ϕ Motors	2	120	9.8
10 HP, 3 ϕ Motors	15	240	28

3 CHECK FREQUENCY RATING OF LOADS ON DEVICE NAMEPLATE

Figure 15-13

The primary side of a transformer must have a rating equal to the supply voltage. Consider each voltage when there is more than one source of voltage available. Use a transformer that has primary taps when there is a variation in the supply voltage. The transformer must have a kVA capacity of at least 10% greater than that required by the loads.

A three-phase full-load current conversion table is used to determine the kVA capacity of 3ϕ circuits when the load rating is given in amperes. **See Figure 15-14.**

THREE-PHASE FULL-LOAD CURRENTS*				
kVA	208 V	240 V	480 V	600 V
3	8.3	7.2	3.6	2.9
4	12.5	10.8	5.4	4.3
6	16.6	14.4	7.2	5.8
9	25	21.6	10.8	8.6
15	41	36	18	14.4
22	62	54	27	21.6
30	83	72	36	28
45	124	108	54	43

* in A

Figure 15-14

To calculate the kVA capacity of a 3ϕ transformer when voltage and current are known, apply the formula:

$$kVA_{CAP} = E \times 1.732 \times \frac{I}{1000}$$

where

kVA_{CAP} = transformer capacity (in kVA)

E = voltage (in V)

1.732 = constant (for 3ϕ power)

I = current (in A)

1000 = constant

Example: Calculating 3ϕ Transformer kVA Capacity

What is the kVA capacity of a 3ϕ transformer used in a 240 V circuit that has loads of 25 A, 30 A, and 8 A connected to its secondary?

$$kVA_{CAP} = E \times 1.732 \times \frac{I}{1000}$$

$$kVA_{CAP} = 240 \times 1.732 \times \frac{63}{1000}$$

$$kVA_{CAP} = \frac{26{,}188}{1000}$$

$$kVA_{CAP} = \mathbf{26.188\ kVA}$$

A 30 kVA transformer is selected for the application because it is the next size available with a capacity of at least 10% greater than that required by the loads.

Determining Single-Phase Transformer Current Draw

A transformer has a voltage ready to be used. However, no current flows and no power is used if no load is connected to the transformer. Current flows and electrical power is consumed when loads are connected to the available voltage source. The amount of current flow depends on the power rating of the load. The higher the power rating, the larger the amount of current flow. Likewise, the lower the power rating, the smaller the amount of current flow. Knowing the amount of current flowing in a circuit is necessary because the conductor size is based on the amount of expected current flow. To calculate transformer current draw when kVA capacity and voltage are known, apply the formula:

$$I = kVA_{CAP} \times \frac{1000}{E}$$

where

I = current draw (in A)

kVA_{CAP} = transformer capacity (in kVA)

1000 = constant (to convert VA to kVA)

E = voltage (in V)

Example: Calculating 1ϕ Transformer Current Draw

What is the current draw of a 1ϕ, 41 kVA, 120 V rated transformer when fully loaded?

$$I = kVA_{CAP} \times \frac{1000}{E}$$

$$I = 41 \times \frac{1000}{120}$$

$$I = \frac{41{,}000}{120}$$

$$I = \mathbf{341\ A}$$

Determining Three-Phase Transformer Current Draw

The current draw of a 3ϕ transformer is calculated similarly to the current draw of a 1ϕ transformer. The only difference is the addition of the constant (1.732) for 3ϕ power. To calculate current draw of a 3ϕ transformer when kVA capacity and voltage are known, apply the formula:

$$I = kVA_{CAP} \times \frac{1000}{E \times 1.732}$$

where

I = current (in A)

kVA_{CAP} = transformer capacity (in kVA)

1000 = constant

E = voltage (in V)

$1.732 = (\sqrt{3})$

Example: Calculating 3ϕ Transformer Current Draw

What is the current draw of a 3ϕ, 30 kVA, 480 V rated transformer when fully loaded?

$$I = kVA_{CAP} \times \frac{1000}{E \times 1.732}$$

$$I = 30 \times \frac{1000}{480 \times 1.732}$$

$$I = \frac{30,000}{831}$$

$$I = \mathbf{36.1\ A}$$

Note: When 3ϕ problems are calculated, the following values* may be substituted to eliminate one mathematical step:

For 208 V × 1.732, use 360

For 230 V × 1.732, use 398

For 240 V × 1.732, use 416

For 440 V × 1.732, use 762

For 460 V × 1.732, use 797

For 480 V × 1.732, use 831

$*\sqrt{3} = 1.732$

Transformer Standard Ambient Temperature Compensation

Temperature rise in a transformer is the temperature of the windings above the existing ambient temperature. Transformer nameplates list their maximum temperature rise. Transformer normal ambient temperature is 40°C.

A transformer must be derated if the ambient temperature exceeds 40°C. Transformer derating charts are used to derate transformers in high ambient temperatures. **See Figure 15-15.** To calculate the derated kVA capacity of a transformer operating at a higher-than-normal ambient temperature condition, apply the formula:

kVA = rated kVA × maximum load

where

kVA = derated transformer capacity (in kVA)

rated kVA = manufacturer transformer rating (in kVA)

maximum load = maximum transformer loading (in %)

Example: Calculating Standard Transformer Derating

What is the derated kVA value of a 30 kVA rated transformer installed in an ambient temperature of 50°C?

kVA = rated kVA × maximum load

kVA = 30 × .92

kVA = **27.6 kVA**

TRANSFORMER DERATINGS	
Maximum Ambient Temperature (°C)	Maximum Transformer Loading (%)
40	100
45	96
50	92
55	88
60	81
65	80
70	76

Figure 15-15

Transformer Special Ambient Temperature Compensation

Standard ambient temperature is the average temperature of the air that cools a transformer over a 24-hour period. Standard ambient temperature assumes that maximum temperature does not exceed 10°C above average ambient temperature.

A transformer is derated above the standard values when the maximum temperature exceeds the average temperature by more than 10°C. A transformer is derated by 1½% for each 1°C above 40°C when the maximum ambient temperature exceeds 10°C above the average temperature. **See Figure 15-16.**

For example, a 30 kVA rated transformer installed in an ambient temperature of 50°C (maximum temperature exceeds ambient temperature by 10°C) is derated to 25.5 kVA. The transformer is derated by 1½% for each degree above 40°C. A 50°C ambient temperature is 10°C above 40°C so the transformer is derated 15% (10°C × 1½% = 15%).

kVA = rated kVA × maximum load

kVA = 30 × .85 (15% derating)

kVA = **25.5 kVA**

Single-Phase Residential Transformer Connections

Electricity is used in residential applications (one-family, two-family, and multifamily dwellings) to provide energy for lighting, heating, cooling, cooking, running motors, etc. The electrical service to dwellings is normally 1ϕ, 120/240 V. The low voltage (120 V) is used for general-purpose receptacles and general lighting. The high voltage (240 V) is used for heating, cooling, cooking, etc.

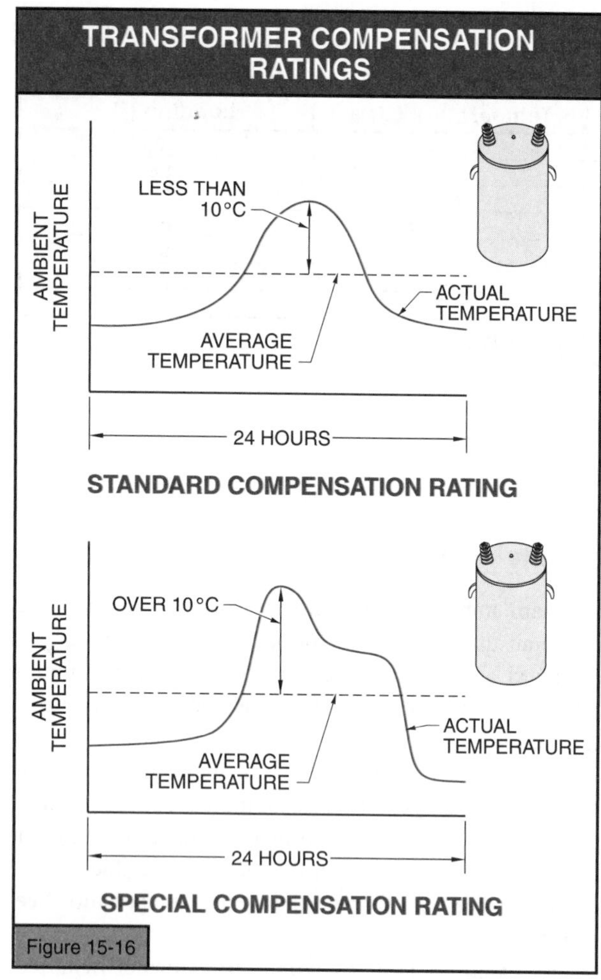

Figure 15-16

Distribution substations contain large utility transformers.

Residential electrical service may be overhead or lateral. *Overhead service* is electrical service in which service-entrance conductors are run from the utility pole through the air and to the dwelling. *Lateral service* is electrical service in which service-entrance conductors are run underground from the utility service to the dwelling. **See Figure 15-17.**

Three-Phase Transformer Connections

Three 1ϕ transformers may be connected to develop 3ϕ voltage. The three transformers may be connected in a wye or delta configuration. A *wye configuration* is a transformer connection that has one end of each transformer coil connected together. The remaining end of each coil is connected to the incoming power lines (primary side) or used to supply power to the load(s) (secondary side). A *delta configuration* is a transformer connection that has each transformer coil connected end-to-end to form a closed loop. Each connecting point is connected to the incoming power lines or used to supply power to the load(s). The voltage output and type available for the load(s) is determined by whether the transformer is connected in a wye or delta configuration. **See Figure 15-18.**

Three-Phase, Delta-to-Delta Connections

Three transformers may be connected in a delta-to-delta connection. A delta-to-delta transformer connection is used to supply 3ϕ voltage on the secondary. In a delta-to-delta connection, each transformer is connected end-to-end. **See Figure 15-19.**

The advantage of a delta-to-delta connection is that if one transformer is disabled, the other two may be used in an open-delta connection for emergency power. The rating of the open-delta bank is 57.7% of the original three transformer bank, but 3ϕ power is available until repairs are made.

One of the delta transformers is center-tapped to supply both 3ϕ voltage and 1ϕ voltage. Single-phase voltage at 120/240 V is available when the transformer is center-tapped. However, because only one transformer is tapped, the transformer that is tapped carries all of the 1ϕ, 120/240 V load and ⅓ of the 3ϕ, 240 V load. The other two transformers each carry ⅓ of the 3ϕ, 240 V load. For this reason, this connection should be used in applications that require a large amount of 3ϕ power and a small amount of 1ϕ power.

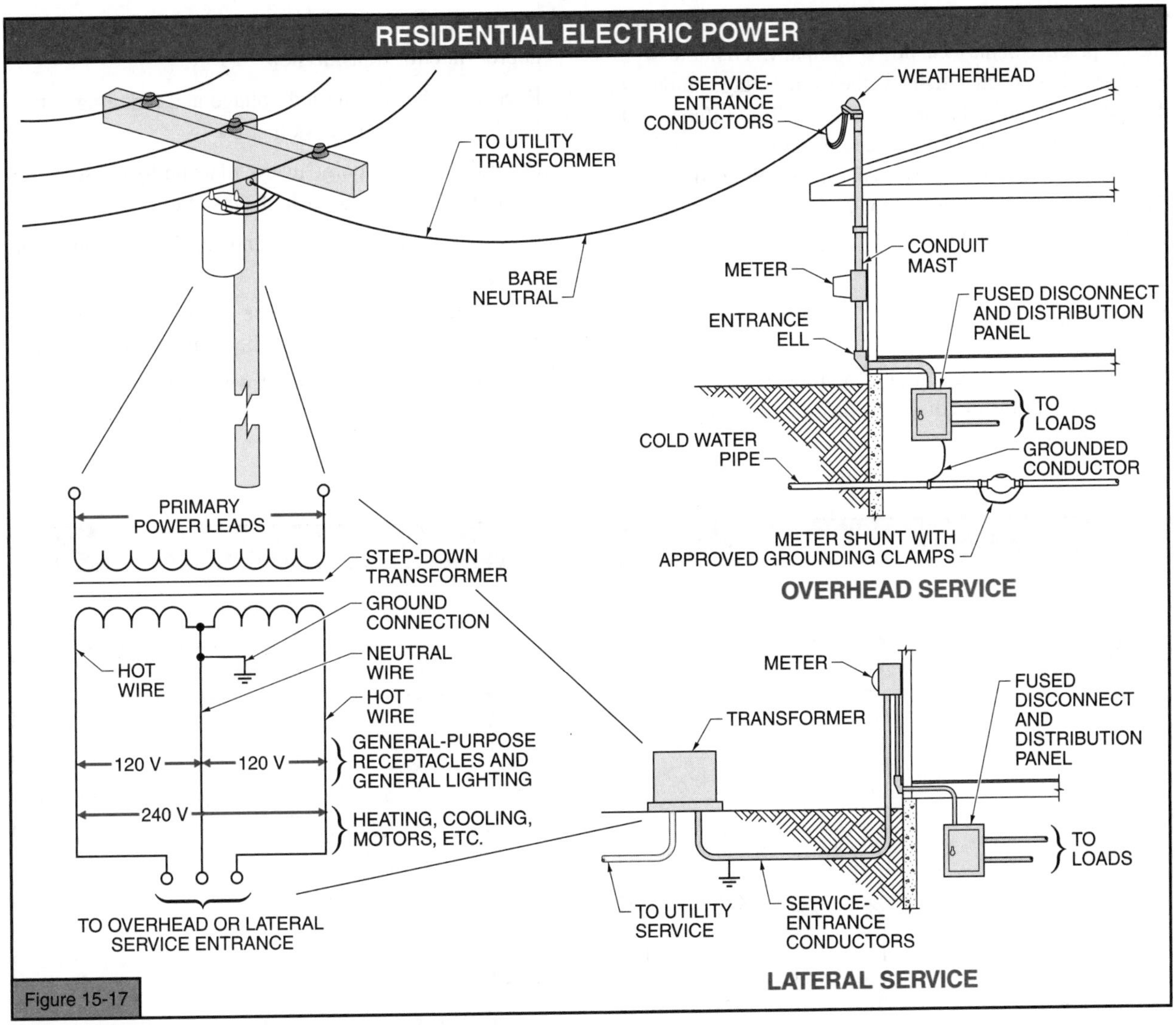

RESIDENTIAL ELECTRIC POWER
TO UTILITY TRANSFORMER
BARE NEUTRAL
SERVICE-ENTRANCE CONDUCTORS
WEATHERHEAD
METER
CONDUIT MAST
ENTRANCE ELL
FUSED DISCONNECT AND DISTRIBUTION PANEL
TO LOADS
COLD WATER PIPE
GROUNDED CONDUCTOR
METER SHUNT WITH APPROVED GROUNDING CLAMPS
OVERHEAD SERVICE
PRIMARY POWER LEADS
STEP-DOWN TRANSFORMER
GROUND CONNECTION
NEUTRAL WIRE
HOT WIRE
HOT WIRE
120 V
120 V
GENERAL-PURPOSE RECEPTACLES AND GENERAL LIGHTING
240 V
HEATING, COOLING, MOTORS, ETC.
TO OVERHEAD OR LATERAL SERVICE ENTRANCE
METER
TRANSFORMER
FUSED DISCONNECT AND DISTRIBUTION PANEL
TO LOADS
TO UTILITY SERVICE
SERVICE-ENTRANCE CONDUCTORS
LATERAL SERVICE
Figure 15-17

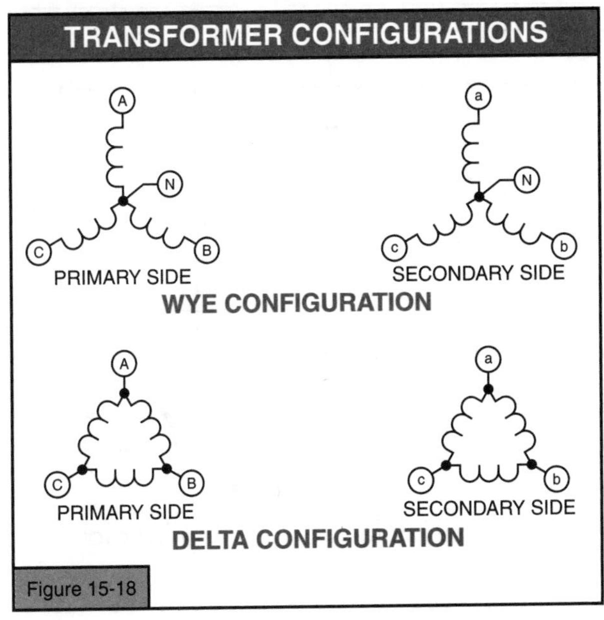

TRANSFORMER CONFIGURATIONS
A
N
C
B
PRIMARY SIDE
a
N
c
b
SECONDARY SIDE
WYE CONFIGURATION
A
C
B
PRIMARY SIDE
a
c
b
SECONDARY SIDE
DELTA CONFIGURATION
Figure 15-18

Three-Phase, Wye-to-Wye Connections

Three transformers may be connected in a wye-to-wye connection. A wye-to-wye transformer connection is used to supply both 1ϕ and 3ϕ voltage. In a wye-to-wye transformer connection, the ends of each transformer are connected together. **See Figure 15-20.**

The advantage of a wye-connected secondary is that the 1ϕ power draw may be divided equally over the three transformers. Each transformer carries ⅓ of the 1ϕ and 3ϕ power if the loads are divided equally. A disadvantage of a wye-to-wye connection is that interference with telephone circuits may result.

Delta-to-Wye/Wye-to-Delta Connections

Transformers may also be connected in a delta-to-wye or wye-to-delta connection. The connection used depends on the incoming supply voltage, the requirements of the loads, and the practice of the local power company. A delta-to-wye transformer connection delivers the same voltage output as the wye-to-wye transformer connection. The difference is the primary is supplied from a delta system. A wye-to-delta transformer connection delivers the same voltage output as the delta-to-delta transformer connection. The difference is that the primary is supplied from a wye system.

Substations

Substations serve as a source of voltage transformation and control along the distribution system. Their function includes:

- Receiving voltage generated and increasing it to a level appropriate for transmission
- Receiving the transmitted voltage and reducing it to a level appropriate for customer use
- Providing a safe point in the distribution system for disconnecting the power in the event of problems
- Providing a place to adjust and regulate the outgoing voltage
- Providing a convenient place to take measurements and check the operation of the distribution system
- Providing a switching point where different connections may be made between various transmission lines

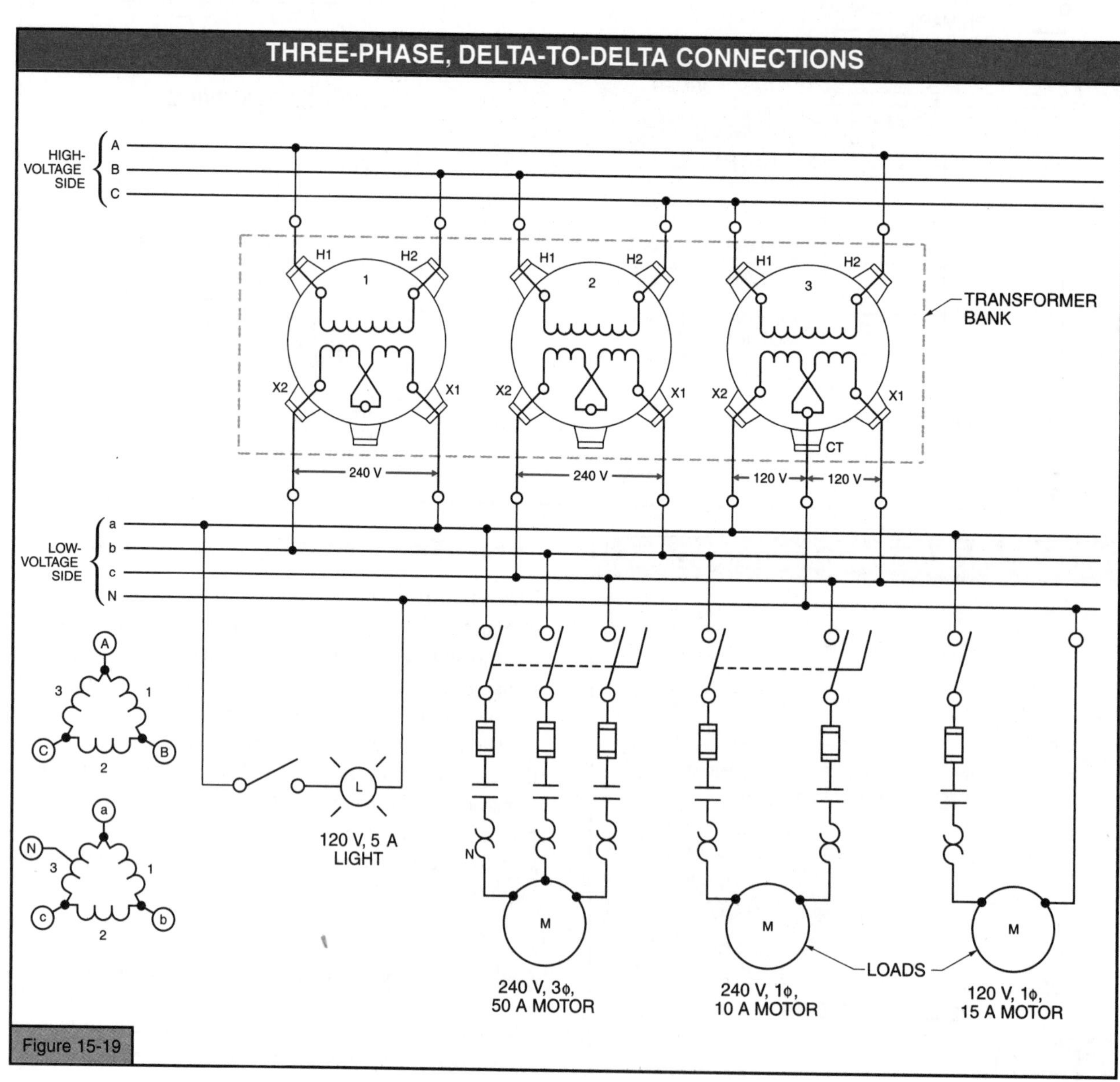

Figure 15-19

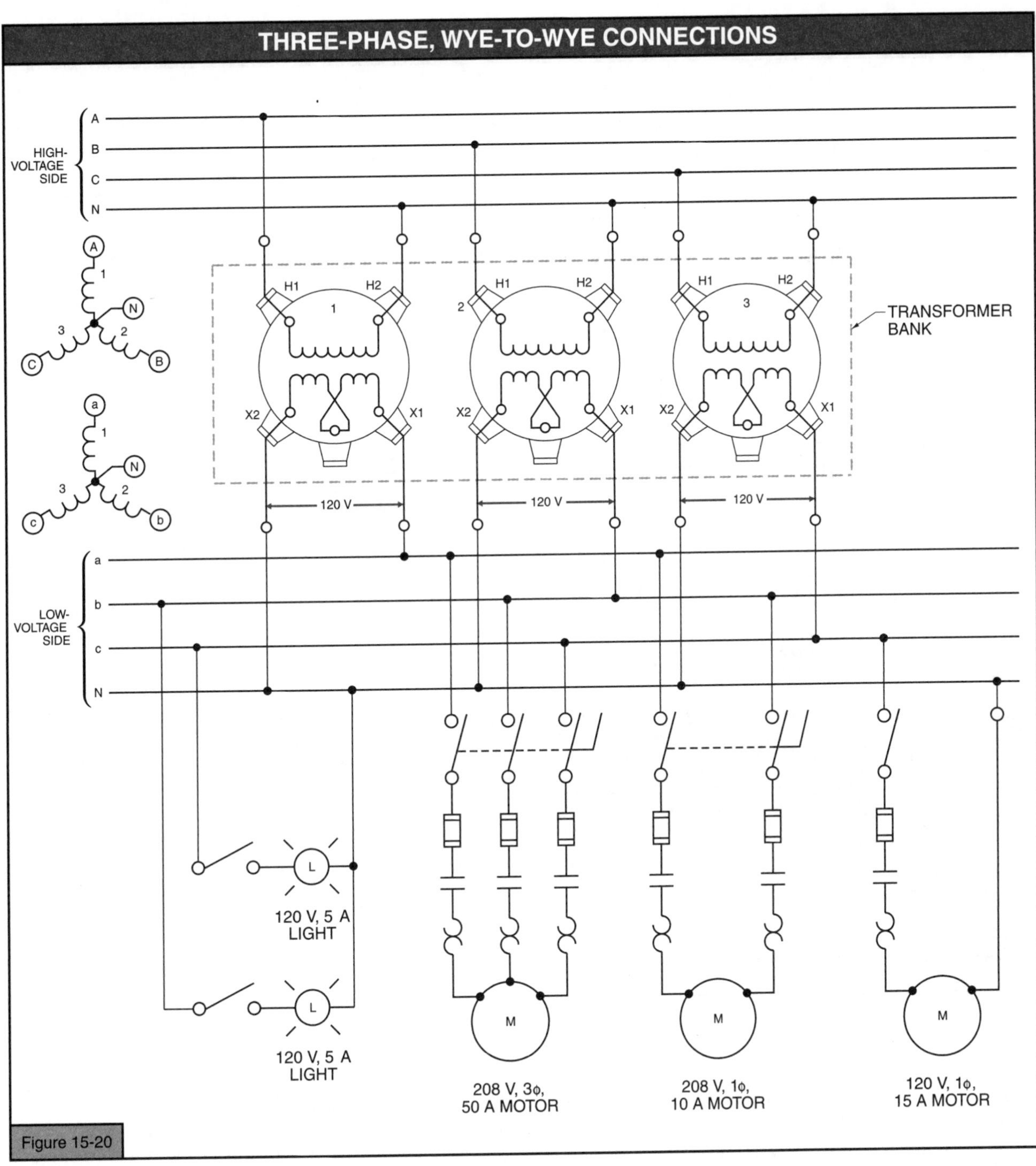

Figure 15-20

Substations have three main sections: primary switchgear, transformer, and secondary switchgear sections. **See Figure 15-21.** Depending on the function of the substation (step-up or step-down voltage), the primary or secondary switchgear section may be the high-voltage or low-voltage section. In a step-up substation, the primary switchgear section is the low-voltage section and the secondary switchgear section is the high-voltage section. In a step-down substation, the primary switchgear section is the high-voltage section and the secondary switchgear section is the low-voltage section. The substation sections normally include breakers, junction boxes, and interrupter switches.

Substations may be entirely enclosed in a building or totally in the open, as in the case of outdoor substations located along a distribution system. The location for a substation is generally selected so that the station is as near as possible to the area to be served.

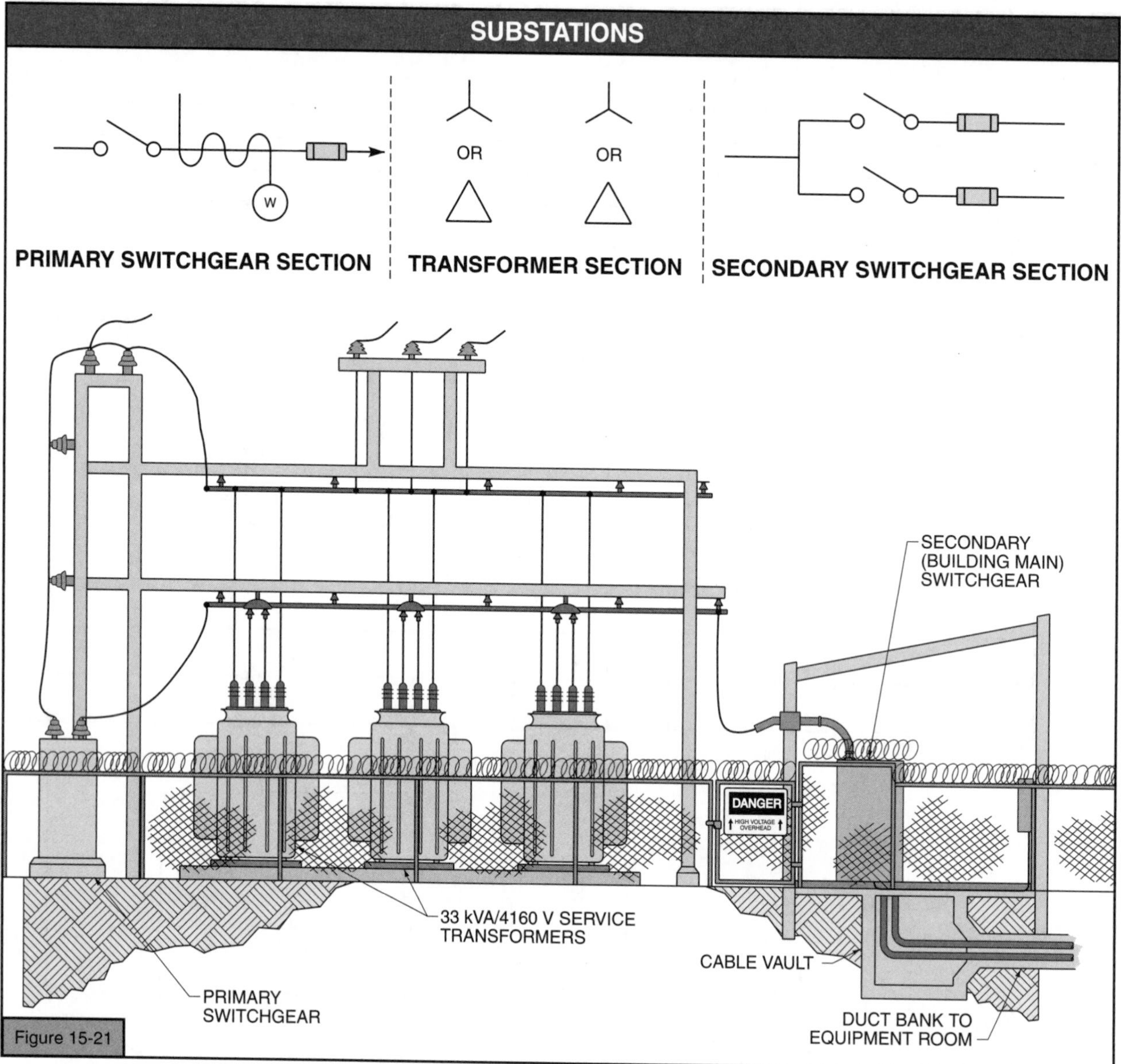

Figure 15-21

Substations can be built to order or purchased from factory-built, metal-enclosed units. The purchased units are unit substations. A unit substation offers standardization and flexibility for future changes when quick replacements are needed.

Transformer Load Balancing

The loads connected to a transformer should be connected so that the transformer is as electrically balanced as possible. Electrical balance occurs when loads on a transformer are placed so that each coil of the transformer carries the same amount of current. **See Figure 15-22.**

Transformer Tap Connections

Transformer taps are connecting points that are provided along the transformer coil. Taps are available on some transformers to correct for excessively high or low voltage conditions. The taps are located on the primary side of the transformer. Standard taps are provided for 2% and 5% of rated primary voltage. **See Figure 15-23.**

For example, if a transformer has a 480 V primary rating and the available supply voltage is 504 V, the primary should be connected to the 5% above-normal tap. This ensures that the secondary voltage is correct even when the primary voltage is high.

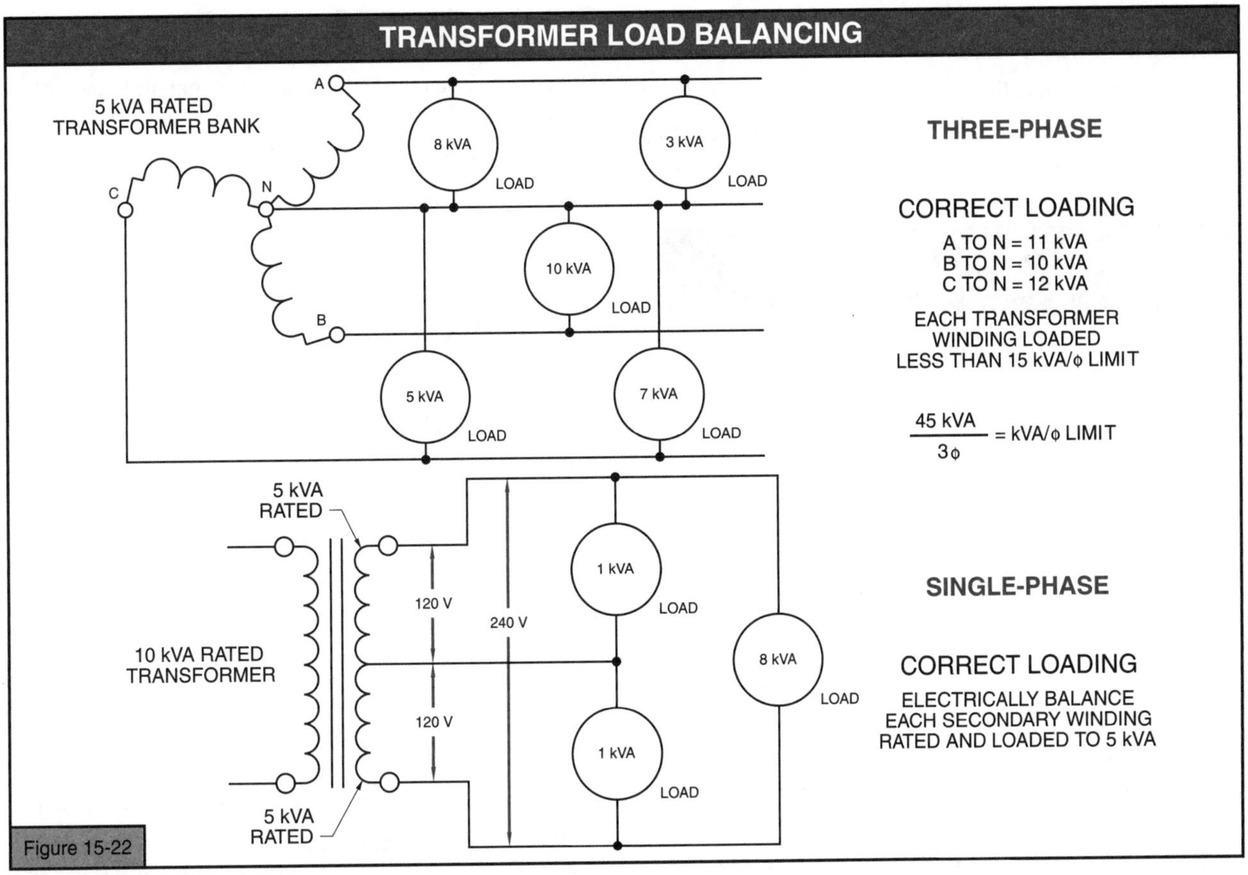

Figure 15-22

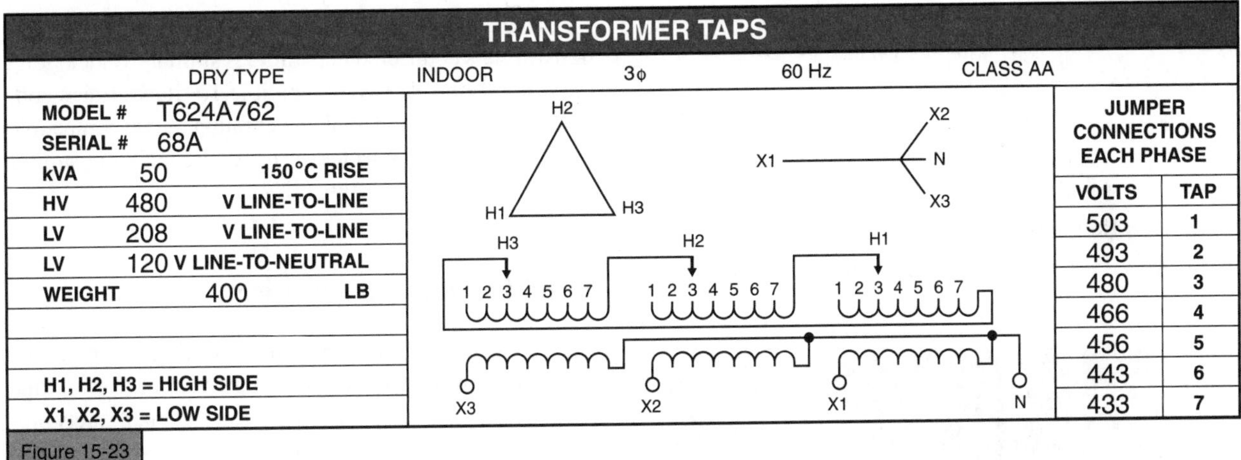

TRANSFORMER TAPS

DRY TYPE | INDOOR | 3ϕ | 60 Hz | CLASS AA

MODEL #	T624A762	
SERIAL #	68A	
kVA	50	150°C RISE
HV	480	V LINE-TO-LINE
LV	208	V LINE-TO-LINE
LV	120	V LINE-TO-NEUTRAL
WEIGHT	400	LB

H1, H2, H3 = HIGH SIDE
X1, X2, X3 = LOW SIDE

JUMPER CONNECTIONS EACH PHASE

VOLTS	TAP
503	1
493	2
480	3
466	4
456	5
443	6
433	7

Figure 15-23

Single-Phase Transformer Parallel Connections

Additional power is required when the capacity of a transformer is insufficient for the power requirements of the load(s). Additional power may be obtained by changing the overloaded transformer to a larger size (higher kVA rating) or adding a second transformer in parallel with the overloaded transformer. The best and most efficient method is to replace the overloaded transformer with a larger one. However, in some applications, it is easier to add a second transformer in parallel. These include systems where extra power is needed only temporarily or a larger transformer is not available.

Single-phase transformers may be connected in parallel as long as certain conditions are met. These conditions include:

- Primary and secondary voltage ratings are identical.
- Frequencies are the same.
- Tap settings are identical.
- Impedance of either transformer is within ±7% (93% to 107%) of the other. **See Figure 15-24.**

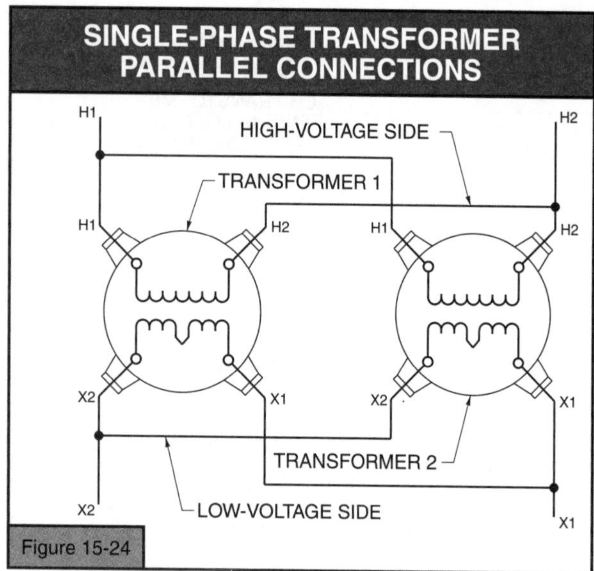

Figure 15-24

The total power rating of two compatible, 1ϕ transformers connected in parallel is equal to the sum of the individual power ratings. To calculate the total power rating of two 1ϕ transformers connected in parallel, apply the formula:

$$kVA_T = kVA_1 + kVA_2$$

where

kVA_T = Total rating of transformer combination (in kVA)

kVA_1 = Rating of transformer 1 (in kVA)

kVA_2 = Rating of transformer 2 (in kVA)

Transformers are often placed on concrete pads, with barriers in place to protect the transformer from damage.

Example: Calculating Total Power Rating—Parallel-Connected Transformers

What is the total output rating of two compatible 1ϕ, 5 kVA transformers connected in parallel?

$$kVA_T = kVA_1 + kVA_2$$

$$kVA_T = 5 + 5$$

$$kVA_T = \mathbf{10\ kVA}$$

Three-Phase Transformer Parallel Connections

Like 1ϕ transformers, 3ϕ transformers may also be connected in parallel. The conditions that must be met to connect 3ϕ transformers in parallel include:

- Primary and secondary voltage ratings are identical.
- Frequencies are the same.
- Tap settings are identical.
- Impedance of either transformer is within ±7% (93% to 107%) of the other.
- Angular displacement of transformer banks is the same. For example, both banks must have a 0°, 30°, or 180° angular displacement. Standard angular displacements are 0° for wye-to-wye or delta-to-delta connected banks and 30° for wye-to-delta or delta-to-wye connected banks. **See Figure 15-25.**

Calculating the total power rating of two compatible, 3ϕ transformers connected in parallel is similar to calculating the power rating of two compatible 1ϕ transformers connected in parallel. The total power rating of two compatible 3ϕ transformers connected in parallel equals the sum of the individual power ratings (kVA).

Transformer Installation

Extreme care must be taken when working around transformers because of the high voltage present. Ensure that proper protective equipment is used and all plant safety procedures are followed. All transformer installations should follow National Electrical Code® (NEC®) and National Fire Protection Association (NFPA) requirements. For example, NEC® Section 450.21 covers the installation of indoor dry-type transformers. **See Figure 15-26.** National Fire Protection Association (NFPA) Section 420.10 covers the installation of all transformers. Following these requirements protects persons working around the transformer, helps prevent fires, and reduces the chance of an electrical shock. Always follow NEC® and NFPA requirements when installing transformers.

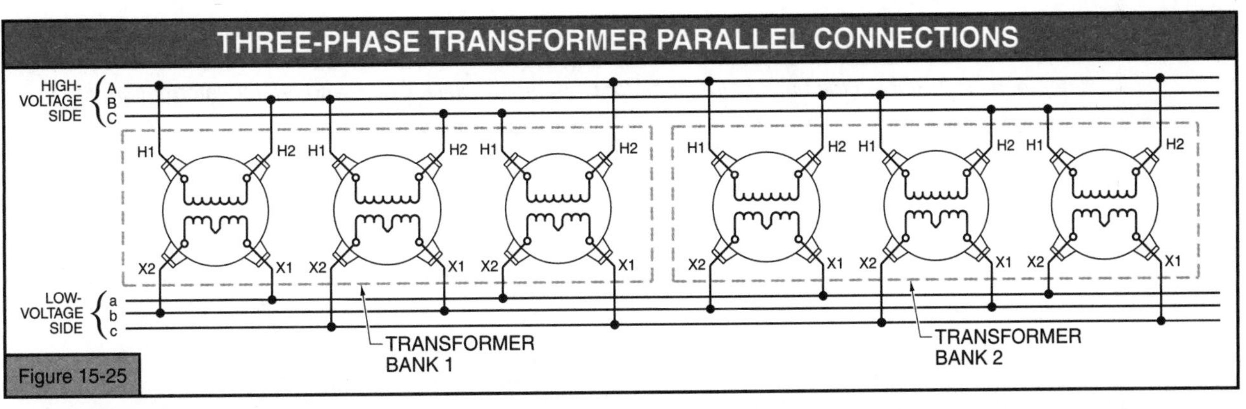

Figure 15-25

DRY-TYPE TRANSFORMERS—INSTALLED INDOORS

CORE
CASE
COILS
TERMINAL ARRANGEMENT

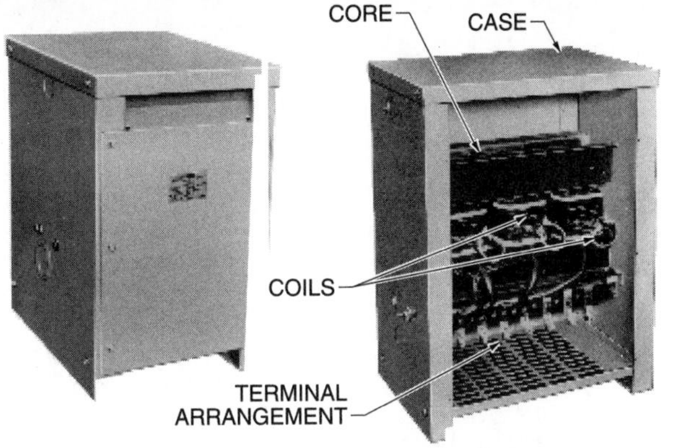

ABB Power T&D Company Inc.

Dry-type transformers may be installed indoors when they are rated at not more than 112½ kVA and not more than 600 V, and the transformer is completely enclosed except for ventilating openings.

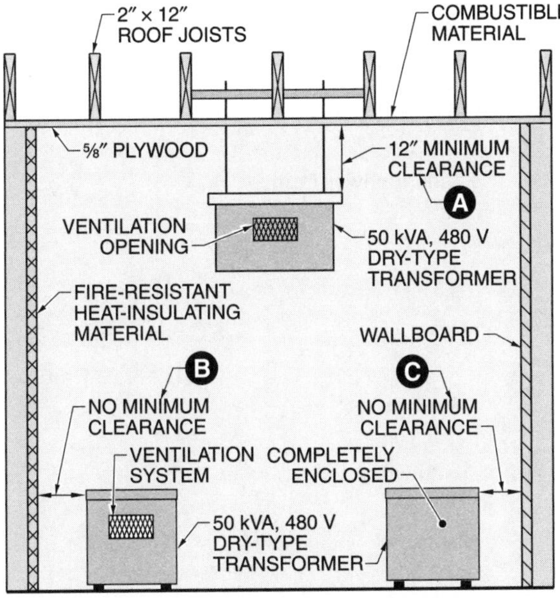

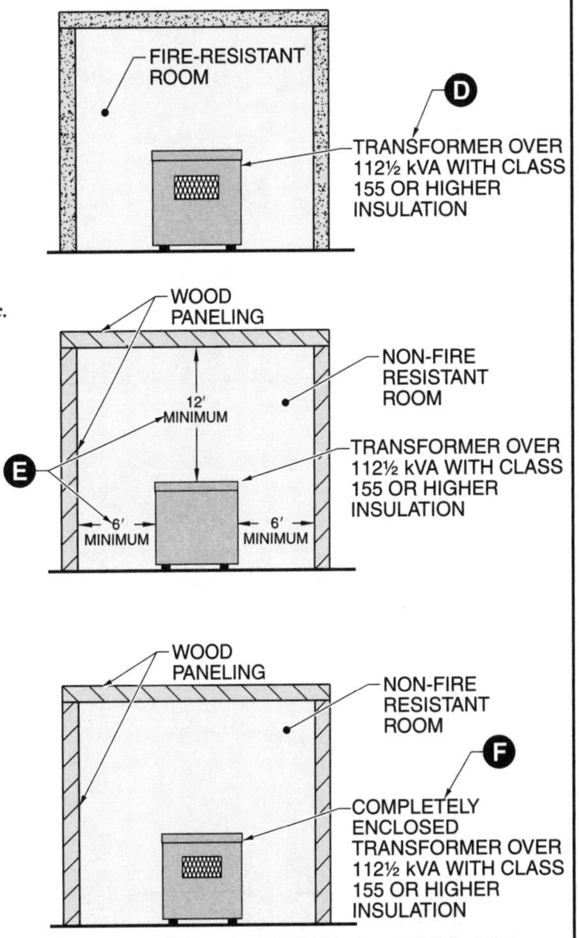

FIRE-RESISTANT ROOM
D
TRANSFORMER OVER 112½ kVA WITH CLASS 155 OR HIGHER INSULATION

WOOD PANELING
NON-FIRE RESISTANT ROOM
12′ MINIMUM
E
6′ MINIMUM
6′ MINIMUM
TRANSFORMER OVER 112½ kVA WITH CLASS 155 OR HIGHER INSULATION

2″ × 12″ ROOF JOISTS
COMBUSTIBLE MATERIAL
⅝″ PLYWOOD
12″ MINIMUM CLEARANCE
A
VENTILATION OPENING
50 kVA, 480 V DRY-TYPE TRANSFORMER
FIRE-RESISTANT HEAT-INSULATING MATERIAL
WALLBOARD
B
C
NO MINIMUM CLEARANCE
NO MINIMUM CLEARANCE
VENTILATION SYSTEM
COMPLETELY ENCLOSED
50 kVA, 480 V DRY-TYPE TRANSFORMER

WOOD PANELING
NON-FIRE RESISTANT ROOM
F
COMPLETELY ENCLOSED TRANSFORMER OVER 112½ kVA WITH CLASS 155 OR HIGHER INSULATION

A. 450.21 (A) Dry-type transformers installed indoors and rated at 112½ kVA or less shall be a minimum of 12″ from combustible material.

B. 450.21 (A) The 12″ minimum is not required when transformers are separated from combustible material by fire-resistant, heat-insulating material.

C. 450.21 (A) The 12″ minimum is not required when the transformer is 600 V or less and completely enclosed.

D. 450.21 (B) Dry-type transformers over 112½ kVA rating shall be installed in fire-resistant rooms.

E. 450.21 (B) Exception 1. Transformers with Class 155 or higher insulation are permitted to be installed in a non-fire-resistant room provided they are separated from combustible material by a minimum of 6′ horizontally and 12′ vertically.

F. 450.21 (B) Exception 2. Transformers with Class 155 or higher insulation and completely enclosed except for ventilation openings are permitted to be enclosed in a non-fire-resistant room.

Figure 15-26

Refer to the CD-ROM in the back of the book for Chapter 15 Quick Quiz® and related information.

Electric Motors

An *electric motor* is a rotating output device that converts electrical power into a rotating, mechanical force. Electric motors include AC motors, DC motors, and AC/DC motors. Different motors are used for different applications. The motor used for a given application depends on the driven load, environmental conditions, and available power.

Many factors must be considered when selecting, wiring, and maintaining electric motors. Satisfying the electrical needs of a motor requires installing the motor according to the manufacturer's wiring diagram. Additionally, motors shall be installed per Article 430 of the NEC®.

Satisfying the mechanical needs of a motor requires that the motor selected meets the intended specifications of enclosures, bearings, frame size, and insulation. These specifications are normally established by the original equipment manufacturer but must be understood when selecting any replacement part(s). The electrician must read the manufacturer's motor nameplate information to select the correct replacement motor.

Motors are used to perform work. The amount of work a motor performs depends on the type, speed, and size of the load to be driven. The driven load type is based on the application. Motors are used to drive fans and compressors, move devices and materials, rotate parts and tools, and mix products.

A motor is normally available that best fits the load requirements for any given application. Work is economically performed for a sustained period of time when the correct motor is used. However, work may be performed, but the motor may not operate economically or last as long as it was designed to, if the wrong motor is used for an application.

Work

Work (W) is applying a force over a distance. **See Figure 16-1.** *Force (F)* is any cause that changes the position, motion, direction, or shape of an object. Work is done when a force overcomes a resistance. *Resistance (R)* is any force that tends to hinder the movement of an object. No work is produced if an applied force does not cause motion.

The amount of work produced is determined by multiplying the force that must be overcome by the distance through which it acts. Thus, work is measured in pound-feet (lb-ft). To calculate the amount of work produced, apply the formula:

$W = F \times D$

where

W = work (in lb-ft)

F = force (in lb)

D = distance (in ft)

Example: Calculating Work

How much work is required to carry a 15 lb bag of groceries vertically from street level to the third floor of a building 25′ above street level?

$W = F \times D$

$W = 15 \times 25$

$W = \mathbf{375\ lb\text{-}ft}$

Resistance must be overcome to perform work. More work is required if the groceries are heavier, the distance longer, or a combination of the two. Less work is required if the groceries are lighter, the distance shorter, or a combination of the two. For example, the amount of work required to carry a 20 lb bag of groceries vertically from street level to the fourth floor of a building 36′ above street level is 720 lb-ft (20 × 36 = 720).

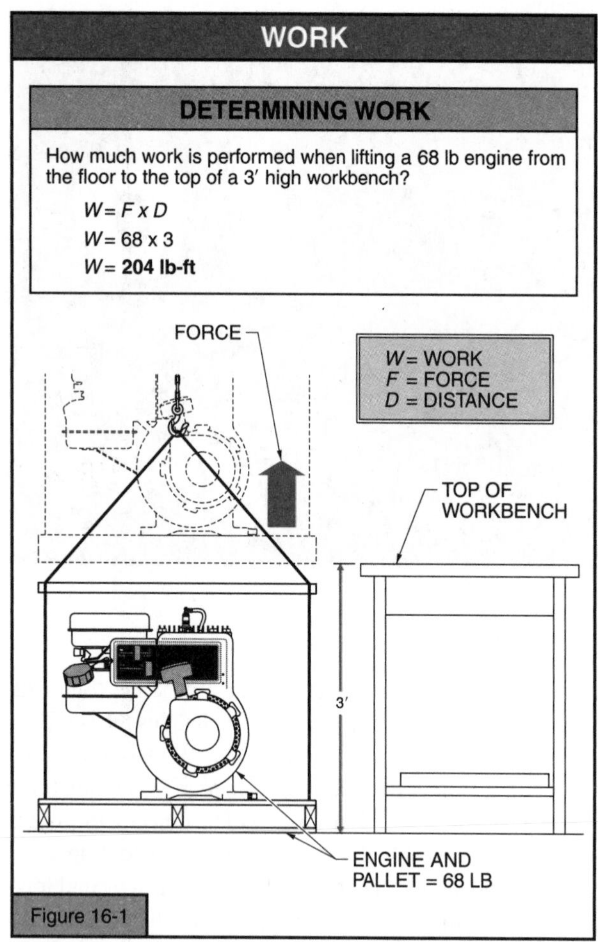

Figure 16-1

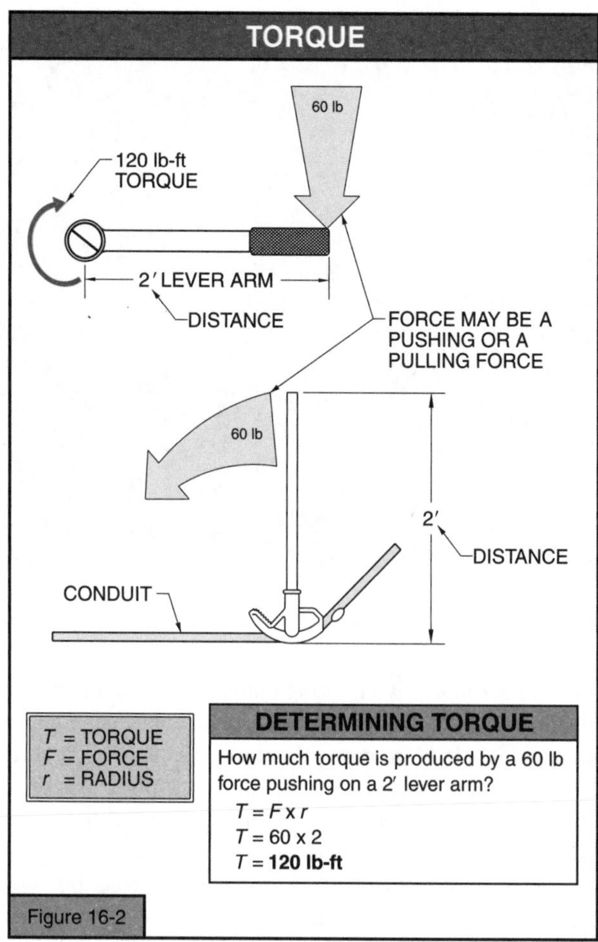

Figure 16-2

Torque

Torque is the force that causes an object to rotate. Torque consists of a force acting on a distance. **See Figure 16-2.** Torque is measured in pound-feet (lb-ft). Torque may exist even though no movement occurs. For example, an engine head bolt may have 50 lb-ft of torque and yet no movement occurs. To calculate torque, apply the formula:

$T = F \times r$

where

T = torque (in lb-ft)

F = force (in lb)

r = radius (in ft)

Example: Calculating Torque

What is the torque produced by a 50 lb force pushing on a 4′ lever arm?

$T = F \times r$

$T = 50 \times 4$

$T = \mathbf{200\ lb\text{-}ft}$

Work is done if the amount of torque produced is large enough to cause movement. More torque is produced if the force is larger, the lever arm longer, or a combination of the two. Less torque is produced if the force is smaller, the lever arm shorter, or a combination of the two. For example, the torque produced by a 70 lb force pushing on a 4′ lever arm is 280 lb-ft (70 × 4 = 280 lb-ft).

Motor Torque

Motor torque is the force that produces or tends to produce rotation in a motor. A motor must produce enough torque to start and keep the load moving for the motor to operate the load connected to it. The torque required to operate a load from initial startup to final shutdown is considered when determining the type and size motor required for an application. A motor connected to a load produces locked rotor torque (LRT), pull-up torque (PUT), full-load torque (FLT), and breakdown torque (BDT). **See Figure 16-3.**

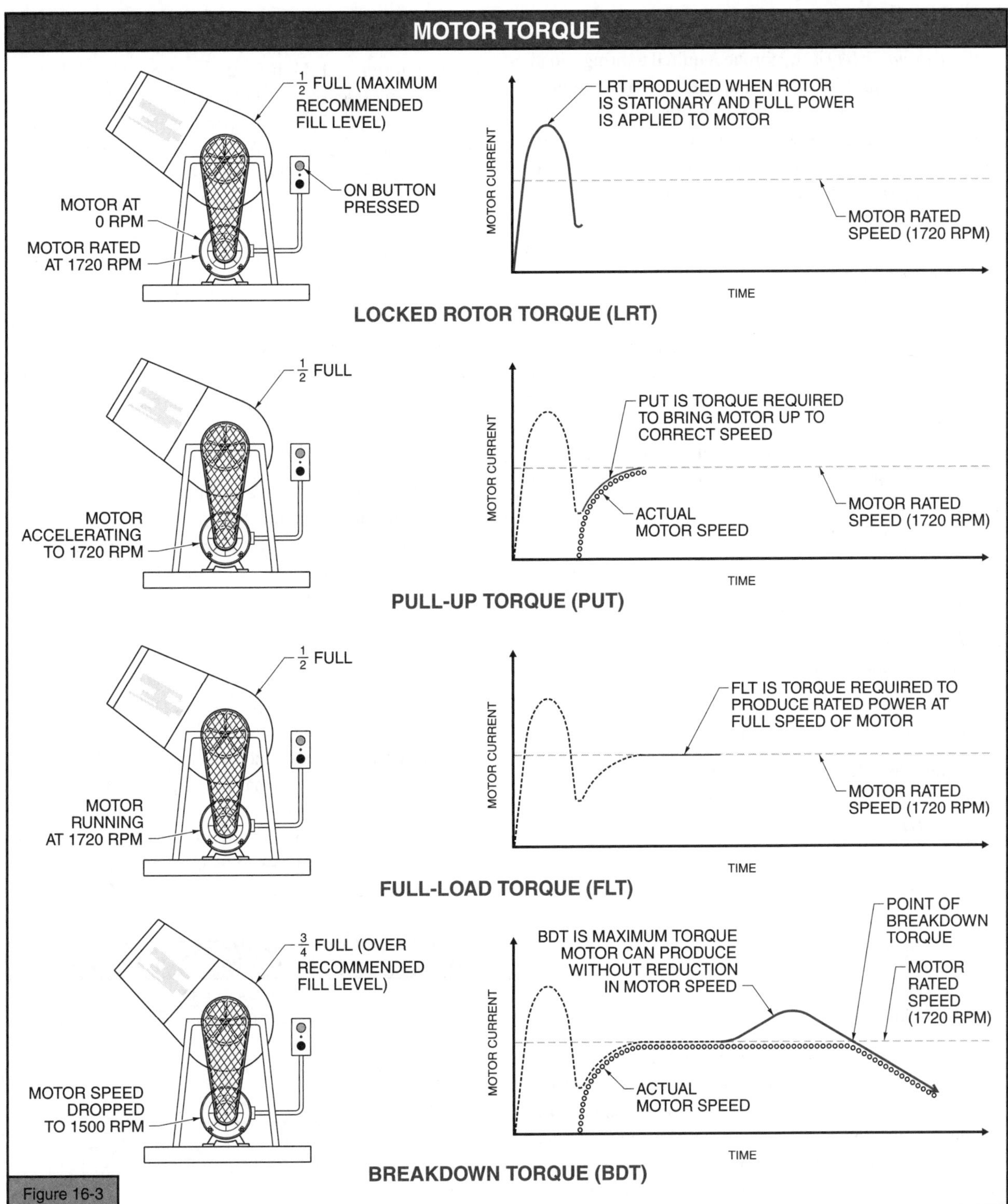

Figure 16-3

Locked Rotor Torque

Many loads require a higher torque to start them moving than to keep them moving. For this reason, a motor must produce a higher torque when starting the load. *Locked rotor torque (LRT)* is the torque a motor produces when the rotor is stationary and full power is applied to the motor. Locked rotor torque is also referred to as breakaway or starting torque. *Starting torque* is the torque required to start a motor. Starting torque is normally expressed as a percentage of full-load torque.

Pull-Up Torque

Pull-up torque (PUT) is the torque required to bring a load up to its rated speed. Pull-up torque is brief if a motor is properly sized for the load. In a motor that does not have sufficient pull-up torque, the locked rotor torque may start the load turning but the pull-up torque does not bring it up to rated speed. Pull-up torque is also referred to as accelerating torque.

Full-Load Torque

Full-load torque (FLT) is the torque required to produce the rated power at full speed of the motor. The amount of torque a motor produces at rated power and full speed (full-load torque) is found by using a conversion chart. **See Appendix.**

A straightedge is placed along the two known quantities and the unknown quantity is read on the third line when using a horsepower to torque conversion chart. To calculate motor FLT, apply the formula:

$$T = \frac{HP \times 5252}{rpm}$$

where

T = torque (in lb-ft)

HP = horsepower

5252 = constant ($33{,}000 \div 2\pi = 5252$)

rpm = revolutions per minute

Example: Calculating Motor FLT

What is the FLT of a 15 HP motor operating at 1725 rpm?

$$T = \frac{HP \times 5252}{rpm}$$

$$T = \frac{15 \times 5252}{1725}$$

$$T = \frac{78{,}780}{1725}$$

$$T = \mathbf{45.67\ lb\text{-}ft}$$

A motor that is fully loaded produces full-load torque. A motor that is underloaded produces less than full-load torque. A motor that is overloaded must produce more than full-load torque to keep the load operating at the motor's rated speed. For example, a 15 HP motor operating at 1725 rpm develops 45.67 lb-ft of torque at full speed. The 15 HP motor produces an output of 15 HP if the load requires 45.67 lb-ft of torque at 1725 rpm. However, the 15 HP motor produces an output of 7.5 HP if the load to which the motor is connected requires only half as much torque (22.83 lb-ft) at 1725 rpm. The 15 HP motor draws less current (and power) from the power lines and operates at a lower temperature when producing 7.5 HP.

However, the motor must produce an output of 30 HP if the 15 HP motor is connected to a load that requires twice as much torque (91.34 lb-ft) at 1725 rpm. The 15 HP motor draws more current (and power) from the power lines and operates at a higher temperature. The 15 HP motor automatically disconnects from the power lines before any permanent damage is done to the motor if the overload protection device is sized correctly.

Breakdown Torque

Breakdown torque (BDT) is the maximum torque a motor can produce without an abrupt reduction in motor speed. As the load on a motor shaft increases, the motor produces more torque. As the load continues to increase, the point at which the motor stalls is reached. This point is the breakdown torque.

Motor Torque Classification

All motors produce torque but not all motors produce the same torque characteristics. Motors are classified by NEMA according to their electrical characteristics. Motor torque characteristics vary with the classification of the motor. Motors are classified as Class A through Class F. Classes B, C, and D are most common.

Class B motors are the most widely used motors. Class B motors increase the starting torque of the motor 150% higher than the motor's full-load torque. As a Class B motor accelerates, pull-up torque can reach 240% higher than the motor's full-load torque. A Class C motor increases the starting torque of the motor by 225% and may have a pull-up torque as high as 200%. A Class D motor increases the starting torque of the motor 275%. Although a Class D motor increases the starting torque the most, a Class D motor's pull-up torque never increases above the starting torque value. **See Figure 16-4.**

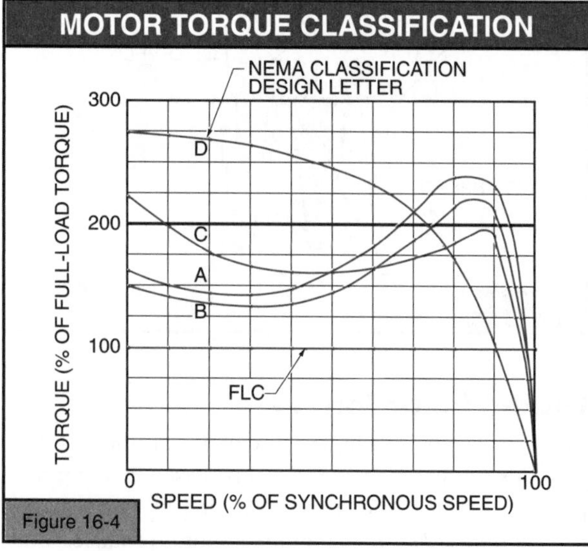

Figure 16-4

Horsepower

Electrical power is rated in horsepower or watts. *Horsepower (HP)* is a unit of power equal to 746 W or 33,000 lb-ft per minute (550 lb-ft per second). A *watt (W)* is a unit of measure equal to the power produced by a current of 1 A across a potential difference of 1 V. A watt is 1/746 of 1 HP and is the base unit of electrical power. Motor power is rated in horsepower or watts. **See Figure 16-5.** Motor horsepower may be calculated when voltage, current, and efficiency are known or when speed and torque are known.

Horsepower is used to measure the energy produced by an electric motor while doing work. To calculate the horsepower of a motor when voltage, current, and efficiency are known, apply the formula:

$$HP = \frac{I \times E \times Eff}{746}$$

where

HP = horsepower

I = current (in A)

E = voltage (in V)

Eff = efficiency

746 = constant to convert watts to horsepower

Example: Calculating Horsepower—Voltage, Current, and Efficiency Known

What is the horsepower of a 115 V motor pulling 8 A and having 82% efficiency?

$$HP = \frac{I \times E \times Eff}{746}$$

$$HP = \frac{8 \times 115 \times .82}{746}$$

$$HP = \frac{754.4}{746}$$

$$HP = \mathbf{1.01\ HP}$$

The horsepower of a motor determines the size of the load the motor can operate and the speed at which the load turns. To calculate the horsepower of a motor when speed and torque are known, apply the formula:

$$HP = \frac{rpm \times T}{5252}$$

where

HP = horsepower

rpm = revolutions per minute

T = torque (in lb-ft)

5252 = constant (33,000 ÷ 2π = 5252)

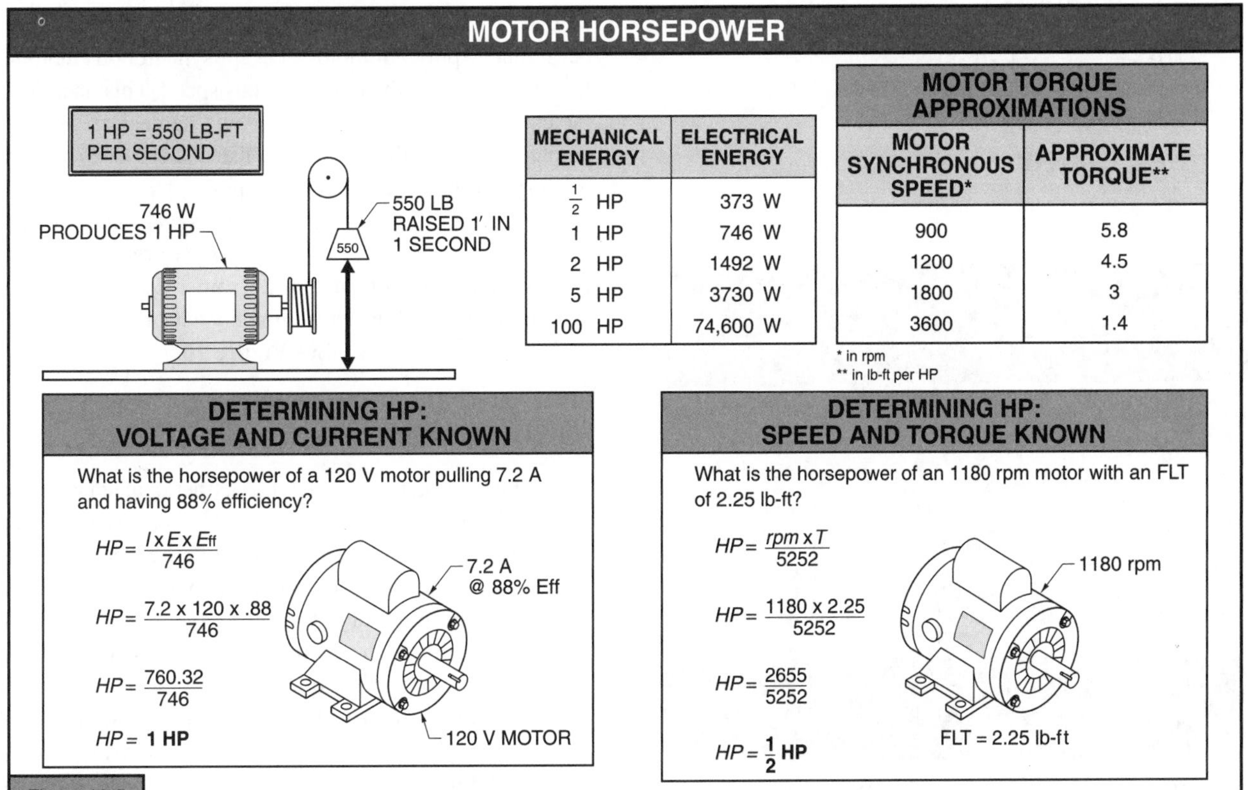

Figure 16-5

Example: Calculating Horsepower—Speed and Torque Known

What is the horsepower of a 1725 rpm motor with an FLT of 3.1 lb-ft?

$$HP = \frac{rpm \times T}{5252}$$

$$HP = \frac{1725 \times 3.1}{5252}$$

$$HP = \frac{5347.5}{5252}$$

$$HP = \mathbf{1.02\ HP}$$

Formulas used to determine torque and horsepower are for theoretical values. An additional 15% to 40% capability may be required to start a given load when applied to a specific application. Loads that are hard to start require a high rating. To increase the rating, the calculated theoretical value is multiplied by 1.15 (15%) to 1.4 (40%).

Example: Calculating Horsepower with Increased Output Capability

What is the horsepower of a 1725 rpm motor with an FLT of 3.1 lb-ft with an added 30% output capability?

$$HP = \frac{rpm \times T}{5252} \times \%$$

$$HP = \frac{1725 \times 3.1}{5252} \times 1.30$$

$$HP = \frac{5347.5}{5252} \times 1.30$$

$$HP = 1.02 \times 1.30$$

$$HP = \mathbf{1.326\ HP}$$

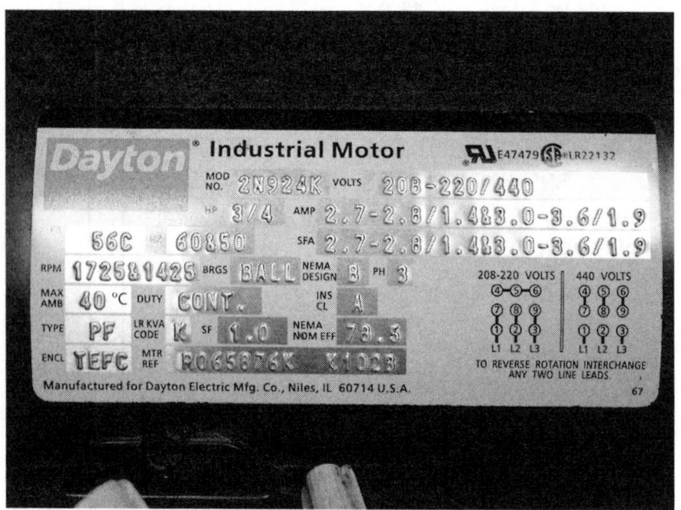

An electric motor typically has a nameplate affixed to the motor housing that contains data used for installation and troubleshooting purposes.

Motor Speed, Torque, and Horsepower Relationship

A motor's operating speed, torque, and horsepower rating determine the work a motor can produce. These factors are interrelated when applied to driving a load. **See Figure 16-6.** Speed and horsepower are proportional if torque remains constant. An increase in horsepower is required to maintain a constant torque if speed increases. A decrease in horsepower is required to maintain a constant torque if speed decreases.

Torque and horsepower are proportional if speed remains constant. An increase in horsepower is required to maintain a constant speed if torque increases. A decrease in horsepower is required to maintain a constant speed if torque decreases.

Horsepower remains constant if torque and speed vary simultaneously but in opposite directions. Horsepower remains constant if torque is increased and speed is reduced. Horsepower also remains constant if torque is decreased and speed is increased.

Motor Speed

Synchronous speed is the theoretical speed of a motor based on the motor's number of poles and the line frequency. *Line frequency* is the number of cycles of supplied voltage per second. The speed of a motor is measured in revolutions per minute (rpm). All motors, except synchronous motors, turn at speeds less than synchronous speed. This operating (actual) speed is listed on the motor's nameplate. For example, a motor with a nameplate-listed speed of 1725 rpm has a synchronous speed of 1800 rpm.

Slip is the difference between the synchronous speed and actual speed of a motor. Slip is measured in percent. For example, the slip of a 1725 rpm motor (1800 rpm synchronous speed) is approximately 4% (1800 − 1725 = 75 ÷ 1800 = .041 = 4%). **See Figure 16-7.**

Single-Phase Motors

Single-phase motors are used when a fractional horsepower motor drive is required or where no 3ϕ power is available. Single-phase large horsepower motors are not normally used because they are inefficient compared to 3ϕ motors and cost more to operate. With the exception of the universal motor, all 1ϕ motors must have an auxiliary means for developing starting torque.

Note: Always refer to the manufacturer's wiring diagram. Although there are many similarities in 1ϕ motors, minor differences among manufacturers have led to hundreds of different wiring diagrams.

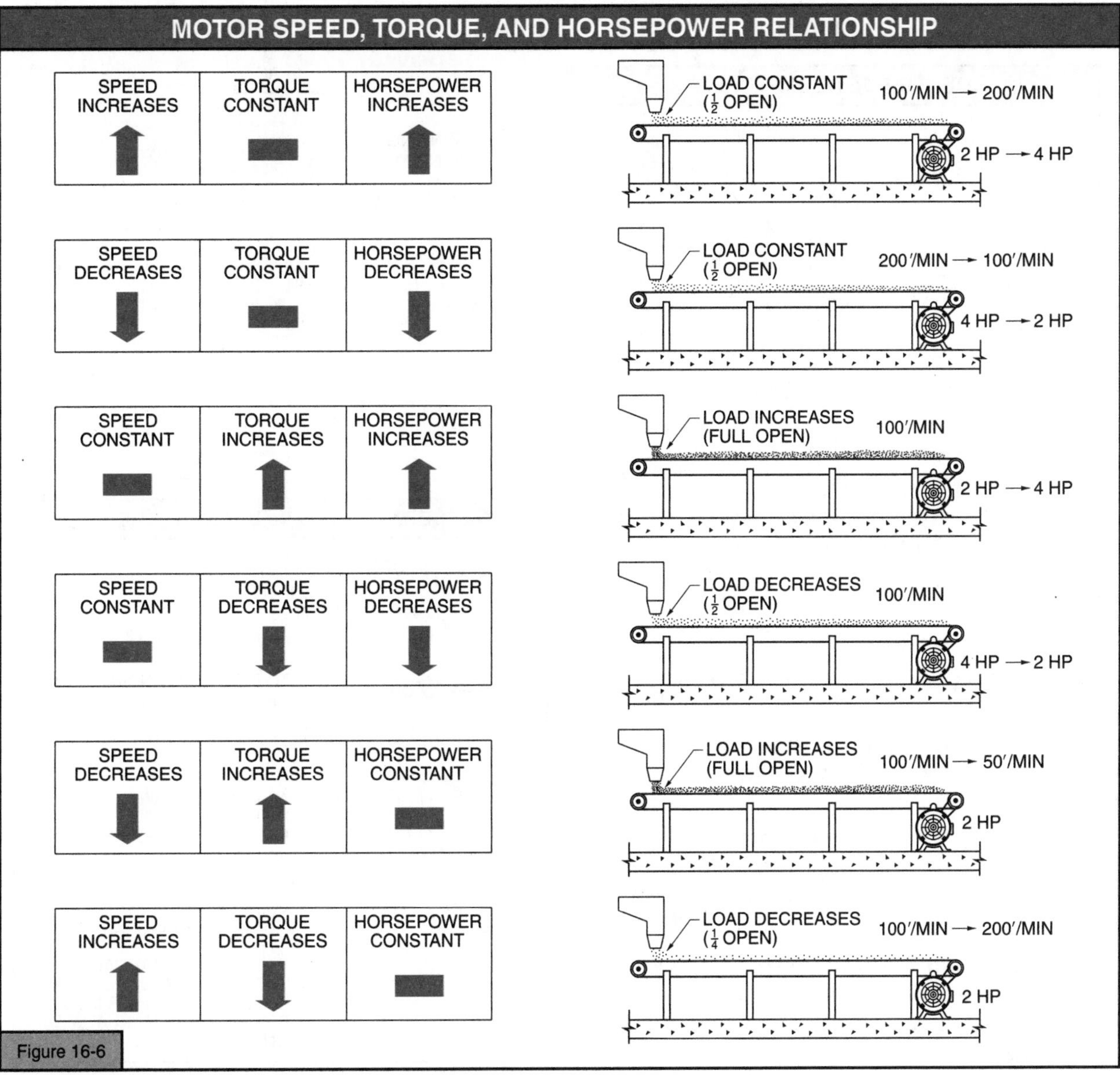

Figure 16-6

MOTOR SPEED*

Poles	Synchronous Speed	With 4% Slip	Actual Speed
2	3600	144	3456
4	1800	72	1728
6	1200	48	1152
8	900	36	864

* in rpm

Figure 16-7

Split-Phase Motors

A *split-phase motor* is a 1ϕ, AC motor that includes a running winding (main winding) and a starting winding (auxiliary winding). A split-phase motor is an AC motor of fractional horsepower, normally 1⁄20 HP to ½ HP.

A split-phase motor has a rotating part (rotor), a stationary part consisting of the running winding and starting winding (stator), a centrifugal switch, and normally an overload switch. **See Figure 16-8.** The running winding and the starting winding are connected in parallel when starting. The running winding normally consists of a heavy, insulated copper wire and the starting winding consists of a fine, insulated copper wire.

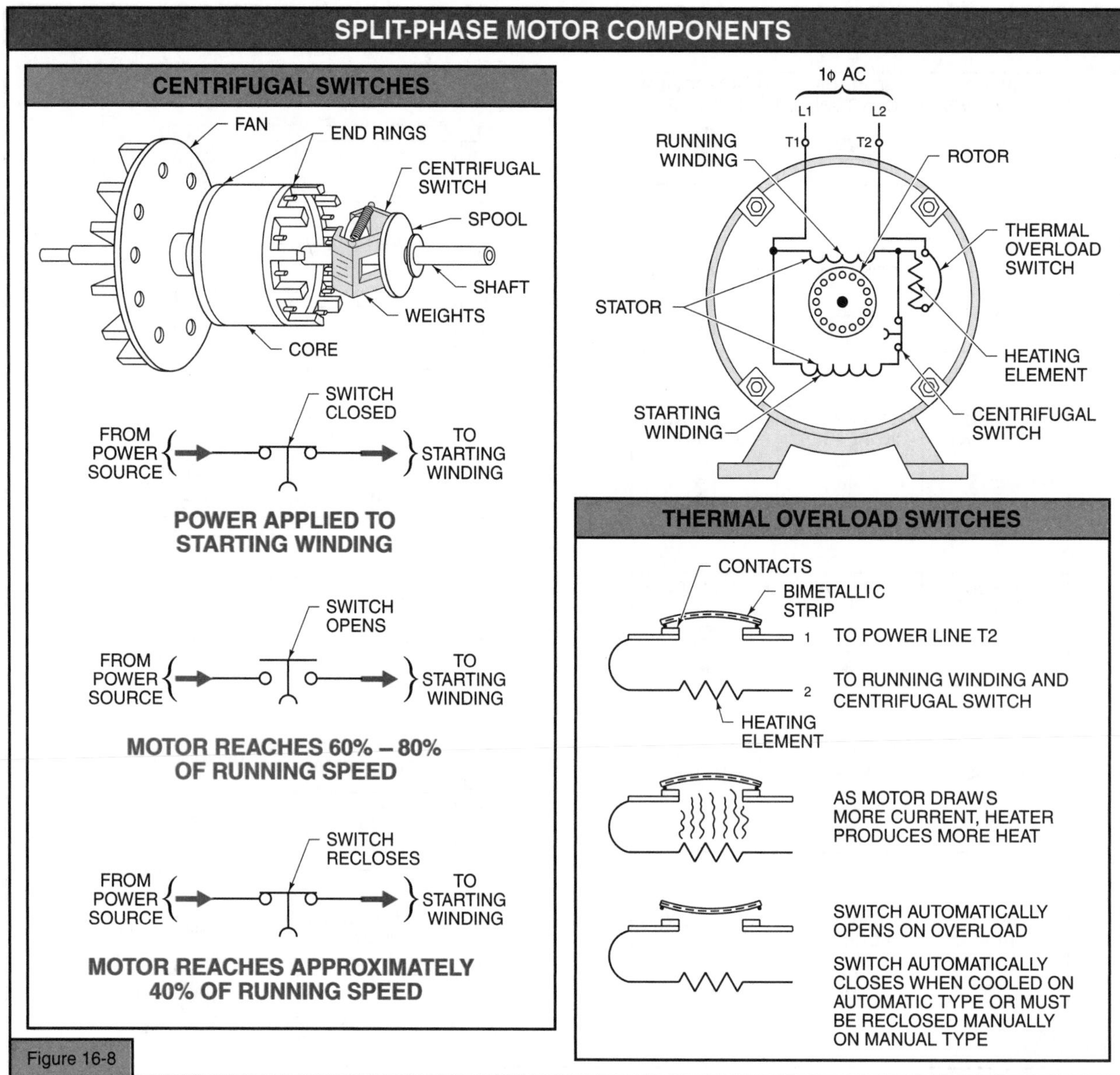

Figure 16-8

The centrifugal switch is located inside the motor, and its purpose is to disconnect the starting winding at about 75% of full-load speed. This allows the motor to operate on the running winding only. The centrifugal switch recloses at approximately 40% of the motor speed when the motor is turned OFF (power removed).

In addition to a centrifugal switch, many motors also contain a thermal overload switch. A *thermal overload switch* is a switch that operates its contacts when a preset temperature is reached. A thermal overload switch protects the motor windings from burnout. A thermal overload switch is activated by high temperatures. It automatically removes the starting and running windings from the power source.

In normal operation, the thermal overload switch is closed. As more current than normal is allowed to pass through the motor windings, the switch begins to heat. At a preset temperature, the switch opens. This automatically removes the motor windings from power. The motor windings begin to cool when removed from power. An automatic thermal overload switch closes as it cools, and the motor is automatically restarted. A manual thermal overload switch must be manually reclosed to restart the motor.

Split-phase motors produce low starting torque (75% to 200% of full-load torque), have a constant speed, and are reversible. Split-phase motors are used in moderate torque applications such as small pumps, blowers, fans, machine tools, and oil burners. **See Figure 16-9.**

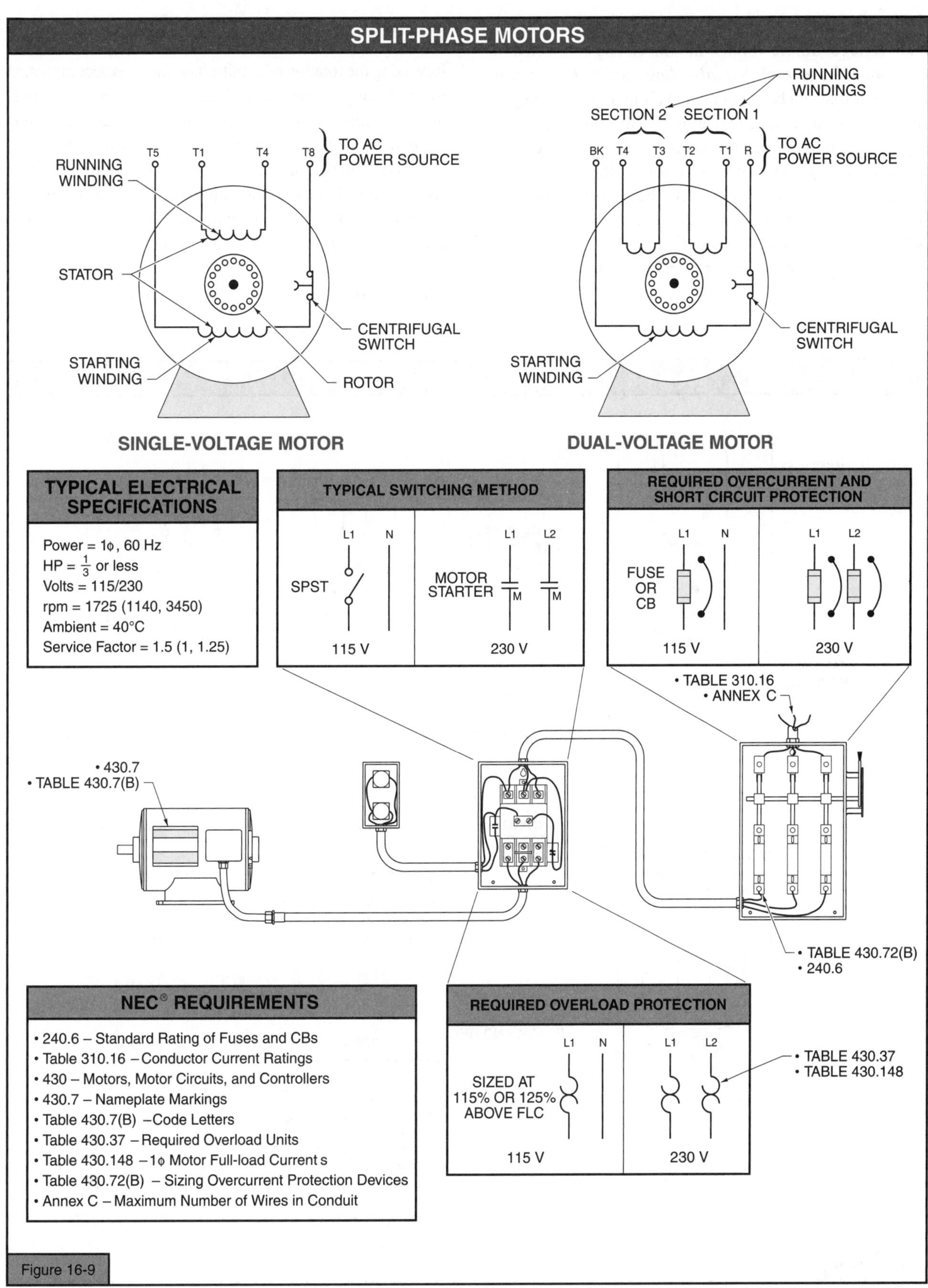
SPLIT-PHASE MOTORS
RUNNING WINDING
T5
T1
T4
T8
TO AC POWER SOURCE
STATOR
CENTRIFUGAL SWITCH
STARTING WINDING
ROTOR
SINGLE-VOLTAGE MOTOR
RUNNING WINDINGS
SECTION 2
SECTION 1
BK
T3
T2
R
DUAL-VOLTAGE MOTOR
TYPICAL ELECTRICAL SPECIFICATIONS
Power = 1ϕ, 60 Hz
HP = 1/3 or less
Volts = 115/230
rpm = 1725 (1140, 3450)
Ambient = 40°C
Service Factor = 1.5 (1, 1.25)
TYPICAL SWITCHING METHOD
L1
N
L2
SPST
MOTOR STARTER
M
115 V
230 V
REQUIRED OVERCURRENT AND SHORT CIRCUIT PROTECTION
FUSE OR CB
• TABLE 310.16
• ANNEX C
• 430.7
• TABLE 430.7(B)
• TABLE 430.72(B)
• 240.6
NEC® REQUIREMENTS
• 240.6 – Standard Rating of Fuses and CBs
• Table 310.16 – Conductor Current Ratings
• 430 – Motors, Motor Circuits, and Controllers
• 430.7 – Nameplate Markings
• Table 430.7(B) – Code Letters
• Table 430.37 – Required Overload Units
• Table 430.148 – 1ϕ Motor Full-load Currents
• Table 430.72(B) – Sizing Overcurrent Protection Devices
• Annex C – Maximum Number of Wires in Conduit
REQUIRED OVERLOAD PROTECTION
SIZED AT 115% OR 125% ABOVE FLC
• TABLE 430.37
• TABLE 430.148

Figure 16-9

Split-phase motors are either single-voltage or dual-voltage. Single-voltage motors can be connected to only one voltage level. A *dual-voltage, split-phase motor* is a motor that can be connected to either of two voltages. In a dual-voltage, split-phase motor, the running winding is split into two sections. All the windings receive the same voltage when wired for low-voltage operation. Wiring for high-voltage connects the running windings in series. The starting winding is connected in parallel with only one running winding to provide proper voltage distribution. Always refer to nameplate or manufacturer's wiring diagrams.

Reversing Split-Phase Motors

Reversing the rotation of a split-phase motor is accomplished by interchanging the leads of the starting or running windings. The manufacturer's wiring diagram is used to determine the exact wires to interchange. **See Figure 16-10.** An electrician can measure the resistance of the starting winding and running winding to determine which leads are connected to which windings if the manufacturer's information is not available. The running winding is made of a heavier gauge wire than the starting winding, so the running winding shows a much lower resistance than the starting winding.

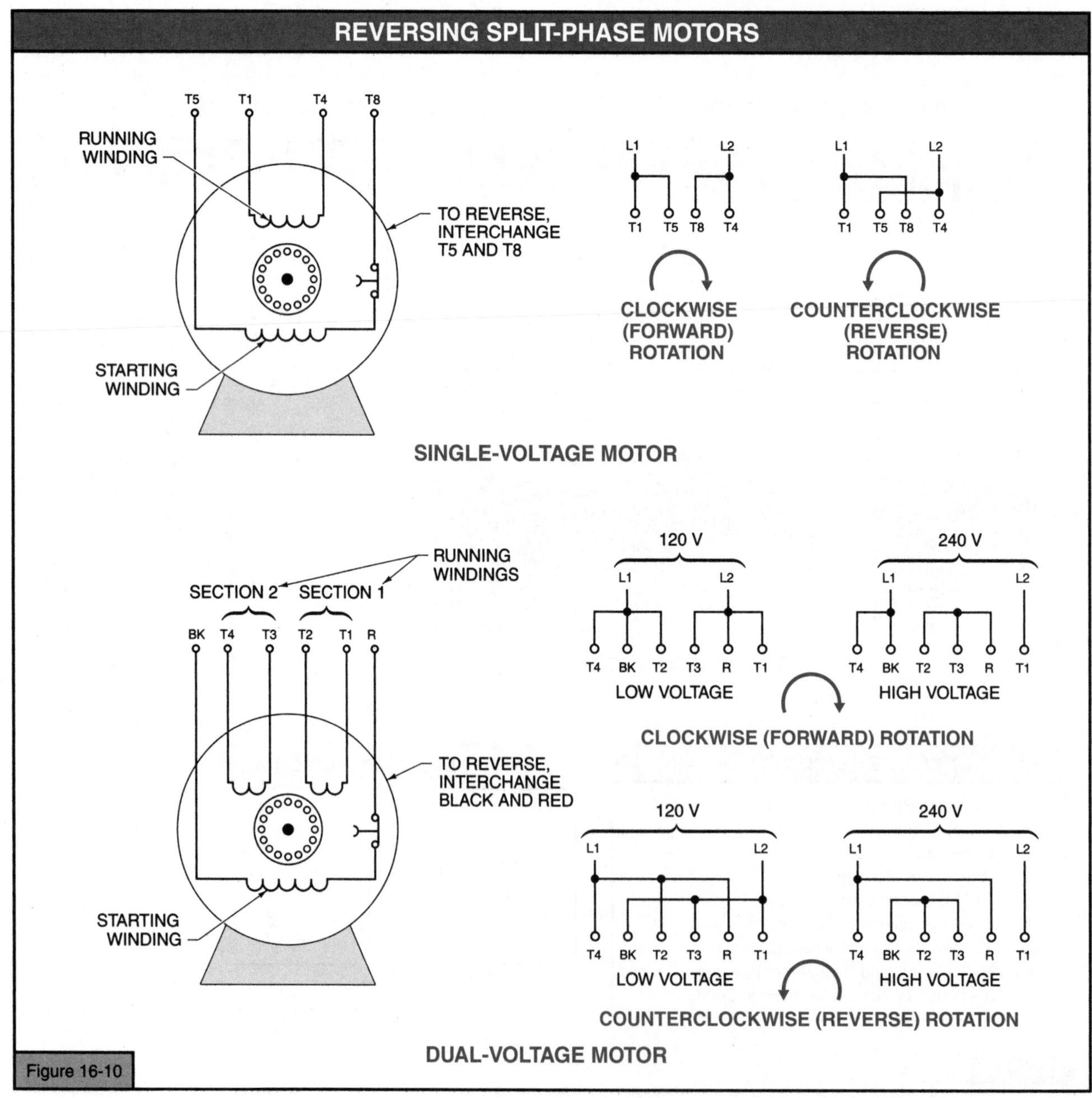

Figure 16-10

Capacitor Motors

A *capacitor motor* is a 1ϕ, AC motor that includes a capacitor in addition to the running and starting windings. A *capacitor* is an electric device designed to store electrical energy by means of an electrostatic field. Capacitor motors are manufactured in sizes ranging from ⅛ HP to 10 HP. The construction of a capacitor motor is similar to that of a split-phase motor. The difference is that a capacitor is connected in series with the starting winding. The addition of a capacitor in the starting winding gives capacitor motors more torque than split-phase motors. Capacitor motors include capacitor-start, capacitor-run, and capacitor start-and-run motors.

Capacitor-Start Motors

A capacitor-start motor operates much the same as a split-phase motor in that it uses a centrifugal switch that operates at approximately 75% of full speed. The starting winding and the capacitor are removed when the circuit opens. The capacitor used in the starting winding gives the capacitor-start motor a high starting torque. **See Figure 16-11.**

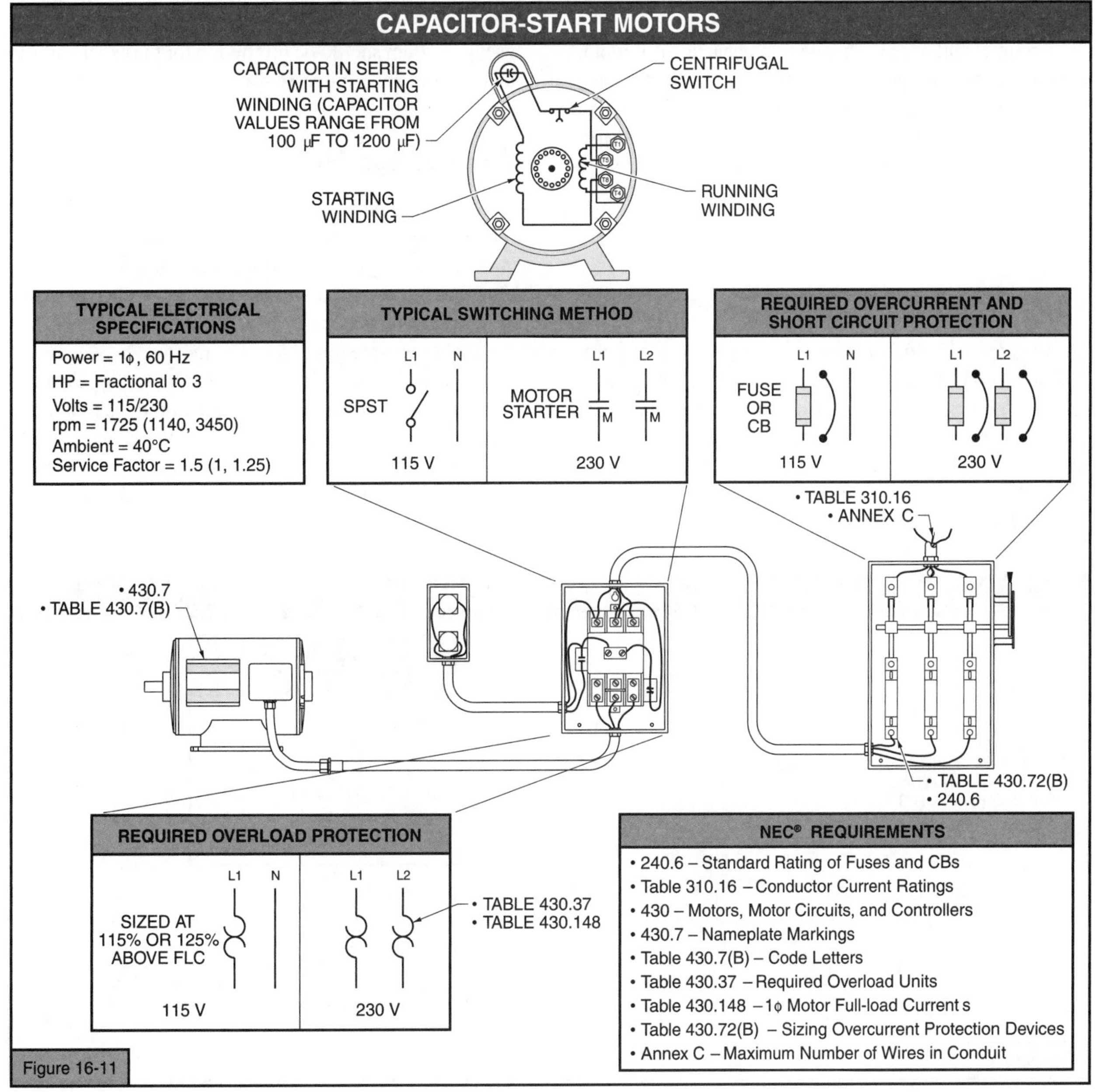

Figure 16-11

Capacitor-start motors have a higher starting torque than split-phase motors (200% to 350% of full-load torque), have a constant speed, and are reversed by interchanging starting winding connections. Capacitor-start motors are the most common capacitor motors and are used in higher starting torque applications such as refrigerator compressors, air compressors, pumps, power tools, washing machines, and air conditioners.

Capacitor-Run Motors

A *capacitor-run motor* is a capacitor motor that has the starting winding and capacitor connected in series at all times. A lower-value capacitor is used in a capacitor-run motor than in a capacitor-start motor because the capacitor remains in the circuit at full speed. This gives a capacitor-run motor medium starting torque (50% to 100% of full-load torque) and somewhat higher running torque than a capacitor-start motor. Capacitor-run motors are constant-speed motors that are reversible. Common uses include high running torque applications such as pumps, blowers, coolers, compressors, and conveyors. **See Figure 16-12.**

Capacitor Start-and-Run Motors

A *capacitor start-and-run motor* is a capacitor motor that has high- and low-value capacitors connected in parallel with each other and in series with the starting winding to provide a very high starting torque. The centrifugal switch disconnects the high-value capacitor at about 75% full-load speed, leaving the low-value capacitor in the circuit. **See Figure 16-13.**

The net result is that a capacitor start-and-run motor combines the advantages of a capacitor-start motor (high starting torque) with the advantages of a capacitor-run motor (high running torque). In each case, the loads and the amount of torque (starting and running) determine which capacitor motor to use. The torque of a capacitor start-and-run motor is between 350% to 450% of full-load torque. The motors have constant speed and are reversible. Common uses include high starting and running torque applications such as air conditioner compressors, conveyors, grinders, and pumps.

Reversing Capacitor Motors

The direction of rotation of a capacitor motor may be changed by reversing the connections to the starting or running windings. Whenever possible, the manufacturer's wiring diagram should be referred to for the exact wires to interchange. **See Figure 16-14.**

Three-Phase Motors

Three-phase motors are the most common motors used in industrial applications. Three-phase motors are used in applications ranging from fractional horsepower to over 500 HP. They are used in most applications because they are simple in construction, require little maintenance, and cost less to operate than 1ϕ or DC motors.

The most common 3ϕ motor used in most applications is the induction motor. An *induction motor* is a motor that has no physical electrical connection to the rotor. Induction motors have no brushes that wear or require maintenance. Current in the rotor is induced by the rotating magnetic field of the stator.

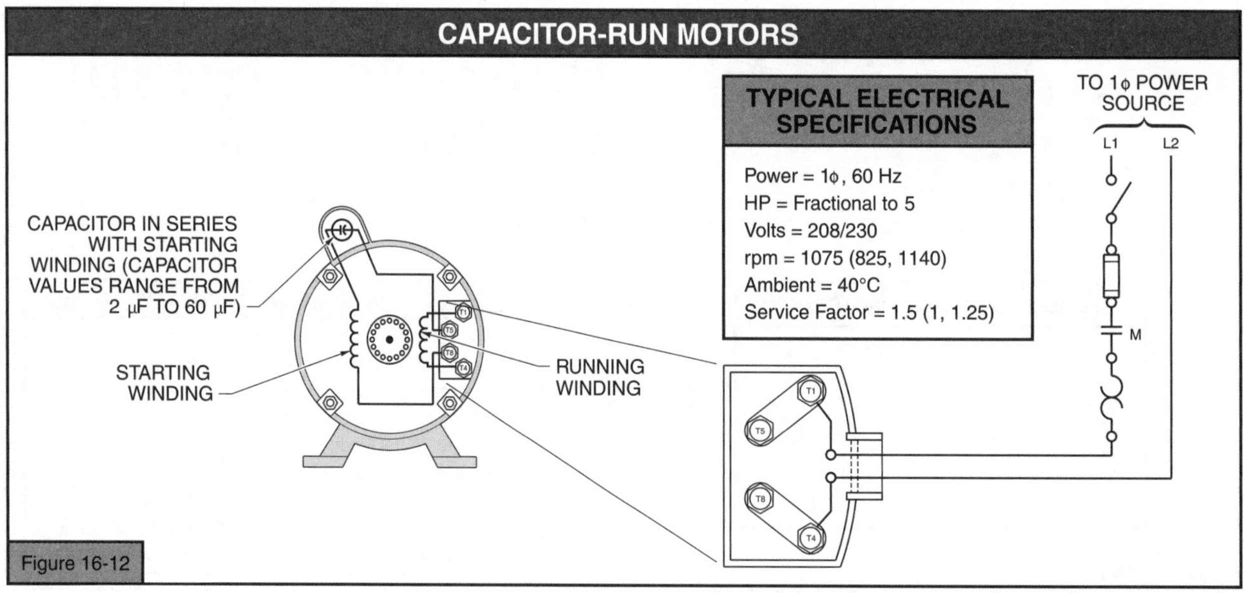

Figure 16-12

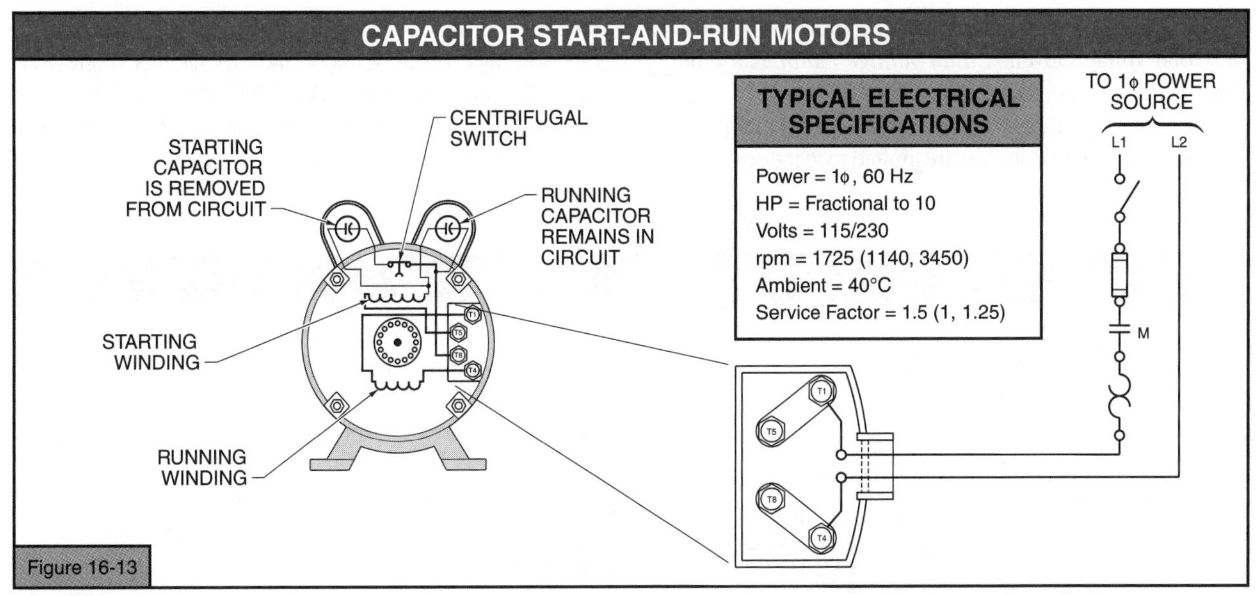

Figure 16-13

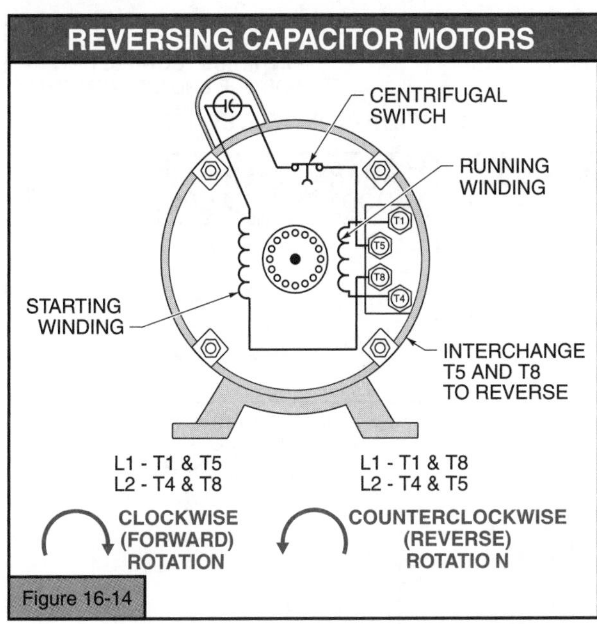

Figure 16-14

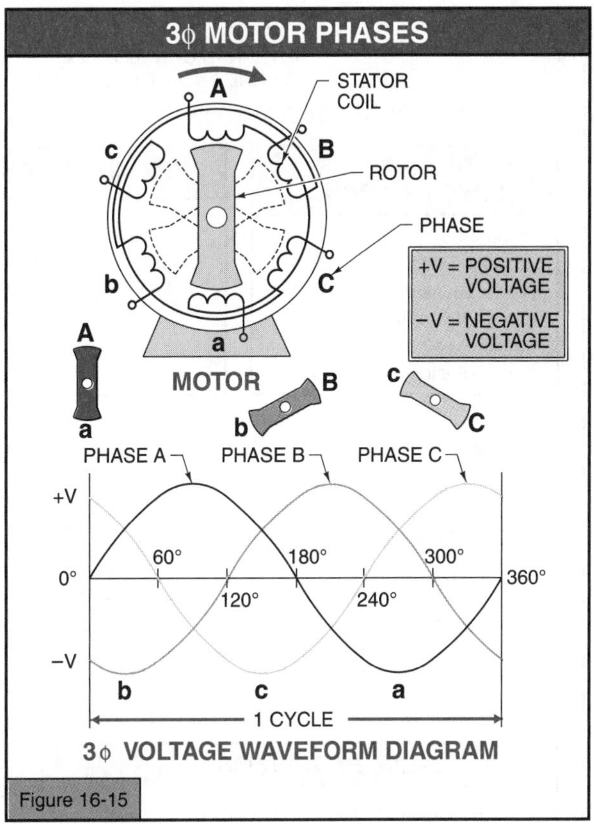

Figure 16-15

The *stator* is the stationary part of an AC motor. The *rotor* is the rotating part of an AC motor. A rotating magnetic field is set up automatically in the stator when the motor is connected to 3φ power. The coils in the stator are connected to form three separate windings (phases). Each phase contains one-third of the total number of individual coils in the motor. These composite windings or phases are the A phase, B phase, and C phase. **See Figure 16-15.**

Each phase is placed in the motor so that it is 120° from the other phases. A rotating magnetic field is produced in the stator because each phase reaches its peak value 120° apart from the other phases. Three-phase motors are self-starting and do not require an additional starting method because of the rotating magnetic field in the motor.

Single-Voltage, 3φ Motor Construction

To develop a rotating magnetic field in the motor, the stator windings must be connected to the proper voltage level. This voltage level is determined by the manufacturer and stamped on the motor nameplate. Three-phase motors are designed as either single-voltage motors or dual-voltage motors. A *single-voltage motor* is a motor that operates at only one voltage level. A *dual-voltage motor* is a motor that operates at more than one voltage level.

Single-voltage motors are less expensive to manufacture than dual-voltage motors, but are limited to locations having the same voltage as the motor. Common single-voltage, 3ϕ motor ratings are 230 V, 460 V, and 575 V. Other single-voltage, 3ϕ motor ratings are 200 V, 208 V, and 220 V. All 3ϕ motors are wired so the phases are connected in either a wye (Y) or delta (D) configuration. **See Figure 16-16.**

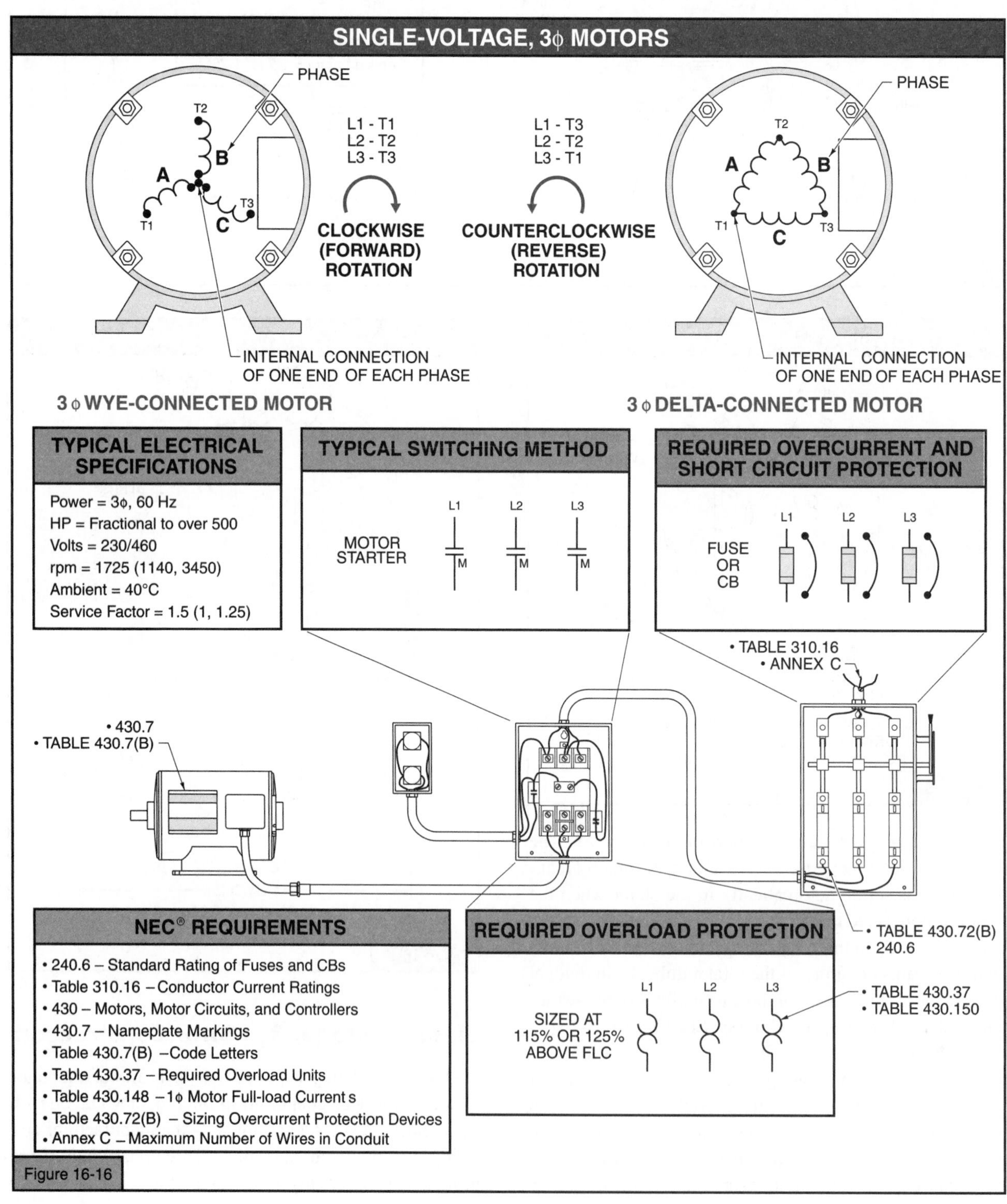

Figure 16-16

In a wye configuration, one end of each of the three phases is internally connected to the other phases. The remaining end of each phase is brought out externally and connected to the power lines. The leads which are brought out externally are labeled terminals one (T1), two (T2), and three (T3). When connected, terminals T1, T2, and T3 are matched to the 3ϕ power lines labeled line one (L1), two (L2), and three (L3). For the motor to operate properly, the 3ϕ power lines supplying power to the motor must have the same voltage and frequency rating as the motor.

In a delta configuration, each winding is wired end-to-end to form a completely closed-loop circuit. At each point where the phases are connected, leads are brought out externally to form terminals one (T1), two (T2), and three (T3). These terminals, like those of a wye-connected motor, are attached to power lines one (L1), two (L2), and three (L3). The 3ϕ power lines supplying power to a delta-connected motor must have the same voltage and frequency rating as the motor.

Single-voltage 3ϕ motors have good starting torque (100% to 275% of full-load torque), constant speed, and are reversible. Three-phase motors are the standard motor used in most industrial applications that require moderate to hard starting, such as pumps, blowers, conveyors, and machine tools.

Dual-Voltage, 3ϕ Motor Construction

Most 3ϕ motors are constructed so that they may be connected for either of two voltages. Manufacturing motors for two voltages enables the same motor to be used with two different power line voltages. The normal dual-voltage rating of industrial motors is 230/460 V. Always check the motor nameplate for proper voltage ratings.

The higher voltage is preferred when a choice between voltages is available. A motor uses the same amount of power and gives the same horsepower output for either high or low voltage, but as the voltage is doubled (230 V to 460 V), the current is cut in half. Using a reduced current enables the use of a smaller wire size which creates a savings on installation.

A wiring diagram is used to show the terminal numbering system for a dual-voltage, 3ϕ, wye-connected motor. Nine leads are brought out of the motor. These leads are marked T1 through T9 and may be externally connected for either of the two voltages. The terminal connections for high and low voltage are shown below the wiring diagram. This information is normally provided on the motor nameplate. **See Figure 16-17.**

The nine leads are connected in either series (high voltage) or parallel (low voltage). To connect a wye configuration for high voltage, L1 is connected to T1, L2 to T2, L3 to T3, and T4 is tied to T7, T5 to T8, and T6 to T9. This connects the individual coils in phases A, B, and C in series, each coil receiving 50% of the line-to-neutral point voltage. The neutral point equals the internal connecting point of all three phases.

To connect a wye motor for low voltage, L1 is connected to T1 and T7, L2 to T2 and T8, L3 to T3 and T9, and T4, T5, and T6 are tied together. This connects the individual coils in phases A, B, and C in parallel so each coil receives 100% of the line-to-neutral point voltage.

A wiring diagram is used to show the terminal numbering system for a dual-voltage, 3ϕ, delta-connected motor. The leads are marked T1 through T9 and a terminal connection diagram is provided for wiring high- and low-voltage operations. The nine leads are connected in either series or parallel for high or low voltage. **See Figure 16-18.** In the high-voltage configuration, the coils are wired in series. In the low-voltage configuration, the coils are wired in parallel to distribute the voltage according to the individual coil ratings.

Reversing 3ϕ, Wye- and Delta-Connected Motors

Reversing the direction of rotation of 3ϕ, wye- and delta-connected motors is accomplished by interchanging any two of the three main power lines to the motor. Although any two lines may be interchanged, the industry standard is to interchange L1 and L3. This standard is true for all 3ϕ motors including three, six, and nine lead wye- and delta-connected motors. Regardless of the type of 3ϕ motor, L1 is connected to T1, L2 to T2, and L3 to T3 for forward rotation. L1 is connected to T3, L2 to T2, and L3 to T1 for reverse rotation. If a 3ϕ motor has more than three leads coming out, these leads are connected according to the motor's wiring diagram.

Interchanging L1 and L3 is a standard for safety reasons. When first connecting a motor, the direction of rotation is not normally known until the motor is started. It is common practice to temporarily connect the motor to determine the direction of rotation before making permanent connections. Motor lead temporary connections are not taped. By always interchanging L1 and L3, L2 can be permanently connected to T2, creating an insulated barrier between L1 and L3.

DC Motors

DC motors are normally used where the load requires an adjustable speed and simple torque control. Common applications for DC motors include printing presses, cranes, elevators, shuttle cars, and automobile starters.

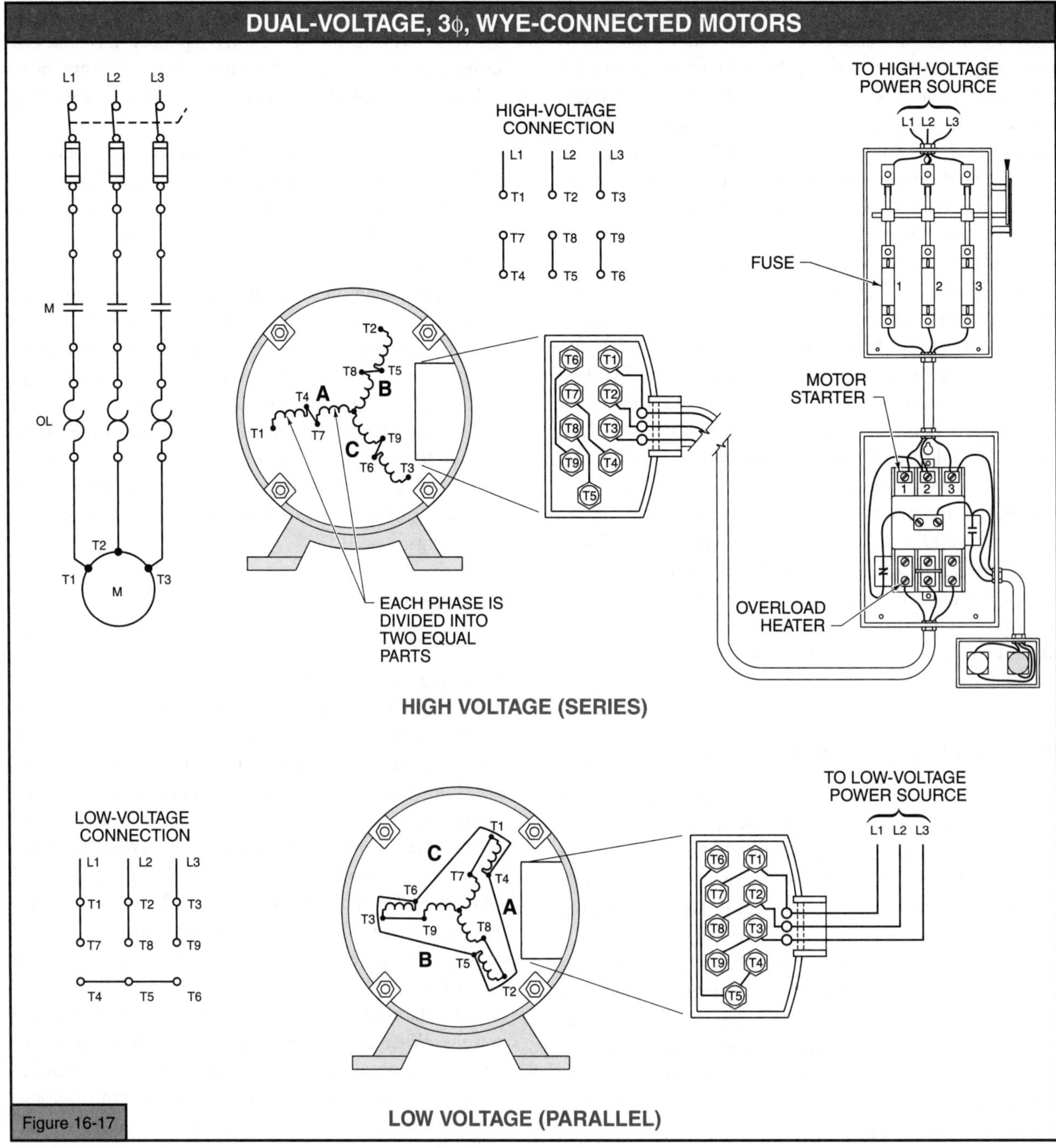

Figure 16-17

A DC motor consists of a field circuit and an armature circuit. The field circuit consists of stationary windings or permanent magnets that provide a magnetic field around the armature. The armature circuit consists of laminated steel slots connected to the shaft and terminated at the commutator. When windings are used in the field circuit, power is applied to the windings through wires that are brought out from the windings. When permanent magnets are used for the field circuit, no power is applied to the windings, so no wires are brought out from the field circuit.

Power is applied to the armature circuit through brushes that ride on the commutator as the armature rotates. The brushes are connected to power through wires that are brought out from the motor.

DC motors include series, shunt, compound, and permanent-magnet motors. The main difference between the different types of DC the motors is in the way in which the field coil and armature coil circuits are wired. The field coil and armature coil circuits may be wired in either a series, parallel, or series-parallel connection.

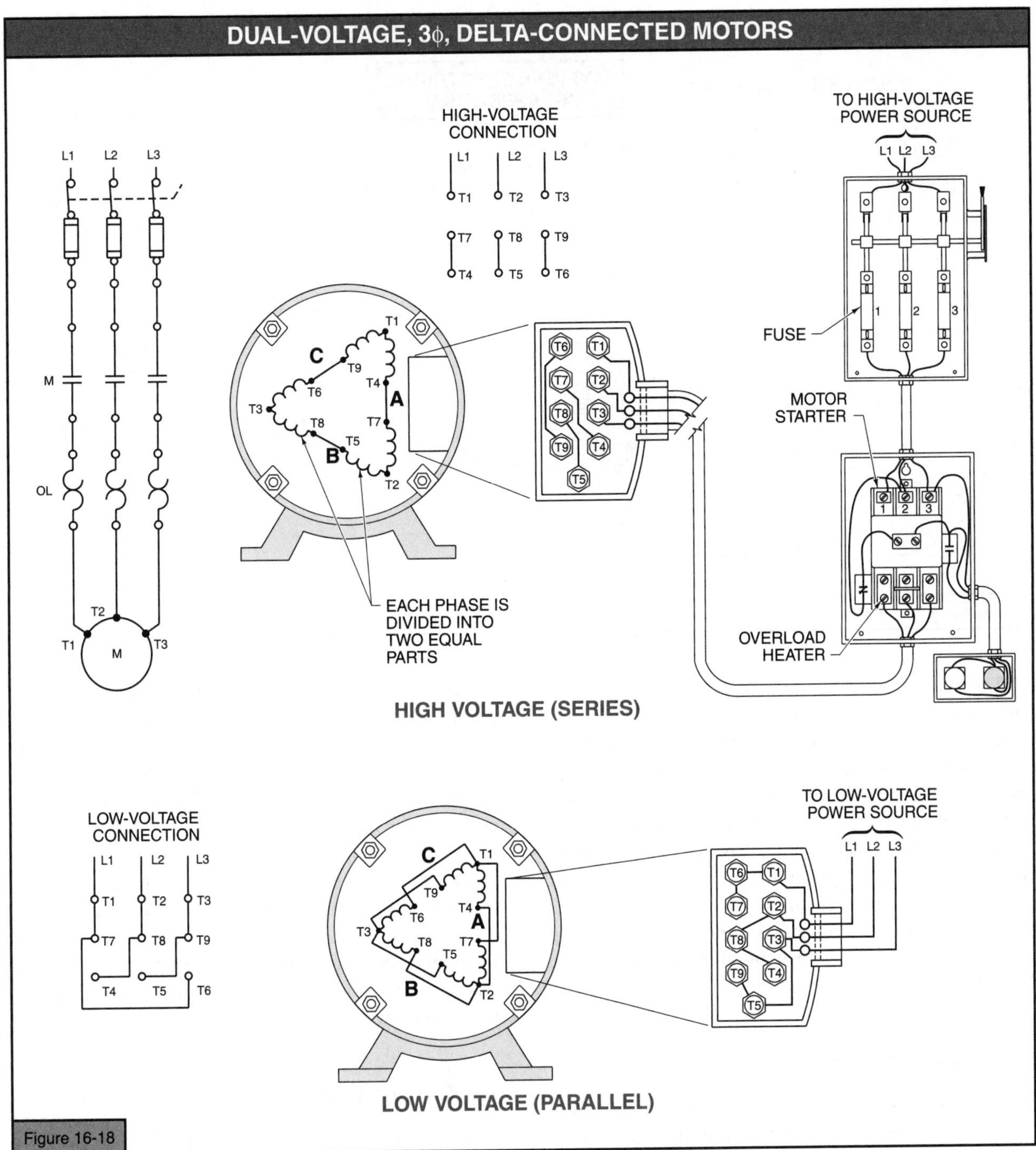

Figure 16-18

DC Series Motors

A *DC series motor* is a DC motor that has the series field coils connected in series with the armature. The series field coils are composed of a few turns of heavy gauge wire. The series coil wires are marked S1 and S2. The armature wires are marked A1 and A2. **See Figure 16-19.**

A DC series motor has a high starting torque (400% to 450% of full-load torque), variable speed, and is reversible. This allows the motor to start very heavy loads. The speed of a DC series motor increases as the load decreases. A DC series motor develops high starting torque because the same current that passes through the armature also passes through the field. If the armature calls for more current (developing more torque), this current also passes through the field, increasing the field strength. DC series motors are used in very high starting torque applications such as hoists, bridges, and traction equipment.

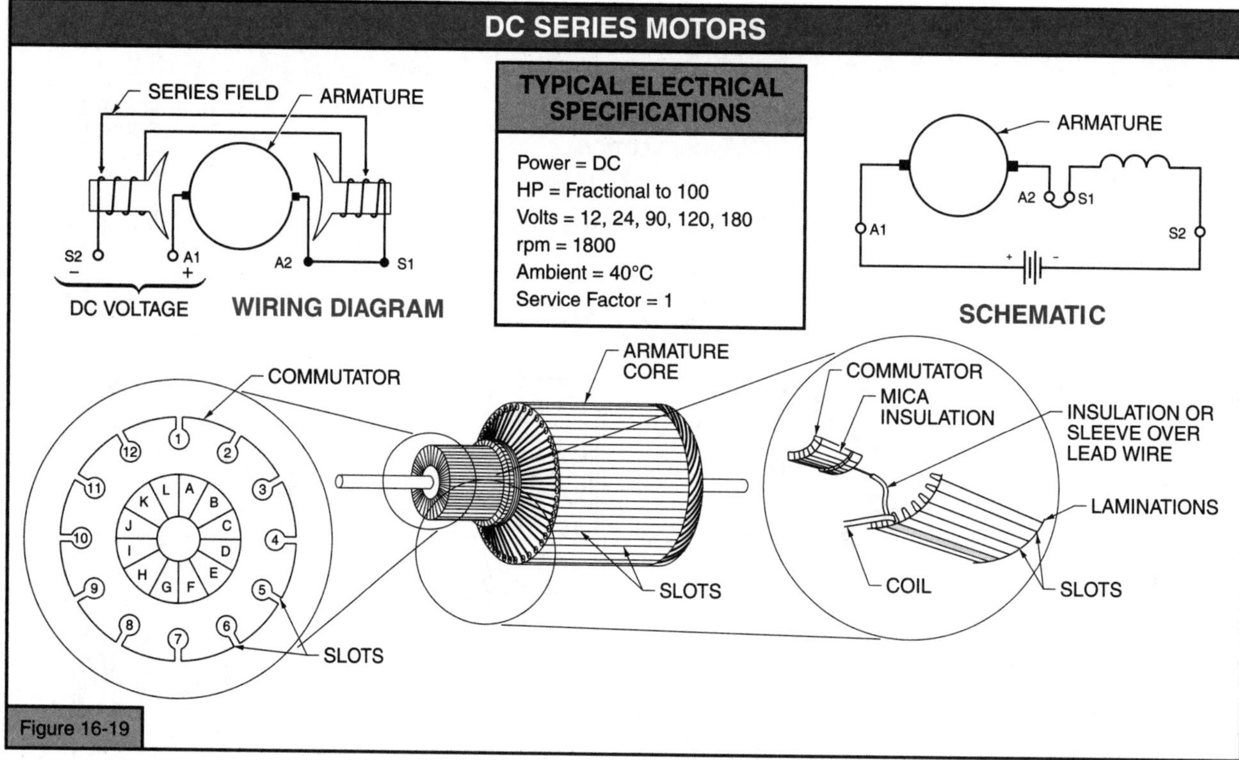

Figure 16-19

The ability to draw more current is an advantage as long as a load is applied to the motor because the load tends to hold the motor speed under control. The variable speed characteristic allows the motor to keep increasing speed after the load is moved. The motor throws itself apart at uncontrolled speeds if left unchecked. For this reason, it is necessary to positively connect DC series motors to loads through couplings or gears that do not allow slip. Belt drives should not be used on DC series motors. In small DC series motors, brush friction, bearing friction, and winding loss may provide sufficient load to hold the speed to a safe level.

DC Shunt Motors

A *DC shunt motor* is a DC motor that has the field connected in parallel with the armature. This produces an essentially constant field strength and motor speed. The windings extending from the shunt field are marked F1 and F2. **See Figure 16-20.** The shunt field can be either connected to the same power supply as the armature (self-excited) or to another power supply (separately-excited).

When DC shunt motors are separately excited, the motor can be speed controlled by varying the field current. Field speed control can be accomplished by inserting external resistances in series with the shunt field circuits. As resistance is increased in the field circuit, field current is reduced and the speed is increased. Conversely, as resistance is removed from the shunt field, field current is increased and the speed is decreased. By selecting the proper controller, the motor may be set for a specific speed control range. Standard DC shunt field voltages are 100 V, 150 V, 200 V, 240 V, and 300 VDC.

DC shunt motors have good starting torque (125% to 250% of full-load torque), are variable from 0 rpm to full speed, and are reversible. DC shunt motors are used in applications that require good speed control, such as woodworking machines, metalworking machines, elevators, and centrifugal pumps.

DC Compound Motors

A *DC compound motor* is a DC motor that has the series field connected in series with the armature and the shunt field connected in parallel with the series field and armature. **See Figure 16-21.** This arrangement combines the characteristics of both the DC series and DC shunt motor. DC compound motors have high starting torque (300% to 400% of full-load torque), fairly good speed torque characteristics at rated load, and are reversible. Only large bidirectional DC compound motors are normally built because of the complicated circuits needed to control compound motors.

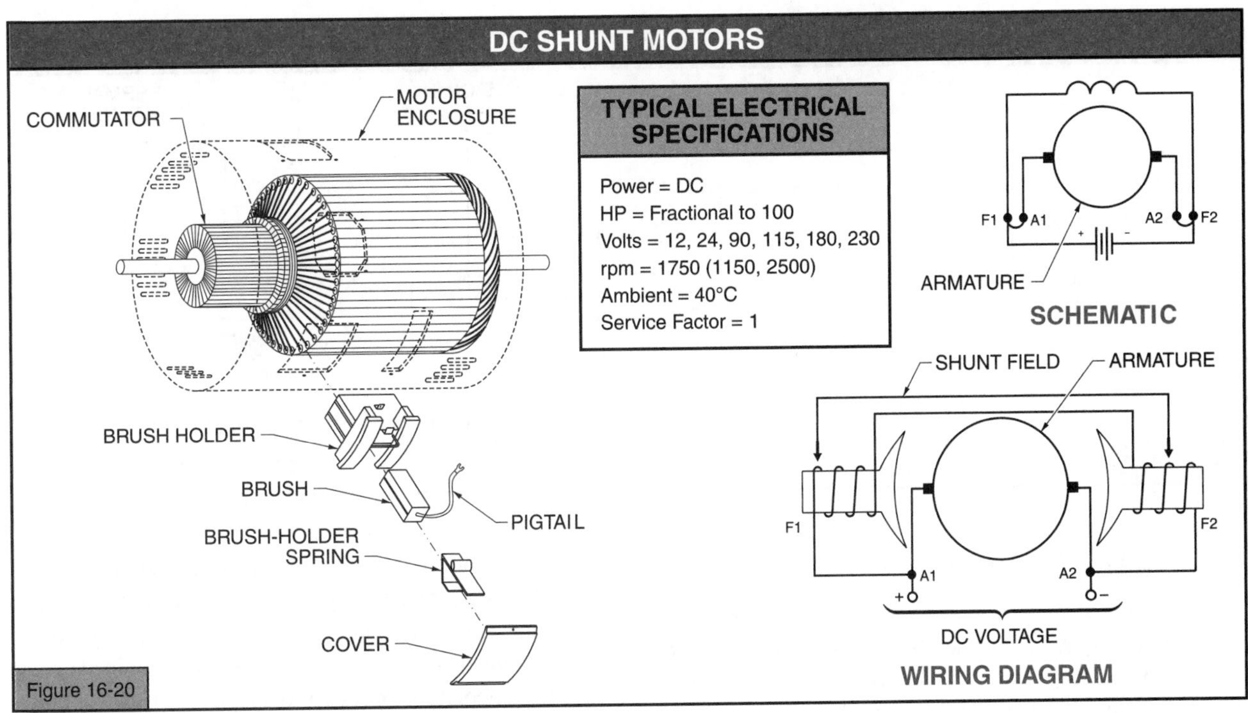

Figure 16-20

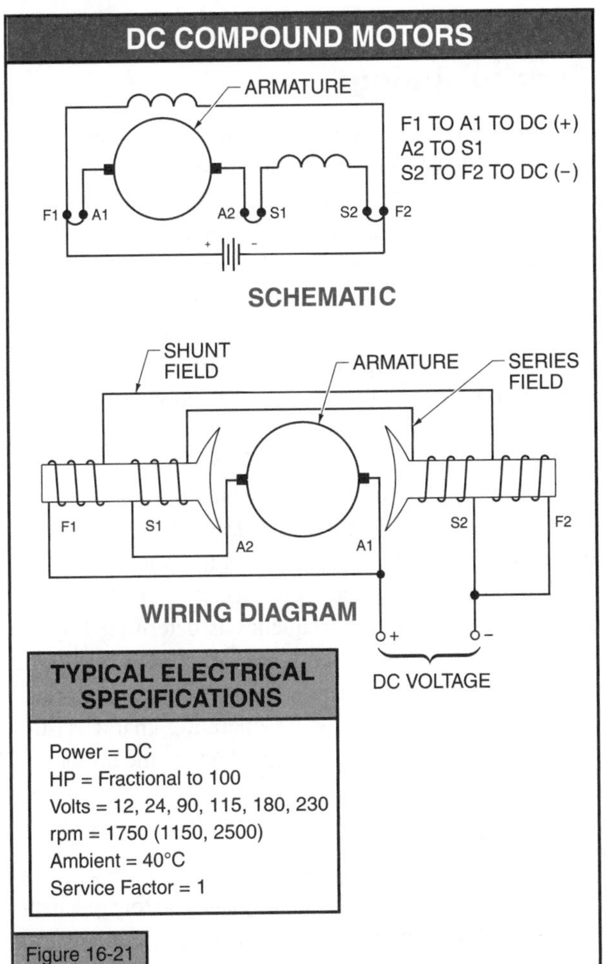

Figure 16-21

Some smaller DC motors may be slightly compounded to improve starting characteristics. DC compound motors are used in applications in which starting torque is higher and speed must be controlled such as elevators, hoists, conveyors, and printing presses. These applications need a fairly high starting torque and a reasonably constant running speed.

DC Permanent-Magnet Motors

A *DC permanent-magnet motor* is a DC motor that has only armature connections and no field connections. Permanent magnet motors use magnets, not a coil of wire, for the field winding. A DC permanent-magnet motor has molded magnets mounted into a steel shell. The permanent magnets are the field coils. DC power is supplied only to the armature. **See Figure 16-22.**

A DC permanent-magnet motor is used in automobiles to control power seats, power windows, and windshield wipers. They produce relatively high torque at low speeds and provide some dynamic (self-) braking when removed from power.

Lubricate motors used in standard operating conditions according to manufacturer's recommendations. Increase the frequency of lubrication by 50% for motors used in severe operating conditions (high temperature, dirty/dusty environments, excess moisture, etc.).

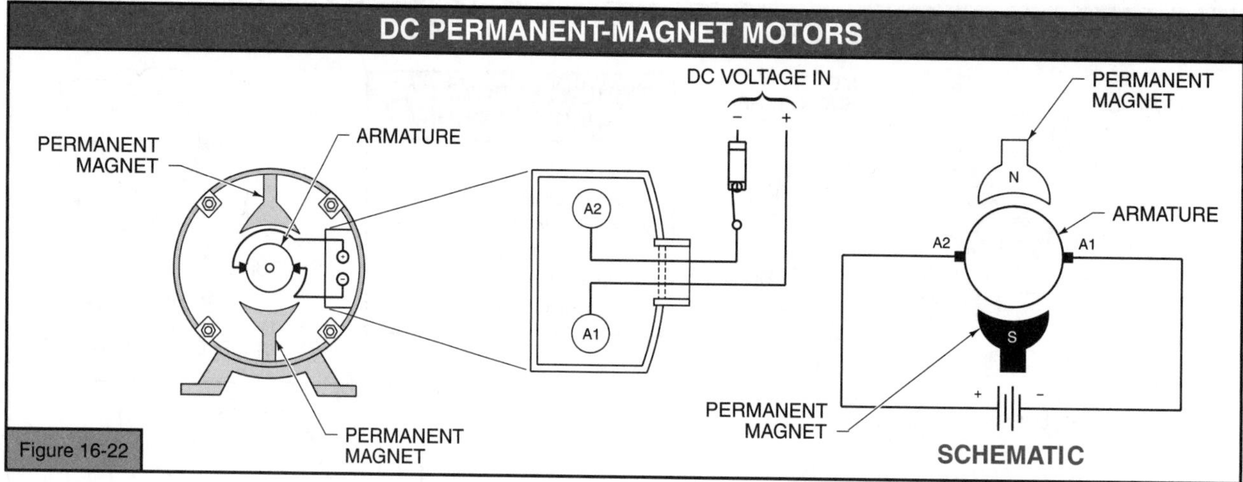

Figure 16-22

Reversing DC Motors

The direction of rotation of DC series, shunt, and compound motors may be reversed by reversing the direction of the current through the field without changing the direction of the current through the armature, or by reversing the direction of the current through the armature, but not both. The industrial standard is to reverse the current through the armature. The direction of rotation of a DC permanent-magnet motor is reversed by reversing the direction of the current through the armature only, since there are no field connections available.

In a DC compound motor, the series and shunt field relationship to the armature must be left unchanged. The shunt must be connected in parallel and the series field in series with the armature. Reversal is accomplished by reversing the armature connections only. If the motor has commutating pole windings, these windings are considered a part of the armature circuit and the current through them must be reversed when the current through the armature is reversed. Commutating windings (interpoles) are used to prevent sparking at the brushes in some DC motors.

Motor Nameplate Data

Electric motors are used to produce work. In order for a motor to safely produce work as required for the expected life of the motor, the motor's electrical, operating, environmental, and mechanical ratings must be considered. The electrical, operating, environmental, and mechanical ratings of a motor are listed on the motor nameplate. Since the nameplate has limited space to convey this information, most information is abbreviated or coded to save space. In addition to the written information, most motor nameplates also include the motor's wiring diagram.

Understanding the abbreviated and coded information provided on a motor nameplate is required when selecting, installing, and troubleshooting electric motors. For motors that are already in service, the information provided on the motor nameplate is often the only information available. **See Figure 16-23.**

Electrical Ratings

Electrical ratings included on most motor nameplates include a voltage rating, current rating, frequency rating, and power rating. A motor's nameplate voltage rating is the optimal voltage that should be connected to the motor for best operating performance. The actual voltage applied to the motor usually varies somewhat from the nameplate rated voltage. Any variation in applied voltage from the nameplate rated voltage changes the operating characteristics of a motor and must be considered when installing and troubleshooting the motor.

The nameplate current rating of a motor is only reached when the motor is fully loaded. Since the load on most motors varies, the actual current draw of a motor varies from the nameplate rated current as the motor operating conditions change and the motor ages.

A motor's nameplate frequency is determined by the manufacturer and is either 50 Hz or 60 Hz. Applying any frequency to a motor other than the nameplate rated frequency changes the motor's operating characteristics (speed and output torque) and can damage the motor.

Power Rating

All motors convert electrical energy into rotating mechanical energy. The amount of rotating mechanical energy produced by a motor determines the amount of work the motor can perform. *Power* is rate of doing work or using energy. The two most common methods of rating motor power are watts (W) and horsepower (HP). Motors manufactured in the United States (or designed for the U.S. market) that are ⅙ HP or greater are typically rated in HP, and motors that are less than ⅙ HP are typically rated in watts. Motors manufactured in Europe are usually rated in kilowatts (kW), regardless of size.

Phase Rating

AC motors are either single-phase or three-phase. DC motors do not have a phase rating. Three-phase motors are preferred over single-phase motors because they can deliver more power at less operating cost and draw less current from the power lines. Less current draw means smaller conductors (wire) and smaller motor control devices can be used.

Code Letter

When a motor is started at full line voltage, the motor draws a much higher current than when the motor is running at rated speed. The amount of starting current is a function of the motor size (in HP) and the motor design characteristics. The amount of current a motor draws when starting is indicated on the motor nameplate by a code letter. **See Figure 16-24.**

The list of motor nameplate code letters begins with the letter "A" and ends with "V." The closer the nameplate listed code letter is to "A," the lower the motor's starting current. Likewise, the closer the nameplate listed code letter is to "V," the higher the motor starting current. For example, a motor with a listed code letter of "G" has a lower starting current than a motor with a listed code letter of "H." Most motors have a code letter in the H to N range.

Operating Ratings

Motors are rated for the type of operation where they will be used. Operating ratings cover ratings for the motor's usage, service factor, speed, operating time rating (duty cycle), and efficiency. These ratings are displayed on the motor nameplate and are critical in performing installation and troubleshooting procedures.

Usage Rating

Motors are rated for general-purpose usage or specific-purpose usage. Motors that are rated for general-purpose usage are used in a wide range of applications and for mechanical loads such as conveyors, machine tools, and belt-driven equipment. They are also used with reciprocating pumps and for moving-air applications.

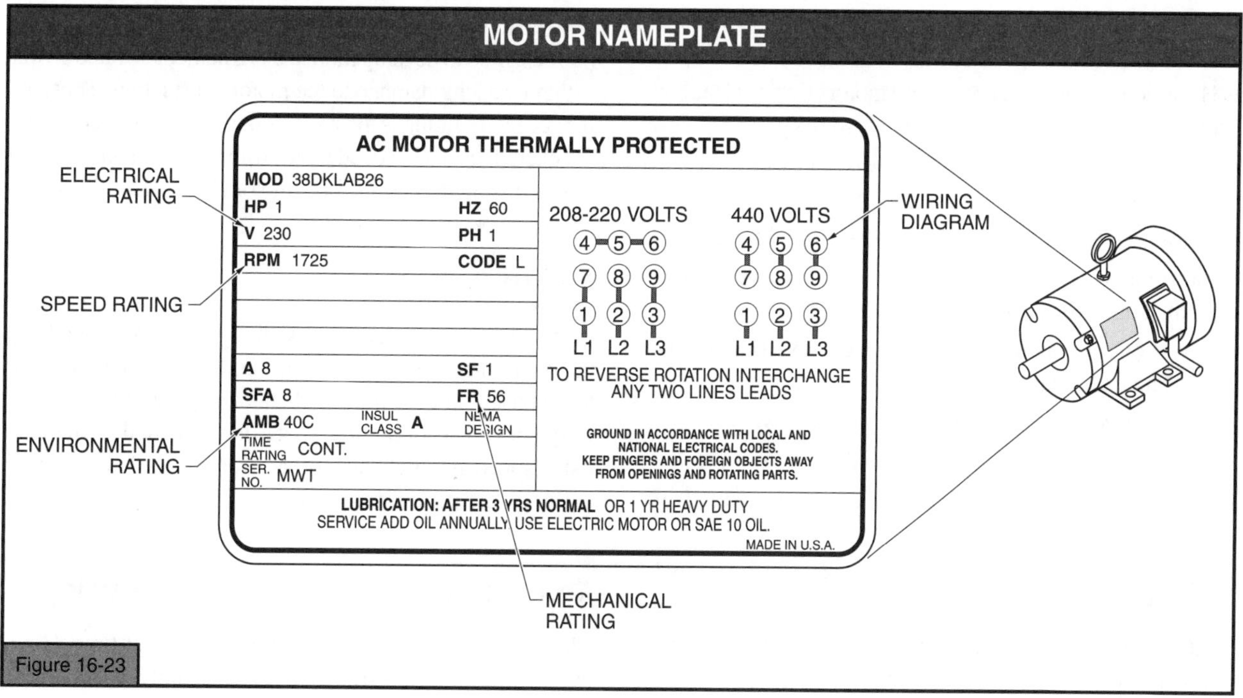

Figure 16-23

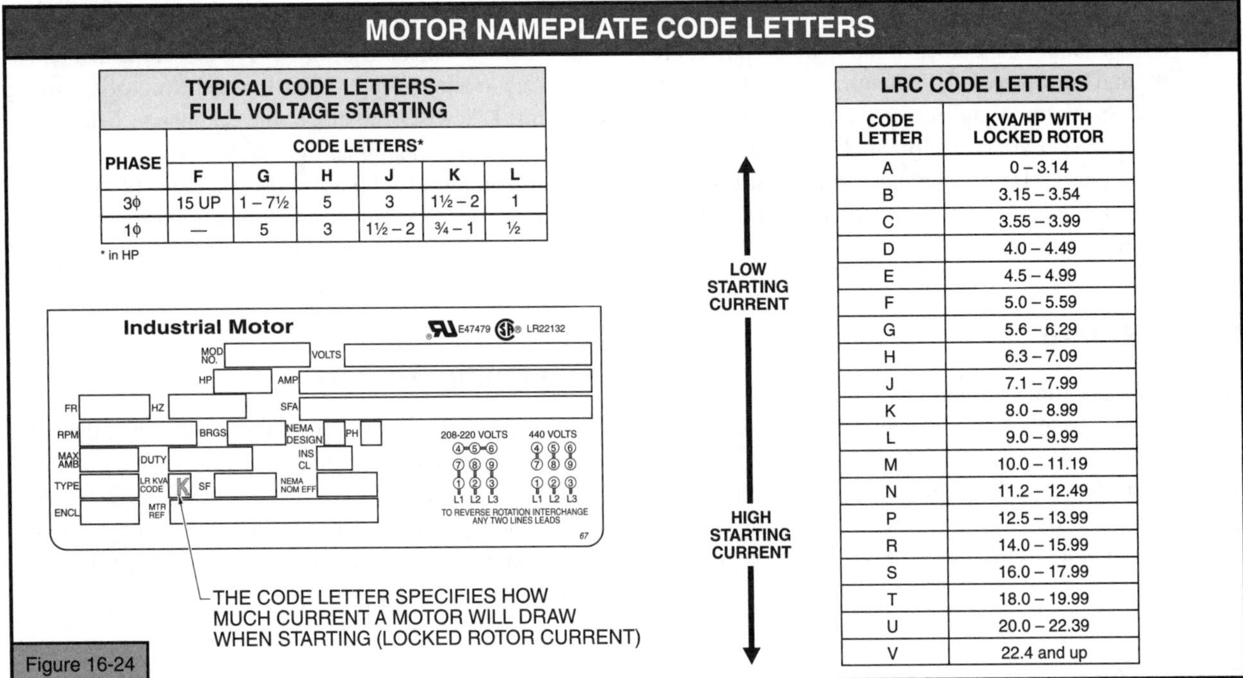

MOTOR NAMEPLATE CODE LETTERS

TYPICAL CODE LETTERS—FULL VOLTAGE STARTING

PHASE	CODE LETTERS*					
	F	G	H	J	K	L
3φ	15 UP	1 – 7½	5	3	1½ – 2	1
1φ	—	5	3	1½ – 2	¾ – 1	½

* in HP

LRC CODE LETTERS

CODE LETTER	KVA/HP WITH LOCKED ROTOR
A	0 – 3.14
B	3.15 – 3.54
C	3.55 – 3.99
D	4.0 – 4.49
E	4.5 – 4.99
F	5.0 – 5.59
G	5.6 – 6.29
H	6.3 – 7.09
J	7.1 – 7.99
K	8.0 – 8.99
L	9.0 – 9.99
M	10.0 – 11.19
N	11.2 – 12.49
P	12.5 – 13.99
R	14.0 – 15.99
S	16.0 – 17.99
T	18.0 – 19.99
U	20.0 – 22.39
V	22.4 and up

Figure 16-24

Specific-purpose motors are designed for applications in which the motor's usage is defined by the specific application, such as the following:

- Washdown rated motors (for food, beverage, and chemical plants)
- Submersible pumps (sump pumps, drainage pumps, water wells, and septic systems)
- Other pumps (for wastewater treatment, water treatment, and water recirculation systems)
- Hazardous locations, severe-duty motors (in dry cleaning plants, paint factories, and grain elevators)
- Instantly reversible motors (for hoists, gates, cranes, and mechanical doors)
- Extra high torque motors (for hard-starting loads, etc.)
- Farm duty/agricultural rated (having added protection against dust, dirt, and chemicals)
- Irrigation (motors that resist corrosion from high moisture and chemical environments)
- Auger drive (for augers and drilling systems)
- HVAC (heating/ventilating/air conditioning)
- Inverter duty rated (motors designed to be controlled by variable frequency drives)
- Pool (for swimming pools, water parks, and whirlpool hot tubs)
- AC/DC vacuum (for commercial vacuum systems, carwash, and sprayer/fogger systems)

Service Factor Rating

A motor attempts to drive a load, even if the load exceeds the motor's power rating. A motor's service factor rating indicates whether the motor can safely handle an overloaded condition. A nameplate service rating of 1 (or no listed rating) indicates the motor is not designed to safely handle an overloaded condition above the motor's rated power (in HP or kW). A nameplate service rating higher than 1 indicates the motor was designed to develop more than its nameplate rated power without causing damage to the motor's insulation. Thus, the service factor rating is the motor's margin of overload rating. For example, a 10 HP rated motor with a service factor of 1.15 can be operated as an 11.5 HP motor ($10 \times 1.15 = 11.5$).

Speed Rating

Motors have two speed ratings, the synchronous speed and the operating speed. The synchronous speed of an AC motor is based on the number of stator poles and the applied frequency. The operating speed is the actual nameplate listed speed at which the motor develops rated horsepower at rated voltage and frequency.

Operating Time Rating (Duty Cycle)

Most motors can be operated for any length of time. However, some motors are designed to operate only for short time periods. Motors designed to operate for unlimited time periods are marked "CONT" (continuous) on the motor nameplate, or have no designation. Motors designed to operate for intermittent time periods before being turned off and allowed to cool down are marked "INTER" on the nameplate, or have a time rating listed. Typical intermittent-duty motor time ratings are listed as 5, 15, 30, or 60 minutes. The amount of time a motor can deliver its rated horsepower is listed on the motor nameplate as TIME RATING, DUTY, or DUTY CYCLE.

Intermittent rated motors are used in applications such as waste disposal systems (garbage disposals), electric hoists, gate opening motors, and other applications in which the motor need only be turned on for short time periods to meet the application requirements.

Efficiency Rating

Motor efficiency is a measure of the effectiveness with which a motor converts electrical energy to mechanical energy. It is the ratio of motor power output to supply power input. All motors take more power to operate than they can produce because of power losses within the motor. Power losses occur because of losses from friction and heat within the motor.

Motor manufacturers produce both standard and energy-efficient motors. Energy-efficient motors are more expensive than standard motors but cost less to operate. Energy-efficient motors are built more ruggedly than standard motors and include larger rotor bars and increased laminations. This reduces losses and increases torque. Since energy-efficient motors are designed to run cooler, there is improved insulation and bearing life. Energy-efficient motors can endure excessive voltage and current, overload, and unbalance conditions better than standard motors. Likewise, energy-efficient motors operate better on nonsinusoidal waveforms and are better suited for use with variable frequency drives, which distort sine waves.

Environmental Ratings

Environmental ratings include the temperature rating and the insulation rating of a motor. Motors must operate under their rated temperature conditions. Operating a motor in an ambient condition that is too hot or with insulation that is not rated for the application could cause premature failure and damage to the motor and related equipment.

Temperature Rating

Ambient temperature is the temperature of the air surrounding a piece of equipment. *Temperature rise* is the difference between a motor's winding temperature when running and the ambient temperature. The temperature rise produced at full load does not harm a motor as long as the ambient temperature does not exceed the nameplate listed temperature.

Higher temperatures caused by either an increase in the ambient temperature or by overloading the motor damages the motor's insulation. Typically, for every 10°F above the rated temperature limit of a motor, the motor's life is cut in half. Heat destroys insulation, and the higher the heat, the greater and faster the damage occurs.

Most motors have a maximum ambient temperature rating of 40°C (104°F). A motor without a rating is typically rated at 40°C. Motors designed to operate in higher ambient temperatures should have additional cooling or have a higher rating (55°C nameplate rating).

Insulation Rating

Insulation breakdown is the main cause of motor failure. Motor insulation is rated according to its thermal breakdown resistance. The four motor insulation classes are Class A, Class B, Class F, and Class H. Class A is the least common insulation in use. Class F is the most commonly used motor insulation. Class H is the best rated and should be used in any application in which a motor drive is used to operate the motor.

Mechanical Ratings

Mechanical ratings on a motor nameplate include frame size rating, design rating, enclosure type, and motor bearings. Following the guidelines on mechanical ratings when installing or servicing a motor results in improved efficiency and lower operating cost.

Frame Size Rating

All motors have a frame to protect the working parts of the motor and provide a means of mounting. Motor frames follow standard ratings that are used to designate the physical size and measurements of the motor. Using standardized dimensions aids in mounting the motor and allows for interchangeability among different motor manufacturers.

Frame size follows standards established by either the National Electrical Manufacturers Association (NEMA) or International Electrotechnical Commission (IEC). Dimensionally, NEMA standards are expressed in English units and IEC standards are expressed in metric units. Frame size is listed on the motor nameplate as FR. For both NEMA and IEC, the larger the frame size number, the larger the motor.

Design Rating

Motor design ratings represent the torque characteristics of a motor. Motor design ratings may be listed as Designs A, B, C, D, or E. Designs B, C, and D are the most common motor design ratings. **See Figure 16-25.** The design rating listed on motor nameplates is used mostly to indicate the amount of starting torque a motor has in comparison to its running current. A motor's starting and running torque characteristics indicate the applications where the motor can be used.

COMMON MOTOR DESIGN RATINGS

Motor Design Rating Letter	Motor Starting Torque*	Motor Application Usage
B	100 – 200	Standard industrial duty motor application
C	200 – 250	Hard starting loads (compressors, blowers, etc.)
D	Over 250	Frequent, sudden loads (punch presses, die stamping, etc.)

* % of running torque

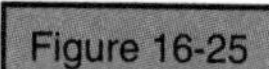
Figure 16-25

Enclosure Type

Motors include an enclosure to protect the motor and the motor's working parts from the outside environment as well as to protect individuals from the electrical and rotating parts of the motor. Protecting individuals from the electrical and rotating parts of a motor can be accomplished using a basic enclosure. However, protecting a motor from all of the environments in which a motor must operate in requires more than a basic enclosure. Environments in which a basic enclosure is not suitable include the following:

- Wet locations (rain, snow, sleet, washdown, etc.)
- Dirty locations (dirt, noncombustible dust, etc.)
- Oily locations (lubricants, cutting oils, coolants, etc.)
- Corrosive locations (salt, chlorine, fertilizers, chemicals, etc.)
- Extremely low- and high-temperature locations (ranging from below zero to above boiling temperatures)
- Gas and other hazardous locations (liquid and solid)
- Combustible dust locations (grain and chemical)

The type of motor enclosure is listed on the motor nameplate as the ENCL rating. Typical motor enclosure ratings include the following:

- ODP (open dripproof) for use in clean, dry, nonhazardous locations
- TENV (totally enclosed nonventilated) for use in moist, dirty, nonhazardous locations
- TEFC (totally enclosed fan-cooled) for use in the same locations as TENV but can tolerate high temperatures
- TEAO (totally enclosed air over) for high nonhazardous-temperature locations

Hazardous location motors include a hazardous location listing specifying the hazardous classification/class/group rating (such as a Class 1, Group D rating).

Bearings

Motor bearings are either sleeve or ball bearings. Both sleeve bearings and ball bearings are used with different size motors ranging from fractional horsepower to hundreds of horsepower. Sleeve bearings are used where a low noise level is important, such as fan and blower motors. Ball bearings are used where higher load capacity is required or periodic lubrication is not practical. The bearing used on a motor is listed on the motor nameplate as BALL or SLEEVE.

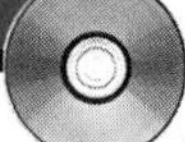

Refer to the CD-ROM in the back of the book for Chapter 16 Quick Quiz® and related information.

17 Resistance, Inductance, and Capacitance

Basic Circuit Elements

Electrical circuits range from simple to complex. Simple electrical circuits may include only a power supply, switch, and load. The power supply may be a DC, AC, DC to AC, or AC to DC power supply. The switch starts or stops the flow of current in the circuit. Switches may be mechanical or solid-state. A *load* is any device that converts electrical energy to light, heat, motion, or sound. Common loads include lamps (electrical energy to light), heating coils (electrical energy to heat), motors (electrical energy to motion), and speakers (electrical energy to sound). **See Figure 17-1.**

A flashlight is an example of a simple electrical circuit. A flashlight includes a power supply (batteries), switch, and a load (lamp). A digital clock is an example of a more complex electrical circuit. A digital clock that plugs into 115 VAC includes a step-down transformer, rectifier (AC to DC) circuit, filter circuits, digital logic circuits, LED displays, and manual input controls (time setting switches).

In a flashlight circuit, the only opposition to current flow is the lamp. A high-wattage lamp offers less resistance (allowing a higher current flow) and a low-wattage lamp offers more resistance (allowing a smaller current flow). Ohm's law and the power formula may be directly applied for determining voltage (E), current (I), resistance (R), and power (P), because the lamp is a resistive load and the power supply (batteries) is pure DC.

Ohm's law and the power formula cannot always be directly applied to more complex DC circuits and most AC circuits. For example, the digital clock circuit includes inductance (step-down transformer) and capacitance (capacitors) in addition to resistance (resistors). Inductance (L) and capacitance (C) affect electrical circuits differently than only resistance (R). Thus, resistance, inductance, and capacitance must all be considered when analyzing a circuit that includes more than just a resistive load. **See Figure 17-2.**

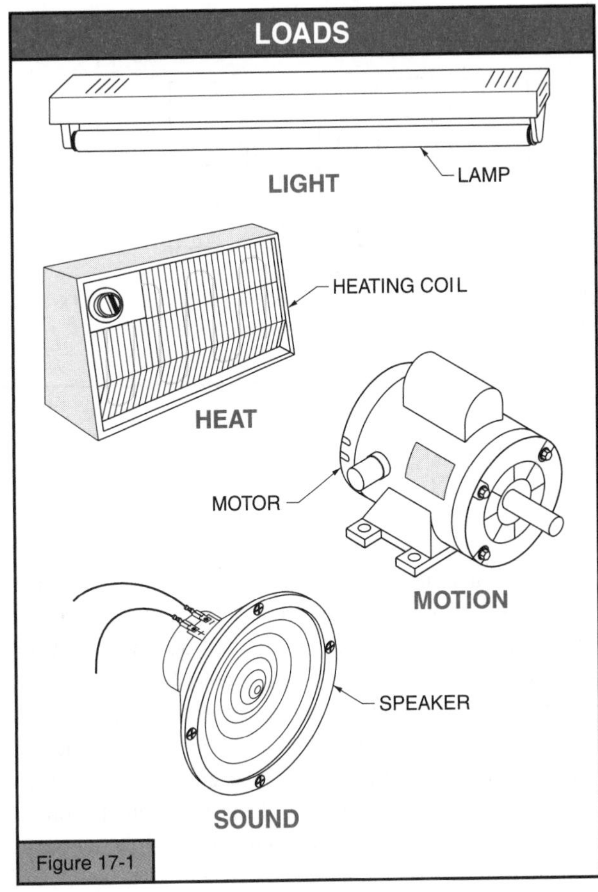

Figure 17-1

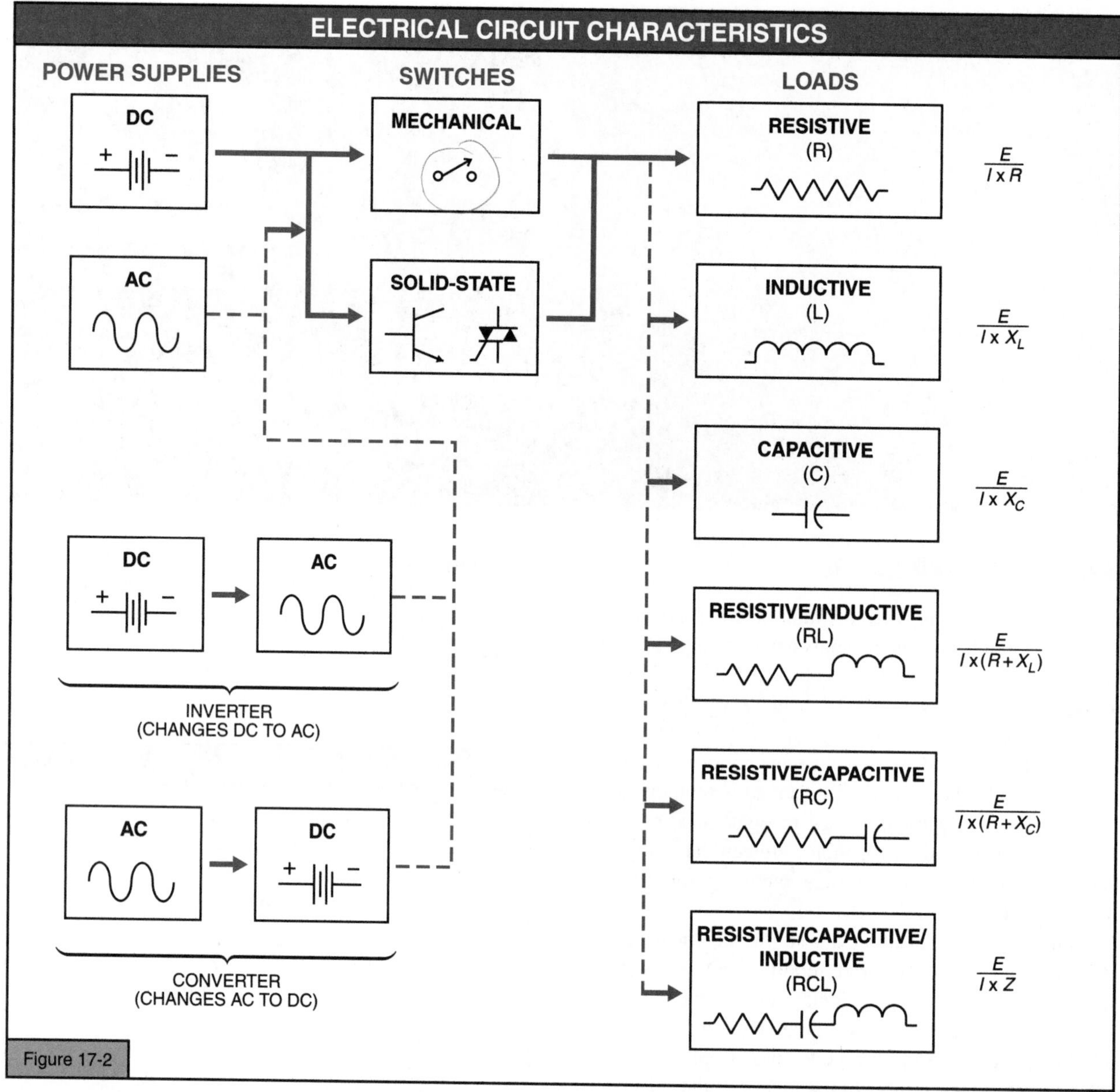

Figure 17-2

Resistive Circuits

A *resistive circuit* is a circuit that contains only resistance. Resistive circuits are the simplest circuits. A circuit that controls a heating element is an example of a resistive circuit. In a resistive circuit, the electrical characteristics of the circuit are the same regardless of whether an AC or DC voltage source is used. **See Figure 17-3.**

In a resistive circuit, voltage, current, resistance, and power are easily calculated. Ohm's law and the power formula may be applied if any two electrical quantities are known (or measured). All electrical circuits include some resistance, because all conductors, switch contacts, connections, loads, etc. have resistance.

Conductor Resistance

All electrical circuits use conductors to carry electricity to different parts of the circuit. A *conductor* is a material that has very little resistance and permits electrons to move through it easily. All conductors have some resistance. Conductor resistance should be kept to a value low enough that the conductor has little or no effect on circuit operation. Conductor resistance is minimized by limiting the temperature in the circuit and by using the correct size, length, and material. The resistance of any conductor (or material) with a uniform cross-sectional area is based on:

- Material Temperature—For metals, the higher the temperature, the greater the resistance.

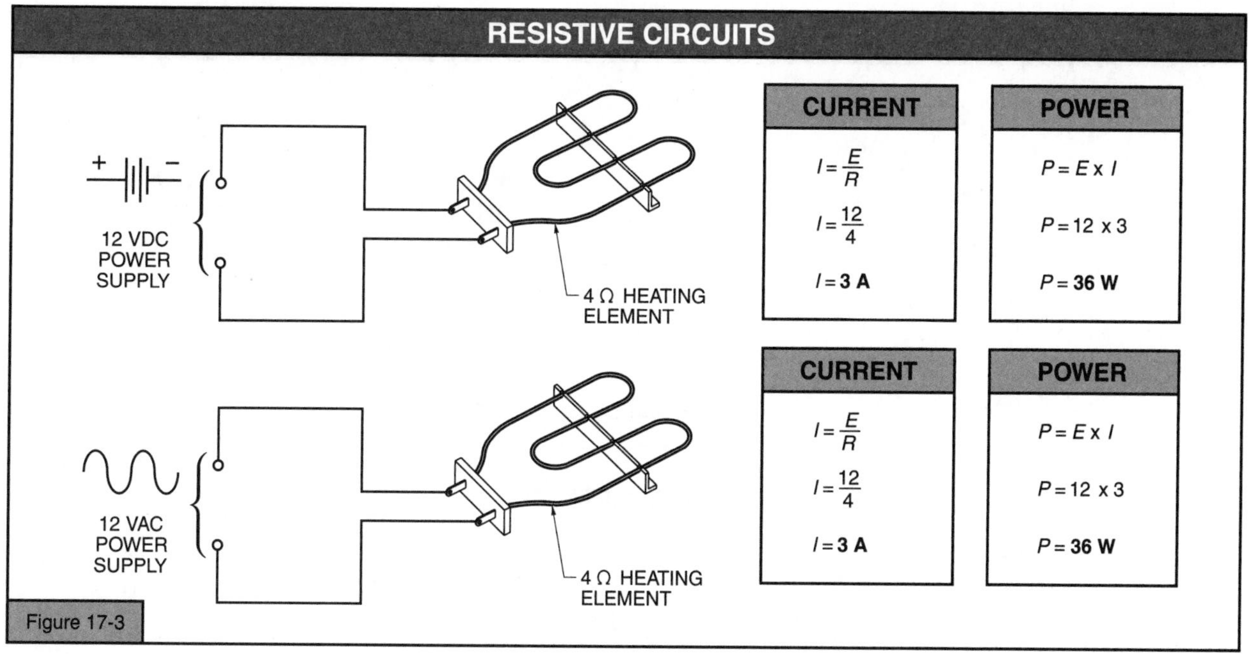

Figure 17-3

- Conductor Size (cross-sectional area)—The smaller the conductor, the greater the resistance.
- Conductor Length—The longer the wire, the greater the resistance.
- Material—Copper has less resistance than aluminum. Aluminum has less resistance than iron.

As the area of the conductor increases, its resistance decreases. As the length of the conductor (current path) increases, its resistance increases. Thus, the resistance of a conductor is directly proportional to its length and inversely proportional to its cross-sectional area.

Resistivity is the ability of a conductor having a specific length and cross-sectional area to offer resistance. Resistivity is used to compare the resistance characteristics of different materials. Materials with high resistivity are poor conductors (good insulators) and materials with low resistivity are good conductors (poor insulators).

As the temperature of a conductor increases, the free electrons between the atoms move faster and their vibrations increase. For this reason, more collisions occur between the free electrons drifting along the conductor and the atoms of the material. As the number of collisions increases, the flow of current decreases. Thus, an increase in temperature affects the resistance of the conductor and the circuit. For example, the resistance of copper at 100°C is about 43% greater than its resistance at 0°C. In general, resistance increases proportionally with an increase in temperature for most conductor materials. However, in most nonmetals, resistance decreases with an increase in temperature.

Cold and Hot Resistance

Cold resistance is the resistance of a component when operating current is not passing through the device. *Hot resistance* is the actual (true) resistance of a component when operating current is passing through the device. Hot resistance is the actual resistance that determines the amount of current flow. However, the hot resistance of a component cannot be measured because ohmmeters cannot be used in an operating circuit.

The resistance measured in some components when no power is applied (cold resistance) is not the same as the actual resistance of the component when power is connected and the circuit is operating. An incandescent lamp is an example of a component that changes resistance once power is applied. The lamp radiates light because the current flowing in the circuit raises the filament temperature to a white-hot level. The physical properties of most materials is such that their resistance increases with a increase in temperature. **See Figure 17-4.**

Although some circuits include only resistance, most circuits include other electrical characteristics in addition to resistance. For example, any circuit that includes a motor, solenoid, bell, alarm, speaker, fluorescent lamp, HID (high-intensity discharge) lamp, transformer, mechanical relay, motor starter, contactor, or an electronic circuit, includes inductance and/or capacitance in addition to resistance. Inductance and capacitance, like resistance, affect how a circuit operates and the actual amount of power the circuit uses. In some circuits, inductance and/or capacitance has more effect on current flow than resistance.

COLD AND HOT RESISTANCES

60W

16.4 Ω

SCREW BASE

CENTER CONTACT

COLD RESISTANCE MEASURED WITH POWER OFF

COLD RESISTANCE

.5 A

120 V

60W

TO 120 VAC POWER SUPPLY

$R = \frac{E}{I}$

$R = \frac{120}{.5}$

$R = \mathbf{240}\ \Omega$

HOT RESISTANCE CALCULATED FROM VOLTAGE AND CURRENT MEASUREMENTS

HOT RESISTANCE

Figure 17-4

Toroidal inductors are used for smoothing and energy storage in inductive input filter circuits.

Inductance

Inductance (L) is the property of an electric device that opposes a change in current due to its ability to store electrical energy in a magnetic field. The opposition to a change in current is a result of the energy stored in the magnetic field of a coil. Inductance is normally stated in henrys (H), millihenrys (mH), or microhenrys (μH).

Any time current flows through a conductor, a magnetic field is produced around the conductor. A conductor wrapped into a coil produces a strong magnetic field around the coil any time current flows through the coil. Direct current flow produces a constant magnetic field around the coil. Alternating current flow produces an alternating magnetic field around the coil. **See Figure 17-5.**

In a DC circuit, once the magnetic field is built, it remains at its maximum potential until the circuit (switch) is opened. Once the circuit is opened, the magnetic field collapses. In an AC circuit, the magnetic field is continuously building and collapsing until the circuit is opened.

In an AC circuit, the magnetic field also changes direction with each change in sine wave alternation.

The unit of inductance, the henry, is named after the American physicist Joseph Henry. Henry's experiments with electromagnets led to the development of the telegraph.

Magnetic Field Strength

The amount of inductance produced by a coil depends on the strength of the magnetic field produced by the coil. The strength of the magnetic field depends on the number of turns of wire, spacing between turns, core used, and wire size (gauge). **See Figure 17-6.**

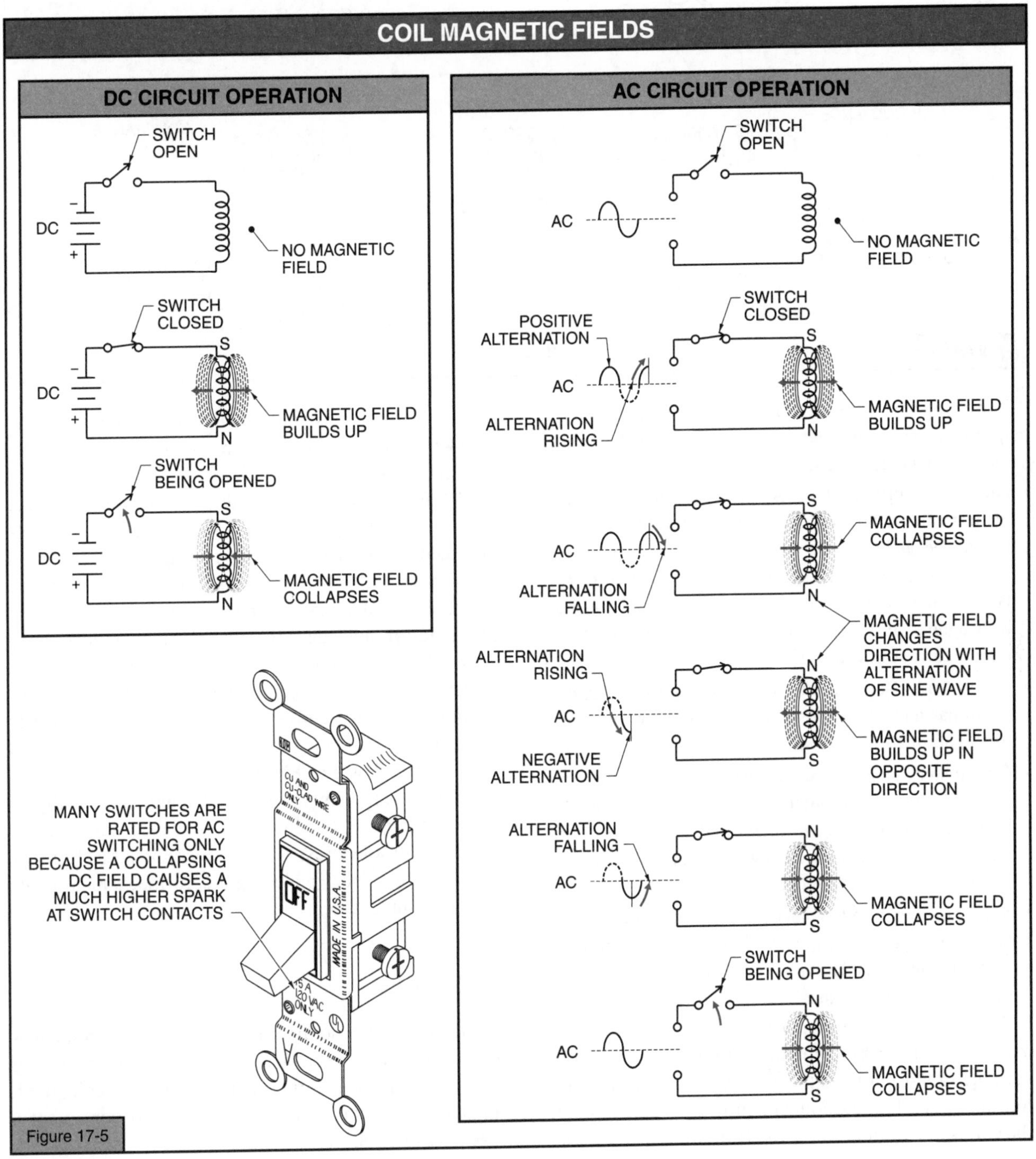

COIL MAGNETIC FIELDS

DC CIRCUIT OPERATION

SWITCH OPEN
DC
NO MAGNETIC FIELD
SWITCH CLOSED
DC
S
N
MAGNETIC FIELD BUILDS UP
SWITCH BEING OPENED
DC
S
N
MAGNETIC FIELD COLLAPSES

MANY SWITCHES ARE RATED FOR AC SWITCHING ONLY BECAUSE A COLLAPSING DC FIELD CAUSES A MUCH HIGHER SPARK AT SWITCH CONTACTS

AC CIRCUIT OPERATION

SWITCH OPEN
AC
NO MAGNETIC FIELD
POSITIVE ALTERNATION
SWITCH CLOSED
AC
ALTERNATION RISING
S
N
MAGNETIC FIELD BUILDS UP
AC
ALTERNATION FALLING
S
N
MAGNETIC FIELD COLLAPSES
MAGNETIC FIELD CHANGES DIRECTION WITH ALTERNATION OF SINE WAVE
ALTERNATION RISING
AC
NEGATIVE ALTERNATION
N
S
MAGNETIC FIELD BUILDS UP IN OPPOSITE DIRECTION
ALTERNATION FALLING
AC
N
S
MAGNETIC FIELD COLLAPSES
SWITCH BEING OPENED
AC
N
S
MAGNETIC FIELD COLLAPSES

Figure 17-5

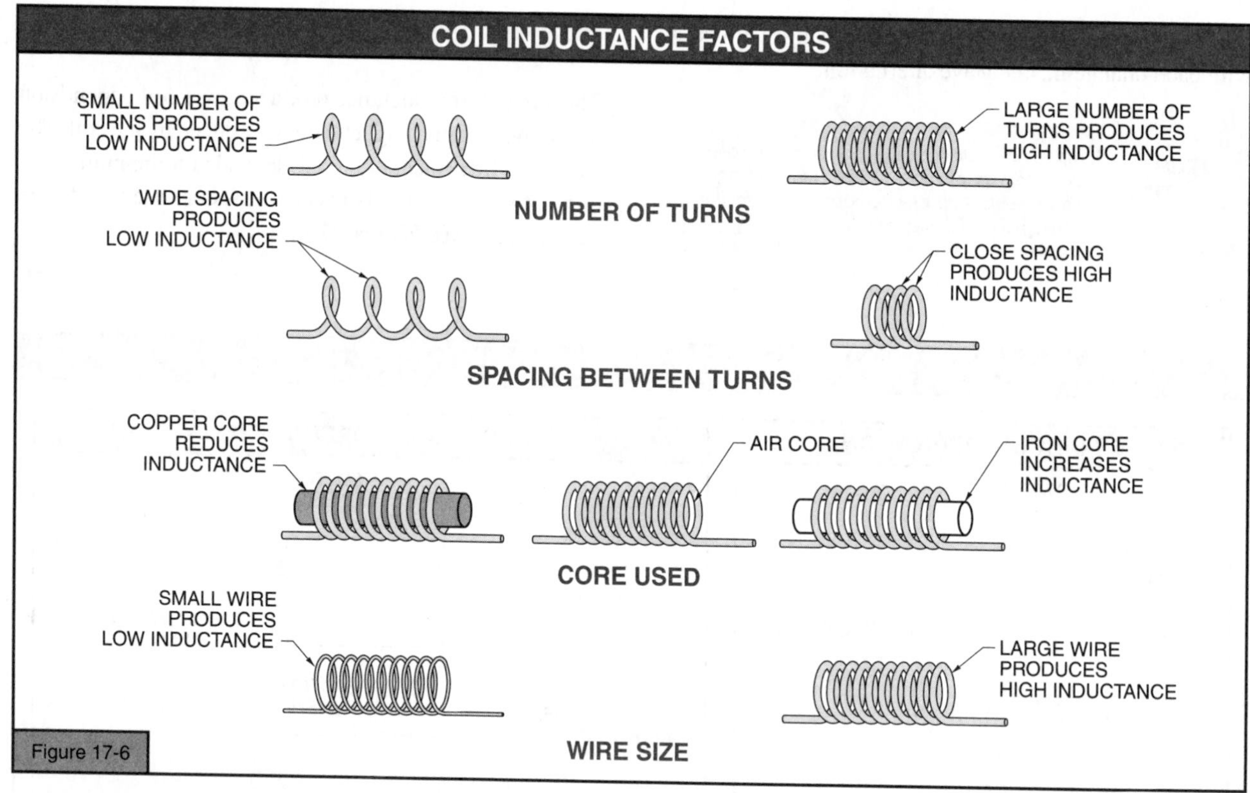

Figure 17-6

Anything that increases the strength of an inductor's magnetic field increases inductance. Factors that determine a coil's inductance include:

- Number of turns in the coil—The greater the number of turns, the higher the inductance. The inductance is greater because more counterelectromotive force (reverse voltage produced by the magnetic lines of force) is induced. Counterelectromotive force (CEMF) pushes back against the supply voltage and attempts to keep the circuit current flow at zero.
- Length and spacing of the coil—The more area (length) a coil has and/or the closer the spacing between turns, the higher the inductance.
- Core material and the relative permeability of the core material—Ferrous (containing iron) materials are the best material to use for magnetic fields. A copper core weakens the magnetic field and reduces inductance. A ferrous core increases inductance.
- Size of the wire used to make the coil—Large wire sizes allow more current to flow, which produces a stronger magnetic field. The larger the wire (and higher current), the higher the inductance.

The number of turns, size of the coil, and core material are fixed by the manufacturer. Thus, the only variable in most coils is the amount of current flowing through the coil. The amount of current flowing through the coil is determined by the total circuit resistance. The higher the total resistance, the lower the circuit current. Likewise, the lower the total resistance, the higher the circuit current.

A component may have inductance by design or due to individual components used to make it operate. Coils designed to have specific values of inductance are referred to as inductors, coils, or chokes. These components are intentionally added into a circuit to take advantage of the properties of inductance that they produce. Inductors are generally added into a circuit to control the amount of current flowing in a circuit or a branch of the circuit. For example, a ballast (coil) is added to fluorescent and HID lamp circuits to provide a high-voltage starting surge and limit current flow. **See Figure 17-7.**

Other components are not designed specifically to have inductance, but include inductance anyway. For example, all solenoids, motors, generators, transformers, and any component including a coil (mechanical relay, etc.) have inductance.

Inductors

Inductors operate on the same basic electrical principles of magnetism as transformers. The difference between an inductor and a transformer is the number of coils used. Inductors use one coil, transformers use two or more coils. **See Figure 17-8.**

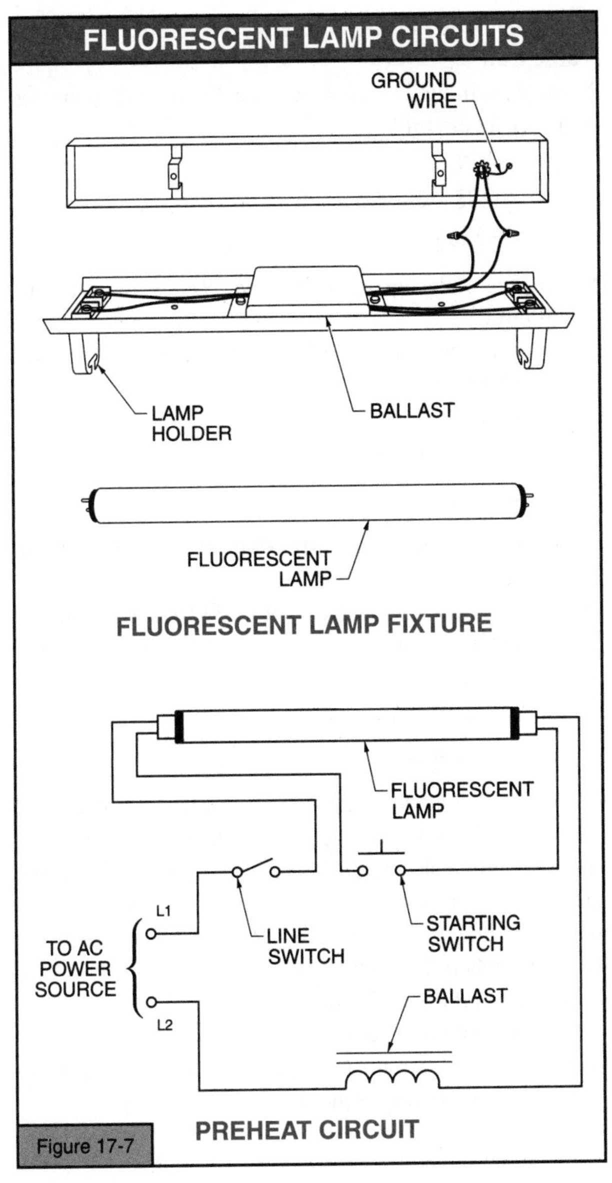

Figure 17-7

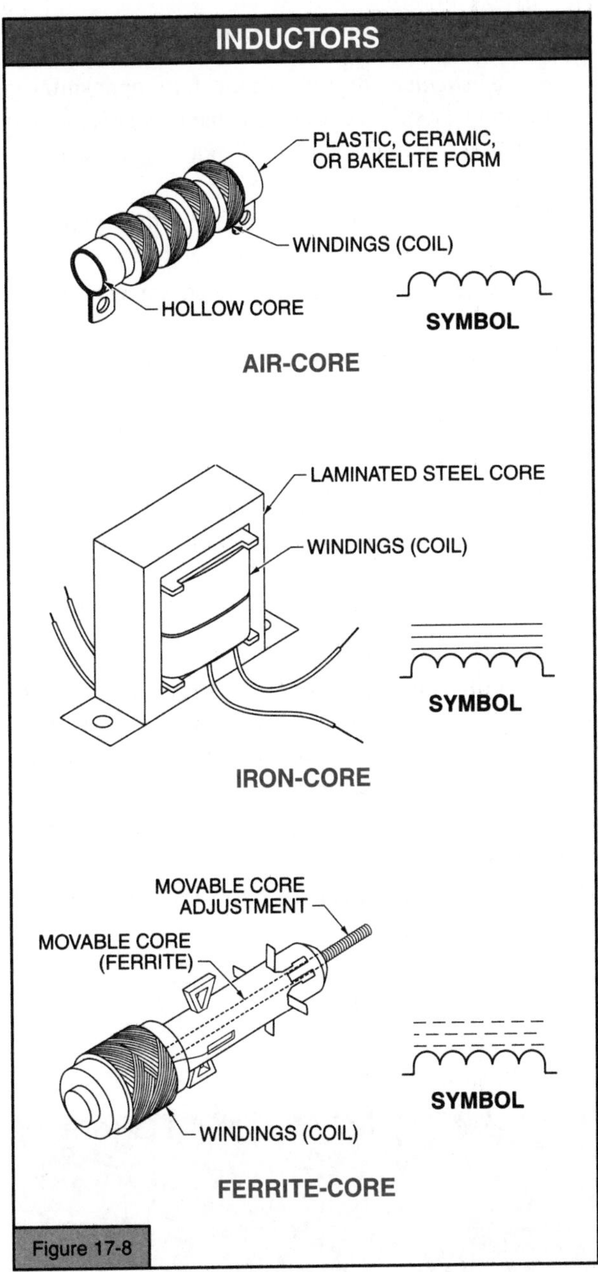

Figure 17-8

The three basic inductors include air-core, iron-core, and ferrite-core. Air-core inductors consist of a coil (copper wire) wrapped around a form (plastic, ceramic, or Bakelite) with no material in the middle. Air-core inductors do not change their inductance value (as do iron-core and ferrite-core) with a change in current. Air-core inductors are used in high frequency equipment circuits, such as radios (AM and FM), TVs, and other transmitter/receiver circuits.

Iron-core inductors consist of a coil (copper wire) wound around a laminated steel core. Iron-core inductors are commonly used in circuits that operate at standard AC power frequencies (50 Hz and 60 Hz) and at audio frequencies (20 Hz to 20 kHz).

Ferrite-core inductors consist of a core made up of ceramic materials (ferrite). *Ferrite* is a chemical compound consisting of powdered iron oxide and ceramic. Ferrite-core inductors are available with a movable ferrite core that can be adjusted for different circuit tuning applications. As the core is moved out of the coil, the air gap reduces the strength of the magnetic field and the amount of inductance. Likewise, as the core is moved into the coil, the air gap is replaced by the core and inductance is increased. Ferrite-core inductors are used to tune many electronic circuits.

AC Circuit Inductive Reactance

Inductive reactance (X_L) is an inductor's opposition to alternating current. Like resistance, inductive reactance is measured in ohms. The amount of inductive reactance in a circuit depends on:

- The amount of inductance (in henrys) of the coil (inductor). Inductance is normally a fixed amount.
- The frequency of the current. Frequency may be a fixed or variable amount.

To calculate inductive reactance for AC circuits, apply the formula:

$X_L = 2\pi fL$

where

X_L = inductive reactance (in Ω)

2π = 6.28 (indicates circular motion that produces the AC sine wave)

f = applied frequency (in Hz)

L = inductance (in H)

The formula indicates that the higher the frequency or greater the inductance, the higher the inductive reactance. Likewise, the lower the frequency or lesser the inductance, the lower the inductive reactance. **See Figure 17-9.**

Example: Calculating Inductive Reactance of a Low Value Coil

What is the inductive reactance of an 8 mH coil connected to a 50 kHz circuit?

$X_L = 2\pi fL$

$X_L = 6.28 \times 50{,}000 \times .008$

$X_L = \mathbf{2512\ \Omega}$

Fluke Corporation

Insulation testers are used for taking resistance measurements.

Example: Calculating Inductive Reactance of a High Value Coil

What is the inductive reactance of a 20 mH coil connected to a 50 kHz circuit?

$X_L = 2\pi fL$

$X_L = 6.28 \times 50{,}000 \times .02$

$X_L = \mathbf{6280\ \Omega}$

Example: Calculating Inductive Reactance at Low Frequency

What is the inductive reactance of a 100 mH coil connected to a 1 kHz circuit?

$X_L = 2\pi fL$

$X_L = 6.28 \times 1000 \times .1$

$X_L = \mathbf{628\ \Omega}$

Example: Calculating Inductive Reactance at High Frequency

What is the inductive reactance of a 100 mH coil connected to a 100 kHz circuit?

$X_L = 2\pi fL$

$X_L = 6.28 \times 100{,}000 \times .1$

$X_L = \mathbf{62{,}800\ \Omega}$

Inductive reactance can take the place of resistance when using Ohm's law because inductive reactance is, in effect, AC resistance. This can be done if voltage and current measurements are taken on a circuit in which frequency is known (or measured). Inductive reactance can be found when the voltage across and current through a coil is measured. **See Figure 17-10.**

To calculate inductive reactance when voltage across a coil (E_L) and current through a coil (I_L) are known (or measured), apply the formula:

$$X_L = \frac{E_L}{I_L}$$

where

X_L = inductive reactance (in Ω)

E_L = voltage across coil (in V)

I_L = current through coil (in A)

Example: Calculating Inductance Reactance—Voltage and Current Measured

What is the inductive reactance of a circuit that contains a coil connected to a 120 VAC power source and has 80 mA of current flow?

$$X_L = \frac{E_L}{I_L}$$

$$X_L = \frac{120}{.08}$$

$X_L = \mathbf{1500\ \Omega}$

RESISTIVE AND INDUCTIVE AC CIRCUITS

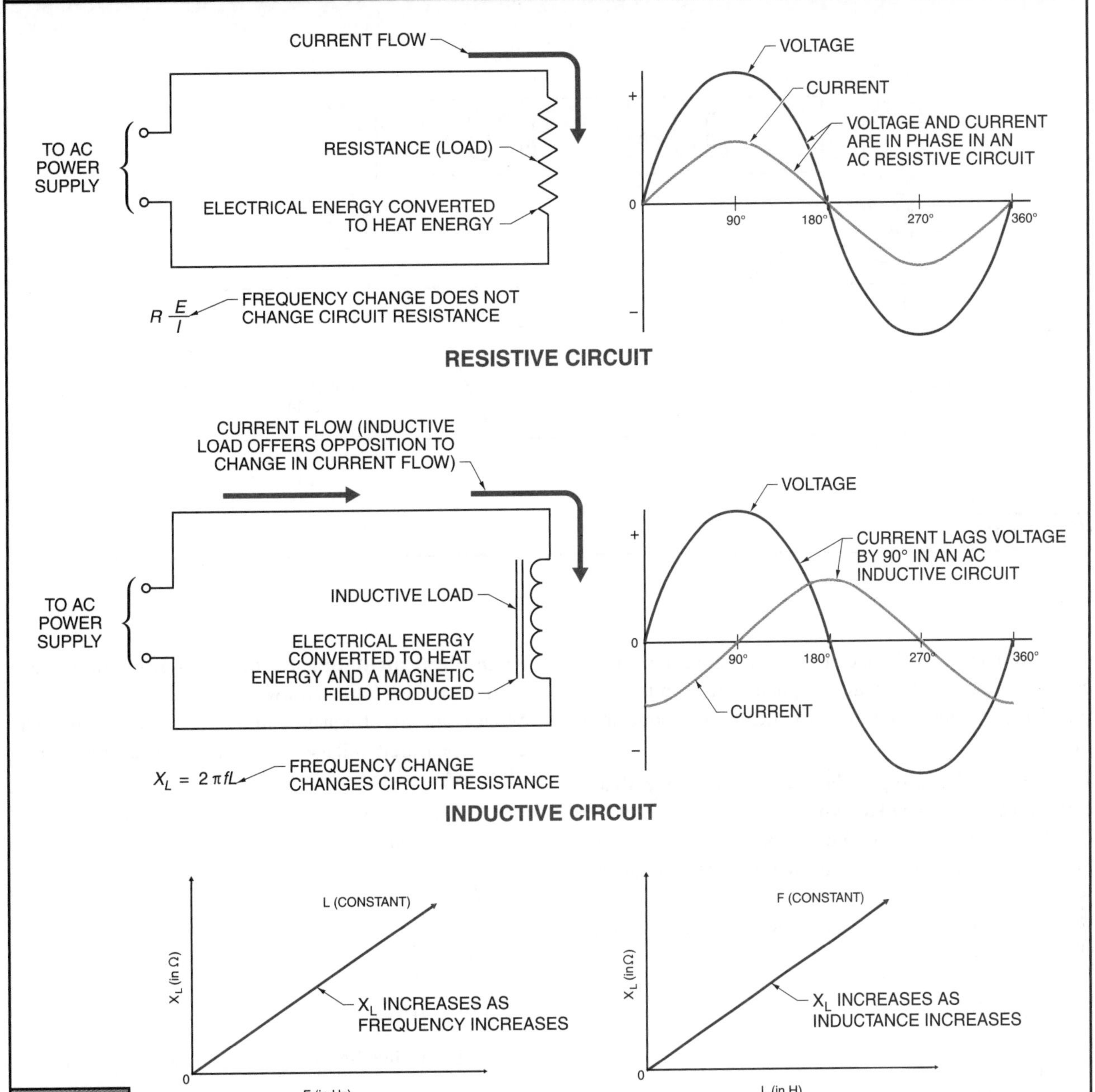

CURRENT FLOW
TO AC POWER SUPPLY
RESISTANCE (LOAD)
ELECTRICAL ENERGY CONVERTED TO HEAT ENERGY
$R = \frac{E}{I}$ — FREQUENCY CHANGE DOES NOT CHANGE CIRCUIT RESISTANCE
VOLTAGE
CURRENT
VOLTAGE AND CURRENT ARE IN PHASE IN AN AC RESISTIVE CIRCUIT
+ 0 − 90° 180° 270° 360°

RESISTIVE CIRCUIT

CURRENT FLOW (INDUCTIVE LOAD OFFERS OPPOSITION TO CHANGE IN CURRENT FLOW)
TO AC POWER SUPPLY
INDUCTIVE LOAD
ELECTRICAL ENERGY CONVERTED TO HEAT ENERGY AND A MAGNETIC FIELD PRODUCED
$X_L = 2\pi fL$ — FREQUENCY CHANGE CHANGES CIRCUIT RESISTANCE
VOLTAGE
CURRENT LAGS VOLTAGE BY 90° IN AN AC INDUCTIVE CIRCUIT
CURRENT
+ 0 − 90° 180° 270° 360°

INDUCTIVE CIRCUIT

L (CONSTANT)
X_L (in Ω)
X_L INCREASES AS FREQUENCY INCREASES
0
F (in Hz)

F (CONSTANT)
X_L (in Ω)
X_L INCREASES AS INDUCTANCE INCREASES
0
L (in H)

Figure 17-9

Inductive reactance may be calculated when the frequency and inductance are known. Also, the third variable may be found if any two factors (inductive reactance, frequency, or inductance) are known. To calculate any one of the unknown quantities, apply the formulas:

$$X_L = 2\pi fL,\ L = \frac{X_L}{2\pi f},\ \text{or}\ f = \frac{X_L}{2\pi L}$$

where

X_L = inductive reactance (in Ω)

2π = 6.28 (indicates circular motion that produces the AC sine wave)

f = applied frequency (in Hz)

L = inductance (in H)

TECH FACT *An inductor placed in the power line of an automobile radio allows DC to pass and blocks any high-frequency AC, such as ignition noise and static, from entering the radio.*

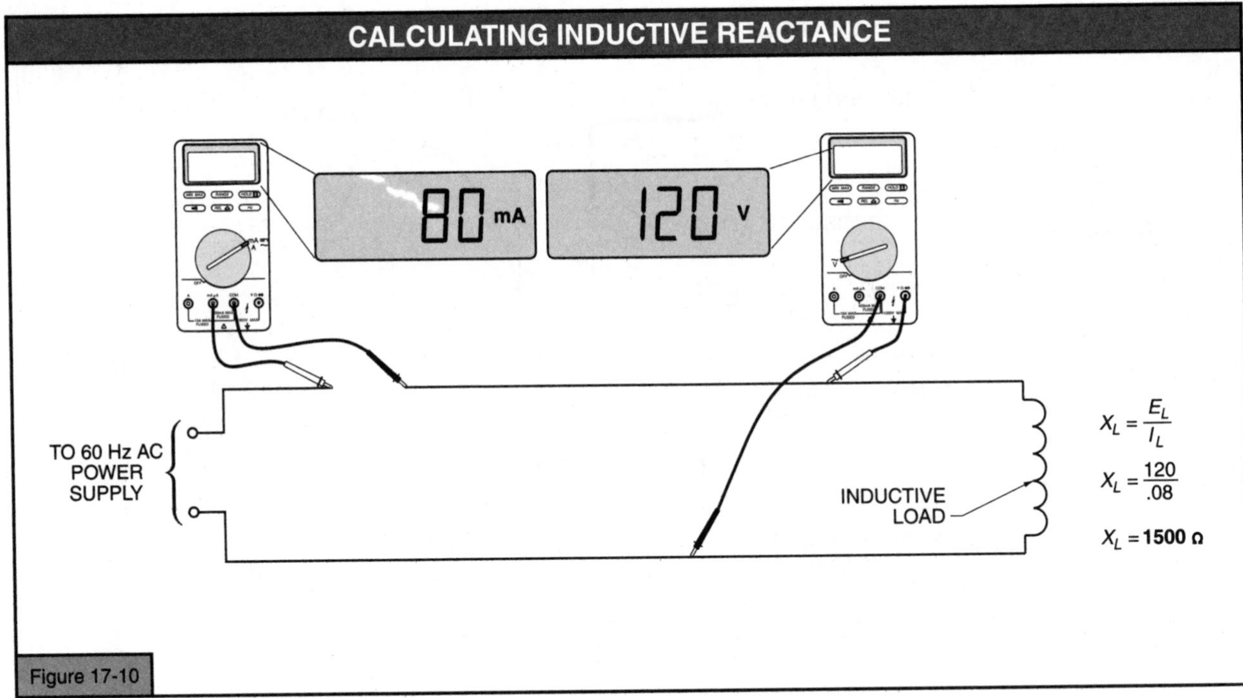

Figure 17-10

A voltmeter and ammeter may be connected into a circuit to measure the circuit's applied voltage and current flow. These values may be used to calculate the inductance of the coil.

Example: Calculating Inductance—Inductive Reactance and Frequency Known

What is the inductance of a coil when the voltage in the circuit is 24 VAC, the current is 60 mA, and the frequency is 60 Hz?

1. Calculate inductive reactance using measured voltage and current.

$$X_L = \frac{E_L}{I_L}$$

$$X_L = \frac{24}{.06}$$

$$X_L = \mathbf{400\ \Omega}$$

2. Calculate inductance based on inductive reactance and frequency.

$$L = \frac{X_L}{2\pi f}$$

$$L = \frac{400}{6.28 \times 60}$$

$$L = \frac{400}{376.8}$$

$$L = \mathbf{1.06\ H}$$

Example: Calculating Frequency—Inductance and Inductive Reactance Known

What is the line frequency in a circuit containing a 1 H coil, a measured voltage of 24 V, and a measured current of 76 mA?

1. Calculate inductive reactance based on measured voltage and current.

$$X_L = \frac{E_L}{I_L}$$

$$X_L = \frac{24}{.076}$$

$$X_L = \mathbf{316\ \Omega}$$

2. Calculate line frequency based on inductive reactance.

$$f = \frac{X_L}{2\pi L}$$

$$f = \frac{316}{6.28 \times 1}$$

$$f = \mathbf{50.3\ Hz}$$

Note: Line frequency may be changed (or adjusted) intentionally as it is in most electronic communication equipment or it may be changed when a piece of equipment is used in a different country. For example, the standard line frequency in the U.S. is 60 Hz. The standard line frequency in Europe is 50 Hz. Such circuit changes must be considered when designing (or using) equipment overseas.

DC Circuit Inductive Reactance

At zero frequency (DC circuits), there is no opposition due to inductance. In DC circuits, a coil has low resistance and no inductive reactance. Current flow through a coil connected to DC is limited only by the resistance of the wire. The inductor has no resistance (X_L) except the small resistance of the wire. Thus, for a steady direct current without any change in current, inductive reactance is normally considered to be zero. The amount of resistance is less with a large wire or less turns. The amount of resistance is greater with a small wire or more turns. **See Figure 17-11.**

Current flow through a coil connected to AC is limited by the wire resistance and the inductive reactance. In AC circuits, a coil has low resistance and high inductive reactance. This reduces current flow in the circuit and produces a voltage drop across the coil.

There are many circuit applications that use inductors. For example, circuits use inductors where it is desired to have high resistance (X_L) to alternating current flow but little resistance (X_L) to direct current flow. Inductors are also used in circuits to produce more opposition to a high frequency alternating current compared with low frequency alternating current.

Series Inductive Reactance

The inductive reactance of a circuit containing more than one coil is found by first determining the total inductance of the circuit and then calculating the inductive reactance. The inductive reactance of three coils connected in series is calculated by adding the inductance of each coil to find the total inductance and then calculating the inductive reactance. **See Figure 17-12.**

Example: Calculating Inductive Reactance—Coils Connected in Series

What is the inductive reactance of a circuit containing coils of 4 mH, 6 mH, and 20 mH connected to a 1000 Hz AC supply?

1. Calculate total inductance of coils connected in series.

$L_T = L_1 + L_2 + L_3 + ...$

$L_T = 4 + 6 + 20$

$L_T = 30 \text{ mH}$

2. Calculate inductive reactance based on total inductance.

$X_L = 2\pi fL$

$X_L = 6.28 \times 1000 \times .03$

$X_L = \mathbf{188.4\ \Omega}$

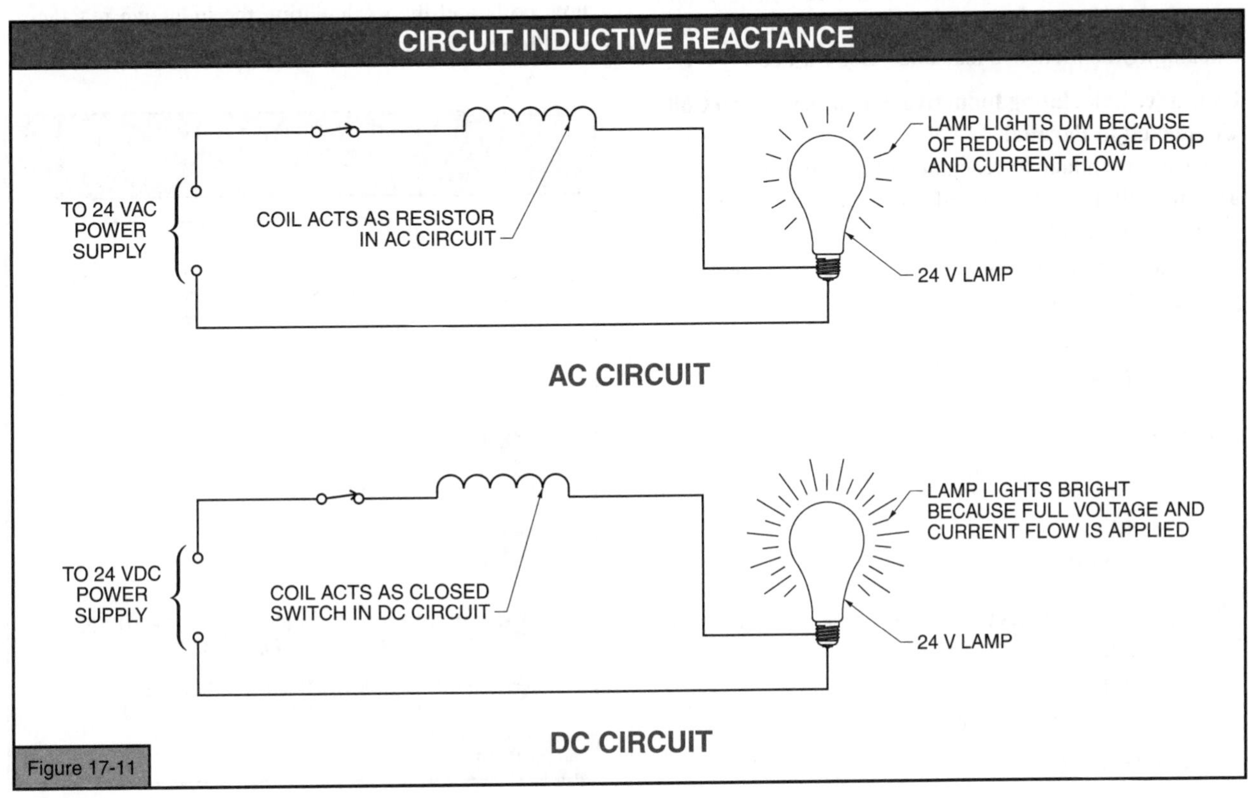

Figure 17-11

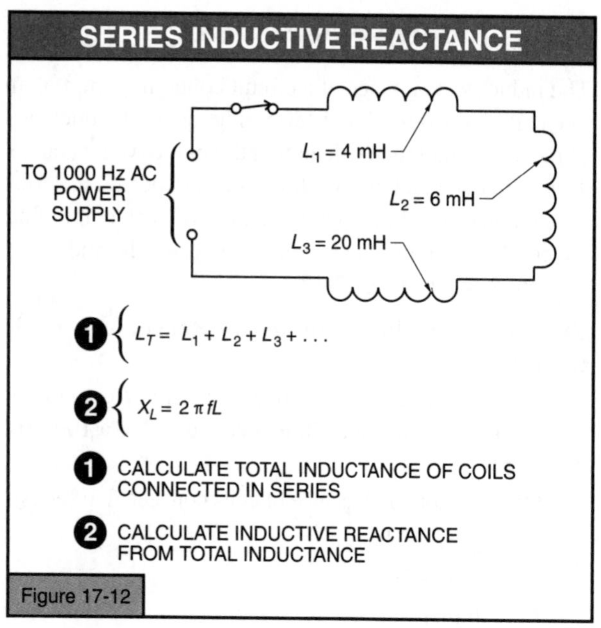

Figure 17-12

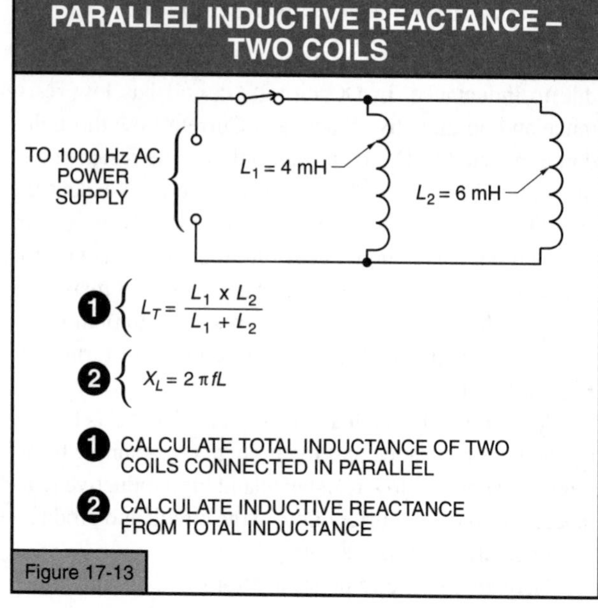

Figure 17-13

Parallel Inductive Reactance

The total inductance in a circuit containing parallel-connected coils is less than the smallest coil value. The total inductance of two coils connected in parallel is found by applying the same formula for calculating the resistance of two loads connected in parallel. The inductive reactance in a circuit containing two parallel-connected coils is found by calculating the total inductance of the two coils and then calculating the inductive reactance. **See Figure 17-13.**

Example: Calculating Inductive Reactance—Two Coils Connected in Parallel

What is the inductive reactance of a circuit containing a 4 mH coil and a 6 mH coil connected in parallel to a 1000 Hz AC supply?

1. Calculate the total inductance of the two coils connected in parallel.

$$L_T = \frac{L_1 \times L_2}{L_1 + L_2}$$

$$L_T = \frac{4 \times 6}{4 + 6}$$

$$L_T = \frac{24}{10}$$

$$L_T = \mathbf{2.4\ mH}$$

2. Calculate the inductive reactance based on the total reactance.

$$X_L = 2\pi fL$$

$$X_L = 6.28 \times 1000 \times .0024$$

$$X_L = \mathbf{15.072\ \Omega}$$

The total inductance of three coils connected in parallel is found by applying the same formula for calculating the resistance of three loads connected in parallel. The total circuit inductance decreases if coils are added in parallel and increases if coils are removed. The inductive reactance in a circuit containing three parallel-connected coils is found by calculating the total inductance of the three coils and then calculating the inductive reactance. **See Figure 17-14.**

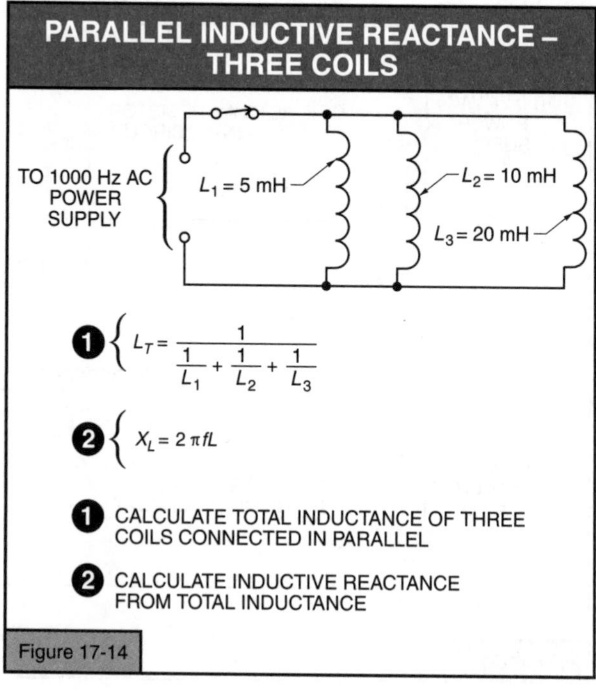

Figure 17-14

Example: Calculating Inductive Reactance—Three Coils Connected in Parallel

What is the inductive reactance of a circuit containing three parallel-connected coils with inductances of 5 mH, 10 mH, and 20 mH connected to a 1000 Hz supply?

1. Calculate total inductance of three coils connected in parallel.

$$L_T = \frac{1}{\frac{1}{L_1}+\frac{1}{L_2}+\frac{1}{L_3}}$$

$$L_T = \frac{1}{\frac{1}{5}+\frac{1}{10}+\frac{1}{20}}$$

$$L_T = \frac{1}{.20+.10+.05}$$

$$L_T = \frac{1}{.35}$$

$$L_T = \mathbf{2.86\ mH}$$

2. Calculate inductive reactance based on circuit total inductance.

$$X_L = 2\pi f L$$

$$X_L = 6.28 \times 1000 \times .00286$$

$$X_L = \mathbf{17.9608\ \Omega}$$

Capacitance

Capacitance (C) is the ability of a component or circuit to store energy in the form of an electrical charge. A capacitor is an electric device specifically designed to store a charge of energy. The unit of capacitance is the farad (F). The farad is a unit that is too large for most electrical/electronic applications. Thus, capacitance (and capacitor values) are normally stated in microfarads (μF) or picofarads (pF).

A capacitor consists of two conductors (plates) separated by an insulator (dielectric), which allows an electrostatic charge to be developed. The electrostatic charge becomes a source of stored energy. The strength of the charge depends on the applied voltage, size of the conductors, and quality of the insulation. The dielectric may be air, paper, oil, ceramic, mica, or any other nonconducting material. **See Figure 17-15.**

The closer the two plates are placed together, the greater the charge on them. A capacitor is charged when its leads are connected to a DC voltage source. The positive terminal of the voltage source attracts electrons from the plate connected to it and the negative terminal of the voltage source repels an equal number of electrons into the negative plate. Charging (current flow) continues until the voltage across the charged plates is equal to the applied voltage. At this time the capacitor is fully charged.

Once charged, the voltage source may be disconnected from the capacitor. The stored energy remains between the two capacitor plates. Thus, a capacitor has the ability to store the amount of charge equal to the applied charging voltage. The capacitor holds this charge until something is connected between the two leads. At this time the capacitor discharges its stored energy. Once discharged, the capacitor must be recharged.

Capacitors are rated for voltage and capacitance. The voltage rating indicates the maximum voltage that can safely be applied to the capacitor leads. The amount of capacitance a capacitor has is determined by:

- The area of the plates. The greater the plate area, the higher the capacitance value.
- The spacing between the plates. The closer the plates, the higher the capacitance.
- The dielectric used. The better the dielectric material, the higher the capacitance value.
- Connection arrangement. Capacitors are connected in parallel to increase total circuit capacitance.

Capacitors

Capacitors may be fixed or variable. A *fixed capacitor* is a capacitor that has one value of capacitance. A *variable capacitor* is a capacitor that varies in capacitance value. Fixed capacitors are classified according to the dielectric used. Fixed capacitors include mica, paper, ceramic, and electrolytic capacitors. Mica, paper, and ceramic capacitors are not polarized. These capacitors may be connected in a circuit with either terminal lead connected to the positive side of the circuit. Electrolytic capacitors are polarized. They must be connected in a circuit with the positive terminal lead connected to the positive side of the circuit. **See Figure 17-16.**

Mica Capacitors

Mica capacitors use mica as the dielectric. Mica capacitors are made by stacking mica sheets between a metal foil (tinfoil or aluminum sheets). The strips of tinfoil are connected to create one plate and the mica sheets are connected to create the other plate. Mica capacitors are small in size and have low capacitance ratings. Although mica capacitors have low capacitance values, they have high-voltage ratings. For this reason, mica capacitors are often used in high-voltage circuits. Mica capacitors commonly range from 1 pF to about .1 μF.

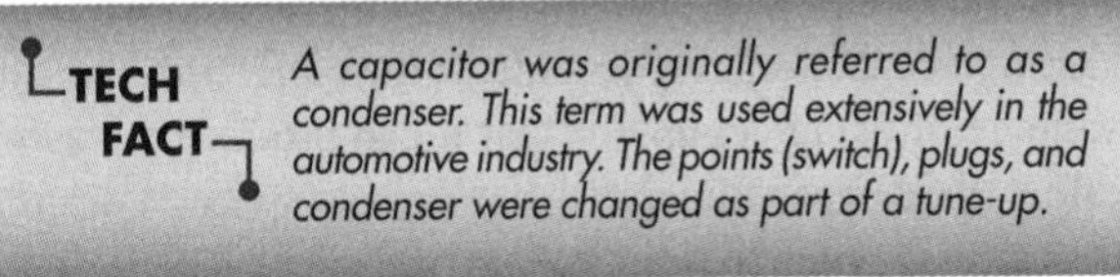

A capacitor was originally referred to as a condenser. This term was used extensively in the automotive industry. The points (switch), plugs, and condenser were changed as part of a tune-up.

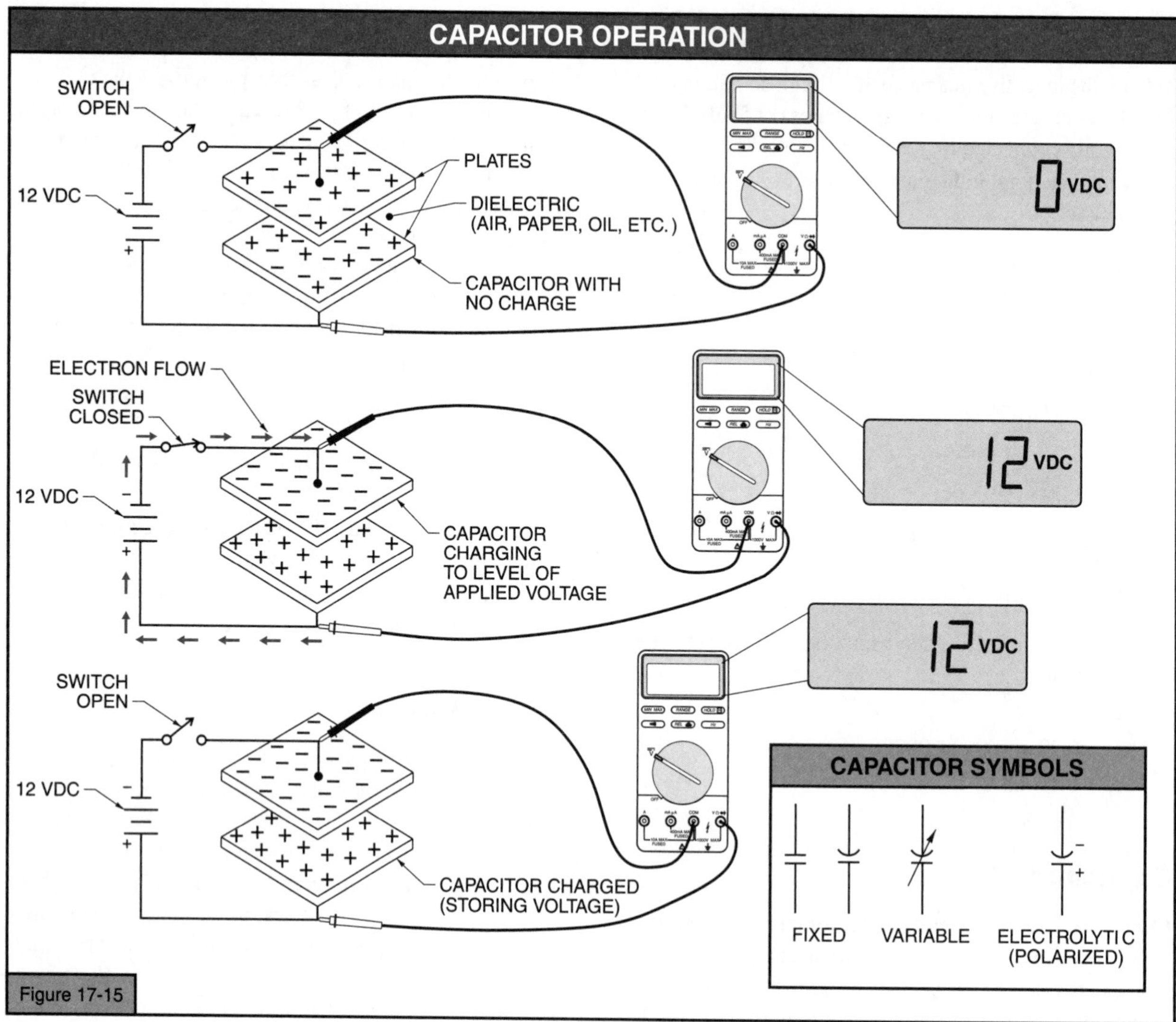

Figure 17-15

Paper (Plastic) Capacitors

Paper capacitors use wax paper or plastic film as the dielectric. In newer capacitors, mylar has replaced the paper film. Paper capacitors are made of flat strips of metal foil separated by a dielectric rolled into a compact cylinder. To prevent moisture problems, they are normally sealed in wax. Paper capacitors have low capacitance ratings from .001 μF to 1 μF. Their voltage ratings are normally less than 600 V.

Ceramic Capacitors

Ceramic capacitors use ceramic as the dielectric. The advantage of using ceramic is that it can produce the same effect as a paper dielectric, but in less space. Ceramic capacitors commonly range from 500 pF to .01 μF.

Electrolytic Capacitors

Electrolytic capacitors use a paste as the dielectric. Electrolytic capacitors provide more capacitance for their size than any other type of capacitor. For this reason, electrolytic capacitors are used when a very high capacitance is required. Electrolytic capacitors may have 100,000 μF of capacitance or more.

Variable Capacitors

Variable capacitors normally use air or mylar as the dielectric and include movable and stationary metal plates. The capacitance is maximum when the movable plates are fully meshed (but not touching) with the stationary plates. Moving the plates apart lowers the capacitance. Variable capacitors are used as tuning capacitors in radio receivers. The capacitance is varied to tune in different stations.

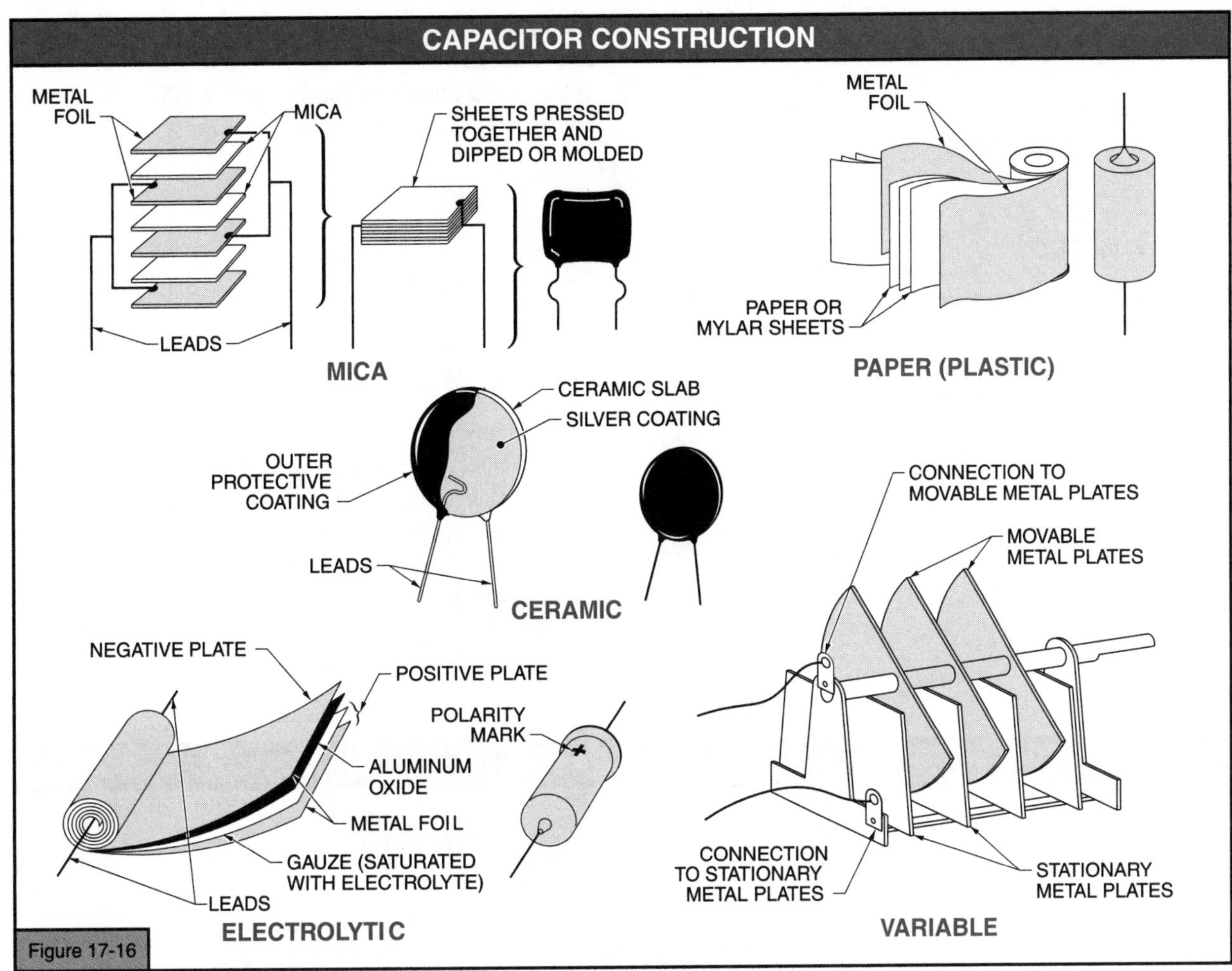

Figure 17-16

Capacitive Reactance

The charges on the plates of a capacitor reverse with each change in the applied voltage polarity when a capacitor is connected to an AC power supply. The plates are alternately charged and discharged. Capacitors offer opposition to the flow of current in a circuit when connected to an AC power supply. *Capacitive reactance* (X_C) is the opposition to current flow by a capacitor. Like inductive reactance (X_L), capacitive reactance is expressed in ohms. **See Figure 17-17.** To calculate capacitive reactance, apply the formula:

$$X_C = \frac{1}{2\pi fC}$$

where

X_C = capacitive reactance (in Ω)

2π = 6.28 (indicates circular motion that produces the AC sine wave)

f = applied frequency (in Hz)

C = capacitance (in F)

The formula for capacitive reactance shows that the higher the frequency or the greater the capacitance, the lower the capacitive reactance. Likewise, the lower the frequency or the lower the capacitance, the higher the capacitive reactance.

Example: Calculating Capacitive Reactance—High Value Capacitor

What is the capacitive reactance of a 50 μF capacitor connected to a 60 Hz supply?

$$X_C = \frac{1}{2\pi fC}$$

$$X_C = \frac{1}{6.28 \times 60 \times .00005}$$

$$X_C = \frac{1}{.01884}$$

$$X_C = \mathbf{53.08\ \Omega}$$

Example: Calculating Capacitive Reactance—Low Value Capacitor

What is the capacitive reactance of a 10 μF capacitor connected to a 60 Hz supply?

$$X_C = \frac{1}{2\pi fC}$$

$$X_C = \frac{1}{6.28 \times 60 \times .00001}$$

$$X_C = \frac{1}{.0038}$$

$$X_C = \mathbf{263.16\ \Omega}$$

Example: Calculating Capacitive Reactance—High Frequency

What is the capacitive reactance of a 100 μF capacitor connected to a 10,000 Hz supply?

$$X_C = \frac{1}{2\pi fC}$$

$$X_C = \frac{1}{6.28 \times 10{,}000 \times .0001}$$

$$X_C = \frac{1}{6.28}$$

$$X_C = \mathbf{.159\ \Omega}$$

Example: Calculating Capacitive Reactance—Low Frequency

What is the capacitive reactance of a 100 μF capacitor connected to a 1000 Hz supply?

$$X_C = \frac{1}{2\pi fC}$$

$$X_C = \frac{1}{6.28 \times 1000 \times .0001}$$

$$X_C = \frac{1}{.628}$$

$$X_C = \mathbf{1.59\ \Omega}$$

Capacitive reactance, like inductive reactance is, in effect, AC resistance. For this reason, it may take the place of resistance when using Ohm's law. This may be done if voltage and current measurements are taken on a circuit in which frequency is known (or measured). **See Figure 17-18.**

To calculate capacitive reactance when voltage across the capacitor (E_C) and current through the capacitor (I_C) are known, apply the formula:

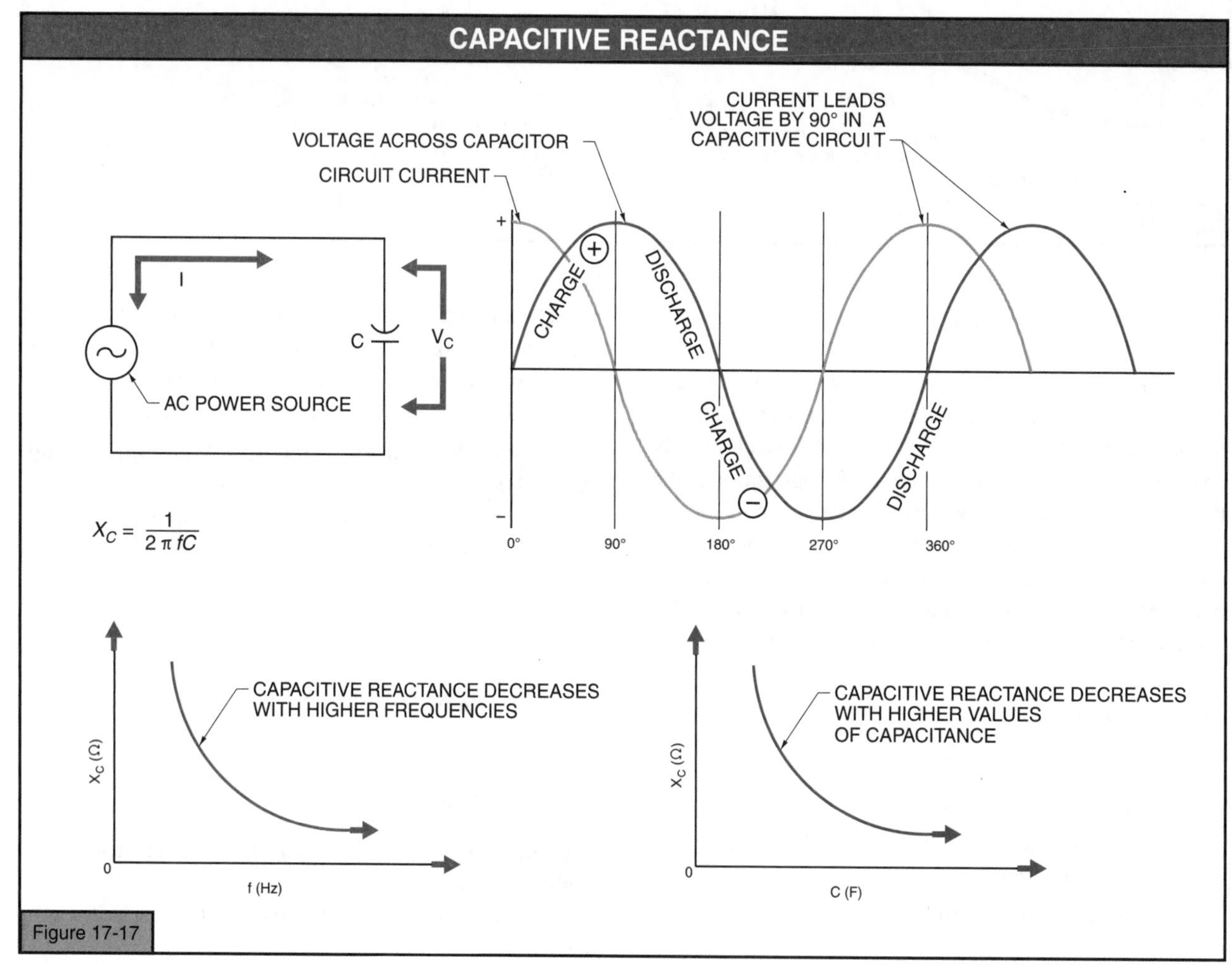

Figure 17-17

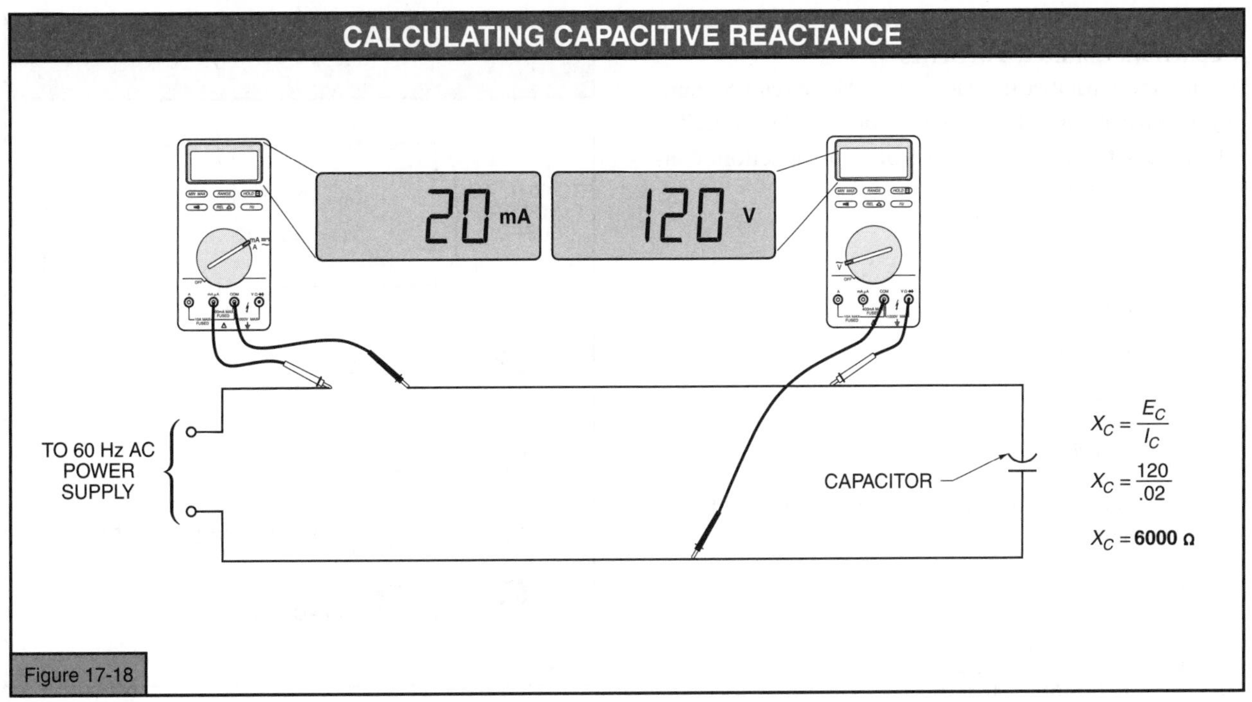

Figure 17-18

$$X_C = \frac{E_C}{I_C}$$

where

X_C = capacitive reactance (in Ω)

E_C = voltage across capacitor (in V)

I_C = current through capacitor (in A)

Example: Calculating Capacitive Reactance—Voltage and Current Known

What is the capacitive reactance of a 120 VAC circuit with 20 mA of current flow?

$$X_C = \frac{E_C}{I_C}$$

$$X_C = \frac{120}{.02}$$

$$X_C = \mathbf{6000\ \Omega}$$

Series Capacitive Reactance

The capacitive reactance of a circuit containing capacitors connected in series is found by determining the total capacitance of the circuit and then calculating the capacitive reactance based on the total capacitance. The total capacitance of two capacitors connected in series is determined by using the same formula used to calculate total resistance of two resistors connected in parallel. **See Figure 17-19.**

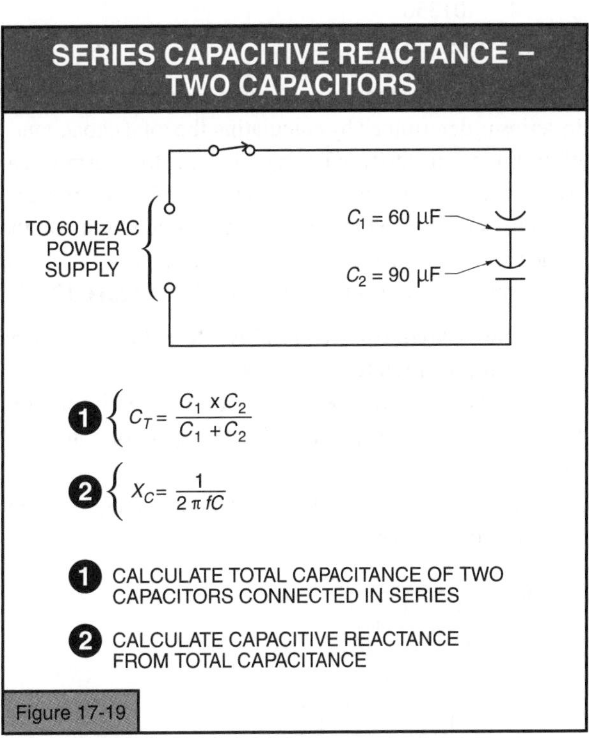

Figure 17-19

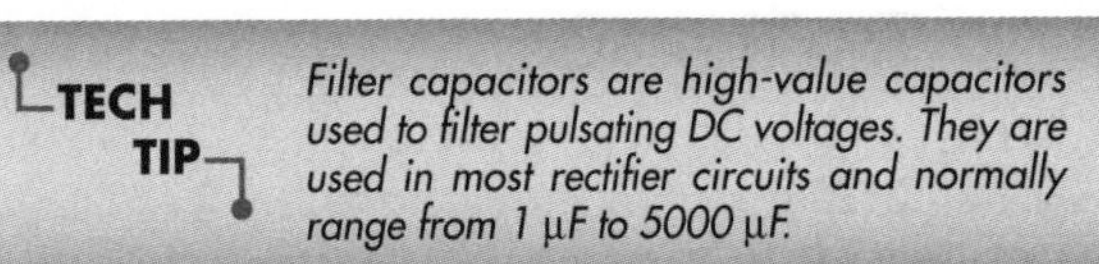

TECH TIP

Filter capacitors are high-value capacitors used to filter pulsating DC voltages. They are used in most rectifier circuits and normally range from 1 μF to 5000 μF.

Example: Calculating Capacitive Reactance—Two Capacitors Connected in Series
What is the capacitive reactance of a 60 Hz circuit containing a 60 μF and a 90 μF capacitor connected in series?

1. Calculate total capacitance of two capacitors connected in series.

$$C_T = \frac{C_1 \times C_2}{C_1 + C_2}$$
$$C_T = \frac{60 \times 90}{60 + 90}$$
$$C_T = \frac{5400}{150}$$
$$C_T = \mathbf{36\ \mu F}$$

2. Calculate capacitive reactance based on circuit total capacitance.

$$X_C = \frac{1}{2\pi fC}$$
$$X_C = \frac{1}{6.28 \times 60 \times .000036}$$
$$X_C = \frac{1}{.01356}$$
$$X_C = \mathbf{73.75\ \Omega}$$

The capacitive reactance of three capacitors connected in series is determined by calculating the total capacitance of the three capacitors and using this value to determine capacitive reactance. The total capacitance of three (or more) capacitors connected in series is determined by using the same formula used to calculate total resistance of three (or more) resistors connected in parallel. **See Figure 17-20.**

Example: Calculating Capacitive Reactance—Three Capacitors Connected in Series
What is the capacitive reactance of a 60 Hz circuit containing a 200 μF, 200 μF, and 400 μF capacitor connected in series.

1. Calculate total capacitance of three capacitors connected in series.

$$C_T = \frac{1}{\frac{1}{C_1} + \frac{1}{C_2} + \frac{1}{C_3}}$$
$$C_T = \frac{1}{\frac{1}{200} + \frac{1}{200} + \frac{1}{400}}$$
$$C_T = \frac{1}{.005 + .005 + .0025}$$
$$C_T = \frac{1}{.0125}$$
$$C_T = \mathbf{80\ \mu F}$$

SERIES CAPACITIVE REACTANCE – THREE CAPACITORS

TO 60 Hz AC POWER SUPPLY

C_1 = 200 μF

C_2 = 200 μF

C_3 = 400 μF

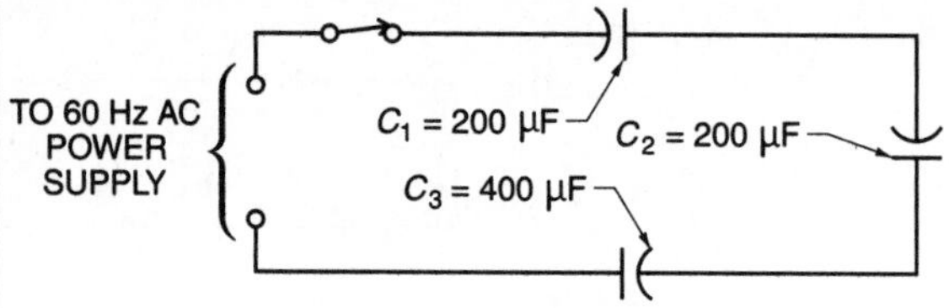

1 $C_T = \frac{1}{\frac{1}{C_1} + \frac{1}{C_2} + \frac{1}{C_3}}$

2 $X_C = \frac{1}{2\pi fC}$

1 CALCULATE TOTAL CAPACITANCE OF THREE CAPACITORS CONNECTED IN SERIES

2 CALCULATE CAPACITIVE REACTANCE FROM TOTAL CAPACITANCE

Figure 17-20

2. Calculate capacitive reactance based on circuit total capacitance.

$$X_C = \frac{1}{2\pi fC}$$
$$X_C = \frac{1}{6.28 \times 60 \times .00008}$$
$$X_C = \frac{1}{.03}$$
$$X_C = \mathbf{33.33\ \Omega}$$

Parallel Capacitive Reactance

The capacitive reactance of capacitors connected in parallel is determined by calculating the total capacitance of the parallel-connected capacitors and using this value to calculate capacitive reactance. The total capacitance of a parallel circuit is determined by adding the value of the individual capacitors. **See Figure 17-21.**

Example: Calculating Capacitive Reactance—Capacitors Connected in Parallel
What is the total capacitive reactance of a 60 Hz circuit containing a 50 μF, 75 μF, and a 25 μF capacitor connected in parallel?

1. Calculate total capacitance of capacitors connected in parallel.

$$C_T = C_1 + C_2 + C_3 + ...$$
$$C_T = 50 + 75 + 25$$
$$C_T = \mathbf{150\ \mu F}$$

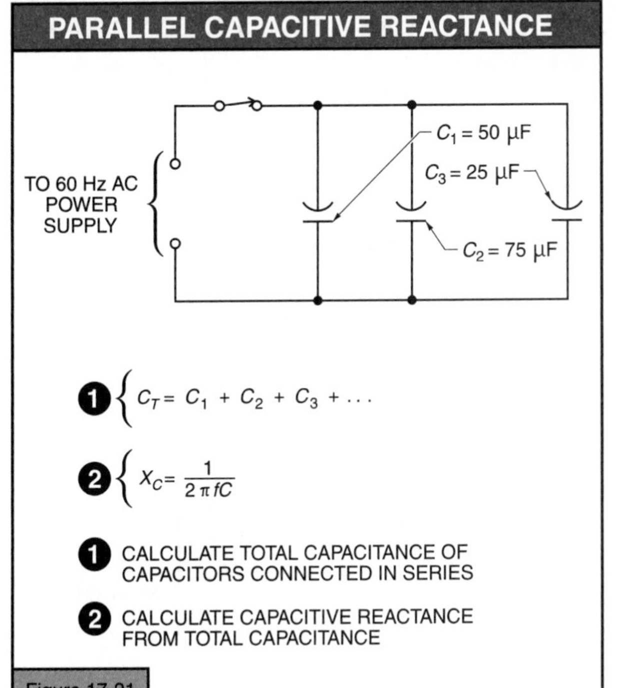

Figure 17-21

2. Calculate capacitive reactance based on circuit total capacitance.

$$X_C = \frac{1}{2\pi fC}$$

$$X_C = \frac{1}{6.28 \times 60 \times .00015}$$

$$X_C = \frac{1}{.0565}$$

$$X_C = \mathbf{17.7\ \Omega}$$

Impedance

Inductors and capacitors are electrical energy-storing devices that are used in AC and DC applications such as timing circuits, surge protection, filtering circuits, fluorescent and HID lamp circuits, 1ϕ motors, motor control circuits, etc. Inductors are used to store a current charge and capacitors are used to store a voltage charge. Inductors and capacitors, when used together, each contribute to the circuit operation. For example, a capacitor and inductor may be used in a filter circuit to smooth a pulsating direct current. **See Figure 17-22.**

In AC circuits, resistance, inductive reactance, and capacitive reactance all limit current flow. Most AC circuits contain all three oppositions to current flow. The exact behavior of current in an AC circuit depends on the amount of resistance, inductive reactance, and capacitive reactance.

Impedance (Z) is the total opposition of any combination of resistance, inductive reactance, and capacitive reactance offered to the flow of alternating current. Impedance, like inductive reactance and capacitive reactance, is measured in ohms. **See Figure 17-23.**

In circuits containing only resistance, current and voltage are normally in-phase. The phase angle between voltage and current is zero (or close to zero).

In circuits containing resistance and inductive reactance (RL), resistance and inductive reactance oppose current flow. RL circuit impedance is the combined effect of the resistance and inductive reactance. For this reason, current lags voltage by a phase angle between 0° and 90°. The more inductive reactance in the circuit, the greater the phase shift (closer to 90°).

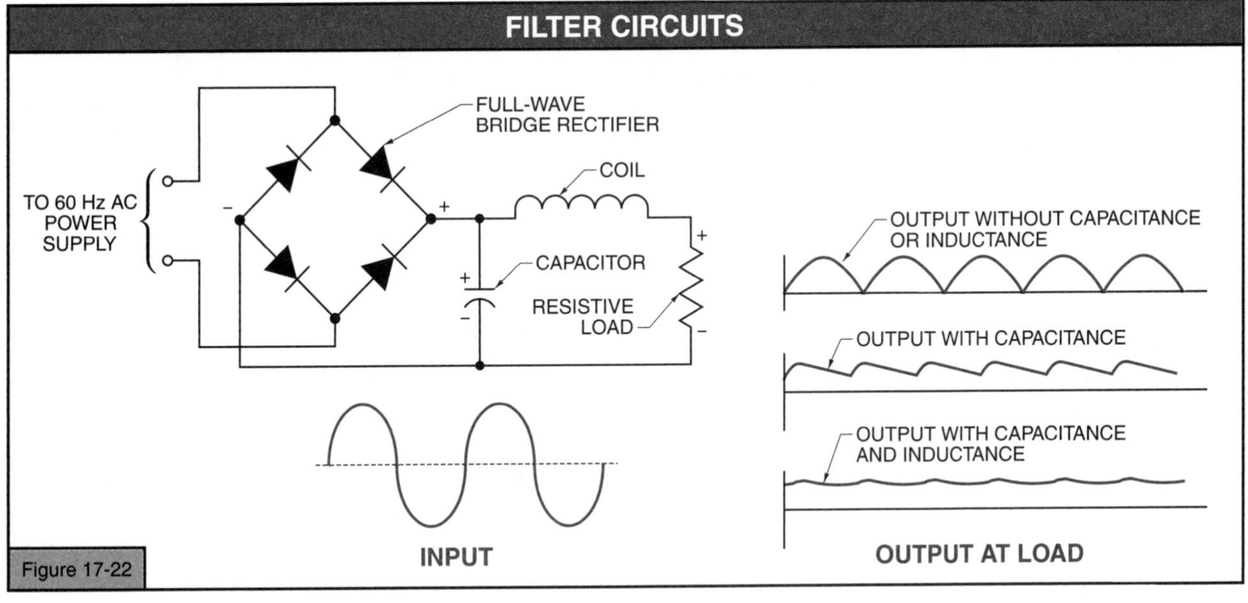

Figure 17-22

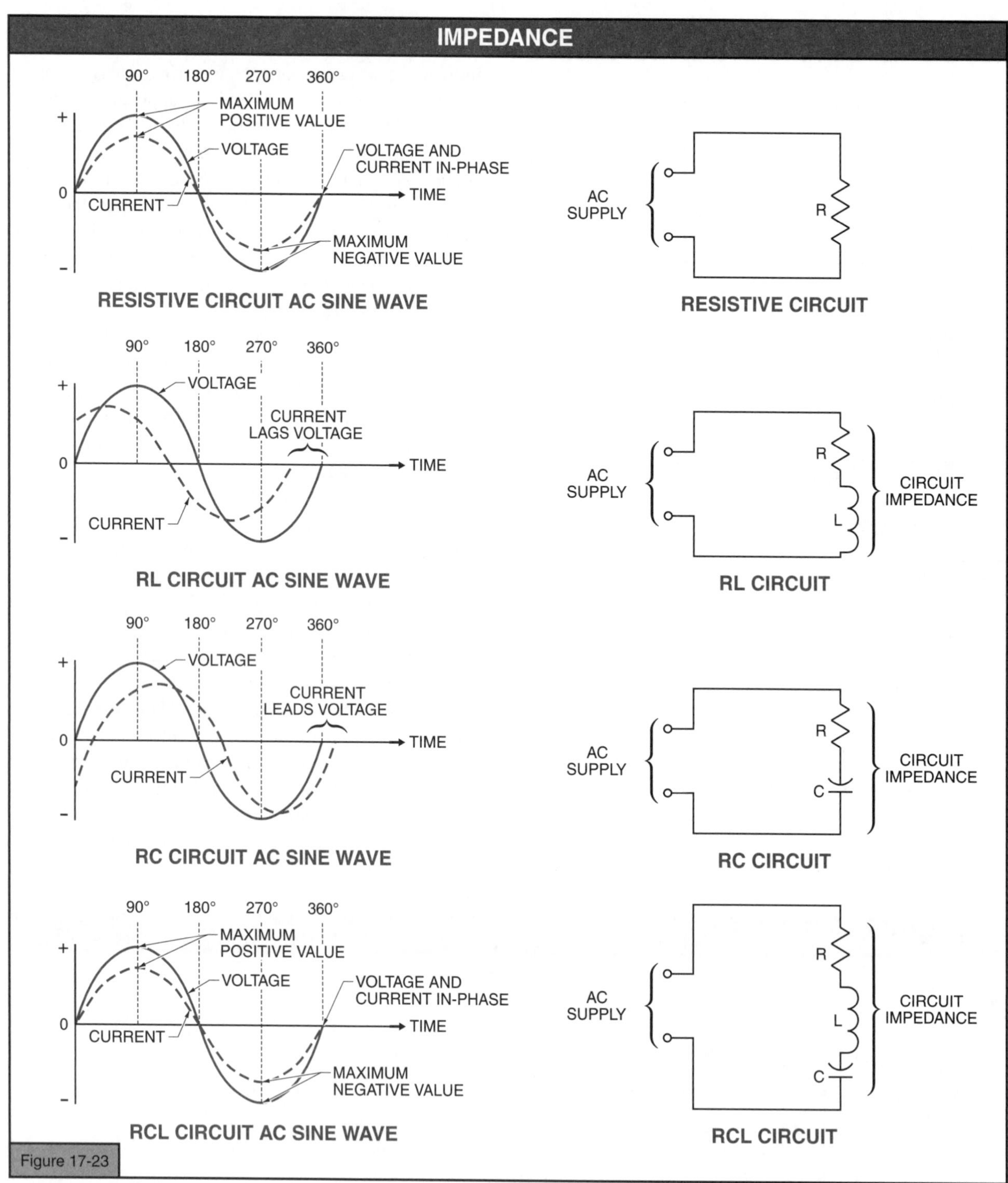

Figure 17-23

In circuits containing resistance and capacitive reactance (RC), resistance and capacitive reactance oppose current flow. RC circuit impedance is the combined effect of the resistance and capacitive reactance. For this reason, current leads voltage by a phase angle between 0° and 90°. The more capacitive the circuit, the greater the phase shift (closer to 90°).

In circuits containing resistance, inductive reactance, and capacitive reactance (RCL), resistance, inductive reactance, and capacitive reactance oppose current flow. RCL circuit impedance is the combined effects of the resistance, inductive reactance, and capacitive reactance. For this reason, the lagging current caused by inductance tends to cancel the leading current caused by capacitance because they are in direct opposition to each other. The total current flow in a circuit that includes resistance, inductive reactance, and capacitive reactance is determined by the applied voltage and total circuit impedance.

Impedance may be substituted for resistance in Ohm's law because impedance represents all of the opposition to current flow in a circuit. Ohm's law may be used to determine any one value when the other two are known. To determine impedance, current, or voltage, apply the formulas:

$$Z = \frac{E}{I}, I = \frac{E}{Z}, \text{ or } E = I \times Z$$

where

Z = impedance (in Ω)

E = voltage (in V)

I = current (in A)

Example: Calculating Impedance—Voltage and Current Known

What is the impedance of a 120 VAC circuit that has 5 mA of current flow?

$$Z = \frac{E}{I}$$

$$Z = \frac{120}{.005}$$

$$Z = \mathbf{24,000\ \Omega}$$

TECH FACT *Speakers are typical impedance loads. Speakers have relatively low impedance (2 Ω to 8 Ω) throughout the audio frequency range.*

Example: Calculating Current—Voltage and Impedance Known

What is the current in a 24 VAC circuit that has an impedance of 15,000 Ω?

$$I = \frac{E}{Z}$$

$$I = \frac{24}{15,000}$$

$$I = \mathbf{.0016\ A}$$

Example: Calculating Voltage—Current and Impedance Known

What is the voltage of an AC circuit that has an impedance of 30,000 Ω and 400 μA of current flow?

$$E = I \times Z$$

$$E = .0004 \times 30,000$$

$$E = \mathbf{12\ V}$$

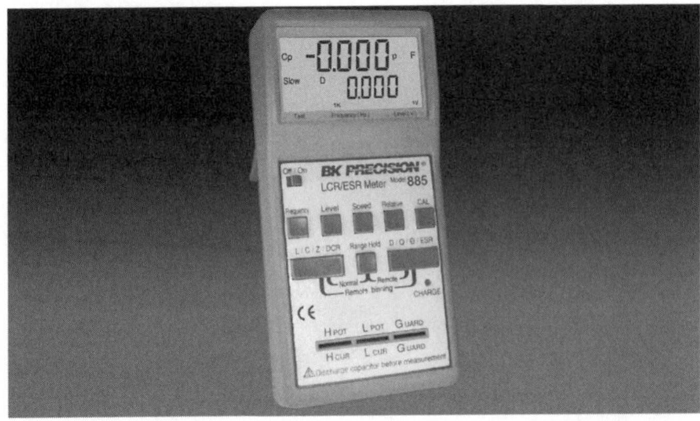

B&K Precision

Inductance-capacitance-resistance (LCR) meters are used to take impedance, inductance, capacitance, and resistance measurements.

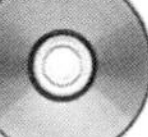

Refer to the CD-ROM in the back of the book for Chapter 17 Quick Quiz® and related information.

Circuit Requirements

Basic Circuits

Electrical and electronic circuits are used to produce work. To produce work, electrical circuits must include a component (load) that converts electrical energy into some other usable form of energy (light, heat, sound, motion), a source of electricity (battery, generator, solarcell, etc.), and a method of controlling the flow of electricity (switch). In addition, a circuit must also include a protection device (fuse, circuit breaker, etc.) to ensure that the circuit operates safely and within its electrical limits. For example, a basic circuit includes a lamp, battery, switch, and fuse. **See Figure 18-1.**

In this circuit, the lamp is the load that converts the electrical energy into light energy. The battery is a chemical source of electricity. The switch determines when the electric current flows. The fuse limits the amount of current (amperes) that is allowed to flow in the circuit.

The lamp could be replaced with a different load, such as a heating element, motor, or horn. The battery could be replaced with a generator and the fuse with a circuit breaker. Additional control devices could also be added to improve the circuit. For example, a rheostat (variable resistor) could be added in series with the switch. A rheostat is used to control the voltage applied to the lamp. The higher the voltage applied to the lamp, the greater the illumination. The lower the voltage applied to the lamp, the less the illumination. The voltage across the rheostat is reduced and the voltage across the lamp is increased when the rheostat resistance is decreased. The voltage across the rheostat is increased and the voltage across the lamp is decreased when the rheostat resistance is increased. **See Figure 18-2.**

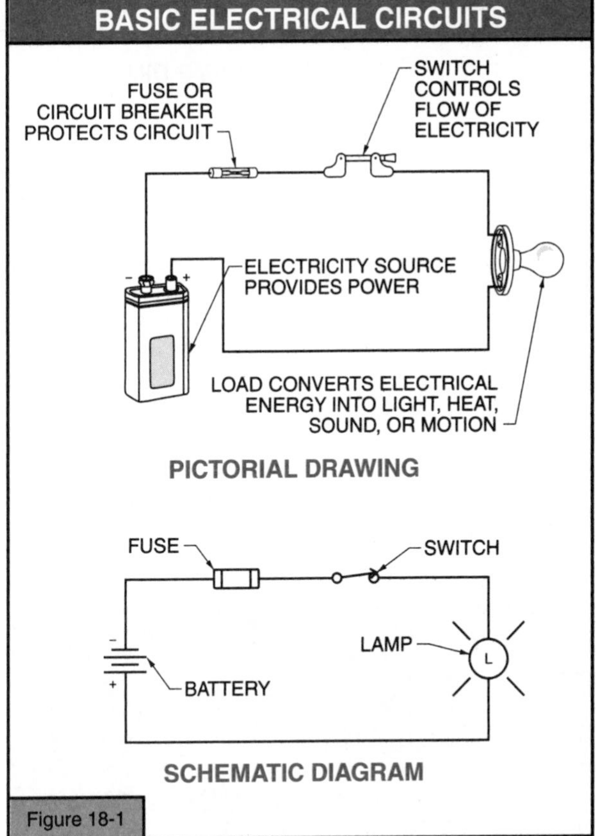

Figure 18-1

TECH FACT *The voltage rating of a fuse, circuit breaker, or switch can be higher than the circuit's actual voltage, but never lower.*

RHEOSTAT VOLTAGE CONTROL

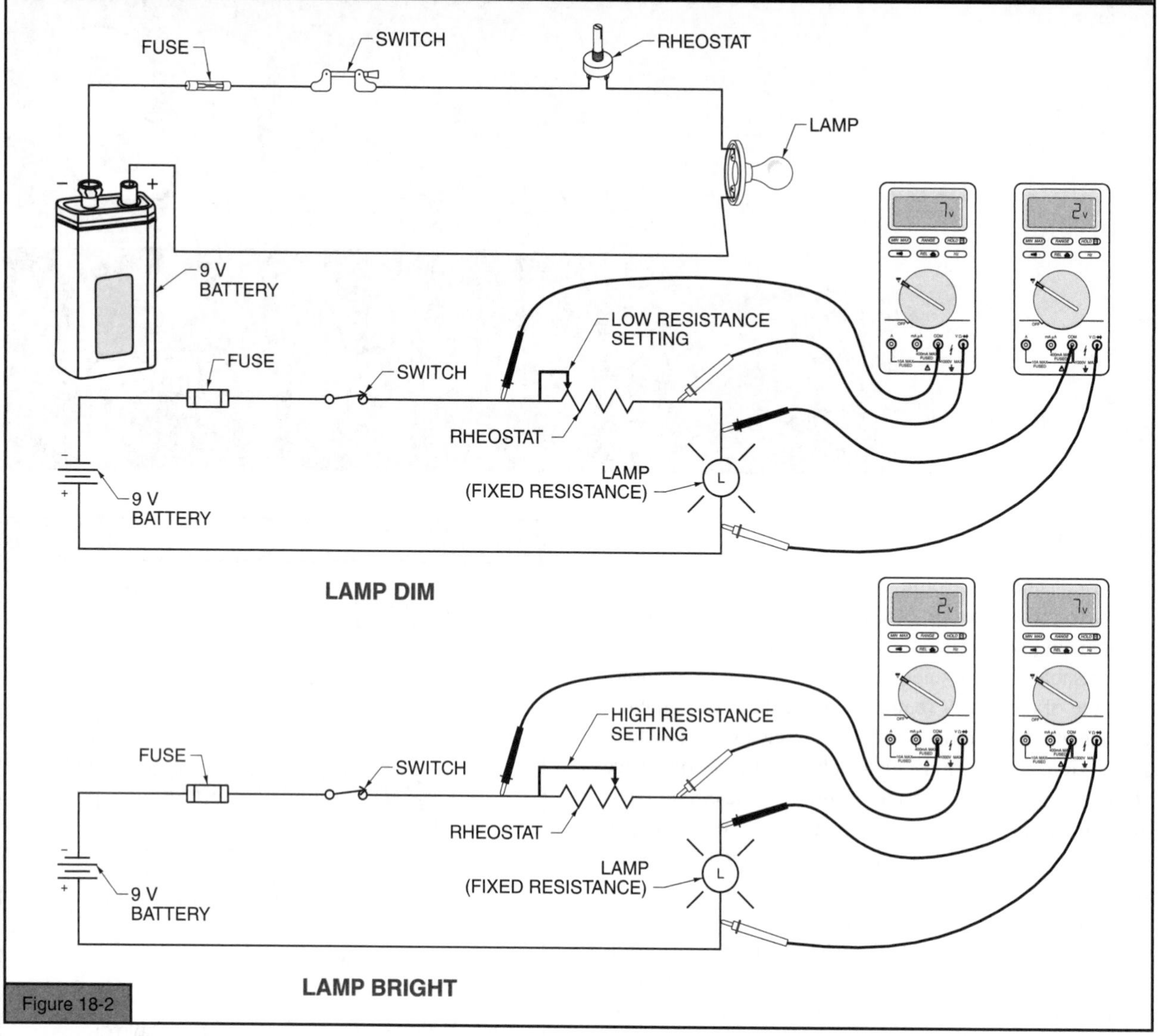

FUSE
SWITCH
RHEOSTAT
LAMP
9 V BATTERY
7 V
2 V
LOW RESISTANCE SETTING
FUSE
SWITCH
RHEOSTAT
LAMP (FIXED RESISTANCE)
9 V BATTERY
LAMP DIM
2 V
7 V
HIGH RESISTANCE SETTING
FUSE
SWITCH
RHEOSTAT
LAMP (FIXED RESISTANCE)
9 V BATTERY
LAMP BRIGHT

Figure 18-2

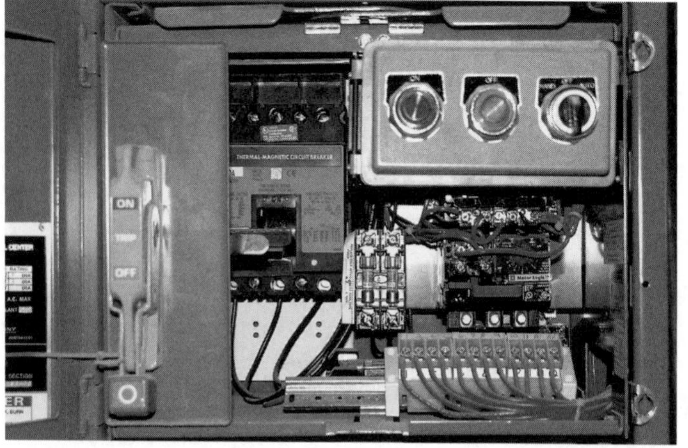

Types of switches used in electric panels include pushbuttons and selectors.

The power source must be able to deliver enough power to operate the load(s). The power source must be able to deliver increased power if the number of loads is increased or if the power requirements of the load(s) are increased (high wattage lamp). In any circuit, the switch(es) and connecting wire must be rated high enough to safely carry the increased power. The switches and wire must be rated to carry the circuit's increased current because increased power in a fixed voltage circuit produces higher current. To protect the circuit (wire, load, switch, and power source), the fuse must be rated at a current level below the level that causes equipment damage. In general, a fuse is sized for a current value approximately 125% higher than the load's rated current value. The rating of the wire and switch must be higher than the rating of the fuse. **See Figure 18-3.**

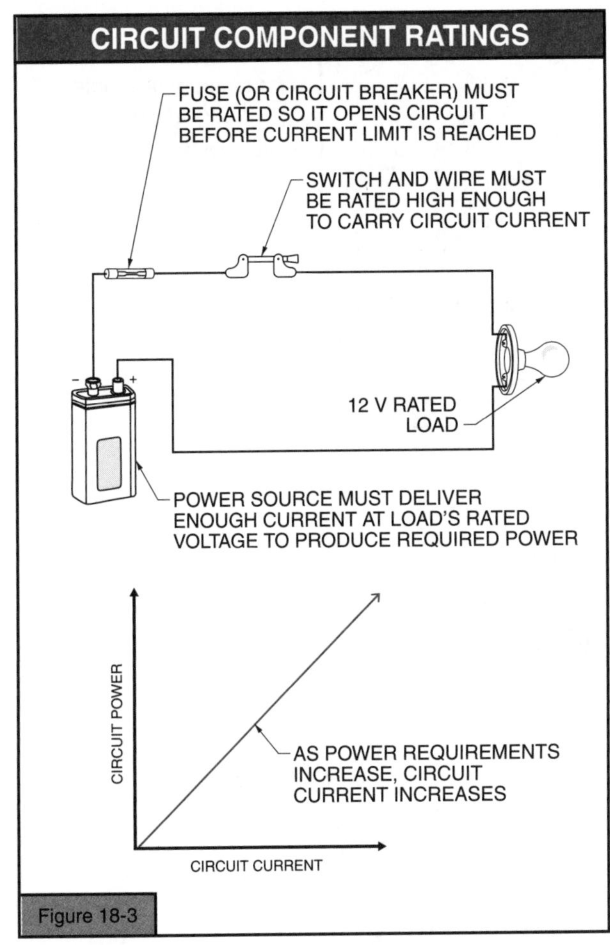

Figure 18-3

Complex Circuits

All electrical circuits begin as basic circuits. In a basic electrical circuit, electricity is produced, protected, controlled, and used to produce work. Basic circuits are improved by adding additional protection (ground fault detection, etc.), monitoring of circuit parameters (voltage, current, power, temperature, etc.), improved control of loads (dimming of lamps, control of motor speed, etc.), and interconnecting basic circuits into systems. A *system* is a combination of components interconnected to perform work or meet a specific need.

A *network* is a system of computers, terminals, and databases connected by communication lines. In a network, inputs such as temperature switches, limit switches, smoke detectors, pushbuttons, etc., and outputs such as motors, lamps, solenoids, alarms, etc., are interconnected to computers (PCs), programmable controllers (PLCs), and other electronic control and monitoring devices. Computers allow the gathering, monitoring, controlling, and displaying of system data. Computers are interconnected through communication lines to form different types and levels of networks. **See Figure 18-4.**

On a plant floor, inputs and outputs are connected to an input/output network. An *input/output network* is a network that allows analog and digital inputs to send information in the system. An *analog input* is an input that sends a continuously changing variable into the system. Temperature, pressure, flow, and level are common analog inputs. ON/OFF switches such as pushbuttons, limit switches, toggle switches, etc. are common digital inputs. A *digital input* is an input that is either ON or OFF. An *output* is a device that is used to produce work, light, heat, sound, or display information. Outputs use the majority of electrical power in a system because they include lamps, motors, and heating elements. The input/output network includes computers and/or PLCs to control the flow of input data and send control signals to the outputs.

A *production control network* is a network used to provide increased control of an input/output network. Production control networks use computers to link numerous machines, work cells, processes, and production lines. Inputs and outputs from different points in the input/output network are monitored. The production control network is used to control process, time, units produced, measure variables, display information, check quality, and other production management functions.

The production control network is connected to the plant information network. A *plant information network* is a network that allows information to flow up from the plant floor and down from engineering, planning and scheduling, quality control, management, and outside sources. The computers, servers, and mainframes of the plant information network help maintain inventories, control cash flow, monitor fire and security systems, generate reports, store/retrieve information, assign workforce, order parts, schedule training, etc.

A properly designed, installed, maintained, and operated system or network can increase production, ensure quality, increase flexibility, and produce documentation. As complexity is increased, additional problems and potential faults may occur. The more complex the circuit, the higher the required level of maintenance and operation.

Load Power Requirements

A load is an electric device that converts electrical energy into other forms of energy. **See Figure 18-5.** Properly operating loads can last for many years. The problem is normally easy to identify when a load fails to operate. For example, it is easy to see that a lamp is out, no heat is being produced, water is not being pumped, there is no sound, etc.

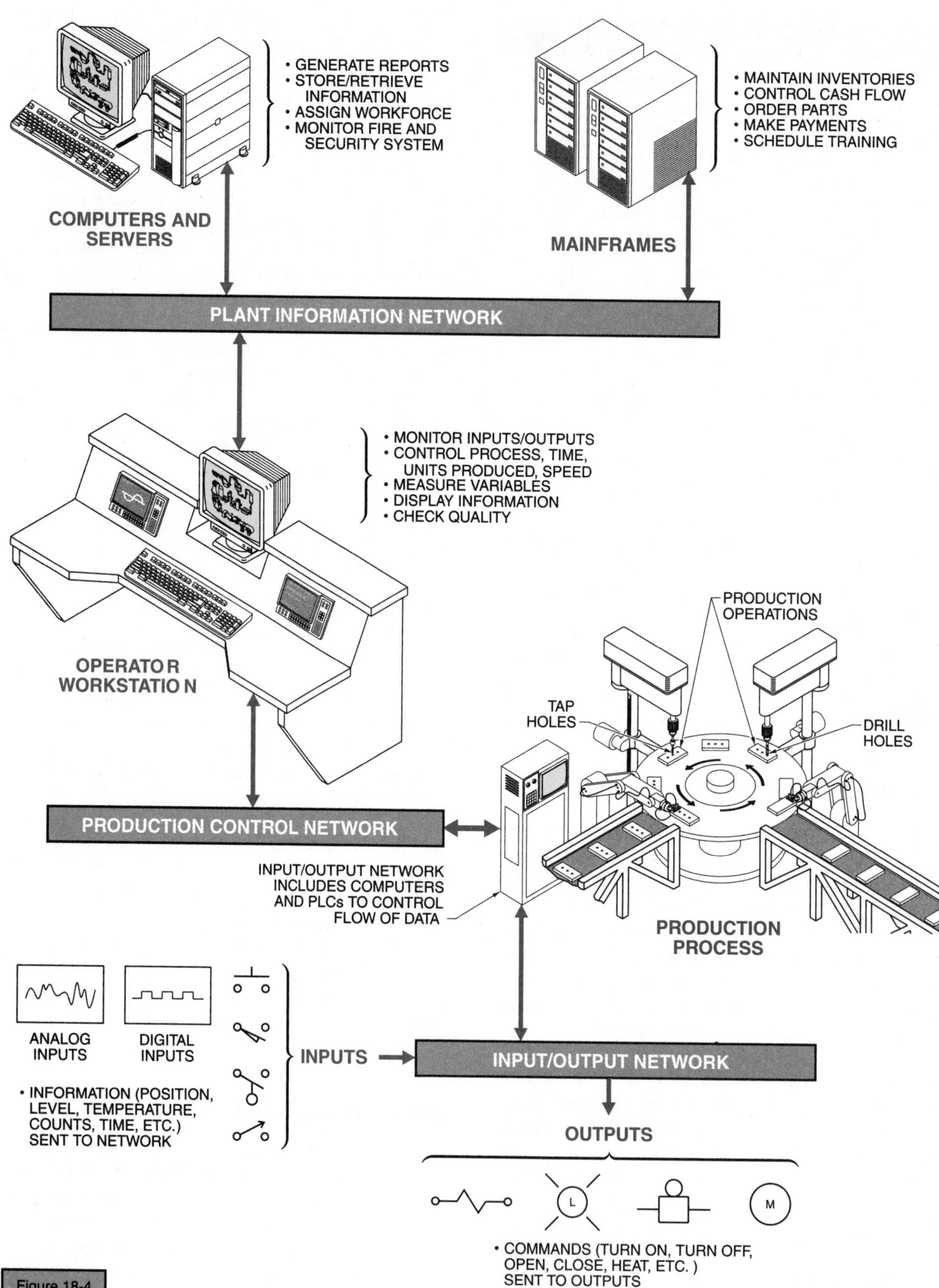
• GENERATE REPORTS
• STORE/RETRIEVE INFORMATION
• ASSIGN WORKFORCE
• MONITOR FIRE AND SECURITY SYSTEM
COMPUTERS AND SERVERS
• MAINTAIN INVENTORIES
• CONTROL CASH FLOW
• ORDER PARTS
• MAKE PAYMENTS
• SCHEDULE TRAINING
MAINFRAMES
PLANT INFORMATION NETWORK
• MONITOR INPUTS/OUTPUTS
• CONTROL PROCESS, TIME, UNITS PRODUCED, SPEED
• MEASURE VARIABLES
• DISPLAY INFORMATION
• CHECK QUALITY
OPERATOR WORKSTATION
PRODUCTION OPERATIONS
TAP HOLES
DRILL HOLES
PRODUCTION CONTROL NETWORK
INPUT/OUTPUT NETWORK INCLUDES COMPUTERS AND PLCs TO CONTROL FLOW OF DATA
PRODUCTION PROCESS
ANALOG INPUTS
DIGITAL INPUTS
• INFORMATION (POSITION, LEVEL, TEMPERATURE, COUNTS, TIME, ETC.) SENT TO NETWORK
INPUTS
INPUT/OUTPUT NETWORK
OUTPUTS
L
M
• COMMANDS (TURN ON, TURN OFF, OPEN, CLOSE, HEAT, ETC.) SENT TO OUTPUTS

Figure 18-4

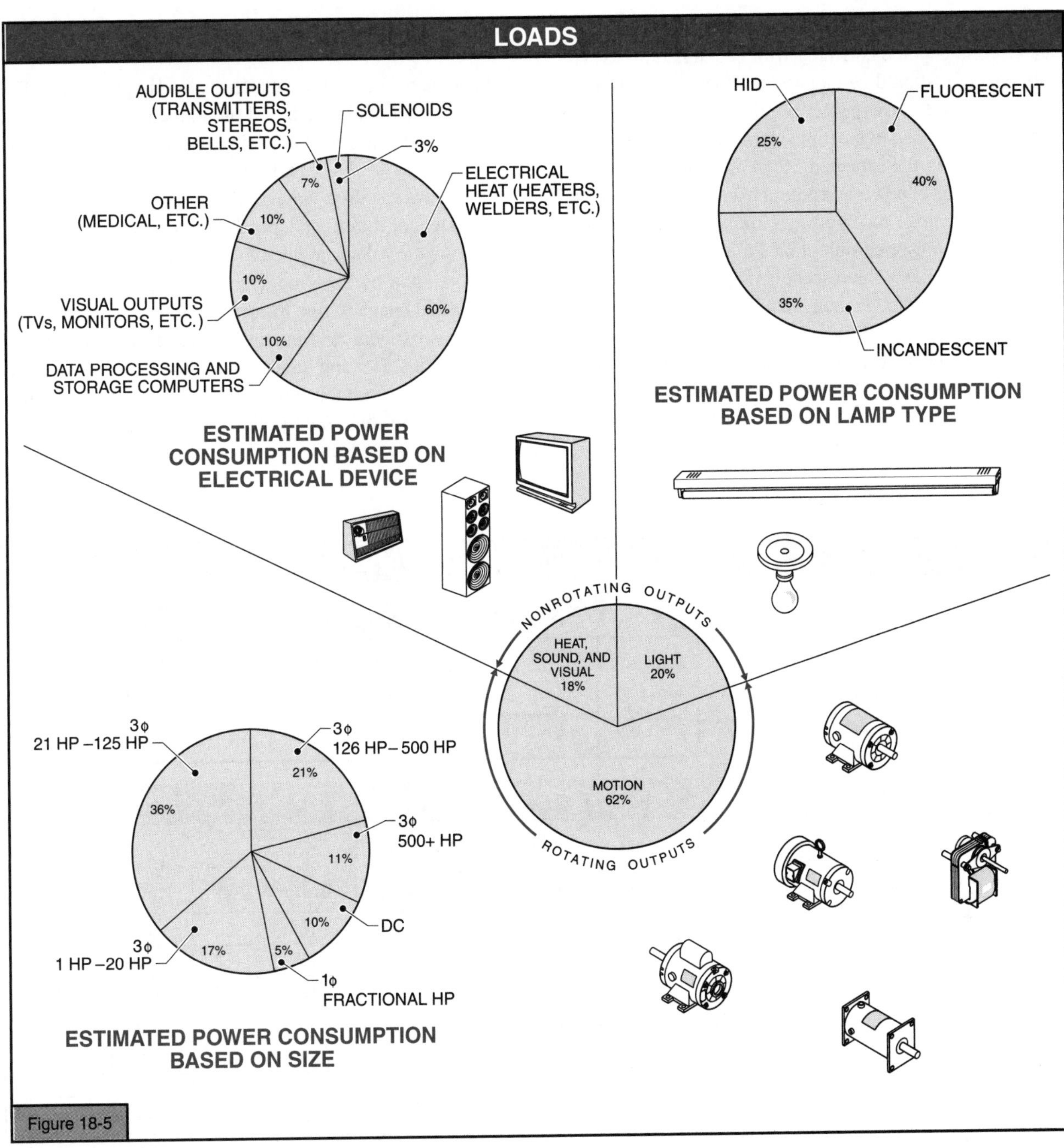

Figure 18-5

A load is normally the starting point of most circuit design functions because the load produces the work required from an electrical circuit. The type and size of load used depends on the application. Application considerations include available power sources, operating conditions, short and long term requirements, and expected results.

The power rating of a load must be high enough to meet the requirements of the application. For example, lamps must produce enough light, heating elements must produce enough heat, alarms must be loud enough, and motors must produce enough torque.

All factors that may exist in a given application must be considered when determining the total power output of a load. For example, lamp output is rated in initial lumens and mean lumens. An *initial lumen* is the amount of light produced when a lamp is new. A *mean lumen* is the average light produced after a lamp has operated for approximately 40% of its rated life. A standard 40 W fluorescent lamp has a mean lumen rating of about 9% less than its initial lumen rating. This indicates that a school classroom or office will have approximately 9% less light after the lamps have reached 40% of their rated life.

Heating elements produce less available heat as the heater coils become corroded with calcium. This is why coffeemakers should be cleaned with vinegar to remove calcium buildup. The power requirements of a load are normally increased to meet the worst case conditions that may occur over the expected life of the load.

The amount of power and type of load must be considered when determining a load's power rating. The load is considered because power consumption varies from load type to type. For example, a fluorescent lamp produces more light per watt than an incandescent lamp of the same wattage rating. Light output may be increased by using a different load type (fluorescent instead of incandescent) without increasing total power required. For any given load (lamp, motor, heater, etc.) of the same type, as the power rating of the load is increased, so is the amount of work delivered.

Power Source Requirements

The source of power must match the load requirements. The ratings of all loads are based on a given applied voltage. The applied voltage includes size (amount of voltage), type (AC, DC, or AC/DC), and phase (1ϕ or 3ϕ). Any change in the power source voltage affects the load's operation. The extent of the effect depends on the load. For example, reducing the voltage applied to a heating element or incandescent lamp by 50% reduces the output of the load but does not damage it. **See Figure 18-6.**

Resistive heating elements are used in many heating applications. Heating elements are rated for a given voltage level. The voltage applied to the heating element may vary because the heating element is basically a resistor used to dissipate heat.

HEATING ELEMENT VOLTAGE VARIATION

LOAD VARIATION		
	APPLIED VOLTAGE	LOAD'S OUTPUT POWER (in W)
LOAD'S RATED VOLTAGE	↑	↑
	↓	↓

PERCENT RATED POWER*

Applied Voltage**	Rated Voltage***										
	110	115	120	208	220	230	240	277	440	460	480
110	100	91	84	28	25	23	21	16	6.2	5.7	5.2
115	109	100	92	31	27	25	23	17	6.7	6.2	5.7
120	119	109	100	33	30	27	25	19	7.4	6.8	6.3
208			300	100	89	82	75	56	22	20	19
220				112	100	91	84	63	25	23	21
230				122	109	109	92	69	27	25	23
240				133	119		100	75	30	27	25
277							133	100	40	36	33
380								188	74	68	63
415									89	81	75
440									100	91	84
460									109	100	92
480									119	109	100

* in W
** in V
*** in %

Figure 18-6

Varying the voltage below the rated voltage of the heating element reduces the heating element output. Reducing the voltage to a resistive heating element control circuit can vary the output of the heating element. The voltage may be reduced anywhere from 0 V to the heater's rated voltage. Increasing the voltage applied to a resistive heating element increases the heating element's output. The voltage must not be increased more than 25% above the heater's rated voltage.

Motor voltage cannot vary over a wide range. Reducing the voltage applied to a motor by 50% can damage the motor. Some loads may have less-than-rated voltage applied to them with no damage while other loads cannot tolerate much of a voltage reduction. The torque output of a motor is reduced when the voltage to a motor is reduced. No work is done and the motor overheats if the torque output of the motor is reduced to the point at which the motor shaft cannot turn the load. **See Figure 18-7.**

An AC motor should generally operate within ±10% of its voltage rating. To ensure long motor life, an AC motor should operate within −10% and +5% of its voltage rating. The voltage to a DC motor may be reduced without damage to the motor. Reducing the voltage to a DC motor varies the speed of the motor and reduces the torque of the motor. The motor overheats if the torque is reduced to the point at which the motor shaft cannot turn the load.

Insulation for conductors used in AC motors can include additional protection against undue wear from motor vibration.

MOTOR VOLTAGE VARIATION

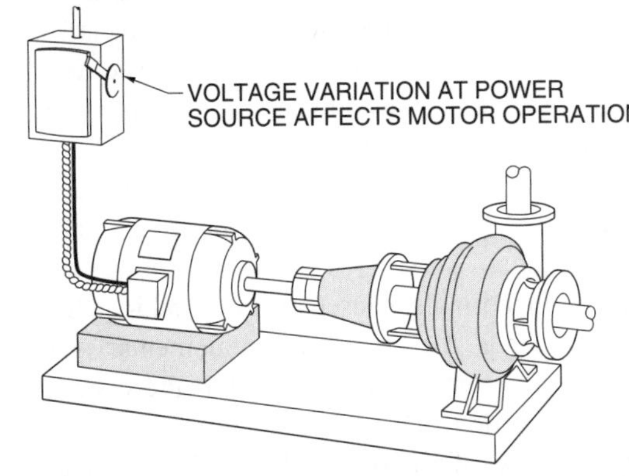

AC MOTOR VOLTAGE VARIATION CHARACTERISTICS		
Performance Characteristics	**10% Above Rated Voltage**	**10% Below Rated Voltage**
Speed	+1%	−1.5%
Motor torque	+20% to +25%	−20% to −25%
Starting current	+10% to +12%	−10% to −12%
Full-load current	−7%	+11%
Motor efficiency	Little change	Little change
Temperature rise	−3°C to −4°C	+6°C to +7°C

DC MOTOR VOLTAGE VARIATION CHARACTERISTICS				
Performance Characteristics	**Voltage 10% Below Rated Voltage**		**Voltage 10% Above Rated Voltage**	
	Shunt	**Compound**	**Shunt**	**Compound**
Speed	−5%	−6%	+5%	+6%
Starting torque	−15%	−15%	+15%	+15%
Current	+12%	+12%	−8%	−8%
Field temperature	Increases	Decreases	Increases	Increases
Armature temperature	Increases	Increases	Decreases	Decreases
Commutator temperature	Increases	Increases	Decreases	Decreases

Figure 18-7

The voltage requirements of lamps vary depending on the type of lamp used. The output power and amount of light is reduced when the voltage to a lamp is reduced. Reducing the voltage to an incandescent lamp reduces the brightness of the lamp. Reducing the voltage also increases the lamp's operating life. The applied voltage must not exceed 10% more than the lamp's rated voltage. **See Figure 18-8.**

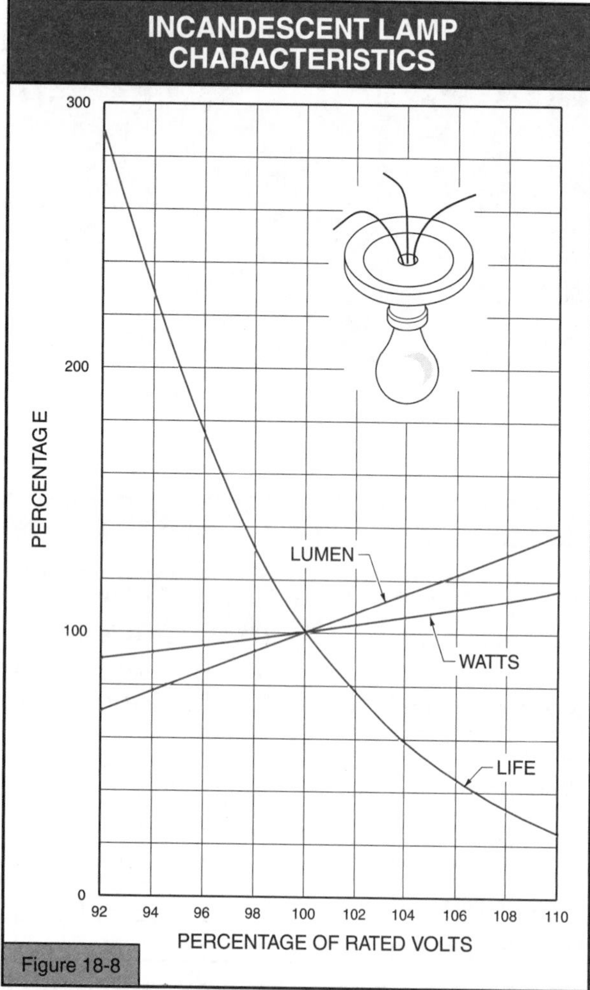

Figure 18-8

Voltage requirements of lamps vary, depending on the type of lamp used.

Although the voltage applied to incandescent lamps may be reduced to any value between 0 V and the lamp's rated voltage, fluorescent and HID lamps must generally operate within ±10% of their voltage ratings. Manufacturers list the voltage rating and range of each lamp type and size. For example, a 100 W metal-halide HID lamp is rated to operate within ±5% of its rated voltage. A 175 W metal-halide lamp is rated to operate within 10% of its rated voltage. **See Figure 18-9.**

Heating elements, motors, and lamps make up a large part of the typical loads in use. Additional loads include alarms, solenoids, appliances, electric tools, computers, printers, etc. Most loads generally operate properly when connected to a power source that is within ±10% of their rated voltage. **See Figure 18-10.**

High and Low Voltages

The voltage in a circuit may be consistently, temporarily, or momentarily too high or too low. Voltage may be consistently too high if a circuit is near the beginning of a power distribution system. Voltage may be consistently too low if a circuit is near the end of a power distribution system.

Voltage may be temporarily too high in the office area of a plant when the production lines are shut down. Voltage may be temporarily too low during certain parts of the day or during a brown-out. A *brown-out* is the deliberate reduction of the voltage level by a power company to conserve power during peak usage times.

Voltage may be momentarily too high or too low at any time. A *voltage surge* is a higher-than-normal voltage that temporarily exists on one or more power line. Voltage surges are normally caused by the switching OFF of heavy electric loads. A *voltage dip* is a momentary low voltage. Voltage dips are normally caused by the switching ON of heavy electric loads.

A voltage dip is often followed by a voltage surge as voltage regulators overadjust for the voltage dip. *Voltage variance* is the difference in voltage between a voltage surge and a voltage dip. The voltage surge following a voltage dip increases the voltage variance. **See Figure 18-11.**

A recording meter or multimeter with a min/max mode is used to measure voltage surges and voltage dips. Input voltage is measured over a 24-hour period when a problem recurs in a piece of equipment. A voltage stabilizer should be added when the voltage fluctuates more than ±8%.

TECH TIP

The response time of a meter is the length of time an input must stay at a new value to record a change. A 100 mS meter rated response time works well for recording power supply surges, dips, and inrush currents.

MANUFACTURER'S LAMP RATING CHART

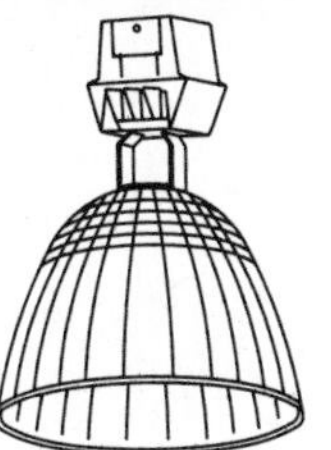

VOLTAGE RANGES (VOLTS)	
A (5%)	B (10%)
126 – 120 – 114	132 – 120 – 108
218 – 208 – 198	229 – 208 – 187
252 – 240 – 228	264 – 240 – 216
291 – 277 – 263	305 – 277 – 249

METAL-HALIDE LAMP CHARACTERISTICS

Lamp Wattage	Voltage Rating	Fuse*	Starting Current*	Operational Current*	Line Input Watts	% Allowable Variation	
						Operating Voltage Range	Lamp Wattage
50	120 277	3 3	.60 .25	.65 .30	72	A	±8
100	120 208 240 277	8 5 3 3	1.15 .66 .58 .50	1.15 .66 .58 .50	129	A	±12
175	120 208 240 277 480	5 3 3 3 3	1.30 .75 .65 .55 .35	1.80 1.05 .90 .80 .45	210	B	±10
250	120 208 240 277 480	8 5 5 3 3	2.10 1.40 1.10 1.00 .60	2.50 1.45 1.25 1.10 .60	285	B	±9
1000	120 208 240 277 480	20 15 10 10 10	8.0 4.6 4.0 3.5 2.0	9.0 5.2 4.5 3.9 2.3	1080	B	±10

* in A

Figure 18-9

Voltage Stabilizers

Most electronic devices are designed to operate at a constant voltage level. A voltage stabilizer is required for any application containing an electronic circuit that requires a constant voltage level. A *voltage stabilizer (regulator)* is a device that provides precise voltage regulation to protect equipment from voltage dips and voltage surges. **See Figure 18-12.**

Most voltage stabilizers hold the output voltage within ±1%. A voltage stabilizer is placed between the incoming power and the load requiring protection when the input voltage varies to ±15%. A voltage stabilizer prevents such problems as loss of memory, incorrect data output, and shutdowns. Voltage stabilizers should be used on computers, local area networks, programmable controllers, X-ray and other medical equipment, radar and sonar equipment, industrial robots, numerical control (NC) machines, air traffic control equipment, photographic equipment, induction heaters, electronic cash registers and banking equipment, electronic security systems, and communication equipment.

GENERAL LOAD ACCEPTABLE AC VOLTAGE RANGES*		
Rating**	**Minimum**	**Maximum**
24	22	26
115	109	121
120	114	126
208	197	218
230	218	242
240	228	252
277	263	291
460	436	484
480	456	504

* in V
** 60 Hz power

Figure 18-10

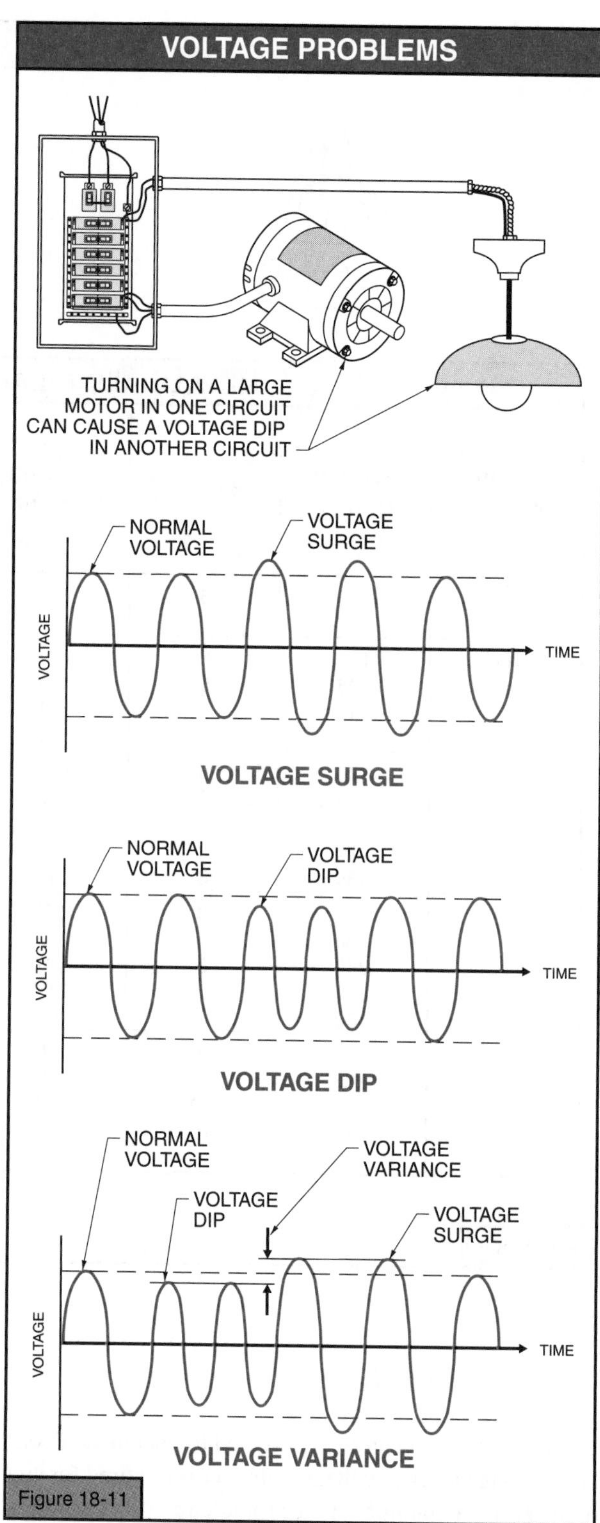

Figure 18-11

Transient Voltages

Most power received from a utility company is changed by the addition of transient voltages on the power lines. A *transient voltage* is a temporary, unwanted voltage in an electrical circuit. Transient voltages occur due to lightning strikes, unfiltered electrical equipment, and electricity being switched ON and OFF. Transient voltages may range in value from a few volts to several thousand volts and may occur at any phase angle on the AC sine wave. Transient voltages on a 120 V power line may reach 5000 V or more.

Transient voltages differ from voltage surges and voltage dips in that transient voltages are normally much larger in magnitude and are very erratic. Lightning is the most common source of transient voltages on outdoor power distribution systems. Lightning induces a surge that travels in two directions from the point of origin. In most cases, the transient voltage is dissipated by the utility company's grounding systems after it has traveled 10 to 20 utility poles (or grounding points). Damage is unlikely if equipment is located farther than 20 utility poles from the lightning strike. Unprotected equipment may be severely damaged if the lightning strike is close to equipment in use.

Transient voltages caused by lightning strikes must be considered in all electrical applications because lightning can strike at almost any place. Low-level transient voltages are produced on indoor power lines when most loads are switched OFF. Transient voltages may be continuously produced when using 1ϕ switched mode power supplies, fluorescent lighting using electronic ballasts, variable-speed motor drives, soft-start motor drives, electronically-controlled welding equipment, and motors that include brushes (DC and universal).

TECH TIP

Power supply surge suppressors are used to prevent damage from spikes and noise problems and fax/modem surge suppressors are used to prevent problems on data networks.

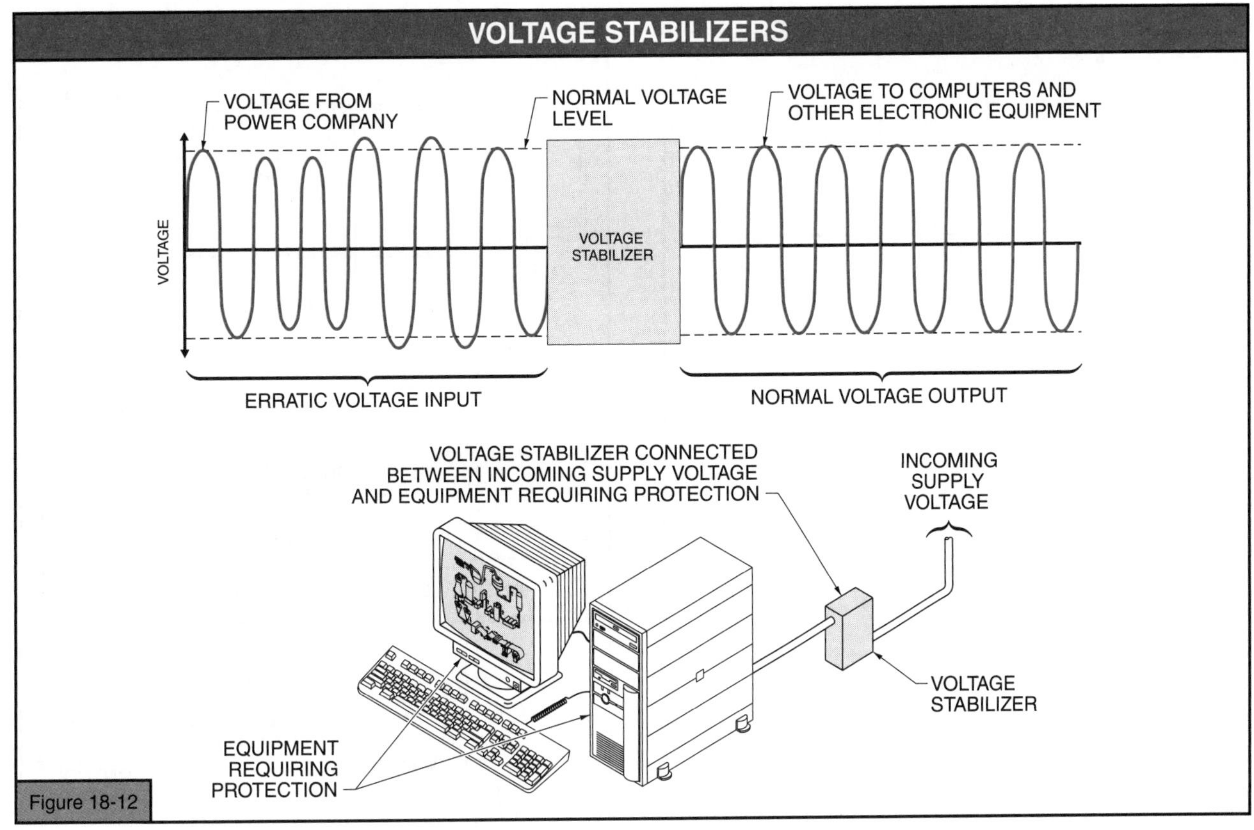

Figure 18-12

Transient voltages include voltage spikes and electrical noise, or both. A *voltage spike* is an increase in voltage (normally several thousand volts) that lasts for a very short time (microseconds to milliseconds). Voltage spikes are shown as sharp peaks on the AC sine wave. *Electrical noise* is any unwanted disturbance that interferes with a signal. Electrical noise distorts the normally pure AC sine wave. Electrical noise is shown as jagged lines on the AC sine wave. An oscilloscope or scopemeter may be used to detect voltage spikes and electrical noise. **See Figure 18-13.**

Electrical noise is produced on power lines from two different points. *Common noise* is noise produced between the ground and hot or the ground and neutral lines. *Transverse noise* is noise produced between the hot and neutral lines. In most electrical systems, common noise occurs more frequently than transverse noise. **See Figure 18-14.**

Common noise is caused by arcing at motor brushes, ground faults, poor grounds, radio transmitters, ignition systems, and the opening of electrical contacts. The opening of electrical contacts generates noise because an arc is produced across the contacts as they are pulled apart. The higher the current in the circuit being opened, the larger and longer the arc. Transverse noise is caused by welders, switched power supplies, and the firing of silicon controlled rectifiers (SCRs) in electrical equipment.

Voltage Spike and Electrical Noise Protection

Computers, electronic circuits, and specialized equipment require protection against voltage spikes and electrical noise. Protection against voltage spikes and electrical noise includes:

- Proper wiring, grounding, and shielding of the power lines. Proper wiring, grounding, and shielding includes the use of a proper grounding system (See NEC® 250) and ensures the proper use of the ground and neutral conductors. The neutral conductor should always carry the normal load current. The ground conductor should only carry fault current. The ground conductor should never carry the normal current drawn by a load. The amount of current carried on the hot, neutral, and ground conductors is measured with a clamp-on ammeter. The current measured on the neutral conductor should equal the current measured on the hot conductor under normal operating conditions. The current measured on the ground conductor should always equal zero under normal operating conditions.
- Proper sizing of neutral and ground conductors.
- Ensuring that all wire connections are tight. Loose connections cause noise and arcing, which may damage computer and other electronic circuits.

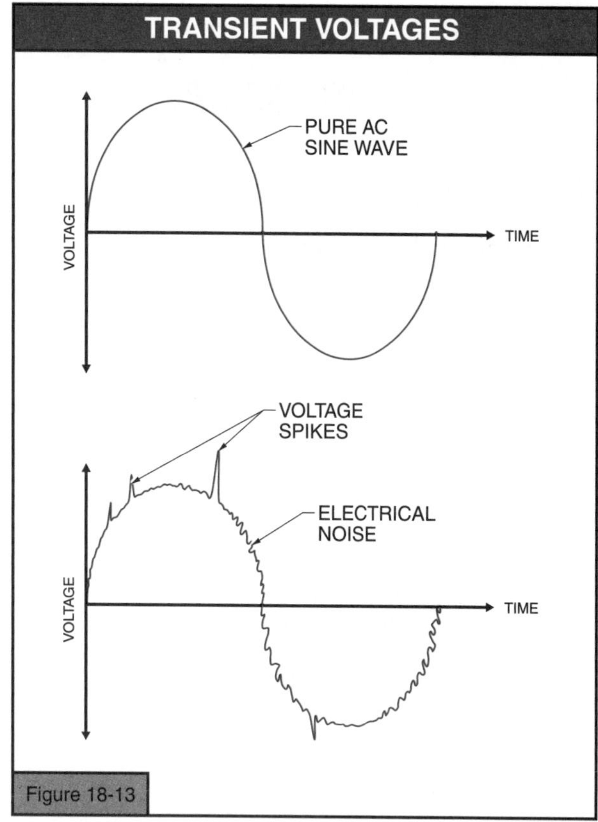

Figure 18-13

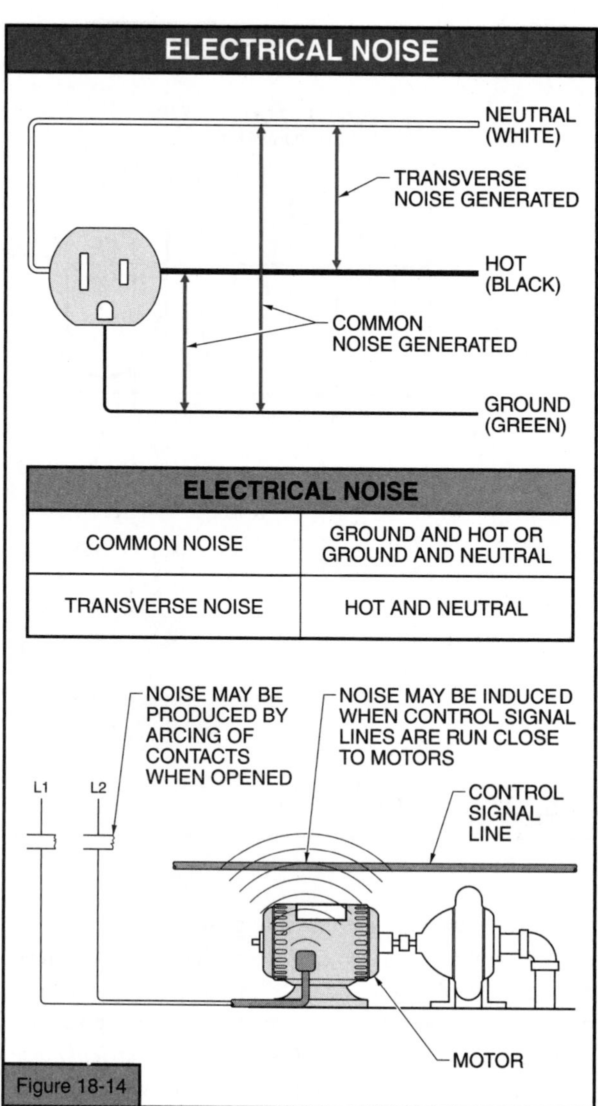

ELECTRICAL NOISE	
COMMON NOISE	GROUND AND HOT OR GROUND AND NEUTRAL
TRANSVERSE NOISE	HOT AND NEUTRAL

Figure 18-14

- Proper selection and application of surge suppressors and noise suppressors. A *surge suppressor* is a protection device that limits voltage surges that may be present on power lines. Surge suppressors limit voltage surges by discharging or bypassing the damaging voltage. Surge suppressors may be installed separately or may be included with a noise suppressor. **See Figure 18-15.**

Large electrical and electronic circuits, such as those used to control tower cranes, require large power supplies.

Power Sources

All electrical and electronic circuits require a power source. The type and level of power required depends on the size and power consumption requirements of the load and circuit. Very small circuits, such as an electric wrist watch, require very little power. In an electric watch, one low-voltage battery (1.5 V) provides enough power to operate the electrical system for a year or more.

A large circuit, such as an electric railway train, requires a large power supply. The power supply required to operate the train must deliver thousands of kilowatts of power. An electric railway train receives power through an overhead wire (catenary), from an electrified third rail, or by a engine (normally diesel) that drives a generator which produces the required electric current.

SURGE AND NOISE SUPPRESSORS

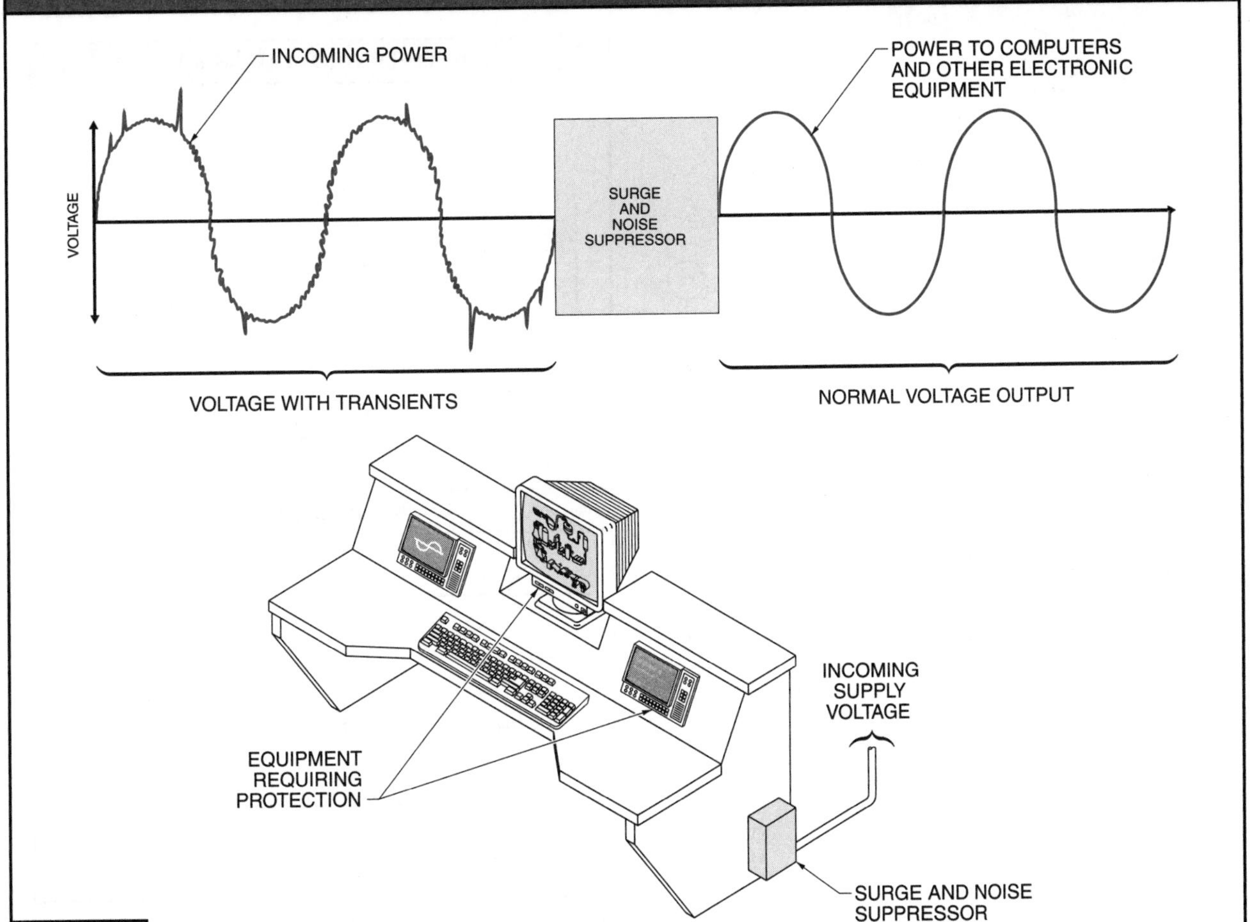

INCOMING POWER
VOLTAGE
SURGE AND NOISE SUPPRESSOR
POWER TO COMPUTERS AND OTHER ELECTRONIC EQUIPMENT
VOLTAGE WITH TRANSIENTS
NORMAL VOLTAGE OUTPUT
INCOMING SUPPLY VOLTAGE
EQUIPMENT REQUIRING PROTECTION
SURGE AND NOISE SUPPRESSOR

Figure 18-15

Electricity may be alternating current (AC) or direct current (DC). AC may be single phase (1ϕ) or three phase (3ϕ). AC is produced by AC generators. DC is produced by batteries, solarcells, thermocouples, DC generators, or rectification of AC. *Rectification* is the process of changing AC into DC. **See Figure 18-16.**

All DC power sources have a definite positive and a negative side. DC is used in applications that require a portable power supply (flashlights, power tools, automobiles, etc.) and in applications using electronic circuits. In most applications using electronic circuits, DC is supplied from a rectified AC supply. Appliances with electronic circuits, such as computers, VCRs, TVs, etc. operate on rectified AC.

AC is produced by an AC generator (alternator). AC does not have a definite positive and negative side because the current reverses its direction of flow twice per cycle. Most AC power comes from power companies that generate and distribute AC power produced by converting coal, oil, natural gas, nuclear energy, or hydroelectric power into electricity.

AC is the most common form of electricity. AC is used in any application in which DC is not required. AC is used exclusively in power distribution systems because it can be connected to a transformer. The transformer changes the voltage level by stepping up or stepping down the voltage.

Generated voltage is stepped up to allow large amounts of power to be transmitted over small wires. Voltage for use is stepped down to provide low and safe voltage levels. AC voltage usages include lighting, heating, and most motor applications.

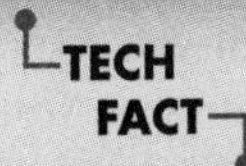

A digital meter displays a negative sign in front of the measured value when the red test lead is connected to the negative potential and the black test lead is connected to the positive potential.

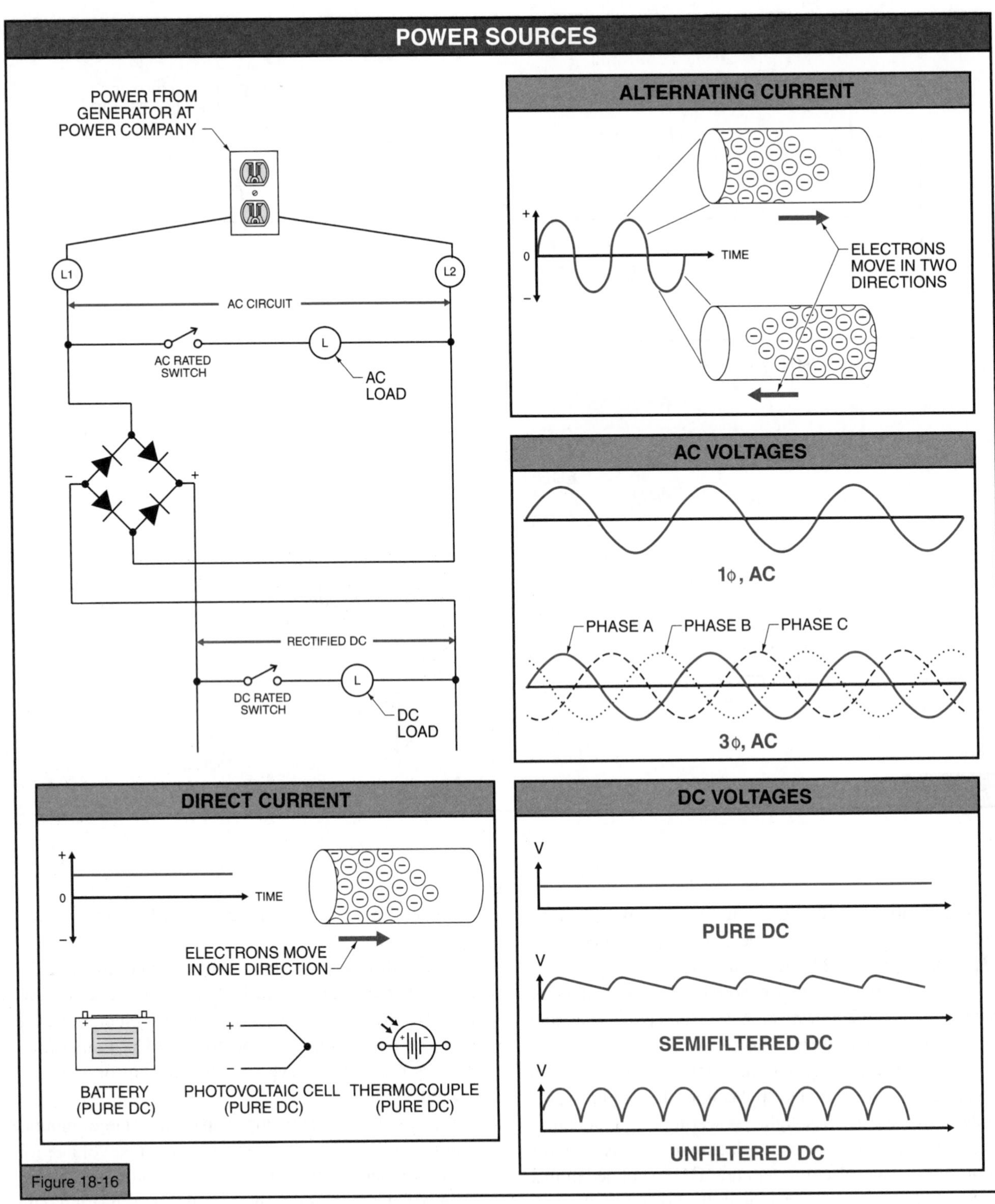

Figure 18-16

AC and DC systems have voltage levels that range from a few millivolts (mV) or less to thousands of volts (kV). AC and DC systems have current levels that range from a few milliamperes (mA) or less to thousands of amperes (kA). The amount of power in a system is calculated by multiplying voltage and current. Most electrical loads have a power rating (100 W lamp, 1200 W heater, 5 HP motor, 25 W speaker, etc.). **See Figure 18-17.**

All electrical circuits must have a source of power. The power source may be chemical (battery), electromagnetic (generator), heat (thermocouple), or solar (photovoltaic cell). The source of power used depends on the application and the amount of power required. For example, a portable calculator requires little power, thus the electrical circuits can be powered from a solarcell. Batteries are used in a golf cart because a golf cart requires more power than can be easily delivered by solarcells. Batteries meet the power requirements and the application needs (portable power source). Generators are used to produce power for common loads used in residential, commercial, and industrial applications because these applications require more power than solarcells or batteries can produce.

LOAD POWER RATINGS

Load	Typical Power*
Central air conditioner (Residential)	4000 – 8000
Window air conditioner	700 – 1500
Dishwasher	1500 – 2500
Refrigerator	250 – 600
Freezer	400 – 600
Television	200 – 400
Waffle iron	900
Computer	500 – 1000
Coffeemaker	500 – 1000
Microwave oven	500 – 900
Electric range	3500 – 8000
Washing machine	1200 – 1700
Can opener	125 – 200
Portable drill	250 – 500
Ceiling fan	300 – 450
Portable fan	200 – 300
Garbage disposer	400 – 900
Hair dryer	600 – 1200
Portable space heater	900 – 1500
Router	750 – 1200
Table saw	900 – 1500
Stereo	300 – 600

* in W

Figure 18-17

Applications such as flashlights, portable electric tools, electric toys, remote control devices (for TVs, VCRs, garage door openers, etc.), portable communication equipment, and entertainment equipment (radios, telephones, etc.) require low-voltage DC power sources. These devices normally use 1.5 V to 9 V DC power sources. The power source can be supplied by one battery (size AAA, AA, C, D, or special), or any combination of series and/or parallel-connected batteries. **See Figure 18-18.**

Applications such as golf carts and electric cars require high-voltage DC power sources. These devices normally use 6 V to 36 V DC power sources. Six volt or 12 V batteries are normally connected in series and/or parallel connections to produce the required power. Batteries are connected in series to increase the total voltage. Batteries are connected in parallel to increase the total current.

Applications that do not require a portable power source, such as fixed lamps, motors, heating elements, etc., normally use an AC voltage source. AC supplied loads are the most common loads used in residential, commercial, and industrial applications.

Chemical Power Sources

A battery produces electricity through the chemical reaction between battery plates and acid. Common battery ratings are 1.5 V, 6 V, 9 V, 12 V, 24 V, and 36 V. Batteries are a common source of electricity, especially in portable equipment.

A battery delivers power for only a fixed amount of operating time. Once the battery power is used, the battery must be replaced (or recharged). Operating time of a battery depends on the load. The more power the load requires, the less operating time. The less power the load requires, the greater the operating time. Thus, a 20 W bulb drains a battery faster than a 10 W bulb. Battery capacity is rated in ampere-hours (Ah). For example, a battery that is rated for 20 Ah can supply 1 A for 20 hr, 2 A for 10 hr, 10 A for 2 hr, or 20 A for 1 hr, etc. **See Figure 18-19.**

Rechargeable batteries are normally used in applications that require large amounts of power. Only rechargeable batteries may be recharged. The most common rechargeable batteries are nickel-cadmium and lead-acid batteries. Nickel-cadmium batteries are used in electric toys, portable shavers, tools, and telephones. Lead-acid batteries are used for high-power applications, such as automobiles, golf carts, forklifts, etc.

Electromagnetic Power Sources

A generator (alternator) produces electricity when magnetic lines of force are cut by a rotating wire coil (rotor). The magnetic lines of force are produced by the magnetic field that is present between the north and south poles of a permanent magnet or electromagnet. An electromagnet can produce a stronger magnetic field than a permanent magnet. **See Figure 18-20.**

As the rotor moves through the magnetic field, electric current flows through the wire coil(s) of the rotor. The voltage produced by a generator depends on the strength of the magnetic field and the speed of rotation of the rotor. The stronger the magnetic field and the faster the rotation, the higher the voltage produced.

Generator output ranges from a few volts to several thousand volts. All generators must be rotated by some force, such as steam or water. Generators are the most common method used to produce electricity. The output of a generator may be connected directly to the load(s), connected to transformers, or connected to a rectifier circuit. In small generator applications, such as portable generators used by emergency crews, the output of the generator is normally connected directly to the load(s), which include lamps, pumps, etc.

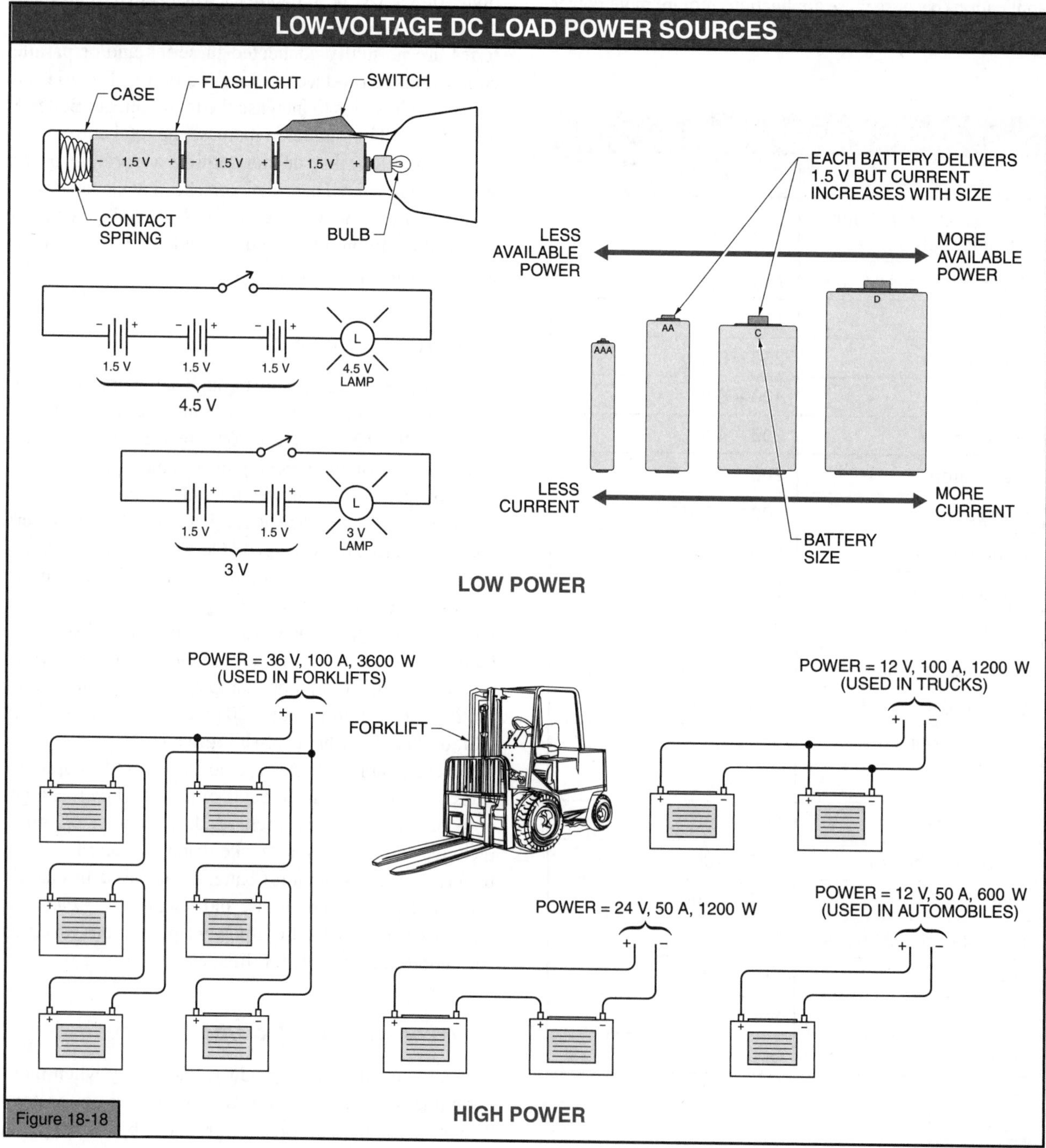

Figure 18-18

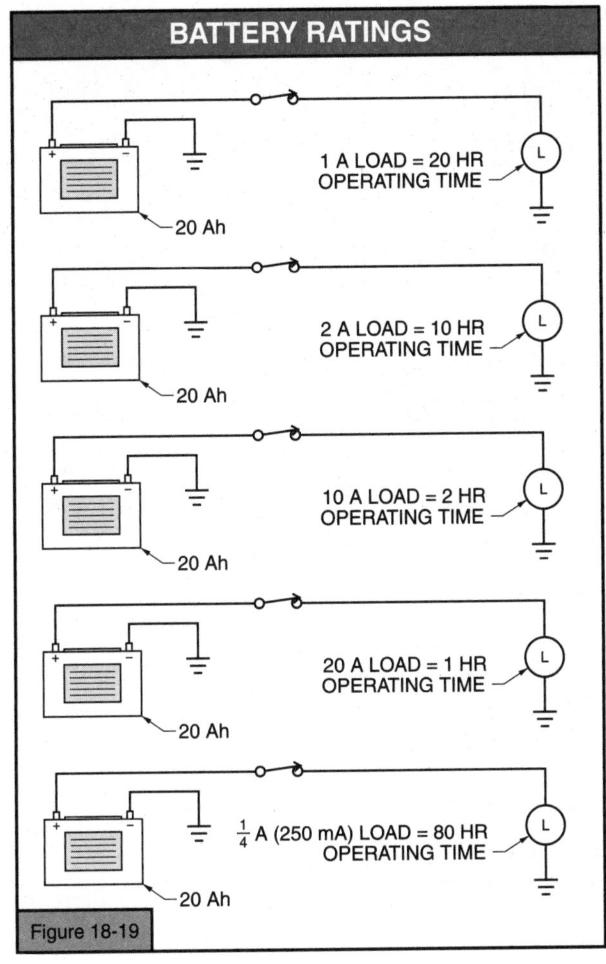

Figure 18-19

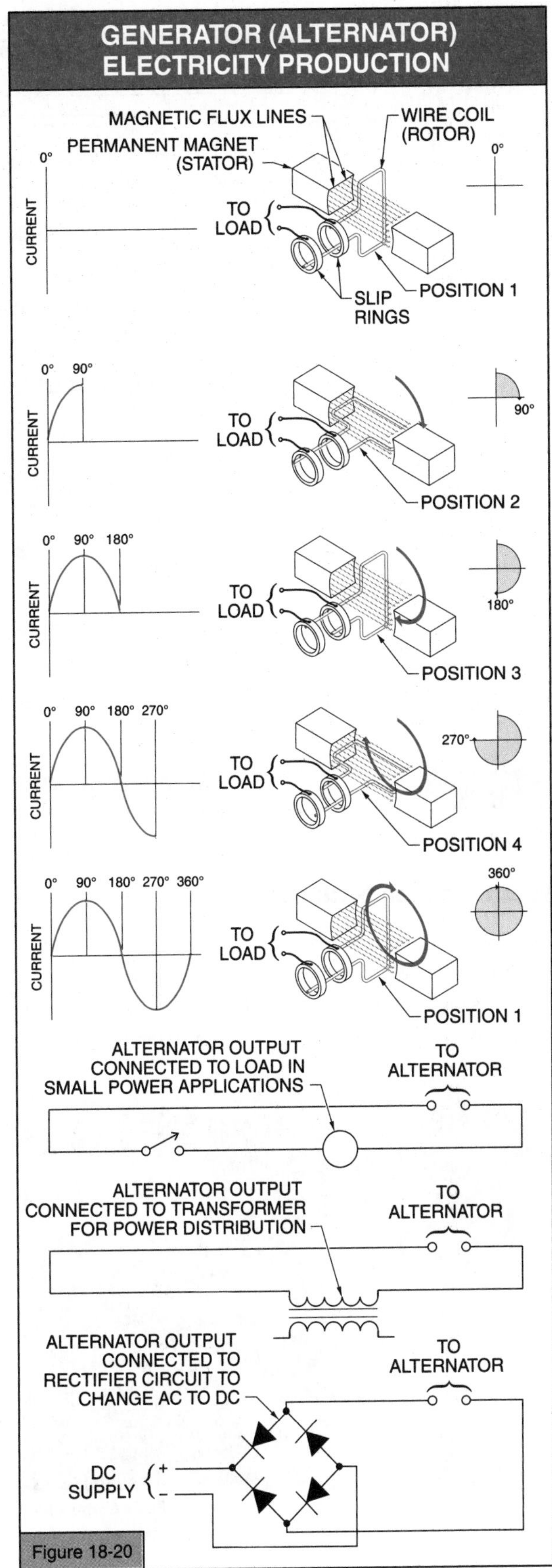

Figure 18-20

In large generator applications, the generator output is connected to transformers. The transformers normally step the voltage up so power can be transmitted at a low current level. As the voltage is increased, the current is reduced for any fixed amount of power. The reduced current allows smaller conductors, etc. to be used to carry the electricity. Step-down transformers step down the voltage to the required level for use by the loads.

In many applications, the generator output is sent to a rectifier circuit so the generated AC may be changed to DC. An automobile is a common application of using an AC generator to deliver DC. An automobile engine is used to drive a generator (alternator). The alternator's output is sent to a rectifier circuit that changes the generated AC into DC. The DC is used to supply the voltage to all the loads in the automobile and to recharge the battery. **See Figure 18-21.**

TECH FACT

Industrial generators are used for peak and standby power systems in locations such as airports, hospitals, and manufacturing and assembly facilities.

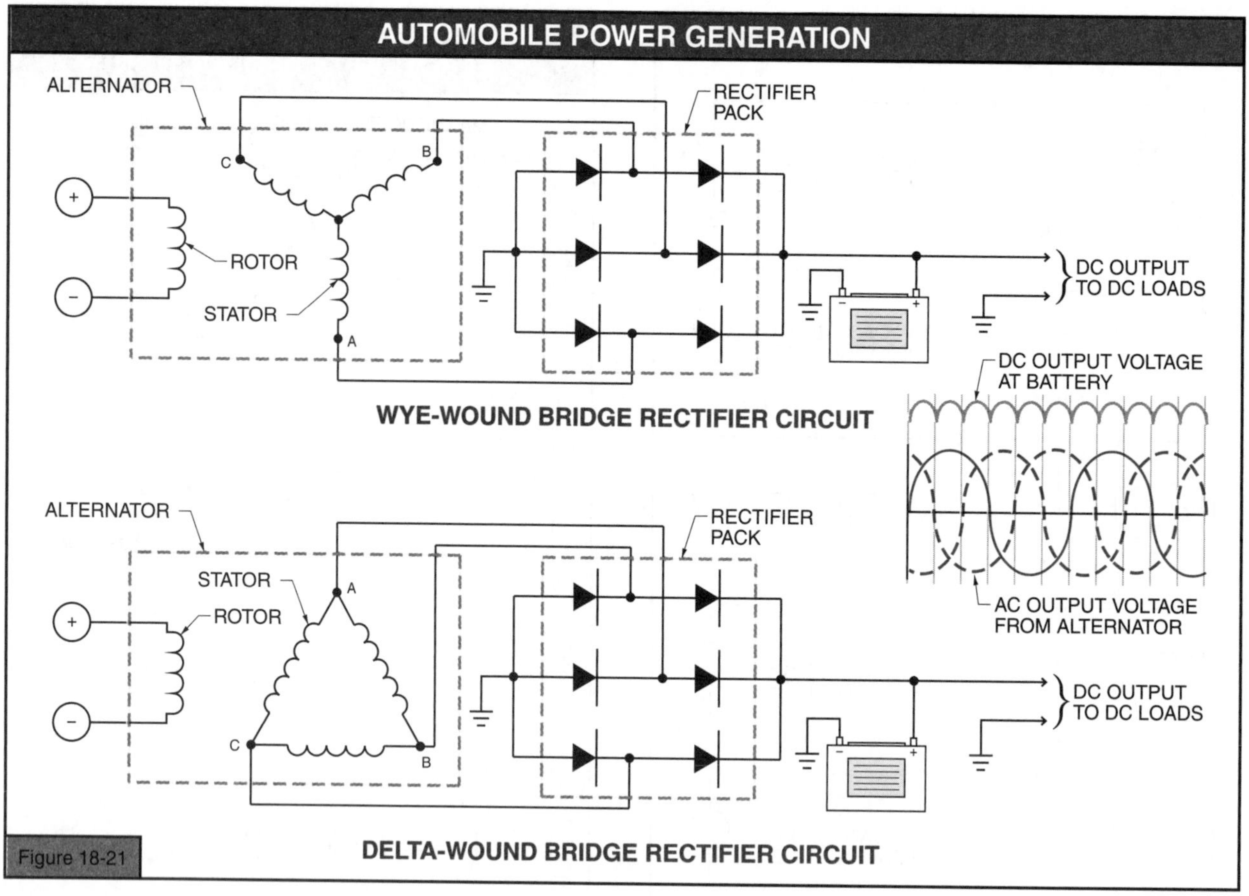

AUTOMOBILE POWER GENERATION

ALTERNATOR
RECTIFIER PACK
ROTOR
STATOR
DC OUTPUT TO DC LOADS
DC OUTPUT VOLTAGE AT BATTERY

WYE-WOUND BRIDGE RECTIFIER CIRCUIT

ALTERNATOR
RECTIFIER PACK
STATOR
ROTOR
AC OUTPUT VOLTAGE FROM ALTERNATOR
DC OUTPUT TO DC LOADS

DELTA-WOUND BRIDGE RECTIFIER CIRCUIT

Figure 18-21

Heat Power Sources

Most applications require a power source that can deliver more than a few millivolts. Some applications require a small voltage source. A thermocouple may be used if the temperature of a solid, gas, or liquid is to be measured.

Fluke Corporation

Thermocouples are used in test instruments that can take temperature measurements.

A *thermocouple* is a device that produces electricity when two different metals (such as iron and constantan) that are joined together are heated. The amount of voltage produced depends on the temperature. The higher the temperature, the higher the voltage produced. The amount of electricity produced is small, normally about 1 mV to 50 mV. Thermocouples are normally used for measuring temperature. The voltage produced by a thermocouple is used to give a temperature measurement when the thermocouple is connected to a temperature meter. The amount of voltage produced by a thermocouple depends on the type of thermocouple used and the temperature at the thermocouple. Manufacturer's specifications list the voltage produced by various thermocouples. **See Figure 18-22.**

Thermocouples may be used in applications other than temperature measurement. For example, a thermocouple heated by a gas pilot light produces a voltage that allows a gas valve to open and let gas through to the main burners of a furnace. An extinguished pilot light provides no heat to the thermocouple. This prevents the thermocouple from generating electrical energy, which closes the electrically-operated gas valve to prevent an explosion. **See Figure 18-23.**

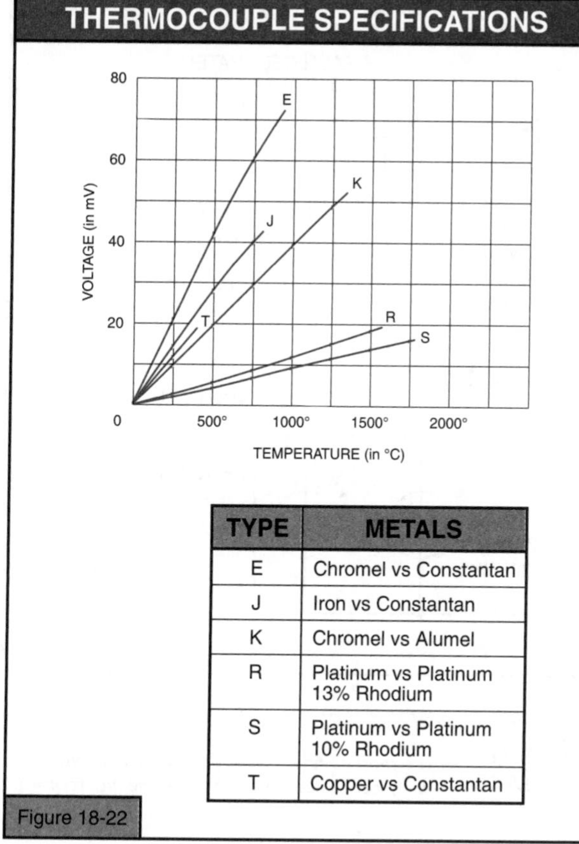

TYPE	METALS
E	Chromel vs Constantan
J	Iron vs Constantan
K	Chromel vs Alumel
R	Platinum vs Platinum 13% Rhodium
S	Platinum vs Platinum 10% Rhodium
T	Copper vs Constantan

Figure 18-22

FURNACE THERMOCOUPLE APPLICATION

PILOT BURNER
PILOT LIGHT
THERMOCOUPLE
MAIN BURNERS
GAS VALVE

Figure 18-23

Solar Power Sources

Electricity can be produced by natural light (sun) and artificial light (lamps). A *photvoltaic (solar) cell* is a device that converts light energy into electrical energy.

Solarcells commonly produce about 1 V, but can be connected to deliver high voltage and current outputs. Connecting solarcells in series increases the total voltage. Connecting solarcells in parallel increases the total current. Increasing either voltage and/or current increases available power. **See Figure 18-24.**

Solarcells are used as a power source for calculators, photographic equipment, and satellite communication systems. Solarcells can also be used to recharge batteries. The battery can be used to deliver more power than may be available at times (no or low light level) from the solarcells alone.

Control Requirements

All electric loads must be controlled. Control circuits used to control loads vary from simple to complex. An ON/OFF switch may be used for simple control of low-power loads. Computers and networks may be used for complex control applications. The application parameters, load size, cost limits, and environment in which the load is to be operated determines the type of control circuit used.

A control switch can be used to directly turn a load ON and OFF when a low-power load is used. A *low-power load* is a load in which the voltage and current rating of the load is less than the voltage and current rating of the switch. Low-power control circuits that directly control loads are used in most residential lighting circuits, low-power power tools, low-power appliances, electric toys, doorbell circuits, etc. **See Figure 18-25.**

In this circuit, the control switch may be any switch that is designed to control the load used. For example, a mechanical switch can control either an AC or DC load. A solid-state triac can control an AC load. A solid-state transistor or SCR can control a DC load. Solid-state switches last longer than mechanical switches.

For example, solid-state switching circuits in automobile ignition systems last 20 or more times longer (more miles) than mechanical switches (points). Solid-state switches are also able to control the voltage to the load. Voltage control is used in circuits that control lamp brightness or the amount of heat output.

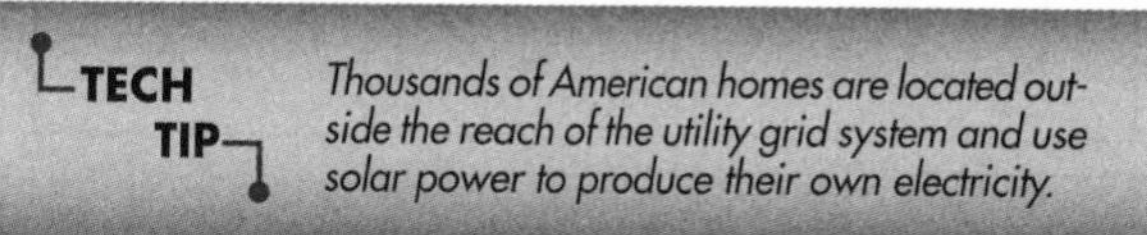

Thousands of American homes are located outside the reach of the utility grid system and use solar power to produce their own electricity.

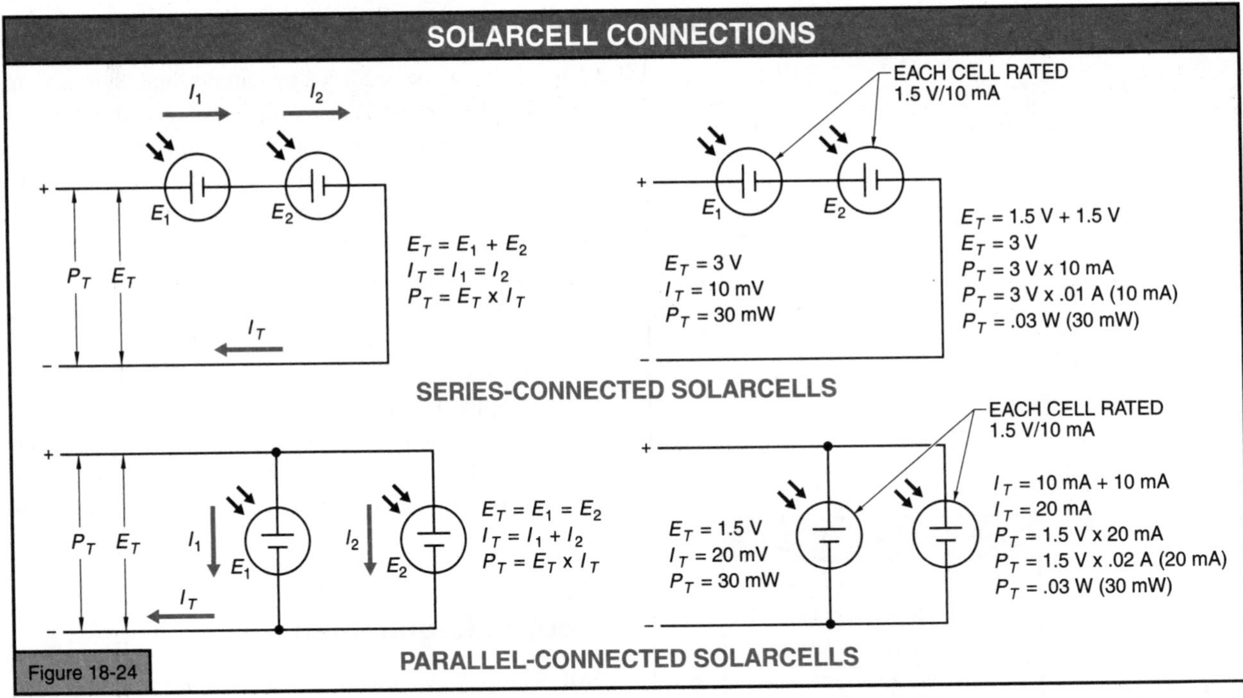

Figure 18-24

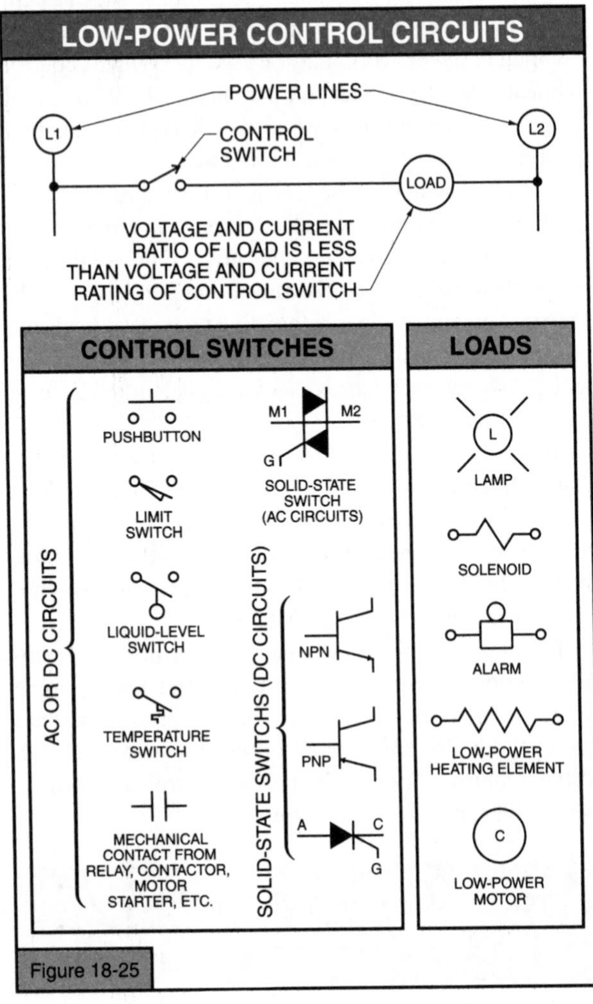

Figure 18-25

TECH FACT *The most common type of solar cell system is a utility-connected system on a residential roof.*

An interface is required when a high-power load is used. A *high-power load* is a load whose voltage and current rating is greater than the voltage and current rating of the switch. An *interface* is a device that allows different levels or types of components to be used together in the same circuit. An interface is used to allow a control switch to control a high-power load. Interfaces include relays, contactors, and motor starters. **See Figure 18-26.**

The control switch is connected to the input of the interface and the load is connected to the output of the interface. In a relay, contactor, and motor starter, the coil is the interface input and the contacts are the output. Relays have low-power rated contacts and are used for switching low-power loads, such as alarms, small solenoids, and very small (mW) motors.

Contactors have high-power rated contacts and are used for switching high-power loads, such as lamps (fluorescent and HID), heating elements, transformers, and small motors (less than ¼ HP) that have built-in overload protection. Motor starters are used for switching most motors over ¼ HP and all motors over 1 HP. Contactors and motor starters are available in sizes that can switch loads ranging from a few amperes to several thousand amperes.

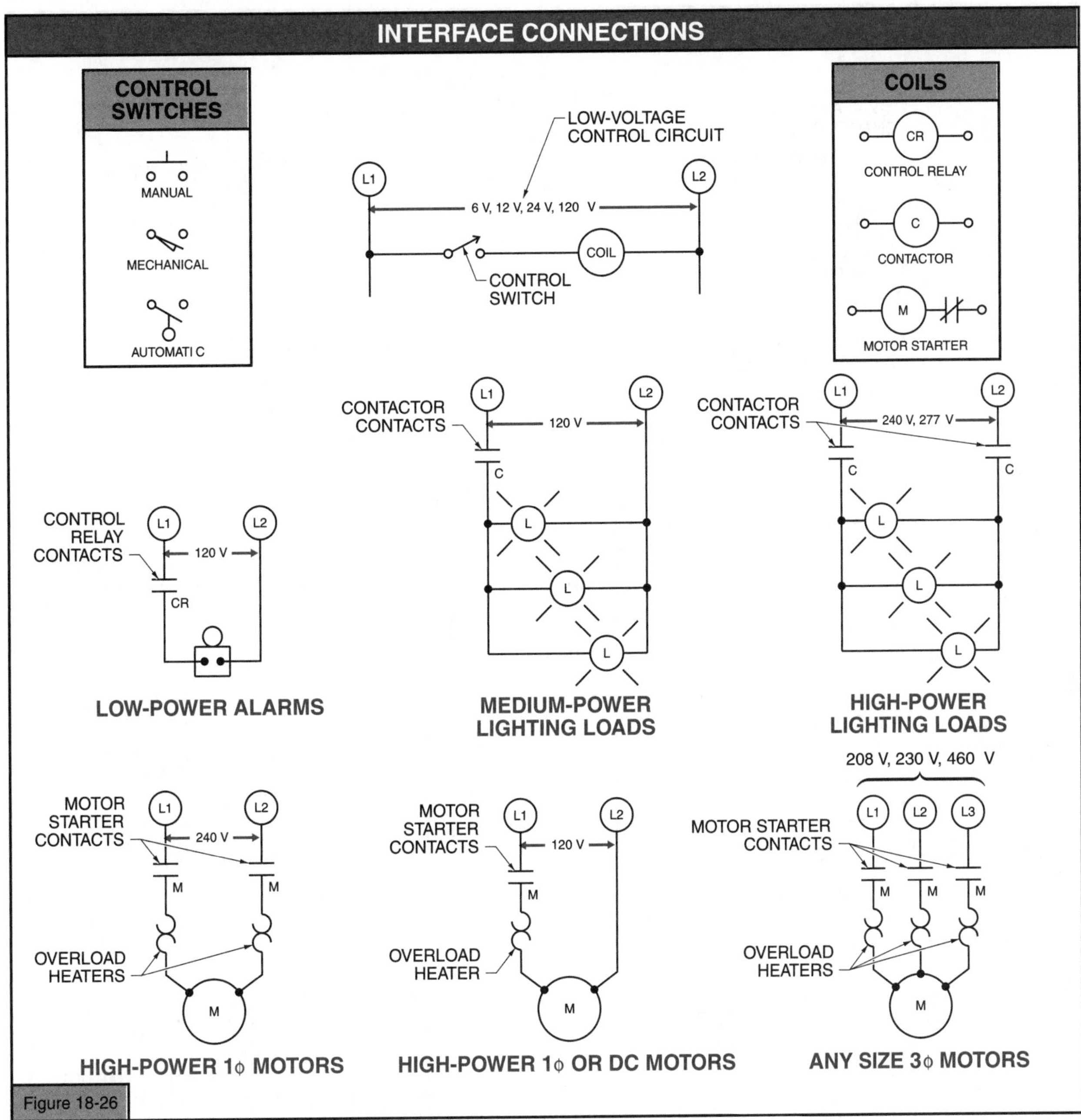

Figure 18-26

Properly applied interfaces allow the smallest control signals to control the highest power loads. For example, a low-power rated temperature switch (thermostat) may be used to control a relay coil. The relay contacts may be used to control a motor starter coil and/or contactor coil. The motor starter coil may be used to control a high-power motor. The contactor coil may be used to control a high-power heating element. **See Figure 18-27.**

TECH FACT *Electromechanical relays are available with silver, silver-cadmium, gold-flashed silver, and tungsten contacts.*

Control Relay Interfaces

A *relay* is a device that controls one electrical circuit by opening and closing contacts in another circuit. Relays may be electromechanical or solid state. An *electromechanical relay* is a switching device that has sets of contacts which are closed by a magnetic force. In an electromechanical relay circuit, a coil opens and closes contacts by a magnetic force that is developed each time the coil is energized. Relay contacts are described by their number of poles, throws, and breaks. **See Figure 18-28.**

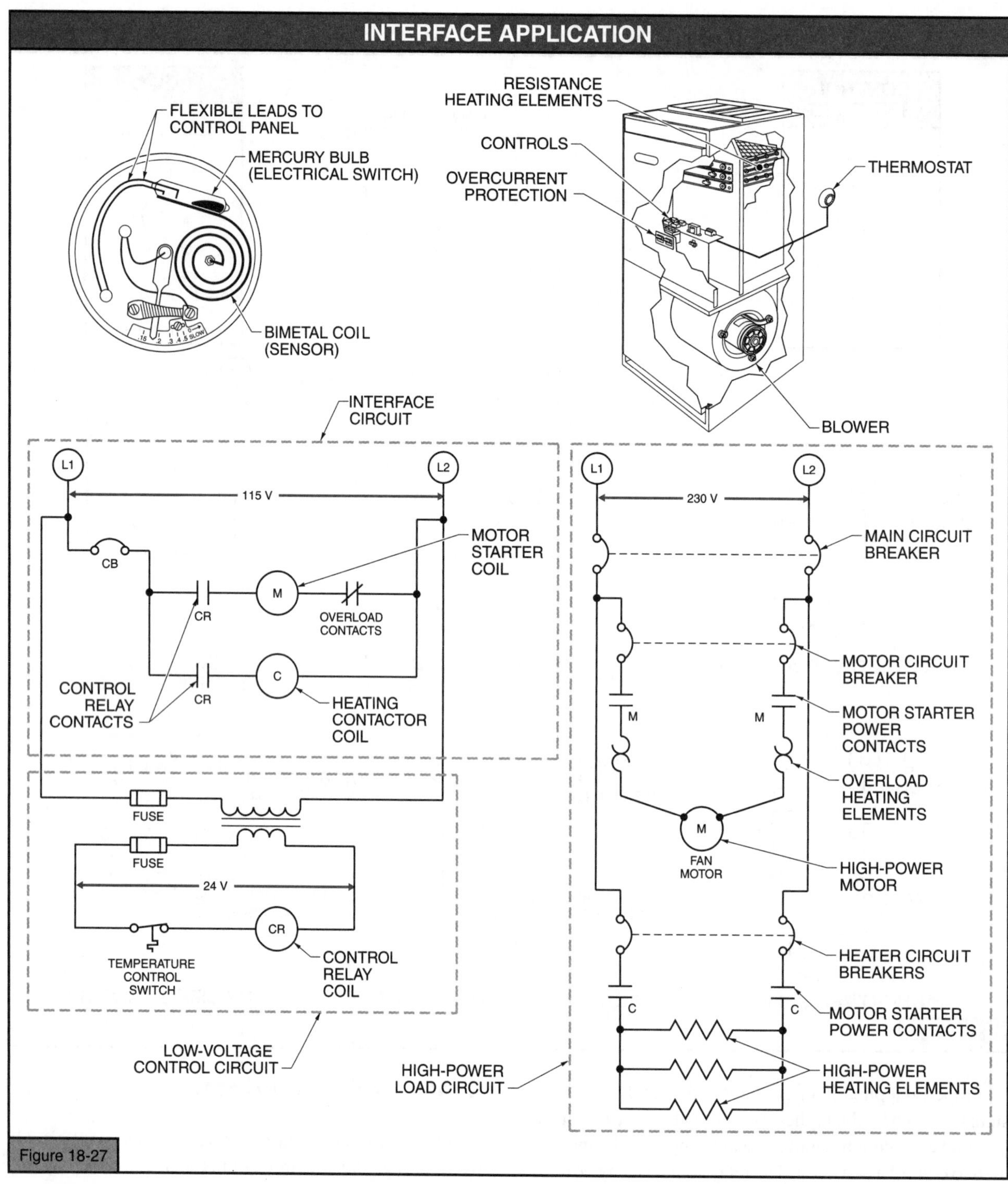

Figure 18-27

A *break* is a place on a contact that opens or closes an electrical circuit. A single-break (SB) contact breaks an electrical circuit in one place. A double-break (DB) contact breaks the electrical circuit in two places. Single-break contacts are normally used when switching low-power devices such as horns and lights. Double-break contacts are used when switching high-power devices such as solenoids.

A *pole* is a completely isolated circuit that a relay can switch. A single-pole contact can carry current through only one circuit at a time. A double-pole contact can carry current through two circuits simultaneously. In a double-pole contact, the two circuits are mechanically connected to open or close simultaneously and are electrically insulated from each other.

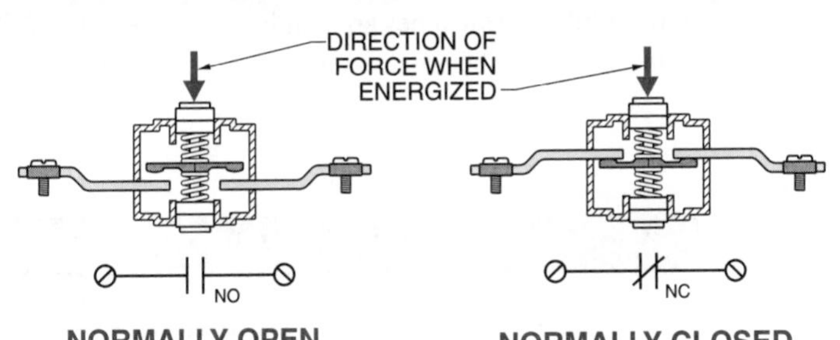

RELAY CONTACT ABBREVIATIONS

Abbreviation	Meaning
SP	Single pole
DP	Double pole
3P	Three pole
ST	Single throw
DT	Double throw
NO	Normally open
NC	Normally closed
SB	Single break
DB	Double break

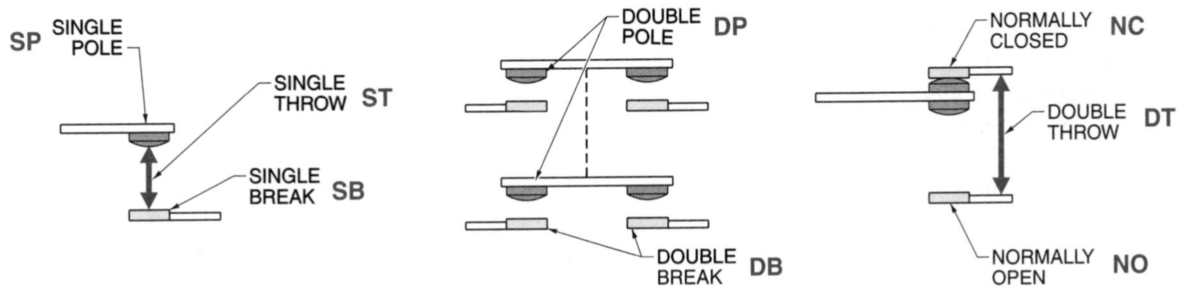

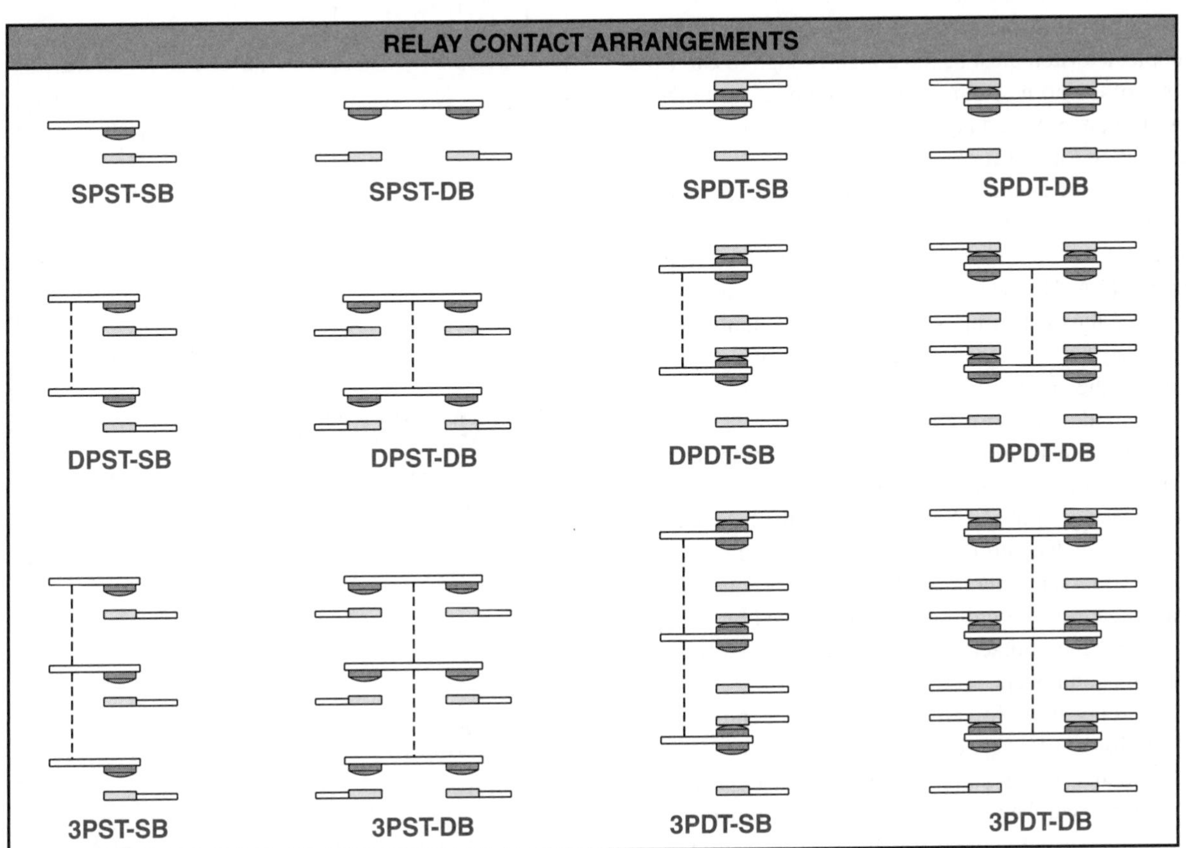

Figure 18-28

The mechanical connection is represented by a dashed line connecting the poles. Relays are available with 1 to 12 poles. A *throw* is the number of closed contact positions per pole. A single-throw contact can control only one circuit. A double-throw contact can control two circuits.

A *solid-state relay* is a relay that uses electronic switching devices in place of mechanical contacts. In a solid-state relay circuit, a coil is replaced by an input circuit and the contacts are replaced with a solid-state switching component (transistor, SCR, or triac).

The input circuit is activated by applying a voltage to the input of the relay that is higher than the specified pickup voltage of the relay. Most solid-state relays have an input voltage range, such as 3 VDC to 32 VDC. The voltage range allows a single solid-state relay to be used with most electronic circuits.

The switching component of a solid-state relay is controlled by the input signal. Most solid-state relays use a thyristor as the output switching component. Thyristors change from their OFF-state (contacts open) to their ON-state (contacts closed) very quickly when their gate switches ON.

A switch must carry the same amount of current as the load when the switch is used to directly control the load. For example, a switch must be rated high enough to carry the current of the horn(s) if the horn switch in an automobile is used to directly start and stop the flow of current to the horn(s). **See Figure 18-29.**

In the horn circuit containing no interface, the horn switch must carry all 16 A (8 A per horn). The wire used to connect the switch must also be rated high enough to carry the 16 A. The switch and wire size is excessively large for wiring into a steering column (space limitations). The circuit may be redesigned to include a relay which is used to reduce the amount of current, switch rating, and wire size required at the horn switch.

A control relay used as an interface reduces the current flowing through that part of the circuit. A smaller wire may be used because only ¼ A flows through the horn switch. The horn switch controls the relay coil, which controls the horn. A solid-state relay may be used as the interface. In a solid-state relay circuit, the coil is replaced by an input circuit and the contacts are replaced with a solid-state switching component.

The relay used for an application depends on the application's life expectancy and electrical and cost requirements. For example, a horn circuit requires that the horn be switched very few times. Electromechanical relays are normally rated for 250,000 operations and solid-state relays are rated for billions of operations. Either relay may be used in the horn circuit because of the small number of switches that occur. However, in a bottling operation that produces thousands of bottles an hour, solid-state switches are required because of the large number of switches that occur.

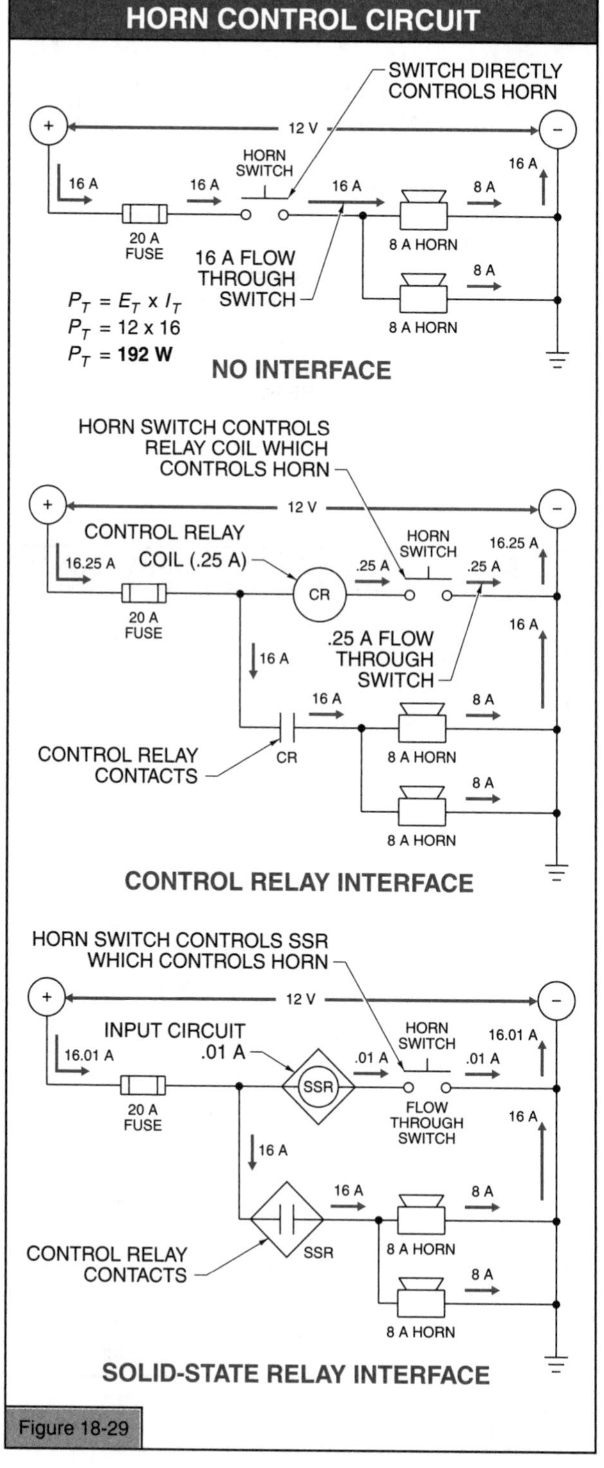

Figure 18-29

Electromechanical Relay Advantages

Advantages of electromechanical relays include:

- Normally have multipole, multithrow contact arrangements
- Contacts can switch AC or DC
- Low initial cost
- Very low contact voltage drop, thus no heat sink is required
- Very resistant to transient voltages
- No OFF-state leakage current through open contacts

Electromechanical Relay Limitations

Limitations of electromechanical relays include:

- Contacts wear, thus have a limited life
- Short contact life when used for rapid switching applications or high current loads
- Generates electromagnetic noise and interference on power lines
- Poor performance when switching high inrush circuits

Solid-State Relay Advantages

Advantages of solid-state relays include:

- Very long life when properly applied
- No contacts to wear
- No contact arcing to generate electromagnetic interference
- Resistant to shock and vibration because they have no moving parts
- Logic compatible to programmable controllers, digital circuits, and computers
- Very fast switching capability
- Different switching modes (zero switching, instant-ON, etc.)

Solid-State Relay Limitations

Limitations of solid-state relays include:

- Normally only one contact available per relay
- Heat sink required due to voltage drop across switch
- Can switch only AC or DC
- OFF-state leakage current when switch is open
- Normally limited to switching only a narrow frequency range such as 40 Hz to 70 Hz

Contactor Interfaces

A *contactor* is a control device that uses a small control current to energize or de-energize the load connected to it. Contactors are relays that have high-current contacts. Contactors control high-power loads and/or many loads on an individual circuit (several loads connected in parallel). Contactors are normally used in high-power lighting circuits. **See Figure 18-30.**

In the lighting circuit, the contactor contacts carry the high current of the loads and the photoelectric switch carries the low current of the contactor coil (normally less than 1 A). This allows the control switch to be rated lower than the loads the switch is controlling. Contactor ratings range from a few amperes to several thousand amperes.

Motor Starter Interfaces

A *motor starter* is an electrically operated switch (contactor) that includes motor overload protection. Motor overload protection is normally provided by heaters or electronic circuits, which are included as part of the motor starter. Motor overload protection protects the motor when the motor is running. **See Figure 18-31.**

Motor overload protection is selected based on the motor's running current listed on the motor nameplate. The current drawn by the motor passes through the overload heating element. The overload heating element heats to a point high enough to open the overload contacts if the motor draws excessive current. Open overload contacts stop the motor because the open overload contact stops the flow of current through the motor starter coil. The motor starter coil stops the flow of current to the motor when the coil is de-energized. The overloads must be reset before the motor can restart. Overloads may be manually reset or automatically reset.

Saftronics Inc.

Motor drives are electronic units designed to control the speed of a motor by using solid-state components.

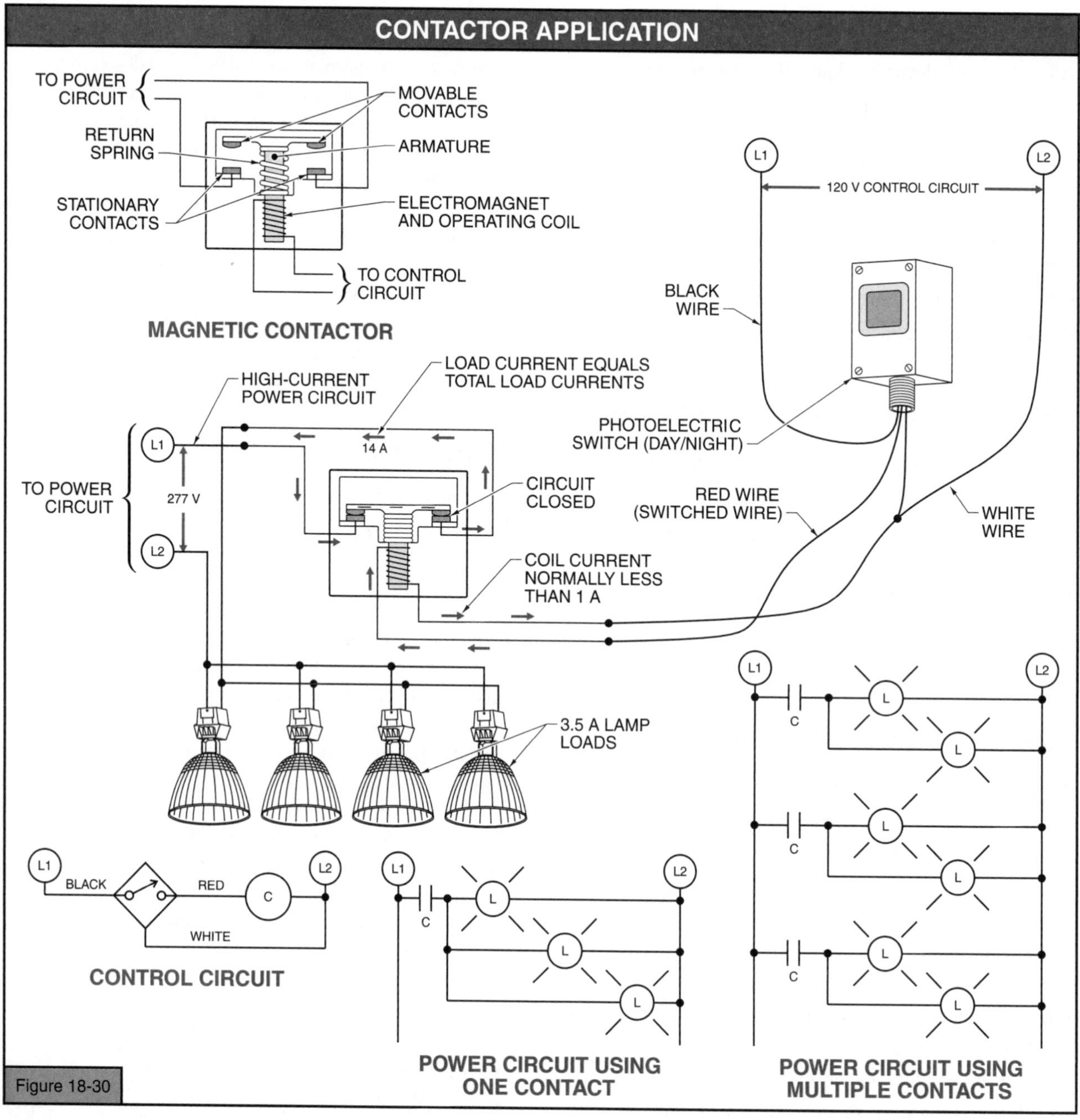

Figure 18-30

In a common start/stop motor control circuit, extra motor starter contacts (holding contacts) are placed in parallel with the start pushbutton. The extra motor starter contacts close, allowing the start pushbutton to be released when the motor starter coil is energized. The current continues to flow through the holding contacts and starter coil. The motor is stopped by pressing the stop pushbutton or if the overload contacts are opened.

Most motor control circuits are powered from step-down transformers to reduce the voltage to the control circuit. A step-down transformer reduces the voltage to the control circuit to a level of 120 V (or less). **See Figure 18-32.**

Complex Circuit Interfacing

In many circuits, the circuit inputs (switches) control the outputs (loads) through computer-controlled circuits. In such systems, the inputs are never connected directly to the loads. The inputs are connected to input modules that change the input signal into a signal that can be sent into a data interface module. A *data interface module* is a device that converts input signals into a signal that can be sent over a network. **See Figure 18-33.**

The input signal from the data interface module is sent into a converter that allows a standard computer, loaded with a control program, to recognize the inputs. The computer program manipulates the input information and sends output signals back over the network, through the converter. The data interface module takes the output signals and energizes output modules that can control lamps, solenoids, motor starters, and other outputs. A printer can also be added to the network to produce hard copy documentation.

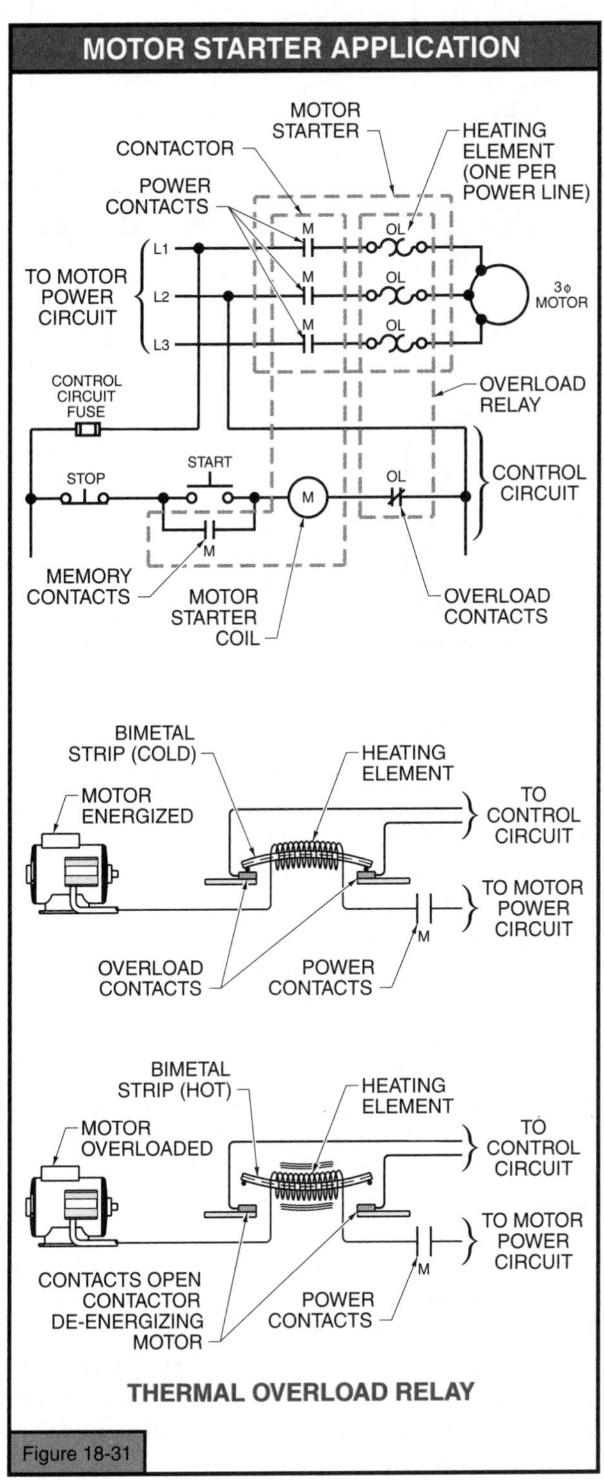

Figure 18-31

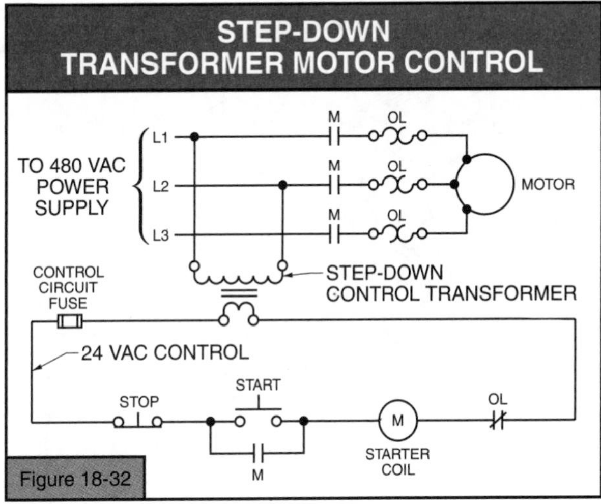

Figure 18-32

Input and output modules are solid-state relays that are designed for many different signal input and output control functions. They are referred to as input and/or output modules. Input and output modules, used along with communication interfaces, allow signals to be sent over telephone and other communication systems. **See Figure 18-34.**

In such complex circuits, care must be taken to ensure each component in the system can communicate with the next component in the system and that they are electrically compatible. Communication is the compatibility between the software (computer programs and procedures) and all parts of the system. Electrical compatibility is the matching of the voltage, current, and power requirements, pin and plug configurations, and size and type of inputs and outputs.

Application software programs are designed to allow circuit control of almost any input and output device. Most industrial software packages use icon-driven software. Icons are picture representations on the computer screen. Icons that show the process and the type of inputs and outputs are used in industrial network systems.

Protection Requirements

All circuits require protection if they are to operate properly and keep fault (overloads, shorts, etc.) damage to a minimum. Circuits are protected from faults by fuses, circuit breakers, overload relays, and electronic protection monitors. Fuses and circuit breakers are designed to protect a circuit from overcurrents and short circuits. Overload relays protect motors from overloads placed on the motor. Electronic protection monitors are devices that are designed to protect the system from over/under voltages, phase loss, phase imbalance, etc.

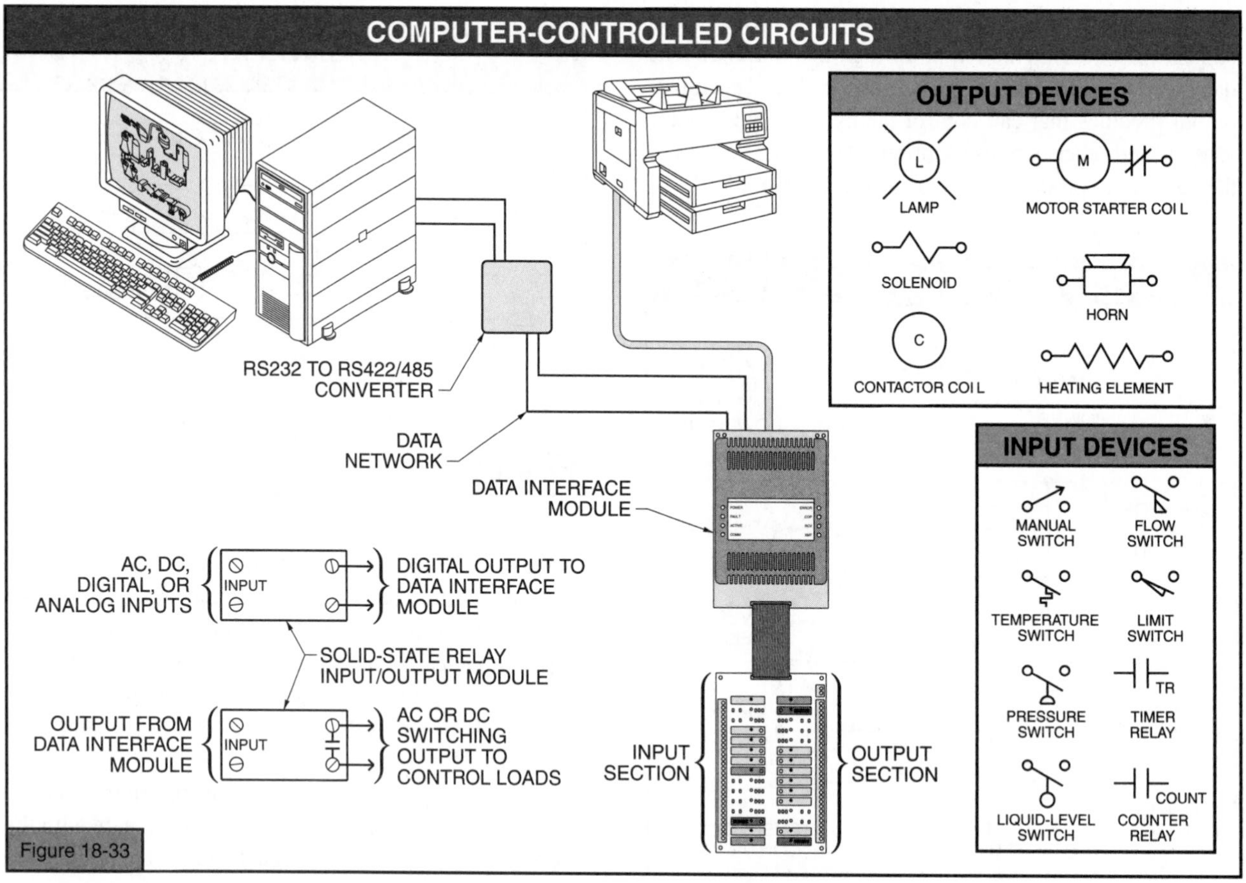

COMPUTER-CONTROLLED CIRCUITS

OUTPUT DEVICES: LAMP, MOTOR STARTER COIL, SOLENOID, HORN, CONTACTOR COIL, HEATING ELEMENT

INPUT DEVICES: MANUAL SWITCH, FLOW SWITCH, TEMPERATURE SWITCH, LIMIT SWITCH, PRESSURE SWITCH, TIMER RELAY (TR), LIQUID-LEVEL SWITCH, COUNTER RELAY (COUNT)

RS232 TO RS422/485 CONVERTER

DATA NETWORK

DATA INTERFACE MODULE

AC, DC, DIGITAL, OR ANALOG INPUTS — INPUT — DIGITAL OUTPUT TO DATA INTERFACE MODULE

SOLID-STATE RELAY INPUT/OUTPUT MODULE

OUTPUT FROM DATA INTERFACE MODULE — INPUT — AC OR DC SWITCHING OUTPUT TO CONTROL LOADS

INPUT SECTION

OUTPUT SECTION

Figure 18-33

System faults, personnel protection, fire hazards, and improper operation are reduced (or eliminated) by proper device installation and maintenance. Proper device installation is accomplished by following codes and standards. See the National Electrical Code® for further information.

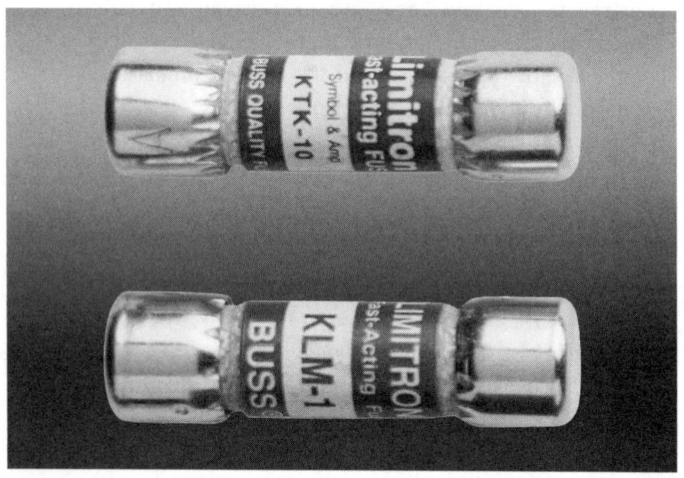

Cooper Bussman, Inc.

Fuses are used to protect electrical circuits from problems such as short circuits and high current overloads.

Monitoring Systems

Electrical loads use electricity to produce work. The electrical power delivered to a load must be the correct type and amount for the load to convert electricity into work. The load operates in a safe, economical, and efficient manner if the delivered power is the correct type and amount. The load operates on whatever power is delivered to it if the delivered power is not correct. The load may not operate at all (very low voltage) or may be destroyed (very high voltage) if the delivered power is completely incorrect. For example, a motor would be destroyed almost immediately if 460 V were applied to a 230 V motor. The motor could operate for a considerable period of time before damage is done if 208 V were applied to a 230 V motor.

Major power problems, such as short circuits and very high overloads are handled by the system's fuses and circuit breakers. Fuses and circuit breakers do not normally take care of problems such as low voltage, high voltage, loss of one power line, and other problems that affect the operation of the load. These problems require special monitors that are designed to detect even slight variations in the supply power.

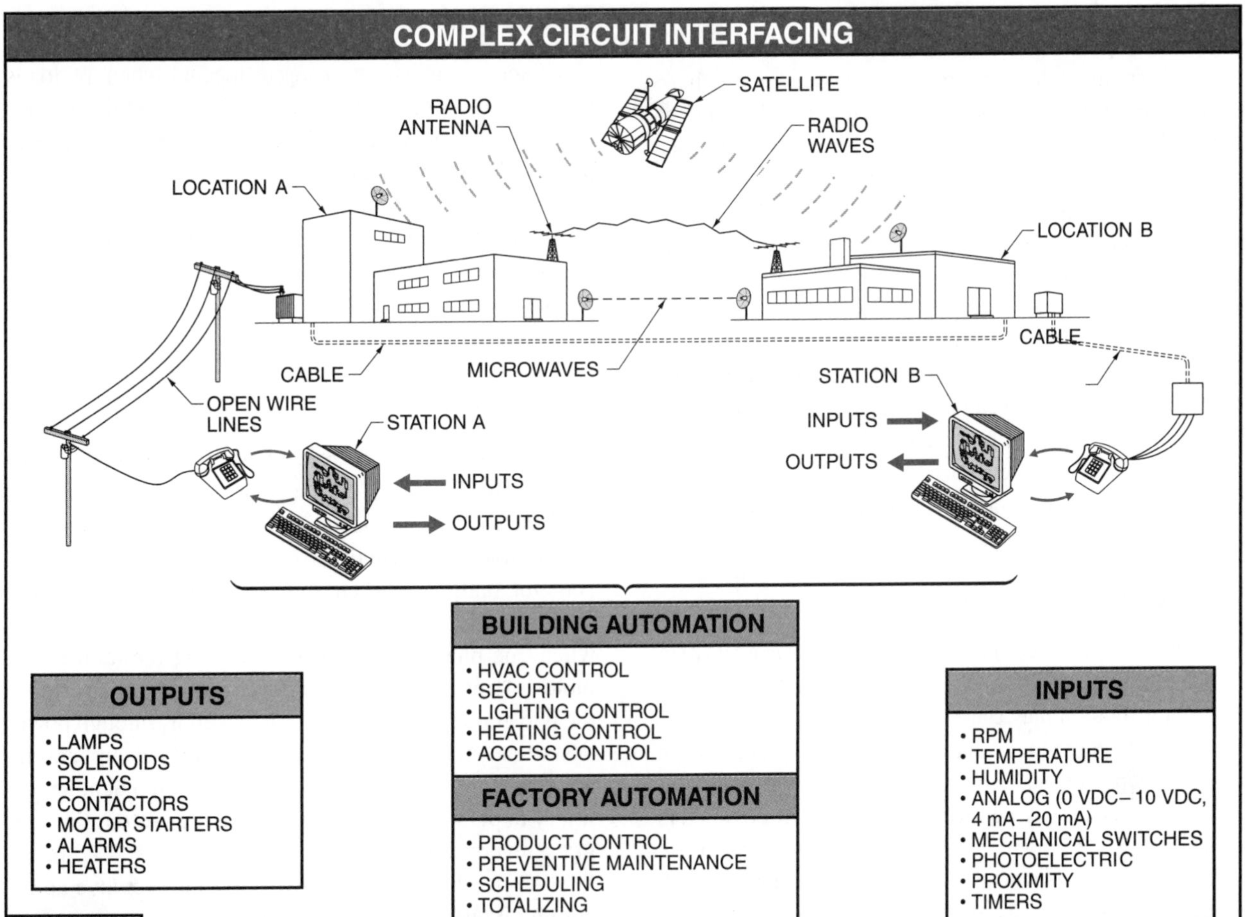

COMPLEX CIRCUIT INTERFACING

SATELLITE
RADIO ANTENNA
RADIO WAVES
LOCATION A
LOCATION B
CABLE
MICROWAVES
CABLE
STATION B
OPEN WIRE LINES
STATION A
INPUTS
OUTPUTS
INPUTS
OUTPUTS

BUILDING AUTOMATION
- HVAC CONTROL
- SECURITY
- LIGHTING CONTROL
- HEATING CONTROL
- ACCESS CONTROL

FACTORY AUTOMATION
- PRODUCT CONTROL
- PREVENTIVE MAINTENANCE
- SCHEDULING
- TOTALIZING

OUTPUTS
- LAMPS
- SOLENOIDS
- RELAYS
- CONTACTORS
- MOTOR STARTERS
- ALARMS
- HEATERS

INPUTS
- RPM
- TEMPERATURE
- HUMIDITY
- ANALOG (0 VDC–10 VDC, 4 mA–20 mA)
- MECHANICAL SWITCHES
- PHOTOELECTRIC
- PROXIMITY
- TIMERS

Figure 18-34

Applying the correct monitor ensures that a fault in the system is detected before expensive problems occur. Monitors may be added to protect the system from many different types of problems that may occur in the main power supply, directly at the electrical load, or in the mechanical load connected to the motor. **See Figure 18-35.**

ELECTRONIC PROTECTION MONITORS

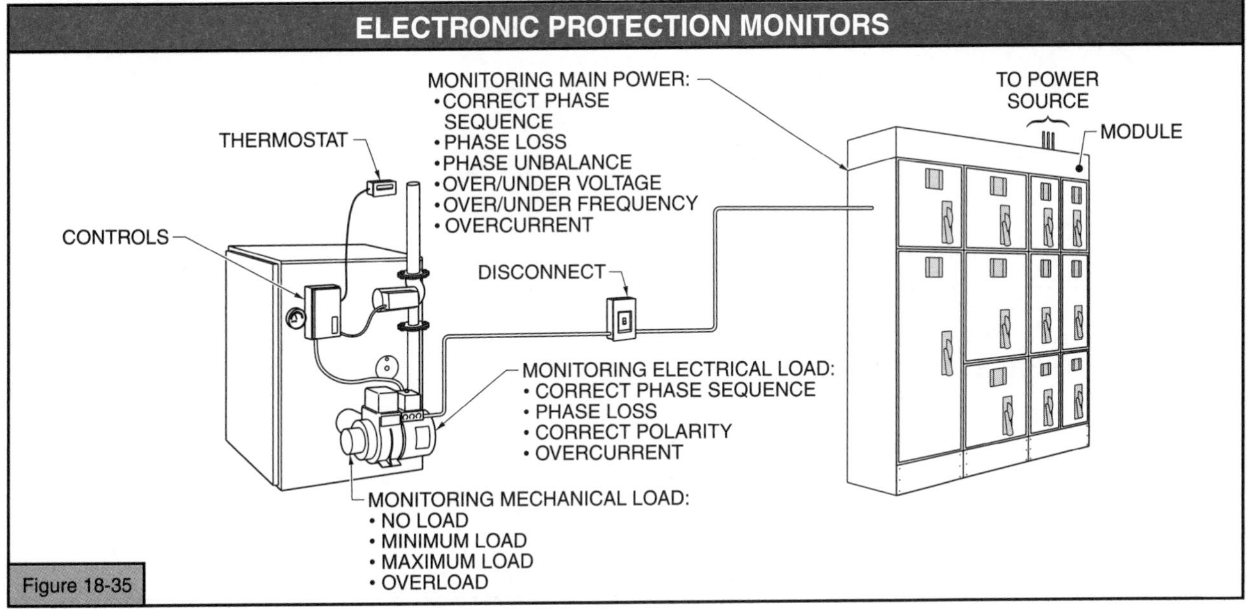

Figure 18-35

The most common power used by large industrial loads is 3ϕ power. Common problems that occur in 3ϕ power systems include improper phase sequence, phase unbalance, phase loss, and phase angle error. Any one of these problems may go undetected by the system's fuses or circuit breakers. **See Figure 18-36.**

Single-phase power is also used for industrial, commercial, and residential loads. Almost all 1ϕ power comes from a 3ϕ power distribution system. Therefore, any problem in the 3ϕ system may also become a problem in the 1ϕ system. One of the most common undetected problems in 1ϕ systems is voltage that is slightly too high or too low. Over time, this problem causes damage to electrical loads, electronic circuits, and other control devices. Any good power distribution system should include power monitors to protect the system from problems that may occur.

Improper Phase Sequence

Changing the sequence of any two phases (phase reversal) in a 3ϕ motor reverses the direction of motor rotation. Reversing motor rotation may cause damage to driven machinery or injury to personnel. Phase reversal may occur when modifications are made to a power distribution system or when maintenance is performed on electrical conductors or switching equipment. The NEC® requires phase reversal protection on all personnel transportation equipment, such as moving walkways, escalators, and ski lifts.

Phase Unbalance

Unbalance of a 3ϕ power circuit occurs when 1ϕ loads are unevenly applied to the circuit. One power line carries more of the load when 1ϕ loads are unevenly applied. A phase unbalance causes 3ϕ motors to run at temperatures above their normal rating. A 3ϕ motor operating in an unbalanced circuit cannot deliver its rated horsepower. Careful attention is given to balancing loads in new 3ϕ power circuit installations.

Phase Loss

The complete loss of one phase in a power circuit (single-phasing) is an extreme case of phase unbalance. A phase loss may cause a motor to start in the opposite direction of rotation and also causes the motor to overheat. A motor, under special circumstances, may rotate in either direction when the two other phase voltages are applied when one phase is lost prior to starting the motor. The motor's temperature increases over a longer time period if the two good phase voltages cannot supply enough power to start the motor. In both cases, the increased temperature may burn out the motor. Damage to driven machinery or injury to personnel is possible because a motor may start in the reverse direction when one phase is lost. Lightning, a fallen tree branch, or a blown fuse may remove an incoming power line and lead to phase loss.

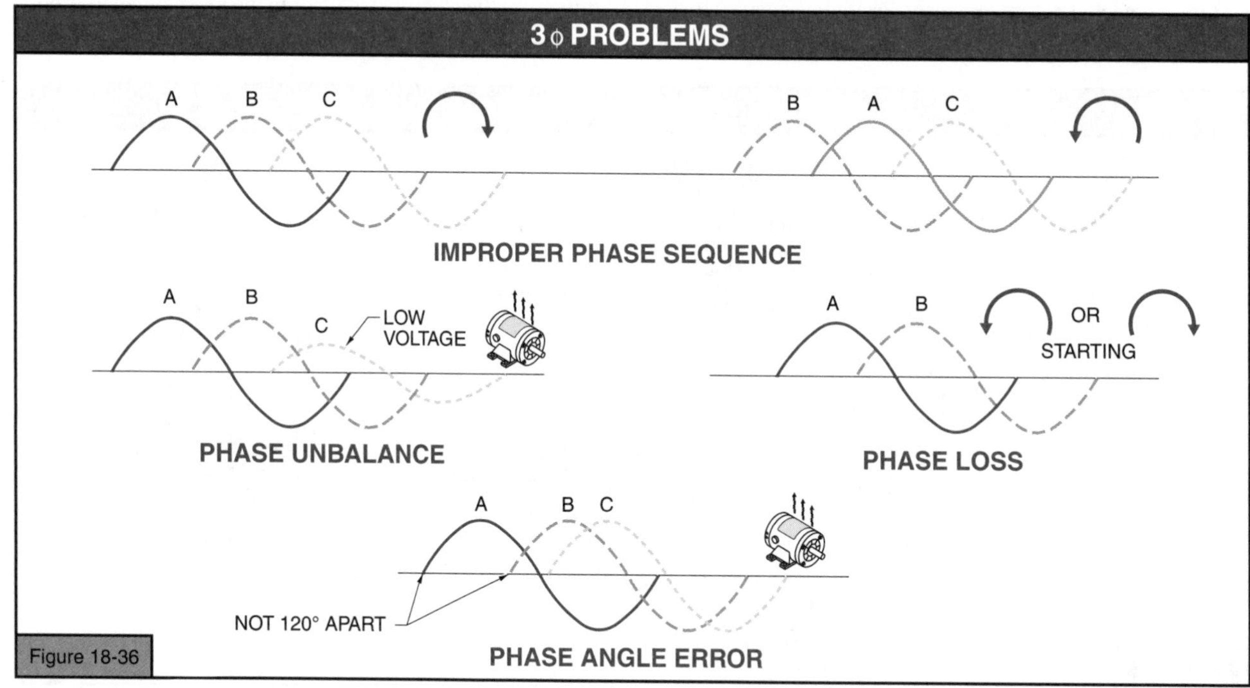

Figure 18-36

Phase Angle Error

In a 3ϕ power circuit, each phase is 120 electrical degrees apart. The phase angle changes if a circuit is unbalanced or if a heavy inductive or capacitive load is placed on one or two of the power lines. A phase angle error occurs when the phase angle changes. A phase angle error causes 3ϕ motors to run at temperatures above their normal ratings. A 3ϕ motor operating in a circuit with a phase angle error cannot deliver its full-rated horsepower.

An over/under voltage power monitor may be used to monitor the control circuit for an over/under voltage condition. The monitor is connected into the circuit so it monitors the voltage out of the control transformer. The monitor operates its relay contacts if there is no problem. The relay's NC contact may be connected to turn ON a warning light when a problem develops. Turning ON a warning light gives a visual indication of the problem. The relay's NO contacts are connected in series with the main power supply line. The contacts open and remove power to the control circuit if the monitor detects a problem. **See Figure 18-37.**

Monitoring Main Power Lines

The power delivered to a system must be the correct type and as clean as possible. Although the power company takes steps to deliver the proper power, outside factors may change it. Lightning or a tree branch may remove one of the three incoming power lines, leading to phase loss and unbalance. Although this problem is taken care of by the power company, expensive work stoppages and equipment damage may occur. A plant may not be able to prevent the external loss of power to the plant, but the plant may monitor the incoming power and take corrective action immediately by monitoring the main power lines. The main power lines may be monitored for correct phase sequence, phase loss, phase unbalance, over/under voltage conditions, over/under frequency conditions, and overcurrent conditions.

A phase loss/phase sequence monitor may be used to monitor the main power lines for a phase loss or the correct phase sequence. The monitor is connected into the circuit to monitor each of the three upstream power lines. The monitor operates its relay contacts if there is no problem. The relay's NO contacts are connected in series with the stop pushbutton. The contacts open and turn OFF any load connected into that part of the circuit if the monitor detects a problem. The relay's NC contacts may be connected to turn ON a warning light when a problem develops. Turning ON a warning light gives a visual indication of which part of the system has a problem. This reduces the time required to find the problem. **See Figure 18-38.**

Monitoring Loads

The power entering a plant is at the end of the power company's distribution system and at the beginning of the in-plant power distribution system. Many problems can occur when the power enters a plant, resulting in incorrect and damaging power being delivered to the loads. Some of these problems are obvious, such as short circuits, overloads, and loss of a phase caused by the blowing of a fuse. Some of these problems are not so obvious, such as over/under voltage, over/under frequencies, and phase asymmetry.

The power company monitors its distribution system from the start (generator) to the end (consumer). Plant power must be monitored from start to end to ensure that power is driven correctly. The electricity in plant systems and electrical loads may be monitored for correct phase sequence, phase loss, contact failure on switching equipment, over/under voltage conditions, correct polarity of all DC circuits, phase reversal in 3ϕ systems, phase unbalance, damaging voltage spikes and electrical noise, dangerous ground faults, and overcurrent conditions.

Conductors

Conductors are a major part of electrical circuits. In general, the NEC® recognizes three types of conductors: copper, aluminum, and copper-clad aluminum. Unless specified, all conductors in the NEC® are considered to be copper. There are better conductors than copper. For example, silver is a better conductor of electricity than copper, however, copper offers the most ampacity at the least cost per foot.

Conductor sizes are expressed in American Wire Gauge (AWG) or circular mils (CM) per Article 110.6. Table 8, Chapter 9, contains some useful information on conductor properties and can also be used to show the relationship between AWG sizes and circular mils. Basically, the AWG is a system for comparing the relative area of a conductor. For the purposes of the NEC®, sizes start at No. 18 (small) and run through No. 1 (larger). As the number value increases, the size of the conductor decreases. Thus, a No. 4 conductor is larger than a No. 8 conductor. After No. 1, the AWG uses the aught sizes. Four sizes are listed using this method: No. 1/0, No. 2/0, No. 3/0, and No. 4/0.

After the No. 4/0 size, the Table switches to the use of kcmils. A *mil* is .001″. A *circular mil* is a measurement used to determine the cross-sectional area of a conductor. The prefix "k" stands for thousands, thus kcmil is equal to thousands of circular mils. As the number of kcmils increases, the size of the conductor increases.

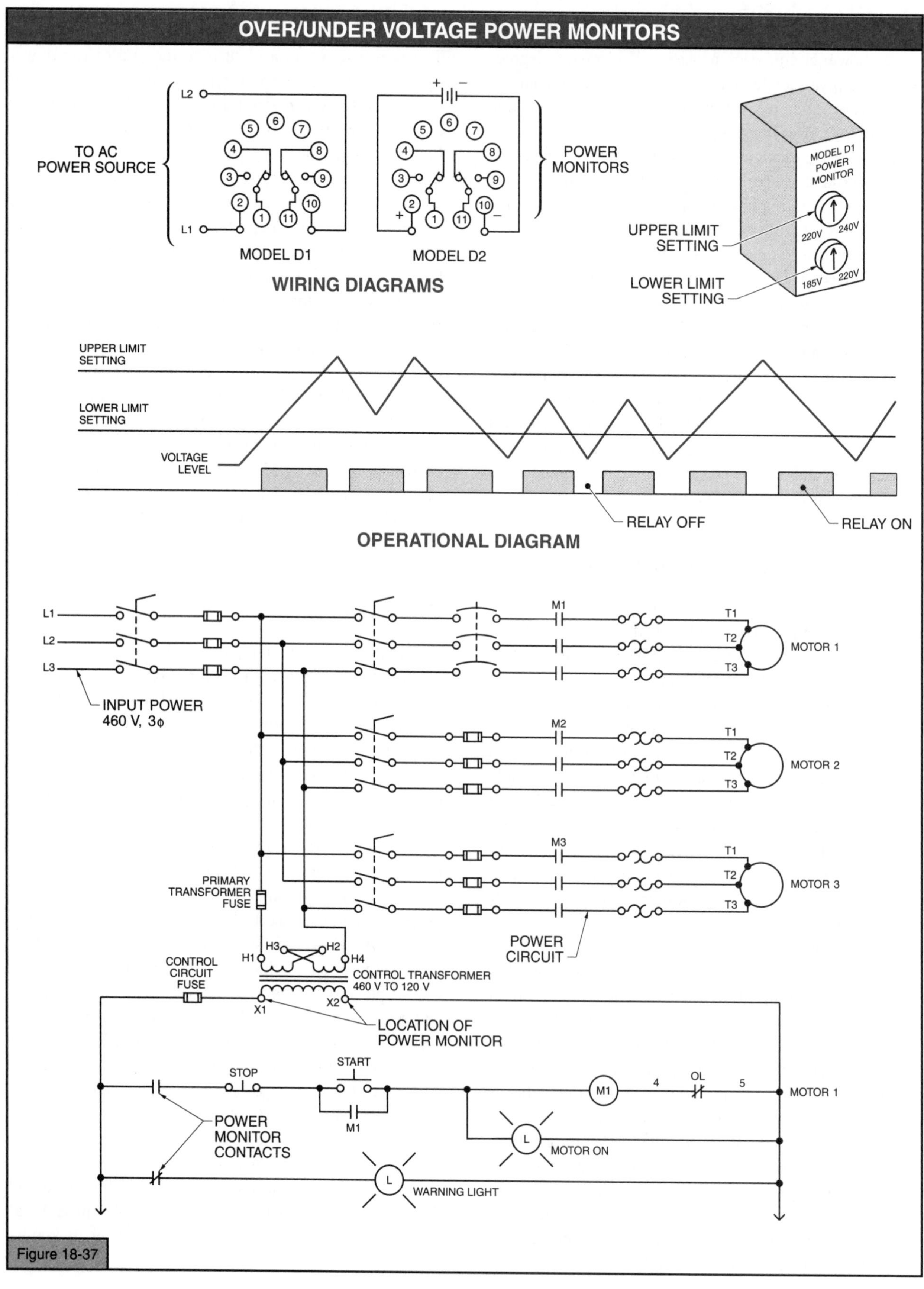
OVER/UNDER VOLTAGE POWER MONITORS
TO AC POWER SOURCE
L2
L1
MODEL D1
MODEL D2
POWER MONITORS
WIRING DIAGRAMS
MODEL D1 POWER MONITOR
UPPER LIMIT SETTING
LOWER LIMIT SETTING
220V
240V
185V
220V
UPPER LIMIT SETTING
LOWER LIMIT SETTING
VOLTAGE LEVEL
RELAY OFF
RELAY ON
OPERATIONAL DIAGRAM
L1
L2
L3
INPUT POWER 460 V, 3ϕ
M1
T1
T2
T3
MOTOR 1
M2
MOTOR 2
M3
MOTOR 3
PRIMARY TRANSFORMER FUSE
POWER CIRCUIT
H1
H3
H2
H4
CONTROL CIRCUIT FUSE
CONTROL TRANSFORMER 460 V TO 120 V
X1
X2
LOCATION OF POWER MONITOR
STOP
START
M1
4
OL
5
MOTOR 1
MOTOR ON
POWER MONITOR CONTACTS
WARNING LIGHT
Figure 18-37

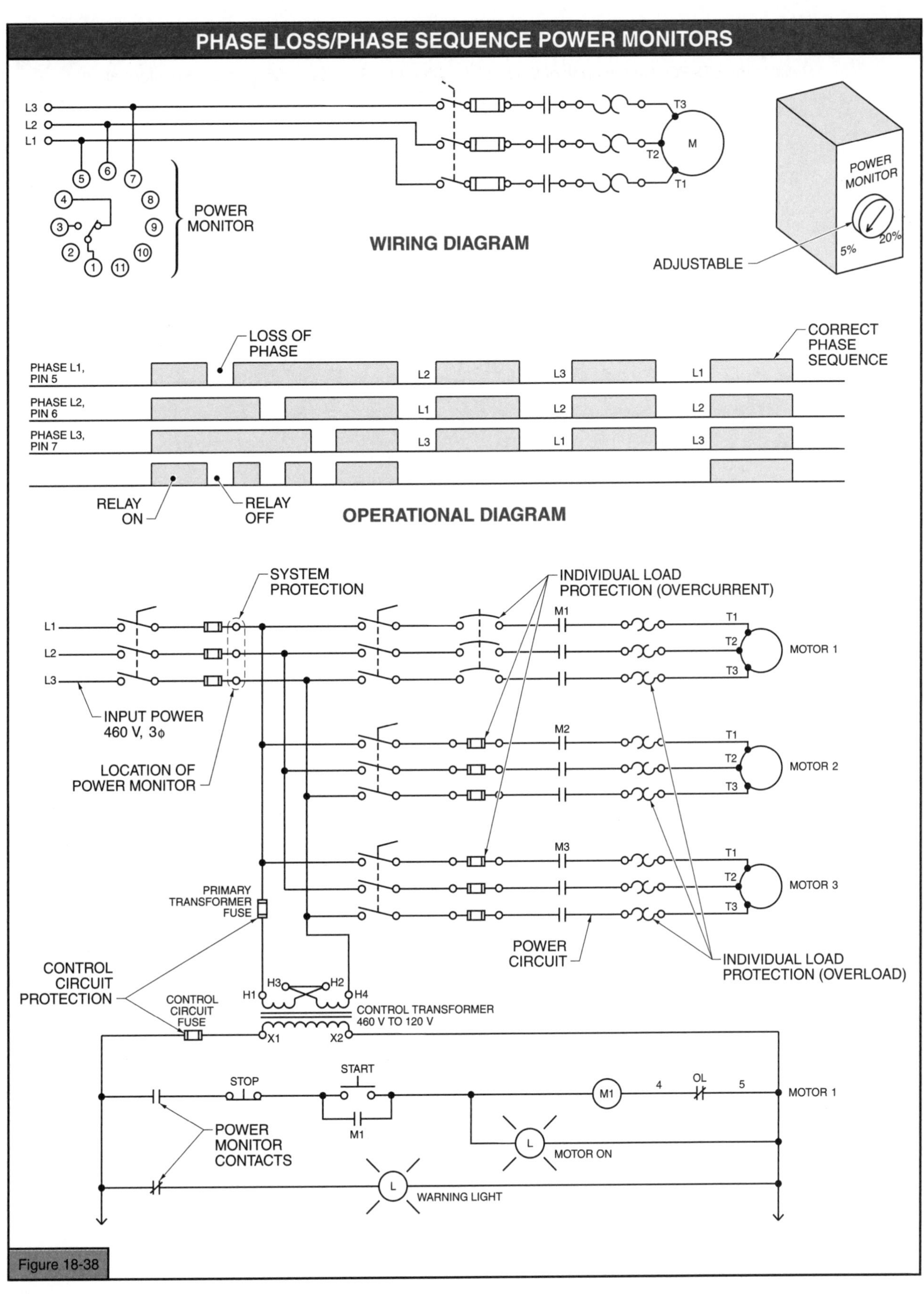
PHASE LOSS/PHASE SEQUENCE POWER MONITORS
POWER MONITOR
WIRING DIAGRAM
ADJUSTABLE
5%
20%
LOSS OF PHASE
CORRECT PHASE SEQUENCE
PHASE L1, PIN 5
PHASE L2, PIN 6
PHASE L3, PIN 7
RELAY ON
RELAY OFF
OPERATIONAL DIAGRAM
SYSTEM PROTECTION
INDIVIDUAL LOAD PROTECTION (OVERCURRENT)
INPUT POWER 460 V, 3ϕ
LOCATION OF POWER MONITOR
MOTOR 1
MOTOR 2
MOTOR 3
PRIMARY TRANSFORMER FUSE
POWER CIRCUIT
INDIVIDUAL LOAD PROTECTION (OVERLOAD)
CONTROL CIRCUIT PROTECTION
CONTROL CIRCUIT FUSE
CONTROL TRANSFORMER 460 V TO 120 V
STOP
START
POWER MONITOR CONTACTS
MOTOR ON
WARNING LIGHT

Figure 18-38

Refer to the CD-ROM in the back of the book for Chapter 18 Quick Quiz® and related information.

19 Residential Circuits

Electrical Circuits

Electricity is used to produce heat, light, sound, and motion in residences, shopping centers, restaurants, businesses, manufacturing facilities, etc. Electricity must be properly distributed and used to safely and efficiently serve the needs of individuals. Electrical systems primarily used in homes, apartments, and condominiums are generally referred to as residential systems. Electrical systems primarily used in hospitals, office buildings, schools, hotels, garages, and government buildings are generally referred to as commercial systems. Electrical systems primarily used to convert raw materials and manufacture products are generally referred to as industrial systems.

Residential, commercial, and industrial systems often include the same basic types of circuits, although each circuit may differ in size (number of components and total power), voltage level (115 V, 230 V, 460 V, etc.), voltage type (AC, DC, 1ϕ, 3ϕ), and usage. For example, residential, commercial, and industrial locations all include general lighting circuits and small appliance circuits (general 115 V outlets). Likewise, some circuits are installed to perform the same basic function, but with different power requirements. For example, residential circuits must include circuits designed to power the appliances that cook food. Restaurants must also include circuits for cooking food on a larger scale than residential buildings. Industries that produce food products must include circuits for preparing and cooking raw materials used in food processing on a very large scale.

Residential, commercial, and industrial circuits make up a large portion of the electrical circuits used to deliver power from the power company and produce work. However, other types of electrical circuits exist. For example, electrical circuits used in automobiles, electric vehicles, boats, planes, trains, spacecraft, and electric toys are generally referred to as mobile electrical circuits.

Residential Circuits

Residential circuits primarily include lamp control (lighting) circuits, general-purpose receptacle (small appliance) circuits, and special-purpose receptacle (ranges, ovens, room heaters, water heaters, air conditioners, pools, spas, hot tubs, etc.) circuits.

Lamp Control

Lamps are controlled by different switches and circuits. The switch used depends on the method by which the lamp is operated. Switches that are manually operated (manual switches), operated at a set time (timers), or operated when motion is detected (motion sensors) are used to control lamps. The different switches apply or disconnect power to a lamp using different methods. The circuit used to control a lamp depends on the application. For example, a two-way switch is used for a lamp that is controlled from one location. A three-way switch is used for a lamp that is controlled from two different locations. A four-way switch is used for a lamp that is controlled from three different locations. A rotary dimmer is used to control the voltage applied to a lamp. See **Figure 19-1.**

A strip gauge is a mark on the back of a switch that identifies the amount of insulation that must be removed from a wire so the wire can be inserted into the switch.

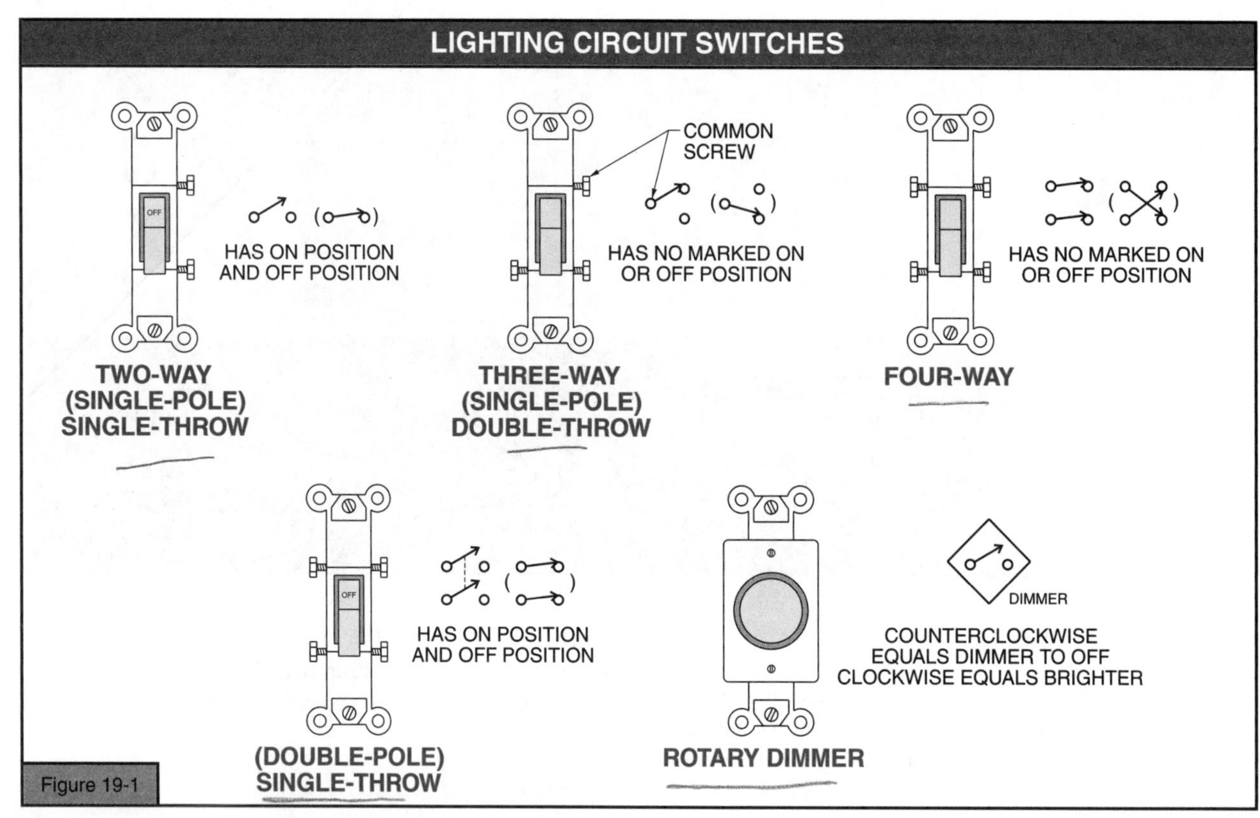

Figure 19-1

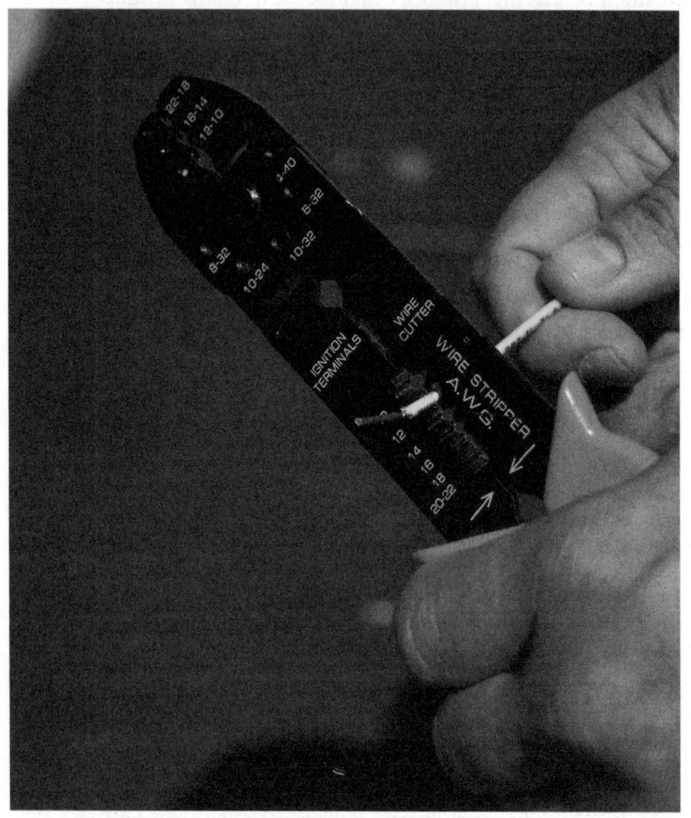

Wire-stripping tools are used to strip insulation from conductors when connecting switches to electrical circuits.

Switches

A *switch* is a device that is used to start, stop, or redirect the flow of current in an electrical circuit. The disconnect operation and rating of a switch are considered when the switch is used to control a lamp. The switch must be rated higher than (or equal to) the voltage and current rating of the lamp(s) the switch is controlling. The switch must disconnect the hot power line in circuits of 120 V (or less) that use one hot and one neutral line. The switch must disconnect both hot lines in circuits that use two hot lines (240 V, etc.). The voltage may arc over the switch contacts if the switch is rated for a voltage less than the applied voltage. The switch contacts are destroyed and may weld together if the switch is rated for a current less than the applied current.

All lighting circuit switches have their ampere and voltage rating stamped on them. They may also have a horsepower rating included. In general, the higher the ampere and voltage rating, the more expensive the switch. For example, a switch rated at 15 A, 120/277 V costs more than a switch rated at 15 A, 120 V. The 277 V rating allows the switch to be used on 277 V fluorescent lamp circuits. **See Figure 19-2.**

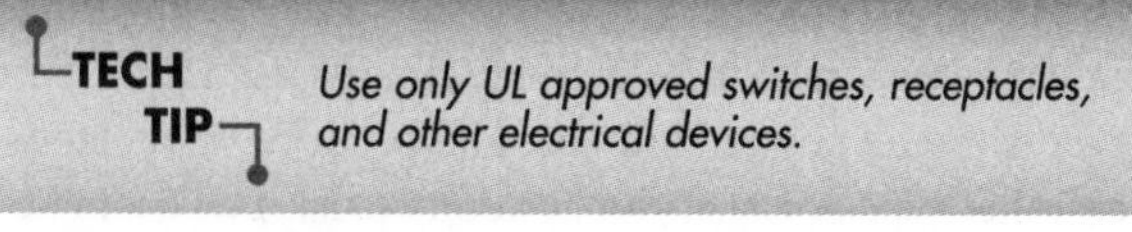
TECH TIP *Use only UL approved switches, receptacles, and other electrical devices.*

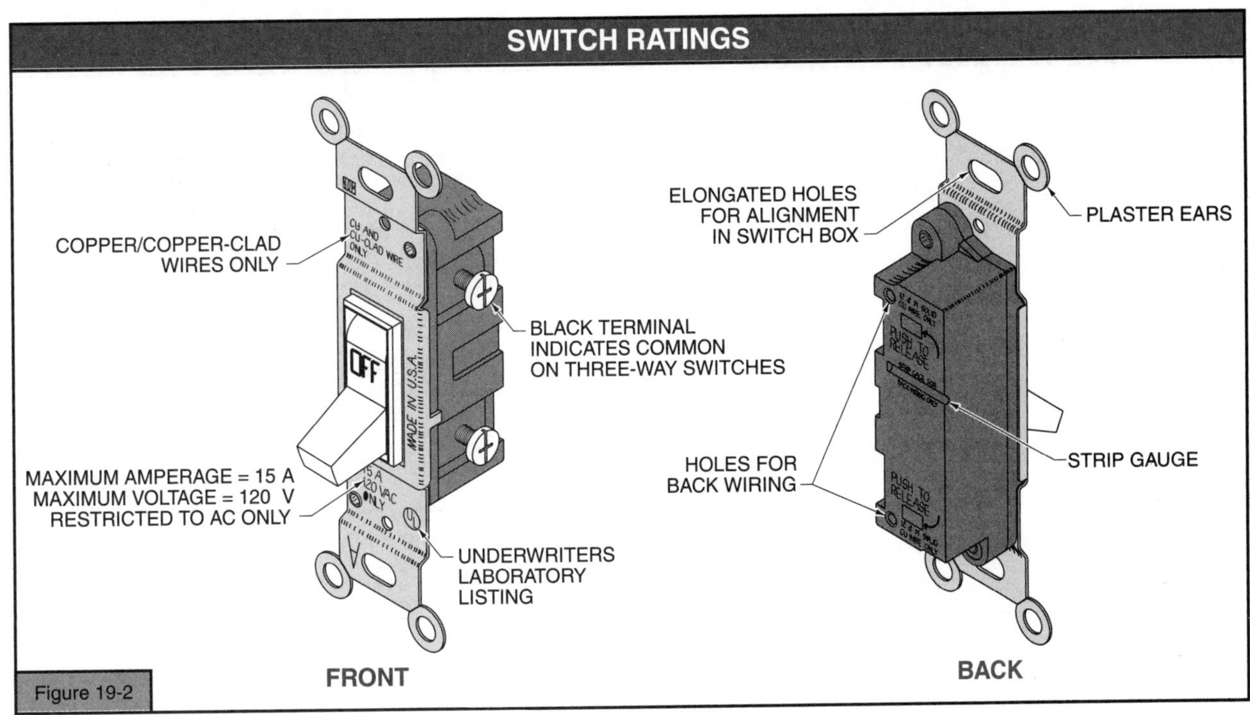

Figure 19-2

Lighting circuit switches are rated as either AC or AC/DC. DC is harder to switch than AC because DC is always flowing at full value. The making (closing) and breaking (opening) of a circuit creates a strong arc during the switching operation when current is flowing at full value. To reduce damage, the contacts of a DC switch are widely gapped and the switching operation is snap-acting to avoid excessive pitting or welding of the switch contacts.

AC operates between zero and full voltage 120 times per second on standard 60 Hz power lines. The voltage at the switch contacts is constantly changing in value. The oscillating characteristic of AC helps extinguish the arc developed at the switch contacts. This allows the use of a slower switching action. AC switches are quieter and lower in cost than DC switches. AC switches are used when wiring most residential lighting circuits.

The switch used to control a lamp depends on the number of different locations the lamp must be controlled from. The three standard manually-operated switches used to control most ON/OFF lighting circuits are two-way, three-way, and four-way switches. **See Figure 19-3.**

Two-Way Switches

A *two-way switch* is a single-pole, single-throw (SPST) switch. A two-way switch may be placed in one of two different positions. The switch allows current to flow through it in the ON position and does not allow current to flow through it in the OFF position. Two-way switches have ON and OFF position markings on them because they have distinct ON and OFF positions. Two-way switches are used to control a lamp from one location.

Three-Way Switches

A *three-way switch* is a single-pole, double-throw (SPDT) switch. A three-way switch has one common terminal and two traveler terminals. The common terminal is darker in color to distinguish it from the two traveler terminals. The switch connects the common terminal to the top traveler terminal in one position and connects the common terminal to the bottom traveler terminal in the other position. A three-way switch does not have a designated ON and OFF position because the common terminal is always connected to one of the traveler terminals. Three-way switches never have ON and OFF position markings on them. Two three-way switches are used to control a lamp from two different locations.

Four-Way Switches

A *four-way switch* is a double-pole, double-throw (DPDT) switch that changes the electrical connections inside the switch from straight to diagonal. Four-way switches do not have ON and OFF position markings on them. Four-way switches are used with three-way switches to control a lamp from three locations. When controlling a lamp from three locations, the four-way switch is placed between the two three-way switches.

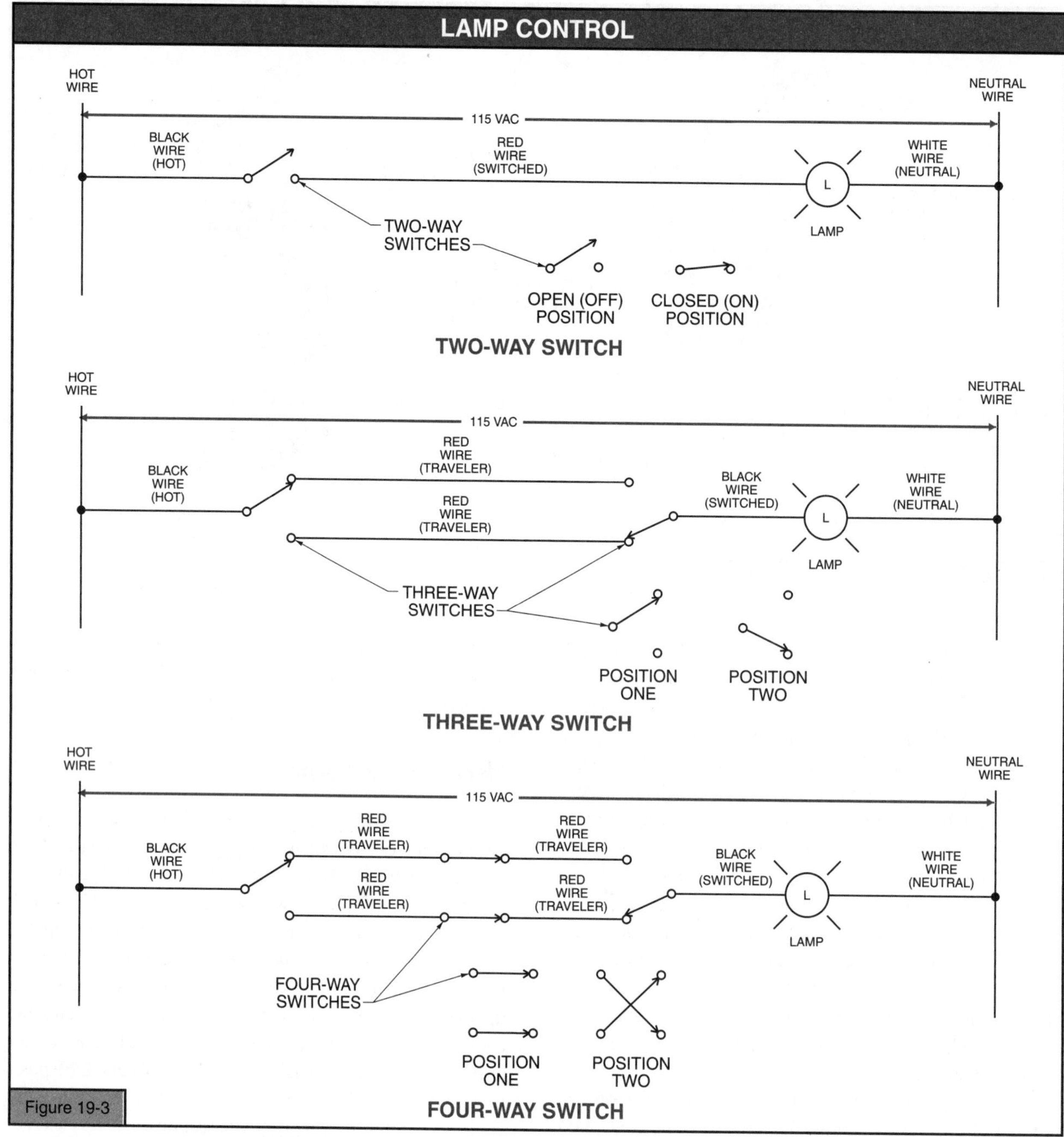

Figure 19-3

Dimmers

A dimmer switch is required to vary the brightness of a lamp. A *dimmer switch* is a switch that changes the lamp brightness by changing the voltage applied to the lamp. Dimmer switches are used to control light levels in living rooms, dining rooms, display cases, and other areas in which a variable light level is desired. A two-way dimmer switch is used if the lamp that is controlled requires only one switch. **See Figure 19-4.**

A two-way dimmer switch is connected into a circuit in the same manner as a standard two-way switch. The only difference is the dimmer switch includes a knob (or lever) that is used to set the brightness in addition to an ON/OFF position. Most dimmer switches adjust the lamp through the full range of brightness.

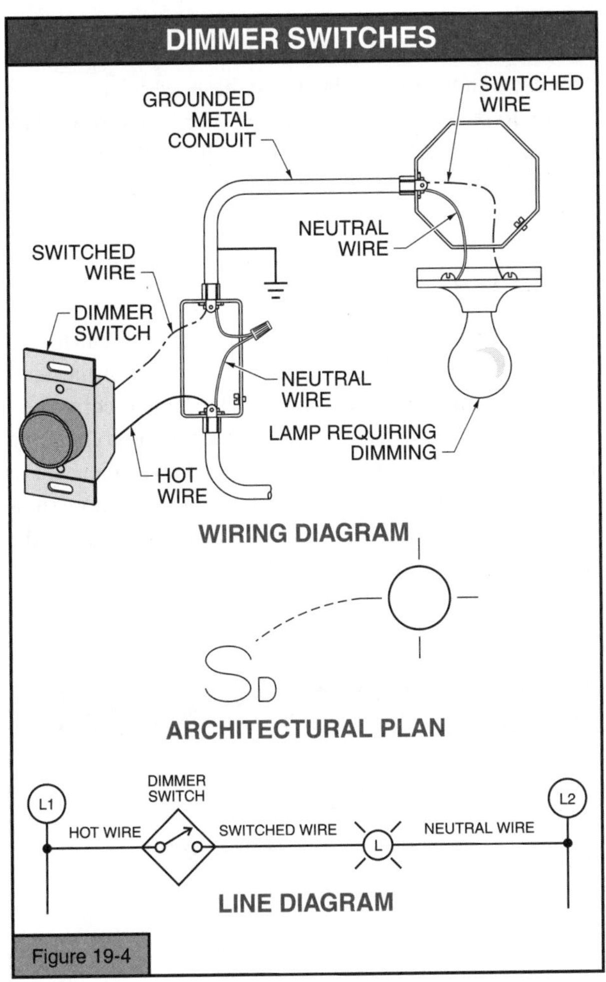

Figure 19-4

A three-way dimmer switch is used if a lamp that is controlled from two locations requires brightness control. A three-way dimmer switch is connected into a circuit in the same manner as a standard three-way switch. A standard three-way switch is used for the other three-way switch in the circuit. In a circuit using a three-way dimmer switch, either three-way switch is used to turn the lamp ON or OFF. Only the dimmer switch sets the brightness of the lamp.

Only two-way and three-way dimmer switches are available because any switching circuit that includes a four-way switch also includes two three-way switches. In this case, a three-way dimmer switch replaces one of the standard three-way switches.

Motion Sensors

A *motion sensor* is an infrared sensing device that detects the movement of a temperature variance and automatically switches when the movement is detected. This change is normally caused by the movement of a person, but also occurs during a fire. Motion sensors are used in parking lots, walkways, hallways, stairs, restrooms, building entrances and exits, and for general room security.

Motion sensors are connected in the same manner as a two-way switch. The switch includes a black wire that is connected to the hot wire and a red wire that is connected to the lamp. White wire is also included that is connected to the neutral wire. The sensor requires a hot and neutral wire to be connected to the switch at all times to power the sensing circuit 24 hours a day. **See Figure 19-5.**

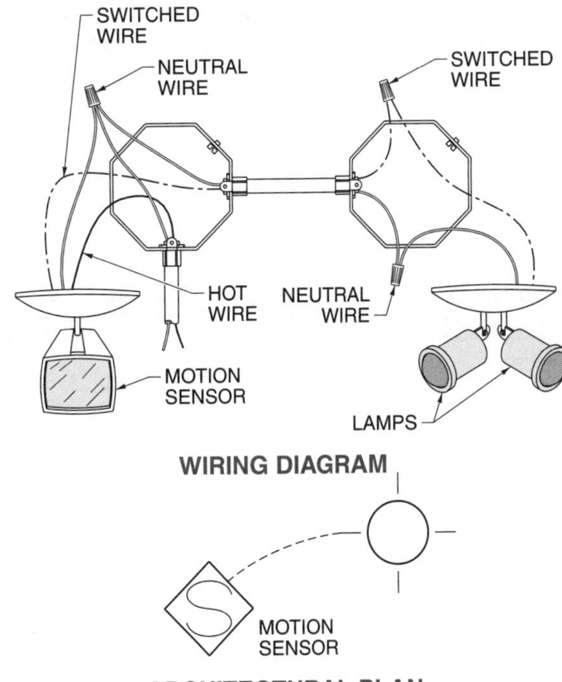

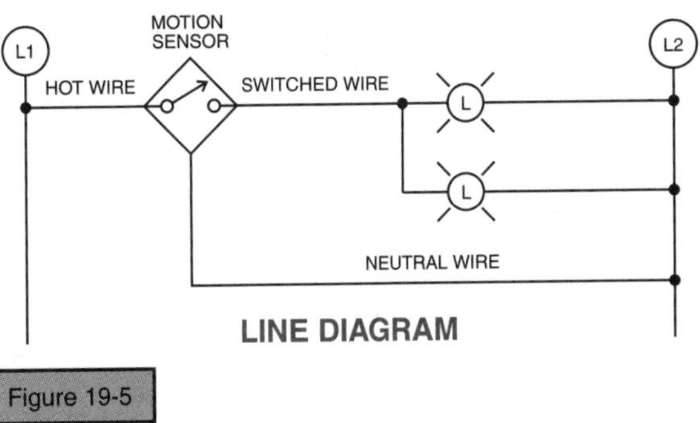

Figure 19-5

TECH FACT

Although voltages are normally stated as 115 V or 120 V, the voltage delivered to residential buildings may safely fluctuate between 114 V and 126 V.

Lamp Fixture Mounting

Lamp fixtures are mounted using one of several different methods. The method used depends on the weight of the lamp fixture. The required mounting parts (mounting strap and locknuts) are normally included as part of the lamp fixture parts. **See Figure 19-6.**

For lighter loads (less than 5 lb), the lamp fixture is held with two mounting screws that hold the fixture base to a mounting strap. The mounting strap is attached by screws to the threaded box ears (on the lamp fixture box). For loads less than 10 lb, the mounting strap is held to a stud by a locknut. The stud is screwed into a hanger that spans the joists between the location where the lamp fixture box is mounted. The fixture is then held to the mounting strap by two screw nuts.

Lamp fixtures between 10 lb and 20 lb are mounted on a hickey that has been screwed onto the stud. A *hickey (reducer)* is a mounting device that screws onto the threaded fixture stud and takes a threaded nipple. The nipple extends through the fixture base, which allows the fixture to be mounted by a locknut. For loads over 20 lb, a threaded adapter is added which allows the locknut to help hold both the lamp fixture and fixture base.

Lamp Control Circuits

A switch (or switches) is connected into a lamp circuit to turn the lamp ON and OFF. The switch used depends on the number of different locations from which the lamp is controlled. The number of different locations a lamp is controlled from is determined by Code requirements and convenience. For example, the NEC® requires a stairway of six or more steps to have a switch that controls the stairway lamp from the top and bottom of the steps per 210.70(A).

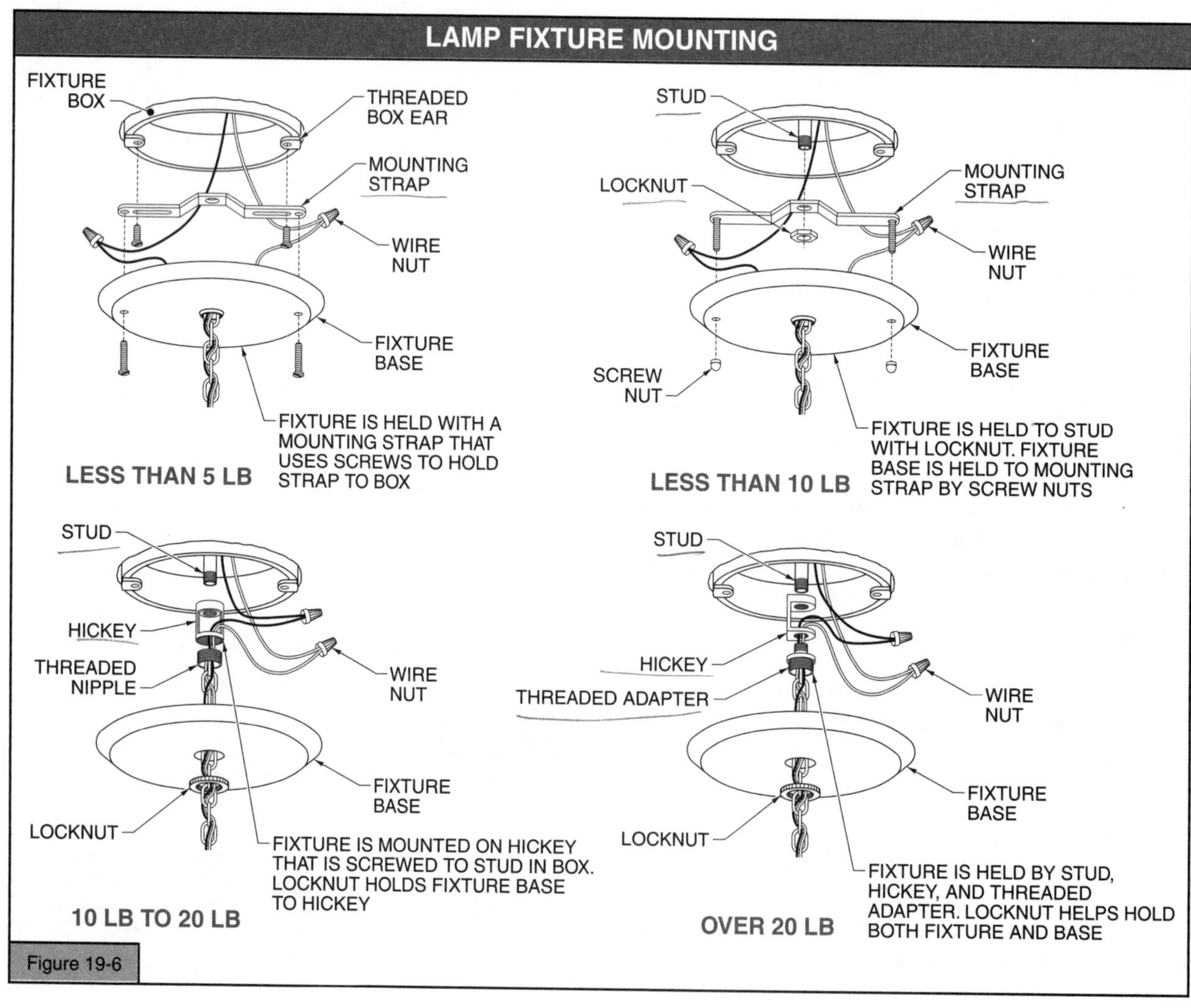

Figure 19-6

Lamps may be controlled from any number of locations. However, lamps are commonly controlled from one, two, or more than two locations.

One Location Lamp Control

Controlling a lamp from one location is the most common lighting circuit. **See Figure 19-7.** A two-way switch is used when controlling a lamp from one location. The switch is connected between the hot power line and the lamp. The circuit is completed and the lamp is turned ON when the switch is placed in the ON position. The switch opens the circuit and the lamp is turned OFF when the switch is placed in the OFF position.

Colored wires are used when connecting a lighting circuit. A black wire is normally used for the hot power line, a white wire is used for the neutral power line, and a red wire is normally used for the switched power line. There is always a voltage between the black and white wires in the circuit. There is power between the red and white wires only when the switch is in the ON position. In this circuit, the metal conduit serves as the system ground and no ground wire is used.

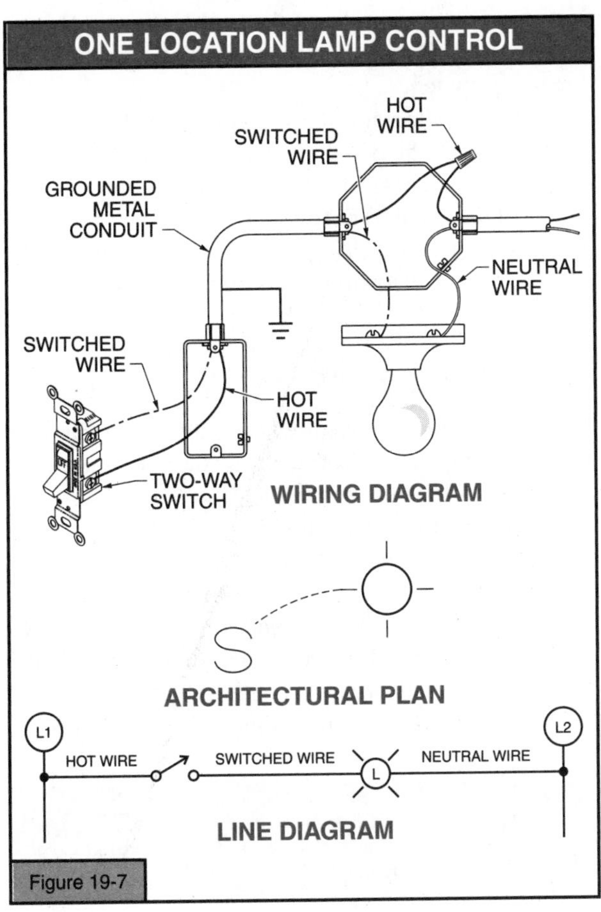

Figure 19-7

Lamps are connected in parallel if more than one lamp is controlled by a switch. The total number of lamps connected in parallel is insignificant as long as the total current required by all the lamps does not exceed the current rating of the switch.

Two Location Lamp Control

In some applications, it is necessary to control a lamp from two separate locations. For example, a lamp that is used to light a staircase must have a switch at the top and bottom of the staircase. Likewise, rooms that have two entrances (or exits) require a switch that controls the lights at each location. Two three-way switches are used when controlling a lamp from two locations. **See Figure 19-8.**

In this circuit, the common point of one three-way switch is connected directly to the hot power supply at all times. The common point of the other three-way switch is connected directly to the lamp at all times. The traveler terminals of both three-way switches are connected together. A black wire is used to connect the hot power line to the first three-way switch. Red wires are used for the travelers. The wire connecting the second three-way switch to the lamp is normally red or black with a color tape marking to distinguish it from the others.

In this circuit, a green (or bare) wire is used for the ground because nonmetallic cable is used. The ground wire is required to ground all non-current carrying metal to earth. The conduit may be used to maintain a solid ground connection if metal conduit is used.

Leviton Manufacturing Co., Inc.

Switch plates for residential use are available in different configurations and colors.

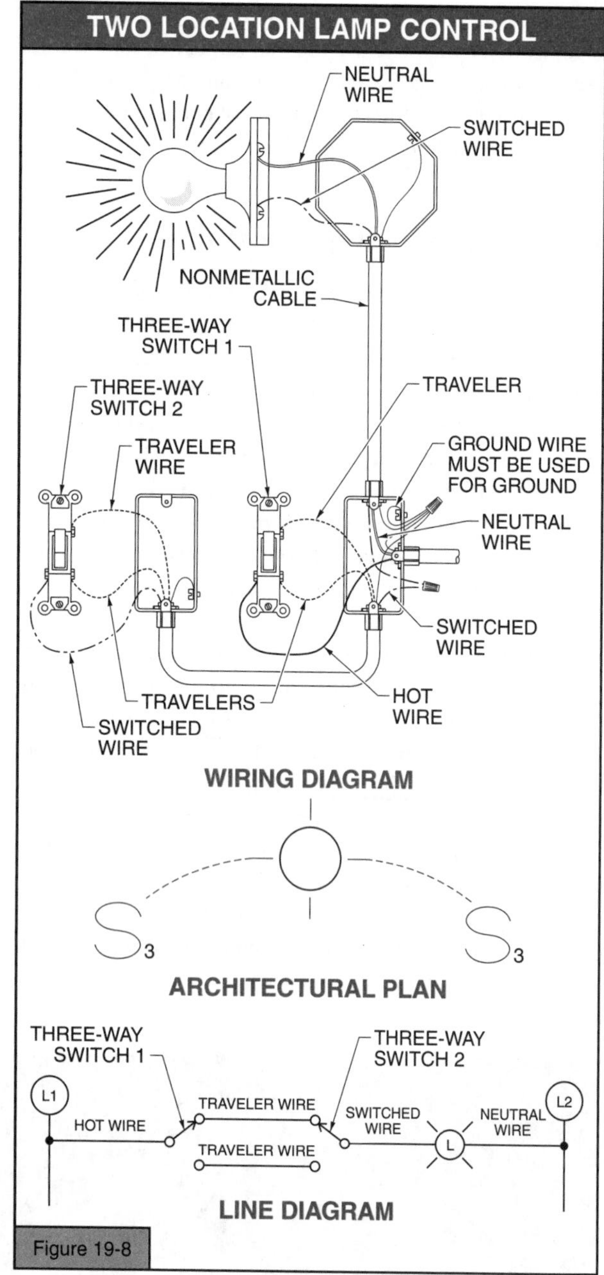

Figure 19-8

The line diagram of two three-way switches controlling a lamp is the same regardless of the actual wiring diagram used to connect the individual components. The actual wiring diagram varies based on the incoming power supply that is brought into the circuit. In this circuit, the incoming power may be brought into the circuit in any one of three different locations. The power may be brought into the box that contains the lamp, the box that contains three-way switch 1, or the box that contains three-way switch 2. **See Figure 19-9.**

The point where the power is brought into a circuit is insignificant from an electrical standpoint. The circuit is wired the same and operates the same regardless of the point where the power is brought into the circuit. However, the number of individual wires required in each conduit may vary. In general, it makes no difference where the power is brought into a circuit as long as the total number of wires in any one conduit run does not exceed the allowable number of wires for that size conduit. A larger size conduit is required if the number of wires exceeds the allowable number of wires for that size conduit. The circuit should be checked for alternative wiring arrangements before a large conduit is used.

More Than Two Location Lamp Control

Some applications require a lamp to be controlled from more than two locations. For example, a room that contains three entrances requires a switch to turn ON the lamp from any of the three entrances. A four-way switch is placed between two three-way switches when controlling a lamp from three locations. **See Figure 19-10.**

The four-way switch is always connected between two three-way switches. Any number of four-way switches may be connected into a circuit, but they all must be connected between the two three-way switches. Red wire is used to connect the traveler wires because the four-way switch is connected between the two three-way switches.

In the line diagram, the four-way switch is always placed between the two three-way switches. The wiring diagram also shows the four-way switch placed in the middle of the circuit. The four-way switch may be placed in any position in the wiring diagram as long as it is wired between the travelers of the two three-way switches.

Multiple Lamp Circuits

Switching circuits may be used to control more than one lamp. The lamps are connected in parallel when more than one lamp is used. **See Figure 19-11.**

Any number of lamps may be connected in parallel as long as the total current drawn by the parallel circuit does not exceed the current limit of the switches and wire. For example, a 100 W incandescent light bulb draws about .88 A at 115 V. Ten 100 W incandescent light bulbs connected in parallel draw about 8.8 A. A standard 15 A lighting circuit safely switches the ten 100 W bulbs.

TECH FACT *The NEC® requires that wire connections and splices be contained inside an approved metal or plastic box.*

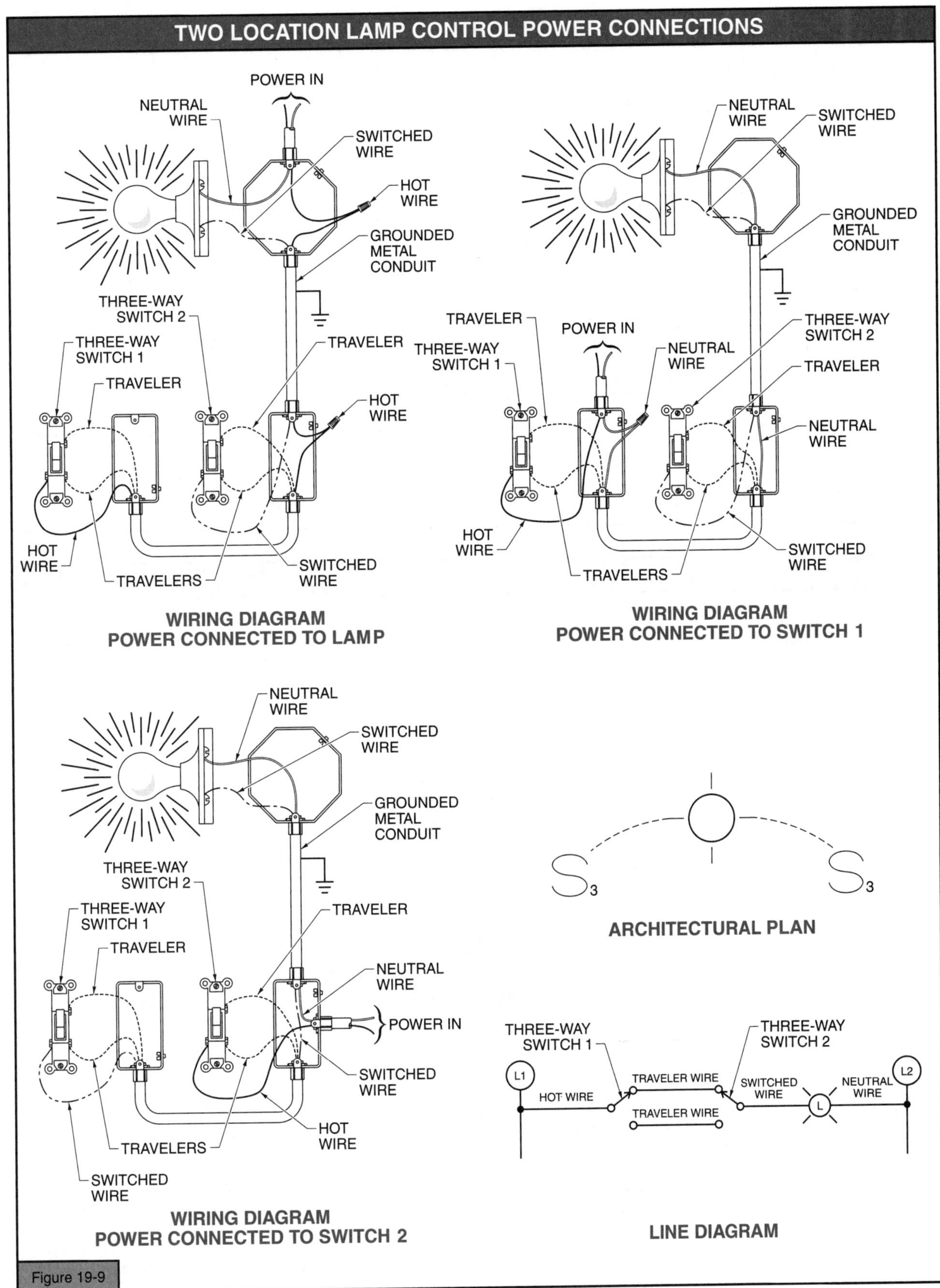
TWO LOCATION LAMP CONTROL POWER CONNECTIONS
POWER IN
NEUTRAL WIRE
SWITCHED WIRE
HOT WIRE
GROUNDED METAL CONDUIT
THREE-WAY SWITCH 2
THREE-WAY SWITCH 1
TRAVELER
TRAVELER
HOT WIRE
HOT WIRE
TRAVELERS
SWITCHED WIRE
WIRING DIAGRAM POWER CONNECTED TO LAMP
NEUTRAL WIRE
SWITCHED WIRE
GROUNDED METAL CONDUIT
TRAVELER
POWER IN
THREE-WAY SWITCH 1
NEUTRAL WIRE
THREE-WAY SWITCH 2
TRAVELER
NEUTRAL WIRE
HOT WIRE
SWITCHED WIRE
TRAVELERS
WIRING DIAGRAM POWER CONNECTED TO SWITCH 1
NEUTRAL WIRE
SWITCHED WIRE
GROUNDED METAL CONDUIT
THREE-WAY SWITCH 2
THREE-WAY SWITCH 1
TRAVELER
TRAVELER
NEUTRAL WIRE
POWER IN
SWITCHED WIRE
HOT WIRE
TRAVELERS
SWITCHED WIRE
WIRING DIAGRAM POWER CONNECTED TO SWITCH 2
S3
S3
ARCHITECTURAL PLAN
THREE-WAY SWITCH 1
THREE-WAY SWITCH 2
L1
L2
HOT WIRE
TRAVELER WIRE
TRAVELER WIRE
SWITCHED WIRE
NEUTRAL WIRE
L
LINE DIAGRAM

Figure 19-9

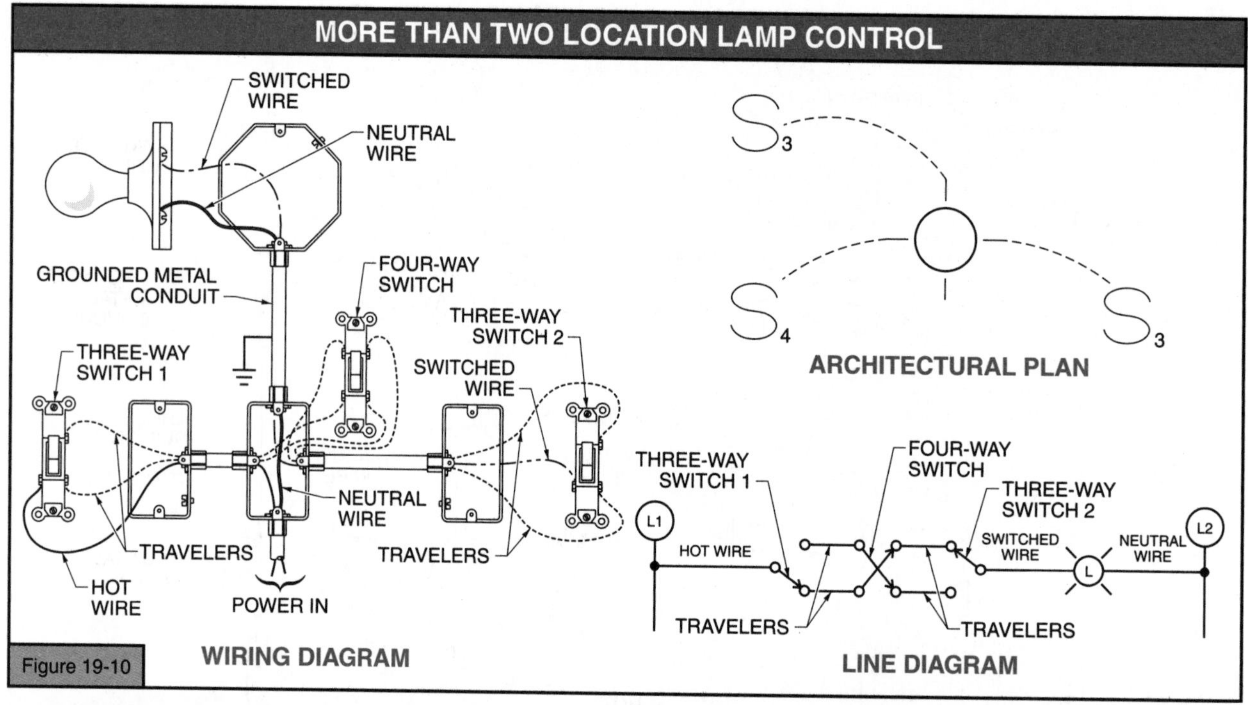

Figure 19-10

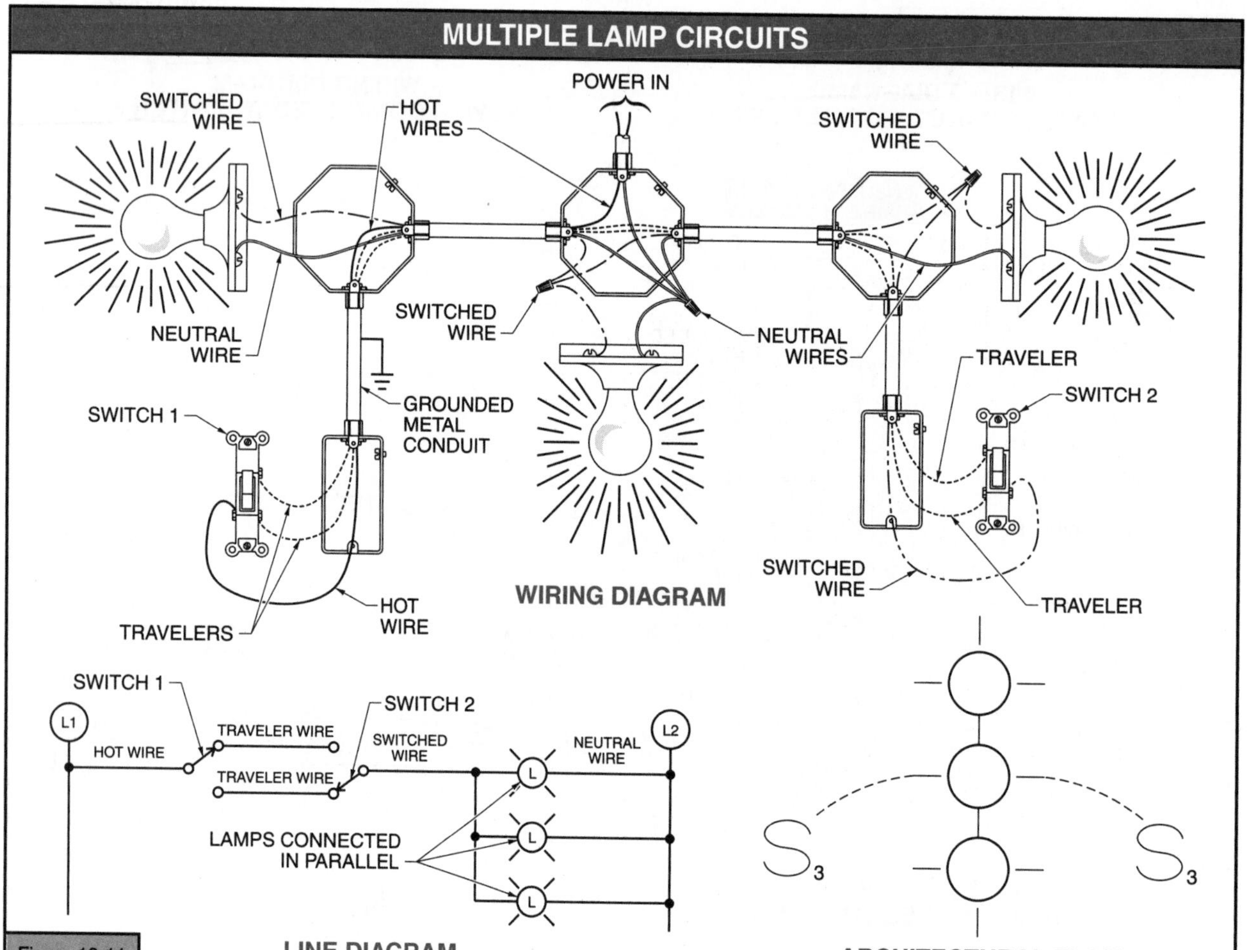

MULTIPLE LAMP CIRCUITS

POWER IN

SWITCHED WIRE

HOT WIRES

SWITCHED WIRE

NEUTRAL WIRE

SWITCHED WIRE

NEUTRAL WIRES

TRAVELER

SWITCH 2

SWITCH 1

GROUNDED METAL CONDUIT

SWITCHED WIRE

TRAVELER

TRAVELERS

HOT WIRE

WIRING DIAGRAM

SWITCH 1

SWITCH 2

L1

HOT WIRE

TRAVELER WIRE

TRAVELER WIRE

SWITCHED WIRE

NEUTRAL WIRE

L2

LAMPS CONNECTED IN PARALLEL

LINE DIAGRAM

ARCHITECTURAL PLAN

Figure 19-11

Circuits should not be operated to their maximum when determining the total number of lamps to connect into a circuit. This is because the person changing the lamps normally has no idea of the limits of the circuit. The circuit current increases if a 100 W lamp is replaced with a higher wattage bulb. Do not load a lighting circuit more than 75% of its maximum ampere rating. A clamp-on ammeter is used to measure the current in a lighting circuit. Measure the current at the main incoming hot power line to take a total (all lamps in parallel) current reading.

Switching Combinations

In many rooms, several different lighting circuits are often used to control the lights. Multiple lighting circuits are used to create different lighting effects and/or save energy. Rooms that use multiple lighting circuits include kitchens, family rooms, bathrooms, and garages. Multiple lighting circuits are also used for outdoor lighting applications.

Lighting circuits that include multiple switching circuits follow the same basic wiring as individual lighting circuits, except the individual circuits normally share the same conduit and boxes. For example, three individual switches may be connected to control three individual lamps. **See Figure 19-12.**

In this circuit, switch 1 controls lamp 1, switch 2 controls lamp 2, and switch 3 controls lamp 3. At any time, all three lamps may be turned OFF and/or ON, or any combination of the three lamps may be ON and/or OFF.

This circuit is often used in bathrooms where one switch may control a ceiling lamp, another switch may control a mirror lamp(s), and a third switch may control a shower lamp or exhaust fan.

Two-way and three-way switches may be used to control two lighting circuits. **See Figure 19-13.** In this circuit, one lamp is controlled from only one location and the other lamp is controlled from two locations. Switch 1 controls lamp 1, and switches 2 and 3 control lamp 2. Lamp 1 can only be controlled by switch 1. However, lamp 2 can be controlled by either switch 1 or switch 2. This circuit is often used in an area, such as a kitchen, where the main room lights need to be controlled from two locations but additional lighting need only be controlled at one location.

TECH FACT

A T rating is a rating given to switches that are capable of handling the severe overloading created by a tungsten load as the switch is closed.

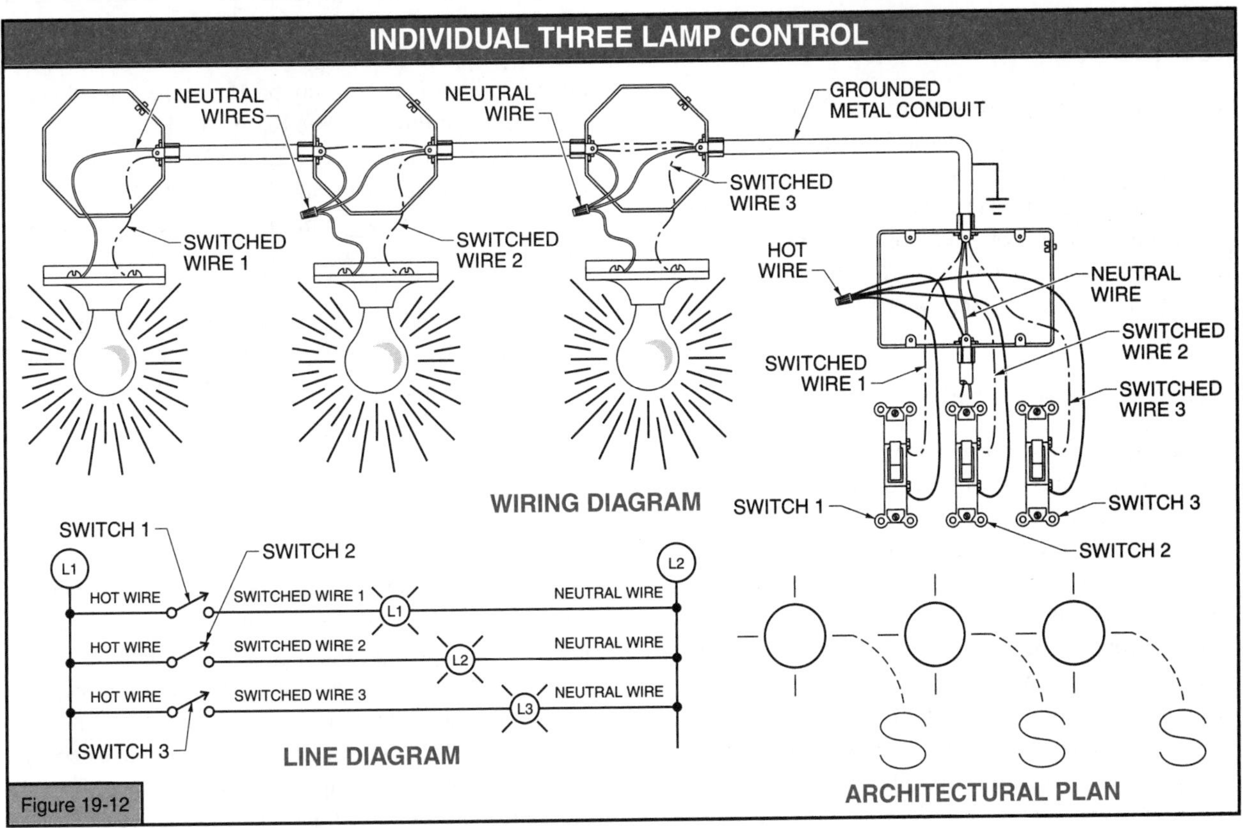

Figure 19-12

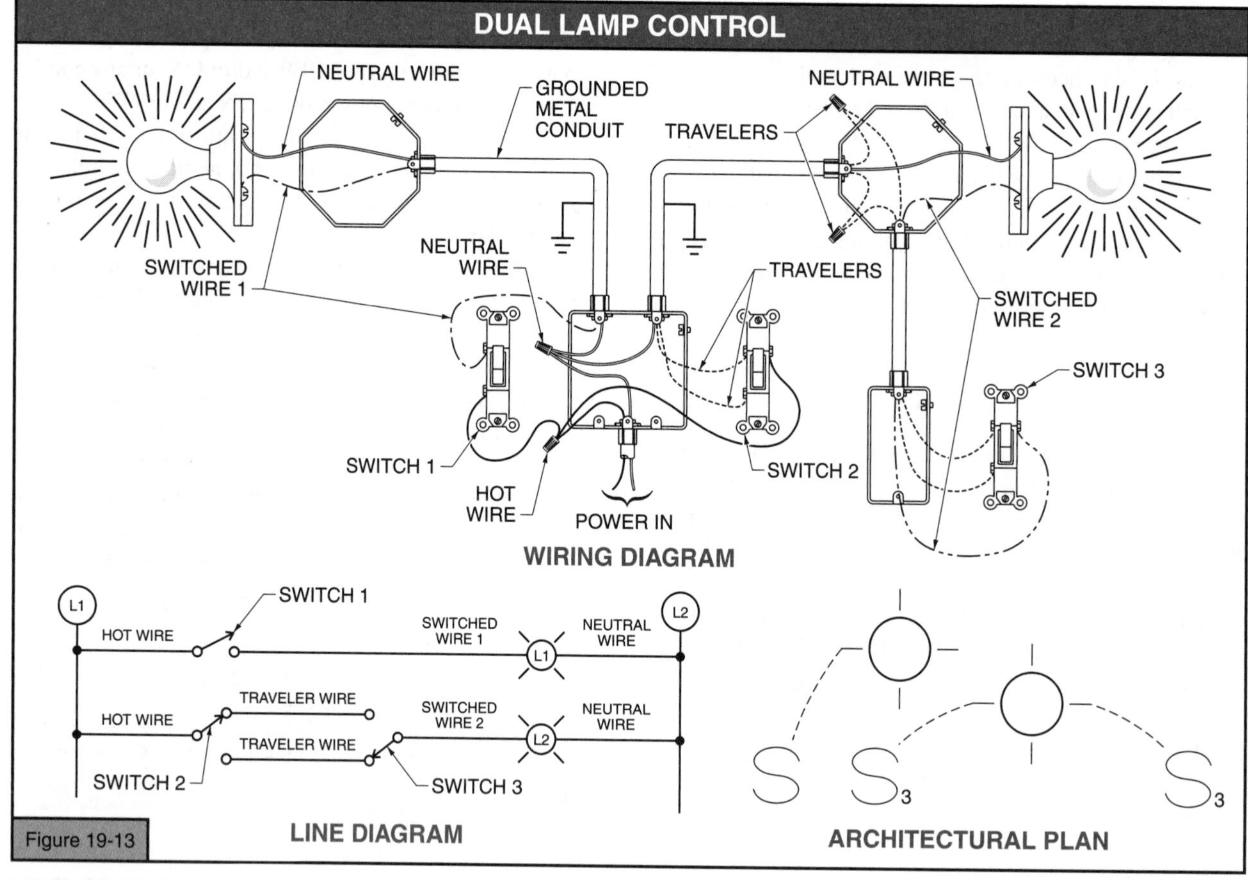

Figure 19-13

Receptacles

A *receptacle (outlet)* is a device used to connect equipment with a cord and plug to an electrical system. The most common 120 V receptacle is the duplex receptacle. A *duplex receptacle* is a receptacle that has two spaces for connecting two different plugs. The three basic receptacles are standard, isolated ground, and GFCI (ground fault circuit interrupter) receptacles. **See Figure 19-14.**

Standard Receptacles

Standard receptacles are the most common receptacles used in general residential wiring. Standard receptacles include a long (neutral) slot, a short (hot) slot, and a U-shaped ground hole. Wires are attached to the receptacle at screw terminals or push-in fittings. A connecting tab between the two hot and two neutral screw terminals provides an electrical connection between the terminals. This electrical connection allows for both terminals to be powered when one wire is connected to either terminal.

Receptacles are marked with ratings for maximum voltage and amperes. Standard receptacles are marked 15 A, 125 V or 20 A, 125 V. The 125 V, 20 A receptacles are normally required for laundry and kitchen circuits. Most other 125 V receptacles may be 15 A (check local codes for exact requirements). **See Figure 19-15.**

Receptacles marked CU or COPPER are used with solid copper wire. Receptacles marked CU-CLAD ONLY are used with copper-coated aluminum wire. Only receptacles marked CO/ALR may be used with solid aluminum wire.

When wiring a 120 V duplex receptacle, the white wire is connected to the silver-colored screw, the black (or red) wire is connected to the brass-colored screw, and the green (or bare) wire is connected to the green screw. **See Figure 19-16.**

Isolated Ground Receptacles

In a standard receptacle, the receptacle ground is connected to the common grounding system when the receptacle is installed in a metal outlet box. The common grounding system normally includes all metal wiring, boxes, conduit, water pipes, and the noncurrent-carrying metal parts of most electrical equipment. The receptacle ground becomes part of the larger grounding system when a piece of electrical equipment is plugged into the receptacle.

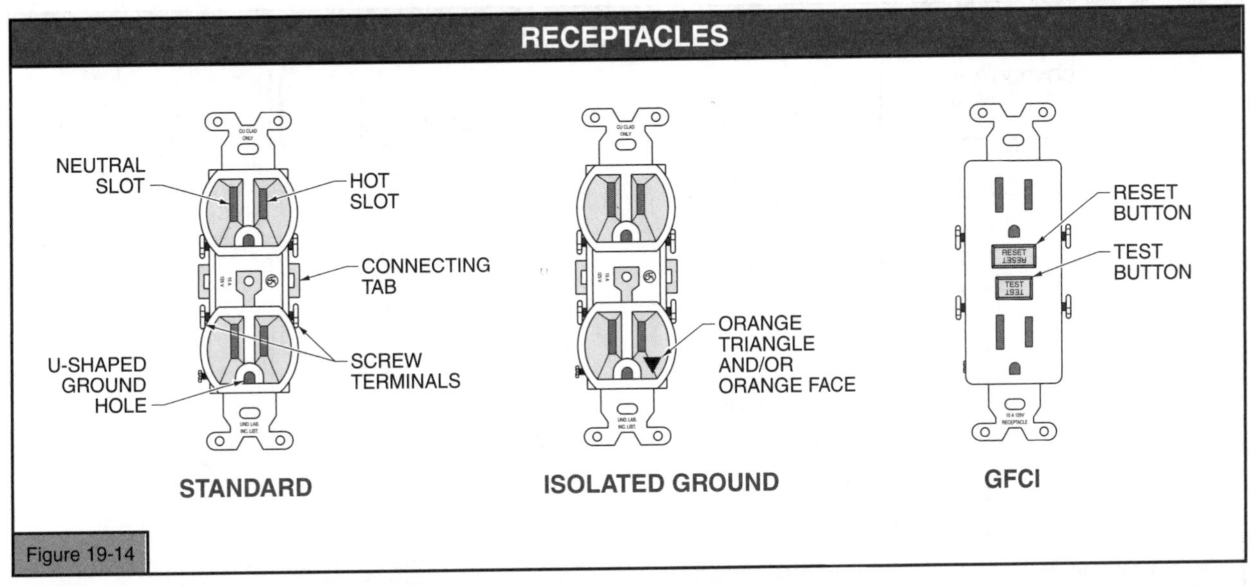

Figure 19-14

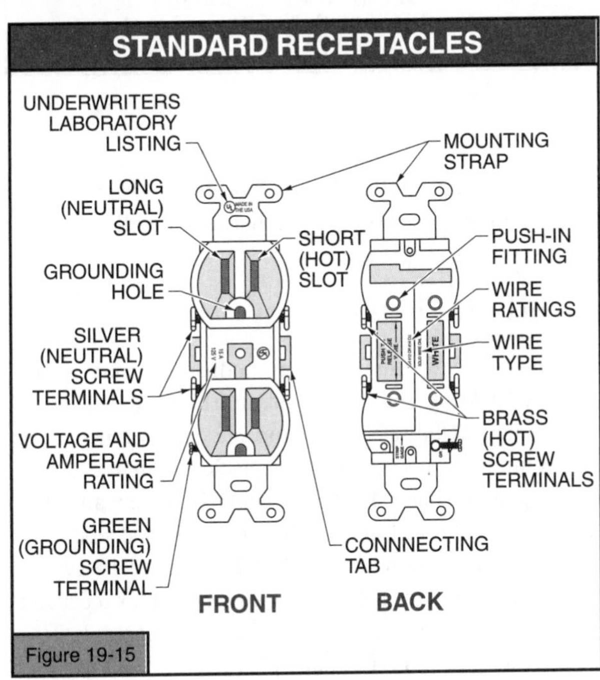

Figure 19-15

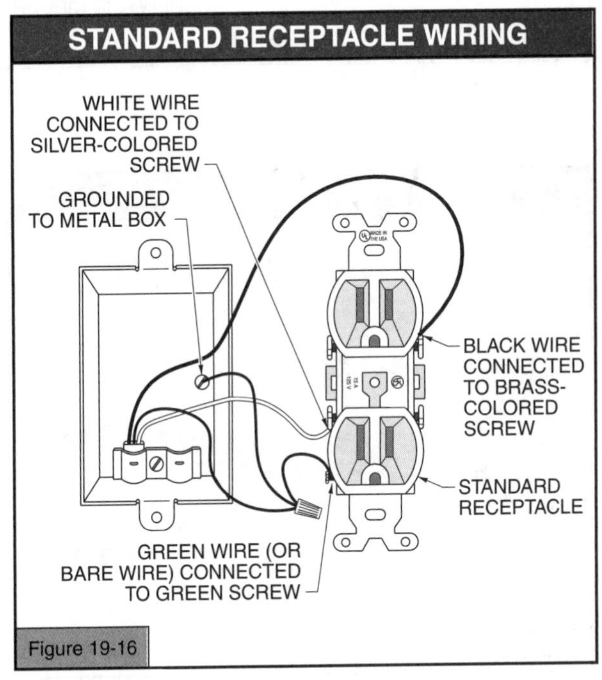

Figure 19-16

The common grounding system may act as a large antenna and conduct electrical noise. This electrical noise may cause interference in computer, medical, security, military, and communication equipment.

An isolated grounded receptacle is used to minimize problems in sensitive applications or in areas of high electrical noise. An *isolated grounded receptacle* is an electrical receptacle that minimizes electrical noise by providing a separate grounding path for each receptacle. Isolated grounding receptacles are identified by an orange triangle on the face of the receptacle and/or an orange-colored faceplate. A separate ground conductor is run with the circuit conductors in an isolated grounding system. The grounding system must be installed per NEC® Article 250. **See Figure 19-17.**

Split-Wired Receptacles

A *split-wired receptacle* is a standard receptacle that has had the tab between the two brass-colored (hot) terminal screws removed. The tab between the two silver-colored terminals is not removed. Split-wired receptacles are used to provide a switched circuit or two separate circuits at the same outlet. **See Figure 19-18.**

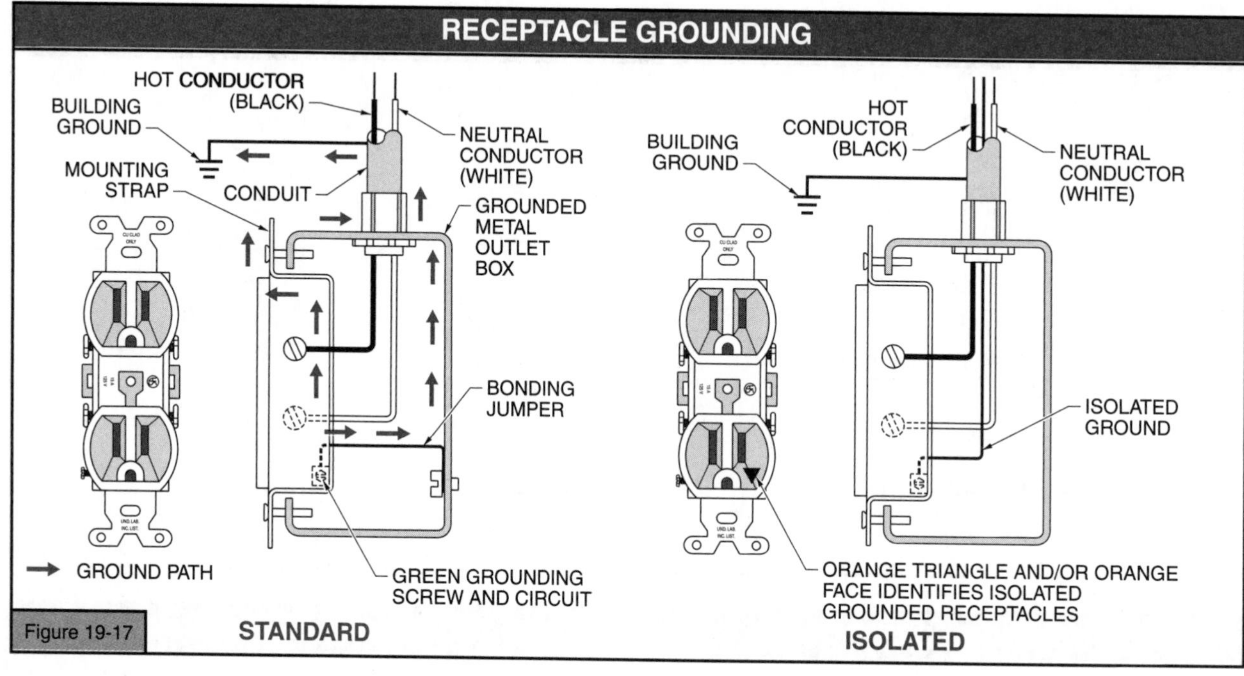

Figure 19-17

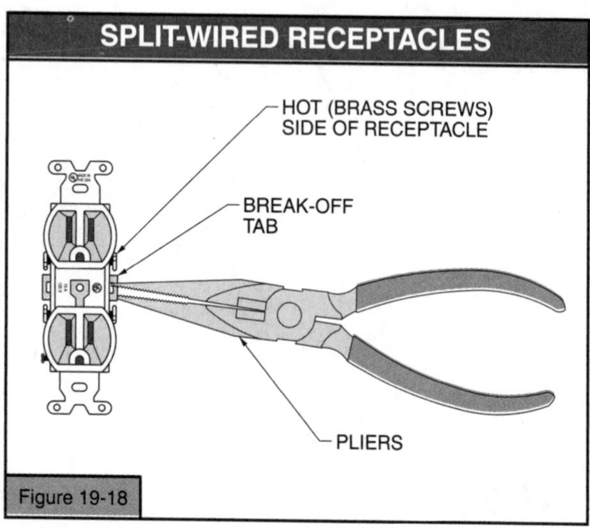

Figure 19-18

GFCI Receptacles

A *ground-fault circuit interrupter* (GFCI) is an electric device which protects personnel by detecting ground faults and quickly disconnecting power from the circuit. A ground fault is any amount of current above the level that may deliver a dangerous shock. GFCIs provide greater protection than standard and isolated grounded receptacles. Per 210.8(A), dwelling units shall have GFCI protection in bathrooms, garages, outdoor areas, crawl spaces, unfinished basements, kitchen countertops, and within 6′ of wet bar sinks. Per 210.8(B), GFCI protection is required in bathrooms and on rooftops of other than dwelling units.

Switched Circuits

Rooms of occupancy, such as bedrooms, may have a switched lighting (split-wired) circuit. This switched lighting circuit may include a built-in lamp and/or a switched receptacle circuit. With a switched receptacle circuit, the lamp(s) is plugged into one side of the receptacle and a wall switch controls the plugged-in lamp. The nonswitched side of the receptacle remains hot at all times. **See Figure 19-19.**

In this circuit, the top receptacle of the duplex receptacle is controlled by a switch and the bottom receptacle is hot at all times. A lamp is normally plugged into the top receptacle and a clock (or other appliance) is plugged into the bottom receptacle. Although there is no standard for which receptacle (top or bottom) is switched, one is selected and made the standard for the entire dwelling.

Two Separate Circuits

In some locations, such as a kitchen countertop, two high-wattage appliances may have to be connected to the same circuit. The circuit may be overloaded if both sides of a receptacle are wired to the same branch circuit. A standard duplex receptacle may be split-wired to two separate branch circuits. **See Figure 19-20.**

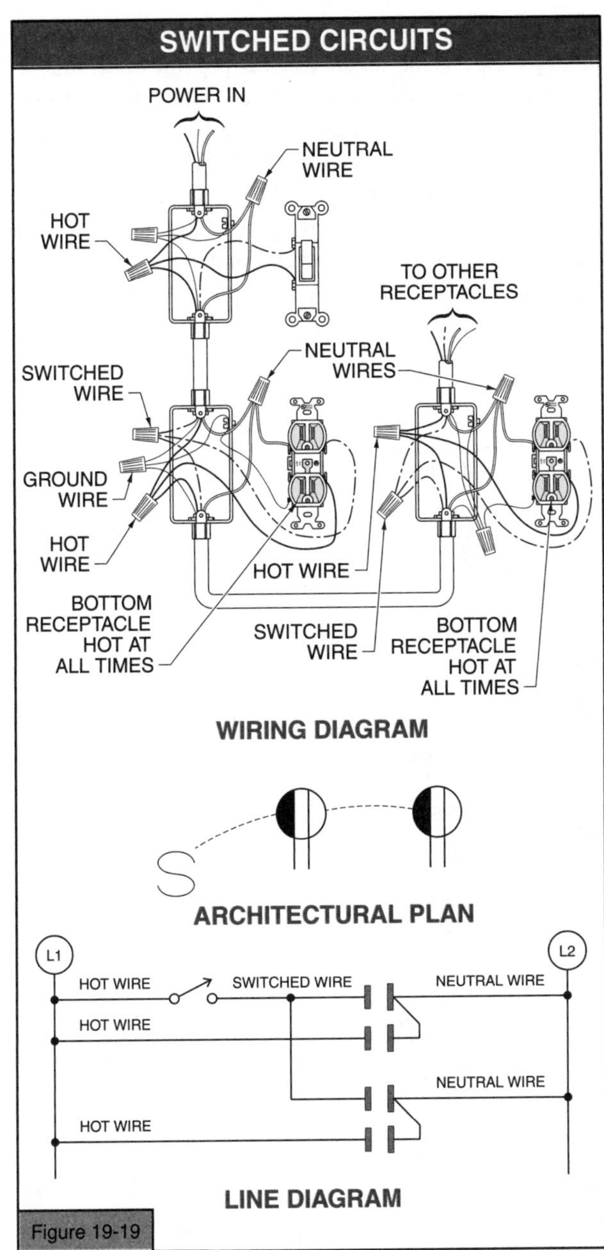

Figure 19-19

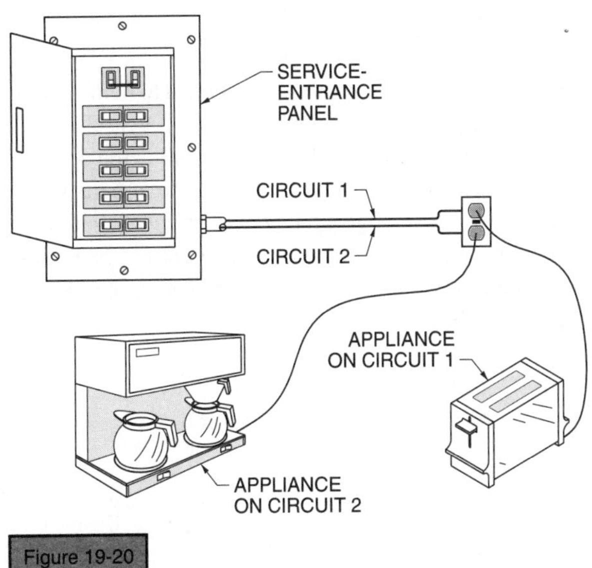

Figure 19-20

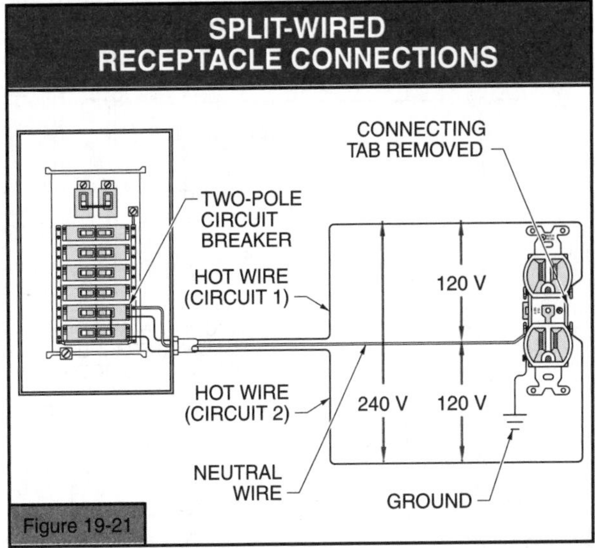

Figure 19-21

Three conducting wires and a ground wire are used when wiring a split-wired receptacle to two different branch circuits. A hot wire from each branch circuit is connected to the two brass-colored screws on the side from which the connecting tab was removed. A neutral wire is connected to one of the silver-colored screws. A green (or bare) ground wire is connected to the green terminal screw. A two-pole circuit breaker is used to ensure that both sides of the receptacle are disconnected when there is an overload on either side. A two-pole circuit breaker includes a mechanical link between the two individual breakers so that they are both always ON or OFF. **See Figure 19-21.**

Receptacle boxes must be large enough to house the receptacle and the conductors that are routed to the receptacle.

Grounding

Residential electrical circuits are grounded to help prevent fires and to prevent personnel that may touch electrical equipment from getting a shock. A *ground* is the connection of all exposed noncurrent-carrying metal parts to the earth. The purpose of grounding is to provide a safe path for a fault current to flow. A complete ground path must be maintained when installing switches, lamps, appliances, and receptacles.

When using metal conduit (rigid or electrical metallic tubing (EMT)), the metal conduit provides a grounded metal pathway back to the service panel. When using nonmetallic cable (Romex), a bare or green insulated copper wire serves as the ground path. The ground wire is connected to all receptacles and metal boxes to provide a continuous pathway for short-circuit current. The ground wire may be connected to each component (component grounding), or to each box using a pigtail or ground clip. **See Figure 19-22.**

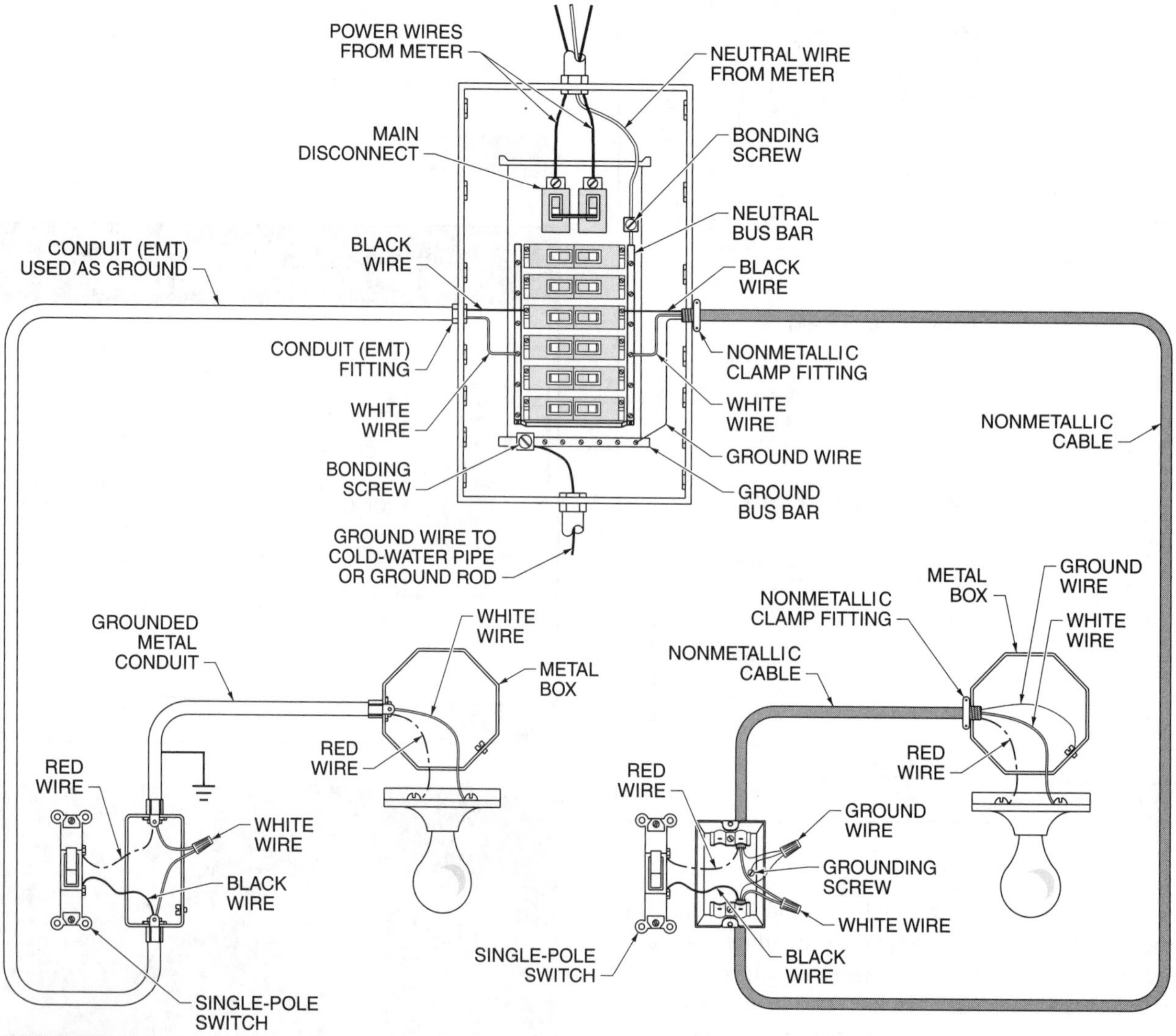

Figure 19-22

Refer to the CD-ROM in the back of the book for Chapter 19 Quick Quiz® and related information.

20 Commercial Circuits

Commercial Circuits

Commercial buildings such as retail shops, restaurants, apartment buildings, hotels, motels, gas stations, bakeries, sports complexes, etc. require high levels of power for lighting, motors, heating elements, and other loads. This high energy demand requires a greater number of individual circuits than residential buildings. Different size power distribution systems are used depending on the amount of power required and the type of loads to be controlled. Relatively low voltage levels (208 V–240 V) are delivered for small commercial applications (gas stations, restaurants, etc.) by the local utility. Medium commercial voltage levels (460 V–480 V) are delivered for mid-sized commercial applications (small office buildings, etc.).

Large commercial applications (colleges, small businesses, hotels, etc.) require voltages of over 1000 V to be delivered and distributed to different transformer banks throughout the building. The high voltage reduces the conductor size required to deliver high amounts of power.

The high voltage is reduced at a substation. A *substation* is an assemblage of equipment installed for switching, changing, or regulating the voltage of electricity. The substations contain transformers and secondary switches that distribute the low-voltage levels (208 V–480 V) to feeder panels, other secondary transformers, and branch-circuit panels. **See Figure 20-1.**

NEC® Branch Circuit Voltage Limits

The exact voltage level used to operate different loads in commercial applications depends on the load, the given application, and the NEC®. For safety reasons, the NEC® places restrictions on the voltage level of some locations. For example, subsection 210.6(A) of the NEC® limits the voltage level to 120 V between conductors that supply lighting fixtures, cords, and plugs that connect loads up to 1440 VA (volt-amperes), and motors up to ¼ HP in dwelling units and guest rooms of hotels, motels, and similar rooms of occupancies. For this reason, most lighting is standard 120 V incandescent or fluorescent fixtures. Less chance exists for an individual to accidentally contact a fatal voltage when the voltage is limited to a low level.

However, it is not practical to limit all circuits to a 120 V level. This is because the lower the voltage, the higher the current for any given power (wattage) rating. Increasing the voltage level to reduce the current level allows for a more practical installation because conductor size, conduit size, and switching equipment size is determined by the amount of current. Thus, higher voltage levels are allowed in commercial areas in which electrical circuits are serviced by qualified maintenance persons. For example, subsection 210.6(C) of the NEC® allows voltage levels up to 277 V for incandescent lamps (equipped with mogul-base screw shell lampholders), fluorescent lamps, and HID lamps used in commercial applications. Such lighting fixtures should not be serviced by nonqualified persons.

TECH FACT — *Never assume that conductors are properly marked or color coded. Take voltage measurements to ensure that voltage is OFF before working on a circuit.*

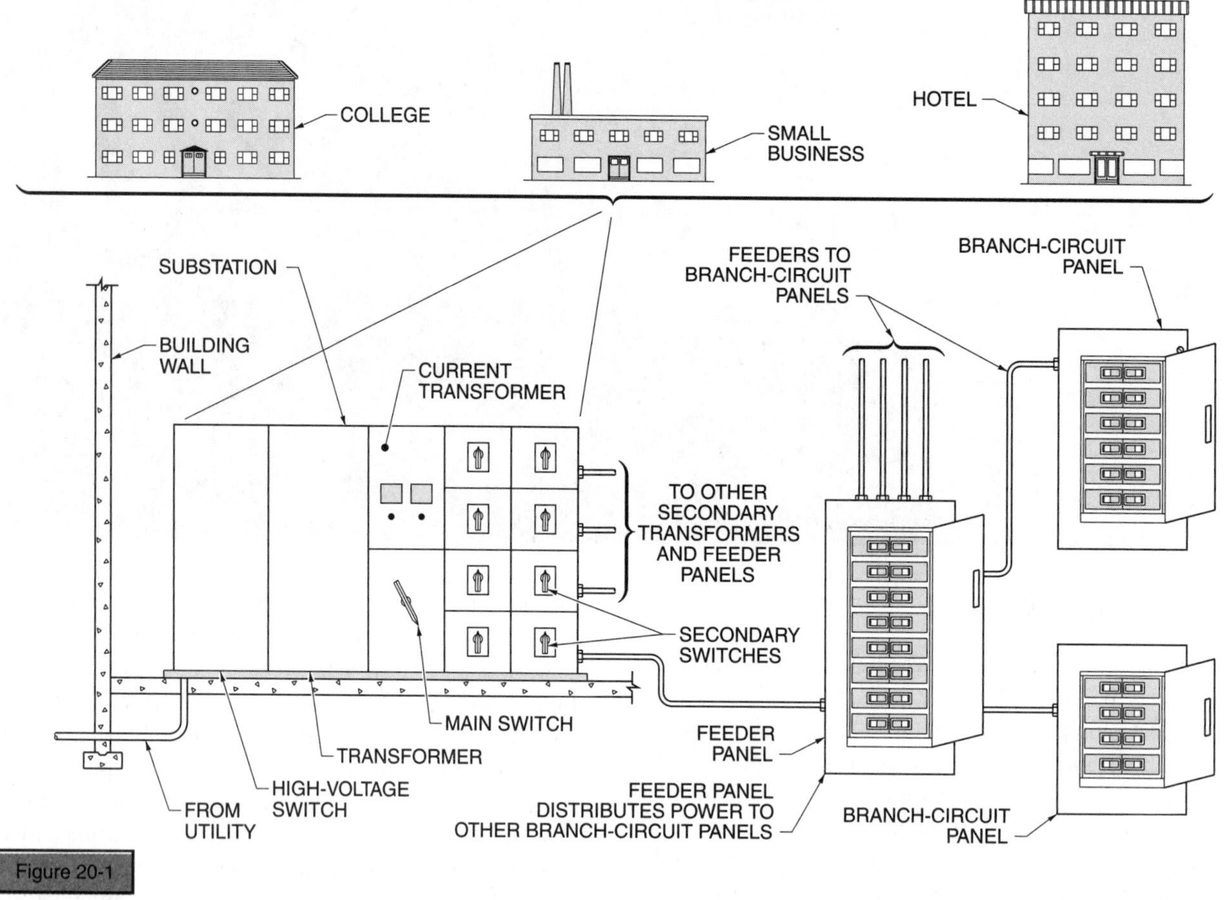

Figure 20-1

Distribution Systems

Electrical power companies generate and distribute high-voltage 3ϕ power. However, commercial customers use low-voltage 3ϕ and 1ϕ power. For this reason, power companies use transformers to deliver power to customers through different distribution systems. Power companies primarily use 1ϕ transformers connected in different combinations to deliver the correct power to customers. Three transformers (except in an open-delta system) are normally interconnected for delivering 3ϕ voltage to commercial customers. The exact voltage level is determined by customer needs and the utility company's existing system.

Commercial power needs include the same needs as residential (1ϕ) and industrial (3ϕ) systems. Apartment buildings and offices require lighting and low-voltage 1ϕ small appliance circuits that are similar to residential buildings except on a much larger scale. Other commercial buildings, such as bakeries and restaurants, require about the same lighting and high-voltage 1ϕ and 3ϕ large appliance circuits as some industrial systems except on a smaller scale. Still other commercial systems, such as outdoor sports stadiums, require a great number of large scale lighting circuits.

The distribution system used varies based on the application, but a few common systems are used for most commercial applications. Common commercial distribution systems include 120/240 V, 1ϕ, 3-wire; 120/208 V, 3ϕ, 4-wire; 120/240 V, 3ϕ, 4-wire; and 277/480 V, 3ϕ, 4-wire services.

120/240 V, 1ϕ, 3-Wire Service

A 120/240 V, 1ϕ, 3-wire service is used to supply power to customers that require 120 V and 240 V, 1ϕ power. This service provides 120 V, 1ϕ; 240 V, 1ϕ; and 120/240 V, 1ϕ circuits. The neutral wire is grounded, therefore it should not be fused or switched at any point. **See Figure 20-2.**

A 120/240 V, 1ϕ, 3-wire service is commonly used for interior wiring for lighting and small appliance use. For this reason, 120/240 V, 1ϕ, 3-wire service is the primary service used to supply most residential buildings. This service is also used for small commercial applications, although a large power panel is used (or there are additional panels).

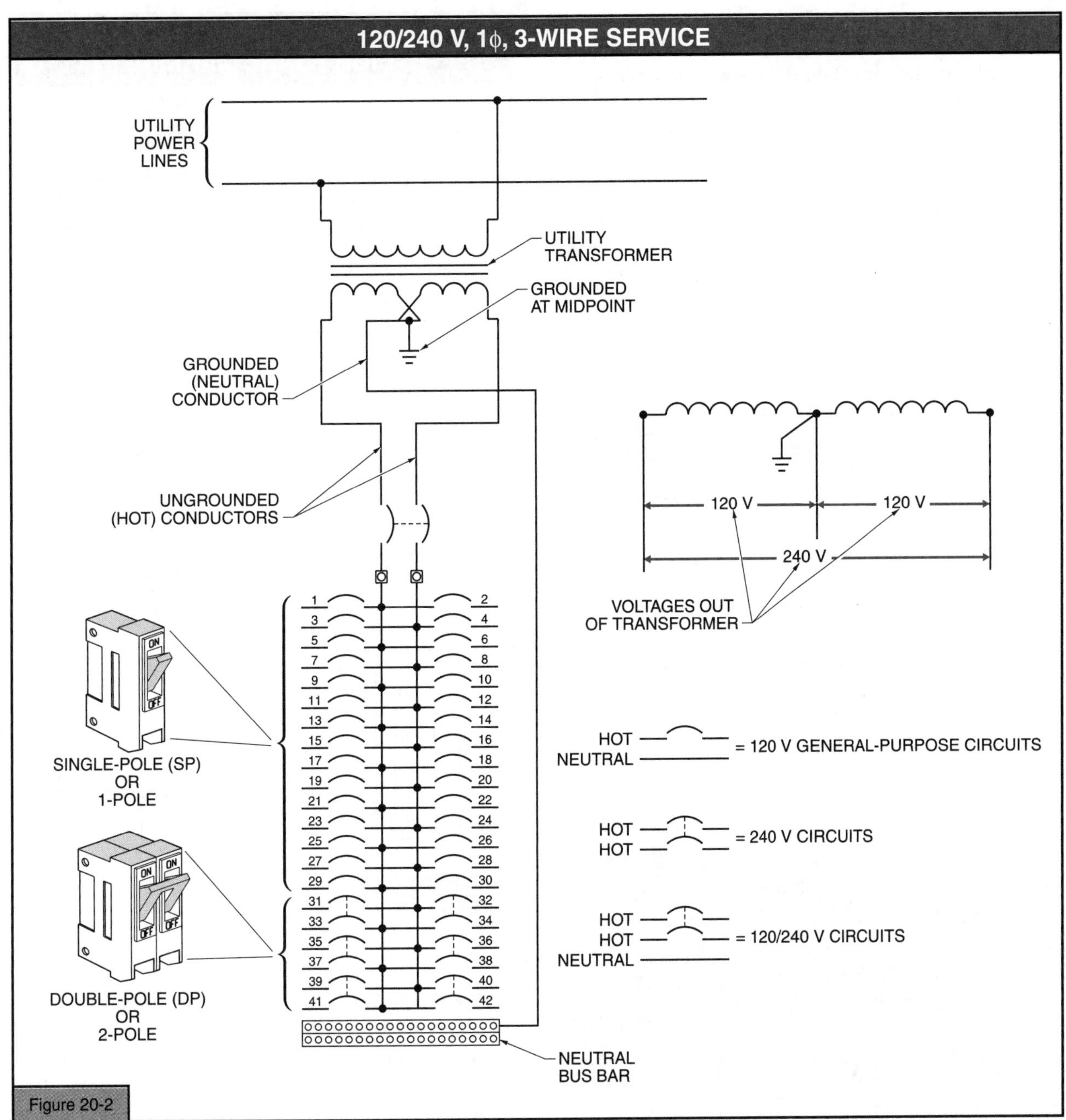

Figure 20-2

In commercial wiring, small appliance usage includes more than typical residential small appliance usage. This is because commercial appliance circuits include high-power devices such as copying machines, commercial refrigerators, washers, etc., and different appliances such as hot tubs and saunas in hotels, automatic ice melting machines, heating elements embedded in the concrete of entranceways, etc. Other commercial small appliance usage includes office computers, printers, motors less than 5 HP, cooking equipment, security equipment, large entertainment systems, hospital equipment, etc. In addition to different appliances, commercial circuits often require duplication because there may be hundreds of rooms or offices at one location.

NEC® Phase Arrangement and High Phase Markings

Three-phase circuits include three individual ungrounded (hot) power lines. These power lines are referred to as phases A (L1), B (L2), and C (L3). Phases A, B, and C are connected to a switchboard or panelboard according to

NEC® subsection 408.3(E). The phases must be arranged A, B, C from front to back, top to bottom, and left to right, as viewed from the front of the switchboard or panelboard. **See Figure 20-3.**

The high leg must be phase B and colored orange (or clearly marked) when the switchboard or panelboard is fed from a 120/240 V, 3ϕ, 4-wire service. This is because in this system, there are 195 V between phase B and the neutral (grounded conductor). This is considered an unreliable source of power because 195 V is too high for standard 115 V or 120 V loads and too low for standard 230 V or 240 V loads.

All ungrounded (hot) conductors read a voltage between the conductor and a ground point (grounded conduit, green ground wire, etc.)

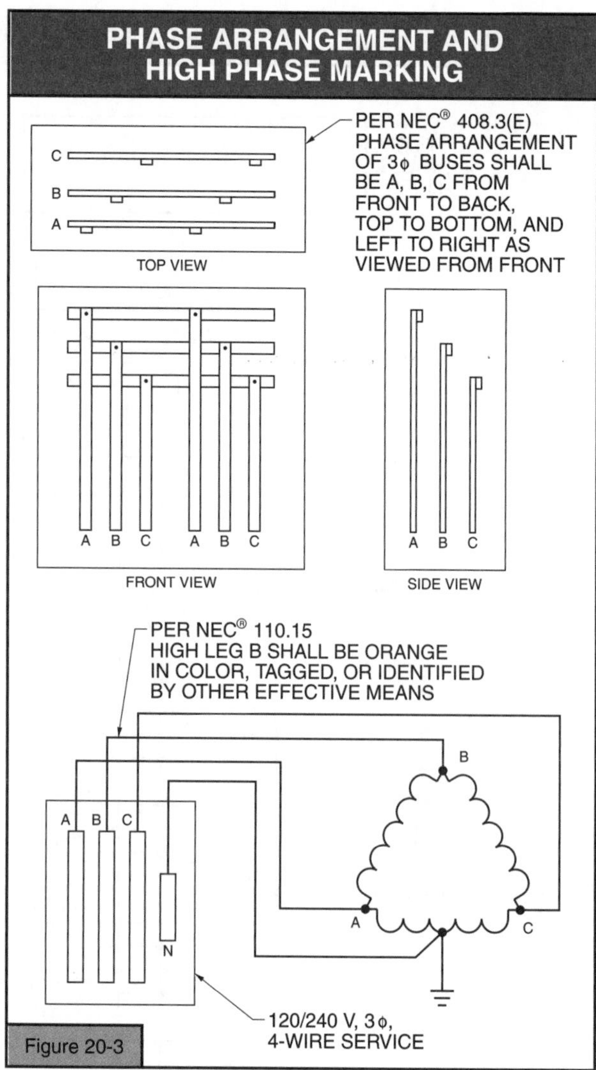

Figure 20-3

120/208 V, 3ϕ, 4-Wire Service

A 120/208 V, 3ϕ, 4-wire service is used to supply customers that require a large amount of 120 V, 1ϕ power; 208 V, 1ϕ power; and low-voltage 3ϕ power. This service includes three ungrounded (hot) lines and one grounded (neutral) line. Each hot line has 120 V to ground when connected to the neutral line. **See Figure 20-4.**

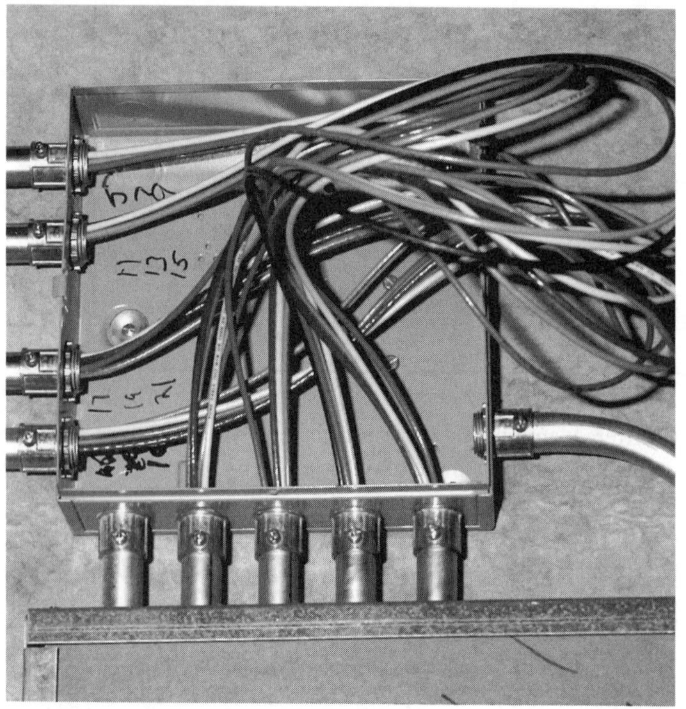

Four-wire service includes three ungrounded (hot) lines and one grounded (neutral) line.

A 120/208 V, 3ϕ, 4-wire service is used to provide large amounts of low voltage (120 V, 1ϕ) power. The 120 V circuits should be balanced to equally distribute the power among the three hot lines. This is accomplished by alternately connecting the 120 V circuits to the power panel so each phase (A to N, B to N, C to N) is divided among the loads (lamps, receptacles, etc.). Likewise, 208 V, 1ϕ loads such as 208 V lamps and heating appliances should also be balanced between phases (A to B, B to C, C to A).

Three-phase loads, such as heating elements designed to be connected to 3ϕ power, can be connected to phases A, B, and C. Three-phase motors may also be connected to the 3ϕ power lines. However, the applied 3ϕ voltage to the motor is only 208 V and may present a problem when using some 3ϕ motors. This is because 208 V is at the lowest end of the 3ϕ voltage range. Any voltage variation on an AC induction motor affects motor operation.

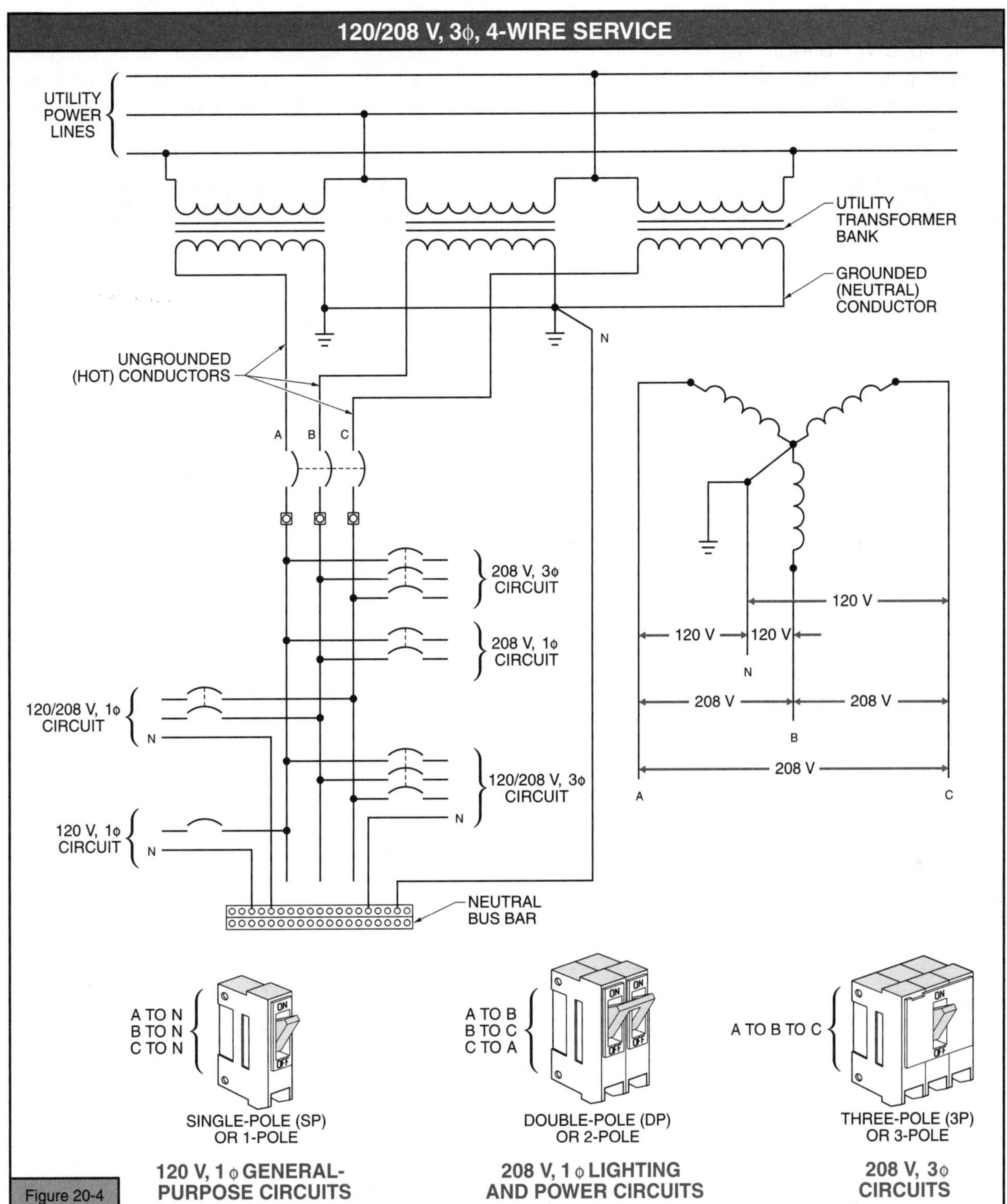

Figure 20-4

120/240 V, 3ϕ, 4-Wire Service

A 120/240 V, 3ϕ, 4-wire service is used to supply customers that require large amounts of 3ϕ power with some 120 V and 240 V, 1ϕ power. This service supplies 1ϕ power delivered by one of the three transformers and 3ϕ power delivered by using all three transformers. The 1ϕ power is provided by center tapping one of the transformers. **See Figure 20-5.**

Because only one transformer delivers all of the 1ϕ power, this service is used in applications that require mostly 3ϕ power or 240 V, 1ϕ power and some 120 V, 1ϕ power. However, large power-consuming electric loads such as air compressors and pumps with motors over 1 HP, and electric heating elements are connected to 3ϕ power. Thus, in many commercial applications, the total amount of 1ϕ power used is small when compared to the total amount of 3ϕ power used. Each transformer may be center tapped if large amounts of 1ϕ power are required.

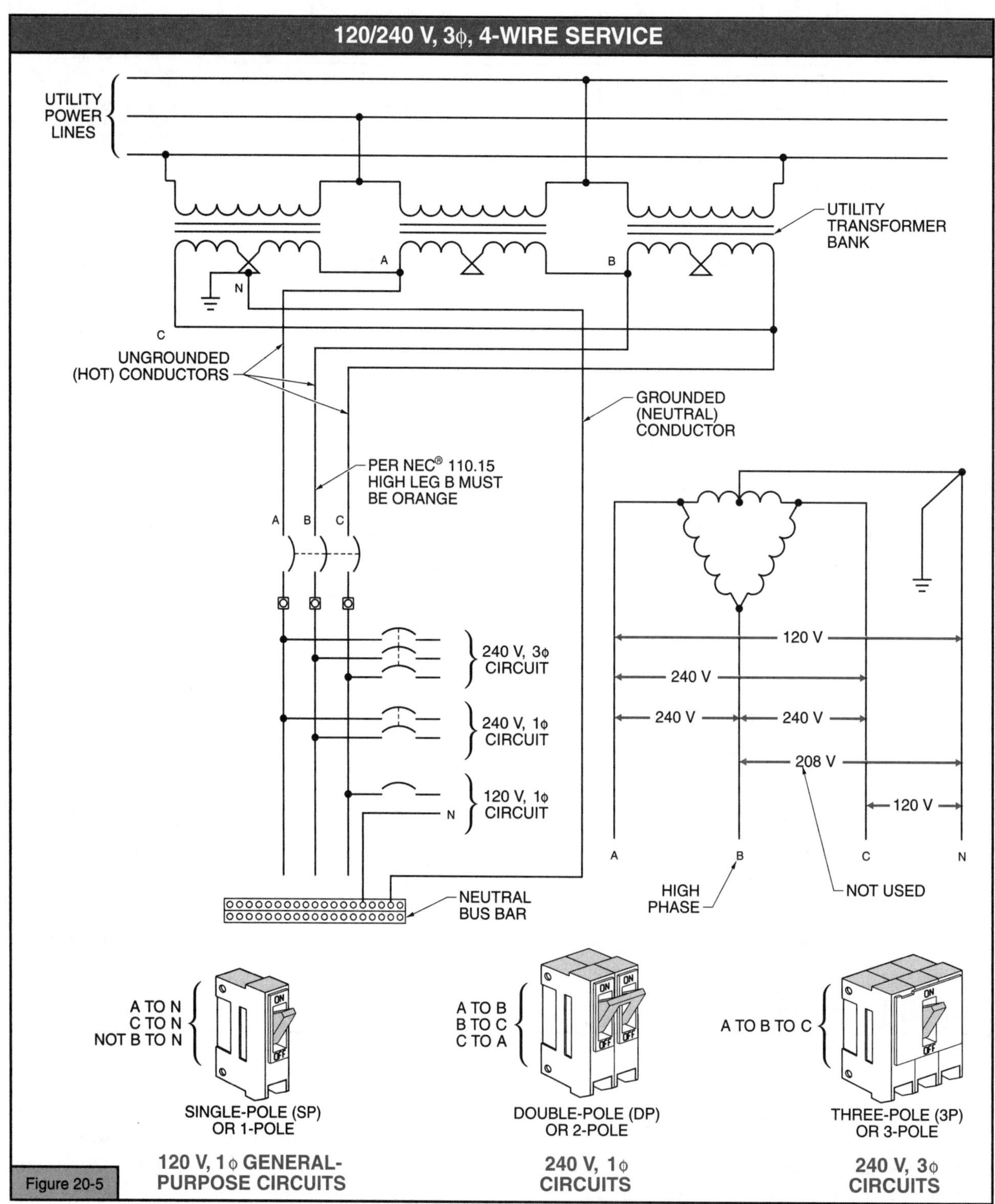

Figure 20-5

277/480 V, 3ϕ, 4-Wire Service

The 277/480 V, 3ϕ, 4-wire service is the same as the 120/208 V, 3ϕ, 4-wire service except the voltage levels are higher. This service includes three ungrounded (hot) lines and one grounded (neutral) line. Each hot line has 277 V to ground when connected to the neutral or 480 V when connected between any two hot (A to B, B to C, or C to A) lines. **See Figure 20-6.**

This service provides 277 V, 1ϕ or 480 V, 1ϕ power, but not 120 V, 1ϕ power. For this reason, 277/480 V, 3ϕ, 4-wire service is not used to supply 120 V, 1ϕ general

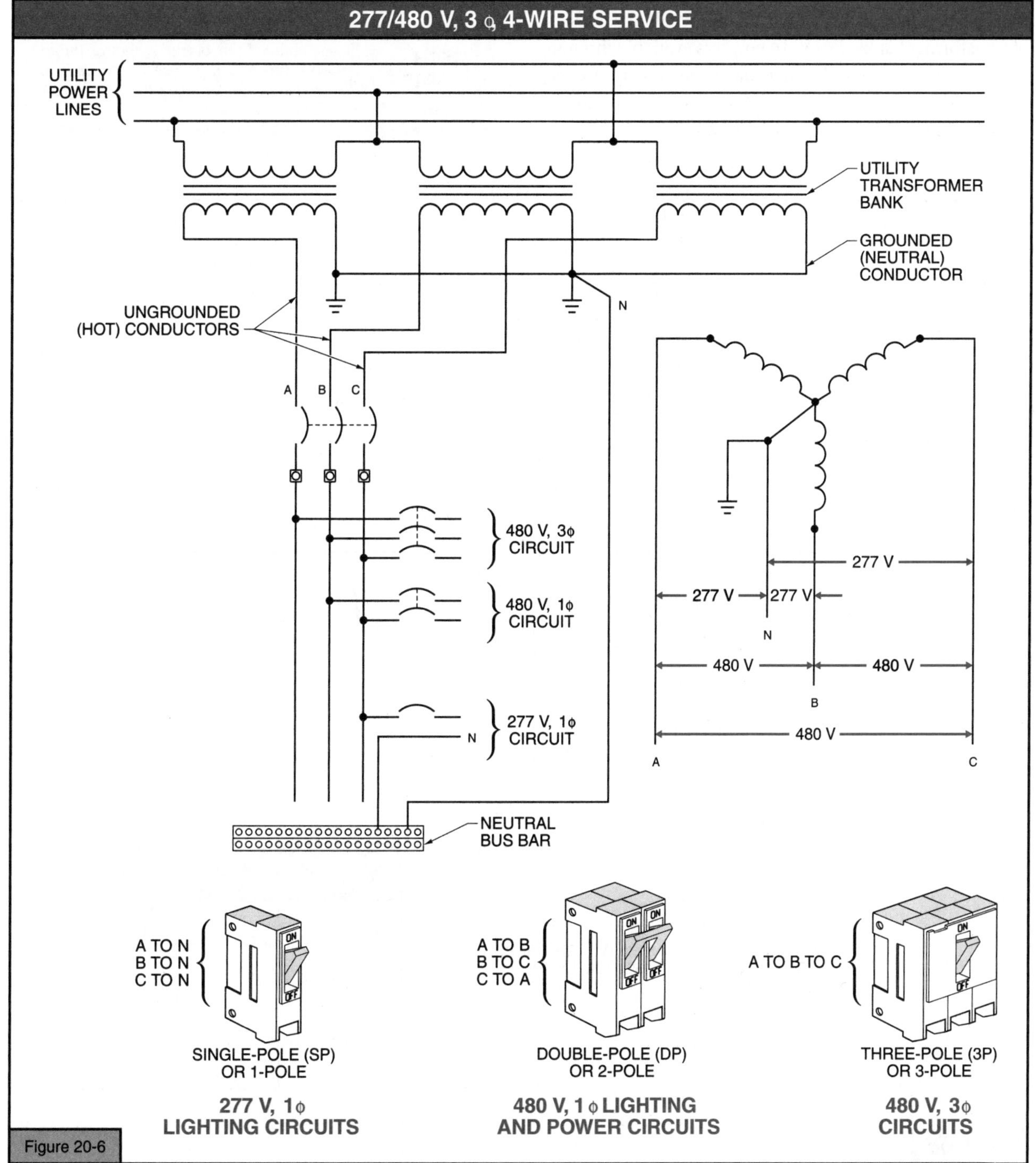

Figure 20-6

lighting and appliance circuits. However, this service can be used to supply 277 V and 480 V, 1ϕ lighting circuits. Such high-voltage lighting circuits are used in commercial fluorescent and HID (high-intensity discharge) lighting circuits.

A system that cannot deliver 120 V, 1ϕ power appears to have limited use. However, in many commercial applications (sport complexes, schools, offices, parking lots, etc.), lighting is a major part of the electrical system. Because large commercial applications include several sets of transformer banks, 120 V, 1ϕ power is available through other transformers. Additional transformers can also be connected to the 277/480 V, 3ϕ, 4-wire service to reduce the voltage to 120 V, 1ϕ. **See Figure 20-7.**

A typical 120 V, 250 W high-pressure sodium lamp has an operating current of 2.70 A; a 240 V, 250 W lamp draws 1.36 A; and a 480 V, 250 W lamp draws .65 A. Thus, the higher the rated lamp voltage, the less operating current required. A different number of 250 W lamps may be connected with other lamps of the same wattage rating to a 15 A circuit. The number varies based on the applied voltage. **See Figure 20-8.**

TECH TIP *Never connect the primary neutral (high-voltage side) to the secondary neutral (low-voltage side) because the transformer may burn out.*

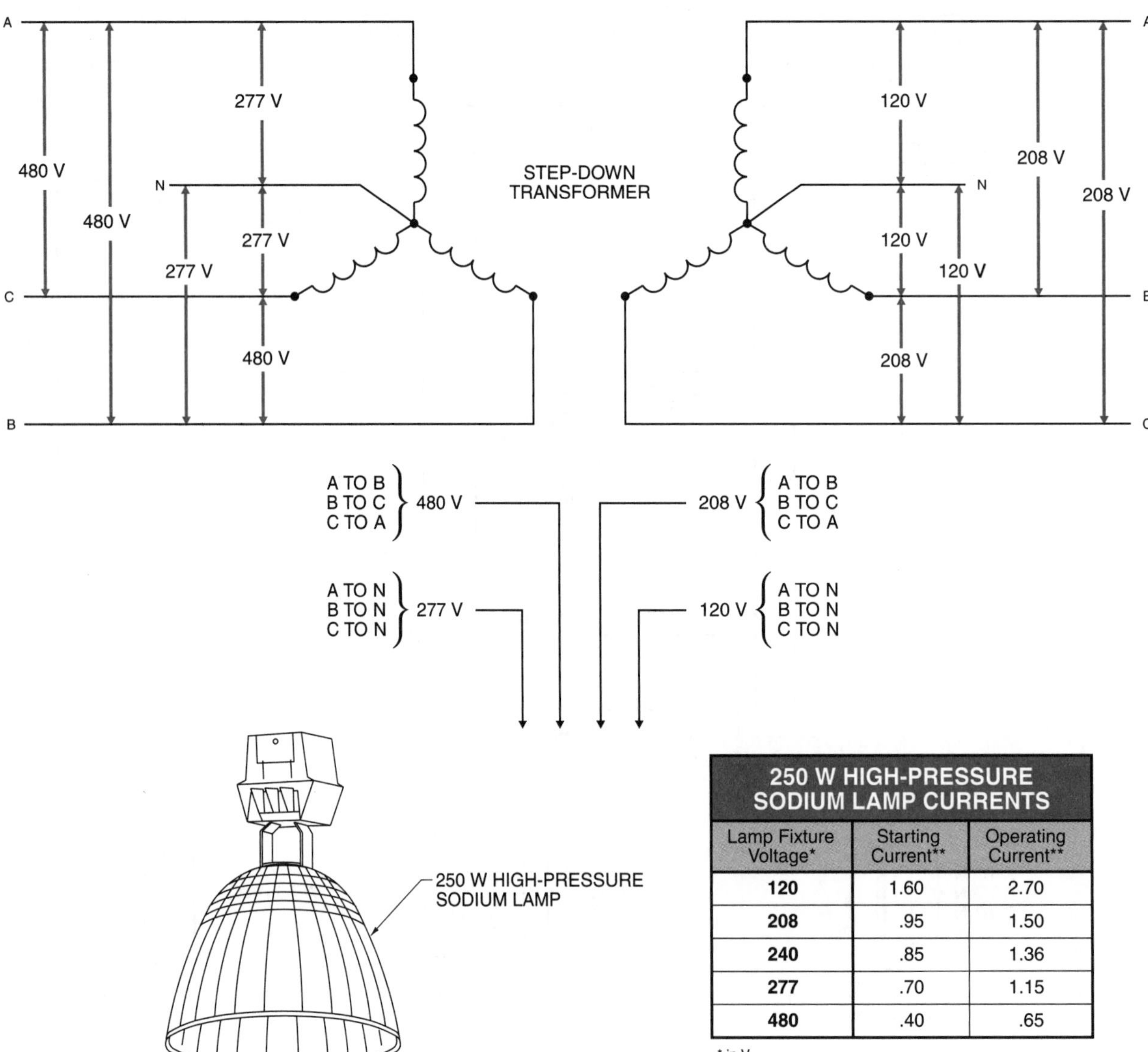

250 W HIGH-PRESSURE SODIUM LAMP CURRENTS

Lamp Fixture Voltage*	Starting Current**	Operating Current**
120	1.60	2.70
208	.95	1.50
240	.85	1.36
277	.70	1.15
480	.40	.65

* in V
** in A

Figure 20-7

250 W HIGH-PRESSURE SODIUM LAMP CIRCUITS			
Lamps on Circuit	Operating Voltage*	Total Circuit Current**	Total Wattage***
5	120	13.5	1250
10	208	15	2500
11	240	14.96	2750
13	277	14.95	3250
23	480	14.95	5750

* in V
** in A
*** in W

Figure 20-8

Voltage Variation Effects on AC Motors

All motors are designed to operate at a given (nameplate rated) voltage. A motor operates if the voltage varies below or above the rated voltage, but there is a change in the motor's performance. Normally, the applied voltage should be within ±10% of the motor's rated voltage. **See Figure 20-9.**

A motor delivers less torque and operates at a higher temperature when connected to a voltage source lower than the motor's rated voltage. However, this does not adversely effect the motor unless the motor is operating in an application that requires it to deliver full-rated power. In such applications, the motor size and/or the delivered voltage must be increased.

TECH FACT *The voltage on the secondary side of a transformer begins to drop if the transformer is overloaded. The higher the overload, the greater the voltage drop.*

INDUCTION MOTOR PERFORMANCE CHANGES*		
Characteristic	Change in Design Voltage	
	110%	90%
Torque Starting and maximum running	Increase 21%	Decrease 19%
Speed** Synchronous Full load Percent slip	 No change Increase 1% Decrease 17%	 No change Decrease 1.5% Increase 23%
Efficiency Full load ¾ load ½ load	 Increase .5 to 1 Little change Decrease 1 to 2	 Decrease 2 Little change Increase 1 to 2
Power Factor Full load ¾ load ½ load	 Decrease 3 Decrease 4 Decrease 5 to 6	 Increase 1 Increase 2 to 3 Increase 4 to 5
Current Starting Full load	 Increase 10% to 12% Decrease 7%	 Decrease 10% to 12% Increase 11%
Temperature Rise	Decrease 3°C to 4°C	Increase 6°C to 7°C
Maximum Overload Capacity	Increase 21%	Decrease 19%
Magnetic Noise	Slight increase	Slight decrease

* 2-, 4-, 6-, and 8-pole motors
** speed of AC induction motors varies directly with frequency

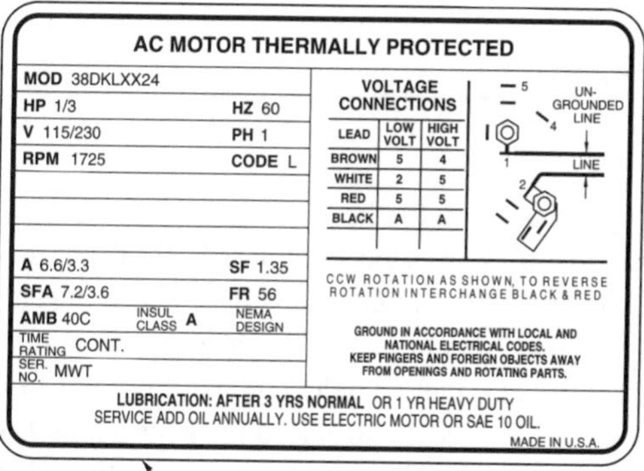

Figure 20-9

Voltage Variation Effects on Heating Elements

Voltage is applied to heating elements to produce heat. Single-phase voltage is used for small heating applications and 3ϕ voltage is used for large commercial applications. **See Figure 20-10.**

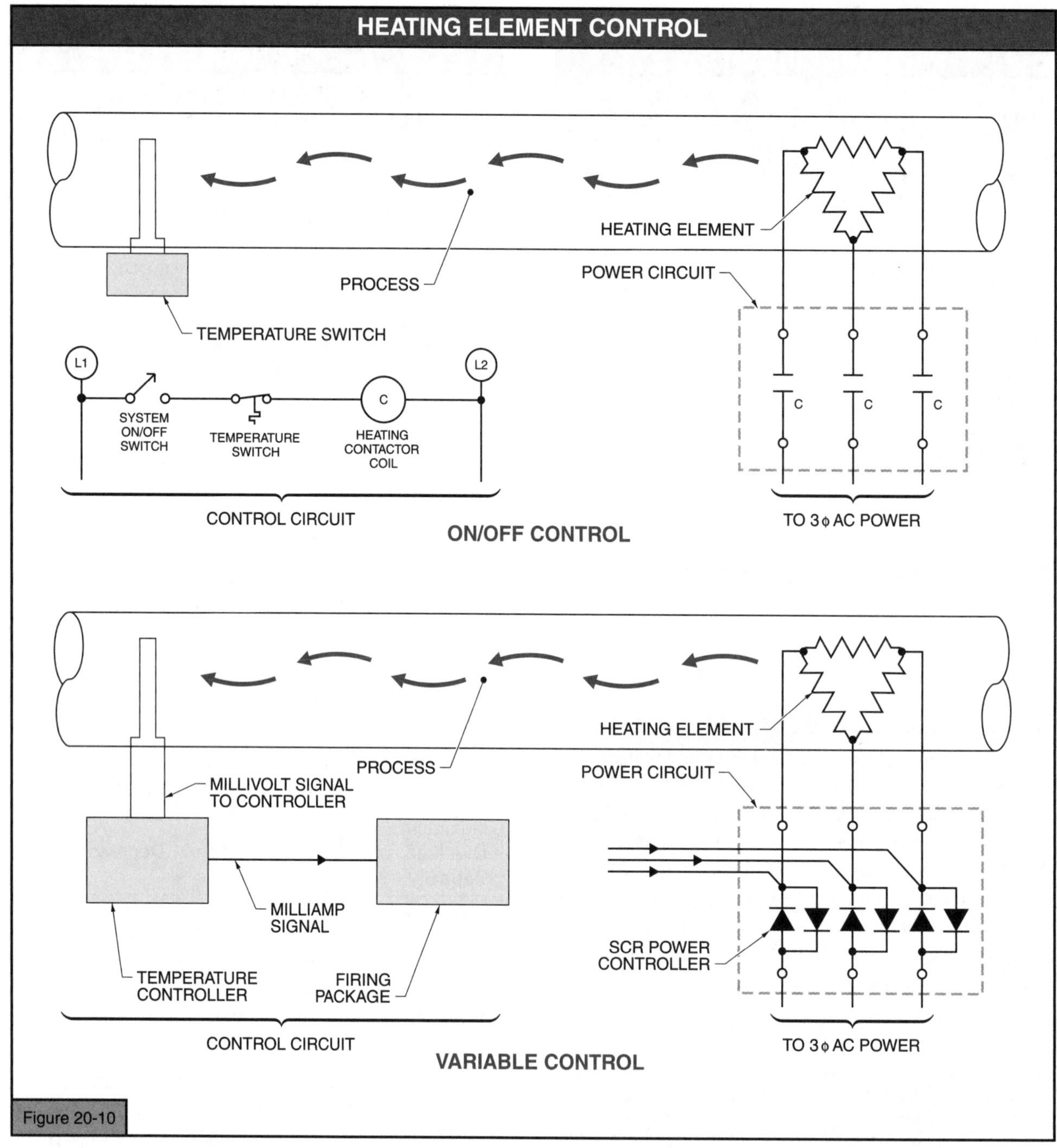

Figure 20-10

Any variation in applied voltage to a heating element increases or decreases the amount of heat produced. However, unlike motors, varying the voltage to heating elements is often done intentionally. The heating element output (amount of heat) can be reduced, and thus controlled, by reducing the applied voltage. Reducing the applied voltage to heating elements does not harm the heating elements. An SCR (silicon controlled rectifier) circuit may be used to set and maintain the amount of heat output.

Voltage Variation Effects on Lamps

The effects of voltage variations on a lamp depend on the lamp used. The voltage on incandescent lamps can vary from 0 V to 10% higher than the lamp's rated voltage. As the voltage is reduced, the lamp's output is reduced (lamp dims). Adjusting the voltage applied to an incandescent lamp is the basic principle of incandescent lamp dimmers.

Unlike incandescent lamps, HID lamps must be operated within a closer voltage range. Low-wattage bulbs (up to 200 W) should have an applied voltage within ±5% of the lamp fixture's rating. Lamps over 200 W should have an applied voltage within ±10% of the lamp fixture's rating.

Fluorescent lamps are available in many different styles. In general, fluorescent lamps should have an applied voltage that is within ±5% of the lamp fixture's rating. Check manufacturer's specifications for voltage variations greater or less than 5%.

Conductor Identification–Color Coding

Conductors (wires) are covered with an insulating material that is available in different colors. The advantage of using different colors on conductors is that the function of each conductor can be easily determined. Some colors have definite meaning. For example, the color green always indicates a conductor used for grounding. Other colors may have more than one meaning depending on the circuit. For example, a red conductor may be used to indicate a hot wire in a 230 V circuit or switched wire in a 115 V circuit. **See Figure 20-11.**

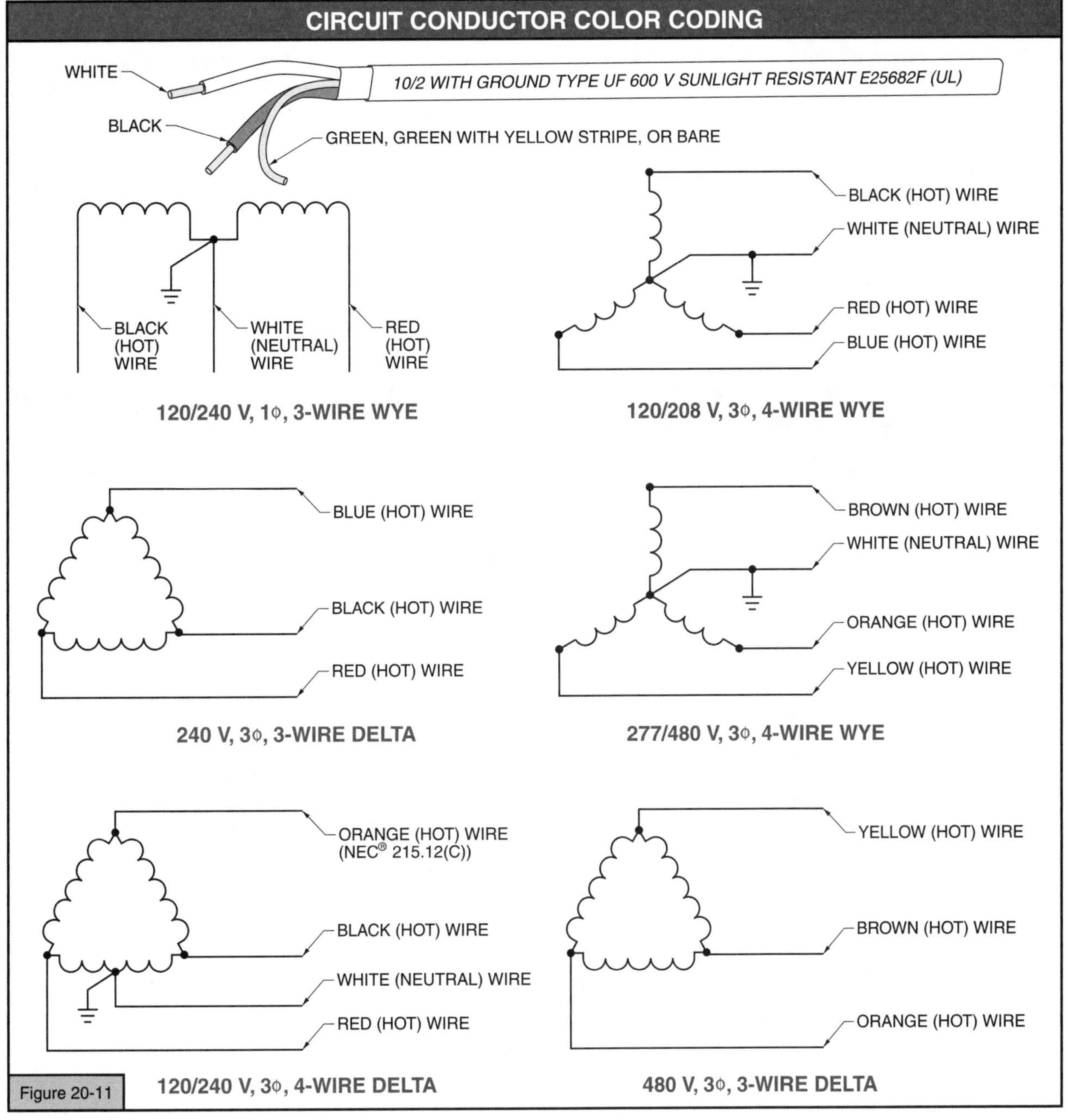

Figure 20-11

Conductor color coding makes balancing loads among the different phases easier and aids when troubleshooting. For this reason, conductor color coding is also used in applications that do not require every conductor to be color coded. Always use standard colors (green, white, black, etc.) where applicable.

Green or green with a yellow stripe is the standard color for a grounding conductor in the electrical field. A solid green conductor is the most common color used for grounding conductors. Green is used to indicate a grounding conductor regardless of the voltage level (115 V, 230 V, 460 V, etc.) or circuit (1ϕ or 3ϕ). A *grounding conductor* is a conductor that does not normally carry current, except during a fault (short circuit).

The color white or natural gray is used for the neutral (grounded circuit) conductor. A *neutral conductor* is a current-carrying conductor that is intentionally grounded. Neutral conductors carry current from one side of a load (lamp, heating element, motor, etc.) to ground. Neutral conductors are connected directly to loads and never connected through fuses, circuit breakers, or switches. The conductor colors green (or green with yellow stripe) and white (or natural gray) are required per NEC® Article 210.5.

Electrical circuits include ungrounded (hot) conductors in addition to grounding and neutral conductors. An *ungrounded conductor* is a current-carrying conductor that is connected to loads through fuses, circuit breakers, and switches. Ungrounded conductors can be any color other than white, natural gray, green, or green with yellow stripe. Black is the most common color used for ungrounded conductors. Red, blue, orange, and yellow are also used for ungrounded conductors. Although such colors as red, blue, orange, and yellow are used to indicate a hot conductor, the exact color used to indicate different hot conductors (A [L1], B [L2], and C [L3] in a 3ϕ system) may vary. The one exception to this is listed in NEC® Article 215.8. This Section states that in a 4-wire delta-connected secondary system, the higher voltage phase should be colored orange (or clearly marked) because it is too high for low voltage 1ϕ power and too low for high voltage 1ϕ power. **See Figure 20-12.**

Most conductors are made from either copper or aluminum. With the exception of silver, copper has the least resistance to the flow of electrons of any metal used for conductors.

CONDUCTOR COLOR CODING COMBINATIONS

Voltage*	Circuit	Conductor Colors
120	1ϕ, 2-wire with ground	One black (hot wire), one white (neutral wire), and one green (ground wire)
120/240	1ϕ, 3-wire with ground	One black (one hot wire), one red (other hot wire), one white (neutral wire), and one green (ground wire)
120/208	3ϕ, 4-wire wye with ground	One black (phase 1 hot wire), one red (phase 2 hot wire), one blue (phase 3 hot wire), one white (neutral wire), and one green (ground wire)
240	3ϕ, 3-wire delta with ground	One black (phase 1 hot wire), one red (phase 2 hot wire), one blue (phase 3 hot wire), and one green (ground wire)
120/240	3ϕ, 4-wire delta with ground	One black (first low phase hot wire), one red (second low phase hot wire), one orange (high phase leg wire), one white (neutral wire), and one green (ground wire)
277/480	3ϕ, 4-wire wye with ground	One brown (phase 1 hot wire), one orange (phase 2 hot wire), one yellow (phase 3 hot wire), one white (neutral wire), and one green (ground wire)
480	3ϕ, 3-wire delta with ground	One brown (phase 1 hot wire), one orange (phase 2 hot wire), one yellow (phase 3 hot wire), and one green (ground wire)

* in V

Figure 20-12

Conductor Identification—Wire Markers

Hot conductors in a 3ϕ system may be identified with other types of markings in addition to color coding. For example, a 3ϕ system may use three black conductors for each of the three hot (A, B, and C) conductors and identification markings for each conductor.

One band of colored (red, yellow, etc.) electrician's tape may be placed around the line one (L1) black conductor when marking 3ϕ hot conductors. Likewise, two bands of colored electrician's tape may be placed around the line two (L2) black conductor, and three bands of colored electrician's tape may be placed around the line three (L3) black conductor.

Wire markers may be used in place of tape bands. A *wire marker* is a preprinted peel-off marker designed to adhere when wrapped around a conductor. Wire markers resist moisture, dirt, oil, etc. and can be used to identify conductors of the same color that have different meanings. For example, the three hot black conductors (L1, L2, and L3) in a 3ϕ system can each be marked with a different numbered wire marker. Wire markers can still be used even when different colored conductors are used. Using wire markers in addition to color coding further clarifies the meanings of the conductors. **See Figure 20-13.**

Receptacles

Power is delivered to convenient locations by receptacles (outlets). Receptacles have different configurations so only matching plugs can be connected to them. This system prevents the connection of equipment to the wrong source of power. For example, a 125 V, 15 A plug cannot be connected to a 125/250 V, 50 A receptacle. Likewise, a 125/250 V plug cannot be connected to a 125 V, 15 A receptacle. **See Figure 20-14.**

Receptacles and plugs may be nonlocking or locking. Nonlocking plugs fit into nonlocking receptacles in a firm but nonlocking manner. Locking plugs fit into locking receptacles by firmly inserting and twisting to lock in place. Once locked, locking plugs cannot be pulled out without first twisting them. Because appliances and machines come in different voltage and current sizes, plugs and receptacles come in many different sizes. **See Appendix.**

Commercial Lighting Systems

Outdoor commercial lighting is required for parking lots, entranceways, sports complexes, airports, buildings, piers, yards, storage areas, etc. Outdoor commercial lighting systems may use low- or high-pressure sodium, metal-halide, or mercury-vapor lamps in voltages of 120 V, 208 V, 240 V, 277 V, or 480 V. Power ratings range from 35 W to above 1500 W. Indoor commercial lighting is required for all general rooms of occupancy, halls, restrooms, work areas, etc. Indoor commercial lighting systems may use incandescent, halogen, or fluorescent lamps in voltages of 120 V, 208 V, 220 V, 240 V, or 277 V.

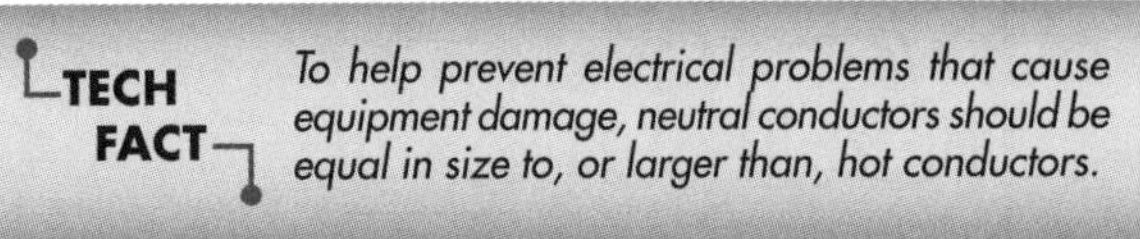

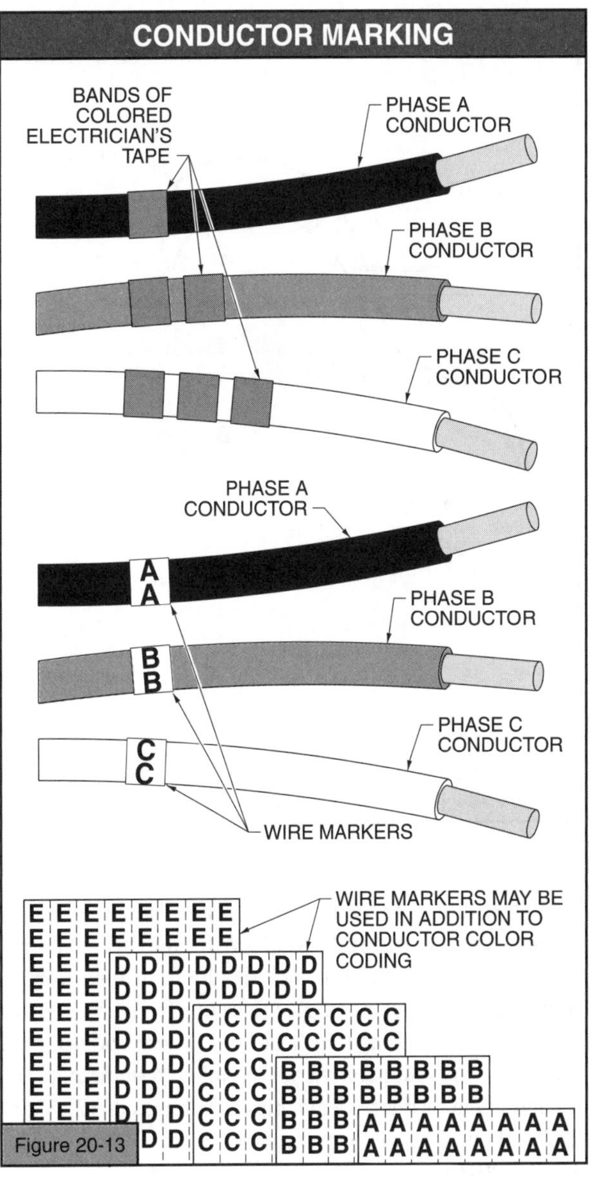

Figure 20-13

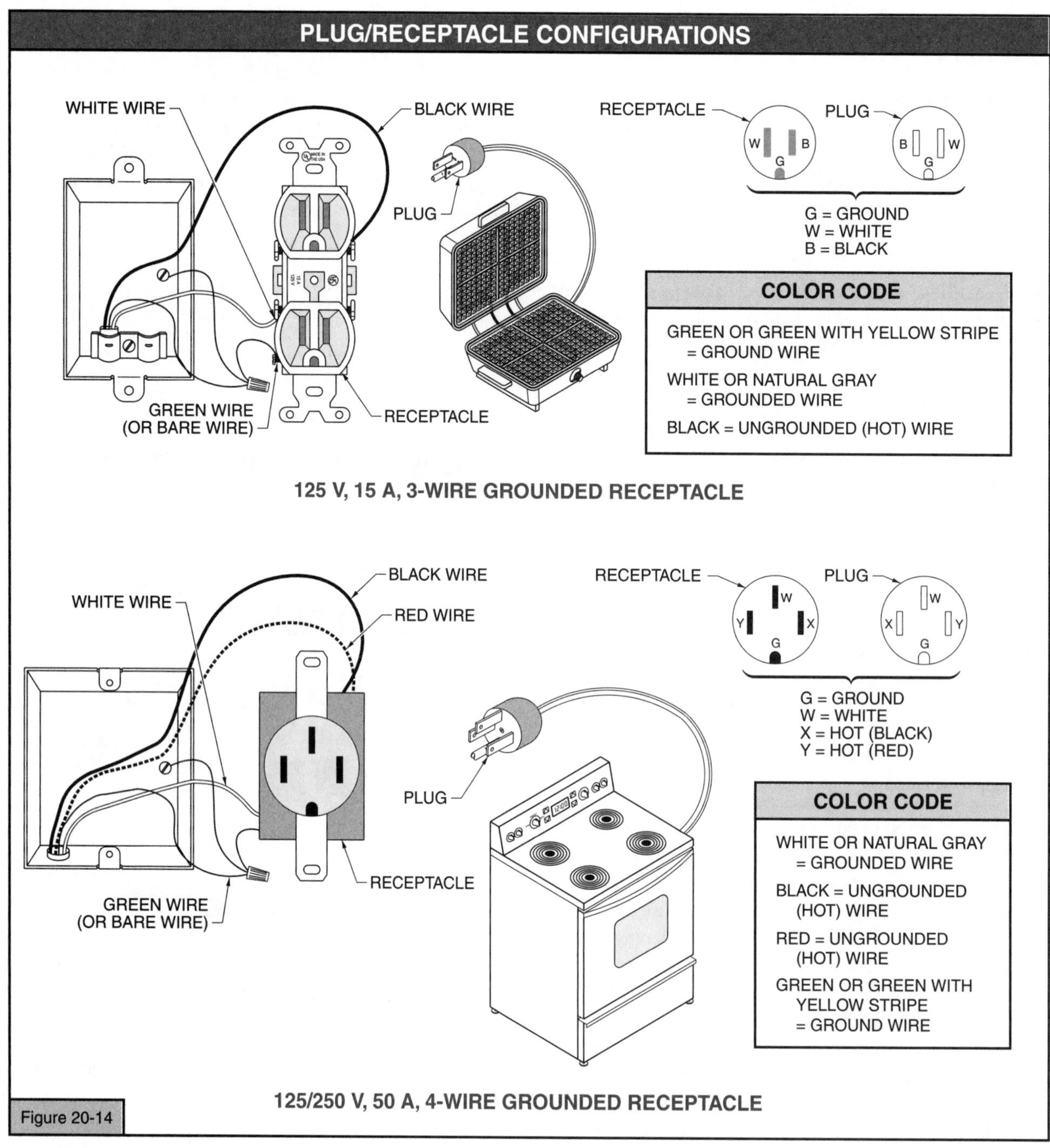

Figure 20-14

Power ratings range from almost any size up to 300 W for incandescent and 15 W to 215 W for fluorescent lamps. Commercial lighting also includes lighting for information (signs and advertising), effect (stage lighting, black lights, etc.), and function (heat lamps, germicidal lamps, etc.). Informational commercial lighting systems may use incandescent or fluorescent lamps in voltages of 120 V and 277 V. Power ratings vary based on size of the sign. **See Figure 20-15.**

Lamps are used to produce light. A *lamp* is an output device that converts electrical energy into light. The amount of light a lamp produces is expressed in lumens. A *lumen (lm)* is the unit used to measure the total amount of light produced by a light source. The more lumens produced by a light source per watt of power used, the more efficient the light source is in producing light. For example, a standard 40 W incandescent lamp produces about 480 lm (or 12 lm/W), and a standard 40 W fluorescent lamp produces about 3100 lm (or 77.5 lm/W). Thus, a 40 W fluorescent lamp source is more light-efficient than a 40 W incandescent lamp. The amount of lumens produced per watt varies depending on the wattage rating of the lamp. Manufacturer's specification sheets are used when determining exact energy efficiency for a given lamp. For example, a 100 W standard high-pressure sodium lamp produces 7800 lm or 78 lm/W. **See Figure 20-16.**

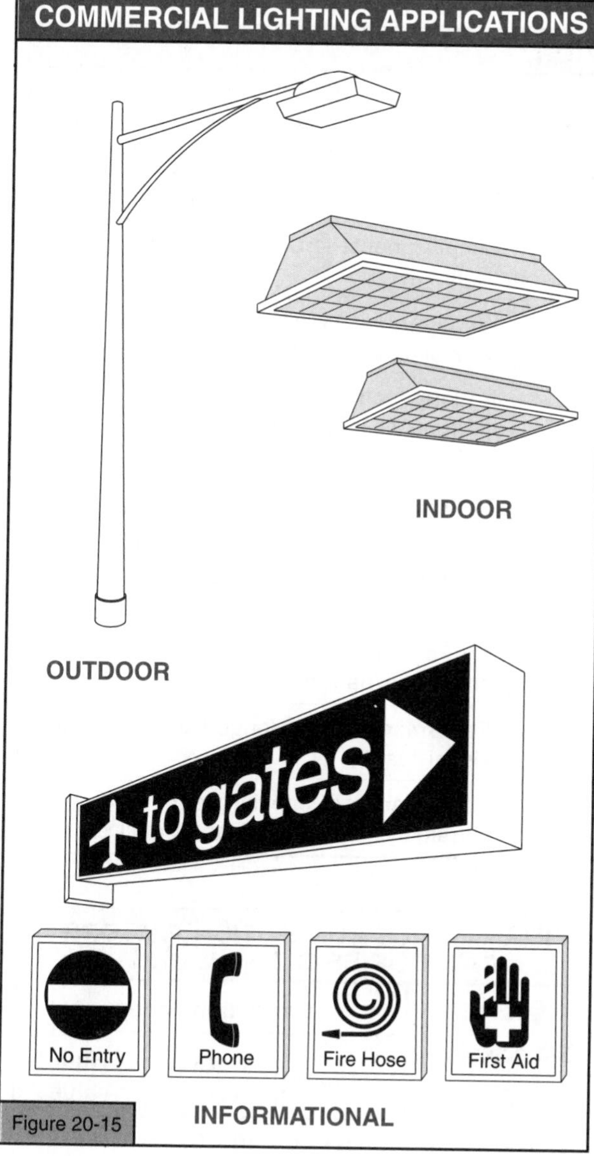

Figure 20-15

TECH TIP *Lamp poles should be spaced at a maximum of 2 times their mounting height from the edge of an area to be lit and 4 times their mounting height from any adjacent lamp pole.*

STANDARD HIGH-PRESSURE SODIUM LAMP CHARACTERISTICS

Lamp Wattage	Initial Lumens	Lumens per Watt
50	3400	68
70	5600	80
100	7800	78
150	14,000	93
175	16,000	91
250	20,000	80
400	40,000	100
1000	117,000	117
1500	155,000	103
1650	177,000	107

Figure 20-16

The light produced by a light source causes illumination. *Illumination* is the effect that occurs when light falls on a surface. The unit of measure of illumination is the footcandle. A *footcandle (fc)* is the amount of light produced by a lamp (lumens) divided by the area that is illuminated. The amount of footcandles required for any given area varies depending on the application. **See Figure 20-17.**

Additional factors may be considered when determining the required amount of footcandles for a given application. For example, the size of the object to be viewed, the speed of a moving object, the reflectiveness of the surrounding surfaces, and the age of the viewer all affect the amount of total light required. Small objects are harder to see and require more light than large objects. The eye requires time to adjust and view an object when the object is moving. Fast moving objects require more light than slow moving objects.

In general, older people need more light for proper viewing. The surface on which an object is viewed must also be considered when determining the required amount of light. For example, the words on a textbook page are placed on a white paper background which has a reflectance of approximately 80%.

Brightness is the perceived amount of light reflecting from an object. Brightness depends on the amount of light falling on an object and the reflecting ability of the object. The object appears to have a uniform brightness if it is uniform and equally illuminated. However, most objects are not uniform nor equally illuminated so they appear to have varying degrees of brightness (shadows). Brightness is described as "perceived" because the eye is constantly adjusting to the amount of brightness. For example, when walking into a dark room, many objects cannot be seen. However, as the eyes adjust to the darkness, objects start appearing in more detail.

Likewise, glare takes place when there is too much brightness. Glare is reduced when the brightness is reduced and/or the eye reduces (squinting) the amount of light received. Proper lamp spacing, number of lamps, and size (wattage) determines the evenness of the brightness. *Contrast* is the ratio of brightness between different objects. For example, the black-on-white letters on a printed page have a high contrast. However, black-on-gray letters have a low contrast. High contrast is helpful when distinguishing outlines and small objects. Contrast between surfaces is sharpened by greater brightness and/or contrasting surfaces. **See Figure 20-18.**

TECH FACT *Fluorescent lamps have been used for commercial and industrial lighting applications since 1938.*

RECOMMENDED LIGHT LEVELS

ILLUMINATION

Illumination Category	Ranges of Illuminances*			Activity
	Low	Average	High	
A	2	3	5	Public spaces with dark surroundings
B	5	7.5	10	Simple orientation for short temporary visits
C	10	15	20	Working spaces where visual tasks are only occasionally performed
D	20	30	50	Performance of visual tasks of high contrast or large size
E	50	75	100	Performance of visual tasks of medium contrast or small size
F	100	150	200	Performance of visual tasks of low contrast or very small size
G	200	300	500	Performance of visual tasks of low contrast and very small size over a prolonged period
H	500	750	1000	Performance of prolonged and exacting visual tasks
I	1000	1500	2000	Performance of special visual tasks of extremely low contrast and small size

* in fc

ILLUMINANCE CORRECTION FACTORS — CATEGORIES A THROUGH C

Characteristics	Correction Factor		
	Low Value	Average Value	High Value
Occupant's Age	Under 40	40 – 55	Over 55
Surface Reflectances	Reflective	Average	Absorbent

ILLUMINANCE CORRECTION FACTORS — CATEGORIES D THROUGH I

Characteristics	Correction Factor		
	Low Value	Average Value	High Value
Worker's Age	Under 40	40 – 55	Over 55
Speed and/or accuracy	Less important	Important	More important

Figure 20-17

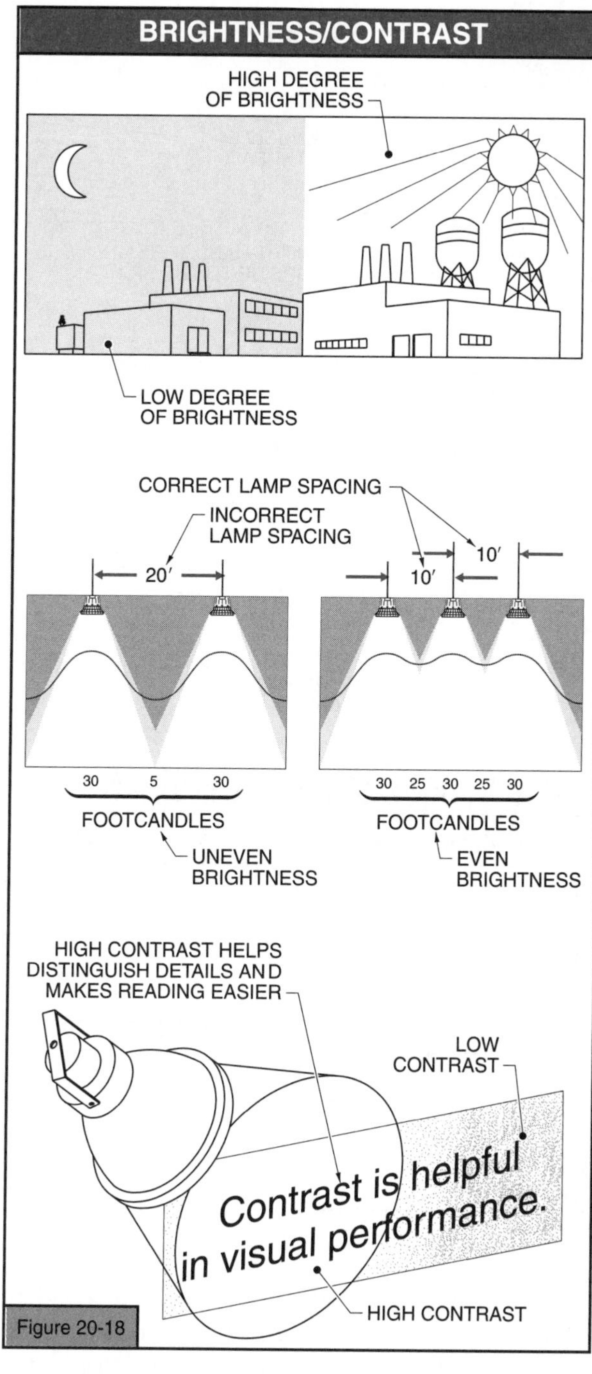

Figure 20-18

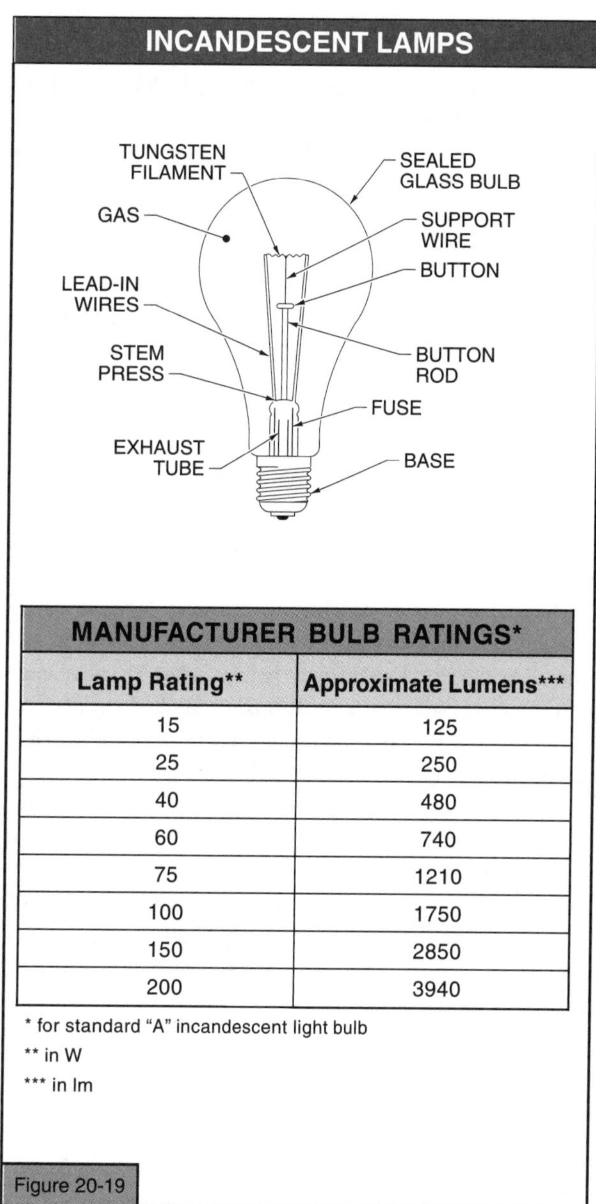

MANUFACTURER BULB RATINGS*	
Lamp Rating**	Approximate Lumens***
15	125
25	250
40	480
60	740
75	1210
100	1750
150	2850
200	3940

* for standard "A" incandescent light bulb
** in W
*** in lm

Figure 20-19

Incandescent Lamps

Incandescent lamps are the most widely used lamps in the world. An *incandescent lamp* is an electric lamp that produces light by the flow of current through a tungsten filament inside a gas-filled, sealed glass bulb. Incandescent lamps have a low initial cost and are simple to install and service. These lamps have a lower electrical efficiency and shorter life than other lamps. **See Figure 20-19.**

An incandescent lamp operates as current flows through the filament. A *filament* is a conductor with a resistance high enough to cause the conductor to heat. The filament glows white-hot and produces light. The filament is made of tungsten which limits the current to a safe operating level.

Inrush current is higher than operating current because the filament has a low resistance when cold. This high inrush current, when first switched ON, is the major cause of most incandescent lamp failures. The air inside the bulb is removed before the bulb is sealed. This prevents oxidation of the filament. The filament burns out quickly if oxygen is present. A gas mixture of nitrogen and argon is placed inside most incandescent bulbs to increase the life of the lamp.

Incandescent Lamp Bases

The base of an incandescent lamp holds the lamp firmly in the socket and connects electricity from the outside circuit to the filament. The ends of the filament are brought out to the base. The base simplifies the replacement of an incandescent lamp. Most incandescent lamp bases are threaded with a knuckle thread. A *knuckle thread* is a rounded thread normally rolled from sheet metal and used in various forms for electric bulbs and bottle caps. The standard incandescent lamp base has right-hand threads that allow removal of the lamp by turning counterclockwise (CCW) and installation by turning clockwise (CW).

The base of a lamp is used to hold the lamp in place and to connect the bulb to power. The size of the bulb and its current rating normally determine the type of base used. The two basic base configurations are bayonet and screw bases.

Small lamps are fitted with bayonet bases. A *bayonet base* is a bulb base that has two pins located on opposite sides. The pins slide into slots located in the bulb's socket. The advantage of a bayonet base is that it firmly holds the bulb. Bayonet base bulbs are used in applications that have high vibration, such as sewing machines and automobile lighting.

Screw bases are used in low-vibration applications. The medium screw and mogul screw are the most common bases used for general lighting service lamps. The medium screw base is normally used for bulbs up to 300 W. The mogul screw base is normally used for bulbs over 300 W. The miniature and candelabra bases are used for low-wattage bulbs, particularly the decorative types. **See Figure 20-20.**

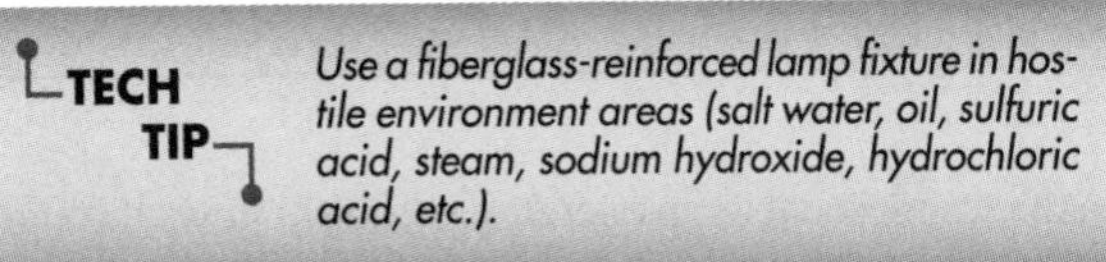
TECH TIP

Use a fiberglass-reinforced lamp fixture in hostile environment areas (salt water, oil, sulfuric acid, steam, sodium hydroxide, hydrochloric acid, etc.).

Incandescent Lamp Bulb Shapes

Bulb sizes and shapes are determined by the end use of the lamp. For any given wattage rating, the larger the bulb, the greater the cost. Large bulbs produce more light over time because the large bulb produces less blacking of the glass. This is because there is a large area over which the vaporized tungsten burning off the filament can dissipate. Bulb shape and size is determined by economics, esthetics, and performance.

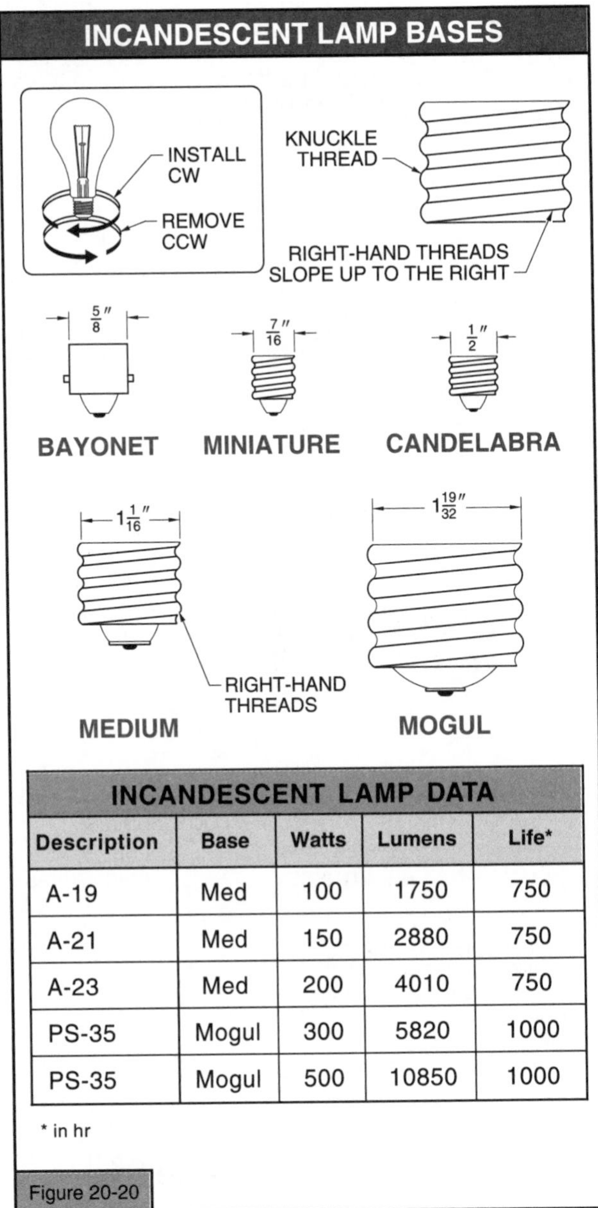

INCANDESCENT LAMP DATA

Description	Base	Watts	Lumens	Life*
A-19	Med	100	1750	750
A-21	Med	150	2880	750
A-23	Med	200	4010	750
PS-35	Mogul	300	5820	1000
PS-35	Mogul	500	10850	1000

* in hr

Figure 20-20

Incandescent lamp bulbs are available in a variety of shapes and sizes. **See Figure 20-21.** Bulb shapes are designated by letters. The most common bulb shape is the "A" bulb. The "A" bulb is used for most residential and commercial indoor lighting applications. The "A" bulb is normally available in sizes from 15 W to 200 W.

Two common variations of the "A" bulb are the "PS" and the "P" bulbs. The "PS" bulb has a longer neck than the "A" bulb. The "PS" bulb is available in sizes from 150 W to 2000 W and is used for commercial and industrial lighting applications. The "P" bulb has a shorter neck than the "A" bulb and is used in indoor lighting applications that require a smaller bulb than the "A" bulb.

Other bulbs are used for decorative or special-purpose applications. The "C" bulb is designed to withstand moderate vibration and is used in sewing, washing, and other machines. The "B," "F," and "G" bulbs are used for decorative applications and are normally of low wattage. They are used for lighting applications in which the bulb is clearly seen, such as chandeliers, lamps, and mirrors in washrooms.

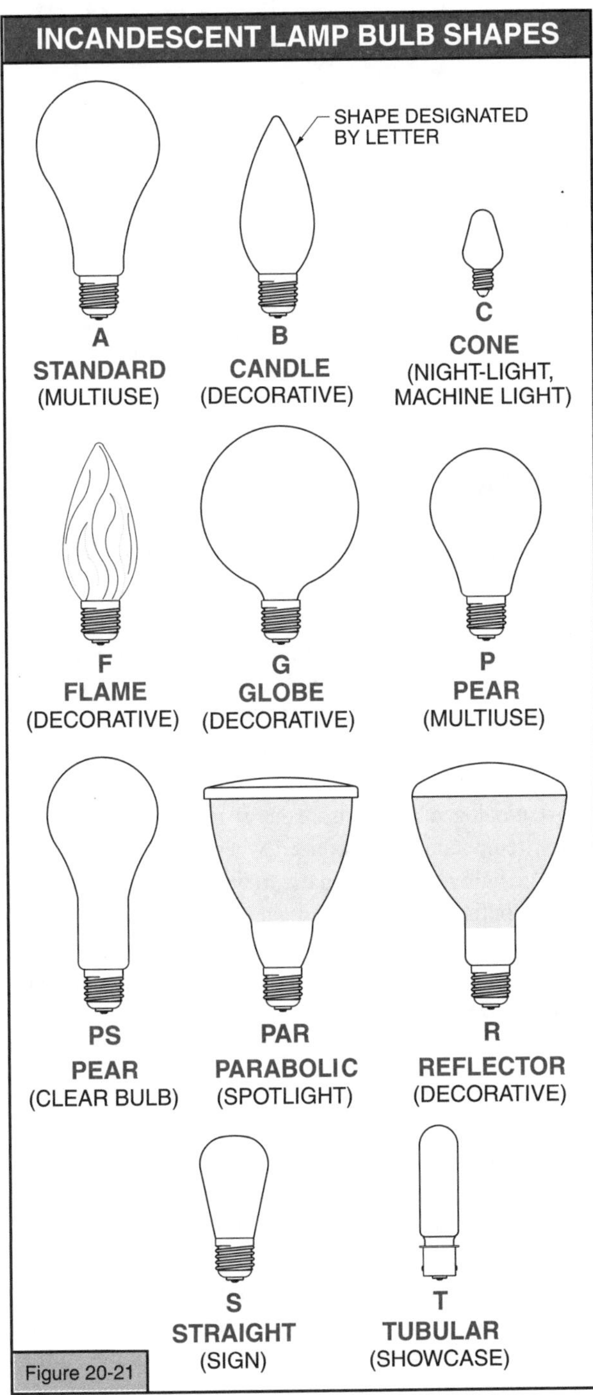

Figure 20-21

The "PAR" bulb is normally used as an outdoor spotlight. Floodlights normally use a reflector "R" bulb (indoors), or "PAR" bulb (outdoors). The "PAR" bulb is preferred outdoors because its shape allows for a more watertight seal. The "R" bulb is also used as a decorative light. Its high wattage is useful for applications such as showroom displays and advertising boards. A *spotlight* is a lamp that casts intensive light in a localized area. A *floodlight* is a lamp that casts general light over a large area. **See Figure 20-22.**

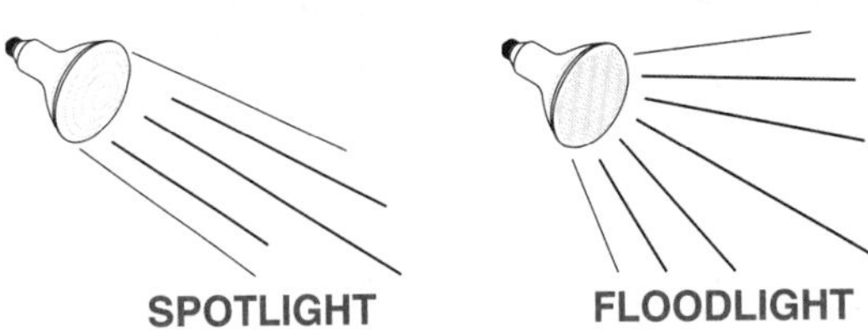

FLOODLIGHT BEAM DESCRIPTIONS

Beam Spread*	Type
10 – 18	1
18 – 29	2
29 – 46	3
46 – 70	4
70 – 100	5
100 – 130	6
130 and up	7

* in degrees

Figure 20-22

The "S" bulb is a straight bulb that is normally used in sign applications. The "T" bulb is a tubular decorative light that is used on display cabinets. "T" bulbs containing bayonet bases are used in high vibration applications such as automobile lighting.

Incandescent Lamp Sizes and Ratings

The size of an incandescent lamp is determined by the diameter of the lamp's bulb. The diameter of a bulb is expressed in eighths of an inch (1⁄8″). For example, an A-21 bulb is 21-eighths (21⁄8″) or 2⅝″ in diameter at the bulb's maximum dimension. A scale calibrated in eighths is used to measure the size of a lamp. The scale is placed at the largest diameter of the bulb. **See Figure 20-23.**

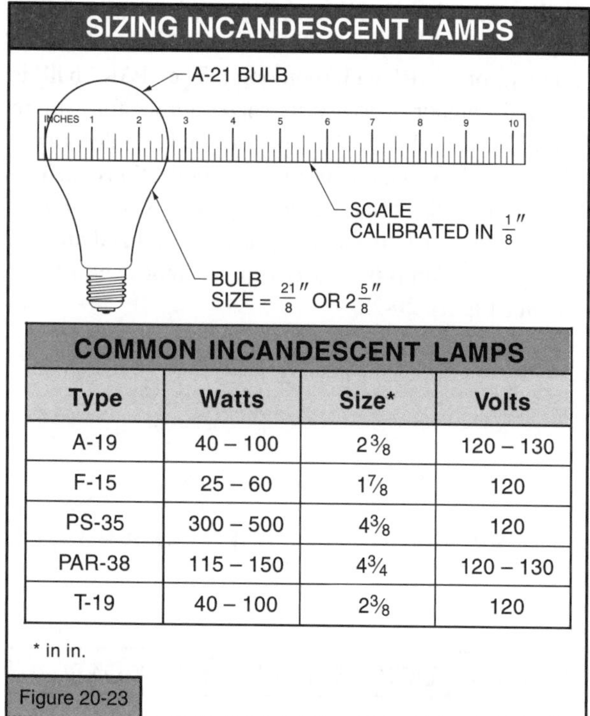

COMMON INCANDESCENT LAMPS

Type	Watts	Size*	Volts
A-19	40 – 100	2⅜	120 – 130
F-15	25 – 60	1⅞	120
PS-35	300 – 500	4⅜	120
PAR-38	115 – 150	4¾	120 – 130
T-19	40 – 100	2⅜	120

* in in.

Figure 20-23

The power rating of incandescent lamp bulbs is given in watts. A *watt (W)* is a unit of measure equal to the power produced by a current of 1 A across a potential difference of 1 V.

To calculate power, apply the formula:

$P = E \times I$

where

P = power (in W)

E = voltage (in V)

I = current (in A)

Example: Calculating Power

How many watts are produced by a 120 V bulb pulling .5 A?

$P = E \times I$

$P = 120 \times .5$

$P = \mathbf{60\ W}$

The higher the wattage rating, the larger the amount (lumens) of light produced. However, the amount of light produced by an incandescent lamp is not directly proportional to the wattage rating. The higher the wattage rating, the more lumens per watt output delivered by the bulb. **See Figure 20-24.**

Tungsten-Halogen Lamps

A *tungsten-halogen lamp* is an incandescent lamp filled with a halogen gas (iodine or bromine). The gas combines with tungsten evaporated from the filament as the lamp burns. Tungsten-halogen lamps are used for display lighting, outdoor lighting, and in photocopy machines because they produce a large amount of light instantly. A tungsten-halogen lamp lasts about twice as long as a standard incandescent lamp.

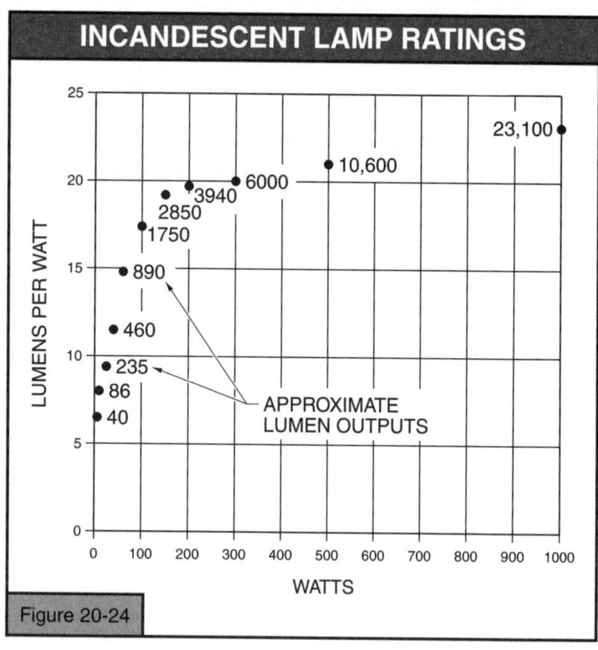

Figure 20-24

Tungsten-halogen lamp wattages range from 15 W to 1500 W. However, because they are not as widely used as the standard incandescent lamp, they are not available in as wide a selection of bulb sizes and wattages. For example, incandescent lamps are available in standard wattages of 2, 3, 6, 7, 10, 13, 15, 18, 20, 25, 30, 33, 40, 45, 50, etc. Tungsten-halogen lamps are available in standard wattages of 15, 25, 50, etc.

Care must be taken when replacing a burned-out tungsten-halogen lamp. Ensure that all power is OFF because a high temperature is produced when the lamp is turned ON. The bulb is made of quartz to withstand the high temperature of the lamp. High-wattage tungsten-halogen lamps are available in "T" or "PAR" bulbs. The replacement cost is about three times the cost of an incandescent lamp. **See Figure 20-25.**

Fluorescent Lamps

A *fluorescent lamp* is a low-pressure discharge lamp in which ionization of mercury vapor transforms ultraviolet energy generated by the discharge into light. The bulb contains a mixture of inert gas (normally argon and mercury vapor) which is bombarded by electrons from the cathode. This provides ultraviolet light. The ultraviolet light causes the fluorescent material on the inner surface of the bulb to emit visible light.

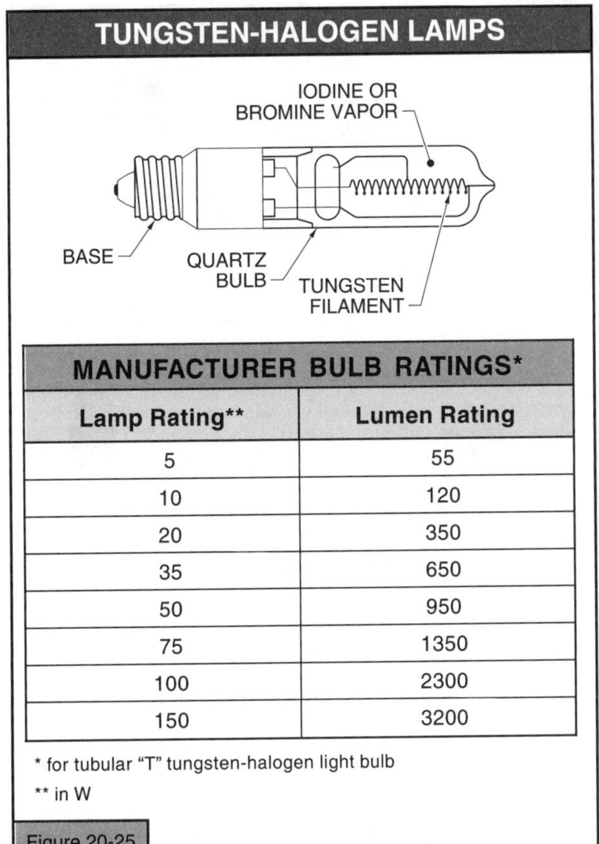

MANUFACTURER BULB RATINGS*	
Lamp Rating**	**Lumen Rating**
5	55
10	120
20	350
35	650
50	950
75	1350
100	2300
150	3200

* for tubular "T" tungsten-halogen light bulb

** in W

Figure 20-25

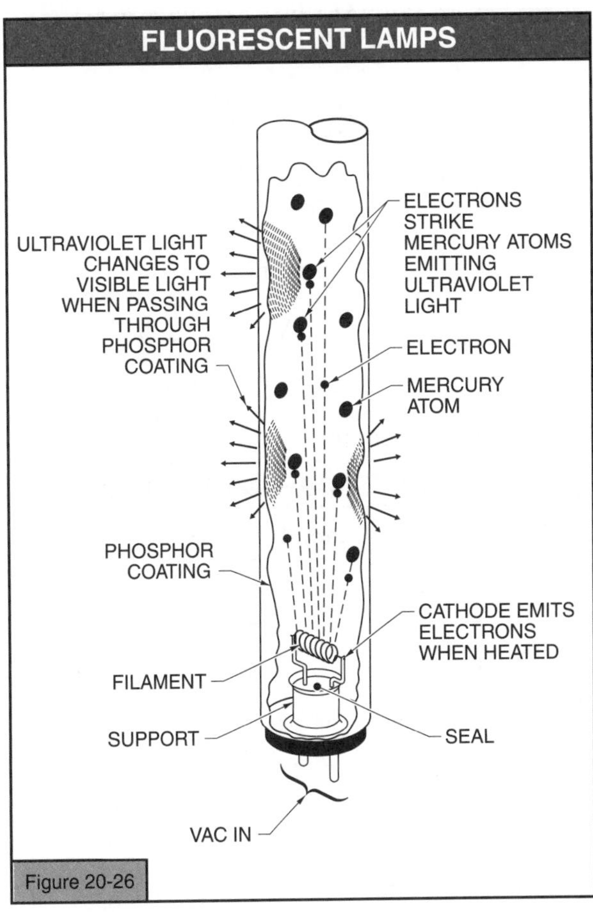

Figure 20-26

To improve energy efficiency and color rendering, some bulbs are double coated. In the double-coated bulb, a layer of rare earth phosphors (red, blue, and green) are added to a layer of conventional phosphors. This combination makes the bulb more expensive, but produces more vivid color rendering. Manufacturers offer colored bulbs that may be used for a desired effect.

A fluorescent lamp consists of a cylindrical glass tube (bulb) sealed at both ends. Each end (base) includes a cathode that supplies the current to start and maintain the arc. A *cathode* is a tungsten coil coated with electron-emissive material which releases electrons when heated.

The cathode in each end is constructed to allow conduction in either direction. Conduction in either direction allows the lamp to be operated on alternating current. The cathode is normally made of coiled tungsten wire and coated with a material that emits electrons. The cathode emits electrons as long as power is applied. **See Figure 20-26.**

Standard fluorescent lamps vary in diameter from ⅝″ to 2⅛″ and in length from 6″ to 96″. The overall diameter of the lamp is designated by a number that indicates eighths of an inch. The most common size is the 40 W lamp, which is designated T-12 and has a length of 48″. The T-12 is 1½″ in diameter ($12 \times \frac{1}{8}'' = 1\frac{1}{2}''$). **See Figure 20-27.**

Fluorescent lamp holders are designed for various types of lamp bases. The base is used to support the lamp and provide the required electrical connections. Fluorescent lamps that use preheat or rapid-start cathodes require four electrical contacts (pins). The four connections are made by using a bi-pin base at either end. Three standard types of bi-pin bases are miniature bi-pin, medium bi-pin, and mogul bi-pin. Instant-start fluorescent lamps require a single pin at each end of the lamp. **See Figure 20-28.**

A *compact fluorescent lamp* is a fluorescent lamp that has a smaller diameter than a conventional fluorescent lamp and a folded bulb configuration. Compact fluorescent lamps provide high light output in much smaller sizes than conventional fluorescent lamps.

Compact fluorescent lamps are available with no ballast or with an integrated ballast. Compact fluorescent lamps containing no ballasts have standard fluorescent lamp bases and are connected to standard fluorescent lamp circuits. Compact fluorescent lamps with an integrated ballast are available with standard incandescent lamp bases and are connected to standard incandescent lamp circuits.

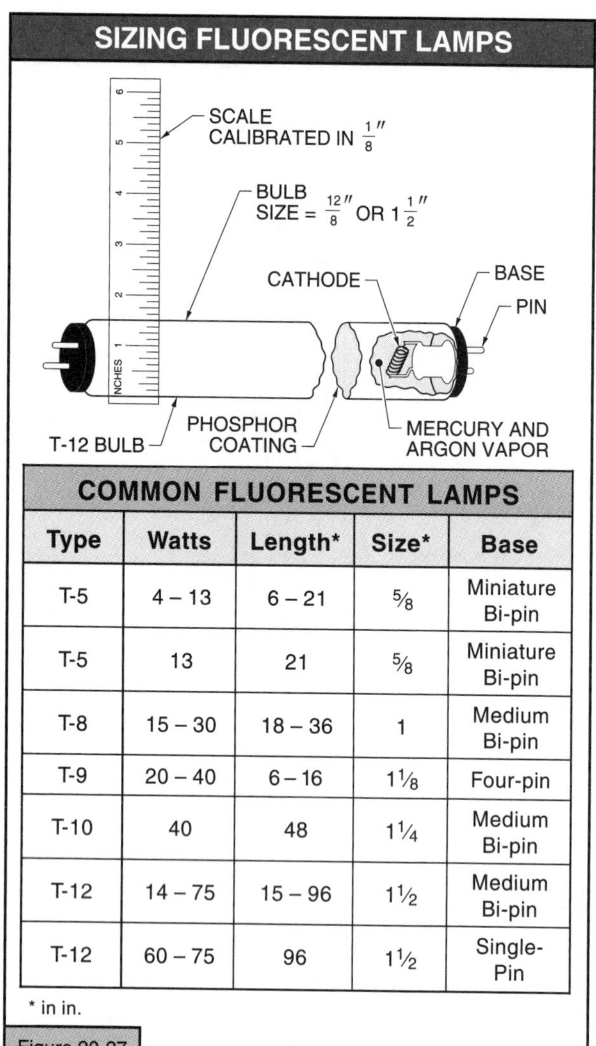

COMMON FLUORESCENT LAMPS

Type	Watts	Length*	Size*	Base
T-5	4 – 13	6 – 21	5/8	Miniature Bi-pin
T-5	13	21	5/8	Miniature Bi-pin
T-8	15 – 30	18 – 36	1	Medium Bi-pin
T-9	20 – 40	6 – 16	1 1/8	Four-pin
T-10	40	48	1 1/4	Medium Bi-pin
T-12	14 – 75	15 – 96	1 1/2	Medium Bi-pin
T-12	60 – 75	96	1 1/2	Single-Pin

* in in.

Figure 20-27

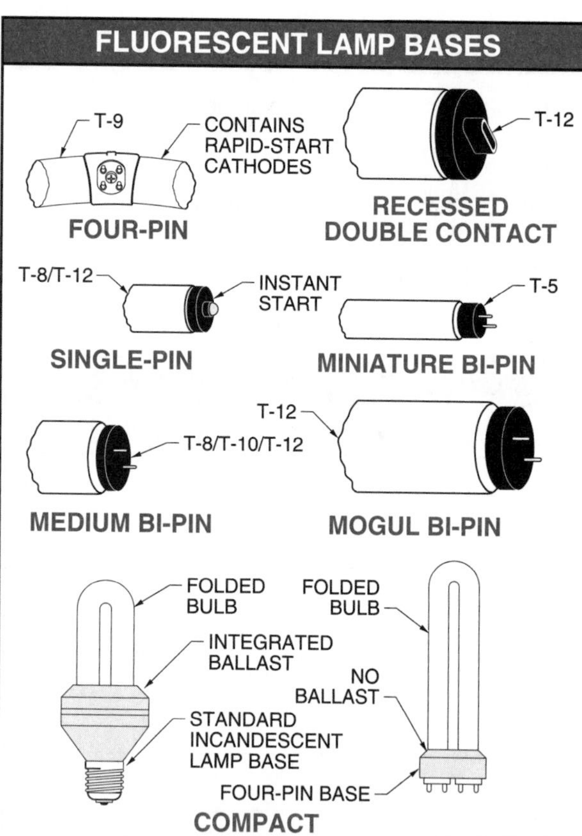

FLUORESCENT LAMP DATA

Description	Base	Watts	Lumens	Life*
T-12 (Preheat)	Med bi-pin	20	1250	9000
T-12/HO (High output)	RDC	35	1650	9000
T8 (Octron)	Med bi-pin	32	2850	20,000
T-12 (Rapid start)	Med bi-pin	34	2650	20,000
T-12 (Rapid start)	Med bi-pin	40	3150	20,000
T-12 (Slimline)	Single pin	40	3000	9000
BX (Long twin tube)	Single end	39	3150	20,000
T-12/HO (High output)	RDC	60	4300	12,000
T-12/VHO (Very high output)	RDC	110	6400	10,000
T-12 (Slimline)	Single pin	75	6150	12,000

* in hr

Figure 20-28

Compact fluorescent lamps are available in a wide variety of sizes and output ratings. Compact fluorescent lamps provide increased energy savings compared to incandescent lamps of comparable light output. Compact fluorescent lamps have an average rated life of 10,000 hr. For this reason, they require less replacement than incandescent lamps. Compact fluorescent lamps are more expensive than incandescent lamps, but, due to their long life, they can provide the same lumens as incandescent lamps at nearly one-fourth the cost.

Fluorescent bulbs do not start well in cold climate conditions because the electrons released by the heated cathode are thermally released. The colder the bulb, the longer the time required to heat the cathode and release electrons into the mercury vapor. Even after the bulb is turned ON, it delivers less light in cold climate conditions as low temperatures affect the mercury-vapor pressure inside the bulb. The colder the bulb, the less the light output.

Standard indoor fluorescent lamps are designed to deliver a peak light output under approximately 75°F ambient temperature conditions. **See Figure 20-29.**

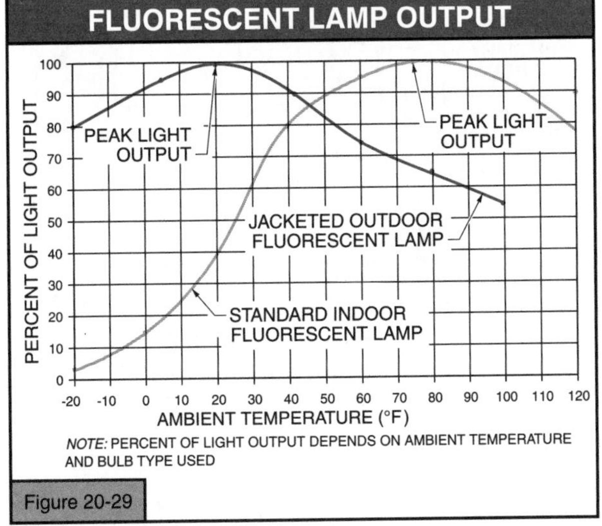

Figure 20-29

The 75°F ambient temperature is considered to be room temperature for most fluorescent lamp installations. However, if some temperature other than the standard room temperature is used to rate the lamp, the manufacturer normally lists a luminaire temperature. *Luminaire temperature* is the temperature at which a lamp delivers its peak light output.

Moderate changes in ambient temperature (50°F to 105°F) have little effect (less than 10%) on light output. Temperatures lower than 50°F or higher than 105°F have a greater effect on light output.

To help hold the heat, outdoor rated fluorescent lamps include an outer glass jacket. The jacket helps hold in heat which shifts the peak output to a lower ambient temperature point. Jacketed lamps are available with different peak output points. Jacketed lamps are recommended for many outdoor applications and certain indoor applications such as freezer warehouses, subways, and tunnels where cold and windy conditions exist.

Fluorescent lamp bulbs that produce different shades of colors are available. The shade of color produced by a fluorescent bulb helps produce a certain environment. Fluorescent bulbs are available for cool, moderate, and warm environments.

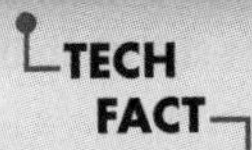

The laminated steel cores of magnetic ballasts for fluorescent and HID lamps produce a hum. Use solid-state electronic ballasts in quiet locations such as libraries, hospitals, etc.

Cool Environments

Cool white or light white fluorescent bulbs produce a pale blue-green whiteness. They are used to enhance blues and greens. This bulb is suggested for high lighting areas, such as industrial and commercial work spaces. It is also used in marketing displays to make people and merchandise look more attractive and natural.

Moderate Environments

White fluorescent bulbs produce a very pale yellow-whiteness. White fluorescent bulbs may also be labeled as natural or daylight. They are used to produce a light environment midway between the blueness of cool white and the yellowness of warm white. White fluorescent bulbs are used in areas where a more neutral effect is desired. For example, white fluorescent bulbs are used in supermarkets, libraries, classrooms, and where fluorescent lighting is mixed with incandescent lighting.

Warm Environments

Warm, white fluorescent bulbs produce a more yellow-whiteness. They are used to give a warm feeling and enhance brighter colors. This bulb type is suggested for low lighting levels. For example, restaurants may use warm, white fluorescent bulbs to help set a particular mood they wish to create.

Preheat Circuits

A *preheat circuit* is a fluorescent lamp-starting circuit that heats the cathode before an arc is created. Preheat circuits are used in some low-wattage lamps (4 W to 20 W) and are common in desk lamps.

A preheat circuit uses a separate starting switch that is connected in series with the ballast. The electrodes in a preheat circuit require a few seconds to reach the proper temperature to start the lamp after the starting switch is pressed. **See Figure 20-30.**

The opening of the starting switch breaks the path of current flow through the starting switch. This leaves the gas in the lamp as the only path for current to travel. A high-voltage surge is produced by the collapsing magnetic field as the starting switch is no longer connected to power. This high-voltage surge starts the flow of current through the gas in the lamp. Once started, the ballast limits the flow of current through the lamp.

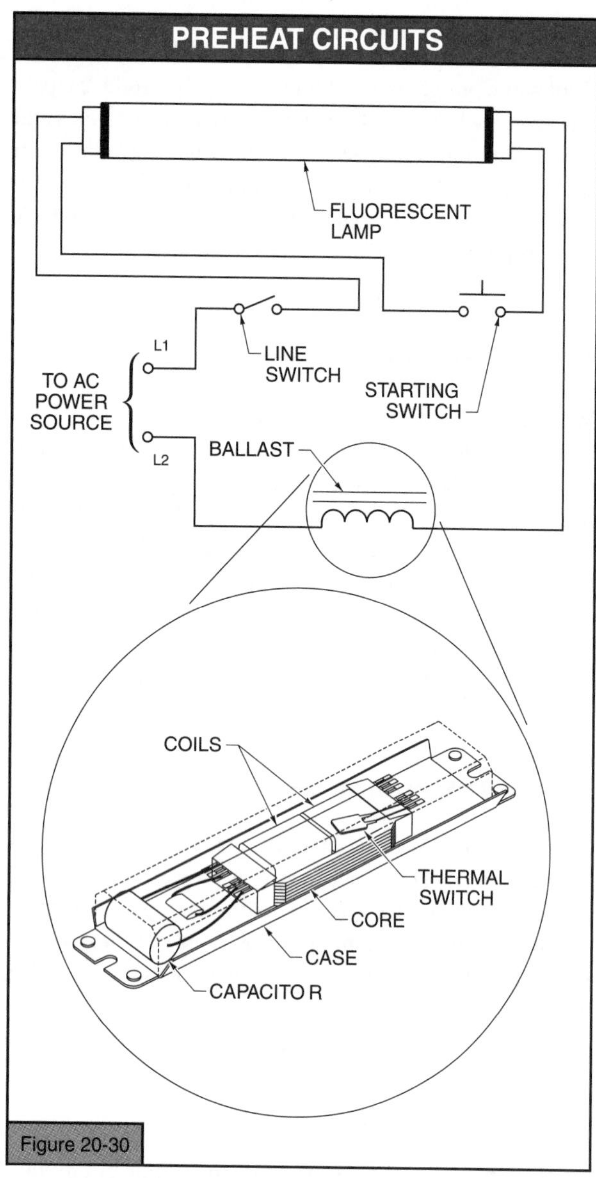

Figure 20-30

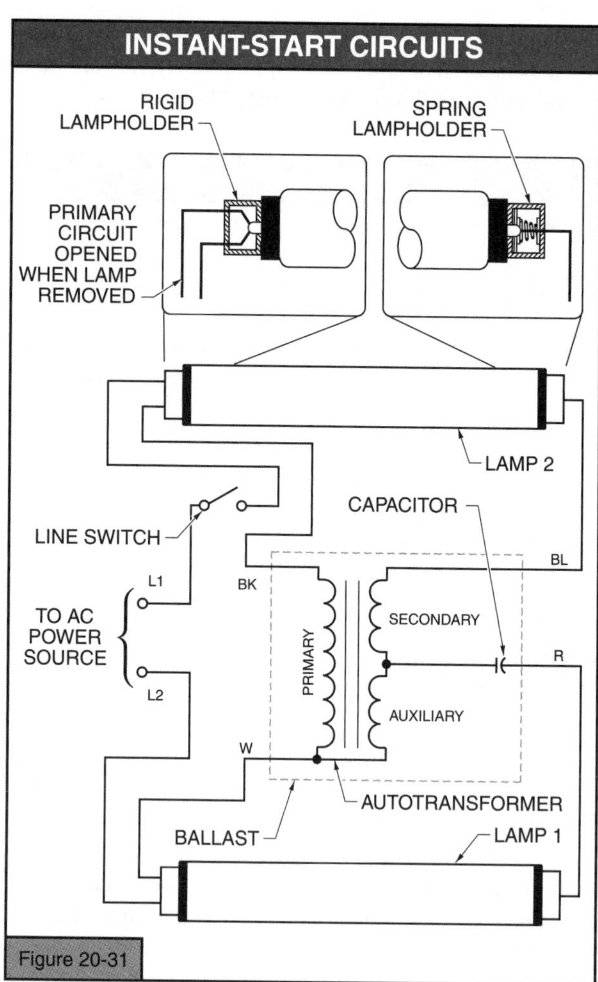

Figure 20-31

Instant-Start Circuits

An *instant-start circuit* is a fluorescent lamp-starting circuit that provides sufficient voltage to strike an arc instantly. The instant-start circuit was developed to eliminate the starting switch and overcome the starting delay of preheat circuits. An arc strikes without preheating the cathodes when a high enough voltage is applied across a fluorescent lamp. **See Figure 20-31.**

The high initial voltage requires a large autotransformer as an integral part of the ballast. The autotransformer delivers an instant voltage of 270 V to 600 V, depending on the bulb size and voltage rating. The larger the bulb, the higher the voltage. Instant-start lamps require only one pin at each base because no preheating of the electrode is required.

A safety circuit is used with an instant-start lamp to prevent electrical shock due to the high starting voltage. The base pin acts as a switch, interrupting the circuit to the ballast when the lamp is removed.

The lamp is replaced by pushing the lamp into the spring lampholder at the high-voltage end of the fixture and then inserting it into the rigid lampholder at the low-voltage end of the fixture. Both lamp ends must be in place before current can flow through the ballast winding. The circuit does not operate when the lamp is not in place.

Rapid-Start Circuits

A *rapid-start circuit* is a fluorescent lamp-starting circuit that has separate windings to provide continuous heating voltage on the lamp cathodes. The lamp starting time is reduced because the cathodes are continuously heated. The rapid-start circuit brings the lamp to full brightness in about two seconds. A rapid-start circuit is the most common circuit used in fluorescent lighting.

A rapid-start circuit uses lamps that have short, low-voltage electrodes which are automatically preheated by the lamp ballast. The rapid-start ballast preheats the cathodes by means of a heater winding. The heater winding continues to provide current to the lamp after ignition. **See Figure 20-32.**

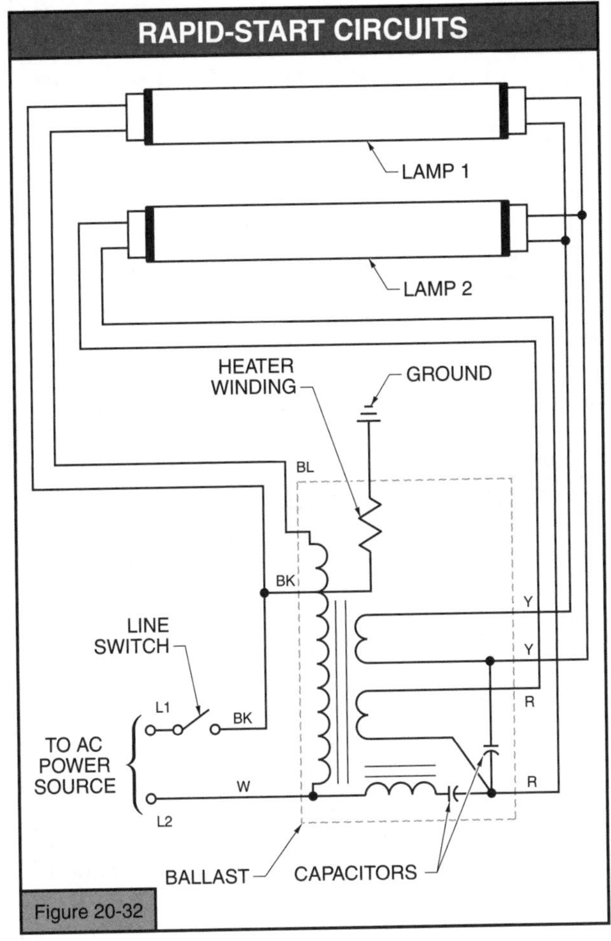

Figure 20-32

High-Intensity Discharge Lamps

A *high-intensity discharge (HID) lamp* is a lamp that produces light from an arc tube. **See Figure 20-33.** An *arc tube* is the light-producing element of an HID lamp. The arc tube contains metallic and gaseous vapors and electrodes. An arc is produced in the tube between the electrodes. The arc tube is enclosed in a bulb which may contain phosphor or a diffusing coating that improves color rendering, increases light output, and reduces surface brightness.

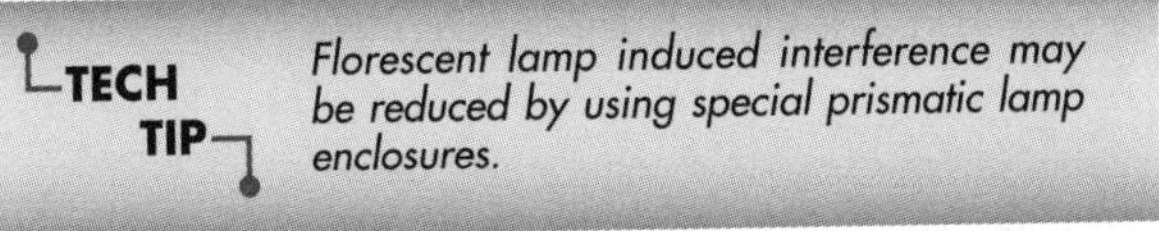

Florescent lamp induced interference may be reduced by using special prismatic lamp enclosures.

HID lamps include low-pressure sodium, mercury-vapor, metal-halide, and high-pressure sodium lamps. All HID lamps are electric discharge lamps. An *electric discharge lamp* is a lamp that produces light by an arc discharged between two electrodes. High vapor pressure is used to convert a large percentage of the energy produced into visible light. Arc tube pressure for most HID lamps is normally from one to eight atmospheres. HID lamps provide an efficient, long-lasting source of light and are used for street, parking lot, and general lighting applications.

HID Lamp Color Rendering

Color rendering is the appearance of a color when illuminated by a light source. For example, a red color may be rendered light, dark, pinkish, or yellowish depending on the light source under which it is viewed. Color rendering of HID lamps varies depending on the lamp used.

Low-pressure sodium lamps produce yellow to yellow-orange light. The color of the lamp distorts the true color of objects viewed under the lamp. The yellow light is produced by the sodium in the low-pressure sodium lamp. Low-pressure sodium lamps are normally not used where the appearance of people and colors are important. The yellow light produces severe color distortion on most light-colored objects.

Mercury-vapor lamps produce a light yellow to white light. Mercury-vapor lamps with clear bulbs have poor rendering of reds. Blue colors appear purplish with most other colors appearing normal. Phosphor-coated bulbs improve color rendering. Blue colors viewed under phosphor-coated bulbs still have a slight purplish hue and yellow colors take on a greenish overtone.

Metal-halide lamps produce a light yellow to white light. Metal-halide lamps produce good overall color rendering. Red colors appear slightly muted with some pinkish overtones.

High-pressure sodium lamps produce a golden white light. High-pressure sodium lamps have good color rendering with all colors being clearly distinguishable. However, reds, greens, blues, and violets are muted.

HID Lamp Selection

Consider using low-pressure sodium lamps first for an application because they are the most efficient. Low-pressure sodium lamps are good for outdoor lighting installations. The yellow-orange color is acceptable for street, highway, parking lot, and floodlight applications.

HID LAMPS

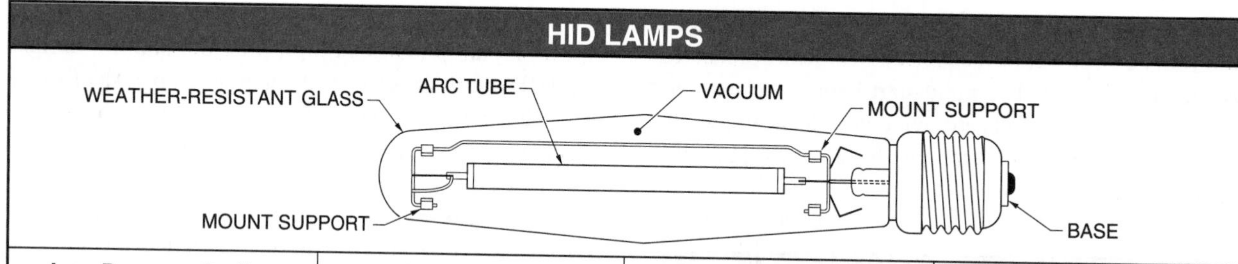

Low-Pressure Sodium	Mercury-Vapor	Metal-Halide	High-Pressure Sodium
• Uses sodium vapor under low pressure to produce light • 6 min to 12 min start time • 4 sec to 12 sec restart time • 190 lm/W to 200 lm/W • Produces yellow to yellow-orange light • 1800 hr rated bulb life*	• Uses mercury vapor to produce light • 5 min to 6 min start time • 3 min to 5 min restart time • 50 lm/W to 60 lm/W • Produces white light with blue colors appearing purplish and yellow colors having a greenish overtone • 16,000 hr to 24,000 hr rated bulb life*	• Uses mercury vapor with metal halides to produce light • 2 min to 5 min start time • 10 min to 15 min restart time • 80 lm/W to 125 lm/W • Produces white light with red colors slightly muted • 3000 hr to 20,000 hr rated bulb life*	• Uses sodium vapor under high pressure to produce light • 3 min to 4 min start time • 30 sec to 60 sec restart time • 65 lm/W to 115 lm/W • Produces golden white light • 7500 hr to 14,000 hr rated bulb life*

* rated bulb life depends on usage, bulb size, and cycle time

HID LAMP DATA

Lamp	Description	Base	Watts	Lumens	Life*
Low-pressure sodium	T-16 clear	Bayonet	18	1570	16,000
	T-16 clear	Bayonet	35	4000	16,000
	T-16 clear	Bayonet	55	6655	16,000
	T-21 clear	Bayonet	90	11,095	16,000
	T-21 clear	Bayonet	135	19,140	16,000
Mercury-vapor	E-23½ white	Mogul	100	4200	24,000
	E-28 white	Mogul	175	8600	24,000
	E-28 white	Mogul	250	12,100	24,000
	E-37 or BT-37 white	Mogul	400	22,500	24,000
	BT-56 white	Mogul	1000	63,000	24,000
High-pressure sodium	B-17 clear	Medium	35	2250	24,000
	E-23½ clear	Mogul	50	4000	24,000
	E-23½ clear	Mogul	70	6400	24,000
	B-17 clear	Medium	100	9500	24,000
	B-17 clear	Medium	150	16,000	24,000
	E-23½ clear	Mogul	150	16,000	24,000
	E-18 clear	Mogul	250	30,000	24,000
	E-25 clear	Mogul	1000	140,000	24,000
Metal-halide	ED-17 coated	Medium	70	4800	10,000
	ED-17 coated	Medium	150	12,200	10,000
	E-28/BT-28 coated	Mogul	175	14,000	10,000
	E-28/BT-28 coated	Mogul	250	20,500	10,000
	E-37/BT-37 coated	Mogul	400	34,000	20,000
	BT-56 coated	Mogul	1000	110,000	12,000

* in hr

Figure 20-33

The long start-up time is not a problem for outdoor lighting because the lights are normally turned ON by a photoelectric cell at dusk. Photoelectric cell switches may be adjusted to turn ON at different light levels. Low-pressure sodium lamps are also used for some indoor applications such as warehouse lighting and other areas where color distortion is not critical to an operation.

Consider using metal-halide lamps when the color rendering of low-pressure sodium lamps is not acceptable for the application. Metal-halide lamps are used for sport, street, highway, parking lot, and floodlight applications. Metal-halide lamps are normally specified for most sport, indoor, and outdoor lighting applications.

Consider using mercury-vapor lamps when the initial installation cost is of major importance. Mercury-vapor lamps are used in all types of applications. Use metal-halide lamps when true color appearance is important such as in a car dealer's lot.

Consider using high-pressure sodium lamps when lower operating cost is important and some color distortion is acceptable. High-pressure sodium lamps may be used in parking lots, street lighting, shopping centers, exterior buildings, and most storage areas.

Low-Pressure Sodium Lamps

A *low-pressure sodium lamp* is an HID lamp that operates at a low vapor pressure and uses sodium as the vapor. A low-pressure sodium lamp has a U-shaped arc tube. The arc tube has both electrodes located at the same end. The arc tube is placed inside a glass bulb and contains a mixture of neon, argon, and sodium-metal. On start-up, an arc is discharged through the neon, argon, and sodium-metal. As the sodium-metal heats and vaporizes, the amber color of sodium is produced. **See Figure 20-34.**

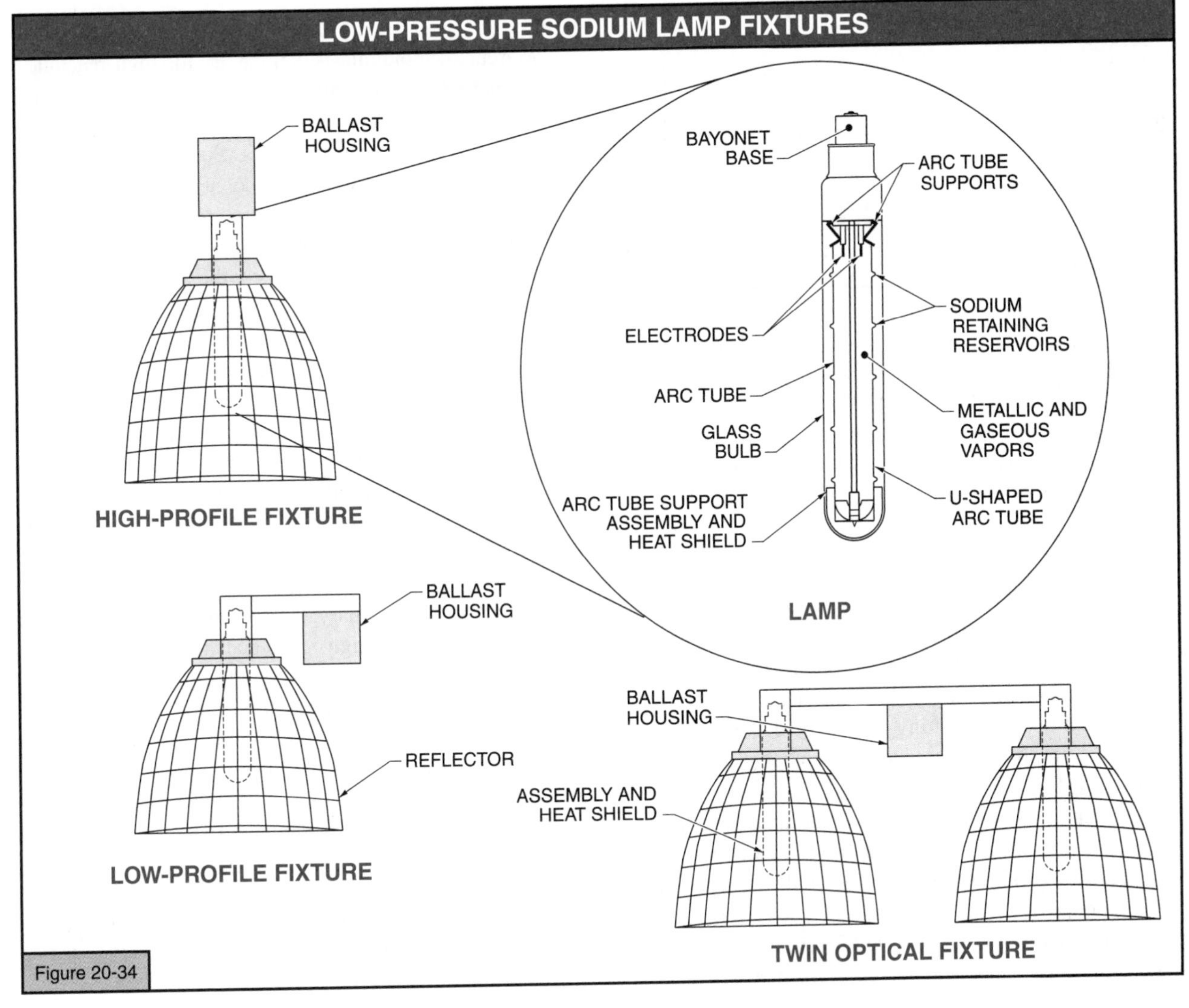

Figure 20-34

A low-pressure sodium lamp is named for its use of sodium inside the arc tube. This lamp has the highest efficiency rating of any lamp. Some low-pressure sodium lamps deliver up to 200 lm/W of power. This is 10 times the output of an incandescent lamp.

Low-pressure sodium lamps must be operated on a ballast that is designed to meet the lamp's starting and running requirements. Low-pressure sodium lamps do not have a starting electrode or ignitor. The ballast must provide an open-circuit voltage of approximately three to seven times the lamp's rated voltage to start and sustain the arc. **See Figure 20-35.**

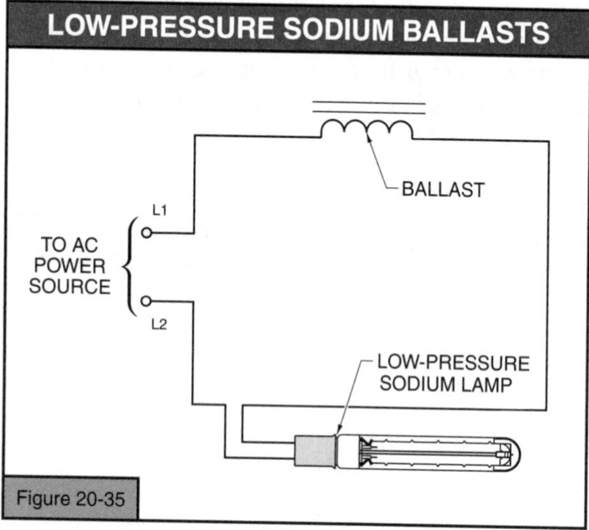

Figure 20-35

Low-pressure sodium lamp fixtures are available in a wide variety of lamp, ballast, and reflector configurations. Lamp fixtures include high-profile, low-profile, and twin optical fixtures. The variety of fixture configurations provides a wide range of light distribution patterns. The pattern chosen is based on the light requirements of the application.

Mercury-Vapor Lamps

A *mercury-vapor lamp* is an HID lamp that produces light by an electrical discharge through mercury vapor. Mercury-vapor lamps are used for general lighting applications. Phosphor coating is added to the inside of the bulb to improve color-rendering characteristics. **See Figure 20-36.**

A mercury-vapor lamp contains a starting electrode and two main electrodes. An electrical field is set up between the starting electrode and one main electrode when power is first applied to the lamp. The electrical field causes current to flow and an arc to strike. Current flows between the two main electrodes as the heat vaporizes the mercury.

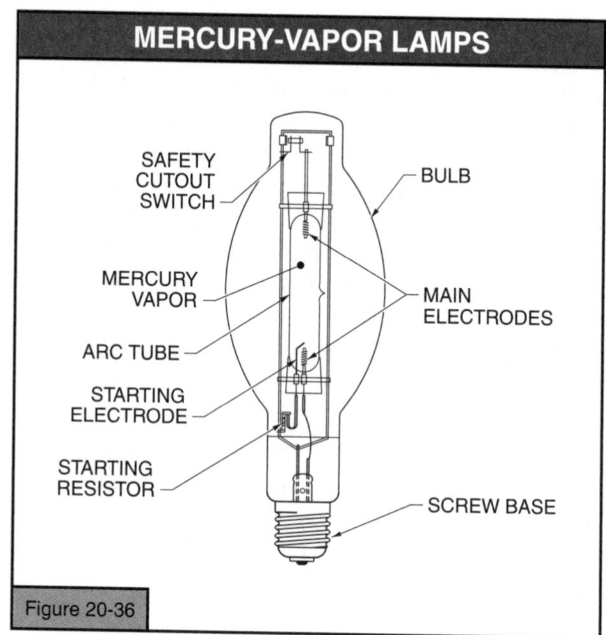

Figure 20-36

Mercury-vapor ballasts include reactor, high-reactance autotransformer, constant-wattage autotransformer, and two-winding, constant-wattage ballasts. The ballast used normally depends on economics. **See Figure 20-37.** A *reactor ballast* is a ballast that connects a coil (reactor) in series with the power line leading to the lamp. Reactor ballasts are used when the incoming supply voltage meets the starting voltage requirements of the lamp. This is common when the incoming supply voltage is 240 V or 277 V. Both 240 V and 277 V mercury-vapor lamps are standard. A capacitor is added to some reactor ballasts to improve the power factor.

Reactor ballasts cost the least of all mercury-vapor ballasts but have poor lamp-wattage regulation. They should only be used when line voltage regulation is good because a 5% change in line voltage produces a 10% change in lamp wattage in a reactor ballast.

A *high-reactance autotransformer ballast* is a ballast that uses two coils (primary and secondary) to regulate both voltage and current. High-reactance autotransformer ballasts are used when the incoming supply voltage does not meet the starting requirements of the lamp. Incoming voltages of 115 V, 208 V, and 460 V require a voltage change to the lamp. The voltage is regulated to within ±5% of the bulb's rating because a small percentage change in voltage results in a large percentage change in light output.

A *constant-wattage autotransformer ballast* is a high-reactance autotransformer ballast with a capacitor added to the circuit. The capacitor improves the power factor. A constant-wattage autotransformer ballast is the most commonly used ballast.

MERCURY-VAPOR BALLASTS

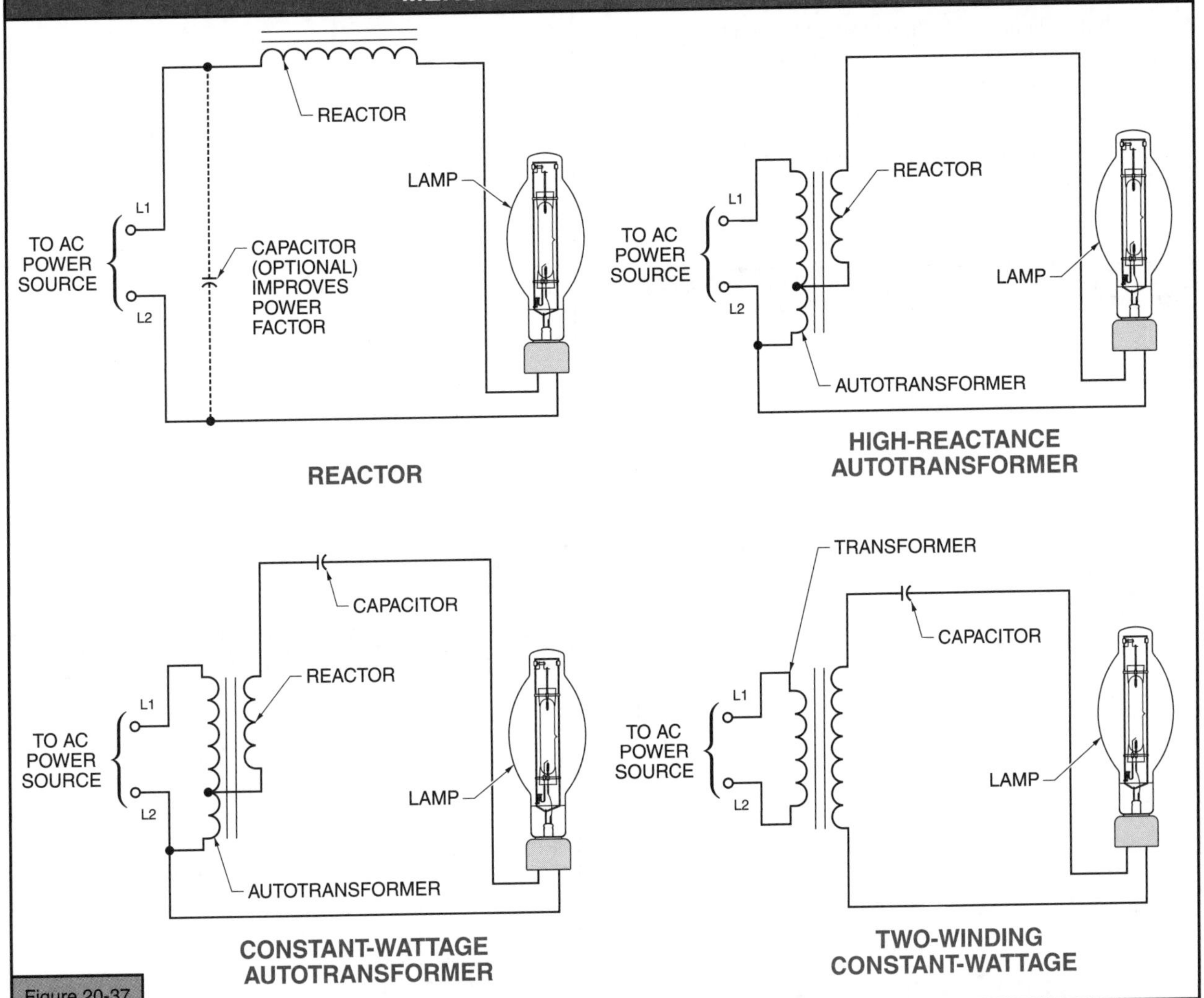

REACTOR

HIGH-REACTANCE AUTOTRANSFORMER

CONSTANT-WATTAGE AUTOTRANSFORMER

TWO-WINDING CONSTANT-WATTAGE

Figure 20-37

A *two-winding constant-wattage ballast* is a ballast that uses a transformer which provides isolation between the primary and secondary circuits. A two-winding constant-wattage ballast has excellent lamp-wattage regulation. A 13% change in line voltage produces only a 2% to 3% change in lamp wattage.

Metal-Halide Lamps

A *metal-halide lamp* is an HID lamp that produces light by an electrical discharge through mercury vapor and metal halide in the arc tube. A *metal halide* is an element (normally sodium and scandium iodide) which is added to the mercury in small amounts. A metal-halide lamp produces more lumens per watt than a mercury-vapor lamp. **See Figure 20-38.**

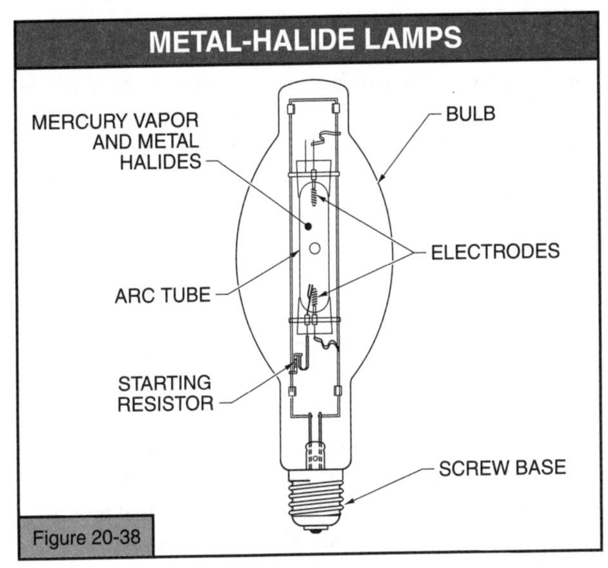

Figure 20-38

The light produced by a metal-halide lamp does not produce as much color distortion as a mercury-vapor lamp. A metal-halide lamp is an efficient source of white light. It has a shorter bulb life than the other HID lamps.

A metal-halide ballast uses the same basic circuit as the constant-wattage autotransformer mercury-vapor ballast. The ballast is modified to provide high starting voltage required by metal-halide lamps. **See Figure 20-39.**

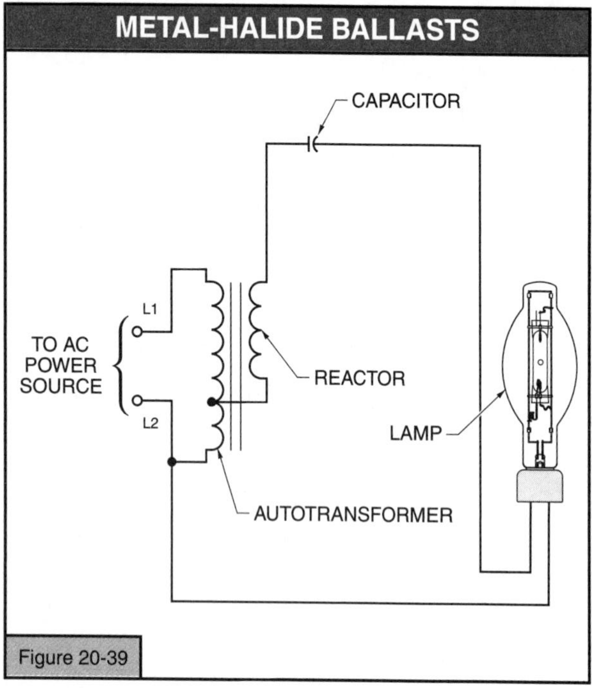

Figure 20-39

High-Pressure Sodium Lamps

A *high-pressure sodium lamp* is an HID lamp that produces light when current flows through sodium vapor under high pressure and high temperature. A high-pressure sodium lamp is a more efficient lamp than a mercury-vapor or metal-halide lamp. The light produced from a high-pressure sodium lamp appears as a golden-white color. **See Figure 20-40.**

A high-pressure sodium lamp is constructed with a bulb and an arc tube. The arc tube is made of ceramic to withstand high temperature. The bulb is made of weather-resistant glass to prevent heat loss and protect the arc tube.

High-pressure sodium lamps do not have a starting electrode. The ballast must deliver a voltage pulse high enough to start and maintain the arc. This voltage pulse must be delivered every cycle and must be 4000 V to 6000 V for 1000 W lamps and 2500 V to 4000 V for smaller lamps. The starter (ignitor) is the device inside the ballast that produces the high starting voltage. **See Figure 20-41.**

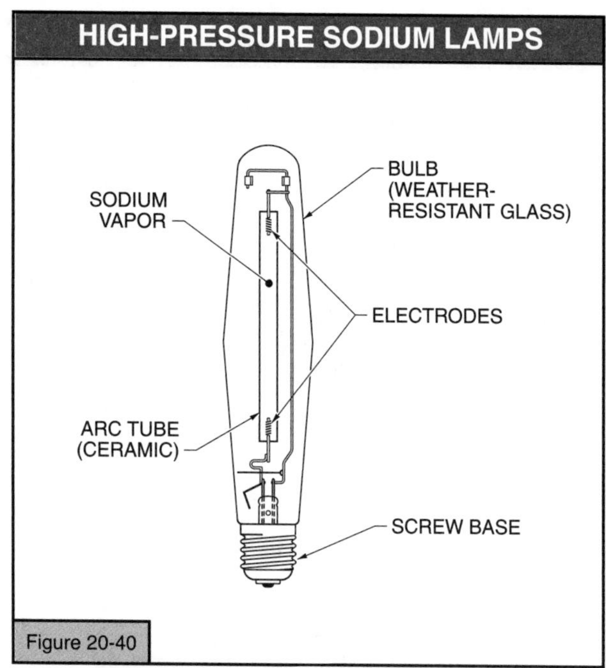

Figure 20-40

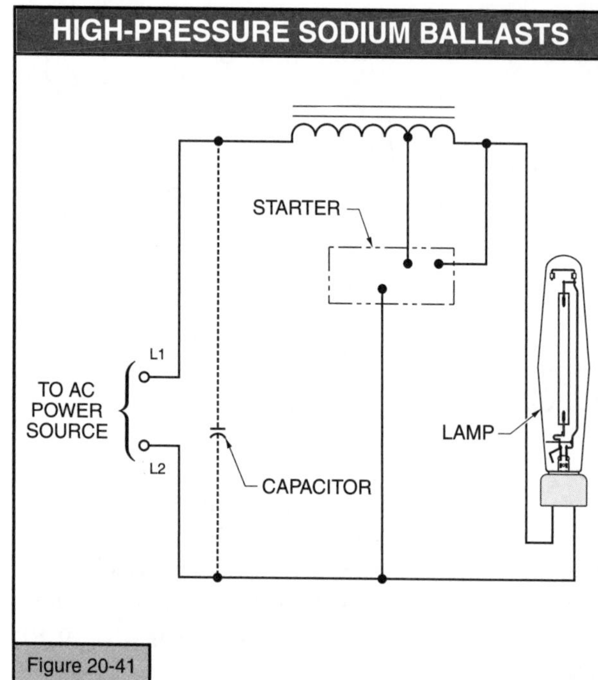

Figure 20-41

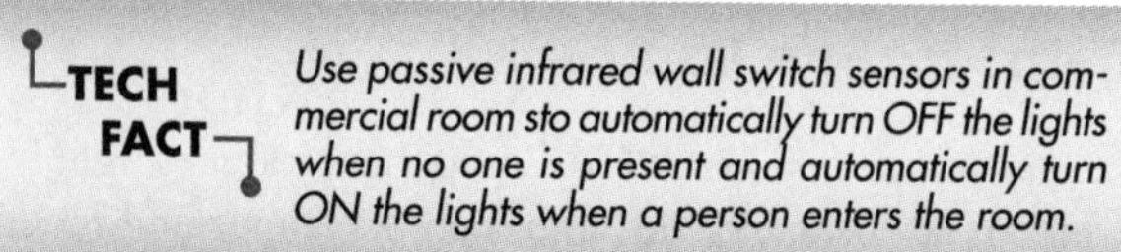

Use passive infrared wall switch sensors in commercial room sto automatically turn OFF the lights when no one is present and automatically turn ON the lights when a person enters the room.

A high-pressure sodium ballast is similar to a mercury-vapor reactor ballast. The main difference is the added starter. The reactor ballast is used where the input voltage meets the lamp's requirement. A transformer or autotransformer is added to the ballast circuit when the incoming voltage does not meet the lamp's requirements.

HID Floodlight Lamp Pole Placement

Most manufacturers recommend that HID lamp poles be spaced not more than 2 times their mounting height from any adjacent pole location. **See Figure 20-42.**

The width of the area to be lit should not exceed 2 times the mounting height of the lamps when lighting an area from one side. The width of the area to be lit should not exceed 4 times the mounting height of the lamps when lighting an area from two sides. The width of the area to be lit should not exceed 2 times the mounting height of the lamps on either side of the pole when lighting an area from the center. Aimed lamps should be directed at a point ⅔ to ¾ the distance across the area to be lit.

Philips Lighting

Most lamp manufacturers recommend that high-intensity discharge (HID) lamp poles be spaced no more than two times their mounting height from any adjacent pole location.

HID LAMP POLE PLACEMENT

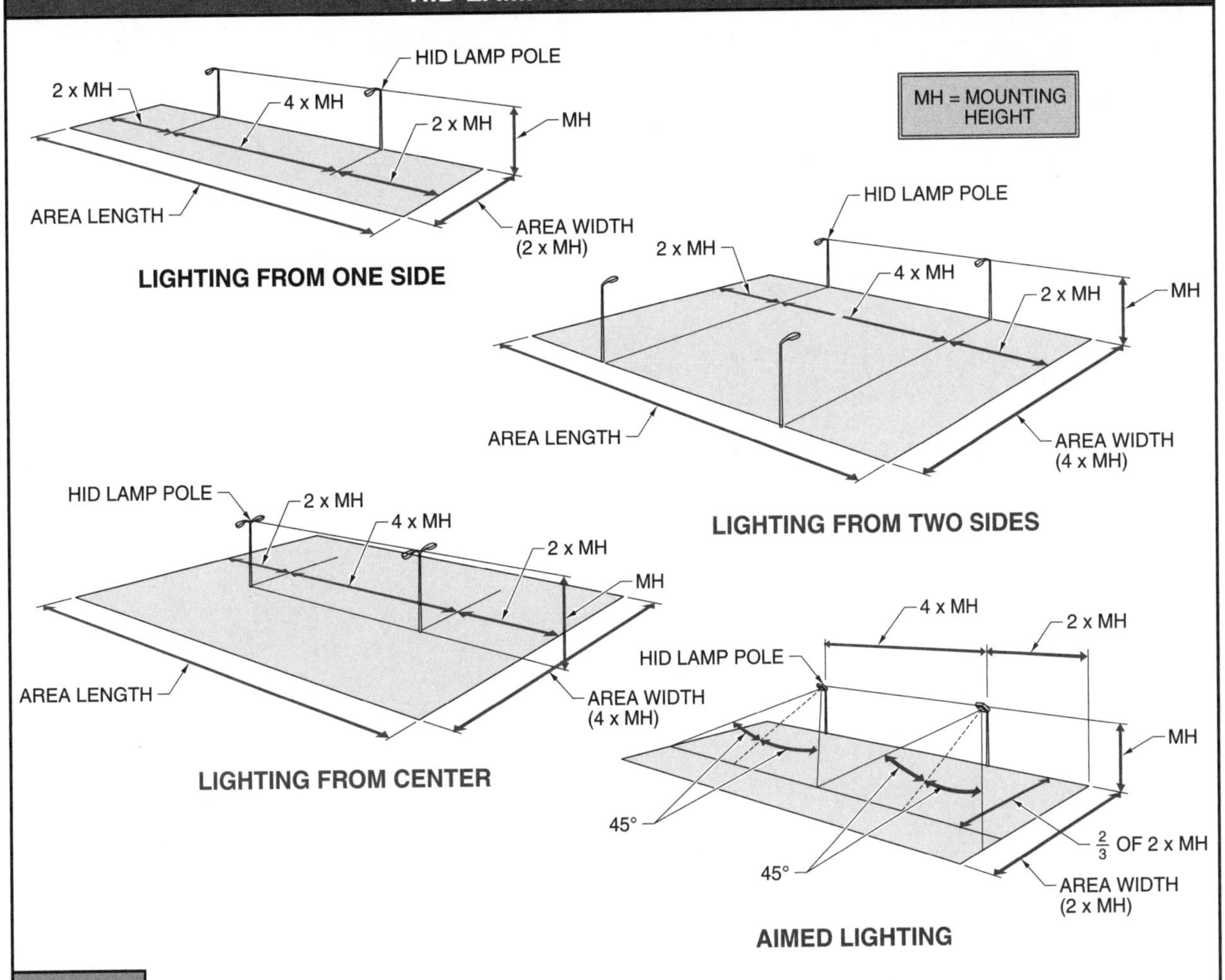

MH = MOUNTING HEIGHT

HID LAMP POLE
2 x MH
4 x MH
2 x MH
MH
AREA LENGTH
AREA WIDTH (2 x MH)

LIGHTING FROM ONE SIDE

HID LAMP POLE
2 x MH
4 x MH
2 x MH
MH
AREA LENGTH
AREA WIDTH (4 x MH)

LIGHTING FROM TWO SIDES

HID LAMP POLE
2 x MH
4 x MH
2 x MH
MH
AREA LENGTH
AREA WIDTH (4 x MH)

LIGHTING FROM CENTER

4 x MH
2 x MH
HID LAMP POLE
MH
45°
45°
$\frac{2}{3}$ OF 2 x MH
AREA WIDTH (2 x MH)

AIMED LIGHTING

Figure 20-42

Refer to the CD-ROM in the back of the book for Chapter 20 Quick Quiz® and related information.

Industrial Circuits

Industrial electrical circuits are circuits that are used to convert raw materials into finished products. Industrial circuits vary in size and complexity depending on the type of industry. Large industrial electrical circuits, such as power generating, mining, food processing, chemical processing, paper and pulp manufacturing, steel and plastics manufacturing, water treatment and purification, consumer product manufacturing, oil and gas drilling and refineries, etc., require very high levels of power. High levels of power are used for lighting, heating, transporting raw materials (oil and gas pipelines, etc.), converting raw materials into finished products (manufacturing and assembling), and numerous other tasks that must be performed in the production of goods. Small industrial electrical circuits, such as small product assembly, electronic circuit board assembly, packaging, etc., require less power but highly complex and specialized electrical circuits.

Most industrial circuits, regardless of their size or type, have similarities to residential, commercial, and other industrial circuits. For example, all industrial circuits include basic residential and commercial electrical circuits to provide branch circuit (general-purpose, 115 VAC, 1ϕ outlets) and 115 VAC lighting circuits. These circuits are the same 15 A and 20 A circuits as those used in residential applications, but are much larger in the total number required and are distributed over numerous panelboards located throughout a plant. **See Figure 21-1.**

Industrial lighting circuits are also the same basic lighting circuits used in commercial lighting. Like commercial lighting, fluorescent lighting is used for office areas and HID lighting is used for outdoor areas. HID lighting is also used for most industrial production and storage area lighting.

Industrial heating, ventilating, and air conditioning (HVAC) systems are basically the same as residential and commercial systems except for the size and number of units used. Residential HVAC systems normally include only one heating and air conditioning unit. Commercial HVAC systems divide the total area into zones, with one large unit normally controlling a given zone. Zones may be individual office areas, individual stores, or individual rooms. Industrial HVAC systems are basically the same as the commercial systems used in office areas. Production areas include heating units, but normally do not include air conditioning units. Industrial storage areas may include little heat and no air conditioning or may include very large refrigeration systems for storing perishable goods.

Although all residential, commercial, and industrial circuits include some motors, motors are normally the largest part of industrial electrical circuits. Motors are included at all points along assembly lines, production lines, material handling systems, etc. The motors range in size from fractional HP to 500 HP (or more). Motors must be controlled for them to deliver the required work. For this reason, motor control is a major part of any industrial circuit.

TECH FACT *If a ladder is required when working on or around electric wiring and circuits, a wood or fiberglass ladder must be used. A metal ladder can conduct electricity and become part of an electrical circuit.*

ELECTRICAL CIRCUIT COMPARISON			
Circuits	**Residential**	**Commercial**	**Industrial**
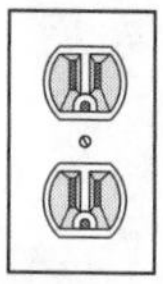 **General-Purpose Receptacles**	Part of branch circuit	Total circuits divided over as many panels as required	Large part of office areas and total plant system. Circuits divided over many areas
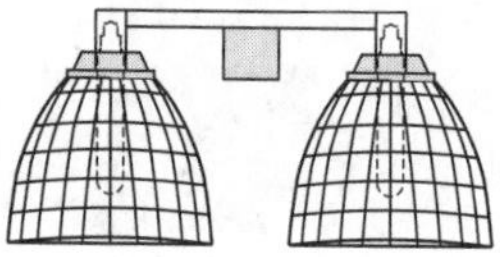 **Lighting**	Generally incandescent	Generally fluorescent indoors and HID outdoors	Generally fluorescent in offices and HID in indoor production areas and all outdoor areas
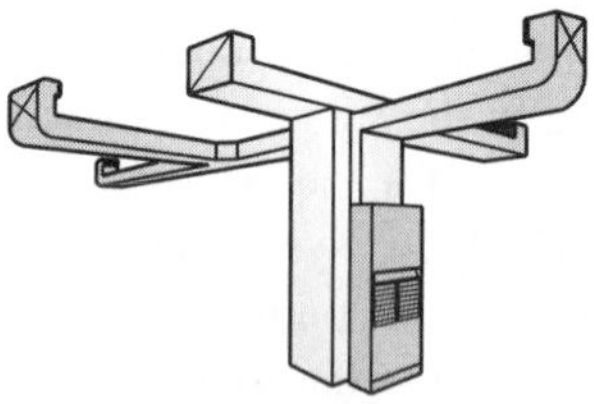 **HVAC**	Normally one heating and cooling unit	Large units used to control a zone (shop, store, office area). Number of units increases as building size increases	Office areas same as commercial application. Production areas heated but may not be air conditioned. Non-air conditioned areas normally include large fans for cooling and ventilation
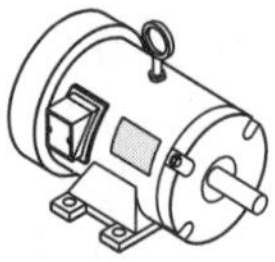 **Motors**	HVAC and sump pump	HVAC, pumps, air compressors, etc.	Normally largest part of total system with motors ranging to 500 HP or more
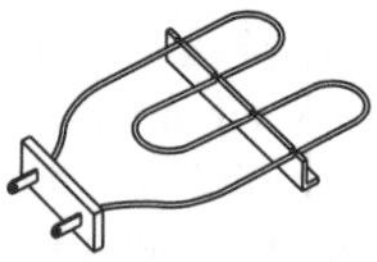 **Heating Elements**	Used only in residences containing electric heat	Normally used only if high heat is required for general heating or for special application	Often major part of industrial process

Figure 21-1

Electric heat is often used in applications that require heat for the production of goods. Electric heat is used to separate minerals, melt metals, refine crude oil, bring products up to the required temperature before processing, keep pipes from freezing, sterilize areas, shrink wrap materials, etc. Electric heat is commonly used because it is easy to control the temperature output of electric heating elements.

Level of Control

The level of circuit control required varies for different applications. For example, the level of control required for controlling a motor may vary from simple manual ON/OFF control to automatically controlling and displaying hundreds of circuit parameters. **See Figure 21-2.**

A manual motor starter is used in a basic motor control circuit. The manual motor starter starts, stops, and provides overload protection for the motor. The advantage of this basic motor control circuit is that it is inexpensive and simple to operate, install, and maintain.

A variable frequency drive is used to provide advanced motor control functions. A *variable frequency drive (VFD)* is a solid-state programmable control module used to control an AC motor. A VFD may control a motor's speed, acceleration time, deceleration time, and provide overload protection, display operating parameters, and perform many other functions. A VFD is programmed to meet almost any application requirement.

In addition to controlling the motor, VFDs are programmed to detect and display many different system faults. Fault detection normally includes power loss, under and over voltage, overcurrent, short circuit, phase loss, etc.

Most industrial circuits include levels of control from basic manual control to controls that include motor drives and programmable controllers.

Industrial Applications

Most industrial circuits include a system for moving a product along a line and performing some function to the product. For example, a conveyor is used to move the product along an assembly or production line. Conveyors are controlled by motors. Motors are controlled by motor control circuits. The product is modified or changed at different points along the line. For example, a product may be sent through a shrink-wrap process before being sent to a warehouse or other process. **See Figure 21-3.**

In the shrink-wrap application, the product is moved by conveyors to a wrapping machine that applies a shrink-wrap to the product. This process includes numerous controls for controlling the required motors, heating elements, and other loads. The infeed conveyor, shrink-wrap motor, heating elements, and outfeed conveyor motor controls are normally grouped together and connected to a common operator station. **See Figure 21-4.**

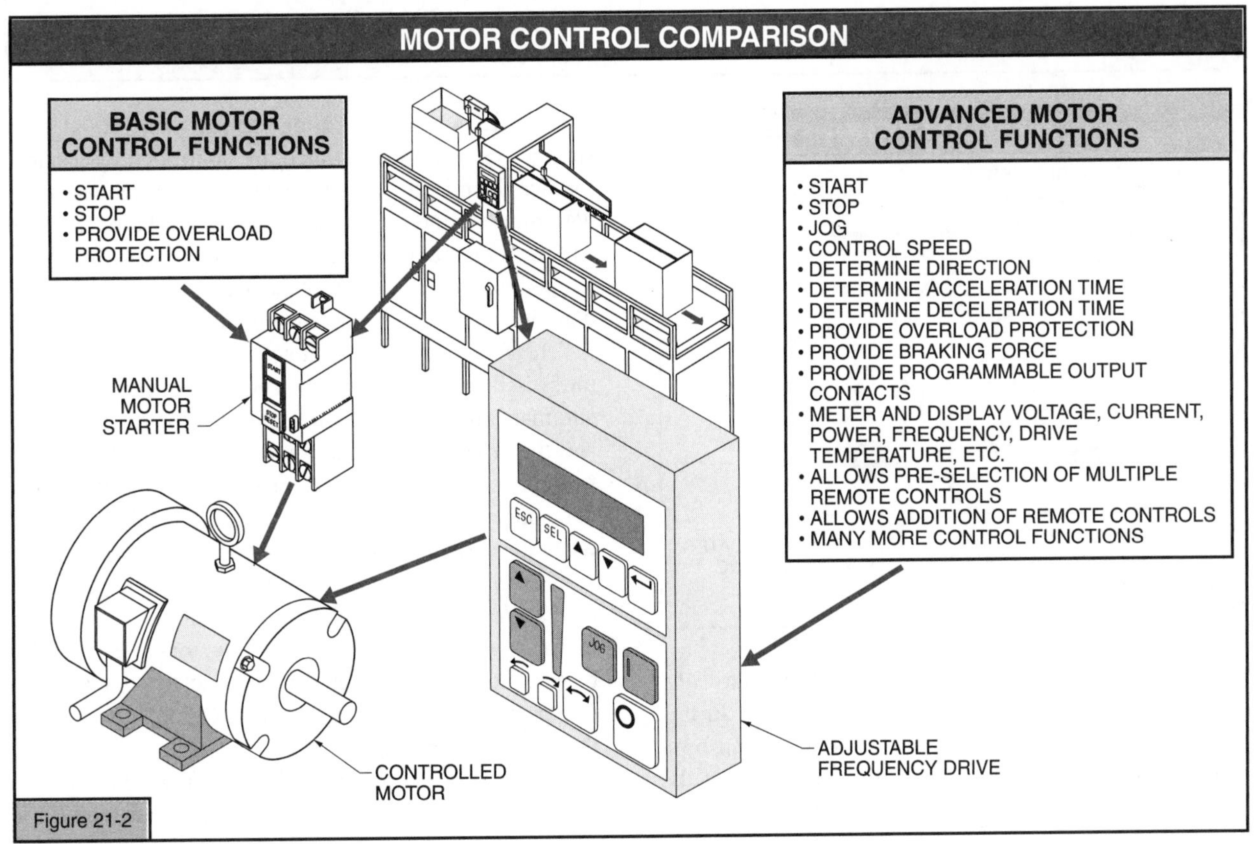

Figure 21-2

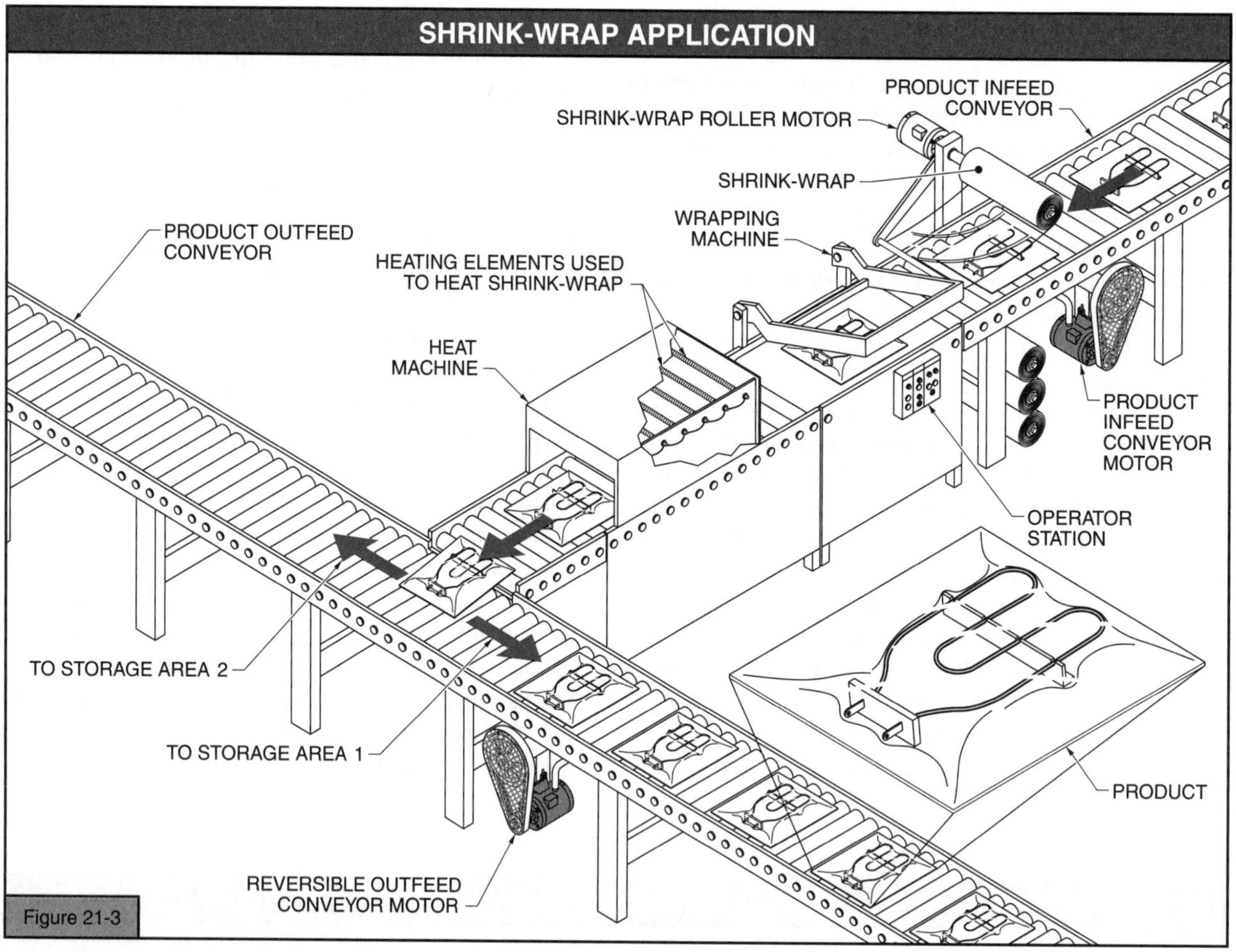

SHRINK-WRAP APPLICATION

Figure 21-3

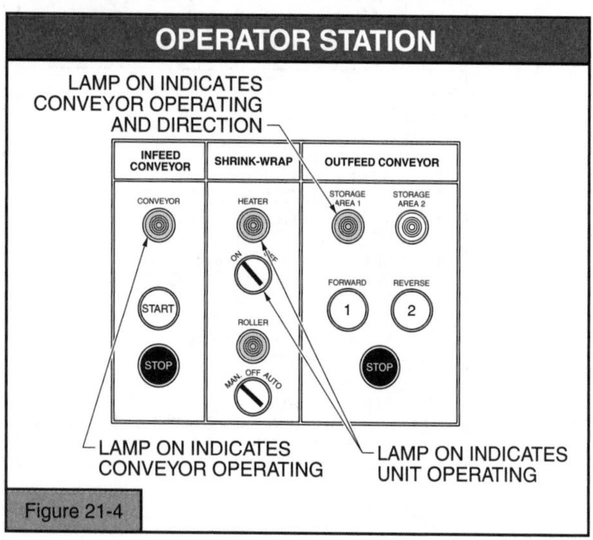

Figure 21-4

The operator station may be designed to control the loads in many different ways (levels of control). In this example, the infeed conveyor motor is controlled by a standard start/stop pushbutton station. Included at the pushbutton station is a lamp that indicates when the motor is running. The heating elements used to heat the shrink-wrap after it is applied are controlled by a selector switch and a temperature switch. The selector switch determines if the heating elements are to be operational and the temperature switch automatically maintains a set temperature. Also included at the operator station is a lamp that indicates when the heating elements are energized.

The shrink-wrap roller motor is controlled by a selector switch and a limit switch. The limit switch is used to automatically control the motor if the product is detected and the selector switch is in the automatic position. The manual position of the selector switch allows the motor to be operated manually. A lamp is included that indicates when the shrink-wrap roller motor is operating.

The outfeed conveyor motor is controlled by a standard forward/reverse/stop pushbutton station. The forward motor direction sends the product to storage area 1. The reverse motor direction sends the product to storage area 2. Lamps are used to indicate which direction (storage area) the motor is operating.

A power circuit and control circuit are required to operate the shrink-wrap circuit. The power circuit includes the 230 V,

3ϕ heating elements, 230 V, 1ϕ shrink-wrap roller motor, 230 V, 3ϕ infeed motor, 230 V, 3ϕ outfeed motor, and primary (high-voltage) side of the control transformer. Also included in the power circuit are the required disconnects, protection devices (fuses and CBs), contactor (heating element), and motor starters (each motor). **See Figure 21-5.**

The control circuit includes the contactor coil, motor starter coils, lamps, selector switches, pushbuttons, temperature switch, and limit switch. Also included in the control circuit is the transformer secondary (low-voltage) side, fuse, and motor starter overload contacts (OLs).

Line (Ladder) Diagrams

All trades and professions have a certain language that enables the rapid and efficient transfer of information and ideas. This language may include symbols, drawings or diagrams, words, phrases, or abbreviations. Line (ladder) diagrams use industrial electrical symbols to provide the information necessary to understand the operation of any electrical control circuit. Industrial electrical symbols illustrate the electrical devices within a circuit. **See Appendix.**

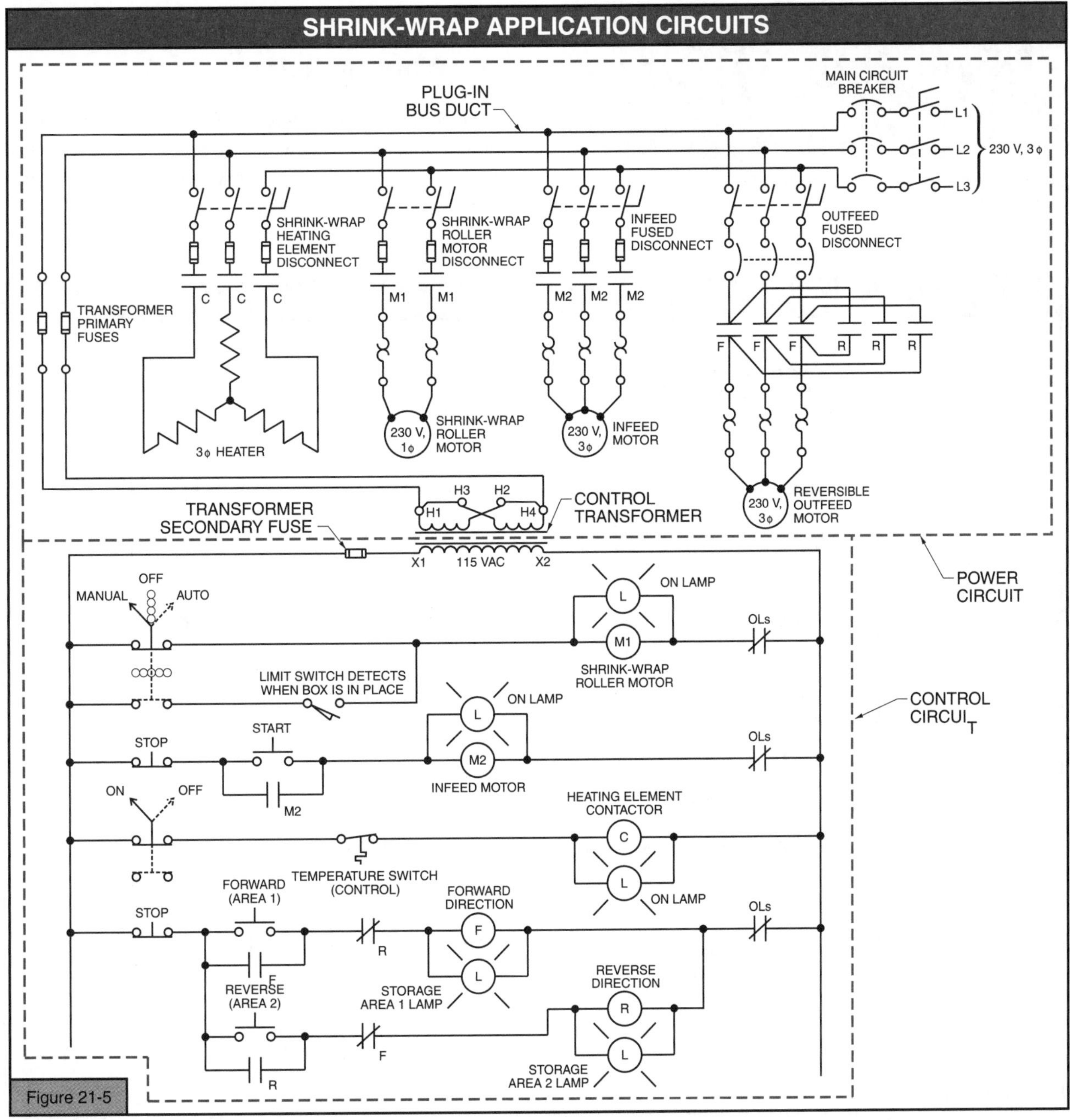

Figure 21-5

Control language is communicating by using line (ladder) diagrams. A *line (ladder) diagram* is a diagram which shows, with single lines and symbols, the logic of an electrical circuit or system of circuits and components. A line diagram indicates the location of electrical devices within a control circuit. **See Figure 21-6.**

Line diagrams are used when connecting and troubleshooting electrical switchgear and panels in industrial facilities.

Circuit design and modifications to existing circuits are possible using line diagrams. A line diagram shows the power source (transformer, etc.) and current flow paths through the various parts of the circuit, such as pushbuttons, contacts, coils, and overloads. A line diagram shows the circuitry necessary for the operation of the components in the power circuit (motors, etc.) and control circuit (indicating lamps, etc.). A line diagram is not intended to show the physical relationship of the various devices in the control cabinet. Line diagrams are much easier to draw than pictorial drawings, power circuits, and wiring diagrams.

Line Diagram Rules

The electrical industry has established a universal set of symbols and rules on how line diagrams (circuits) are laid out. An electrician establishes a working practice common to all electricians by applying these standards.

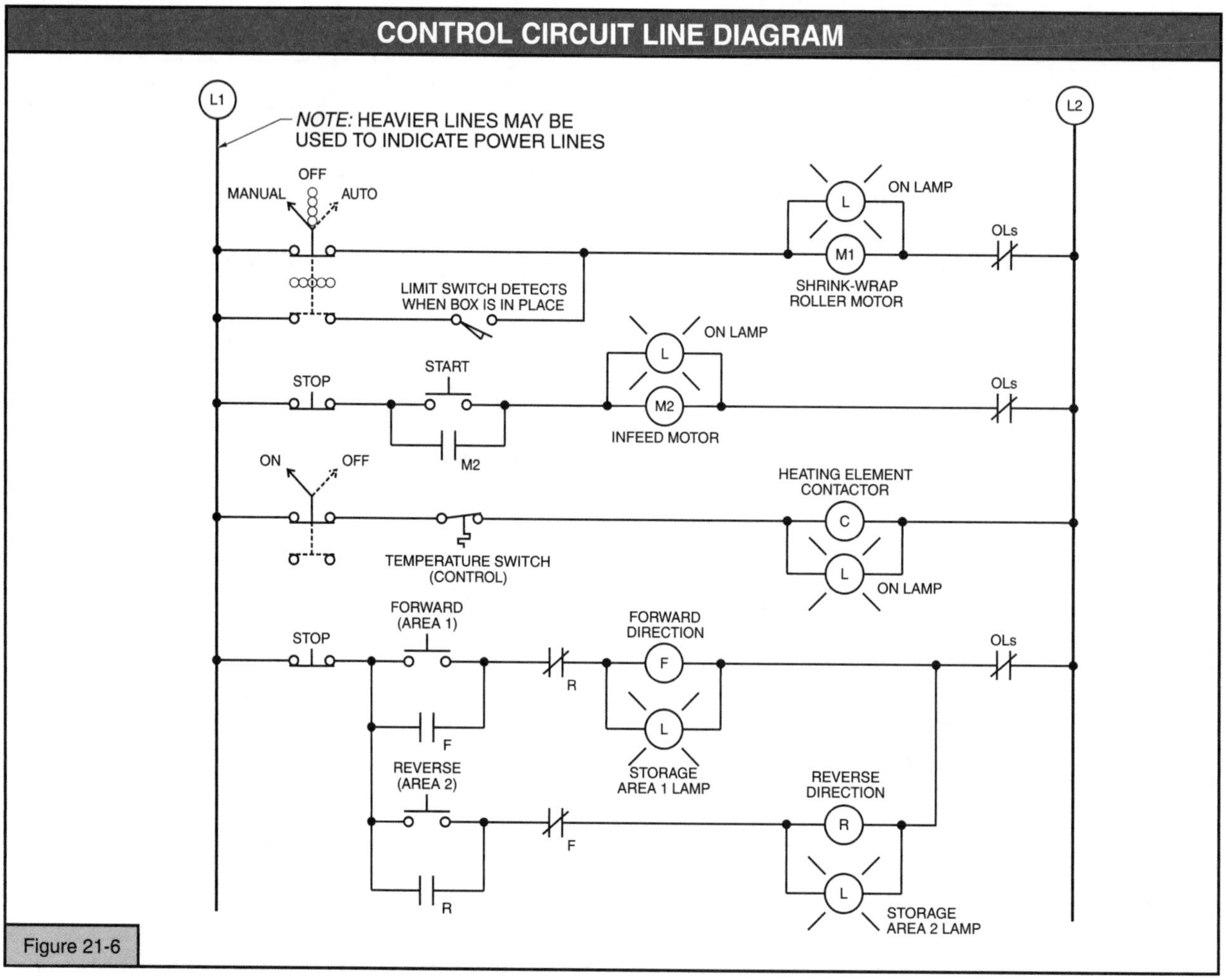

CONTROL CIRCUIT LINE DIAGRAM

L1
L2
NOTE: HEAVIER LINES MAY BE USED TO INDICATE POWER LINES
OFF
MANUAL
AUTO
ON LAMP
L
M1
OLs
LIMIT SWITCH DETECTS WHEN BOX IS IN PLACE
SHRINK-WRAP ROLLER MOTOR
ON LAMP
L
START
STOP
M2
OLs
INFEED MOTOR
M2
ON
OFF
HEATING ELEMENT CONTACTOR
C
L
ON LAMP
TEMPERATURE SWITCH (CONTROL)
FORWARD (AREA 1)
STOP
FORWARD DIRECTION
F
OLs
R
L
F
REVERSE (AREA 2)
STORAGE AREA 1 LAMP
REVERSE DIRECTION
R
F
L
R
STORAGE AREA 2 LAMP

Figure 21-6

Establishing standards in line diagrams also aids when troubleshooting or changing a standard line diagram to a PLC line diagram. Using a line diagram when troubleshooting helps to identify the circuit's potential test points for taking circuit measurements. Standard line diagrams are also used to develop PLC line diagrams. PLC line diagrams are used to develop and enter a program into a PLC. **See Figure 21-7.**

The standard line diagram of three start and three stop pushbuttons is used to show the control of a motor starter coil. This line diagram is used to connect the wiring diagram when hard-wiring the pushbuttons to the motor starter. *Hard-wiring* is the physical connection of electrical components.

In a PLC wiring diagram, the components are not hard-wired together. Instead, all control circuit inputs are connected to the input section of the PLC and all control circuit outputs are connected to the output section of the PLC. Inputs include all control circuit devices that are used to switch (start or stop) the flow of current. Inputs include all switches (pushbuttons, temperature switches, etc.) and overload contacts. Outputs include all control circuit devices that use the circuit's current to change the electrical power into some other form of energy such as light, linear motion, etc. Outputs include lamps, contactor coils, motor starter coils, solenoids, etc.

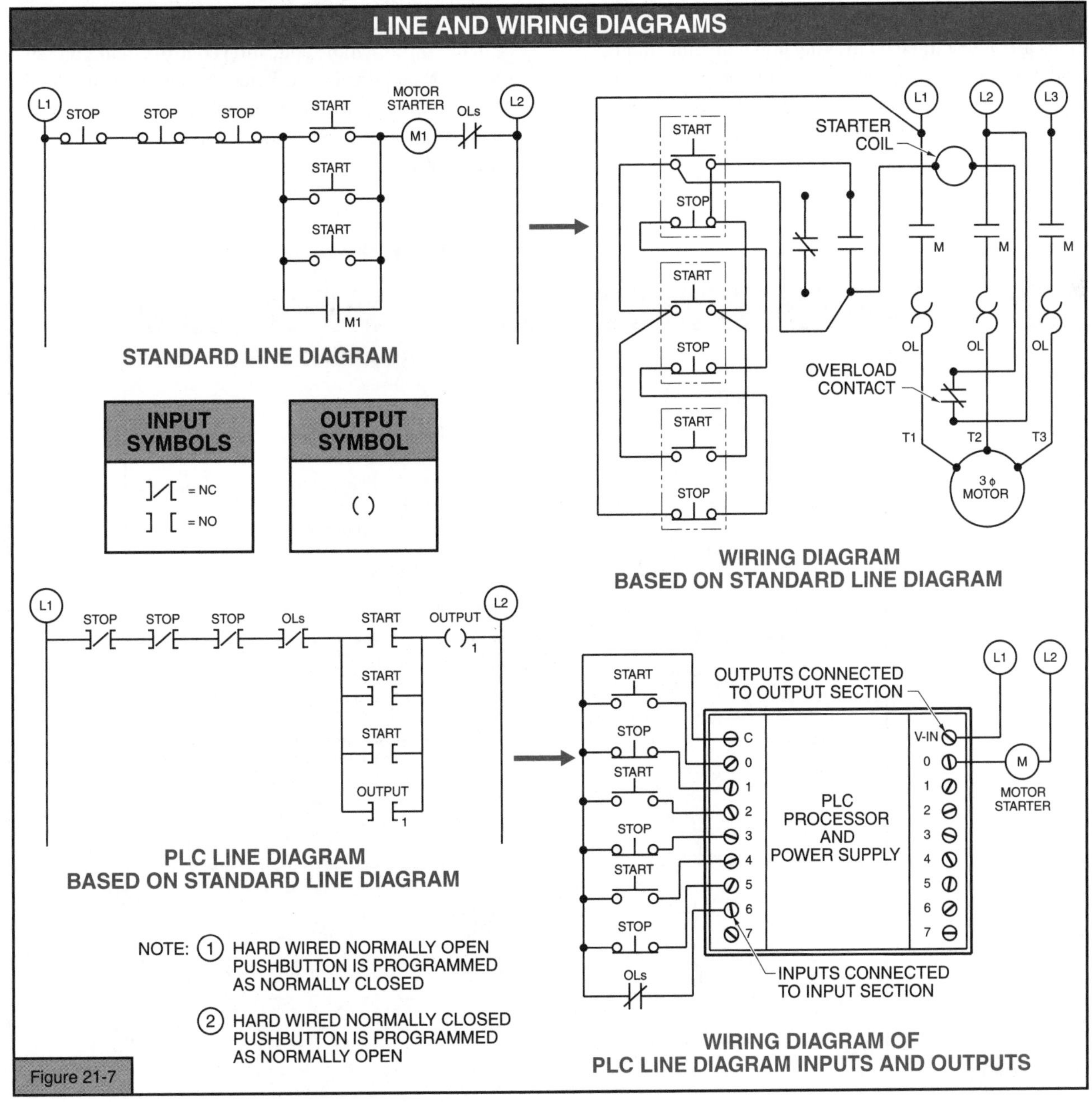

Figure 21-7

Line Number References

Each line in a line diagram should be numbered starting with the top line and reading down. Numbering each line simplifies the understanding of the function of the circuit. The importance of this numbering system becomes clear as circuits become more complex and lines are added. Many industrial line diagrams include hundreds of lines in the control circuit. **See Figure 21-8.**

In this circuit, line 1 contains the shrink-wrap heating element ON lamp. Line 2 contains the manual position of the selector switch, M1 starter coil, and the M1 overload contact. Line 3 contains the automatic position of the selector switch and the limit switch. Line 4 contains the infeed motor ON lamp. The rest of the lines are determined by moving down the line diagram and assigning a line number to any component that is not already assigned to a line. A line must include at least one component. The component may be a load (lamp) or a switch (M2 contact).

Numerical Cross-Reference System (NO Contacts)

Numerical cross-reference systems are required to trace the action of a circuit in complex line diagrams. Common rules help to quickly simplify the understanding of complex circuit operation.

Relays, contactors, and magnetic motor starters normally have more than one set of auxiliary contacts. These contacts may appear at several different locations in the line diagram. Numerical cross-reference systems quickly identify the location and type of contacts controlled by a given device. A numerical cross-reference system consists of numbers in parenthesis to the right of the line diagram. NO contacts are represented by line numbers. **See Figure 21-9.**

In the shrink-wrap application control circuit, pressing the start pushbutton in line 5 energizes motor starter coil 2 (M2). Energizing M2 starts the infeed motor in the power circuit and also closes the normally open contact in line 6 of the control circuit.

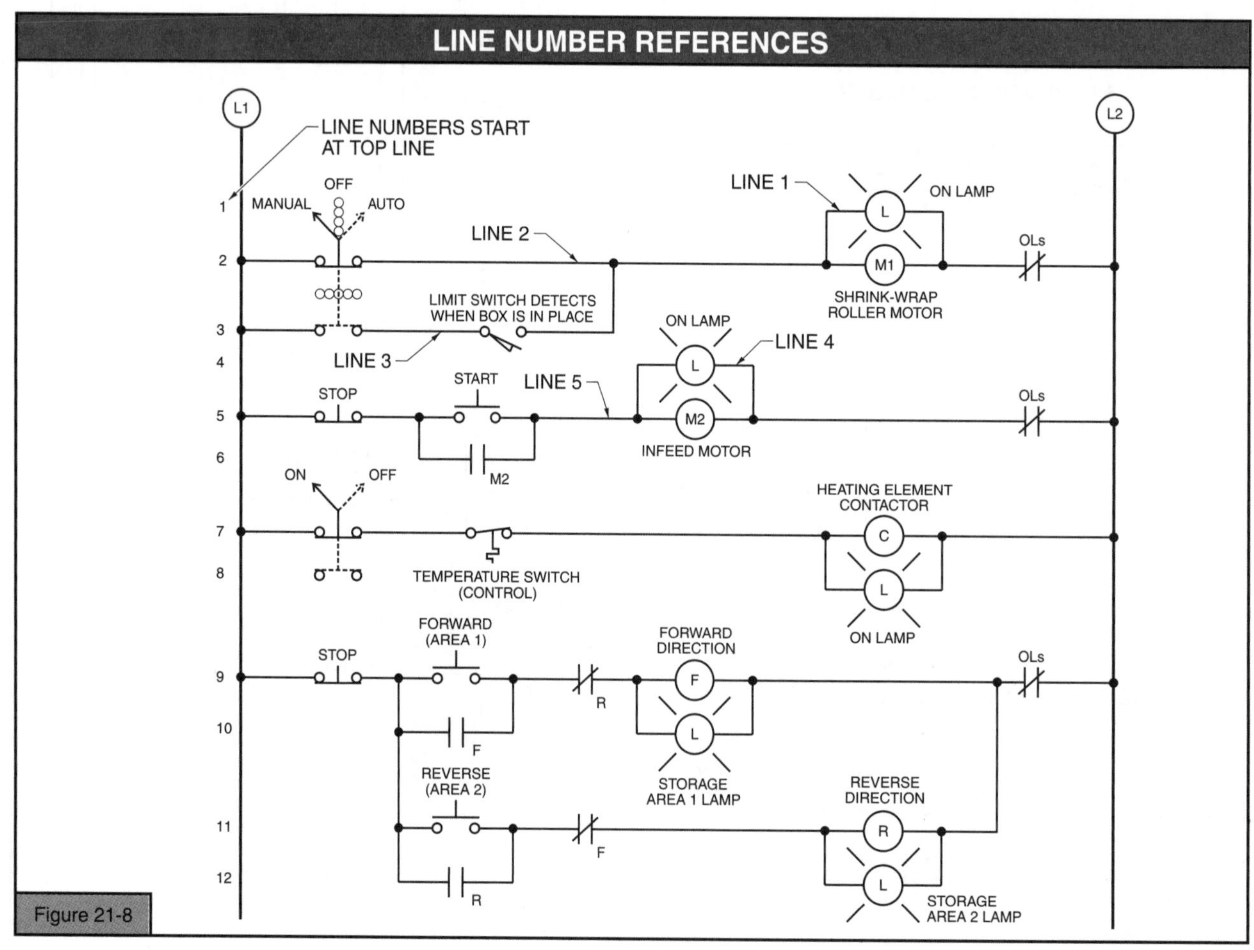

Figure 21-8

The start pushbutton may be released when the normally open contact in line 6 closes because the M2 closed contact allows the current to flow around the start pushbutton. Even though the M2 normally closed contact is located in the next line (line 6), the contact could be located in any line of the control circuit. Thus, the exact location of the contact is easily determined by adding the line number of the contact's location to the right of the coil.

The numerical cross-reference system allows the simplification of complex line diagrams. Each NO contact must be clearly marked because each set of NO contacts are numbered according to the line in which they appear.

Numerical Cross-Reference System (NC Contacts)

In addition to NO contacts, there are also NC contacts in a circuit. To differentiate between NO contacts and NC contacts, NC contacts are indicated by an underlined number. In the shrink-wrap application control circuit, a forward and reversing motor starter is used to control the outfeed motor. Interlocking is provided to protect the system from having the forward starter coil and reverse starter coil energized at the same time. For example, the normally closed forward contact in line 11 opens when the forward starter coil is energized. No power can be delivered to the reversing coil any time the normally closed forward contact is opened, even if the reverse pushbutton is pressed. For this reason, both normally open and normally closed contacts are used when forwarding and reversing a motor. Normally open contacts are used for adding memory, and normally closed contacts are used for interlocking. All normally closed contacts are underlined in the numerical cross-reference system.

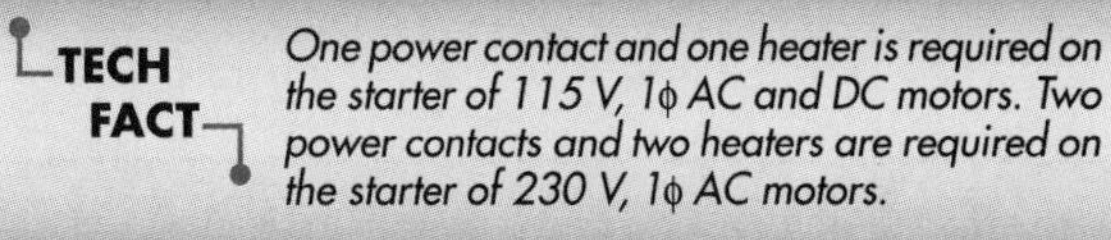
TECH FACT

One power contact and one heater is required on the starter of 115 V, 1ϕ AC and DC motors. Two power contacts and two heaters are required on the starter of 230 V, 1ϕ AC motors.

NUMERICAL CROSS-REFERENCE SYSTEM

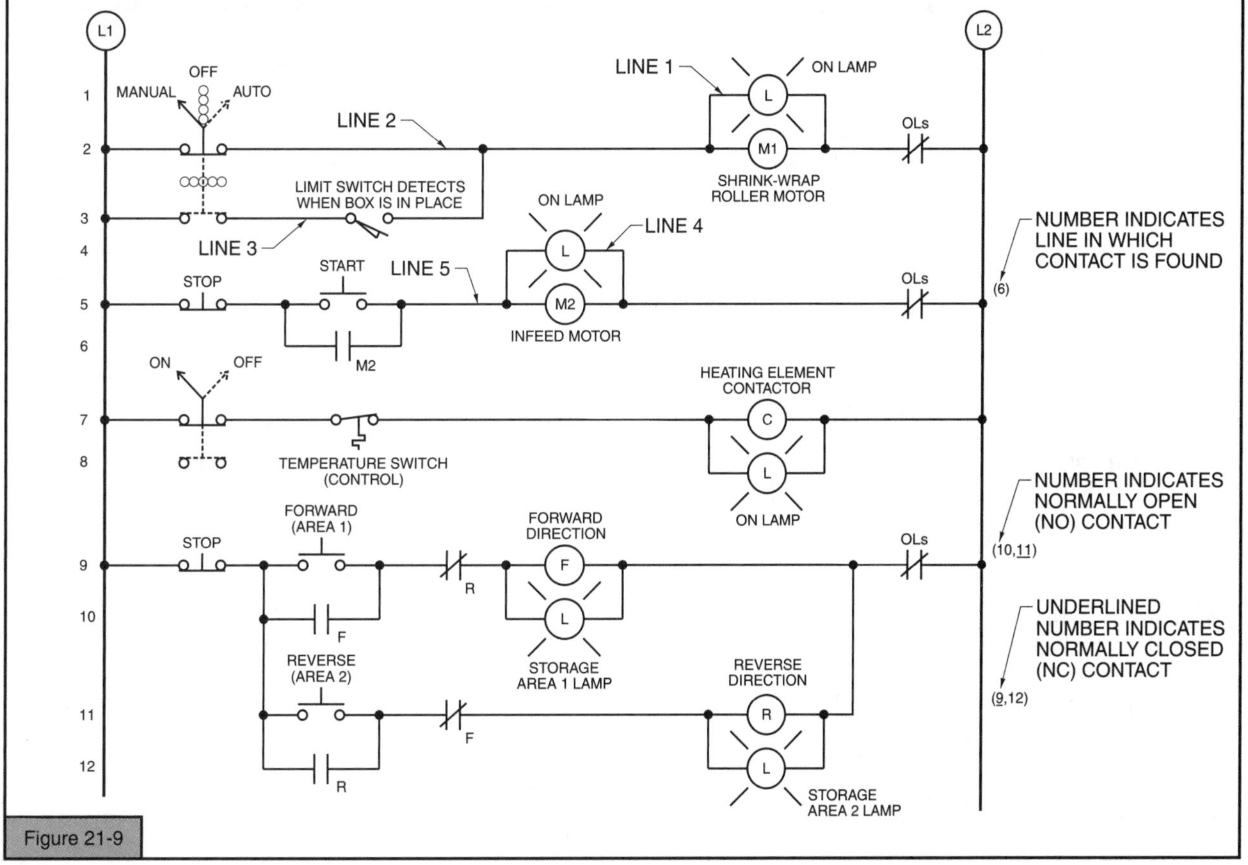

Figure 21-9

Wire-Reference Numbers

Each wire in a control circuit is assigned a reference point on a line diagram to keep track of the different wires that connect the components in the circuit. Each reference point is assigned a wire-reference number. Wire-reference numbers are normally assigned from the top left to the bottom right. This numbering system can apply to any control circuit, such as single-station, multistation, or reversing circuits. **See Figure 21-10.**

Any wire that is always connected to a point is assigned the same number. The wires that are assigned a number vary from 2 to the number required by the circuit. In the shrink-wrap application control circuit, 17 total wire numbers are used. Any wire that is prewired when the component is purchased is normally not assigned a wire-reference number. Different numbering systems may be used. The exact numbering system varies for each manufacturer or design engineer. One common method used is to circle the wire-reference numbers. Circling the wire-reference numbers helps separate them from other numbering systems.

Manufacturer's Terminal Numbers

Manufacturers of relays, timers, counters, etc. include numbers on the terminal connection points of their devices. The terminal numbers are used to identify and separate the different component parts (coil, NC contacts, etc.) included on the individual devices. Manufacturer's terminal numbers are often added to a line diagram after the specific equipment to be used in the control circuit is identified. **See Figure 21-11.**

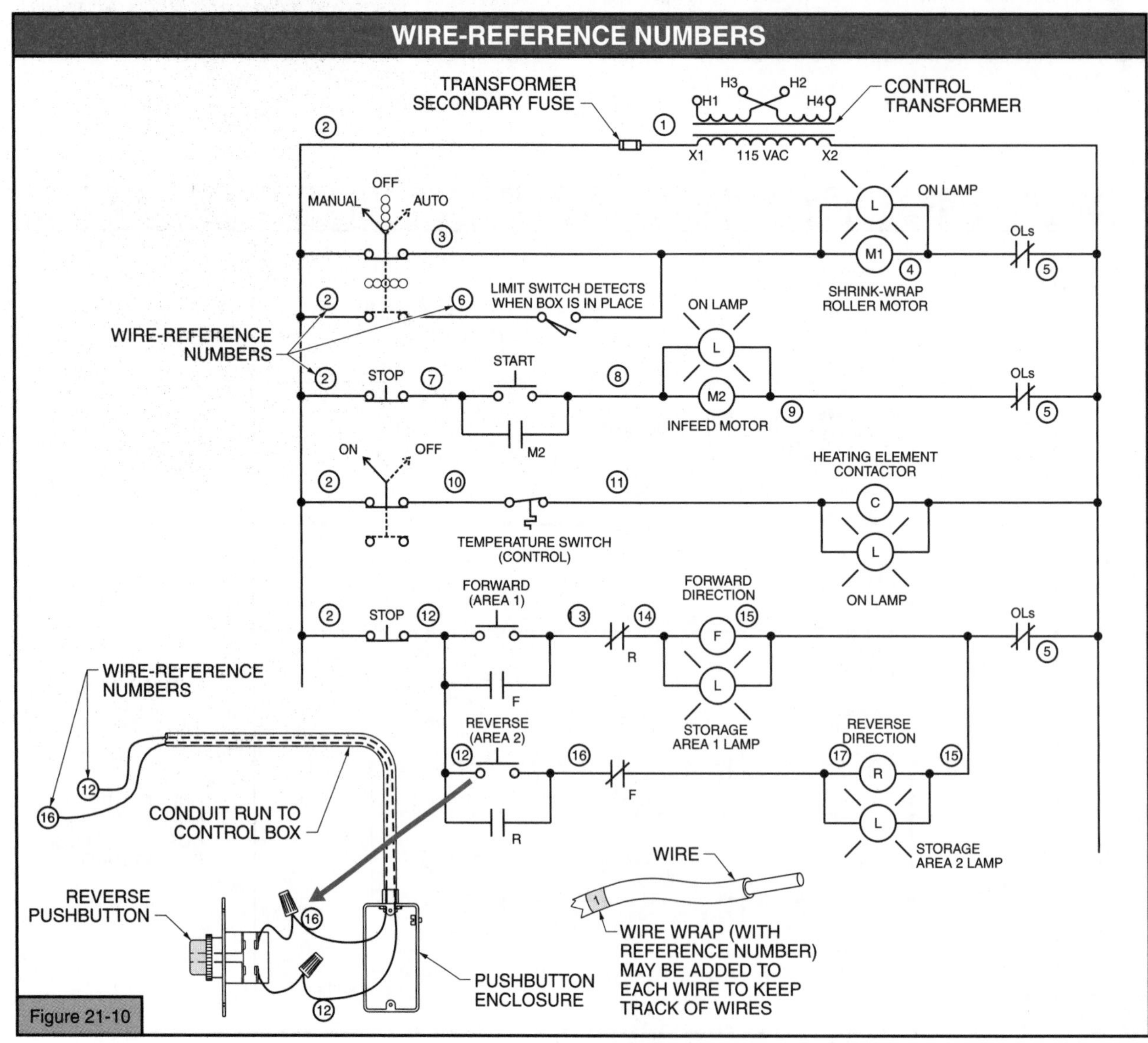

Figure 21-10

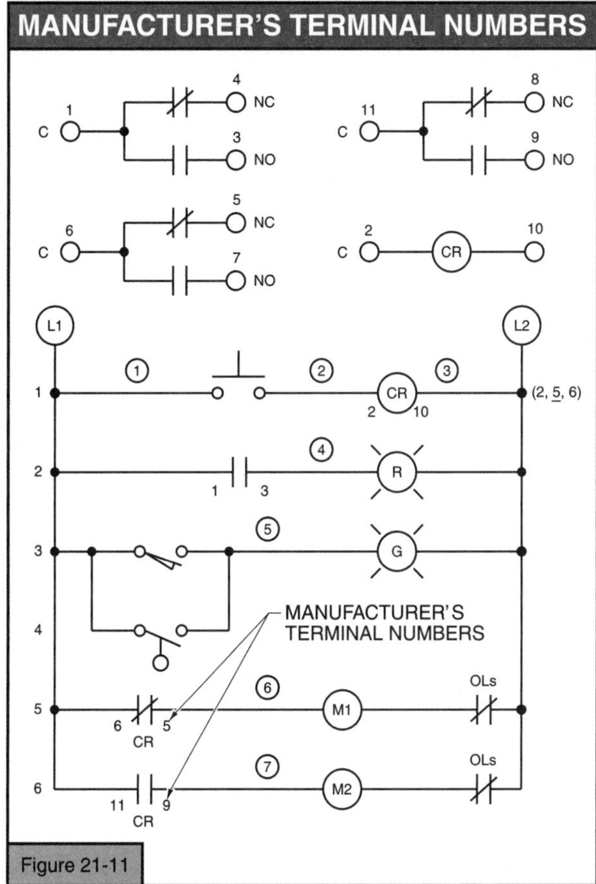

Figure 21-11

In this circuit, line numbers, numerical cross-reference numbers, wire-reference numbers, and manufacturer's terminal numbers are all added to the line diagram. Each numbering system performs a designated function, and together they clarify the circuit's operation and wiring.

The line numbers and numerical cross-reference numbers are used when reading the circuit's logic and operation. Wire-reference and manufacturer's terminal numbers are used when installing and troubleshooting the circuit. For example, a voltmeter could be connected between wires (wire-reference numbers) marked 1 and 3 to measure the control circuit's voltage.

Motor Circuit Operation

An industrial electrical circuit consists of many small individual parts that are interconnected to perform work in a predetermined logical order. The shrink-wrap application includes motors for moving the product along and applying the shrink-wrap, heating elements for heating the shrink-wrap, and indicator lamps that indicate to the operator the status (ON or OFF) of each load. The loads (motors and heating elements) are placed in the high-voltage (230 VAC) power circuit, and the control switches and indicator lamps are placed in the low-voltage (115 VAC) control circuit. The high voltage is used to operate the loads because they require high levels of power (voltage and current). The control circuit components do not require high power, so they are operated at a low voltage. The control voltage is normally 24 VDC, 24 VAC, 115 VAC, or lower when solid-state control components are used.

Each motor is turned ON and OFF through a magnetic motor starter. The shrink-wrap roller motor is a 230 VAC motor, so a 1ϕ motor starter is used. A 1ϕ motor starter includes one power contact and one overload heater (or other current-sensing device) for 115 VAC or DC motor operation, and two power contacts and two overload heaters for 230 VAC operation (such as the shrink-wrap roller motor). **See Figure 21-12.**

The infeed motor requires a motor starter with three power contacts and three overload heaters. The outfeed motor requires a forward motor starter with three power contacts and a reversing motor starter with three power contacts. The forward starter and reversing starter share the same three overload heaters because the heaters/overload contacts can protect the motor from an overload in either direction. This type of starter is normally included as one interconnected unit (reversing starter).

The shrink-wrap circuit motor starters, contactors, and control transformer are located in a control panel. The control panel may be part of the operator station, but is normally a separate panel. In the case of several loads (motors, heating elements, etc.) operating as a system (packaging/wrap line), the individual units are normally grouped in one common control center (motor control center) even if non-motor loads (heating elements) are included. **See Figure 21-13.**

The individual motor control components and disconnects are normally placed in individual enclosures that are interconnected by conduit if a motor control center (MCC) is not used. Power to each individual disconnect panel is delivered through conduit or a plug-in busway. **See Figure 21-14.**

A *busway* is a metal-enclosed distribution system of busbars available in prefabricated sections. Prefabricated fittings, tees, elbows, and crosses simplify the connecting and reconnecting of the distribution system. By bolting sections together, the electrical power is available at many locations and throughout the system.

TECH FACT — *Measure the voltage drop across the contacts when the motor is operating to check the condition of the power contacts on the starter. The greater the voltage drop, the poorer the contacts.*

MOTOR CIRCUIT OPERATION

TO SHRINK-WRAP ROLLER MOTOR DISCONNECT

MAGNETIC MOTOR STARTER

MOTOR STARTER COIL (M1)

③ ④ TO TERMINALS 3 AND 4 IN CONTROL PANEL

NORMALLY OPEN AUXILIARY CONTACTS (NOT USED IN SHRINK-WRAP ROLLER MOTOR APPLICATION)

HEATERS (TWO USED FOR 230 V, 1φ MOTORS)

④ ⑤ TO TERMINALS 4 AND 5 IN CONTROL PANEL

OVERLOAD CONTACTS (OLs)

TO SHRINK-WRAP ROLLER MOTOR

PICTORIAL DRAWING

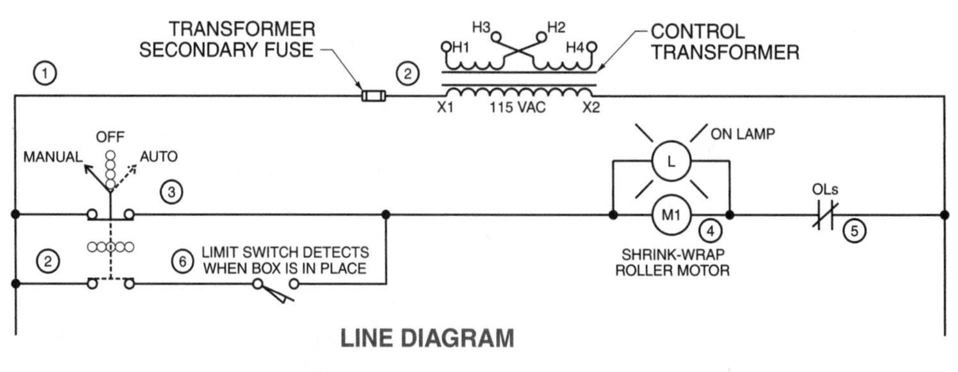

LINE DIAGRAM

Figure 21-12

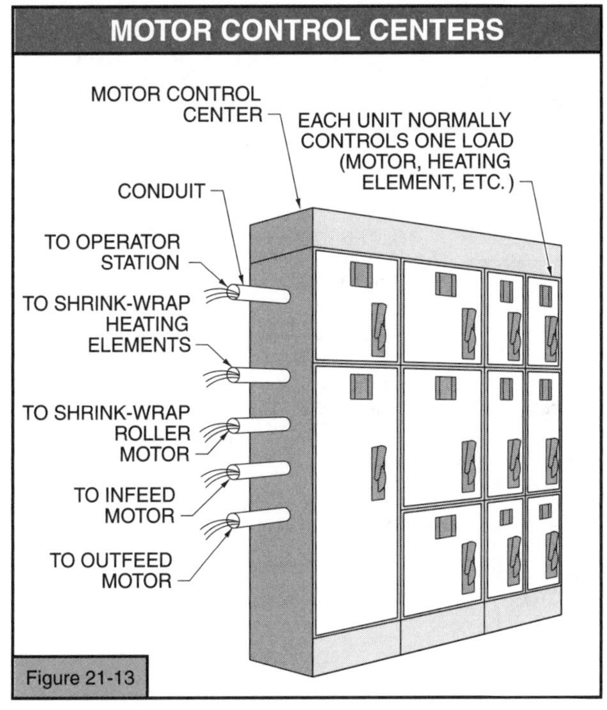

Figure 21-13

Heater Circuit Operation

A *heating element* is a conductor that offers enough resistance to produce heat when connected to an electrical power supply. The lower the resistance, the greater the heat produced. The greater the resistance, the lower the heat produced. The major difference between most heating elements is in their power rating, the material they are made from, and their physical construction. Heating element power ratings determine the amount of heat produced. The material they are made from and their physical construction determines the location where they may be used.

Heating element control circuits range from simple to complex. A simple control circuit is used to turn the heating element completely ON or OFF. For example, in the shrink-wrap application, the circuit that controls the heating element contactor is controlled by a temperature switch that turns ON the heating element any time the heat is below the setting of the temperature switch. A more complex control circuit would allow the heating element to produce a varying amount of heat output. A complex control circuit normally allows the heating element to produce a proportional amount of heat output in response to the difference between the set temperature and the actual temperature. The greater the temperature difference, the higher the heat produced. The lesser the temperature difference, the lower the heat produced.

A heating element control circuit may turn ON and OFF a heating element directly or indirectly. In the direct-control method, the temperature control switch is connected to the heating element. **See Figure 21-15.** The control switch must have an ampere and voltage rating equal to or higher than the rating of the heating element when connected directly to the heating element. For this reason, direct control is used only when the heating element is rated for low power, normally up to about 1500 W.

Indirect heating element control is used on high-rated heating elements, such as the shrink-wrap application. In the indirect-control method, the temperature control switch controls a contactor, which connects the heating element to the power supply. In this method, the temperature control switch is rated for the ampere and voltage of the contactor coil. The contactor contacts are used to switch the high current and voltage of the heating element. **See Figure 21-16.**

An ON/OFF temperature control circuit is the simplest heating element control circuit. The heating element is either ON or OFF. In the shrink-wrap application, an industrial 3ϕ heating element is used to deliver heat to shrink the wrap. In its simplest form, the heating element may be controlled by a manually-operated ON/OFF switch, such as an ON/OFF selector switch. The heating element contactor is energized when the switch is placed in the ON position. The contactor is used to connect the heating element to the 3ϕ power supply when energized. The heating element produces heat as long as the switch is in the ON position. The contactor is de-energized and power is removed from the heating element when the switch is placed in the OFF position.

A manually-operated heating element is often used to manually add heat to a given area. A manually-operated circuit would not work well (or be inefficient) in the shrink-wrap application. A temperature switch is added to automatically maintain the set temperature. The temperature switch is set for the required temperature and the heating element is automatically turned ON and OFF. The heating element is turned fully ON or OFF when using a temperature switch.

Electric Motor Drives

Magnetic motor starters are being replaced in many applications by electric motor drives. An *electric motor drive* is an electronic device that controls the direction, speed, torque, and other operating functions of an electric motor in addition to providing motor protection and monitoring functions. Electric motor drives are either AC drives or DC drives, with AC drives being the most common. AC drives are also referred to as adjustable-speed drives, variable-frequency drives, or inverters. **See Figure 21-17.**

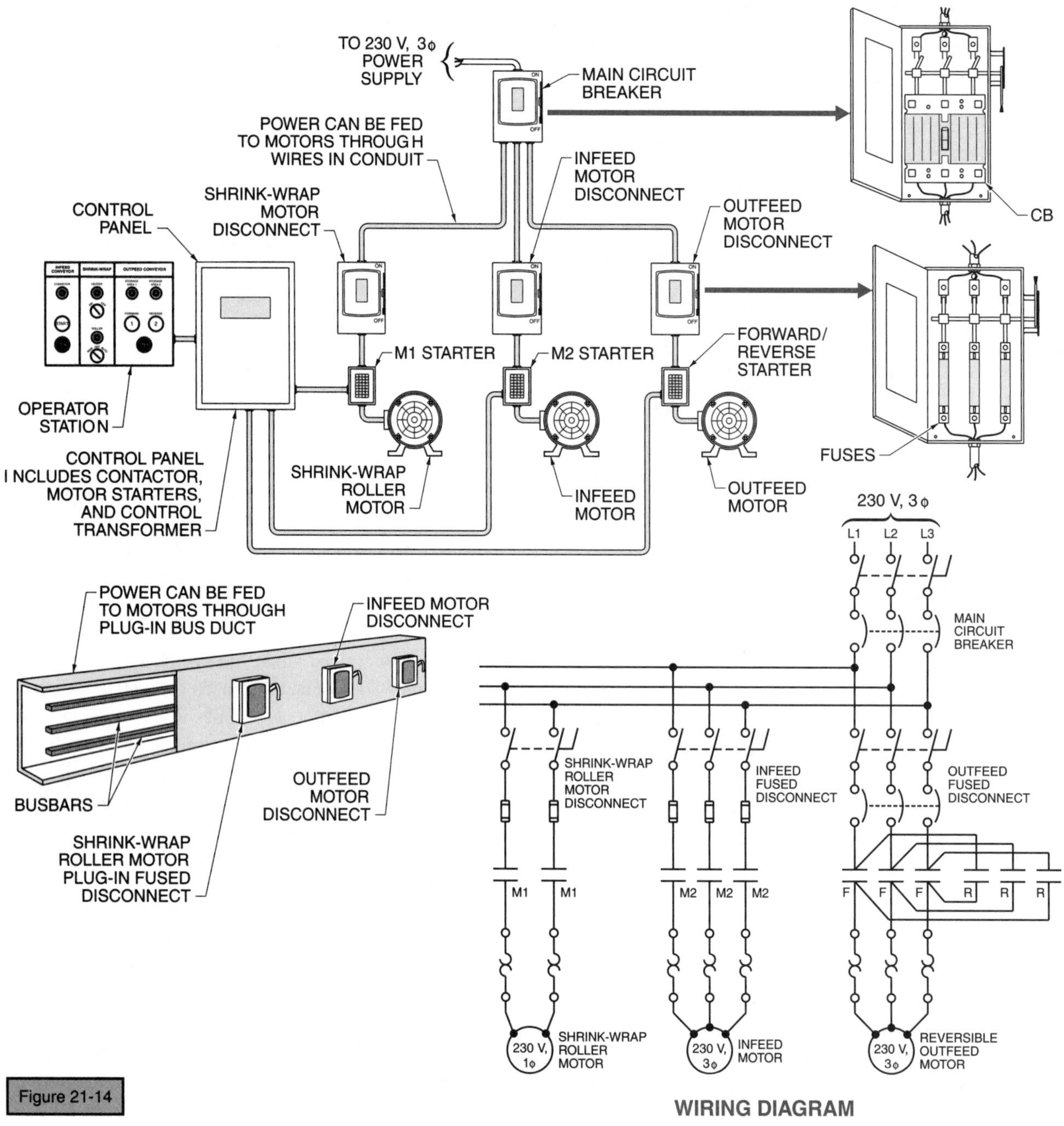

TO 230 V, 3φ POWER SUPPLY
MAIN CIRCUIT BREAKER
POWER CAN BE FED TO MOTORS THROUGH WIRES IN CONDUIT
INFEED MOTOR DISCONNECT
SHRINK-WRAP MOTOR DISCONNECT
CONTROL PANEL
OUTFEED MOTOR DISCONNECT
CB
M1 STARTER
M2 STARTER
FORWARD/ REVERSE STARTER
OPERATOR STATION
CONTROL PANEL INCLUDES CONTACTOR, MOTOR STARTERS, AND CONTROL TRANSFORMER
SHRINK-WRAP ROLLER MOTOR
INFEED MOTOR
OUTFEED MOTOR
FUSES
230 V, 3φ
L1 L2 L3
POWER CAN BE FED TO MOTORS THROUGH PLUG-IN BUS DUCT
INFEED MOTOR DISCONNECT
MAIN CIRCUIT BREAKER
BUSBARS
OUTFEED MOTOR DISCONNECT
SHRINK-WRAP ROLLER MOTOR PLUG-IN FUSED DISCONNECT
SHRINK-WRAP ROLLER MOTOR DISCONNECT
INFEED FUSED DISCONNECT
OUTFEED FUSED DISCONNECT
M1 M1
M2 M2 M2
F F F R R R
230 V, 1φ
SHRINK-WRAP ROLLER MOTOR
230 V, 3φ
INFEED MOTOR
230 V, 3φ
REVERSIBLE OUTFEED MOTOR
WIRING DIAGRAM

Figure 21-14

HEATING ELEMENT DIRECT CONTROL

L1
L2
SUPPLY VOLTAGE
HIGH CURRENT PATH
TEMPERATURE CONTROL SWITCH
HEATING ELEMENT

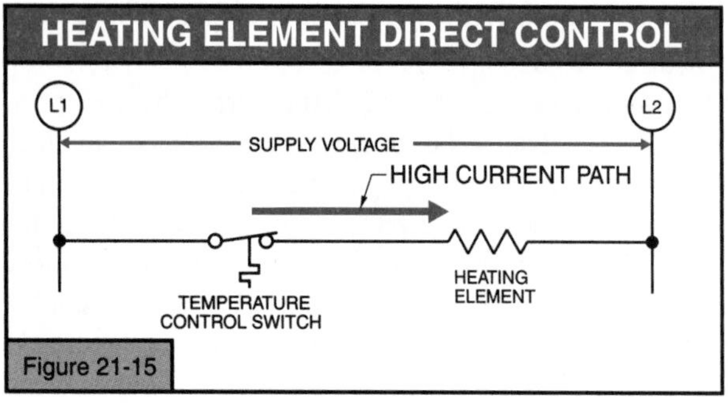

Figure 21-15

TECH TIP

An electric motor drive is programmed for a specific application and controlled by inputs and outputs specific to the application. The complexity of an electric motor drive application may require more than one technician or qualified person to complete a test. A qualified person must be able to verify that the electric motor drive is working properly in relation to the entire process, and must also be able to make suggestions for optimizing the electric motor drive, such as slowing down a mixing motor on a tank in a batch process.

HEATING ELEMENT INDIRECT CONTROL
L1
L2
L3
SHRINK-WRAP HEATING ELEMENT DISCONNECT
C
C
C
3φ HEATER
POWER CIRCUIT
WYE-CONNECTED HEATING ELEMENT
DELTA-CONNECTED HEATING ELEMENT
L1
L2
ON
OFF
HEATING ELEMENT CONTACTOR
C
L
TEMPERATURE CONTROL SWITCH
ON LAMP
CONTROL CIRCUIT

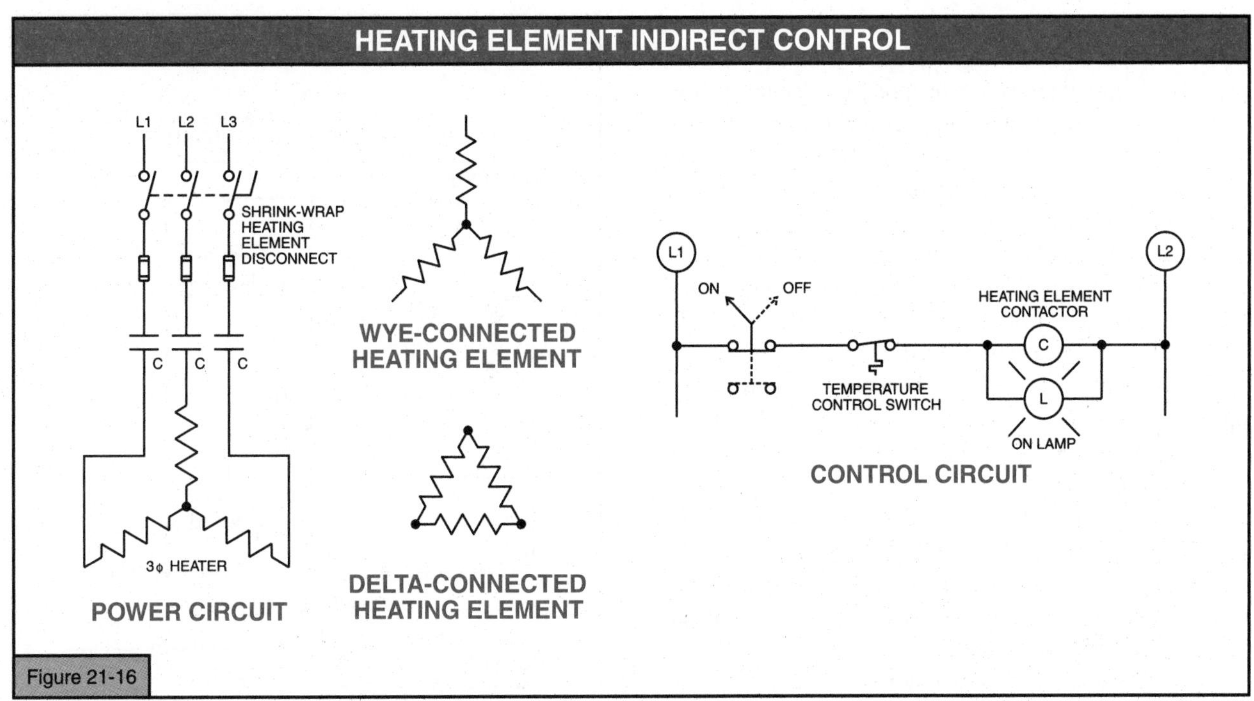

Figure 21-16

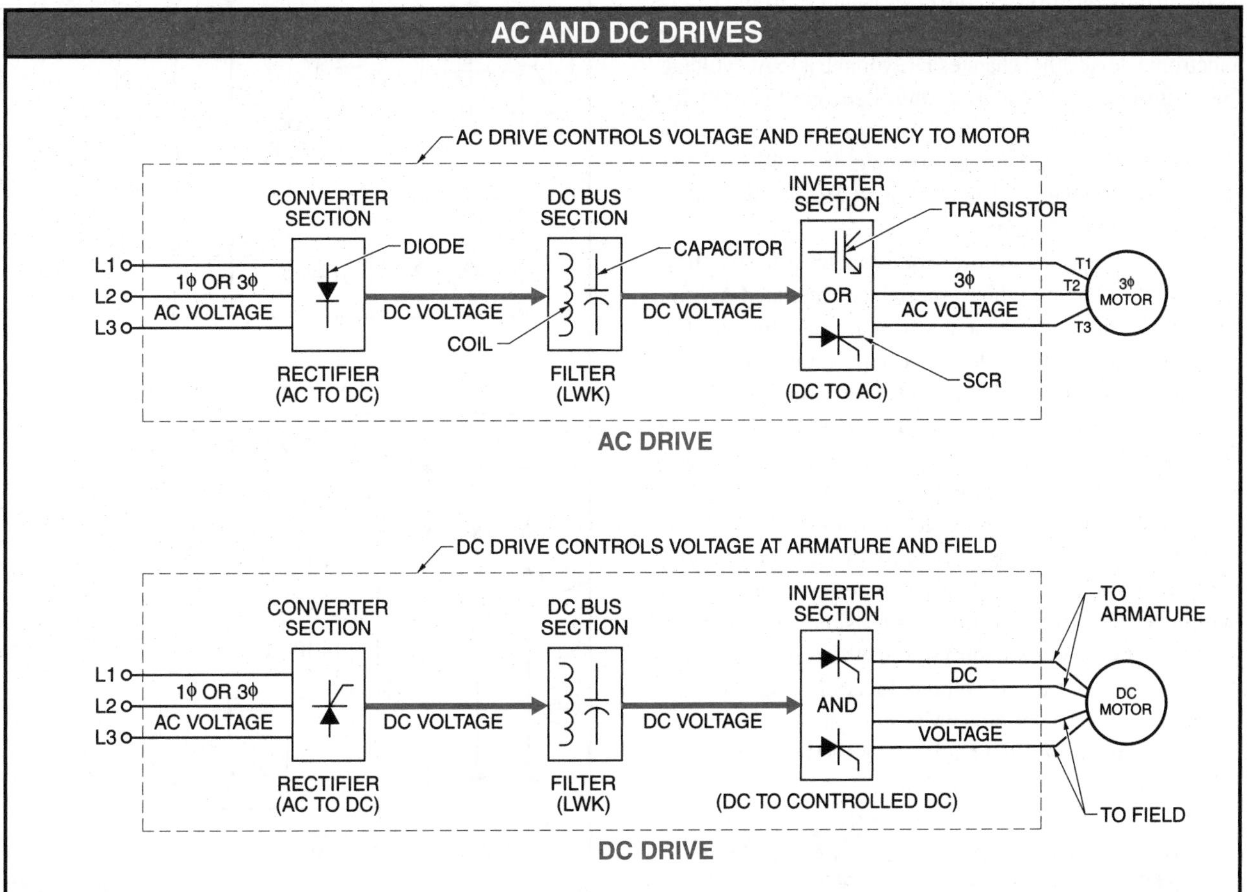

AC AND DC DRIVES
AC DRIVE CONTROLS VOLTAGE AND FREQUENCY TO MOTOR
CONVERTER SECTION
DC BUS SECTION
INVERTER SECTION
TRANSISTOR
DIODE
CAPACITOR
L1
L2
L3
1φ OR 3φ
AC VOLTAGE
DC VOLTAGE
DC VOLTAGE
OR
3φ
AC VOLTAGE
T1
T2
T3
3φ MOTOR
COIL
RECTIFIER (AC TO DC)
FILTER (LWK)
(DC TO AC)
SCR
AC DRIVE
DC DRIVE CONTROLS VOLTAGE AT ARMATURE AND FIELD
CONVERTER SECTION
DC BUS SECTION
INVERTER SECTION
TO ARMATURE
L1
L2
L3
1φ OR 3φ
AC VOLTAGE
DC VOLTAGE
DC VOLTAGE
AND
DC
VOLTAGE
DC MOTOR
RECTIFIER (AC TO DC)
FILTER (LWK)
(DC TO CONTROLLED DC)
TO FIELD
DC DRIVE

Figure 21-17

Electric motor drives perform the same functions as motor starters, but also vary motor speed, reverse motors, provide additional protection features, and display operating information, and can be interfaced with other electrical equipment. Electric motor drives are used to control any size motor from fractional horsepower to hundreds of horsepower.

AC drives control and monitor motor speed by converting incoming AC voltage to DC voltage and then converting the DC voltage back to a variable-frequency AC voltage. To change the speed of a motor, electric motor drives vary the frequency (Hz) to the motor. A standard 60 Hz AC motor operates at full speed when connected to 60 Hz, at half speed when connected to 30 Hz, and at one-quarter speed when connected to 15 Hz.

DC drives control speed by controlling the DC voltage to a motor by varying the amount of voltage and current on the field and armature of the motor. AC power can also be connected to DC drives because the incoming AC voltage is converted to DC voltage.

The section of an electric motor drive that coverts AC voltage to DC voltage is the converter section, and the section that converts the DC voltage back to AC voltage is the inverter section. A *converter* is an electronic device that changes AC voltage into DC voltage. An *inverter* is an electronic device that changes DC voltage into AC voltage. **See Figure 21-18.** Silicon-controlled rectifiers (SCRs) or diodes are used in the converter section to convert the incoming AC voltage into DC voltage. In addition to converting AC voltage into DC voltage, SCRs can also control the level of the DC voltage.

AC drives use the inverter section switches to switch the DC voltage ON and OFF to reproduce simulated AC sine waves. The simulated AC sine waves are actually pulsating DC voltages at different levels. The pulsating DC voltage is used to operate AC motors and control the speed of the motors by varying the voltage and frequency applied to the motors. AC drives control the voltage and frequency sent to a motor by switching the DC voltage ON and OFF at the correct moments. Electronic switches (SCRs and transistors) are used because of the fast switching required to simulate AC sine waves. A microprocessor circuit located within an electric motor drive controls the switching.

Inverters are classified by the method used to change the frequency of the DC voltage. Inverter designs include variable voltage inverters (VVI), current source inverters (CSI), and pulse width modulated (PWM) inverters.

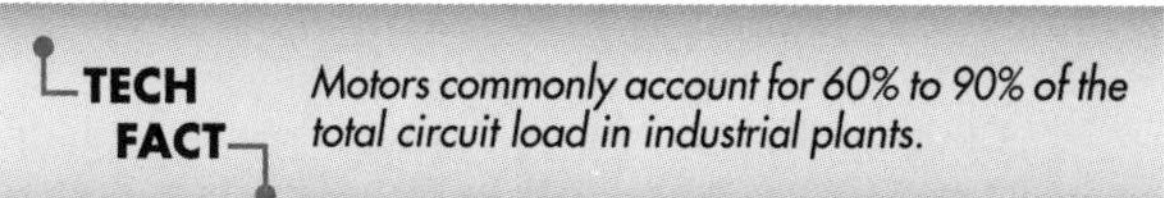

Motors commonly account for 60% to 90% of the total circuit load in industrial plants.

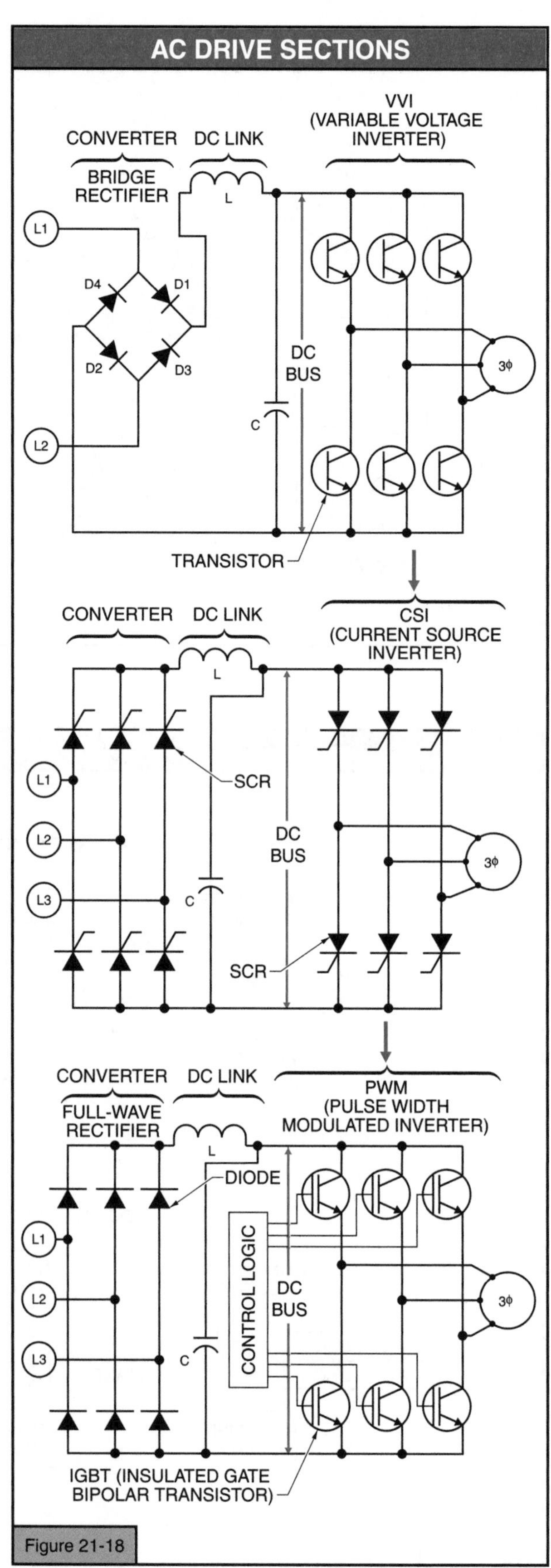

Figure 21-18

Switch Functions

All switches serve some function in an electrical circuit. The primary function of any switch is to start, stop, or redirect the flow of current in a circuit. Switches are typically designated by the function the switch performs such as on, off, start, stop, emergency stop, up, down, left, right, forward, reverse, fast, slow, or reset. To perform a function, switches are wired into an electrical circuit as single-function switches or dual-function switches. **See Figure 21-19.**

A *single-function switch* is a switch that performs only one switching function such as start or stop. When performing one function, one position of the switch performs the designated function, such as starting the motor, and the other switch position performs no function at all.

TECH FACT — *Maintain a standard on all machines when adding pilot indicating lamps in a control circuit. For example, red normally means stop, OFF, close, and green normally means start, ON, open.*

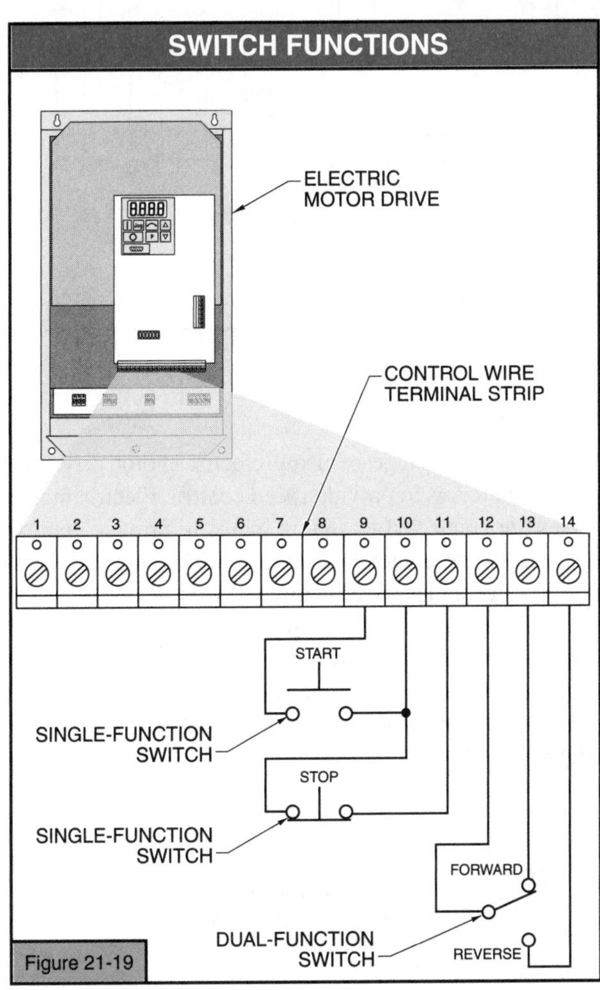

Figure 21-19

A *dual-function switch* is a switch that performs two different switching functions such as forward or reverse. When performing two functions, one position of the switch is used for one function, such as turning the motor clockwise (forward), and the other position of the switch is used for the second function, such as turning the motor counterclockwise (reverse).

Two-Wire Motor Control

Control circuits are often referred to by the number of conductors used in the control circuit, such as two-wire or three-wire control. *Two-wire control* is an input control for an electric motor drive requiring two conductors to complete a circuit. *Three-wire control* is an input control for an electric motor drive requiring three conductors to complete a circuit.

Two-wire control has two wires leading from a control device to a starter. **See Figure 21-20.** A control device can be a thermostat, float switch, or other device. Two-wire motor control circuits are used when motors are required to operate automatically from such control devices as liquid level switches, pressure switches, or temperature switches.

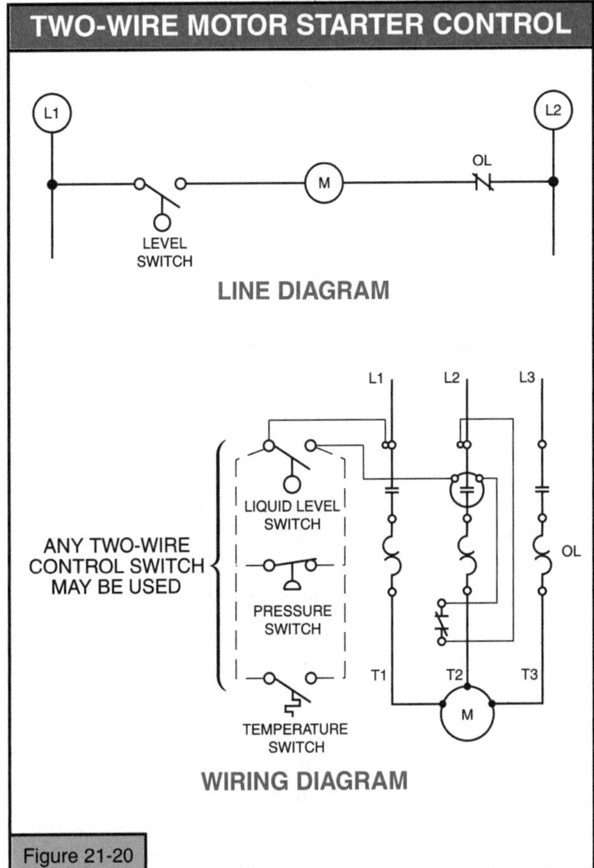

Figure 21-20

Two-wire control is also used when using an electric motor drive to control a motor. Electric motor drives automatically reduce the voltage to control switch circuits. Typically, the control voltage used by electric motor drives is low voltage DC (24 VDC). **See Figure 21-21.**

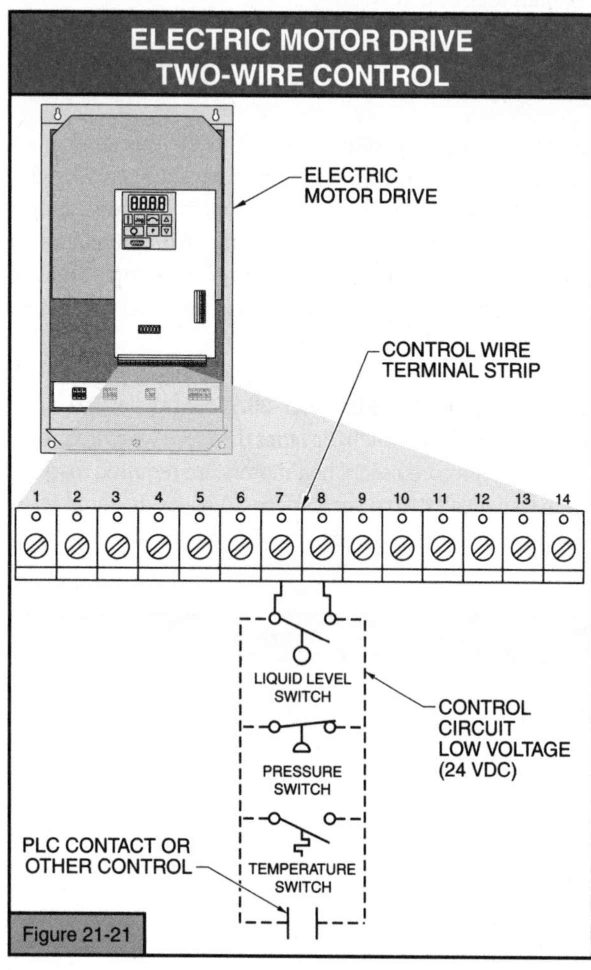

Figure 21-21

Three-Wire Motor Control

Three-wire control has three wires leading from the control device to a starter to complete the circuit. Auxiliary contacts are added to starters to give memory to three-wire control circuits that use pushbuttons. When a motor starter coil (M) is energized, the coil causes the normally open contacts (NO) to close and remain closed (memory) until the coil is de-energized. *Memory* is a control function that keeps a motor running after the start pushbutton is released. Memory circuits are also known as holding or sealing circuits. When a memory circuit is ON, the circuit remains ON until turned OFF, and remains OFF until the circuit is turned back ON. **See Figure 21-22.**

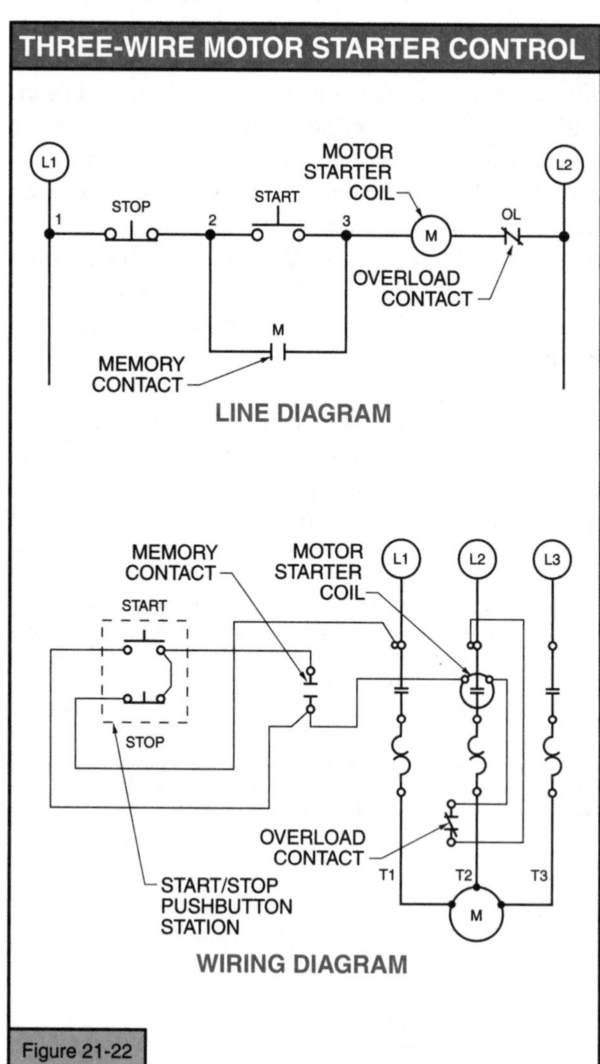

Figure 21-22

Analog Inputs

One major advantage of using electric motor drives to control motors is to provide speed control. Electric motor drives are connected to analog inputs to provide control over a full range of motor speeds. An *analog signal* is a type of input signal to an electric motor drive that can be either varying voltage or varying current. Analog input devices connected to electric motor drives include potentiometers, variable voltage inputs (commonly 0 VDC to 10 VDC), or variable current inputs (commonly 4 mA to 20 mA). **See Figure 21-23.**

Electric motor drive parameter settings and external control signals may allow a drive to automatically start a motor after an input power failure.

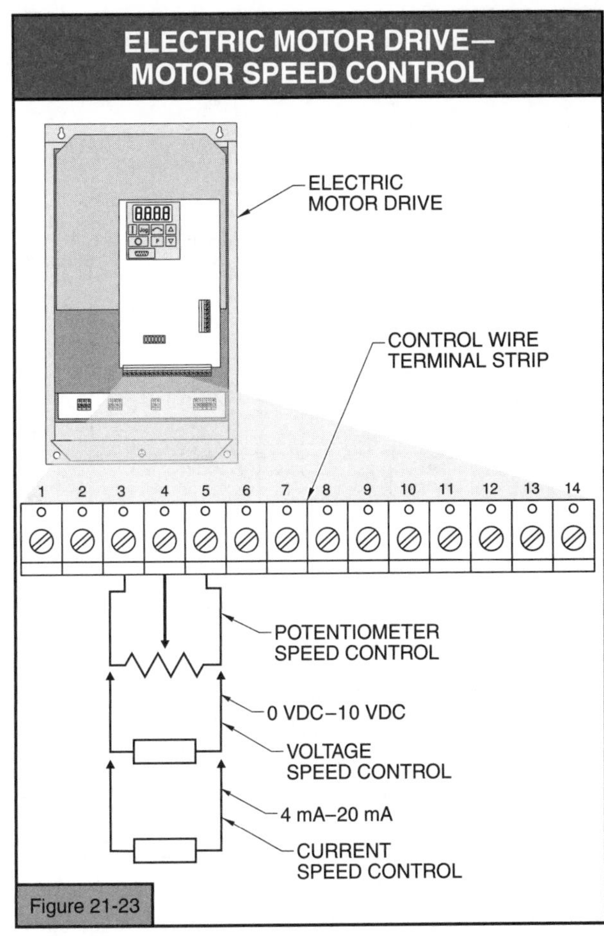

Figure 21-23

Common Control Circuits

Control circuit requirements vary with different applications. Each application requires a circuit that meets its particular requirements. The requirements of a circuit include the electrical requirements of the load(s), the control circuit requirements, safety requirements, and operator requirements. The requirements of the loads are that the power source delivers the proper amount of voltage and current and the correct type of power (AC or DC). The requirements of the control circuit are that the loads are properly turned ON and OFF at the correct time and that all control functions (speed, reversing, etc.) are met. The safety requirements are that the equipment is properly installed (per NEC®, local codes, etc.), and that the equipment is of the correct size and type for the application. Operator requirements are that the correct type of controls are used (pushbutton, selector switch, etc.), the amount of circuit feedback to the operator meets requirements (indicating lamps, alarms, etc.), and that the operator knows how to safely and properly operate the equipment. Although individual circuit requirements vary, common control circuits are used in many applications.

Start/Stop Stations Controlling Magnetic Starters

A load is often required to be started and stopped from more than one location. **See Figure 21-24.** In this circuit, the magnetic motor starter may be started or stopped from two locations. Additional stop pushbuttons are connected in series (NOR logic) with the existing stop pushbuttons.

Additional start pushbuttons are connected in parallel (OR logic) with the existing start pushbuttons. Pressing any one of the start pushbuttons (PB3 or PB4) causes coil M1 to energize. This causes auxiliary contacts (memory contacts) M1 to close, adding memory to the circuit until coil M1 is de-energized. Coil M1 may be de-energized by pressing stop pushbuttons PB1 or PB2, by an overload which activates the overload contacts, or by a loss of voltage to the circuit. In the case of an overload, the overload must be removed and the circuit overload devices reset before the circuit returns to normal starting condition if manual reset overloads are used. The overload contacts automatically reclose if automatic reset overloads are used.

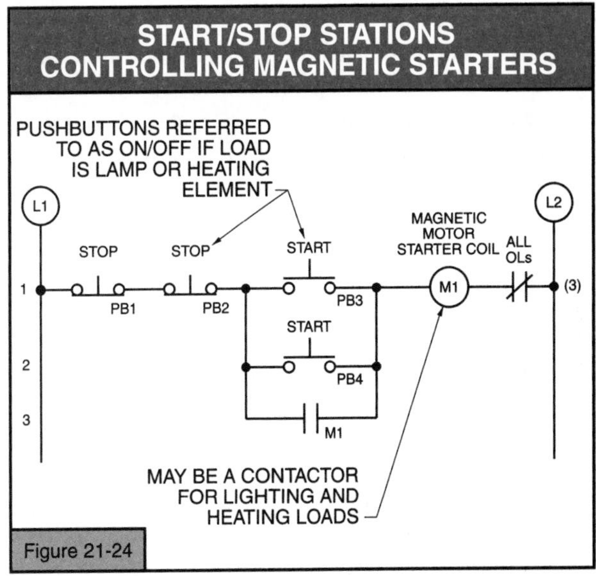

Figure 21-24

Two Magnetic Starters Operated by Two Start/Stop Stations with Common Emergency Stop

In almost all electrical circuits, several devices are run off a common supply voltage. Two start/stop stations may be used to control two separate magnetic motor starter coils (or contactor coils) with a common emergency stop protecting the entire circuit. **See Figure 21-25.**

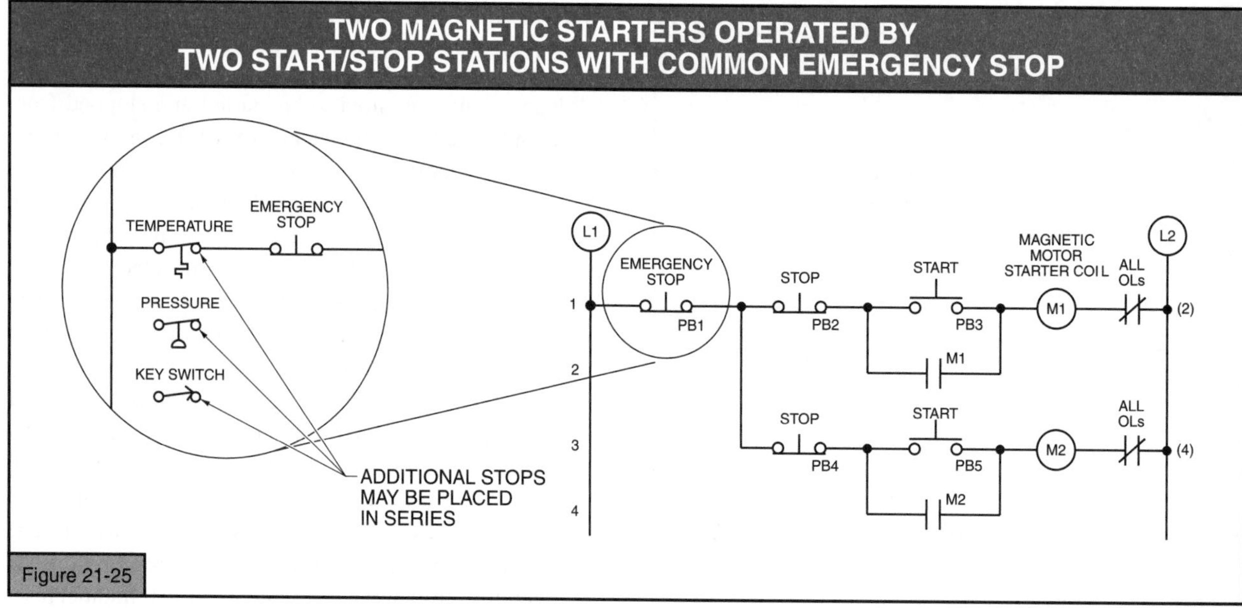

Figure 21-25

In this circuit, pressing start pushbutton PB3 causes coil M1 to energize and seal in auxiliary contacts M1. Pressing start pushbutton PB5 causes coil M2 to energize and seal in auxiliary contacts M2. Once the entire circuit is operational, emergency stop pushbutton PB1 can shut down the entire circuit or the individual stop pushbuttons (PB2 or PB4) can de-energize the coils in their respective circuits. Each circuit is overload protected and does not affect the other when one magnetic motor starter experiences a problem. For added protection, additional normally closed stop pushbuttons (key switch, temperature switch, etc.) may be connected in series with the emergency stop pushbutton.

Start/Stop Station Controlling Two or More Magnetic Starters

Steel mills, paper mills, bottling plants, and canning plants are industrial applications which require simultaneous operation of two or more motors. In these industrial applications, products and/or materials are spread out over great lengths. These products and/or materials must be started together to prevent product or material separation, stretching, breaking, or jam up. To accomplish this, two motors may be started almost simultaneously from one location. **See Figure 21-26.**

In this circuit, pressing start pushbutton PB2 energizes coil M1 and seals in both sets of auxiliary contacts M1. *Note:* It is acceptable to have more than one set of auxiliary contacts controlled by one coil. The first set of M1 contacts (line 2) provides memory for the start pushbutton and completes the circuit to energize coil M1 when both sets of contacts close.

The second set of M1 contacts (line 3) completes the circuit to coil M2, energizing coil M2. The motors associated with these magnetic motor starters start almost simultaneously because both coils energize almost simultaneously. Pressing the stop pushbutton breaks the circuit (line 1), de-energizing coil M1. Both sets of auxiliary contacts are deactivated when coil M1 drops out. The motors associated with these magnetic motor starters stop almost simultaneously because both coils de-energize almost simultaneously.

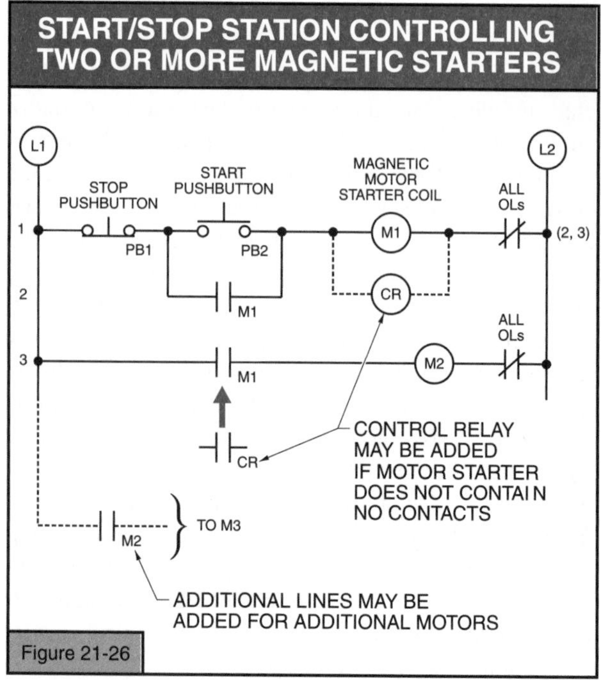

Figure 21-26

An overload in magnetic motor starter M2 affects only the operation of coil M2. The entire circuit stops because de-energizing coil M1 also affects both sets of auxiliary contacts M1. This protection is used where a machine, such as an industrial drill, is damaged if the cooling liquid pump shuts OFF while the drill is still operating.

Pressure Switch with Pilot Light Indicating Device Activation

Pilot lights are manufactured in a variety of colors, shapes, and sizes to meet the needs of industry. The illumination of these lights signals to an operator that any one of a number of events is taking place. A pilot light may be used with a pressure switch (or any other automatic switch) to indicate when a device is activated. **See Figure 21-27.**

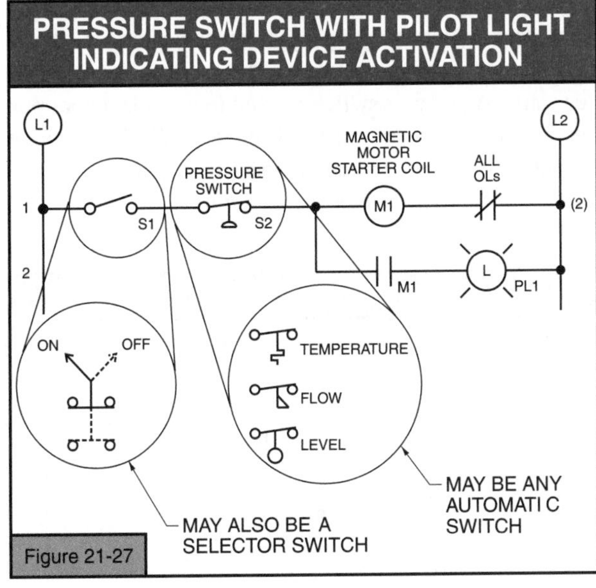

Figure 21-27

In this circuit, pressure switch S2 has automatic control over the circuit when switch S1 is closed. Pressure switch S2 closes and activates coil M1, which controls the magnetic starter of the compressor motor, starting the compressor when the pressure to switch S2 drops. At the same time, contacts M1 close and pilot light PL1 turns ON. The compressor continues to run and the pilot light stays ON as long as the motor runs. Coil M1 de-energizes and the magnetic motor starter drops out, stopping the compressor motor when pressure builds sufficiently to open pressure switch S2. The pilot light shuts OFF because contact M1 controlled by coil M1 opens. The pilot light is ON only when the compressor motor is running. This circuit may be used in a garage to let the operator know when the air compressor is ON or OFF.

A pilot light my also be used with a start/stop station to indicate when a device is activated. **See Figure 21-28.** In this circuit, pressing start pushbutton PB2 energizes coil M1, causing auxiliary contacts M1 to close. Closing contacts M1 provides memory for start pushbutton PB2 and maintains an electrical path for the pilot light. The pilot light stays ON as long as coil M1 is energized. Pressing stop pushbutton PB1 de-energizes coil M1, opening contacts M1 and turning OFF the pilot light. An overload in this circuit also de-energizes coil M1, opening contacts M1 and turning OFF the pilot light. This circuit may be used as a positive indicator that some process is taking place. The process may be in a remote location such as in a well or in another building.

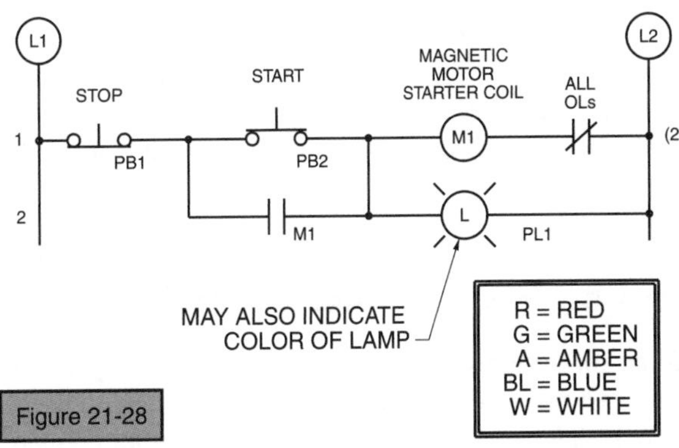

Figure 21-28

Start/Stop Station with Pilot Light Indicating No Device Activation

Pilot lights may be used to show when an operation is stopped as well as when it is started. NOT logic is used in a circuit to show that an operation has stopped. NOT logic is established by placing one set of NC contacts in series with a device. **See Figure 21-29.**

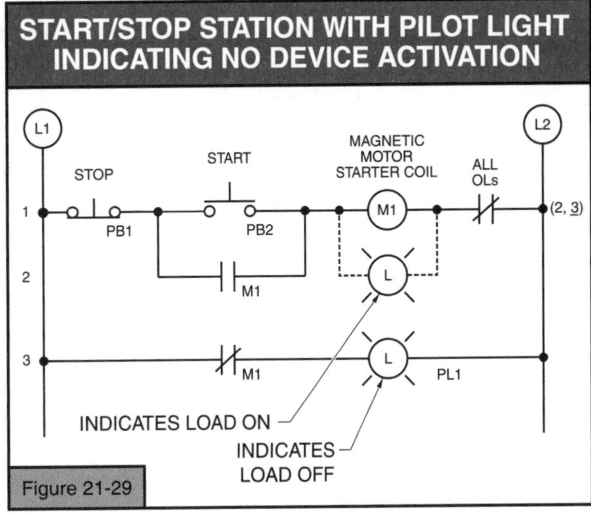

Figure 21-29

In this circuit, pressing start pushbutton PB2 energizes coil M1, causing both sets of auxiliary contacts M1 to energize. Normally open contacts M1 (line 1) close, providing memory for PB2, and normally closed contacts M1 (line 3) open, disconnecting pilot light PL1 from the line voltage, causing the light to turn OFF. Pressing stop pushbutton PB1 de-energizes coil M1, causing both sets of contacts to return to their normal positions. Normally open contacts M1 (line 2) return to their normally open position, and normally closed contacts M1 (line 3) return to their normally closed position, reconnecting the pilot light to the line voltage and causing it to turn ON. The pilot light is ON only when the coil to the magnetic motor starter is OFF. A bell or siren could be substituted for the pilot light to serve as an audible warning device. This circuit may be used to monitor critical operating procedures such as a cooling pump for a nuclear reactor. The pilot light, bell, or siren immediately calls attention to the fact that the process has been stopped if the cooling pump stops.

Pushbutton Sequence Control

Conveyor systems often require one conveyor system to feed cartons or other materials onto another conveyor system. A control circuit is required to prevent the pileup of material on the second conveyor if the second conveyor is stopped. A sequence control circuit prevents the first conveyor from continuing to operate unless the second conveyor has started and is running. **See Figure 21-30.**

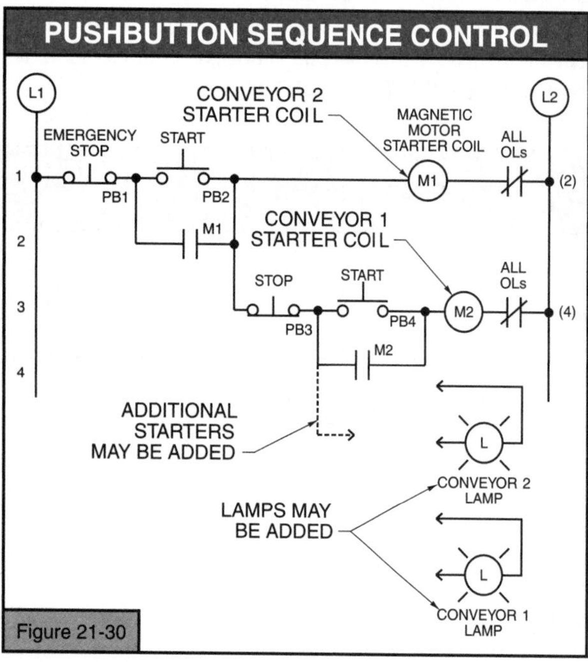

Figure 21-30

In this circuit, pressing start pushbutton PB2 energizes coil M1 and causes auxiliary contacts M1 to close. With auxiliary contacts M1 closed, PB2 has memory and provides an electrical path to allow coil M2 to be energized when start pushbutton PB4 is pressed.

With start pushbutton PB4 pressed, coil M2 energizes and closes contacts M2, providing memory for start pushbutton PB4 so both conveyors run. Conveyor 1 (coil M2) cannot start unless conveyor 2 (coil M1) is energized. Both conveyors shut down if an overload occurs in the circuit with coil M1 or if emergency stop pushbutton PB1 is pressed. Only conveyor 1 shuts down if conveyor 1 (coil M2) experiences an overload. A problem in conveyor 1 does not affect conveyor 2.

Selector Switch Jogging

Jogging is the frequent starting and stopping of a motor for short periods of time. Jogging is used to position materials by moving the materials small distances each time the motor starts. A selector switch is used to provide a common industrial jog/run circuit. **See Figure 21-31.** The selector switch (2-position switch) is used to manually open or close a portion of the electrical circuit.

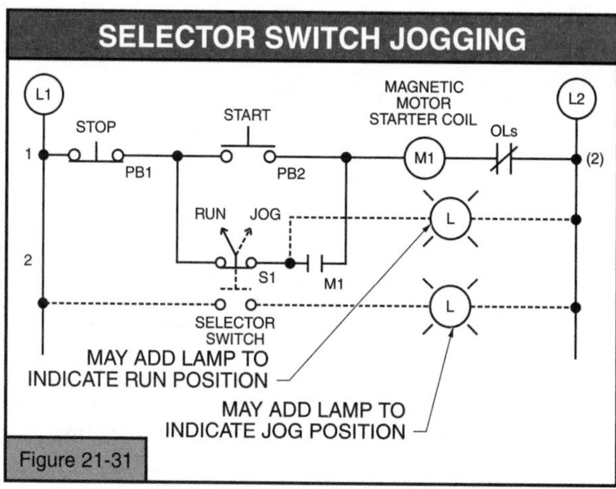

Figure 21-31

In this circuit, the selector switch determines if the circuit is a jog circuit or run circuit. Pressing start pushbutton PB2 with selector switch S1 in the open (jog) position energizes coil M1, causing the magnetic motor starter to operate. Releasing start pushbutton PB2 de-energizes coil M1, causing the magnetic motor starter to stop. Pressing start pushbutton PB2 with selector switch S1 in the closed (run) position energizes coil M1, closing auxiliary contacts M1, providing memory so that the magnetic starter operates and continues to operate until stop pushbutton PB1 is pressed.

Coil M1 de-energizes and all circuit components return to their original condition when stop pushbutton PB1 is pressed. The overloads may also open the circuit and must be reset after the overload is removed to return the circuit to normal operation. This circuit may be found where an operator may run a machine continuously for production, but may stop it at any time for small adjustments or repositioning.

Starting and Stopping in Forward and Reverse with Indicator Lights

Operators are often required to know the direction of rotation of a motor at a given moment. An example is a motor controlling a crane which raises and lowers a load. The line diagram of a crane control circuit is capable of indicating through lights the direction the motor is operating. By the electrician adding nameplates, these lights could indicate up and down directions of the hoist. **See Figure 21-32.**

In this circuit, pressing the momentary contact forward pushbutton causes the normally open and normally closed contacts to move simultaneously. The normally open contacts close, energizing coil F while this pushbutton is depressed. Coil F causes the memory contacts F to close and the normally closed electrical interlock to open, isolating the reversing circuit. The forward pilot light turns ON when holding contacts F are closed. The NC contacts of the forward pushbutton open and isolate the reversing coil R for the period of time the pushbutton is depressed.

Pushing the momentary contact reverse pushbutton causes the normally open and normally closed contacts to move simultaneously. The opening of the normally closed contacts de-energizes coil F. With coil F de-energized, the memory contacts F1 open and the electrical interlock F closes. The closing of the normally open contacts energizes coil R. Coil R causes the holding contacts R1 to close and the normally closed electrical interlock to open, isolating the forward circuit. The reverse pilot light turns ON when the memory contacts R close. Pushing the stop pushbutton with the motor running in either direction stops the motor and causes the circuit to return to its normal position.

Overload protection for the circuit is provided by the heater coils. Operation of the overload contacts break the circuit, opening the overload contacts. The motor cannot be restarted until the overloads are reset and the forward or reverse pushbutton is pressed.

This circuit provides protection against low voltage or a power failure. A loss of voltage de-energizes the circuit and hold-in contacts F or R open. This design prevents the motor from starting automatically after the power returns.

TECH FACT *High ambient temperature affects the life of an electric motor drive. Do not install an electric motor drive in any location that exceeds the rated ambient-temperature limit of the drive.*

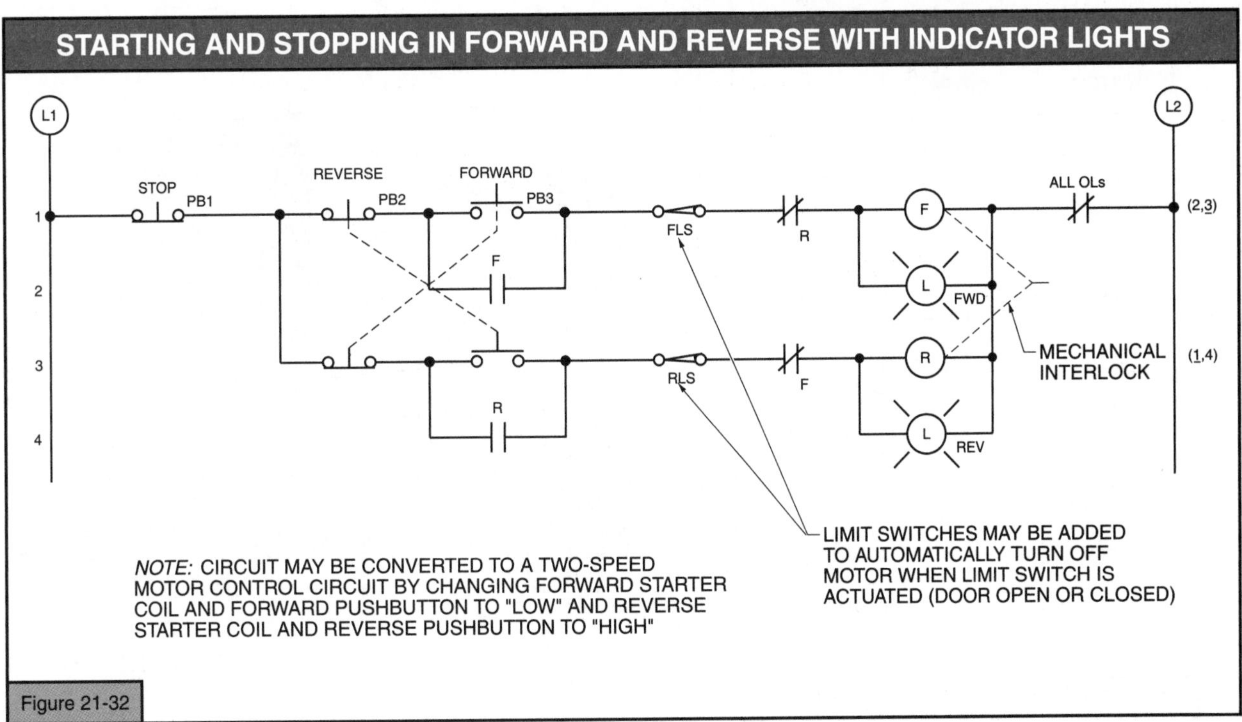

Figure 21-32

Programmable Logic Controllers

A *programmable logic controller (PLC)* is a solid-state control device that can be programmed to automatically control electrical systems in residential, commercial, and industrial facilities. PLCs can be used to control heating and air conditioning (HVAC), security, building automation, process control, assembly, and process electrical systems. Although in the past PLCs were used exclusively in large industrial processes and manufacturing facilities, PLCs are now used to control electrical systems in furnaces, air conditioners, car washes, sporting complexes, transportation systems, bakeries, and automated food production, and almost any other electrical system that has more than a few inputs (switches and sensors) and outputs (motors, lights, and solenoids).

PLCs are manufactured in a range of sizes, from nano and micro sizes that control a few input and output terminals, to large sizes that control thousands of input and output terminals. PLC system diagrams can be viewed using software with integrated programming devices, handheld programming devices, operator interface panels, and desktop or portable computers. **See Figure 21-33.**

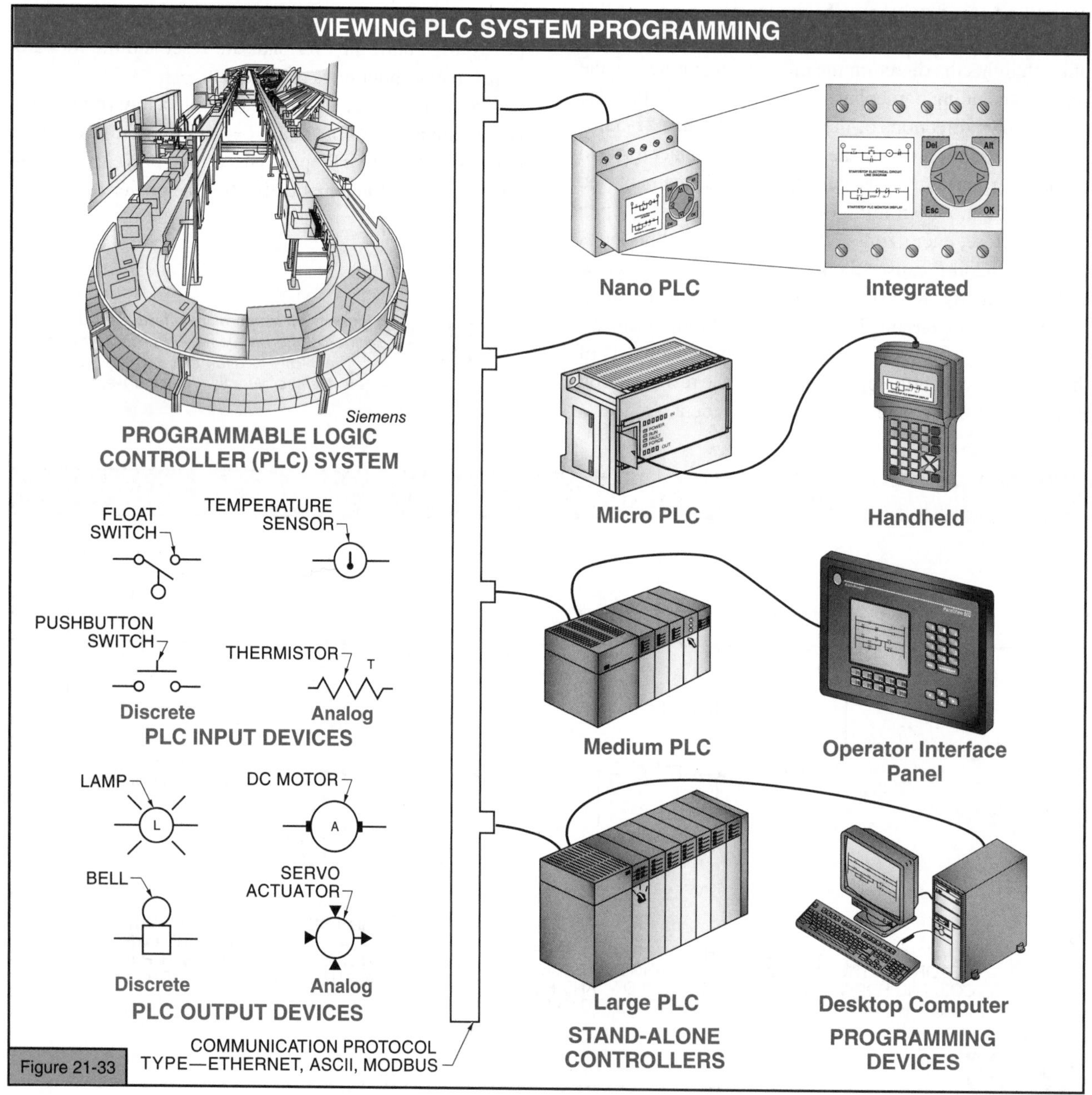

Figure 21-33

PLC Programming Symbols

PLC programs are designed using PLC programming software. PLC programming software uses various symbols, letters, and numbers to designate devices and components. Devices such as input relays, timers, and counters each have a symbol and an identifying address (assigned values and numbers).

When designing a PLC program for controlling a circuit, device and component symbols are selected and displayed on a computer screen as the electrical circuit is developed. The symbols are typically selected from a tool palette that is displayed on the screen. Basic symbols used to program a PLC include normally open inputs, normally closed inputs, and standard outputs. **See Figure 21-34.**

Input devices such as pushbuttons, limit switches, and pressure switches are drawn as generic, normally open or normally closed contacts. Unlike standard electrical symbols, there is no way to distinguish which type of switch (pressure, temperature, or limit) is being used by looking at a generic, normally open or normally closed PLC input device symbol. However, when programming a PLC contact, text can be added to a generic symbol. For example, the words "level switch" can be added to a generic symbol to indicate a level switch.

Additional text can be added to provide even more information about a device. For example, the phrase "overflow level switch 1—tank 3" can be added to a generic symbol for a level switch to indicate that the switch is used for overflow detection in tank 3.

Standard output components include lamps, solenoids, motor starters, and alarms that are used to perform work or give an indication of circuit operation. When using standard electrical symbols, output components have specific symbols that easily identify an output component. As with input devices, generic symbols are used to represent output components, and words can be added to identify a component when programming PLCs.

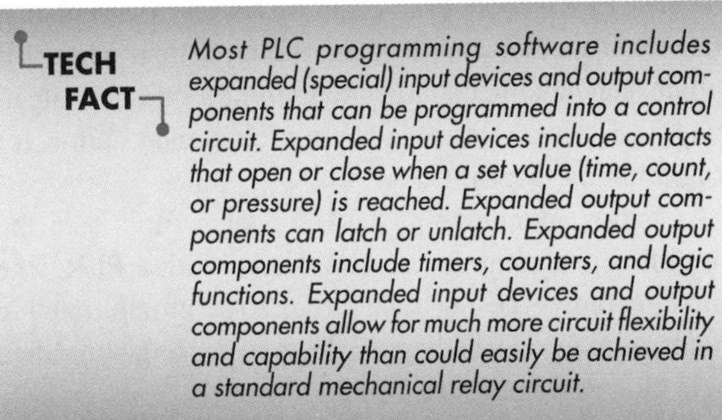
TECH FACT

Most PLC programming software includes expanded (special) input devices and output components that can be programmed into a control circuit. Expanded input devices include contacts that open or close when a set value (time, count, or pressure) is reached. Expanded output components can latch or unlatch. Expanded output components include timers, counters, and logic functions. Expanded input devices and output components allow for much more circuit flexibility and capability than could easily be achieved in a standard mechanical relay circuit.

PLC PROGRAMMING SYMBOLS

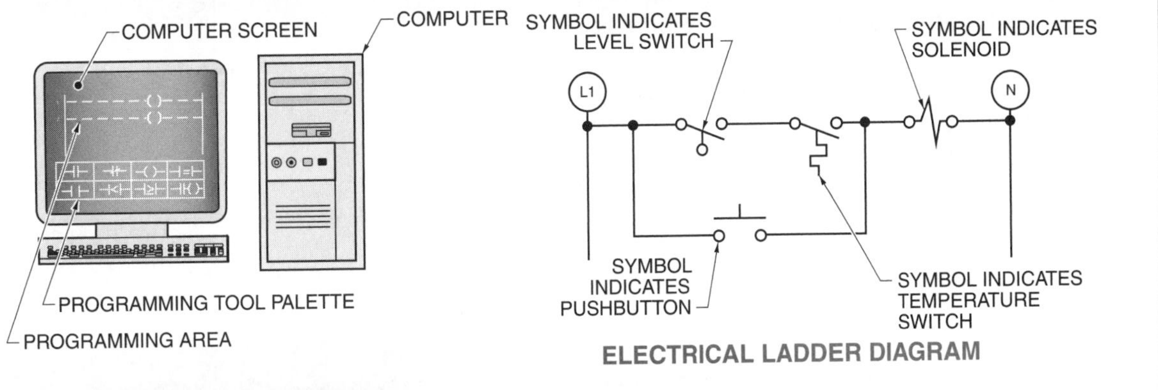

BASIC PROGRAMMING SYMBOLS

Component	Symbol
Input device (normally open)	—| |—
Input device (normally closed)	—|/|—
Standard output	—( )—

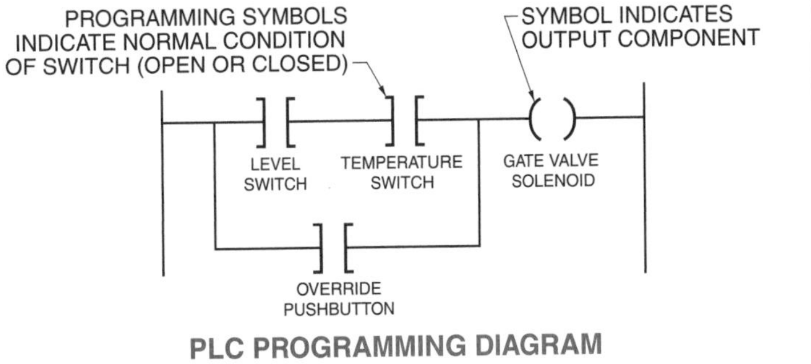

Figure 21-34

Programmable Logic Relays (PLRs)

A programmable relay is a nano PLC that is designed to perform basic control functions using either the programming keypad/display window on the PLR and/or simplified programming software on a computer. Although a PLR can be referred to as a nano PLC or micro PLC, a distinction is usually made between PLRs and PLCs because the simplicity of a PLRs allows them to be used in areas that require operation of a programmed circuit by a person with minimal electrical knowledge. Examples of PLRs applications include use in areas such as HVAC controls, security systems, automated sprinkling systems, and basic machine control.

Most PLRs allow both digital (ON/OFF) and analog inputs to be connected to their input section. The PLR outputs can be either solid-state transistor or mechanical-relay types. Internal timers, counters, and relays allow for programming flexibility.

A major advantage of a PLR over a PLC is in the simplified programming software used with a PLR. The software allows the electrical circuit to be programmed in standard line (ladder) diagram format or using simplified logic blocks. **See Figure 21-35.**

Once the electrical circuit is programmed, a pretest can be run using the simulator mode included with the PLR software. The simulation software allows the circuit to be tested directly from the line (ladder) diagram or logic block diagram, or from a screen that shows the PLR, PLR inputs, and PLR outputs in a pictorial form. When using the simulator mode with a line diagram or logic block diagram, the path of current flow is indicated by a change in the color of the connecting wires when current is flowing through them. When using the simulator in the pictorial screen, the outputs are highlighted when they are ON and the inputs are highlighted when they are closed. **See Figure 21-36.**

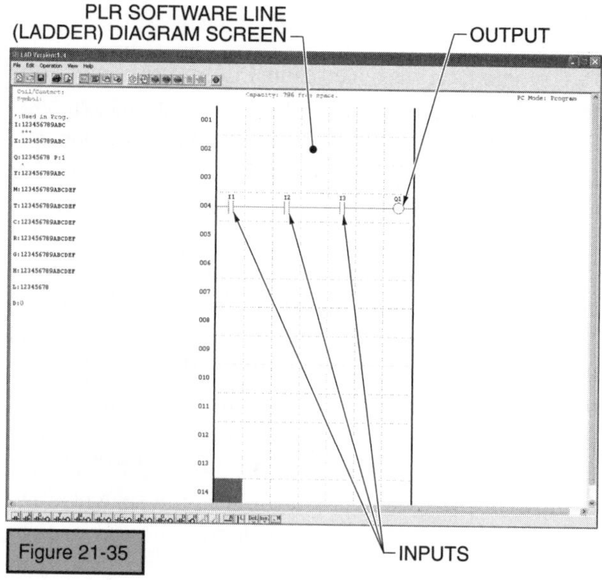

Figure 21-35

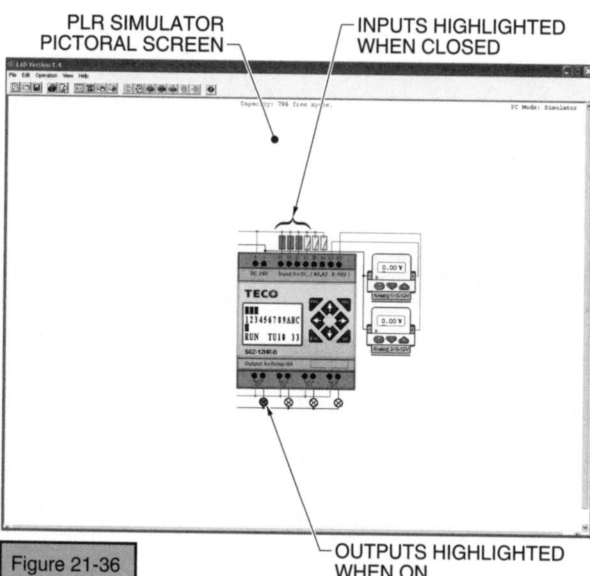

Figure 21-36

Refer to the CD-ROM in the back of the book for Chapter 21 Quick Quiz® and related information.

22 Fluid Power Circuits

Fluid Power

Fluid power is the transmission and control of energy by means of a pressurized fluid. A *fluid* is a liquid or a gas that can move and change shape without separating when under pressure. Pressurized fluid multiplies force simply and efficiently. Fluid power is used to move, cut, bend, punch, align, and control applications requiring forces ranging from several pounds to several hundred thousand pounds. Fluid power may use liquid (hydraulics) or air (pneumatics) to produce work.

Hydraulics

Hydraulics is the transmission of energy using a liquid (normally oil). A *liquid* is a state of matter that has a definite volume but not a definite shape. The flow rate in a hydraulic system is easily regulated because a liquid is relatively noncompressible. All hydraulic systems are closed systems because the fluid is returned, filtered, and reused. Hydraulic systems are preferred for applications requiring great forces. Forces of a few pounds to several hundred thousand pounds can be delivered by hydraulic cylinders, motors, and rotary actuators. Hydraulic systems are standard for most applications requiring a high force in a small area. **See Figure 22-1.**

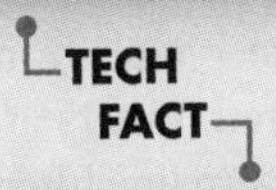

When a new hydraulic system is installed, clean the intake pump strainer after 25 hr of operation and again after 50 hr of operation. After the first two cleanings, clean every 600 hr of operation.

Hydraulic system applications include forklifts, forming and stamping machines, clamping, pile driving, industrial robots, farm equipment, mining equipment, most high-force industrial machining operations, and power steering systems for automobiles, trains, ships, and airplanes. In industrial machining operations, the hydraulic force can be directly used to perform work on the product. Such work includes clamping, bending, forming, punching, feeding, and ejecting parts, as well as moving worktables in and out of machine and clamping tools.

Force Transmission

Force transmission is accomplished by pressurizing the oil and directing it throughout the hydraulic system. Filtered oil from a reservoir flows into the hydraulic pump. The hydraulic pump develops flow as the pump rotates. The pump is normally driven by an electric motor or internal combustion engine. Electric motors are commonly used in industrial applications in which the hydraulic unit is fixed in one place. Internal combustion engines are used in mobile applications, such as earthmovers, forklifts, farm equipment, mining equipment, and automobiles. **See Figure 22-2.**

Drive Motors

Electric motors are the most common motor used to drive hydraulic pumps. The shaft of an electric motor is coupled to the shaft of a hydraulic pump directly using a motor coupling, or indirectly using belts. In many hydraulic systems, an electric motor is used to drive several different pumps simultaneously. Different pump sizes are used to deliver different flow volumes to different parts of the circuit.

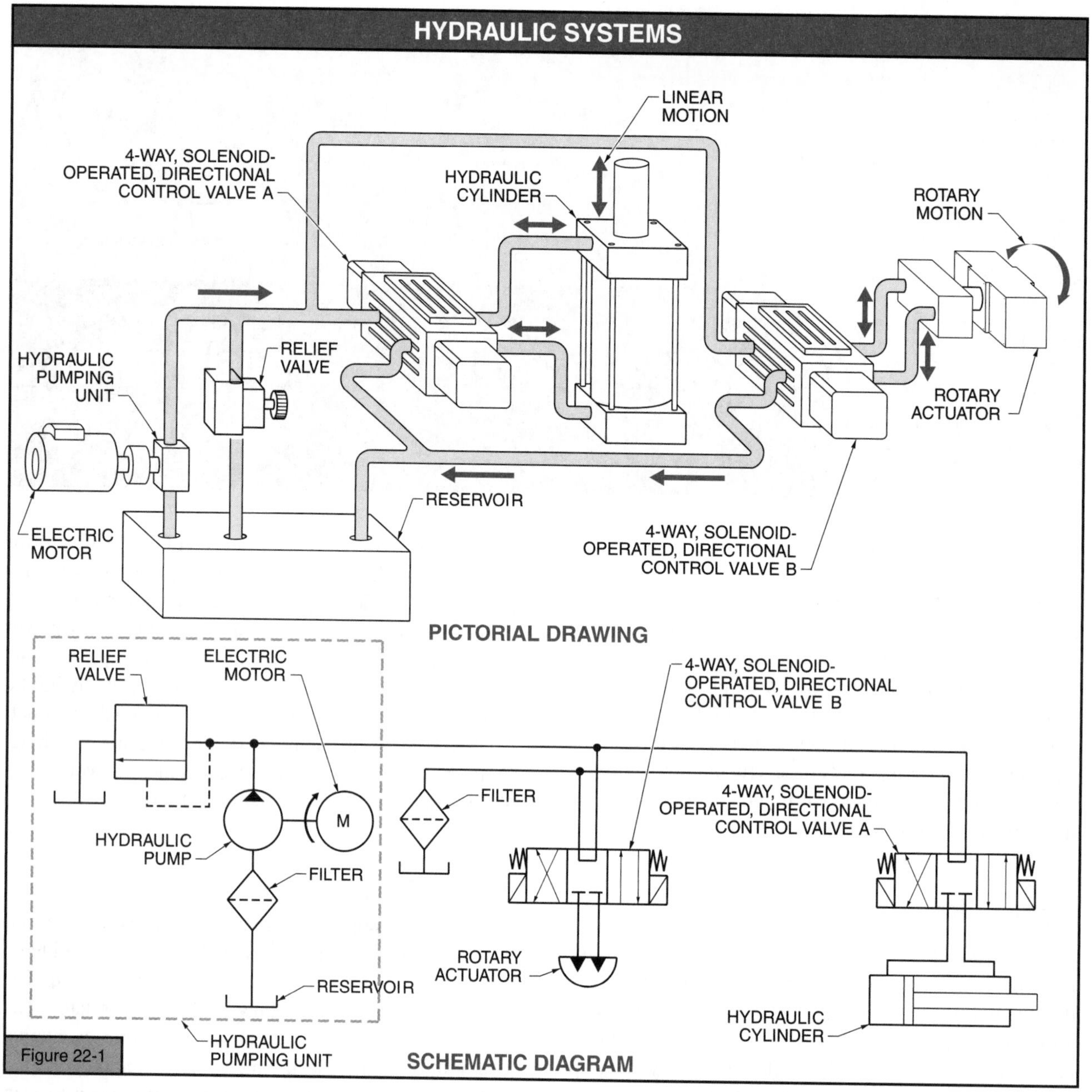

HYDRAULIC SYSTEMS
LINEAR MOTION
4-WAY, SOLENOID-OPERATED, DIRECTIONAL CONTROL VALVE A
HYDRAULIC CYLINDER
ROTARY MOTION
RELIEF VALVE
HYDRAULIC PUMPING UNIT
ROTARY ACTUATOR
RESERVOIR
ELECTRIC MOTOR
4-WAY, SOLENOID-OPERATED, DIRECTIONAL CONTROL VALVE B
PICTORIAL DRAWING
RELIEF VALVE
ELECTRIC MOTOR
4-WAY, SOLENOID-OPERATED, DIRECTIONAL CONTROL VALVE B
FILTER
M
4-WAY, SOLENOID-OPERATED, DIRECTIONAL CONTROL VALVE A
HYDRAULIC PUMP
FILTER
ROTARY ACTUATOR
RESERVOIR
HYDRAULIC CYLINDER
HYDRAULIC PUMPING UNIT
SCHEMATIC DIAGRAM
Figure 22-1

The electric motor is turned ON and OFF using a motor starter. The motor starter is normally controlled by a standard start/stop pushbutton station. Three-phase motors are preferred because they are more energy-efficient than 1ϕ motors. Single-phase motors are used if 3ϕ power is not available. Regardless of the motor used, the motor control circuit remains the same. The motor power circuit depends on the exact motor type and voltage used. **See Figure 22-3.**

TECH FACT *Buildup on the intake pump strainer causes cavitation (inability of a pump to draw a full charge of oil) and other pump problems.*

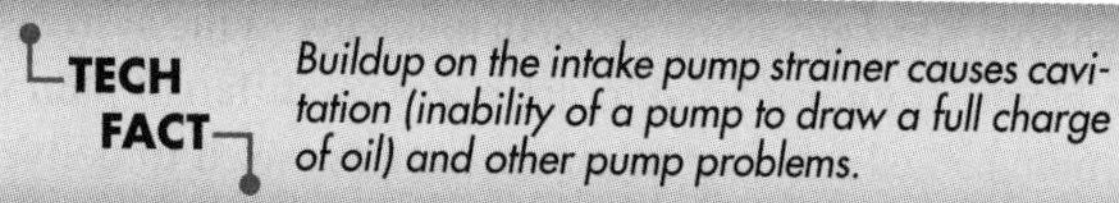

Hydraulic Pumps

A hydraulic pump creates a vacuum on its inlet port. The vacuum causes oil to be drawn into the pump. The pump traps the oil and forces it through its outlet port under pressure. **See Figure 22-4.**

A strainer is normally connected to the pump's inlet port. The inlet strainer prevents dirt from entering and damaging the hydraulic pump. The inlet strainer is normally mounted in a horizontal position to help prevent air from entering the system as the oil level is lowered. Filters are also commonly added to the return lines. Return line filters help remove particles before they enter the reservoir.

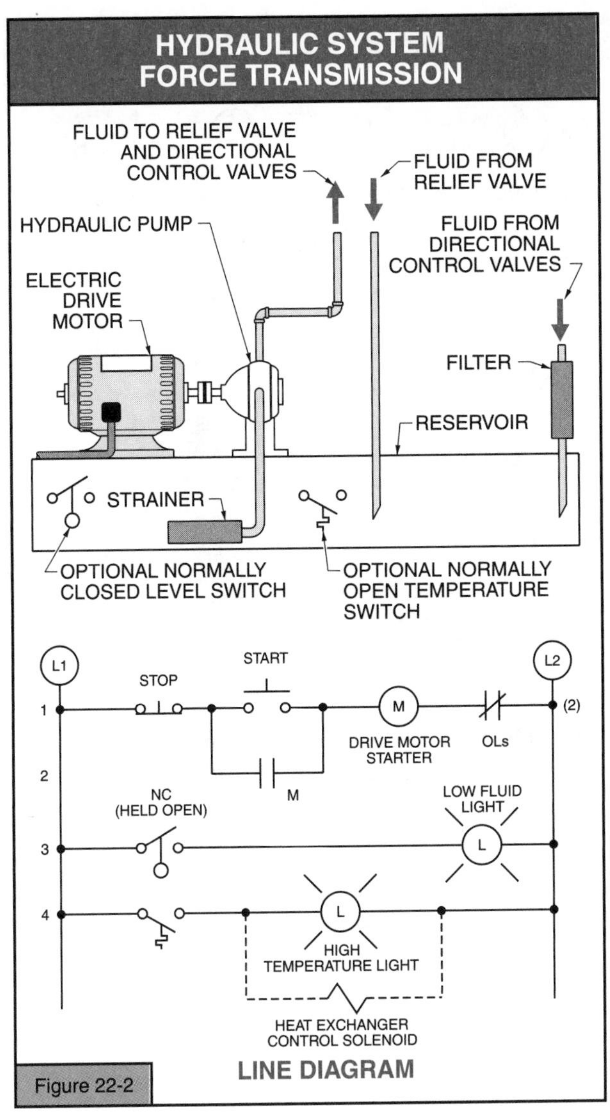

Figure 22-2

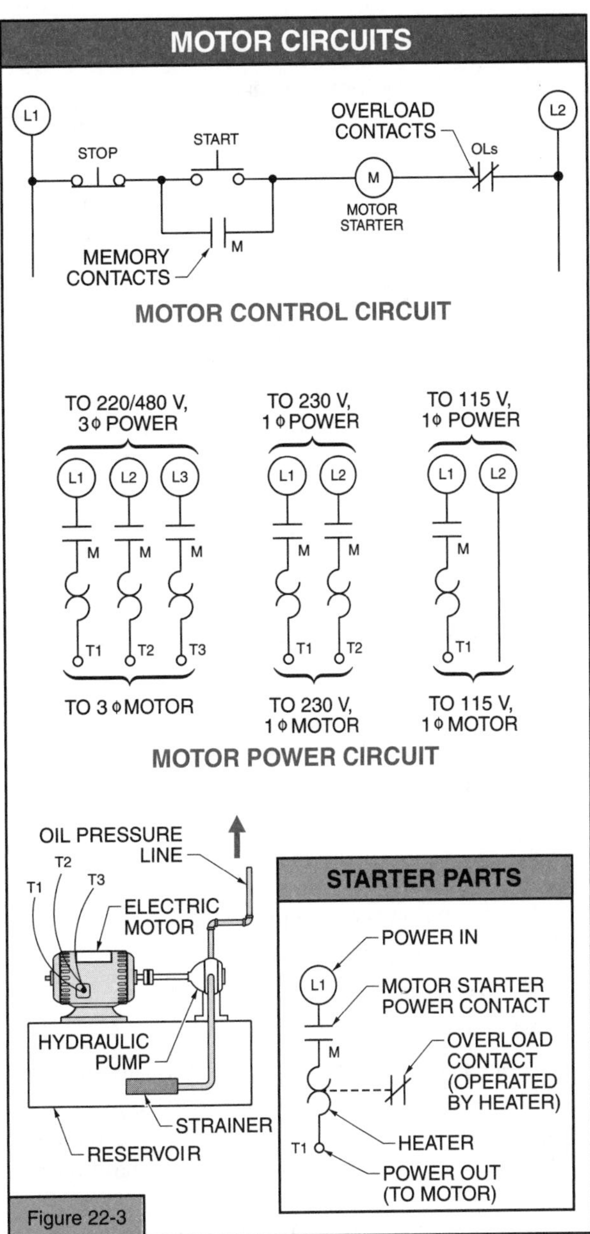

Figure 22-3

Hydraulic pumps may be classified as positive displacement or variable displacement. A *positive-displacement pump* is a pump that delivers a definite quantity of fluid for every stroke, revolution, or cycle. Positive-displacement pumps are the most common pumps used in fluid power applications.

The primary purpose of a positive-displacement pump is to produce flow, not pressure. The output flow of a positive-displacement pump is directly proportional to the speed of the pump shaft. For example, if the speed of the shaft increases by 25%, the output flow increases by 25%.

The amount of pump flow is stated in gallons per minute (gpm). Although a pump's output can be increased by increasing the speed of the pump, pump life is reduced when operated at high speeds. For this reason, a large pump operating at a slow speed is selected for an application over a small pump operating at a fast speed.

Pressure in a fluid power system is produced when the fluid flow is restricted. The pressure produced is directly proportional to the resistance placed against the fluid flow. Resistance to the flow of fluid is produced by the forces placed on the cylinders and other circuit components. The higher the resistance, the higher the pressure. Likewise, the lower the resistance, the lower the pressure.

The amount of force required to operate a pump depends on the amount of flow (in gpm) and pressure (in psi) at which the pump must operate. To calculate the required motor horsepower when an electric motor is used to produce the force, apply the formula:

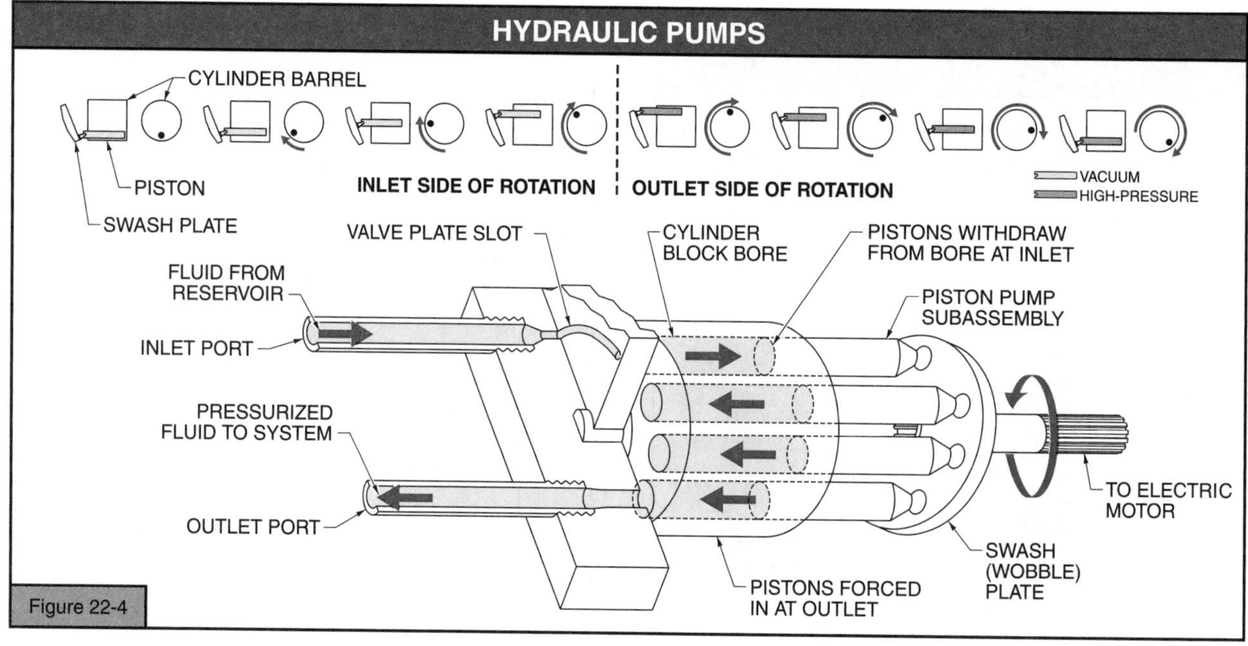

Figure 22-4

$$HP = \frac{Q \times P}{1714 \times Eff}$$

where

HP = required motor horsepower

Q = pump flow rate (in gpm)

P = pump pressure (in psi)

1714 = constant

Eff = motor efficiency (in %)

Motor efficiency is the measure of the effectiveness with which a motor converts electrical energy to mechanical energy. Motor efficiency varies with the type and the size of the motor. The larger the motor, the higher the efficiency. **See Figure 22-5.**

An 80% to 85% motor efficiency is used when determining motor size if the actual efficiency of a motor is not known. The 80% efficiency number produces a slightly higher horsepower requirement, but allows for greater overloading of the motor that may develop as additional requirements are placed on the fluid power system and/or the motor's efficiency is reduced over time from wear and improper ventilation. For this reason, an 80% efficiency number is used when calculating motor horsepower.

TECH TIP

Select a 25% larger cylinder and a 25% higher system pressure than is mathematically required to move the load when determining cylinder size and system pressure.

Example: Calculating Required Horsepower

What is the required horsepower of a motor that drives a pump that must deliver 10 gpm when the system pressure is 1000 psi?

$$HP = \frac{Q \times P}{1714 \times Eff}$$

$$HP = \frac{10 \times 1000}{1714 \times .8}$$

$$HP = \frac{10{,}000}{1371.2}$$

$$HP = \mathbf{7.29\ HP}$$

Although the required horsepower is 7.29 HP, a 7.29 HP motor is not available as a standard motor. The next larger standard motor size is selected when a nonstandard size motor is required for an application. For this application, a 7½ HP standard motor is selected. **See Figure 22-6.**

Tables show the required motor horsepower for pumps that must deliver flow rates of 1 gpm to 100 gpm at system pressures of 100 psi to 5000 psi. **See Appendix.** The required motor horsepower must be increased if the system pressure or flow rate is increased. This must be considered when selecting a motor because the motor size is normally not increased after the system is installed. After the system is installed, an increase in required horsepower results in overloading the motor. An overloaded motor is determined by measuring the current at the motor with a clamp-on ammeter. The current drawn by a motor should not exceed the current rating listed on the motor nameplate when the fluid power system is in normal operation.

TYPICAL MOTOR EFFICIENCIES		
HP	Standard Motor*	Energy-Efficient Motor*
1	76.5	84.0
1½	78.5	85.5
2	79.9	86.5
3	80.8	88.5
5	83.1	88.6
7½	83.8	90.2
10	85.0	90.3
15	86.5	91.7
20	87.5	92.4
25	88.0	93.0
30	88.1	93.1
40	89.3	93.6
50	90.4	93.7
75	90.8	95.0
100	91.6	95.4
125	91.8	95.8
150	92.3	96.0
200	93.3	96.1
250	93.6	96.2
300	93.8	96.5

* in %

Figure 22-5

Pressure Relief Valves

A *pressure relief valve* is a valve that limits the amount of pressure in a fluid power system. Pressure relief valves are used as safety valves to prevent damage that may be caused by excessively high pressure. Pressure relief valves are 2-way, normally-closed valves that are set to open at a given pressure. Pressure relief valves are closed and no fluid is allowed through the valve when the pressure in the system is below the setting of the valve. The valve opens and relieves the pressure in the system when the pressure in the system increases to the setting of the valve (or higher). The relief valve closes when the pressure in the system drops below the relief valve setting. **See Figure 22-7.**

Cylinders

Cylinders are the most common fluid power actuators. A *cylinder* is a device that converts fluid energy into a linear mechanical force. An *actuator* is a device that converts fluid energy into mechanical motion. Actuators can produce linear motion (cylinders), rotary motion (fluid motors), or partial rotary motion (rotary actuators).

Fluid power cylinders are used in applications requiring pushing, pulling, lifting, clamping, tilting, blocking, etc. The area of the cylinder and the pressure in the fluid power system determine the force that the cylinder produces. The total force of a fluid power cylinder is determined by calculating the area of the cylinder and then applying the formula to determine total force. To calculate total force of a fluid power cylinder when extending, apply the procedure:

1. Calculate area of cylinder.

   $A = \pi r^2$

   where

   A = area (in sq in.)

   $\pi = 3.14$

   r^2 = radius squared

2. Calculate total force.

   $TF = P \times A$

   where

   TF = total force (in lb)

   P = pressure (in sq in.)

   A = area (in sq in.)

Example: Calculating Cylinder Extending Force

What is the total extending force of a fluid power cylinder that has a 4″ bore operating at 2000 psi of system pressure?

1. Calculate area of cylinder.

   $A = \pi r^2$

   $A = 3.14 \times 2 \times 2$

   $A = 12.56$ sq in.

2. Calculate total force.

   $TF = P \times A$

   $TF = 2000 \times 12.56$

   $TF =$ **25,120 lb**

TECH FACT

Types of hydraulic actuators include single-acting, external return, single-acting spring return, telescopic, double-acting double-ended rod, and tandem cylinder.

MOTOR RATINGS*		
Classification	Rating	Size
Milli	W	1, 1.5, 2, 3, 5, 7.5, 10, 15, 25, 35
Fractional	HP	1/20, 1/12, 1/8, 1/6, 1/4, 1/3, 1/2, 3/4
Full	HP	1, 1 1/2, 2, 3, 5, 7 1/2, 10, 15, 20, 25, 30, 40, 50, 60, 75, 100, 125, 150, 200, 250, 300
Full-special order	HP	350, 400, 450, 500, 600, 700, 800, 900, 1000, 1250, 1500, 1750, 2000, 2250, 2500, 3000, 3500, 4000, 4500, 5000, 5500, 6000, 7000, 8000, 9000, 10,000, 11,000, 12,000, 13,000, 14,000, 15,000, 16,000, 17,000, 18,000, 19,000, 20,000, 22,500, 30,000, 32,500, 35,000, 37,500, 40,000, 45,000, 50,000

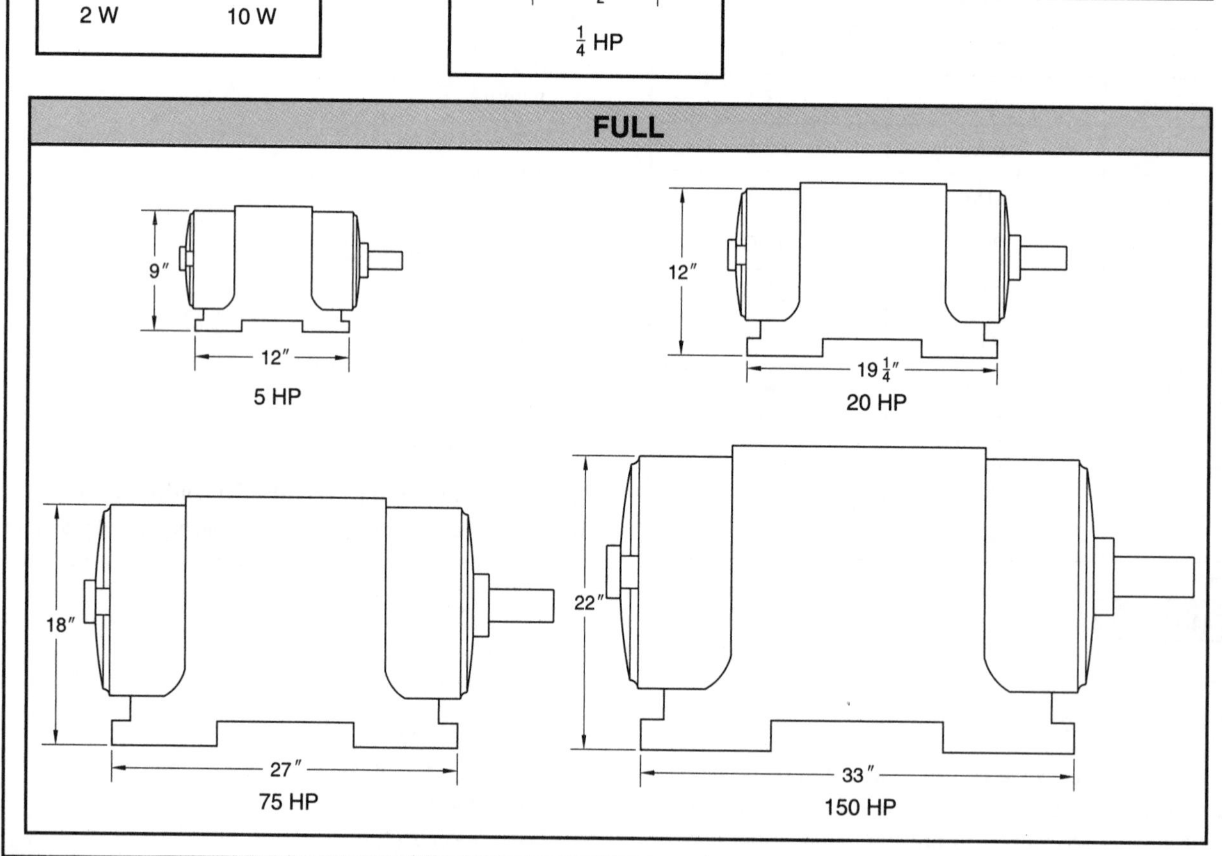

* dimensions are for typical motor sizes and vary by manufacturers

Figure 22-6

The area of the rod must be considered when determining the total force a cylinder can produce when retracting. The reason for this is that the rod area reduces the total (effective) area at which the fluid can push against the piston when retracting. **See Figure 22-8.** The area of the rod is subtracted from the total area of the cylinder when calculating the total force of a retracting cylinder. To calculate the area of a cylinder when retracting, apply the procedure:

1. Calculate area of cylinder.

   $A = \pi r^2$

   where

   A = area (in sq in.)

   $\pi = 3.14$

   r^2 = radius of cylinder squared

2. Calculate area of rod.

   $A_R = \pi r^2$

   where

   A_R = area of rod (in sq in.)

   $\pi = 3.14$

   r^2 = radius of rod squared

3. Calculate effective area of cylinder when retracting.

   $A_E = A - A_R$

   where

   A_E = effective area of cylinder (after subtracting rod area)

   A = area of cylinder (in sq in.)

   A_R = area of rod (in sq in.)

4. Calculate total retracting force.

   $TF = P \times A$

   where

   TF = total force (in lb)

   P = pressure (in sq in.)

   A = area (in sq in.)

Example: Calculating Cylinder Retracting Force

What is the total retracting force of a fluid power cylinder with a 2″ bore and a ½″ rod operating at 1000 psi of system pressure?

1. Calculate area of cylinder.

   $A = \pi r^2$

   $A = 3.14 \times 1 \times 1$

   $A = 3.14$ sq in.

2. Calculate area of rod.

   $A_R = \pi r^2$

   $A_R = 3.14 \times (.25 \times .25)$

   $A_R = 3.14 \times .0625$

   $A_R = .1962$ sq in.

3. Calculate effective area of cylinder when retracting.

   $A_E = A - A_R$

   $A_E = 3.14 - .1962$

   $A_E = 2.9438$ sq in.

4. Determine total retracting force.

   $TF = P \times A$

   $TF = 1000 \times 2.9438$

   $TF =$ **2943.9 lb**

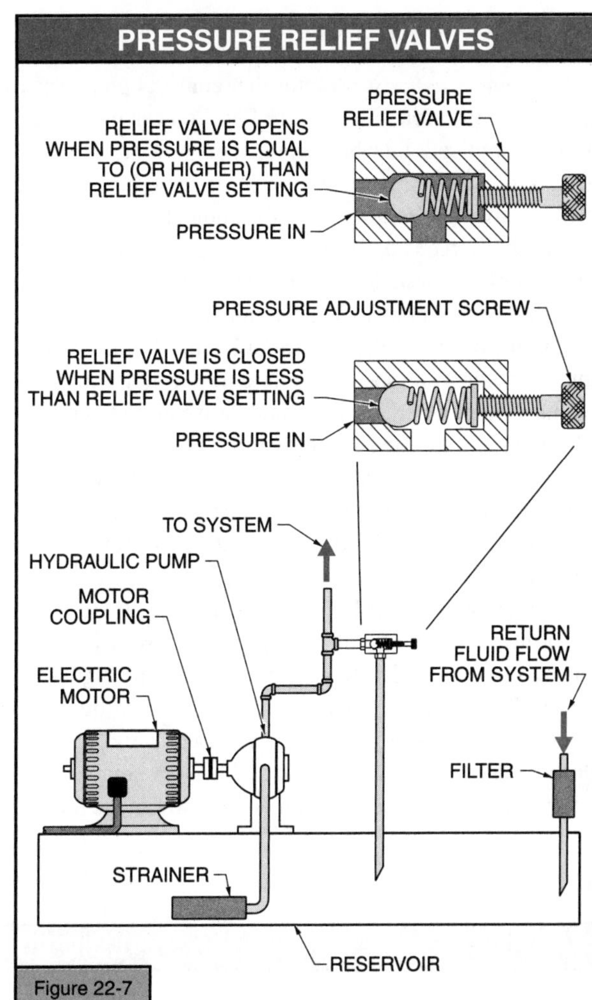

Figure 22-7

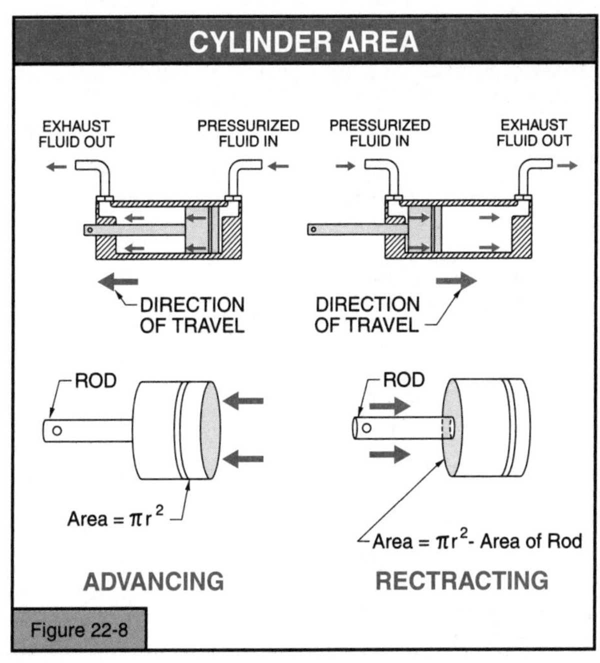

Figure 22-8

Fluid Power Motors

A *fluid power motor* is a motor that converts pressurized fluid force into rotating motion. Fluid power motors are used in applications where electric motors are not suitable. Fluid power motor applications include operating in explosive environments, high ambient temperatures, totally submerged applications, and applications in which the motor is often stalled. No damage is done to the source of power or components when a fluid power motor stalls.

Fluid power motors are variable speed and variable torque motors. The speed of a fluid power motor depends on the volume of pressurized fluid delivered to the motor. The volume is adjusted by increasing or decreasing the flow rate into the motor. The flow rate is set by a flow control valve. The higher the flow rate, the faster the motor rotates. The lower the flow rate, the slower the motor rotates. **See Figure 22-9.**

The amount of fluid pressure determines the amount of torque. The amount of fluid pressure is set by a pressure control valve. The higher the pressure, the higher the torque. The lower the pressure, the lower the torque.

Most fluid power motors are designed to run in either direction. The direction of rotation of a fluid power motor is reversed by reversing the direction of fluid flow in the motor.

Rotary Actuators

A *rotary actuator (fluid power oscillator)* is an actuator that moves back and forth over a fixed arc that is less than one complete revolution (360°). The actuator may move any number of degrees for which it is designed. Common movements include 15°, 30°, 45°, 90°, 120°, and 180°.

Rotary actuators are used for clamping parts, opening and closing gates, loading parts into a machine, unloading or dumping parts, transferring parts, and applications that require a turning force. **See Figure 22-10.**

The advantage of using a rotary actuator is that it can be used to move very small (light weight), and very large (heavy) loads. For example, rotary actuators are used on ships to open and close hatches, on railroads for controlling valves and dumping freight cars, and in industry for lifting, lowering, moving, pulling, pushing, and transferring parts of all sizes.

Directional Control Valves

A *directional control valve* is a valve that connects, disconnects, or directs the fluid pressure from one part of a circuit to another. Directional control valves are identified by the number of positions, ways, and types of actuators the valve contains.

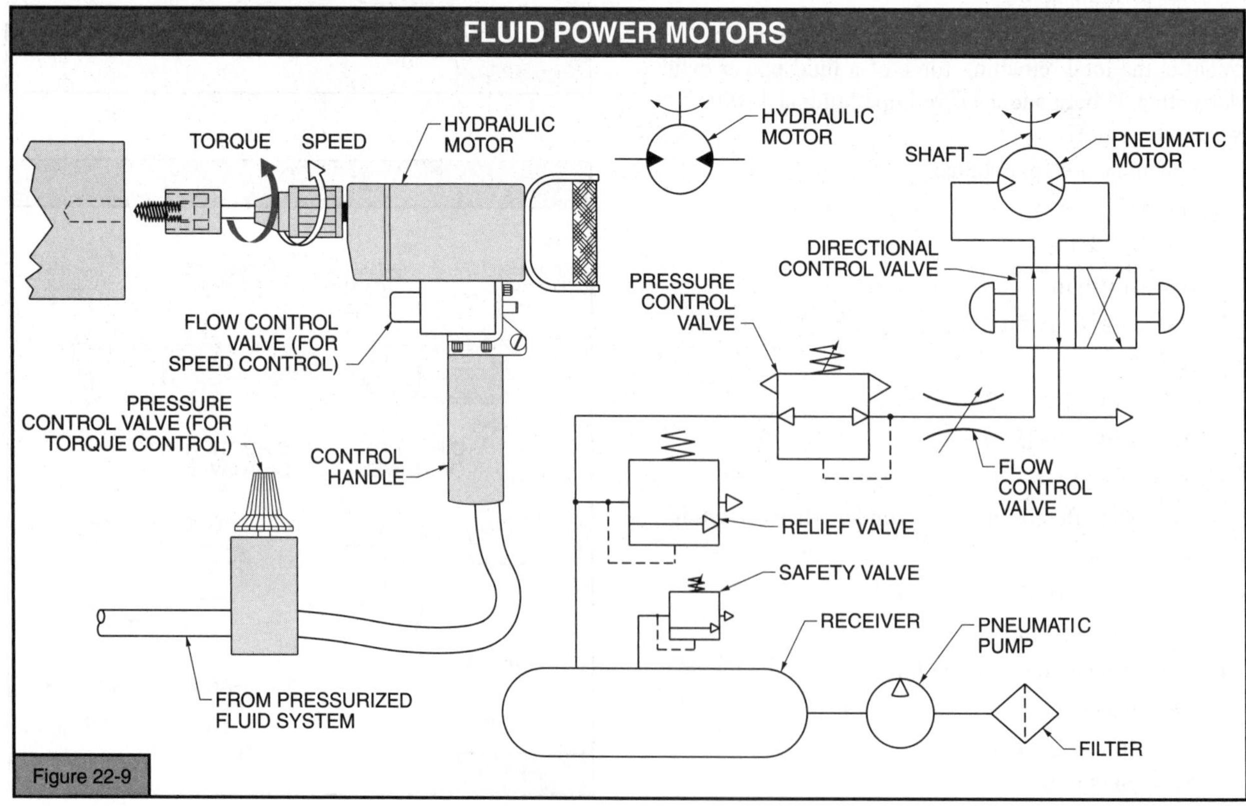

Figure 22-9

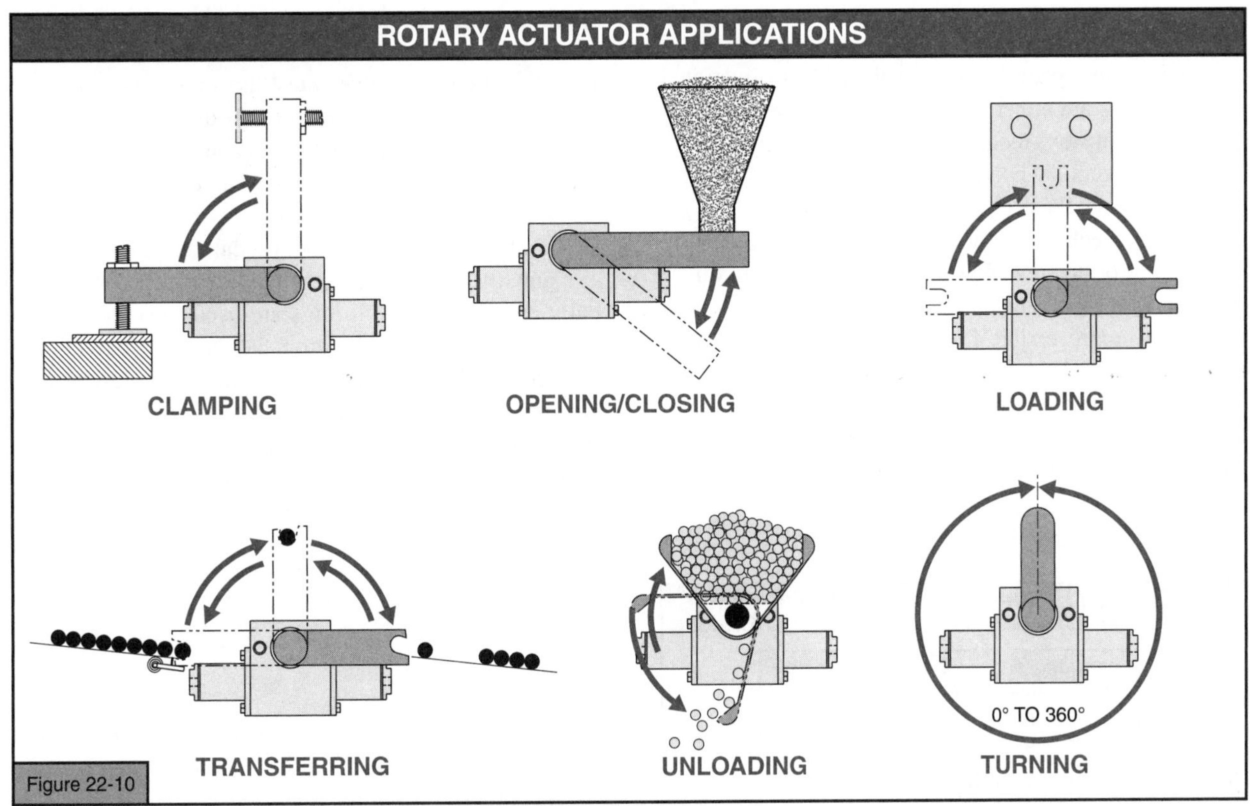

Figure 22-10

Positions

A directional control valve is placed in different positions to start, stop, or change the direction of fluid flow. A *position* is the number of positions within the valve in which the spool is placed to direct fluid through the valve. A directional control valve normally has two or three positions. Some directional control valves have four or more positions.

Two-position valves are used to place a hydraulic or pneumatic cylinder in the fully extended or fully retracted position. Three-position valves are used to place a hydraulic cylinder in the fully extended position, fully retracted position, or any position between fully extended and fully retracted. Four- (or more) position valves are used in special applications. **See Figure 22-11.**

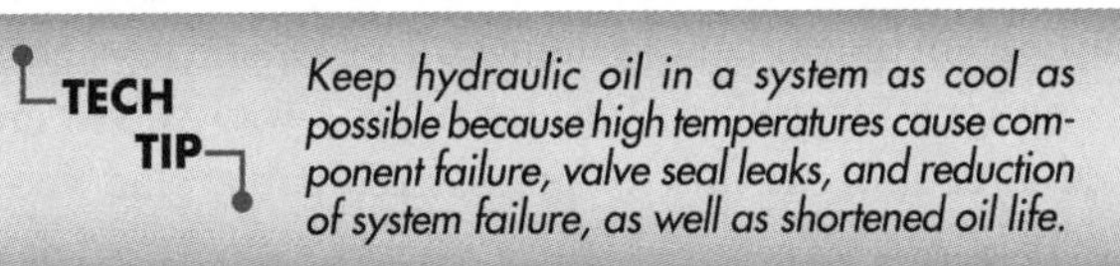
TECH TIP *Keep hydraulic oil in a system as cool as possible because high temperatures cause component failure, valve seal leaks, and reduction of system failure, as well as shortened oil life.*

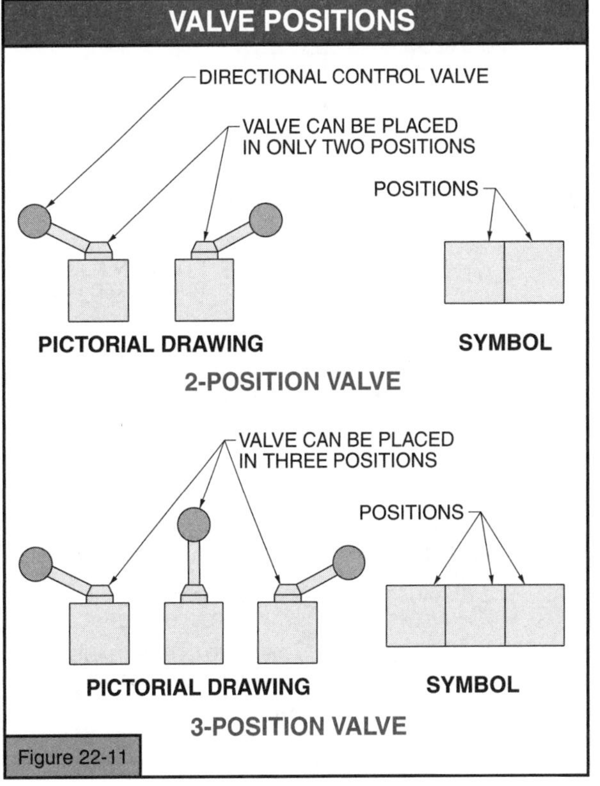

Figure 22-11

Ways

A *way* is a flow path through a valve. Most directional control valves are either 2-way, 3-way, or 4-way valves. The number of ways required depends on the application. Two-way directional control valves have two main ports that allow or stop the flow of fluid. Two-way valves are used as shutoff, check, and quick-exhaust valves. Three-way valves allow or stop fluid flow or exhaust. Three-way valves are used to control single-acting cylinders, fill-and-drain tanks, and nonreversible fluid power motors. Four-way directional control valves have four (or five) main ports that change fluid flow from one port to another. Four-way valves are used to control the direction of double-acting cylinders and reversible fluid motors. **See Figure 22-12.**

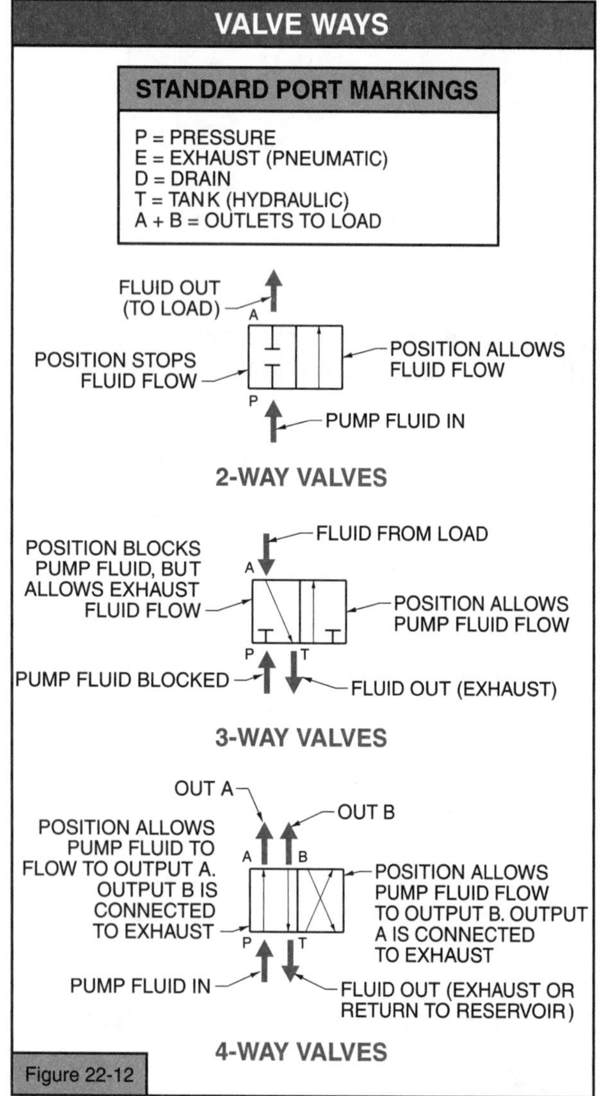

Figure 22-12

Valve Actuators

Directional control valves must have a means to change the valve position. An actuator is a device that changes the position of a valve. Directional control valve actuators include pilots, manual levers, mechanical levers, springs, and solenoids. **See Figure 22-13.**

Pushbutton, lever, and foot (pedal) actuators are used to manually operate valves. Solenoids are used to electrically operate valves. Springs are used to set a valve in a "normal" position. Actuators can be used individually or in combination. For example, many solenoid-operated valves also include a manual operator. The manual operator is normally included so that the valve can be manually tested when troubleshooting the circuit, and to allow the valve to be switched when there is no electrical power in the circuit.

Normally Closed and Normally Open Valves

A *normally closed (NC) valve* is a valve that does not allow pressurized fluid to flow out of the valve in the spring-actuated position. A *normally open (NO) valve* is a valve that allows pressurized fluid to flow out of the valve in the spring-actuated position. Two-way and 3-way valves may be either normally open or normally closed. The valve must have a spring position to be normally open or normally closed. **See Figure 22-14.**

Normally closed and normally open are used in the fluid power and electrical fields. The terms have the opposite meaning in each field. For example, normally closed in the fluid power field means that no fluid pressure flows in the valve's normal position. Normally closed in the electrical field means that electrical pressure flows in the switch's normal position. Care must be taken not to confuse the meaning of normally closed and normally open when working on fluid power equipment that is electrically controlled.

Two-Way Valve Applications

Two-way valves are used to start and stop the flow of a liquid or gas. Two-way valves have two ports located on the valve. One port is connected to the pressurized fluid and the other port is the outlet port. The pressurized fluid may be air (pneumatic system), oil (hydraulic system), or any type of product. For example, 2-way valves are used to control the flow of pressurized paint, glue, food products, gas, etc.

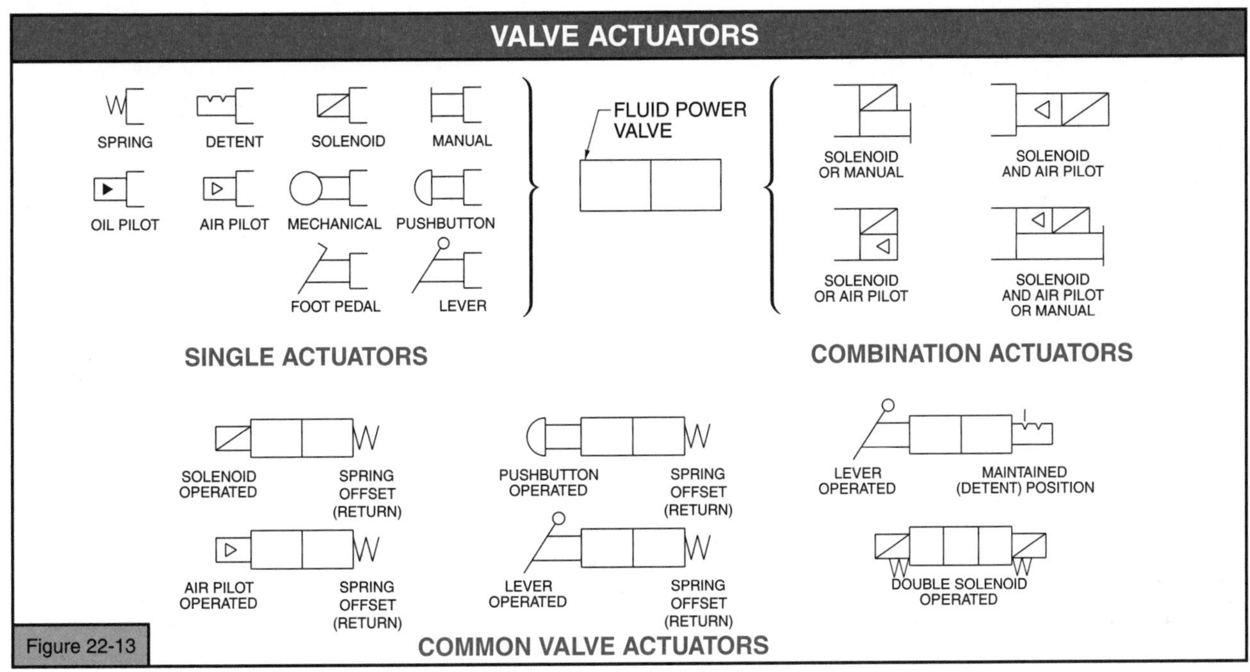

Figure 22-13

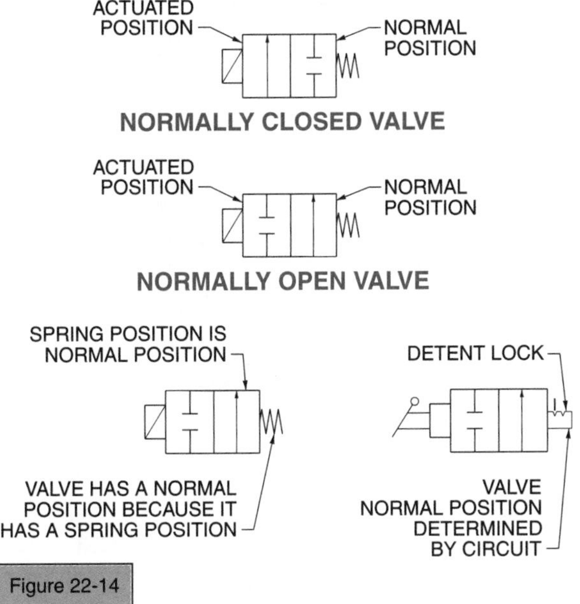

Figure 22-14

Two-way valves that include a spring position are either normally closed or normally open. A normally closed valve does not allow the pressurized fluid to flow through it in the normal (spring) position. **See Figure 22-15.**

The 2-way valve is a solenoid-operated, spring-return valve. The spring position is the normal position and the valve is in the spring position any time the solenoid is not energized. The valve changes position and allows the pressurized fluid to flow when the solenoid is energized.

Two-way valves may also be normally open. A normally-open valve allows the pressurized fluid to flow through it in the normal (spring) position. **See Figure 22-16.**

The 2-way valve is a solenoid-actuated, spring-return valve. The spring position is the normal position and the valve is in the spring position any time the solenoid is not energized. The valve changes position and does not allow the pressurized fluid to flow when the solenoid is energized. The valve changes back to the spring position and the pressurized fluid is allowed to flow through the valve when the solenoid is not energized.

Four-way valves are used to control the flow of liquid in applications such as hydraulic cranes.

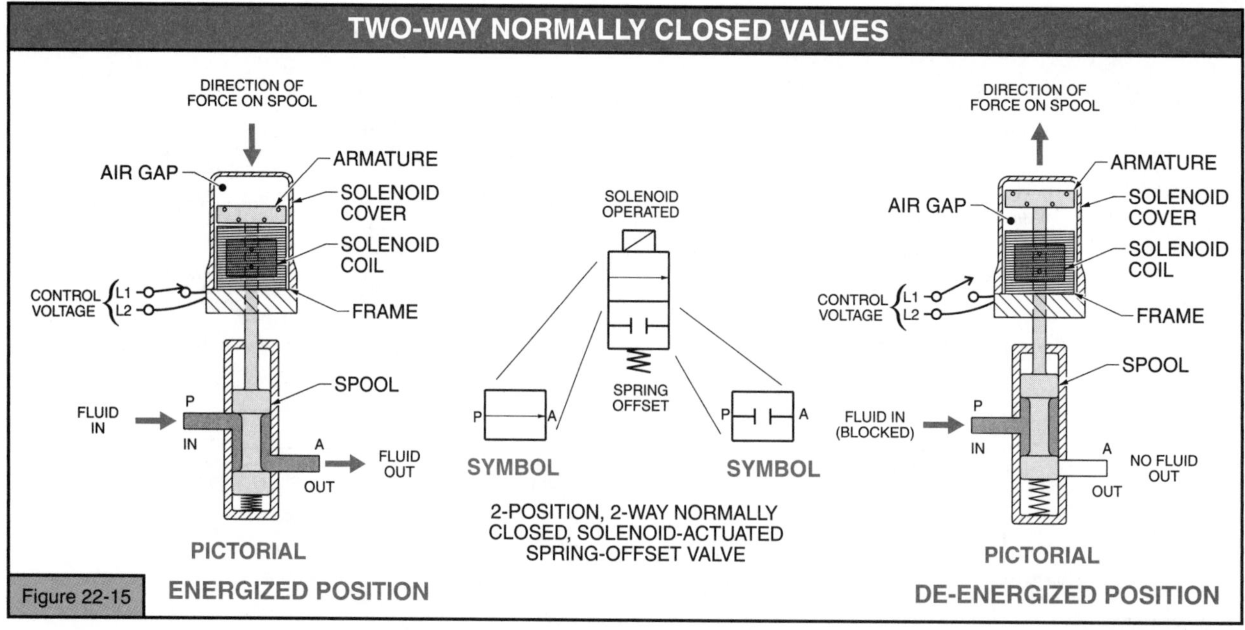

Figure 22-15

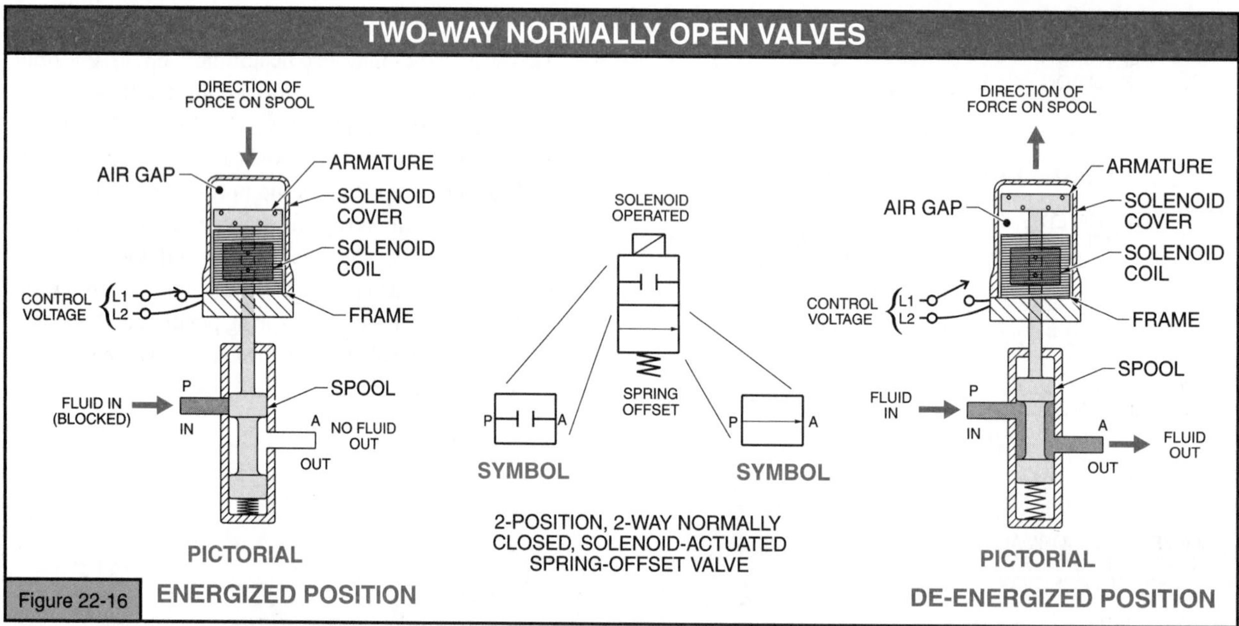

Figure 22-16

Solenoid-actuated, 2-way valves are commonly used in applications that require the starting and stopping of product flow. For example, a 2-way valve may be used to control the flow of product into a container. The pressurized product (food, etc.) is connected to the valve. The valve is opened to dispense the product and closed to stop the flow. The product container may be detected by a photoelectric or proximity switch. Once detected, a one-shot timer is used to energize the solenoid for a preset amount of time. **See Figure 22-17.**

Three-Way Valve Applications

Three-way valves are used to start and stop the flow of a liquid or gas and also allow for the exhaust (or return to reservoir) of the compressed fluid. Three-way valves have three ports located on the valve. One port is connected to the pressurized fluid, another port is the outlet port, and the third port is the exhaust (air) or reservoir (oil) port. The pressurized fluid is normally air or oil. Unlike 2-way valves, 3-way valves are generally not used to control any type of product flow. This is because the flow of a product is normally only started or stopped, not returned to a tank. Three-way valves are generally used to control single-acting cylinders and nonreversible fluid power motors. A *single-acting cylinder* is a cylinder that requires pressurized fluid in only one port, such as a spring-return cylinder, spring-advance cylinder, or cylinder in which the weight of the load is used to move the cylinder in one direction only.

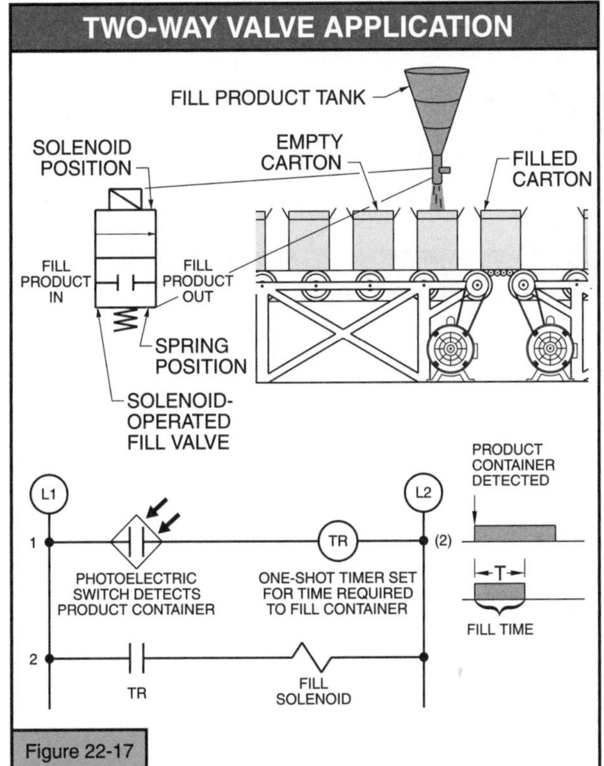

Figure 22-17

Like 2-way valves, 3-way valves that include a spring position are either normally closed or normally open. A normally-closed valve does not allow the pressurized fluid to flow through it in the normal (spring) position. **See Figure 22-18.**

The 3-way valve is a solenoid-actuated, spring-return valve. The spring position is the normal position and the valve is in the spring position any time the solenoid is not energized. The valve changes position and allows the pressurized fluid to flow when the solenoid is energized. The valve returns to the spring position and connects the outlet port to the exhaust (or tank port) when the solenoid is not energized. This allows any air in the cylinder to exhaust (pneumatic system), or any oil in the cylinder to return to reservoir (hydraulic system).

Three-way valves may also be normally open. A normally-open valve allows the pressurized fluid to flow through it in the normal (spring) position. **See Figure 22-19.**

The 3-way valve is a solenoid-actuated, spring-return valve. The spring position is the normal position and the valve is in the spring position any time the solenoid is not energized. The valve changes position and does not allow the pressurized fluid to flow when the solenoid is energized. The pressurized fluid from the circuit is connected to the outlet (exhaust) port. The valve returns to the spring position and the pressurized fluid is allowed to flow through the valve when the solenoid is de-energized.

Solenoid-actuated, 3-way valves are commonly used in applications that include a spring-return cylinder. For example, a 3-way valve can be used to control a spring-return cylinder that is used to clamp a part in place. The pressurized fluid (air or oil) is used to advance the cylinder and clamp the part under high force. The spring in the cylinder is used to return the cylinder. The advantage of using a spring-return cylinder is that only one fluid power line must be run to the cylinder. **See Figure 22-20.**

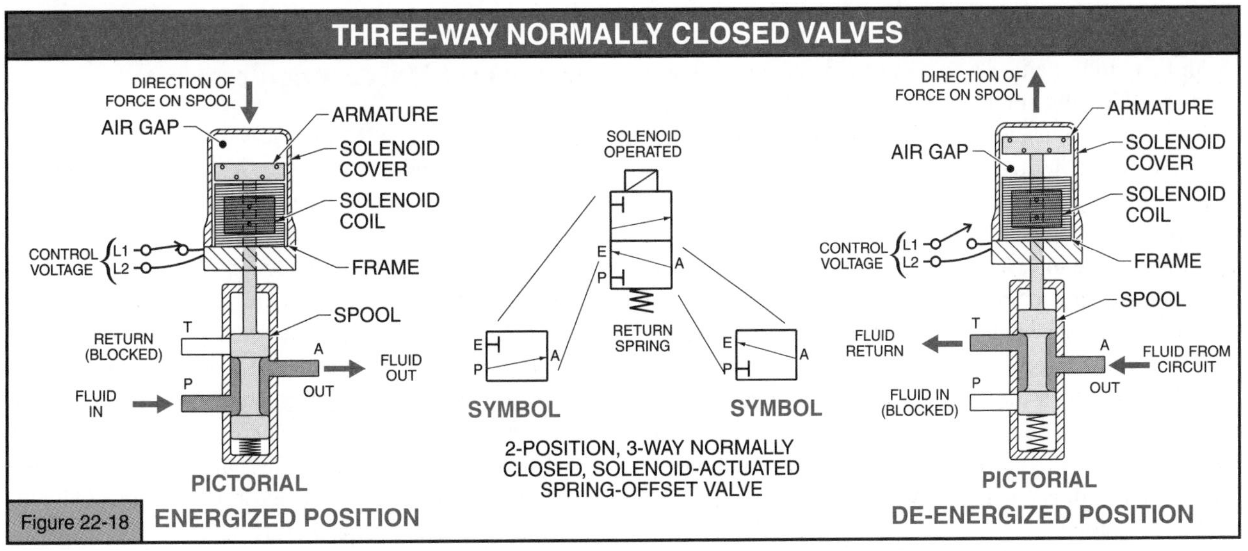

Figure 22-18

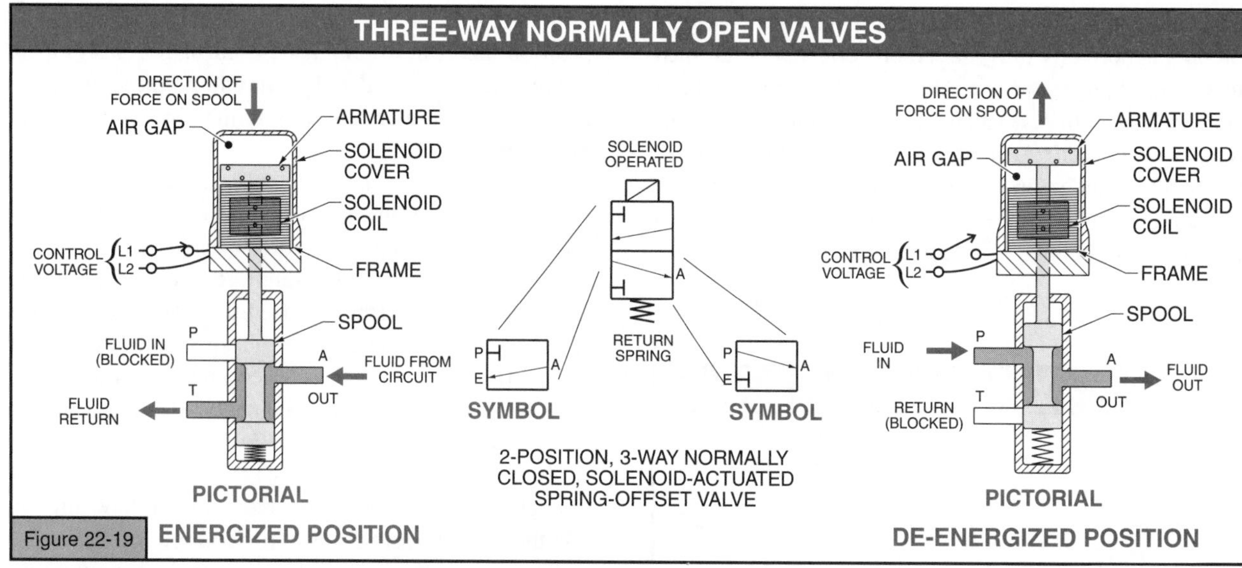

Figure 22-19

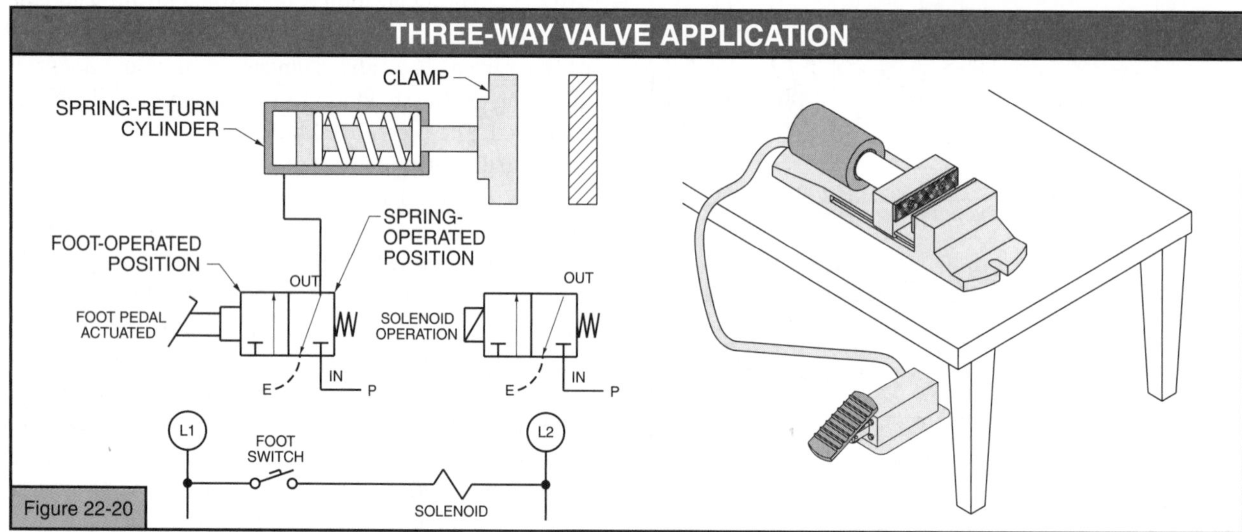

Figure 22-20

In this circuit, either a foot-operated valve or a solenoid-actuated valve controlled by a foot switch can be used. The advantage of using the foot-operated valve is that no electricity is required to operate the circuit. The advantage of using a solenoid-actuated valve is that additional circuit control is easily added. For example, a limit switch may be added to the electrical circuit to prevent the cylinder from being operated unless a guard is in place. The guard activates the limit switch and the limit switch is connected in series with the foot switch.

Likewise, a pressure switch could also be added to prevent the clamp cylinder from advancing unless the required system pressure was reached. A low system pressure could create a dangerous condition by not clamping the part at the required force.

Four-Way Valve Applications

Four-way valves are used to control double-acting cylinders and reversible fluid power motors. A double-acting cylinder is a cylinder that requires fluid pressure to advance and retract the cylinder. A fluid power motor is a device that converts fluid energy into rotary mechanical motion. Four-way valves have four (or five) ports located on the valve. One port is connected to the pressurized air, two ports are used as the outlet ports, and one (or two) ports are used as the exhaust ports. The two outlet ports are connected to a double-acting cylinder or reversible fluid power motor. Only one tank (drain) port is normally used if the valve is controlling oil. Either one exhaust (single exhaust) or two exhaust (dual exhaust) ports are used if the valve is controlling air. **See Figure 22-21.**

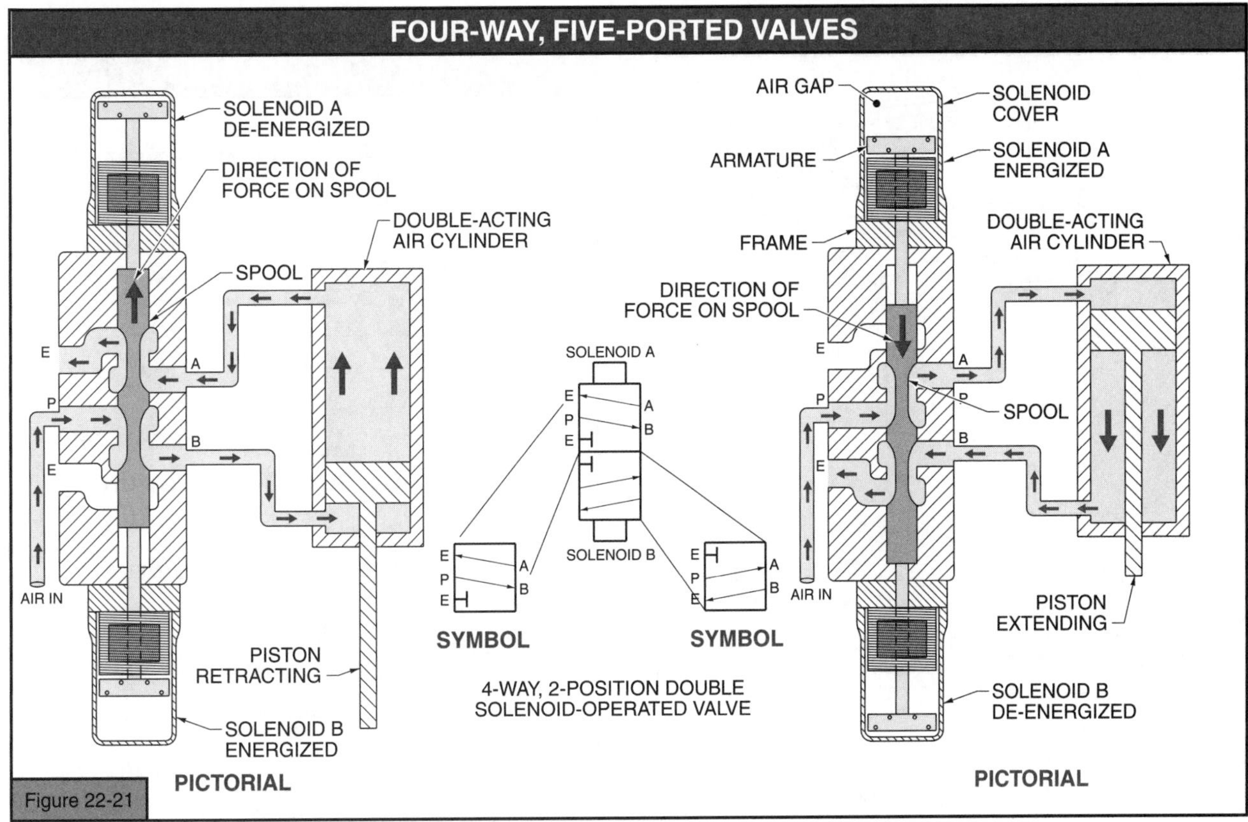

Figure 22-21

Unlike 2-way and 3-way valves, 2-position, 4-way valves do not have a normally open or normally closed position. This is because the pressurized fluid is always flowing out from one of the two outlet ports. The valve changes the outlet port that the pressurized fluid is flowing out of, but does not stop it. A 3-position valve must be used if an application requires the fluid not to flow out of a 4-way valve in one position of the valve. In a 3-position valve, the center position is normally used to stop the flow of fluid through the valve. **See Figure 22-22.**

In this application, a 4-way valve is used to control two double-acting cylinders. A 2-position valve is used if the cylinders are to be either fully retracted or fully extended. A 3-position valve is used if the cylinders are to be stopped in any position between fully extended and fully retracted. The 3-position valve includes a center position that blocks the flow of fluid to both sides of the cylinder. In a hydraulic system, the cylinders are held in whatever position they are in when the valve switches to the center position. The valve switches to the center position any time both solenoids are de-energized. The valve is a spring-centered valve. Spring-centered valves work with oil because oil is considered to be noncompressible. An air-operated application would not work well because air is compressible and the weight on the cylinders could allow the cylinders to move.

Flow Control Valves

A *flow control valve* is a valve used to control the volume of fluid that flows in a part of the circuit. By controlling the flow, the cylinder speed or the fluid motor speed is controlled. Liquid flow rate is normally expressed in gpm. One gallon is equal to 231 cu in. Air flow is normally expressed in cubic feet per minute (cfm). Most flow control valves include a built-in check valve. A *check valve* is a valve that allows the fluid to flow in one direction only. **See Figure 22-23.**

When a flow control valve includes a check valve, the valve can control fluid flow in one direction and allow full flow in the other direction. To control the amount of fluid flowing through the valve, the valve includes a flow adjustment. The more the valve is opened, the greater the flow through the valve. The more the valve is closed, the less the flow through the valve. **See Figure 22-24.**

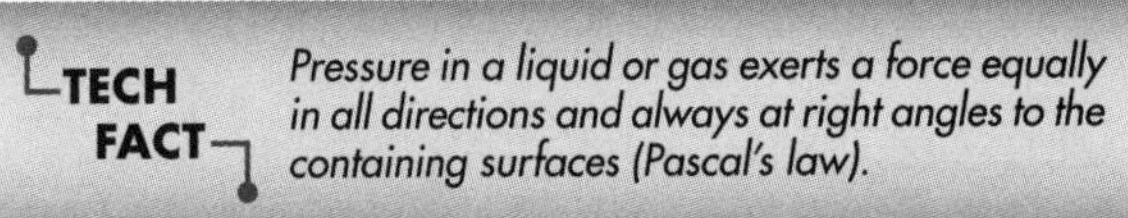

Pressure in a liquid or gas exerts a force equally in all directions and always at right angles to the containing surfaces (Pascal's law).

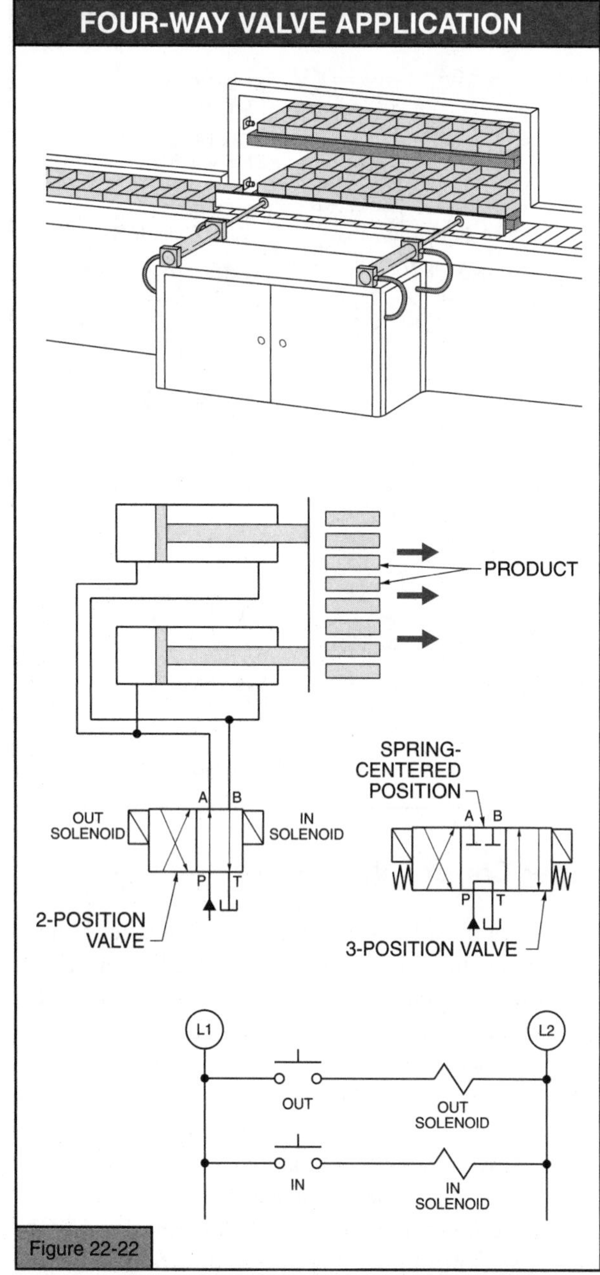

Figure 22-22

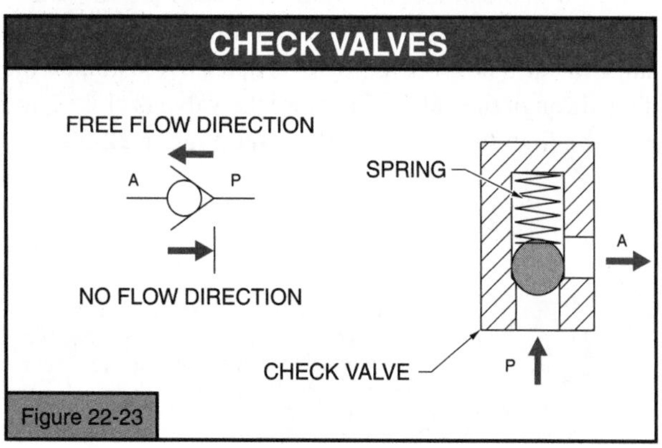

Figure 22-23

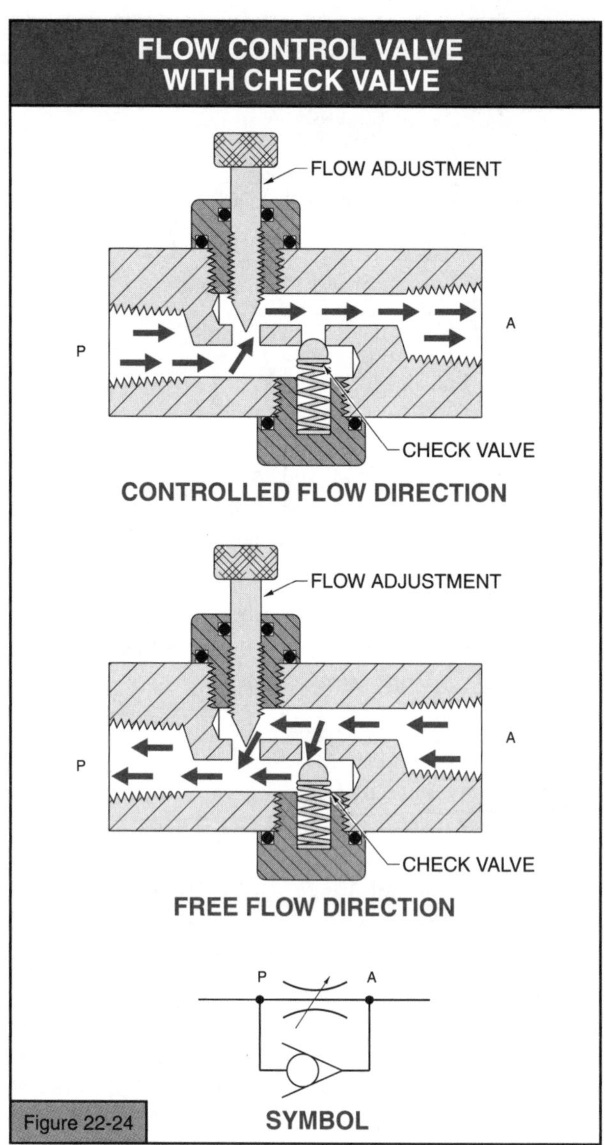

Figure 22-24

Pneumatics

Pneumatics is the transmission of energy using air. Air is the mixture of gases that are present in the atmosphere. Air has the properties of a fluid. Unlike molecules of a solid or liquid, air molecules are not freely attracted to each other and are easily compressible. Pneumatic systems are open systems because the air is exhausted to the atmosphere after it is used. **See Figure 22-25.**

Pneumatic systems are preferred for applications requiring low forces. The advantages of using pneumatic systems in low-force applications include relatively low cost, high speeds, and small leaks that do not shut down the operation. Examples of pneumatic applications include pneumatic wrenches, air brakes, dental drills, air clamps, injection molding machines, automatic gates and doors, and impact guns.

PNEUMATIC SYSTEMS

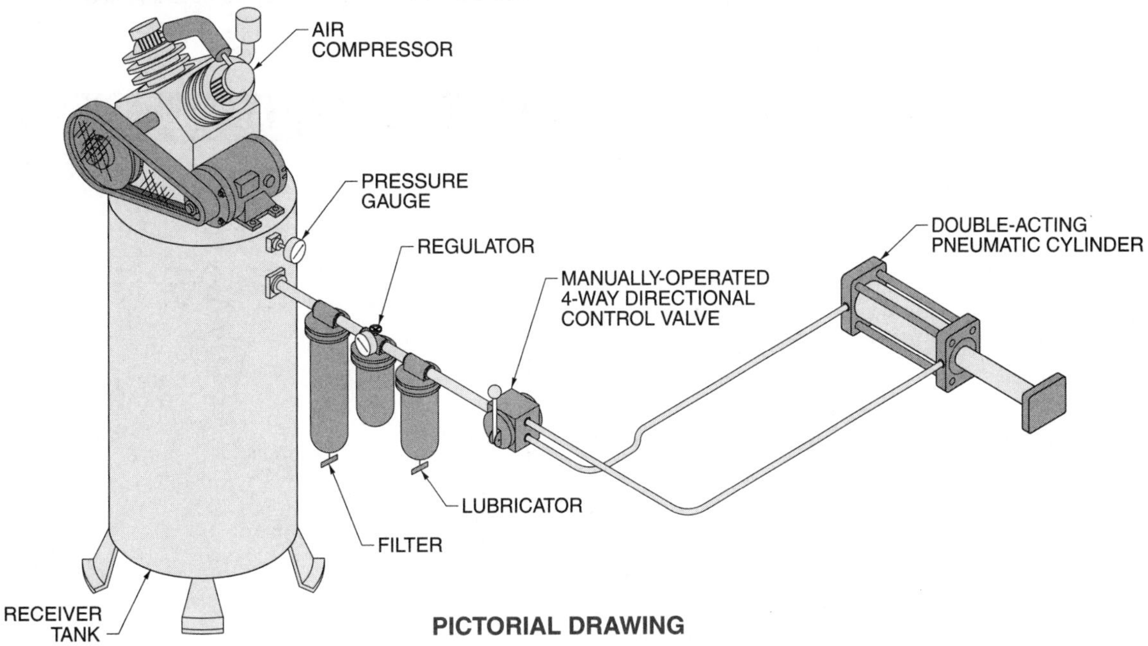

AIR COMPRESSOR
PRESSURE GAUGE
REGULATOR
MANUALLY-OPERATED 4-WAY DIRECTIONAL CONTROL VALVE
DOUBLE-ACTING PNEUMATIC CYLINDER
LUBRICATOR
FILTER
RECEIVER TANK

PICTORIAL DRAWING

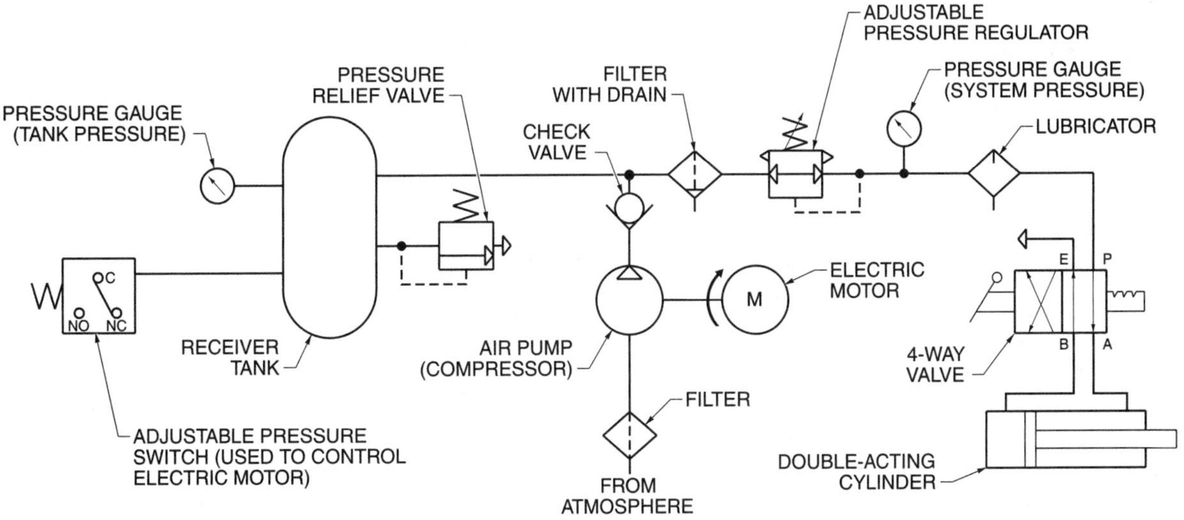

ADJUSTABLE PRESSURE REGULATOR
PRESSURE RELIEF VALVE
FILTER WITH DRAIN
PRESSURE GAUGE (SYSTEM PRESSURE)
PRESSURE GAUGE (TANK PRESSURE)
CHECK VALVE
LUBRICATOR
E
P
C
NO
NC
ELECTRIC MOTOR
M
RECEIVER TANK
AIR PUMP (COMPRESSOR)
B
A
4-WAY VALVE
FILTER
ADJUSTABLE PRESSURE SWITCH (USED TO CONTROL ELECTRIC MOTOR)
FROM ATMOSPHERE
DOUBLE-ACTING CYLINDER

SCHEMATIC DIAGRAM

Figure 22-25

Pneumatic systems are also used in wet locations or in areas that may contain an explosive atmosphere. For example, pneumatic systems are used in many mining operations to reduce the hazards of sparks and shock that may be present in an electrical system.

TECH FACT *The absolute pressure of a confined body of gas varies inversely with its volume, provided its temperature remains constant (Boyle's law).*

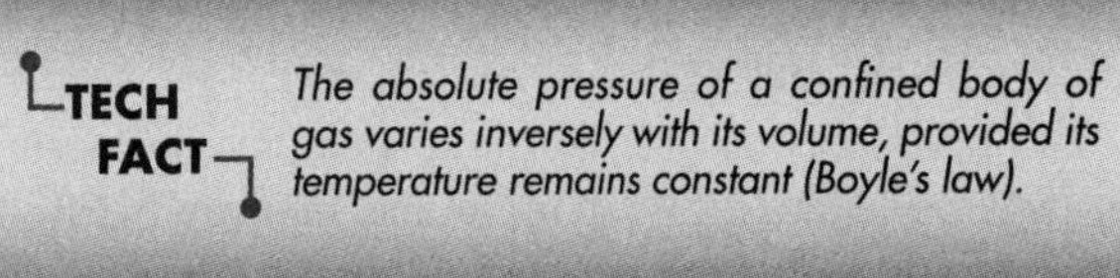

Air Compressors

An *air compressor (pump)* is a device that increases the pressure of atmospheric air to the desired pressure of the receiver tank. Air from the atmosphere flows into the inlet of the air compressor. The air compressor is normally controlled by an electric motor. The electric motor is controlled by a pressure switch. The pressure switch is connected to the receiver tank. The pressure switch energizes the motor starter that controls the electric motor driving the compressor when the pressure in the receiver tank drops below the setting of the pressure switch. The normally closed contacts on the pressure switch are used because the switch is in the normal position any time pressure is below the setting of the switch.

Pneumatic system piping is designed to minimize air loss between the receiver tank located near the compressor and the point of use. A 10% pressure loss is allowable under normal operating conditions. Pneumatic system piping is selected for its mechanical and noncorrosive properties rather than for its ability to withstand pressure because 250 psi is normally the maximum pressure used in a pneumatic system. Long distribution piping systems (and pressure drop) may be reduced by installing more than one compressor at convenient locations.

Fluid Conditioners

Atmospheric air contains some moisture and dirt. Some dirt can be removed by adding a filter on the inlet side of the compressor. Air filters are also normally located in the pressure line before components and immediately preceding power tools. The filters protect against failure from large particles as well as gradual failure from small particles that increase wear in pneumatic components. Filters must be cleaned or replaced before they become clogged. A clogged filter reduces compressor pumping efficiency.

Compressed air contains moisture. The presence of moisture in a pneumatic system can cause problems such as rust, pressure drops, and lubricant removal from surfaces. To condition the compressed air, the air should pass through an air filter, a pressure regulator, and a lubricator before it is delivered to components using the compressed air. Filters are used to remove dirt, dust, and moisture from the air. Lubricators are used to add a small amount of oil to the compressed air. Lubrication increases efficiency and helps maintain the sliding surfaces inside cylinders, valves and air tools. A regulator is added to set the pressure of the system. **See Figure 22-26.**

A filter/regulator/lubricator unit is commonly located at each workstation to condition and regulate air from the main supply line. Porous metal filters combined with deflectors use centrifugal force to separate out moisture from the air. The moisture is periodically removed from the filter by means of a drain cock located in the bottom of the filter.

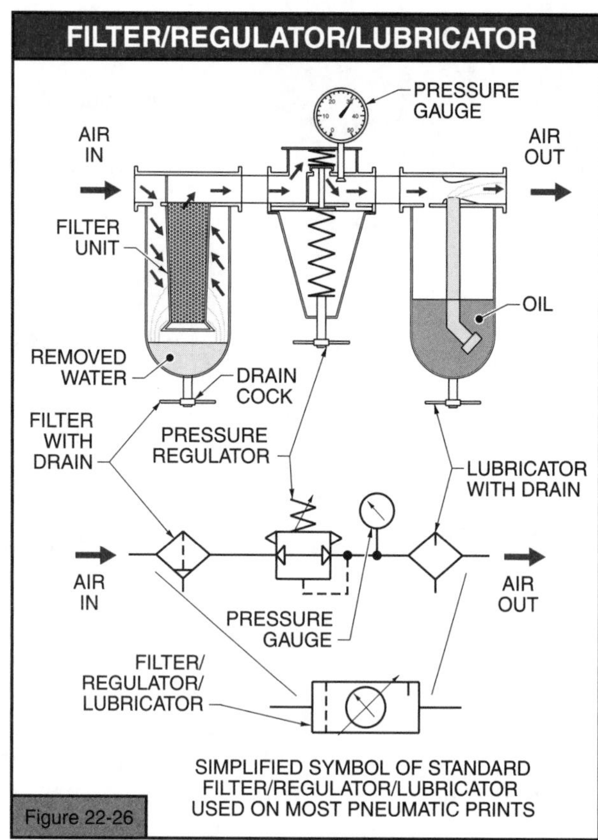

Figure 22-26

Pneumatic Circuits

After air is compressed, filtered, lubricated, and regulated, the air is used to produce work similarly to hydraulic oil. The symbols for most of the components used in a pneumatic circuit are basically the same as those used in a hydraulic circuit. Generally, the difference between hydraulic and pneumatic circuits is that a pneumatic circuit requires meter-out devices on cylinders to prevent lunging when the load resistance is reduced or removed. This occurs because air is compressible and expands to fill the available space provided. In addition, no return lines are required because air can be released directly into the atmosphere. **See Figure 22-27.**

In this circuit, two double-acting cylinders are used to form a part by making two different bends on the part. Four-way directional control valves are required because the cylinders are double-acting. To add more control to the circuit, the 4-way valves are solenoid operated. The solenoids can be controlled manually by placing the selector switch in the manual position and using the pushbuttons. The solenoids can also be controlled automatically by placing the selector switch in the automatic position and allowing the timers to control the solenoids.

In the automatic position, timer 1 is used to delay the start of the operation. The operation starts 5 sec after the selector switch is placed in the automatic position because timer 1 is set for 5 sec. A start-up alarm sounds during the 5 sec period. This alarm is added to warn operators that the circuit is automatically going to start.

All five timers are energized when the selector switch is placed in the automatic position. The timers can be individually set for any required time setting. In this application, each timer is set 5 sec longer than the other. This allows the circuit to automatically change which solenoid is energized every 5 sec. Timer 5 does not operate any solenoids. It is used to automatically restart the circuit after both cylinders have completed their operation.

In this circuit, the alarm energizes when the selector switch is placed in automatic position. The initial bend cylinder extends 5 sec after the selector switch is placed in the automatic position (solenoid 1 energizes), and the alarm de-energizes. The initial bend cylinder retracts 5 sec after it extends (solenoid 1 de-energizes and solenoid 2 energizes). The finish bend cylinder extends 5 sec after the initial bend cylinder retracts (solenoid 2 de-energizes and solenoid 3 energizes). The finish bend cylinder retracts 5 sec after it extends (solenoid 3 de-energizes and solenoid 4 energizes). Timer 5 times out in 5 sec and opens the normally-closed TR5 contact. The opening of the normally-closed TR5 contact resets the circuit and the operation automatically restarts with the alarm being energized.

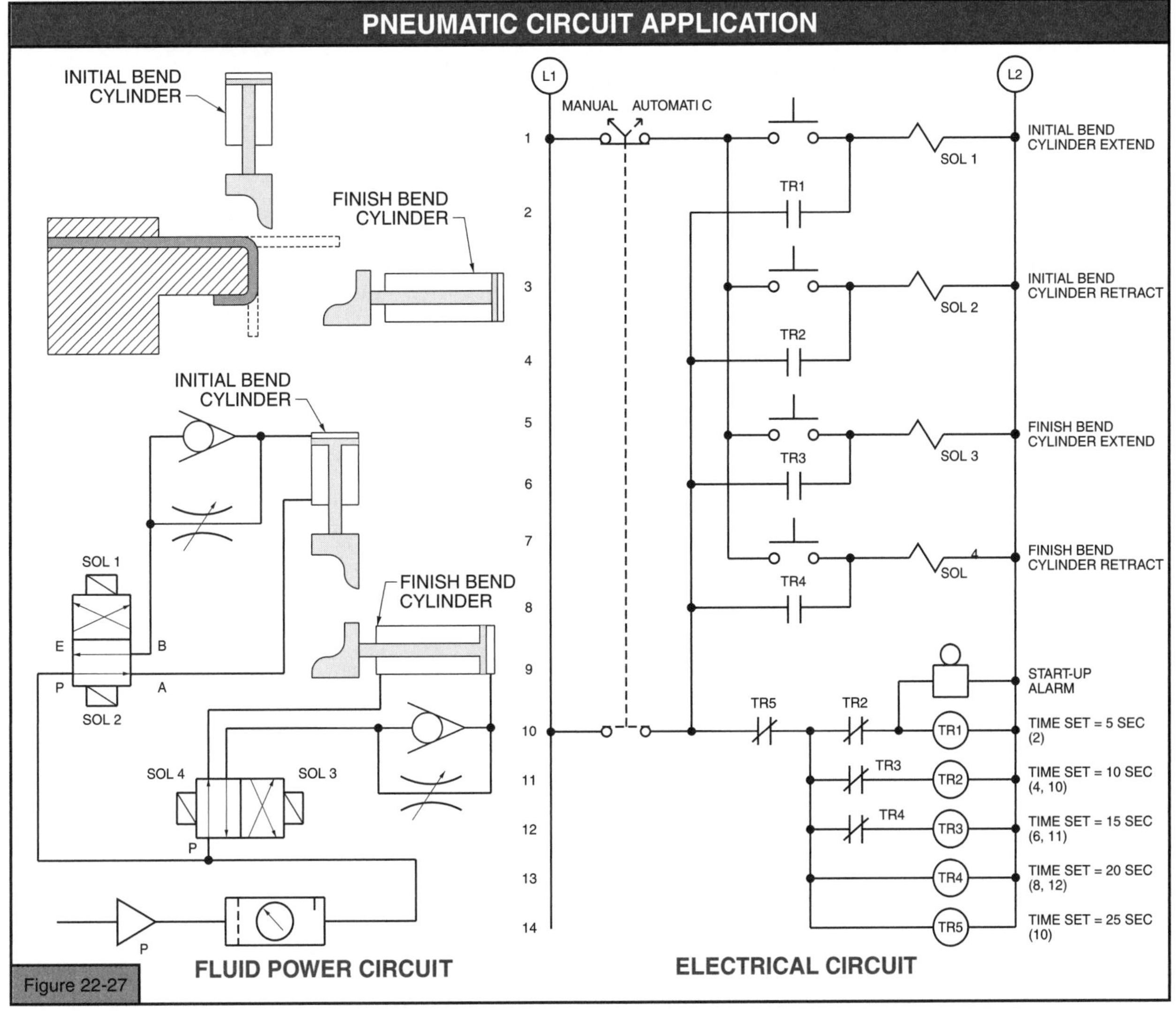

Figure 22-27

Refer to the CD-ROM in the back of the book for Chapter 22 Quick Quiz® and related information.

23 Audio Systems

Sound

Sound is energy that consists of pressure vibrations in the air. Sound is produced by pressure vibrations that originate from a vibrating object. The vibrations travel outward in waves. Vibrating objects produce pressure vibrations which cause recurring compression and rarefaction of air. *Compression* is an area of increased pressure in a sound wave produced when a vibrating object moves outward. *Rarefaction* is an area of reduced pressure in a sound wave produced when a vibrating object moves inward. Sound is heard when the vibrations are picked up by a person's ears. **See Figure 23-1.**

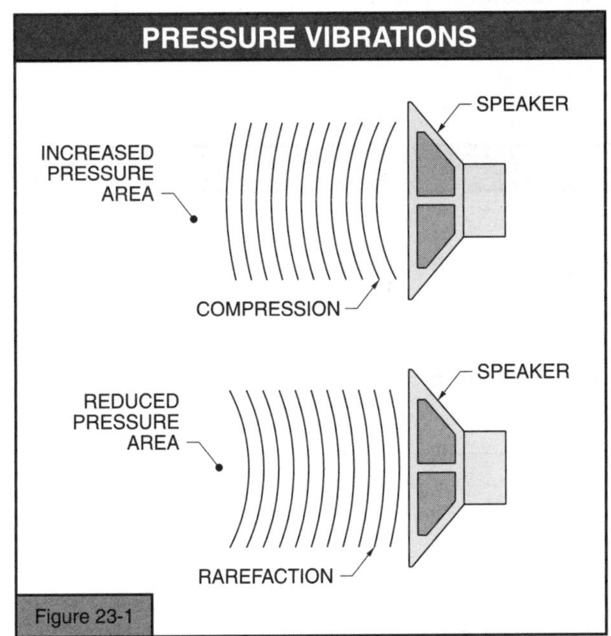

Figure 23-1

Sound waves take the shape of a sine wave. A *sine wave* is a symmetrical waveform that contains 360°. A sine wave reaches its peak positive value at 90°, returns to 0 at 180°, increases to its peak negative value at 270°, and returns to 0 at 360°. **See Figure 23-2.** A sine wave is a pure tone (fundamental) with no harmonics. A *harmonic* is an overtone whose frequency is a multiple of the fundamental frequency.

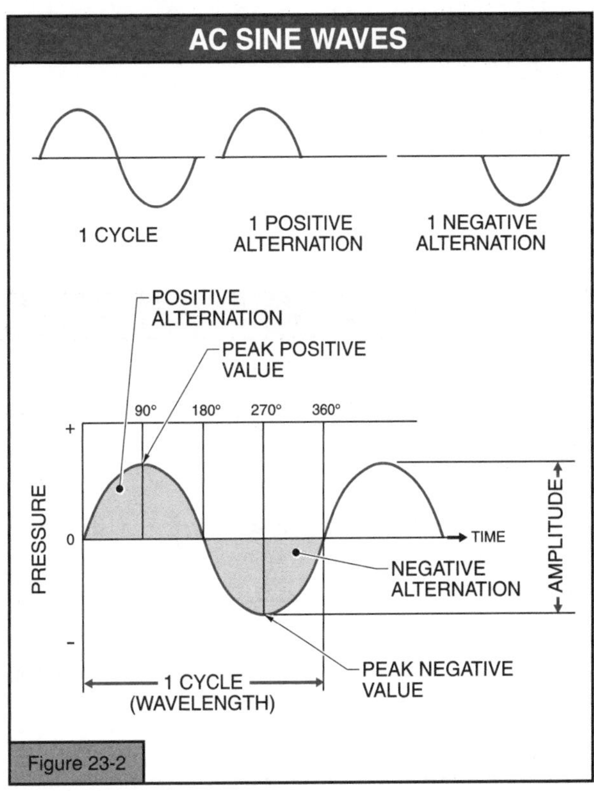

Figure 23-2

A *cycle* is one complete positive and negative alternation of a wave form. A sine wave has one positive alternation and one negative alternation per cycle. A single complete cycle of a sound pressure wave consists of one half-cycle of compression of air molecules, followed by one half-cycle of rarefaction of air molecules. Loud sounds compress and rarefy air molecules to a greater extent than soft sounds.

Frequency

Frequency (f) is the number of air pressure fluctuation cycles produced per second. Sound consists of various frequencies. Frequency is measured in hertz (Hz). *Hertz* is the international unit of frequency and is equal to one cycle per second (cps). **See Figure 23-3.** To calculate frequency, apply the formula:

$$f = \frac{1}{T}$$

where

f = frequency (in Hz)

1 = constant

T = period (in sec)

Example: Calculating Sound Wave Frequency

What is the frequency of a waveform that requires .2 sec to complete one cycle?

$$f = \frac{1}{T}$$

$$f = \frac{1}{.2}$$

$$f = \mathbf{5\ Hz}$$

The faster an object vibrates, the higher the frequency produced. The slower an object vibrates, the lower the frequency produced. As the frequency increases, the wavelength decreases. *Wavelength* is the distance covered by one complete cycle of a given frequency sound as it passes through the air. Sound waves travel through the air at a speed of 1130 ft/sec at sea level at a temperature of 59°F. The speed of sound is independent of frequency. To calculate wavelength, apply the formula:

$$W = \frac{s}{f}$$

where

W = wavelength

s = speed of sound (in ft/sec)

f = frequency (in cps)

Example: Calculating Wavelength

What is the wavelength of a 5000 Hz sound wave traveling at 1130 ft/sec?

$$W = \frac{s}{f}$$

$$W = \frac{1130}{5000}$$

$$W = \mathbf{.226\ ft}$$

Low frequencies produce deep, bass sounds that vibrate at several hundred vibrations per second. High frequencies produce high, shrill sounds that vibrate at several thousand vibrations per second. A sound with a 10 Hz frequency vibrates at 10 cps. Every second, the air particles move up and down 10 times.

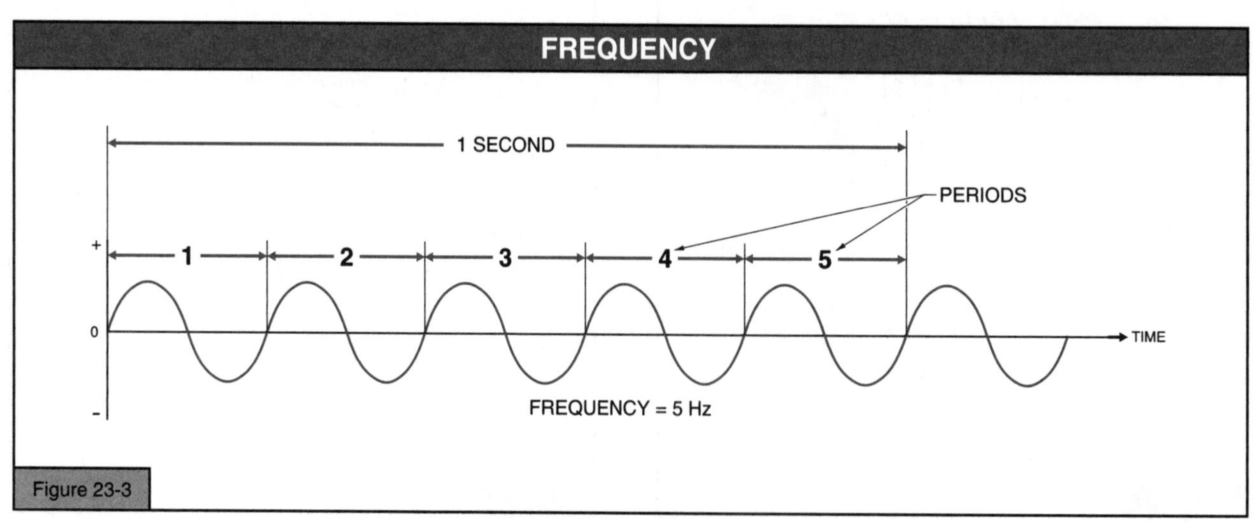

Figure 23-3

A person with good hearing can hear sounds with frequencies ranging from about 20 Hz to 20 kHz. As a person ages, the upper limit of their hearing decreases. Alarms, horns, and bells normally operate at one frequency. Sirens normally operate at a frequency that varies slightly. Speakers operate at frequencies within the human hearing range by converting electrical energy into vibrations (sound waves). The *frequency spectrum* is the range of all possible frequencies. The *audio spectrum* is the part of the frequency spectrum that humans can hear (20 Hz–20 kHz). **See Figure 23-4.**

Period (T) is the time required to produce one complete cycle of a sound wave. The period of a sound wave is expressed in seconds. To calculate the period of a sound wave, apply the formula:

$$T = \frac{1}{f}$$

where

T = period (in sec)

1 = constant

f = frequency (in cps)

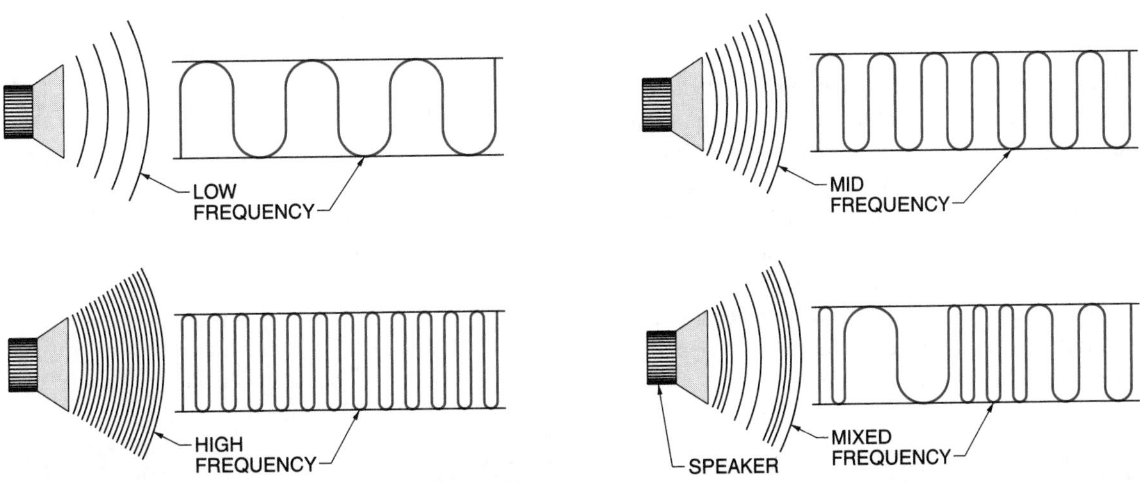

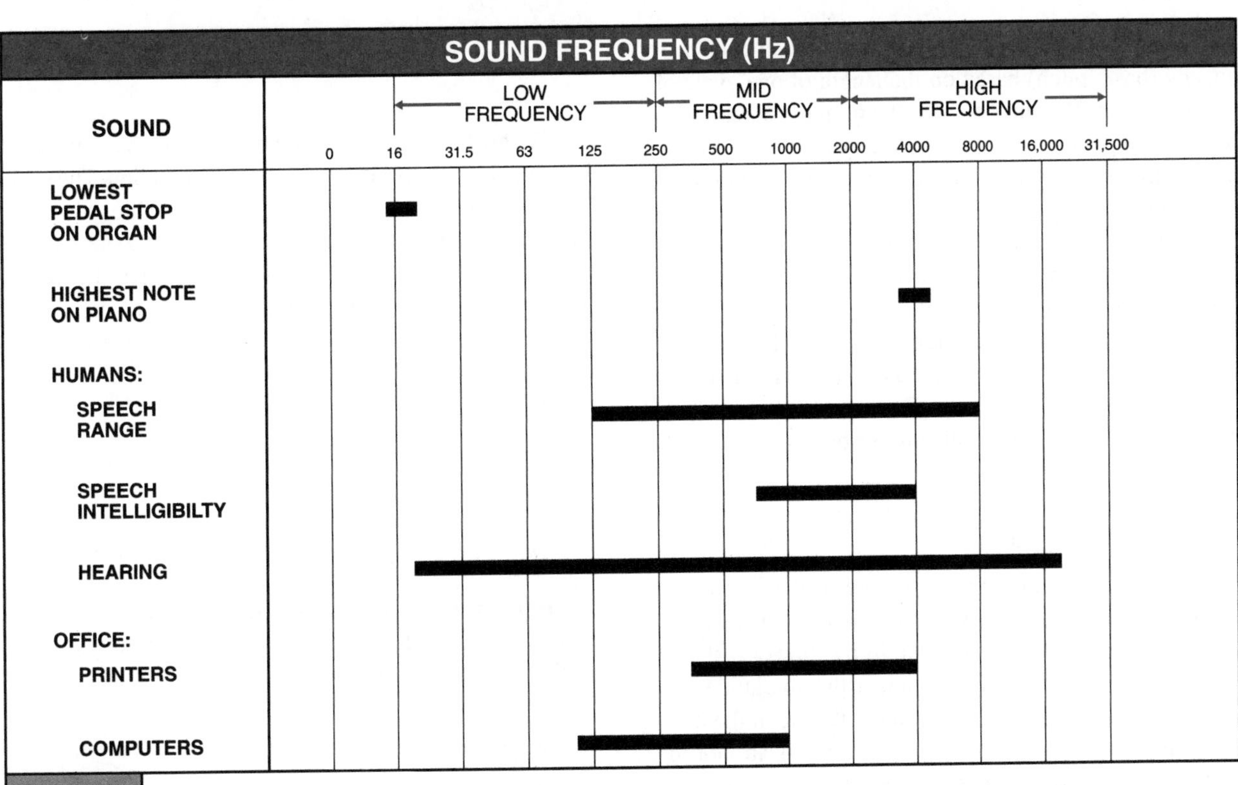

Figure 23-4

Example: Calculating Sound Wave Period
What is the period of a 2000 Hz sound wave?

$$T = \frac{1}{f}$$

$$T = \frac{1}{2000}$$

$$T = \mathbf{.0005\ sec}$$

Pitch

All sounds have the three basic aspects of pitch (fundamental frequency), timbre (overtone structure), and sound intensity (volume). *Pitch* is the highness or lowness of a sound. Frequency corresponds to the musical attribute of pitch. For example, the higher the frequency, the higher the perceived pitch of the sound. Pitch is determined by the speed of the back-and-forth motion as air molecules are compressed and rarefied. The sound has a recognizable musical pitch if the compressions and rarefactions occur regularly between 20 and 20,000 times per second. Sound without a definite pitch occurs if air molecules move back-and-forth without a discernible pattern. Generally, unpitched sounds are referred to as noise.

Timbre

Any single-pitched sound possesses a fundamental frequency (basic pitch) based on the length of one cycle of the waveform. Within the cycle, there may be subcycles (overtones) of higher frequencies mixed with the fundamental frequency. The presence and relative strengths of the frequencies determine the characteristics (tone) of the sound. The presence and relative strengths of the overtones of a sound determines whether the sound is bright, dull, shrill, mellow, rough, or smooth.

The fundamental frequency of a sound is normally much louder than its overtones. The sound is muted and dull if the overtones are quiet. The sound is bright if the high overtones are loud relative to the others.

Sound Intensity

Sound intensity (volume) is a measure of the amount of energy flowing in a sound wave. *Amplitude* is the distance that a vibrating object moves from its position of rest as it vibrates. **See Figure 23-5.** The larger the amplitude, the louder the sound. Sound waves that fluctuate in the air a small amount produce little sound. Eardrums are not vibrated much and the sound is considered soft. Sound waves that fluctuate in the air a large amount produce loud sounds. Eardrums are vibrated a great deal and the sound is considered loud. Eardrums may be damaged if the waves are too large (excessive sound level). The peak-to-peak value (amplitude) of a sine wave is the height of the waveform. Sine waves are specified as peak, peak-to-peak, average, or root-mean-square.

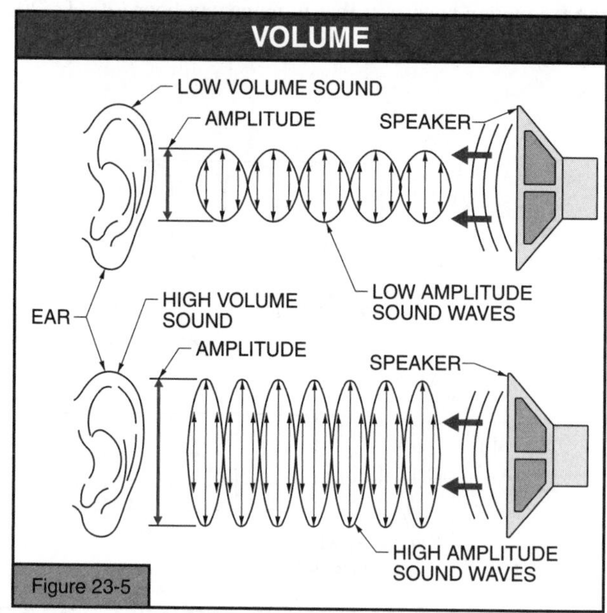

Figure 23-5

The *peak value* (V_{max}) of a sine wave is the maximum value of either the positive or negative alternation. The positive and negative alternation are equal in a sine wave. **See Figure 23-6.** The *peak-to-peak value* ($V_{p\text{-}p}$) of a sine wave is the value measured from the maximum positive alternation to the maximum negative alternation. Peak-to-peak value is twice the peak value. To calculate peak-to-peak value (in volts, decibels, etc.), apply the formula:

$$V_{p\text{-}p} = 2 \times V_{max}$$

where

$V_{p\text{-}p}$ = peak-to-peak value

2 = constant

V_{max} = peak value

Example: Calculating Peak-to-Peak Value
What is the peak-to-peak value of a sine wave that has a positive peak alternation of 40 dB?

$$V_{p\text{-}p} = 2 \times V_{max}$$

$$V_{p\text{-}p} = 2 \times 40$$

$$V_{p\text{-}p} = \mathbf{80\ dB}$$

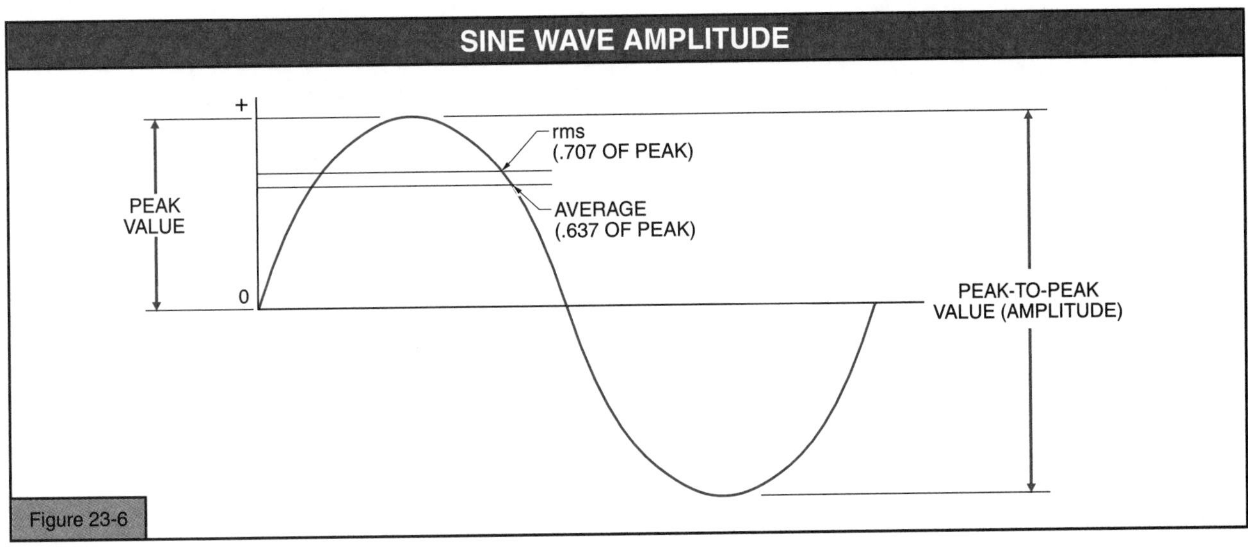

Figure 23-6

The *average value* (V_{avg}) of a sine wave is the mathematical mean of all instantaneous voltage values in the sine wave. Average value is equal to .637 of the peak value of a sine wave. To calculate average value (in volts, decibels, etc.), apply the formula:

$V_{avg} = V_{max} \times .637$

where

V_{avg} = average value

V_{max} = peak value

.637 = constant

Example: Calculating Average Value

What is the average value of a waveform that has a positive peak alternation of 50 dB?

$V_{avg} = V_{max} \times .637$

$V_{avg} = 50 \times .637$

V_{avg} = **31.85 dB**

Root-mean-square (rms) value is a mathematical expression equal to .707 times the peak value of a waveform. Rms is used to describe the level of an audio signal. Rms values are similar to average values. The average value of a signal does not vary much even with great variations in the peak value of a waveform. The peak values may vary wildly without a major effect on the average value of a waveform. The rms value corresponds closely with the sensitivity of a person's ears to sound energy.

The loudness and harmonic content of a sound is shown when the amplitude of the sound is plotted against time. **See Figure 23-7.** The height of the wave represents the sound's loudness (volume) and the number of times it crosses the horizontal axis while maintaining a recognizably repeating shape represents the fundamental frequency. The wave's overall shape corresponds to its harmonic content (timbre).

The *decibel (dB)* is the unit used to measure the intensity (volume) of sound. A decibel is 1⁄10 of a Bel (unit named after Alexander Graham Bell). A *Bel* is the logarithm of an electric, acoustic, or other power ratio. The decibel scale is logarithmic, which means that for every 3 dB increase in sound level, the power level must be doubled. For any given sound, the loudness doubles for every increase of 10 dB or is reduced by ½ for every decrease of 10 dB. For example, the sound level of normal conversation is one-half the sound level of a hair dryer. **See Figure 23-8.** Sound waves lose intensity as they spread outward. Thus, the loudness of a sound decreases as the distance it travels increases.

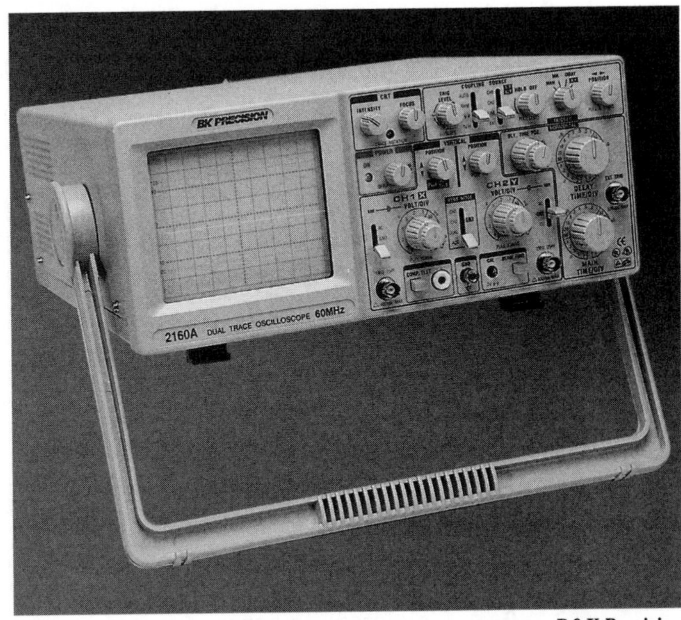

B&K Precision

Oscilloscopes are test instruments that display electronic test results through the use of frequency.

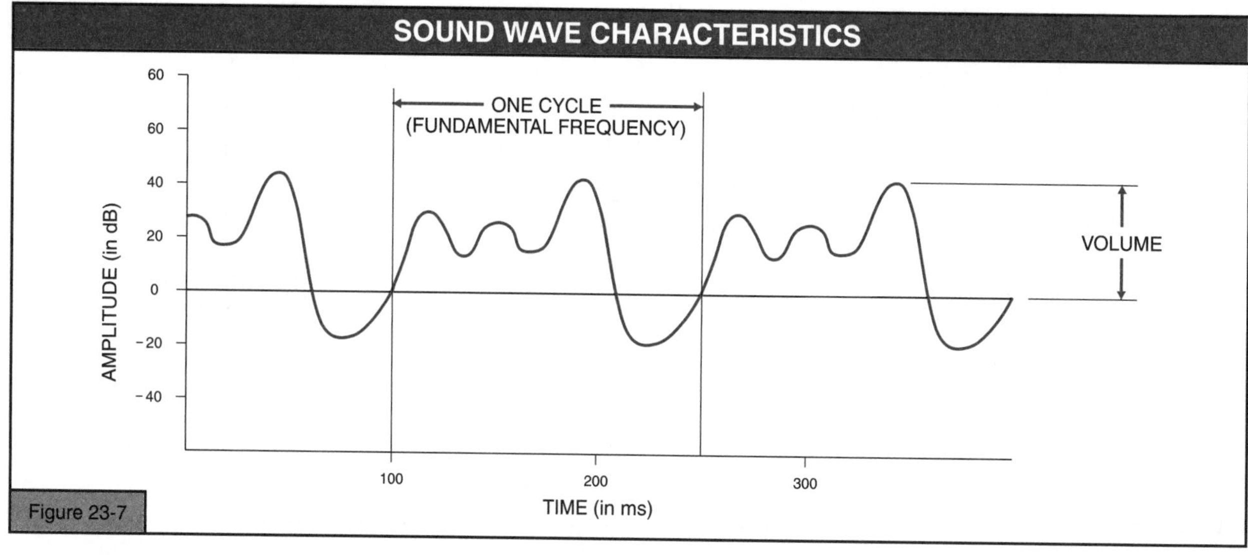

Figure 23-7

SOUND LEVELS

Decibel	Example	Loudness
180	Rocket engine	Deafening
160	Jet engine	
150	Explosion	
140	Loud rock music	
130	Air raid siren (pain threshold)	Very loud
120	Thunder	
110	Chain saw	
100	Subway	
90	Heavy truck traffic	
80	Vacuum cleaner	Loud
70	Busy street	
60	Hair dryer	Moderate
50	Normal conversation	
40	Running refrigerator	
30	Quiet conversation	Faint
20	Quiet living room	
10	Whisper	Very faint
0	Intolerably quiet	

Figure 23-8

The decibel level of a sound wave represents a ratio of the level of sound to a reference level (normally 0 dB). Zero dB represents the absolute faintest sound that a normal human ear can hear. As the sound intensity increases, the decibel level increases. The sound becomes painful if the level of sound continues to increase. A sound intensity of 130 dB is considered the pain threshold in the human ear.

The human ear is not equally sensitive over its entire frequency range. For example, a 20 Hz sound at 80 dB seems as loud as a 1 kHz sound at 20 dB. This is because at 20 Hz the human ear is not very sensitive to sound. A normal human ear is most sensitive to sounds between 3 kHz and 4 kHz. It normally takes more sound pressure at low and very high frequencies for a person to perceive a sound to be equally as loud as a sound at 1 kHz. Sound sensitivity decreases with age. The hearing of old women drops to an upper range of about 10 kHz, and old men to about 8 kHz. Children and women tend to be more sensitive to sound than adult men.

Caution: The sound levels that may cause permanent hearing damage vary with individuals and with the length of exposure to the sound. According to OSHA, 30 min is all a human ear can take at 110 dB. The human ear can take 2 hr at 100 dB or 4 hr at 95 dB before it begins to be permanently damaged.

Phase

Phase is the time relationship of a sine (or sound) wave to a known time reference. Phase is expressed in degrees. One complete cycle of a sine wave equals 360°. In sound systems, phase must be controlled because it affects how sounds add together. Sine waves add algebraically when they mix in the air.

Sine waves add to form a sine wave of twice the level of the original if the sine waves are in phase. Sine waves that are 90° out-of-phase add to form a sine wave that is 1.414 times higher than the original sine waves. Sine waves that are 180° out-of-phase totally cancel each other. **See Figure 23-9.**

Audio Systems

An *audio system* is an arrangement of electronic components that is designed to amplify sound. Audio systems amplify sound by converting it into electrical energy, increasing the power of the electrical energy by electronic means, and converting the more powerful electrical energy back into sound. Almost all households have some type of audio system. Audio systems amplify sound to help individuals hear something better, make sound louder for artistic purposes, enable people to hear sound in remote locations, and reproduce recorded or broadcast sound. Audio systems include clock radios, portable systems, shelf systems, separate component systems, home theater systems, intercoms, public address, and sophisticated concert theater sound systems.

A *clock radio* is a small electric device that keeps time and has a tuner to receive broadcast signals. A *portable audio system* is a small, lightweight electric device used to reproduce and amplify sound at outdoor locations. Portable audio systems may include an amplifier, tuner, CD player, tape player, and speakers in one portable enclosure. Portable systems also include digital audio players. A *digital audio player* is an electric device that stores and plays digital music files. Most digital audio players are portable and can be connected to car and home stereo systems. Digital audio players may include tuners and/or microphones.

Digital audio players are commonly known as MP3 players due to the popularity of the MP3 file format. However, many other file formats are compatible with digital audio players. These formats include Windows Media® Audio (WMA), Waveform Audio (WAV), Advanced Audio Coding (AAC), and Ogg Vorbis.

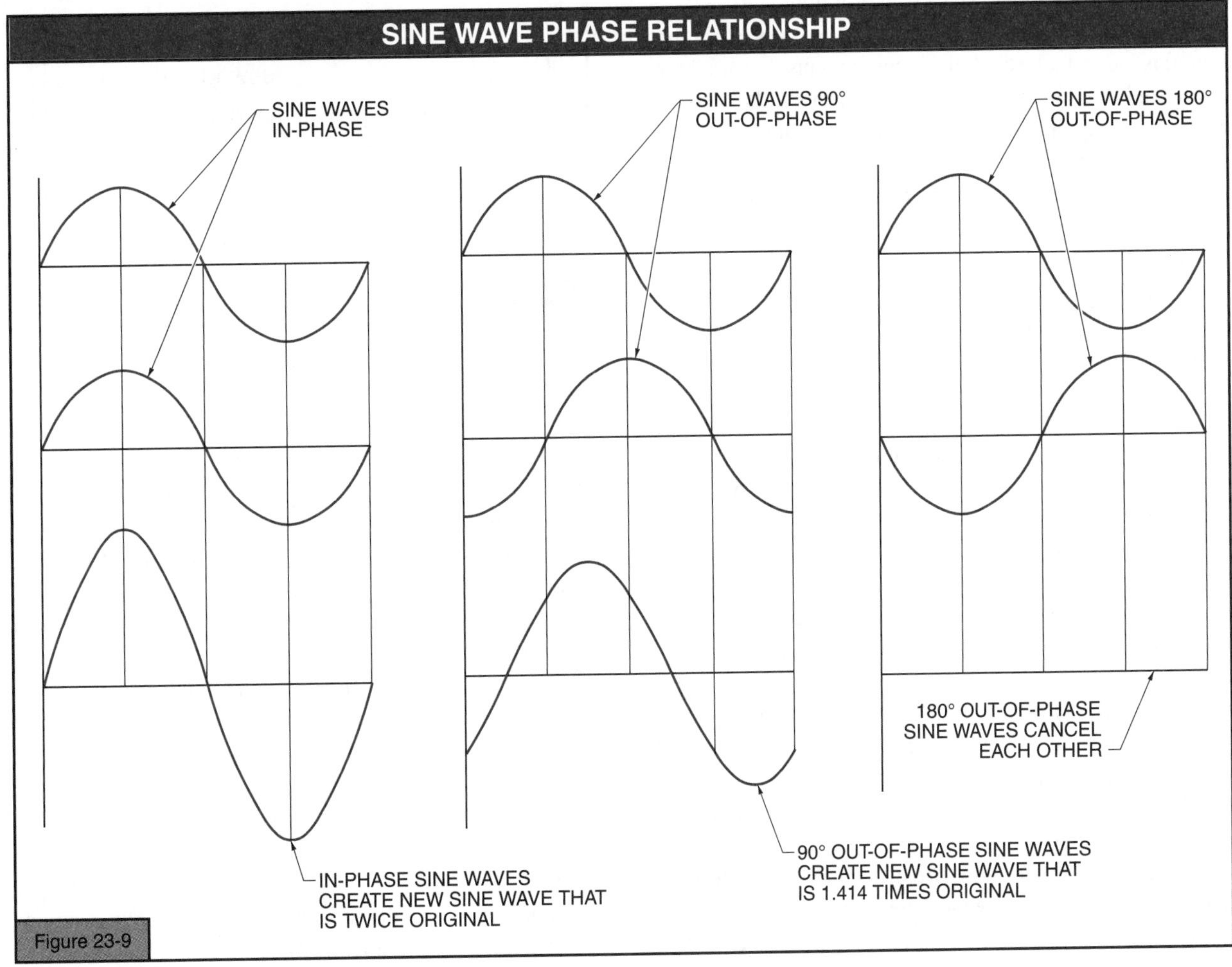

Figure 23-9

Basic digital audio players include a data/power port, microprocessor, digital signal processor, memory, playback controls, display, and audio port. The data/power port transfers data between the digital audio player and a computer and enables power to be supplied to the digital audio player.

The microporcessor monitors user input through playback controls, displays information about the current song, and sends directions to the digital signal processor. The digital signal processor accesses the digital music file from memory and sends it to the amplifier. The digital signal processor also converts the digital information into sound waves.

The playback controls allow a user to play/pause, fast-forward, rewind, skip a song, and change the volume of the audio output. The display visually indicates system operation based on user input. The audio port enables the connection of headphones or speaker jacks.

A *shelf system* is a complete audio system that is designed to play the various audio formats in a minimum of space. Shelf systems contain all electronic circuitry in one enclosure and are designed for offices, apartments, dorm rooms, etc. or anywhere space is a premium.

A *separate component system* is an audio system that contains individual, stand-alone components that are interconnected to allow the playing of the various audio formats. Separate component systems require a greater space than portable or shelf systems, but the individual components may be selected based on the user's preferences.

A *home theater system* is a system that integrates a high-quality video source with multichannel audio electronics. The objective of a home theater system is to reproduce cinema-quality video and sound. An audio-video (AV) receiver is one of the main components of a home theater system. AV receivers are used to amplify sound from various audio sources to speakers and route video signals from various sources to a television. AV receivers offer a wide variety of inputs and features, in addition to being the drivers for surround-sound speaker systems.

Surround sound is the concept of expanding reproduced sound characteristics from one dimension (mono) to three dimensions. Surround sound systems reproduce the sound characteristics of a realistic audio environment. The electronic circuitry in a receiver or amplifier decodes the surround information in the source signal to produce the left, right, center, front, and back aspect of the original sound environment. Surround sound is only available on amplifiers and receivers that contain surround-sound circuitry. **See Figure 23-10.**

Surround-sound systems are classified with a number, such as "3.0 surround," which means that there are three audio channels extracted from a stereo source. The digit in front of the decimal point indicates the number of extracted audio channels, and the number after the decimal point indicates the number of low-frequency channels. Surround-sound systems include 3.0, 4.0, 4.1, 5.1, 6.1, 7.1, and 10.2.

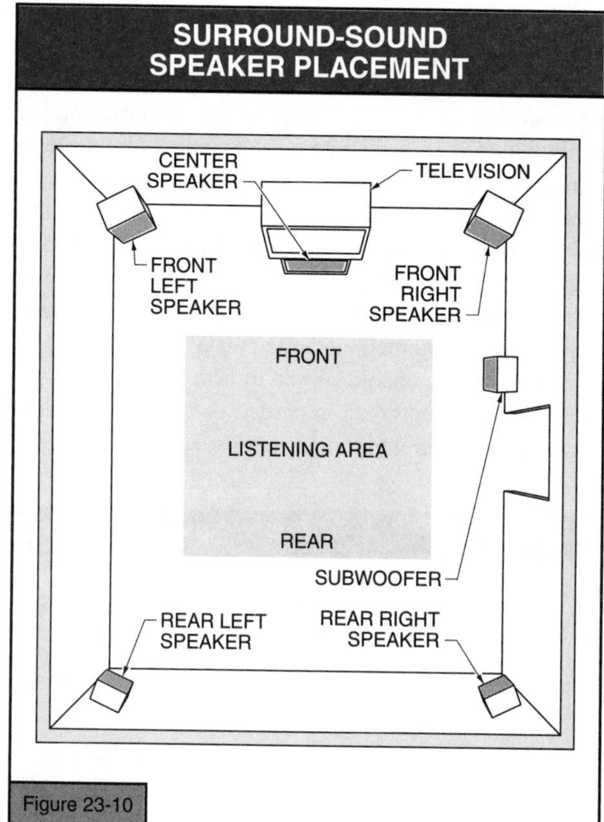

Figure 23-10

An *intercom* is an audio system that allows users to verbally communicate from different stations, monitor certain stations, and play music throughout a building or at certain stations. Intercom systems include a microphone, speaker, and controls at each station (such as a room or location). Intercom systems may also include a master unit, which can control the various features of the system along with other audio systems.

A radio/audio/intercom system combines the functions of chimes, intercoms, and audio systems into one master unit. The master unit is used to control the various systems. **See Figure 23-11.** With radio/audio/intercom systems, a piece of equipment can be used to serve more than one function. Radio/audio/intercom systems may include the following features:

- Room-to-room intercom that allows for calling any or all stations at one time from any station

- Audio monitoring of specific stations
- Remote station privacy settings to shut off intercom to specific rooms
- Hands-free call answering
- Music muting or intercom override
- FM/AM radio
- CD and audio inputs

A *public address (PA) system* is an electronic amplification system used as a communication system in public areas. PA systems are used to reinforce sound so that individuals can hear sounds clearly at a distance from the source. A basic PA system consists of microphones, a mixer/amplifier, and loudspeakers. The mixer combines audio signals from various audio sources, modifies the tone, and controls the overall level of the signal sent to an amplifier. Basic PA systems are often used in areas such as school auditoriums and churches. PA systems are also known as sound-reinforcement systems.

A *concert theater sound system* is a large audio system used to project high-quality sound for a theater. Concert theater sound systems are normally high-power systems that are tailored for the specific theater. Concert theater sound systems may contain many speakers and can be used to amplify live performance sounds.

Audio systems may contain amplifiers, preamplifiers, receivers, tuners, tape decks, turntables, compact disc (CD) players, speakers, and microphones. The electronic circuitry for these devices may be integrated into a single cabinet, such as in a portable or shelf system, or may be placed in stand-alone devices such as in the components in a separate component system. These devices enable the reproduction of sounds from different formats such as broadcast signals (tuner), cassette tapes (tape decks), live sound (microphone), records (turntables), and compact discs (compact disc players). **See Figure 23-12.**

Most audio systems, with the exception of clock radios, are stereo systems. A *stereo system* is an audio system that uses two independent channels that are routed to a pair of speakers situated to the right and left of the listener. Stereo systems reproduce the sound characteristics of the three-dimensional space in which the recorded performance originally took place. Stereo systems sound bigger, brighter, fuller, and more realistic than a mono (one-channel) system.

Amplifiers

An *amplifier* is an electric device that turns medium-level audio signals into strong audio signals that are sent to speakers. An amplifier is designed to increase the current, voltage, or power of an applied signal. Some amplifiers have built-in preamplifiers. A *preamplifier* is an electric device that accepts low-level audio signals like those from microphones and turntables and amplifies them to medium-level audio signals that are sent to an amplifier. Consumer amplifiers are often integrated with tuners, tape players, compact disc players, and/or speakers to form a shelf or portable system. A *receiver* is a tuner and amplifier contained in the same enclosure. A *tuner* is an electric device designed to receive radio broadcast signals and convert them into electrical signals that can be amplified. **See Figure 23-13.**

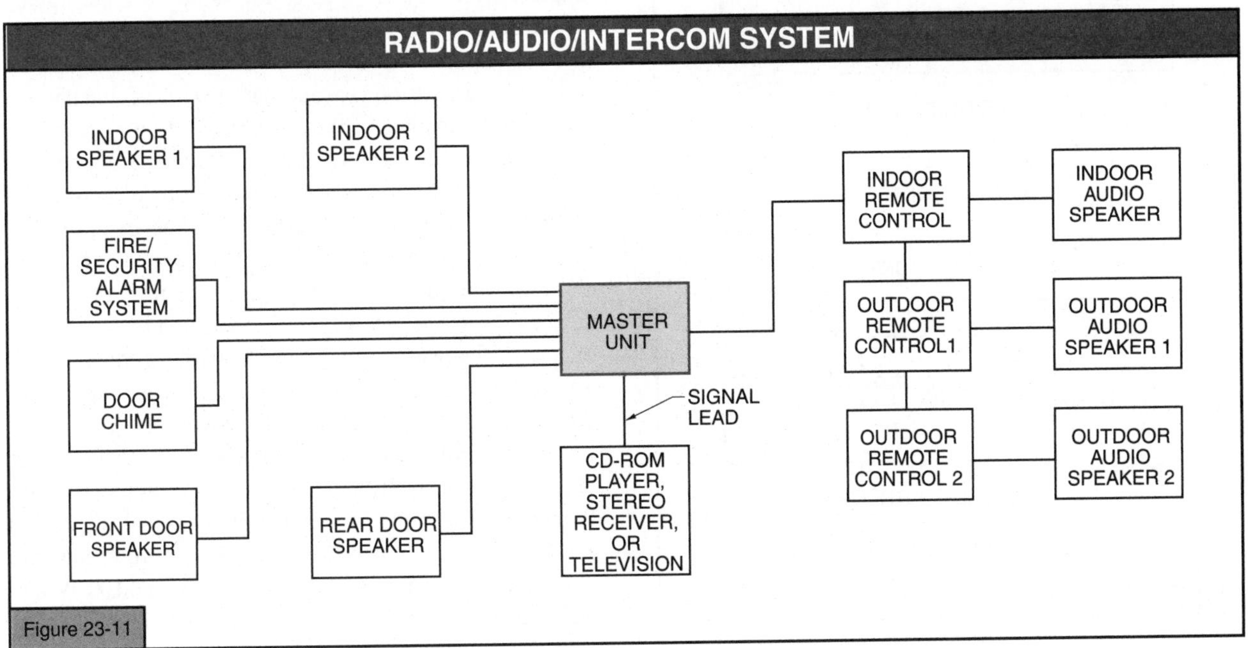

Figure 23-11

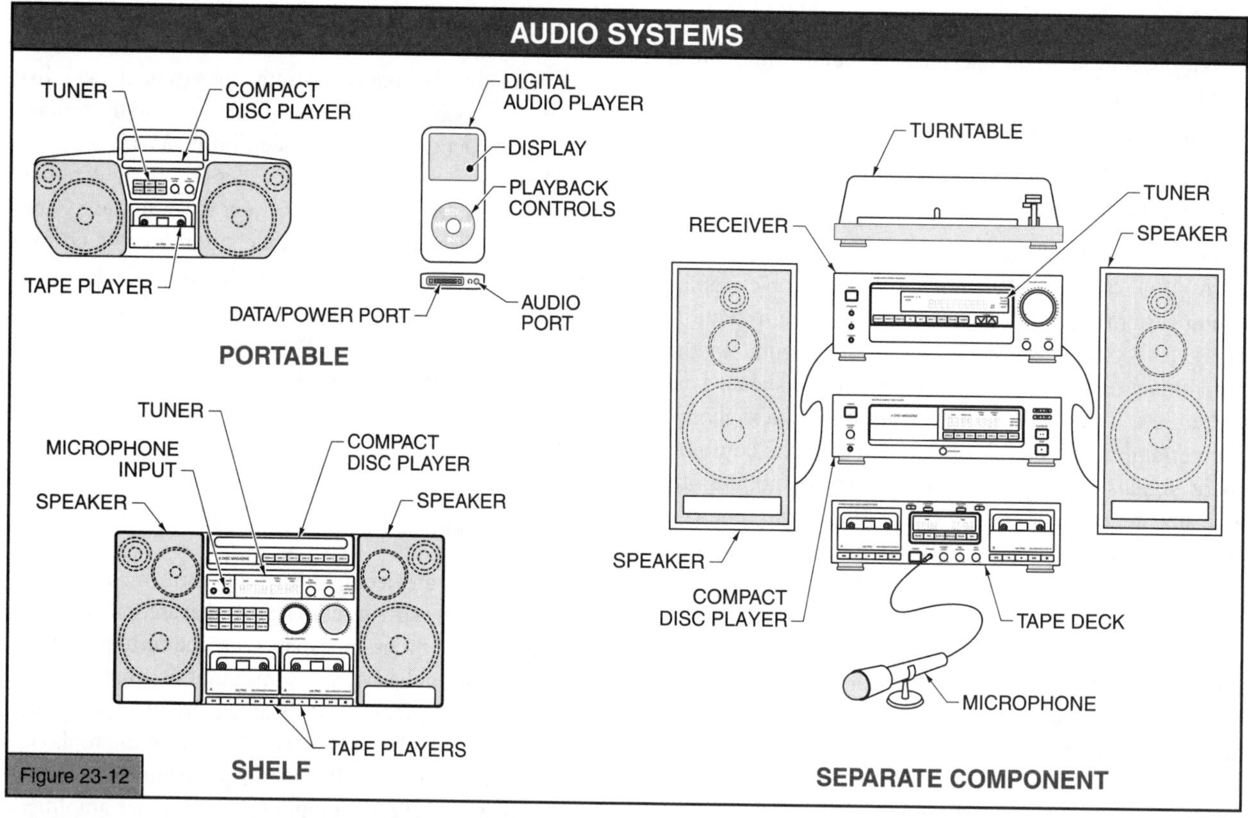

Figure 23-12

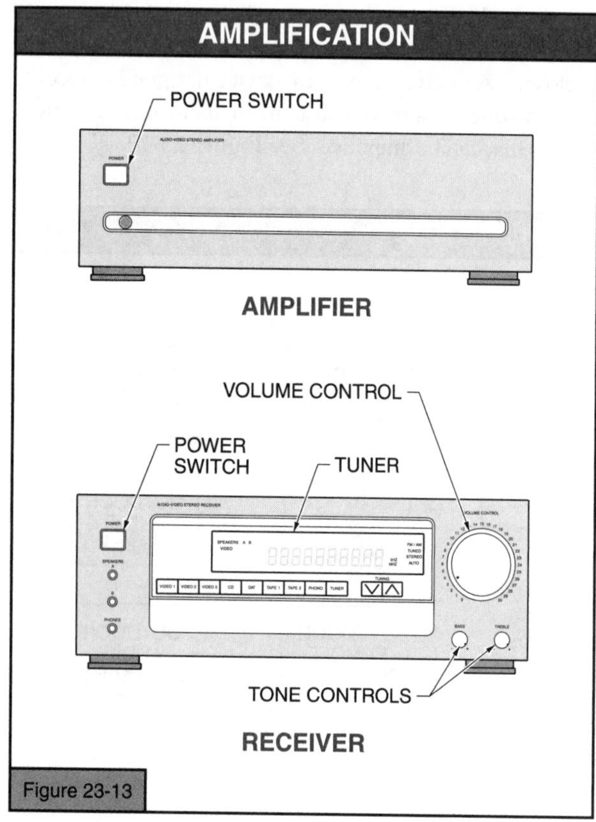

Figure 23-13

An amplifier, powered by alternating current (AC) or direct current (DC), strengthens the signal applied to it. An amplifier sends a strong electrical signal to the speakers. The electrical signal from the amplifier becomes vibrations in the speakers, making sound that is heard. By turning the amplifier volume control up or down, the electrical signals sent to the speakers are strengthened or weakened, making them vibrate more vigorously or gently. An amplifier may be described by its power (number of watts of signal it can produce), frequency response, and distortion.

Amplifier power is the measure of power going into an amplifier to make it work. The more watts of power, the louder the sound the amplifier and speakers can produce. Power (wattage) is normally listed on the back of almost any electronic device. *Wattage* is the number of watts a device uses to operate. A typical amplifier may use 100 W of electrical power to run its circuits while producing only 4 W of sound signal sent to the speakers. The more power sent to the speakers, the louder the sound.

Portable radios normally have 1 W to 30 W amplifiers. Shelf systems normally have 10 W to 100 W amplifiers. A 40 W amplifier has enough power to drive high-quality speakers and fill a room with distortion-free sound. Auditorium and multi-speaker public address systems require 100 W and higher amplifiers to feed quality sound to hundreds of ears.

For stereo sound, amplifiers require twice the power, such as 50 W for the left channel and 50 W for the right channel. Most power goes into the bass sounds. The high-frequency sounds require only a few watts for clear reproduction. The bass sounds require large amounts of power for clear reproduction.

Speakers create sound by generating pressure waves in the air. The molecules in the air are packed together more tightly in certain areas (peaks) and are spread slightly farther apart in other areas (troughs). The peaks and troughs are not stationary but move through the air at about 1130 ft/sec, which is the speed of sound. While the waves move at a constant speed, they are not evenly spaced. Hundreds or thousands may pass a stationary point every second when many are packed close together. The number passing a stationary point every second may be one, two, or a few dozen when they are spaced farther apart.

Bandwidth is the range of frequencies that a device can accept within tolerable limits. Different audio devices have different frequency responses due to the limitations of their technology. Amplifiers, speakers, tape decks, CD/DVD players, microphones, and turntables all have specific, non-overlapping uses. Therefore, their performance requirements are different.

For high-quality sound reproduction, an amplifier and speaker that gives the full range of frequency response (20 Hz–20 kHz) is required. Frequency response is given as a range of frequencies and a plus or minus decibel level. The smoother and flatter the frequency response, the better the amplifier. For example, a ±1 dB variation at the high and low frequencies indicates a good amplifier. A low-quality amplifier may start to drop off at 10 kHz. **See Figure 23-14.**

Amplifier distortion is often referred to as total harmonic distortion (THD). Total harmonic distortion is a measure of how much an amplifier exaggerates or colors the sound going through it. THD is given as a percentage, such as 0.05% THD. The lower the value, the better quality amplifier.

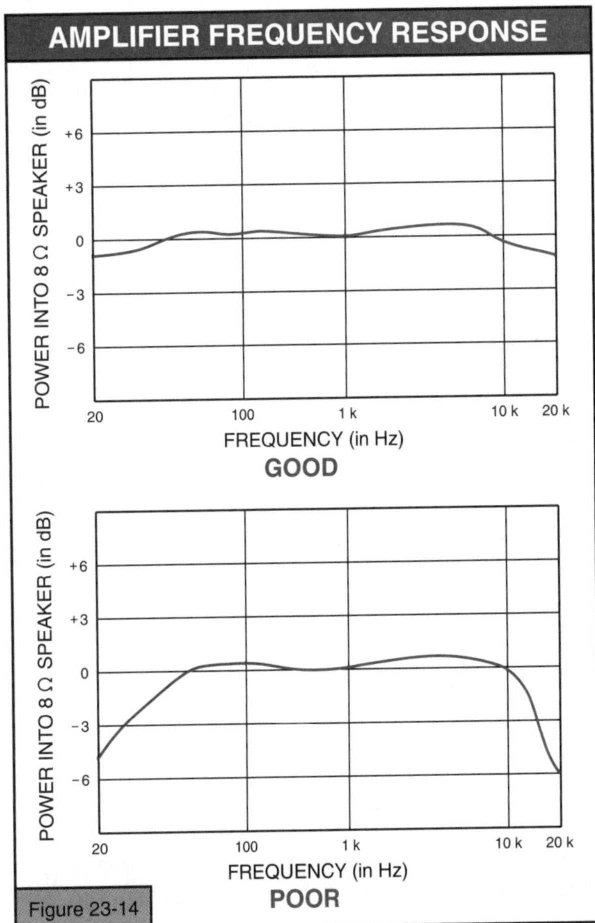

Figure 23-14

Impedances

Impedance is a measure of how much an input or an output resists the signal traveling through it. Impedance is expressed in ohms (Ω). Impedance is the sum of resistance and reactance. Resistance in a circuit is always constant regardless of the frequency passing through the circuit. Reactance is frequency-dependent. However, at low frequencies like those in the audio spectrum (20 Hz to 20 kHz), most impedances act like resistances and are treated in the same way.

Microphones have outputs and speakers have inputs. Almost all other audio devices have inputs and outputs. A *signal source* is an audio device that provides a signal to another audio device. A *load* is a device that receives a signal from the source. A speaker is a load to an amplifier. Both sources and loads have impedances.

Audio devices normally have two impedances, one for their input and one for their output. The lower the impedance, the less the opposition to current flow. The higher the impedance, the greater the opposition to current flow. An ideal amplifier has extremely high input impedance (preferably infinite) and extremely low output impedance (preferably zero). The input impedance is kept as high as possible to prevent the load from overloading the source. The connection looks like a short circuit to the source when the input impedance of the load falls too low. An extremely high input impedance looks like an open circuit to the source driving the load and does not require much of a signal. A low source output impedance is required to maximize the voltage transfer from the output of the source to the input of the load.

For the transfer of a signal to occur with a minimum of loss, the source impedance should be much lower than the load impedance. Only 50% of the voltage reaches the load if the two impedances are identical. For this reason, almost all audio devices have a very low output impedance and a very high input impedance. Input impedances commonly range from 1 kΩ to 200 kΩ. Output impedances commonly range from 200 Ω to 2 Ω.

The output of an amplifier (source) has an impedance that matches the impedance of the speakers (load) to carry the signal with maximum efficiency. Most speakers designed for residential use have 8 Ω impedances. Thus, most stereo amplifiers for residential use have 8 Ω speaker outputs because in driving a speaker, the prime concern is that power transfer is maximized. Some amplifiers have outputs for speakers with different impedances. Small speakers may have 4 Ω impedances and a few speakers may have 2 Ω or 16 Ω impedances. Versatile amplifiers may have several outputs, one for 4 Ω, 8 Ω, and 16 Ω impedances.

Connecting devices in parallel lowers the total load impedance seen by the source. The total load impedance can fall below the minimum load impedance the amplifier can handle, causing the amplifier's output section to blow a fuse when multiple speakers are connected to one output.

Amplifier Connections

Compact disc/DVD players, tuners, turntables, VCRs, digital audio players, and tape decks are connected to a receiver or amplifier to enable the reproduction of sounds from CD/DVD, broadcast, record, video cassette tape, digital audio player, and audio cassette tape formats. Ensure that the power switch is OFF before connecting any device to a receiver or amplifier. An RCA cable is commonly used to interface audio devices. RCA cables consist of two conductors in a protective insulation. The ends of the cable consist of a metal ring and pin (plug) that is connected to the jacks on the back of an audio device. **See Figure 23-15.**

RCA cables are normally grouped in sets of two. RCA cables are available with red and white plugs. The white plug is used to connect the left channel and the red plug is used to connect the right channel. RCA plugs should be fully inserted into the jacks. A loose connection may cause noise. An antenna is connected to the back of a receiver or tuner to improve the reception of AM and FM broadcasts.

TECH FACT *Environmental conditions such as wind, temperature, and humidity have an effect on how sound waves travel outdoors.*

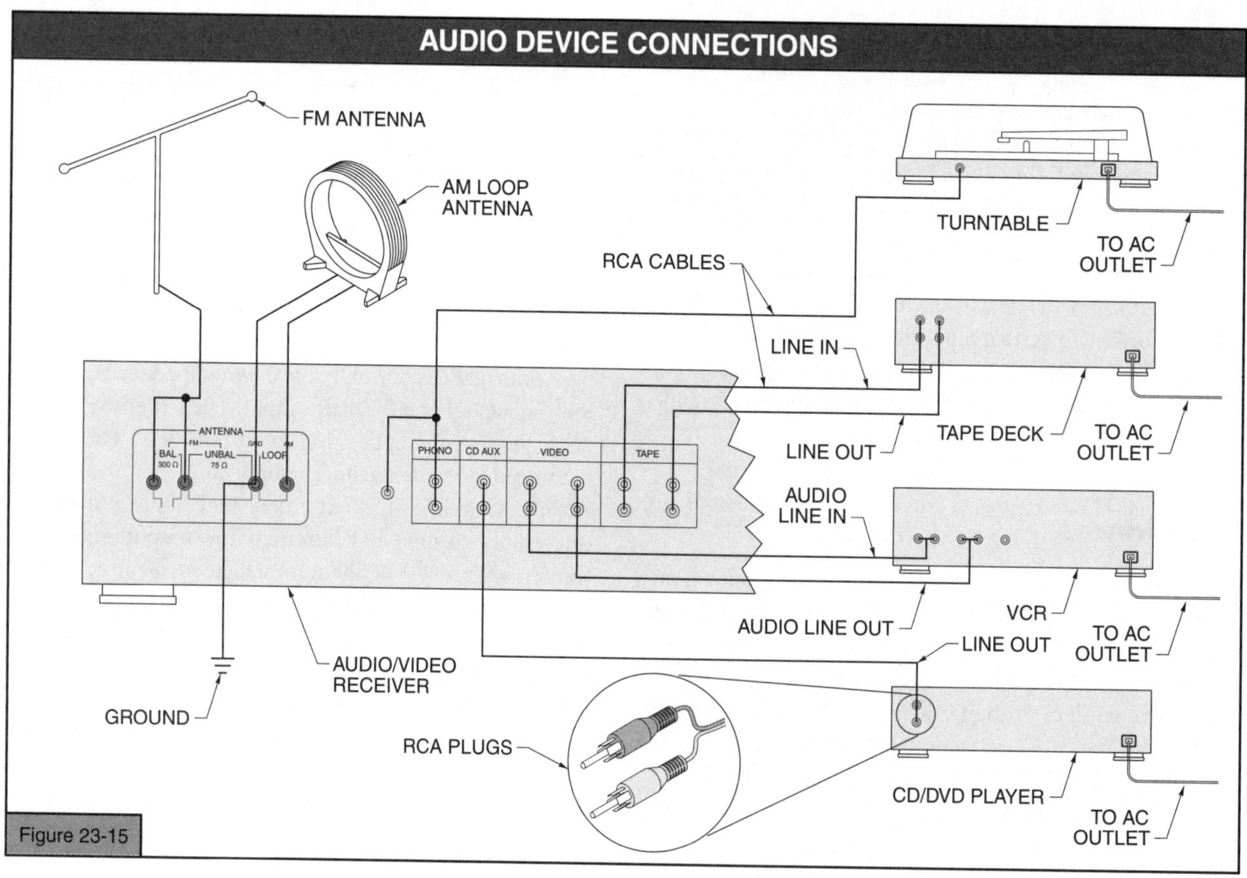

Figure 23-15

Tape Decks

A *tape deck (tape player)* is a device that converts electrical signals to magnetic impulses and transfers these impulses onto tape. Tape decks allow the recording and playback of sound on magnetic cassette tapes. Tape decks are connected to amplifiers or receivers to record and play cassette tapes or may be an integral part of a portable or shelf system. **See Figure 23-16.**

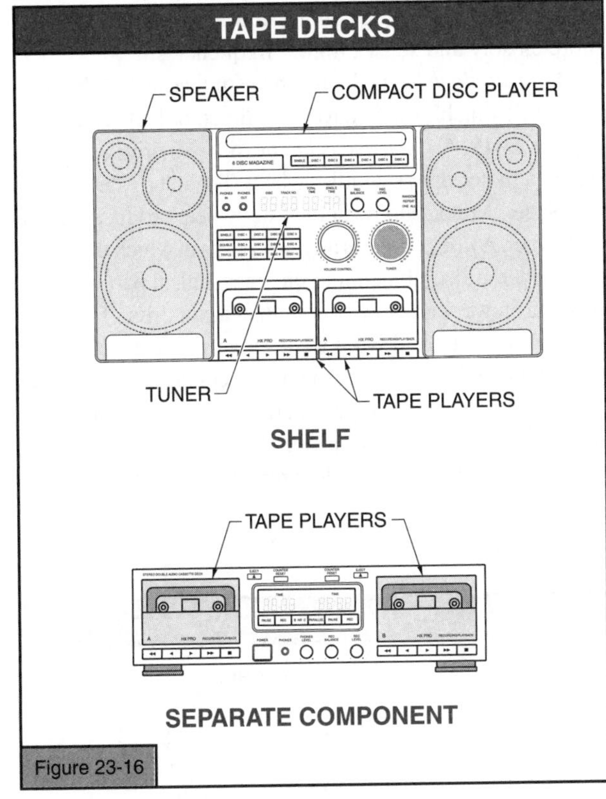

Figure 23-16

A tape deck contains record, playback, and erase heads. The heads may be different devices or combined on one device. A *record head* is an electromagnet that converts electrical vibrations into magnetic vibrations. A ribbon of magnetizable tape slides over the record head and is magnetized. The sound vibrations are changed to electrical vibrations inside the tape deck. The electrical vibrations are changed to magnetic vibrations and stored on the tape by the record head. **See Figure 23-17.**

A *playback head* is an electromagnet which senses the changing magnetism on a magnetic tape and converts it into a weak electrical signal. As a tape is played, it slides over the playback head, which converts the stored magnetic vibrations into an electrical signal that is sent to an amplifier or receiver, where it is amplified and sent to speakers. An *erase head* is an electromagnet that erases all magnetic vibrations on a tape.

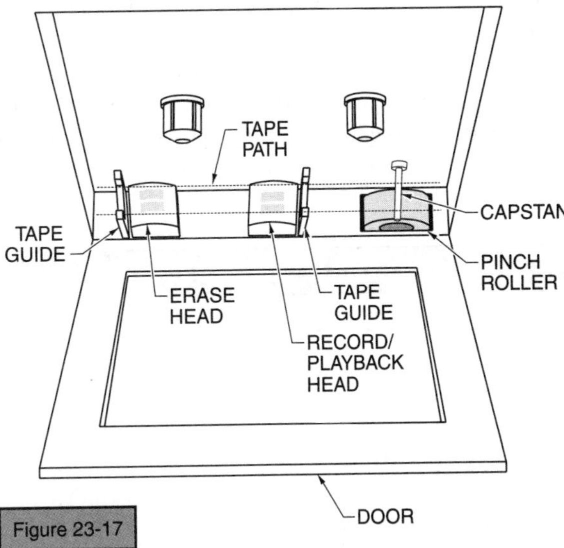

Figure 23-17

The advantage of tape decks is that cassette tapes may be recorded and erased repeatedly. The disadvantages of tape decks are that the vibrations may get jumbled if the tape machine does not pull the tape smoothly over the record and playback heads and the repeated recording and erasing of cassette tapes decreases the sound quality of the recorded material. Cassette tapes may have magnetic vibrations already on them that leave noise when the tape is played. The tape may stretch and cause the vibrations to stretch and be played at the wrong frequency. The tape deck motor may not run at a constant speed due to slipping, loose belts, or pulleys. This may cause the pitch of the music to waver.

Turntables

A *turntable* is a record player without an amplifier. A *record* is a vinyl disk that contains small grooves which contain recorded sounds (music, audio, etc.). A record is made when electrical vibrations are sent to an electromagnet that wiggles a needle as it scrapes the grooves in a record. Every sound vibration is turned into an electrical vibration. The electrical vibrations are turned into needle vibrations that leave a tiny vibration in the groove of the record. Turntables may be integrated into the same cabinet as other audio devices or may be a separate audio component. **See Figure 23-18.**

A *platter* is the flat disk that rotates a record. A *tone arm* is the device that holds the cartridge. The tone arm transfers the electrical signal from the cartridge to the amplifier. A *cartridge* is a vibration sensor that holds a needle. The cartridge sends the electrical signal from the needle to the tone arm. The needle sits in the record groove. As each vibration vibrates the needle, a tiny electric current is generated. The electric current is sent to an amplifier where it is turned into sound vibrations. Most cartridges are magnetic. A magnetic cartridge contains a coil of wire and a magnet that produces the electrical signal from the vibrating needle. Magnetic cartridges produce weak signals that are similar to a microphone.

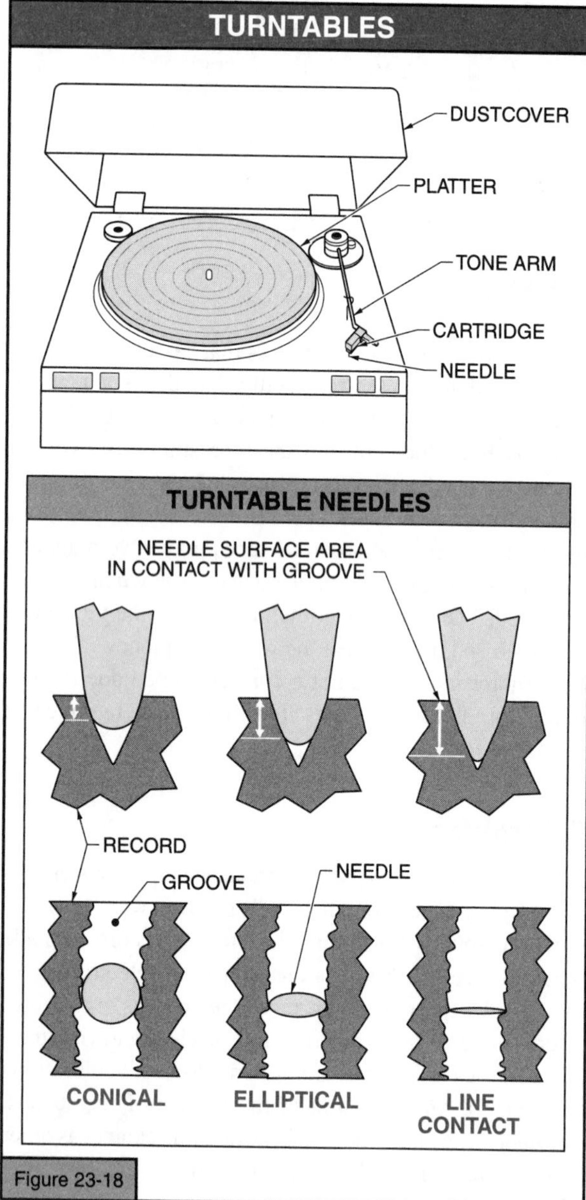

Figure 23-18

TECH FACT *While traditional turntables use direct contact between a needle and the playing surface of a record, high-tech turntables are available that use a laser to contact (read) the grooves in a record without causing the grooves of the record to experience any wear.*

Record needles (styli) are made of diamond and are available in conical, elliptical, and line contact styles. A conical tip needle is the most common. A conical tip needle is the broadest needle. It misses some musical detail and wears the record grooves fairly rapidly. Elliptical tip needles are more expensive, ride deeper in the record groove, reproduce more of the high frequencies, and cause less wear than conical tip needles. Line contact needles are flatter than elliptical needles and ride deeper in the groove. This allows more surface area to contact the groove which causes less wear. A line contact needle follows the groove more closely and senses higher frequencies.

A *turntable drive* is the mechanism that makes the platter spin. Turntables may have rim, direct, or belt drives. **See Figure 23-19.** A *rim drive* is an inexpensive drive system in which a motor drives an idler wheel. The idler wheel drives the platter. Minor bumps in the idler wheel produce rumble and flutter. A *direct drive* is a drive system in which the motor and platter are combined into one mechanical part. A direct drive has fewer parts to wear out than a rim drive. The highest quality turntable drive is a belt-drive system. A *belt drive* is a turntable drive that has the motor connected to an inner platter through a belt and pulley arrangement. The inner platter is used to turn the outer platter. Belt-drive turntables have excellent speed control and low rumble and flutter.

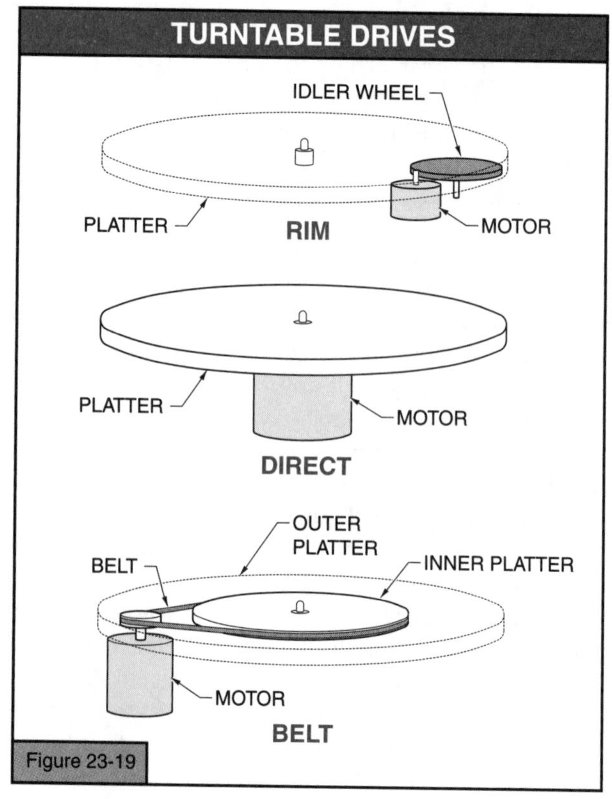

Figure 23-19

Compact Disc Players

A *compact disc (CD) player* is an electric device that plays music stored on a compact disc (CD) by means of reflecting light from a laser beam. Audio information (sound) is digitally recorded in stereo on the surface of a compact disc in the form of microscopic pits and flats. A lens arrangement focuses the laser beam from the CD player onto the CD surface. As the CD rotates, the laser beam, which is altered by the pits and flats, is reflected from the CD through a lens and optical system. The signal is used to reproduce the digitally recorded sound. The reflected laser beam is converted into 1s and 0s, which are then converted into an analog audio signal that is sent to an amplifier and speakers. CDs yield extraordinary sound quality and can be played thousands of times without deterioration. **See Figure 23-20.**

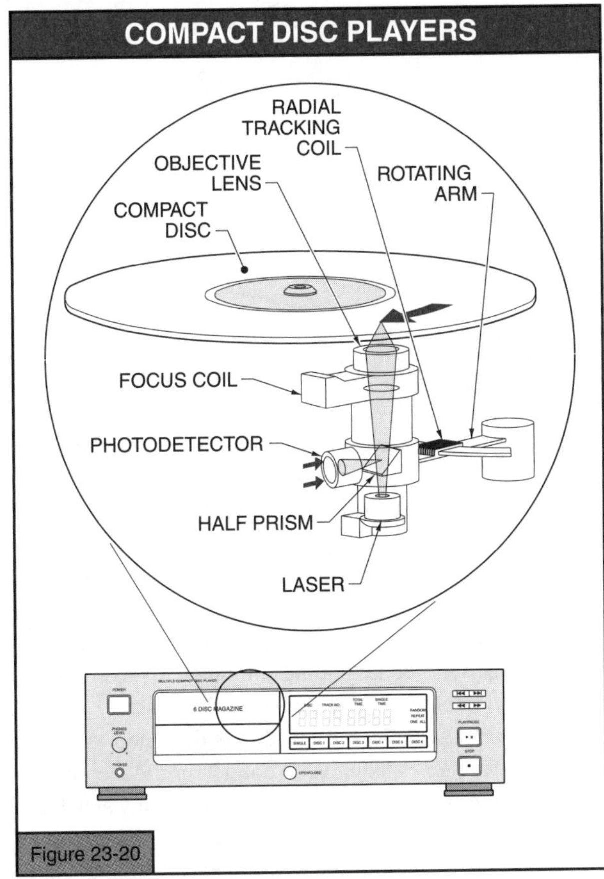

Figure 23-20

Some compact discs have been digitally recorded in a studio while others are copies of analog tape recordings. The Society of Professional Audio Recording Studios (SPARS) has devised a three-digit code using As (analog) and Ds (digital) to identify a CD's source. The first letter identifies the original recording as analog or digital, the second letter identifies whether the original recording was mixed to a digital or an analog recorder, and the third letter identifies whether the mixdown was made into a digital or an analog recording for the CD laser master. For example, the code AAD indicates that the source was recorded on an analog tape, mixed to an analog tape recorder, and transferred to a digital master.

Speakers

A *speaker* is an electric device that converts electrical signals into sound waves. A speaker consists of a coil of wire (voice coil) attached to a paper or polypropylene cone. A vibrating electric current passing through the voice coil creates a magnetic field that attracts or repels a large permanent magnet built into the rear of the speaker. This moves the cone back and forth, which causes the air around the speaker to pressurize and depressurize, producing sound waves. High-frequency signals cause the speaker cone to vibrate rapidly and produce high-frequency sounds. Low-frequency signals cause the speaker cone to vibrate slowly and produce low-frequency sounds.

About 95% of all speakers used in audio systems are electrodynamic speakers. An *electrodynamic speaker* is a speaker that uses a magnet and a coil of wire to move the cone to produce sound. The parts of an electrodynamic speaker include the frame, permanent magnet, voice coil, iron core, bobbin, dust cap, spider, and suspension. **See Figure 23-21.**

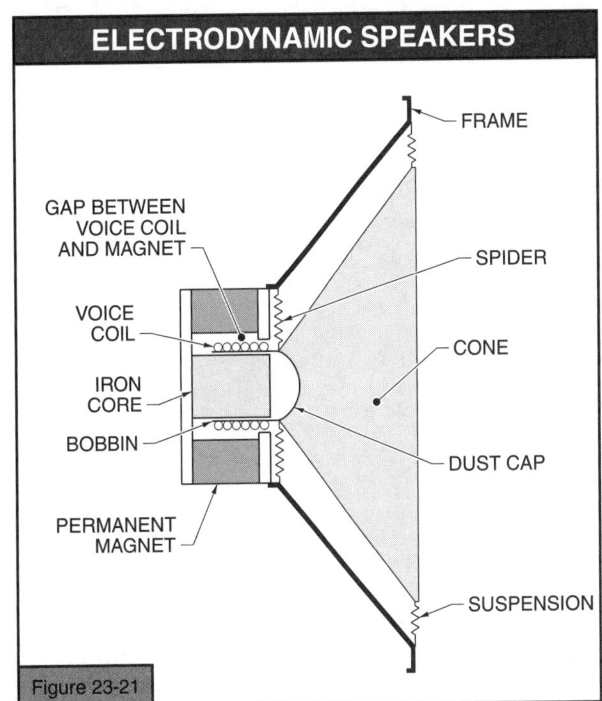

Figure 23-21

The frame holds all the components together and provides the means to fasten the speaker to an enclosure or panel. The frame has cutouts in the back so air can circulate freely around the speaker cone. The diameter of the frame ranges from 1″ to over 15″. Generally, the larger the size, the more volume the speaker can produce. Large speakers are normally used to reproduce bass sounds and small speakers are normally used to reproduce midrange and treble sounds.

As audio signals are continuously fed to the speaker, the varying magnetic field causes the voice coil to move back and forth over the iron core of the permanent magnet. A *bobbin* is a small paper, plastic, or metal cylinder around which the voice coil is wound. The bobbin is attached to the cone. The bobbin moves the cone and causes it to vibrate when the voice coil moves due to the electrical signal applied to it. A dust cap is placed over the bobbin to form the inside center of the cone and keep dust and debris from entering the small gap between the voice coil and the permanent magnet. The inner edge of the cone, voice coil, and bobbin are attached to the frame through a flexible rubber or paper element (spider). The outer edge of the cone is attached to the frame through a flexible rubber, foam, or paper element (suspension).

Several different speakers are used to reproduce sounds in the frequency range between 20 Hz and 20 kHz because no individual speaker can accurately reproduce all sounds in the complete range. Speaker types include tweeters, midranges, woofers, and full-range speakers. The frequency range overlaps to some extent across speaker types. **See Figure 23-22.**

SPEAKERS

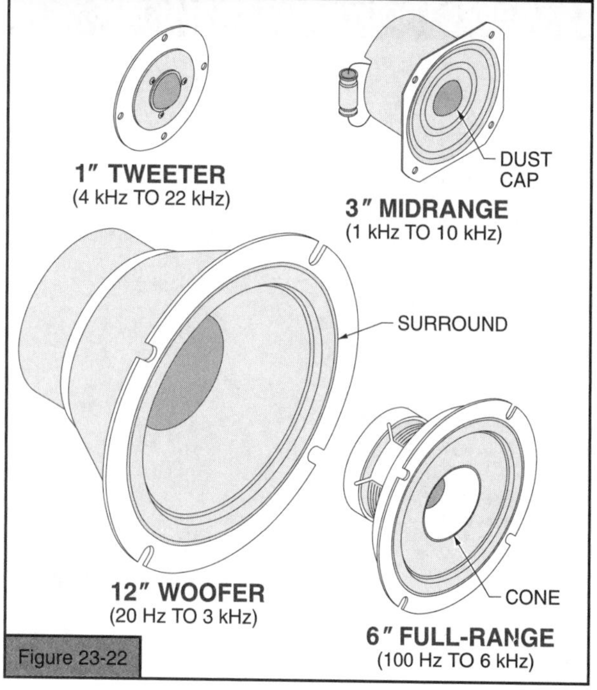

1″ TWEETER
(4 kHz TO 22 kHz)
DUST CAP
3″ MIDRANGE
(1 kHz TO 10 kHz)
SURROUND
12″ WOOFER
(20 Hz TO 3 kHz)
CONE
6″ FULL-RANGE
(100 Hz TO 6 kHz)

Figure 23-22

A *tweeter* is a small speaker that is designed to reproduce high frequencies between 4 kHz and 22 kHz. Tweeters are normally under 2″ in diameter. Tweeters use a paper or cloth cone, a plastic or metal dome, or a piezoelectric element and diaphragm. Tweeters have the most narrow sound dispersion. *Dispersion* is the scattering of sound waves. Sound waves produced by tweeters travel in a narrow corridor through the air. Tweeters are often outfitted with horns and baffles to help disperse the sound. **See Figure 23-23.**

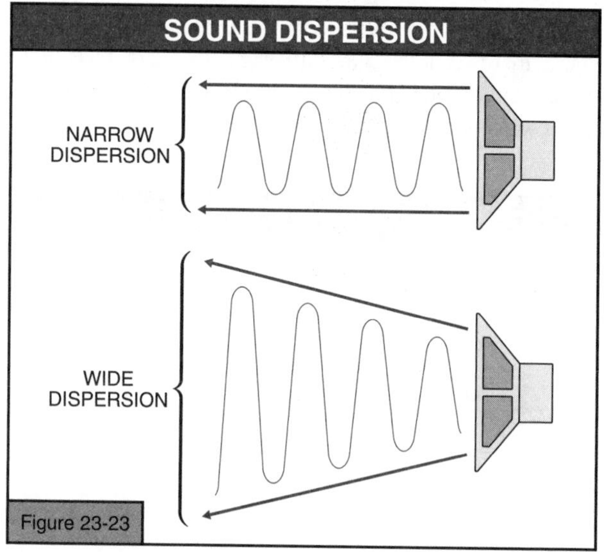

Figure 23-23

A midrange is the most common speaker used to produce sound. A *midrange* is a speaker designed to reproduce sounds from about 1 kHz to 10 kHz. A midrange speaker reproduces most of the vocal and instrumental range, except for very high and low frequency sounds. Midrange speakers are available in sizes from about 3″ to 8″ in diameter. Most midrange speakers use paper, cloth, or polypropylene cones. Some midrange speakers use a small plastic dome to increase sound dispersion.

A *woofer* is a large speaker that is designed to reproduce low-frequency sounds. Woofers reproduce sounds from about 20 Hz to 3 kHz. Woofers can be any size from about 4″ in diameter to over 15″ in diameter. Woofers are commonly 10″ to 12″ in diameter in residential speaker systems. Felt paper and polypropylene are common woofer cone materials.

Large quantities of air must be moved to realistically reproduce bass sounds. The lower the sound (frequency), the greater the amount of air that must be moved. For this reason, woofers are large speakers that are designed to move the large volume of air necessary to reproduce the long wavelengths of bass sounds.

A *full-range speaker* is a speaker designed to adequately reproduce most of the audio spectrum. Full-range speakers normally represent a compromise over a system using individual tweeter, midrange, and woofer speakers. No single speaker can accurately reproduce the entire range of human hearing. Full-range speakers are found in radios and TV sets. They are normally midsized so they reproduce low and high sounds adequately while reproducing midrange sounds (most voice and music tones) excellently. Full-range speakers reproduce sounds from about 100 Hz to 6 kHz.

Speaker Frequency Response

Speakers work over a certain frequency range. The *frequency response* is the variation of the speaker output over a frequency range. A speaker colors a sound if it reproduces certain frequencies to a greater extent while ignoring others. Ideally, a speaker should reproduce all frequencies relatively equally. Frequency response is normally illustrated in graph form plotting output level against frequency. **See Figure 23-24.**

A frequency response graph shows the sonic range that speakers can reproduce and the accuracy to which the frequencies can be reproduced relative to the original recording. The speaker output varies plus or minus some decibel value around the midrange value. A good quality speaker has a frequency response graph that varies slightly over the audio frequency range with a slight drop off in output below 20 Hz and above 20 kHz. A poor quality speaker has a frequency response graph that varies widely over the audio frequency range with a rapid drop off in output around 20 Hz and 20 kHz.

A *crossover network* is an electrical device used to limit the frequencies sent to a particular speaker in a multiple-speaker system. A crossover network is required because it is inefficient to send low-range signals to a tweeter and high-range signals to a woofer. Neither produce any sound, yet both use some power.

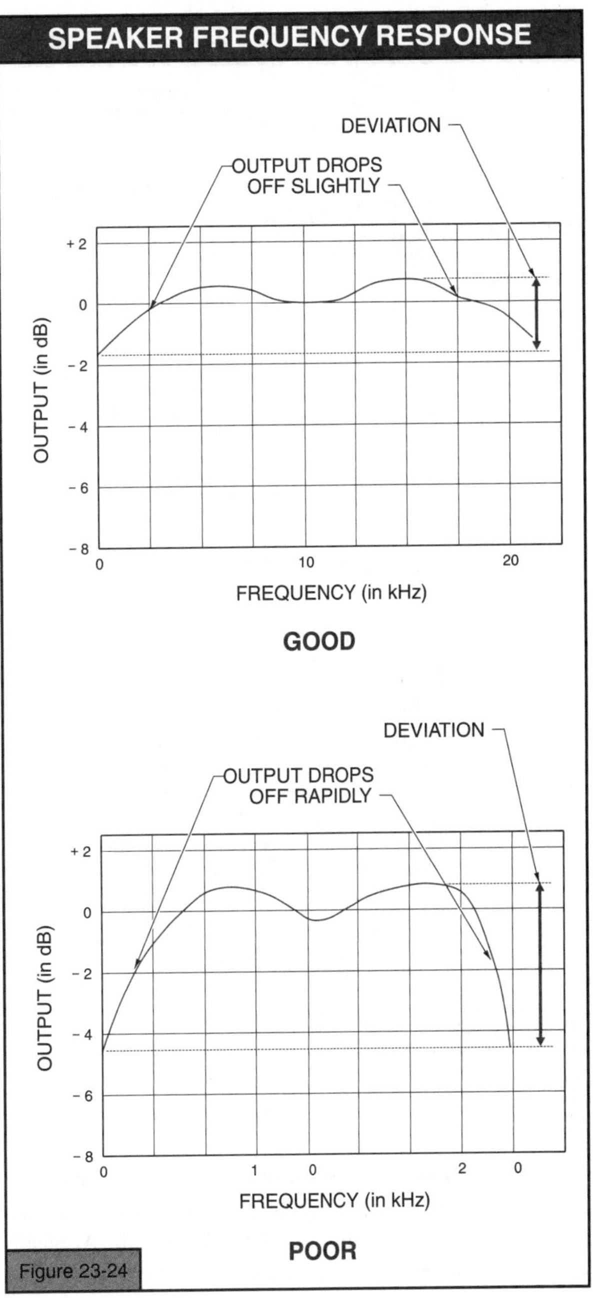

Figure 23-24

Speaker Enclosures

A speaker that is not mounted in or on anything does not operate at maximum efficiency nor produce the best sound reproduction. For this reason, all speakers are designed to be placed inside an enclosure or cabinet.

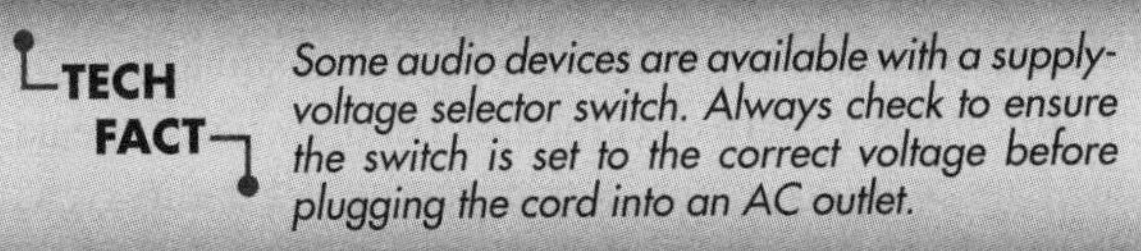

Some audio devices are available with a supply-voltage selector switch. Always check to ensure the switch is set to the correct voltage before plugging the cord into an AC outlet.

Sound waves are emitted from the front and back of the speaker cone. Pressure waves from the back of a bare speaker can flow around the side of the speaker and fill the low-pressure area at the front of the speaker. This results in total or partial cancellation of the sound waves. The speaker is less efficient and distortion may result. *Distortion* is any unintentional change in the characteristics of an audio signal. The two most popular speaker enclosures are acoustic suspension and ported reflex enclosures. **See Figure 23-25.**

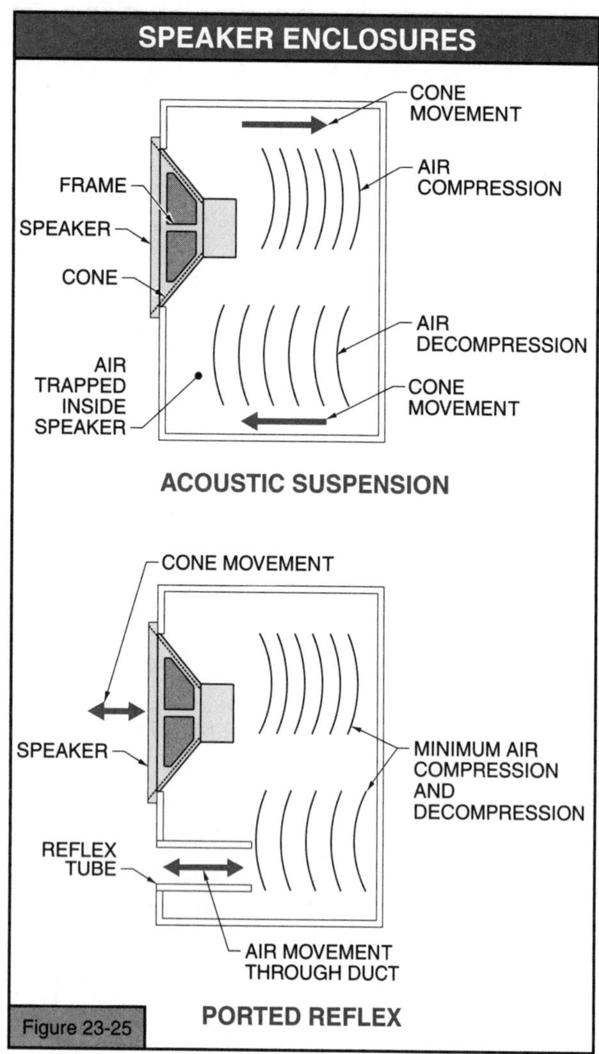

Figure 23-25

In an acoustic suspension enclosure, the speaker enclosure is sealed. The movement of the speaker acts as a piston to compress and decompress the air. Acoustic suspension enclosures increase the stiffness of the speaker. This improves bass response, but sacrifices some speaker efficiency. The amplifier has to deliver more power to produce the same volume.

In a ported reflex enclosure, a small hole (normally two to three inches in diameter) is cut into the enclosure. In most ported reflex enclosures, a tube is inserted in the hole. The tube acts as an air duct, providing a partial vent for the compressed and decompressed air. Speaker cone motion emits sound from the port. Low-frequency air vibrations within the port are in-phase with cone motion and output is increased when the port/enclosure combination is properly constructed (tuned). The bass efficiency of a ported reflex enclosure can be 50% to 100% greater than that of an acoustic suspension enclosure. The cross-sectional area and length of the duct together with the enclosure volume determine the tuned frequency of the enclosure.

Speaker Power Ratings

Speakers are measured by the amount of power (wattage) they can handle. Speakers (especially woofers) often carry a peak and an average wattage rating. A *peak power rating* is the maximum power the speaker can take for a very short time. An *average power rating* is the amount of power a speaker can accept continuously without damage. The speaker voice coil may overheat and deform if the peak power rating is exceeded for more than a few seconds. This may cause permanent damage to the speaker. A typical peak power rating for 10″ to 12″ woofers is 50 W to 500 W. Midrange and tweeter cone speakers are smaller and cannot easily dissipate excess heat caused by extreme wattages. For this reason, midrange speakers have a peak power rating of 50 W to 100 W, and tweeters have a peak power rating of 40 W to 50 W.

Speaker Distortion

Distortion is any unintentional change in the characteristics of an audio signal. Distortion is created in the electronic circuitry of tuners, amplifiers, turntables, and tape decks. Distortion normally appears as a fuzzy sound. Speakers also contribute distortion. Speaker distortion includes harmonic distortion, noise, transient response, and clipping.

Harmonic distortion is the presence of frequencies in the speaker output which are not in the electrical input. These signals are caused by imperfect speaker and suspension behavior and are especially prominent when a speaker is driven to very high volume levels. Proper design of the speaker enclosure significantly reduces harmonic distortion.

Speaker noise is a raspy sound caused by a damaged component, such as a torn cone. Pieces of acoustic batting used to muffle the sound inside a speaker cabinet may create noise when they fall on the cone and vibrate as the cone moves.

Transient response is the time delay from when a speaker cone moves from rest to when it responds to a sharp electrical pulse. The transient response (how fast the speaker reacts to make sound) depends on the stiffness of the cone and suspension and the enclosure design. A stiff cone and suspension resist movement, so the speaker may respond slower and cause a delay in the reproduced sound.

Clipping is a sharp clicking that occurs when a speaker cone cannot move as far as required by the audio signal. The speaker is not able to accurately reproduce the sound when clipping occurs. Damage may result if the speaker is left operating in this condition because the cone, bobbin, or suspension may be torn. Clipping is most prevalent in woofers, where the cone must move a great distance to produce a high volume level at very low frequencies.

Speaker Connections

A speaker becomes a load when it is connected to an amplifier or receiver. The speaker draws power to produce work. The amount of power a speaker draws depends on its power rating and impedance. *Speaker impedance* is the load value (in Ω) that the speaker presents to an amplifier or receiver. Speakers normally have an impedance value of 8 Ω. Low-impedance speakers (4 Ω or less) may cause problems with certain amplifiers and receivers.

Speakers added in parallel decrease the circuit resistance and increase the circuit current. **See Figure 23-26.** For example, if two 8 Ω speakers are connected in parallel, their combined resistance is 4 Ω. The amplifier or receiver must deliver twice the power to the speaker circuit to operate the speakers. Always check the manufacturer's recommended connections before adding any extra speakers to a system.

PARALLEL-CONNECTED SPEAKERS

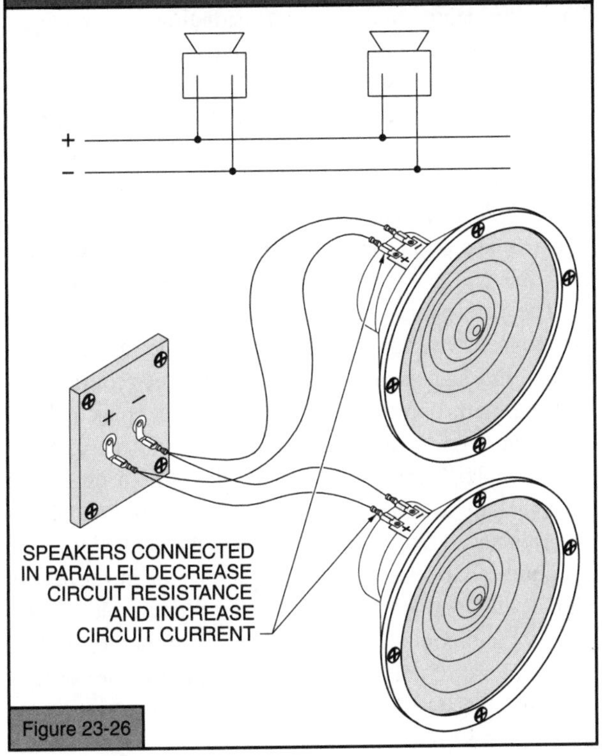

+
–
SPEAKERS CONNECTED IN PARALLEL DECREASE CIRCUIT RESISTANCE AND INCREASE CIRCUIT CURRENT

Figure 23-26

Caution: Ensure the amplifier is turned OFF before connecting or disconnecting speakers to an amplifier or a receiver. Some amplifiers and receivers can be damaged by short circuits or lack of circuits while connecting or disconnecting speaker wires.

Speakers normally have a plus (red) and minus (black) connector for the two wires leading to them. Similarly, the amplifier or receiver has plus (red) and minus (black) terminals for the two wires coming from it. A speaker is connected by removing approximately ½″ of insulation from the ends of the speaker wire and connecting one speaker wire to the plus (red) terminal on the speaker and connecting the same wire to the plus (red) terminal on the amplifier or receiver. One end of the other speaker wire is connected to the minus (black) terminal on the speaker. The other end is connected to the minus (black) terminal on the amplifier or receiver. **See Figure 23-27.**

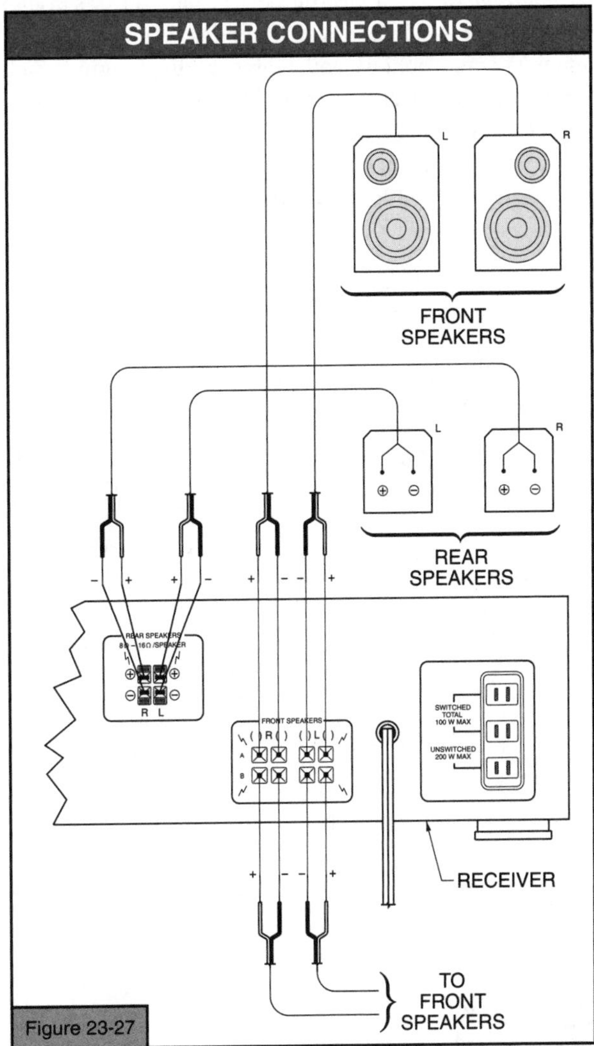

Figure 23-27

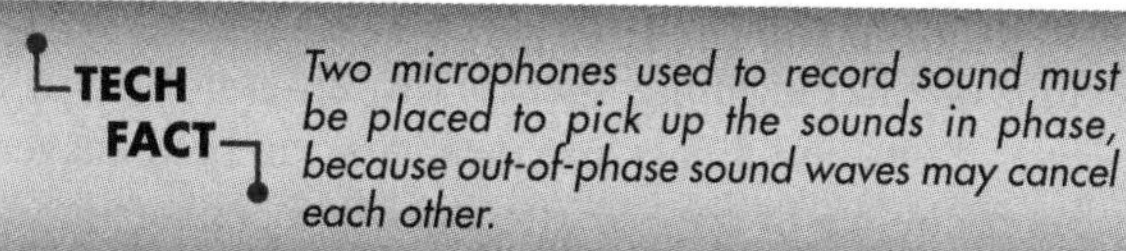

Two microphones used to record sound must be placed to pick up the sounds in phase, because out-of-phase sound waves may cancel each other.

Although a speaker works if the wires are swapped, the speakers must be in-phase because a positive electrical vibration sent to an amplifier or receiver output should make both speakers move forward together (in-phase). One speaker moves forward while the other speaker moves away when the amplifier pumps out a plus vibration, if the wires are switched on one speaker. The speaker moving away creates a negative sound pressure that cancels the positive sound pressure of the other speaker. The result is weaker sound. Out-of-phase speakers give a hollow, thin sound.

Microphones

A *microphone* is an electric device that converts sound vibrations to weak electrical vibrations. Sound vibrations enter the microphone and vibrate a diaphragm which creates a weak electric current. The weak electric current varies at the same rate as the original sound. This weak electric current is sent to the input of an amplifier, receiver, or tape deck. **See Figure 23-28.**

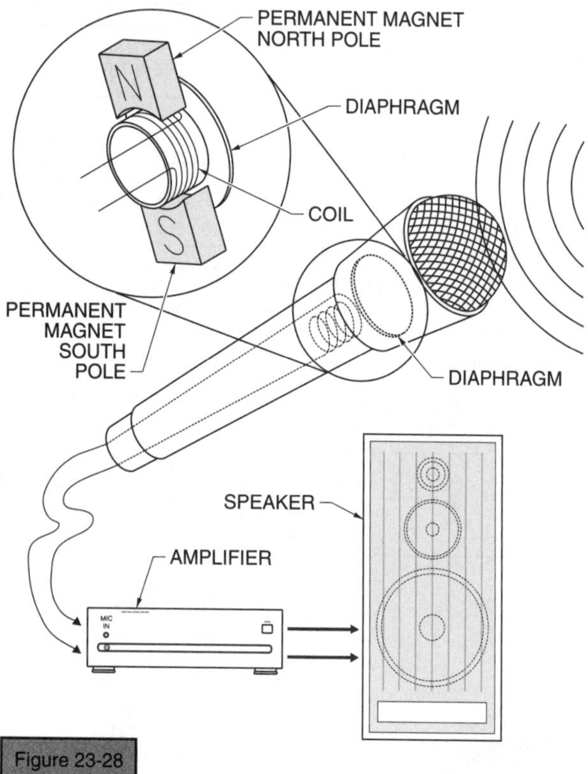

Figure 23-28

The sound quality of a microphone is measured by its frequency response. High-quality microphones have a wide frequency response which indicates that they are equally sensitive to low-, medium-, and high-frequency sounds. Microphones are often described using a microphone frequency response graph. The graph shows how sensitive a microphone is at various frequencies. **See Figure 23-29.**

A good quality microphone has a relatively straight line from 20 Hz to 20 kHz. This indicates that the microphone is equally sensitive to all (or most) frequencies across the audio spectrum. Normal microphones are flat over a certain range. Peaks and valleys in the graph indicate that the microphone is more sensitive to some frequencies and less sensitive to others. A 10 dB dip in the graph line represents a 10 dB drop in power when the microphone reproduces that frequency. The more peaks and valleys in the graph line, the less desirable the microphone is because it colors the sound.

An *electret condenser microphone* is a microphone in which sound waves vibrate a diaphragm in the microphone next to a charged capacitor. A battery-powered electric circuit in the microphone senses the movement of the capacitor's electrical charge and sends these electrical vibrations to an amplifier, a receiver, or a tape deck. A *dynamic microphone* is a microphone that senses sound using a magnet and a coil of wire. Dynamic microphones require no batteries.

Microphones can carry their signals to an amplifier, a receiver, or a tape deck through balanced or unbalanced lines. Residential and semiprofessional audio equipment uses unbalanced lines such as a patch cord. A *patch cord* is a cable that consists of two wires: a thin center conductor that carries the signal and a woven shield wire that encircles it. Patch cords are inexpensive wires that normally have two-conductor plugs on the ends, such as a mini, phone, or RCA plug. In an unbalanced line, the shield is a ground wire that keeps interference (hum and buzz) from entering the cable and distorting the signal. **See Figure 23-30.**

The microphones, plugs, and wires for unbalanced lines are inexpensive and simple to maintain. They are adequate for carrying signals the short distance between components and amplifiers or receivers. Unbalanced lines pick up hum, radio interference, and noise from nearby motors. The longer the cable, the more interference picked up. Unbalanced lines pick up less hum when kept to a maximum of 10′.

A balanced line has three wires in the cable. One wire is connected to ground, one wire carries the in-phase (hot signal), and one wire carries the anti-phase (cold signal). The hot and cold signals are the same signal, only 180° out-of-phase of each other. Two signals that are exactly the same and are 180° out-of-phase with each other cancel each other when added together.

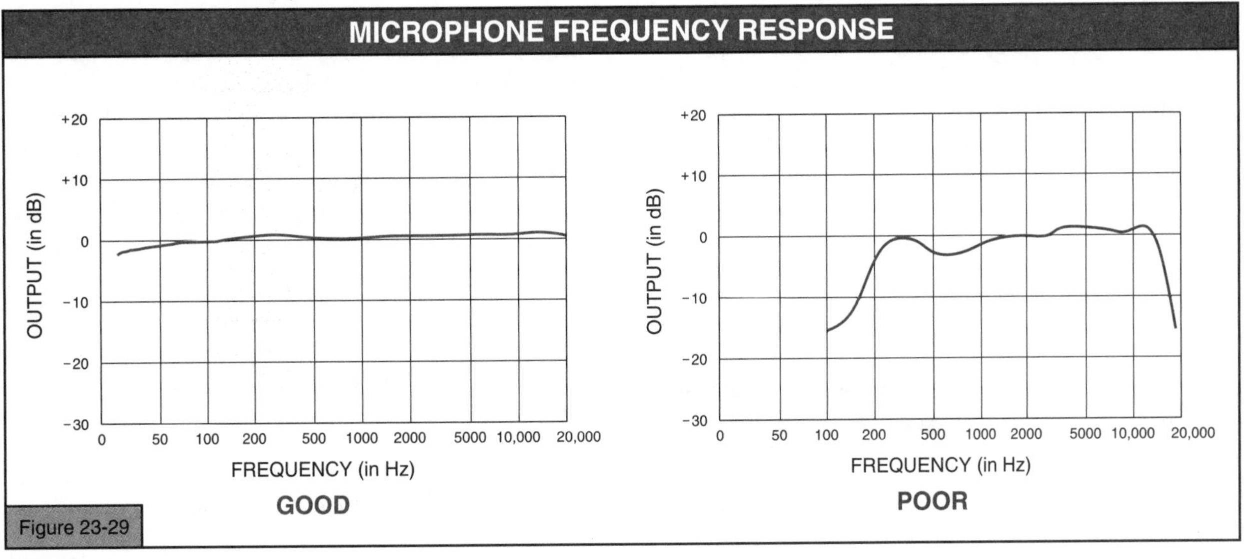

Figure 23-29

CONDUCTOR PLUGS

Plug	Line	Equipment
MINI	Unbalanced	Audiotape recorders, VCRs, small portable equipment
PHONE	Unbalanced	Audiotape recorders, VCRs, most school AV equipment, and musical equipment
RCA	Unbalanced	Some VCRs, nearly all turntables, tape decks, DVD players, and CD players
XLR or A3M	Balanced	Professional VCRs, most microphone mixers and other high-quality audio equipment, nearly all good microphones

Figure 23-30

Noise accumulates equally in the hot and cold signal paths as an audio signal travels down a balanced line. The two signals are subtracted from each other electronically when they arrive at their destination. The similar signals (noise) get removed because they are found equally in both wires. The audio signals get boosted because subtracting out-of-phase signals is the same as adding in-phase signals.

The plug at the end of a balanced cable has three pins and a canon (referred to as XLR or A3M) connector. Balanced lines can carry signals 100′ or more and pick up very little electrical noise. Most professional and industrial audio equipment uses balanced lines.

The various audio devices use different plugs and adapters to connect to different pieces of equipment. Portable equipment normally uses mini plugs because of their small size. Large equipment often uses 1/4″ phone plugs and jacks because they are sturdy and inexpensive. Most connections from machine to machine use RCA plugs and jacks. Professional audio equipment generally uses XLR plugs and jacks. Balanced lines with XLR plugs are connected to devices with unbalanced inputs through the use of a line matching transformer. **See Figure 23-31.** A *line matching transformer* is an adapter containing a transformer that changes the three-wire balanced system to a two-wire unbalanced system.

Signal Levels

An *audio signal* is an electrical representation of a sound in the form of fluctuating voltage or current. Many different signals exist in an audio system. *Signal level* is the amplitude or strength of an audio signal. Signal level is expressed in decibels. Different audio devices are designed to operate with specific signal levels. Some audio devices operate with high signals and some operate with low signals. Level standards have been developed so most equipment is compatible. Standard operating levels are microphone, instrument, line, and speaker levels.

Microphone level is the signal level produced by a standard microphone. Microphone level is the lowest signal level. The diaphragm of a microphone must be sensitive enough to respond to the most minute variations in air pressure to pick up subtle sonic details. Because of the small scale of the diaphragm's motion, the levels it generates are very low. Thus, a microphone's standard operating level ranges from –60 dB to –40 dB. This is why microphones are so susceptible to noise in long cable runs and require good shielding to reduce the chance of stray noise distorting the signal. Turntables are other devices that produce low signal levels.

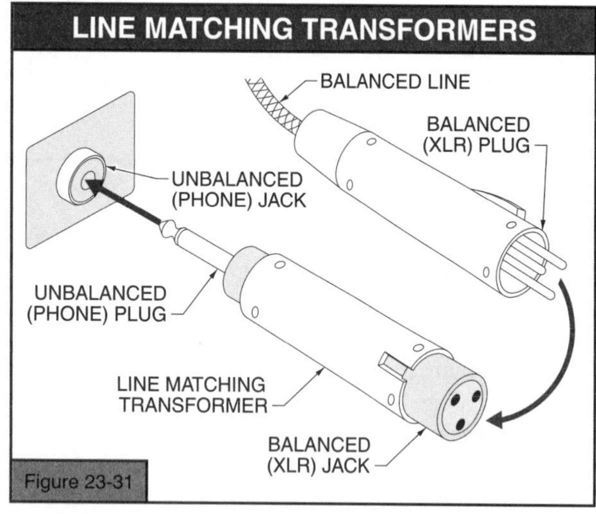

Figure 23-31

Motorola Inc.

The base of a cordless telephone attaches to a phone jack through a standard telephone wire and converts electrical signals into FM radio signals.

Instrument level is the signal level produced by instruments such as guitars and old keyboards. Instrument level ranges from –60 dB to +4 dB.

Line level is the signal level produced by audio devices such as tape decks, PA systems, tuners, VCRs, and other devices. Line levels are normally –10 dB. Line-level devices must not be connected to the supersensitive mic in on a VCR, camcorder, or tape deck because the sound produced is distorted and raspy.

Speaker levels vary with the volume settings of the amplifier or receiver. Never connect a speaker output to a line-level input. This can melt down the input of the line-level device.

Audio System Maintenance

Preventive maintenance and proper cleaning is the best protection against audio system problems. Audio circuits do not normally go bad. The moving mechanical parts such as switches, plugs, motors, wheels, speakers, and pushbuttons are the cause of most problems. Dust and dirt buildup is the major cause of a decrease in sound quality and device failure.

Audio system devices should be cleaned periodically with a soft cloth moistened lightly with water. Never use scouring powder or solvent such as alcohol or benzene when cleaning audio devices.

Tape decks become dirty when tape moves over the record and playback heads. This is due to the tape carrying dirt and flaking oxide, which builds up and creates a gap between the heads and the tape. This gap causes a muffled sound to be recorded and/or played back from the tape.

The audio heads of tape decks should be cleaned when the sound becomes weak or muffled, the high frequencies suddenly disappear, or the tape squeaks or sticks as it passes through the device. The heads may have been knocked out of alignment or worn out if numerous cleanings do not improve the sound quality. Audio heads normally last about 2000 hr to 3000 hr. As they wear, they lose their high-frequency response, muffling the sound. A tape deck that eats tapes normally has a dirty or sticky capstan or pinch roller.

Use a tape head cleaner or manually clean the heads and tape path with a cotton swab and solvent. Tape head cleaners are required for cleaning car tape players and other hard-to-reach tape decks. Manual head-cleaning kits contain swabs and a bottle of fluid which are used to manually clean the tape heads. **See Figure 23-32.** To clean a tape deck using a manual head-cleaning kit, apply the procedure:

1. Turn OFF the tape deck.

2. Remove the tape from the machine.
3. Vacuum dust.
4. Clean the record/playback head and the capstan by dipping a swab in cleaning fluid and rubbing it across the parts.
5. Clean the erase head.
6. Clean tape guide(s).
7. Clean the pinch roller. Rotate the roller with a finger while pressing the swab against the face of the roller. Let all parts dry before reassembling and putting a tape in the tape deck.

Audio tape heads are electromagnets that are magnetized by varying electric current. The heads normally become demagnetized when the current is shut OFF. Long usage or very loud recording signals may cause a small amount of residual magnetism to remain on the tape heads and any metal parts near the tape path. The residual magnetism can gradually erase a tape each time it is played. The tape gradually loses high-frequency response and gains hiss. This is avoided by occasionally demagnetizing the tape heads. A tape head demagnetizer is used to demagnetize tape heads. **See Figure 23-33.** To demagnetize tape heads, apply the procedure:

1. Turn OFF the tape deck.
2. Remove the tape from the machine.
3. Turn ON the tape head demagnetizer approximately 4′ away from the tape deck. Move the tape head demagnetizer very slowly and smoothly whenever it is in the vicinity of the tape deck. Quick jerky motions can induce, rather than eliminate, a magnetic charge.
4. Move the tape head demagnetizer probe close to the record/playback head, erase head, tape guides, and capstan. Hold the probe by each for 1 second and slowly remove when finished.
5. Hold the tape head demagnetizer approximately 4′ away from the tape deck and turn OFF. This is required because the tape head demagnetizer sends out a pulse of magnetism when turned OFF. This could be left on the heads when the demagnetizer is turned OFF too close to the device.

A *cassette tape* is a plastic film covered with a magnetic powder. Anything that nicks, wrinkles, or scrapes the magnetic powder off the film reduces the quality of the recorded sound. Heat and dirt are the most detrimental to cassette tapes. Never leave a cassette tape on a hot radiator, hot car, sunny window sill, dusty workshop, etc. Store cassette tapes in their plastic cases when not in use to keep out dust and dirt.

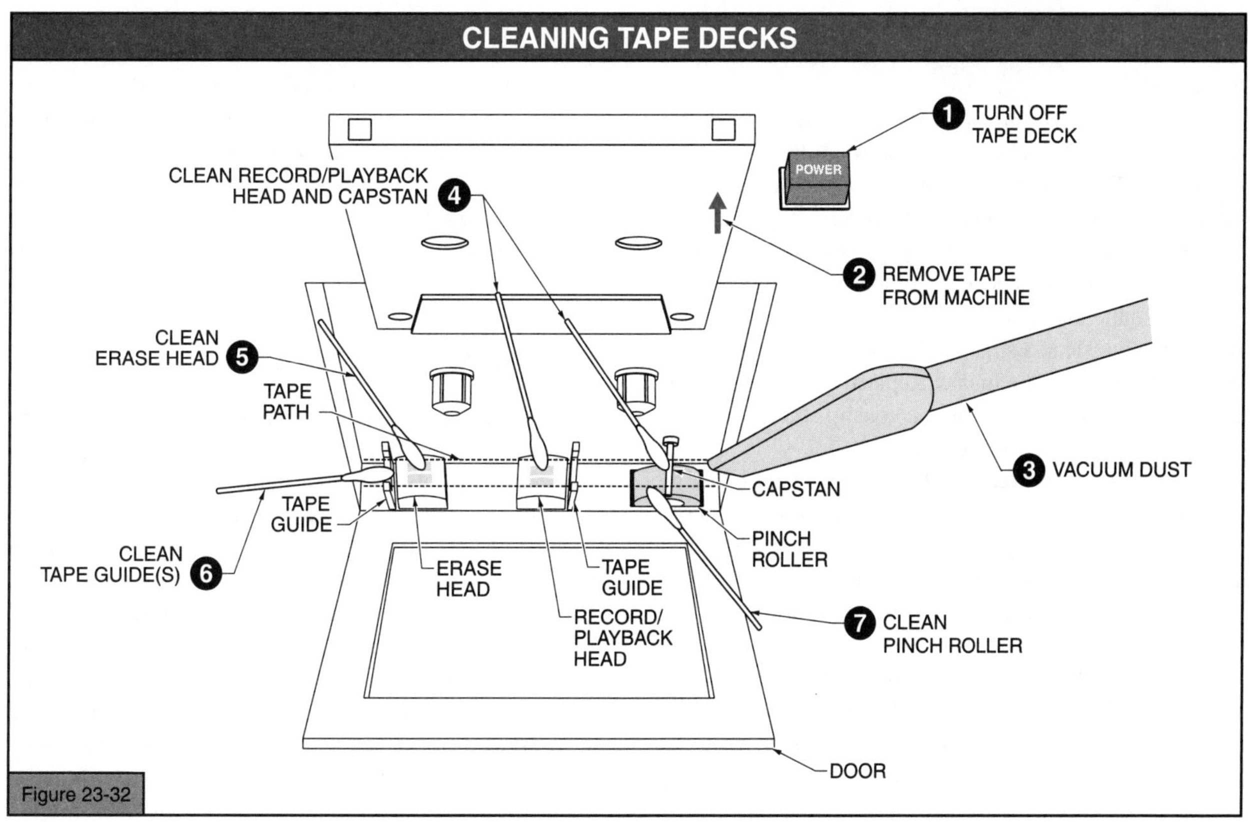

Figure 23-32

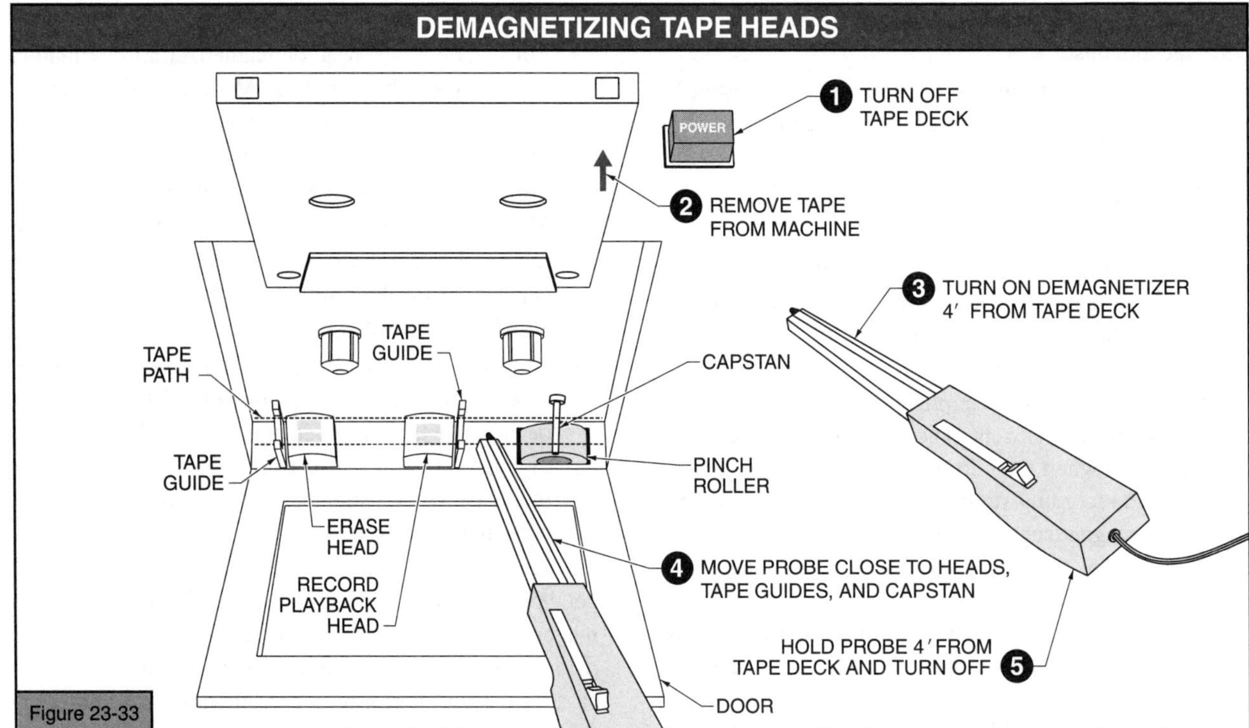

Figure 23-33

Store cassette tapes at average temperature and humidity (about 70°F and 50% or less humidity). Keep cassette tapes away from magnetic fields such as speakers, amplifiers, transformers, magnets, and electric motors. Remove the safety tab on any cassette tape that must not be erased.

In a direct-drive turntable, the motor directly drives the turntable platter. There are no belts, pulleys, or rubber wheels to slip or wear out. For this reason, direct-drive turntables have fewer maintenance problems than belt-drive turntables. Belt-drive turntable speed is reduced or made uneven when the belts get old, stretched, or slippery.

A turntable may have a stretched drive belt if the turntable plays music at a slightly slower speed or if there are distorted sounds in the music. Replace the belt if it feels loose or looks stretched. Apply a nonslip solvent to clean glazed belts if they appear shiny.

Records are very delicate. Records are kept clean by keeping the turntable platter clean, sweeping the record in the direction of the grooves with a fiber brush or velvet pad each time the record is played, and handling records by their edges or label only. Wash a record only with record cleaner.

Store records vertically and away from sun and heat. Store records in their paper, polyethylene, or rice paper sleeves. Discard the plastic record cover wrap. Its static electricity attracts dust and may shrink, warping the record.

Compact disc and DVD players reproduce sound by means of a reflected laser beam. Error-correction circuitry in CD and DVD players covers momentary lapses in a signal. For this reason, CDs and DVDs play if scratched and smeared with fingerprints. Covered-up sound is not true sound and the compensation process adds distortion. Bad scratches and dirt may overload the error-correction circuitry, causing a CD or DVD to skip, jump ahead, or eject. Keep CDs and DVDs in their cases to protect them. Handle the discs by their edges only and clean by gently wiping from the center to the edge of the disc with a soft lint-free cloth.

Troubleshooting Audio Components

Sound may be distorted at its source, in the cables, or when recorded. Plug headphones into the last device in the audio chain when troubleshooting audio components. This is normally the tape deck that is recording the sound. The problem is heard from the tape deck if anything is wrong along the way. Begin troubleshooting by working toward the source if there is no sound at the tape deck.

Ground all sources if all signals have hum. This is accomplished by plugging all devices into the same multiple AC outlet so the grounding wires in the electric cords all go to the same place. Ensure that all volume controls share the load (amplifiers, tape decks, etc.). One volume control should not be excessively high while another is low.

A dirty or corroded connection between devices may cause a sound to crackle or pop in and out. The dirty connection is found by wiggling or twisting the plugs until the sound is restored. A corroded plug or jack may be polished with a pencil eraser or fine sandpaper. Flip any switches that affect the sound back and forth several times or rotate volume or tone controls up and down a few times. This removes oxide from the switch and volume control contacts. Contact cleaner may be used to clean switches or volume controls if they are not restored by switching or rotating.

A microphone that sounds weak or has hum or hiss may not be the correct impedance. The microphone may be defective or of poor quality. The battery may be weak if it is an electret condenser microphone. The problem is in the microphone, cable, plug, or input if a single microphone does not work. There may be a loose or broken wire if wiggling the wire near the microphone and plug produces a crackling sound.

Turntable needle life is approximately 500 hr with most diamond needles. Diamond needles may chip especially if used heavily. This leaves a diamond stub that does not play well. Change the needle if a turntable sounds scratchy or the needle skips. Many turntable needles can be pulled out and a new one inserted into the cartridge. A needle that skips may also indicate an unlevel turntable platter. Use a pocket torpedo level to set the platter level.

Ensure the ground wire of a turntable is secure at the back of the turntable and amplifier or receiver because a turntable that has a faint hum or buzz may have a loose ground wire. Ensure the turntable is not placed too close to an amplifier, high-intensity lamp, motor, or other device that radiates electrical interference which can pass through the electrical wiring and into the amplifier. Move the turntable farther from the speakers if it rumbles or place it on a more solid support. Buy and install an AC line filter if hum and buzz are still a problem. An *AC line filter* is a device that absorbs electrical interference from motors, etc. before these signals can travel down the power line.

Move a compact disc player farther from the speakers if it skips easily. The speaker sound may vibrate the disc as it plays. Clean the laser (refer to manufacturer's instructions). It may be smudged or dusty.

Check the amplifier, receiver, or source if speakers buzz or rattle, or if a sound is heard when there is no music playing. Remove the grille and tighten the screws that hold the speakers to the cabinet. The screws loosen with time and vibration. Gently and evenly push the midrange and woofer cones in if a speaker produces a buzz or rattle. The cone should move freely without scraping. The speaker may be warped and require replacement if it scrapes. A speaker cabinet may have a blown tweeter, woofer, or midrange if the speaker cabinet lacks highs, lows, or midranges. Cover the other speakers to determine if the remaining speaker makes any sound.

Device Safeguards

Certain safeguards should be observed when using electric devices. Safeguards ensure long and safe device operation. Read all safety and operating instructions before operating any electric device. Keep safety and operating instructions for future reference.

Never use an electric device near water such as bathtubs, kitchen sinks, laundry tubs, wet basements, or swimming pools. An electric device should be placed on a cart or stand that is recommended by the manufacturer. The cart or stand should be moved with care. Quick stops, excessive force, and uneven surfaces may cause the cart or stand to overturn. Electric devices should be mounted to a wall or ceiling only as recommended by the manufacturer.

Place electric devices in a location that does not interfere with proper ventilation. For example, an electric device should not be placed on a bed, sofa, rug, or similar surface that may block the ventilation openings. Electric devices should not be placed in a bookcase or cabinet that may impede the flow of air through the ventilation openings. The device should be placed away from heat sources such as radiators, heat registers, stoves, amplifiers, or other devices that produce heat. Electric devices should be connected to a power supply of the type described in the operating instructions or as marked on the device. Any device equipped with a polarized plug must be used with a polarized receptacle. The polarization or ground of a plug should not be defeated. Contact a qualified electrician if the receptacle being used is not polarized. Power-supply cords should be routed so that they are not likely to be walked on or pinched by items placed on or against them.

An outdoor antenna should be located away from power lines. Ensure the antenna system is grounded if an outdoor antenna is connected to a receiver or tuner. Grounding is recommended if FM reception is impaired by noise. To ground, connect a thick polyvinyl insulated wire to the GND terminal on the receiver or tuner and attach the other end to a metal water pipe or grounding electrode. This provides some protection against voltage surges and built up static charges. Article 810 of the NEC® provides information on proper grounding of the mast and supporting structure, grounding of the lead-in wire to an antenna discharge unit, size of grounding conductors, location of antenna-discharge unit, connection to grounding electrodes, and requirements for the grounding electrode. **See Figure 23-34.**

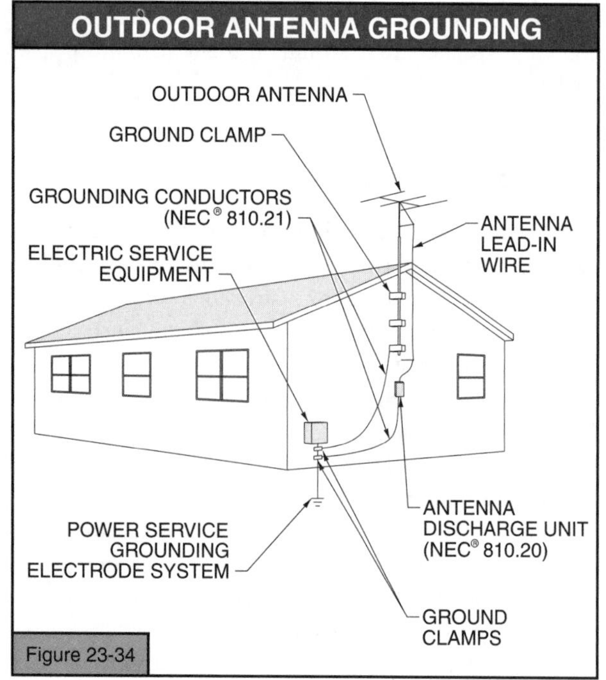

Figure 23-34

The power cord of an electric device should be unplugged from the outlet when unused for a long period of time. Care should be taken so that objects do not fall and liquids are not spilled into the enclosure. The device should be serviced by qualified service personnel when the power-supply cord or the plug has been damaged, objects have fallen or liquid has spilled into the device, the device has been exposed to rain, has been dropped, or the enclosure damaged.

The user should not attempt to service the device beyond that described in the operating instructions. All other servicing should be referred to qualified service personnel. Do not overload wall outlets and extension cords because this can result in fire or electric shock. Ensure the service technician has used replacement parts specified by the manufacturer or parts having the same characteristics as the original part when replacement parts are required. Unauthorized substitutions may result in fire, electric shock, or other hazards. Ask the service technician to perform safety checks to determine that the device is in proper operating condition after completing any service or repairs.

Refer to the CD-ROM in the back of the book for Chapter 23 Quick Quiz® and related information.

24 Electronic Control Devices

Solid-State Controls

Solid-state controls are devices that use electronic components such as transistors, SCRs, and triacs to switch and control the flow of current in a circuit. Advantages of solid-state controls include:

- Long life relative to electromechanical devices
- High reliability when properly installed and used
- No contact arcing
- High resistance to shock and vibration due to no moving parts
- Very fast switching relative to mechanical switches

Solid-state controls are used in many applications that previously used mechanical switches. For example, SCRs and triacs are now used to control motors. The use of solid-state controls to operate motors eliminates electrical contacts that may arc and cause a potential explosion hazard. Solid-state controls allow motors to be started and stopped in a controlled manner. Solid-state controls can soft start and soft stop a motor. Soft starting and soft stopping is the applying and removal of voltage to and from the motor in a controlled manner. Soft starts and soft stops reduce mechanical stress on motors, shafts, gear boxes, belts, and products being moved. **See Figure 24-1.**

Semiconductor Theory

All matter consists of an organized collection of atoms. An *atom* is the smallest particle that an element can be reduced to and still maintain the properties of that element. The three fundamental particles contained in atoms are protons, neutrons, and electrons. Protons and neutrons make up the nucleus, and electrons whirl around the nucleus in orbits or shells.

The *nucleus* is the heavy, dense center of an atom which contains protons and neutrons and has a positive electrical charge. A *proton* is a particle with a positive charge of 1 unit that exists in the nucleus of an atom. A *neutron* is a neutral particle, with a mass approximately the same as a proton, that exists in the nucleus of an atom. The nucleus is surrounded by one or more electrons. An *electron* is a negatively charged particle in an atom. Electrons whirl around the nucleus at great speeds in shells. Each shell can hold a specific number of electrons. The innermost shell can hold two electrons. The second shell can hold eight electrons. The third shell can hold 18 electrons, etc. The shells are filled starting with the inner shell and working outward, so when the inner shells are filled with as many electrons as they can hold, the next shell is started. Electrons and protons have equal amounts of opposite charges. There are as many electrons as there are protons in an atom, which leaves the atom electrically neutral.

Valence Electrons

Most elements do not have a completed outer shell with the maximum allowable number of electrons. *Valence electrons* are electrons in the outermost shell of an atom. Valence electrons determine the conductive or insulative value of a material. Conductors normally have only one or two valence electrons in their outer shell. **See Figure 24-2.** Insulators normally have several electrons in their outer shell which is either almost or completely filled with electrons. Semiconductor materials fall between the low resistance offered by a conductor and the high resistance offered by an insulator. Semiconductors are made from materials that have four valence electrons.

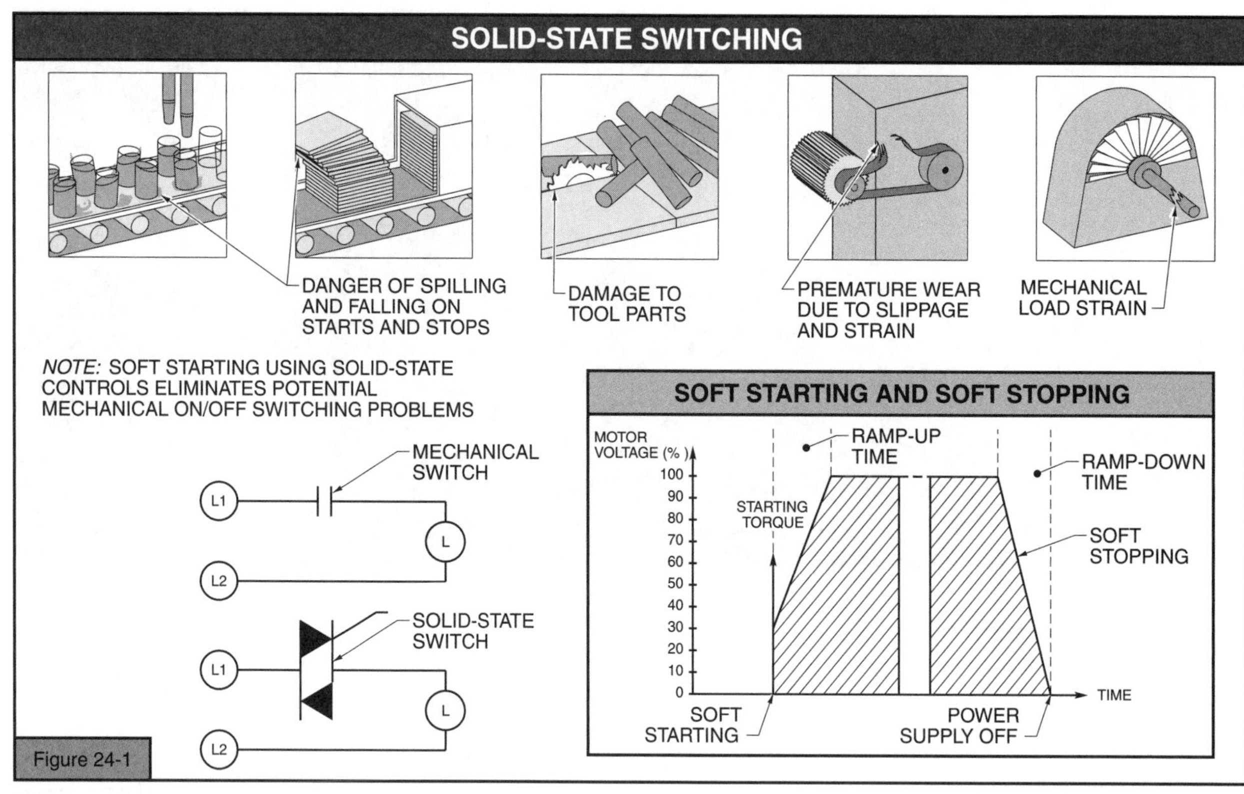

Figure 24-1

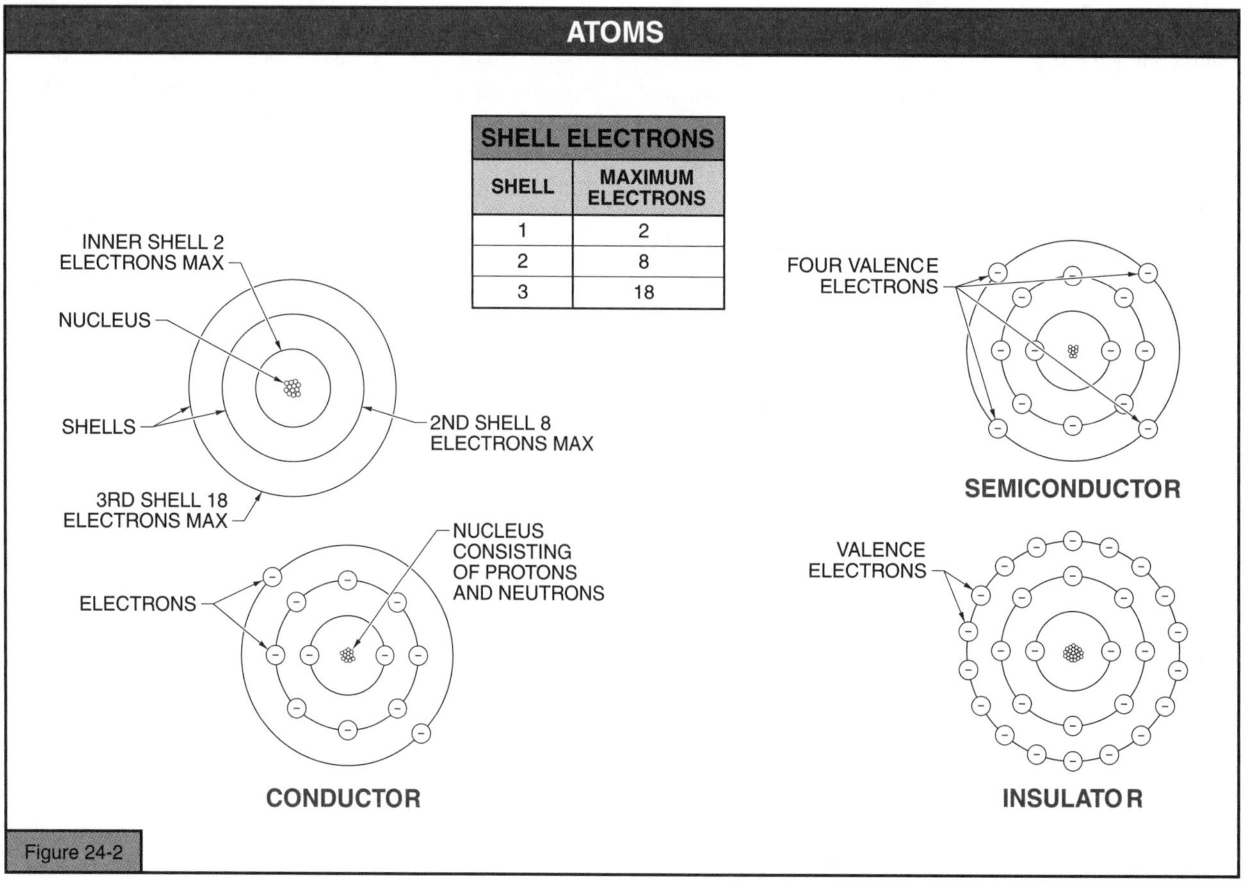

ATOMS

SHELL ELECTRONS	
SHELL	MAXIMUM ELECTRONS
1	2
2	8
3	18

INNER SHELL 2 ELECTRONS MAX
NUCLEUS
SHELLS
2ND SHELL 8 ELECTRONS MAX
3RD SHELL 18 ELECTRONS MAX
ELECTRONS
NUCLEUS CONSISTING OF PROTONS AND NEUTRONS

CONDUCTOR

FOUR VALENCE ELECTRONS

SEMICONDUCTOR

VALENCE ELECTRONS

INSULATOR

Figure 24-2

Doping

The basic material used in most semiconductor devices is either germanium or silicon. In their natural state, germanium and silicon are pure crystals. These pure crystals do not have enough free electrons to support a significant current flow. To prepare these crystals for use as a semiconductor device, their structure must be altered to permit significant current flow.

Doping is the process by which the crystal structure is altered. In doping, some of the atoms in the crystal are replaced with atoms from other elements. The addition of new atoms in the crystal creates N-type material and P-type material.

N-Type Material

N-type material is material created by doping a region of a crystal with atoms from an element that has more electrons in its outer shell than the crystal. Adding these atoms to the crystal results in more free electrons. Free electrons (carriers) support current flow. Current flows from negative to positive through the crystal when voltage is applied to N-type material. The material is N-type material because electrons have a negative charge. **See Figure 24-3.**

Elements commonly used for creating N-type material are arsenic, bismuth, and antimony. The quantity of doping material used ranges from a few parts per billion to a few parts per million. By controlling these small quantities of impurities in a crystal, the manufacturer controls the operating characteristics of the semiconductor.

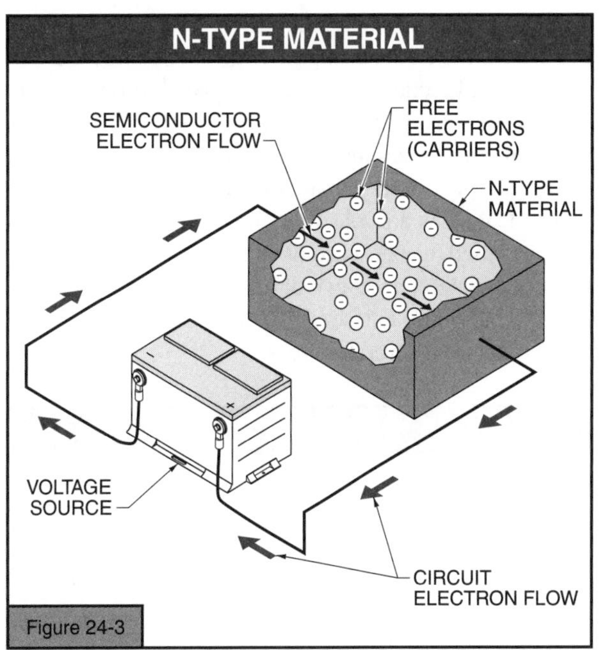

Figure 24-3

P-Type Material

P-type material is material with empty spaces (holes) in its crystalline structure. To create P-type material, a crystal is doped with atoms from an element that has fewer electrons in its outer shell than the crystal.

Holes are the missing electrons in the crystal structure of a semiconductor. The holes are represented as positive charges. In P-type material, the holes act as carriers. The holes are filled with free electrons when voltage is applied, and the free electrons move from negative potential to positive potential through the crystal. **See Figure 24-4.** Movement of the electrons from one hole to the next makes the holes appear to move in the opposite direction. Hole flow is equal to and opposite of electron flow. Typical elements used for doping a crystal to create P-type material are gallium, boron, and indium.

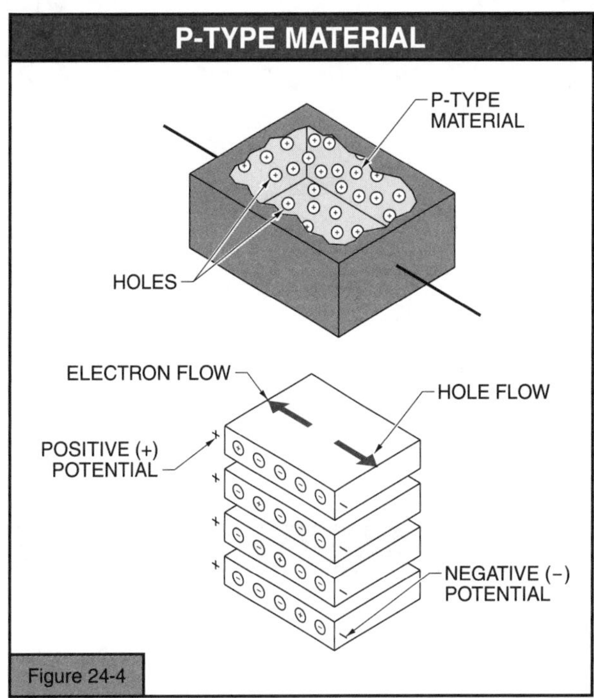

Figure 24-4

Semiconductor Devices

Semiconductor devices are devices in which electrical conductivity is between that of a conductor (high conductivity) and that of an insulator (low conductivity). The electrical conductivity of a semiconductor device is nearly as great as the conductivity of a metal at high temperatures and is nearly absent at low temperatures.

Semiconductor devices are often mounted on a PC board. A *PC board* is an insulating material such as fiberglass or phenolic with conducting paths laminated to one or both sides of the board. PC boards provide electrical paths of sufficient size to ensure a reliable electronic circuit. **See Figure 24-5.** *Pads* are small round conductors to which component leads are soldered. *Traces (foils)* are conducting paths used to connect components on a PC board. They are used to interconnect two or more pads. A *bus* is a large trace extending around the edge of a circuit board to provide conduction from several sources.

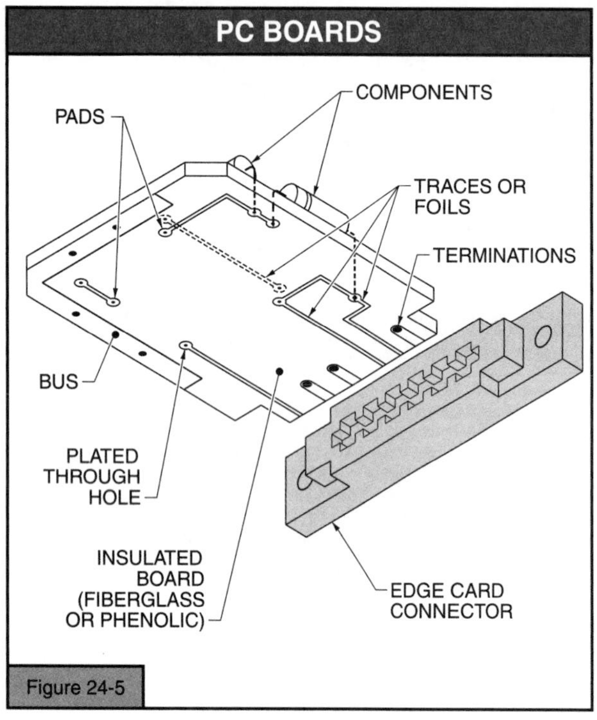

Figure 24-5

An *edge card* is a PC board with multiple terminations (terminal contacts) on one end. Most edge cards have terminations made from copper, which is the same material as the traces. In some instances, the terminations are gold plated, allowing for the lowest possible contact resistance. An edge card connector allows the edge card to be connected to the system's circuitry with the least amount of hardware.

Solid-state devices are normally mounted on one side of a PC board. **See Figure 24-6.** In some cases where space is a premium, components may be mounted on both sides of the PC board. Device leads extend through the board and are connected to the pads, traces, and bus with solder.

The higher the current in the circuit, the greater the heat produced in the solid-state device. A *heat sink* is a device that conducts and dissipates heat away from a component. Heat sinks consist of a piece of metal used to dissipate the heat produced by solid-state devices mounted on it. For proper operation, the heat sink must be mounted so that the heat can be pulled away from it. Heat can be removed by natural convection and/or forced-air cooling using a fan motor.

Diodes

A *diode* is a semiconductor device that offers very high opposition to current flow in one direction and very low opposition to current flow in the opposite direction. This is made possible by the doping process, which creates N-type material and P-type material on the same component. The N-type and P-type materials exchange carriers at the junction of the two materials, creating a thin depletion region. The depletion region responds rapidly to voltage changes. The operating characteristics of a specific diode are illustrated through the use of a diode characteristic curve. **See Figure 24-7.**

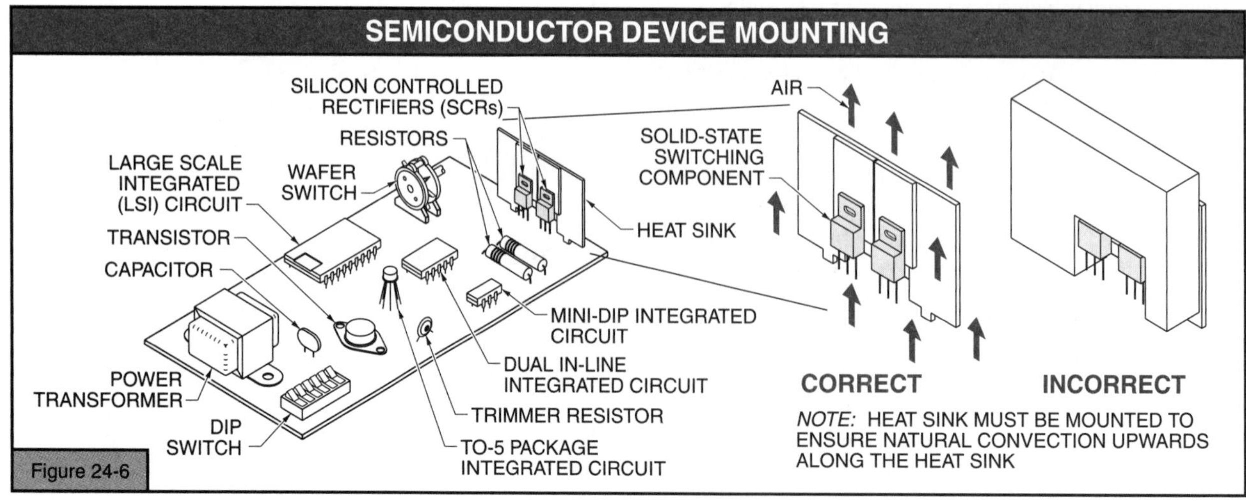

Figure 24-6

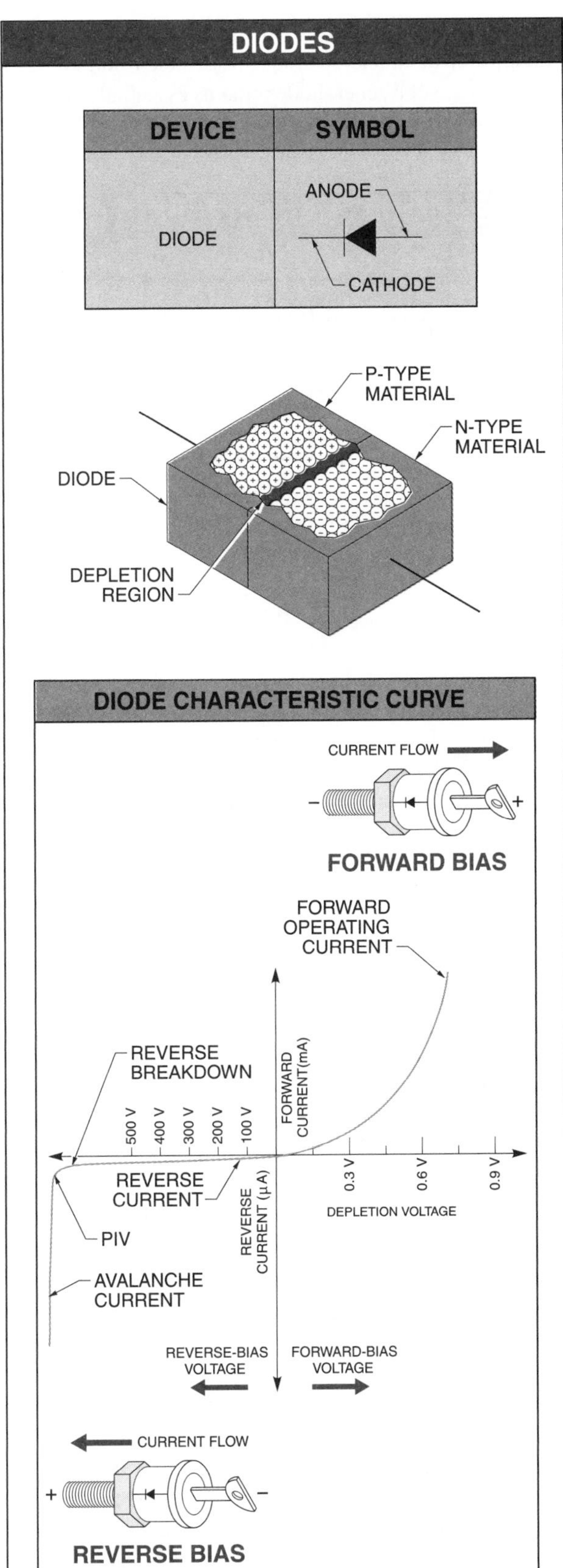

Figure 24-7

The action occurring in the depletion region either blocks current flow or passes current when voltage is applied to the diode. *Forward-bias voltage* is the application of the proper polarity to a diode. Forward bias results in forward current. *Reverse-bias voltage* is the application of the opposite polarity to a diode. Reverse bias results in a very small reverse current (normally 1 mA).

Peak inverse voltage (PIV) is the maximum reverse bias voltage that a diode can withstand. The PIV ratings for most diodes used in industry range from a few volts to several thousand volts. The diode breaks down and passes current freely if the reverse bias applied to the diode exceeds its PIV rating. *Avalanche current* is current passed when a diode breaks down. Avalanche current can destroy diodes. Diodes with the correct voltage rating must be used to avoid avalanche current. **See Figure 24-8.**

Zener Diodes

A *zener diode* is a silicon PN junction that differs from a standard (rectifier) diode in that it operates in the reverse breakdown region. A *PN junction* is the area on a semiconductor material between the P-type and N-type material. A zener diode acts as a voltage regulator either by itself or in conjunction with other semiconductor devices. A zener diode symbol differs from a standard diode symbol in that the normally vertical cathode line is bent slightly at each end. Standard diodes normally conduct in forward bias and can be destroyed if the reverse voltage or bias is exceeded. A zener diode is often referred to as an avalanche diode because the zener diode normally operates in reverse breakdown.

Most cellular telephones contain printed circuit boards with small electronic components mounted on one side.

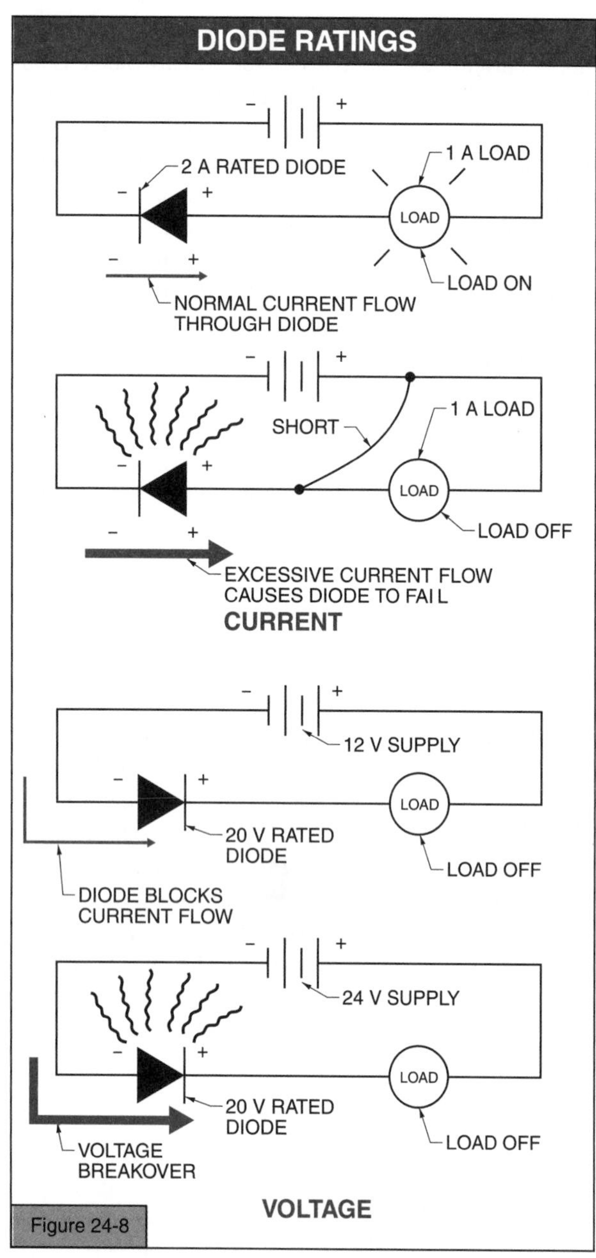

Figure 24-8

Zener Diode Operation

The forward breakover voltage and current characteristics are similar to a standard diode when a source voltage is applied to a zener diode in the forward direction. **See Figure 24-9.** The current of a zener diode remains low until the reverse voltage reaches reverse breakdown (zener breakdown) when a source voltage is applied to the zener diode in the reverse direction. The zener diode conducts heavily (avalanches) at zener breakdown. Reverse current flow through a zener diode must be limited by a resistor or other device to prevent diode destruction. The maximum current that may flow through a zener diode is determined by diode size. Like the forward voltage drop of a standard diode, the reverse voltage drop or zener voltage of a zener diode remains essentially constant despite large current fluctuations.

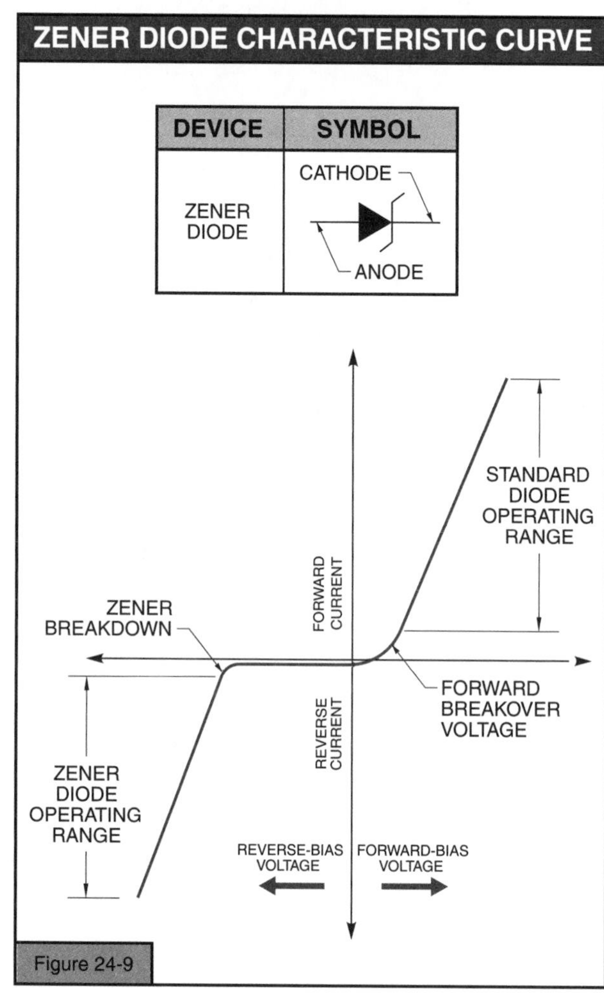

Figure 24-9

A zener diode is used as a voltage regulator to maintain a constant voltage across the load even when load current fluctuates. A zener diode is capable of being a constant voltage source because of the resistance changes that take place within the PN junction. **See Figure 24-10.** The resistance of the PN junction remains high and produces leakage current in the microampere range when a source of voltage is applied to the zener diode in the reverse direction. However, as the reverse voltage is increased, the PN junction reaches a critical voltage and the zener diode avalanches. As the avalanche voltage is reached, the normally high resistance of the PN junction drops to a low value and the current increases rapidly. The current is normally limited by a circuit resistor or load resistance.

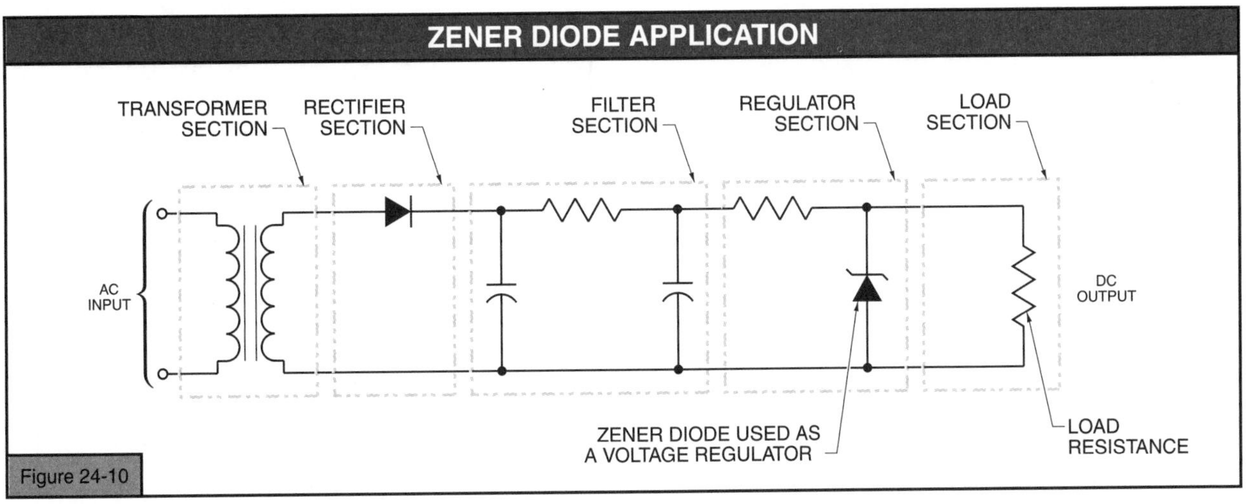

Figure 24-10

TECH FACT *Correct heat sink selection and mounting is confirmed by taking an actual temperature measurement under worst-case conditions and comparing the reading to the listed manufacturer's allowable temperature chart.*

Transistors

A *transistor* is a three-terminal device that controls current through the device depending on the amount of voltage applied to the base. Transistors may be PNP or NPN transistors. A *PNP transistor* is a transistor formed by sandwiching a thin layer of N-type material between two layers of P-type material. An *NPN transistor* is a transistor formed by sandwiching a thin layer of P-type material between two layers of N-type material. **See Figure 24-11.** Transistor terminals are the emitter (E), base (B), and collector (C). The emitter, base, and collector are located in the same place for both symbols. The only difference is the direction in which the emitter arrow points. In both cases, the arrow points from the P-type material toward the N-type material. Transistors are bipolar devices. A *bipolar device* is a device in which both holes and electrons are used as internal carriers for maintaining current flow.

Transistor Terminals

Transistors are manufactured with two or three leads extending from their case. A transistor's outline (TO and SP) number is used as a reference when a specific-shaped transistor must be used. **See Figure 24-12.** TO and SP numbers are determined by individual manufacturers. *Note:* The bottom view of transistor TO-3 shows only two leads (terminals). Frequently, transistors use the metal case as the collector-pin lead.

Spacing can also be used to identify transistor leads. Normally, the emitter and base leads are close together and the collector lead is farther away. The base lead is normally in the middle. A transistor with an index pin must be viewed from the bottom. An *index pin* is a metal extension from the transistor case. The leads are identified in a clockwise direction from the index pin. For example, the leads on TO-5 are identified as E, B, and C. The emitter is closest to the index pin. Refer to a transistor manual or to manufacturer's specification sheets for detailed information on transistor construction and identification.

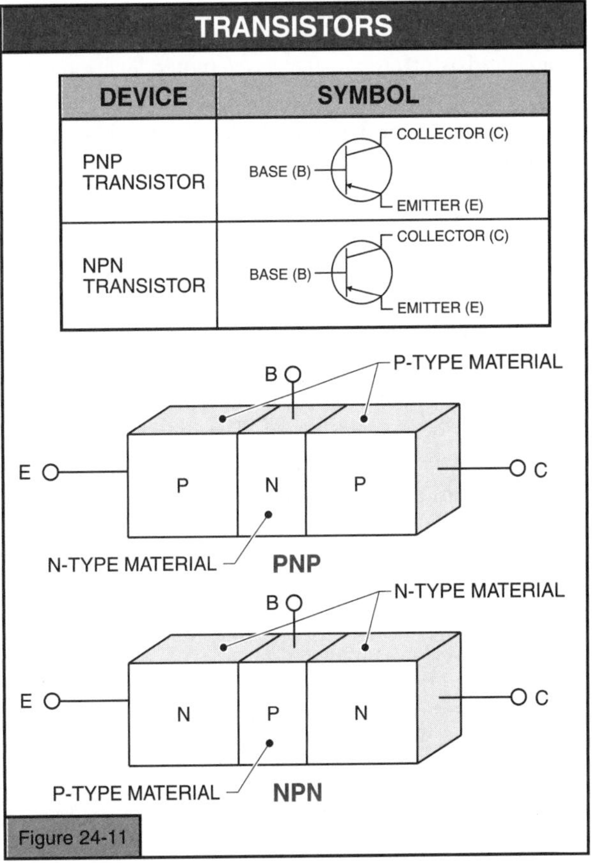

Figure 24-11

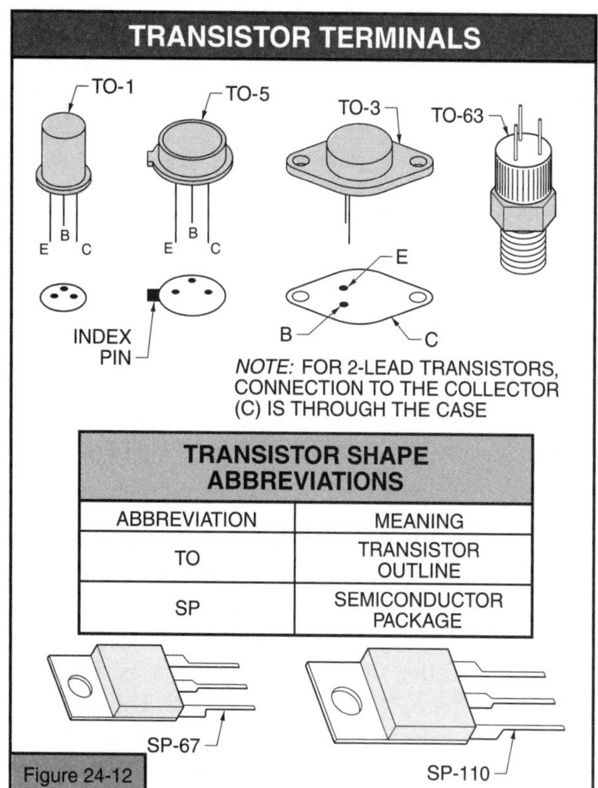

TRANSISTOR SHAPE ABBREVIATIONS	
ABBREVIATION	MEANING
TO	TRANSISTOR OUTLINE
SP	SEMICONDUCTOR PACKAGE

Figure 24-12

Biasing Transistor Junctions

In any transistor circuit, the base/emitter junction must always be forward biased and the base/collector junction must always be reverse biased. **See Figure 24-13.** The external voltage (bias voltage) is connected so the positive terminal connects to the P-type material (base) and the negative terminal connects to the N-type material (emitter). This arrangement forward biases the base/emitter junction. Current flows from the emitter to the base. The action that takes place is the same as the action that occurs for a forward-biased semiconductor diode.

In any transistor circuit, the base/collector junction must always be reverse biased. The external voltage is connected so the negative terminal connects to the P-type material (base) and the positive terminal connects to the N-type material (collector). This arrangement reverse biases the base/collector junction. Only a very small current (leakage current) flows in the external circuit. The action that takes place is the same as the action that occurs for a semiconductor diode with reverse bias applied.

TECH TIP *Optimum thermal connection beetween a heat sink and mounted solid-state component requires that a thermally conductive compound (thermal paste) be placed between the two metal surfaces.*

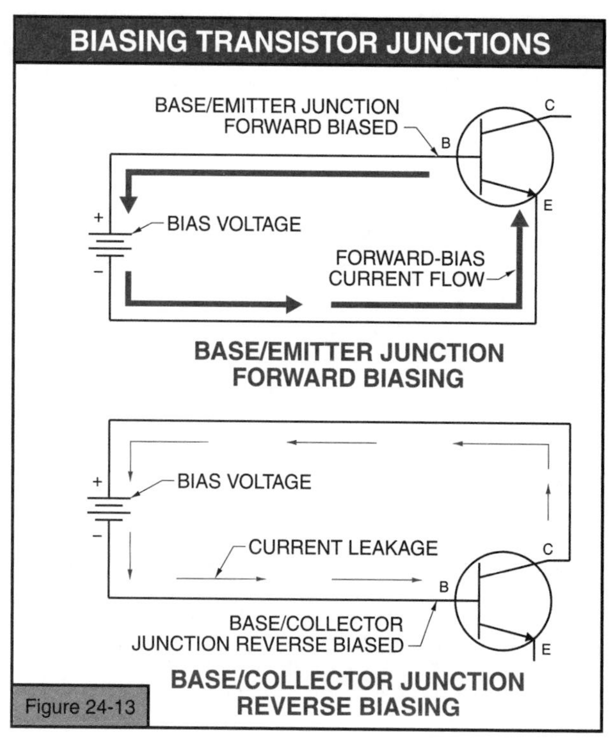

Figure 24-13

Transistor Current Flow

Individual PN junctions of a transistor may be used in combination with two bias arrangements. **See Figure 24-14.** The base/emitter junction is forward biased while the base/collector junction is reverse biased. This circuit arrangement results in an entirely different current path than the path that occurs when the individual circuits are biased separately.

The forward bias of the base/emitter circuit causes the emitter to inject electrons into the depletion region between the emitter and the base. Because the base is less than .001″ thick for most transistors, the more positive potential of the collector pulls the electrons through the thin base. As a result, the greater percentage (95%) of the available free electrons from the emitter passes directly through the base (I_C) into the N-type material, which is the collector of the transistor.

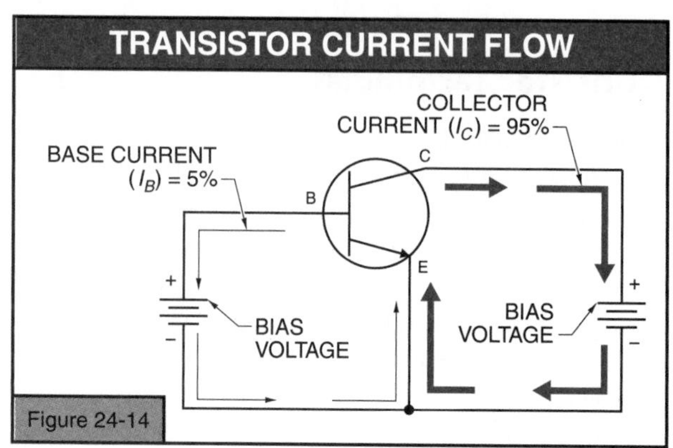

Figure 24-14

The base current (I_B) is a critical factor in determining the amount of current flow in a transistor because the forward-biased junction has a very low resistance and could be destroyed by heavy current flow. Therefore, the base current must be limited and controlled.

Transistors as DC Switches

Transistors were mainly developed to replace mechanical switches. Transistors have no moving parts and can switch ON and OFF quickly. Mechanical switches have two conditions: open and closed or ON and OFF. The switch has a very high resistance when open and a very low resistance when closed.

A transistor may be operated like a switch. For example, a transistor may be used to turn a pilot light ON or OFF. **See Figure 24-15.** In this circuit, the resistance between the collector (C) and the emitter (E) is determined by the current flow between the base (B) and emitter (E). The collector/emitter resistance is high, like that of an open switch, when no current flows between B and E. The pilot light does not glow because there is no current flow.

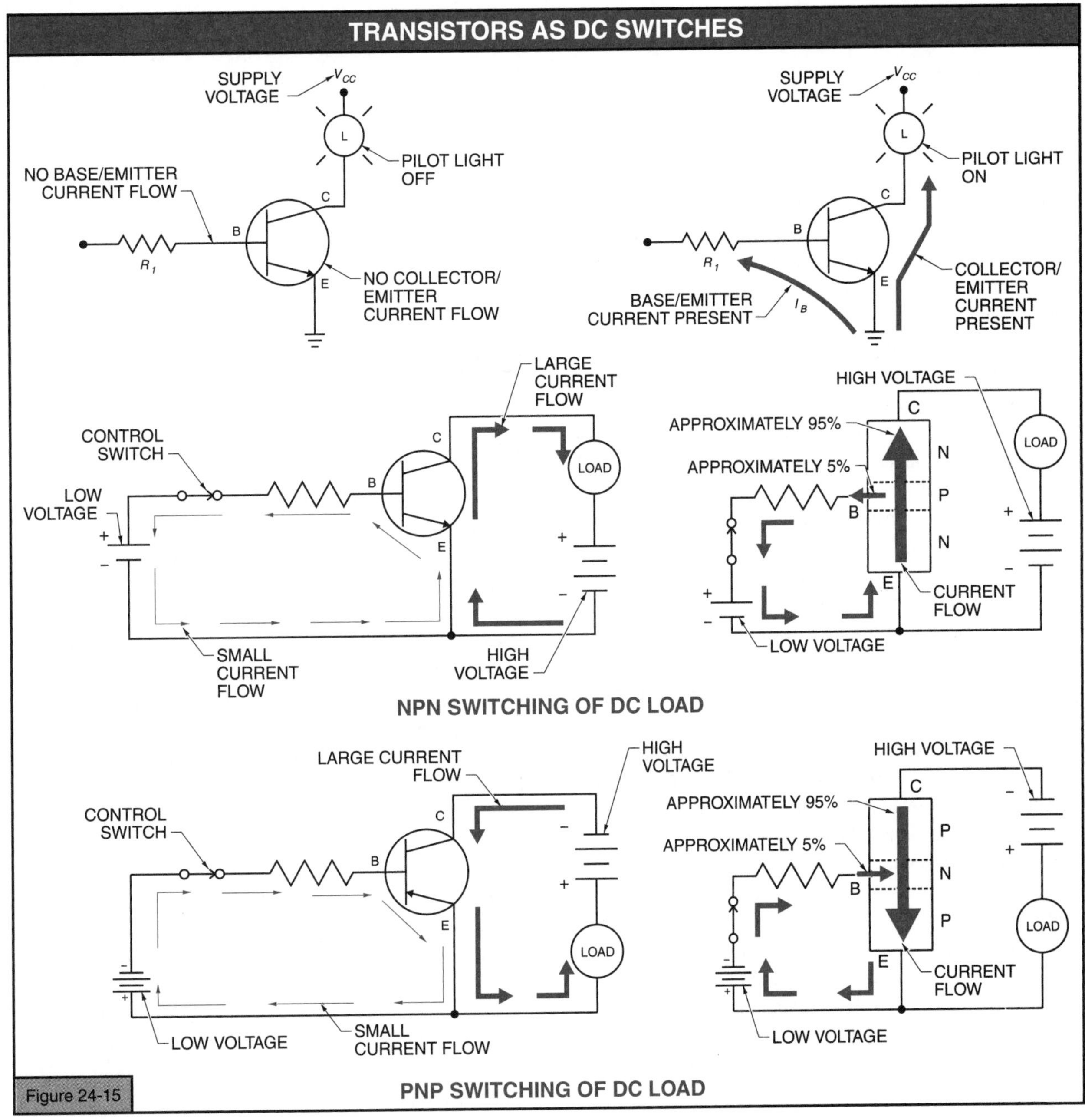

Figure 24-15

The collector/emitter resistance is reduced to a very low value, like that of a closed switch, if a small current flows between B and E. This switches ON the pilot light. A transistor switched ON is normally operating in the saturation region. The *saturation region* is the maximum current that can flow in the transistor circuit. At saturation, the collector resistance is considered zero and the current is limited only by the resistance of the load.

The resistance of the pilot light is the only current-limiting device in the circuit when the circuit reaches saturation. The transistor is operating in the cutoff region when the transistor is switched OFF. The *cutoff region* is the point at which the transistor is turned OFF and no current flows. At cutoff, all the voltage is across the open switch (transistor) and the collector/emitter voltage is equal to the supply voltage V_{CC}.

Transistor Amplifiers

Transistors may be used as AC amplification devices as well as DC switching devices. *Amplification* is the process of taking a small signal and increasing its size. In control systems, transistor AC amplifiers are used to increase small signal currents and voltages so they can do useful work. Amplification is accomplished by using a small signal to control the energy output from a large source, such as a power supply.

The primary objective of an amplifier is to produce gain. *Gain* is a ratio of the amplitude of the output signal to the amplitude of the input signal. Gain is a ratio of output to input and has no unit of measure, such as volts or amps, attached to it. Gain is used to describe current gain, voltage gain, and power gain. In each case, the output is compared to the input.

A single amplifier may not provide enough gain to increase the amplitude for the output signal needed. In such a case, two or more amplifiers can be used to obtain the gain required. *Cascaded amplifiers* are two or more amplifiers connected to obtain the required gain. For many amplifiers, gain is in the hundreds and even thousands.

The three basic transistor amplifiers are the common-emitter, common-base, and common-collector. **See Figure 24-16.** Each amplifier is named after the transistor connection that is common to both the input and the load. For example, the input of a common-emitter amplifier is across the base and emitter, while the load is across the collector and emitter. Thus, the emitter is common to the input and load.

The four main classes of operation for an amplifier are designated by the letters A, B, AB, and C. In each case, the letter is a reference to the level of an amplifier operation in relation to the cutoff region. The cutoff region is the point at which all collector current is stopped by the absence of base current.

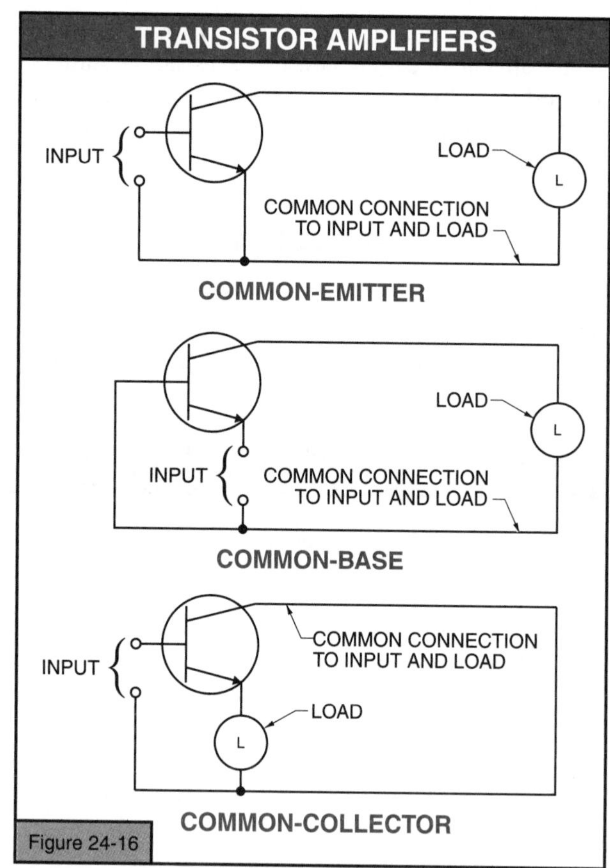

Figure 24-16

Silicon Controlled Rectifiers (SCRs)

A *silicon controlled rectifier (SCR)* is a three-terminal semiconductor thyristor that is normally an open circuit until a signal applied to the gate switches it to the conducting state in one direction. A *thyristor* is a solid-state switching device that switches current ON by a quick pulse of control current. **See Figure 24-17.** The three terminals are the anode, cathode, and gate. The anode and cathode of the SCR are similar to the anode and cathode of an ordinary semiconductor diode.

The gate serves as the control point for the SCR. The SCR differs from an ordinary semiconductor diode in that it does not pass significant current, even when forward biased, unless the anode voltage equals or exceeds the forward breakover voltage. *Forward breakover voltage* is the voltage required to switch the SCR into a conductive state. The SCR switches ON and becomes highly conductive when forward breakover voltage is reached. The SCR is unique because the gate current is used to reduce the level of breakover voltage necessary for the SCR to conduct.

Low-current SCRs can operate with an anode current of less than 1 mA. High-current SCRs can handle load currents in the hundreds of amperes. The size of an SCR increases with an increase in its current rating.

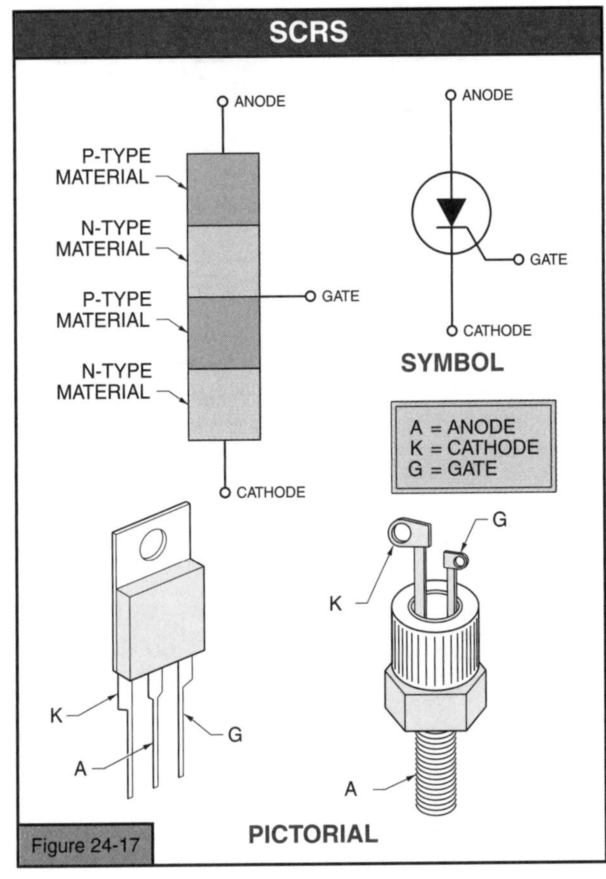

Figure 24-17

SCR Operation

An SCR operates like a regular diode in reverse bias. An SCR characteristic curve shows the operating characteristics of an SCR. **See Figure 24-18.** With a reverse-bias voltage applied, there is a small current flow in the SCR until avalanche current is reached. After avalanche current is reached, the current increases dramatically and may cause damage if thermal runaway begins.

With a forward-bias voltage applied, there is also a small forward leakage current (forward blocking current). This current stays relatively constant until the forward breakover voltage is reached. At this point, the current increases rapidly and is often referred to as the forward avalanche region. In the forward avalanche region, the resistance of the SCR is very low. The SCR acts like a closed switch and the current is limited only by the external load resistance. A short in the load circuit of an SCR can destroy the SCR if overload protection is not adequate.

An SCR operates much like a mechanical switch. An SCR is either ON or OFF. The SCR is ON (fires) when the applied voltage is above the forward breakover voltage (V_{BRF}). The SCR remains ON as long as the current stays above the holding current. *Holding current* is the minimum current necessary for an SCR to continue conducting. An SCR returns to its OFF state when voltage across the SCR drops to a value too low to maintain the holding current.

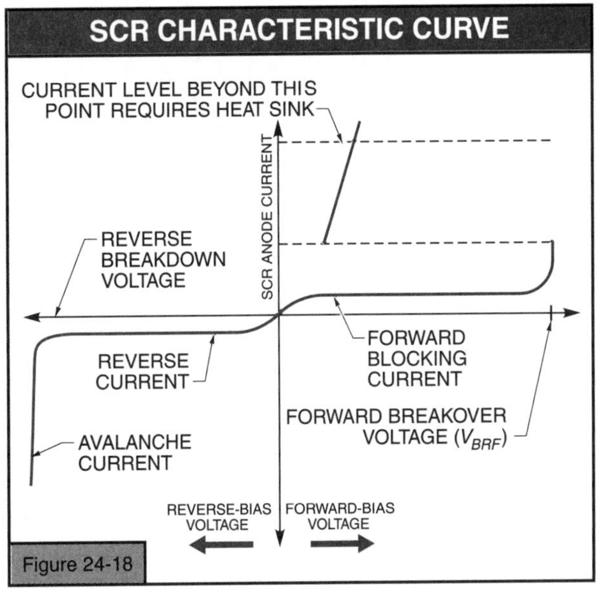

Figure 24-18

The value of forward breakover voltage can be reduced when the gate is forward biased and current begins to flow in the gate/cathode junction. Increasing values of forward-bias voltage can be used to reduce the amount of forward breakover voltage (V_{BRF}) necessary to switch the SCR into conduction.

Once the SCR has been switched ON by the gate current, the gate current loses control of the SCR forward current. Even if the gate current is completely removed, the SCR remains ON until the anode voltage has been removed. **See Figure 24-19.** The SCR also remains ON until the anode voltage has been significantly reduced to a level where the current is not large enough to maintain the proper level of holding current.

Triacs

A *triac* is a three-terminal semiconductor thyristor that is triggered into conduction in either direction by a small current to its gate. Triacs were developed to provide a means for producing improved controls for AC power. Triacs are available in a variety of packaging arrangements. Triacs can handle a wide range of current and voltages but normally have relatively low-current capabilities compared to SCRs. Triacs are normally limited to less than 50 A and cannot replace SCRs in high-current applications.

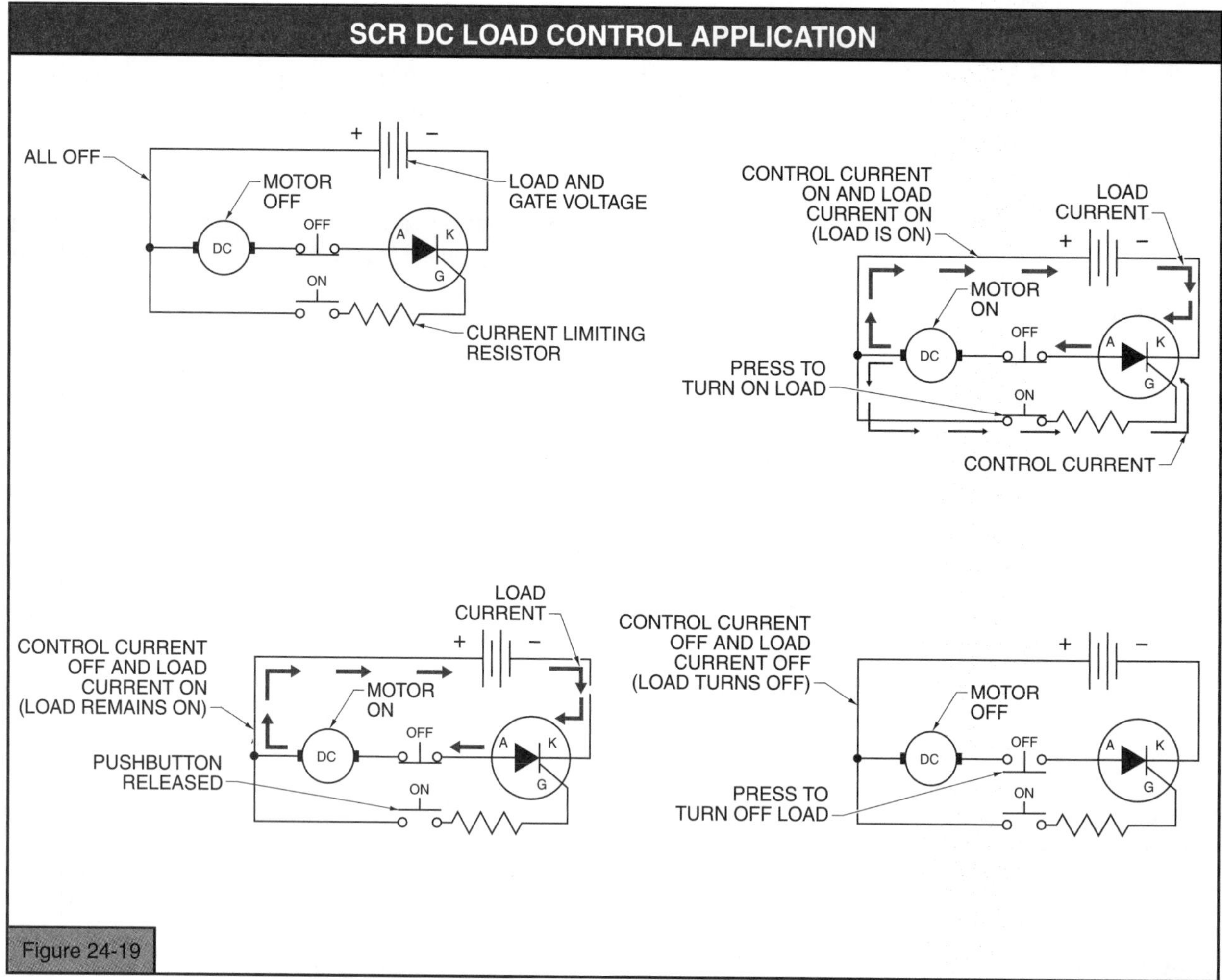

Figure 24-19

The terminals of a triac are the gate, main terminal 1 (MT1), and main terminal 2 (MT2). There is no designation of anode and cathode. Current may flow in either direction through MT1 and MT2. MT2 is the case- or metal-mounting tab to which the heat sink is attached if required. A triac may be considered two NPN switches sandwiched together on a single N-type material wafer. **See Figure 24-20.**

Triac Operation

A triac blocks current in either direction between MT1 and MT2. A triac may be triggered into conduction in either direction by a momentary pulse in either direction supplied to the gate. A triac operates like a pair of SCRs connected in a reverse parallel arrangement. The triac conducts if the appropriate signal is applied to the gate.

A triac characteristic curve shows the characteristics of a triac when triggered into conduction. **See Figure 24-21.** The triac remains OFF until the gate is triggered. The trigger circuit pulses the gate and turns ON the triac, allowing current to flow. The trigger circuit can be designed to produce a pulse that varies at any point in the positive or negative half-cycle. Therefore, the average current supplied to the load may vary.

An advantage of a triac is that virtually no power is wasted by being converted to heat. Heat is generated when current is impeded, not when current is switched OFF. A triac is either fully ON or fully OFF. It never partially limits current. Triacs do not have a reverse breakdown condition of high voltage and high current, such as those found in diodes and SCRs. The triac turns ON if the voltage across the triac becomes excessive. The triac can conduct a reasonably high current when turned ON.

TECH FACT

Triacs and SCRs are referred to as thyristors. The word thyristor is derived from the Greek word "thyra," which means door, indicating that they can be either open or closed.

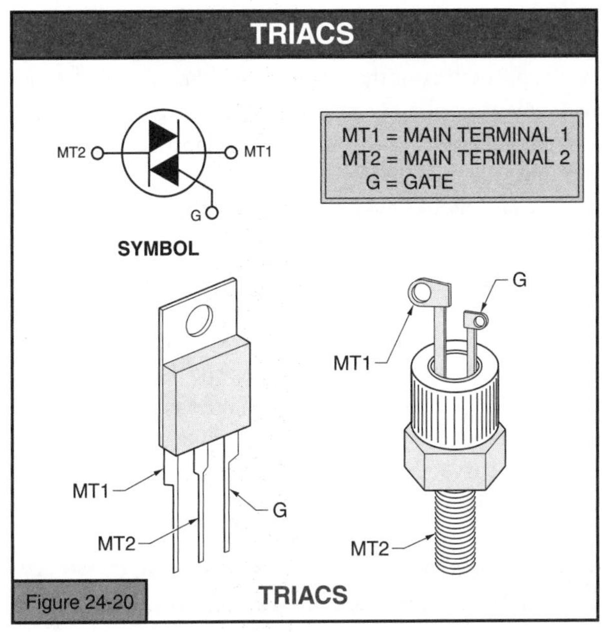

Figure 24-20 TRIACS

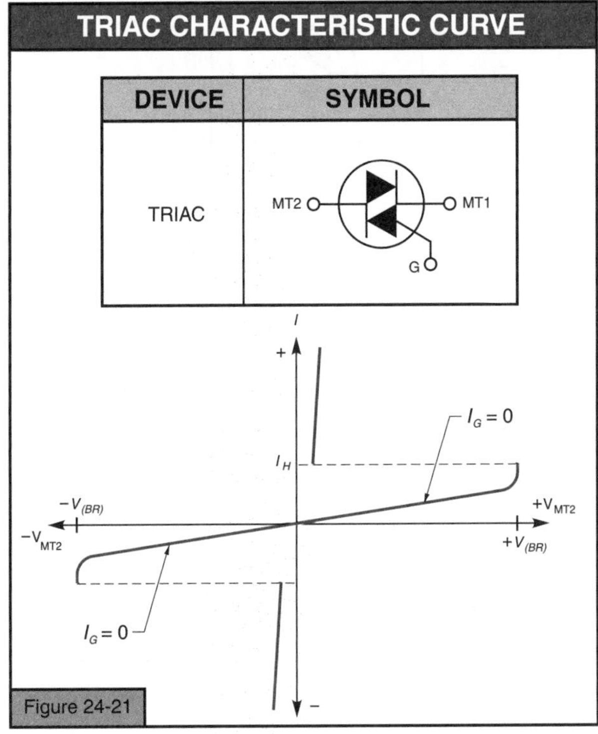

Figure 24-21

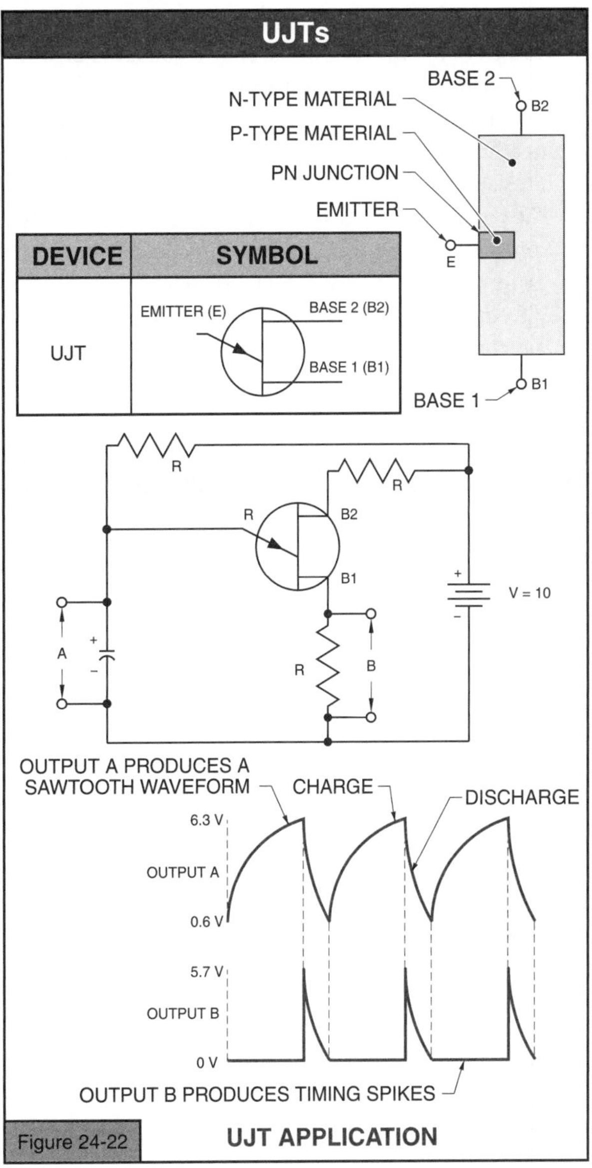

Figure 24-22 UJT APPLICATION

Unijunction Transistors (UJTs)

A *unijunction transistor (UJT)* is a transistor consisting of N-type material with a region of P-type material doped within the N-type material. The N-type material functions as the base and has two leads, base 1 (B1) and base 2 (B2). The lead extending from the P-type material is the emitter (E). **See Figure 24-22.**

A UJT is used primarily as a triggering device because it serves as a step-up device between low-level signals and SCRs and triacs. Outputs from photocells, thermistors, and other transducers can be used to trigger UJTs. UJTs are also used in oscillators, timers, and voltage/current-sensing applications.

UJT Biasing

In normal operation, B1 is negative, and a positive voltage is applied to B2. The internal resistance between B1 and B2 divides at the emitter (E), with approximately 60% of the resistance between E and B1. The remaining 40% of resistance is between E and B2. The net result is an internal voltage split. This split provides a positive voltage at the N-type material of the emitter junction, creating an emitter junction that is reverse biased. As long as the emitter voltage remains less than the internal voltage, the emitter junction remains reverse biased, even at a very high resistance.

The junction of a UJT is forward biased when the emitter voltage is greater than the internal value. This rapidly drops the resistance between E and B1 to a very low value. A UJT characteristic curve shows the dramatic change in voltage due to this resistance change. **See Figure 24-23.**

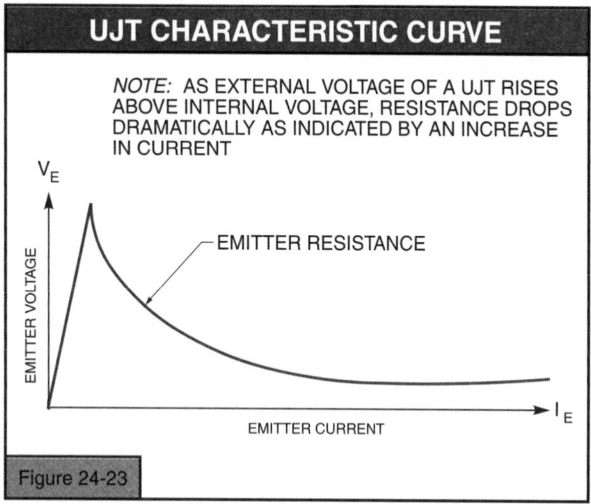

Figure 24-23

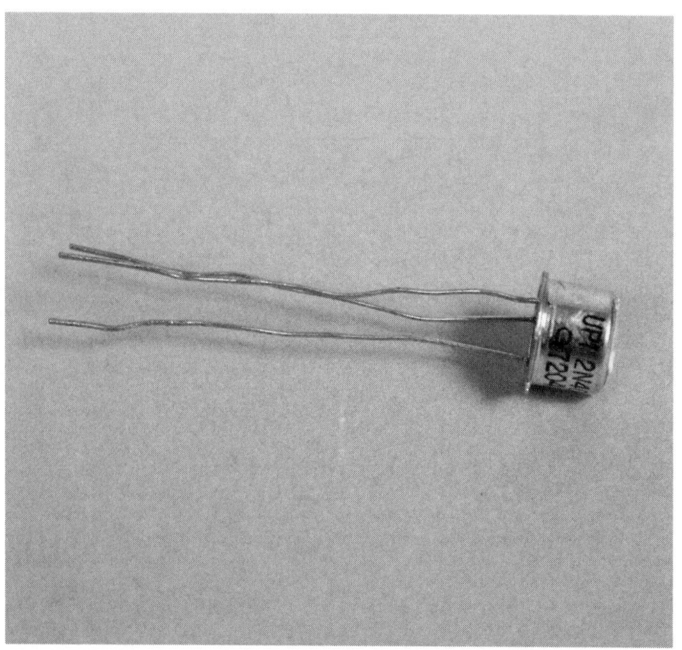
Transistors are devices used to amplify DC voltage or small AC signals.

Diacs

A *diac* is a thyristor that triggers in either direction when its breakover voltage is exceeded. Unlike a transistor, the two junctions are heavily and equally doped. Each junction is almost identical to the other.

A diac acts like two zener diodes that are series connected in opposite directions. The diac is used primarily as a triggering device. It accomplishes this through the use of its negative resistance characteristic. *Negative resistance characteristic* is the characteristic that current decreases with an increase of applied voltage. The diac has negative resistance because it does not conduct current until the voltage across it reaches breakover voltage. A diac characteristic curve shows the operating characteristics of a diac. **See Figure 24-24.**

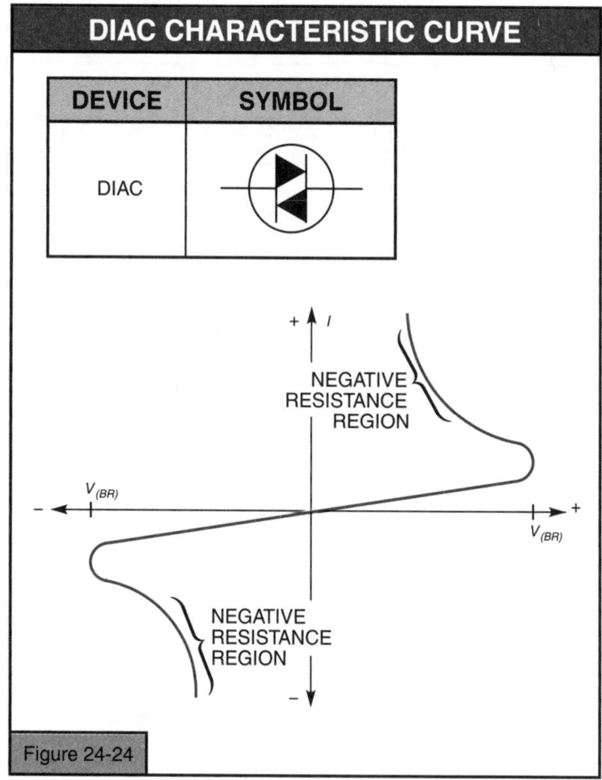

Figure 24-24

A diac rapidly switches from a high-resistance state to a low-resistance state when a positive or negative voltage reaches the breakover voltage. Because the diac is a bidirectional device, it is ideal for controlling triacs, which are also bidirectional.

A diac may be used to control the brightness of a lamp. **See Figure 24-25.** During the beginning of each half cycle of the applied AC voltage, the triac is in the OFF state and no voltage is applied to the lamp. However, the AC voltage is also applied across the resistor/capacitor circuit connected in parallel with the triac. This applied voltage charges the capacitor. The diac conducts when the voltage across the capacitor rises to the breakover voltage of the diac. The capacitor voltage is then discharged through the diac to the triac gate. The voltage at the triac gate triggers the triac into conduction and voltage is applied to the lamp. This sequence of events is repeated for each half cycle of the AC sine wave.

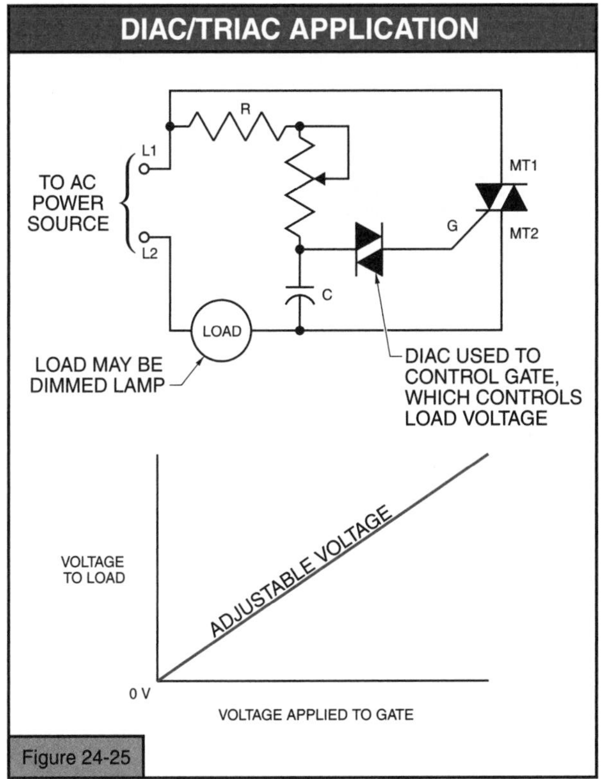

Figure 24-25

The variable resistor may be adjusted to control the brightness of the lamp by determining when on the AC sine wave the triac is allowed to start conducting. A decrease in resistance allows the capacitor to charge faster and the breakover voltage of the diac is reached earlier. This allows the triac to triggered into conductance sooner in the AC sine wave. A higher voltage is applied to the lamp when the triac is allowed to conduct sooner in the AC sine wave. Likewise, if the resistance is increased, the triggering of the triac occurs later in the cycle, and less voltage is applied to the lamp.

Electronic Switching

Electric loads may be controlled by using mechanical or electronic (solid-state) switches. Mechanical switches have the advantage of being able to switch both AC and DC loads. Mechanical switches normally include several contacts (NO and NC). However, mechanical switches are relatively slow when compared to solid-state switches and have a limited life span (normally 500,000 operations).

Solid-state switches can switch circuits very quickly and have a long life span (normally several billion operations). However, solid-state switches are generally limited to switching AC or DC loads only (depending on the type used). The main disadvantage to solid-state switching is that all solid-state switches have a voltage drop across them, thus require heat sinks when switching high currents. **See Figure 24-26.**

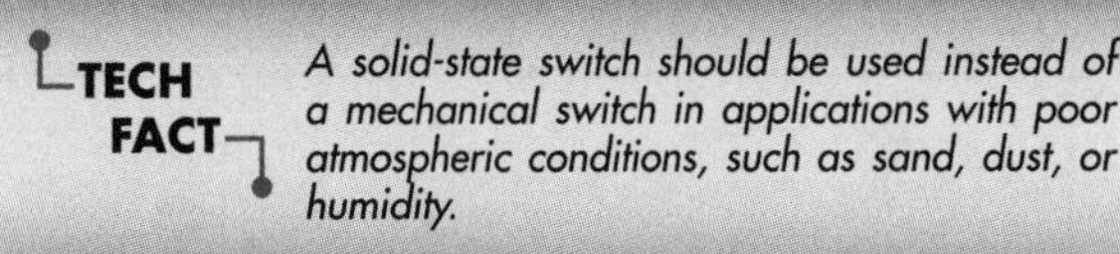

A solid-state switch should be used instead of a mechanical switch in applications with poor atmospheric conditions, such as sand, dust, or humidity.

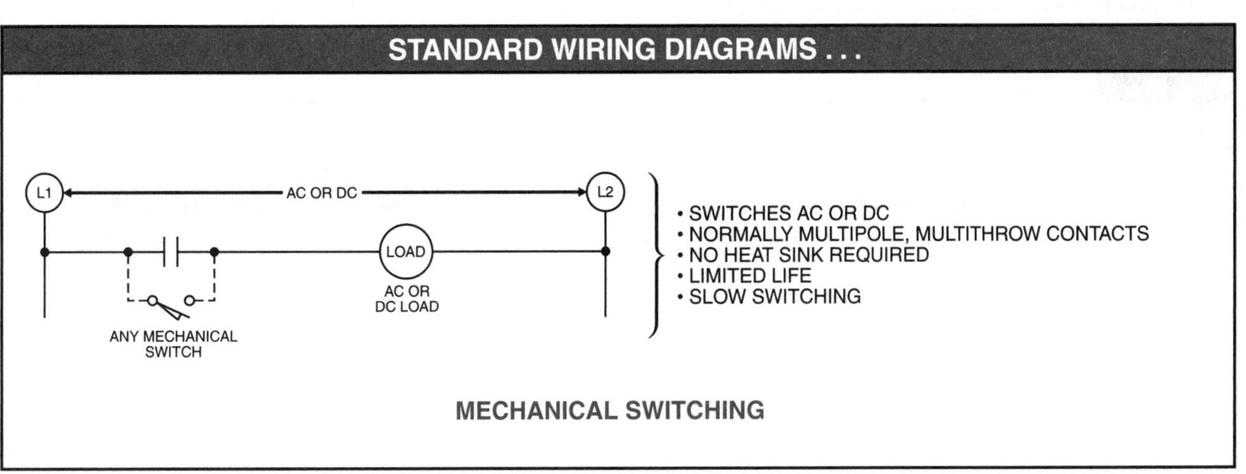

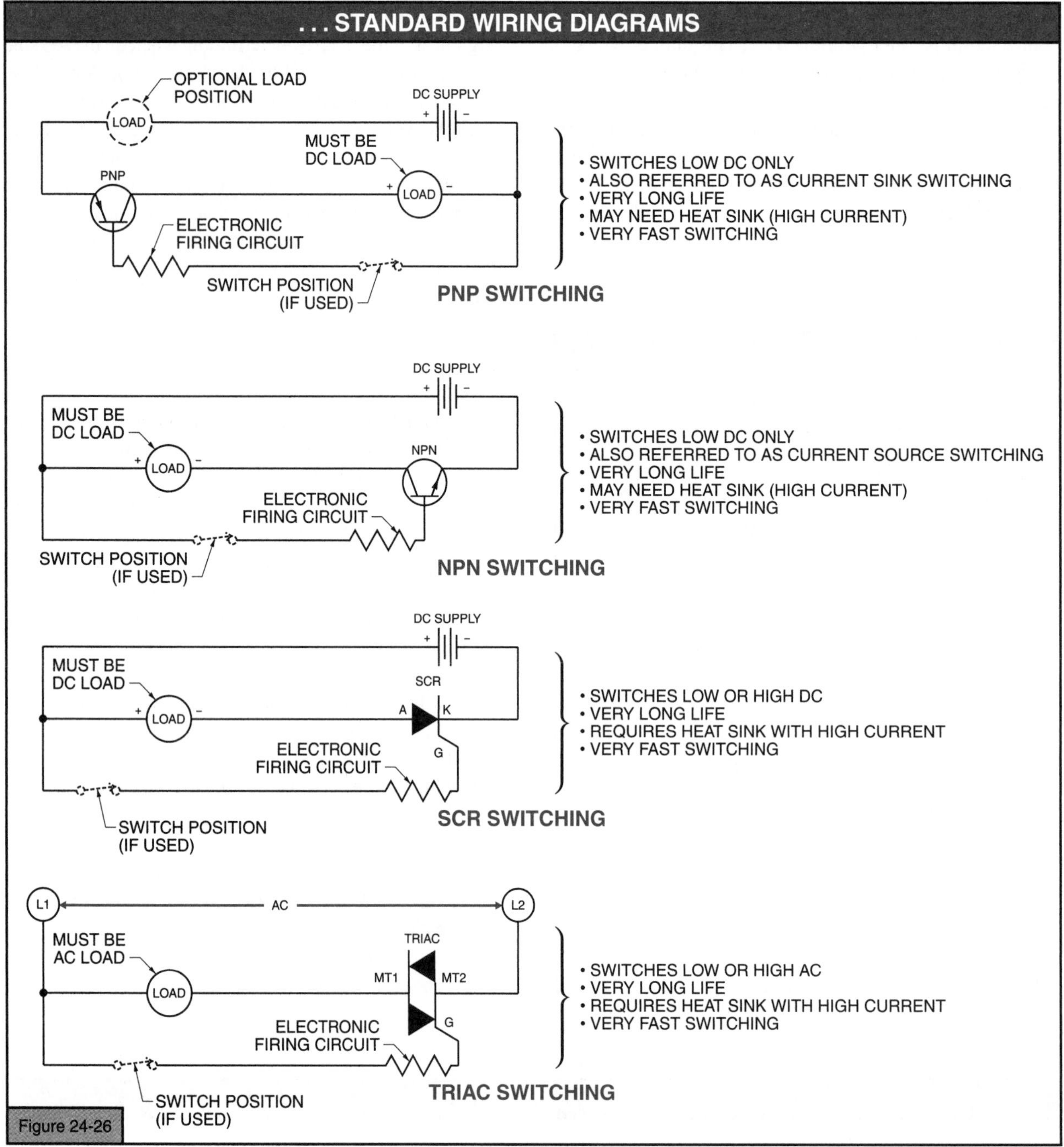

Figure 24-26

Refer to the CD-ROM in the back of the book for Chapter 24 Quick Quiz® and related information.

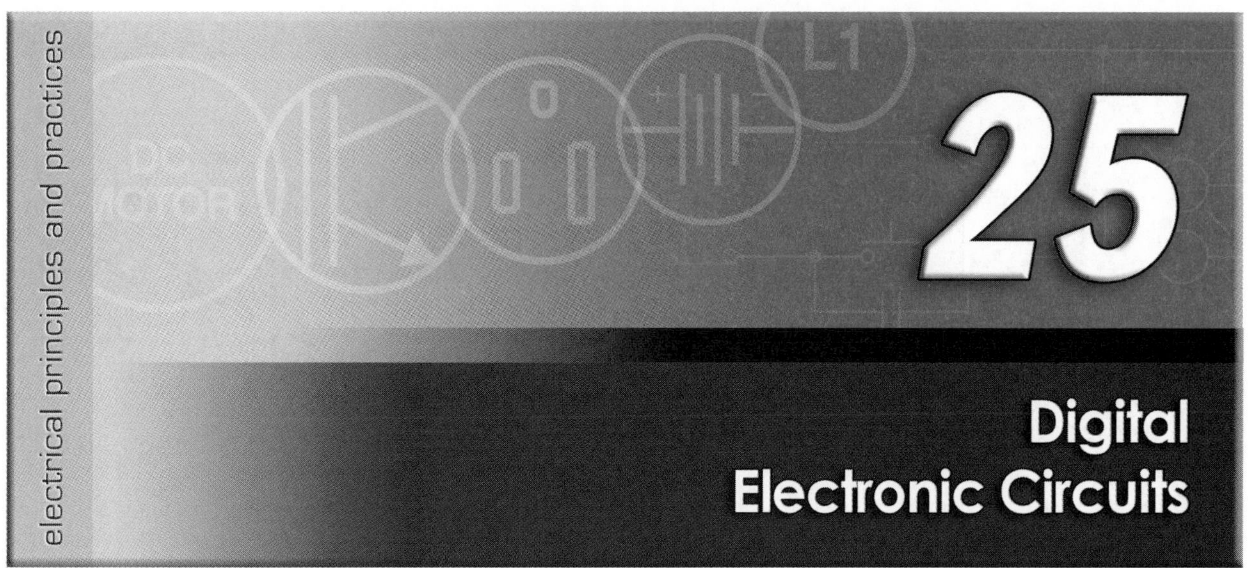

Digital Electronic Circuits

Electronic devices use electronic components and circuits to produce sound, perform mathematical and logical functions, amplify signals, control operations, or increase operating speed. Today, almost all electronic systems use solid-state components that have no moving parts.

Electronic systems are either analog or digital. *Analog* is the output (display) of circuits with continuously changing quantities that may have any value between the defined limits. Quantities such as temperature, speed, voltage, current, and frequency are common analog signals. *Digital* is the output (display) of specific quantities that change in discrete steps, such as the seconds, minutes, and hours displayed on a digital watch. **See Figure 25-1.**

Early electronic circuits were analog. Today, most electronic circuits are digital. The development of digital electronics parallels the development of computers. Computer development required small and fast devices. These requirements led to the development of integrated circuits (ICs). Developments in ICs led to higher quality and faster computers as well as to higher quality electronic circuits in televisions, microwaves, telephones, and almost all other devices that used electronic circuitry. Improved ICs have also created additional uses of electronic circuits in the medical, military, communication, transportation, security, space, and other fields.

Integrated Circuits

Integrated circuits (ICs) are electronic devices in which all components (transistors, diodes, and resistors) are contained in a single package or chip. ICs contain numerous circuits that are microscopically photoetched in layers on a silicon base. ICs are used to process and store information in computers, calculators, programmable logic controllers (PLCs), video games, digital watches, and almost all other electronic circuits. The two basic categories of ICs are digital and linear ICs. **See Figure 25-2.**

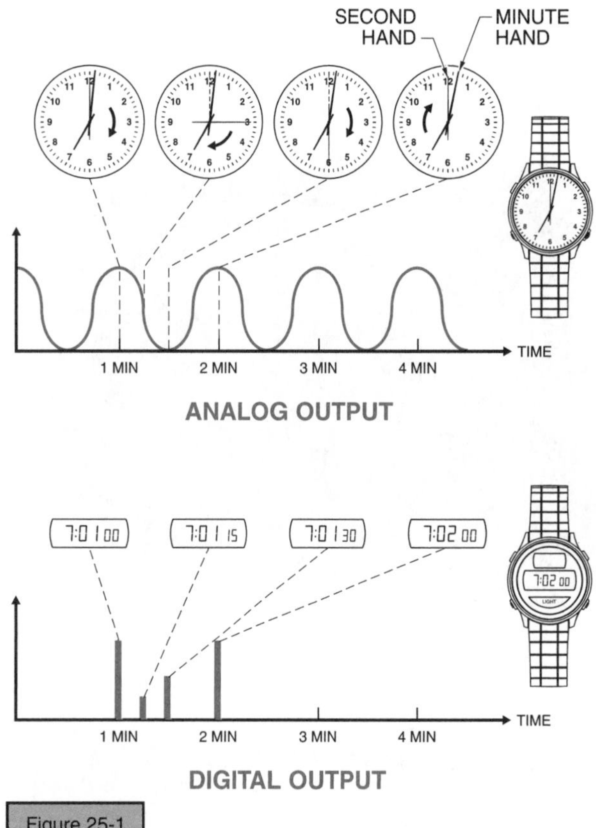

Figure 25-1

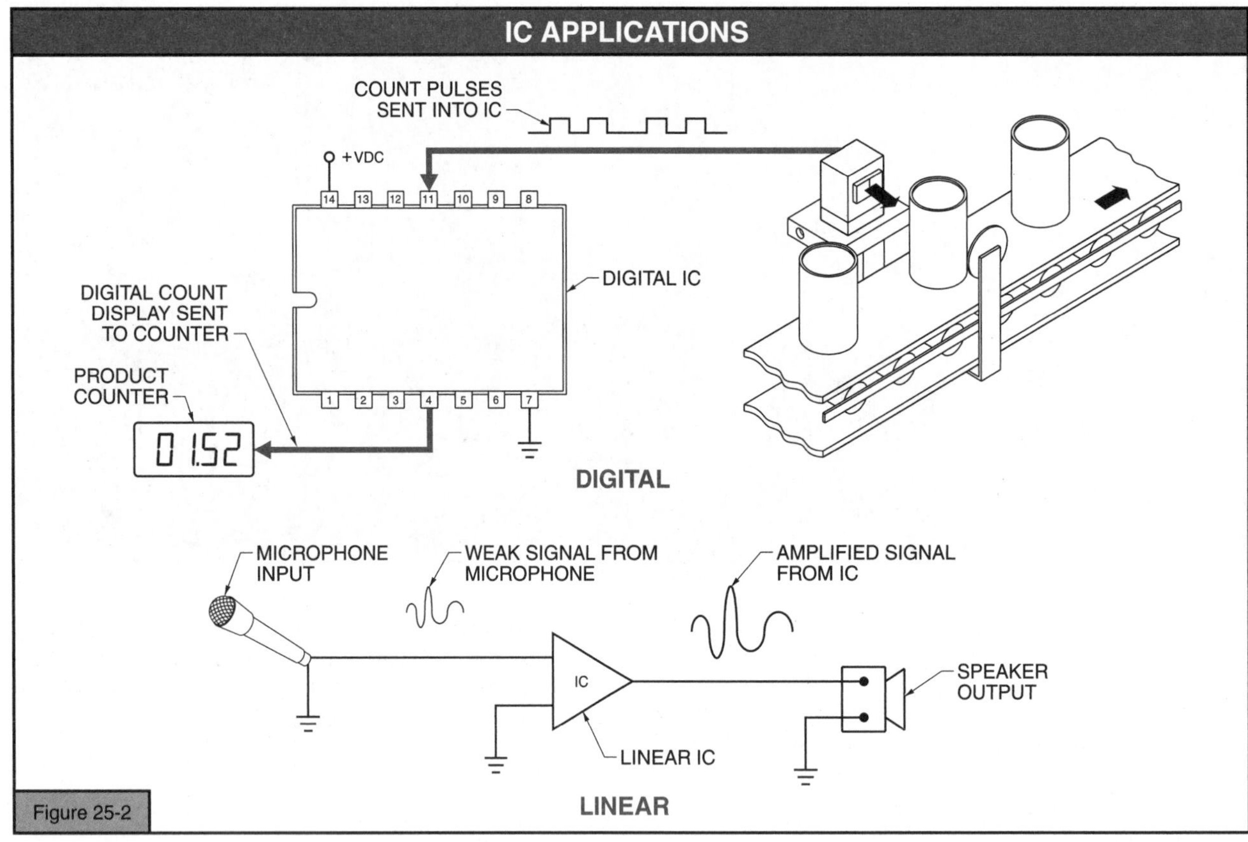

Figure 25-2

Electronic circuits contain components such as transistors, diodes, and resistors on a single circuit board.

TECH FACT *A short in a digital logic circuit normally results in a permanent logic state (high or low) being applied to one or more gates.*

A *digital IC* is an IC that contains a switching integrated circuit that processes finite (zero and maximum) signal levels. Digital ICs are used to perform functions that require logic (AND, NAND, etc.), mathematics, or other decision functions. Digital ICs are used in computers, calculators, and almost all other types of electronic circuits. Digital ICs are the most widely used IC.

A *linear IC* is an IC that contains an amplifying integrated circuit that produces an output signal that is proportional to the applied input signal. Linear ICs are used to provide amplification and regulation functions. Linear ICs are used in radios, amplifiers, televisions, and power supplies. Operational amplifiers (op-amps) are the most common linear IC.

IC Packages

The circuitry contained inside an IC must be connected to a power supply and other electronic components. ICs are powered and interconnected with other components through their pins. Most ICs are mounted on a printed circuit (PC) board with their pins soldered to the copper traces on the PC board. ICs are available in different packages and pin configurations. **See Figure 25-3.**

IC PACKAGES

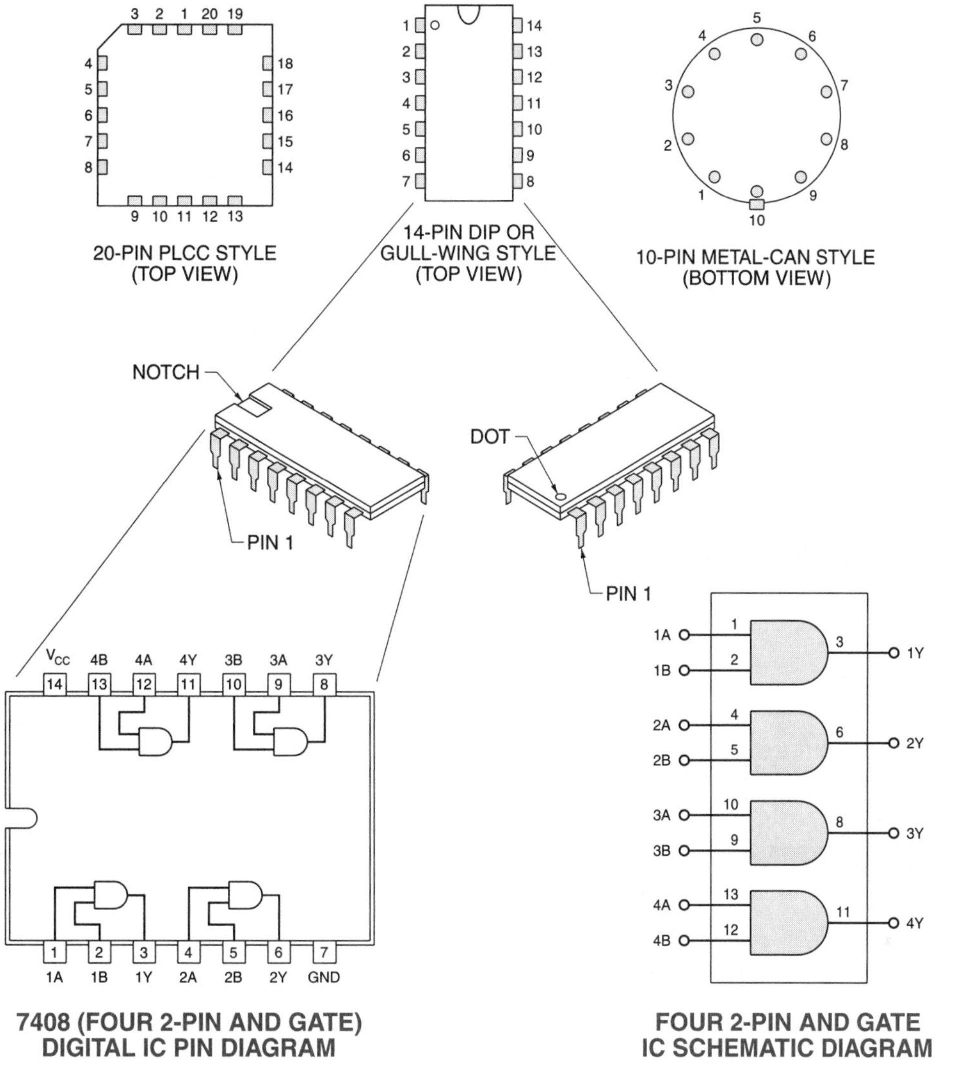

20-PIN PLCC STYLE (TOP VIEW)
14-PIN DIP OR GULL-WING STYLE (TOP VIEW)
10-PIN METAL-CAN STYLE (BOTTOM VIEW)
NOTCH
DOT
PIN 1
PIN 1
VCC 4B 4A 4Y 3B 3A 3Y
1A 1B 1Y 2A 2B 2Y GND
1A
1B
2A
2B
3A
3B
4A
4B
1Y
2Y
3Y
4Y
7408 (FOUR 2-PIN AND GATE) DIGITAL IC PIN DIAGRAM
FOUR 2-PIN AND GATE IC SCHEMATIC DIAGRAM

Figure 25-3

The more complex the internal circuitry of an IC, the greater the number of pins required. For example, a basic logic function (AND, OR, etc.) IC circuit may have only 8 or 14 pins. A microprocessor IC circuit used in a personal computer may have 64 pins.

Truth Tables

A *truth table* is a table used to describe the output condition of a logic gate or combination of gates for every possible input condition. A truth table normally lists the inputs as letters such as A, B, C, etc. The output is listed as letters such as X or Y. Inputs are listed on the left side of the truth table and the output is listed on the right side. Most digital truth tables use a 0 to represent a low (no signal) and a 1 to represent a high (signal present). A 1 in the output of an AND gate indicates that the output is ON when both inputs are ON. **See Figure 25-4.**

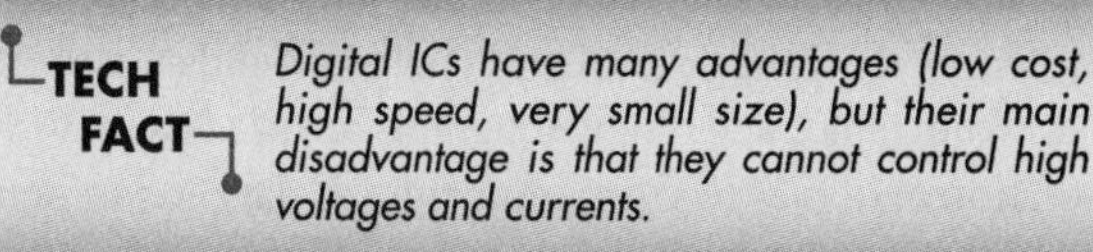

TECH FACT *Digital ICs have many advantages (low cost, high speed, very small size), but their main disadvantage is that they cannot control high voltages and currents.*

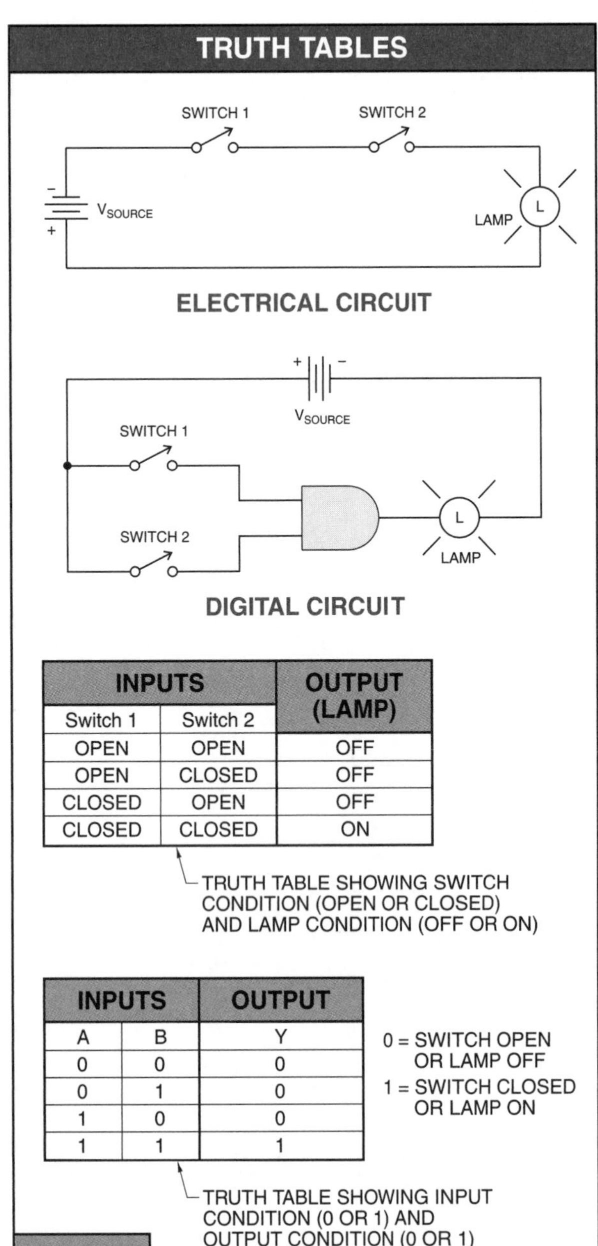

INPUTS		OUTPUT (LAMP)
Switch 1	Switch 2	
OPEN	OPEN	OFF
OPEN	CLOSED	OFF
CLOSED	OPEN	OFF
CLOSED	CLOSED	ON

INPUTS		OUTPUT
A	B	Y
0	0	0
0	1	0
1	0	0
1	1	1

Figure 25-4

Timing Diagrams

A *timing diagram* is a diagram that shows graphically how the output of a circuit changes in response to input changes. Timing diagrams show the high (1) and low (0) levels of a digital signal as it changes relative to time. The 1s and 0s may represent numbers, control signals, symbols, letters, or other types of signals. On a timing diagram, the vertical axis shows the voltage level and the horizontal axis shows time. Truth tables and timing diagrams are used to show the operation of digital ICs in manufacturer's data and training manuals. **See Figure 25-5.**

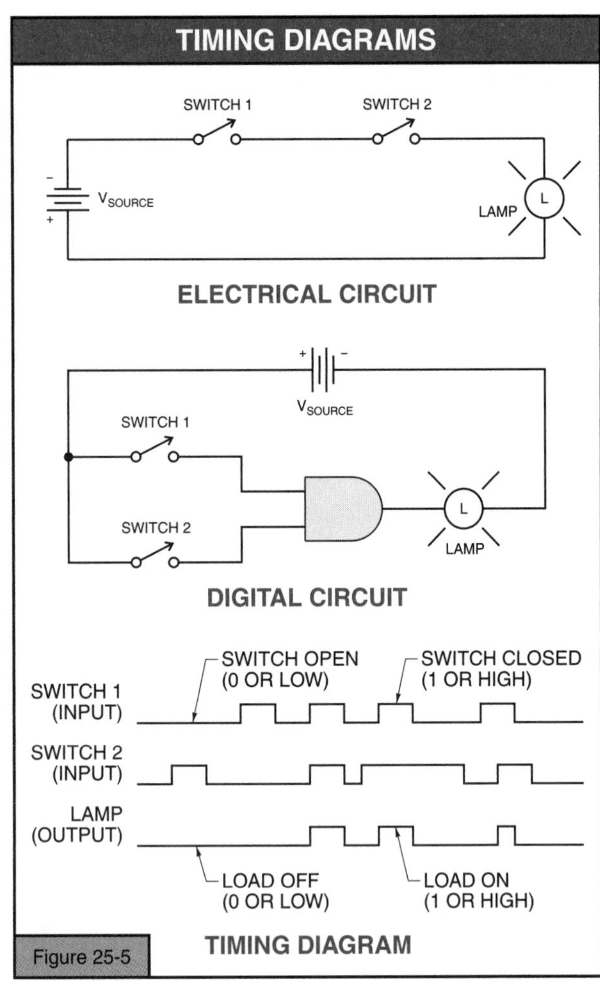

Figure 25-5

In this circuit, switch 1 and switch 2 must be closed before the lamp is energized. Switch 1 and 2 are the inputs and the lamp is the output. The timing diagram shows that the output is ON (1) only when both inputs are present (1). The timing diagram also shows that the output is OFF (0) anytime one or both inputs are absent (0).

Digital signals often require precise timing. Special clock and timing circuits are used to produce clock waveforms. The waveform is used to trigger the digital signals at precise intervals. Clock signals are used in digital circuits to synchronize the changing of data in various parts of a digital circuit. **See Figure 25-6.** In this circuit, five clock pulses are transmitted to the receiving unit when the enable signal is present for 5 μs. The enable signal allows the circuit to output data. Clock pulses are used in digital flip-flops, counters, shift registers, multiplexers, and many other circuits requiring synchronization.

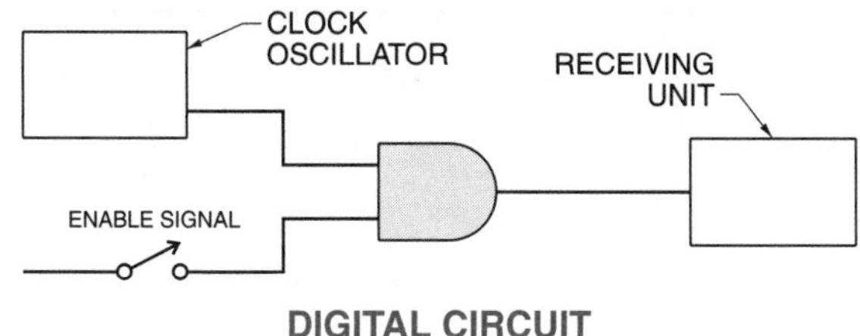

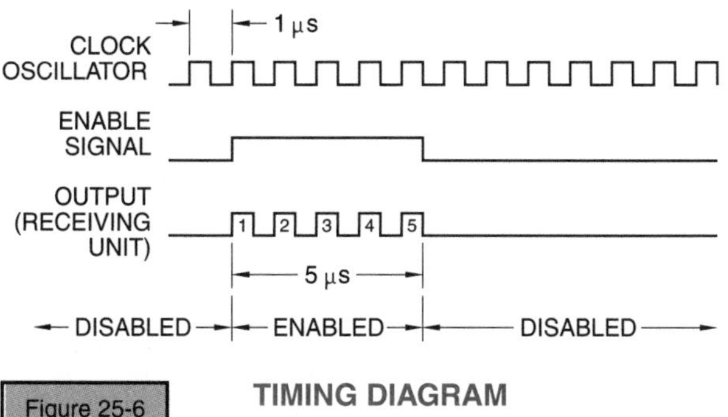

Figure 25-6

Logic Gates

ICs are manufactured in a variety of logic circuits. These logic circuits may include basic logic functions such as AND gates, or specialized circuits such as counters. Most digital logic operations are performed by one of the basic logic gates or by a combination of basic logic gates.

The five basic logic gates are the AND, OR, NOT (inverter), NOR, and NAND gates. The basic AND, OR, NOR, and NAND gates have two or more inputs and one output. The NOT (inverter) gate has one input and one output. In addition to the five basic gates, an exclusive OR (XOR) gate is also available. An exclusive OR gate has two inputs and one output. **See Figure 25-7.**

Each logic gate is designed to operate on binary signals. A *binary signal* is a signal that has only two states. A binary signal is either high (1) or low (0). A high signal is normally 5 V, but may range from 2.4 V to 5 V. A low signal is normally 0 V, but may range from 0 V to .8 V. A high signal may represent an ON, yes, or true condition. A low signal may represent an OFF, no, or false condition. For example, a light may be ON or OFF, a valve may be open or closed, and a motor may be running or stopped.

TECH TIP

Use a logic probe to determine the basic logic level (high or low) in a digital circuit. Use a logic analyzer to determine complex digital problems.

AND Gates

An *AND gate* is a logic gate that provides logic level 1 only if all inputs are at logic level 1. AND gates are normally referred to by the number of inputs they have, such as two-input AND gate, three-input AND gate, etc. AND gates may be used in any circuit that uses AND circuit logic. An example of AND logic is a control circuit that requires two or more switches to be closed before a load can be energized. **See Figure 25-8.**

In this circuit, an operator cannot turn ON a pressure pump unless the ON/OFF switch is closed and the foot switch is closed. Using a foot switch in this application allows the operator's hands to be free for holding or moving an object inside the machine. An electrical circuit that connects the two switches in series could be used to control the 115 VAC pump. The disadvantage to this circuit is that the control switches are at the same voltage level (115 VAC) as the load. In addition to being at the same voltage level, the control switches must also be rated high enough to carry the current of the load.

The circuit operates in the same manner if a digital AND logic gate is used. The difference is that lower-rated control switches may be used because they only have to be rated for the voltage of the logic gate. This voltage rating is normally 5 VDC. This also allows the use of smaller wires connecting the control switches (18 AWG or smaller). A DC to AC solid-state relay (SSR) is used as the interface between the DC control circuit and the 115 VAC power circuit containing the pump motor.

Integrated circuits are used to provide a complete circuit function in one small package.

BASIC DIGITAL LOGIC GATES

Device	Boolean Expression	Read As	Symbol	Function/Notes
AND GATE	$Y = A \bullet B$ ($\bullet$ = AND)	Y = A AND B	INPUT A, B; OUTPUT Y	To provide logic level 1 only if all inputs are at logic level 1. HIGH = 1 LOW = 0
OR GATE	$Y = A + B$ ($+$ = OR)	Y = A OR B	INPUT A, B; OUTPUT Y	To provide logic level 1 if one or more inputs are at logic level 1. HIGH = 1 LOW = 0
NOT (INVERTER) GATE	$Y = \overline{A}$ ($-$ = NOT)	Y = NOT A	INPUT A; OUTPUT Y BUBBLE MEANS INVERTED	To provide an output that is the opposite of the input. HIGH = 1 LOW = 0
NOR GATE	$Y = \overline{A+B}$	Y = NOT (A OR B)	INPUT A, B; OUTPUT Y	To provide logic level 0 if one or more inputs are at logic level 1. HIGH = 1 LOW = 0
NAND GATE	$Y = \overline{A \bullet B}$	Y = NOT (A AND B)	INPUT A, B; OUTPUT Y	To provide logic level 0 only if all inputs are at logic level 1. HIGH = 1 LOW = 0
XOR (EXCLUSIVE OR GATE)	$Y = A \oplus B$ ($\oplus$ = EXCLUSIVE)	Y = A BUT NOT B OR Y = B BUT NOT A	INPUT A, B; OUTPUT Y MEANS EXCLUSIVE	To provide logic level 1 only if one, but not both of the inputs are at logic level 1. HIGH = 1 LOW = 0

TRUTH TABLES

AND

INPUTS		OUTPUT
A	B	Y
0	0	0
0	1	0
1	0	0
1	1	1

OR

INPUTS		OUTPUT
A	B	Y
0	0	0
0	1	1
1	0	1
1	1	1

NOT

INPUT	OUTPUT
A	Y
0	1
1	0

NOR

INPUTS		OUTPUT
A	B	Y
0	0	1
0	1	0
1	0	0
1	1	0

NAND

INPUTS		OUTPUT
A	B	Y
0	0	1
0	1	1
1	0	1
1	1	0

XOR

INPUTS		OUTPUT
A	B	Y
0	0	0
0	1	1
1	0	1
1	1	0

Figure 25-7

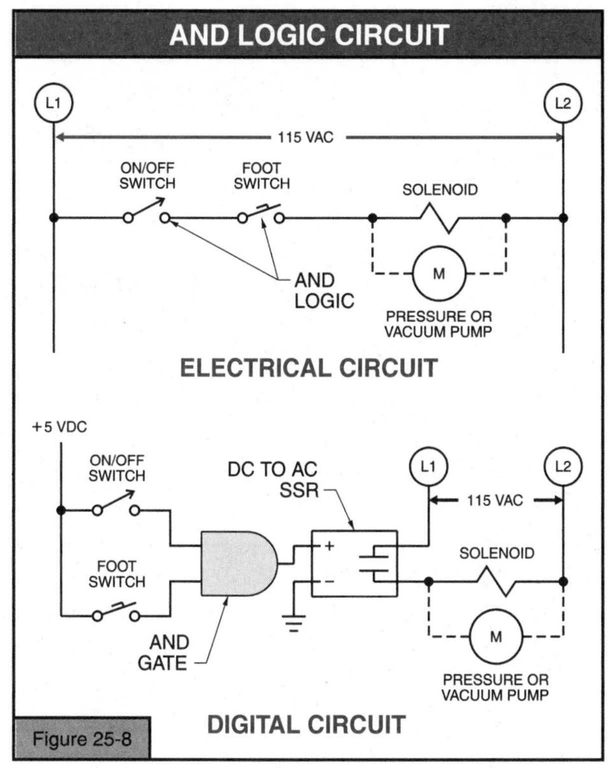

Figure 25-8

OR Gates

An *OR gate* is a logic gate that provides logic level 1 if one or more inputs are at logic level 1. Like AND gates, OR gates are normally referred to by the number of inputs they have such as two-input OR gate, three-input OR gate, etc. OR gates may be used in any circuit that uses OR circuit logic. An example of OR logic is a control circuit that requires any one or more switches to be closed before a load can be energized. **See Figure 25-9.**

In this circuit, pressing either pushbutton energizes the doorbell. An SSR is used as the interface between the digital circuit and the 24 VAC doorbell circuit because the doorbell draws more current and normally operates at a higher voltage than the digital IC gate could deliver.

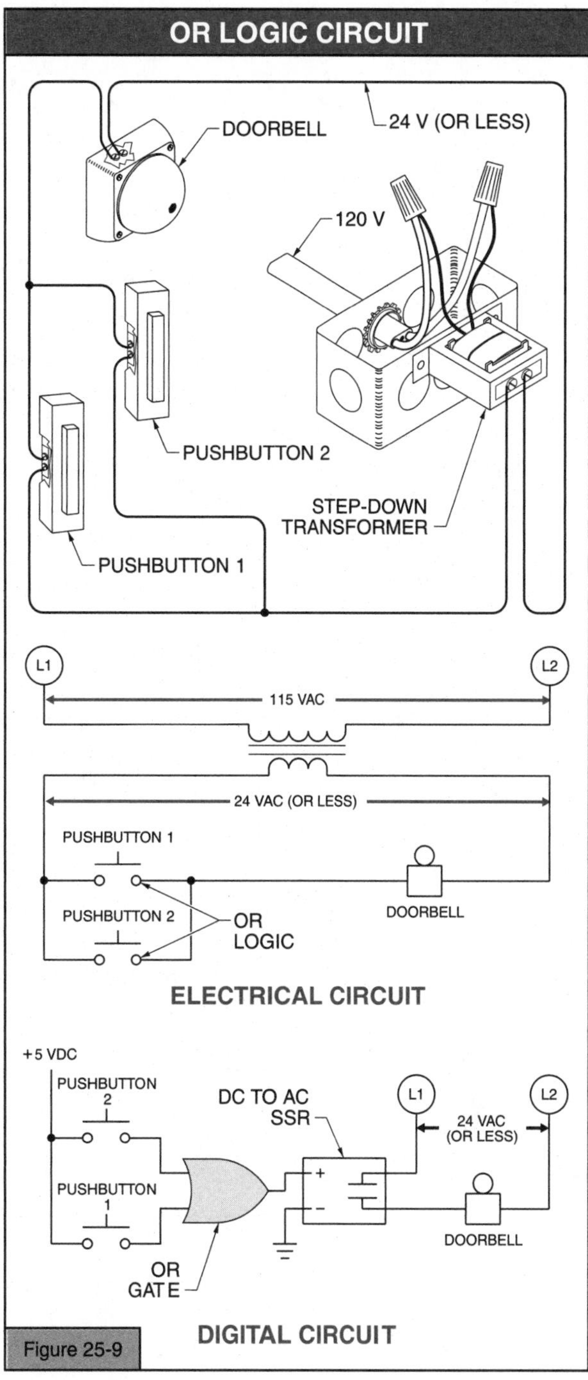

Figure 25-9

NOT (Inverter) Gates

A *NOT (inverter) gate* is a logic gate that provides an output that is the opposite of the input. Inverters are used in circuits any time a signal must be changed to its opposite state. An example of an inverter circuit is when a normally closed (NC) switch needs to operate as a normally open (NO) switch in the circuit, or a NO switch needs to operate as a NC switch. The inverter is used to change the function of the switch. **See Figure 25-10.**

In this circuit, a NC liquid level switch could be made to operate as a NO switch by adding an inverter into the digital circuit. Without inverters, NO and NC control switches are required to perform circuit operations. NO switches are commonly used to start or allow the flow of current in a circuit. NC switches are commonly used to stop or remove the flow of current in a circuit. Any NO or NC switch may be changed to its opposite operating function by adding an inverter.

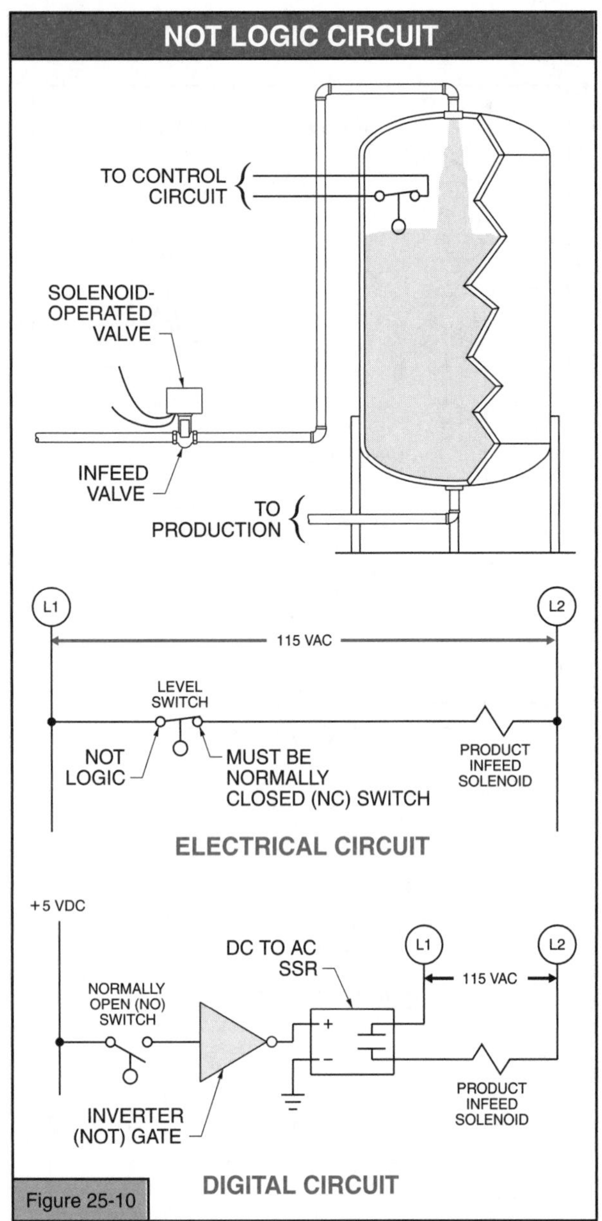

Figure 25-10

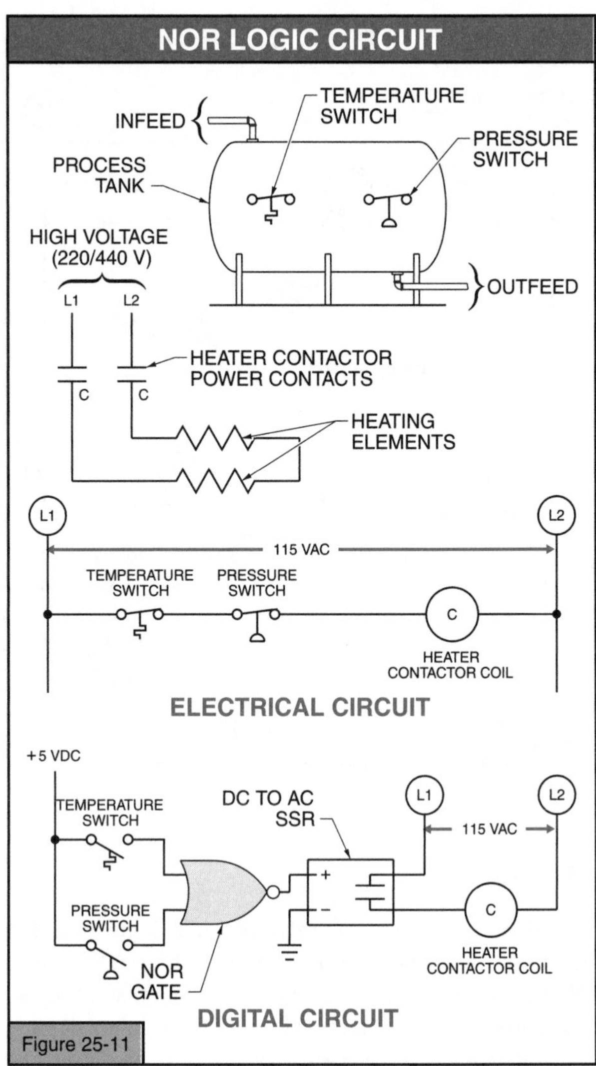

Figure 25-11

In this circuit, this function may be accomplished by using NC switches in the electrical circuit or NO switches and a NOR gate in the digital circuit. In the electrical circuit, the temperature and pressure switches are used to control the heater contactor coil. The heater contactor coil is used to control the heating elements. In the digital circuit, the temperature and pressure switches are used to control an SSR. The SSR is used to control the heater contactor coil.

NOR Gates

A *NOR gate* is a logic gate that provides logic level 0 if one or more inputs are at logic level 1. NOR gates are used in applications that require an output to turn OFF if any one of the input conditions is above a set limit. The condition above the set limit of an input may be excessive temperature, pressure, liquid level, current level, voltage level, counts, or any other input condition. An example of NOR logic is when a heater must be turned OFF if the temperature or pressure in a system is above the set limits. **See Figure 25-11.**

NAND Gates

A *NAND gate* is a logic gate that provides logic level 0 only if all inputs are at logic level 1. NAND gates are used in applications that require an output to turn OFF only if all the input conditions are met. An example of NAND logic is in the control of a solenoid-operated valve that controls the flow of product that is critical to an operation. **See Figure 25-12.**

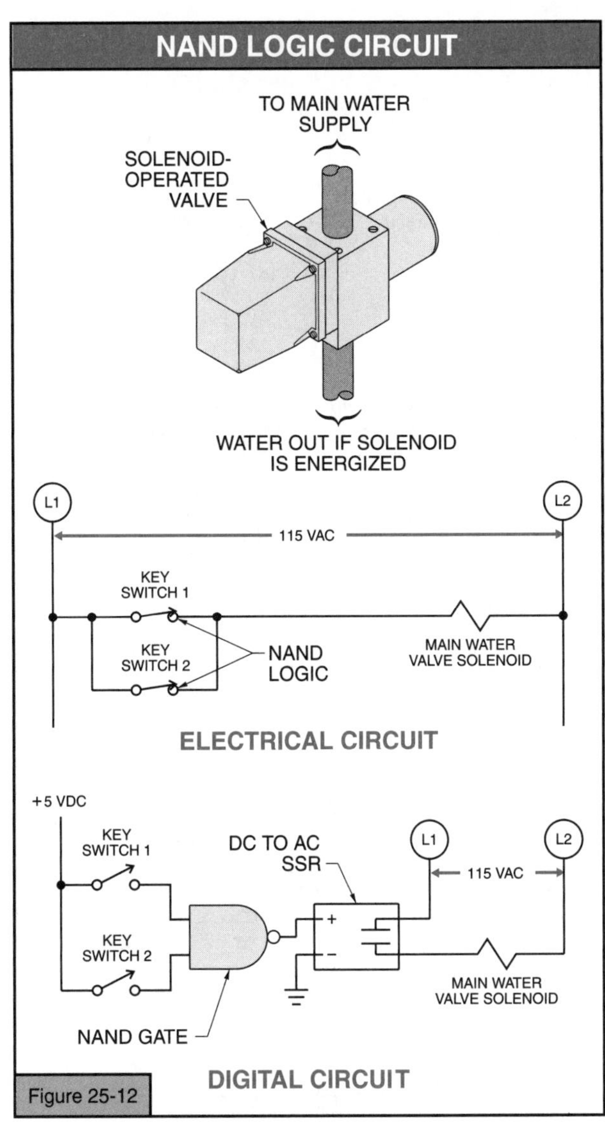

Figure 25-12

TECH TIP *ICs very rarely go bad on their own. Most of the time, some outside source (voltage-surge, short, etc.) causes ICs to fail. Always check for additional circuit or system problems when replacing a bad IC.*

In this circuit, the solenoid-operated valve controls the flow of water though a pipe. The water may be flowing to a fire sprinkler system, cooling system, process system, or other vital operation. Dangerous problems may develop if the flow of water is stopped. At times, the flow may have to be stopped. By using NAND circuit logic, more than one switch must be activated to stop the flow. For example, if two key-operated switches are used, both switches have to be activated before the flow is stopped. An operator may have one key and a supervisor may have the other.

This same logic may be applied to other circuits that require more than one switch activation. For example, the solenoid may be used to control an electrically-operated prison door lock or any other electrically-operated lock.

XOR (Exclusive OR) Gates

An *XOR (exclusive OR) gate* is a logic gate that provides logic level 1 only if one, but not both of the inputs are at logic level 1. XOR gates are used in applications in which the flow of two or more products into a common point must be controlled. An example of the need for exclusive OR logic is when two products that cannot be mixed must both use a common pipe. **See Figure 25-13.**

In this circuit, ammonia and bleach tanks use a common pipe at different times of the operation. Ammonia and bleach are two products that should not be mixed together. Exclusive OR logic is used in the electrical circuit to ensure that the ammonia and bleach are never mixed. Switches with multiple contacts are required to develop exclusive OR logic in an electrical circuit. For example, the tank and bleach control switches would both have to have three contacts each for proper circuit operation. To reduce the number of required contacts on the switches, control relays could be added into the circuit to provide more switching contacts. There is no need to have multiple switch contacts or additional relays when an XOR gate is used.

Digital circuit logic is commonly used to control functions in applications such as HVAC control units.

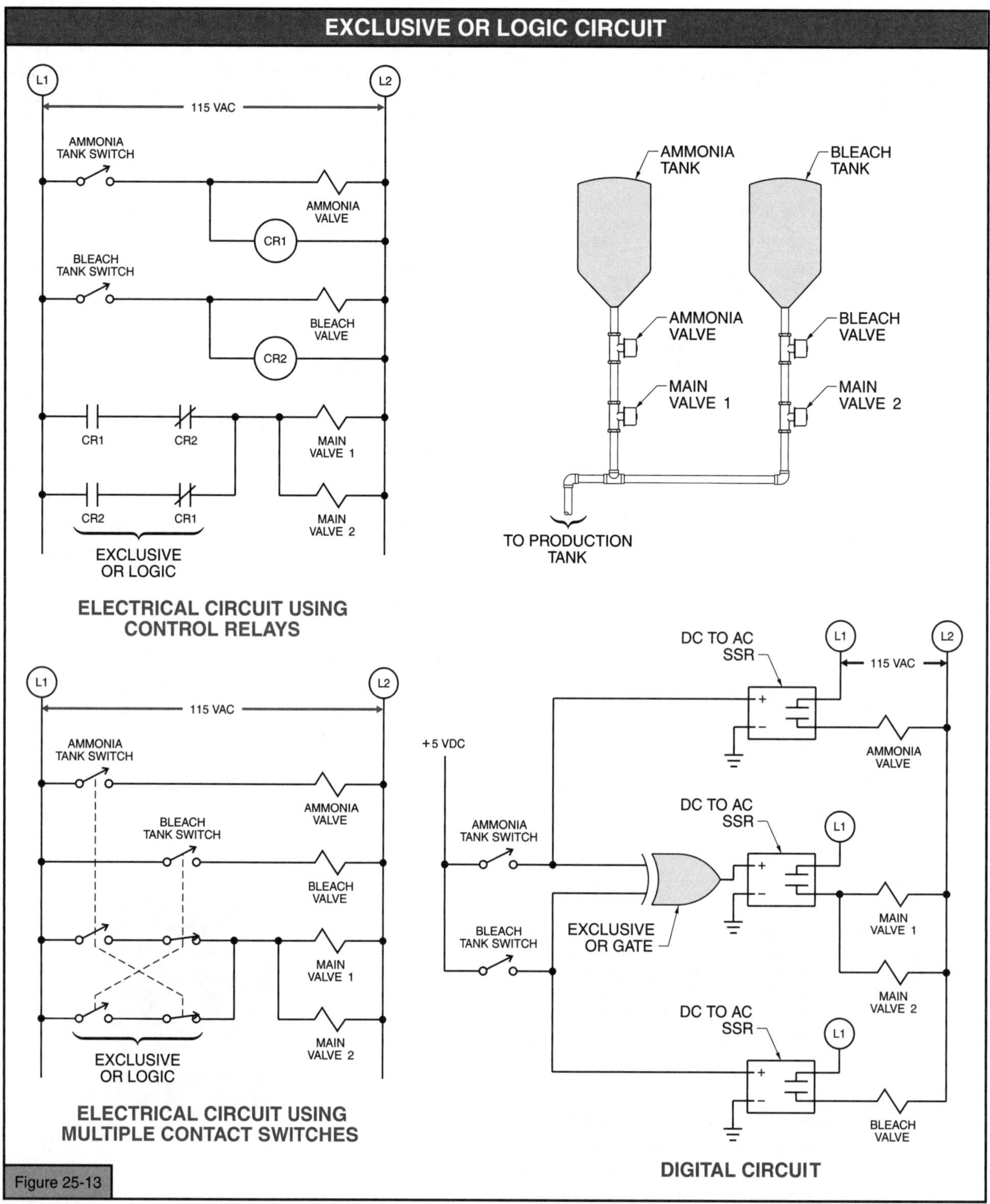

Figure 25-13

Combination Logic

In addition to using basic logic gates alone, any number of logic gates may be combined to produce the required logic function. One of the most common gates used to change the output logic of a circuit is an inverter (NOT gate). Inverters are used to convert a basic gate such as an AND, OR, NAND, or NOR gate to another logic function. Inverters may be added to the output, inputs, or inputs and output of other gates. **See Figure 25-14.**

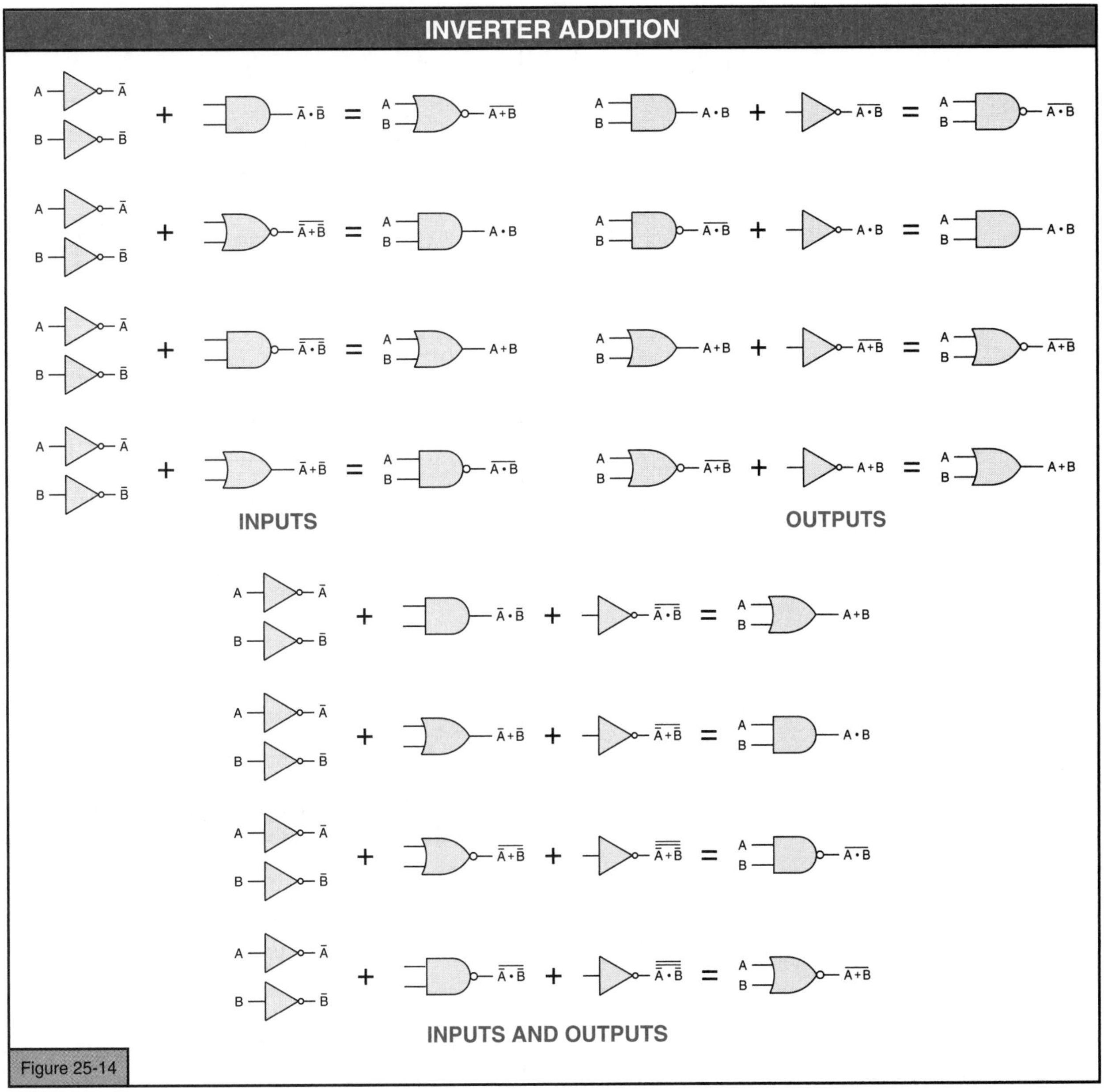

INVERTER ADDITION

Figure 25-14

Pull-Up and Pull-Down Resistors

Digital circuits last a relatively long time if they are properly applied, installed, and used. Basic considerations must be given when designing a digital circuit to prevent problems before they occur. This includes deciding what to do with gates that are not used and ensuring that the gate output is not overloaded.

Digital inputs are designed to be connected to either a high (1) or a low (0). A high is equal to the input supply voltage and a low is equal to ground potential. An input is floating when it is not connected to the supply voltage (high) or ground (low). A *floating input* is a digital input that is not a true high or low at all times. A floating input should not occur in a digital circuit because they cause intermittent problems in circuit operation, such as false counts into a counter and false input signals.

A pull-up resistor may be used to prevent unused inputs from floating. A *pull-up resistor* is a resistor that has one side connected to the power supply at all times and the other side connected to the unused gate(s). **See Figure 25-15.**

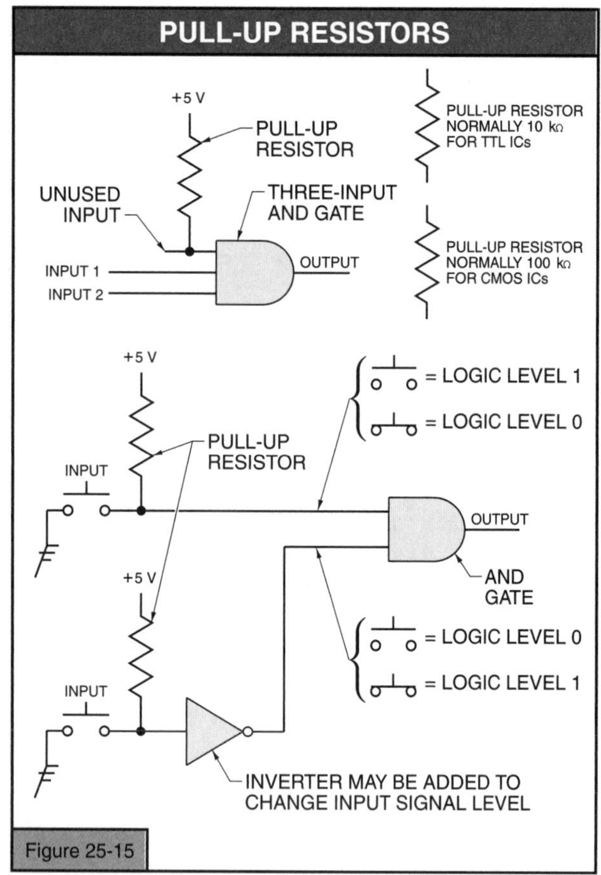

Figure 25-15

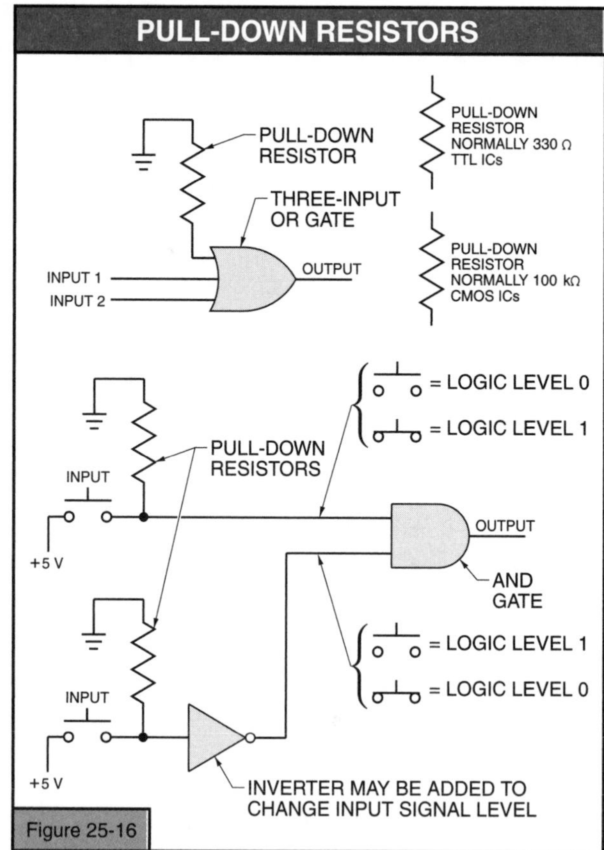

Figure 25-16

A pull-up resistor may also be used to prevent a switch contact from floating. The switch is connected to the gate and the pull-up resistor when using a pull-up resistor with a switch. The other side of the pull-up resistor is connected to the positive side of the power supply.

A pull-down resistor may also be used in addition to using a pull-up resistor. A *pull-down resistor* is a resistor that has one side connected to ground at all times and the other side connected to the unused gate(s). **See Figure 25-16.**

A pull-down resistor may be used to prevent a switch contact from floating. The switch is connected to the gate and the pull-down resistor when using a pull-down resistor with a switch. The other side of the pull-down resistor is connected to ground.

Tying Inputs Together

Logic gate inputs may be tied together and connected to a pull-up or pull-down resistor if more than one input gate is not used. In addition to using pull-up and pull-down resistors, input gates that are not used may also be tied together. **See Figure 25-17.**

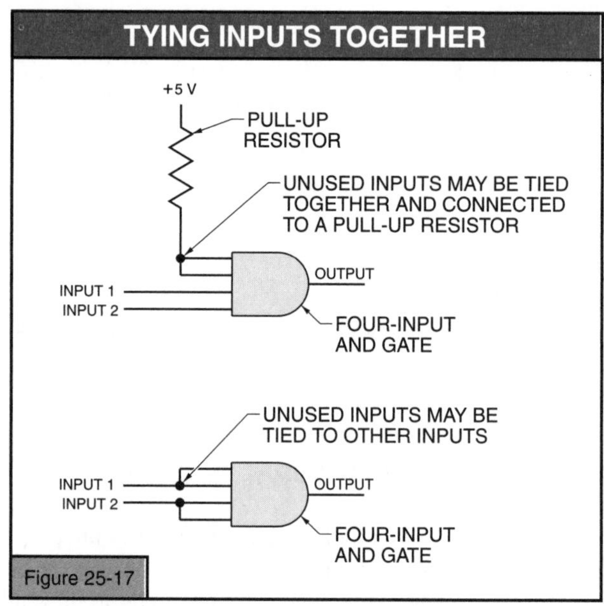

Figure 25-17

Fan-Out

The output of a digital gate can only supply a fixed amount of power that can be used to drive additional logic gates or loads. The amount of power a gate can deliver is determined by the gate's fan-out. *Fan-out* is the number of loads that can be driven by the output of a logic gate. Fan-out is normally determined for a standard logic gate in a family (TTL or CMOS) driving other gates in the same family. The circuit operates erratically because the output voltage drops if the driving gate is connected to more gates than the fan-out number specifies. Interfaces such as transistors and SSRs may be used when a higher or different power output is required from a digital circuit. **See Figure 25-18.**

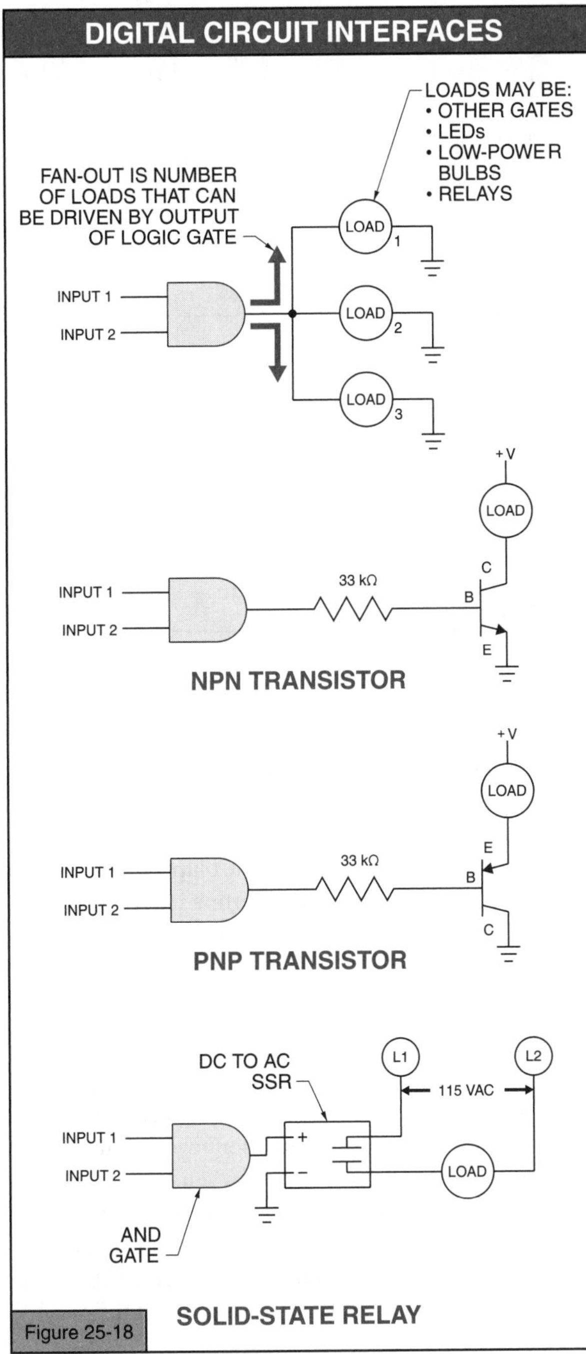

Figure 25-18

Logic Families

Manufacturers of digital chips have developed several families of digital ICs. These families are groups of ICs that are designed to work together. For example, two major family groups are the TTL and CMOS families of ICs. Each family includes many of the same digital gates (AND, NOR, etc.) but have different operating specifications. The specification of input voltage and current, output voltage and current, power consumption, noise margin level, operating speed, and other operating conditions vary from family to family. The prevailing characteristics of a logic family are speed of operation and power usage.

TTL Family

TTL (transistor-transistor logic) ICs use transistors for inputs and outputs. The TTL family is commonly referred to as the 7400 series. The 74 identifies commercially available TTL ICs. ICs preceded with a 54, such as 5400, are normally military TTL ICs that are functionally equivalent to the 7400 series. The supply voltage for TTL ICs is 5.0 VDC, ±.25 V.

CMOS Family

CMOS ICs use complementary metal-oxide semiconductors. The CMOS family of ICs was developed to use less power than TTL ICs. The CMOS family of ICs may be used in battery-operated electronic devices. The first CMOS family was the 4000 series. Later came the 74C00 series and the 74HC00 series. The supply voltage for CMOS ICs ranges from 3 VDC to 18 VDC.

Operational Amplifier ICs

Operational amplifiers (op-amps) are the most common linear ICs. An *op-amp* is a general-purpose, high-gain, DC or AC amplifier. Op-amps are used in most digital circuits requiring an analog amplification of a weak signal. For example, weak audio or data signals must be amplified to be useful. In addition to being used for amplification, op-amps may also be used to perform other electronic functions. An op-amp has five leads attached to the amplifier. Two are input leads, one is an output lead, and two are power supply leads. **See Figure 25-19.**

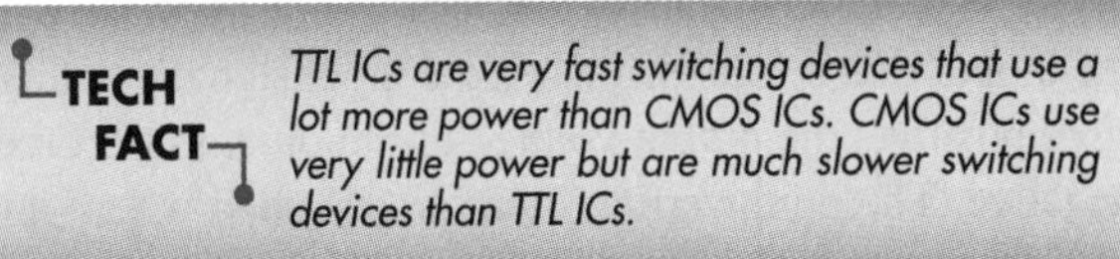

TTL ICs are very fast switching devices that use a lot more power than CMOS ICs. CMOS ICs use very little power but are much slower switching devices than TTL ICs.

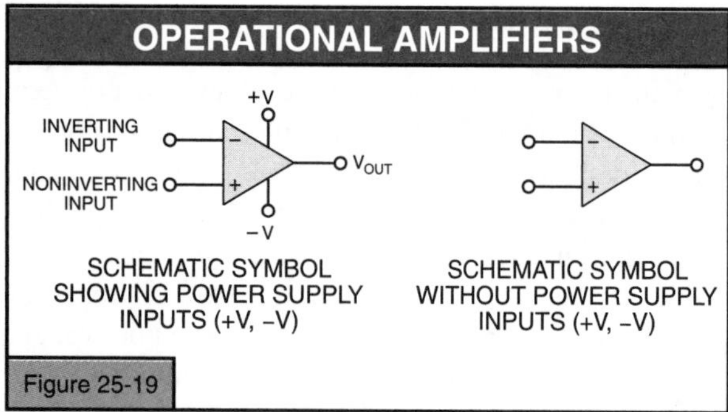

Figure 25-19

An op-amp IC chip may be made to perform addition, subtraction, multiplication, division, and other math functions by connecting a few components to it. An op-amp's main use today is amplifying small signals to levels that can be used for the control of other devices. The main advantages of an op-amp include:

- A very high input impedance (resistance). Op-amps do not draw a lot of power from the input source. This feature is required when amplifying very weak input signals.
- A very high gain (thousands of times). One op-amp can do the job of dozens of individual transistors.
- A very low output impedance. A high output impedance reduces the amplified output.

Power Supply Inputs

The power supply leads deliver power to the op-amp. The op-amp consists of many transistors and resistors. The internal transistors, resistors, and circuitry must be powered to operate. Normally, the power needed by an op-amp is ±15 V. Although the power supply leads are sometimes shown, it is common practice to omit the power supply leads in schematic diagrams.

Although an op-amp can produce very high amplification, it cannot have an output voltage that exceeds the supply voltage. If the supply voltage is ±15 V, the maximum amplified output voltage is normally between ±13 V to ±14 V. This output voltage is somewhat less than the supply voltage and is due to internal losses.

The input terminals of an op-amp are labeled minus (–) and plus (+). The minus input is an inverting input. A positive voltage applied to the inverting input is amplified and exits the op-amp as a negative voltage. If the input signal is a sine wave and is connected to the inverting input, the output signal is a sine wave that is 180° out-of-phase with the input. This means that when the input signal's voltage rises positive, the output voltage of the op-amp becomes negative. The noninverting input is grounded when only using the inverting input. **See Figure 25-20.**

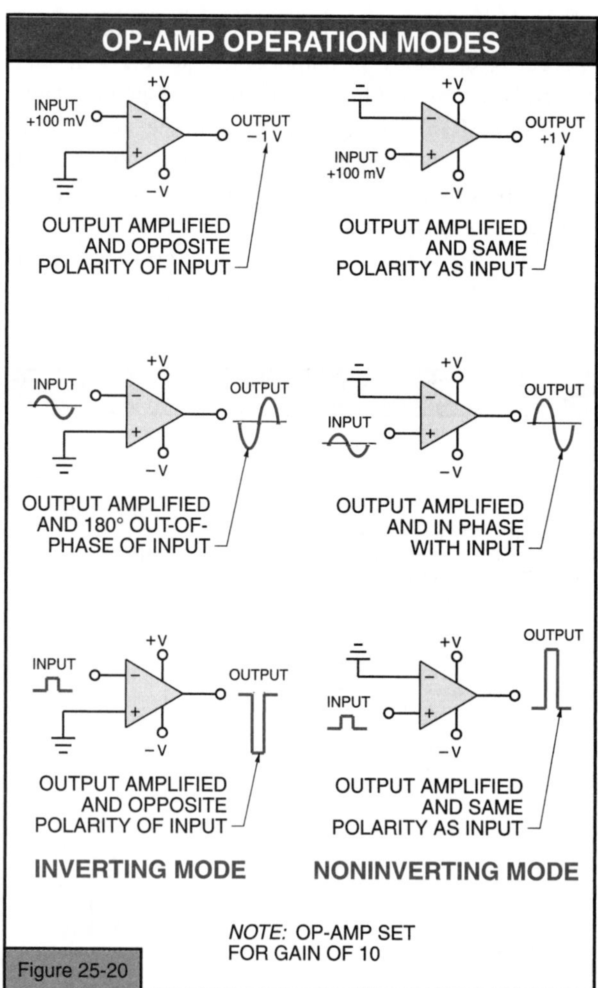

Figure 25-20

The plus input is the noninverting input. A signal applied to the noninverting input of an op-amp is amplified and reproduced at the output. If the input signal is positive, the output signal is positive. If the input signal is negative, the output signal is negative. If the signal applied to the noninverting input is a sine wave, the output of the op-amp is in phase with the input. The inverting input is grounded when only using the noninverting input.

Differential Inputs

The inverting and noninverting inputs may be powered simultaneously. No input is grounded when both inputs are powered. The inverting and noninverting inputs are referred to as differential inputs when both inputs are powered. The op-amp amplifies only the difference between the inverting input and the noninverting input. If the inverting input has +4 V applied to it and the noninverting input has +7 V applied to it, the op-amp amplifies +3 V. **See Figure 25-21.**

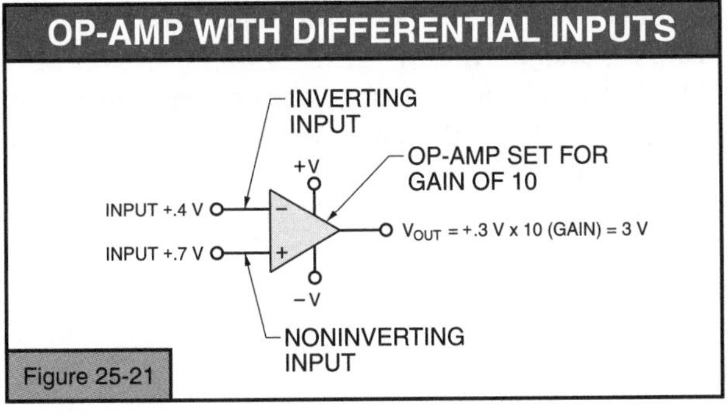

Figure 25-21

Op-Amp Gain

Gain is the ratio of the amplitude of the output signal to the amplitude of the input signal. Op-amps may have gains of 500,000 or more, no gain, or controlled gain. Gain in an op-amp is controlled by adding external resistors to provide closed-loop feedback. **See Figure 25-22.**

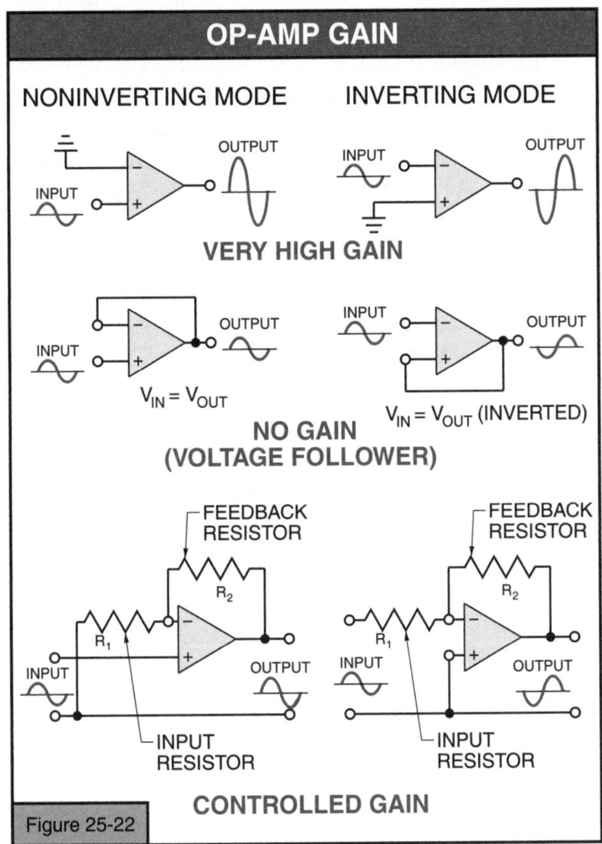

Figure 25-22

In the very high gain mode, a very small change in voltage on either input results in a very large change in the output voltage. This circuit is far too sensitive and unstable for most applications. The gain is normally reduced to a much lower level because such a high gain produces a very sensitive and unstable circuit. Gain is reduced and the circuit is stabilized by feeding back some of the output signal to one of the inputs through a resistor.

In the no gain mode, the output is connected to the input that does not have the input signal applied to it. In the no gain mode, voltage out equals voltage in. For this reason, this circuit is often referred to as a voltage follower. Although there is no gain in this mode, the op-amp is used as an isolation amplifier. An isolation amplifier has a very high input impedance (resistance) and does not produce a loading effect on the input signal.

In the controlled gain mode, the amount of gain is controlled by the value of two resistors (R_1 and R_2). The two resistors are the input and feedback resistors. To calculate voltage gain in an inverting op-amp, apply the formula:

$$A_v = \frac{R_2}{R_1}$$

where

A_v = voltage amplification

R_1 = input resistor (in Ω)

R_2 = feedback resistor (in Ω)

Example: Calculating Voltage Gain—Inverting Op-Amp

What is the voltage gain of an inverting op-amp circuit when the input resistor equals 10 kΩ and the feedback resistor equals 90 kΩ?

$$A_v = \frac{R_2}{R_1}$$

$$A_v = \frac{90,000}{10,000}$$

$$A_v = \mathbf{9}$$

To calculate the voltage gain of a noninverting op-amp, apply the formula:

$$A_v = \frac{R_1 + R_2}{R_1}$$

where

A_v = voltage amplification

R_1 = input resistor (in Ω)

R_2 = feedback resistor (in Ω)

TECH FACT *CMOS integrated circuits are very sensitive to static electricity. Store CMOS integrated circuits in the manufacturer's antistatic packages until ready for use. Never touch the pins of CMOS integrated circuits.*

Example: Calculating Voltage Gain—Noninverting Op-Amp

What is the voltage gain of a noninverting op-amp circuit when the input resistor equals 25 kΩ and the feedback resistor equals 100 kΩ?

$$A_v = \frac{R_1 + R_2}{R_1}$$

$$A_v = \frac{25,000 + 100,000}{25,000}$$

$$A_v = \frac{125,000}{25,000}$$

$$A_v = \mathbf{5}$$

The gain of an op-amp may be calculated when the input and output voltage is known or measured. Op-amp gain is calculated by dividing the output voltage by the input voltage. To calculate op-amp gain when input and output voltage is known, apply the formula:

$$A_G = \frac{V_O}{V_I}$$

where

A_G = amplifier gain

V_O = output voltage (in V)

V_I = input voltage (in V)

Example: Calculating Op-Amp Gain—Input and Output Voltages Known

What is the gain of an op-amp when the measured input voltage is 2 mV and the output voltage is 8 V?

$$A_G = \frac{V_O}{V_I}$$

$$A_G = \frac{8}{.002}$$

$$A_G = \mathbf{4000}$$

Op-Amp Applications

Op-amps are used in applications that require a small signal to be amplified so that the output signal can drive a higher power load. For example, an op-amp can take the signal from a photoconductive cell, thermocouple, gas detector, or smoke detector and amplify it. **See Figure 25-23.**

The op-amp can directly control the load if the output voltage of the op-amp is large enough and the current draw of the load is within the op-amp's limits. An interface is required if the load requires more power for operation than the op-amp can safely deliver. The most common interface is a solid-state relay (SSR). The input of the SSR is connected to the op-amp and the output of the SSR is used to drive the load.

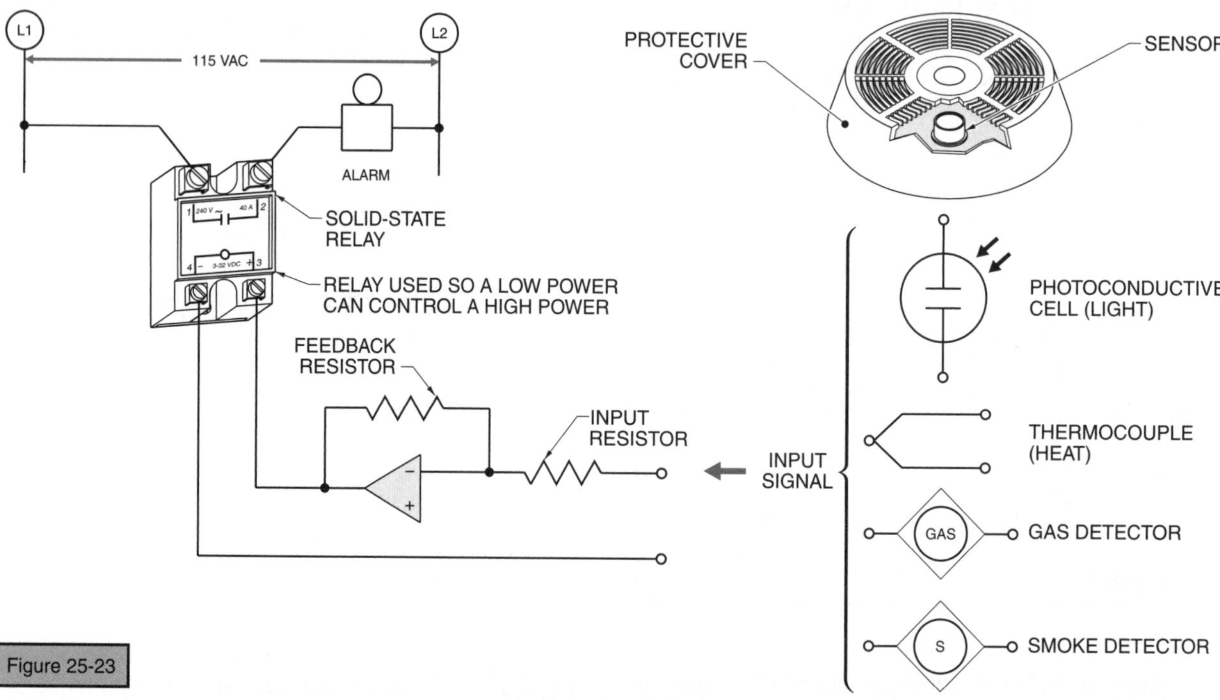

Figure 25-23

Refer to the CD-ROM in the back of the book for Chapter 25 Quick Quiz® and related information.

CONTENTS

COMPARISON OF NUMBERING SYSTEMS				
Decimal	Binary	Octal	Hexadecimal	BCD
0	000	0	0	0000
1	001	1	1	0001
2	010	2	2	0010
3	011	3	3	0011
4	100	4	4	0100
5	101	5	5	0101
6	110	6	6	0110
7	111	7	7	0111
8	1000	10	8	1000
9	1001	11	9	1001
10	1010	12	A	0001 1000
11	1011	13	B	0001 0001
12	1100	14	C	0001 0010
13	1101	18	D	0001 0011
14	1110	16	E	0001 0100

BINARY NUMBERING SYSTEM									
Binary				Decimal	Binary				Decimal
8s	4s	2s	1s		8s	4s	2s	1s	
0	0	0	0	0	1	0	0	0	8
0	0	0	1	1	1	0	0	1	9
0	0	1	0	2	1	0	1	0	10
0	0	1	1	3	1	0	1	1	11
0	1	0	0	4	1	1	0	0	12
0	1	0	1	5	1	1	0	1	13
0	1	1	0	6	1	1	1	0	14
0	1	1	1	7	1	1	1	1	15

STANDARD NEMA RATINGS OF AC CONTACTORS, 60 HZ						
Size	8 Hr Open Rating (A)	Power*				
		3ϕ			1ϕ	
		200	230	230/460	115	230
00	9	1½	1½	2	⅓	1
0	18	3	3	5	1	2
1	27	7½	7½	10	2	3
2	45	10	15	25	3	7½
3	90	25	30	50	–	–
4	135	40	50	100	–	–
5	270	75	100	200	–	–
6	540	150	200	400	–	–
7	810	–	300	600	–	–
8	1215	–	450	900	–	–
9	2250	–	800	1600	–	–

* in volts

ENGLISH TO METRIC EQUIVALENTS

		Unit	Metric Equivalent
LENGTH		mile	1.609 km
		rod	5.029 m
		yard	.9144 m
		foot	30.48 cm
		inch	2.54 cm
AREA $A = l \times w$		square mile	2.590 k^2
		acre	.405 hectacre, 4047 m^2
		square rod	25.293 m^2
		square yard	.836 m^2
		square foot	.093 m^2
		square inch	6.452 cm^2
VOLUME $V = l \times w \times t$		cubic yard	.765 m^3
		cubic foot	.028 m^3
		cubic inch	16.387 cm^3
CAPACITY — WATER, FUEL, ETC.	*U.S. liquid measure*	gallon	3.785 l
		quart	.946 l
		pint	.473 l
		gill	118.294 ml
		fluidounce	29.573 ml
		fluidram	3.697 ml
		minim	.061610 ml
VEGETABLES, GRAIN, ETC.	*U.S. dry measure*	bushel	35.239 l
		peck	8.810
		quart	1.101 l
		pint	.551 l
DRUGS	*British imperial liquid and dry measure*	bushel	.036 m^3
		peck	.0091 m^3
		gallon	4.546 l
		quart	1.136 l
		pint	568.26 cm^3
		gill	142.066 cm^3
		fluidounce	28.412 cm^3
		fluidram	3.5516 cm^3
		minim	.059194 cm^3
MASS AND WEIGHT — COAL, GRAIN, ETC.	*avoirdupois*	short ton	.907 t
		long ton	1.016 t
		pound	.454 kg
		ounce	28.350 g
		dram	1.772 g
		grain	.0648 g
GOLD, SILVER, ETC.	*troy*	pound	.373 kg
		ounce	31.103 g
		pennyweight	1.555 g
		grain	.0648 g
DRUGS	*apothecaries'*	pound	.373 kg
		ounce	31.103 g
		dram	3.888 g
		scruple	1.296 g
		grain	.0648 g

METRIC TO ENGLISH EQUIVALENTS

	Unit	English Equivalent		
LENGTH	kilometer	.62 mi		
	hectometer	109.36 yd		
	dekameter	32.81′		
	meter	39.37″		
	decimeter	3.94″		
	centimeter	.39″		
	millimeter	.039″		
AREA A = l x w	square kilometer	.3861 sq mi		
	hectacre	2.47 A		
	are	119.60 sq yd		
	square centimeter	.155 sq in.		
VOLUME V = l x w x t	cubic centimeter	.061 cu in.		
	cubic decimeter	61.023 cu in.		
	cubic meter	1.307 cu yd		
CAPACITY		***cubic***	***dry***	***liquid***
WATER, FUEL, ETC.	kiloliter	1.31 cu yd		
	hectoliter	3.53 cu ft	2.84 bu	
VEGETABLES, GRAIN, ETC.	dekaliter	.35 cu ft	1.14 pk	2.64 gal.
	liter	61.02 cu in.	.908 qt	1.057 qt
DRUGS	cubic decimeter	61.02 cu in.	.908 qt	1.057 qt
	deciliter	6.1 cu in.	.18 pt	.21 pt
	centiliter	.61 cu in.		338 fl oz
	milliliter	.061 cu in.		.27 fl dr
MASS AND WEIGHT	metric ton	1.102 t		
COAL, GRAIN, ETC.	kilogram	2.2046 lb		
	hectogram	3.527 oz		
GOLD, SILVER, ETC.	dekagram	.353 oz		
	gram	.035 oz		
DRUGS	decigram	1.543 gr		
	centigram	.154 gr		
	milligram	.015 gr		

ELECTRICAL/ELECTRONIC ABBREVIATIONS/ACRONYMS . . .			
Abbr/ Acronym	**Meaning**	**Abbr/ Acronym**	**Meaning**
A	Ammeter; Ampere; Anode; Armature	DPST	Double-Pole, Single-Throw
AC	Alternating Current	DS	Drum Switch
AC/DC	Alternating Current; Direct Current	DT	Double-Throw
A/D	Analog to Digital	DVM	Digital Voltmeter
AF	Audio Frequency	EMF	Electromotive Force
AFC	Automatic Frequency Control	F	Fahrenheit; Fast; Forward; Fuse; Farad
Ag	Silver	FET	Field-Effect Transistor
ALM	Alarm	FF	Flip-Flop
AM	Ammeter; Amplitude Modulation	FLC	Full-Load Current
AM/FM	Amplitude Modulation; Frequency Modulation	FLS	Flow Switch
ARM.	Armature	FLT	Full-Load Torque
Au	Gold	FM	Frequency Modulation
AU	Automatic	FREQ	Frequency
AVC	Automatic Volume Control	FS	Float Switch
AWG	American Wire Gauge	FTS	Foot Switch
BAT.	Battery (electric)	FU	Fuse
BCD	Binary-Coded Decimal	FWD	Forward
BJT	Bipolar Junction Transistor	G	Gate; Giga; Green; Conductance
BK	Black	GEN	Generator
BL	Blue	GRD	Ground
BR	Brake Relay; Brown	GY	Gray
C	Celsius; Capacitance; Capacitor, Coulomb	H	Henry; High Side of Transformer; Magnetic Flux
CAP.	Capacitor	HF	High Frequency
CB	Circuit Breaker; Citizen's Band	HP	Horsepower
CC	Common-Collector; Configuration	Hz	Hertz
CCW	Counterclockwise	I	Current
CE	Common-Emitter Configuration	IC	Integrated Circuit
CEMF	Counter-Electromotive Force	INT	Intermediate; Interrupt
CKT	Circuit	IOL	Instantaneous Overload
cmil	Circular Mil	IR	Infrared
CONT	Continuous; Control	ITB	Inverse Time Breaker
CPS	Cycles Per Second	ITCB	Instantaneous Trip Circuit Breaker
CPU	Central Processing Unit	J	Joule
CR	Control Relay	JB	Junction Box
CRM	Control Relay Master	JFET	Junction Field-Effect Transistor
CT	Current Transformer	K	Kilo; Cathode
CW	Clockwise	kWh	kilowatt-hour
D	Diameter; Diode; Down	L	Line; Load; Coil; Inductance
D/A	Digital to Analog	LB-FT	Pounds Per Foot
DB	Dynamic Braking Contactor; Relay	LB-IN.	Pounds Per Inch
DC	Direct Current	LC	Inductance-Capacitance
DIO	Diode	LCD	Liquid Crystal Display
DISC.	Disconnect Switch	LCR	Inductance-Capacitance-Resistance
DMM	Digital Multimeter	LED	Light-Emitting Diode
DP	Double-Pole	LRC	Locked Rotor Current
DPDT	Double-Pole, Double-Throw	LS	Limit Switch

. . . ELECTRICAL/ELECTRONIC ABBREVIATIONS/ACRONYMS			
Abbr/ Acronym	Meaning	Abbr/ Acronym	Meaning
LT	Lamp	RF	Radio Frequency
M	Motor; Motor Starter; Motor Starter Contacts	RH	Rheostat
MAX.	Maximum	rms	Root-Mean-Square
MB	Magnetic Brake	ROM	Read-Only Memory
MCS	Motor Circuit Switch	rpm	Revolutions Per Minute
MEM	Memory	RPS	Revolutions Per Second
MED	Medium	S	Series; Slow; South; Switch; Second; Siemen
MIN	Minimum	SCR	Silicon-Controlled Rectifier
MMF	Magnetomotive Force	SEC	Secondary
MN	Manual	SF	Service Factor
MOS	Metal-Oxide Semiconductor	1 PH; 1ϕ	Single-Phase
MOSFET	Metal-Oxide Semiconductor Field-Effect Transistor	SOC	Socket
MTR	Motor	SOL	Solenoid
N; NEG	North; Negative; Number of Turns	SP	Single-Pole
NC	Normally Closed	SPDT	SIngle-Pole, Double-Throw
NEUT	Neutral	SPST	Single-Pole, Single-Throw
NO	Normally Open	SS	Selector Switch
NPN	Negative-Positive-Negative	SSW	Safety Switch
NTDF	Nontime-Delay Fuse	SW	Switch
O	Orange	T	Tera; Terminal; Torque; Transformer
OCPD	Overcurrent Protection Device	TB	Terminal Board
OHM	Ohmmeter	3 PH; 3 ϕ	Three-Phase
OL	Overload Relay	TD	Time-Delay
OZ/IN.	Ounces Per Inch	TDF	Time-Delay Fuse
P	Peak; Positive; Power; Power Consumed	TEMP	Temperature
PB	Pushbutton	THS	Thermostat Switch
PCB	Printed Circuit Board	TR	Time-Delay Relay
PH	Phase	TTL	Transistor-Transistor Logic
PLS	Plugging Switch	U	Up
PNP	Positive-Negative-Positive	UCL	Unclamp
POS	Positive	UHF	Ultrahigh Frequency
POT.	Potentiometer	UJT	Unijunction Transistor
P-P	Peak-to-Peak	UV	Ultraviolet; Undervoltage
PRI	Primary Switch	V	Violet; Volt
PS	Pressure Switch	VA	Voltampere
PSI	Pounds Per Square Inch	VAC	Volts Alternating Current
PUT	Pull-Up Torque	VDC	Volts Direct Current
Q	Transistor; Quality Factor	VHF	Very High Frequency
R	Radius; Red; Resistance; Reverse	VLF	Very Low Frequency
RAM	Random-Access Memory	VOM	Volt-Ohm-Milliammeter
RC	Resistance-Capacitance	W	Watt; White
RCL	Resistance-Inductance-Capacitance	w/	With
REC	Rectifier	X	Low Side of Transformer
RES	Resistor	Y	Yellow
REV	Reverse	Z	Impedance

FLUID POWER ABBREVIATIONS/ACRONYMS			
Abbr/Acronym	**Meaning**	**Abbr/Acronym**	**Meaning**
BTU	British Thermal Unit	IN-LB	Inch Pound
C	Degrees Celsius	INT	Internal
CC	Closed Center (valves)	I/O	Input/Output
CCW	Counterclockwise	IPM	Inches Per Minute
CFM	Cubic Feet Per Minute	IPS	Inches Per Second
CFS	Cubic Feet Per Second	LB	Pound
CIM	Cubic Inches Per Minute	MAX	Maximum
COM	Common	MIN	Minimum
CPM	Cycles Per Minute	NC	Normally Closed
CPS	Cycles Per Second	NO	Normally Open
CW	Clockwise	NPT	National Pipe Thread
CYL	Cylinder	OC	Open Center (valves)
D	Drain	OZ	Ounce
DIA	Diameter	PO	Pilot Operated
EXT	External	PRES or P	Pressure
F	Degrees Fahrenheit	PSI	Pounds Per Square Inch
FL	Fluid	PSIA	PSI Absolute
FPM	Feet Per Minute	PSIG	PSI Gauge
FS	Full Scale	PT	Pint
FT	Foot	QT	Quart
FT-LB	Foot Pound	R	Radius
GA	Gauge	RMS	Root Mean Square
GAL	Gallon	RPM	Revolutions Per Minute
GPM	Gallons Per Minute	RPS	Revolutions Per Second
HP	Horsepower	SOL	Solenoid
Hz	Hertz	T	Torque; Thrust; Tank
ID	Inside Diameter	VAC	Vacuum
IN	Inches	VI	Viscosity Index
		VISC	Viscosity

HAZARDOUS LOCATIONS		
Class	**Group**	**Material**
I	A	Acetylene
	B	Hydrogen, butadiene, ethylene oxide, propylene oxide
	C	Carbon monoxide, ether, ethylene, hydrogen sulfide, morpholine, cyclopropane
	D	Gasoline, benzene, butane, propane, alcohol, acetone, ammonia, vinyl chloride
II	E	Metal dusts
	F	Carbon black, coke dust, coal
	G	Grain dust, flour, starch, sugar, plastics
III	No groups	Wood chips, cotton, flax, nylon

NON-LOCKING WIRING DEVICES

2-POLE, 3-WIRE

WIRING DIAGRAM	NEMA ANSI	RECEPTACLE CONFIGURATION	RATING
	5-15 C73.11		15 A 125 V
	5-20 C73.12		20 A 125 V
	5-30 C73.45		30 A 125 V
	5-50 C73.46		50 A 125 V
	6-15 C73.20		15 A 250 V
	6-20 C73.51		20 A 250 V
	6-30 C73.52		30 A 250 V
	6-50 C73.53		50 A 250 V
	7-15 C73.28		15 A 277 V
	7-20 C73.63		20 A 277 V
	7-30 C73.64		30 A 277 V
	7-50 C73.65		50 A 277 V

4-POLE, 4-WIRE

WIRING DIAGRAM	NEMA ANSI	RECEPTACLE CONFIGURATION	RATING
	18-15 C73.15		15 A 3ϕY 120/208 V
	18-20 C73.26		20 A 3ϕY 120/208 V
	18-30 C73.47		30 A 3ϕY 120/208 V
	18-50 C73.48		50 A 3ϕY 120/208 V
	18-60 C73.27		60 A 3ϕY 120/208 V

3-POLE, 3-WIRE

WIRING DIAGRAM	NEMA ANSI	RECEPTACLE CONFIGURATION	RATING
	10-20 C73.23		20 A 125/250 V
	10-30 C73.24		30 A 125/250 V
	10-50 C73.25		50 A 125/250 V
	11-15 C73.54		15 A 3ϕ 250 V
	11-20 C73.55		20 A 3ϕ 250 V
	11-30 C73.56		30 A 3ϕ 250 V
	11-50 C73.57		50 A 3ϕ 250 V

3-POLE, 4-WIRE

WIRING DIAGRAM	NEMA ANSI	RECEPTACLE CONFIGURATION	RATING
	14-15 C73.49		15 A 125/250 V
	14-20 C73.50		20 A 125/250 V
	14-30 C73.16		30 A 125/250 V
	14-50 C73.17		50 A 125/250 V
	14-60 C73.18		60 A 125/250 V
	15-15 C73.58		15 A 3ϕ 250 V
	15-20 C73.59		20 A 3ϕ 250 V
	15-30 C73.60		30 A 3ϕ 250 V
	15-50 C73.61		50 A 3ϕ 250 V
	15-60 C73.62		60 A 3ϕ 250 V

LOCKING WIRING DEVICES

2-POLE, 3-WIRE

WIRING DIAGRAM	NEMA ANSI	RECEPTACLE CONFIGURATION	RATING
	ML2 C73.44		15 A 125 V
	L5-15 C73.42		15 A 125 V
	L5-20 C73.72		20 A 125 V
	L6-15 C73.74		15 A 250 V
	L6-20 C73.75		20 A 250 V
	L6-30 C73.76		30 A 250 V
	L7-15 C73.43		15 A 277 V
	L7-20 C73.77		20 A 277 V
	L8-20 C73.79		20 A 480 V
	L9-20 C73.81		20 A 600 V

3-POLE, 4-WIRE

WIRING DIAGRAM	NEMA ANSI	RECEPTACLE CONFIGURATION	RATING
	L14-20 C73.83		20 A 125/250 V
	L14-30 C73.84		30 A 125/250 V
	L15-20 C73.85		20 A 3ϕ 250 V
	L15-30 C73.86		30 A 3ϕ 250 V
	L16-20 C73.87		20 A 3ϕ 480 V
	L16-30 C73.88		30 A 3ϕ 480 V
	L17-30 C73.89		30 A 3ϕ 600 V

3-POLE, 3-WIRE

WIRING DIAGRAM	NEMA ANSI	RECEPTACLE CONFIGURATION	RATING
	ML3 C73.30		15 A 125/250 V
	L10-20 C73.96		20 A 125/250 V
	L10-30 C73.97		30 A 125/250 V
	L11-15 C73.98		15 A 3ϕ 250 V
	L11-20 C73.99		20 A 3ϕ 250 V
	L12-20 C73.101		20 A 3ϕ 480 V
	L12-30 C73.102		30 A 3ϕ 480 V
	L13-30 C73.103		30 A 3ϕ 600 V

4-POLE, 4-WIRE

WIRING DIAGRAM	NEMA ANSI	RECEPTACLE CONFIGURATION	RATING
	L18-20 C73.104		20 A 3ϕY 120/208 V
	L18-30 C73.105		30 A 3ϕY 120/208 V
	L19-20 C73.106		20 A 3ϕY 277/480 V
	L20-20 C73.108		20 A 3ϕY 347/600 V

4-POLE, 5-WIRE

WIRING DIAGRAM	NEMA ANSI	RECEPTACLE CONFIGURATION	RATING
	L21-20 C73.90		20 A 3ϕY 120/208 V
	L22-20 C73.92		20 A 3ϕY 277/480 V
	L23-20 C73.94		20 A 3ϕY 347/600 V

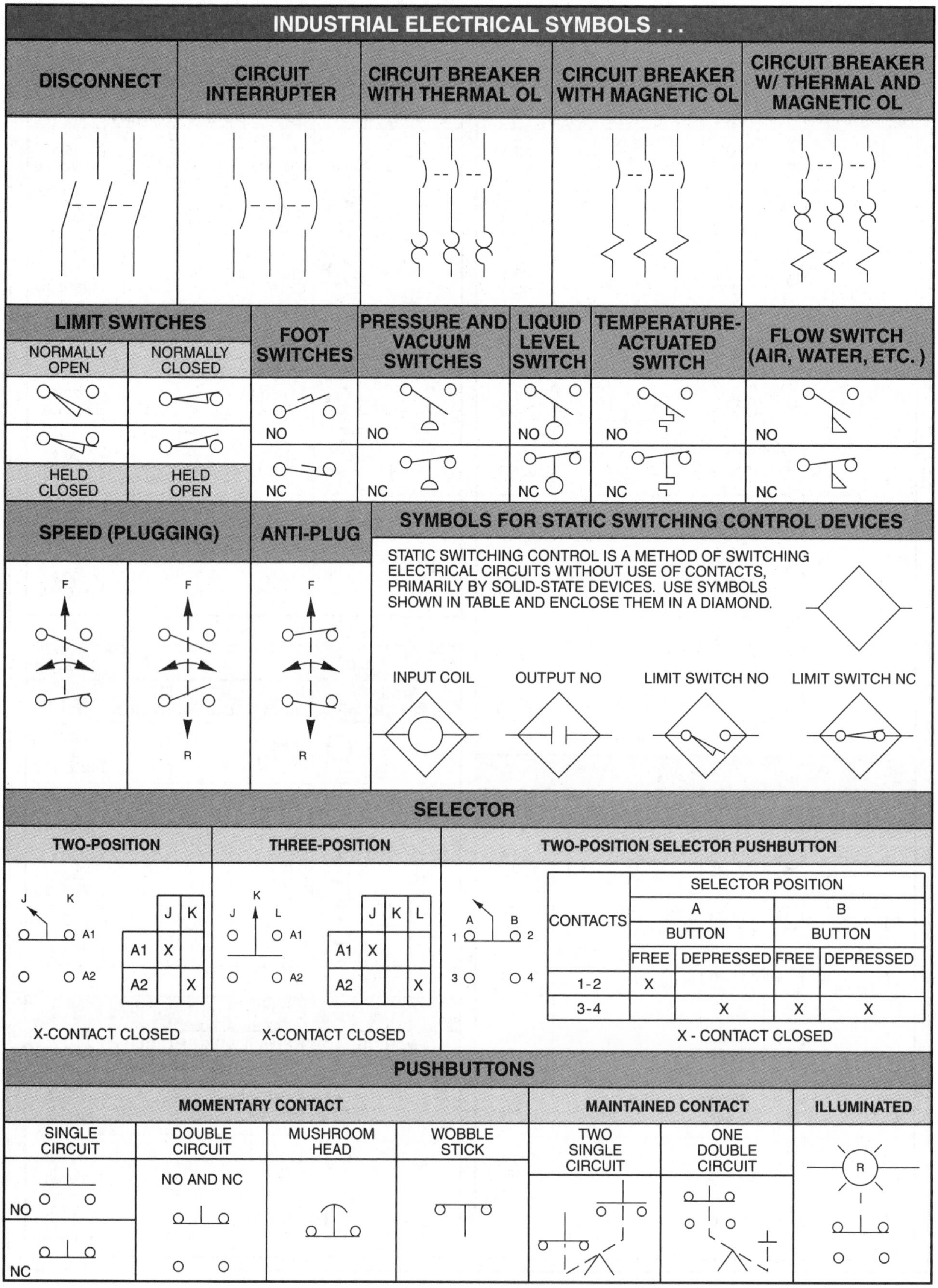

INDUSTRIAL ELECTRICAL SYMBOLS . . .

DISCONNECT	CIRCUIT INTERRUPTER	CIRCUIT BREAKER WITH THERMAL OL	CIRCUIT BREAKER WITH MAGNETIC OL	CIRCUIT BREAKER W/ THERMAL AND MAGNETIC OL

LIMIT SWITCHES		FOOT SWITCHES	PRESSURE AND VACUUM SWITCHES	LIQUID LEVEL SWITCH	TEMPERATURE-ACTUATED SWITCH	FLOW SWITCH (AIR, WATER, ETC.)
NORMALLY OPEN	NORMALLY CLOSED	NO	NO	NO	NO	NO
HELD CLOSED	HELD OPEN	NC	NC	NC	NC	NC

SPEED (PLUGGING)		ANTI-PLUG
F	F R	F R

SYMBOLS FOR STATIC SWITCHING CONTROL DEVICES

STATIC SWITCHING CONTROL IS A METHOD OF SWITCHING ELECTRICAL CIRCUITS WITHOUT USE OF CONTACTS, PRIMARILY BY SOLID-STATE DEVICES. USE SYMBOLS SHOWN IN TABLE AND ENCLOSE THEM IN A DIAMOND.

INPUT COIL | OUTPUT NO | LIMIT SWITCH NO | LIMIT SWITCH NC

SELECTOR

TWO-POSITION

J K A1 A2

	J	K
A1	X	
A2		X

X-CONTACT CLOSED

THREE-POSITION

K J L A1 A2

	J	K	L
A1	X		
A2			X

X-CONTACT CLOSED

TWO-POSITION SELECTOR PUSHBUTTON

A B 1 2 3 4

CONTACTS	SELECTOR POSITION			
	A		B	
	BUTTON		BUTTON	
	FREE	DEPRESSED	FREE	DEPRESSED
1-2	X			
3-4		X	X	X

X - CONTACT CLOSED

PUSHBUTTONS

MOMENTARY CONTACT				MAINTAINED CONTACT		ILLUMINATED
SINGLE CIRCUIT	DOUBLE CIRCUIT	MUSHROOM HEAD	WOBBLE STICK	TWO SINGLE CIRCUIT	ONE DOUBLE CIRCUIT	R
NO NC	NO AND NC					

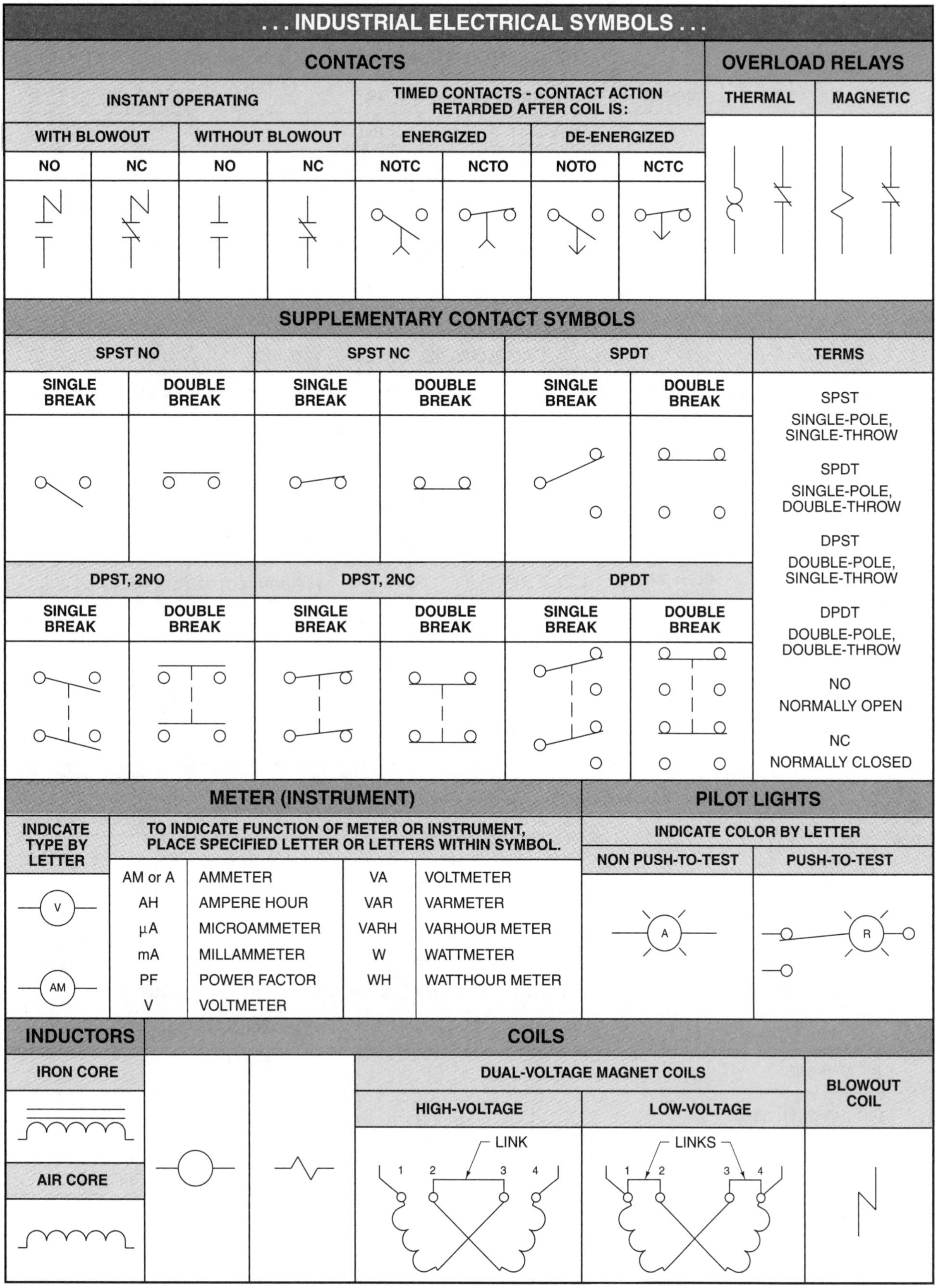

. . . INDUSTRIAL ELECTRICAL SYMBOLS . . .

CONTACTS

INSTANT OPERATING				TIMED CONTACTS - CONTACT ACTION RETARDED AFTER COIL IS:			
WITH BLOWOUT		WITHOUT BLOWOUT		ENERGIZED		DE-ENERGIZED	
NO	NC	NO	NC	NOTC	NCTO	NOTO	NCTC

OVERLOAD RELAYS

THERMAL	MAGNETIC

SUPPLEMENTARY CONTACT SYMBOLS

SPST NO		SPST NC		SPDT	
SINGLE BREAK	DOUBLE BREAK	SINGLE BREAK	DOUBLE BREAK	SINGLE BREAK	DOUBLE BREAK
DPST, 2NO		**DPST, 2NC**		**DPDT**	
SINGLE BREAK	DOUBLE BREAK	SINGLE BREAK	DOUBLE BREAK	SINGLE BREAK	DOUBLE BREAK

TERMS

SPST
SINGLE-POLE, SINGLE-THROW

SPDT
SINGLE-POLE, DOUBLE-THROW

DPST
DOUBLE-POLE, SINGLE-THROW

DPDT
DOUBLE-POLE, DOUBLE-THROW

NO
NORMALLY OPEN

NC
NORMALLY CLOSED

METER (INSTRUMENT)

INDICATE TYPE BY LETTER: V, AM

TO INDICATE FUNCTION OF METER OR INSTRUMENT, PLACE SPECIFIED LETTER OR LETTERS WITHIN SYMBOL.

AM or A	AMMETER	VA	VOLTMETER
AH	AMPERE HOUR	VAR	VARMETER
μA	MICROAMMETER	VARH	VARHOUR METER
mA	MILLAMMETER	W	WATTMETER
PF	POWER FACTOR	WH	WATTHOUR METER
V	VOLTMETER		

PILOT LIGHTS

INDICATE COLOR BY LETTER

NON PUSH-TO-TEST	PUSH-TO-TEST
A	R

INDUCTORS

IRON CORE

AIR CORE

COILS

DUAL-VOLTAGE MAGNET COILS

HIGH-VOLTAGE	LOW-VOLTAGE
LINK; 1 2 3 4	LINKS; 1 2 3 4

BLOWOUT COIL

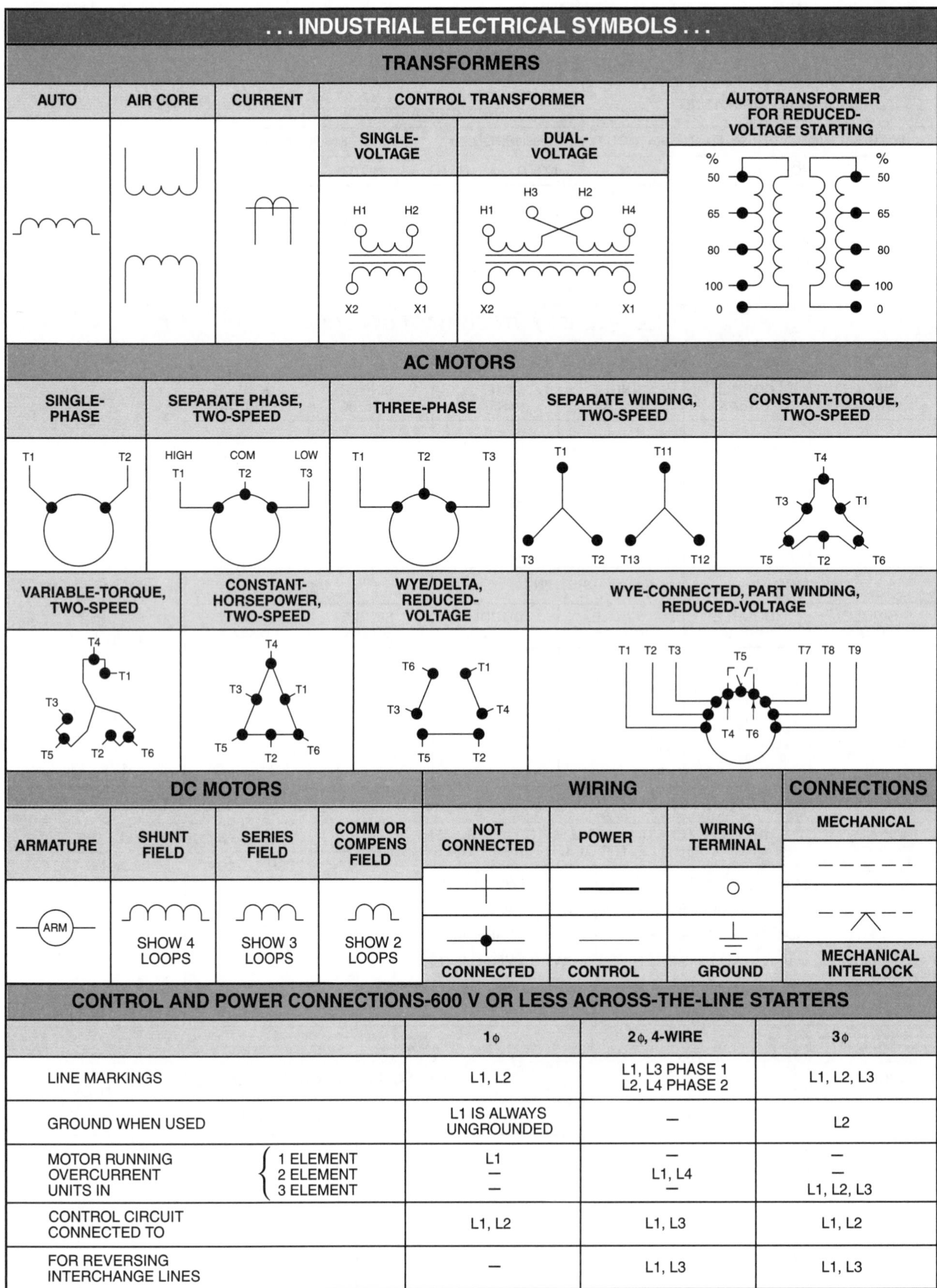

CONTROL AND POWER CONNECTIONS-600 V OR LESS ACROSS-THE-LINE STARTERS

		1φ	2φ, 4-WIRE	3φ
LINE MARKINGS		L1, L2	L1, L3 PHASE 1 L2, L4 PHASE 2	L1, L2, L3
GROUND WHEN USED		L1 IS ALWAYS UNGROUNDED	—	L2
MOTOR RUNNING OVERCURRENT UNITS IN	1 ELEMENT	L1	—	—
	2 ELEMENT	—	L1, L4	—
	3 ELEMENT	—	—	L1, L2, L3
CONTROL CIRCUIT CONNECTED TO		L1, L2	L1, L3	L1, L2
FOR REVERSING INTERCHANGE LINES		—	L1, L3	L1, L3

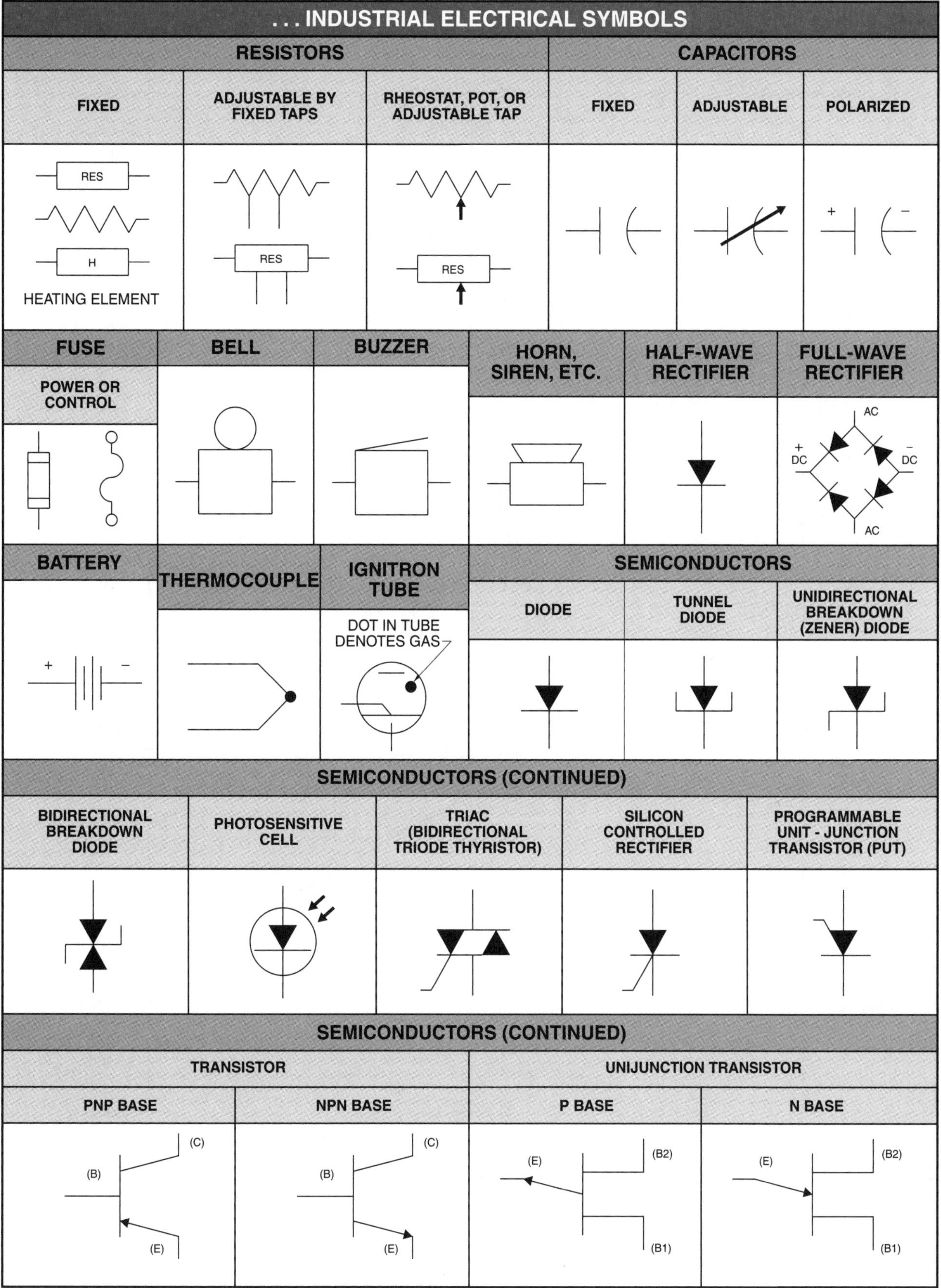

. . . INDUSTRIAL ELECTRICAL SYMBOLS
RESISTORS
CAPACITORS
FIXED
ADJUSTABLE BY FIXED TAPS
RHEOSTAT, POT, OR ADJUSTABLE TAP
FIXED
ADJUSTABLE
POLARIZED
RES
H
HEATING ELEMENT
RES
RES
+
−
FUSE
BELL
BUZZER
HORN, SIREN, ETC.
HALF-WAVE RECTIFIER
FULL-WAVE RECTIFIER
POWER OR CONTROL
AC
+
DC
−
DC
AC
BATTERY
THERMOCOUPLE
IGNITRON TUBE
SEMICONDUCTORS
DIODE
TUNNEL DIODE
UNIDIRECTIONAL BREAKDOWN (ZENER) DIODE
DOT IN TUBE DENOTES GAS
+
−
SEMICONDUCTORS (CONTINUED)
BIDIRECTIONAL BREAKDOWN DIODE
PHOTOSENSITIVE CELL
TRIAC (BIDIRECTIONAL TRIODE THYRISTOR)
SILICON CONTROLLED RECTIFIER
PROGRAMMABLE UNIT - JUNCTION TRANSISTOR (PUT)
SEMICONDUCTORS (CONTINUED)
TRANSISTOR
UNIJUNCTION TRANSISTOR
PNP BASE
NPN BASE
P BASE
N BASE
(C)
(B)
(E)
(C)
(B)
(E)
(E)
(B2)
(B1)
(E)
(B2)
(B1)

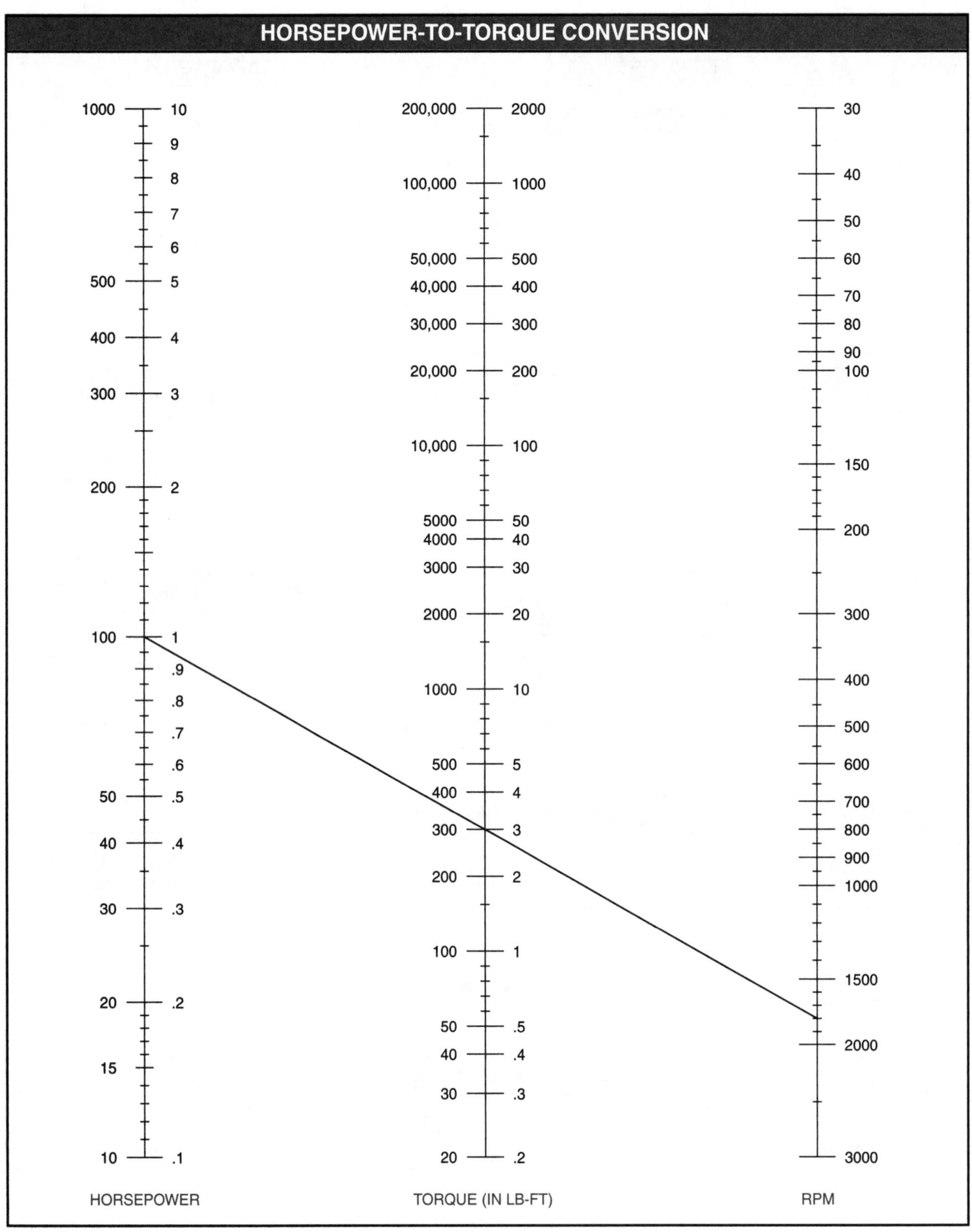
HORSEPOWER-TO-TORQUE CONVERSION
1000
10
9
8
7
6
500
5
400
4
300
3
200
2
100
1
.9
.8
.7
.6
50
.5
40
.4
30
.3
20
.2
15
10
.1
200,000
2000
100,000
1000
50,000
500
40,000
400
30,000
300
20,000
200
10,000
100
5000
50
4000
40
3000
30
2000
20
1000
10
500
5
400
4
300
3
200
2
100
1
50
.5
40
.4
30
.3
20
.2
30
40
50
60
70
80
90
100
150
200
300
400
500
600
700
800
900
1000
1500
2000
3000
HORSEPOWER
TORQUE (IN LB-FT)
RPM

MOTOR HORSEPOWER											
Pump Flow*	Pump Pressure**										
	100	250	500	750	1000	1250	1500	2000	3000	4000	5000
1	.07	.18	.36	.54	.72	.91	1.09	1.45	2.18	2.91	3.64
2	.14	.36	.72	1.09	1.45	1.82	2.18	2.91	4.37	5.83	7.29
3	.21	.54	1.09	1.64	2.18	2.73	3.28	4.37	6.56	8.75	10.93
4	.29	.72	1.45	2.18	2.91	3.64	4.37	5.83	8.75	11.66	14.58
5	.36	.91	1.82	2.73	3.64	4.55	5.46	7.29	10.93	14.58	18.23
8	.58	1.45	2.91	4.37	5.83	7.29	8.75	11.66	17.50	23.33	29.17
10	.72	1.82	3.64	5.46	7.29	9.11	10.93	14.58	21.87	29.17	36.46
12	.87	2.18	4.37	6.56	8.75	10.93	13.12	17.50	26.25	35.00	43.75
15	1.09	2.73	5.46	8.20	10.93	13.67	16.40	21.87	32.81	43.75	54.69
20	1.45	3.64	7.29	10.93	14.58	18.23	21.87	29.17	43.75	58.34	72.92
25	1.82	4.55	9.11	13.67	18.23	22.79	27.34	36.46	54.69	72.92	91.16
30	2.18	5.46	10.93	16.40	21.87	27.34	32.81	43.75	65.63	87.51	109.39
35	2.55	6.38	12.76	19.14	25.52	31.90	38.28	51.05	76.57	102.10	127.62
40	2.91	7.29	14.58	21.87	29.17	36.46	43.75	58.34	87.51	116.68	145.85
45	3.28	8.20	16.40	24.61	32.81	41.02	49.22	65.63	98.45	131.27	164.08
50	3.64	9.11	18.23	27.34	36.46	45.58	54.69	72.92	109.39	145.85	182.32
55	4.01	10.20	20.05	30.08	40.11	50.13	60.16	80.22	120.33	160.44	200.55
60	4.37	10.93	21.87	32.81	43.75	54.69	65.63	87.51	131.27	175.02	218.78
65	4.74	11.85	23.70	35.55	47.40	59.25	71.10	94.80	142.21	189.61	237.01
70	5.10	12.76	25.52	38.28	51.05	63.81	76.57	102.10	153.13	204.20	255.25
75	5.46	13.67	27.36	41.02	54.69	68.37	82.04	109.39	164.08	218.78	273.48
80	5.83	14.58	29.17	43.75	58.34	72.92	87.51	116.68	175.02	233.37	291.71
90	6.56	16.40	32.81	49.22	65.63	82.04	98.45	131.27	196.90	262.54	328.17
100	7.29	18.23	36.46	54.69	72.92	91.16	109.39	145.85	218.78	291.71	364.64

* in gpm

** pump pressure in psi (efficiency assumed to be 80%)

AC/DC FORMULAS				
To Find	DC	AC		
		1φ, 115 or 220 V	1φ, 208, 230, or 240 V	3φ – All Voltages
I, HP known	$\frac{HP \times 746}{E \times E_{ff}}$	$\frac{HP \times 746}{E \times E_{ff} \times PF}$	$\frac{HP \times 746}{E \times E_{ff} \times PF}$	$\frac{HP \times 746}{1.73 \times E \times E_{ff} \times PF}$
I, kW known	$\frac{kW \times 1000}{E}$	$\frac{kW \times 1000}{E \times PF}$	$\frac{kW \times 1000}{E \times PF}$	$\frac{kW \times 1000}{1.73 \times E \times PF}$
I, kVA known		$\frac{kVA \times 1000}{E}$	$\frac{kVA \times 1000}{E}$	$\frac{kVA \times 1000}{1.763 \times E}$
kW	$\frac{I \times E}{1000}$	$\frac{I \times E \times PF}{1000}$	$\frac{I \times E \times PF}{1000}$	$\frac{I \times E \times 1.73 \times PF}{1000}$
kVA		$\frac{I \times E}{1000}$	$\frac{I \times E}{1000}$	$\frac{I \times E \times 1.73}{1000}$
HP (output)	$\frac{I \times E \times E_{ff}}{746}$	$\frac{I \times E \times E_{ff} \times PF}{746}$	$\frac{I \times E \times E_{ff} \times PF}{746}$	$\frac{I \times E \times 1.73 \times E_{ff} \times PF}{746}$

E_{ff} = efficiency

HORSEPOWER FORMULAS				
To Find	Use Formula	Example		
		Given	Find	Solution
HP	$HP = \frac{I \times E \times E_{ff}}{746}$	240 V, 20 A, 85% E_{ff}	HP	$HP = \frac{I \times E \times E_{ff}}{746}$ $HP = \frac{20\ A \times 240\ V \times 85\%}{746}$ $HP =$ **5.5**
I	$I = \frac{HP \times 746}{E \times E_{ff} \times PF}$	10 HP, 240 V, 90% E_{ff}, 88% PF	I	$I = \frac{HP \times 746}{E \times E_{ff} \times PF}$ $I = \frac{10\ HP \times 746}{240\ V \times 90\% \times 88\%}$ $I =$ **39 A**

VOLTAGE DROP FORMULAS – 1φ, 3φ					
Phase	To Find	Use Formula	Example		
			Given	Find	Solution
1φ	VD	$VD = \frac{2 \times R \times L \times I}{1000}$	240 V, 40 A, 60 L, .764 R	VD	$VD = \frac{2 \times R \times L \times I}{1000}$ $VD = \frac{2 \times .764 \times 60 \times 40}{1000}$ $VD =$ **3.67 V**
3φ	VD	$VD = \frac{2 \times R \times L \times I}{1000}$ .866	208 V, 110 A, 75 L, .194 R, .866 multiplier	VD	$VD = \frac{2 \times R \times L \times I}{1000} \times .866$ $VD = \frac{2 \times .194 \times 75 \times 110}{1000} \times .866$ $VD =$ **2.77 V**

$^{*}\frac{\sqrt{3}}{2} = .866$

POWERS OF 10						
1×10^{4}	=	10,000	=	$10 \times 10 \times 10 \times 10$		Read ten to the fourth power
1×10^{3}	=	1000	=	$10 \times 10 \times 10$		Read ten to the third power or ten cubed
1×10^{2}	=	100	=	10×10		Read ten to the second power or ten squared
1×10^{1}	=	10	=	10		Read ten to the first power
1×10^{0}	=	1	=	1		Read ten to the zero power
1×10^{-1}	=	.1	=	1/10		Read ten to the minus first power
1×10^{-2}	=	.01	=	$1/(10 \times 10)$ or 1/100		Read ten to the minus second power
1×10^{-3}	=	.001	=	$1/(10 \times 10 \times 10)$ or 1/1000		Read ten to the minus third power
1×10^{-4}	=	.0001	=	$1/(10 \times 10 \times 10 \times 10)$ or 1/10,000		Read ten to the minus fourth power

UNITS OF ENERGY

Energy	Btu	ft lb	J	kcal	kWh
British thermal unit	1	777.9	1.056	0.252	2.930×10^{-4}
Foot-pound	1.285×10^{-3}	1	1.356	3.240×10^{-4}	3.766×10^{-7}
Joule	9.481×10^{-4}	0.7376	1	2.390×10^{-4}	2.778×10^{-7}
Kilocalorie	3.968	3.086	4.184	1	1.163×10^{-3}
Kilowatt-hour	3.413	2.655×10^{6}	3.6×10^{6}	860.2	1

UNITS OF POWER

Power	W	ft lb/s	HP	kW
Watt	1	0.7376	$.341 \times 10^{-3}$	0.001
Foot-pound/sec	1.356	1	$.818 \times 10^{-3}$	1.356×10^{-3}
Horsepower	745.7	550	1	0.7457
Kilowatt	1000	736.6	1.341	1

STANDARD SIZES OF FUSES AND CBs

NEC® 240.6 (a) lists standard ampere ratings of fuses and fixed-trip CBs as follows:
15, 20, 25, 30, 35, 40, 45,
50, 60, 70, 80, 90, 100, 110,
125, 150, 175, 200, 225,
250, 300, 350, 400, 450,
500, 600, 700, 800,
1000, 1200, 1600,
2000, 2500, 3000, 4000, 5000, 6000

VOLTAGE CONVERSIONS

To Convert	To	Multiply By
rms	Average	.9
rms	Peak	1.414
Average	rms	1.111
Average	Peak	1.567
Peak	rms	.707
Peak	Average	.637
Peak	Peak-to-peak	2

AC MOTOR CHARACTERISTICS						
Motor Type 1ϕ	Typical Voltage	Starting Ability (Torque)	Size (HP)	Speed Range (rpm)	Cost*	Typical Uses
Shaded-pole	115 V, 230 V	Very low 50% to 100% of full load	Fractional ½ HP to ⅓ HP	Fixed 900, 1200, 1800, 3600	Very low 75% to 85%	Light-duty applications such as small fans, hair dryers, blowers, and computers
Split-phase	115 V, 230 V	Low 75% to 200% of full load	Fractional ⅓ HP or less	Fixed 900, 1200, 1800, 3600	Low 85% to 95%	Low-torque applications such as pumps, blowers, fans, and machine tools
Capacitor-start	115 V, 230 V	High 200% to 350% of full load	Fractional to 3 HP	Fixed 900, 1200, 1800	Low 90% to 110%	Hard-to-start loads such as refrigerators, air compressors, and power tools
Capacitor-run	115 V, 230 V	Very low 50% to 100% of full load	Fractional to 5 HP	Fixed 900, 1200, 1800	Low 90% to 110%	Applications that require a high running torque such as pumps and conveyors
Capacitor start-and-run	115 V, 230 V	Very high 350% to 450% of full load	Fractional to 10 HP	Fixed 900, 1200, 1800	Low 100% to 115%	Applications that require both a high starting and running torque such as loaded conveyors
3ϕ Induction	230 V, 460 V	Low 100% to 175% of full load	Fractional to over 500 HP	Fixed 900, 1200, 3600	Low 100%	Most industrial applications
Wound rotor	230 V, 460 V	High 200% to 300% of full load	½ HP to 200 HP	Varies by changing resistance in rotor	Very high 250% to 350%	Applications that require high torque at different speeds such as cranes and elevators
Synchronous	230 V, 460 V	Very low 40% to 100% of full load	Fractional to 250 HP	Exact constant speed	High 200% to 250%	Applications that require very slow speeds and correct power factors

* based on standard 3ϕ induction motor

DC AND UNIVERSAL MOTOR CHARACTERISTICS						
Motor Type	Typical Voltage	Starting Ability (Torque)	Size (HP)	Speed Range (rpm)	Cost*	Typical Uses
DC Series	12 V, 90 V, 120 V, 180 V	Very high 400% to 450% of full load	Fractional to 100 HP	Varies 0 to full speed	High 175% to 225%	Applications that require very high torque such as hoists and bridges
Shunt	12 V, 90 V, 120 V, 180 V	Low 125% to 250% of full load	Fractional to 100 HP	Fixed or adjustable below full speed	High 175% to 225%	Applications that require better speed control than a series motor such as woodworking machines
Compound	12 V, 90 V, 120 V, 180 V	High 300% to 400% of full load	Fractional to 100 HP	Fixed or adjustable	High 175% to 225%	Applications that require high torque and speed control such as printing presses, conveyors, and hoists
Permanent-magnet	12 V, 24 V, 36 V, 120 V	Low 100% to 200% of full load	Fractional	Varies from 0 to full speed	High 150% to 200%	Applications that require small DC-operated equipment such as automobile power windows, seats, and sun roofs
Stepping	5 V, 12 V, 24 V	Very low** .5 to 5000 oz/in.	Size rating is given as holding torque and number of steps	Rated in number of steps per sec (maximum)	Varies based on number of steps and rated torque	Applications that require low torque and precise control such as indexing tables and printers
AC/DC Universal	115 VAC, 230 VAC, 12 VDC, 24 VDC, 36 VDC, 120 VDC	High 300% to 400% of full load	Fractional	Varies 0 to full speed	High 175% to 225%	Most portable tools such as drills, routers, mixers, and vacuum cleaners

* based on standard 3ϕ induction motor
** torque is rated as holding torque

OVERCURRENT PROTECTION DEVICES						
Motor Type	Code Letter	FLC (%)				
		Motor Size	TDF	NTDF	ITB	ITCB
AC*	—	—	175	300	150	700
AC*	A	—	150	150	150	700
AC*	B – E	—	175	250	200	700
AC*	F – V	—	175	300	250	700
DC	—	⅛ to 50 HP	150	150	150	150
DC	—	Over 50 HP	150	150	150	175

* full voltage and resistor starting

ANNEALED SOLID COPPER WIRE CHARACTERISTICS						
AWG Gauge	Diameter (mils)	Cross-Sectional Area		Resistance*		Weight†
		Circular mils	Square inches	25°C	65°C	
0000	460.0	212,000.0	0.166	0.0500	0.0577	641.0
000	410.0	168,000.0	0.132	0.0630	0.0727	508.0
00	365.0	133,000.0	0.105	0.0795	0.0917	403.0
0	325.0	106,000.0	0.0829	0.100	0.116	319.0
1	289.0	83,700.0	0.0657	0.126	0.146	253.0
2	258.0	66,400.0	0.0521	0.159	0.184	201.0
3	229.0	52,600.0	0.0413	0.201	0.232	159.0
4	204.0	41,700.0	0.0328	0.253	0.292	126.0
5	182.0	33,100.0	0.0260	0.319	0.369	100.0
6	162.0	26,300.0	0.0206	0.403	0.465	79.5
7	144.0	20,800.0	0.0164	0.508	0.586	63.0
8	128.0	16,500.0	0.0130	0.641	0.739	50.0
9	114.0	13,100.0	0.0103	0.808	0.932	39.6
10	102.0	10,400.0	0.00815	1.02	1.18	31.4
11	91.0	8230.0	0.00647	1.28	1.48	24.9
12	81.0	6530.0	0.00513	1.62	1.87	19.8
13	72.0	5180.0	0.00407	2.04	2.36	15.7
14	64.0	4110.0	0.00323	2.58	2.97	12.4
15	57.0	3260.0	0.00256	3.25	3.75	9.86
16	51.0	2580.0	0.00203	4.09	4.73	7.82
17	45.0	2050.0	0.00161	5.16	5.96	6.20
18	40.0	1620.0	0.00128	6.51	7.51	4.92
19	36.0	1290.0	0.00101	8.21	9.48	3.90
20	32.0	1020.0	0.000802	10.4	11.9	3.09
21	28.5	810.0	0.000636	13.1	15.1	2.45
22	25.3	642.0	0.000505	16.5	19.0	1.94
23	22.6	509.0	0.000400	20.8	24.0	1.54
24	20.1	404.0	0.000317	26.2	30.2	1.22
25	17.9	320.0	0.000252	33.0	38.1	0.970
26	15.9	254.0	0.000200	41.6	48.0	0.769
27	14.2	202.0	0.000158	52.5	60.6	0.610
28	12.6	160.0	0.000126	66.2	76.4	0.484
29	11.3	127.0	0.0000995	83.4	96.3	0.384
30	10.0	101.0	0.0000789	105.0	121.0	0.304
31	8.9	79.7	0.0000626	133.0	153.0	0.241
32	8.0	63.2	0.0000496	167.0	193.0	0.191
33	7.1	50.1	0.0000394	211.0	243.0	0.152
34	6.3	39.8	0.0000312	266.0	307.0	0.120
35	5.6	31.5	0.0000248	335.0	387.0	0.0954
36	5.0	25.0	0.0000196	423.0	488.0	0.0757
37	4.5	19.8	0.0000156	533.0	616.0	0.0600
38	4.0	15.7	0.0000123	673.0	776.0	0.0476
39	3.5	12.5	0.0000098	848.0	979.0	0.0377
40	3.1	9.9	0.0000078	1230.0	1420.0	0.0299

* in Ω/1000′
† in lb/1000′

CHEMICAL ELEMENTS

Name	Symbol	Valence Electrons	Atomic Weight*	Atomic Number	Name	Symbol	Valence Electrons	Atomic Weight*	Atomic Number
Actinium	Ac	2	[227]	89	Neon	Ne	8	20.183	10
Aluminum	Al	3	26.9815	13	Neptunium	Np	2	[237]	93
Americium	Am	2	[243]	95	Nickel	Ni	2	58.71	28
Antimony	Sb	5	121.75	51	Niobium	Nb	1	92.906	41
Argon	Ar	8	39.948	18	Nitrogen	N	5	14.0067	7
Arsenic	As	5	74.9216	33	Nobelium	No	2	[255]	102
Astatine	At	7	[210]	85	Osmium	Os	2	190.2	76
Barium	Ba	2	137.34	56	Oxygen	O	6	15.9994	8
Berkelium	Bk	2	[247]	97	Palladium	Pd	—	106.4	46
Beryllium	Be	2	9.0122	4	Phosphorus	P	5	30.9738	15
Bismuth	Bi	5	208.980	83	Platinum	Pt	1	195.09	78
Boron	B	3	10.811	5	Plutonium	Pu	2	[244]	94
Bromine	Br	7	79.909	35	Polonium	Po	6	[210]	84
Cadmium	Cd	2	112.40	48	Potassium	K	1	39.102	19
Calcium	Ca	2	40.08	20	Praseodymium	Pr	2	140.907	59
Californium	Cf	2	[251]	98	Promethium	Pm	2	[145]	61
Carbon	C	4	12.01115	6	Protactinium	Pa	2	[231]	91
Cerium	Ce	2	140.12	58	Radium	Ra	2	[226]	88
Cesium	Cs	1	132.905	55	Radon	Rn	8	[222]	86
Chlorine	Cl	7	35.453	17	Rhenium	Re	2	186.2	75
Chromium	Cr	1	51.996	24	Rhodium	Rh	1	102.905	45
Cobalt	Co	2	58.9332	27	Rubidium	Rb	1	85.47	37
Copper	Cu	1	63.54	29	Ruthenium	Ru	1	101.07	44
Curium	Cm	2	[247]	96	Samarium	Sm	2	150.35	62
Dysprosium	Dy	2	162.50	66	Scandium	Sc	2	44.956	21
Einsteinium	Es	2	[254]	99	Selenium	Se	6	78.96	34
Erbium	Er	2	167.26	68	Silicon	Si	4	28.086	14
Europium	Eu	2	151.96	63	Silver	Ag	1	107.870	47
Fermium	Fm	2	[257]	100	Sodium	Na	1	22.9898	11
Fluorine	F	7	18.9984	9	Strontium	Sr	2	87.62	38
Francium	Fr	1	[223]	87	Sulfur	S	6	32.064	16
Gadolinium	Gd	2	157.25	64	Tantalum	Ta	2	180.948	73
Gallium	Ga	3	69.72	31	Technetium	Tc	2	[97]	43
Germanium	Ge	4	72.59	32	Tellurium	Te	6	127.60	52
Gold	Au	1	196.967	79	Terbium	Tb	2	158.924	65
Hafnium	Hf	2	178.49	72	Thallium	Tl	3	204.37	81
Helium	He	2	4.0026	2	Thorium	Th	2	232.038	90
Holmium	Ho	2	164.930	67	Thulium	Tm	2	168.934	69
Hydrogen	H	1	1.00797	1	Tin	Sn	4	118.69	50
Indium	In	3	114.82	49	Titanium	Ti	2	47.90	22
Iodine	I	7	126.9044	53	Tungsten	W	2	183.85	74
Iridium	Ir	2	192.2	77	Unnilennium	Une	2	[266]	109
Iron	Fe	2	55.847	26	Unnilhexium	Unh	2	[263]	106
Krypton	Kr	8	83.80	36	Unniloctium	Uno	—	[265]	108
Lanthanum	La	2	138.91	57	Unnilpentium	Unp	2	[262]	105
Lawrencium	Lr	2	[256]	103	Unnilquadium	Unq	2	[261]	104
Lead	Pb	4	207.19	82	Unnilseptium	Uns	2	[262]	107
Lithium	Li	1	6.939	3	Uranium	U	2	238.03	92
Lutetium	Lu	2	174.97	71	Vanadium	V	2	50.942	23
Magnesium	Mg	2	24.312	12	Xenon	Xe	8	131.30	54
Manganese	Mn	2	54.9380	25	Ytterbium	Yb	2	173.04	70
Mendelevium	Md	2	[258]	101	Yttrium	Y	2	88.905	39
Mercury	Hg	2	200.59	80	Zinc	Zn	2	65.37	30
Molybdenum	Mo	1	95.94	42	Zirconium	Zr	2	91.22	40
Neodymium	Nd	2	144.24	60					

* a number in brackets indicates the mass number of the most stable isotope

Glossary

A

abbreviation: A letter or combination of letters that represents a word.

accumulator: A device that stores fluid under pressure.

AC line filter: A device that absorbs electrical interference from motors, etc. before these signals can travel down the power line.

acronym: A word formed from the first letter(s) of a compound term.

AC sine wave: A symmetrical waveform that contains 360 electrical degrees.

actuator: 1. A device that moves the spool of a valve. **2.** A device that converts fluid energy into mechanical motion.

AC voltage: Voltage that reverses its direction of flow at regular intervals.

adjustable core coil: A coil with a means for varying the magnetic field.

aerial cable: Cable suspended in the air on poles or other overhead structures.

air compressor (pump): A device that increases the pressure of atmospheric air to the desired pressure of the receiver tank.

air core coil: A coil with an air core that develops a magnetic field when current passes through the coil.

air-core inductor: A coil of wire wrapped around a hollow core.

air muffler: A device that reduces the noise that is made when air passes through a restriction.

alternating current (AC): Current that reverses its direction of flow at regular intervals.

alternating current motor: A motor that operates on AC.

alternation: Half of a cycle.

ambient temperature: The temperature of the air surrounding a piece of equipment.

American National Standards Institute (ANSI): A national organization that helps identify industrial and public needs for standards.

ampere (A): The number of electrons passing a given point in one second.

amplification: The process of taking a small signal and increasing its size.

amplifier: An electric device that turns medium-level audio signals into strong audio signals that are sent to speakers.

amplifier power: The measure of power going into an amplifier to make it work.

amplitude: The distance that a vibrating object moves from its position of rest as it vibrates.

analog: The output (display) of circuits with continuously changing quantities that may have any value between the defined limits.

analog display: An electromechanical device that indicates readings by the mechanical motion of a pointer.

analog input: An input that sends a continuously changing variable into the system.

analog signal: A type of input signal to an electric motor drive that can be either varying voltage or varying current.

AND gate: A logic gate that provides logic level 1 only if all inputs are at logic level 1.

apparent power: The product of the voltage and current in a circuit calculated without considering the phase shift that may be present between the voltage and current in the circuit.

appliance (heater) plug: A plug used to power appliances that produce heat, such as electric grills, roasters, broilers, waffle irons, and large coffeemakers.

arc blast: An explosion that occurs when the air surrounding electrical equipment becomes ionized and conductive.

arc blast hood: An eye and face protection device that consists of a flame-resistant hood and face shield.

arc flash: An extremely high-temperature discharge produced by an electrical fault in the air.

architectural symbol: A symbol used on floor plans, site (plot) plans, and detail drawings to show the location of electrical and other devices.

arc tube: The light-producing element of an HID lamp.

area: The number of unit squares equal to the surface of an object.

arithmetic: The branch of mathematics that involves the computation (addition, subtraction, multiplication, and division) of positive real numbers.

armor-clad cable: A cable that has an outer cover made of flexible galvanized steel.

atom: The smallest particle that an element can be reduced to and still maintain the properties of that element.

audio signal: An electrical representation of a sound in the form of fluctuating voltage or current.

audio spectrum: The part of the frequency spectrum that humans can hear (20 Hz–20 kHz).

audio system: An arrangement of electronic components that is designed to reproduce sound.

autotransformer: A transformer that changes voltage level using the same common coil for both the primary and secondary sides.

avalanche current: Current passed when a diode breaks down.

average power rating: The amount of power a speaker can accept continuously without damage.

average value: The mathematical mean of all instantaneous voltage values in the sine wave.

B

backbone cable: A conductor (copper or fiber optic) used between telecommunications closets or floor distribution terminals and equipment rooms within buildings.

back-to-back bend: A bend that consists of two 90° bends on one piece of conduit.

bandwidth: The range of frequencies that a device can accept within tolerable limits.

bar graph: A graph composed of segments that function as an analog pointer.

base: The repeated factor that is multiplied to obtain a power.

battery: A DC voltage source that converts chemical energy to electrical energy.

bayonet base: A bulb base that has two pins located on opposite sides.

bel: The logarithm of an electrical, acoustical, or other power ratio.

belt drive: A turntable drive that has the motor connected to an inner platter through a belt and pulley arrangement.

bench testing: Testing performed when equipment under test is brought to a designated service area.

binary coded decimal (BCD) system: A coding system that represents each digit from 0 through 9 as a four-bit binary number.

binary numbering system: A base two numbering system that uses only two digits, 0 and 1.

binary signal: A signal that has only two states.

bipolar device: A device in which both holes and electrons are used as internal carriers for maintaining current flow.

bit: Each digit (0 or 1) of a binary number.

bobbin: A small paper, plastic, or metal cylinder around which the voice coil is wound.

branch circuit: The portion of an electrical circuit between the circuit breaker or fuse and the receptacles, lamps, and loads connected in the circuit.

break: A place on a contact that opens or closes an electrical circuit.

breakdown torque (BDT): The maximum torque a motor can produce without an abrupt reduction in motor speed.

brightness: The perceived amount of light reflecting from an object.

British thermal unit (Btu): The amount of heat required to raise the temperature of 1 lb of water 1°F.

brown-out: The deliberate reduction of the voltage level by a power company to conserve power during peak usage times.

bus: A large trace extending around the edge of a circuit board to provide conduction from several sources.

busway: A metal-enclosed distribution system of busbars available in prefabricated sections.

byte: A group of 8 bits (10001101, etc.).

C

cable: Two or more conductors grouped together within a common protective cover and used to connect individual components.

calorie: The amount of heat required to raise 1 g of water 1°C.

Canadian Standards Association (CSA): A Canadian organization similar to UL that tests equipment and products to ensure they meet national standards.

capacitance: The ability to store energy in the form of an electrical charge.

capacitive circuit: A circuit in which current leads voltage.

capacitive reactance: The opposition to current flow by a capacitor.

capacitor: An electric device designed to store electrical energy by means of an electrostatic field.

capacitor motor: A 1ϕ, AC motor that includes a capacitor in addition to the running and starting windings.

capacitor-run motor: A capacitor motor that has the starting winding and capacitor connected in series at all times.

capacitor start-and-run motor: A capacitor motor that has high- and low-value capacitors connected in parallel with each other and in series with the starting winding to provide a very high starting torque.

cartridge: A vibration sensor that holds a needle.

cascaded amplifiers: Two or more amplifiers connected to obtain the required gain.

cassette tape: A plastic film covered with a magnetic powder.

cathode: A tungsten coil coated with electron-emissive material which releases electrons when heated.

causes: Actions that produce effects.

cell: A unit that produces electricity at a fixed voltage and current level.

check valve: A valve that allows the fluid to flow in one direction only.

circle: A plane figure generated about a centerpoint.

circuit breaker: An OCPD with a mechanical mechanism that may manually or automatically open the circuit when an overload condition or short circuit occurs.

circular mil: A measurement used to determine the cross-sectional area of a conductor.

circumference: The boundary of a circle.

cladding: A layer of glass or other transparent material surrounding the fiber core and used as optical insulation.

clamp-on ammeter: A meter that measures current in a circuit by measuring the strength of the magnetic field around a conductor.

class: The hazardous location category that refers to the generic hazardous material present.

clipping: A sharp clicking that occurs when a speaker cone cannot move as far as required by the audio signal.

clock radio: A small electric device that keeps time and has a tuner to receive broadcast signals.

code: A regulation or minimum requirement.

coefficient: A factor that is multiplied by the base number.

coil: A winding consisting of insulated conductors arranged to produce magnetic flux.

cold resistance: The resistance of a component when operating current is not passing through the device.

cold solder joint: A poor electrical and mechanical connection caused by insufficient heat during soldering or connecting parts moving after the solder is applied but before it has cooled.

color rendering: The appearance of a color when illuminated by a light source.

common noise: Noise produced between the ground and hot or the ground and neutral lines.

communication cable: Cable (usually copper or fiber optic) used to transmit data from one location (or device) within a system to another location.

compact disc (CD) player: An electric device that plays music stored on a compact disc (CD) by means of reflecting light from a laser beam.

compact fluorescent lamp: A fluorescent lamp that has a smaller diameter than a conventional fluorescent lamp and a folded bulb configuration.

complementary metal-oxide semiconductor (CMOS) ICs: A group of ICs that employ MOS transistors.

compound: A combination of the atoms of two or more elements.

compound proportion: A proportion in which some terms are products of two variables.

compression: An area of increased pressure in a sound wave produced when a vibrating object moves outward.

concert theater sound system: A large audio system used to project high-quality sound for a theater.

conductor: Material that has very little resistance and permits electrons to move through it easily.

conduit: A hollow pipe used to protect conductors.

constant: A number with a fixed value.

constant-wattage autotransformer ballast: A high-reactance autotransformer ballast with a capacitor added to the circuit.

contactor: A control device that uses a small control current to energize or de-energize the load connected to it.

contacts: The conducting part of a switch that operates with another conducting part to make or break a circuit.

contact tachometer: A device that measures the rotational speed of an object through direct contact of the tachometer tip with the object to be measured.

contact thermometer: An instrument that measures temperature at a single point.

continuous quantities: The numbers resulting from measurements that cannot be directly counted.

contrast: The ratio of brightness between different objects.

control circuit transformer: A transformer used to provide power for a machine or system control circuit.

control switch: A switch that controls the flow of current in a circuit.

control valve: A valve that sets the flow or pressure of fluid or air in the circuit.

convection: Heat transfer that occurs when currents circulate between warm and cool regions of a fluid (air).

conventional current flow: Current flow from positive to negative.

converter: An electronic device that changes AC voltage into DC voltage.

copper-clad aluminum: A conductor that has copper bonded to an aluminum core.

copper loss: Loss caused by the resistance of the copper wire to the flow of current.

cord: Two or more conductors grouped together and used to deliver power to a load by means of a plug.

crimped connection: A connection in which two wires are held together by crimping them in a specially designed fitting.

crossover network: An electrical device used to limit the frequencies sent to a particular speaker in a multiple speaker system.

cube root: A number which has three equal roots.

current: The amount of electrons flowing through an electrical circuit.

current transformer: A transformer that measures the amount of current flowing through a wire by measuring the strength of the magnetic field around the wire.

cutoff region: The point at which the transistor is turned OFF and no current flows.

cycle: One complete positive and negative alternation of a wave form.

cylinder: A device that converts fluid energy into a linear mechanical force.

D

data interface module: A device that converts input signals into a signal that can be sent over a network.

DC compound motor: A DC motor that has the series field connected in series with the armature and the shunt field connected in parallel with the series field and armature.

DC permanent-magnet motor: A DC motor that has only armature connections and no field connections.

DC series motor: A DC motor that has the series field coils connected in series with the armature.

DC shunt motor: A DC motor that has the field connected in parallel with the armature.

DC voltage: Voltage that flows in one direction only.

decibel (dB): The unit used to measure the intensity (volume) of sound.

decimal: A number expressed in base 10.

decimal fraction: A fraction with a denominator of 10, 100, 1000, etc.

decimal point: The period at the left of a proper decimal number or the period that separates the parts of a mixed decimal number.

delta configuration: A transformer connection that has each transformer coil connected end-to-end to form a closed loop.

denominator: A number in a fraction that indicates the number of equal parts into which a unit is divided.

detent actuator: A device that automatically holds a valve in its last position.

diac: A thyristor that triggers in either direction when its breakover voltage is exceeded.

diameter: The distance from circumference to circumference through the centerpoint of a circle.

digital: The output (display) of specific quantities that change in discrete steps, such as the seconds, minutes, and hours displayed on a digital watch.

digital audio player: A device that stores and plays digital music files.

digital display: An electronic device that displays readings as numerical values.

digital IC: An IC that contains a switching integrated circuit that processes finite (zero and maximum) signal levels.

digital input: An input that is either ON or OFF.

digital logic gate: A circuit that performs a special logic operation such as AND, OR, NOT, NOR, NAND, and exclusive OR.

digital logic probe: A special DC voltmeter that detects the presence or absence of a signal.

dimmer switch: A switch that changes the lamp brightness by changing the voltage applied to the lamp.

diode: A semiconductor device that offers very high opposition to current flow in one direction and very low opposition to current flow in the opposite direction.

direct current (DC): Current that flows in only one direction.

direct drive: A turntable drive system in which the motor and platter are combined into one mechanical part.

directional control valve: A valve that connects, disconnects, or directs the fluid pressure from one part of a circuit to another.

direct proportion: A statement of equality between two ratios in which the first of four terms divided by the second equals the third divided by the fourth.

disconnect switch (disconnect): A switch that disconnects electrical circuits from motors and machines.

discrete: Refers to unconnected elements that constitute a separate entity.

dispersion: The scattering of sound waves.

distortion: Any unintentional change in the characteristics of an audio signal.

division: The hazardous location category that refers to the probability that a hazardous material is present.

doping: The process by which the crystal structure of a semiconductor is altered.

double offset (saddle) bend: A bend that consists of four 45° offset bends on one piece of conduit.

dual-function switch: A switch that performs two different switching functions such as forward or reverse.

dual-voltage motor: A motor that operates at more than one voltage level.

dual-voltage, split-phase motor: A motor that can be connected for either of two voltages.

duplex receptacle: A receptacle that has two spaces for connecting two different plugs.

dynamic microphone: A microphone that senses sound using a magnet and a coil of wire.

E

earmuff: An ear protection device worn over the ears.

earplug: An ear protection device made of moldable rubber, foam, or plastic and inserted into the ear canal.

eddy-current loss: Loss caused by the induced currents that are produced in metal parts that are being magnetized.

edge card: A PC board with multiple terminations (terminal contacts) on one end.

effects: The results of actions or causes.

electret condenser microphone: A microphone in which sound waves vibrate a diaphragm in the microphone next to a charged capacitor.

electric arc: A discharge of electric current across an air gap.

electric discharge lamp: A lamp that produces light by an arc discharged between two electrodes.

electricity: The movement of electrons from atom to atom.

electric motor: A rotating output device that converts electrical power into a rotating, mechanical force.

electric motor drive: An electronic device that controls the direction, speed, torque, and other operating functions of an electric motor in addition to providing motor protection and monitoring functions.

electrical enclosure: A housing that protects wires and equipment and prevents personnel injury by accidental contact with a live circuit.

electrical noise: Any unwanted disturbance that interferes with a signal.

electrical resistance: Opposition to the flow of electrons through any material.

electrodynamic speaker: A speaker that uses a magnet and a coil of wire to move the cone to produce sound.

electromagnetism: The magnetic field produced when electricity passes through a conductor.

electromechanical relay: A switching device that has sets of contacts which are closed by a magnetic force.

electron: A negatively charged particle in an atom.

electron current flow: Current flow from negative to positive.

element: A substance that cannot be chemically broken down and contains atoms of only one variety.

energy: The capacity to do work.

envelope: A pictorial representation of what is occurring inside a valve for one given position.

erase head: A electromagnet that erases all magnetic vibrations on a tape.

even number: Any number that can be divided by 2 an exact number of times.

exclusive OR gate: A logic gate that provides a logic level 1 only if one, but not both of the inputs are at logic level 1.

exponent: A number that indicates the number of times a base is raised to a power (number of times used as a factor).

extremes: The two outer numbers of a proportion.

F

face shield: An eye and face protection device that covers the entire face with a plastic shield and is used for protection from flying objects.

factors: Two or more numbers multiplied together to give a product.

fan-out: The number of loads that can be driven by the output of a logic gate.

ferrite: A chemical compound consisting of powdered iron oxide and ceramic.

ferrite-core inductor: A coil of wire wrapped around a ceramic material (ferrite).

ferromagnetic materials: Materials, such as soft iron, that are easily magnetized.

fiber optics: The method of using light to transport information from one location to another location through thin filaments of glass or plastic.

filament: A conductor with a resistance high enough to cause the conductor to heat.

filter: A device that removes solid contaminants from a fluid power system.

filter with drain: A device that removes contaminants and moisture from a fluid power system.

first aid: Help for a victim immediately after an injury and before professional medical help arrives.

fixed capacitor: A capacitor that has one value of capacitance.

fixed displacement fluid power motor: A fluid power motor that provides rotary motion and drives a load in one direction.

fixed displacement pump: A pump that develops a fixed amount of flow for a given pump speed.

fixed resistor: A resistor with a set value, such as 100 Ω.

fixture splice: A splice that connects two wires in which one of the wires is smaller than the other, or one wire is stranded and the other is solid.

flash protection boundary: The distance at which PPE is required to prevent burns when an arc occurs.

floating input: A digital input that is not a true high or low at all times.

floodlight: A lamp that casts general light over a large area.

flow control valve: A valve used to control the volume of fluid that flows in a part of the circuit.

flow switch: A switch that senses the movement of a fluid.

fluid: A liquid or a gas that can move and change shape without separating when under pressure.

fluid actuator: A device that converts fluid power to mechanical power.

fluid conditioner: A device that maintains clean fluid in the system.

fluid power: The transmission and control of energy by means of a pressurized fluid.

fluid power motor: A motor that converts pressurized fluid force into rotating motion.

fluid power oscillator: A device that moves in fixed increments each time fluid pressure is applied.

fluid pump: A pump that converts mechanical power to hydraulic or pneumatic power by pushing the fluid into the system.

fluorescent lamp: A low-pressure discharge lamp in which ionization of mercury vapor transforms ultraviolet energy generated by the discharge into light.

foot actuator: An actuator that changes the valve position manually by using a person's foot.

footcandle (fc): The amount of light produced by a lamp (lumens) divided by the area that is illuminated.

foot switch: A control switch that is operated by a person's foot.

force: Any cause that changes the position, motion, direction, or shape of an object.

forced-air cooled transformers: Transformers that use a fan to move air over the transformer.

forward-bias voltage: The application of the proper polarity to a diode.

forward breakover voltage: The voltage required to switch the SCR into a conductive state.

fossil fuel: A fuel formed by plant and animal remains taken from the ground.

four-way switch: A double-pole, double-throw (DPDT) switch that changes the electrical connections inside the switch from straight to diagonal.

fraction: A part of a whole unit or number.

frequency: The number of air pressure fluctuation cycles produced per second.

frequency response: The variation of the speaker output over a frequency range.

frequency spectrum: The range of all possible frequencies.

full-load torque (FLT): The torque required to produce the rated power at full speed of the motor.

full-range speaker: A speaker designed to adequately reproduce most of the audio spectrum.

fuse: An OCPD with a fusible link that melts and opens the circuit when an overload condition or short circuit occurs.

G

gain: A ratio of the amplitude of the output signal to the amplitude of the input signal.

gas: A state of matter that is fluid, has a relatively low density, and is highly compressible.

general wiring: The wiring used to connect electrical components in a circuit.

ghost voltage: A voltage that appears on a meter that is not connected to a circuit.

graph: A diagram that shows the continuous relationship between two or more variables.

ground: The connection of all exposed noncurrent-carrying metal parts to the earth.

ground-fault circuit interrupter (GFCI): An electric device which protects personnel by detecting ground faults and quickly disconnecting power from the circuit.

grounding: The connection of all exposed noncurrent-carrying metal parts to earth.

grounding conductor: A conductor that does not normally carry current, except during a fault (short circuit).

H

hard-wiring: The physical connection of electrical components.

harmonic: An overtone whose frequency is a multiple of the fundamental frequency.

harmonic distortion: The presence of frequencies in the speaker output which are not in the electrical input.

hasp: A multiple-lockout/tagout device.

heat: Thermal energy.

heat exchanger: A device that removes heat from the fluid in a hydraulic system.

heating element: A conductor that offers enough resistance to produce heat when connected to an electrical power supply.

heat sink: A device that conducts and dissipates heat away from a component.

heavy-duty plug: A plug that is used on high-wattage appliances and devices that operate on 230 V or 460 V.

hertz: The international unit of frequency, which is equal to one cycle per second.

hexadecimal numbering system: A base 16 numbering system that uses digits 0, 1, 2, 3, 4, 5, 6, 7, 8, 9 and the first six letters of the alphabet, A, B, C, D, E, and F.

hickey (reducer): A mounting device that screws onto the threaded fixture stud and takes a threaded nipple.

high-intensity discharge (HID) lamp: A lamp that produces light from an arc tube.

high-power load: A load whose voltage and current rating is greater than the voltage and current rating of the switch.

high-pressure sodium lamp: An HID lamp that produces light when current flows through sodium vapor under high pressure and high temperature.

high-reactance autotransformer ballast: A ballast that uses two coils (primary and secondary) to regulate both voltage and current.

holding current: The minimum current necessary for an SCR to continue conducting.

holes: The missing electrons in the crystal structure of a semiconductor.

horsepower (HP): A unit of power equal to 746 W or 33,000 lb-ft per minute (550 lb-ft per second).

hot resistance: The actual (true) resistance of a component when operating current is passing through the device.

hydraulics: The transmission of energy using a liquid (normally oil).

hysteresis loss: Loss caused by magnetism that remains (lags) in a material after the magnetizing force has been removed.

I

illumination: The effect that occurs when light falls on a surface.

impedance: The total opposition of any combination of resistance, inductive reactance, and capacitive reactance offered to the flow of alternating current.

improper fraction: A fraction that has a numerator larger than its denominator.

incandescent lamp: An electric lamp that produces light by the flow of current through a tungsten filament inside a gas-filled, sealed glass bulb.

index: The small figure to the left of the radical sign ($\sqrt{}$).

index pin: A metal extension from the transistor case.

inductance: The property of an electric device that opposes a change in current due to its ability to store electrical energy in a magnetic field.

induction motor: A motor that has no physical electrical connection to the rotor.

inductive circuit: A circuit in which current lags voltage.

inductive reactance: An inductor's opposition to alternating current.

infrared meter: A meter that measures heat energy by measuring the infrared energy that a material emits.

infrared region: The region of the spectrum with wavelengths slightly longer than the color red.

initial lumen: The amount of light produced when a lamp is new.

in-line ammeter: A meter that measures current in a circuit by inserting the meter in series with the component(s) under test.

in-phase: The state when voltage and current reach their maximum amplitude and zero level simultaneously.

input/output network: A network that allows analog and digital inputs to send information in the system.

instant-start circuit: A fluorescent lamp-starting circuit that provides sufficient voltage to strike an arc instantly.

instrument level: The signal level produced by instruments such as guitars and old keyboards.

insulator: A material that has a very high resistance and resists the flow of electrons.

integrated circuits (ICs): Electronic devices in which all components (transistors, diodes, and resistors) are contained in a single package or chip.

intensity: The level of brightness.

intercom: An audio system that allows users to verbally communicate from different stations (locations/rooms), monitor certain stations, and play music throughout a building or at certain stations.

interface: A device that allows different levels or types of components to be used together in the same circuit.

inverse proportion: A statement of equality between two ratios in which an increase in one quantity results in a proportional decrease in the other related quantity.

inverse square law: States that the amount of illumination on a surface varies inversely with the square of the distance from the light source.

inverter: An electronic device that changes DC voltage into AC voltage.

invisible light: The portion of the electromagnetic spectrum on either side of the visible light spectrum.

iron core coil: A coil with an iron core added to increase the strength of the magnetic field.

iron-core inductor: A coil of wire wrapped around a laminated steel core.

isolated grounded receptacle: An electrical receptacle that minimizes electrical noise by providing a separate grounding path for each receptacle.

J

jogging: The frequent starting and stopping of a motor for short periods of time.

junction field-effect transistor (JFET): A device in which output current is controlled by the voltage on the input.

K

kinetic energy: The energy of motion.

knee pad: A rubber, leather, or plastic pad strapped onto the knees for protection.

knuckle thread: A rounded thread normally rolled from sheet metal and used in various forms for electric bulbs and bottle caps.

L

lamp: An output device that converts electrical energy into light.

lateral service: Electrical service in which service-entrance conductors are run underground from the utility service to the dwelling.

law of charges: A law that states that opposite charges attract and like charges repel.

level switch: A switch that measures and responds to the level of a material.

lever actuator: An actuator that changes the valve position using a lever.

light: That portion of the electromagnetic spectrum which produces radiant energy.

light emitting diode (LED): A diode that emits light when forward current is applied.

limit switch: A switch that converts mechanical motion into an electrical signal.

linear IC: An IC that contains an amplifying integrated circuit that produces an output signal that is proportional to the applied input signal.

linear scale: A scale that is divided into equally spaced segments.

line frequency: The number of cycles of supplied voltage per second.

line (ladder) diagram: A diagram which shows, with single lines and symbols, the logic of an electrical circuit or system of circuits and components.

line level: The signal level produced by audio devices such as tape decks, PA systems, tuners, VCRs, and other devices.

line matching transformer: An adapter containing a transformer that changes the three-wire balanced system to a two-wire unbalanced system.

liquid: A state of matter that has a definite volume but not a definite shape.

liquid-immersed/forced-air cooled transformer: A transformer that uses a liquid (oil) and fans to cool the transformer.

liquid-immersed/self-cooled transformer: A transformer that uses a liquid (refined oil or synthetic oil) to help cool the transformer windings.

load: 1. Any device that converts electrical energy to light, heat, motion, or sound. **2.** A device that receives a signal from the source.

local area network (LAN): A communication network that spans a relatively small area.

locked rotor torque (LRT): The torque a motor produces when the rotor is stationary and full power is applied to the motor.

lockout: The process of removing the source of electrical power and installing a lock which prevents the power from being turned ON.

lockout devices: Lightweight enclosures that allow the lockout and tagout of standard control devices.

lowest common denominator (LCD): The smallest number into which the denominators of a group of two or more fractions divides an exact number of times.

low-power load: A load in which the voltage and current rating of the load is less than the voltage and current rating of the switch.

low-pressure sodium lamp: An HID lamp that operates at a low vapor pressure and uses sodium as the vapor.

lubricator: A device that adds lubrication to pneumatic systems.

lubricator with drain: A device that adds lubrication and removes moisture from a pneumatic system.

lumen (lm): The unit used to measure the total amount of light produced by a light source.

luminaire temperature: The temperature at which a lamp delivers its peak light output.

M

magnet: A device that attracts iron and steel because of the molecular alignment of its material.

magnetic flux: The invisible lines of force that make up the magnetic field.

magnetism: A force that interacts with other magnets and ferromagnetic materials.

manual shutoff valve: A valve that is manually operated to shut OFF or apply fluid power.

mathematics: The science of numbers and their operations, interrelations, and combinations.

matter: Anything that has mass and occupies space.

mean lumen: The average light produced after a lamp has operated for approximately 40% of its rated life.

means: The two inner numbers of a proportion.

megohmmeter: A device that detects insulation deterioration by measuring high resistance values under high test voltage conditions.

memory: A control function that keeps a motor running after the start pushbutton is released.

mercury-vapor lamp: An HID lamp that produces light by an electrical discharge through mercury vapor.

metal halide: An element (normally sodium and scandium iodide) which is added to the mercury in small amounts.

metal-halide lamp: An HID lamp that produces light by an electrical discharge through mercury vapor and metal halide in the arc tube.

microphone: An electric device that converts sound vibrations to weak electrical vibrations.

microphone level: The signal level produced by a standard microphone.

midrange: A speaker designed to reproduce sounds from about 1 kHz to 10 kHz.

mixed decimal number: A decimal number that has a whole number and a decimal number separated by a decimal point.

molecule: The smallest particle that a compound can be reduced to and still possess the chemical properties of the original compound.

motion sensor: An infrared sensing device that detects the movement of a temperature variance and automatically switches when the movement is detected.

motor: A machine that develops torque (rotating mechanical force) on a shaft which is used to produce work.

motor efficiency: The measure of the effectiveness with which a motor converts electrical energy to mechanical energy.

motor starter: An electrically-operated switch (contactor) that includes motor overload protection.

motor torque: The force that produces or tends to produce rotation in a motor.

multimeter: A meter that is capable of measuring two or more quantities.

mutual inductance: The effect of one coil inducing a voltage into another coil.

N

NAND gate: A logic gate that provides a logic level 0 only if all inputs are at logic level 1.

National Electrical Manufacturers Association (NEMA): A national organization that assists with information and standards concerning proper selection, ratings, construction, testing, and performance of electrical equipment.

National Fire Protection Association (NFPA): A national organization that provides guidance in assessing the hazards of the products of combustion.

negative charge: An electrical charge produced when there are more electrons than normal.

negative resistance characteristic: The characteristic that current decreases with an increase of applied voltage.

neon test light: A test light that has a bulb that is filled with neon gas and uses two electrodes to ionize the gas (excite the atoms).

network: A system of computers, terminals, and databases connected by communication lines.

neutral conductor: A current-carrying conductor that is intentionally grounded.

neutron: A neutral particle, with a mass approximately the same as a proton, that exists in the nucleus of an atom.

nibble: A group of 4 bits (1001, 0011, etc.).

nonlinear scale: A scale that is divided into unequally spaced segments.

nonmetallic cable: A tough, plastic-sheathed cable that is normally made of a moisture-resistant, flame-retardant material.

non-positive displacement pump: A pump that circulates or transfers fluid.

non-time delay fuse (NTDF): A fuse that may detect an overcurrent and open the circuit almost instantly.

NOR gate: A logic gate that provides a logic level 0 if one or more inputs are at logic level 1.

normally closed (NC) valve: A valve that does not allow pressurized fluid to flow out of the valve in the spring-actuated position.

normally open (NO) valve: A valve that allows pressurized fluid to flow out of the valve in the spring-actuated position.

NOT gate: A logic gate that provides an output that is the opposite of the input.

NPN transistor: A transistor that is formed by sandwiching a thin layer of P-type material between two layers of N-type material.

NTC thermistor: A thermistor that has a decreasing resistance value with an increase in temperature.

N-type material: Material created by doping a region of a crystal with atoms from an element that has more electrons in its outer shell than the crystal.

nuclear fission: The splitting of atomic nuclei, resulting in the release of large amounts of energy.

nucleus: The heavy, dense center of an atom which contains protons and neutrons and has a positive electrical charge.

numerator: A number in a fraction that indicates how many of these parts are used.

O

Occupational Safety and Health Administration (OSHA): A federal agency that requires all employers to provide a safe environment for their employees.

octal numbering system: A base eight numbering system that uses only digits 0, 1, 2, 3, 4, 5, 6, and 7.

odd number: Any number that cannot be divided by 2 an exact number of times.

OFF-delay timer: A timer that delays for a predetermined time period when the timer no longer receives a signal to stay activated or is turned OFF before the timer contacts return to their normal condition.

offset bend: A bend that consists of two 45° angles on one piece of conduit.

ohmmeter: A device that is used to measure the amount of resistance in a component (or circuit) that is not powered.

Ohm's law: The relationship between voltage, current, and resistance in a circuit.

ON-delay timer: A timer that delays for a predetermined time after receiving a signal to activate or turn ON before the timer's contacts are changed from their normal condition.

op-amp: A general-purpose, high-gain, DC or AC amplifier.

OR gate: A logic gate that provides a logic level 1 if one or more inputs are at logic level 1.

oscilloscope: An instrument that displays an instantaneous voltage.

output: A device that is used to produce work, light, heat, sound, or display information.

overcurrent: A condition that exists on an electrical circuit when the normal load current is exceeded.

overcurrent protection device (OCPD): A disconnect switch with circuit breakers (CBs) or fuses added to provide overcurrent protection of the switched circuit.

overhead service: Electrical service in which service-entrance conductors are run from the utility pole through the air and to the dwelling.

overload: The condition that occurs when circuit current rises above the normal current level at which the load and/or circuit is designed to operate.

P

pads: Small round conductors to which component leads are soldered.

palm button actuator: A device that is manually operated by a person's hand to change the valve position.

parallel connection: A connection that has two or more components connected so that there is more than one path for current flow.

patch cord: A cable that consists of two wires: a thin center conductor that carries the signal and a woven shield wire that encircles it.

PC board: An insulating material such as fiberglass or phenolic with conducting paths laminated to one or both sides of the board.

peak inverse voltage (PIV): The maximum reverse bias voltage that a diode can withstand.

peak load: The maximum output required of a transformer.

peak power rating: The maximum power the speaker can take for a very short time.

peak-to-peak value: The value measured from the maximum positive alternation to the maximum negative alternation.

peak value: The maximum value of either the positive or negative alternation.

percentage: A fraction that has a denominator of 100.

period: The time required to produce one complete cycle of a sound wave.

permanent magnets: Magnets that hold their magnetism for a long period of time.

personal protective equipment (PPE): Clothing and/or equipment used by a technician to reduce the possibility of injury in the work area.

phase: The time relationship of a sine (or sound) wave to a known time reference.

phase shift: The state when voltage and current in a circuit do not reach their maximum amplitude and zero level simultaneously.

photoconductive diode: A diode that conducts current when energized by light.

photo tachometer: A device that measures the speed of an object without direct contact with the object.

phototransistor: A transistor that controls the amount of current flowing through the emitter/base junction based on the amount of light.

photovoltaic cell (solarcell): A device that converts solar energy into electrical energy.

pigtail splice: A splice that consists of twisting two wires together.

pilot actuator: An actuator that changes the valve position with air or oil force.

pilot light: A light that provides illumination or gives a visual indication of the presence or absence of power in part of a circuit.

pitch: The highness or lowness of a sound.

plant information network: A network that allows information to flow up from the plant floor and down from engineering, planning and scheduling, quality control, management, and outside sources.

platter: The flat disk that rotates a record.

playback head: An electromagnet which senses the changing magnetism on a magnetic tape and converts it into a weak electrical signal.

plot plan: an aerial view of one building lot and provides specific information about the lot.

plug: The device at the end of a cord that connects the device to the electrical power supply by means of a receptacle.

pneumatics: The transmission of energy using air.

PN junction: The area on a semiconductor material between the P-type and N-type material.

PNP transistor: A transistor formed by sandwiching a thin layer of N-type material between two layers of P-type material.

polarity: The positive (+) or negative (–) state of an object.

polarized plug: A plug in which one blade is wider than the other blade.

pole: A completely isolated circuit that a relay can switch.

portable audio system (boombox): A small, lightweight electric device that contains an amplifier, tuner, tape player, and speakers in one portable enclosure.

position: The number of positions within the valve in which the spool is placed to direct fluid through the valve.

positive charge: An electrical charge produced when there are fewer electrons than normal.

positive displacement pump: A pump that delivers a definite quantity of fluid for every stroke, revolution, or cycle.

potential energy: Stored energy a body has due to its position, chemical state, or condition.

power: **1.** The product of a repeated factor. **2.** The rate of doing work or using energy.

power factor: The ratio of true power used in an AC circuit to apparent power delivered to the circuit.

power formula: The relationship between power (P), voltage (E), and current (I) in an electrical circuit.

power source: A device that converts various forms of energy into electricity.

preamplifier: An electric device that accepts low-level audio signals like those from microphones and turntables and amplifies them to medium-level audio signals that are sent to an amplifier.

preheat circuit: A fluorescent lamp-starting circuit that heats the cathode before an arc is created.

pressure gauge: A device used to indicate pressure at given points in the system.

pressure-reducing valve: A valve that reduces the pressure in one part of a circuit.

pressure relief valve: A valve that limits the amount of pressure in a fluid power system.

pressure switch: A switch that opens or closes contacts in response to pressure changes in media such as air, water, or oil.

primary coil (input side): The coil of a transformer to which the voltage is connected.

primary division: A division with a listed value.

prime number: A number that can be divided an exact number of times, only by itself and the number 1.

produced electricity: The electricity produced either by magnetism (generator), chemicals (battery), light (photocell), heat (thermocouple), or pressure (crystals).

production control network: A network used to provide increased control of an input/output network.

programmable logic controller (PLC): A solid-state control device that can be programmed to automatically control electrical systems in residential, commercial, and industrial facilities.

proper decimal number: A decimal number that has no whole numbers.

proper fraction: A fraction that has a numerator smaller than its denominator.

proportion: An expression indicating that two ratios are equal.

protective helmet: A hard hat that is used in the workplace to prevent injury from the impact of falling and flying objects, and from electrical shock.

proton: A particle with a positive electrical charge of one unit that exists in the nucleus of an atom.

proximity switch: A switch that detects the presence or absence of an object without touching the object.

PTC thermistor: A thermistor that has an increasing resistance value with an increase in temperature.

P-type material: Material with empty spaces (holes) in its crystalline structure.

public address (PA) system: An electronic amplification system used as a communication system in public areas.

pull-down resistor: A resistor that has one side connected to ground at all times and the other side connected to the unused gate(s).

pull-up resistor: A resistor that has one side connected to the power supply at all times and the other side connected to the unused gate(s).

pull-up torque (PUT): The torque required to bring a load up to its rated speed.

pushbutton: A switch that makes (NO) or breaks (NC) a circuit when manually pressed.

R

radicand: The number that appears under the radical sign.

rapid-start circuit: A fluorescent lamp-starting circuit that has separate windings to provide continuous heating voltage on the lamp cathodes.

rarefaction: An area of reduced pressure in a sound wave produced when a vibrating object moves inward.

ratio: The relationship between two quantities or terms.

reactor ballast: A ballast that connects a coil (reactor) in series with the power line leading to the lamp.

real number: Any number associated with a point on a line.

receiver: A tuner and amplifier contained in the same enclosure.

receptacle (outlet): A device used to connect equipment with a cord and plug to an electrical system.

receptacle tester: A device that is plugged into a standard receptacle to determine if the receptacle is properly wired and energized.

record: A vinyl disk that contains small grooves which contain recorded music.

record head: An electromagnet that converts electrical vibrations into magnetic vibrations.

rectangle: A four-sided figure with opposite sides equal and four 90° angles.

rectification: The process of changing AC into DC.

rectifier: A device that converts AC voltage to DC voltage by allowing the voltage and current to move in only one direction.

reducing fractions: The process of changing a fraction from one fractional form to another.

relation: How much larger or smaller one term is in comparison with another term.

relay: A device that controls one electrical circuit by opening and closing contacts in another circuit.

resistance: **1.** The opposition to current flow. **2.** Any force that tends to hinder the movement of an object.

resistive circuit: A circuit that contains only resistance.

resistivity: The resistance of a conductor having a specific length and cross-sectional area.

resistor: A device that limits the current flowing in an electronic circuit.

reverse-bias voltage: The application of the opposite polarity to a diode.

reversible fluid power motor: A fluid power motor that can drive a load in two directions.

rim drive: An inexpensive drive system in which a motor drives an idler wheel.

ring conductor: The second wire in a pair of wires.

ripcord: A cord included in a cable that aids in removing the outer jacket.

root: The number which must be multiplied by itself a given number of times to obtain the original number.

root-mean-square (effective) value: 1. The value that produces the same amount of heat in a pure resistive circuit as a DC of the same value. **2.** A mathematical expression equal to .707 times the peak value of a waveform.

rosin-core solder: Electrical solder that contains a rosin core.

rotary actuator (fluid power oscillator): An actuator that moves back and forth over a fixed arc that is less than one complete revolution (360°).

rotor: The rotating part of an AC motor.

rounding: The process of obtaining a number to a required degree of accuracy.

rubber insulating matting: A floor covering that provides technicians protection from electrical shock when working on live electrical circuits.

S

safety glasses: An eye protection device with special impact-resistant glass or plastic lenses, reinforced frames, and side shields.

saturation region: The maximum current that can flow in the transistor circuit.

scope: A device that gives a visual display of voltages.

scopemeter: A combination oscilloscope and digital multimeter.

secondary coil (output side): The coil of a transformer in which the voltage is induced.

secondary division: A division that divides primary divisions in halves, thirds, fourths, fifths, etc.

selector switch: A switch with an operator that is rotated (instead of pushed) to activate the electrical contacts.

self-air cooled transformer: Transformer that dissipates heat through the air surrounding the transformer.

semiconductor: An electronic device that has electrical conductivity between that of a conductor (high conductivity) and that of an insulator (low conductivity).

semiconductor devices: Devices in which electrical conductivity is between that of a conductor (high conductivity) and that of an insulator (low conductivity).

separate component system: An audio system that contains individual, stand-alone components that are interconnected to allow the playing of the various audio formats.

sequence valve: A valve that directs fluid from one part of a circuit to another part only after the pressure has reached a set value.

series connection: A connection that has two or more components connected so there is only one path for current flow.

series/parallel connection: A combination of series- and parallel-connected components.

shelf system: A complete audio system that is designed to play the various audio formats in a minimum of space.

short circuit: Overcurrent that leaves the normal current-carrying path by going around the load and back to the power source or ground.

signal level: The amplitude or strength of an audio signal.

signal source: An audio device that provides a signal to another audio device.

silicon controlled rectifier (SCR): A three-terminal semiconductor thyristor that is normally an open circuit until a signal applied to the gate switches it to the conducting state in one direction.

sine wave: A symmetrical waveform that contains 360°.

single-acting cylinder: A cylinder that requires pressurized fluid in only one port, such as a spring-return cylinder, spring-advance cylinder, or cylinder in which the weight of the load is used to move the cylinder in one direction only.

single-function meter: A meter capable of measuring and displaying only one quantity.

single-function switch: A switch that performs only one switching function such as start or stop.

single-voltage motor: A motor that operates at only one voltage level.

slip: The difference between the synchronous speed and actual speed of a motor.

solarcell: A voltage source that converts light energy to electrical energy.

solder: An alloy of tin (Sn) and lead (Pb).

soldered connection: A connection that joins conductors by heat to make a strong electrical and mechanical connection.

soldering: The process of joining metals by heat to make a strong electrical and mechanical connection.

solderless connection: A connection that joins conductors without using heat to make a strong electrical and mechanical connection.

solenoid: An electric output device that converts electrical energy into a linear, mechanical force.

solid: A state of matter that has a definite volume and shape.

solid-state controls: Devices that use electronic components such as transistors, SCRs, and triacs to switch and control the flow of current in a circuit.

solid-state relay: A relay that uses electronic switching devices in place of mechanical contacts.

sound: Energy that consists of pressure vibrations in the air.

sound intensity (volume): A measure of the amount of energy flowing in a sound wave.

speaker: An electric device that converts electrical signals into sound waves.

speaker impedance: The load value (in Ω) that the speaker presents to an amplifier or receiver.

speaker noise: A raspy sound caused by a damaged component, such as a torn cone.

speed: The rate at which an object is moving.

split-phase motor: A 1ϕ, AC motor that includes a running winding (main winding) and a starting winding (auxiliary winding).

split-wired receptacle: A standard receptacle that has the tap between the two brass-colored (hot) terminal screws removed.

spotlight: A lamp that casts intensive light in a localized area.

spring actuator: An actuator that automatically returns a valve to a set position.

square: A four-sided figure with all sides equal and four 90° angles.

square foot: A measurement that contains 144 sq in. (12″ × 12″ = 144 sq in.).

square inch: A measurement of 1″ × 1″.

square root: A number which, when multiplied by itself, gives the original number.

standard: An accepted reference or practice.

standard tap splice: A splice that connects two wires together when one wire is to remain unbroken.

starting torque: The torque required to start a motor.

static electricity: An electrical charge at rest.

stator: The stationary part of an AC motor.

step-down transformer: A transformer in which the secondary coil has fewer turns of wire than the primary coil.

step-up transformer: A transformer in which the secondary coil has more turns of wire than the primary coil.

stereo system: An audio system that uses two independent channels that are routed to a pair of speakers situated to the right and left of the listener.

strobe tachometer: A device that uses a flashing light to measure the speed of a moving object.

subdivision: A division that divides secondary divisions in halves, thirds, fourths, fifths, etc.

substation: An assemblage of equipment installed for switching, changing, or regulating the voltage of electricity.

sulfidation: The formation of film on the contact surface.

superscript: A small number above and to the right of the base.

surge protector: An electrical device that continuously monitors incoming power and automatically clamps any excess voltage that could cause damage.

surge suppressor: An electrical device that limits voltage surges that may be present on power lines.

surround sound: The concept of expanding reproduced sound characteristics from one dimension (mono) to three dimensions.

sweep: The movement of the displayed trace across the scope screen.

switch: A device that is used to start, stop, or redirect the flow of current in an electrical circuit.

switchboard: a freestanding distribution cabinet in commercial and industrial buildings.

symbol: A graphic element that represents an operation, a quantity, or unit.

synchronous speed: The theoretical speed of a motor based on the motor's number of poles and the line frequency.

system: A combination of components interconnected to perform work or meet a specific need.

T

tagout: The process of placing a danger tag on the source of electrical power which indicates that equipment may not be operated until the danger tag is removed.

tape deck (tape player): A device that converts electrical signals to magnetic impulses and transfers these impulses onto tape.

tapped core coil: A coil with taps for selecting different magnetic field strengths.

tapped resistor: A resistor that contains fixed tap points of different resistances.

temperature: The measurement of the intensity of heat.

temperature rise: The difference between a motor's winding temperature when running and the ambient temperature.

temperature switch: A switch that responds to temperature changes.

temporary magnets: Magnets that lose their magnetism as soon as the magnetizing force is removed.

test light: A test instrument with a bulb that is connected to two test leads to give a visual indication when voltage is present in a circuit.

thermal conductivity: The property of a material to conduct heat in the form of thermal energy.

thermal overload switch: A switch that operates its contacts when a preset temperature is reached.

thermistor: A device that changes resistance with a change in temperature.

thermocouple: A device that produces electricity when two different metals (such as iron and constantan) that are joined together are heated.

three-way switch: A single-pole, double-throw (SPDT) switch.

three-wire control: An input control for an electric motor drive requiring three conductors to complete a circuit.

throw: The number of closed contact positions per pole.

thyristor: A solid-state switching device that switches current ON by a quick pulse of control current.

time delay fuse (TDF): A fuse that may detect and remove a short circuit almost instantly, but allow small overloads to exist for a short period of time.

timer: A control device that uses a preset time period as part of the control function.

timing diagram: A diagram that shows graphically how the output of a circuit changes in response to input changes.

tinning: The process of applying solder to a clean soldering iron tip to prevent corrosion on the tip.

tip conductor: The first wire in a pair of wires.

tone arm: The turntable device that holds the cartridge.

torque: The force that causes an object to rotate.

trace: **1.** A reference point/line that is visually displayed on the face of the scope screen. **2.** Conducting paths used to connect components on a PC board.

transformer: An electric device that uses electromagnetism to change voltage from one level to another or to isolate one voltage from another.

transformer taps: Connecting points that are provided along the transformer coil.

transient response: The time delay from when a speaker cone moves from rest to when it responds to a sharp electrical pulse.

transient voltage: A temporary, unwanted voltage in an electrical circuit.

transistor: A three-terminal device that controls current through the device depending on the amount of voltage applied to the base.

transistor-transistor logic (TTL) ICs: A broad family of ICs that employ a two-transistor arrangement.

transverse noise: Noise produced between the hot and neutral lines.

triac: A three-terminal semiconductor thyristor that is triggered into conduction in either direction by a small current to its gate.

triangle: A three-sided figure with three interior angles.

true power: The actual power used in an electrical circuit expressed in watts (W).

truth table: A table used to describe the output condition of a logic gate or combination of gates for every possible input condition.

tuner: An electric device designed to receive radio broadcast signals and convert them into electrical signals that can be amplified.

tungsten-halogen lamp: An incandescent lamp filled with a halogen gas (iodine or bromine).

tunnel diode: A diode designed so that the current flowing through the diode decreases with an increase in applied voltage for a specified range of forward voltage.

turntable: A record player without an amplifier.

turntable drive: The mechanism that makes the platter spin.

tweeter: A small speaker that is designed to reproduce high frequencies between 4 kHz to 22 kHz.

twisted conductors: Conductors that are intertwined at a constant rate.

two-way switch: A single-pole, single-throw (SPST) switch.

two-winding constant-wattage ballast: A ballast that uses a transformer which provides isolation between the primary and secondary circuits.

two-wire control: An input control for an electric motor drive requiring two conductors to complete a circuit.

U

ultraviolet region: The region of the spectrum with wavelengths just short of the color violet.

Underwriters Laboratories Inc. (UL): An independent organization that tests equipment and products to see if they conform to national codes and standards.

ungrounded conductor: A current-carrying conductor that is connected to loads through fuses, circuit breakers, and switches.

unijunction transistor (UJT): A transistor consisting of N-type material with a region of P-type material doped within the N-type material.

V

valence electrons: Electrons in the outermost shell of an atom.

valence shell: The outermost shell of an atom and contains the electrons that form new compounds.

variable: A quantity that may change.

variable capacitor: A capacitor that varies in capacitance value.

variable displacement fluid power motor: A fluid power motor that drives a load in one direction and provides a variable output.

variable displacement pump: A pump in which the amount of flow is manually adjusted or automatically adjusted without changing the pump speed.

variable frequency drive (VFD): A solid-state programmable control module used to control an AC motor.

variable (adjustable) resistor: A resistor with a set range of values, such as 0 Ω to 1000 Ω.

visible light: The portion of the electromagnetic spectrum to which the human eye responds.

voltage: The amount of electrical pressure in a circuit.

voltage dip: A momentary low voltage.

voltage indicator: A test instrument that indicates the presence of voltage when the test tip touches, or is near, an energized hot conductor or energized metal part.

voltage spike: An increase in voltage (normally several thousand volts) that lasts for a very short time (microseconds to milliseconds).

voltage stabilizer (regulator): A device that provides precise voltage regulation to protect equipment from voltage dips and voltage surges.

voltage surge: A higher-than-normal voltage that temporarily exists on one or more power line.

voltage tester: An electrical test instrument that indicates the approximate voltage amount and type of voltage (AC or DC) in a circuit by the movement of a pointer (and vibration on some models).

voltage variance: The difference in voltage between a voltage surge and a voltage dip.

W

watt (W): A unit of measure equal to the power produced by a current of 1 A across a potential difference of 1 V.

wattage: The number of watts a device uses to operate.

wavelength: The distance covered by one complete cycle of a given frequency sound as it passes through the air.

way: A flow path through a valve.

Western Union splice: A splice that connects two wires that may be placed under a lot of mechanical strain.

whole numbers (integers): Numbers that have no fractional or decimal parts.

wire: Any individual conductor.

wire marker: A preprinted peel-off marker designed to adhere when wrapped around a conductor.

wire nut: A solderless plastic connector that uses a tapered metal coil spring to twist wires together.

wire stripper: A tool that is designed to properly remove insulation from small gauge (normally AWG sizes No. 10–No. 22) wires.

woofer: A large speaker that is designed to reproduce low-frequency sounds.

word: A group of one or more bytes (1111000011110000, etc.) that is transmitted or stored as a unit.

work: Applying a force over a distance.

wrap-around bar graph: A bar graph that displays a fraction of the full range on the graph.

wye configuration: A transformer connection that has one end of each transformer coil connected together.

X

XOR (exclusive OR) gate: A logic gate that provides logic level 1 only if one, but not both of the inputs are at logic level 1.

Z

zener diode: A silicon PN junction that differs from a standard diode in that it operates in the reverse breakdown region.

Index

A

B

R

S

W

X

Z

USING THE *Electrical Principles and Practices* CD-ROM

Before removing the CD-ROM from the protective sleeve, please note that the book cannot be returned for refund or credit if the CD-ROM sleeve seal is broken.

System Requirements

The *Electrical Principles and Practices* CD-ROM is designed to work best on a computer meeting the following hardware/software requirements:

- Intel® Pentium® (or equivalent) processor
- Microsoft® Windows® XP, 2000, NT®, Me, 98, 98 SE, or 95 operating system
- 128 MB of free available system RAM (256 MB recommended)
- 90 MB of available disk space
- 1024 × 768 24-bit (millions of colors) color display or better
- Sound output capability and speakers
- CD-ROM drive
- Internet Explorer™ 3.0 or Netscape® 3.0 or later browser software and an Internet connection

Opening Files

Insert the CD-ROM into the computer CD-ROM drive. Within a few seconds, the home screen will be displayed allowing access to all features of the CD-ROM. Information about the usage of the CD-ROM can be accessed by clicking on USING THIS CD-ROM. The Quick Quizzes®, Illustrated Glossary, Math Formulas, Media Clips, and Reference Material can be accessed by clicking on the appropriate button on the home screen. Clicking on the American Tech web site button (www.go2atp.com) accesses information on related educational products. Unauthorized reproduction of the material on this CD-ROM is strictly prohibited.

iii

The Turkish Cookbook

Regional Recipes and Stories

NUR İLKIN AND SHEILAH KAUFMAN

First published in 2010 by

INTERLINK BOOKS
An imprint of Interlink Publishing Group, Inc.
46 Crosby Street, Northampton, Massachusetts 01060
www.interlinkbooks.com

Book design by Juliana Spear

Photographs on pages iii, xvii, 331, and 346–347 courtesy of Gulhan Corman; pages iv, v, xxvii, and 174 courtesy of Sheilah Kaufman; pages vi, viii, xiv, and 285 (top) courtesy of the Turkish Culture and Tourism Office. Additional area photography © www.dreamstime.com: page x: Adlin Hisyamuddin; p. 1 (top): Tevfik Ozakut (bottom) Kevin Knuth; p. 53 (top): Sinan Durdu (bottom): Çiğdem Metin; p. 115 (top): Jokerproduction (bottom): Lu2006; p. 171 (top): Yaroslav (bottom): Syrsln; p. 229 (top): Ali Rıza Yıldız (bottom): Anibal Trejo; p. 233: Jokerproduction; p. 261 (top): Radist (bottom): Jokerproduction; p. 285 (bottom): Can Yilmaz.

Library of Congress Cataloging-in-Publication Data
İlkin, Nur.
The Turkish cookbook : regional recipes and stories / by Nur İlkin and Sheilah Kaufman.
p. cm.
Includes bibliographical references and index.
ISBN 978-1-56656-793-0
1. Cookery, Turkish. I. Kaufman, Sheilah. II. Title.
TX725.T8I433 2009
641.59561—dc22 2009042698

Printed and bound in China